ANNUAL REVIEW OF ENTOMOLOGY

Production Editor Teri Hyde
Subject Indexer Bruce Tracy

ANNUAL REVIEW OF ENTOMOLOGY

VOLUME 45, 2000

MAY R. BERENBAUM, *Editor*
University of Illinois, Urbana-Champaign

RING T. CARDÉ, *Associate Editor*
University of California, Riverside

GENE E. ROBINSON, *Associate Editor*
University of Illinois, Urbana-Champaign

www.AnnualReviews.org science@annurev.org 650-493-4400

ANNUAL REVIEWS
4139 El Camino Way • P.O. Box 10139 • Palo Alto, California 94303-0139

ANNUAL REVIEWS
Palo Alto, California, USA

International Standard Serial Number: 0066-4170
International Standard Book Number: 0-8243-0145-5
Library of Congress Catalog Card Number: A56-5750

Typeset by Impressions Book and Journal Services, Inc., Madison, WI
Printed and Bound in the United States of America

Preface

Each year the outgoing editorial board member is asked to write the preface for the next volume of the *Annual Review of Entomology*. By this simple mechanism, the volume published in the year 2000 falls to me. The *Annual Review of Entomology* is a highly successful publication, as outlined by Editor May Berenbaum and Board Member Michael Irwin in their prefaces to Volumes 42 and 43. This is due, in part, to the excellent management by past editors and board members, supported by the leadership and staff of the Annual Reviews. In the past six years, I have gained some insight into the process that makes this volume an annual success and into the changes that will insure its continued vigor. In 1993, when I first joined the board, Tom Miller, Frank Radovsky and Vince Resh had been the editors for a group total of 50 years. Tom Miller set the tone of courtesy and respect, Frank Radovsky was a fountain of ideas for topics, especially in the area of medical entomology, and Vince Resh had a long and seamless memory of the past flow of reviews. By 1998, my last full year as a board member, all three of these mentors had retired as editors. The new cast of editors, May Berenbaum, Gene Robinson and Ring Cardé, have now begun to place their own stamp of originality, scope and fairness on the Annual Review and the process by which topics and authors are selected. The process of selection seems simple enough. Our charge is to identify and evaluate topics ripe for review, along with potential authors. Suggestions for topics come from a variety of sources, including editors, committee members, foreign correspondents, previous authors, and volunteering prospective authors. The goal of the selection process is to provide a wide community of biologists with useful and insightful syntheses.

As entomology departments across the United States are facing challenges to their existence, the *Annual Review of Entomology* faces no such threat and, in fact, continues to unite us. We can look to the health of the *Annual Review of Entomology* as a model for success. The process of selecting topics for review is, in effect, an ongoing, evolving definition of our field and our role in the biological sciences. The continuing self-assessment inherent in the process of the reviews themselves provides benchmarks for specific, and more delimited areas. The *Annual Review of Entomology*, then, provides an aid in defining the field of entomology and its role in the greater community of biologists in the future.

Entomology, by its broad and most correct definition, is the study of insects. The spectacularly diverse insects are a major feature of most terrestrial ecosystems, making up about 75% of all animal species. Given such statistics, it follows that understanding insects and their roles in ecosystems and communities is essential to understanding local and global ecologies. Further, as the dominant forms of animal life sharing this planet, insects and humans often influence each other's welfare. Insects as vectors of pathogens and pests in agricultural systems represent a continual source of friction between us. Experiences with pests can make us forget that insects also enhance human interactions with the natural world, for example as pollinators and biological control agents. Moreover, insects as a widespread and diverse form of life provide biological sciences with model systems for research at all levels of biological integration. Studies of insects continue to make significant contributions to our understanding of how life, in general, works. The *Annual Review of Entomology* provides a single site for showcasing all these faces of Entomology.

DIANA WHEELER
For the Editorial Committee

CONTENTS

Related Articles

From the ***Annual Review of Ecology and Systematics,*** Volume 30, 1999:

Streams in Mediterranean Climate Regions: Abiotic Influences and Biotic Responses to Predictable Seasonal Events, Avital Gasith and Vincent H. Resh

Choosing the Appropriate Scale of Reserves for Conservation, Mark W. Schwartz

Conspecific Sperm and Pollen Precedence and Speciation, Daniel J. Howard

Using Phylogenetic Approaches for the Analysis of Plant Breeding System Evolution, Stephen G. Weller and Ann K. Sakai

Evolution of Diversity in Warning Color and Mimicry Polymorphisms, Shifting Balance, and Speciation, James Mallet and Mathieu Joron

The Relationship Between Productivity and Species Richness, R. B. Waide, M. R. Willig, C. F. Steiner, G. Mittelbach, L. Gough, S. I. Dodson, G. P. Juday, and R. Parmenter

Analysis of Selection on Enzyme Polymorphisms, Walter F. Eanes

Polymorphism in Systematics and Comparative Biology, John J. Wiens

Physical-Biological Coupling in Streams: The Pervasive Effects of Flow on Benthic Organisms, David D. Hart and Christopher M. Finelli

Full of Sound and Fury: The Recent History of Ancient DNA, Robert K. Wayne, Jennifer A. Leonard, and Alan Cooper

Do Plant Populations Purge Their Genetic Load? Effects of Population Size and Mating History on Inbreeding Depression, D. L. Byers and D. M. Waller

Gene Flow and Introgression from Domesticated Plants into Their Wild Relatives, Norman C. Ellstrand, Honor C. Prentice, and James F. Hancock

Resistance of Hybrid Plants and Animals to Herbivores Pathogens, and Parasites, Robert S. Fritz, Catherine Moulia, and George Newcombe

Evolutionary Computation: An Overview, Melanie Mitchell and Charles E. Taylor

From the ***Annual Review of Genetics,*** Volume 33, 1999:

History of Plant Population Genetics, Robert W. Allard

Mechanisms of mRNA Surveillance in Eukaryotes, P. Hilleren and R. Parker

Genetics of Chemotaxis and Thermotaxis in the Nematode Caenorhabditis elegans, Ikue Mori

From the ***Annual Review of Microbiology,*** Volume 54, 2000:

The Adaptive Mechanisms of Trypansoma brucei for Sterol Homeostasis in Its Different Life-Cyle Environments, Isabelle Coppens and Pierre J. Courtoy
The Development of Genetic Tools for Dissecting the Biology of Maleria Parasites, Andrew P. Waters

From the ***Annual Review of Phytopathology,*** Volume 37, 1999:

The Caenorhabditis elegans Genome: A Guide in the Post Genomics Age, David McK Bird, Charles H. Opperman, Steven J. M. Jones, and David L. Baillie
Phytoalexins: What We Have Learned After 60 Years? Ray Hammerschmidt
Natural Genomic and Antigenic Variation in Whitefly-Transmitted Geminiviruses (Begomoviruses), B. D. Harrison and D. J. Robinson

ANNUAL REVIEWS is a nonprofit scientific publisher established to promote the advancement of the sciences. Beginning in 1932 with the *Annual Review of Biochemistry,* the Company has pursued as its principal function the publication of high-quality, reasonably priced *Annual Review* volumes. The volumes are organized by Editors and Editorial Committees who invite qualified authors to contribute critical articles reviewing significant developments within each major discipline. The Editor-in-Chief invites those interested in serving as future Editorial Committee members to communicate directly with him. Annual Reviews is administered by a Board of Directors, whose members serve without compensation.

ANNUAL REVIEWS OF

Anthropology
Astronomy and Astrophysics
Biochemistry
Biomedical Engineering
Biophysics and Biomolecular Structure
Cell and Developmental Biology
Earth and Planetary Sciences
Ecology and Systematics
Energy and the Environment
Entomology
Fluid Mechanics
Genetics
Immunology
Materials Science
Medicine
Microbiology
Neuroscience
Nuclear and Particle Science
Nutrition
Pharmacology and Toxicology
Physical Chemistry
Physiology
Phytopathology
Plant Physiology and Plant Molecular Biology
Political Science
Psychology
Public Health
Sociology

SPECIAL PUBLICATIONS
Excitement and Fascination of Science, Vols. 1, 2, 3, and 4

Annu. Rev. Entomol. 2000. 45:1–54

THE CURRENT STATE OF INSECT MOLECULAR SYSTEMATICS: A Thriving Tower of Babel

Michael S. Caterino[1,2], Soowon Cho[1,3], and Felix A. H. Sperling[1,4]

[1]*Insect Molecular Systematics Laboratory, Division of Insect Biology, Department of Environmental Science, Policy and Management, University of California, Berkeley, California 94720-3112*

[2]*Department of Entomology, The Natural History Museum, Cromwell Road, London, SW7 5BD, UK; e-mail: histerid@nhm.ac.uk*

[3]*Department of Agricultural Biology, Chungbuk National University, Cheongju 361-763, Korea; e-mail: soowon@trut.chungbuk.ac.kr*

[4]*Department of Biological Sciences, CW-405 Biological Sciences Centre, University of Alberta, Edmonton, Alberta, T6G 2E9, Canada; e-mail: felix.sperling@ualberta.ca*

Key Words mtDNA, rDNA, phylogeny, evolution, DNA standards

■ **Abstract** Insect molecular systematics has undergone remarkable recent growth. Advances in methods of data generation and analysis have led to the accumulation of large amounts of DNA sequence data from most major insect groups. In addition to reviewing theoretical and methodological advances, we have compiled information on the taxa and regions sequenced from all available phylogenetic studies of insects. It is evident that investigators have not usually coordinated their efforts. The genes and regions that have been sequenced differ substantially among studies and the whole of our efforts is thus little greater than the sum of its parts. The cytochrome oxidase I, 16S, 18S, and elongation factor-1α genes have been widely used and are informative across a broad range of divergences in insects. We advocate their use as standards for insect phylogenetics. Insect molecular systematics has complemented and enhanced the value of morphological and ecological data, making substantial contributions to evolutionary biology in the process. A more coordinated approach focused on gathering homologous sequence data will greatly facilitate such efforts.

INTRODUCTION

The 16 years that have elapsed since Berlocher's comprehensive 1984 *Annual Review* treatment of insect molecular systematics have witnessed sweeping changes (55, 413, 440), including the advent of the polymerase chain reaction (PCR; 342), the development of automated sequencing, and greatly increased computational power. As a result, accessibility of molecular techniques has improved to the point where the majority of phylogenetic studies involve a molecular component, and DNA sequence data is being generated with an ease and economy inconceivable 15 years ago. Furthermore, the development of phylogenetic theory and methodology has kept apace

0066–4170/00/0107–0001/$14.00

and we are able to make excellent use of these massive quantities of new data. This introduction would seem to paint a rosy picture of the field; this is by and large an accurate depiction. However, as our title implies, we feel there is just cause for a critical assessment of the overarching coherence of insect systematists' efforts and that there is room for substantial improvement.

The story of the Tower of Babel serves as an appropriate metaphor for what we perceive to be a lack of coordination across the breadth of insect systematics. In this Old Testament account, the people of Babylon were thwarted in their attempts to build a tower to heaven because of increasing linguistic incompatibilities. The tower was abandoned, incomplete, due to a collectively incoherent array of verbiage. For insect systematics, the entire enterprise will not likely be abandoned. However, inconsistency among systematists in their choice of phylogenetic markers has led to a situation in which the whole of insect molecular systematics is only negligibly greater than the sum of its parts. This situation stands in contrast to that of the plant and vertebrate systematic communities. In these two cases a small number of generally accepted markers has been singled out as standards, with the result that global phylogenies for these groups are emerging from the amassed data. The chloroplast ribulose-1,5 biphosphate carboxylase (*rbcL*) data set for plants now includes nearly 5000 taxa (GenBank, March 1999) while over 7000 (nonhuman) vertebrate sequences are available for the cytochrome *b* (*cyt*b) gene (215; also GenBank, March 1999). In addition to the significance for the systematics of these diverse groups, such large data sets offer unparalleled opportunities for the study of molecular evolution. Such studies, through reciprocal illumination, promise to continue to enlighten systematic analyses (319).

There is much cause for optimism and satisfaction regarding the state of insect molecular systematics. In this review we accordingly survey some of the major advances of the past decade or so. However, we also graphically illustrate the pervasive discontinuities among studies, which render our collective efforts incompatible and incoherent. While we hope that pointing out this perceived problem will lead to its remediation, we also offer some suggestions that would allow us to bridge the gaps among existing works and build a solid foundation for the future.

WHAT HAS BEEN DONE?

Taxa

Here we review progress toward the goal of understanding insect phylogenetic relationships. In general, it seems that efforts to resolve relationships at lower taxonomic levels have been relatively successful. Given sufficient sampling, most studies that set out to understand relationships among subspecies, species, and species groups offer significant new insights, if not complete resolution. This relative success is a function of the effort exerted at lower levels as well as the fact that many markers now in widespread use are highly informative around the species level, and were initially developed for this purpose. At higher taxonomic levels, however, few studies have included sufficient samples for doing more than sketching rough phylogenetic outlines.

Phylogenetic relationships within family-level taxa have, for the most part, been resolved where examined. This generalization, of course, depends heavily on the age and diversity of the taxon, and families are by no means comparable across insects. Nonetheless, there have been notable successes across a range of groups, including mosquitoes (38), vespid wasps (431), braconid wasps (35, 136), and tiger beetles (519). The macrolepidoptera have been particularly well studied, with the Noctuoidea and Bombycoidea serving as the proving grounds for development of several promising single-copy nuclear loci (99, 153, 168, 326, 404). These data, as well as the nuclear ribosomal (rDNA) and mitochondrial DNA (mtDNA) data of Weller et al (530, 533) and Weller & Pashley (531), have allowed several solid inferences to be made regarding the diversification of the taxonomically difficult Noctuoidea.

Molecular phylogenetic studies among major lineages within insect orders number far fewer, although there have been some important findings. The most thoroughly studied group at this level has been the Hemiptera (including Homoptera). Taken together, the 18S rDNA studies of Wheeler et al (542), Campbell et al (74, 75), Sorensen et al (449), and Von Dohlen & Moran (523) have firmly established the basic structure of a global Hemiptera phylogeny, in which the "Homoptera" are clearly paraphyletic. While a few other published studies have made promising inroads within other ordinal-level groups, particularly the Orthoptera (161, 162), Hymenoptera (136, 544, 546), and Lepidoptera (548), none have achieved the comprehensive phylogenetic picture obtained for Hemiptera.

As far as studies among the orders and major lineages of insects are concerned, the work of Whiting et al (547) is in a class by itself. Their analysis examined relationships among 85 taxa, including multiple members of all the holometabolous orders on the basis of 18S and 28S rDNA as well as morphology. Though this study emphasized the Holometabola, the thorough taxon sampling (including 20 non-holometabolous taxa) and diverse character set have led to the most complete picture of insect phylogeny to date. Aside from the placement of Strepsiptera, their overall conclusions regarding holometabolous relationships are largely consistent with classical hypotheses. The idea that the Strepsiptera may be closely related to the Diptera, however intriguing, cannot be considered strongly supported at present (80, 238). The "Strepsiptera problem" has nonetheless been very successful in attracting attention to the problems of insect higher phylogeny.

Despite significantly advancing our understanding of phylogeny, the impact of molecular systematics on insect taxonomy has been minimal. Molecular techniques have proven very useful for identification of particular genotypes (e.g. 45, 279, 471). However, this work typically relies on existing taxonomy. The use of molecular techniques in actually creating or modifying formal nomenclature has been limited. There have now been a few species descriptions based at least in part on molecular data (33, 350, 381), and the recent suggestion that identities of old type specimens may be established using molecular data bears exploration (504). However, with few exceptions (notably 449) higher-level molecular studies have not addressed primarily taxonomic concerns.

Markers

The database of insect sequences is growing rapidly, as evidenced by our bibliography. GenBank now boasts nearly 100,000 insect sequences. Although 80,000 of those are *Drosophila* sequences, a substantial fraction of them have been used in phylogenetic studies. Systematic studies of insects have now examined around 40 protein-coding genes, all of the major ribosomal RNA genes (both mitochondrial and nuclear) as well as numerous non-coding regions. This number continues to grow as work on additional loci, most of which are nuclear protein-coding, make the jump from *Drososphila* studies to more general application. The diversity of available markers has unquestionably furthered the cause of insect molecular systematics. However, this plethora of markers also brings with it the risk of pluralism, in that the use of different estimators among studies makes comparison and synthesis difficult or impossible.

The most commonly sequenced regions in insect systematics are mtDNA and nuclear rDNA. As contiguous pieces of DNA, these two classes also lend themselves to easy comparison. Therefore this review deals primarily with these two classes of markers. However, our point is no less true for nuclear protein–coding genes and we briefly discuss these as well. In Figures 1 and 2 we have mapped the mitochondrial and nuclear ribosomal fragments sequenced for all published insect systematic studies that we have been able to find, as well as some unpublished data from GenBank. While the popularity of these regions would seem at first to refute our point, a quick glance at these figures shows that there is substantial incompatibility among studies. There exist a few commonly used regions for both mitochondrial and ribosomal genes but in no cases are these consistent across all taxa. In many cases the same gene has been sequenced across several studies, but these regions often do not overlap with regions sequenced for related taxa (e.g. see 28S for Diptera in Figure 2).

For mitochondrial DNA, the most frequently sequenced genes are cytochrome oxidase I (COI), COII and 16S rDNA, with 12S rDNA not far behind. Of these, COII has been sequenced over the widest variety of taxa, with homologous sequences available for nearly all orders. COI has been sequenced in as many different groups. However, due to its length, the specific region chosen varies from study to study. A region of the 16S gene corresponding roughly to positions 12900–13400 (relative to *Drosophila yakuba;* 103) has been consistently sequenced across most taxa with the exception of Lepidoptera. COIII, NADH dehydrogenase 5 (ND5) and *cyt*b have been sequenced in a few studies each. ND2, ND4, and ND1 have received only isolated attention, and the remaining mitochondrial genes have largely been ignored. Unfortunately, it is not possible to offer much justification for these patterns. In general, it seems that most of the protein-coding genes perform approximately equally well, given similar length fragments. Far too few comparative studies examining rates among genes and among taxa have been done for any clear pattern of rate differences

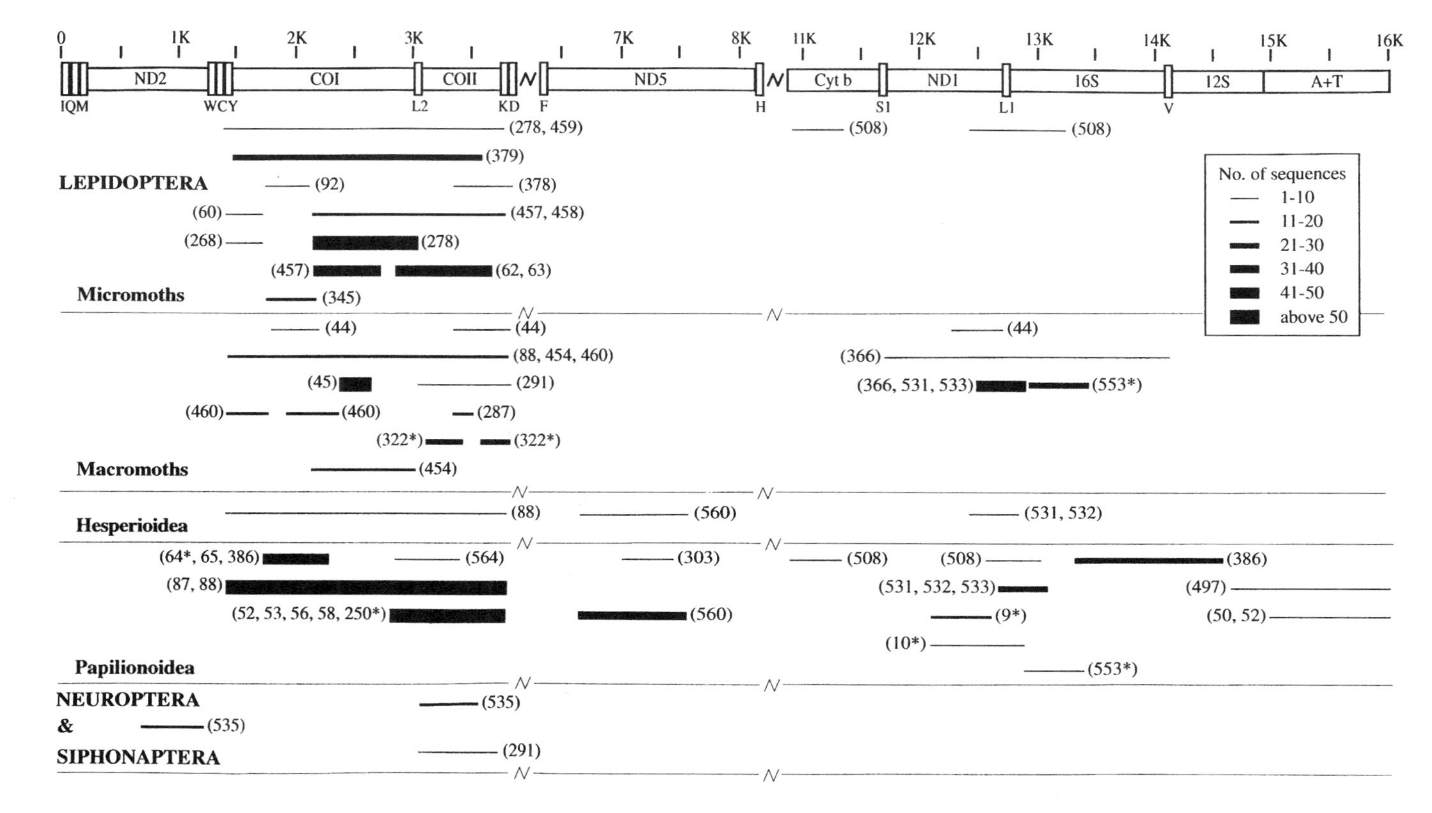

Figure 1 (*a*) Mitochondrial regions sequenced in molecular systematic studies of hexapods. This figure treats all systematically oriented studies that we were able to find published before March 15, 1999. Studies that used only previously published data (i.e. GenBank data) were not included. A study was deemed systematic if the data gathered were used to build a phylogenetic tree, normally involving more than a single species. The length of each bar indicates the size of the region sequenced and the thickness of the bar reflects the number of sequences published for that taxon, for that region. Studies utilizing the same region (plus or minus a few base pairs) were lumped. The positions of the bars are determined by the endpoints. (*continued*)

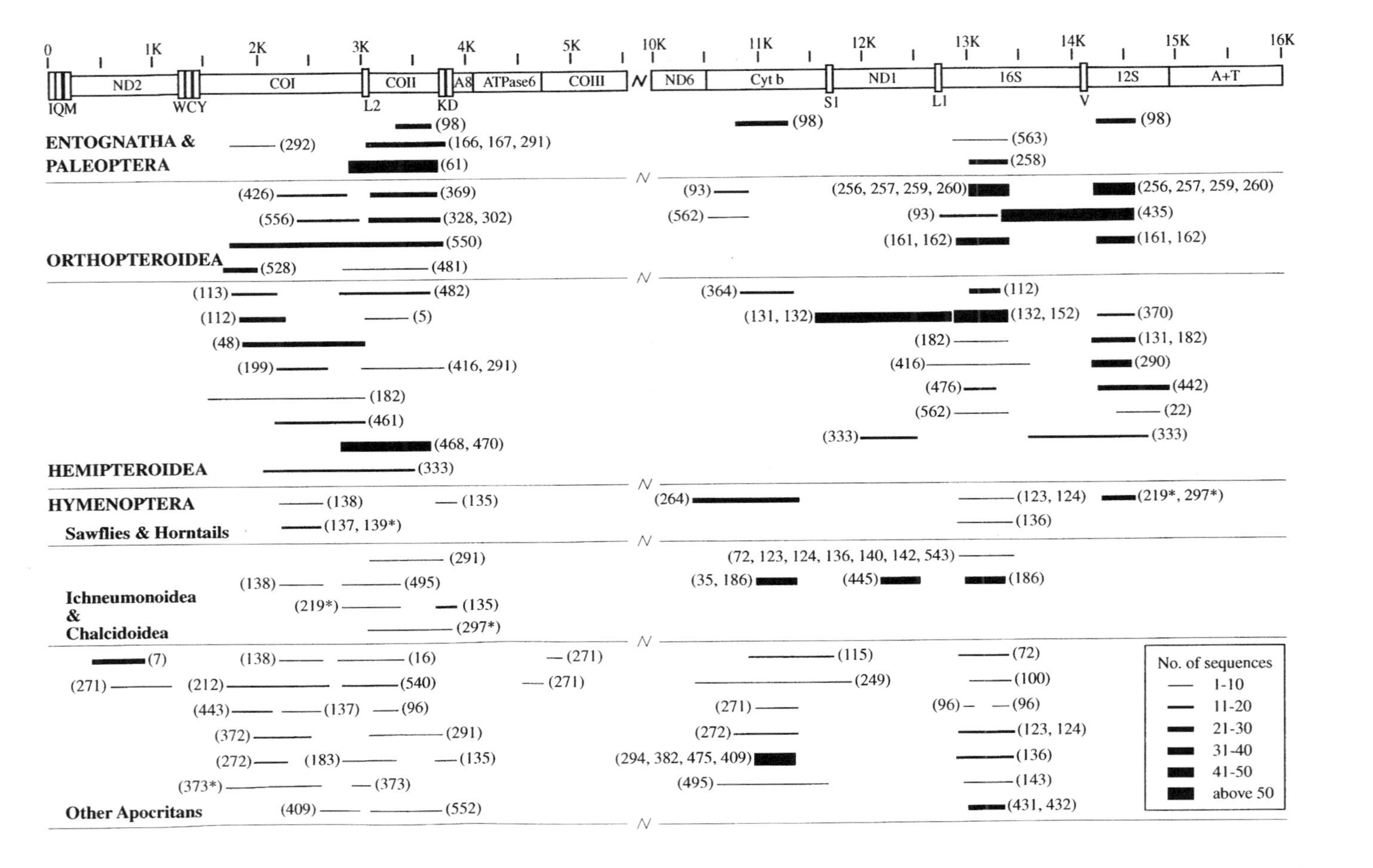

Figure 1 (*b*) (*continued*) Therefore, in length-variable regions (control region, rDNA) the length of the figured bar may not exactly represent the length of the fragment sequenced. Asterisks after the references indicate that the data used in the study are not presently available for inspection, either in the original publication or from GenBank.

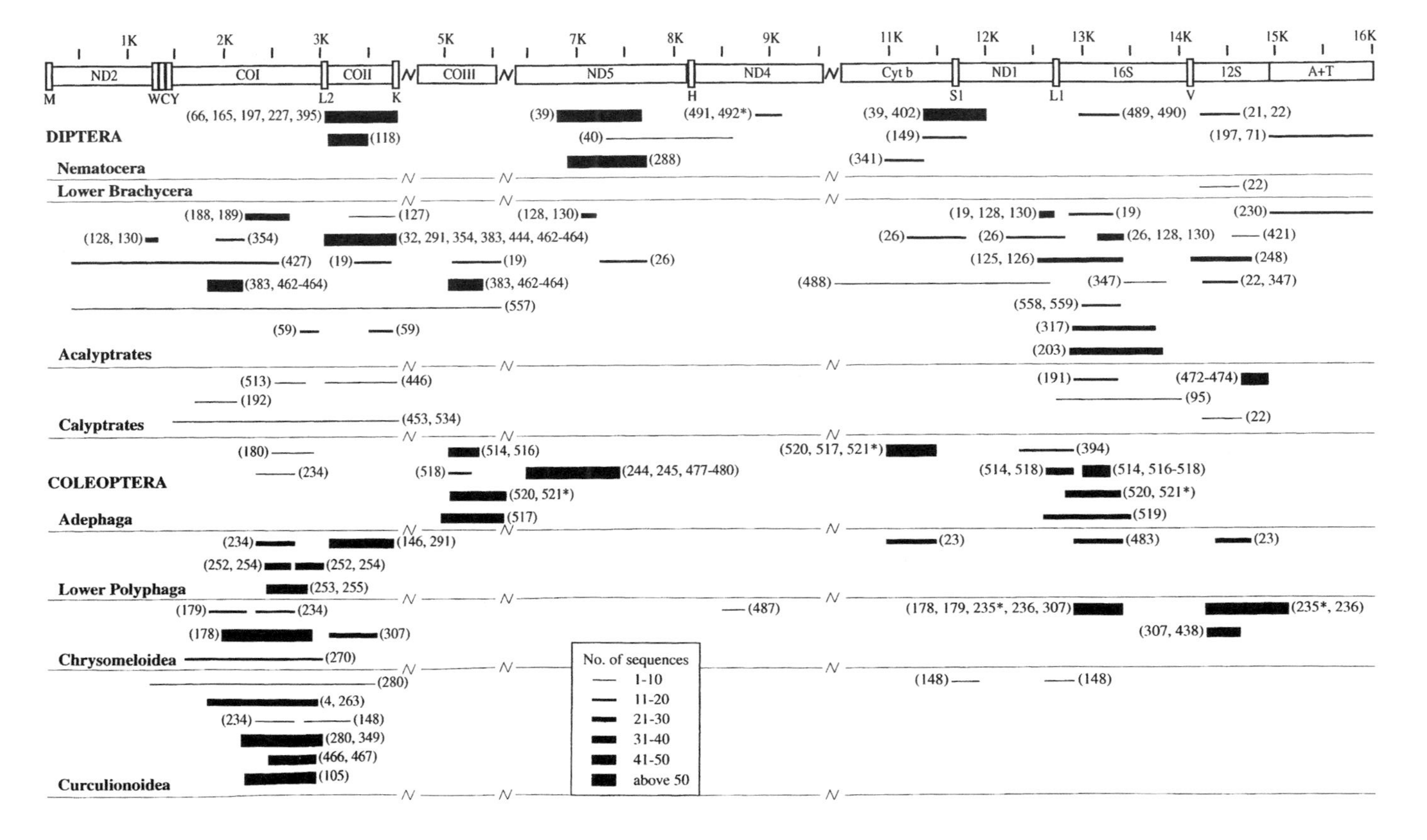

Figure 1 (*c*) (*continued*) In addition to the studies cited here, complete mitochondrial sequences are available for the following six insects: *Locusta migratoria* (164), *Drosophila yakuba* (103), *D. melanogaster* (289), *Anopheles gambiae* (31), *A. quadrimaculatus* (327), and *Apis mellifera* (114). Note that the regions illustrated in each figure are not identical. *(a)* Mitochondrial regions sequenced in Lepidoptera, Siphonaptera and Neuroptera. *(b)* Mitochondrial regions sequenced in Ametabola, Hemimetabola, and Hymenoptera. *(c)* Mitochondrial regions sequenced in Diptera and Coleoptera.

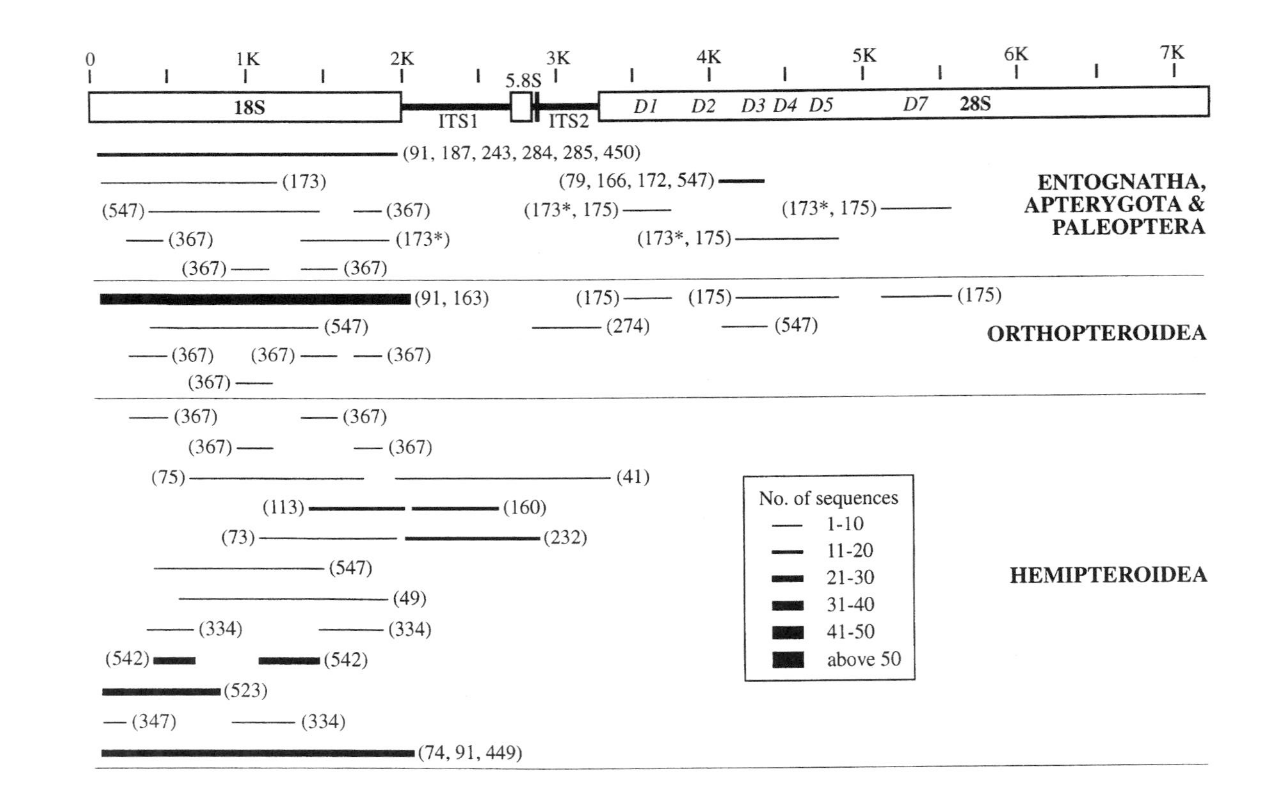

0
1K
2K
3K
4K
5K
6K
7K
18S
ITS1
5.8S
ITS2
D1
D2
D3
D4
D5
D7
28S
(91, 187, 243, 284, 285, 450)
(173)
(79, 166, 172, 547)
(547)
(367)
(173*, 175)
(173*, 175)
(367)
(173*)
(173*, 175)
(367)
(367)
ENTOGNATHA, APTERYGOTA & PALEOPTERA
(91, 163)
(175)
(175)
(175)
(547)
(274)
(547)
(367)
(367)
(367)
(367)
ORTHOPTEROIDEA
(367)
(367)
(367)
(367)
(75)
(41)
(113)
(160)
(73)
(232)
(547)
(49)
(334)
(334)
(542)
(542)
(523)
(347)
(334)
(74, 91, 449)
No. of sequences
1-10
11-20
21-30
31-40
41-50
above 50
HEMIPTEROIDEA

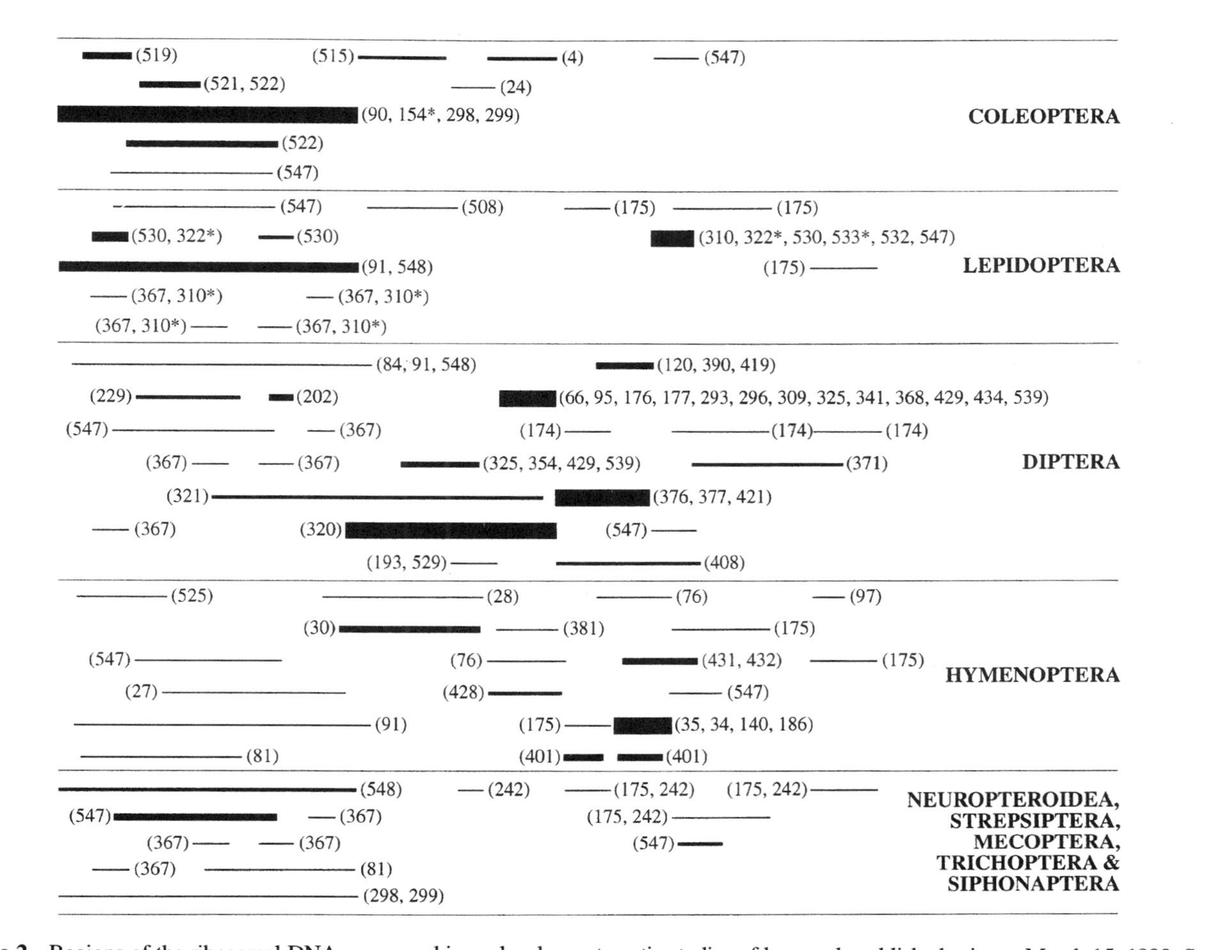

Figure 2 Regions of the ribosomal DNA sequenced in molecular systematic studies of hexapods published prior to March 15, 1999. See Figure 1 for conditions regarding the studies included.

among loci and among regions within those loci to have emerged (but see 440). In fact, the variation in regions that have been chosen presents an obstacle to a synthesis of such patterns. Given this relative ignorance, it would be advisable to examine substantially more sequence than is typically done. We have examined the phylogenetic performance of various subsets of our *Papilio* COI–COII data (88) relative to the complete, combined sequences. No commonly sequenced subsets found the same tree reconstructed by the complete data set (which was almost entirely congruent with independent elongation factor-1 alpha [*EF-1α*] sequences; 403). Sequencing larger fragments (more than 1 kb) increases the likelihood that regions of different levels of variation will be spanned, giving resolution for a greater range of divergences.

The picture for nuclear rDNA is one of greater consistency. Sequencing much or all of the 18S rRNA gene has become standard for most higher-level studies, although there are exceptions: The internal transcribed spacer (ITS) and 28S regions predominate in studies of Diptera and Hymenoptera. Generally, complete 18S analyses have supported relationships that are largely congruent with previous hypotheses based on morphology (e.g. 449, 547, 548). The 28S data sets that have been generated have typically been much smaller (less than 500 bp) and comparison of results for this gene is difficult. Given satisfactory results from the 18S data and a substantial existing database, it seems advisable to concentrate future efforts in this region.

Phylogenetic studies utilizing nuclear protein-coding genes are far fewer in number than studies of either mtDNA or nuclear rDNA (see Table 1). However, a few loci are coming into wider use. Of these *EF-1α* has been the most popular. Its sequences have proven very useful for studies among species groups and genera within (sub)families (35, 99, 326, 403), and support relationships congruent with mitochondrial COI-II in *Papilio* (87, 403). However, amino acids in *EF-1α* are highly conserved and third codon positions provide essentially all the phylogenetic information. The studies of Regier & Schultz (405) and Roger et al (415) found that amino acid variation may be informative at the interordinal level and above. Besides *EF-1α*, no single copy nuclear genes have been widely applied, though several of those listed in Table 1 show significant promise (particularly *ddc, g6pdh, hb, per, pepck* and *wg*). The alcohol dehydrogenase (*adh*) locus has been useful in Drosophilidae, but it has not been applied outside of this group. Though nuclear protein-coding genes present a seemingly formidable array of barriers to their simple application (e.g. multiple copies, potentially large introns), these barriers are increasingly surmountable (55).

MOLECULAR SYSTEMATIC METHODS

In this section we review prevailing trends in data generation for molecular systematics. The sections are organized to reflect steps in the design phase of a molecular systematic study. These include taxon sampling strategies, choice of appropriate markers, and analytical issues. All of these steps have been reviewed to a greater or

TABLE 1 Nuclear protein-coding loci that have been used in molecular systematic studies of insects. Taxa are listed in order of decreasing number of existing sequences: Di = Diptera; Le = Lepidoptera; Tr = Trichoptera; Me = Mecoptera; Si = Siphonaptera; Hy = Hymenoptera; He = Hemiptera; Co = Coleoptera; Or = Orthoptera; Th = Thysanura; Cm = Collembola; Ar = Archaeognatha; Bl = Blattodea. For lists of additional potential nuclear loci see (169) and (55).

Locus	Sequences	Taxa	References
α-amylase (*amy*)	38	Di, Co	221, 246, 356, 436
Acetylcholine esterase (*achE*)	21	Di	19, 127
Actin	5	Hy, Le	201, 304
Alcohol dehydrogenase (*adh*)	157	Di	2, 3, 19, 70, 106, 107, 134, 189, 207, 247, 308, 352, 354, 408, 417, 418, 420, 499, 500, 512, 529
Arylphorin (hexamerin family)	12	Le	69, 241, 437
Cecropin family	36	Di	119, 399
Chorion genes: *s18*, *s15*, and *s19*	5	Di	282, 283, 311
Cpnl-1 (anonymous nuclear DNA)	8	Or	110, 510
Dopa decarboxylase (*ddc*)	57	Le	153, 168
Elongation factor-1-alpha (*EF-1α*)	196	Le, Hy, He, Cm, Ar, Bl	35, 99, 166, 168, 326, 333, 403, 405, 526
EF-1α (copy 2)	1	Hy	117
Esterase (*est*)	31	Di	17, 39
Glycerol-3-phosphate dehydrogenase (*gpdh* or *g3pdh*)	39	Di	25, 275, 503, 536, 537
Glycerol-6-phosphate dehydrogenase (*g6pdh*)	13	Di, Le, Th	451
Guanylate cyclase	13	Di	181
Globin family	33	Di	196, 204, 205

continued

TABLE 1 *continued* Nuclear protein-coding loci that have been used in molecular systematic studies of insects. Taxa are listed in order of decreasing number of existing sequences: Di = Diptera; Le = Lepidoptera; Tr = Trichoptera; Me = Mecoptera; Si = Siphonaptera; Hy = Hymenoptera; He = Hemiptera; Co = Coleoptera; Or = Orthoptera; Th = Thysanura; Cm = Collembola; Ar = Archaeognatha; Bl = Blattodea. For lists of additional potential nuclear loci see (169) and (55).

Locus	Sequences	Taxa	References
Histone H1	8	Di	206, 505
Histone H4	7	Hy	201
Hunchback (*hb*)	39	Di	19, 127, 448, 494
Kruppel	3	Di	448
Luciferase	2	Co	355, 502
Lysozyme intron	16	Le	286
Myosin alkali light chain (*Mlc1*) intron	30	Di	102
Nullo	7	Di	71
Opsin	28	Hy, Di	15, 85[1], 86, 306, 388
Period (*per*)	111	Di, Le	111, 189, 190, 267, 346, 374, 375, 404
Phosphoglucose isomerase (*Pgi*)	2	Or	262
Phosphoenolpyruvate carboxykinase (*PEPCK*)	12	Le, Tr, Di, Me, Si	171
Prune	10	Di	439
Resistance to dieldrin (*Rdl*)	17	Co	4
Retrotransposon reverse transcriptase	30	Hy	314
Ribosomal protein 49 (*rp49*)	9	Di	400
Serendiptiy-α	7	Di	71
Cu, Zn - superoxide dismutase (*sod*)	30	Di	25, 277

Sodium channel para locus[1]	27	Le, He, Bl, Co	1, 312, 496
Snail	3	Di	448
Timeless	1	Di	359
Triosephosphate isomerase (*Tpi*)	39	Di	213, 276
Vestigial (*vg*)	5	Di	127
White	15	Di	38
Wingless (*wg*)	57	Le, Di	19, 58, 56, 127
Xanthine dehydrogenase (*xdh*)	37	Di	493
Yolk protein 1 (*yp1*)	55	Di	261, 380
Yolk protein 2 (*yp2*)	40	Di	220, 228
Zeste	24	Di	220

[1] Amino acid sequence only

lesser degree in the past few years. We, therefore, rest heavily on the shoulders of our predecessors (55, 57, 223, 225, 413, 440, 486) and attempt to focus on very recent as well as insect-specific developments.

Sampling

One of the key elements in designing a molecular systematic study is selecting ingroup and outgroup taxa that maximize the investigator's chances of accurately answering the question being posed. While this issue has proven rather contentious, most studies agree that sampling can significantly affect phylogenetic inference. The primary point of contention has been whether better phylogenetic accuracy is attained by densely sampling the taxon of interest (e.g. 195, 222, 223) despite the fact that some of the sampled taxa may be extraneous to the question at hand, or whether more scant sampling may provide better estimates (e.g. 265). The underlying agreement between these opinions is that it is important to sample the diversity of the group of interest evenly. It is well known that very short internal branches are difficult to reconstruct, and that long terminal branches tend to attract each other, at least under the parsimony criterion (157, 486). Therefore, when possible these long branches should be broken up by the addition of taxa or eliminated when not directly relevant to the study. Dense taxon sampling is likely to accomplish the first of these solutions and simulation studies have supported this as the most generally reliable strategy (195, 222). Graybeal's (195) study further showed that addition of taxa to a problem may improve phylogenetic accuracy faster than addition of characters (i.e. more sequence). Beyond topological accuracy, taxon sampling is likely to affect the probability of correctly identifying the root of a tree (422) and the estimation of rates of evolution on the tree (412). Thus, the optimal sampling strategy is inseparable from the goal of the study.

Concerns regarding branch lengths are especially relevant to insect molecular systematics, because major lineages of insects inherently have long branches. Particularly in higher-level studies, i.e. among tribes, families, and orders of insects, thorough taxonomic sampling will be of the utmost importance. Including single examples of highly divergent taxa in such studies is as likely as not to cause problems with phylogeny estimation. This point also serves to amplify our theme: ultimately coordinated efforts across many lower-level studies are more likely to solve higher-level problems in insect systematics than sparse sampling across a diverse array of taxa.

Markers

The modern systematist has a wide variety of techniques at his or her disposal. While DNA sequencing has justifiably become the method of choice for most molecular systematic studies, it should not be forgotten that many alternatives may be more appropriate or more practical for certain applications. The central guiding principle should be the question at hand. The importance of clearly outlining the objectives of a study is paramount. Sequencing is generally the most appropriate for studies at interspecific levels and higher (14, 225). However, questions of intraspecific population structure, species limits, and species diagnosis often can be effectively addressed using time-honored techniques like allozyme electrophoresis, restriction fragment length polymorphisms (RFLP), or any of a variety of newer DNA-based techniques such as single-stranded conformational polymorphisms (SSCP), amplified fragment length polymorphism (AFLP), or random amplified polymorphic DNA (RAPD). The best strategy for certain problems will often use one or more of these in combination with sequencing. Excellent discussions of the problem of choosing markers for a study can be found in Hillis et al (225: Chapters 2 and 12). To their recommendations we would mainly add that consideration be given to maximizing compatibility with previous work. Choosing similar loci, primers, and analytical methods will ensure that a study adds to a body of accumulating knowledge, thus accomplishing more than merely solving the problem at hand. While in some cases it may seem that the objectives of the study are at odds with this goal—that popular markers may not be completely appropriate to the question—some simulation studies have shown that individual markers may be informative across broad ranges of divergence (396). Furthermore, the more knowledge we have of particular loci, the more likely we are to be able to develop weighting schemes or models that make optimal use of observed variation.

DNA Sequences

It is obvious that DNA sequencing has become the dominant technique for generating comparative molecular data. The many desirable properties of sequence data have been reviewed at length elsewhere (14, 225). We reiterate two of these properties here: *(a)* the inherent comparability of sequence data across studies—the "connectibility" of Avise (14)—and *(b)* a unique insight into the evolutionary processes driving the diversification of DNA itself. Automated sequencing facilities are accessible to most biologists within a reasonable budget, and appropriate markers have been developed for most evolutionary applications.

mtDNA Mitochondrial DNA is by far the most widely used of DNA regions for insects as well as for animals in general (225). For detailed discussions of the general properties of mtDNA, see the reviews of Avise (13, 14), Harrison (209) and Simon et al (440). The popularity of mtDNA markers derives in large part from its relative ease of isolation and amplification, even from marginally preserved specimens. It is also amenable to straightforward analyses, individuals typically being homoplasmic and part of a strictly dichotomous lineage of haplotypes. Heteroplasmy has been reported only infrequently (e.g. 528). One of the main developments in recent work on mtDNA is the documentation of paralogous nuclear copies of mitochondrial genes, or pseudogenes (482, 565). These pseudogenes may seriously confound phylogenetic analyses. Given our near ignorance regarding their frequency and distribution in insects, Zhang & Hewitt's precautions (565) should be kept in mind whenever one is working with mtDNA.

A synthesis of the information content of various mitochondrial loci is difficult for the reasons presented in the preceding section. Very few studies have explicitly compared the resolving power of different mitochondrial loci over a particular set of taxa. In those that have, topological incongruence is common (e.g. 26, 88, 520), sometimes significantly so (e.g. 19). However, no clear patterns among loci have become apparent. What is clear is the degree to which variation depends on structural constraints in both protein coding (88, 295, 440) and rRNA genes (163). Thus, individual genes do not exhibit an overall rate of evolution, but comprise both conserved and variable regions. Variation in the mitochondrial control region (= AT rich region) has been reviewed (71, 497, 566, 567) and has been found to be problematic for analyses at any level. Additional studies of variation patterns within other mitochondrial genes would be very useful.

Nuclear DNA Due to the recognition that mitochondrial gene trees may represent only a partial and potentially biased view of organismal phylogeny, numerous nuclear genes have been developed for arthropod phylogenetics (55, 56, 169, 170; see also 194, 315). Nuclear ribosomal RNA genes have been especially widely used owing to their abundance within a genome and concomitant ease of amplification and sequencing. The nuclear ribosomal repeat (an array containing the 18S, 5.8S, and 28S genes, separated by transcribed and nontranscribed spacers) are mosaics of highly conserved and variable regions (224). While the conserved regions have been informative for resolving relationships at higher taxonomic levels, alignment of the variable regions is generally problematic (315, 524, 555). With sufficient time, concerted evolution effectively homologizes the many copies of nuclear rDNA within a genome (224), but some genomes have been shown to retain a considerable diversity of paralogous sequences (67, 498). Nonetheless, rDNA sequences have proven highly informative for phylogenetics. A compilation of sequences and secondary structures has recently been posted to the World Wide Web (http://rrna.uia.ac.be/ssu/; 122, 507).

Nuclear protein-coding genes exhibiting a wide range of evolutionary rates have recently been assessed for phylogenetic utility. These loci show great promise for resolving relationships at levels too conserved to be easily examined with mitochondrial proteins. The inherent rate variation among codon positions in protein coding genes suggests that many of these loci may be effective for more recent divergences as well (56, 99, 326, 403). Furthermore, the relatively unbiased nucleotide composition of most nuclear loci compared to mtDNA should lead to a higher saturation threshold. However, working with nuclear genes poses potential difficulties with heterozygosity, and low copy number may hinder amplification. In addition, many nuclear genes may contain large introns that necessitate reverse transcriptase PCR (RT-PCR; e.g. 153), and many purportedly single-copy loci may be represented by one or more paralogues in other groups (for examples in *EF-1*α see 117, 527).

Noncoding Markers Seeking variation, which may often be elusive, systematists and population biologists are delving increasingly into the vast noncoding portions of the genome. Among the significant insights gained in this effort has been the extent to which noncoding regions comprise repetitive, or satellite, DNA: as much as 45% of the entire genome in some cases (387). Because much of this satellite DNA is localized in the centromeric regions of chromosomes, and may exhibit highly conserved higher order structure, it is becoming apparent that satellite DNA plays an important role in chromosome structure and function (384, 506). Interestingly, the quantity of minisatellite DNA in the genome appears to vary with reproductive strategy. For example, parthenogenic *Bacillus* walking sticks harbor only 10% to 30% of the total satellite DNA of close bisexual relatives (305).

Microsatellites, named for the very short repeat units of which they are composed (dinucleotide repeat units being the most common), have come into widespread use for genetic analysis at or below the species level (397). Their generally high mutation rate ensures high levels of polymorphism (101; but see 430) and they have therefore been particularly useful for examining relationships among individuals and breeding groups within insect populations. Development of microsatellite loci for a particular organism is rather labor intensive, but there is some indication that loci may be more broadly applicable than originally thought (151). In addition, methods have been developed recently to screen for microsatellite loci without creating a complete genomic library (147). Nonetheless, it appears that microsatellite loci may be substantially more difficult to develop in some insects, such as butterflies (318), than in others. Particularly exciting applications of microsatellites have focused on testing hypotheses of kin selection and behavior as they covary with individual relatedness in social insects (e.g. 150, 357). A surprising result of these studies is that our ability to discern relationships in many cases exceeds that of the insects themselves and, therefore, behaviors predicted by kin selection are not realized (6, 447). Microsatellites have also provided fascinating insights into sperm competition and paternity (109), the origins of sociality (94), and population structure (281). A recent study of the four species of the *Drosophila melanogaster* species complex (a phylogenetic enigma on par with the gorilla-chimpanzee-human trichotomy) indicates excellent promise for the resolution of interspecific relationships using multiple microsatellites (208).

Increasingly, introns within single-copy nuclear protein-coding loci are being applied to systematic studies. Though these noncoding regions are generally perceived to be highly variable relative to the coding portions of their host genes (365), this is, in fact, a controversial point. Moriyama & Powell (339) found lower average polymorphism in noncoding introns than among silent sites in adjacent coding regions (across 24 loci). However, Villablanca et al (511) have demonstrated significantly higher levels of variation in introns than in either isozymes or mitochondrial DNA for invasive Mediterranean fruit fly (*Ceratitis capitata*) populations. The disagreement may simply be a result of the more variable introns having been developed and used by systematists. It should, nonetheless, be emphasized that higher variability may not be a general property of introns. Palumbi (365) offers a selection of potentially useful intron loci and primers. In addition to Villablanca et al's (511) study of the population structure of a recently invading species, introns have been successfully applied to problems of native population structure in yucca moths (286) and interspecific phylogeny in *Spodoptera* (Noctuidae; 1).

Gene Arrangements

As large amounts of DNA data have accumulated it has become apparent that the organization of the genome itself may contain substantial phylogenetic information. The value of mitochondrial gene rearrangements in arthropods was first pointed out by Boore et al (46) in a study that appeared to support monophyly of the chelicerates, crustaceans, hexapods, and myriapods to the exclusion of the Onychophora. Such rearrangements are likely quite rare and thus would seem to be excellent potential characters, though some parallelisms have been shown (e.g. 141, 324).

Allozymes

Berlocher's (37) review of insect molecular systematics was written during the heyday of enzyme electrophoresis and this technique accordingly dominated his treatment. The intervening years have seen a steady decline in the application of allozymes to systematic problems at all levels. Although now used infrequently for the reconstruction of interspecific phylogenies (but see 155, 307, 313, 353, 461), allozymes have maintained a foothold in studies of population structure and species limits (e.g. 218, 361, 467). The essential assumption that enzyme homology could be inferred from electrophoretic mobility has been criticized, but techniques such as sequential electrophoresis have been developed to deal with this problem (37, 410) and allozymes remain an inexpensive and effective method for the generation of multi-locus, diallelic (codominant) data sets (14, 343, 501).

RFLP

As sequencing becomes common, data generated via restriction enzymes has largely disappeared from phylogenetic studies of insects, although the technique is still commonly applied in plant studies. Provided that the restriction sites are mapped, however, they still offer a quick alternative for studying sequence variation across broad cross sections of the genome, and whole mitochondrial digestions have proven very useful (e.g. 452, 456) and reasonably accurate (87) for phylogeny reconstruction. The most common use for restriction enzymes in systematics is for the diagnosis of otherwise cryptic taxa. In this application, diagnostic digestion profiles are sought within amplified DNA fragments (most frequently mtDNA) either with sequencing (e.g. 217, 453, 458, 459) or without sequencing (e.g. 184, 200, 441) first to identify diagnostic sites directly.

RAPD

RAPDs (arbitrarily primed PCR, or AP-PCR; 538) burst onto the systematic stage in 1991 (551) and immediately achieved wide use. The sudden access to variability in otherwise monomorphic or in previously unstudied taxa assured RAPDs a large audience. However, many weaknesses of the RAPD technique quickly became apparent (42, 215). Most obviously problematic is the assumption that comigrating PCR products represent homologous pieces of DNA, particularly given that RAPD markers are expected largely to come from repetitive regions of the genome (215). A more insidious problem is the difficulty of replicating results under even slightly different PCR conditions. Unpredictable variation in specimen preservation will also affect results, as will endosymbionts, parasites and phoretics. The confident use of RAPDs requires, as an absolute minimum, scrupulous standardization of all conditions of collection, extraction, and amplification; sample sizes sufficient to confirm consistency; and some check of band homology, whether through sequencing, SSCP (e.g. 433, 509), or hybridization experiments (e.g. 104). Given these cautions, however, RAPDs can provide a quick and inexpensive method of generating comparative genetic data. Their value in the diagnosis of pest host races, biotypes, and species has been considerable (216, 389).

SSCP (and DSCP)

Among the techniques that identify sequence polymorphism without sequencing, SSCP (226) and its cousin double-stranded conformational polymorphism (DSCP; 8) are becoming increasingly popular. These techniques rely on differences in DNA secondary structures owing to variation in primary sequence and consequent variation in rates of electrophoretic mobility. Under appropriate conditions, SSCP is sensitive to single base pair differences (214). In entomological applications, SSCP is typically used in conjunction with PCR amplification of a particular fragment for the purpose of species diagnosis (e.g. 226, 476). SSCP also provides a quick method of screening for polymorphisms among populations and species that can then be sequenced for phylogenetic purposes (286). A particularly interesting application employs SSCP as an alternative to cloning for the isolation of nuclear alleles (358).

AFLP

AFLP is a recently developed technique that, like RAPDs, quickly locates widely dispersed genetic polymorphisms among organisms (see 43 for a recent review). Techniques for isolating insect DNA for AFLP applications have been evaluated in a recent study by Reineke et al (406). Despite apparent promise for entomological applications, AFLP has not been widely applied in studies of insect systematics (but see 316).

ANALYTICAL ISSUES

Phylogenetic Methods

Methods for reconstructing phylogenies have been reviewed at length, most cogently by Swofford et al (486). A few advances that are especially pertinent to insect phylogenetics, however, are worth highlighting. Sequence alignment constitutes the primary assessment of homology among sequence data sets and has accordingly figured prominently in discussions of phylogenetic methodology. In most studies of protein-coding genes, alignment proves straightforward as long as amino acid sequence is conserved (though introns pose potential problems). However, the variable regions of ribosomal genes may be substantially more difficult. Methods developed to deal with this problem include incorporating information on phylogeny (323, 541) and secondary structure (266) during alignment.

Many analytical advances derive from our increasing ability to incorporate knowledge of molecular evolution in tree reconstruction. While some have argued that it is best to let the data itself guide our analyses (e.g. 55), this goal can be achieved through many analytical techniques. Insect DNA exhibits many potentially confounding features (e.g. AT bias in mtDNA; rate heterogeneity in rDNA evolution). The ability to accommodate, as well as gain a deeper understanding of, these biases in the process of tree reconstruction is an important strength of model-based methods such as maximum likelihood and distance methods. Because model-based methods can account for unobserved substitutions, these techniques may also extend the range of phylogenetic utility of particular loci.

The tremendous increase in computational power available to the systematist has itself been an important advance. The ease of implementation of many methods through programs like Phylogenetic Analysis Using Parsimony (PAUP; 484, 485), the Phylogenetic Inference Package (PHYLIP; 159), MacClade (301), and Phylogenetic Analysis by Maximum Likelihood (PAML; 561), has been an important factor in broadening our analytical perspectives. Continued disagreement over the merits of various approaches is in part rendered moot by the ease of performing multiple analyses and comparing the results. Indeed, concordance among analyses has been considered strong support for tree topologies (116).

Gene Trees Versus Species Trees

Phylogenetic analysis of a particular locus produces a so-called gene tree which may or may not agree with the species phylogeny (14, 145, 300). Although this concern may have been overly emphasized (57), incongruence among gene trees is expected under certain circumstances: horizontal transfer (including hybridization), lineage sorting (deep coalescence), and gene duplication coupled with extinction (144, 300). Due to their smaller effective population sizes, mitochondrial genes may more reliably track short internodes than nuclear autosomal genes for recent divergences (330). However, it should be kept in mind that in species with polygynous mating systems and sex-biased dispersal (e.g. female philopatry), mtDNA may exhibit longer coalescence times than nuclear alleles do (231, 331). Specific methods have been proposed to accommodate a variety of such complications under the view that a species tree may be considered to be a cloud of gene histories (300). Reconciliation methods for inferring species trees from multiple gene trees have been implemented in the program GeneTree (363).

To Combine or Not to Combine

There has been much discussion of whether independent data sets bearing on phylogenetic relationships of the same taxa should be analyzed separately or combined and analyzed simultaneously (20, 68, 121, 127, 239, 268, 329, 348). Combining data sets can give misleading results if there is heterogeneity among data sets with respect to some property that affects phylogeny estimation (121). Opinions about combining data range from "rarely" (329) to "always" (268). In practice, however, the trend is toward conditional data combination based on levels of congruence among data partitions, and Cunningham (116) has shown that congruence and phylogenetic accuracy are generally correlated. Combinability is most commonly assessed by the incongruence length difference (ILD) test (equivalent to the partition homogeneity test; 156). In addition, Reed & Sperling (403) have developed a graphical method of detecting incongruence. While occasional incongruence may be expected, it is not generally clear how to treat incongruent partitions. Ballard et al (23) recently outlined a compatible evidence (CE) algorithm for accommodating otherwise incongruent partitions. Similarly, Wiens (549) proposed a method for analyzing partially combinable partitions. Even in cases where a simultaneous analysis is to be preferred on theoretical grounds, it is valuable to study incongruence and support contributed by various partitions (19, 403, 408).

One aspect of combinability that has only begun to be discussed is how to assemble the phylogenetic branches we are each working on to reconstruct a "supertree" of life. One approach is that of the Maddisons' Tree of Life (http://phylogeny.arizona.edu/tree/phylogeny.html) in which independent studies are essentially grafted together at their roots. Sanderson et al (423) called for a database specifically to house phylogenetic trees as well as the data used to generate them. This idea has since developed into TreeBASE (http://herbaria.harvard.edu/treebase/; 335, 424) which, its authors hope, will serve as a development platform for methods of combining partially overlapping data sets for producing supertrees (425).

USING PHYLOGENIES: COMPARATIVE METHODS

As reflected by the emphasis of this review, much of molecular systematists' effort has been devoted to exploring the methodological aspects of molecular phylogenetics. Massive quantities of sequence data have provided both obstacles and opportunities. Methods for reconstructing and assessing confidence in molecular trees have steadily improved and emphasis has begun to shift toward putting these trees to work. The power of a phylogenetic perspective has become widely apparent and systematists, as well as nonsystematists, are using molecular trees to address a wide variety of evolutionary questions. This broader awareness of phylogenetic issues has spurred substantial discussion and development of methods for testing evolutionary scenarios in light of phylogenetic hypotheses (see 51, 211, 301). Such methods include mapping ecological traits onto a phylogeny, testing for evolutionary correlations among trait changes, and comparing phylogenies among symbiotic organisms. Comparative methods are highly diverse in their applications and properties and are themselves in need of a comprehensive review. In this section we illustrate some evolutionary problems that have been particularly well-served by a molecular phylogenetic approach.

Organismal Evolution

Molecular phylogenies have provided the means for examining many longstanding issues in insect behavior and ecology. For example, the principle of coevolution (broadly defined) has now been examined from many perspectives. Through phylogenetic investigations it has been shown that associations between organisms have, in many cases, persisted through considerable periods of diversification. Significant phylogenetic concordance has been found between pocket gophers and their lice (199, 362), between aphids and their endosymbiotic bacteria (332, 334), and between some herbivorous insects and their host plants (e.g. 155, 414) to name a few examples. By examining relationships among social insects and their relatives, molecular trees have shed light on the origins of sociality in several different groups, including thrips (94) and wasps (431, 432), as well as the evolution of caste diversity within certain social groups (112, 468, 470). Molecular phylogenies have enabled tests of Emery's "rule" of social parasitism (that social parasites are closely related to their hosts), with controversial results. Baur et al's (29, 30) work on myrmecine ants seems to support this rule, though it does not appear to be the case in polistine vespids (82, 83, 100) or in allodapine bees (294). Other interesting ecological applications have included studying the evolution of Mullerian mimicry in *Heliconius* butterflies (53, 54), gall induction by aphids (469), host plant breadth in swallowtail butterflies (455), and the relationship between defensive chemistry and aposematic coloration in tiger beetles (517). Behavioral traits are well-suited to investigation via molecular phylogenies and significant advances have been made regarding the evolution of mating systems (in crickets: 210; weevils: 349, 351; and walking sticks: 305, 426), parental investments in *Drosophila* (383) and nest construction in social bees and wasps (e.g. 360). Finally, robust molecular phylogenies have provided the framework necessary for studying the factors underlying adaptive radiations (19, 185, 261)

Molecular Evolution

Primarily through studies of *Drosophila melanogaster* and a few close relatives, insect molecular phylogenies promise profound, though as yet largely unrealized, insight into molecular evolution. Species of *Drosophila* exhibit surprising variation in characters such as genome size, karyotype, gene linkage, and nucleotide and codon biases. This database provides an unparalleled resource for studying changes in organisms' genetic makeup. Powell & DeSalle (393) illustrate several examples of the types of insights that *Drosophila* studies can offer, including lability of introns and transposable elements and changes in chromosome number and arrangement. Robertson et al (411) also review what is known about *mariner* transposon distribution and behavior. Additional phylogenetic studies are required to assess the direction, order, and frequency of changes in nucleotide biases, gene rearrangements, and overall genomic economy. While several papers have offered intriguing glimpses of explanations of some of these (e.g. 338–340), a true synthesis will require a broader comparative approach. A particularly fruitful application of molecular phylogenetics has been the examination of relationships among proteins in multigene families. This has followed the recognition that much of the protein coding portion of the genome derives from duplication and specialization of existing loci (392). Several insect systems have helped to illustrate the generality of this phenomenon, including the globin gene family in midges (196), cytochrome P450 detoxification enzymes in Papilionidae (36) and especially immune system–associated proteins (240). Finally, molecular systematics and the study of development are undergoing considerable reciprocal illumination as we gain insight into the ontogenetic functions of many phylogenetically useful loci (e.g. *wg:* 56; *ddc:* 153, 269).

Conservation

Undoubtedly, one of the major changes in the past few years has been the decline in the number of species on the planet to study. It is well worth considering the contributions molecular systematics has made toward biodiversity preservation. Moritz (336) recognizes two basic roles for studies of molecular data in conservation efforts: understanding phylogeographic and demographic patterns. Good examples of both approaches exist for endangered or potentially endangered arthropods (e.g. 10, 50, 398, 516). These studies offer significant insight into the population structures of their respective species. Intraspecific phylogenies also offer the potential of estimating population sizes (158) and rates of population growth or decline (344) via coalescence theory, although with a coalescence approach caution is warranted (337). In addressing serious issues such as conservation we should be careful not to overstate the abilities of molecular techniques (47). One study in particular illustrates the limitations. Legge et al (287) applied allozymes and mitochondrial sequences to a study of the genetic distinctness of Cryan's buckmoth (*Hemileuca* sp.), a host-plant–specific saturniid. Despite a unique ability to survive on its host, they found no genetic differences from more widespread species. This case may be anomalous, but at this point it stands with only a few studies of threatened insects. More work in this area is needed.

OTHER ISSUES

Preserving Specimens for Molecular Study

A substantial amount of discussion has centered on how best to preserve nucleic acids for molecular study. Comparative studies on insect DNA, in addition to our own experiences, have shown that ultracold freezing ($-80°$ C) of live specimens is by far the most effective method (133, 391, 407). Single-copy nuclear genes are easily amplified from such frozen specimens. Proteins for allozyme analysis are preserved as well. Preservation with ethanol (100%), as well as fast-drying methods using silica gel, critical-point drying, or air-drying in a dessicator, have also proven to be effective, although they are questionable for long-term storage. General guidelines for preserving specimens for phylogenetic study have been developed (89, 225, 237). Huber (237) also emphasized the need for voucher specimens to be available, identifiable, and traceable to collection locality (country, state/province, city, or GPS coordinates) and date. Ideally, some part of the specimen should remain frozen (or otherwise preserved) to permit replication of the entire experiment, if necessary. There has been some discussion regarding museums' roles in molecular systematics (e.g. 545) and we agree there is a need for development of a general policy among collections with respect to frozen collections, voucher deposition, and destructive sampling, especially given the possibility that museums may house species that no longer exist in nature.

Ancient DNA

Given the difficulties in extracting DNA from extant species, the likelihood of getting amplifiable DNA from fossils would seem very low. However, so-called ancient DNA has been the subject of much scientific and popular discussion. Initial (prepublication) reports inspired the "bioscifi" book and subsequent 1992 movie "Jurassic Park". This preceded a flurry of papers on DNA from amber-entombed insects including a 30-million-year-old (MYO) termite (129), a 30 MYO bee (78) and a 120 MYO weevil (77). However, the veracity of these reports has been seriously challenged (11, 12, 198, 233, 465, 525) and the challenges have themselves been challenged (18, 385). The recent successful characterization of Neanderthal DNA based on multiple controls (273) has set new standards for the field (108). Hopefully, the insect studies will stand the test.

CONCLUSIONS

In this review we have tried to forge a cohesive view of insect systematics as a field driven by questions of organismal evolution. If "nothing in biology makes sense except in light of evolution," it is the job of systematists to provide this light. Coherence in insect systematics will ultimately depend on having a large database of homologous data. Currently, exploring a variety of markers is advantageous. However, direct comparisons among them should be requisite. It is fantasy to think that we will eventually fill in the gaps through random sequencing and that our studies will grow together and eventually fuse. It is necessary that we consciously work toward this goal.

How can we accomplish this fusion? We believe that our goal should be to identify a relatively small set of markers to serve as standards for comparison. At this point, four sequencing markers stand out as both well-surveyed and informative across a range of divergences: the mitochondrial COI and 16S genes, and the nuclear 18S and *EF-1*α genes. We have reservations about *EF-1*α: It is already known to be problematic for some groups due to multiple copies. However, it may be impossible to identify a nuclear protein-coding gene that is truly single copy across the whole of insect diversity. Those that have been tested need to be examined over a broader selection of taxa. While offering such concrete recommendations may seem premature, even presumptuous, we do not expect that ours will be the final word on the subject. We hope that the ideal of standards, at least, is palatable.

Ultimately, the essence of science is repeatability, and in its interest we also take this opportunity to reiterate some basics. We had surprising difficulty identifying the regions and taxa sequenced for many studies (Figures 1 and 2). These basic data need to be made clear in all published works, including loci, position numbers relative to some standard (e.g. *D. yakuba* for mtDNA, *D. melanogaster* for nearly everything else), primers used, number of specimens sequenced, exactly which sequences were used for phylogenetic analyses, etc. Posting sequences to GenBank (or EMBL/DDBJ) has become a prerequisite for publication in certain journals (e.g. *Molecular Phylogenetics and Evolution*); this practice must be a universal standard. Any publication for which the data are not available for inspection must be regarded as incomplete. Molecular systematists have been particularly lax regarding voucher specimens from their work. Being able to verify specimen identifications is absolutely essential. All of these considerations are the bare minimum to ensure the repeatability of molecular phylogenetic studies. We also endorse and encourage participation in communal efforts to summarize phylogenetic data and hypotheses such as the Tree of Life and TreeBASE.

In summary, the field of insect molecular systematics is flourishing. But the time for deciding whether we do, in fact, constitute a single community with a common goal is now. Many signs point toward fragmentation and factionalization, and this trend must be reversed. In this review we have tried to identify the points of commonality among our varied efforts and we hope that these will serve as the foundation for new standards. Only through such coordinated efforts shall we cease to babble and build our own tower.

ACKNOWLEDGMENTS

We thank Anthony Cognato, Joshua Herbeck, James Kruse, Marilyn Myers, and Daniel Rubinoff for providing helpful feedback on this manuscript. This work was supported in part by NSF-PEET and California Agriculture Experiment Station grants to F.A.H.S.

LITERATURE CITED

1. Adamczyk JJ, Silvain JF, Pashley Prowell D. 1996. Intra- and interspecific DNA variation in a sodium channel intron in *Spodoptera* (Lepidoptera: Noctuidae). *Ann. Entomol. Soc. Am.* 89:812–21
2. Albalat R, Gonzalez-Duarte R. 1993. *Adh* and *adh*-dup sequences of *Drosophila lebanonensis* and *Drosophila immigrans:* interspecies comparisons. *Gene* 126(2): 171–78
3. Anderson CL, Carew EA, Powell JR. 1993. Evolution of the *adh* locus in the *Drosophila willistoni* group the loss of an intron and shift in codon usage. *Mol. Biol. Evol.* 10(3):605–18
4. Andreev D, Breilid H, Kirkendall L, Brun LO, ffrench-Constant RH. 1998. Lack of nucleotide variability in a beetle pest with extreme inbreeding. *Insect Mol. Biol.* 7(2):197–200
5. Aoki S, Von Dohlen CD, Kurosu U, Ishikawa, H. 1997. Migration to roots by first-instar nymphs, and not by alates, in the gall aphid *Clydesmithia canadensis. Naturwissenschaften* 84:35–36
6. Arévalo E, Strassmann JE, Queller DC. 1998. Conflicts of interest in social insects: male production in two species of *Polistes. Evolution* 52:797–805
7. Arias MC, Sheppard WS. 1996. Molecular phylogenetics of honey bee subspecies (*Apis mellifera* L.) inferred from mitochondrial DNA sequence. *Mol. Phylogenet. Evol.* 5:557–66
8. Atkinson L, Adams ES. 1997. Double-strand conformation polymorphism (DSCP) analysis of the mitochondrial control region generates highly variable markers for population studies in a social insect. *Insect Mol. Biol.* 6:369–76
9. Aubert J, Barascud B, Descimon H, Michel F. 1996. Systématique moléculaire des *Argynnes* (Lepidoptera: Nymphalidae). *C. R. Acad. Sci. III* 319:647–51
10. Aubert J, Barascud B, Descimon H, Michel F. 1997. Ecology and genetics of interspecific hybridization in the swallowtails, *Papilio hospiton* Géné and *P. machaon* L., in Corsica (Lepidoptera: Papilionidae). *Biol. J. Linn. Soc.* 60:467–92
11. Austin JJ, Smith AB, Thomas RH. 1997. Palaeontology in a molecular world: the search for authentic ancient DNA. *Trends Ecol. Evol.* 12(8):303–6
12. Austin JJ, Ross AJ, Smith AB, Fortey RA, Thomas RH. 1997. Problems of reproducibility—does geologically ancient DNA survive in amber-preserved insects? *Proc. R. Soc. London Ser. B.* 264:467–74
13. Avise JC. 1987. Intraspecific phylogeography: the mitochondrial DNA bridge between population genetics and systematics. *Annu. Rev. Ecol. Syst.* 18:489–522
14. Avise JC. 1994. *Molecular Markers, Natural History and Evolution.* New York: Chapman & Hall. 511 pp.
15. Ayala FJ, Chang BS, Hartl DL. 1993. Molecular evolution of the *rh3* gene in *Drosophila. Genetica* 92(1):23–32
16. Ayala FJ, Wetterer JK, Longino JT, Hartl DL. 1996. Molecular phylogeny of *Azteca* ants (Hymenoptera: Formicidae) and the colonization of *Cecropia* trees. *Mol. Phylogenet. Evol.* 5(2):423–28
17. Babcock CS, Anderson WW. 1996. Molecular evolution of the sex-ratio inversion complex in *Drosophila pseudoobscura:* analysis of the esterase-5 gene region. *Mol. Biol. Evol.* 13(2):297–308
18. Bada JL, Wang XYS, Hamilton H. 1999. Preservation of key biomolecules in the fossil record: current knowledge and future challenges. *Philos. Trans. R. Soc. London Ser. B.* 354(1379):77–86
19. Baker RH, DeSalle R. 1997. Multiple sources of character information and the phylogeny of Hawaiian drosophilids. *Syst. Biol.* 46(4):654–73

20. Baker RH, Yu X, DeSalle R. 1998. Assessing the relative contribution of molecular and morphological characters in simultaneous analysis trees. *Mol. Phylogenet. Evol.* 9(3):427–36
21. Ballard JWO. 1994. Evidence from 12S ribosomal RNA sequences resolves a morphological conundrum in *Austrosimulium* (Diptera: Simuliidae). *J. Aust. Entomol. Soc.* 33(2):131–35
22. Ballard JWO, Ballard O, Olsen GJ, Faith DP, Odgers WA, et al. 1992. Evidence from 12S ribosomal RNA sequences that onychophorans are modified arthropods. *Science* 258(5086):1345–48
23. Ballard JWO, Thayer MK, Newton AF, Grismer ER. 1998. Data sets, partitions, and characters: philosophies and procedures for analyzing multiple data sets. *Syst. Biol.* 47:367–96
24. Barciszewska MZ, Gawronski A, Szymanski M, Erdmann VA. 1995. The primary structure of *Harpalus rufipes* 5S ribosomal RNA: a contribution for understanding insect evolution. *Mol. Biol. Rep.* 21:165–67
25. Barrio E, Ayala FJ. 1997. Evolution of the *Drosophila obscura* species group inferred from the *gpdh* and *sod* genes. *Mol. Phylogenet. Evol.* 7:79–93
26. Barrio E, Latorre A, Moya A. 1994. Phylogeny of the *Drosophila obscura* species group deduced from mitochondrial DNA sequences. *J. Mol. Evol.* 39(5):478–88
27. Baur A, Buschinger A, Zimmermann FK. 1993. Molecular cloning and sequencing of 18S rDNA gene fragment from six different species. *Soc. Insects* 40:325–35
28. Baur A, Buschinger A, Zimmermann FK. 1993. Internal transcribed spacer (ITS-1) in ants (Hymenoptera: Formicidae): high homologies within tribes, high diversity between tribes. *GenBank.* http://www.ncbi.nlm.nih.gov
29. Baur A, Chalwatzis N, Buschinger A, Zimmermann FK. 1995. Mitochondrial DNA sequences reveal close relationships between social parasitic ants and their host species. *Curr. Genet.* 28:242–47
30. Baur A, Sanetra M, Chalwatzis N, Buschinger A, Zimmermann FK. 1996. Sequence comparisons of the internal transcribed spacer region of ribosomal genes support close relationships between parasitic ants and their respective host species (Hymenoptera: Formicidae). *Insectes Soc.* 43(1):53–67
31. Beard CB, Hamm DM, Collins FH. 1993. The mitochondrial genome of the mosquito *Anopheles gambiae:* DNA sequence, genome organization, and comparisons with mitochondrial sequences of other insects. *Insect Mol. Biol.* 2:103–24
32. Beckenbach AT, Wei YW, Liu H. 1993. Relationships in the *Drosophila obscura* species group inferred from mitochondrial cytochrome oxidase II sequences. *Mol. Biol. Evol.* 10(3):619–34
33. Bellows TS, Perring TM, Gill RJ, Headrick DH. 1994. Description of a species of *Bemisia* (Homoptera: Aleyrodidae). *Ann. Entomol. Soc. Am.* 87(2):195–206
34. Belshaw R, Fitton M, Herniou E, Gimeno C, Quicke DLJ. 1998. A phylogenetic reconstruction of the Ichneumonoidea (Hymenoptera) based on the D2 variable region of 28S ribosomal RNA. *Syst. Entomol.* 23:109–23
35. Belshaw R, Quicke DLJ. 1997. A molecular phylogeny of the Aphidiinae (Hymenoptera: Braconidae). *Mol. Phylogenet. Evol.* 7:281–93
36. Berenbaum MR, Favret C, Schuler MA. 1996. On defining "key innovations" in an adaptive radiation: cytochrome P450s and Papilionidae. *Am. Nat.* 148(Suppl.):S139–55
37. Berlocher SH. 1984. Insect molecular systematics. *Annu. Rev. Entomol.* 29:403–33
38. Besansky NJ, Fahey GT. 1997. Utility of the white gene in estimating phylogenetic relationships among mosquitoes (Diptera: Culicidae). *Mol. Biol. Evol.* 14(4):442–54
39. Besansky NJ, Lehmann T, Fahey GT, Fontenille D, Braack LEO, et al. 1997. Patterns

of mitochondrial variation within and between African malaria vectors, *Anopheles gambiae* and *An. arabiensis,* suggest extensive gene flow. *Genetics* 147(4): 1817–28
40. Besansky NJ, Powell JR, Caccone A, Hamm DM, Scott JA, Collins FH. 1994. Molecular phylogeny of the *Anopheles gambiae* complex suggests genetic introgression between principal malaria vectors. *Proc. Natl. Acad. Sci. USA* 91(15):6885–88
41. Beuning LL, Wu E, Murphy P, Charles J, Morris BAM. 1997. A comparison of the ITS region of four mealybug species. *GenBank.* http://www.ncbi.nlm.nih.gov
42. Black WC. 1993. PCR with arbitrary primers: approach with care. *Insect Mol. Biol.* 2:1–5
43. Blears MJ, De Grandis SA, Lee H, Trevors JT. 1998. Amplified fragment length polymorphism (AFLP): a review of the procedure and its applications. *J. Ind. Microbiol. Biotechnol.* 21(3):99–114
44. Bogdanowicz SM, Schaefer PW, Harrison RG. 1998. MtDNA variation among worldwide populations of gypsy moths. *GenBank.* http://www.ncbi.nlm.nih.gov
45. Bogdanowicz SM, Walner WE, Bell J, Odell TM, Harrison RG. 1993. Asian gypsy moths (Lepidoptera: Lymantriidae) in North America: evidence from molecular data. *Ann. Entomol. Soc. Am.* 86:710–15
46. Boore JL, Collins TM, Stanton D, Daehler LL, Brown WM. 1995. Deducing the pattern of arthropod phylogeny from mitochondrial DNA rearrangements. *Nature* 376(6536):163–65
47. Bossart JL, Pashley Prowell D. 1998. Genetic estimates of population structure and gene flow: limitations, lessons and new directions. *Trends Ecol. Evol.* 13: 202–6
48. Boulding EG. 1998. Molecular evidence against phylogenetically distinct host races of the pea aphid (*Acyrthosiphon pisum*). *Genome* 41(6):769–75
49. Bourgoin T, Steffen-Campbell JD, Campbell BC. 1997. Molecular phylogeny of Fulgoromorpha (Insecta, Hemiptera, Archaeorrhyncha). The enigmatic Tettigometridae: evolutionary affiliations and historical biogeography. *Cladistics* 13:207–24
50. Brookes MI, Graneau YA, King P, Rose OC, Thomas CD, Mallet JLB. 1997. Genetic analysis of founder bottlenecks in the rare British butterfly *Plebejus argus. Conserv. Biol.* 11:648–61
51. Brooks DR, McLennan DA. 1991. *Phylogeny, Ecology, and Behavior: A Research Program in Comparative Biology.* Chicago: Univ. Chicago Press. 434 pp.
52. Brower AVZ. 1994. Phylogeny of *Heliconius* butterflies inferred from mitochondrial DNA sequences (Lepidoptera: Nymphalidae). *Mol. Phylogenet. Evol.* 3:159–74
53. Brower AVZ. 1994. Rapid morphological radiation and convergence among races of the butterfly *Heliconius erato* inferred from patterns of mitochondrial DNA evolution. *Proc. Natl. Acad. Sci. USA* 91:6491–95
54. Brower AVZ. 1996. Parallel race formation and the evolution of mimicry in *Heliconius* butterflies: a phylogenetic hypothesis from mitochondrial DNA sequences. *Evolution* 50:195–221
55. Brower AVZ, DeSalle R. 1994. Practical and theoretical considerations for choice of a DNA sequence region in insect molecular systematics, with a short review of published studies using nuclear gene regions. *Ann. Entomol. Soc. Am.* 87:702–16
56. Brower AVZ, DeSalle R. 1998. Patterns of mitochondrial versus nuclear DNA sequence divergence among nymphalid butterflies: the utility of *wingless* as a source of characters for phylogenetic inference. *Insect Mol. Biol.* 7:73–82
57. Brower AVZ, DeSalle R, Vogler AP. 1996. Gene trees, species trees, and systematics: a cladistic perspective. *Annu. Rev. Ecol. Syst.* 27:423–50
58. Brower AVZ, Egan MG. 1997. Cladistics of *Heliconius* butterflies and relatives (Nymphalidae: Heliconiiti): the phyloge-

netic position of *Eueides* based on sequences from mtDNA and a nuclear gene. *Proc. R. Soc. London Ser. B* 264:969–77

59. Brown JM, Abrahamson WG, Way PA. 1996. Mitochondrial DNA phylogeography of host races of the goldenrod ball gallmaker, *Eurosta solidaginis* (Diptera: Tephritidae). *Evolution* 50(2):777–86
60. Brown JM, Leebens-Mack JH, Thompson JN, Pellmyr O, Harrison RG. 1997. Phylogeography and host association in a pollinating seed parasite *Greya politella* (Lepidoptera: Prodoxidae). *Mol. Ecol.* 6(3):215–24
61. Brown JM, McPeek MA, May ML. 1998. Phylogeny of the North American *Enallagma* based on variation in mitochondrial cytochrome oxidase sequences and morphology. *GenBank.* http://www.ncbi.nlm.nih.gov
62. Brown JM, Pellmyr O, Thompson JN, Harrison RG. 1994. Mitochondrial DNA phylogeny of the Prodoxidae (Lepidoptera: Incurvarioidea) indicates rapid ecological diversification of yucca moths. *Ann. Entomol. Soc. Am.* 87:795–802
63. Brown JM, Pellmyr O, Thompson JN, Harrison RG. 1994. Phylogeny of *Greya* (Lepidoptera: Prodoxidae), based on nucleotide sequence variation in mitochondrial cytochrome oxidase I and II: congruence with morphological data. *Mol. Biol. Evol.* 11:128–41
64. Brunton CFA. 1998. The evolution of ultraviolet patterns in European *Colias* butterflies (Lepidoptera, Pieridae): a phylogeny using mitochondrial DNA. *Heredity* 80:611–16
65. Brunton CFA, Hurst GDD. 1998. Mitochondrial DNA phylogeny of brimstone butterflies (genus *Gonepteryx*) from the Canary Islands and Madeira. *Biol. J. Linn. Soc.* 63:69–79
66. Brust RA, Ballard JWO, Driver F, Hartley DM, Galway NJ, Curran J. 1998. Molecular systematics, morphological analysis, and hybrid crossing identify a third taxon, *Aedes (Halaedes) wardangensis* sp.nov., of the *Aedes (Halaedes) australis* species group. *Can. J. Zool.* 76:1236–46
67. Buckler ES, Ippolito A, Holtsford TP. 1997. The evolution of ribosomal DNA: divergent paralogues and phylogenetic implications. *Genetics* 145:821–32
68. Bull JJ, Huelsenbeck JP, Cunningham CW, Swofford DL, Waddell PJ. 1993. Partitioning and combining data in phylogenetic analysis. *Syst. Biol.* 42:384–97
69. Burmester T, Massey HC, Zakharkin SO, Benes H. 1998. The evolution of hexamerins and the phylogeny of insects. *J. Mol. Evol.* 47(1):93–108
70. Caccone A, Garcia BA, Powell JR. 1996. Evolution of the mitochondrial DNA control region in the *Anopheles gambiae* complex. *Insect Mol. Biol.* 5(1):51–59
71. Caccone A, Moriyama EN, Gleason JM, Nigro L, Powell JR. 1996. A molecular phylogeny for the *Drosophila melanogaster* subgroup and the problem of polymorphism data. *Mol. Biol. Evol.* 13(9):1224–32
72. Cameron SA. 1993. Multiple origins of advanced eusociality in bees inferred from mitochondrial DNA sequences. *Proc. Natl. Acad. Sci. USA* 90:8687–91
73. Campbell BC. 1993. Congruent evolution between whiteflies (Homoptera: Aleyrodidae) and their bacterial endosymbionts based on respective 18S and 16S rDNAs. *Curr. Microbiol.* 26:129–32
74. Campbell BC, Steffen-Campbell JD, Gill RJ. 1994. Evolutionary origin of whiteflies (Hemiptera: Sternorrhyncha: Aleyrodidae) inferred from 18S rDNA sequences. *Insect Mol. Biol.* 3:73–88
75. Campbell BC, Steffen-Campbell JD, Sorensen JT, Gill RJ. 1995. Paraphyly of Homoptera and Auchenorrhyncha inferred from 18S rDNA nucleotide sequences. *Syst. Entomol.* 20:175–94
76. Campbell BC, Steffen-Campbell JD, Werren JH. 1993. Phylogeny of the *Nasonia* species complex (Hymenoptera: Pteromalidae) inferred from an internal transcribed

spacer (ITS2) and 28S rDNA sequences. *Insect Mol. Biol.* 2:225–37
77. Cano RJ, Poinar HN, Pieniazek NJ, Acra A, Poinar GO Jr. 1993. Amplification and sequencing of DNA from a 120–135-million-year-old weevil. *Nature* 363(6429): 536–38
78. Cano RJ, Poinar HN, Roubik DW, Poinar GO Jr. 1992. Enzymatic amplification and nucleotide sequencing of portions of the 18S rRNA gene of the bee *Proplebeia dominicana* (Apidae: Hymenoptera) isolated from 25–40 million year old Dominican amber. *Med. Sci. Res.* 20:619–22
79. Carapelli A, Frati F, Fanciulli PP, Dallai R. 1995. Genetic differentiation of six sympatric species of *Isotomurus:* Is there any difference in their microhabitat preference? *Eur. J. Soil Biol.* 31:87–99
80. Carmean D, Crespi BJ. 1995. Do long branches attract flies? *Nature* 373(6516): 666
81. Carmean D, Kimsey LS, Berbee ML. 1992. 18S rDNA sequences and the holometabolous insects. *Mol. Phylogenet. Evol.* 1:270–78
82. Carpenter JM. 1997. Phylogenetic relationships among European *Polistes* and the evolution of social parasitism (Hymenoptera: Vespidae). *Mem. Mus. Natl. Hist. Nat.* 173:135–61
83. Carpenter JM, Strassmann JE, Turillazzi S, Hughes CR, Solis CR, Cervo R. 1993. Phylogenetic relationships among paper wasp social parasites and their hosts (Hymenoptera: Vespidae). *Cladistics* 9:129–46
84. Carreno RA, Barta JR. 1998. Small subunit ribosomal RNA genes of tabanids and hippoboscids: evolutionary relationships and comparison with other Diptera. *J. Med. Entomol.* 35(6):1002–6
85. Carulli JP, Chen DM, Stark WS, Hartl DL. 1994. Phylogeny and physiology of *Drosophila* opsins. *J. Mol. Evol.* 38(3):250–62
86. Carulli JP, Hartl DL. 1992. Variable rates of evolution among *Drosophila* opsin genes. *Genetics* 132(1):193–204
87. Caterino MS, Reed RD, Kuo MM, Sperling FAH. 1999. Getting to the root of a difficult phylogenetic problem: the phylogeny of Papilionidae. *GenBank.* http://www.ncbi.nlm.nih.gov
88. Caterino MS, Sperling FAH. 1999. *Papilio* phylogeny based on mitochondrial cytochrome oxidase I and II genes. *Mol. Phylogenet. Evol.* 11(1):122–37
89. Catzeflis FM. 1991. Animal tissue collections for molecular genetics and systematics. *Trends Ecol. Evol.* 6:168
90. Chalwatzis N, Baur A, Stetzer E, Kinzelbach R, Zimmermann FK. 1995. Strongly expanded 18S rRNA genes correlated with a peculiar morphology in the insect order of Strepsiptera. *Zoology* 98:115–26
91. Chalwatzis N, Hauf J, Van De Peer Y, Kinzelbach R, Zimmermann FK. 1996. 18S ribosomal RNA genes of insects: primary structure of the genes and molecular phylogeny of the Holometabola. *Ann. Entomol. Soc. Am.* 89(6):788–803
92. Chang WXZ, Tabashnik BE, Artelt B, Malvar T, Balester V, et al. 1997. Mitochondrial DNA sequence variation among geographic strains of diamondback moth (Lepidoptera: Plutellidae). *Ann. Entomol. Soc. Am.* 90(5):590–95
93. Chapco W, Martel RKB, Kuperus WR. 1997. Molecular phylogeny of North American band-winged grasshoppers (Orthoptera: Acrididae). *Ann. Entomol. Soc. Am.* 90:555–62
94. Chapman TW, Crespi B. 1998. High relatedness and inbreeding in two species of haplodiploid eusocial thrips (Insecta: Thysanoptera) revealed by microsatellite analysis. *Behav. Ecol. Sociobiol.* 43(4–5):301–6
95. Chen X, Li S, Aksoy S. 1999. Concordant evolution of a symbiont with its host insect species: molecular phylogeny of genus *Glossina* and its bacteriome-associated endosymbiont, *Wigglesworthia glossinidia. J. Mol. Evol.* 48:49–58
96. Chenuil A, McKey DB. 1996. Molecular phylogenetic study of a myrmecophyte

symbiosis: Did *Leonardoxa*/ant associations diversify via cospeciation? *Mol. Phylogenet. Evol.* 6:270–86
97. Chenuil A, Solignac M, Bernard M. 1997. Evolution of the large-subunit ribosomal RNA binding site for protein L23/25. *Mol. Biol. Evol.* 14(5):578–88
98. Chippindale PT, Davé VK, Whitmore DH, Robinson JV. 1999. Phylogenetic relationships of North American damselflies of the genus *Ischnura* (Odonata: Coenagrionidae) based on sequences of three mitochondrial genes. *Mol. Phylgenet. Evol.* 11(1):110–21
99. Cho S, Mitchell A, Regier JC, Mitter C, Poole RW, et al. 1995. A highly conserved nuclear gene for low-level phylogenetics: Elongaton factor-1α recovers morphology-based tree for heliothine moths. *Mol. Biol. Evol.* 12:650–56
100. Choudhary M, Strassmann JE, Queller DC, Turillazzi S, Cervo R. 1994. Social parasites in polistine wasps monophyletic: implications for sympatric speciation. *Proc. R. Soc. London Ser. B* 257(1348):31–35
101. Choudhary M, Strassmann JE, Solis CR, Queller DC. 1993. Microsatellite variation in a social insect. *Biochem. Genet.* 31(1–2):87–96
102. Clark AG, Leicht BG, Muse SV. 1996. Length variation and secondary structure of introns in the *Mlc1* gene in six species of *Drosophila. Mol. Biol. Evol.* 13(3):471–82
103. Clary DO, Wolstenholme DR. 1985. The mitochondrial DNA molecule of *Drosophila yakuba:* nucleotide sequence, gene organization, and genetic code. *J. Mol. Evol.* 22:252–71
104. Cognato AI, Rogers SO, Teale SA. 1995. Species diagnosis and phylogeny of the *Ips grandicollis* group (Coleoptera: Scolytidae) using random amplified polymorphic DNA. *Ann. Entomol. Soc. Am.* 88(4):397–405
105. Cognato AI, Sperling FAH. 2000. Phylogeny of *Ips* DeGeer species (Coleoptera: Scolytidae) inferred from mitochondrial cytochrome oxidase I sequence. *Mol. Phylogenet. Evol.* In press
106. Cohn VH, Moore GP. 1988. Organization and evolution of the alcohol dehydrogenase gene in *Drosophila. Mol. Biol. Evol.* 5(2):154–66
107. Cohn VH, Thompson MA, Moore GP. 1984. Nucleotide sequence comparison of the *Adh* gene in 3 drosophilids. *J. Mol. Evol.* 20:31–37
108. Cooper A, Wayne R. 1998. New uses for old DNA. *Curr. Opin. Biotechnol.* 9:49–53
109. Cooper G, Miller PL, Holland PWH. 1996. Molecular genetic analysis of sperm competition in the damselfly *Ischnura elegans* (Vander Linden). *Proc. R. Soc. London Ser. B* 263(1375):1343–49
110. Cooper SJB, Hewitt GM. 1993. Nuclear DNA sequence divergence between parapatric subspecies of the grasshopper *Chorthippus parallelus. Insect Mol. Biol.* 2(3):185–94
111. Costa R, Peixoto AA, Thackeray JR, Dalgleish R, Kyriacou CP. 1991. Length polymorphism in the threonine-glycine-encoding repeat region of the period gene in *Drosophila. J. Mol. Evol.* 32(3):238–46
112. Crespi BJ, Carmean DA, Mound LA, Worobey M, Morris D. 1998. Phylogenetics of social behavior in Australian gall-forming Thrips: evidence from mitochondrial DNA sequence, adult morphology and behavior, and gall morphology. *Mol. Phylogenet. Evol.* 9:163–80
113. Crespi BJ, Carmean DA, Vawter L, Von Dohlen C. 1996. Molecular phylogenetics of Thysanoptera. *Syst. Entomol.* 21:79–87
114. Crozier RH, Crozier YC. 1993. The mitochondrial genome of the honeybee *Apis mellifera:* complete sequence and genome organization. *Genetics* 133: 97–117
115. Crozier RH, Dobric N, Imai HT, Graur D, Cornuet JM, Taylor RW. 1995. Mitochondrial DNA sequence evidence on the phylogeny of Australian jack-jumper ants of the *Myrmecia pilosula* complex. *Mol. Phylogenet. Evol.* 4:20–30
116. Cunningham CW. 1997. Is congruence

between data partitions a reliable predictor of phylogenetic accuracy? Empirically testing an iterative procedure for choosing among phylogenetic methods. *Syst. Biol.* 46:464–78

117. Danforth BN, Ji S. 1998. Elongation factor-1 alpha occurs as two copies in bees: implications for phylogenetic analysis of *EF-1α* sequences in insects *Mol. Biol. Evol.* 15:225–35
118. Danoff-Burg JA, Conn JE. 1997. Character congruence in *Anopheles* mosquito phylogenetics. *GenBank.* http://www.ncbi.nlm.nih.gov
119. Date A, Sata Y, Takahata N, Chigusa SI. 1998. Evolutionary history and mechanism of the *Drosophila* cecropin gene family. *Immunogenetics* 47(6):417–29
120. Depaquit J, Perrotey S, Lecointre G, Tillier A, Tillier S, et al. 1998. Molecular systematics of Phlebotominae: a pilot study. Paraphyly of the genus *Phlebotomus. C. R. Acad. Sci. III* 321(10):849–55
121. De Queiroz A, Donoghue MJ, Kim J. 1995. Separate versus combined analysis of phylogenetic evidence. *Annu. Rev. Ecol. Syst.* 26:657–81
122. De Rijk P, Caers A, Van de Peer Y, De Wachter R. 1999. Database on the structure of large ribosomal subunit RNA. *Nucleic Acids Res.* 27(1):174–78
123. Derr JN, Davis SK, Woolley JB, Wharton RA. 1992. Reassessment of the 16S rRNA nucleotide sequence from members of the parasitic Hymenoptera. *Mol. Phylogenet. Evol.* 1(4):338–41
124. Derr JN, Davis SK, Woolley JB, Wharton RA. 1992. Variation and the phylogenetic utility of the large ribosomal subunit of mitochondrial DNA from the insect order Hymenoptera. *Mol. Phylogenet. Evol.* 1(2):136–47
125. DeSalle R. 1992. The origin and possible time of divergence of the Hawaiian Drosophilidae: evidence from DNA sequences. *Mol. Biol. Evol.* 9(5):905–16
126. DeSalle R. 1992. The phylogenetic relationships of flies in the family Drosophilidae deduced from mtDNA sequences. *Mol. Phylogenet. Evol.* 1(1):31–40
127. DeSalle R, Brower AVZ. 1997. Process partitions, congruence, and the independence of characters: inferring relationships among closely related Hawaiian *Drosophila* from multiple gene regions. *Syst. Biol.* 46(4):751–64
128. DeSalle R, Freedman T, Prager EM, Wilson AC. 1987. Tempo and mode of sequence evolution in mitochondrial DNA of Hawaiian *Drosophila. J. Mol. Evol.* 26(1–2):157–64
129. DeSalle R, Gatesy J, Wheeler W, Grimaldi D. 1992. DNA sequences from a fossil termite in Oligo-Miocene amber and their phylogenetic implications. *Science* 257(5078):1933–36
130. DeSalle R, Templeton AR. 1992. The mtDNA genealogy of closely related *Drosophila silvestris. J. Hered.* 83(3):211–16
131. Dietrich CH, Fitzgerald SJ, Holmes JL, Black WC, Nault LR. 1998. Reassessment of *Dalbulus* leafhopper (Homoptera: Cicadellidae) phylogeny based on mitochondrial DNA sequences. *Ann. Entomol. Soc. Am.* 91(5):590–97
132. Dietrich CH, Whitcomb RF, Black WC. 1997. Phylogeny of the grassland leafhopper genus *Flexamia* (Homoptera: Cicadellidae) based on mitochondrial DNA sequences. *Mol. Phylogenet. Evol.* 8:139–49
133. Dillon N, Austin AD, Bartowsky E. 1996. Comparison of preservation techniques for DNA extraction from hymenopterous insects. *Insect Mol. Biol.* 5:21–24
134. Dorit RL, Ayala FJ. 1995. *Adh* evolution and the phylogenetic footprint. *J. Mol. Evol.* 40(6):658–62
135. Dowton M. 1999. Relationships among the cyclostome braconid (Hymenoptera: Braconidae) subfamilies inferred from a mitochondrial tRNA gene rearrangement. *Mol. Phylogenet. Evol.* 11(2):283–87
136. Dowton M, Austin AD. 1994. Molecular phylogeny of the insect order Hymenop-

tera: Apocritan relationships. *Proc. Natl. Acad. Sci. USA* 91(21):9911–15

137. Dowton M, Austin AD. 1994. Multiple origins of parasitism in the wasps. *GenBank.* http://www.ncbi.nlm.nih.gov
138. Dowton M, Austin AD. 1995. Increased genetic diversity in mitochondrial genes is correlated with the evolution of parasitism in the Hymenoptera. *J. Mol. Evol.* 41:958–65
139. Dowton M, Austin AD. 1997. Evidence for AT-transversion bias in wasp (Hymenoptera: Symphyta) mitochondrial genes and its implications for the origin of parasitism. *J. Mol. Evol.* 44(4):398–405
140. Dowton M, Austin AD. 1997. Phylogenetic relationships among the microgastroid wasps (Hymenoptera: Braconidae): Combined analysis of 16S and 28S rDNA genes. *GenBank.* http://www.ncbi.nlm.nih.gov
141. Dowton M, Austin AD, 1999. Evolutionary dynamics of a mitochondrial rearrangement "hot spot" in the Hymenoptera. *Mol. Biol. Evol.* 16(2):298–309
142. Dowton M, Austin AD, Antolin MF. 1998. Evolutionary relationships among the Braconidae (Hymenoptera: Ichneumonoidea) inferred from partial 16S rDNA gene sequences. *Insect Mol. Biol.* 7:129–50
143. Dowton M, Austin AD, Dillon N, Bartowsky E. 1997. Molecular phylogeny of the apocritan wasps: the Proctotrupomorpha and Evaniomorpha. *Syst. Entomol.* 22:245–55
144. Doyle JJ. 1992. Gene trees and species trees: molecular systematics as one-character taxonomy. *Syst. Bot.* 17:144–63
145. Doyle JJ. 1997. Trees within trees: genes and species, molecules and morphology. *Syst. Biol.* 46:537–53
146. Emerson BC, Wallis GP. 1995. Phylogenetic relationships of the *Prodontria* (Coleoptera: Scarabaeidae: Melolonthinae), derived from sequence variation in the mitochondrial cytochrome oxidase II gene. *Mol. Phylogenet. Evol.* 4:433–47
147. Ender A, Schwenk K, Staedler T, Streit B, Schierwater B. 1996. RAPD identification of microsatellites in *Daphnia. Mol. Ecol.* 5:437–41
148. Erney SJ, Pruess KP, Danielson SD, Powers TO. 1996. Molecular differentiation of alfalfa weevil, *Hypera postica,* strains (Coleoptera: Curculionidae). *Ann. Entomol. Soc. Am.* 89(6):804–11
149. Esseghir S, Ready PD, Killick-Kendrick R, Ben-Ismail R. 1997. Mitochondrial haplotypes and phylogeography of *Phlebotomus* vectors of *Leishmania major. Insect Mol. Biol.* 6(3):211–25
150. Evans JD. 1996. Competition and relatedness between queens of the facultatively polygynous ant *Myrmica tahoensis. Anim. Behav.* 51:831–40
151. Ezenwa VO, Peters JM, Zhu Y, Arévalo E, Hastings MD, et al. 1998. Ancient conservation of trinucleotide microsatellite loci in polistine wasps. *Mol. Phylogenet. Evol.* 10(2):168–77
152. Fang QQ, Black WC, Blocker HD, Whitcomb RF. 1993. A phylogeny of new world *Deltocephalus*-like leafhopper genera based on mitochondrial 16S ribosomal DNA sequences. *Mol. Phylogenet. Evol.* 2:119–31
153. Fang QQ, Cho S, Regier JC, Mitter C, Matthews M, et al. 1997. A new nuclear gene for insect phylogenetics: dopa decarboxylase is informative of relationships within Heliothinae (Lepidoptera: Noctuidae). *Syst. Biol.* 46:269–83
154. Farrell BD. 1998. "Inordinate fondness" explained: Why are there so many beetles? *Science* 281(5376):555–59
155. Farrell BD, Mitter C. 1998. The timing of insect-plant diversification: Might *Tetraopes* (Coleoptera: Cerambycidae) and *Asclepias* (Asclepiadaceae) have co-evolved? *Biol. J. Linn. Soc.* 63:553–77
156. Farris JS, Kallersjo M, Kluge AG, Bult C. 1995. Constructing a significance test for incongruence. *Syst. Biol.* 44:570–72
157. Felsenstein J. 1978. Cases in which parsimony or compatibility methods will be

positively misleading. *Syst. Zool.* 27:401–10
158. Felsenstein J. 1992. Estimating effective population size from samples of sequences: inefficiency of pairwise and segregating sites as compared to phylogenetic estimates. *Genet. Res.* 59:139–47
159. Felsenstein J. 1993. *PHYLIP (Phylogeny Inference Package)* version 3. 5c. Dep. Genet., Univ. Wash., Seattle
160. Fenton B, Malloch G, Germa F. 1998. A study of variation in rDNA its regions shows that two haplotype coexists within a single aphid genome. *Genome* 41:337–45
161. Flook PK, Rowell CHF. 1997. The phylogeny of the Caelifera (Insecta, Orthoptera) as deduced from mtrRNA gene sequences. *Mol. Phylogenet. Evol.* 8:89–103
162. Flook PK, Rowell CHF. 1997. The effectiveness of mitochondrial rRNA gene sequences for the reconstruction of the phylogeny of an insect order (Orthoptera). *Mol. Phylogenet. Evol.* 8:177–92
163. Flook PK, Rowell CHF. 1998. Inferences about orthopteroid phylogeny and molecular evolution from small subunit nuclear ribosomal DNA sequences. *Insect Mol. Biol.* 7:163–78
164. Flook PK, Rowell CHF, Gellissen G. 1995. The sequence, organization, and evolution of the *Locusta migratoria* mitochondrial genome. *J. Mol. Evol.* 41:928–41
165. Foley DH, Bryan JH, Yeates D, Saul A. 1998. Evolution and systematics of *Anopheles:* insights from a molecular phylogeny of Australasian mosquitoes. *Mol. Phylogenet. Evol.* 9(2):262–75
166. Frati F, Carapelli A. 1998. An assessment of the value of nuclear and mitochondrial genes in elucidating the origin and evolution of *Isotoma klovstadi. GenBank.* http://www.ncbi.nlm.nih.gov
167. Frati F, Simon C, Sullivan J, Swofford DL. 1997. Evolution of the mitochondrial cytochrome oxidase II gene in Collembola. *J. Mol. Evol.* 44:145–58
168. Friedlander TP, Horst KR, Regier JC, Mitter C, Peigler RS, Fang QQ. 1998. Two nuclear genes yield concordant relationships within Attacini (Lepidoptera: Saturniidae). *Mol. Phylogenet. Evol.* 9:131–40
169. Friedlander TP, Regier JC, Mitter C. 1992. Nuclear gene sequenes for higher level phylogenetic analysis: 14 promising candidates. *Syst. Biol.* 41:483–90
170. Friedlander TP, Regier JC, Mitter C. 1994. Phylogenetic information content of five nuclear gene sequences in animals: initial assessment of character sets from concordance and divergence studies. *Syst. Biol.* 43:511–25
171. Friedlander TP, Regier JC, Mitter C, Wagner DL. 1996. A nuclear gene for higher level phylogenetics: Phosphoenolpyruvate carboxykinase tracks Mesozoic-age divergences within Lepidoptera (Insecta). *Mol. Biol. Evol.* 13:594–604
172. Friedrich M. 1995. *GenBank.* http://www.ncbi.nlm.nih.gov
173. Friedrich M, Tautz D. 1995. Ribosomal DNA phylogeny of the major extant arthropod classes and the evolution of myriapods. *Nature* 376(6536):165–67
174. Friedrich M, Tautz D. 1997. Evolution and phylogeny of the Diptera: a molecular phylogenetic analysis using 28S rDNA sequences. *Syst. Biol.* 46(4):674–98
175. Friedrich M, Tautz D. 1997. An episodic change of rDNA nucleotide substitution rate has occurred during the emergence of the insect order Diptera. *Mol. Biol. Evol.* 14:644–53
176. Fritz GN. 1998. Sequence analysis of the rDNA internal transcribed spacer 2 of five species of South American human malaria mosquitoes *DNA Seq.* 18:215–21
177. Fritz GN, Conn J, Cockburn A, Seawright J. 1994. Sequence analysis of the ribosomal RNA internal transcribed spacer 2 from populations of *Anopheles nuneztovari* (Diptera: Culicidae). *Mol. Biol. Evol.* 11(3):406–16
178. Funk DJ. 1999. Molecular systematics of cytochrome oxidase I and 16S from *Neo-*

chlamisus leaf beetles and the importance of sampling. *Mol. Biol. Evol.* 16:67–82
179. Funk DJ, Futuyma DJ, Orti G, Meyer A. 1995. Mitochondrial DNA sequences and multiple data sets: a phylogenetic study of phytophagous beetles (Chrysomelidae: *Ophraella*). *Mol. Biol. Evol.* 12:627–40
180. Galian J, De la Rua P, Serrano J, Juan C, Hewitt GM. 1998. Phylogenetic relationships in Iberian Scaritini (Coleoptera, Carabidae) inferred from mitochondrial DNA sequences and karyotype analysis. *GenBank.* http://www.ncbi.nlm.nih.gov
181. Garcia BA, Caccone A, Mathiopoulos KD, Powell JR. 1996. Inversion monophyly in African anopheline malaria vectors. *Genetics* 143(3):1313–20
182. Garcia BA, Powell JR. 1998. Phylogeny of species of *Triatoma* (Hemiptera: Reduviidae) based on mitochondrial DNA sequences. *J. Med. Entomol.* 35:232–38
183. Garnery L, Cornuet JM, Solignac M. 1992. Evolutionary history of the honey bee *Apis mellifera* inferred from mitochondrial DNA analysis. *Mol. Ecol.* 1(3):145–54
184. Gasparich GE, Silva JG, Ho-Yoen H, McPheron BA, Steck GJ, Sheppard WS. 1997. Population structure of mediterranean fruit fly and implications for worldwide colonization patterns. *Ann. Entomol. Soc. Am.* 90(6):790–97
185. Gillespie RG, Croom HB, Palumbi SR. 1994. Multiple origins of a spider radiation in Hawaii. *Proc. Natl. Acad. Sci. USA* 91:2290–94
186. Gimeno C, Belshaw RD, Quicke DLJ. 1997. Phylogenetic relationships of the Alysiinae/Opiinae (Hymenoptera: Braconidae) and the utility of cytochrome b, 16S and 28S D2 rRNA. *Insect Mol. Biol.* 6(3):273–84
187. Giribet G, Ribera C. 1998. The position of arthropods in the animal kingdom: a search for a reliable outgroup for internal arthropod phylogeny. *Mol. Phylogenet. Evol.* 9(3):481–88
188. Gleason JM, Caccone A, Moriyama EN, White KP, Powell JR. 1997. Mitochondrial DNA phylogenies for the *Drosophila obscura* group. *Evolution* 51(2):433–40
189. Gleason JM, Griffith EC, Powell JR. 1998. A molecular phylogeny of the *Drosophila willistoni* group: Conflicts between species concepts? *Evolution* 52(4):1093–103
190. Gleason JM, Powell JR. 1997. Interspecific and intraspecific comparisons of the *period* locus in the *Drosophila willistoni* sibling species. *Mol. Biol. Evol.* 14(7):741–53
191. Gleeson DM, Howitt RLJ, Newcomb RD. 1998. The phylogenetic position of the New Zealand batfly, *Mystacinobia zelandica* inferred from mitochondrial 16S ribosomal DNA sequence data. *GenBank.* http://www.ncbi.nlm.nih.gov
192. Gleeson DM, Sarre S. 1997. Mitochondrial DNA variability and geographic origin of the sheep blowfly, *Lucilia cuprina* (Diptera: Calliphoridae), in New Zealand. *Bull. Entomol. Res.* 87(3):265–72
193. Grau R, Bachmann L. 1997. The evolution of intergenic spacers of the 5S rDNA genes in the *Drosophila obscura* group: Are these sequences suitable for phylogenetic analyses? *Biochem. Syst. Ecol.* 25(2):131–39
194. Graybeal A. 1994. Evaluating the phylogenetic utility of genes: a search for genes informative about deep divergences among vertebrates. *Syst. Biol.* 43:174–93
195. Graybeal A. 1998. Is it better to add taxa or characters to a difficult phylogenetic problem? *Syst. Biol.* 47: 9–17
196. Gruhl M, Kao WY, Bergtrom G. 1997. Evolution of orthologous intronless and intron-bearing globin genes in two insect species. *J. Mol. Evol.* 45:499–508
197. Guillemaud T, Pasteur N, Rousset F. 1997. Contrasting levels of variability between cytoplasmic genomes and incompatibility types in the mosquito *Culex pipiens. Proc. R. Soc. London Ser. B* 264(1379):245–51
198. Gutierrez G, Marin A. 1998. The most ancient DNA recovered from an amber-preserved specimen may not be as ancient as it seems. *Mol. Biol. Evol.* 15:926–29
199. Hafner MS, Sudman PD, Villablanca FX, Spradling TA, Demastes JW, Nadler SA.

1994. Disparate rates of molecular evolution in cospeciating hosts and parasites. *Science* 265(5175):1087–90

200. Hall HG, Smith DR. 1991. Distinguishing African and European honeybee matrilines using amplified mitochondrial DNA. *Proc. Natl. Acad. Sci. USA* 88(10):4548–52
201. Hamelin E, Bigot YYB, Rouleux F, Renault S, Periquet G. 1995. Fast cloning and sequencing of histone H4 and actin messenger RNA fragments and their uses as control. *GenBank.* http://www.ncbi.nlm.nih.gov
202. Han H-Y, McPheron BA. 1994. Phylogenetic study of selected tephritid flies (Diptera: Tephritidae) using partial sequences of the nuclear 18S ribosomal DNA. *Biochem. Syst. Ecol.* 22(5):447–57
203. Han H-Y, McPheron BA. 1997. Molecular phylogenetic study of Tephritidae (Diptera) using partial sequences of the mitochondrial 16S ribosomal DNA. *Mol. Phylogenet. Evol.* 7(1):17–32
204. Hankeln T, Amid C, Weich B, Niessing J, Schmidt ER. 1998. Molecular evolution of the globin gene cluster E in two distantly related midges, *Chironomus pallidivittatus* and *C. thummi thummi*. *J. Mol. Evol.* 46(5):589–601
205. Hankeln T, Friedl H, Ebersberger I, Martin J, Schmidt ER. 1997. A variable intron distribution in globin genes of *Chironomus:* evidence for recent intron gain. *Gene* 205(1–2):151–60
206. Hankeln T, Schmidt ER. 1993. Divergent evolution of an "orphon" histone gene cluster in *Chironomus*. *J. Mol. Biol.* 234(4):1301–7
207. Haring E, Hagemann S, Lankinen P, Pinsker W. 1998. The phylogenetic position of *Drosophila eskoi* deduced from P element and *Adh* sequence data. *Hereditas* 128(3):235–44
208. Harr B, Weiss S, David JR, Brem G, Schlötterer C. 1998. A microsatellite-based multilocus phylogeny of the *Drosophila melanogaster* species complex. *Curr. Biol.* 8:1183–86
209. Harrison RG. 1989. Animal mitochondrial DNA as a genetic marker in population and evolutionary biology. *Trends Ecol. Evol.* 4:6–11
210. Harrison RG, Bogdanowicz SM. 1995. Mitochondrial DNA phylogeny of North American field crickets: perspectives on the evolution of life cycles, songs, and habitat associations. *J. Evol. Biol.* 8:209–32
211. Harvey PH, Pagel MD. 1991. *The Comparative Method in Evolutionary Biology.* New York: Oxford Univ. Press. 239 pp.
212. Hasegawa E, Tinaut A, Ruano F. 1998. Molecular phylogeny of Formicini: relationships of *Rossomyrmex* and *Polyergus* (Hymenoptera: Formicidae). *GenBank.* http://www.ncbi.nlm.nih.gov
213. Hasson E, Wang IN, Zeng LW, Kreitman M, Eanes WF. 1998. Nucleotide variation in the triosephosphate isomerase (*Tpi*) locus of *Drosophila melanogaster* and *Drosophila simulans*. *Mol. Biol. Evol.* 15(6):756–69
214. Hayashi K. 1991. PCR-SSCP: a simple and sensitive method for detection of mutations in the genomic DNA. *PCR Method Appl.* 1:34–38
215. Haymer DS. 1994. Random amplified polymorphic DNAs and microsatellites: What are they, and can they tell us anything we don't already know? *Ann. Entomol. Soc. Am.* 87:717–22
216. Haymer DS, McInnis DO. 1994. Resolution of populations of the mediterranean fruit fly at the DNA level using random primers for the polymerase chain reaction. *Genome* 37:244–48
217. He M, Haymer DS. 1997. Polymorphic intron sequences detected within and between populations of the oriental fruit fly. *Ann. Entomol. Soc. Am.* 90(6):825–31
218. Heinze J, Lipski N, Schlehmeyer K, Hölldobler B. 1995. Colony structure and reproduction in the ant, *Leptothorax acervorum*. *Behav. Ecol.* 6:359–67
219. Herre EA, Machado CA, Bermingham E, Nason JD, Windsor DM, et al. 1996. Molecular phylogenies of figs and their

pollinator wasps. *J. Biogeogr.* 23(4):521–30

220. Hey J, Kliman RM. 1993. Population genetics and phylogenetics of DNA sequence variation at multiple loci within the *Drosophila melanogaster* species complex. *Mol. Biol. Evol.* 10(4):804–22
221. Hickey DA, Benkel BF, Boer PH, Genest Y, Abukashawa S, Ben-David G. 1987. Enzyme-coding genes as molecular clocks: the molecular evolution of animal alpha-amylases. *J. Mol. Evol.* 26(3):252–56
222. Hillis DM. 1996. Inferring complex phylogenies. *Nature* 383(6596):130–31
223. Hillis DM. 1998. Taxonomic sampling, phylogenetic accuracy, and investigator bias. *Syst. Biol.* 47:3–8
224. Hillis DM, Dixon MT. 1991. Ribosomal DNA molecular evolution and phylogenetic inference. *Q. Rev. Biol.* 66:411–54
225. Hillis DM, Moritz C, Mabel BK, eds. 1996. *Molecular Systematics,* Sunderland, MA: Sinauer. 655 pp. 2nd ed.
226. Hiss RH, Norris DE, Dietrich CH, Whitcombe RF, West DF, et al. 1994. Molecular taxonomy using single-strand conformation polymorphism (SSCP) analysis of mitochondrial ribosomal DNA genes. *Insect Mol. Biol.* 3:171–82
227. Ho C-M, Liu Y-M, Wei Y-H, Hu S-T. 1995. Gene for cytochrome c oxidase subunit II in the mitochondrial DNA of *Culex quinquefasciatus* and *Aedes aegypti* (Diptera: Culicidae). *J. Med. Entomol.* 32(2):174–80
228. Ho K-F, Craddock EM, Piano F, Kambysellis MP. 1996. Phylogenetic analysis of DNA length mutations in a repetitive region of the Hawaiian *Drosophila* yolk protein gene *Yp2*. *J. Mol. Evol.* 43(2):116–24
229. Hoeben P, Daniel LJ, Ma J, Drew RAI. 1996. The *Bactrocera* (*Notodacus*) *xanthoides* (Broun) species complex (Diptera: Tephritidae): comparison of 18S rRNA sequences from Fiji, Tonga and Vanuatu specimens suggest two distinct strains. *Aust. J. Entomol.* 35(1):61–64
230. Hoeben P, Ma J. 1997. *GenBank.* http://www.ncbi.nlm.nih.gov
231. Hoelzer GA. 1997. Inferring phylogenies from mtDNA variation: mitochondrial gene trees versus nuclear gene trees revisited. *Evolution* 51:622–26
232. Honda JY, Nakashima Y, Yanase T, Kawarabata T, Hirose Y. 1998. Use of the internal transcribed spacer (ITS-1) region to infer *Orius* (Hemiptera: Anthocoridae) species phylogeny. *Appl. Entomol. Zool.* 33(4):567–71
233. Hoss M, Jaruga P, Zastawny TH, Dizdaroglu M, Pääbo S. 1996. DNA damage and DNA sequence retrieval from ancient tissues. *Nucleic Acids Res.* 24:1304–7
234. Howland DE, Hewitt GM. 1995. Phylogeny of the Coleoptera based on mitochondrial cytochrome oxidase I sequence data. *Insect Mol. Biol.* 4:203–15
235. Hsiao TH. 1994. Molecular techniques for studying systematics and phylogeny of Chrysomelidae. In *Novel Aspects of the Biology of the Chrysomelidae,* ed. PH Jolivet, ML Cox, E Petitpierre. The Netherlands: Kluwer
236. Hsiao TH. 1996. Studies of interactions between alfalfa weevil strains, *Wolbachia* endosymbionts and parasitoids by DNA based techniques. In *The Ecology of Agricultural Pests: Biochemical Approaches,* ed. WOC Symondson, JE Liddell, pp. 51–71. London: Chapman & Hall
237. Huber JT. 1998. The importance of voucher specimens, with practical guidelines for preserving specimens of the major invertebrate phyla for identification. *J. Natl. Hist.* 32:367–85
238. Huelsenbeck JP. 1997. Is the Felsenstein zone a fly trap? *Syst. Biol.* 46:69–74
239. Huelsenbeck JP, Bull JJ, Cunningham CW. 1996. Combining data in phylogenetic analysis. *Trends Ecol. Evol.* 11:152–58
240. Hughes AL. 1998. Protein phylogenies provide evidence of a radical discontinuity between arthropod and vertebrate immune systems. *Immunogenetics* 47:283–96
241. Hughes AL. 1999. Evolution of the artho-

pod prophenoloxidase/hexamerin protein family. *Immunogenetics* 49(2):106–14
242. Hwang UW, Kim W, Tautz D, Friedrich M. 1998. Molecular phylogenetics at the Felsenstein zone: approaching the Strepsiptera problem using 5. 8S and 28S rDNA sequences. *Mol. Phylogenet. Evol.* 9:470–80
243. Hwang UW, Lee BH, Kim W. 1995. Sequences of the 18S rDNAs from two collembolan insects: shorter sequences in the V4 and V7 regions. *Gene* 154(2):293–94
244. Imura Y, Su ZH, Osawa S. 1997. Morphology and molecular phylogeny of some tibetan ground beetles belonging to the subgenera *Neoplesius* and *Eocechenus* (Coleoptera: Carabidae). *Elytra* 25(1):231–45
245. Imura Y, Zhou HZ, Okamoto M, Su ZH, Osawa S. 1999. Phylogenetic relationships of some Chinese beetles belonging to the subgenera *Neoplesius, Pagocarabus* and *Aristocarabus* (Coleoptera, Carabidae) based on mitochondrial ND5 gene sequences *Elytra.* In press
246. Inomata N, Tachida H, Yamazaki T. 1997. Molecular evolution of the *amy* multigenes in the subgenus *Sophophora* of *Drosophila. Mol. Biol. Evol.* 14(9):942–50
247. Jeffs PS, Holmes EC, Ashburner M. 1994. The molecular evolution of the alcohol dehydrogenase and alcohol dehydrogenase-related genes in the *Drosophila melanogaster* species subgroup. *Mol. Biol. Evol.* 11(2):287–304
248. Jenkins TM, Basten CJ, Anderson WW. 1996. Mitochondrial gene divergence of Colombian *Drosophila pseudoobscura. Mol. Biol. Evol.* 13(9):1266–75
249. Jermiin LS, Crozier RH. 1994. The cytochrome b region in the mitochondrial DNA of the ant *Tetraponera rufonigra:* sequence divergence in Hymenoptera may be associated with nucleotide content. *J. Mol. Evol.* 38:282–94
250. Jiggins CD, Davies N. 1998. Genetic evidence for a sibling species of *Heliconius charithonia* (Lepidoptera: Nymphalidae). *Biol. J. Linn. Soc.* 64:57–67
251. Johns GC, Avise JC. 1998. A comparative summary of genetic distances in the vertebrates from the mitochondrial cytochrome b gene. *Mol. Biol. Evol.* 15(11):1481–90
252. Juan C, Ibrahim KM, Oromi P, Hewitt GM. 1996. Mitochondrial DNA sequence variation and phylogeography of *Pimelia* darkling beetles on the Island of Tenerife (Canary Islands). *Heredity* 77:589–98
253. Juan C, Oromi P, Hewitt GM. 1995. Mitochondrial DNA phylogeny and sequential colonization of Canary Islands by darkling beetles of the genus *Pimelia* (Tenebrionidae). *Proc. R. Soc. London Ser. B* 261(1361):173–80
254. Juan C, Oromi P, Hewitt GM. 1996. Phylogeny of the genus *Hegeter* (Tenebrionidae, Coleoptera) and its colonization of the Canary Islands deduced from cytochrome oxidase I mitochondrial DNA sequences. *Heredity* 76:392–403
255. Juan C, Oromi P, Hewitt GM. 1997. Molecular phylogeny of darkling beetles from the Canary Islands: comparison of inter island colonization patterns in two genera. *Biochem. Syst. Ecol.* 25:121–30
256. Kambhampati S. 1995. A phylogeny of cockroaches and related insects based on DNA sequence of mitochondrial ribosomal RNA genes. *Proc. Natl. Acad. Sci. USA* 92:2017–20
257. Kambhampati S. 1996. Phylogenetic relationship among cockroach families inferred from mitochondrial 12S rRNA gene sequence. *Syst. Entomol.* 21:89–98
258. Kambhampati S, Charlton RE. 1999. Phylogenetic relationship among *Libellula, Ladona* and *Plathemis* (Odonata: Libellulidae) based on DNA sequence of mitochondrial 16S rRNA gene. *Syst. Entomol.* 24: 37–49
259. Kambhampati S, Kjer KM, Thorne BL. 1996. Phylogenetic relationship among termite families based on DNA sequence of

mitochondrial 16S ribosomal RNA gene. *Insect Mol. Biol.* 5:229–38
260. Kambhampati S, Luykx P, Nalepa CA. 1996. Evidence for sibling species in *Cryptocercus punctulatus,* the wood roach, from variation in mitochondrial DNA and karyotype. *Heredity* 76:485–96
261. Kambysellis MP, Ho KF, Craddock EM, Piano F, Parisi M, Cohen J. 1995. Pattern of ecological shifts in the diversification of Hawaiian *Drosophila* inferred from a molecular phylogeny. *Curr. Biol.* 5(10):1129–39
262. Katz LA. 1997. Characterization of the phosphoglucose isomerase gene from crickets: an analysis of phylogeny, amino acid conservation and nucleotide composition. *Insect Mol. Biol.* 6:305–18
263. Kelley ST, Farrell BD. 1998. Is specialization a dead end? The phylogeny of host use in *Dendroctonus* bark beetles. *Evolution* 52(6):1731–43
264. Kerdelhue C, Le Clainche I, Rasplus JY. 1999. Molecular phylogeny of the *Ceratosolen* species pollinating *Ficus* of the subgenus *Sycomorus* sensu stricto: biogeographical history and origins of the species—specificity breakdown cases. *Mol. Phylogenet. Evol.* 11(3):401–14
265. Kim J. 1996. General inconsistency conditions for maximum parsimony: effects of branch lengths and increasing numbers of taxa. *Syst. Biol.* 45:363–74
266. Kjer KM. 1995. Use of rRNA secondary structure in phylogenetic studies to identify homologous positions: an example of alignment and data presentation from the frogs. *Mol. Phylogenet. Evol.* 4(3):314–30
267. Kliman RM, Hey J. 1993. DNA sequence variation at the *period* locus within and among species of the *Drosophila melanogaster* complex. *Genetics* 133(2):375–87
268. Kluge AG. 1998. Total evidence or taxonomic congruence: cladistics or consensus classification. *Cladistics* 14:151–58
269. Koch PB, Keys DN, Rocheleau T, Aronstein K, Blackburn M, et al. 1998. Regulation of dopa decarboxylase expression during colour pattern formation in wild-type and melanic tiger swallowtail butterflies. *Development* 125(12):2303–13
270. Kopf A, Rank NE, Roininen H, Julkunen-Tiitto R, Pasteels JM, Tahvaninen J. 1998. The evolution of host-plant use and sequestration in the leaf beetle genus *Phratora* (Coleoptera: Chrysomelidae). *Evolution* 52:517–28
271. Koulianos S, Crozier RH. 1997. Mitochondrial sequence characterisation of Australian commercial and feral honeybee strains (*Apis mellifera* L.), in the context of the species worldwide. *J. Aust. Entomol. Soc.* 36(4):359–64
272. Koulianos S, Schmid-Hempel P. 1998. Phylogenetic relationships among bumble bees (*Bombus* Latreille) inferred from mitochondrial cytochrome b and cytochrome oxidase I sequences. *GenBank.* http://www.ncbi.nlm.nih.gov
273. Krings M, Stone A, Schmitz RW, Krainitzki H, Stoneking M, Pääbo S. 1997. Neandertal DNA sequences and the origin of modern humans. *Cell* 90:19–30
274. Kuperus WR, Chapco W. 1994. Usefulness of internal transcribed spacer regions of ribosomal DNA in melanopline (Orthoptera: Acrididae) systematics. *Ann. Entomol. Soc. Am.* 87:751–54
275. Kwiatowski J, Krawczyk M, Jaworski M, Skarecky D, Ayala FJ. 1997. Erratic evolution of glycerol-3-phosphate dehydrogenase in *Drosophila, Chymomyza* and *Ceratitis. J. Mol. Evol.* 44(1):9–22
276. Kwiatowski J, Krawczyk M, Kornacki M, Bailey K, Ayala FJ. 1995. Evidence against the exon theory of genes derived from the triose-phosphate isomerase gene. *Proc. Natl. Acad. Sci. USA* 92(18):8503–6
277. Kwiatowski J, Skarecky D, Bailey K, Ayala FJ. 1994. Phylogeny of *Drosophila* and related genera inferred from the nucleotide sequence of the Cu, Zn *Sod* gene. *J. Mol. Evol.* 38(5):443–54
278. Landry B, Powell JA, Sperling FAH. 1999. Systematics of the *Argyrotaenia franciscana* (Lepidoptera: Tortricidae) species

group: evidence from mitochondrial DNA. *Ann. Entomol. Soc. Am.* 92(1):40–46

279. Langor DW, Sperling FAH. 1995. Mitochondrial DNA variation and identification of bark weevils in the *Pissodes strobi* species group in western Canada (Coleoptera: Curculionidae). *Can. Entomol.* 127:895–911
280. Langor DW, Sperling FAH. 1997. Mitochondrial DNA sequence divergence in weevils of the *Pissodes strobi* species complex (Coleoptera: Curculionidae). *Insect Mol. Biol.* 6:255–65
281. Lanzaro GC, Toure YT, Carnahan J, Zheng L, Dolo G, et al. 1998. Complexities in the genetic structure of *Anopheles gambiae* populations in West Africa as revealed by microsatellite DNA analysis. *Proc. Natl. Acad. Sci. USA* 95:14260–65
282. Lecanidou R, Rodakis GC, Eickbush TH, Kafatos FC. 1986. Evolution of the silk moth chorion gene superfamily: gene families CA and CB. *Proc. Natl. Acad. Sci. USA* 83(17):6514–18
283. Leclerc RF, Regier JC. 1994. Evolution of chorion gene families in Lepidoptera: characterization of 15 cDNAs from the gypsy moth *J. Mol. Evol.* 39(3):244–54
284. Lee BH, Hwang UW, Kim W, Park KH, Kim JT. 1995. Systematic position of cave Collembola, *Gulgastrura reticulosa* (Insecta) based on morphological characters and 18S rDNA nucleotide sequence analysis. *Mem. Biospeol.* 22:83–90
285. Lee BH, Hwang UW, Kim W, Park KH, Kim JT. 1995. Phylogenetic study of the suborder Arthropleona (Collembola) based on morphological characters and 18S rDNA sequence analysis. *Pol. Pismo Entomol.* 64(1–4):261–77
286. Leebens-Mack J, Pellmyr O, Brock M. 1998. Host specificity and the genetic structure of two yucca moth species in a yucca hybrid zone. *Evolution* 52:1376–82
287. Legge JT, Roush R, DeSalle R, Vogler AP, May B. 1996. Genetic criteria for establishing evolutionary significant units in Cryan's buckmoth. *Conserv. Biol.* 10:85–98
288. Lehmann T, Besansky NJ, Hawley WA, Fahey TG, Kamau L, Collins FH. 1997. Microgeographic structure of *Anopheles gambiae* in western Kenya based on mtDNA and microsatellite loci. *Mol. Ecol.* 6(3):243–53
289. Lewis DL, Farr CL, Kaguni LS. 1995. *Drosophila melanogaster* mitochondrial DNA: completion of the nucleotide sequence and evolutionary comparisons. *Insect Mol. Biol.* 4(4):263–278
290. Liu D. 1996. *GenBank.* http://www.ncbi.nlm.nih.gov
291. Liu H, Beckenbach AT. 1992. Evolution of the mitochondrial cytochrome oxidase II gene among 10 orders of insects. *Mol. Phylogenet. Evol.* 1:41–52
292. Logan JA. 1999. Extraction, PCR and sequencing of a 440 base pair region of the mitochondrial cytochrome oxidase I gene from two species of acetone-preserved damselflies. *Environ. Entomol.* 28(2): 143–47
293. Lounibos LP, Wilkerson RC, Conn JE, Hribar LJ, Fritz GN, Danoff-Burg JA. 1998. Morphological, molecular, and chromosomal discrimination of cryptic *Anopheles* (*Nyssorhynchus*) from South America. *J. Med. Entomol.* 35(5):830–38
294. Lowe RM, Crozier RH. 1997. The phylogeny of bees of the socially parasitic Australian genus *Inquilina* and their *Exoneura* hosts (Hymenoptera: Anthophoridae). *Insectes Soc.* 44:409–14
295. Lunt DH, Zhang D-X, Szymura JM, Hewitt GM. 1996. The insect cytochrome oxidase I gene: evolutionary patterns and conserved primers for phylogenetic studies. *Insect Mol. Biol.* 5:153–65
296. Ma Y, Xu J, Qu F, Zheng Z. 1998. Sequence differences of ribosomal DNA internal transcribed spacer 2 between *Anopheles sinensis* and *Anopheles anthropophagus*. *GenBank.* http://www.ncbi.nlm.nih.gov
297. Machado CA, Herre EA, McCafferty S,

Bermingham E. 1996. Molecular phylogenies of fig pollinating and non-pollinating wasps and the implications for the origin and evolution of the fig–fig wasp mutualism. *J. Biogeogr.* 23(4):531–42
298. Maddison DR, Baker MD, Ober KA. 1999. A preliminary phylogenetic analysis of 18S ribosomal DNA of carabid beetles (Coleoptera). *Boll. Mus Reg. Sci. Nat. Torino.* 1998:229–50
299. Maddison DR, Baker MD, Ober KA. 1999. Phylogeny of carabid beetles as inferred from 18S ribosomal DNA (Coleoptera: Carabidae). *Syst. Entomol.* 24(2):103–38
300. Maddison WP. 1997. Gene trees in species trees. *Syst. Biol.* 46:523–36
301. Maddison WP, Maddison DR. 1992. *MacClade: Analysis of Phylogeny and Character Evolution.* Sunderland, MA: Sinauer. Ver. 3. 0
302. Maekawa K, Kitade O, Matsumoto T. 1999. Molecular phylogeny of orthopteroid insects based on mitochondrial cytochrome oxidase II gene. *Zool. Sci.* 16:175–84
303. Makita H, Shinkawa T, Ohta K, Nakazawa T. 1996. Phylogeny of Papilionidae butterflies inferred from mitochondrial ND5 gene sequences. *Zool. Sci.* 13(Suppl.):41
304. Mange A, Prudhomme JC. 1999. Comparison of *Bombyx mori* and *Helicoverpa armigera* cytoplasmic actingenes provides clues to the evolution of actin genes in insects. *Mol. Biol. Evol.* 16(2):165–72
305. Mantovani B, Tinti F, Bachmann L, Scali V. 1997. The Bag320 satellite DNA family in *Bacillus* stick insects (Phasmatodea): different rates of molecular evolution of highly repetitive DNA in bisexual and parthenogenetic taxa. *Mol. Biol. Evol.* 14:1197–205
306. Mardulyn P, Cameron SA. 1999. The major opsin in bees (Insecta: Hymenoptera): a promising nuclear gene for higher level phylogenetics. *Mol. Phylogenet. Evol.* 12:168–76
307. Mardulyn P, Milinkovitch MC, Pasteels JM. 1997. Phylogenetic analyses of DNA and allozyme data suggest that *Gonioctena* leaf beetles (Coleoptera: Chrysomelidae) experienced convergent evolution in their history of host-plant family shifts. *Syst. Biol.* 46:722–47
308. Marfany G, Gonzalez-Duarte R. 1993. Characterization and evolution of the Adh genomic region in *Drosophila guanche* and *Drosophila madeirensis. Mol. Phylogenet. Evol.* 2(1):13–22
309. Marinucci M, Romi R, Mancini P, DiLuca M, Severini C. 1999. Sequence analysis of the ribosomal DNA in seven palearctic species of the *Anopheles maculipennis* complex. *GenBank.* http://www.ncbi.nlm.nih.gov
310. Martin JA, Pashley DP. 1992. Molecular systematic analysis of butterfly family and some subfamily relationships (Lepidoptera: Papilionoidea). *Ann. Entomol. Soc. Am.* 85:127–39
311. Martinez-Cruzado JC. 1990. Evolution of the autosomal chorion cluster in *Drosophila.* IV. The Hawaiian *Drosophila:* rapid protein evolution and constancy in the rate of DNA divergence. *J. Mol. Evol.* 31(5):402–23
312. Martinez-Torres D, Devonshire AL, Williamson MS. 1997. Molecular studies of knockdown resistance to pyrethroids: cloning of domain II sodium channel gene sequences from insects. *Pestic. Sci.* 51(3):265–70
313. Matsuoka N, Hosoya T, Hamaya T, Abe A. 1998. Phylogenetic relationships among four species of stag beetles (Coleoptera: Lucanidae) based on allozymes. *Comp. Biochem. Physiol. B* 119:401–6
314. McAllister BF, Werren JH. 1997. Phylogenetic analysis of a retrotransposon with implications for strong evolutionary constraints on reverse transcriptase *Mol. Biol. Evol.* 14:69–80
315. McHugh D. 1998. Deciphering metazoan phylogeny: the need for additional molecular data. *Am. Zool.* 38(6):859–66
316. McMichael M, Pashley Prowell D. 1999. Differences in amplified fragment-length polymorphisms in fall armyworm (Lepidoptera: Noctuidae) host strains. *Ann. Entomol. Soc. Am.* 92(2):175–81

317. McPheron BA, Han H-Y. 1997. Phylogenetic analysis of North American *Rhagoletis* (Diptera: Tephritidae) and related genera using mitochondrial DNA sequence data. *Mol. Phylogenet. Evol.* 7(1):1–16
318. Meglécz E, Solignac M. 1998. Microsatellite loci for *Parnassius mnemosyne*. *Hereditas* 128:179–80
319. Meyer A. 1994. Shortcomings of the cytochrome b gene as a molecular marker. *Trends Ecol. Evol.* 9:278–80
320. Miller BR, Crabtree MB, Savage HM. 1996. Phylogeny of fourteen *Culex* mosquito species, including the *Culex pipiens* complex, inferred from the internal transcribed spacers of ribosomal DNA. *Insect Mol. Biol.* 5(2):93–107
321. Miller BR, Crabtree MB, Savage HM. 1997. Phylogenetic relationships of the *Culicomorpha* inferred from 18S and 5. 8S ribosomal DNA sequences (Diptera: Nematocera). *Insect Mol. Biol.* 6(2): 105–14
322. Miller JS, Brower AVZ, DeSalle R. 1997. Phylogeny of the neotropical moth tribe Josiini (Notodontidae: Dioptinae): comparing and combining evidence from DNA sequences and morphology. *Biol. J. Linn. Soc.* 60:297–316
323. Mindell DP. 1991. Aligning DNA sequences: homology and phylogenetic weighting. In *Phylogenetic Analysis of DNA Sequences,* ed. MM Miyamoto, J Cracraft, p. 73–89. New York: Oxford Univ. Press. 358 pp.
324. Mindell DP, Sorenson MD, Dimcheff DE. 1998. Multiple independent origins of mitochondrial gene order in birds. *Proc. Natl. Acad. Sci. USA* 95:10693–97
325. Ming Y. 1995. Characterization of the rDNA of the genus *Rhagoletis*. *GenBank.* http://www.ncbi.nlm.nih.gov
326. Mitchell A, Cho S, Regier JC, Mitter C, Poole RW, Matthews M. 1997. Phylogenetic utility of elongation factor-1α in Noctuoidea (Insecta: Lepidoptera): the limits of synonymous substitution. *Mol. Biol. Evol.* 14:381–90
327. Mitchell SE, Cockburn AF, Seawright JA. 1993. The mitochondrial genome of *Anopheles quadrimaculatus* species A: complete nucleotide sequence and gene organization. *Genome* 36(6):1058–73
328. Miura T, Maekawa K, Kitade O, Abe T, Matsumoto T. 1998. Phylogenetic relationships among subfamilies in higher termites (Isoptera: Termitidae) based on mitochondrial COII gene sequences. *Ann. Entomol. Soc. Am.* 91:515–23
329. Miyamoto MM, Fitch WM. 1995. Testing species phylogenies and phylogenetic methods with congruence. *Syst. Biol.* 44:64–76
330. Moore WS. 1995. Inferring phylogenies from mtDNA variation: mitochondrial gene trees versus nuclear gene trees. *Evolution* 49:718–26
331. Moore WS. 1997. Mitochondrial gene trees versus nuclear gene trees, a reply to Hoelzer. *Evolution* 51:627–29
332. Moran NA, Baumann P, Von Dohlen C. 1994. Use of DNA sequences to reconstruct the history of the association between members of the Sternorrhyncha (Homoptera) and their bacterial endosymbionts. *Eur. J. Entomol.* 91:79–83
333. Moran NA, Kaplan ME, Gelsey MJ, Murphy TG, Scholes EA. 1999. Phylogenetics and evolution of the aphid genus *Uroleucon* based on mitochondrial and nuclear DNA sequences. *Syst. Entomol.* 24:85–93
334. Moran NA, Von Dohlen CD, Baumann P. 1995. Faster evolutionary rates in endosymbiotic bacteria than in cospeciating insect hosts. *J. Mol. Evol.* 41:727–31
335. Morell V. 1996. TreeBASE: the roots of phylogeny. *Science* 273:569
336. Moritz C. 1994. Applications of mitochondrial DNA analysis in conservation: a critical review. *Mol. Ecol.* 3:401–11
337. Moritz C. 1996. Use of molecular phylogenies in conservation. In *New Uses for New Phylogenies,* ed. PH Harvey, AJL Brown, JM Smith, S Nee, p. 203–14. Oxford: Oxford Univ. Press. 349 pp.
338. Moriyama EN, Hartl DL. 1993. Codon

usage bias and base composition of nuclear genes in *Drosophila. Genetics* 134:847–58
339. Moriyama EN, Powell JR. 1996. Intraspecific nuclear DNA variation in *Drosophila. Mol. Biol. Evol.* 13:261–77
340. Moriyama EN, Powell JR. 1997. Codon usage bias and tRNA abundance in *Drosophila. J. Mol. Evol.* 45:514–23
341. Mukabayire O, Boccolini D, Lochouarn L, Fontenille D, Besansky NJ. 1999. Mitochondrial and ribosomal internal transcribed spacer (ITS2) diversity of the African malaria vector *Anopheles funestus. Mol. Ecol.* 8(2):289–97
342. Mullis K, Faloona F, Scharf S, Saiki R, Horn G, Erlich H. 1986. Specific enzymatic amplification of DNA in-vitro: the polymerase chain reaction. *Cold Spring Harbor Symp. Quant. Biol.* 51:263–74
343. Murphy RW, Sites JW, Buth DG, Haufler CH. 1996. Proteins: isozyme electrophoresis. In *Molecular Systematics,* ed. DM Hillis, C Moritz, BK Mable, p. 51–120. Sunderland, MA: Sinauer. 655 pp. 2nd ed.
344. Nee S, Holmes EC, Rambaut A, Harvey PH. 1996. Inferring population history from molecular phylogenies. In *New Uses for New Phylogenies,* ed. PH Harvey, AJL Brown, JM Smith, S Nee. Oxford: Oxford Univ. Press
345. Newcomb RD, Gleeson DM. 1998. Pheromone evolution within the genera *Ctenopseustis* and *Planotortrix* (Lepidoptera: Tortricidae) inferred from a phylogeny based on cytochrome oxidase I gene variation. *Biochem. Syst. Ecol.* 26(5):473–85
346. Nielsen J, Peixoto AA, Piccin A, Costa R, Kyriacou CP, Chalmers D. 1994. Big flies, small repeats: the "Thr-Gly" region of the *period* gene in Diptera. *Mol. Biol. Evol.* 11(6):839–53
347. Nigro L, Grapputo A. 1993. Evolution of the mitochondrial ribosomal RNA in the oriental species subgroups of *Drosophila. Biochem. Syst. Ecol.* 21(1):79–83
348. Nixon KC, Carpenter JM. 1996. On simultaneous analysis. *Cladistics* 12:221–41
349. Normark BB. 1996. Phylogeny and evolution of parthenogenetic weevils of the *Aramigus tessellatus* species complex (Coleoptera: Curculionidae: Naupactini): evidence from mitochondrial DNA sequences. *Evolution* 50:734–45
350. Normark BB, Lanteri AA. 1996. *Aramigus uruguayensis* (Coleoptera: Curculionidae), a new species based on mitochondrial DNA and morphological characters. *Entomol. News* 107(5):311–16
351. Normark BB, Lanteri AA. 1998. Incongruence between morphological and mitochondrial-DNA characters suggests hybrid origins of parthenogenetic weevil lineages (genus *Aramigus*). *Syst. Biol.* 47:475–94
352. Nurminsky DI, Moriyama EN, Lozovskaya ER, Hartl DL. 1996. Molecular phylogeny and genome evolution in the *Drosophila virilis* species group: duplications of the alcohol dehydrogenase gene. *Mol. Biol. Evol.* 13(1):132–49
353. Nyman T, Roininen H, Vuorinen JA. 1998. Evolution of different gall types in willow-feeding sawflies (Hymenoptera: Tenthredinidae). *Evolution* 52:465–74
354. O'Grady PM, Clark JB, Kidwell MG. 1998. Phylogeny of the *Drosophila saltans* species group based on combined analysis of nuclear and mitochondrial DNA sequences. *Mol. Biol. Evol.* 15(6):656–64
355. Ohmiya Y, Ohba N, Toh H, Tsuju FI. 1995. Cloning, expression and sequence analysis of cDNA for the luciferases from the Japanese fireflies: *Pyrocoelia miyako* and *Hotaria parvula. Photochem. Photobiol.* 62:309–13
356. Okuyama E, Shibata H, Tachida H, Yamazaki T. 1996. Molecular evolution of the 5′-flanking regions of the duplicated Amy genes in *Drosophila melanogaster* species subgroup. *Mol. Biol. Evol.* 13(4):574–83
357. Oldroyd BP, Clifton MJ, Wongsiri S, Rinderer TE, Sylvester HA, Crozier RH. 1997. Polyandry in the genus *Apis,* particularly *Apis andreniformis. Behav. Ecol. Sociobiol.* 40:17–26
358. Ortí G, Hare MP, Avise JC. 1997. Detec-

tion and isolation of nuclear haplotypes by PCR-SSCP. *Mol. Ecol.* 6:575–80

359. Ousley A, Zafarullah K, Chen Y, Emerson M, Hickman L, Sehgal A. 1998. Conserved regions of the timeless (*tim*) clock gene in *Drosophila* analyzed through phylogenetic and functional studies. *Genetics* 148(2): 815–25
360. Packer L. 1991. The evolution of social behavior and nest architecture in sweat bees of the subgenus *Evylaeus* (Hymenoptera: Halictidae): a phylogenetic approach. *Behav. Ecol. Sociobiol.* 29:153–60
361. Packer L, Taylor JS, Savignano DA, Bleser CA, Lane CP, Sommers LA. 1998. Population biology of an endangered butterfly, *Lycaeides melissa samuelis* (Lepidoptera; Lycaenidae): genetic variation, gene flow, and taxonomic status. *Can. J. Zool.* 76(2):320–29
362. Page RDM. 1996. Temporal congruence revisited: comparison of mitochondrial DNA sequence divergence in cospeciating pocket gophers and their chewing lice. *Syst. Biol.* 45:151–67
363. Page RDM. 1998. GeneTree: comparing gene and species phylogenies using reconciled trees. *Bioinformatics* 14:(9):819–20
364. Page RDM, Lee PLM, Becher SA, Griffiths R, Clayton DH. 1998. A different tempo of mitochondrial DNA evolution in birds and their parasitic lice. *Mol. Phylogenet. Evol.* 9:276–93
365. Palumbi SR. 1996. Nucleic acids II: the polymerase chain reaction. In *Molecular Systematics,* ed. DM Hillis, C Moritz, BK Mable. Sunderland, MA: Sinauer. 2nd ed.
366. Pashley DP, Ke LD. 1992. Sequence evolution in mitochondrial ribosomal and ND-1 genes in Lepidoptera: implications for phylogenetic analyses. *Mol. Biol. Evol.* 6:1061–75
367. Pashley DP, McPheron BA, Zimmer EA. 1993. Systematics of holometabolous insect orders based on 18S ribosomal RNA. *Mol. Phylogenet. Evol.* 2:132–42
368. Paskewitz SM, Wesson DM, Collins FH. 1993. The internal transcribed spacers of ribosomal DNA in five members of the *Anopheles gambiae* species complex. *Insect Mol. Biol.* 2(4):247–57
369. Passamonti M, Mantovani B, Scali V. 1998. *GenBank.* http://www.ncbi.nlm.nih.gov
370. Paterson AM, Wallis GP, Wallis LJ, Gray RD. 1999. Seabird and louse coevolution: complex histories revealed by sequence data and reconciliation analysis. *Syst. Biol.* In press
371. Pawlowski J, Szadziewski R, Kmieciak D, Fahrni J, Bittar G. 1996. Phylogeny of the infraorder Culicomorpha (Diptera: Nematocera) based on 28S RNA gene sequences. *Syst. Entomol.* 21(2):167–78
372. Pedersen BV. 1996. A phylogenetic analysis of cuckoo bumblebees (*Psithyrus* Lepeletier) and bumblebees (*Bombus* Latreille) inferred from sequences of the mitochondrial gene cytochrome oxidase I. *Mol. Phylogenet. Evol.* 5(2):289–97
373. Pedersen BV. 1996. On the phylogenetic position of the Danish strain of the black honeybee (the Laeso bee), *Apis mellifera mellifera* L. (Hymenoptera: Apidae) inferred from mitochondrial DNA sequences. *Entomol. Scand.* 27(3):241–50
374. Peixoto AA, Costa R, Wheeler DA, Hall JC, Kyriacou CP. 1992. Evolution of the Threonine-Glycine repeat region of the *period* gene in the *melanogaster* species subgroup of *Drosophila. J. Mol. Evol.* 35(5):411–19
375. Peixoto AA, Hennessy JM, Townson I, Hasan G, Rosbash M, et al. 1998. *Proc. Natl. Acad. Sci. USA* 95(8):4475–80
376. Pelandakis M, Higgins DG, Solignac M. 1991. Molecular phylogeny of the subgenus *Sophophora* of *Drosophila* derived from large subunit of ribosomal RNA sequences. *Genetica* 84(2):87–94
377. Pelandakis M, Solignac M. 1993. Molecular phylogeny of *Drosophila* based on ribosomal RNA sequences. *J. Mol. Evol.* 37(5):525–43
378. Pellmyr O, Leebens-Mack J. 1998. Herbi-

vores and molecular clocks as tools in plant biogeography. *Biol. J. Linn. Soc.* 63(3):367–78

379. Pellmyr O, Leebens-Mack J, Huth CJ. 1996. Non-mutualistic yucca moths and their evolutionary consequences. *Nature* 380(6570):155–56
380. Piano F, Craddock EM, Kambysellis MP. 1997. Phylogeny of the island populations of the Hawaiian *Drosophila grimshawi* complex: evidence from combined data. *Mol. Phylogenet. Evol.* 7(2):173–84
381. Pinto JD, Stouthamer R, Platner GR. 1997. A new cryptic species of *Trichogramma* (Hymenoptera: Trichogrammatidae) from the Mojave Desert of California as determined by morphological, reproductive and molecular data. *Proc. Entomol. Soc. Wash.* 99(2):238–47
382. Pirounakis K, Koulianos S, Schmid-Hempel P. 1998. Genetic variation among European populations of *Bombus pascuorum* from mitochondrial DNA sequence data. *Eur. J. Entomol.* 95:27–33
383. Pitnick S, Markow TA, Spicer GS. 1995. Delayed male maturity is a cost of producing large sperm in *Drosophila. Proc. Natl. Acad. Sci. USA* 92:10614–18
384. Plohl M, Mestrovic N, Bruvo B, Ugarkovic D. 1998. Similarity of structural features and evolution of satellite DNAs from *Palorus subdepressus* (Coleoptera) and related species. *J. Mol. Evol.* 46:234–39
385. Poinar HN, Hoess M, Bada JL, Pääbo S. 1996. Amino acid racemization and the preservation of ancient DNA. *Science* 272(5263):864–66
386. Pollock DD, Watt WB, Rashbrook VK, Iyengar EV. 1998. Molecular phylogeny for *Colias* butterflies and their relatives (Lepidoptera: Pieridae). *Ann. Entomol. Soc. Am.* 91:524–31
387. Pons J, Bruvo B, Juan C, Petitpierre E, Plohl M, Ugarkovic D. 1997. Conservation of satellite DNA in species of the genus *Pimelia* (Tenebrionidae, Coleoptera). *Gene* 205(1–2):183–90
388. Popp MP, Grisshammer R, Hargrave PA, Smith WC. 1996. Ant opsins: sequences from the Saharan silver ant and the carpenter ant. *Invertebr. Neurosci.* 1:323–29
389. Pornkulwat S, Skoda SR, Thomas GD, Foster JE. 1998. Random amplified polymorphic DNA used to identify genetic variation in ecotypes of the European corn borer (Lepidoptera: Pyralidae). *Ann. Entomol. Soc. Am.* 91:719–25
390. Porter CH, Collins FH. 1996. Phylogeny of nearctic members of the *Anopheles maculipennis* species group derived from the D2 variable region of 28S ribosomal RNA. *Mol. Phylogenet. Evol.* 6(2):178–88
391. Post RJ, Flook PK, Millest AL. 1993. Methods for the preservation of insects for DNA studies. *Biochem. Syst. Ecol.* 21:85–92
392. Powell JR. 1997. *Progress and Prospects in Evolutionary Biology: The Drosophila Model.* New York: Oxford Univ. Press. 562 pp.
393. Powell JR, DeSalle R. 1995. *Drosophila* molecular phylogenies and their uses. *Evol. Biol.* 28:87–138
394. Prueser F, Mossakowski D. 1998. Low substitution rates in mitochondrial DNA in mediterranean carabid beetles. *Insect Mol. Biol.* 7:121–28
395. Pruess KP, Adams BJ, Parsons TJ, Zhu X, Powers TO. 1999. Phylogenetic relationships of black flies inferred from the mitochondrial cytochrome oxidase II gene. *GenBank.* http://www.ncbi.nlm.nih.gov
396. Purvis A, Quicke DLJ. 1997. Building phylogenies: Are the big easy? *Trends Ecol. Evol.* 12(2):49–50
397. Queller DC, Strassmann JE, Hughes CR. 1993. Microsatellites and kinship. *Trends Ecol. Evol.* 8:285–88
398. Ramirez MG, Froehlig JL. 1997. Minimal genetic variation in a coastal dune arthropod: the trapdoor spider *Aptostichus simus* (Cyrtaucheniidae). *Conserv. Biol.* 11(1): 256–59
399. Ramos-Onsins S, Aguade M. 1998. Molecular evolution of cecropin multigene family

in *Drosophila:* functional genes vs. pseudogenes. *Genetics* 150(1):157–71

400. Ramos-Onsins S, Segarra C, Rozas J, Aguade M. 1998. Molecular and chromosomal phylogeny in the *obscura* group of *Drosophila* inferred from sequences of the rp49 gene region. *Mol. Phylogenet. Evol.* 9(1):33–41
401. Rasplus J-Y, Kerdelhue C, Le Clainche I, Mondor G. 1998. Molecular phylogeny of fig wasps: Agaonidae are not monophyletic. *C. R. Acad. Sci. III* 321(6):517–27
402. Ready PD, Day JC, Souza AAD, Rangel EF, Davies CR. 1997. Mitochondrial DNA characterization of populations of *Lutzomyia whitmani* (Diptera: Psychodidae) incriminated in the peri-domestic and silvatic transmission of *Leishmania* species in Brazil. *Bull. Entomol. Res.* 87(2):187–95
403. Reed RD, Sperling FAH. 1999. Interaction of process partitions in phylogenetic analysis: an example from the swallowtail butterfly genus *Papilio. Mol. Biol. Evol.* 16:286–97
404. Regier JC, Fang QQ, Mitter C, Peigler RS, Friedlander TP, Solis AM. 1998. Evolution and phylogenetic utility of the *period* gene in Lepidoptera. *Mol. Biol. Evol.* 15:1172–82
405. Regier JC, Shultz JW. 1997. Molecular phylogeny of the major arthropod groups indicates polyphyly of crustaceans and a new hypothesis for the origin of hexapods. *Mol. Biol. Evol.* 14:902–13
406. Reineke A, Karlovsky P, Zebitz CPW. 1998. Preparation and purification of DNA from insects for AFLP analysis. *Insect Mol. Biol.* 7(1):95–99
407. Reiss RA, Schwert DP, Ashworth AC. 1995. Field preservation of Coleoptera for molecular genetic analyses. *Environ. Entomol.* 24:716–19
408. Remsen J, DeSalle R. 1998. Character congruence of multiple data partitions and the origin of the Hawaiian Drosophilidae. *Mol. Phylogenet. Evol.* 9(2):225–35
409. Reyes SG, Cooper SJB, Schwarz MP. 1999. Species phylogeny of the bee genus *Exoneurella* Michener (Hymenoptera: Apidae: Allodapini): evidence from molecular and morphological data sets. *Ann. Entomol. Soc. Am.* 92(1):20–29
410. Richardson BJ, Baverstock PR, Adams M. 1986. *Allozyme Electrophoresis: A Handbook for Animal Systematics and Population Studies.* Orlando, FL: Academic
411. Robertson HM, Soto-Adames FN, Walden KKO, Avancini RMP, Lampe DJ. 1998. The mariner transposons of animals: horizontally jumping genes. In *Horizontal Gene Transfer,* ed. M Syvanen, C Kado, pp. 268–84. London: Chapman & Hall
412. Robinson M, Gouy M, Gautier C, Mouchiroud D. 1998. Sensitivity of the relative-rate test to taxonomic sampling. *Mol. Biol. Evol.* 15:1091–98
413. Roderick GK. 1996. Geographic structure of insect populations: gene flow, phylogeography, and their uses. *Annu. Rev. Entomol.* 41:325–52
414. Roderick GK. 1997. Herbivorous insects and the Hawaiian silversword alliance: coevolution or cospeciation? *Pac. Sci.* 51:440–49
415. Roger AJ, Sandblom O, Doolittle WF, Philippe H. 1999. An evaluation of elongation factor 1-alpha as a phylogenetic marker for eukaryotes *Mol. Biol. Evol.* 16(2):218–33
416. Rouhbakhsh D, Lai C-Y, Von Dohlen CD, Clark MA, Baumann L, et al. 1996. The tryptophan biosynthetic pathway of aphid endosymbionts (*Buchnera*): genetics and evolution of plasmid-associated anthranilate synthase (trpEG) within the Aphididae. *J. Mol. Evol.* 42:414–21
417. Rowan RG, Hunt JA. 1987. Phylogenetic comparisons of the *Adh* gene from five species of Hawaiian *Drosophila. Genetics* 116(1 Part 2):S23
418. Rowan RG, Hunt JA. 1991. Rates of DNA change and phylogeny from the DNA sequences of the alcohol dehydrogenase gene for five closely related species of Hawaiian *Drosophila. Mol. Biol. Evol.* 8(1):49–70
419. Ruiz Linares A, Hancock JM, Dover GA.

1991. Secondary structure constraints on the evolution of *Drosophila* 28S ribosomal RNA expansion segments. *J. Mol. Biol.* 219(3):381–90

420. Russo CAM, Takezaki N, Nei M. 1995. Molecular phylogeny and divergence times of *Drosophilid* species. *Mol. Biol. Evol.* 12(3):391–404
421. Ruttkay H, Solignac M, Sperlich D. 1992. Nuclear and mitochondrial ribosomal RNA variability in the *obscura* group of *Drosophila. Genetica* 85(2):131–38
422. Sanderson MJ. 1996. How many taxa must be sampled to identify the root node of a large clade? *Syst. Biol.* 45:168–73
423. Sanderson MJ, Baldwin BG, Bharathan G, Campbell CS, Von Dohlen C, et al. 1993. The growth of phylogenetic information and the need for a phylogenetic data base. *Syst. Biol.* 42(4):562–68
424. Sanderson MJ, Donoghue MJ, Piel W, Eriksson T. 1994. TreeBASE: a prototype database of phylogenetic analyses and an interactive tool for browsing the phylogeny of life. *Am. J. Bot.* 81(6):183
425. Sanderson MJ, Purvis A, Henze C. 1998. Phylogenetic supertrees: assembling the trees of life. *Trends Ecol. Evol.* 13(3): 105–9
426. Sandoval C, Carmean DA, Crespi BJ. 1998. Molecular phylogenetics of sexual and parthenogenetic *Timema* walkingsticks. *Proc. R. Soc. London Ser. B* 265(1396):589–95
427. Satta Y, Takahata N. 1990. Evolution of *Drosophila* mitochondrial DNA and the history of the *melanogaster* subgroup. *Proc. Natl. Acad. Sci. USA* 87:9558–62
428. Schilthuizen M, Nordlander G, Stouthamer R, Van Alphen JJM. 1998. Morphological and molecular phylogenetics in the genus *Leptopilina* (Hymenoptera: Eucoilidae). *Syst. Entomol.* 23:253–64
429. Schlötterer C, Hauser M-T, Von Haeseler A, Tautz D. 1994. Comparative evolutionary analysis of rDNA ITS regions in *Drosophila. Mol. Biol. Evol.* 11(3):513–22
430. Schlötterer C, Ritter R, Harr B, Brem G. 1998. High mutation rate of a long microsatellite allele in *Drosophila melanogaster* provides evidence for allele-specific mutation rates. *Mol. Biol. Evol.* 15(10):1269–74
431. Schmitz J, Moritz RFA. 1998. Molecular phylogeny of Vespidae (Hymenoptera) and the evolution of sociality in wasps. *Mol. Phylogenet. Evol.* 9:183–91
432. Schmitz J, Moritz RFA. 1998. Sociality and the rate of rDNA sequence evolution in wasps (Vespidae) and honeybees. *J. Mol. Evol.* 47:606–12
433. Schreiber DE, Garner KJ, Slavicek JM. 1997. Identification of three randomly amplified polymorphic DNA-polymerase chain reaction markers for distinguishing Asian and North American gypsy moths (Lepidoptera: Lymantriidae). *Ann. Entomol. Soc. Am.* 90(5):667–74
434. Severini C, Silvestrini F, Mancini P, La Rosa G, Marinucci M. 1996. Sequence and secondary structure of the rDNA second internal transcribed spacer in the sibling species *Culex pipiens* L. and *Cx. quinquefasciatus* Say (Diptera: Culicidae). *Insect Mol. Biol.* 5:181–86
435. Shaw KL. 1996. Sequential radiations and patterns of speciation in the Hawaiian cricket genus *Laupala* inferred from DNA sequences. *Evolution* 50:237–55
436. Shibata H, Yamazaki T. 1995. Molecular evolution of the duplicated *Amy* locus in the *Drosophila melanogaster* species subgroup: Concerted evolution only in the coding region and an excess of nonsynonymous substitutions in speciation. *Genetics.* 141(1):223–36
437. Shimada T, Kurimoto Y, Kobayashi M. 1995. Phylogenetic relationship of silkmoths inferred from sequence data of the arylphorin gene. *Mol. Phylogenet. Evol.* 4:223–34
438. Silvain J-F, Delobel A. 1998. Phylogeny of West African *Caryedon* (Coleoptera: Bruchidae): congruence between molecular and morphological data. *Mol. Phylogenet. Evol.* 9:533–41
439. Simmons GM, Kwok W, Matulonis P, Ven-

katesh T. 1994. Polymorphism and divergence at the *prune* locus in *Drosophila melanogaster* and *D. simulans*. *Mol. Biol. Evol.* 11(4):666–71

440. Simon C, Frati F, Beckenbach A, Crespi B, Liu H, Flook P. 1994. Evolution, weighting, and phylogenetic utility of mitochondrial gene sequences and a compilation of conserved polymerase chain reaction primers. *Ann. Entomol. Soc. Am.* 87:651–701
441. Simon C, McIntosh C, Deniega J. 1993. Standard restriction fragment analysis of the mitochondrial genome is not sensitive enough for phylogenetic analysis or identification of 17-year periodical cicada broods (Hemiptera: Cicadidae): the potential for a new technique. *Ann. Entomol. Soc. Am.* 86:228–38
442. Simon C, Nigro L, Sullivan J, Holsinger K, Martin A, et al. 1996. Large differences in substitutional pattern and evolutionary rate of 12S ribosomal RNA genes. *Mol. Biol. Evol.* 13:923–32
443. Sipes SD. 1997. The female of *Diadasia afflictula* unveiled and verified using molecular markers (Hymenoptera: Apidae). *GenBank.* http://www.ncbi.nlm.nih.gov
444. Smith JJ, Bush GL. 1997. Phylogeny of the genus *Rhagoletis* (Diptera: Tephritidae) inferred from DNA sequences of mitochondrial cytochrome oxidase II. *Mol. Phylogenet. Evol.* 7(1):33–43
445. Smith PT, Kambhampati S, Völkl W, Mackauer M. 1999. A phylogeny of aphid parasitoids (Hymenoptera: Braconidae: Aphidiinae) inferred from mitochondrial NADH 1 dehydrogenase gene sequence. *Mol. Phylogenet. Evol.* 11(2):236–45
446. Smith SM, Fuerst P, Meckelenburg KL. 1996. Mitochondrial DNA sequence of cytochrome oxidase II from *Calliphora erythrocephala:* Evolution of blowflies. *Ann. Entomol. Soc. Am.* 89(1):28–36
447. Solis CR, Hughes CR, Klingler CJ, Strassmann JE, Queller DC. 1998. Lack of kin discrimination during wasp colony fission. *Behav. Ecol.* 9:172–76
448. Sommer RJ, Retzlaff M, Goerlich K, Sander K, Tautz D. 1992. Evolutionary conservation pattern of zinc-finger domains of *Drosophila* segmentation genes. *Proc. Natl. Acad. Sci. USA* 89:10782–86
449. Sorensen JT, Campbell BC, Gill RJ, Steffen-Campbell JD. 1995. Non-monophyly of Auchenorrhyncha ("Homoptera"), based upon 18S rDNA phylogeny: eco-evolutionary and cladistic implications within pre-Heteropterodea Hemiptera (S.L.) and a proposal for new monophyletic suborders. *Pan-Pac. Entomol.* 71:31–60
450. Soto-Adames FN. *Phylogeny of the families of Collembola based on the 18S rRNA gene.* PhD diss. Champaign-Urbana: Univ. Ill.
451. Soto-Adames FN, Robertson HM, Berlocher SH. 1994. Phylogenetic utility of partial DNA sequences of G6pdh at different taxonomic levels in Hexapoda with emphasis on Diptera. *Ann. Entomol. Soc. Am.* 87:723–36
452. Sperling FAH. 1993. Mitochondrial DNA phylogeny of the *Papilio machaon* species group (Lepidoptera: Papilionidae). *Mem. Entomol. Soc. Can.* 165:233–42
453. Sperling FAH, Anderson GS, Hickey DA. 1994. A DNA-based approach to the identification of insect species used for postmortem interval estimation. *J. Forensic Sci.* 39(2):418–27
454. Sperling FAH, Byers R, Hickey D. 1996. Mitochondrial DNA sequence variation among pheromotypes of the dingy cutworm, *Feltia jaculifera* (Gn.) (Lepidoptera: Noctuidae). *Can. J. Zool.* 74:2109–17
455. Sperling FAH, Feeny P. 1996. Umbellifer and composite feeding in *Papilio:* phylogenetic frameworks and constraints on caterpillars. In *Swallowtail Butterflies: Their Ecology and Evolutionary Biology,* ed. JM Scriber, Y Tsubaki, RC Lederhouse. Gainesville, FL: Scientific
456. Sperling FAH, Harrison RG. 1994. Mitochondrial DNA variation within and between species of the *Papilio machaon*

group of swallowtail butterflies. *Evolution* 48:408–22

457. Sperling FAH, Hickey DA. 1994. Mitochondrial DNA sequence variation in the spruce budworm species complex (*Choristoneura:* Lepidoptera). *Mol. Biol. Evol.* 11:656–65
458. Sperling FAH, Hickey DA. 1995. Amplified mitochondrial DNA as a diagnostic marker for species of conifer-feeding *Choristoneura* (Lepidoptera: Tortricidae). *Can. Entomol.* 127(3):277–88
459. Sperling FAH, Landry J-F, Hickey DA. 1995. DNA-based identification of introduced ermine moth species in North America (Lepidoptera: Yponomeutidae). *Ann. Entomol. Soc. Am.* 88:155–62
460. Sperling FAH, Raske AG, Otvos IS. 1999. Mitochondrial DNA sequence variation among populations and host races of *Lambdina fiscellaria* (Gn.) (Lepidoptera: Geometridae). *Insect Mol. Biol.* 8(1):97–106
461. Sperling FAH, Spence JR, Andersen NM. 1997. Mitochondrial DNA, allozymes, morphology, and hybrid compatibility in *Limnoporus* water striders (Heteroptera: Gerridae): Do they all track species phylogenies? *Ann. Entomol. Soc. Am.* 90:401–15
462. Spicer GS. 1995. Phylogenetic utility of the mitochondrial cytochrome oxidase gene. Molecular evolution of the *Drosophila buzzatii* species complex. *J. Mol. Evol.* 41(6):749–59
463. Spicer GS, Jaenike J. 1996. Phylogenetic analysis of breeding site use and alpha-amanitin tolerance within the *Drosophila quinaria* species group. *Evolution* 50(6):2328–37
464. Spicer GS, Pitnick S. 1996. Molecular systematics of the *Drosophila hydei* subgroup as inferred from mitochondrial DNA sequences. *J. Mol. Evol.* 43(3):281–86
465. Stankiewicz BA, Poinar HN, Briggs DEG, Evershed RP, Poinar GO Jr. 1998. Chemical preservation of plants and insects in natural resins. *Proc. R. Soc. London Ser. B* 265(1397):641–47
466. Stauffer C, Lakatos F, Hewitt GM. 1997. The phylogenetic relationships of seven European *Ips* (Scolytidae, Ipinae) species. *Insect Mol. Biol.* 6:233–40
467. Stauffer C, Zuber M. 1998. *Ips amitinus* var. *montana* (Coleoptera, Scolytidae) is synonymous to *Ips amitinus:* a morphological, behavioural and genetic re-examination. *Biochem. Syst. Ecol.* 26:171–83
468. Stern DL. 1994. A phylogenetic analysis of soldier evolution in the aphid family Hormaphididae. *Proc. R. Soc. London Ser. B* 256(1346):203–9
469. Stern DL. 1995. Phylogenetic evidence that aphids, rather than plants, determine gall morphology. *Proc. R. Soc. London Ser. B* 260(1357):85–89
470. Stern DL. 1998. Phylogeny of the tribe Cerataphidini (Homoptera) and the evolution of the horned soldier aphids. *Evolution* 52:155–65
471. Stern DL, Aoki S, Kurosu U. 1997. Determining aphid taxonomic affinities and life cycles with molecular data: a case study of the tribe Cerataphidini (Hormaphididae: Aphidoidea: Hemiptera). *Syst. Entomol.* 22:81–96
472. Stevens J, Wall R. 1996. Species, sub-species and hybrid populations of the blowflies *Lucilia cuprina* and *Lucilia sericata. Proc. R. Soc. Lond. Ser. B* 263:1335–41
473. Stevens J, Wall R. 1997. The evolution of ectoparasitism in the genus *Lucilia* (Diptera: Calliphoridae). *Int. J. Parasitol.* 27(1):51–59
474. Stevens J, Wall R. 1997. Genetic variation in populations of the blowflies *Lucilia cuprina* and *Lucilia sericata* (Diptera: Calliphoridae). Random amplified polymorphic DNA analysis and mitochondrial DNA sequences. *Biochem. Syst. Ecol.* 25(2):81–97
475. Stone GN, Cook JM. 1998. The structure of cynipid oak galls: patterns in the evolution of an extended phenotype. *Proc. R. Soc. London Ser. B* 265:979–88

476. Stothard JR, Yamamoto Y, Cherchi A, Garcia AL, Valenta SAS, et al. 1998. A preliminary survey of mitochondrial sequence variation in Triatominae (Hemiptera: Reduviidae) using polymerase chain reaction-based single strand conformational polymorphism (SSCP) analysis and direct sequencing. *Bull. Entomol. Res.* 88:553–60
477. Su Z-H, Ohama T, Okada TS, Nakamura K, Ishikawa R, Osawa S. 1996. Phylogenetic relationships and evolution of the Japanese Carabinae ground beetles based on mitochondrial ND5 gene sequences. *J. Mol. Evol.* 42:124–29
478. Su Z-H, Ohama T, Okada TS, Nakamura K, Ishikawa R, Osawa S. 1996. Geography-linked phylogeny of the *Damaster* ground beetles inferred from mitochondrial ND5 gene sequences. *J. Mol. Evol.* 42:130–34
479. Su Z-H, Tominaga O, Ohama T, Kajiwara E, Ishikawa R, et al. 1996. Parallel evolution in radiation of *Ohomopterus* ground beetles inferred from mitochondrial ND5 gene sequences. *J. Mol. Evol.* 43:662–71
480. Su Z-H, Tominaga O, Okamoto M, Osawa S. 1998. Origin and diversification of hindwingless *Damaster* ground beetles within the Japanese islands as deduced from mitochondrial ND5 gene sequences (Coleoptera, Carabidae). *Mol. Biol. Evol.* 15(8):1026–39
481. Sunnucks P, Driver F, Brown WV, Carver M, Hales DF, Milne WM. 1997. Biological and genetic characterization of morphologically similar *Therioaphis trifolii* (Monell) (Hemiptera: Aphididae) with different host utilization. *Bull. Entomol. Res.* 87(4):425–36
482. Sunnucks P, Hales DF. 1996. Numerous transposed sequences of mitochondrial cytochrome oxidase I-II in aphids of the genus *Sitobion* (Hemiptera: Aphididae). *Mol. Biol. Evol.* 13:510–24
483. Suzuki H. 1997. Molecular phylogenetic studies of Japanese fireflies and their mating systems (Coleoptera: Cantharoidea) *Tokyo Metro. Univ. Bull. Natl. Hist.* 3:1–53
484. Swofford DL, 1993. *PAUP: Phylogenetic Anaysis Using Parsimony.* Washington, DC: Smithson. Inst. Ver. 3. 1. 1
485. Swofford DL. 1998. *PAUP*: Phylogenetic Anaysis Using Parsimony.* Sunderland, MA: Sinauer. Ver. 4.0b1
486. Swofford DL, Olsen GJ, Waddell PJ, Hillis DM. 1996. Phylogenetic inference. In *Molecular Systematics,* ed. DM Hillis, C Moritz, BK Mable. Sunderland, MA: Sinauer. 2nd ed.
487. Szalanski AL, Powers TO. 1996. Molecular diagnostics of three *Diabrotica* (Coleoptera: Chrysomelidae) pest species. *J. Kans. Entomol. Soc.* 69:260–66
488. Tamura K. 1992. The rate and pattern of nucleotide substitution in *Drosophila* mitochondrial DNA. *Mol. Biol. Evol.* 9(5):814–25
489. Tang J, Pruess K, Cupp EW, Unnasch TR. 1996. Molecular phylogeny and typing of blackflies (Diptera: Simuliidae) that serve as vectors of human or bovine onchocerciasis. *Med. Vet. Entomol.* 10(3):228–34
490. Tang J, Pruess K, Unnasch TR. 1996. Genotyping North America black flies by means of mitochondrial ribosomal RNA sequences. *Can. J. Zool.* 74(1):39–46
491. Tang J, Toe L, Back C, Unnasch TR. 1995. Mitochondrial alleles of *Simulium damnosum* sensu lato infected with *Onchocerca volvulus. Int. J. Parasitol.* 25(10):1251–54
492. Tang J, Toe L, Back C, Zimmerman PA, Pruess K, Unnasch TR. 1995. The *Simulium damnosum* species complex: phylogenetic analysis and molecular identification based upon mitochondrially encoded gene sequences. *Insect Mol. Biol.* 4(2):79–88
493. Tarrio R, Rodriguez-Trelles F, Ayala FJ. 1998. New *Drosophila* introns originate by duplication. *Proc. Natl. Acad. Sci. USA* 95(4):1658–62
494. Tautz D, Nigro L. 1998. Microevolutionary divergence pattern of the segmentation gene *hunchback* in *Drosophila. Mol. Biol. Evol.* 15(11):1403–11
495. Taylor DB, Peterson RD, Szalanski AL,

Petersen JJ. 1997. Mitochondrial DNA variation among *Muscidifurax* spp. (Hymenoptera: Pteromalidae), pupal parasitoids of filth flies (Diptera). *Ann. Entomol. Soc. Am.* 90:814–24

496. Taylor MFJ, Heckel DG, Brown TM, Kreitman ME, Black B. 1993. Linkage of pyrethroid insecticide resistance to a sodium channel locus in the tobacco budworm. *Insect Biochem. Mol. Biol.* 23(7):763–75
497. Taylor MFJ, McKechnie SW, Pierce N, Kreitman M. 1993. The lepidopteran mitochondrial control region: structure and evolution. *Mol. Biol. Evol.* 10:1259–72
498. Telford MJ, Holland PWH. 1997. Evolution of 28S ribosomal DNA in chaetognaths: duplicate genes and molecular phylogeny. *J. Mol. Evol.* 44:135–44
499. Thomas RH, Hunt JA. 1991. The molecular evolution of the alcohol dehydrogenase locus and the phylogeny of Hawaiian *Drosophila. Mol. Biol. Evol.* 8(5):687–702
500. Thomas RH, Hunt JA. 1993. Phylogenetic relationships in *Drosophila:* a conflict between molecular and morphological data. *Mol. Biol. Evol.* 10(2):362–74
501. Thorpe JP, Solé-Cava AM. 1994. The use of allozyme electrophoresis in invertebrate systematics. *Zool. Scr.* 23(1):3–18
502. Toh H. 1991. Sequence analysis of firefly luciferase family reveals a conservative sequence motif. *Protein Seq. Data Anal.* 4:111–18
503. Tominaga H, Narise S. 1995. Sequence evolution of the *Gpdh* gene in the *Drosophila virilis* species group. *Genetica* 96(3):293–302
504. Townson H, Harbach RE, Callan TA. 1999. DNA identification of musuem specimens of the *Anopheles gambiae* complex: an evaluation of PCR as a tool for resolving the formal taxonomy of sibling species complexes. *Syst. Entomol.* 24:95–100
505. Trieschmann L, Schulze E, Schulze B, Grossbach U. 1997. The histone H1 genes of the dipteran insect, *Chironomus thummi,* fall under two divergent classes and encode proteins with distinct intranuclear distribution and potentially different functions. *Eur. J. Biochem.* 250(1):184–96
506. Ugarkovic D, Podnar M, Plohl M. 1996. Satellite DNA of the red flour beetle *Tribolium castaneum:* comparative study of satellites from the genus *Tribolium. Mol. Biol. Evol.* 13:1059–66
507. Van de Peer Y, Caers A, De Rijk P, De Wachter R. 1999. Database on the structure of small ribosomal subunit RNA. *Nucleic Acids Res.* 27(1):179–83
508. Vane-Wright RI, Raheem DC, Cieslak A, Vogler AP. 1999. Evolution of the mimetic African swallowtail butterfly *Papilio dardanus:* molecular data confirm relationships with *P. phorcas* and *P. constantinus. Biol. J. Linn. Soc.* 66(2):215–29
509. Vaughn TT, Antolin MF. 1998. Population genetics of an opportunistic parasitoid in an agricultural landscape. *Heredity* 80:152–62
510. Vazquez P, Cooper SJB, Gosalvez J, Hewitt GM. 1994. Nuclear DNA introgression across a Pyrenean hybrid zone between parapatric subspecies of the grasshopper *Chorthippus parallelus. Heredity* 73(4):436–43
511. Villablanca FX, Roderick GK, Palumbi SR. 1998. Invasion genetics of the mediterranean fruit fly: variation in multiple nuclear introns. *Mol. Ecol.* 7:547–60
512. Villarroya A, Juan E. 1991. *Adh* and phylogenetic relationships of *Drosophila lebanonensis* (*Scaptodrosophila*). *J. Mol. Evol.* 32(5):421–28
513. Vincent S. 1997. Partial sequence of the subunit I cytochrome oxidase b mitochondrial gene from European blowflies of interest in forensic sciences. *GenBank.* http://www.ncbi.nlm.nih.gov
514. Vogler AP, DeSalle R. 1993. Phylogeographic patterns in coastal North American tiger beetles (*Cicindela dorsalis* Say) inferred from mitochondrial DNA sequences. *Evolution* 47(4):1192–202
515. Vogler AP, DeSalle R. 1994. Evolution and phylogenetic information content of the

ITS-1 region in the tiger beetle *Cicindela dorsalis*. *Mol. Biol. Evol.* 11:392–405
516. Vogler AP, DeSalle R, Assmann T, Knisley CB, Schultz TD. 1993. Molecular population genetics of the endangered tiger beetle *Cicindela dorsalis* (Coleoptera: Cicindelidae). *Ann. Entomol. Soc. Am.* 86:142–52
517. Vogler AP, Kelley KC. 1998. Covariation of defensive traits in tiger beetles (genus *Cicindela*): A phylogenetic approach using mtDNA. *Evolution* 52:529–38
518. Vogler AP, Knisley BC, Glueck SB, Hill JM, DeSalle R. 1993. Using molecular and ecological data to diagnose endangered populations of the puritan tiger beetle *Cicindela puritana*. *Mol. Ecol.* 2:375–83
519. Vogler AP, Pearson DL. 1996. A molecular phylogeny of the tiger beetles (Cicindelidae): congruence of mitochondrial and nuclear rDNA data sets. *Mol. Phylogenet. Evol.* 6:321–38
520. Vogler AP, Welsh A. 1997. Phylogeny of North American *Cicindela* tiger beetles inferred from multiple mitochondrial DNA sequences. *Mol. Phylogenet. Evol.* 8(2):225–35
521. Vogler AP, Welsh A, Barraclough TG. 1998. Molecular phylogeny of the *Cicindela maritima* (Coleoptera: Cicindelidae) group indicates fast radiation in western North America. *Ann. Entomol. Soc. Am.* 91:185–94
522. Vogler AP, Welsh A, Hancock JM. 1997. Phylogenetic analysis of slippage-like sequence variation in the V4 rRNA expansion segment in tiger beetles (Cicindelidae). *Mol. Biol. Evol.* 14(1):6–19
523. Von Dohlen CD, Moran NA. 1995. Molecular phylogeny of the Homoptera: a paraphyletic taxon. *J. Mol. Evol.* 41:211–23
524. Waegele JW, Stanjek G. 1995. Arthropod phylogeny inferred from partial 12S rRNA revisited: monophyly of the Tracheata depends on sequence alignment. *J. Zool. Syst. Evol. Res.* 33(2):75–80
525. Walden KK, Robertson HM. 1997. Ancient DNA from amber fossil bees? *Mol. Biol. Evol.* 14(10):1075–77
526. Walldorf U, Hovemann BT. 1990. *Apis mellifera* cytoplasmic elongation factor 1 alpha (EF-1α) is closely related to *Drosophila melanogaster* EF-1α. *FEBS Lett.* 267:245–49
527. Walldorf U, Hovemann BT, Bautz EKF. 1985. F1 and F2: two similar genes regulated differently during development of *Drosophila melanogaster. Proc. Natl. Acad. Sci. USA* 82(17):5795–99
528. Walton C, Butlin RK, Monk KA. 1997. A phylogeny for grasshoppers of the genus *Chitaura* (Orthoptera: Acrididae) from Sulawesi, Indonesia, based on mitochondrial DNA sequence data. *Biol. J. Linn. Soc.* 62:365–82
529. Watabe H, Bachmann L, Haring E, Sperlich D. 1997. Taxonomic and molecular studies on *Drosophila sinobscura* and *D. hubeiensis,* two sibling species of the *D. obscura* group. *J. Zool. Syst. Evol. Res.* 35(2):81–94
530. Weller SJ, Friedlander TP, Martin JA, Pashley DP. 1992. Phylogenetic studies of ribosomal RNA variation in higher moths and butterflies (Lepidoptera: Ditrysia). *Mol. Phylogenet. Evol.* 1:312–37
531. Weller SJ, Pashley DP. 1995. In search of butterfly origins. *Mol. Phylogenet. Evol.* 4:235–46
532. Weller SJ, Pashley DP, Martin JA. 1996. Reassessment of butterfly family relationships using independent genes and morphology. *Ann. Entomol. Soc. Am.* 89:184–92
533. Weller SJ, Pashley DP, Martin JA, Constable JL. 1994. Phylogeny of noctuoid moths and the utility of combining independent nuclear and mitochondrial genes. *Syst. Biol.* 43:194–211
534. Wells JD, Sperling FAH. 1999. Molecular phylogeny of *Chrysomya albiceps* and *C. rufifaces*. *J. Med. Entomol.* 36(3):222–26
535. Wells MM, Henry CS. 1998. Songs, reproductive isolation and speciation in cryptic species of insects: a case study using green lacewings. In *Endless Forms: Species And Speciation,* ed. D Howard, SH Berlocher,

p. 217–33. New York: Oxford Univ. Press. 470 pp.

536. Wells RS. 1996. Nucleotide variation at the *gpdh* locus in the genus *Drosophila. Genetics* 143(1):375–84
537. Wells RS. 1996. Excessive homoplasy in an evolutionarily constrained protein. *Proc. R. Soc. London Ser. B* 263(1369):393–400
538. Welsh J, McClelland M. 1990. Fingerprinting genomes using PCR with arbitrary primers. *Nucleic Acids Res.* 18:7213–19
539. Wesson DM, Porter CH, Collins FH. 1992. Sequence and secondary structure comparisons of ITS rDNA in mosquitoes. *Mol. Phylogenet. Evol.* 1(4):253–69
540. Wetterer JK, Schultz TR, Meier R. 1998. Phylogeny of fungus-growing ants (tribe Attini) based on mtDNA sequence and morphology. *Mol. Phylogenet. Evol.* 9(1):42–47
541. Wheeler WC, Gladstein DS. 1994. MALIGN: a multiple sequence alignment program. *J. Hered.* 85:417–18
542. Wheeler WC, Schuh RT, Bang R. 1993. Cladistic relationships among higher groups of Heteroptera: congruence between morphological and molecular data sets. *Entomol. Scand.* 24:121–37
543. Whitfield JB. 1997. Molecular and morphological data suggest a single origin of the polydnaviruses among braconid wasps. *Naturwissenschaften* 84(11):502–7
544. Whitfield JB. 1998. Phylogeny and evolution of host-parasitoid interactions in Hymenoptera. *Annu. Rev. Entomol.* 43:129–51
545. Whitfield JB, Cameron SA. 1994. Museum policies concerning specimen loans for molecular systematic research. *Mol. Phylogenet. Evol.* 3:268–70
546. Whitfield JB, Cameron SA. 1998. Hierarchical analysis of variation in the mitochondrial 16S rRNA gene among Hymenoptera. *Mol. Biol. Evol.* 15(12): 1728–43
547. Whiting M, Carpenter JC, Wheeler QD, Wheeler WC. 1997. The Strepsiptera problem: phylogeny of the holometabolous insect orders inferred from 18S and 28S ribosomal DNA sequences and morphology. *Syst. Biol.* 46:1–68
548. Wiegmann BM. 1994. *The earliest radiation of the Lepidoptera: evidence from 18S rDNA.* PhD diss. College Park: Univ. Maryland
549. Wiens JJ. 1998. Testing phylogenetic methods with tree congruence: phylogenetic analysis of polymorphic morphological characters in phrynosomatid lizards. *Syst. Biol.* 47:427–44
550. Willett CS, Ford MJ, Harrison RG. 1997. Inferences about the origin of a field cricket hybrid zone from a mitochondrial DNA phylogeny. *Heredity* 79:484–94
551. Williams JGK, Kubelik AR, Livak KJ, Rafalski JA, Tingey SV. 1990. DNA polymorphisms amplified by arbitrary primers are useful as genetic markers. *Nucleic Acids Res.* 18:6531–36
552. Willis LG, Winston ML, Honda BM. 1992. Phylogenetic relationships in the honeybee (genus *Apis*) as determined by the sequence of the cytochrome oxidase II region of mitochondrial DNA. *Mol. Phylogenet. Evol.* 1(3):169–78
553. Wink M, von Nickisch-Rosenegk E. 1997. Sequence data of mitochondrial 16S rDNA of Arctiidae and Nymphalidae: evidence for a convergent evolution of pyrrolizidine alkaloid and cardiac glycoside sequestration. *J. Chem. Ecol.* 23:1549–68
554. Wink M, Mikes Z, Rheinheimer J. 1997. Phylogenetic relationships in weevils (Coleoptera: Curculionoidea) inferred from nucleotide sequences of mitochondrial 16S rDNA. *Naturwissenschaften* 84:318–21
555. Winnepenninckx B, Backeljau T. 1996. 18S rRNA alignments derived from different secondary structure models can produce alternative phylogenies. *J. Zool. Syst. Evol. Res.* 34:135–43
556. Wirth T, Le Guellec R, Vancassel M, Veuille M. 1998. Molecular and reproductive characterization of sibling species in the european earwig (*Forficula auricularia*). *Evolution* 52(1):260–65

557. Wolstenholme DR, Clary DO. 1985. Sequence evolution of *Drosophila* mitochondrial DNA. *Genetics* 109(4):725–44
558. Xiong B, Kocher TD. 1991. Comparison of mitochondrial DNA sequences of seven morphospecies of black flies (Diptera: Simuliidae). *Genome* 34(2):306–11
559. Xiong B, Kocher TD. 1993. Phylogeny of sibling species of *Simulium venustrum* and *S. verecundum* (Diptera: Simuliidae) based on sequences of the mitochondrial 16S rRNA gene. *Mol. Phylogenet. Evol.* 2(4):293–303
560. Yagi T, Sasaki G, Takebe H. 1999. Phylogeny of Japanese papilionid butterflies inferred from nucleotide sequences of mitochondrial *ND5* gene. *J. Mol. Evol.* 48:42–48
561. Yang Z. 1997. *Phylogenetic Analysis by Maximum Likelihood (PAML)*. Dept. Integr. Biol., Univ. Calif., Berkeley. Ver. 1. 3
562. Yeh W-B, Yang C-T, Hui C-F. 1998. Phylogenetic relationships of the Tropiduchidae-group (Homoptera: Fulgoroidea) of planthoppers inferred through nucleotide sequences. *Zool. Stud.* 37:45–55
563. Yeh W-B, Yang C-T, Kang S-C. 1997. Identification of two sibling species, *Ephemera formosana* and *E. sauteri* (Ephemeroptera: Ephemeridae), based on mitochondrial DNA sequence analysis. *Zhonghua Kunchong* 17:257–68
564. Yokoyama J, Odagiri K, Yokoyama A, Fukuda T. 1998. Phylogenetic relationship of genus *Japonica* subgenus *Yuhbae* (Lepidoptera, Lycaenidae) inferred from mitochondrial DNA sequences. *GenBank.* http://www.ncbi.nlm.nih.gov
565. Zhang D-X, Hewitt GM. 1996. Nuclear integrations: challenges for mitochondrial DNA markers. *Trends Ecol. Evol.* 11:247–51
566. Zhang D-X, Hewitt GM. 1997. Insect mitochondrial control region: a review of its structure, evolution and usefulness in evolutionary studies. *Biochem. Syst. Ecol.* 25:99–120
567. Zhang D-X, Szymura JM, Hewitt GM. 1995. Evolution and structural conservation of the control region of insect mitochondrial DNA. *J. Mol. Evol.* 40:382–91

Annu. Rev. Entomol. 2000. 45:55–81

MEDICINAL MAGGOTS: An Ancient Remedy for Some Contemporary Afflictions

R. A. Sherman[1], M. J. R. Hall[2], and S. Thomas[3]
[1]*Assistant Professor, Department of Medicine, University of California, Irvine, USA; e-mail: rsherman@UCI.edu*
[2]*Head, Veterinary Entomology Programme, Department of Entomology, The Natural History Museum, London, UK; e-mail: mjrh@nhm.ac.uk*
[3]*Director, The Surgical Material Testing Laboratory, Princess of Wales Hospital, Bridgend, UK; e-mail: steve@smtl.co.uk*

Key Words larva, diptera, fly, wound, therapy

■ **Abstract** Certain fly larvae can infest corpses or the wounds of live hosts. Those which are least invasive on live hosts have been used therapeutically, to remove dead tissue from wounds, and promote healing. This medicinal use of maggots is increasing around the world, due to its efficacy, safety and simplicity. Given our low cultural esteem for maggots, the increasing use and popularity of maggot therapy is evidence of its utility. Maggot therapy has successfully treated many types of chronic wounds, but much clinical and basic research is needed still. In this review, the biology of myiasis and the history of maggot therapy are presented, the current status of our understanding and clinical use of medicinal maggots is discussed, and opportunities for future research and applications are proposed.

INTRODUCTION

The use and popularity of maggot therapy—the treatment of wounds with live fly larvae—is increasing in many countries throughout the world. Euphemistically called larval therapy (141), maggot debridement therapy (MDT) (120), and bio-surgery (19), the advantages of maggot therapy include its profound efficacy in removing (debriding) dead (necrotic) tissue, its safety, and its simplicity. These and other advantages have been responsible for the recent revival in the use of maggot therapy.

The end of the twentieth century witnessed in the microbial world the development of antibiotic resistance to some of the most potent antimicrobials yet created. Ironically (or perhaps as a consequence), the end of the century also witnessed the health care community once again embracing the maggot—a creature that thrives in the presence of bacteria, putrefaction, and "filth."

Most risks associated with maggot therapy result from a lack of understanding about fly biology. Occasionally, the entomologist is called upon to consult or

0066-4170/00/0107–0055/$14.00

participate in the treatment. Therefore, a review of this topic is timely and worthwhile.

BIOLOGY AND NATURAL HISTORY

The ancestral habit of blowflies (Diptera: Calliphoridae) was probably that of breeding in carrion, especially vertebrate carcasses (37). Fleshflies (Diptera: Sarcophagidae) had similar saprophagous origins (110). From these origins, there has been a great diversification of the breeding habits of these two families, which now include species that are specialized as breeders in dung, or as parasitoids of insects, earthworms, slugs, snails, and amphibians. Of special importance to this review are the species distributed along what is hypothesized to be one particular evolutionary pathway (33, 145), from saprophages, through species that are facultatively parasitic, developing in the necrotic tissues of animal wounds, to parasites of healthy birds and mammals, including humans (Figure 1). The host location strategies of adult females of the parasitic species (54) are closely related to the strategies used by carrion breeding species to locate dead bodies, both being based on a response to the odors of infection and tissue decay (42).

The primitive, carrion-breeding habit of blowflies has been known and recorded for centuries. For example, reference to this can be found in the *Hortus Sanitatus,* one of the earliest European medical texts, published at Mainz in 1491

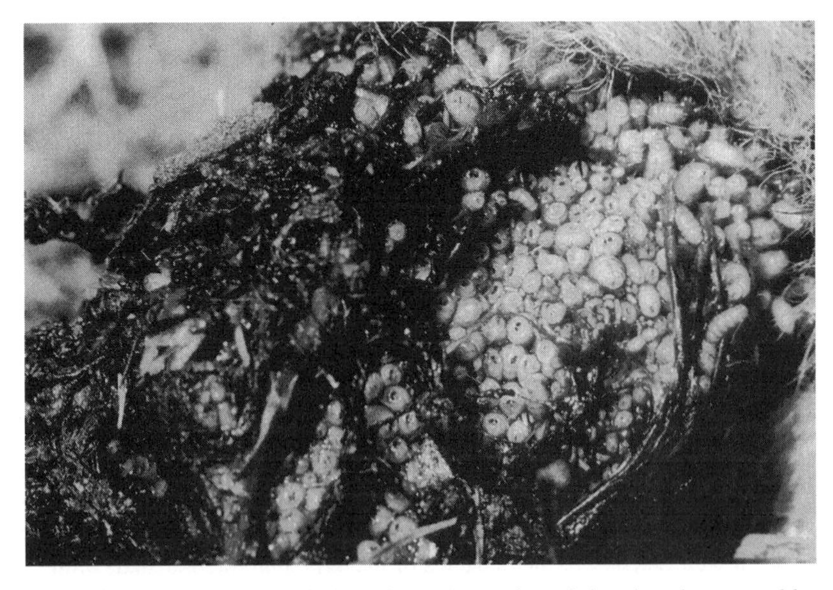

Figure 1 Obligatory myiasis, with deep tissue destruction of sheep's vulva, caused by *Wohlfahrtia magnifica* (Sarcophagidae).

(81). Less well known or recorded are the habits of parasitic species that develop on living hosts, causing the condition known as myiasis. Myiasis was defined by Zumpt (145) as "the infestation of live human and vertebrate animals with dipterous larvae, which, at least for a certain period, feed on the host's dead or living tissue, liquid body-substances, or ingested food." Hall & Wall (57) provide a broad review of myiasis of humans and domestic animals. Based on the relationship with their hosts, flies causing myiasis can be grouped in two categories: obligate parasites that can develop only on live hosts, and facultative parasites that can develop either on carrion or on live hosts. Accidental infestations with fly larvae can sometimes occur, for example when eggs or larvae are inhaled or swallowed inadvertently with food; but these should be considered "pseudomyiasis" (145) rather than true parasitic myiasis.

Maggot therapy grew out of observations of the beneficial effects that resulted from maggot infestations in the wounds of injured soldiers. Essentially, maggot therapy is a carefully controlled, artificially induced myiasis, in which the medical practitioner aims to balance the positive effects of maggot activity on necrotic tissues (benign myiasis) against their potentially negative effects on healthy tissues (malign myiasis). Negative effects might arise from inappropriately using a fly species that feeds preferentially on living tissues, or from introducing too many larvae into a wound. In the latter situation, there could be a risk to healthy tissues if all the necrotic tissues were digested before removal of larvae from a wound. As Stewart (128) cautioned, " . . . [although] fly larvae play an important therapeutic role, they must be utilized with care by an experienced individual."

While many species of fly have been recorded to cause human myiasis (56), only a relatively small number are known to have been used medicinally (Table 1). Most myiasis-causing flies belong to one of three major families: Oestridae, Sarcophagidae, or Calliphoridae (although representatives of other families, such as Muscidae and Phoridae, also are known to cause myiasis; 56). Approximately 150 species of Oestridae are known, and all cause myiasis. Of the approximately 1000 species of Calliphoridae and 2000 species of Sarcophagidae, only about 80 species have been reported to cause myiasis. The biological properties that make different groups within these three families either good or poor candidates for use in maggot therapy are summarized in Table 2.

The Oestridae have no value in maggot therapy: All are obligate parasites, generally with a high degree of host specificity. Likewise, the obligate myiasis-causing Calliphoridae, such as the New World and Old World screwworms (*Cochliomyia hominivorax* and *Chrysomya bezziana*, respectively; 53, 126) are not suitable for maggot therapy because they are truly parasitic, feeding on living tissue.

The Sarcophagidae include two species that are obligate parasites of vertebrates and can infest humans: *Wohlfahrtia magnifica* and *W. vigil*. *W. magnifica* is well recognized as an agent of human myiasis, particularly in eastern Europe (96), Israel (6, 143), and North Africa (30). *W. vigil* (= *W. opaca*; 91) has been reported to cause a furuncular myiasis of infants in North America (26, 41, 59,

TABLE 1 Species of fly used in maggot therapy. Lucilia sericata has been used so commonly that only reference to its first use by Baer 1931 (4) is made.

Family	Species	References
Calliphoridae	*Calliphora vicina*	Teich 1986 (130)
	Chrysomya rufifacies	
	Lucilia caesar	Baer 1931 (4); McLellan 1932 (78)
	Lucilia cuprina	Fine & Alexander 1934 (38)
	Lucilia illustris	Leclercq 1990 (70)
	Lucilia sericata	Baer 1931 (4)
	Phormia regina	Baer 1931 (4)
		Horn et al 1976 (64)
		Horn et al 1976 (64)
		Robinson 1933 (101)
		Reames 1988 (100)
	Protophormia terraenovae	Leclercq 1990 (70)
Sarcophagidae	*Wohlfahrtia nuba*	Grantham-Hill 1933 (44)
Muscidae	*Musca domestica*	

89, 127), but not in Europe. Neither species is suitable for maggot therapy because of their development in healthy tissue. A closely related species, *W. nuba*, was used successfully in wound therapy by Grantham-Hill (44) in the Sudan, suggesting that at least this species and perhaps other saprophagous sarcophagids may have merit in maggot therapy. However, he observed that *W. nuba* would feed on healthy tissues at the edges of a wound if all necrotic tissue was exhausted. *Sarcophaga bullata* has been used successfully in the US (R Sachse, personal communication), but details about its relative risks and benefits are unstudied. Similarly, *Sarcophaga crassipalpis*, found infesting bed sores in an elderly patient in Italy (23) and *Boettcherisca peregrina*, found infesting a parotid gland papillary adenocarcinoma (68), could be candidates for maggot therapy. A major problem with Sarcophagidae is that females deposit larvae rather than eggs. Larvae are much more difficult to disinfect than eggs. It is possible, however, that techniques such as rearing in a germ-free facility (45, 13) could be developed to ensure sterility of the larvae.

The flies most often used in maggot therapy are the facultative calliphorids (Table 1), which all share several advantageous biological properties (Table 2). The most widely used species is the "greenbottle" blowfly, *Lucilia* (= *Phaenicia* of some authors) *sericata*. Weil et al (141) found that larvae of *L. sericata* starved on clean granulation tissue and were, therefore, ideally suited for maggot therapy. Other authors, however, believe that this species may occasionally feed on healthy human tissue (128) or at least produce a local erythema of the surrounding skin due to the action of the larval enzymes (135). Clearly, caution should be used when selecting any maggot for medicinal use.

TABLE 2 Summary of the properties of the vertebrate myiasis-causing members of Oestridae, Sarcophagidae, and Calliphoridae that make them good (✔) or poor (✗) candidates for use in maggot therapy. The principal obligate parasites among the Calliphoridae (*Cochliomyia hominivorax, Chrysomya bezziana,* and *Cordylobia*) and Sarcophagidae (*Wohlfahrtia magnifica*), are not included in this analysis.

Property	Oestridae	Sarcophagidae	Calliphoridae
Invasion of internal organs by larvae?	✗Many species parasitize internal organs	✔Larvae do not usually invade internal organs	✔Larvae do not usually invade internal organs
Larval behavior in cutaneous tissues	✗Larvae in cutaneous tissues usually isolated in furuncles	✔Larvae in cutaneous tissues usually congregate	✔Larvae in cutaneous tissues usually congregate (except Cordylobia)
Rate of development in host	✗Larvae develop slowly in hosts	✔Larvae develop rapidly in hosts	✔Larvae develop rapidly in hosts
Host specificity	✗Species show a high degree of host specificity	✔Species are not generally host specific	✔Species are not generally host specific
In vitro rearing	✗Species are extremely difficult to rear in vitro	✔Species are relatively easy to rear in vitro (except for obligate parasites)	✔Species are relatively easy to rear in vitro (except for obligate parasites)
Egg laying	✗Females can lay eggs or larvae depending on sub-family	✗Females are larviparous, laying first-instar larvae that are relatively difficult to handle and sterilize	✔Females lay eggs that are easy to make sterile (= "germ free")
Food source	✗Larvae feed on living tissues	✔Larvae generally feed on necrotic tissues (except for obligate parasites, which are aggressive feeders on living tissues)	✔Larvae generally feed on necrotic tissues (except for obligate parasites, which are aggressive feeders on living tissues)

Naturally occurring cases of myiasis in humans with facultative parasites are most commonly seen in young children, in the elderly, in the physically or mentally infirm, in cases of personal neglect, or in the setting of high fly density (15, 48, 76, 109). The usual and most prudent medical response to such infestations is removal of the larvae to prevent tissue damage and bacterial infection, even when the maggots are known to belong to a therapeutically useful species (16, 65, 82). Natural infestations with these maggots can have beneficial effects (10, 44) and these benefits can be lost when the larvae are removed (100). Nevertheless, natural infestations are uncontrolled and as such can complicate ongoing medical treatment (1) or be harmful outright (24, 25, 51, 58, 124, 137).

Among the calliphorids not generally used therapeutically, due to their propensity to cause malign myiasis, are *Cochliomyia macellaria* (18, 125), *Chrysomya megacephala* (71, 145) and *Lucilia cuprina* (76). However, with careful management, even these three species might someday have a role in maggot therapy. An apparently nonpathogenic strain of *L. cuprina* has already been used with success (38).

Interestingly, *Lucilia sericata*, the species used most commonly in maggot therapy today, is a serious pest to the sheep industry of the United Kingdom, Europe (55), and New Zealand (131). Called "sheep strike," myiasis in sheep due to *Lucilia* can be fatal in cases of heavy infestation (52).

HISTORY OF MAGGOT THERAPY

Origins to the Present Day

Because several thorough reviews have been published recently (14, 49, 70, 94, 114, 139), this section focuses only on the highlights of maggot therapy history. Some societies have recognized for centuries that the larvae of certain flies can have beneficial effects upon the healing of infected wounds. There is evidence that maggot therapy has been used by aboriginal tribes of Australia (29), the Hill Peoples of Northern Burma (47), and possibly the Mayans of Central America (141). Yet, the beneficial aspects of myiasis have not always been appreciated universally.

Possibly the first written mention of human myiasis is in the Bible, where Job complained,

> . . . My body is clothed with worms and scabs, my skin is broken and festering . . . (87).

Like many surgeons who followed, Ambroise Paré (1509–1590), chief surgeon to Charles IX and Henri III, observed in 1557 at the battle of St. Quentin that maggots frequently infested suppurating wounds (43). Hieronymus Fabricus (35) also described the presence of maggots in wounds. In 1829, Napoleon's surgeon in chief, Baron Dominic Larrey, reported that when maggots developed in wounds sustained in battle, they prevented the development of infection and accelerated healing (69). There is no evidence, however, that Larrey deliberately introduced maggots into his patients' wounds.

The beneficial effects of wound myiasis were noted by the Confederate medical officer Joseph Jones, quoted by Chernin (14):

> I have frequently seen neglected wounds . . . filled with maggots . . . as far as my experience extends, these worms only destroy dead tissues, and do not injure specifically the well parts. I have heard surgeons affirm that a gangrenous wound which has been thoroughly cleansed by maggots heals more rapidly than if it had been left to itself.

According to Baer (4), the Confederate surgeon J Zacharias, may have been the first western physician to intentionally introduce maggots into wounds for the purpose of cleaning or debriding the wound. Baer (4) quotes Zacharias as stating:

> Maggots . . . in a single day would clean a wound much better than any agents we had at our command. . . . I am sure I saved many lives by their use. . . .

Crile & Martin (22) also noted that soldiers whose wounds were infested with maggots did far better than wounded soldiers not infested.

The founder of modern maggot therapy is William Baer (1872–1931), clinical professor of orthopaedic surgery at the Johns Hopkins School of Medicine in Maryland. During the First World War, Baer treated two wounded soldiers who had lain overlooked on the battlefield for a week. Although they had sustained serious injury and their wounds swarmed with maggots, Baer noted that the soldiers had no fever, no evidence of systemic infection, and no pus; instead, they had the "most beautiful pink granulation tissue that one can imagine." Drawing upon his wartime experiences, Baer treated four children with intractable bone infections (osteomyelitis) at the Children's Hospital in Baltimore (3). His initial use of unsterilized maggots was very successful and the wounds healed within six weeks. Encouraged by these results, Baer used the technique more widely. However, several of his patients developed tetanus, and he concluded that "it would be necessary to have sterile [viz. germ free] maggots" (4).

In the absence of any equally effective alternative for the treatment of osteomyelitis or infected soft tissue injuries, the use of maggots spread quickly during the 1930s, particularly in the United States where *Lucilia sericata* larvae were produced by Lederle Corporation (98) and sold for five dollars per 1000 (now equivalent to about $100). By the mid-1930s, Robinson surveyed 947 North American surgeons known to have employed maggot therapy (104). Of the 605 responding surgeons, 91.2 percent expressed a favorable opinion; only 4.4 percent expressed an unfavorable view. The most common complaints raised by surveyed practitioners were the cost of the maggots, the time and effort required to construct the maggot dressings, and the discomfort to patients. Other than Baer's cases of tetanus and one case of erysipelas (141), which were thought to be associated with the use of non-sterile larvae, no other serious adverse reactions were reported.

The early maggot therapy literature describes the successful treatment of chronic or acutely infected wounds, including bone infections (osteomyelitis) (9, 72, 75, 141), abscesses, carbuncles, and leg ulcers (36). Although the larvae were unable to liquify dead bone, they did appear to cleave the pieces of dead bone (sequestra) at their interface with normal bone, leaving behind clean healthy granulation tissue (141). Based on clinical outcomes and wound cultures, Weil and colleagues (141) believed that medicinal maggots treated many soft tissue infections, including *Clostridium welchii* (*Cl. perfringens*). In addition, they reported maggots to be of value in the management of some tumors, including two cases of inoperable breast cancer. More recently, Bunkis et al (10) and Reames et al (100) described the benefits of debridement and odor control resulting from accidental myiasis of head and neck tumors. Seaquist and colleagues (111) also reported benefits from naturally occurring *Phormia regina* myiasis in a malignant lesion; however, this infestation was accompanied by pain.

During the 1930s, attempts to isolate the "maggot active principle" generated several reports of the successful topical application of maggot extracts to promote wound debridement and disinfection (73–75). An injected maggot extract "vaccine" was reportedly successful (73, 75), but was associated with significant systemic reactions, and eventually was abandoned.

These years also marked the beginning of the antibiotic era. By 1940, sulfonamides were already available, and Chain et al (12) had discovered the methods for mass producing Flemming's penicillin. By the mid-1940s, maggot therapy nearly disappeared from use, probably because of (i) the emergence of antibiotics as a readily available alternative to maggot therapy; (ii) the reduced incidence of bone and soft tissue infections, as a consequence of widespread antibiotic use; (iii) improved wound care and aseptic techniques; (iv) improved surgical techniques; (v) the expense of medicinal maggots; (vi) the cumbersome maggot dressings; and (vii) the unacceptability of live maggot dressings, relative to the newer alternatives.

Subsequently, maggot therapy rarely was used, except as a last resort (64, 130).

In 1988, maggot therapy was described by some as being beneficial in modern military and survival medicine (21); while others wrote:

> . . . Fortunately maggot therapy is now relegated to a historical backwater, of interest more for its bizarre nature than its effect on the course of medical science . . . a therapy the demise of which no one is likely to mourn . . . (139).

Meanwhile, an infectious diseases fellow at the University of California was planning clinical trials of maggot therapy for treating pressure ulcers and other chronic wounds. Preliminary evaluation of this study suggested that maggot therapy offered several advantages over other wound treatments currently employed (117–20). By 1995, dozens of patients with pressure ulcers, diabetic foot wounds, and chronic leg ulcers were being treated also at the Biosurgical Research Unit in Bridgend, South Wales (135), and at the Hadassah Hospital maggot therapy center in Jerusalem (85). In 1996, the International Biotherapy Society was founded "to investigate and develop the use of living organisms, or their products, in tissue repair." The society is now one of the sponsors of an annual International Conference on Biotherapy. Thus, the revival of maggot therapy is well under way.

Early Techniques for Producing and Applying Medicinal Larvae

Numerous papers were published in the early 1930s describing techniques for breeding flies (86, 72) and producing sterile larvae (4, 9, 17, 86, 141). Although Livingston (72) and Weil's group (141) claimed some success, sterilization of hatched larvae was found to be virtually impossible. Efforts at sterilizing the egg were more successful, but most of the early methods that destroyed all bacteria also were lethal to the eggs. Egg sterilization usually began with pretreatment in Dakin's solution (dilute sodium hypochlorite, or bleach) followed by immersion in mercuric chloride or formaldehyde. Simmons (121) reported satisfactory sterilization using five percent formalin, one percent sodium hydroxide; yet even his method did not kill all spore-forming bacteria such as *Cl. perfringens* or *Cl. tetanii*.

Once applied to the wound, medicinal maggots were held in place by specially constructed dressings (38), but many were difficult to construct and probably uncomfortable to wear. They consisted of layers of crinoline or gauze, or were made from copper mesh (141), held in place with adhesive tape or, sometimes, Unna's Paste—a mixture of zinc oxide, gelatin, glycerin, and water (67). Self-retaining metal (141) or glass (79) devices were developed to hold wounds open during therapy, allowing drainage of the wound and providing access to the maggots. Ochsenhirt & Komara (88) described a complex technique for intra-oral treatment, involving dentures with tubes through which the larvae were introduced.

Livingston (72) recommended exposing the medicinal maggots to bright light in order to drive them deep into the wound, but this was considered unnecessary by Robinson (102). Robinson suggested that patients' skin surrounding a wound should be protected from larval secretions in order to reduce irritation by the secretions, and to eliminate the tickling sensation caused by the maggots' movements. He also emphasized the need to control the number of larvae applied, proposing that as few as five to six might be sufficient for a fingertip injury, while 500 to 600 might be required for more extensive wounds.

CURRENT STATUS OF MAGGOT THERAPY

During the second half of the twentieth century, maggot therapy was used sporadically, and only as a treatment of last resort for serious and recalcitrant wounds. Life-threatening infections such as temporal mastoiditis (64) and perineal gangrene (130) were treated with maggot therapy, following unsuccessful surgical and antibiotic treatments. More recently, small prospective controlled trials demonstrated benefits from maggot therapy as an early intervention in the treatment of pressure ulcers (117, 120). Maggot therapy is now used as an adjunct to conventional modalities, and not only as an alternative when all else has failed (34, 84, 90, 99, 113, 135). Laboratory investigations have demonstrated that medicinal maggots can be administered to patients concurrently receiving systemic antibiotics without adverse effects on the debriding capability of the larvae (116). However, there is evidence that the residues of some topical treatments, such as hydrogels, may adversely effect larval development (132).

Today, maggot therapy is used to debride many different types of skin and soft tissue wounds: pressure ulcers, venous stasis ulcers (Figure 2), neurovascular ulcers such as diabetic foot wounds, traumatic and post-surgical wounds, and burns. Incidental maggot infestations of necrotic skin tumors have been noted to eliminate odor and even destroy some of the malignant tissue (10, 100). However, even though it may be useful for controlling some of the problems associated with tumor necrosis, maggot therapy is not considered to be a likely cure for cancer.

Maggot therapy candidates generally have non-acute external wounds that have failed one or more courses of conventional treatment (Figures 3, 4, and 5). Often, surgical intervention is risky or not even an option. Wound infections that are life- or limb-threatening are usually not treated with maggot debridement; surgical intervention is faster and thus the treatment of choice in these circumstances. However, maggot therapy has been used successfully as an adjunct to surgery in the treatment of life-threatening necrotizing fasciitis (130). Necrotic

Figure 2 Medicinal maggots applied to a necrotic venous stasis ulcer on the leg.

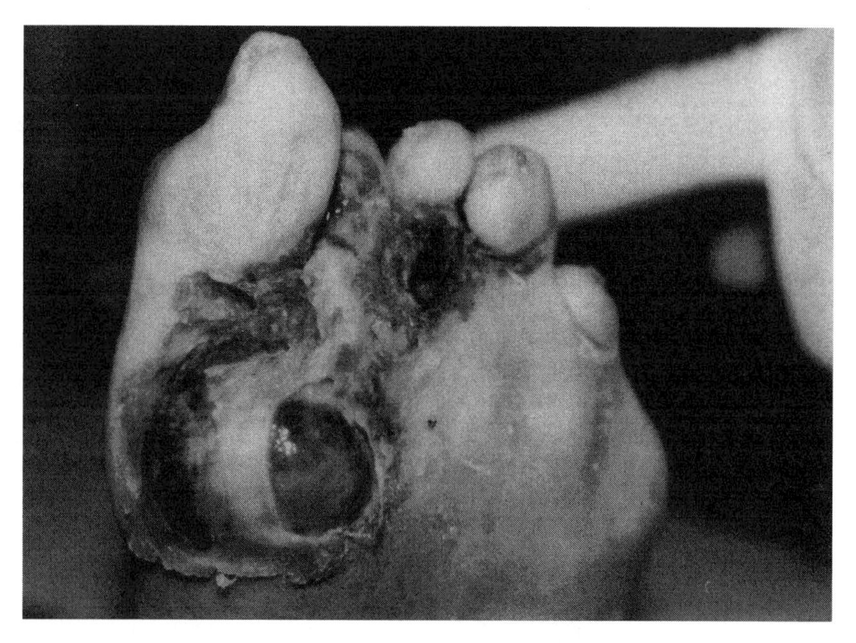

Figure 3 Before maggot therapy, this 72-year-old man received two years of medical and surgical treatment for this chronic foot ulcer. Note underlying skin atrophy and muscle contractions.

bone is not digested by medicinal maggots, although maggot therapy has been used in combination with surgical excision, or as an alternative when the bone infection is minimal. Circulatory obstruction is a relative contraindication; vascular flow that is inadequate to facilitate healing despite debridement is usually an indication for amputation. Concurrent antibiotic administration, age, pregnancy, immobility, and altered mental status are not themselves contraindications to maggot therapy.

Modern maggot therapy dressings, constructed from readily available medical supplies, are designed to contain the maggots within the wound (112, 115, 134, 136). Whenever possible, a tracing of the wound is prepared on a sterile plastic sheet. This template is then used to cut a matching wound-sized hole from a hydrocolloid dressing (a self adhesive wafer with a semipermeable plastic outer surface). This hydrocolloid pad forms the foundation of the dressings: It provides a base to which adhesives can be fixed, it protects the intact skin from irritation by the maggots' proteolytic enzymes, and it protects the patient from sensing the movement of the larvae. Medicinal maggots (also called disinfected or "sterile" maggots) are placed on the wound, with or without a small piece of gauze. The maggots are then covered by a sterile sheet of nylon mesh (pore size of 100 to 400 microns), which is affixed to the surrounding hydrocolloid base.

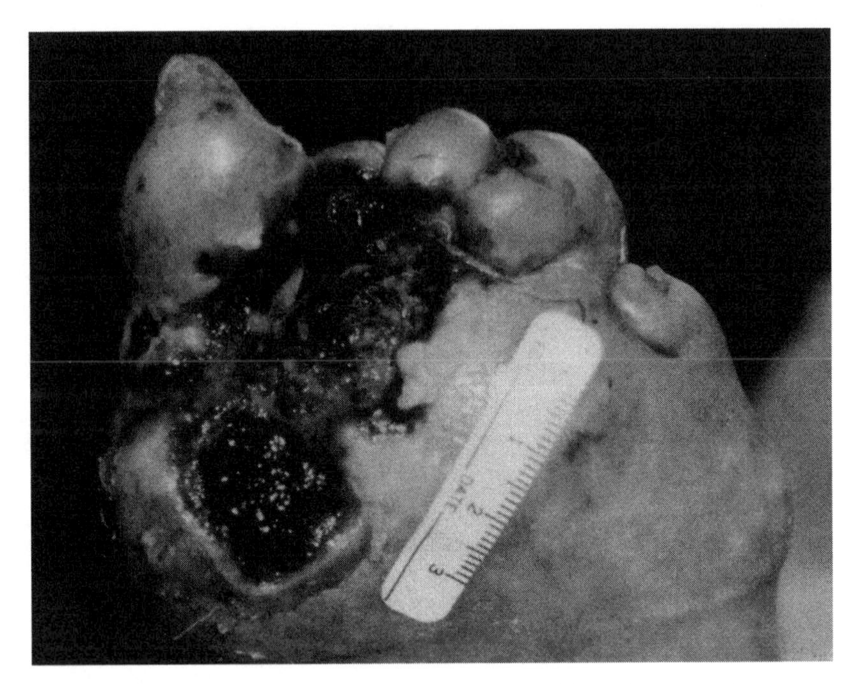

Figure 4 Within just one week, the same wound seen in Figure 3 now has been completely debrided. Healthy, red, vascularized tissue is growing within the wound base.

This dressing allows air to reach the maggots, and facilitates the drainage of the liquified necrotic tissue and serous exudate. The dressing is covered with a simple absorbent pad, that contains the wound exudate. Alternative dressings, using nylon stockings (115) or heat-sealed netting in the shape of a sleeve or bag (135) have been proposed for treating nonplanar wounds such as those on the toes or heels. For patients with skin too fragile to permit the use of adhesive dressings, or patients who are allergic to hydrocolloid dressings, a zinc oxide paste can be used to protect the skin (134) and form a seal with the nylon net, in a manner similar to that described by Jewett (67).

The number of maggots applied to a wound depends on a number of factors, including the age and size of the maggot, the relative amount of necrotic tissue, and most importantly, the size of the wound. In general, five to ten larvae per cm^2 are used. The larvae are removed from the wound one to three days later, simply by detaching the outer dressing and collecting the maggots as they attempt to escape. The wandering larvae should be securely contained, and disposed of in the manner customary for other potentially infectious dressings and waste.

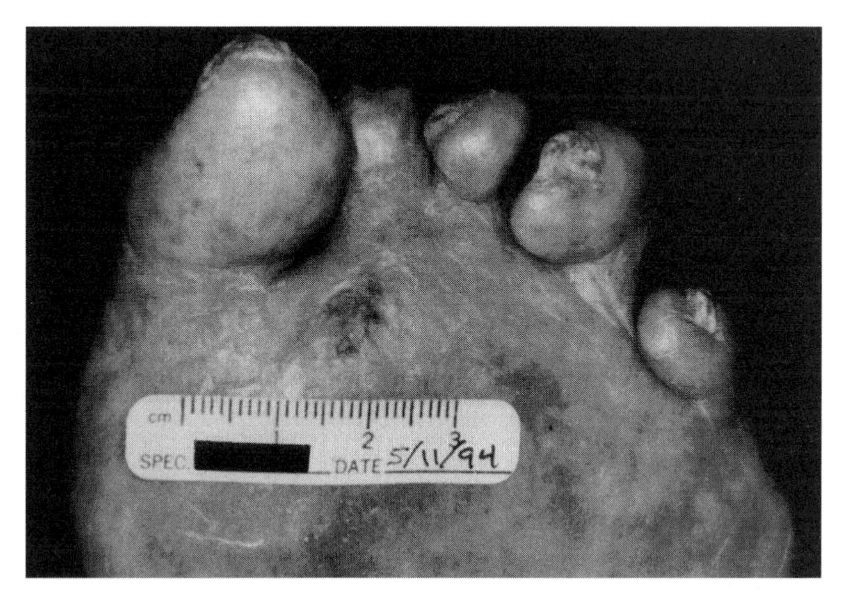

Figure 5 The same wound presented in Figures 3 and 4, here shown one year later. Only a small scar marks the location of the prior chronic wound.

Depending upon the condition of the wound, a fresh batch of larvae or a conventional dressing may be applied.

A review of the literature reveals no significant risks or adverse events resulting from the clinical use of medicinal maggots. The most common patient complaint is physical discomfort. Treatment-associated pain has been reported by six percent of nursing home patients (113). In another series (R Sherman, J Sherman & K Mumcuoglu, unpublished data), pain was reported by 8 of 21 ambulatory patients. Pain was reportedly controlled with oral analgesics, or by removal of the growing maggots. Pain did not cause any patient in either study to terminate therapy. There is some evidence to suggest that maggot-related pain is more common and more severe in patients with ischemic ulcers.

Patient acceptance has been very high (113); psychological distress has been rare.

Theoretically, contamination of wounds with maggot-transmitted microorganisms is a concern but it has not been witnessed since "sterile" larvae became the standard of maggot therapy care in the 1930s. In sheep experimentally infested with16,000 larvae, systemic illness has been associated with ammonia toxicity (52), resulting from the absorption of maggot-secreted ammonia into the blood stream (50). Sheep struck by parasitic *Lucilia cuprina* display a rapid increase in temperature and respiratory rate accompanied by loss of weight and appetite (7).

Ammonia toxicity is theoretically possible in humans, although maggot therapy utilizes far fewer larvae. Fever has been reported in some patients (78), but its etiology has never been explained.

Maggot Therapy in North America

During the 1990s, prospective controlled studies compared maggot therapy to conventional treatments for managing pressure ulcers in spinal cord–injured patients. Maggot therapy debrided wounds faster than any of the non-surgical modalities being used, and it hastened overall wound healing. Necrotic wounds of 5 to 30 cm^2 in surface area (mean = 13.0 cm^2) were debrided in 1.5 weeks with maggot therapy, compared to more than 4 weeks with conventional dressings (120). Wounds that had been enlarging at an average rate of 21.8% of surface area per week before receiving maggot therapy began to heal at a rate of about 20% per week shortly after maggot therapy was initiated (117).

The utility of maggot therapy for treating chronic soft-tissue wounds in elderly nursing home patients is currently being evaluated. Preliminary analysis of the first 28 patients indicates that maggot therapy halted or reversed the progression of pressure ulcers, venous stasis ulcers, and diabetic foot ulcers (113). At the time of this writing, approximately 50 centers in North America are once again including maggot therapy as a treatment option in wound care. Based on sales records, an average of five treatments are administered weekly in the United States. Although most centers in America use maggot therapy only in hospitalized patients, successful therapy in the outpatient setting has provided continuous wound debridement at low cost to patients who choose to remain at home (R Sherman, J Sherman & K Mumcuoglu, unpublished data).

Maggot Therapy in the United Kingdom

Clinical research and application of larval therapy has advanced rapidly in the United Kingdom, where considerable press coverage has encouraged public acceptance. Medicinal maggots are produced in the United Kingdom under the brand name of Larv E™ by the Biosurgical Research Unit, a hospital-based facility in South Wales. In three years, more than 5000 treatments have been distributed to more than 400 centers, including hospitals, specialist clinics and general practitioners. Although the majority of these centers are based in the United Kingdom, some larvae have been distributed also to centers in Belgium, Germany, and Sweden.

The United Kingdom experience with maggot therapy supports that of earlier researchers, in which larvae have been found useful for many types of necrotic wounds including leg ulcers, pressure sores and infected surgical wounds (Table 3). After using maggot therapy on their own patients, 95% of respondents to a survey in the United Kingdom expressed the opinion that this treatment fulfilled a limited or significant role in wound management. On average, wounds were completely debrided with only two applications (range: one to five). Maggot

TABLE 3 Analysis of 100 questionnaires returned by UK purchasers of Larv E(TM) brand medicinal maggots.

Questionnaire Responses	No. of respondents
Wound Types	
Leg ulcers	55
Pressure Ulcers	26
Necrotic toes or feet	8
Surgical wounds	5
Traumatic wounds	2
Sinus	1
Not identified	3
Wound Condition before Treatment	
Sloughy	90
Necrotic	38
Infected	31
Presence of Clinical Infection	
Prior to treatment	32
After treatment	14
Reported Efficacy of Treatment	
Completely debrided	35
Partially debrided	55
Unchanged	6
No data	4
Granulation Tissue Present after Debridement	
Yes	86
No	13
No data	1
Effect on Pain	
Pain increased	33
Pain decreased	7
No change	54
No data	6
Therapist Opinions on the Role of Maggot Therapy in Wound Management	
Major role	62
Limited role	33
No role	0
No view expressed	5

therapy has been used successfully in the treatment of diabetic foot ulcers (99), and in the treatment of wounds infected with antibiotic-resistant bacteria such as MRSA (134). Additionally, maggot therapy has decreased wound odor and pain (34), thereby significantly improving the quality of life for some patients.

Maggot Therapy Elsewhere

The use of maggot therapy elsewhere in Europe has been met with success. It has been used recently in Hungary (90), Germany, Sweden, Belgium, and the Ukraine. BioMonde is now producing medicinal *L. sericata* in Germany, and distributing them to more than 140 hospitals and 50 general practitioners (D Goj, personal communication).

The Maggot Therapy Center of the Hadassah Hospital in Jerusalem has treated more than 70 patients since 1995 (K Mumcuoglu, personal communication), including those with diabetic foot ulcers, venous stasis ulcers, and chronic wounds in the setting of lymphedema and blood disorders (83–95). Patients throughout the country are referred to one of two affiliated hospitals, or one of three geriatric clinics. Maggot therapy is used both in the inpatient and outpatient setting. The center also conducts basic research, in collaboration with other medical center departments, and the Volcani Institute (Bet Dagan, Israel).

At the time of this writing, the Maggot Therapy Research Development Group is under formation in Australia, but is not yet fully operational (B Hughes, personal communication).

Maggot therapy offers several advantages to conventional wound care in rural and tropical regions of the world (20, 118), where highly skilled surgeons, technologically advanced resources, or even electricity may not be readily available. However, we are not aware of maggot therapy now ongoing in any other countries, except perhaps as isolated events.

THE BIOCHEMICAL BASIS OF MAGGOT-INDUCED WOUND HEALING

Medicinal maggots are recognized to have three beneficial effects on wounds: (i) debridement, or elimination of necrotic tissue; (ii) disinfection of the wound through microbial killing; and (iii) promotion of wound healing. All three of these benefits have been observed clinically, and demonstrated in the laboratory.

Wound Debridement

Necrophagous larvae feed on the dead tissue, cellular debris, and serous drainage (exudate) of corpses or necrotic wounds. Extracorporeal digestion by proteolytic enzymes is one mechanism by which wounds are cleaned. Secreted collagenases and trypsin-like and chymotrypsin-like enzymes have been described (3, 8, 11, 39, 60, 61, 77, 95, 138, 140, 144). Debridement is aided by ingestion of the liquified tissue, and also may be enhanced by the maggots' crawling about the wound, probing and macerating it with their mouthhooks (5).

Antimicrobial Activity of Maggot Secretions

The natural habitats of blow fly larvae—corpses, wounds and excrement—abound with bacteria. It has long been realized that the maggots must be able to tolerate, if not eradicate, resident pathogens (61–63). In the 1930s many believed that microbial killing resulted from bacterial ingestion and digestion by the maggots (75, 101, 108). Wound exudate, abundant in response to the maggots, was believed to facilitate the irrigation of bacteria out of wounds (101). The alkalinity of maggot-treated wounds also was believed to be a factor in wound disinfection. Baer (4) first demonstrated that wound fluid was alkaline during maggot therapy, and Messer & McClellan (80) believed that ammonia secreted by the maggots was the cause. Subsequently, ammonia and ammonium derivatives such as ammonium bicarbonate were suggested as the factors responsible for disinfection and wound-healing (106, 107). Stewart's studies (128) led him to conclude that calcium and calcium carbonate produced by the maggots killed bacteria directly, stimulated phagocytosis, and possibly promoted the growth of granulation tissue.

In 1935 Simmons reported that *L. sericata* excretions were antimicrobial (122, 123). By 1957, Pavillard & Wright (92) isolated and partially purified a substance from the secretions of *Phormia terraenovae (= Protophormia terraenovae)* (Calliphoridae) larvae, which killed *Streptococcus pyogenes* and *Streptococcus pneumoniae* (pneumococcus). *Staphylococcus aureus* was much less sensitive, and *Escherichia coli* and *Proteus vulgaris* were both highly resistant to this substance. Injection of their isolate into mice protected the mice from the lethal effects of intraperitoneal injections of pneumococcus. Pavillard & Wright never identified their antimicrobial agent. Recent laboratory studies confirm the antimicrobial activity of medicinal maggot secretions (40), and demonstrate their ability to kill multi-drug resistant strains of *S. aureus* and clinical isolates of pathogenic *Streptococcus* sp. (133).

In 1968, Greenberg (46) showed that metabolic products of *Proteus mirabilis*, a commensal of the larval gut, produced agents that were highly lethal to Gram-positive and Gram-negative bacteria under acidic conditions; he called these agents "mirabilicides." Erdmann & Khalil (32) went on to identify two antibacterial substances, phenylacetic acid and phenylacetaldehyde, produced by *Proteus mirabilis* isolated from the gut of screwworm larvae (*Cochliomyia hominovorax*). Erdmann (31) proposed that the larvae acted like a sterilizing filter for the host's contaminated wound.

Growth-Promoting Activity of Maggot Secretions

The rapid healing of maggot-treated wounds, and the prolific response of healthy granulation tissue, was an exciting clinical observation (4, 101). Although some researchers speculated that the maggots simply facilitated normal healing by removing the debris and infection that was impairing wound healing (108), most believed that the maggots directly stimulated wounds to heal more quickly. Some of the same substances described as antimicrobial were said to also account for the growth-promoting activity of maggots. It was even claimed that the abundant growth of granulation tissue was, at least in part, a response to the physical stimulation of the maggots crawling over the wound (9). Allantoin, ([2,5-Dioxo-4-imidazolidinyl]urea, a product of purine metabolism first isolated from comfry root and still used today in some cosmetics and pharmaceuticals) was highly acclaimed to be responsible for this effect (103). Ammonium bicarbonate (106) and urea, too, were noted to be active constituents. Subsequently, allantoin and urea (105), ammonium bicarbonate (106), and mixtures of calcium carbonate with picric acid (129) were being used in wound healing as alternatives to live maggots, with reported success.

Being a physician, not an entomologist, Livingston (73) expressed in the following manner his opinion that no simple group of chemicals could replace the complex nature of maggot secretions:

> . . . maggots are fly embryos and, therefore, of necessity are rich in complex organic substances which, because of their embryonic nature, are growth stimulating.

Efforts to validate in the laboratory the growth-promoting activity of medicinal maggots continue to this day (97).

FUTURE STATUS OF MAGGOT THERAPY

As the twentieth century comes to a close, maggot therapy is being embraced once again by mainstream western medicine. Maggot therapy is acknowledged to be useful in many clinical situations, and its use is rapidly growing throughout the world. Nevertheless, several questions remain unanswered, and many paths of investigation remain unexplored. The future holds the promise—and the obligation—for significant research in the following areas.

Clinical Studies

A large, prospective, randomized, controlled study is clearly needed to provide scientifically and statistically sound support to the more anecdotal evidence that comprises the majority of maggot therapy literature. However, the very advantage of low cost has made it difficult to find financial support for such clinical trials. Limited experience with maggot therapy for burns and tumors begs for further investigations.

Increasing the Clinical Use of Maggot Therapy

Preliminary experience shows that maggot therapy in the outpatient setting is safe, effective, and can be administered with minimal medical resources and personnel (R Sherman, J Sherman, K Mumcuoglu, unpublished data). Outpatient and inpatient use of maggot therapy is expected to rise, especially for common problematic wounds such as diabetic foot ulcers (84, 99), where efficacy has already been demonstrated. The World Health Organization estimates that by the year 2025, 228 million people in developing countries will suffer from diabetes (142). Land mine morbidity, too, is an increasingly devastating problem in rural, war-ravaged, and developing nations, where access to surgical and medical treatment may be poor (2, 93). Thus, many opportunities exist to extend the use of maggot debridement worldwide.

Optimizing the Treatment

The efficacy of maggot therapy might be improved by pre-inoculating larvae with a non-pathogenic strain of *Proteus mirabilis* in order to prime the gut with antimicrobials. Despite the limited number of fly species used therapeutically (Table 1), other species might be equally or more effective. Particular species might be optimal for treating specific wounds. For example, an otherwise invasive blow fly might be effective in treating tumors with both necrotic and viable tissue. This avenue of investigation remains unexplored.

Optimizing Maggot Production

Simplifying the methods of maggot rearing could lower the costs of producing medicinal maggots. Attempts are already underway at the Biosurgical Research Unit in Wales and BioMonde in Germany (D Goj, personal communication) to maintain colonies of germ-free flies, thereby avoiding the laborious process of egg sterilization.

Improving the Image of Maggot Therapy

The concept that maggot therapy is an antiquated modality, and the anxiety of having live insects roaming over the body, have prevented some care providers and patients from benefiting from maggot therapy. Acceptance of maggot therapy will come about with the publication of well-designed clinical trials, and through education of health care staff and the public.

Studies Of Immune Response To Maggots

Maggot therapy, with its multiple cycles of therapeutic myiasis, represents a repeated challenge to the human host. It has been suggested that an immune response may affect the clinical benefits of treatment and perhaps explain why the maggots applied during the second or third week of therapy may not survive.

Sheep are known to develop an immune response (57), but a corresponding immune response has never been documented in humans. If an immune response did develop, and if it was detrimental to therapy, alternating species might overcome the problem.

Veterinary Research

With rare exception (27, 28), maggot therapy has not been applied to veterinary medicine. Since myiasis can be a serious disease of sheep and other animals (57), many are reluctant to treat animal wounds with blow fly larvae. Recently, however, maggot therapy was proposed as a treatment for work animals (66), and veterinary trials may soon begin (E Iversen, personal communication).

Biochemical Studies

There are many advantages to using pharmaceuticals or extracts, rather than live maggots: (i) a more predictable and uniform product; (ii) easier and perhaps less expensive production and application of the treatment; (iii) the ability to treat areas not normally treated with maggots, such as corneal ulcers or open abdominal wounds; and (iv) better acceptance. If the maggot cannot be replaced by its own products, perhaps it can be enhanced through genetic engineering, based on what we may learn about its biochemistry and mechanisms of action. In addition, understanding the complex biochemical relationship between the maggot and its host advances our understanding of host-parasite ecology in general.

CONCLUSION

Our cultural image of maggots as the antithesis of health and cleanliness hampers the acceptance of live maggots as a medical treatment. The fact that maggot therapy is accepted at all in mainstream western medicine is a testament to its efficacy. Perhaps the therapeutic benefits of maggots will be acknowledged and applied during the twenty-first century for years to come, rather than be forsaken as readily as they were in the past century.

ACKNOWLEDGMENTS

The authors are grateful to John Church, Kosta Mumcuoglu, and Rory Post for reviewing and contributing to our manuscript. Information contributed by Detlev Goj, Bradford Hughes, Eve Iversen, and Rainer Sachse also was used in writing this manuscript. During the writing of this manuscript, RA Sherman received research and salary support from the National Institutes of Health (K08AI 01454) and the University of California (Biotechnology STAR Grant #S96-27).

Visit the Annual Reviews home page at www.AnnualReviews.org.

LITERATURE CITED

1. Abram LJ, Froimson AI. 1987. Myiasis (maggot infection) as a complication of fracture management. *Orthopedics* 10:625–27
2. Anderson N, Palha da Sousa C, Paredes S. 1995. Social cost of land mines in four countries: Afghanistan, Bosnia, Cambodia, and Mozambique. *Br. Med. J.* 311:718–21
3. Baer WS. 1929. Sacro-iliac joint—arthritis deformans—viable antiseptic in chronic osteomyelitis. *Proc. Int. Assembly Inter-state Postgrad. Med. Assoc. North Am.* 371:365–72
4. Baer WS. 1931. The treatment of chronic osteomyelitis with the maggot (larva of the blowfly). *J. Bone Jt. Surg.* 13:438–75
5. Barnard DR. 1977. Skeletal-muscular mechanisms of the larva of *Lucilia sericata* (Meigen) in relation to feeding habit. *Pan-Pac. Entomol.* 53:223–29
6. Baruch E, Godel V, Lazar M, Gold D, Lengy J. 1982. Severe external ophthalmomyiasis due to larvae of *Wohlfahrtia* sp. *Isr. J. Med. Sci.* 18:815–16
7. Broadmeadow M, Gibson JE, Dimmock CK, Thomas RJ, O'Sullivan BM. 1984. The pathogenesis of flystrike in sheep. *Wool Technol. Sheep Breed.* 32:28–32
8. Brookes VJ. 1961. Partial purification of a proteolytic enzyme from an insect, *Phormia regina. Biochim. Biophys. Acta* 46:13–21
9. Buchman J, Blair JE. 1932. Maggots and their use in the treatment of chronic osteomyelitis. *Surg. Gynecol. Obstet.* 55:177–90
10. Bunkis J, Gherini S, Walton RL. 1985. Maggot therapy revisited. *West. J. Med.* 142:554–56
11. Casu RE, Pearson RD, Jarmey JM, Cadogan LC, Riding GA, Tellam RL. 1994. Excretory/secretory chymotrypsin from *Lucilia cuprina:* purification, enzymatic specificity and amino acid sequence deduced from mRNA. *Insect Mol. Biol.* 3:201–11
12. Chain E, Florey HW, Gardner AD, Heatley HG, Jenning MA, Orr-Ewing J, et al. 1940. Penicillin as a chemotherapeutic agent. *Lancet* 2:226–28
13. Charnley AK, Hunt J, Dillon RJ. 1985. The germ-free culture of desert locusts, *Schistocerca gregaria. J. Insect Physiol.* 31:477–85
14. Chernin E. 1986. Surgical maggots. *South. Med. J.* 79:1143–45
15. Chigusa Y, Kirinoki M, Yokoi H, Matsuda H, Okada K, et al. 1996. Two cases of wound myiasis due to *Lucilia sericata* and *L. illustris* (Diptera: Calliphoridae). *Med. Entomol. Zool.* 47:73–76
16. Chigusa Y, Matsumoto J, Kirinoki M, Kawai S, Matsuda H, et al. 1998. A case of wound myiasis due to *Lucilia sericata* (Diptera: Calliphoridae) in a patient suffering from alcoholism and mental deterioration. *Med. Entomol. Zool.* 49:125–27
17. Child FS, Roberts EF. 1931. The treatment of chronic osteomyelitis with live maggots. *NY State J. Med.* 31:937–43
18. Chodosh J, Clarridge JE, Matoba A. 1991. Nosocomial conjunctival ophthalmomyiasis with *Cochliomyia macellaria. Am. J. Ophthalmol.* 111:520–21
19. Church JCT. 1995. Larvatherapy—biosurgery. *Eur. Tissue Repair Soc. Bull.* 2:109–10
20. Church JCT. 1996. The early management of open wounds: shall we use maggots? *East Cent. Afr. J. Surg.* 2:9–12
21. Craig GK. 1988. *U.S. Army Special*

Forces Medical Handbook, pp. 510–12. Boulder, CO: Paladin Press. 600 pp.

22. Crile G, Martin E. 1917. Clinical Congress of Surgeons of North America, "war session," *JAMA* 69:1538–41
23. Cutrupi V, Lovisi A, Bernardi A, Meggio A. 1988. Miasi, considerazioni su di un caso. *Riv. Parassitol.* 3 (=47) (1986):185–88
24. Damsky LJ, Bauer H, Reeber E, Shaw JO, Anselment LA. 1976. Human myiasis by the black blow fly: brief clinical and laboratory observations of three cases. *Minn. Med.* 59:303–5
25. Daniel M, Šrámová H, Zálabská E. 1994. *Lucilia sericata* (Diptera: Calliphoridae) causing hospital-acquired myiasis of a traumatic wound. *J. Hosp. Infect.* 28:149–52
26. Degiusti DL, Zackheim H. 1963. A first report of *Wohlfahrtia vigil* (Walker) myiasis in man in Michigan. *JAMA* 184:782–83
27. Dicke RJ. 1953. Maggot treatment of actinomycosis. *J. Econ. Entomol.* 46:706–7
28. Dixon OHJ. 1933. The treatment of chronic osteomyelitis and other suppurative infections with live maggots (larva of the blow fly). *Vet. Bull.* 27:16–20
29. Dunbar GK. 1944. Notes on the Ngemba tribe of the Central Darling River of Western New South Wales. *Mankind* 3:177–80
30. El Kadery A, El-Begermy MA. 1989. Aural myiasis caused by *Wohlfahrtia magnifica. J. Egypt. Soc. Parasitol.* 19:751–53
31. Erdmann GR. 1987. Antibacterial action of myiasis-causing flies. *Parasitol. Today* 3:214–16
32. Erdmann GR, Khalil SKW. 1986. Isolation and identification of two antibacterial agents produced by a strain of *Proteus mirabilis* isolated from larvae of the screwworm (*Cochliomyia hominivorax*) (Diptera: Calliphoridae). *J. Med. Entomol.* 23:208–11
33. Erzinçlioglu YZ. 1989. The origin of parasitism in blowflies. *Br. J. Entomol. Nat. Hist.* 2:125–27
34. Evans H. 1997. A treatment of last resort. *Nurs. Times* 93:23, 62–65
35. Fabricius Ab Aquapendente H. 1634. *Medicina Practica.* Paris, Clodoveum Cottard IV. 651 pp. Quoted by Goldstein H. 1931. Maggots in the treatment of wound and bone infections. *J. Bone Jt. Surg.* 13:476–78
36. Ferguson LK, McLaughlin CW. 1935. Maggot therapy—a rapid method of removing necrotic tissues. *Am. J. Surg.* 29:72–84
37. Ferrar P. 1987. A guide to the breeding habits and immature stages of Diptera Cyclorrhapha. *Entomonograph* 8:1–907
38. Fine A, Alexander H. 1934. Maggot therapy—technique and clinical application. *J. Bone Jt. Surg.* 16:572–82
39. Fraser A, Ring RA, Stewart RK. 1961. Intestinal proteinases in an insect, *Calliphora vomitoria L. Nature* 4806:999–1000
40. Friedman E, Shaharabany M, Ravin S, Golomb E, Gollop N, et al. 1998. *Partially Purified Antibacterial Agent from Maggots Displays a Wide Range of Antibacterial Activity.* Presented at 3rd Int. Conf. Biotherapy, Jerusalem, Israel
41. Gertson GD, Lancaster WEG, Larson GA, Wheeler GC. 1933. Wohlfartia myiasis in North Dakota: report of two cases. *JAMA* 100:487–88
42. Gill CO. 1982. Microbial interaction with meats. In *Meat Microbiology*, ed. MH Brown, pp. 225–64. London: App. Sci.
43. Goldstein H. 1931. Maggots in the treatment of wound and bone infections. *J. Bone Jt. Surg.* 13:476–78
44. Grantham-Hill C. 1933. Preliminary note on the treatment of infected wounds with the larva of *Wohlfahrtia nuba. Trans. R. Soc. Trop. Med. Hyg.* 27:93–98
45. Greenberg B. 1954. A method for the

sterile culture of housefly larvae, *Musca domestica* L. *Can. Entomol.* 86:527–28

46. Greenberg B. 1968. Model for destruction of bacteria in the midgut of blow fly maggots. *J. Med. Entomol.* 5:31–38
47. Greenberg B. 1973. Flies through history. In *Flies and Disease,* 1:2–18. Princeton, NJ: Princeton Univ. Press. 447 pp.
48. Greenberg B. 1984. Two cases of human myiasis caused by *Phaenicia sericata* (Diptera: Calliphoridae) in Chicago area hospitals. *J. Med. Entomol.* 21:615
49. Grossman J. 1994. Flies as medical allies. *The World and I* 9(10):186–93
50. Groves BA, Bates PG. 1998. Preliminary investigations of plasma ammonia levels in sheep infested with *Lucilia sericata* and their potential in the aging of blowfly lesions in cases of neglect. *Med. Vet. Entomol.* 12:208–10
51. Guarnera EA, Mariluis JC. 1986. Caso humane de miasis cutánea diseminada por *Phaenicia sericata. Bol. Chil. Parasitol.* 41:79–82
52. Guerrini VH. 1988. Ammonia toxicity and alkalosis in sheep infested by *Lucilia cuprina* larvae. *Int. J. Parasitol.* 18:79–81
53. Hall MJR. 1991. Screwworm flies as agents of wound myiasis. In *New World Screwworm: Response to an Emergency,* ed. RDS Branckaert. *World Anim. Rev.* Oct.:8–17 (Spec. issue)
54. Hall MJR. 1995. Trapping the flies that cause myiasis: their responses to host-stimuli. *Ann. Trop. Med. Parasitol.* 89:333–57
55. Hall MJR. 1997. Traumatic myiasis of sheep in Europe: a review. *Parassitologia* 39:409–13
56. Hall MJR, Smith KGV. 1993. Diptera causing myiasis in man. In *Medical Insects and Arachnids,* ed. RP Lane, RW Crosskey, pp. 429–69. London: Chapman & Hall. xv +723 pp.
57. Hall MJR, Wall R. 1995. Myiasis of humans and domestic animals. *Adv. Parasitol.* 35:257–334
58. Hall RD, Anderson PC, Clark DP. 1986. A case of human myiasis caused by *Phormia regina* (Diptera: Calliphoridae) in Missouri, USA. *J. Med. Entomol.* 23:578–79
59. Haufe WO, Nelson WA. 1957. Human furuncular myiasis caused by the flesh fly *Wohlfahrtia opaca* (Coq.) (Sarcophagidae: Diptera). *Can. Entomol.* 89:325–27
60. Hobson RP. 1931. On an enzyme from blow-fly larvae (*Lucilia sericata*) which digests collagen in alkaline solution. *Biochemistry* 25:1458–63
61. Hobson RP. 1931. Studies on the nutrition of blow-fly larvae. I. Structure and function of the alimentary tract. *J. Exp. Biol.* 8:110–23
62. Hobson RP. 1932. Studies on the nutrition of blow-fly larvae. II. Role of the intestinal flora in digestion. *J. Exp. Biol.* 9:128–38
63. Hobson RP. 1932. Studies on the nutrition of blow-fly larvae. IV. The normal role of micro-organisms in larval growth. *J. Exp. Biol.* 9:366–77
64. Horn KL, Cobb AH, Gates GA. 1976. Maggot therapy for subacute mastoiditis. *Arch. Otolaryngol.* 102:377–79
65. Ioli V, Forinod D, Mento G, Catalano A. 1997. Myiase cutanée a *Lucilia sericata. Bull. Soc. Fr. Parasitol.*15:62–66
66. Iversen E. 1996. Methods of treating injuries of work animals. *Buffalo Bull.* 15:34–37
67. Jewett EL. 1933. The use of Unna's paste in the maggot treatment of osteomyelitis. *J. Bone Jt. Surg.* 15:513–515
68. Kani A, Nakamura O, Ono H, Nagase K, Totani T, Morishita T. 1981. A case of myiasis for papillary adenocarcinoma of parotid gland. *Acta Dermatol. Kyoto, Engl. Ed.* 76:173–77
69. Larrey DJ. 1829. Observations on wounds and their complications by erysipelas, gangrene and tetanus etc. pp. 51–52. Paris: Clin. Chir. Transl. EF Rivinus,

1932. p. 34. Philadelphia: Key, Mielke & Biddle
70. Leclercq M. 1990. Utilisation de larves de Diptères—Maggot Therapy—en médecine: historique et actualité. *Bull. Ann. Soc. R. Belge Entomol.* 126:41–50
71. Lee HL, Yong YK. 1991. Human aural myiasis. *Southeast Asian J. Trop. Med. Public Health* 22:274–75
72. Livingston SK. 1932. Maggots in the treatment of chronic osteomyelitis, infected wounds, and compound fractures. An analysis based on the treatment of one hundred cases with a preliminary report on the isolation and use of the active principle. *Surg. Gynecol. Obstet.* 54:702–6
73. Livingston SK. 1936. The therapeutic active principle of maggots—with a description of its clinical application in 567 cases. *J. Bone Jt. Surg.* 18:751–56
74. Livingston SK. 1937. Therapeutics of maggot active principle. *Am. J. Surg.* 35:554–56
75. Livingston SK, Prince LH. 1932. The treatment of chronic osteomyelitis with special reference to the use of the maggot active principle. *JAMA* 98:1143–49
76. Lukin LG. 1989. Human cutaneous myiasis in Brisbane: a prospective study. *Med. J. Aust.* 150:237–40
77. Maseritz IH. 1934. Digestion of bone by larvae of *Phormia regina*. *Arch. Surg.* 28:589–607
78. McClellan NW. 1932. The maggot treatment of osteomyelitis. *Can. Med. Assoc. J.* 27:256–60
79. McKeever DC. 1933. Maggots in treatment of osteomyelitis. A simple inexpensive method. *J. Bone Jt. Surg.* 15:85–93
80. Messer FC, McClellan RH. 1935. Surgical maggots. A study of their functions in wound healing. *J. Lab. Clin. Med.* 20:1219–26
81. Meydenbach J, ed. 1491. *Ortus Sanitatis* (Hortus Sanitatis). Compiled chiefly from the German *Hortus Sanitatis*, J. von Cube. Moguntia. Mainz: Jacobus Meydenbach. 454 leaves.
82. Miller KB, Hribar LJ, Sanders LJ. 1990. Human myiasis caused by *Phormia regina* in Pennsylvania. *J. Am. Podiatr. Med. Assoc.* 80:600–2
83. Mumcuoglu KY, Ingber A, Gilead L, Stessman J, Friedman R, et al. 1999. Maggot therapy for the treatment of intractable wounds. *Int. J. Dermatol.* (In press)
84. Mumcuoglu KY, Ingber A, Stessman J, Friedman R, Schulman H, et al. 1998. Maggot therapy for the treatment of diabetic foot ulcers. *Diabetes Care* 21:2030–31
85. Mumcuoglu KY, Lipo M, Ioffe-Uspensky I, Miller J, Galun R. 1996. Maggot therapy for the treatment of a severe skin infection in a patient with gangrene and osteomyelitis. Int. Congr. Entomol. 20th, Firenze, Italy. p. 772. (Abstr.)
86. Murdoch FF, Smart TL. 1931. A method of producing sterile blowfly larvae for surgical use. *US Nav. Med. Bull.* 29:406–17
87. NY Int. Bible Soc. 1978. *The Holy Bible,* New Int. Version, Job. 7:5
88. Ochsenhirt NC, Komara MA. 1933. Treatment of osteomyelitis of mandible by intraoral maggot-therapy. *J. Dent. Res.* 13:245–46
89. O'Rourke FJ. 1954. Furuncular myiasis due to *Wohlfahrtia vigil* (Walker). *Can. Med. Assoc. J.* 71:146–49
90. Óvári A, Farkas R, Adrián E. 1998. Experiences in wound treatment with sterile maggots in Hungary. *Lege Artis Med* 8:874–79 (In Hungarian)
91. Pape T. 1996. Catalogue of the Sarcophagidae of the world (Insecta: Diptera). *Mem. Entomol. Int.* 8:1–558
92. Pavillard ER, Wright EA. 1957. An antibiotic from maggots. *Nature* 180(4592): 916–17
93. Pearn JH. 1996. Landmines: time for an international ban. *Br. Med. J.* 312:990–91
94. Pechter EA, Sherman RA. 1983. Maggot

therapy: the medical metamorphosis. *Plast. Reconstr. Surg.* 72:567–70
95. Pendola S, Greenberg B. 1975. Substrate-specific analysis of proteolytic enzymes in the larval midgut of *Calliphora vicina. Ann. Entomol. Soc. Am.* 68:341–45
96. Portchinsky IA. 1916. *Wohlfahrtia magnifica*, Schin., and allied Russian species. The biology of this fly and its importance to man and domestic animals. *Mem. Bur. Entomol. Sci. Comm. Cent. Board Land Admin. Agric. Petrograd.* 11:1–108 (In Russian)
97. Prete PE. 1997. Growth effects of *Phaenicia sericata* larval extracts on fibroblasts: mechanism for wound healing by maggot therapy. *Life Sci.* 60:505–10
98. Puckner WA. 1932. New and nonofficial remedies, surgical maggots—Lederle. *JAMA.* 98:401
99. Rayman A, Stansfield G, Woolard T, Mackie A, Rayman G. 1998. Use of larvae in the treatment of the diabetic necrotic foot. *Diabetic Foot* 1:7–13
100. Reames MK, Christensen C, Luce EA. 1988. The use of maggots in wound debridement. *Ann. Plast. Surg.* 21:388–91
101. Robinson W. 1933. The use of blowfly larvae in the treatment of infected wounds. *Ann. Entomol. Soc. Am.* 26:270–76
102. Robinson W. 1934. Suggestions to facilitate the use of surgical maggots in suppurative infections. *Am. J. Surg.* 25:525
103. Robinson W. 1935. Stimulation of healing in non-healing wounds by allantoin occurring in maggot secretions and of wide biological distribution. *J. Bone Jt. Surg.* 17:267–71
104. Robinson W. 1935. Progress of maggot therapy in the United States and Canada in the treatment of suppurative diseases. *Am. J. Surg.* 29:67–71
105. Robinson W. 1937. The healing properties of allantoin and urea discovered through the use of maggots in human wounds. *Annu. Rep. Smithson. Inst.* 1937:451–61
106. Robinson W. 1940. Ammonium bicarbonate secreted by surgical maggots stimulates healing in purulent wounds. *Am. J. Surg.* 47:111–15
107. Robinson W, Baker FL. 1939. The enzyme urease and occurrence of ammonia in maggot infected wounds. *J. Parasitol.* 25:149–55
108. Robinson W, Norwood VH. 1934. Destruction of pyogenic bacteria in the alimentary tract of surgical maggots implanted in infected wounds. *J. Lab. Clin. Med.* 19:581–86
109. Roche S, Cross S, Burgess I, Pines C, Cayley ACD. 1990. Cutaneous myiasis in an elderly debilitated patient. *Postgrad. Med. J.* 66:776–77
110. Schaefer CW. 1979. Feeding habits and hosts of Calyptrate flies (Diptera: Brachycera: Cyclorrhapha). *Entomol. Gen.* 5:193–200
111. Seaquist ER, Henry TR, Cheong E, Theologides A. 1983. *Phormia regina* myiasis in a malignant wound. *Minn. Med.* 66:409–10
112. Sherman RA. 1997. A new dressing design for treating pressure ulcers with maggot therapy. *Plast. Reconstr. Surg.* 100:451–56
113. Sherman RA. 1998. Maggot therapy in modern medicine. *Infect. Med.* 15:651–56
114. Sherman RA, Pechter EA. 1988. Maggot therapy: a review of the therapeutic applications of fly larvae in human medicine, especially for treating osteomyelitis. *Med. Vet. Entomol.* 2:225–30
115. Sherman RA, Tran JM-T, Sullivan R. 1996. Maggot therapy for venous stasis ulcers. *Arch. Dermatol.* 132:254–56
116. Sherman RA, Wyle FA, Thrupp L. 1995. Effects of seven antibiotics on the growth and development of *Phaenicia sericata*

(Diptera: Calliphoridae) larvae. *J. Med. Entomol.* 32:646–49

117. Sherman RA, Wyle F, Vulpe M. 1995. Maggot debridement therapy for treating pressure ulcers in spinal cord injury patients. *J. Spinal. Cord. Med.* 18:71–74
118. Sherman RA, Wyle FA, Vulpe M, Levsen L, Castillo L. 1993. *Utility of Treating Chronic Wounds with Maggot Therapy.* Presented at Annu. Conf. Am. Soc. Trop. Med. Hyg. Atlanta, GA
119. Sherman RA, Wyle FA, Vulpe M, Levsen L, Castillo L. 1993. The utility of maggot therapy for treating pressure sores. *J. Am. Paraplegia Soc.* 16:269 (Abstr.)
120. Sherman RA, Wyle FA, Vulpe M, Wishnow R, Iturrino J, et al. 1991. Maggot therapy for treating pressure sores in spinal cord patients. *J. Am. Paraplegia Soc.* 14:200 (Abstr.)
121. Simmons SW. 1934. Sterilization of blowfly eggs in the culture of surgical maggots for use in the treatment of pyogenic infections. *Am. J. Surg.* 25:140–47
122. Simmons SW. 1935. The bactericidal properties of excretions of the maggot of *Lucilia sericata. Bull. Entomol. Res.* 26:559–63
123. Simmons SW. 1935. A bactericidal principle in excretions of surgical maggots which destroys important etiological agents of pyogenic infections. *J. Bacteriol.* 30:253–67
124. Smart J. 1936. Larvae of *Lucilia sericata* Mg., from a case of aural myiasis reported from Essex (Diptera). *Proc. R. Entomol. Soc. London Ser. A* 11:1
125. Smith DR. Clevenger RR. 1986. Nosocomial myiasis. *Arch. Pathol. Lab. Med.* 110:439–40
126. Spradbery JP. 1994. Screw-worm fly: a tale of two species. *Agric. Zool. Rev.* 6:1–62
127. Stabler RM, Nelson MC, Lewis BL, Berthrong M. 1962. *Wohlfahrtia opaca* myiasis in man in Colorado. *J. Parasitol.* 48:209–10
128. Stewart MA. 1934. The role of *Lucilia sericata* Meig. larvae in osteomyelitis wounds. *Ann. Trop. Med. Parasitol.* 28:445–60
129. Stewart MA. 1934. A new treatment of osteomyelitis. *Surg. Gynecol. Obstet.* 5:155–65
130. Teich S, Myers RAM. 1986. Maggot therapy for severe skin infections. *South. Med. J.* 79:1153–55
131. Tenquist JD, Wright DF. 1976. The distribution, prevalence and economic importance of blowfly strike in sheep. *NZ J. Exp. Agric.* 4:291–95
132. Thomas S, Andrews A. 1999. The effect of hydrogel dressings on maggot development. *J. Wound Care* 8:75–77
133. Thomas S, Andrews A. 1999. The antimicrobial activity of maggot secretions: results of a preliminary study. *J. Tissue Viability* (In press)
134. Thomas S, Andrews A, Jones M. 1998. The use of larval therapy in wound management. *J. Wound Care* 7:521–24
135. Thomas S, Jones M, Shutler S, Jones S. 1996. Using larvae in modern wound management. *J. Wound Care* 5:60–69
136. Thomas S, Jones M, Shutler S, Jones S. 1997. The use of fly larvae in the treatment of wounds. *Nurs. Stand.* 12:54–59
137. Townsend LH Jr, Hall RD. 1976. Human myiasis in Virginia caused by *Phaenicia sericata* (Meigen) (Diptera: Calliphoridae). *Proc. Entomol. Soc. Wash.* 78:113
138. Vistnes LM, Lee R, Ksander GA. 1986. Proteolytic activity of blowfly larvae secretions in experimental burns. *Surgery* 90:835–41
139. Wainwright M. 1988. Maggot therapy—a backwater in the fight against bacterial infection. *Pharm. Hist.* 30:19–26
140. Waterhouse DF, Irzykiewicz H. 1957. An examination of proteolytic enzymes from several insects for collagenase activity. *J. Insect Physiol.* 1:18–22
141. Weil GC, Simon RJ, Sweadner WR. 1933. A biological, bacteriological and clinical study of larval or maggot therapy in the treatment of acute and chronic pyogenic infections. *Am. J. Surg.* 19:36–48

142. WHO. 1998. *World Health Organization Annual Report.* Geneva: WHO
143. Zeltser R, Lustmann J. 1988. Oral myiasis. *Int. J. Oral Maxillofacial Surg.* 17: 288–89
144. Ziffren SE, Heist HE, May SC, Womack NA. 1953. The secretion of collagenase by maggots and its implication. *Ann. Surg.* 153:932–34
145. Zumpt F. 1965. *Myiasis in Man and Animals in the Old World.* London: Butterworths. 267 pp.

Annu. Rev. Entomol. 2000. 45:83–110

LIFE HISTORY AND PRODUCTION OF STREAM INSECTS

Alexander D. Huryn[1] and J. Bruce Wallace[2]

[1]*Department of Biological Sciences, University of Maine, Orono, Maine 04469-5722; e-mail: huryn@maine.maine.edu*

[2]*Department of Entomology, University of Georgia, Athens, Georgia 30602-2202; e-mail: wallace@sparc.ecology.uga.edu*

Key Words aquatic insects, macroinvertebrates, growth, biomass, secondary production, molting, predation, food limitation, disturbance

■ **Abstract** Studies of the production of stream insects are now numerous, and general factors controlling the secondary production of stream communities are becoming evident. In this review we focus on how life-history attributes influence the production dynamics of stream insects and other macroinvertebrates. Annual production of macroinvertebrate communities in streams world-wide ranges from approximately 10^0 to 10^3 g dry mass m^{-2}. High levels are reported for communities dominated by filter feeders in temperate streams. Filter feeding enables the accrual and support of high biomass, which drives the very highest production. Frequently disturbed communities in warm-temperate streams are also highly productive. Biomass accrual by macroinvertebrates is limited in these streams, and production is driven by rapid growth rates rather than high biomass. The lowest production, reported for macroinvertebrate communities of cool-temperate and arctic streams, is due to the constraints of low seasonal temperatures and nutrient or food limitation. Geographical bias, paucity of community-wide studies, and limited knowledge of the effects of biotic interactions limit current understanding of mechanisms controlling stream productivity.

INTRODUCTION

Insects and other macroinvertebrates play a central role in the flow of materials and energy through most terrestrial and benthic freshwater food-webs (reviews in 120, 124, 144). Their quantitative influence on this flow will be largely defined by the magnitude of their production. Production at a given trophic level sets limits for productivity of higher trophic levels, while simultaneously affecting rates of resources removed from lower levels. Details of this interactive process are poorly understood for most macroinvertebrate communities (107). However, it is becoming increasingly apparent that this process is an important determinant of overall ecosystem productivity (62).

0066-4170/00/0107–0083/$14.00

We agree that production "by itself does not tell us much about the influence of invertebrates on ecosystem functioning since it is only one end product of organic matter processing" (39:1217). However, we emphasize that, when combined with information about food-web interactions (sensu 100), a comprehensive knowledge of production will improve understanding of the structure and function of both communities and ecosystems (15, 107). For stream communities in particular, estimates of macroinvertebrate production have recently been used to quantify the consequences of direct consumption and cascading trophic interactions to total stream primary productivity (64, 65), and to indicate the importance of a given food-web link or trophic resource to overall ecosystem function (15, 53).

Secondary Production and Life History

In general use, "secondary production" refers to the formation of animal biomass over time (mass area^{-1} · time^{-1}; review in 8). Annual secondary production, for example, is the sum of all biomass produced by a population during one year, including production remaining at the end of the year and all biomass produced during this period. Losses may include mortality (e.g. disease, parasitism, cannibalism, and predation), loss of tissue reserves (e.g. molting, silk, and starvation), and emigration. The secondary production of a population (P) is the product of the biomass-specific growth rate (g, mass · mass^{-1} · time^{-1}) and population biomass (B, mass · area^{-1}) ($P = g \times B$; review in 8). Numerous methods exist for measuring secondary production, and the relationship between the biomass-specific growth rate and population biomass is implicit to all (7).

The contrasting roles of growth rate and biomass in determining production are conveniently summarized by the P/B ratio. Cohort P/Bs for stream macroinvertebrates usually range from two to eight (7, 148). Annual P/Bs for entire communities, however, range from less than 1–117 because some populations may complete more or less than one cohort each year (8). The annual and daily P/B provides an index allowing comparison of growth rates (i.e. $g = P/B$) among populations and communities, and will be used as such throughout this review (e.g. Tables 1 and 2).

Rates of biomass growth and accrual for any macroinvertebrate population are constrained by "life history"—the temporal pattern of development from egg through adult stages and the duration and abundance of each stage (25). Natality, abundance, individual growth rate, individual biomass, dispersal, and survivorship are all important life-history attributes that together determine levels of production at the population level (7, 8). Because of the close relationship between life-history attributes and production, patterns of production among macroinvertebrate populations and communities must also reflect the relationship between life history and the environmental template (129).

The general objective of this review is to address the relationship between life-history attributes and production by freshwater macroinvertebrate communities.

TABLE 1 Examples of exceptionally low and high growth rates for freshwater macroinvertebrates.

Organism	Growth rate (day^{-1})	Annual P/B	Developmental period	Habitat/location	Source
Low Growth Rates					
Unio tumidus (Unionacea)	0.00001–0.00016 a	0.004–0.057	10+ years?	lake/Hungary	108
Unio tumidus (Unionacea)	0.0004[a]	0.13	12 years	river/England	94
Austropotamobius pallipes (Decapoda)	0.0009[a]	0.32	11 years	stream/England	18
Chironomus (Diptera)	0.0013[a]	0.49	7 years	tundra pools/Alaska	23
Lara avara (Coleoptera)	0.0031–0.0036[b]	1.1–1.3[c]	5–6 years	stream/Oregon	131
Acroneuria lycorias (Plecoptera)	0.0039–0.0043[a]	1.4–1.6	3 years	stream/Michigan	34
Sericostoma personatum (Trichoptera)	0.0066[a]	2.4	3 years	stream/Denmark	72
Philocasca alba (Trichoptera)	0.0079[b]	2.9[c]	3 years	stream/Alberta	93
Hexagenia limbata (Ephemeroptera)	0.0082[d]	3.0[c]	3–4 years	reservoir/Manitoba	44
Euthyplocia hecuba (Ephemeroptera)	0.012–0.014[b]	4.4–5.1[c]	22 months	stream/Costa Rica	137
High Growth Rates					
Diamesa incallida (Diptera)	0.17–0.22[e]	62–80[c]	35–45 days	stream/Germany	96
Baetis spp. (Ephemeroptera)	0.265[a]	97	19 days	stream/Georgia	11
Polypedilum epomis (Diptera)	0.29[e]	106[c]	22 days	stream/Costa Rica	75
Chironomidae (Diptera)	0.33[a]	120	12 days	stream/Arizona	74
Orthocladius calvus (Diptera)	0.56[b]	203[c]	16 days	exp. stream/England	80
Leptohyphes packeri (Ephemeroptera)	0.66[a]	240	12 days	stream/Arizona	74
Polypedilum spp. (Diptera)	0.71	258	7–12 days	stream/Georgia	9

[a] Estimated as annual P/B divided by 365.
[b] Estimated as ln(final mass/initial mass)/developmental period (days).
[c] Approximate annual P/B = daily growth rate × 365; provided as index only (assumes continuous development and exponential growth).
[d] Estimated as "b" using length-frequency data and length-weight equations (10).
[e] Estimated as "d" assuming initial length of 0.5 mm (cf. 63) and length-weight equations (10).

TABLE 2 Examples of exceptionally low and high levels of production for macroinvertebrate communities in streams and rivers. Values are standardized to dry mass (DM) equivalents (147).

Location	Production (g DM · m^{-2} · $year^{-1}$)	Annual P/B	Habitat—organisms	Source
Low Production				
North Carolina	1.2	5.2	headwater stream with four-year detritus exclusion	143
Alaska	~ 0.8 to 2.2[a]	~5.5	tundra stream (unfertilized reach)	57
South Carolina	2 to 4.1	7.5 to 9.4	stream—collector-gatherers, filter-feeders, predators	128
Norway	3.9	~ 4.8	river—high discharge upstream of a weir	41
Minnesota	5.4	4.2	river—collector-gatherers, predators, shredders	79
High Production				
Arizona	121 to 135	78 to 117.4[b]	desert stream—midges, mayflies, collector-gatherers	39, 74
Germany	129	0.7	lake outflow—bivalves dominate biomass	109
North Carolina	169[c]	8.5	river—primarily filter-feeding caddisflies	53
Georgia	200[c]	8.2	river—primarily filter-feeding caddisflies	52
Wales (U.K.)	268	3.0	moat stream—Tubificidae (organic enrichment)	82
West Virginia	612[c]	30.3	below dam—black flies and filter-feeding caddisflies	141
Iceland	~40 to 880	~2.5 to 11	lake outflow—black flies (eight-year study)	45
England	~1000	2.2 to 2.7	pond outflow—black flies (summer cohorts)	154

[a] Six-year range for three dominant invertebrate taxa.
[b] P/B based on zero winter biomass.
[c] Primarily bedrock substrata covered by the hydrophyte *Podostemum ceratophyllum*, which provides habitat structure.

The literature concerning freshwater macroinvertebrate life histories (25, 136, 145) and production (8, 147) is large. Rather than providing a global review of this information, we focus on how selected life-history attributes of different developmental stages of benthic insects [egg, first-instar larva (= "first-instar" hereafter), larva during subsequent instars (= "larva" hereafter), pupa, and adult] and other macroinvertebrates influence their production. We pay particular attention to factors that influence growth rate or biomass because these will have the strongest influence on secondary production. Although examples from both lentic and lotic habitats are included, we focus primarily on streams.

EGGS AND FIRST-INSTAR

Estimates of the biomass of adult insects emerging from streams are generally only about 24% of total annual production (130), and most estimates of adult mortality prior to oviposition exceed approximately 95% (36, 47, 74, 150). These high losses indicate that egg production will be a small fraction of annual production. The few attempts to measure egg production support this assessment (e.g. approximately .03 to 2% of annual production; 22, 99, 153). Although rarely considered in this context, this statistic is important because the egg and first-instar represent the initial investment of biomass that begins the process of production. Other factors directly related to the egg and first-instar that may have important influences on production are mortality, hatching synchrony, dispersal and cannibalism.

Mortality

The initial biomass of a cohort will be strongly influenced by egg mortality. Available information for stream insects indicates that hatching success is usually fairly high, often ranging from 70% to more than 90% (20, 36, 87, 104, 153, 157). Low levels of hatching success (approximately 20%) have also been reported (29). Agents contributing to egg mortality include invertebrate predators and parasitoids (36, 104, 106, 111, 145) and zoosporic fungi (87). Physical disturbances such as flooding can impose high levels of mortality for taxa that oviposit near the water's surface (36), as well as eggs resting among substrata (46, but see 157).

Hatching Synchrony

Eggs of many aquatic insects show synchronized development and hatching (21, 85, 106, 153). Synchronization offers several advantages, including reduction of cannibalism (61, but see 153) and efficient access to food resources available only for a short period (e.g. autumn detritus) (85). Asynchronous egg development, however, is also widespread among aquatic insects (35, 55, 92, 106), and may have an important role in recolonization of streams following scouring floods (157). The pros and cons of hatching synchrony in the context of secondary production have received little attention. Given similar larval abundances of the mosquito *Aedes triseriatus*, cohorts with synchronous development yielded smaller adult females compared with cohorts with asynchronous development (33). The diverse size–structure of larvae in the asynchronous cohorts apparently reduced competition and enhanced larval growth and production (33).

First-Instar

The primary function of the first-instar of many aquatic insects is dispersal, which is distinct from subsequent instars whose function is primarily growth (104, 106). A dispersal stage shortly after hatching is critical for stream insects where access to oviposition sites may not allow optimal placement of eggs (42, 104, 106). Competition near oviposition sites due to clumped distribution of eggs may lead to reduced feeding rates, dispersal, and density-dependent mortality (42, 104). Mortality of first-instars can be high. Willis & Hendricks (153) reported 93% mortality for first-instars of the caddisfly *Hydropsyche slossonae*.

Cannibalism by first-instars has been documented for the Odonata and Trichoptera (61, 104, 153). Willis & Hendricks (153) reported that 10% to 20% of first-instars of *H. slossonae* cannibalize siblings, a statistic that is remarkably similar to their survivorship (approximately 10%). Cannibalism may have important implications for growth and survival of larvae. By eating siblings, first-instars of some insects can double their survival probabilities and proceed to the second instar (153).

LARVA

The egg and first-instar represent the initial investment of biomass for subsequent production, but factors influencing individual growth rates and biomass accrual of larvae following these initial stages are probably the most critical in determining overall secondary production by stream insects. In this section we focus on the extremes of variation in growth (slow versus fast) and biomass (low versus high) of freshwater benthic macroinvertebrates, and show how these interact to determine extreme levels of production (low versus high) by populations and communities.

Slow Growth Rates and Long Life Cycles

To our knowledge, larvae of the xylophagous elmid beetle *Lara avara* have the slowest growth rates (~0.3% to 0.4% day^{-1}) and the longest life cycles (5–6 years) of all stream insects (131; Table 1). The slowest growth rates recorded among all aquatic insects (~0.1% day^{-1}; Table 1), however, are those for larvae of *Chironomus* inhabiting tundra pools in northern Alaska. These larvae require seven years to complete their life cycle (23, 24). Other reports of extremely slow growth rates (<1% day^{-1}) and long life cycles (~2+ years) for stream insects are summarized in Table 1. Although the majority of these are for species from cool-temperate streams, long life cycles are not restricted to higher latitudes. Sweeney et al (137) studied the tropical mayfly *Euthyplocia hecuba* in Costa Rica and found growth rates of <1.4% day^{-1} and a life cycle of 22 months (Table 1). This is the first evidence of a life cycle of more than one year for a tropical mayfly.

Crayfish (Decapoda), particularly those from streams at high latitudes or altitudes, exhibit some of the lowest growth rates and longest life cycles reported for stream macroinvertebrates (Table 1). Examples of low annual P/Bs for crayfish include: 0.3 to 0.4 (18, 151); 0.58 to less than 1.0 (70, 91); or approximately 1.0 (132). Life cycles of more than 10 years have been reported for several species (6, 18, 70, 151).

Without doubt, the slowest-growing and longest-lived stream macroinvertebrates are mollusks (Table 1). The freshwater mussel *Unio*, for example, has annual P/Bs of less than 0.1 and life cycles greater than 10 years in England and Europe (94, 108; Table 1). Although not measured, the P/Bs of the mussel *Margaritifera* are probably even lower because some populations in North American streams have modal ages ranging from 48 to 100 years (134, 140). In contrast, some bivalves such as the Asiatic clam *Corbicula fluminea* and the zebra mussel *Dreissena polymorpha* have higher growth rates and shorter life cycles (three years). Populations of *Corbicula* have annual P/Bs ranging from 0.5 to 1.8 (133), and P/Bs for zebra mussels range from 3.3 to 3.5 (76). Other mollusks that have slow growth rates include the pleurocerid snail *Elimia*, with annual P/Bs ranging from about 0.5 to 1.5 and life cycles of two to three or more years for eight populations in Alabama (66).

Rapid Growth Rates and Short Life Cycles

The highest annual P/B measured for a stream insect is 258 for larvae of the chironomid *Polypedilum* in the Ogeechee River in Georgia (9; Table 1). The bioenergetic significance of this statistic becomes clear if one realizes that it equals a daily P/B of 0.7, indicating that this population replaces its biomass almost daily. This growth rate is among the highest estimated for metazoans, and is more similar to growth rates of microbes than to other stream macroinvertebrates. Relatively rapid growth for other macroinvertebrates, particularly the Ephemeroptera, have also been reported for this system (e.g. annual P/Bs ~60 to 90; 12). Warm temperatures (maximum ~31 to 33° C), a large supply of high-quality food, and small body size contribute to rapid growth rates of mayflies and midges in the Ogeechee River (9), as well as other warm-temperate streams (e.g. 39, 46, 51). However, the effect of such factors is not universal.

Jackson & Sweeney (75) documented growth rates and developmental time for 35 species of Ephemeroptera, Plecoptera, Trichoptera, and Chironomidae in Costa Rican streams (annual range of water temperature = 20–23° C), and showed most taxa had multivoltine life cycles (32 of 35 taxa), overlapping cohorts and complex size structure. The results from this study were surprising, however:

Although food and temperature were apparently not limiting, the highest growth rates measured for insects in these tropical streams (~0.29 day^{-1} for *Polypedilum epomis*; Table 1) were well below those reported from some warm-temperate streams (e.g. ~0.66 day^{-1} for *Leptohyphes packeri*; 0.71 day^{-1} for *Polypedilum* spp; Table 1).

Rapid growth rates and short developmental times are not limited to macroinvertebrates of tropical or warm-temperate streams. A short life cycle (approximately 16 days) and extremely rapid growth rates (~0.56 day^{-1}; Table 1) have been reported for the midge *Orthocladius calvus* in an outdoor recirculating stream in England (80). A short life cycle of 35 to 45 days and a relatively high growth rate (~ 0.17 day^{-1}; Table 1) was also reported for the midge *Diamesa incallida* in a spring stream in Germany (96). These rapid growth rates are particularly surprising given the large terminal size of the larvae of these taxa—9 to 10 mm in length, 0.7 to 0.8 mg dry mass (=DM)—and relatively cool temperatures of their habitats (10.5° C–80; 8° C–96). Nevertheless, comprehensive studies of the growth rates for midges in many cool-temperate streams indicate that these are generally less than approximately 0.05 day^{-1} (17, 63).

Factors Controlling Growth Rates

The maximum and minimum limits for growth rates of freshwater macroinvertebrates are determined by food quality (9, 49, 51, 71, 93, 131), temperature (9, 39, 44, 49, 71, 75, 93), time available for development (24), body size (9, 70, 75, 113, 137, 151), population density (37, 56, 66) and intrinsic genetically based constraints (140). These major factors have been the subject of comprehensive reviews (cf. 30, 81, 136, 140; see also 49). Rather than focusing on these here, we will address the effects of several additional factors that, although probably significant, are generally not considered in the context of production.

Region and Habitat Template The highest growth rates known for any stream insect are reported for the warm-temperate streams, Sycamore Creek and the Ogeechee River (9, 12, 46, 74). The organisms involved (primarily midges and mayflies) not only grow rapidly, but also complete life cycles rapidly because they attain relatively small sizes (e.g. length 4 to 5 mm; 9, 12, 46). Given the habitat template offered by Sycamore Creek and the Ogeechee River, one explanation for the extremely high growth rates may lie in the balance of fitness costs associated with fast growth and short life cycles versus slow growth and long life cycles (97). In habitats where risk of mortality is high during development, growth rate should be maximized and size minimized; where risk of mortality is low, size should be maximized, generally resulting in slower growth rates and longer developmental periods (75).

In Sycamore Creek, mortality due to random flash flooding (see "Larval Production"), and in the Ogeechee River, mortality from dehydration because of frequent stranding of snag communities, has apparently resulted in selection for taxa with rapid growth rates and short life cycles (12, 13, 46, 74). The extreme rates of growth and turnover recorded for these systems are attributable to warm temperatures, abundant food and a harsh disturbance regime. Attributing extremely high growth rates and production to these factors may seem counter-intuitive, but given a similar thermal regime and food supply, with a more benign disturbance regime, selection for taxa with longer life cycles and a larger terminal size may result in lower average growth rates and production (see "Larval Production"; 75).

Body Size and Armor Although large body size is often associated with high fecundity (117), a period of prolonged growth may be required with increased risk of mortality. Perhaps in response, many taxa with extremely slow growth rates (bivalves, snails, crayfish) tend to be large and armored compared with those with rapid growth rates (cf. Table 1). Protection from predation (94, 113, 116) or physical disturbance (140) may allow these groups to reduce mortality and accrue biomass over longer life cycles than most insects (also see 19). Larvae of the exceptionally slow-growing elmid beetle *Lara* provide an important example from the insects (Table 1). These larvae presumably avoid predation by possessing remarkably heavily sclerotized bodies and by remaining concealed within feeding grooves on submerged wood (131).

Presence of Predators The mere presence of a predator within a stream community may have a profound influence on growth rates of prey (27). When exposed to predators, or chemicals released by predators or injured conspecifics, benthic insects typically show reduced movement activity, sheltering behavior (121), or greater nocturnal activity compared to daytime activity (89). Such behaviors reduce time spent feeding, which reduces growth rates (90). Peckarsky et al (103), for example, showed that larvae of the mayfly *Baetis bicaudatus* did not grow in the presence of predators, but in the absence of predators larval mass increased 50 percent during one week. Similarly, data from field observations and experiments predicted that in the presence of predaceous stoneflies and trout, *B. bicaudatus* emerge at a mass 50% smaller compared to larvae reared in the absence of predators (90). However, growth rates do not always decrease in the presence of a predator. The pulmonate snail *Physella* shows higher growth rates and attains a larger size in the presence of their crayfish predators in Oklahoma streams (27).

Few data show how growth rates differ among macroinvertebrate communities subject to different levels of predation, but available examples indicate that predation may have a significant effect. Huryn (65) studied production of macroinvertebrate communities in two streams with high and low levels of predation by fish (i.e. approximately 100% versus 18% of macroinvertebrate production consumed). The community subjected to high predation was characterized by individuals of smaller size and higher growth rates (annual P/B = 6.9) compared to that experiencing low predation (annual P/B = 3.9) (65). Body size is often negatively correlated with growth rate (e.g. 4, 75), so higher growth rates observed in this study (65) may be either an indirect effect of size-selective predation, or a bias toward rapidly growing taxa that attained smaller body size, or both.

Competition Competition can affect growth rates of macroinvertebrates in streams, particularly grazers (56, 37). A tenfold increase in the density of the grazing caddisfly *Glossosoma nigrior* reduced prepupal biomass by approximately 25% (56). High abundance of larvae of the grazing caddisfly *Helicopsyche borealis* following dry winters in a California stream resulted in lower pupal mass (approximately 25% lower) compared to years with wet winters (37). Scouring flows during wet winters reduced both larval abundance and competition for food (37). Growth rates of the pleurocerid snail *Elimia* in six Alabama streams also showed a strong negative relationship with population biomass, suggesting that growth was limited by competition for periphyton (66).

LARVAL PRODUCTION

Production is the product of growth rate and biomass (8). The highest levels of secondary production will therefore occur when growth rates are rapid and biomass is high (8). However, because these factors are often negatively correlated, both parameters are rarely maximized simultaneously (37, 56, 66; see "Competition"). Consequently, high levels of production observed in the field usually result from rapid growth rates plus low biomass, or slow to moderate growth rates plus high biomass. To illustrate the fundamental importance of this simple observation, we provide examples using chironomid production in two North American streams.

Sycamore Creek, Arizona—A Dynamic System Sycamore Creek is a warm temperate, desert stream with annual water temperatures ranging from 5 to 33° C. Sycamore Creek is subject to two to nine flash floods each year that scour substrata and reduce benthic insect abundance by as much as 98% (46). Annual production of Chironomidae is ~58 g dry mass (=DM) m^{-2} of stream bottom (74). Average biomass is relatively low (~0.5 g DM m^{-2}), but larval growth rates are rapid (~ 0.3 day^{-1}), larvae are small (maximum length <5 mm), and development is continuous with approximately 20 or more generations per year (Figure 1). These factors result in an annual P/B of 121.5.

Juday Creek, Indiana—An Inertial System Juday Creek is a temperate forest stream with annual water temperature ranging from 2.5 to 17° C. Floods are infrequent and seasonal. Annual production of Chironomidae is 29.7 g DM m^{-2} (16). Although biomass is high compared to Sycamore Creek (~4.6 g DM m^{-2}), larval growth rates are slower (maximum <0.09 day^{-1}), larvae of the dominant taxa are larger (~7 to 11 mm), and only two to three generations are completed

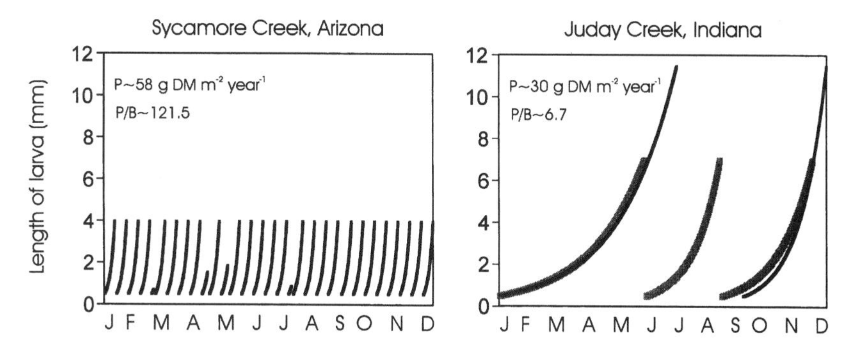

Figure 1 Life cycles of midge larvae in a warm-temperate desert stream prone to random flash flooding (Sycamore Creek), and a cool-temperate stream with more stable flows (Juday Creek). Emergence and oviposition occur at the maximum length indicated by growth trajectories. *Left:* Generalized life cycle for midge assemblage of Sycamore Creek (46, 74). Breaks indicate the effects of random flash floods (cf. 46). *Right:* Life cycles for *Diamesa nivoriunda (black)* and *Cricotopus bicinctus (gray)*. Both taxa are major contributors to midge production in Juday Creek (16, 17).

each year (Figure 1). These factors result in an annual P/B of 6.7.

The comparison of these studies shows that the life history attributes of midges in Sycamore Creek resulted in dynamic control of production, whereas those in Juday Creek resulted in inertial control. In Sycamore Creek, rapid growth rates, short life cycles, and small terminal size of larvae compensated for low biomass and resulted in high production and P/Bs (see "Region and Habitat Template"). In Juday Creek, midges had longer life cycles, slower growth, and larger terminal sizes, yet chironomid production in Juday Creek was similar to that of Sycamore Creek because biomass was approximately nine times higher in this cool-temperate stream. The distinction between dynamic and inertial control of production is particularly evident in these examples. The balance between growth rate and biomass in determining macroinvertebrate production in most highly-productive streams appears to be biased toward inertial control (Table 2).

Extremes of Larval Production

High Production The highest production known for an aquatic insect is 7.4 and 8.8 g carbon m^{-2} day^{-1} for each of two summer cohorts of filter-feeding larvae of the black fly *S. noelleri* in a lake outlet stream (154, Table 2). Minimum annual production was 414.6 g carbon m^{-2}; total annual production probably exceeded 500 g carbon m^{-2} [~1 kg ash-free dry mass (=AFDM) m^{-2}; cf. 88]. This example is extreme, however. More reasonable "extremely high" rates of community-level production range from 100 to 300 g DM m^{-2} $year^{-1}$ (8; Table 2)

What factors contribute to high secondary productivity? Estimates of annual production >100 g DM m^{-2} are usually for filter-feeding insects. Examples include hydropsychid caddisflies (15, 52, 53, 101, 141; Table 2) and black flies (14, 45, 154; Table 2). Some filter-feeding mollusks also show high levels of production (e.g. 26, 109; Table 2). High production by filter-feeders is always due in part to relatively rapid growth rates and high population biomass (e.g. some black flies; Table 2) or to slow to moderate growth rates combined with exceptionally high biomass (e.g. bivalves, hydropsychids, some black flies; Table 2). The astonishingly high production of filter feeders in some lake outlet streams (Table 2) is subsidized by favorable thermal regimes and nutrient-rich plankton exported from the lake, which enhance growth rates and allow for accrual of high levels of population biomass (155; see "Biomass Accrual"). Levels of annual production approaching 100 g DM m^{-2} are also reported for communities of collector-gathering macroinvertebrates in warm-temperate streams with abundant food and a harsh disturbance regime (e.g. 9, 40, 74; see "Region and Habitat Template").

Low Production Estimates of macroinvertebrate production >100 g DM m^{-2} are exceptionally high. But what level of production should be considered exceptionally low? Based on a summary of 58 studies of production for entire macroinvertebrate communities, 40% reported levels of less than 10 g DM m^{-2} $year^{-1}$, and 78% reported levels of less than 50 g DM m^{-2} $year^{-1}$ (8). Only three studies, however, reported levels less than approximately 3 g DM m^{-2} $year^{-1}$. Data available since Benke's (8) review are in accordance with these values (e.g. 64, 65, 109, 142, 143; Table 2).

What factors contribute to low secondary productivity? Obviously, any factor that reduces growth rates or biomass of macroinvertebrates will also reduce their production. Low food quality and quantity, and low temperature can reduce growth rates (see "Factors Controlling Growth Rates"). Reductions in biomass can be caused by many factors, including emigration (60, 67), predation and competition (78, 149), food availability and nutrients (57, 79, 105, 143), disturbance (40, 41, 57, 66, 122, 123), availability of suitable substrata (14, 28, 52, 53, 109, 128), and dissolved oxygen and low stream velocities (128). Several studies suggest that bottom-up control (food limitation) has a critical influence on the production of stream macroinvertebrates. For example, one of the lowest levels of secondary production measured in a stream resulted from a four-year experimental exclusion of terrestrial detritus inputs to a heavily shaded headwater stream in North Carolina (142, 143; Table 2). Exclusion of detritus resulted in a consecutive four-year decline in abundance (−77%), biomass (−79%), and secondary production (−78%) of both primary and secondary consumers in the dominant habitat, which underscores the importance of bottom-up control of productivity (143). Likewise, the addition of nutrients to an arctic tundra stream increased both primary and secondary productivity (105).

FATE OF LARVAL PRODUCTION

Biomass Accrual

Biomass of stream macroinvertebrates accrues whenever gains due to production or immigration exceed losses within a given area (see "Emigration/Immigration"). Accumulating biomass may have simultaneous positive and negative influences on rates of production because of its dual role as both an investment that contributes to future production and a burden because of the greater fraction of available energy needed for its maintenance (19, 66; see "Disturbance"). A survey of 57 studies that estimated biomass for entire macroinvertebrate communities revealed that 68% were less than 5 g DM m^{-2} and 81% were less than 10 g DM m^{-2} (8). Levels of biomass for entire macroinvertebrate communities in streams appear to be generally less than 5 to 10 g AFDM m^{-2}. Estimates approaching 100 g m^{-2} are rare (8, 52, 53, 109).

High levels of biomass accrual often characterize taxa with long life cycles (e.g. more than two years) and overlapping cohorts. In such cases, high biomass can offset relatively low growth rates and result in surprisingly high production (e.g. 23, 151). The annual P/Bs of populations that tend to accrue biomass over successive cohorts are usually less than one (e.g. 23, 24, 66, 94, 151).

Long life cycles are not required for substantial accrual of biomass, however. Populations of filter-feeding macroinvertebrates, most with life cycles of less than one year, attain exceptionally high levels of population biomass. Biomass of filter-feeding larvae of the black fly *S. noelleri* in a lake outlet stream in England, for example, can exceed 80 g carbon m^{-2} (~160 g AFDM m^{-2}; cf. 88) (154; Table 2). This extraordinary biomass accrues in a matter of weeks and is due to larval densities that may exceed one million individuals m^{-2}. Animal populations that maintain such high levels of biomass require a continuous and reliable source of energy. This demand is met by filter feeders in stream habitats with stable substrate, adequate current velocity, and high organic seston concentrations, where they are able to exploit the kinetic energy of flowing water for the delivery of food produced elsewhere (28). Because of these subsidies, filter feeders expend less energy in search of food while supporting high levels of biomass (28, 154; Table 2). For most communities of stream insects, biomass losses are relatively high compared with biomass accrual, as indicated by annual P/Bs that are often more than five and may exceed 100 for some communities (8).

Biomass Loss

Molting and Silk Production Loss of biomass as cast exuviae by stream macroinvertebrates has received little study with regard to its effects on production, although this may be substantial (5). Molting losses for several species of Ephemeroptera and Plecoptera range from 8% to 21% of larval AFDM (see 5), and losses by crayfish can range from 21% to 38% of AFDM at each molt (70). Molting losses for the dobsonfly *Corydalus cornutus* are approximately 10% (22). Molting losses for most other holometabolous insects are expected to be lower, e.g. 2% to 4% for Trichoptera (5, 72, 99). Lower losses for larvae of the Holometabola are expected to be a function of lower proportional losses of biomass at each molt (5, 22, 72, 99) and lower molting frequency (58, 72, but see 22).

Molting losses may have additional consequences for production because of the compounding effect of exponential growth. For example, assuming that a mayfly larva grows at 5% day^{-1} over 150 days, molts 14 times during 15 stadia, and loses 12% of its mass at each molt (cf. selected Plecoptera, Ephemeroptera), its final mass will be 2.5 mg. Given an identical growth rate and period, a caddisfly larva that molts 4 times during 5 stadia and loses 5% of its mass during each molt (cf. selected Trichoptera) will achieve a final mass of 12.3 mg. In the case of the mayfly, approximately 0.7 mg (about 28% of production) of individual biomass will be discarded as exuviae, compared with 0.4 mg (approximately 3% of production) for the caddisfly. Although the mass lost as exuviae is only about two times greater for the mayfly, its final body mass (=individual production) is approximately five times lower than the caddisfly.

The bioenergetic efficiency of biomass accrual by most Holometabola is predicted to exceed that of most Hemimetabola. However, the lack of a hardened exoskeleton in many holometabolous taxa may require substantial expenditure of silk for construction of retreats and protective cases (e.g. Chironomidae, Trichoptera). Silk used in case construction by Trichoptera, for example, requires 12% to 23% of annual production (69, 72). Other similar losses include production of silk for holdfasts by black fly larvae, production of mucus by gastropods (marine example 31), and production of glochidia by mussels (94).

The fate of molted exoskeletons, silk, or mucus produced by stream macroinvertebrates has received little attention. However, in some ecosystems it has been suggested that these are a major flux of carbon. Molted exoskeletons from krill in the Antarctic Ocean may be one of the major sinks of carbon in Antarctic waters (95). Although large numbers of molted exoskeletons are routinely found in drift samples from streams (59, 135), the bioenergetic consequences and fate of these are unknown. Silk produced by hydropsychid caddisflies can function as a substrate for calcium carbonate precipitation, increases travertine deposition rates in hardwater streams (32), and may indirectly affect community structure by adhering and stabilizing sediment particles (B Statzner, personal communication).

Disturbance Many studies have examined the effects of physical disturbance on macroinvertebrate abundance. However, the effects of disturbance on their biomass has received surprisingly little attention, although available information indicates that physical disturbance may have a major effect on production dynamics of stream communities. Five to 50-fold reductions in biomass were measured following floods in an unstable, braided New Zealand stream (122, 123), and flash floods reduced macroinvertebrate biomass by as much as 98% in Sycamore Creek (40, 51). Accrual of biomass following disturbance is rapid in both of these systems, with recovery to pre-disturbance levels occurring in less than three months (40, 51, 122). Streams with low levels of disturbance often have higher biomass than more frequently disturbed streams (112, 114, 115). Channel structures serving as refugia for freshwater mussels provide an excellent example of the potential effect of disturbance on biomass accrual by long-lived macroinvertebrates (139). Beds of the freshwater mussel *Margaritifera* with the largest and oldest populations (modal age ~46 to 100 year) occur in large block-boulder reaches, and it has been hypothesized that such large, relatively immobile substrata protect *Margaritifera* from being buried during 50- to 100-year floods (139).

Removal of biomass by disturbance may affect growth rates via density-dependent competition for food. This effect on growth rate complicates predictions of how disturbance may affect production. Some studies have indicated that a reduction in biomass as a result of disturbance may result in higher growth rates (37, 66). But higher growth rates may not necessarily lead to increased production, compared to pre-disturbance levels, if losses of biomass are severe and frequent relative to the developmental periods of the dominant producers (66). In Sycamore Creek macroinvertebrate biomass tends to decline following extended periods without disturbance (e.g more than 60 to 80 days) because of reduced food quality resulting from cyclical coprophagy (51). Disturbance may thus actually facilitate high food quality and consequently high levels of insect production in this warm-temperate desert stream (40, 51, 74).

Emigration/Immigration Simultaneous losses and additions of macroinvertebrate biomass occur continuously via immigration and emigration of individuals among stream habitats and reaches. Most movements are probably in the form of drifting individuals, although upstream movements are of considerable significance in some geographic regions, particularly those by snails and atyid shrimp (67, 110, 119). The effect of such migrants as a net gain or loss of biomass has received little study. Their effects may be significant, however, especially when interpreting results of studies where the spatial scale of observation is limited (e.g. single habitats; 54). Large larvae of the mayfly *Tricorythodes atratus* in Minnesota, for example, drift from riffles and accumulate in pools, while small larvae remain in riffles (54). This behavior will result in erroneously high (pools) or low (riffles) estimates of production if only one habitat is considered (54). Movements of stream invertebrates into ephemeral floodplain habitats have been regularly documented, but the effects of such movements on community or ecosystem-level processes are rarely quantified (see 68). In one case, migrations from stream channels to forested floodplain wetlands in Virginia resulted in a small input of biomass to the floodplain (less than four percent of the annual mean biomass and less than one percent of the annual invertebrate production on the floodplain, 127).

To our knowledge, quantitative estimates of movements of macroinvertebrate biomass between stream reaches are not available. However, these movements probably have significant effects on reach-level patterns of production. Hinterleitner-Anderson et al (60), for example, showed that fewer larvae of the mayfly *Baetis* drifted from a fertilized reach of an Alaskan river when compared with those drifting from an unfertilized reach. Larvae drifted from the unfertilized (low food resource) reach where per-capita food supply was limiting growth, and accumulated in the fertilized (high food resource) reach downstream (60). This study provides some support for the long-standing hypothesis that drift may represent "excess production" where animals migrate out of stream reaches as a function of resource availability (cf. 2, 146).

Predators may also have important influences on the biomass distribution of macroinvertebrates in streams (125). The mere presence of predatory macroinvertebrates tends to increase behavioral drift of prey, thereby increasing emigration of individuals from patches with predators and immigration into predator-free areas (125). The presence of invertivorous fishes, on the other hand, may result in a reduction in the behavioral drift of macroinvertebrates which should decrease rates of emigration (43). Although the occurrence of such movements within stream channels is well documented (43, 125), their effects on spatial and temporal patterns of macroinvertebrate biomass and production remain unstudied.

Predation and the Allen Paradox The fate of most production by stream macroinvertebrates may be to support production by their predators. High demands on macroinvertebrate production by vertebrate predators have been reported for approximately 50 years (i.e. "Allen Paradox" 1, review in 149). Most recently, Huryn (65) used a comprehensive production budget to show that brown trout (*Salmo trutta*) consumed virtually all macroinvertebrate production in a New Zealand stream (approximately 101% to 110%). Similar rates of prey consumption by salmonids were reported for 11 of 13 streams for which information is available (64, review in 149).

The Allen paradox is not restricted to fish-macroinvertebrate interactions. Predaceous invertebrates in a sand-bottomed stream in Virginia consumed 94% of prey production (126). This estimate did not account for production by prey in the hyporheic zone of the stream, and their inclusion indicated that predator demand was still substantial (66% of prey production; 126). In another study, predaceous invertebrates and larval salamanders consumed an estimated 74% to 83% of macroinvertebrate prey production in two fishless mountain streams in North Carolina (83). Wallace et al (142) measured production by macroinvertebrate predators and prey in one of these same streams for eight years and showed that predator production required essentially all prey production (approximately 100%), and that predator production is probably limited by prey production. These studies together indicate that the effects of predation may result in vanishingly small surpluses of prey production (see "Adult Emergence as a Function of Larval Production").

Although a number of studies provide compelling evidence that predation can be an important sink for production by stream macroinvertebrates (64, 83, 126, 142, 149), such findings are by no means universal (e.g. 65). Under conditions where predator populations are limited by nontrophic factors (e.g. spawning habitat, floods), or in cases where prey are relatively immune to predation (e.g. mussels; 94), consumption of production by predators may be low. The fate of surplus macroinvertebrate production in such systems is unknown, but probably includes losses due to disturbance (see "Disturbances"), emigration (see "Emigration/Immigration"), disease (29, 78), parasitism (138), and emergence (see Section, "Adult Emergence as a Function of Larval Production").

PUPA

The role of the pupal stage in production dynamics of holometabolous stream insects has received little attention. As a nonfeeding stage, the pupa functions as a biomass sink rather than a source of biomass production. Losses of pupal mass occur through respiration, molting, and mortality.

Pupal Respiration and Molting

Metabolic processes of insect pupae rely on energy and nutrients accumulated as larvae. Studies of the terrestrial sphinx moth *Manduca sexta* and darkling beetle *Tenebrio molitor* indicate that losses of wet mass may range from 14% to 50% (98). There are few similar studies for aquatic insects, but the available examples indicate similar losses, including an approximate loss of 19% to 21% (DM) during prepupal to pupal transformation for the caddisfly *Glossosoma nigrior* (56), and an approximate loss of 48% loss (J) during prepupal to adult transformation for females of the dobsonfly *Corydalus cornutus* (22).

Pupal Mortality

The limited data available indicate that mortality of pupae can be substantial. Total adult emergence for a population of the caddisfly *Potamophylax cingulatus* in a Swedish stream accounted for only about ten percent of total pupal biomass (99). Mortality rates of more than 45% were observed for hydropsychid caddisfly pupae in an Ontario stream (118), and Willis & Hendricks (153) showed that pupae of *Hydropsyche slossonae* in a Virginia stream had the second highest mortality rate (76%) of all aquatic life-history stages except the first-instar. Predaceous chironomid larvae have been regularly implicated as agents of high levels of pupal mortality for hydropsychid caddisflies (102, 118, 153). Larvae of the Empididae (Hemerodromiinae) are also known to be predators of caddisfly pupae in South America (77). Substantial pupal mortality may also result from desiccation due to stranding. Stranding apparently resulted in the death of 38% of prepupae of the caddisfly *Neophylax fuscus* in an Ontario stream (86). Mortality directly associated with adult emergence may also be high. Rutherford (118), for example, reported that 34% of pharate adult hydropsychid caddisflies ("swimmers") drowned before molting to the adult stage.

ADULT

The aerial adults of stream insects form links between stream and riparian food webs, and the mortality associated with this process represents the endpoint for much adult production. In most studies, however, the hatching of eggs and emergence of adult insects are often viewed as starting points and endpoints of their life cycles. Because of this limited perspective, little is actually known about quantitative processes that affect the adult stage of stream insects.

Adult Emergence as a Function of Larval Production

A review of data for 18 populations of stream insects in Europe and North America indicated that emerging adult biomass consistently represented approximately 24% of annual production (130). This finding is remarkably similar to the proportion of annual insect production emerging as adult biomass from a warm-temperate desert stream (24% to 29% Diptera, 27% Trichoptera, 2% to 15% Ephemeroptera, mean ~19%; 74). This similarity is particularly striking because annual production reported by Jackson & Fisher (74) and for the stream populations included in the analysis by Statzner & Resh (130) ranged over three orders of magnitude (<1 to >100 g DM m^{-2}).

Fate of Adult Biomass

Adult Growth Tissue growth by adult stream insects, which occurs for selected Odonata, Plecoptera, Coleoptera, Trichoptera, and Diptera, has not been addressed in studies of secondary production. With the exception of black flies, adult growth has been perhaps best documented for the Odonata and Plecoptera. As predators, adult Odonata may increase their mass by 84% to 125% (3, 84). Adult females of the stoneflies *Leuctra* may gain 50% and *Siphonoperla* may gain 25% of their emergence weight through terrestrial feeding (156). Adult males of nemourid stoneflies may double their mass and mature females may gain up to three times the mass of freshly emerged females by feeding on lichens, algae, and pollen (5). Although unstudied, production by aerial adults may provide a further subsidy to riparian food webs that is ultimately dependent on the benthic community in the adjacent stream.

Adult Mortality Mass-balance approaches show that the amount of adult insect biomass returning to streams to oviposit following emergence may be only a small fraction of that emerging from the stream. Jackson & Fisher (74) estimated that of 23.1 g DM m^{-2} of insects emerging from Sycamore Creek each year, only three percent returned to the stream. In a similar study, Gray (47) estimated that of 20 mg DM m^{-2} of insects emerging daily from a prairie stream in Kansas, only one percent returned to the stream. Similar levels of loss have been shown for adults of insects emerging from forested streams, although biomass lost must be inferred from abundance. The abundance of females of the mayfly *Baetis* and the caddisfly *Apatania fimbriata* emerging from a forested German stream, for example, were compared with the abundance of egg masses to estimate approximately 90% to 99% and 82% mortality, respectively (36, 150). Based on the cited studies, it is apparent that most (e.g. > 80%) of the production of emerging adult stream insects, which may represent approximately 20% to 25% of total annual production, is supplied to riparian food webs.

Sources of Adult Mortality Wind, rain, extreme temperatures, and humidity have all been implicated as a source of mortality for aerial adults of aquatic insects (50, 73, 150). However, predation by insectivorous birds and bats is probably most significant. Fisher (38) estimated that a single flycatcher inhabiting the riparian zone of Sycamore Creek during the summer will consume all insect biomass emerging from 1000 m^2 of stream bed, or the total annual secondary production from 200 m^2 of stream bed. Gray (48) estimated that insectivorous birds inhabiting the riparian zone of a prairie stream in Kansas consumed 57% to 87% of emerging aquatic insects on a daily basis. Although these studies should be considered preliminary estimates, they provide some explanation for observations that only a small percentage of adult insects return to oviposit in the streams from which they emerge (36, 47, 74, 150). Terrestrial invertebrate predators, particularly spiders, ants, and ground beetles, are also sources of mortality for emerging aquatic insects (e.g. 50, 59, 150). However, their quantitative effects seem to be insignificant compared with those of vertebrate predators (cf. 38, 48, 152).

SUMMARY AND CONCLUSIONS

The annual production of stream macroinvertebrate communities worldwide ranges over four orders of magnitude ($\sim 10^0$ to 10^3 g DM m^{-2}). Factors contributing to high levels of production appear to be closely related to mode of feeding and life-history attributes of communities in temperate streams. Highly productive communities tend to occur in warm- to cool-temperate streams and tend to be dominated by filter-feeding bivalves, black flies, or caddisflies. Populations of filter-feeders are able to accrue and support high levels of biomass, which drives the very highest levels of production. Communities in warm-temperate streams subject to frequent disturbance also tend to be unusually productive. A harsh disturbance regime combined with a warm thermal regime apparently selects for taxa with short developmental times because of rapid growth rates and small terminal size. Because such life-history attributes tend to result in low biomass accrual, high levels of production are driven by rapid growth rates rather than high levels of biomass. The very lowest levels of production, reported for macroinvertebrate communities in cool-temperate to arctic streams, appear to be attributable to a combination of seasonally low temperatures and nutrient or food limitation rather than specific modes of feeding or life-history attributes.

Most studies of the production biology of stream macroinvertebrates have focused on larval production as driven by growth rate and population biomass. In comparison, the relationship between production and attributes of the egg, first-instar, pupa, and adult stages are essentially unknown. As beginning and end-points of the insect life cycle, it has been suggested that adult mortality may be an important bottleneck for larval production. The few comprehensive studies of the adult stage of stream insects do indicate that mortality before oviposition may be extremely high. Recent models based on stream insects, however, suggest that recruitment above surprisingly low thresholds may actually have little effect on larval production. Strong density-dependent control of abundance is predicted to result in continuous downstream movement of larvae in many stream systems (2, 146). Because production, rather than abundance per se, most directly underlies density-dependent competition for trophic resources, rates of larval production, rather than rates of oviposition by adults, may be the ultimate determinant of larval abundance and distribution for some populations of stream insects. The relationship between larval production and attributes of other life-history stages, both aquatic and terrestrial, deserves further study.

Most studies of benthic macroinvertebrate production either simply report levels of production for selected taxa, emphasize the effects of abiotic factors on production (i.e. habitat, temperature, water chemistry), or estimate the trophic support required to sustain selected taxa. Consequently, the role of food-web interactions in regulating production and its fate is poorly understood. For the few streams for which comprehensive production budgets are available, it is apparent that the fate of most macroinvertebrate production may be to support further production by predators. Extremely high losses to predators (~90% + in some cases) indicate that predation influences production of prey most significantly by removing biomass. However, the effects predators have on growth rates of prey may be as significant as the direct consumption of prey biomass is in determining community production dynamics. The mere presence of a predator within a stream community may influence the feeding behavior of their prey to the degree that an overall decrease in growth rates, and consequently community production, results (but see 27). By removing biomass while simultaneously affecting the growth rates of their prey, predators may have effects on the production of benthic communities that are much greater than currently recognized. Although potentially of great significance, such relationships remain largely unstudied in this context.

Current understanding of factors underlying differences in production among stream macroinvertebrate communities is hampered by the strong geographical bias of the few comprehensive studies, and by the dominance of studies concentrating on the effect of physical rather than biotic factors. Coordinated studies incorporating multiple streams with contrasting disturbance and nutrient regimes in diverse geographic regions are sorely needed for a predictive understanding of the relationship between the habitat template, life history, and secondary production.

Production is an interactive process that has simultaneous effects on multiple trophic levels—the potential for strong biotic control of production clearly exists. However, this potential has been ignored in most studies. Perhaps the major challenge for future studies of stream productivity will be to gain a predictive understanding of the role of biotic interactions (i.e. emigration/immigration, anti-predator behavior, and food-web interaction strength) in controlling quantities of materials and energy flowing through food webs.

ACKNOWLEDGMENTS

We thank John Hutchens and Vivian Butz Huryn for comments on earlier versions of the manuscript, and Ann Hershey for providing access to unpublished data.

LITERATURE CITED

1. Allen KR. 1951. The Horokiwi Stream: a study of a trout population. *NZ Marine Dep. Fish. Bull.* 10
2. Anholt BR. 1995. Density dependence resolves the stream drift paradox. *Ecology* 76:2235–239
3. Anholt BR, Marden JH, Jenkins DM. 1991. Patterns of mass gain and sexual dimorphism in adult dragonflies (Insecta: Odonata). *Can. J. Zool.* 69:1156–63
4. Banse K, Mosher S. 1980. Adult body mass and annual production/biomass relationships of field populations. *Ecol. Monogr.* 50:355–79
5. Beer-Stiller A, Zwick P. 1995. Biometric studies of some stoneflies and a mayfly (Plecoptera and Ephemeroptera). *Hydrobiologia* 299:169–78
6. Belchier M, Edsman L, Sheehy MRJ, Shelton PMJ. 1998. Estimating age and growth in long-lived temperate crayfish using lipofuscin. *Freshw. Biol.* 39:439–46
7. Benke AC. 1984. Secondary production of aquatic insects. In *The Ecology of Aquatic Insects*, ed. VH Resh, DM Rosenberg, pp. 289–322. New York: Praeger
8. Benke AC. 1993. Concepts and patterns of invertebrate production in running waters. *Verh. Int. Ver. Limnol.* 25:15–38
9. Benke AC. 1998. Production dynamics of riverine chironomids: extremely high biomass turnover rates of primary consumers. *Ecology* 79:899–910
10. Benke AC, Huryn AD, Smock LA, Wallace JB. 1999. Length-mass relationships for freshwater macroinvertebrates in North America with particular reference to the southeastern United States. *J. N. Am. Benthol. Soc.* (In press)
11. Benke AC, Jacobi DI. 1986. Growth rates of mayflies in a subtropical river and their implications for secondary production. *J. N. Am. Benthol. Soc.* 5:107–14
12. Benke AC, Jacobi DI. 1994. Production

dynamics and resource utilization of snag-dwelling mayflies in a blackwater river. *Ecology* 75:1219–32
13. Benke AC, Parsons KA. 1990. Modelling black fly production dynamics in blackwater streams. *Freshw. Biol.* 24:167–80
14. Benke AC, Van Arsdall TC Jr, Gillespie DM. 1984. Invertebrate productivity in a subtropical blackwater river: the importance of habitat and life history. *Ecol. Monogr.* 54:25–63
15. Benke AC, Wallace JB. 1997. Trophic basis of production among riverine caddisflies: implications for food web analysis. *Ecology* 78:1132–45
16. Berg MB, Hellenthal RA. 1991. Secondary production of Chironomidae (Diptera) in a north temperate stream. *Freshw. Biol.* 25:497–505
17. Berg MB, Hellenthal RA. 1992. Life histories and growth of lotic chironomids (Diptera: Chironomidae). *Ann. Entomol. Soc. Am.* 85:578–89
18. Brewis JM, Bowler K. 1983. A study of the dynamics of a natural population of the freshwater crayfish, *Austropotamobius pallipes*. *Freshw. Biol.* 13:443–52
19. Brey T, Gage JD. 1997. Interactions of growth and mortality in benthic invertebrate populations: empirical evidence for a mortality growth continuum. *Arch. Fish. Mar. Res.* 45:45–59
20. Brittain JE. 1995. Egg development in Australian mayflies (Ephemeroptera). In *Current Directions in Research on Ephemeroptera*, ed. LD Corkum, JJH Ciborowski, pp. 307–16. Toronto: Can. Sch. Press
21. Brittain JE, Campbell IC. 1991. The effect of temperature on egg development in the Australian mayfly genus *Coloburiscoides* (Ephemeroptera: Coloburiscidae) and its relationship to distribution and life history. *J. Biogeogr.* 18:231–35
22. Brown AV, Fitzpatrick LC. 1978. Life history and population energetics of the dobson fly, *Corydalus cornutus*. *Ecology* 59:1091–108
23. Butler MG. 1982. Production dynamics of some arctic *Chironomus* larvae. *Limnol. Oceanogr.* 27:728–36
24. Butler MG. 1982. A 7-year life cycle for two *Chironomus* species in arctic Alaskan tundra ponds (Diptera: Chironomidae). *Can. J. Zool.* 60:58–70
25. Butler MG. 1984. Life histories of aquatic insects. See Ref. 114a, pp. 24–55. New York: Praeger
26. Cleven E-J, Frenzel P. 1993. Population dynamics and production of *Dreissena polymorpha* (Pallas) in River Seerhein, the outlet of Lake Constance (Obersee). *Arch. Hydrobiol.* 127:395–407
27. Crowl TA, Covich AP. 1990. Predator-induced life-history shifts in a freshwater snail. *Science* 247:949–51
28. Cudney MD, Wallace JB. 1980. Life cycles, microdistribution and production dynamics of six species of net-spinning caddisflies in a large southeastern (U.S.A.) river. *Holarct. Ecol.* 3:169–82
29. Cummins KW, Wilzbach MA. 1988. Do pathogens regulate stream invertebrate populations? *Verh. Int. Ver. Limnol.* 23:1232–43
30. Danks HV. 1992. Long life cycles in insects. *Can. Entomol.* 124:167–87
31. Davies MS, Hawkins SJ, Jones HD. 1990. Mucus production and physiological energetics in *Patella vulgata* L. *J. Molluscan Stud.* 56:499–503
32. Drysdale, RN. 1999. The sedimentological signficance of hydropsychid caddisfly larvae (Order: Trichoptera) in a travertine-depositing stream: Louie Creek, Northwest Queensland, Australia. *J. Sediment. Res.* 69:145–50
33. Edgerly JS, Livdahl TP. 1992. Density-dependent interactions within a complex life cycle: the roles of cohort structure and mode of recruitment. *J. Anim. Ecol.* 61:139–50
34. Eggert SL, Burton TM. 1994. A compar-

ison of *Acroneuria lycorias* (Plecoptera) production and growth in northern Michigan hard- and soft-water streams. *Freshw. Biol.* 32:21–31
35. Elliott JM. 1995. Egg hatching and ecological partitioning in carnivorous stoneflies (Plecoptera). *C.R. Acad. Sci. III.* 318:237–43
36. Enders G, Wagner R. 1996. Mortality of *Apatania fimbriata* (Insecta: Trichoptera) during embryonic, larval and adult stages. *Freshw. Biol.* 36:93–104
37. Feminella JW, Resh VH. 1990. Hydrologic influences, disturbance, and intraspecific competition in a stream caddisfly population. *Ecology* 71:2083–94
38. Fisher SG. 1991. Presidential address: emerging global issues in freshwater ecology. *Bull. N. Am. Benthol. Soc.* 8:235–45
39. Fisher SG, Gray LJ, Grimm NB, Busch DE. 1982. Temporal succession in a desert stream ecosystem following flash flooding. *Ecol. Monogr.* 52:93–110
40. Fisher SG, Gray LJ. 1983. Secondary production and organic matter processing by collector macroinvertebrates in a desert stream. *Ecology* 64:1217–24
41. Fjellheim A, Håvardstun J, Raddum GG, Schnell ØA. 1993. Effects of increased discharge on benthic invertebrates in a regulated river. *Reg. Riv.* 8:179–87
42. Fonseca DM, Hart DD. 1996. Density-dependent dispersal of black fly neonates is mediated by flow. *Oikos* 75:49–58
43. Forester GE. 1994. Influences of predatory fish on the drift dispersal and local density of stream insects. *Ecology* 75:1208–18
44. Giberson DJ, Rosenberg DM. 1994. Life histories of burrowing mayflies (*Hexagenia limbata* and *H. rigida*, Ephemeroptera: Ephemeridae) in a northern Canadian reservoir. *Freshw. Biol.* 32: 501–18
45. Gíslason GM, Gardarsson A. 1988. Long term studies on *Simulium vittatum* Zett. (Diptera: Simuliidae) in the River Laxá, North Iceland, with particular reference to different methods used in assessing population changes. *Verh. Int. Ver. Limnol.* 23:2179–88
46. Gray LJ. 1981. Species composition and life histories of aquatic insects in a lowland Sonoran Desert stream. *Am. Midl. Nat.* 106:229–42
47. Gray LJ. 1989. Emergence production and export of aquatic insects from a tallgrass prairie stream. *Southwest. Nat.* 34:313–18
48. Gray LJ. 1993. Response of insectivorous birds to emerging aquatic insects in riparian habitats in a tallgrass prairie stream. *Am. Midl. Nat.*129:288–300
49. Gresens SE. 1997. Interactive effects of diet and thermal regime on growth of the midge *Pseudochironomus richardsoni* Malloch. *Freshw. Biol.* 38:365–73
50. Gribbin SD, Thompson DJ. 1990. A quantitative study of mortality at emergence in the damselfly *Pyrrhosoma nymphula* (Sulzer) (Zygoptera: Coenagrionidae). *Freshw. Biol.* 24:295–302
51. Grimm NB, Fisher SG. 1989. Stability of periphyton and macroinvertebrates to disturbance by flash floods in a desert stream. *J. N. Am. Benthol. Soc.* 8:293–307
52. Grubaugh JW, Wallace JB. 1995. Functional structure and production of the benthic community in a Piedmont river: 1956–1957 and 1991–1992. *Limnol. Oceanogr.* 40:490–501
53. Grubaugh JW, Wallace JB, Houston ES. 1997. Production of benthic macroinvertebrate communities along a southern Appalachian river continuum. *Freshw. Biol.* 37:581–96
54. Hall RJ, Waters TF, Cook EF. 1980. The role of drift dispersal in production ecology of a stream mayfly. *Ecology* 61:37–43
55. Harker JE. 1997. The role of parthenogenesis in the biology of two species of mayfly (Ephemeroptera). *Freshw. Biol.* 37:287–97
56. Hart DD. 1987. Experimental studies of

exploitative competition in a grazing stream insect. *Oecologia* 73:41–47
57. Harvey CJ, Peterson BJ, Bowden WB, Hershey AE, Miller MC, et al. 1998. Biological responses to fertilization of Oksrukuyik Creek, a tundra stream. *J. N. Am. Benthol. Soc.* 17:190–209
58. Harvey RS, Vannote RL, Sweeney BW. 1980. Life history, developmental processes, and energetics of the burrowing mayfly *Dolania americana*. In *Advances in Ephemeroptera Biology*, ed. JF Flannagan, KE Marshall, pp. 211–230. New York: Plenum
59. Hering D, Plachter H. 1997. Riparian ground beetles (Coleoptera, Carabidae) preying on aquatic invertebrates: a feeding strategy in alpine floodplains. *Oecologia* 111:261–70
60. Hinterleitner-Anderson D, Hershey AE, Schuldt JA. 1992. The effects of river fertilization on mayfly (*Baetis* sp.) drift patterns and population density in an arctic river. *Hydrobiologia* 240:247–58
61. Hopper KR, Crowley PH, Kielman D. 1996. Density dependence, hatching synchrony, and within-cohort cannibalism in young dragonfly larvae. *Ecology* 77:191–200
62. Hunter MD, Price PW. 1992. Playing chutes and ladders: heterogeneity and the relative roles of bottom-up and top-down forces in natural communities. *Ecology* 73:724–32
63. Huryn AD. 1990. Growth and voltinism of lotic midge larvae: patterns across an Appalachian mountain basin. *Limnol. Oceanogr.* 35:339–51
64. Huryn AD. 1996. An appraisal of the Allen Paradox in a New Zealand trout stream. *Limnol. Oceanogr.* 41:243–52
65. Huryn AD. 1998. Ecosystem-level evidence for top-down and bottom-up control of production in a grassland stream system. *Oecologia* 115:173–83
66. Huryn AD, Benke AC, Ward GM. 1995. Direct and indirect effects of regional geology on the distribution, biomass, and production of the freshwater snail *Elimia*. *J. N. Am. Benthol. Soc.* 14:519–34
67. Huryn AD, Denny MW. 1997. A biomechanical hypothesis explaining upstream movements by the freshwater snail *Elimia*. *Funct. Ecol.* 11:472–83
68. Huryn AD, Gibbs KE. 1999. Macroinvertebrates of riparian sedge meadows in Maine: a community structured by river-floodplain interaction. In *Invertebrates in Freshwater Wetlands of North America: Ecology and Management*, ed. DP Batzer, RB Rader, SA Wissinger, pp. 363–82. New York: Wiley & Sons
69. Huryn AD, Wallace JB. 1985. Life history and production of *Goerita semata* Ross (Trichoptera: Limnephilidae) in the southern Appalachian mountains. *Can. J. Zool.* 63:2601–11
70. Huryn AD, Wallace JB. 1987. Production and litter processing by crayfish in an Appalachian stream. *Freshw. Biol.* 18:277–86
71. Iversen TM. 1973. Life cycle and growth of *Sericostoma personatum* Spence (Trichoptera, Sericostomatidae) in a Danish spring. *Entomol. Scand.* 4:323–27
72. Iversen TM. 1980. Densities and energetics of two stream living larval populations of *Sericostoma personatum* (Trichoptera). *Holarct. Ecol.* 3:65–73
73. Jackson JK. 1988. Diel emergence, swarming and longevity of selected adult aquatic insects from a Sonoran Desert stream. *Am. Midl. Nat.* 119:344–52
74. Jackson JK, Fisher SG. 1986. Secondary production, emergence, and export of aquatic insects of a Sonoran Desert stream. *Ecology* 67:629–38
75. Jackson JK, Sweeney BW. 1995. Egg and larval development times for 35 species of tropical stream insects from Costa Rica. *J. N. Am. Benthol. Soc.* 14:115–30
76. Kharchenko TA, Lyashenko AV. 1997. The growth and production of *Dreissena* under conditions of artificial water flows. *Gidrobiolog. Zhur.* 33:3–16 (In Russian)
77. Knutson L, Flint OS. 1979. Do dance

flies feed on caddisflies?—Further evidence (Diptera: Empididae; Trichoptera). *Proc. Entomol. Soc. Wash.* 81:32–33

78. Kohler SL, Wiley MJ. 1997. Pathogen outbreaks reveal large-scale effects of competition in stream communities. *Ecology* 78:2164–76
79. Krueger CC, Waters TF. 1983. Annual production of macroinvertebrates in three streams of different water quality. *Ecology* 64:840–50
80. Ladle M, Cooling AD, Welton JS, Bass JAB. 1985. Studies on Chironomidae in experimental recirculating stream systems. II. The growth, development and production of a spring generation of *Orthocladius (Euorthocladius) calvus* Pinder. *Freshw. Biol.* 15:243–55
81. Lamberti GA, Moore JW. 1984. Aquatic insects as primary consumers. See Ref. 114a, pp. 164 –95. New York: Praeger
82. Lazim NM, Learner MA. 1986. The life-cycle and production of *Limnodrilus hoffmeisteri* and *L. udekemianua* (Tubificidae: Oligochaeta) in the organically enriched Moat-Feeder Stream, Cardiff, South Wales. *Arch. Hydrobiol. Suppl.* 74:200–25
83. Lugthart GJ, Wallace JB. 1992. Effects of disturbance on benthic functional structure and production in mountain streams. *J. N. Am. Benthol. Soc.*11:138–64
84. Marden JH, Fitzhugh GH, Wolf MR. 1998. From molecules to mating success: integrative biology of muscle maturation in a dragonfly. *Am. Zool.* 38:528–44
85. Marten M, Zwick P. 1989. The temperature dependence of embryonic and larval development in *Protonemura intricata* (Plecoptera: Nemouridae). *Freshw. Biol.* 22:1–14
86. Martin ID, Barton DR. 1987. The formation of diapause aggregations by larvae of *Neophylax fuscus* Banks (Trichoptera:Limnephilidae) and their influence on mortality and development. *Can. J. Zool.* 65:2612–18
87. Martin WW. 1991. Egg parasitism by zoosporic fungi in a littoral chironomid community. *J. N. Am. Benthol. Soc.* 10:455–62
88. McCullough DA, Minshall GW, Cushing CE. 1979. Bioenergetics of a stream "collector" organism, *Trichorythodes minutus* (Insecta: Ephemeroptera). *Limnol. Oceanogr.* 24:45–58
89. McIntosh AR, Peckarsky BL. 1996. Differential behavioral responses of mayflies from streams with and without fish to trout odour. *Freshw. Biol.* 35:141–48
90. McPeek MA, Peckarsky BL. 1998. Life histories and the strength of species interactions: combining mortality, growth, and fecundity effects. *Ecology* 79:867–79
91. Mitchell JT, Smock LA. 1991. Distribution, life history, and production of crayfish in the James River, Virginia. *Am. Midl. Nat.* 126:353–63
92. Moreira GRP, Peckarsky BL. 1994. Multiple developmental pathways of *Agnetina capitata* (Plecoptera: Perlidae) in a temperate forest stream. *J. N. Am. Benthol. Soc.* 13:19–29
93. Mutch RA, Pritchard G. 1984. The life history of *Philocasca alba* (Trichoptera: Limnephilidae) in a Rocky Mountain stream. *Can. J. Zool.* 62:1282–88
94. Negus CL. 1966. A quantitative study of growth and production of unionid mussels in the River Thames at Reading. *J. Anim. Ecol.* 35:513–32
95. Nicol S. 1990. The age-old problem of krill longevity. *BioScience* 40:833–36
96. Nolte U, Hoffman T. 1992. Fast life in cold water: *Diamesa incallida* (Chironomidae). *Ecography* 15:25–30
97. Nylin S, Gotthard K. 1998. Plasticity in life-history traits. *Annu. Rev. Entomol.* 43:63–83
98. Odell JP. 1998. Energetics of metamorphosis in two holometabolous insect species: *Manduca sexta* (Lepidoptera:

Sphingidae) and *Tenebrio molitor* (Coleoptera: Tenebrionidae). *J. Exp. Zool.* 280:344–53

99. Otto C. 1975. Energetic relationships of the larval population of *Potamophylax cingulatus* (Trichoptera) in a South Swedish stream. *Oikos* 26:159–69
100. Paine RT. 1980. Food webs: linkage, interaction strength and community infrastructure. *J. Anim. Ecol.* 49:667–85
101. Parker CR, Voshell JR Jr. 1983. Production of filter-feeding Trichoptera in an impounded and a free-flowing river. *Can. J. Zool.* 61:70–87
102. Parker CR, Voshell JR Jr. 1979. *Cardiocladius* (Diptera:Chironomidae) larvae ectoparasitic on pupae of Hydropsychidae (Trichoptera). *Ann. Entomol. Soc. Am.* 8:808–9
103. Peckarsky BL, Cowan CA, Penton MA, Anderson C. 1993. Sublethal consequences of stream-dwelling predatory stoneflies on mayfly growth and fecundity. *Ecology* 74:1836–46
104. Petersen RC Jr, Petersen LB-M. 1988. Compensatory mortality in aquatic populations: its importance for interpretation of toxicant effects. *Ambio* 17:381–86
105. Peterson BJ, Deegan L, Helfrich J, Hobbie JE, Hullar M, et al. 1993. Biological responses of a tundra river to fertilization. *Ecology* 74:653–72
106. Pinder LCV. 1995. Biology of eggs and first-instar larvae. In *The Chironomidae: The Biology and Ecology of Non-Biting Midges*, eds. P Armitage, PS Cranston, LCV Pinder, pp. 87–106. London: Chapman & Hall
107. Polis GA. 1994. Food webs, trophic cascades and community structure. *Aust. J. Ecol.* 19:121–36
108. Ponyi JE. 1992. The distribution and biomass of Unionidae (Mollusca, Bivalvia), and the production of *Unio tumidus* Retzius in Lake Balaton (Hungary). *Arch. Hydrobiol.* 125:245–51
109. Popperl R. 1996. The structure of a macroinvertebrate community in a northern German lake outlet (Lake Belau, Schleswig-Holstein) with special emphasis on abundance, biomass, and secondary production. *Int. Rev. Gesamten Hydrobiol.* 81:183–98
110. Pringle CM. 1997. Exploring how disturbance is transmitted upstream: going against the flow. *J. N. Am. Benthol. Soc.*16:425–381
111. Proctor H, Pritchard G. 1989. Neglected predators: water mites (Acari: Parasitengona: Hydrachnellae) in freshwater communities. *J. N. Am. Benthol. Soc.* 16:425–38
112. Quinn JM, Hickey CW. 1990. Characterisation and classification of benthic invertebrate communities in 88 New Zealand rivers in relation to environmental factors. *NZ J. Mar. Freshw. Res.* 24:387–410
113. Rabeni CF. 1992. Trophic linkages between stream centrarchids and their crayfish prey. *Can. J. Fish. Aquat. Sci.* 49:1714–21
114. Rader RB, Ward JV. 1989. Influence of impoundments on mayfly diets, life histories, and production. *J. N. Am. Benthol. Soc.* 8:64–73

114a. Resh VH, Rosenberg DM, eds. 1984. *The Ecology of Aquatic Insects.* New York: Praeger

115. Robinson CT, Reed LM, Minshall GW. 1992. Influence of flow regime on life history, production, and genetic structure of *Baetis tricaudatus* (Ephemeroptera) and *Hesperoperla pacifica* (Plecoptera). *J. N. Am. Benthol. Soc.* 11:278–89
116. Roell MJ, Orth DJ. 1993. Trophic basis of production of stream-dwelling smallmouth bass, rock bass, and flathead catfish in relation to bait harvest. *Trans. Am. Fish. Soc.* 122:46–62
117. Roff DA. 1992. *The Evolution of Life Histories.* New York: Chapman & Hall
118. Rutherford JE. 1986. Mortality in reared hydropsychid pupae (Trichoptera: Hydropsychidae). *Hydrobiologia* 131: 97–111
119. Schneider DW, Lyons J. 1993. Dynamics

of upstream migration in two species of tropical freshwater snails. *J. N. Am. Benthol. Soc.* 12:3–17

120. Schoenly K, Beaver R, Heumier T. 1992. On the trophic relations of insects: a food web approach. *Am. Nat.* 137:597–638
121. Scrimgeour GJ, Culp JM, Cash KJ. 1994. Anti-predator responses of mayfly larvae to conspecific and predator stimuli. *J. N. Am. Benthol. Soc.* 13:299–309
122. Scrimgeour GJ, Davidson RJ, Davidson JM. 1988. Recovery of benthic macroinvertebrate and epilithic communities following a large flood, in an unstable, braided, New Zealand river. *NZ J. Mar. Freshw. Res.* 22:337–44
123. Scrimgeour GJ, Winterbourn MJ. 1989. Effects of floods on epilithon and benthic macroinvertebrate populations in an unstable New Zealand river. *Hydrobiologia* 171:33–44
124. Seastedt TR, Crossley DA Jr. 1984. The influence of arthropods on ecosystems. *BioScience* 34:157–61
125. Sih A, Wooster DE. 1994. Prey behavior, prey dispersal, and predator impacts on stream prey. *Ecology* 75:1199–207
126. Smith LC, Smock LA. 1992. Ecology of invertebrate predators in a Coastal Plain stream. *Freshw. Biol.* 28:319–29
127. Smock LA. 1994. Movements of invertebrates between stream channels and forested floodplains. *J. N. Am. Benthol. Soc.*13:524–31
128. Smock LA, Gilinsky E, Stoneburner DL. 1985. Macroinvertebrate production in a southeastern United States Blackwater stream. *Ecology* 66:1491–503
129. Southwood TRE. 1988. Tactics, stragtegies and templets. *Oikos* 52:3–18
130. Statzner B, Resh VH. 1993. Multiple-site and -year analyses of stream insect emergence: a test of ecological theory. *Oecologia* 96:65–79
131. Steedman RJ, Anderson NH. 1985. Life history and ecological role of the xylophagous aquatic beetle, *Lara avara* LeConte (Dryopoidea: Elmidae). *Freshw. Biol.* 15:535–46
132. Steltzer RS, Burton TM. 1993. Growth and abundance of the crayfish *Orconectes propinquus* in a hard water and a soft water stream. *J. Freshw. Ecol.* 8:329–40
133. Stites DL, Benke AC, Gillespie DM. 1995. Population dynamics, growth, and production of the Asiatic clam, *Corbicula fluminea*, in a blackwater river. *Can. J. Fish. Aquat. Sci.* 52:425–37
134. Stober QJ. 1972. Distribution and age of *Margaritifera margaritifera* (L.) in a Madison River (Montana, U.S.A.) mussel bed. *Malacologia* 11:343–50
135. Stoneburner DL, Smock LA. 1979. Seasonal fluctuations of macroinvertebrate drift in a South Carolina piedmont stream. *Hydrobiologia* 63:49–56
136. Sweeney BW. 1984. Factors influencing life-history patterns of aquatic insects. See Ref. 114a, pp. 56–100
137. Sweeney BW, Jackson JK, Funk DH. 1995. Semivoltinism, seasonal emergence, and adult size variation in a tropical stream mayfly (*Euthyplocia hecuba*). *J. N. Am. Benthol. Soc.* 14:131–46
138. Vance SA, Peckarsky BL. 1996. The infection of nymphal Baetis caudatus by the mermithid nematode Gasteromermis sp. *Ecol. Entomol.* 21:377–81
139. Vannote RL, Minshall GW. 1982. Fluvial processes and local lithology controlling abundance, structure, and composition of mussel beds. *Proc. Natl. Acad. Sci. USA* 79:4103–7
140. Vannote RL, Sweeney BW. 1980. Geographic analysis of thermal equilibria: a conceptual model for evaluating the effect of natural and modified thermal regimes on aquatic insect communities. *Am. Nat.* 115:667–95
141. Voshell JR. 1985. *Trophic Basis of Production for Macroinvertebrates in the New River below Bluestone Dam.* Blacksburg: Dep. Entomol. VA Polytech. Inst. State Univ.
142. Wallace JB, Eggert SL, Meyer JL, Web-

ster JR. 1997. Multiple trophic levels of a forested stream linked to terrestrial litter inputs. *Science* 277:102–4

143. Wallace JB, Eggert SL, Meyer JL, Webster JR. 1999. Effects of resource limitation on a detrital-based ecosystem. *Ecol. Monogr.* (In press)
144. Wallace JB, Webster JR. 1996. The role of macroinvertebrates in stream ecosystem function. *Annu. Rev. Entomol.* 41:115–39
145. Ward JV. 1991. *Aquatic Insect Ecology: 1. Biology and Habitat.* New York: Wiley
146. Waters TF. 1966. Production rate, population density, and drift of a stream invertebrate. *Ecology* 47:595–604
147. Waters TF. 1977. Secondary production in inland waters. *Adv. Ecol. Res.* 10:91–164
148. Waters TF. 1987. The effect of growth and survival patterns upon the cohort P/B ratio. *J. N. Am. Benthol. Soc.* 6:223–29
149. Waters TF. 1988. Fish production–benthos production relationships in trout streams. *Pol. Arch. Hydrobiol.* 35:545–61
150. Werneke U, Zwick P. 1992. Mortality of the terrestrial adult and aquatic nymphal life stages of Baetis *vernus* and *Baetis rhodani* in the Breitenbach, Germany (Insecta: Ephemeroptera). *Freshw. Biol.* 28:249–55
151. Whitmore N, Huryn AD. 1999. Life history and production of *Paranephrops zealandicus* in a lowland forest stream: implications for the sustainable harvest of a freshwater crayfish. *Freshw. Biol.* (In press)
152. Williams DD, Ambrose LG, Browning LN. 1995. Trophic dynamics of two sympatric species of riparian spiders (Araneae: Tetragnathidae). *Can. J. Zool.* 73:1545–53
153. Willis LD, Hendricks AC. 1992. Life history, growth, survivorship, and production of *Hydropsyche slossonae* in Mill Creek, Virginia. *J. N. Am. Benthol. Soc.* 11:290–303
154. Wotton RS. 1988. Very high secondary production at a lake outlet. *Freshw. Biol.* 20:341–46
155. Wotton RS. 1995. Temperature and lake-outlet communities. *J. Therm. Biol.* 20:121–25
156. Zwick P. 1990. Emergence, maturation and upstream oviposition flights of Plecoptera from the Breitenbach, with notes on the adult phase as a possible control of stream insect populations. *Hydrobiologia* 194:207–23
157. Zwick P. 1996. Variable egg development of *Dinocras* spp. (Plecoptera, Perlidae) and the stonefly seed bank theory. *Freshw. Biol.* 35:81–100

Annu. Rev. Entomol. 2000. 45:111–120

AMINO ACID TRANSPORT IN INSECTS

Michael G. Wolfersberger
Department of Biology, Temple University, Philadelphia, Pennsylvania 19122; e-mail: mgwolf@astro.ocis.temple.edu

Key Words cotransport, potassium, sodium, symport, uniport

■ **Abstract** Most insect cell membranes seem to contain uniporters that facilitate the diffusion of amino acids into and out of the cells. In addition to these passive diffusion systems, all but one of the insect tissues studied to date seem to contain at least one amino acid–cation symport system that allows their cells to accumulate certain amino acids from the extracellular medium. cDNAs encoding three such symporters have very recently been cloned and sequenced. The deduced amino acid sequence of each insect symporter was determined to be homologous to that of symporters mediating the transport of the same or related substrates in mammalian tissues.

INTRODUCTION

Four years ago VF Sacchi and I completed a review of what had been learned about amino acid absorption in insect midgut during the preceding 18 to 20 years (22). It turned out that the vast majority of studies had been conducted on lepidopteran larvae. Recent literature searches have revealed that most studies of amino acid transport in insects continue to focus on larval midgut with Lepidoptera continuing to receive the greatest attention. Studies of amino acid transport in insect tissues other than midgut have been restricted to transport systems for amino acids that function as neurotransmitters. A few papers (11, 15, 25) concerning transport in insects of neurotransmitters derived from amino acids have also been published in recent years. They are not discussed in this review.

Selective membrane permeability mechanisms for water-soluble solutes such as amino acids have long been predicted to be protein mediated (28). However, it is only recently that a few of the expected transport proteins have been identified in insect cell membranes (2, 3, 23). The finite number of transport proteins in a cell membrane plus specific interactions between transporters and solutes lead to flux saturation and inhibition. These properties of solute fluxes through transporters can usually be described by the same equations used to describe the kinetics of enzyme-catalyzed reactions. Therefore, one often sees the affinity of transporters for solutes expressed in terms of K_m and the maximal solute flux expressed in terms of V_{max}. Although $K_{1/2}$ and J_{max} may be more correct, the more familiar terms from enzymology will be used throughout this article.

0066-4170/00/0107-0111/$14.00

AMINO ACID TRANSPORT IN LARVAL LEPIDOPTERAN MIDGUT

State of the Art in March 1995

At the time of our recent review of amino acid absorption in insect midgut (22) it was clear that far more was known about midgut amino acid absorption in lepidopteran larvae than was known about midgut amino acid absorption in any other insect class or developmental stage. Also, it became clear that our knowledge of amino acid absorption in the midgut of *Philosamia cynthia* exceeded that for any other lepidopteran larvae. Numerous publications supported the existence of several different amino acid–alkali metal cation symport systems in larval *P. cynthia* midgut. The system responsible for absorption of most of the monoamino monocarboxylic acid components of common proteins had been the subject of extensive kinetic characterization, and progress had been made toward its isolation. It was therefore somewhat surprising to find that no studies of amino acid transport in larval *P. cynthia* midgut had been published since 1995. It seems that the advantages of being able to rear larvae throughout the year on an artificial diet have finally lured all students of amino acid transport in lepidopteran midgut to *Bombyx mori* or *Manduca sexta.*

Studies conducted up to 1995 had shown that absorption of at least most neutral (zwitterionic at physiological pH) amino acids in midguts of *B. mori* and *M. sexta* larvae seemed to be mediated by an amino acid-K^+ symport system more similar to than different from that characterized in larval *P. cynthia* midgut. This system seemed to be present throughout the larval *B. mori* midgut but showed much higher capacity as well as greater sensitivity to external pH and membrane potential in the posterior portion (22).

Recent Studies on *Bombyx Mori*

Shinbo and associates (24) have since provided independent evidence for amino acid absorption throughout larval *B. mori* midgut using a novel experimental approach. Larvae feeding on artificial diets were quickly frozen then cut crosswise into segments. The total amino acid composition of the midgut contents in each segment plus in samples of diet and fecal pellets was determined quantitatively. The midgut contents of each segment were also analyzed for protease activity. Protease activity was detected in the contents of each segment with the highest specific activity in segments corresponding to the central and early posterior portions of the midgut. The concentration of each amino acid was lower in fecal pellets than in diet. The concentration of each amino acid was also lower in the midgut contents of the first anterior larval segment than in diet. The concentration in the midgut contents of all amino acids, except glycine, decreased in a regular manner between the first or second anterior segment and the last or next-to-last posterior larval segment. From these results Shinbo and associates concluded that protein digestion and absorption of neutral as well as acidic and basic amino acids must begin in the anterior portion of larval *B. mori* midgut and continue throughout the midgut with absorption reaching a maximum in the posterior portion.

The mechanisms by which absorption of amino acids occurs in larval *B. mori* midgut has been the subject of two recent publications (6, 13). Together they demonstrate that uptake of neutral amino acids is mediated by at least two agencies: the previously characterized neutral amino acid–K^+ symport system, and a high-capacity amino acid uniport system. Unlike the symport system, the uniport system is essentially uninfluenced by alkaline external pH or a transmembrane electrical potential difference ($\Delta\Psi$). The uniport system is also less selective than the symport system. Arginine, which is clearly not a substrate of the neutral amino acid-K^+ symport system (22), was more effective than alanine in inhibiting leucine uniport. Considering the affinities and capacities of the two transport systems it was estimated that in the central portion of *B. mori* midgut the majority of leucine uptake would occur through symport at a leucine concentration of less than two mM, but at concentrations greater than 10 mM uniport would be the major mechanism of leucine uptake. Although the capacity of the uniport system in posterior midgut is similar to that in central midgut, the capacity of the symport system in posterior midgut is sufficiently high so that even at a leucine concentration of 15 mM uniport would remain a minor uptake mechanism (13).

Other recent studies of amino acid uptake in larval *B. mori* midgut seem to have been directed toward either elucidating the mode of action of known insecticides or finding specific inhibitors of amino acid uptake that might show potential as insecticides. One of the insecticides studied was ethyl[2-(p-phenoxyphenoxy)ethyl]carbamate (fenoxycarb), a commercially available insect growth regulator that, when applied to lepidopteran larvae, mimics the effects of juvenile hormone (12). Topical application of 60 pg of fenoxycarb per larva was estimated to be sufficient to greatly inhibit growth and prevent 50 percent of the insects from undergoing larval-pupal ecdysis. The initial rate of K^+-gradient driven leucine uptake by brush border membrane vesicles (BBMV) prepared from posterior midguts of larvae treated with 10 ng or 10 μg of fenoxycarb was the same as that in BBMV prepared from posterior midguts of untreated controls. However, the initial rate of K^+-gradient driven leucine uptake by BBMV prepared from posterior midguts of larvae treated with 10 fg of fenoxycarb was almost 75 percent greater than that in BBMV prepared from posterior midguts of untreated controls. The authors admitted that interpretation of these results was not simple (12). However, a direct affect of fenoxycarb on the leucine-K^+ symport system seems unlikely.

The ability of *Bacillus thuringiensis* delta-endotoxins to inhibit amino acid–K^+ symport in larval lepidopteran midgut by increasing the K^+ permeability of the brush border membrane is well documented (22). An extensive reinvestigation of the mechanism of inhibition of leucine uptake in larval *B. mori* midgut by *B. thuringiensis* Cry1Aa toxin (14) showed inhibition of leucine uptake not only in the presence but also in the absence of potassium. Inhibition in the absence of potassium was taken as evidence for a direct interaction between this microbial insecticide and the membrane component mediating leucine uptake, presumed to be primarily the leucine-K^+ symport system. The site of this direct interaction was not determined. However, the noncompetitive nature of the toxin inhibition plus the failure of the toxin to compete with bestatin effectively ruled out the amino acid binding site as the site of toxin-transporter interaction. Bestatin, a leucine derivative widely used as an aminopeptidase inhibitor, had previously been found to be a competitive inhibitor of leucine-K^+ symport in larval *B. mori* midgut (18).

The search for specific inhibitors of amino acid transport in larval *B. mori* midgut recently uncovered a novel candidate: cyclochampedol. This flavone derivative, isolated from tree bark, inhibited leucine uptake in both the presence and absence of potassium. The inhibition was noncompetitive. K_i values were determined to be 0.24 $\pm$ 0.01 mM in the presence and 0.26 $\pm$ 0.02 mM in the absence of potassium (19). Determination of the usefulness of this compound as an insecticide or as a tool for studying insect amino acid transporters awaits further investigation.

Recent Studies on Manduca Sexta

The feasibility of cloning a cDNA encoding a neutral amino acid symporter from larval lepidopteran midgut by expression of insect mRNA in *Xenopus laevis* oocytes was demonstrated by Sacchi and associates in 1995 (21). However, it required three more years and several more collaborators to realize this goal (2). The cloned cDNA contained an open reading frame that encoded a 634 amino acid residue protein. The protein, named KAAT1, showed weak but significant sequence identity with amino acid transporters belonging to the sodium-dependent and chloride-dependent gamma-aminobutyric acid (GABA) transporter superfamily. Like other members of this superfamily, it was predicted to have twelve transmembrane domains.

This protein was functionally characterized by voltage-clamp experiments on cRNA expressing oocytes (2). These oocytes showed potassium or sodium currents in the presence of leucine as well as several other amino acids known to be substrates for the neutral amino acid symport system of larval *M. sexta* midgut. Importantly, external application of L-arginine, L-glutamate, and α-(methylamino)isobutyric acid (MeAIB) failed to elicit potassium currents in the oocytes. Studies of the dependence of current on potassium concentration yielded a K_m for potassium of 32 mM. This is the same value as that determined in studies of leucine-K^+ symport by BBMV prepared from the posterior portion of larval *M. sexta* midgut (29). Studies of the dependence of current on sodium concentration yielded a K_m for sodium of 6 mM. A K_m for sodium is not available from studies of leucine uptake by *M. sexta* BBMV. However, the value from studies of leucine-Na^+ symport by *P. cynthia* BBMV is 2 mM; considerably less than the K_m for potassium (37 mM) in these vesicles (20). Studies of the dependence of potassium current into KAAT1 expressing oocytes on leucine concentration yielded a K_m for leucine of 123 $\pm$ 21 μM. This value is about half that (270 $\pm$ 84 μM) obtained from BBMV studies (29). However, considering the major differences in experimental systems and conditions, these leucine affinities seem reassuringly similar. Leucine-dependent potassium current into KAAT1 expressing oocytes required the presence of specific anions. No significant current was recorded in the presence of potassium gluconate. However, when gluconate was replaced by sulfate, chloride, or thiocynate increasingly larger currents were recorded. Maximum K^+-gradient driven neutral amino acid accumulation by larval *M. sexta* midgut BBMV typically increases with anions in the same manner (7). The dependence of current into KAAT1 expressing oocytes on substrate concentrations suggested a stoichiometry of two K^+ to one Cl^- to one amino acid (2). This stoichiometry nicely explains the previously observed acceleration of uptake by an inside negative $\Delta\Psi$ (6, 22).

Northern analysis revealed the presence of mRNA that hybridized with KAAT1 cDNA in larval *M. sexta* midgut and labial glands but not foregut, hindgut, Malpighian tubules, or nervous tissue (2). In midgut this mRNA was found only in columnar cells. In contrast to leucine-K^+ symport activity (29), mRNA that hybridized with KAAT1 cDNA seemed to be similarly abundant in the anterior and posterior portions of larval *M. sexta* midgut (2). Among explanations one might consider for this unexpected result are the possibilities that (*a*) KAAT1 mRNA is synthesized but not translated in anterior midgut; (*b*) anterior midgut preparation was contaminated with tissue from central midgut where specific leucine-K^+ symport activity is high (29); and (*c*) the KAAT1 cDNA hybridized with mRNA encoding a leucine uniporter and/or an arginine-K^+ symporter, both of which are known to be active in anterior midgut (29).

AMINO ACID TRANSPORT IN LARVAL COLEOPTERAN MIDGUT

Initial studies of leucine uptake by subcellular fractions prepared from frozen midguts of *Leptinotarsa decemlineata* larvae failed to reveal any stimulatory effects of either KCl or NaCl. The results obtained were consistent with leucine uptake by means of facilitated diffusion and provided no evidence for the presence of neutral amino acid–alkali ion symport systems in the midgut of this coleopteran (22). More recent studies of histidine uptake produced essentially the same results as had been obtained for leucine (17). However, studies of methionine and tyrosine uptake yielded more complex and puzzling results.

The maximum initial rate of L-tyrosine uptake by larval Colorado potato beetle midgut BBMV was approximately 70 times greater in the presence of an inwardly directed KSCN gradient and 140 times greater in the presence of an inwardly directed NaSCN gradient than in the absence of salt. However, the salt gradients did not drive any transient L-tyrosine accumulation by the vesicles and less than 40 percent of the tyrosine associated with the vesicles at equilibrium was sensitive to changes in medium osmolarity. Furthermore, the presence of salt resulted in a decrease in tyrosine affinity of the same order of magnitude as the increase in the initial rate of tyrosine uptake (9). These effects on L-tyrosine uptake seemed to be potassium and sodium specific and anion independent because they were also obtained with KCl and NaCl but not LiCl, CsCl, RbCl, $MgCl_2$, or $CaCl_2$ gradients (8). Experiments with KSCN-loaded vesicles showed that an inwardly directed gradient was not required for the stimulatory effects of this salt on L-tyrosine uptake. However, inwardly directed potassium phosphate and potassium gluconate gradients failed to stimulate L-tyrosine uptake. Taken together these results provide evidence that in the absence of sodium or potassium L-tyrosine uptake into larval Colorado potato beetle midgut cells seems to be mediated by a high-affinity, low-capacity (K_m 1.7 μM, V_{max} 0.15 pmol/s/mg protein) uniport system that, in the presence of sodium or potassium, is converted to or overtaken by a (much) higher-capacity and lower-affinity cation-dependent L-tyrosine uniport system. However, even in the presence of a maximally stimulating NaSCN concentration the capacity of this system is only about three percent of that of the neutral amino acid uniport system in larval *B. mori* midgut (13). The failure of externally applied potassium phosphate or potassium gluconate to bring about this transition could possibly be due to the cation-dependent uniport system requiring a higher potassium concentration on the internal side of the membrane than can be achieved by adding salts of very poorly permeating anions to the external medium.

Like tyrosine uptake, methionine uptake by larval Colorado potato beetle midgut BBMV is stimulated by the addition of NaCl or KCl but not $CaCl_2$ or $MgCl_2$. However, unlike tyrosine uptake, methionine uptake is also stimulated by RbCl and to a lesser extent by CsCl and LiCl (17). Again, gradients of the stimulatory salts fail to drive amino acid accumulation, indicating that the salts are modifying the properties of a uniport system rather than energizing a symport system. Even in the presence of 100 mM NaSCN, the methionine uptake system shows unusually low transport capacity and high substrate affinity (V_{max} 0.08 pmol/s/mg protein, K_m 0.07 μM). One might reasonably question the contribution of a transport system with such a low capacity to satisfying the nutritional needs of the larvae.

INSECT GLUTAMATE TRANSPORTERS

The observation that when bathed in a solution containing twenty common amino acids pieces of whole larval *Tenebrio molitor* epidermis accumulated strongly only aspartic and glutamic acids (26) seems to have prompted a more detailed analysis of acidic amino acid uptake by this tissue. McLean & Caveney (16) found that at low substrate concentrations uptake rates in the absence of sodium were less than 10% of those in the presence of sodium. However, the sodium-dependent uptake rate began to decrease with amino acid concentration for glutamate or aspartate concentrations greater than 0.1 mM. At 10 mM substrate, uptake rates were essentially the same in the presence or absence of sodium. The sodium-dependent component of L-glutamate uptake was determined to have a K_m for glutamate of 146 μM, a K_m for Na^+ of 21 mM, and a Na^+ to L-glutamate ratio slightly greater than 2. Lithium substituted poorly for sodium in accelerating glutamate uptake, whereas other monovalent cations were completely ineffective. Both L-aspartate and D-aspartate were competitive inhibitors of L-glutamate-Na^+ symport but D-glutamate was not. The authors speculated that, in light of the role of glutamate as a neuromuscular transmitter in insects, uptake of aspartic and glutamic acids by epidermis is much more important to keeping the plasma levels of these amino acids below synaptic threshold than to supplying epidermal metabolic requirements (16).

Among the studies that contributed to the recognition of glutamate as a neuromuscular transmitter in insects were those showing that electrically stimulated nerve-muscle preparations from adult male cockroaches took up glutamate at twice the rate of nonstimulated preparations, whereas leucine uptake by parallel preparations was not affected by nerve stimulation (5). Evidence for L-glutamate uptake by amino acid-Na^+ symport in excitable insect tissues came from studies on isolated abdominal nerve cords from adult *Periplaneta americana* (4) and abdominal ganglia from pupal *M. sexta* (11). L-Glutamate uptake in both insect preparations seemed to involve both sodium-dependent and sodium-independent components. The K_m for L-glutamate of the higher affinity sodium-dependent component was nearly the same in the two preparations (0.33 mM vs 0.48 mM). GABA and D-glutamate were ineffective inhibitors of L-glutamate uptake by either preparation. L-aspartate was a potent inhibitor of L-glutamate uptake by the *M. sexta* preparation but a weak inhibitor of L-glutamate uptake by the *P. americana* preparation. L-glutamate uptake by cultured cerebral ganglia as well as by abdominal plus thoracic muscle from *P. americana* embryos has since been shown to occur mainly by sodium-dependent processes with essentially the same inhibitor sensitivities as sodium-dependent glutamate uptake by pupal *M. sexta* abdominal ganglia (1).

Donly and associates used the polymerase chain reaction (PCR) with primers based on highly conserved regions of mammalian glutamate transporters to isolate from a late instar *Trichoplusia ni* larval head library a cDNA encoding a Na^+-dependent glutamate transporter (3). The deduced sequence of the 479 amino acids encoded by the isolated cDNA was 37% to 42% identical to the sequences of four human glutamate transporters. Glutamate uptake by cultured insect cells expressing this cDNA was sodium dependent, saturable, and at least fiftyfold more rapid than in cells infected with a baculovirus that lacked the cDNA. The mean K_m of the transporter for glutamate was 39.4 μM; an affinity about one order of magnitude greater than that found in *M. sexta* abdominal ganglia (11) and more than one order of magnitude less than that found in embryonic *P. americana* cerebral ganglia (1). The Na^+ to glutamate stoichiometery was estimated to be two to one. L-aspartate was found to be a more potent inhibitor of sodium-dependent glutamate uptake than L-glutamate was, whereas D-glutamate was ineffective. Expression of transporter mRNA was localized predominantly to caterpillar brain (3).

A cDNA encoding a Na^+-dependent glutamate transporter has been isolated recently from a library prepared from *Drosophila melanogaster* heads (23) using methods essentially identical to those used by Donly and associates (3). The protein encoded by this cDNA contained the same number of residues as the *T. ni* glutamate transporter but slightly more of them were identical to those in similar positions in human glutamate transporters one to four. Glutamate uptake by COS-7 cells expressing this cDNA was sodium-dependent, saturable, and at least threefold more rapid than in vector-transfected cells. cRNA encoding this *D. melanogaster* transporter was also expressed in *X. laevis* oocytes. In this expression system glutamate transport activity was assayed by both radiolabeled substrate uptake and two-electrode voltage-clamp. When expressed in COS-7 cells the mean K_m of the transporter for glutamate was 72 μM, whereas when expressed in oocytes the K_m dropped to 25 μM. The K_m for L-aspartate was nearly the same as that for L-glutamate. However, the K_m for D-aspartate was more than fourfold greater than that for L-aspartate, and the K_m for D-glutamate was at least 200-fold greater than that for L-glutamate. Northern blots showed expression of transporter mRNA was much greater in heads than bodies, whereas in situ hybridizations showed expression almost exclusively in brain and thoracic ganglia of embryos (23). Closer localization was not attempted, but based on the results of earlier locust studies one might expect to find this transporter concentrated in axon terminals and glial cells (27).

SUMMARY AND CONCLUSIONS

At least the more hydrophobic amino acids seem able to cross the membranes of various insect cells by simple diffusion (1, 29). Many insect cell membranes also seem to contain uniporters that facilitate the diffusion of certain amino acids into and out of the cells (8, 11, 13). In addition to these passive diffusion systems all insect tissues studied to date, except *L. decemlineata* midgut, seem to contain at least one amino acid–cation symport system that allows their cells to accumulate certain amino acids from the extracellular medium. cDNAs encoding three such symporters have very recently been cloned and sequenced (2, 3, 23). The deduced amino acid sequence of each insect symporter turned out to be homologous to that of symporters mediating the transport of the same or related substrates in mammalian tissues. If this proves to be generally the case, one can reasonably expect to learn the primary structure of many more insect amino acid transporters in the next few years.

ACKNOWLEDGMENT

Cited work from my laboratory was supported in part by NIH Grant AI30464.

Visit the Annual Reviews home page at www.AnnualReviews.org.

LITERATURE CITED

1. Bermudez I, Botham RP, Beadle DJ. 1988. High- and low-affinity uptake of amino acid transmitters in cultured neurones and muscle cells of the cockroach, *Periplaneta americana*. *Insect Biochem.* 18:249–62
2. Castagna M, Shayakul C, Trotti D, Sacchi VF, Harvey WR, Hediger MA. 1998. Cloning and characterization of a potassium-coupled amino acid transporter. *Proc. Natl. Acad. Sci. USA* 95:5395–400
3. Donly BC, Richman A, Hawkins E, McLean H, Caveney S. 1997. Molecular cloning and functional expression of an insect high-affinity Na^+-dependent glutamate transporter. *Eur. J. Biochem.* 248:535–42
4. Evans PD. 1975. The uptake of L-glutamate by the central nervous system of the cockroach, *Periplaneta americana*. *J. Exp. Biol.* 62:55–67
5. Faeder IR, Salpeter MM. 1970. Glutamate uptake by a stimulated insect nerve muscle preparation. *J. Cell Biol.* 46: 300–7
6. Giordana B, Leonardi MG, Casartelli M, Consonni P, Parenti P. 1998. K^+-neutral amino acid symport of *Bombyx mori* larval midgut: a system operative in extreme conditions. *Am. J. Physiol.* 274:R1361–71
7. Hennigan BB, Wolfersberger MG, Parthasarthy R, Harvey WR. 1993. Cation-dependent leucine, alanine, and phenylalanine uptake at pH 10 in brush border membrane vesicles from larval *Manduca sexta* midgut. *Biochim. Biophys. Acta* 1148:209–15
8. Hong YS, Neal JJ. 1997. Tyrosine transporter in larval *Leptinotarsa decemlineata* midgut brush border membrane. *Insect Biochem. Mol. Biol.* 27:193–200
9. Hong YS, Reuveni M, Neal JJ. 1995. A sodium- and potassium-stimulated tyrosine transporter from *Leptinotarsa decemlineata* midguts. *J. Insect Physiol.* 41:527–33
10. Kingan TG, Hildebrand JG. 1985. γ-Aminobutyric acid in the central nervous system of metamorphosing and mature *Manduca sexta*. *Insect Biochem.* 15:667–75
11. Kingan TG, Hishinuma A. 1987. Transport and metabolism of L-glutamic acid by abdominal ganglia of the hawk moth, *Manduca sexta*. *Comp. Biochem. Physiol. C* 87:9–14
12. Leonardi MG, Cappellozza S, Ianne P, Cappellozza L, Parenti P, Giordana B. 1996. Effects of the topical application of an insect growth regulator (fenoxycarb) on some physiological parameters in the fifth instar larvae of the silkworm *Bombyx mori*. *Comp. Biochem. Physiol. B* 113:361–65
13. Leonardi MG, Casartelli M, Parenti P, Giordana B. 1998. Evidence for a low-affinity, high-capacity uniport for amino acids in *Bombyx mori* larval midgut. *Am. J. Physiol.* 274:R1372–75
14. Leonardi MG, Parenti P, Casartelli M, Giordana B. 1997. *Bacillus thuringiensis* Cry1Aa δ-endo-toxin affects the K^+/amino acid symport in *Bombyx mori* larval midgut. *J. Membr. Biol.* 159:209–17
15. Mbungu D, Ross LS, Gill SS. 1995. Cloning, functional expression, and

pharmacology of a GABA transporter from *Manduca sexta. Arch. Biochem. Biophys.* 318:489–97

16. McLean H, Caveney S. 1993. Na^+-dependent medium-affinity uptake of L-glutamate in insect epidermis. *J. Comp. Physiol. B* 163:297–306
17. Neal JJ, Wu D, Hong YS, Reuveni M. 1996. High affinity transport of histidine and methinonine across *Leptinotarsa decemlineata* midgut brush border membrane. *J. Insect Physiol.* 42:328–35
18. Parenti P, Morandi P, McGivan JD, Consonnic P, Leonardi G, Giordana B. 1997. Properties of the aminopeptidase N from the silkworm midgut (*Bombyx mori*). *Insect Biochem. Molec. Biol.* 27:397–403
19. Parenti P, Pizzigoni A, Hanozet G, Hakim EH, Makmur L, et al. 1998. A new prenylated flavone from *Arctocarrpus champeden* inhibits the K^+-dependent amino acid transport in *Bombyx mori* midgut. *Biochem. Biophys. Res. Commun.* 244:445–48
20. Sacchi VF, Parenti P, Perego C, Giordana B. 1994. Interaction between Na^+ and the K^+-dependent amino acid transport in midgut brush border membrane vesicles from *Philosamia cynthia* larvae. *J. Insect Physiol.* 40:69–74
21. Sacchi VF, Perego C, Magagnin S. 1995. Functional characterization of leucine transport induced in *Xenopus laevis* oocytes injected with mRNA isolated from midguts of lepidopteran larvae (*Philosamia cynthia*). *J. Exp. Biol.* 198:961–66
22. Sacchi VF, Wolfersberger MG. 1996. Amino acid absorption. In *Biology of the Insect Midgut,* ed. MJ Lehane, PF Billingsley, pp. 265–92. London: Chapman & Hall
23. Seal RP, Daniels GM, Wolfgang WJ, Forte MA, Amara SG. 1998. Identification and characterization of a cDNA encoding a neuronal glutamate transporter from *Drosophila melanogaster. Recept. Channel* 6:51–64
24. Shinbo H, Konno K, Hirayama C, Watanabe K. 1996. Digestive sites of dietary proteins and absorptive sites of amino acids along the midgut of the silkworm, *Bombyx mori. J. Insect Physiol.* 42:1129–38
25. Soehnge H, Huang X, Becker M, Whitley P, Conover D, Stern M. 1996. A neurotransmitter transporter encoded by the drosophila inebriated gene. *Proc. Natl. Acad. Sci. USA* 93:13262–67
26. Tomlin E, McLean H, Caveney S. 1993. Active accumulation of glutamate and aspartate by insect epidermal cells. *Insect Biochem. Mol. Biol.* 23:561–69
27. van Marle J, Piek T, Lind A, van Weeren-Kramer J. 1983. Localization of a Na^+-dependent system for glutamate in excitatory neuromuscular junction of the locust *Schistocerca gregaria. Comp. Biochem. Physiol. C* 74:191–94
28. Wolfersberger MG. 1994. Uniporters, symporters and antiporters. *J. Exp. Biol.* 196:5–6
29. Wolfersberger MG. 1996. Localization of amino acid absorption systems in the larval midgut of the tobacco hornworm *Manduca sexta. J. Insect Physiol.* 42:975–82

Annu. Rev. Entomol. 2000. 45:121–150

SOCIAL WASP (HYMENOPTERA: VESPIDAE) FORAGING BEHAVIOR

M. Raveret Richter
Department of Biology, 815 N. Broadway, Skidmore College, Saratoga Springs, New York 12866–1632; e-mail: mrichter@skidmore.edu

Key Words prey, carbohydrate, interactions, recruitment, community

■ **Abstract** Social wasps (Hymenoptera: Vespidae) forage for water, pulp, carbohydrates, and animal protein. When hunting, social wasps are opportunistic generalists and use a variety of mechanisms to locate and choose prey. Individual foragers are influenced by past foraging experience and by the presence of other foragers on resources. A forager's ability to learn odors and landmarks, which direct its return to foraging sites, and to associate cues such as odor or leaf damage with resource availability provide the behavioral foundation for facultative specialization by individual foragers. Social wasps, by virtue of their behavior and numbers, have a large impact on other organisms by consuming them directly. Indirect effects such as disruption of prey and resource depletion may also be important. Community-level impacts are particularly apparent when wasps feed upon clumped prey vulnerable to depredation by returning foragers, or when species with large, long-lived colonies are introduced into island communities. A clearer understanding of these relationships may provide insight into impacts of generalist predators on the evolution of their prey.

OVERVIEW

In both temperate and tropical communities, social wasps (Hymenoptera: Vespidae) are often strikingly abundant. They collect water, plant fibers, and carbohydrates, and hunt arthropod prey or scavenge animal protein (28). Social wasps are generalist foragers, but individuals are able to learn and may specialize by hunting for prey or collecting other resources at specific locations (106). The foraging behavior of generalist social bees, who are able to associate the presence and quality of resources with colors, odors, shapes, and features of the environment, has greatly influenced the evolution of floral characteristics (8, 31). Similarly, the foraging behavior of generalist invertebrate predators such as social wasps, with a comparable proclivity for associative learning, may shape the evolution of their insect prey.

In this review, I summarize research on behavioral mechanisms of social wasp foraging and discuss potential ecological impacts of foraging behavior, particularly with regard to prey foraging. I report what social wasps forage for, and the manner in which they choose and collect these resources. Interactions among wasps, via recruitment of other wasps to resources and by theft of or exclusion from resources, may be important determinants of where and how wasps forage. However, relative to recruitment to resources by eusocial bees, ants, and termites, recruitment to food is poorly developed in social wasps (65). I describe observed patterns of recruitment in these wasps and consider possible relationships between patterns of interactions among wasps and the distribution and abundance of competitors and resources.

Because social wasps are abundant components of their communities, they are likely to have considerable ecological impact (64a). Wasps may exert selective pressures on their prey, and I discuss the behavior of lepidopteran prey in relation to wasp predators. Ecological impacts of social wasps at the community level are further illustrated by examining the consequences of introducing social wasps to new environments and by documenting some examples of the use of social wasps in biological control.

Studies of social bees have been particularly fruitful in illustrating connections among ecological patterns, social systems, and behavioral mechanisms (114). Wasps, who live in a variety of ecological circumstances and exemplify a rich diversity of life histories and social systems (137), offer similar potential for productive studies on the relationship between social systems, foraging behavior, and the distribution and abundance of resources. In addition, studies of the hunting mechanisms and ecological impacts of prey foraging by social wasps may provide insight into the evolutionary importance of generalist foragers. By summarizing and synthesizing past work on the foraging behavior of social wasps from mechanistic, ecological, and evolutionary perspectives, this review is intended to serve as a catalyst for future investigations.

INTRODUCTION

Focus

The behavior and interactions of social wasps at their nests are well studied (reviewed in 112a, 137). Members of the genera *Polistes* (131, 134) and *Mischocyttarus* (58), with unenclosed colonies that are relatively easy to observe, have proven particularly popular and productive research subjects. Studies of these wasps have been a fertile source of ideas about how sociality has evolved and is maintained (5, 44, 59, 135).

Foraging behavior, especially that which takes place away from the nest at often unpredictable locations (57), is less well studied (106). Social wasps forage for water, plant fibers, protein, and carbohydrates (28). Water is imbibed by the wasps and is used in nest cooling (1, 43) and construction (64) as well as for metabolic processes. Plant fibers (pulp) are used for nest construction (133). Protein for developing brood consists of arthropod prey and, for some species, scavenged protein (1). Carbohydrate-rich foods such as nectar, sap, and fruit serve as energy sources (117, 43). This review focuses on the basics of water, pulp, prey and carbohydrate foraging. I discuss the importance of recruitment and interactions of wasps on food and explore the impact of social wasp foraging on communities.

Phylogeny

Wasps are members of the insect order Hymenoptera, comprised of the plant-feeding suborder Symphyta and the suborder Apocrita. Apocrita contains both the parasitoid wasps and the Aculeata: stinging wasps, ants, and bees (30). Parasitoid Apocrita generally oviposit on or in arthropod hosts and may also inject a paralytic poison into hosts through their ovipositors. Aculeate wasps are believed to have evolved from a generalized parasitoid ancestor (30). The shaft of the ovipositor is modified into a sting, and the eggs of aculeate wasps leave the body through an opening at the base of the shaft. The majority of aculeate wasps use their sting to paralyze hosts more or less permanently and then carry them to a crevice or a nest where an egg is laid on the host. However, social aculeate wasps, upon which this review focuses, most often kill their prey by biting rather than stinging, and the sting is used for defense. Placement of the sting on the tip of the highly flexible metasoma enables the wasp to adeptly aim and inject venom into its foes (30).

Social aculeate wasps, with rare exception (80), are all in the family Vespidae, comprised of the subfamilies Euparagiinae, Masarinae, Eumeninae, Stenogastrinae, Polistinae and Vespinae (17). The latter three subfamilies contain eusocial species (17) and are the focus of this review. Sociality expands the options available for finding resources and provisioning the colony. Group living makes possible behavior such as recruitment and division of labor. Different patterns of social organization among wasp species correspond to different possibilities for partitioning the tasks of locating resources and collecting, transporting, dispensing, or storing them. The following sections illustrate ways in which wasps locate, choose, and collect resources.

Water and Pulp Foraging

In addition to imbibing water or passing it along to larvae (69), social wasps mix water with masticated plant fiber in processing material for nest construction and also use water in conjunction with wing fanning in evaporative cooling of the nest (1, 43, 137). Wasps imbibe water at sources such as standing water or rain droplets on vegetation, and water is carried to the nest in the forager's crop. At the nest, the water is regurgitated from the crop and transferred to nestmates (61). The forager may then take off from the surface of the nest and resume water collection in the field (64). Water can be a limiting resource for wasps (53). In areas or during seasons in which fresh water is scarce, social wasp foragers are frequent visitors to sinks and dripping water spigots (MA Raveret Richter, unpublished observations).

Plant fibers (reviewed in 133) such as scrapings from dead branches or weathered, unpainted wood are collected by social wasps and serve as the major structural components of their nests. In plant fiber collection by *Polybia occidentalis,* a forager lands on a piece of wood, regurgitates water from her crop onto the surface of the wood, and scrapes her mandibles across the surface of the wood until she forms a ball of pulp, which she carries in her mandibles as she flies back to her nest. Upon arriving at the nest, foragers typically transfer the pulp to nestmates and may resume foraging for additional loads of pulp (61).

Studies of the collection by *Polybia occidentalis* foragers of water and pulp for use in the task of nest construction (61, 64, 91) have provided many insights into the organization of work in social wasp colonies. The location of water and pulp in the field, at sites such as puddles, ditches, and fence posts, is somewhat more predictable than that of food resources such as prey and nectar (91). *Polybia occidentalis* foragers tend to specialize upon gathering a single type of resource, and if foragers do change from collecting one resource type to another, the switch tends to be within a functionally related group of materials: foragers more frequently switched from collecting one type of nest building material (pulp or water) to another, or one kind of food material (prey or nectar) to another, rather than switching between collecting nest building and food material (91).

In nest construction by *P. occidentalis,* pulp and water are collected by two different groups of foragers and used by a third group, the builders, in nest construction. A preliminary study (62) suggested that the rate of water collection by foragers set the pace for nest construction activity for *P. occidentalis.* However, a more recent and extensive investigation (64) suggests instead that builders respond directly to nest damage in regulating their activity. Jeanne (64) observed that the level of pulp foraging activity was a function of the demand by builders for pulp from foragers. Water foragers adjusted their foraging in response to demands for water by pulp foragers and builders. Thus, in cases where pulp and water supplies are not limited, nest construction activity was not driven by the rate at which water was delivered to the nest, but rather by the nest builders' demand for supplies from foragers.

PREY HUNTING AND PROTEIN SCAVENGING

Solitary and Presocial Wasps

The hunting behavior of a number of solitary sphecid wasps has been studied in depth. These studies provide much of our knowledge of how wasps hunt, and several are briefly presented here as points of reference for the observations that follow on the social species. Many examples of the hunting behavior of solitary wasps are provided by Evans & West Eberhard (30) and Iwata (57). Tinbergen's studies of *Philanthus triangulum* (129) document the responses of hunting wasps to visual and olfactory prey cues. Steiner's (122, 123) comparisons among cricket-hunting *Liris* and fly-hunting *Oxybelus* illustrate the characteristics and variability of hunting and prey-stinging behavior. The majority of solitary sphecid wasps are to some degree host-specific and mass-provision their young with prey such as caterpillars or bees that they have stung and paralyzed (30). Some, such as *Clypeadon laticinctus* (Hymenoptera: Sphecidae) are highly specialized and hunt only a single prey species (4).

Within the Vespidae, the solitary and presocial subfamilies Euparagiinae, Masarinae, and Eumeninae employ a variety of methods for provisioning their young (reviewed in 23). All vespids oviposit before provisioning cells (16), enabling social contact between the mother and her young in cases where the mother progressively provisions her brood (29). In the solitary Euparagiinae, comprising a single genus with nine species, young are mass-provisioned with curculionid beetles (23, 136). Most of these wasps have some degree of specificity, at least at the generic level, in prey choice (30). The subfamily Masarinae is unique among the Vespidae: These wasps feed their young a mixture of pollen and nectar and do not hunt insect prey.

In the solitary to subsocial Eumeninae, where single females nest in burrows, tubes, or free-standing mud nests (30), young are supplied primarily with stung and paralyzed caterpillar prey; a few species provision with curculionid larvae (117). Although the majority of these wasps mass-provision their brood, some are progressive provisioners, particularly at times of prey scarcity (30). The degree of prey specificity in eumenine wasps is variable (23). For example, Jennings and Houseweart found differences among species and between years in the prey specificity of four conspecific eumenine wasps (66). *Ancistrocerus adiabatus* and *Ancistrocerus antilope* each provisioned with only one caterpillar species in a particular year. However, *A. adiabatus* switched from provisioning with only *Nephoteryx* sp. (Lepidoptera) in 1977 to using only *Acleris variana* (Lepidoptera) in 1978. *Ancistrocerus catskill* and *Euodynerus leucomelas* both provisioned their nests with several species of caterpillar. The caterpillar species captured differed between the two seasons, and in neither year were all cells provisioned with a single species of prey, suggesting that these wasps, similar to the social wasps to be described here, were opportunistic predators. The degrees of host specificity and constancy were influenced by changes in prey abundance and distribution in these latter wasps (66). Cowan (23) observed that, in conspecific eumenine nests, one may find one nest provisioned with a variety of caterpillar species, while the adjacent nest contains cells provisioned with a single prey species. He suggests that individual foragers may learn prey searching habits that bring them into contact with a limited range of prey, or that hunting females may return repeatedly to locations where prey are concentrated.

Social Wasps

Prey Specificity Social vespids are opportunistic, generalist prey foragers (43, 106, 117). However, individual social wasp foragers often return to hunt in sites of previous hunting success (104, 108, 125, 126) and may feed repeatedly on the same species of prey, thus functioning individually as facultative specialists (106). The choice of hunting sites can be influenced by prey density: For example, *Polistes jadwigae* and *Polistes chinensis* foragers hunting large *Spodoptera litura* (Lepidoptera: Noctuidae) larvae more frequently searched for and attacked this prey at the highest density study site (89). Thus, as with the Eumeninae described above, both past experience and the current distribution and abundance of prey appear to be important in influencing an individual's choice of prey.

Social wasps use masticated arthropod prey and other animal protein to progressively provision their developing brood. Prey items most commonly include a variety of arthropods such as caterpillars, flies, alate ants, termites (1, 32) spiders (30, 130), bees (15, 56, 73, 79, 94, 130) and other social wasps (2, 78 as in 57, 79a). Many social wasps also scavenge vertebrate and invertebrate carrion, in addition to foraging for arthropod prey. Vespinae that scavenge protein (*Vespa* and species in the *Vespula vulgaris* species group) are notorious pests on human food, making outdoor consumption of food perilous when they are abundant (2).

Occasionally, live vertebrates are fed upon by foraging wasps and blood or tissue is collected. An imperturbable entomologist visiting British Columbia left a *Vespula* forager to chew on his ear, and the wasp drew blood and carried off a drop of the blood in its mouthparts (102). In Israel, *Vespula germanica* chew tissue on the teats of dairy cattle, inducing mastitis in the cows (14). Grant (41) observed hornets consuming young hummingbirds, and Lacey (74) observed *Angiopolybia pallens* preying upon eggs and tadpoles in the foam nests of *Leptodactylus pentadactylus*. One response to such predation is early hatching and escape of the tadpoles. K Warkentin (personal communication) observed escape hatching of eggs of the arboreal tree frog *Agalychnis callidryas* in response to predation by *Polybia rejecta* foragers. Blood or tissue collection has the potential to pose serious problems for animals unable to escape or to protect vulnerable tissue from the onslaught of foraging wasps, which return repeatedly to sites at which they have foraged successfully (106).

Prey Capture and Handling Wasp social organization, the regimen for provisioning brood, and environmental factors such as weather (117) and the abundance and distribution of prey (89) and predators may all influence the hunting strategies of wasps. Foragers may hunt for live prey, return to remains of prey that they have already killed, steal prey captured by other wasps (106), or scavenge carrion (2).

A hunting wasp faces several tasks. She must locate prey and distinguish it from non-food items, capture the prey, and either consume it or prepare it for transport back to the nest. While she is processing the prey, a wasp may need to protect it from other predators. Portions of prey that cannot be carried back to the nest on the first trip may need to be stowed so that they are less vulnerable to theft while the wasp is ferrying multiple loads to the nest from the capture site (109).

Live prey are generally pounced upon by social wasp foragers and killed by biting. Foragers then malaxate (process it into a ball, using the mouthparts to chew and manipulate it) the prey, carry it to the nest, and feed it progressively to their young. If prey are large and require multiple trips to carry back to the colony, a forager bites off pieces of the prey, malaxates them, carries them back to the colony, and returns quickly for the remainder of the kill. Prey remains left untended in the field are subject to theft by other hunting wasps (106) and by ants (109).

Although several authors (reviewed in 93) have documented cases in which foraging social wasps have used their sting when grappling with particularly large and active prey, the sting is generally reserved for defensive use in social wasps (3, 30). However, Olson (93) observed a *Parachartergus fraternus* forager deliver stings to the mid-abdominal segments of a stationary fourth instar *Rothschildia lebeau* (Lepidoptera: Saturniidae) caterpillar. After stinging the caterpillar, the wasp retired to nearby vegetation while the caterpillar became progressively more limp and lost its grip on the vegetation. On two separate occasions, the wasp returned to deliver an additional sting to the caterpillar. When the caterpillar had become motionless and fallen from the leaf, the wasp then began the typical malaxation of the prey. Olson (93) observed a similar incident in which a *P. fraternus* forager stung a 1.5 cm green looper caterpillar and subsequently perched on a nearby leaf while the caterpillar became limp. When the caterpillar hung motionless and was attached to the substrate only by its anal prolegs, the wasp approached and malaxated the caterpillar. Olson (93) suggests that this stinging behavior, similar to the prey stinging of non-social vespids in the Eumeninae (23), may be associated with the small mandible-length–to–body-length ratio in *P. fraternus* relative to that of other eusocial vespid genera; paralyzing the prey might better enable the wasp to process it and carry it to the nest.

Prey handling by social wasp foragers may also be influenced by the presence of allelochemicals in the host plants of their prey. Rayor et al (111) conducted tests in which paper wasp *Polistes dominulus* foragers were presented with a choice between *Pieris napi* (Lepidoptera: Pieridae) caterpillars raised on either cabbage, *Brassica oleracea*, (Brassicaceae), which contains glucosinolates, or wormseed mustard, *Erysimum cheiranthoides* (Brassicaceae), which contains both glucosinolates and cardenolides. Wasps did not detect or were not deterred by the presence of cardenolides in caterpillars raised on *E. cheiranthoides* and showed no preference for feeding on caterpillars reared upon either plant. Thus, feeding on a chemically defended host plant did not afford the caterpillars reduced predation.

Sequestration of allelochemicals from host plants can successfully reduce predation upon caterpillars by generalist hymenopteran predators such as wasps and ants (86). *Mischocyttarus flavitarsis* (Hymenoptera: Vespidae) foragers reject the colored caterpillars of *Uresiphita reversalis* (Lepidoptera: Pyralidae) after superficial contact. Experimental manipulations (86) demonstrate that quinolizidine alkaloids sequestered from the host plant *Genista monspessulana* (Papilionaceae) stored in the cuticle of *U. reversalis* caterpillars are repellent to *M. flavitarsus* foragers.

Chemical analyses demonstrated that the *Pieris napi* caterpillars in Rayor et al's experiments (111) did not sequester cardenolides from *E. cheiranthoides* in their tissues. Thus, when biting the mustard-fed caterpillars, wasps did not encounter high cardenolide concentrations in the epidermis or hemolymph. Upon encountering gut tissue in mustard-fed caterpillars, however, wasps contacted cardenolides. When processing these caterpillars, *P. dominulus* foragers used their mandibles to skillfully excise the gut and its contents from the balls of prey tissue they transported back to their nests. In contrast, wasps did not remove the gut from cabbage-fed caterpillars during processing. Thus, depending upon the host plant they had consumed, caterpillars were processed differently by *P. dominulus* foragers, and it took wasps significantly longer to process caterpillars fed upon wormseed mustard than those fed upon cabbage (111).

Orientation Flights Landmarks are commonly used in insect navigation (19) and help to direct the return of foraging wasps to the remains of prey too large to carry to the nest in a single load. When first departing from large prey (or other sites to which they will return), social wasps, as well as other winged Hymenoptera, perform an orientation flight. While facing the prey (109), feeder (20), nest site (28, 117) or other focal site, the wasps fly back and forth in arcs of gradually increasing radius as they increase their horizontal and vertical distance from the focus. Wasps departing from feeding sites will complete the departure flight by circling high above the ground and then flying away from the feeding site (20).

When a *Polybia sericea* Olivier forager captures prey too large to carry to the nest in one load, the forager, while facing into the wind, departs from the prey capture site flying in side-to-side arcs or circles, successively increasing its horizontal distance from the prey and at the same time ascending from 0.5 to 3.0 m above the prey; the wasp then flies to the nest (109). Wasps return to the prey capture site from downwind, facing the prey. *Polistes jadwigae* Dalla Torre (126) also make orientation flights at sites where they have killed large prey.

Field experiments demonstrate that, upon returning from downwind to the site of a past kill, *P. sericea* (105, 108) and *P. jadwigae* (126) foragers use visual landmarks to direct intensive aerial search of the foraging site; olfactory prey cues direct the forager's landing. Displacement of the visual landmarks that had surrounded a prey capture site when a *P. sericea* forager made orientation flights significantly increased the amount of time it took for the returning forager to relocate and land upon prey remains (108).

Detailed video analyses of spatial patterning in the orientation flight of the yellowjacket *Vespula vulgaris* (20) demonstrates that these wasps orient toward and visually inspect feeders at points located toward the outermost edges of the arcs traced by the wasps during their orientation flights. They suggest that this alignment of inspection points, by providing the wasp with serial images of the goal and associated landmarks viewed at different distances, minimizes the number of views the animal must retain to steer the return to the remembered location by means of image matching, and that wasps' landmark memories are acquired primarily during these aligned inspections. A subsequent study (18) demonstrated that for several individual foragers, the alignment of the inspection points was similar both for the departure flights and the return flights of the wasps, as would be predicted if wasps acquire landmark memories in inspection flights, and use these remembered images for orientation during the return flight.

Use of Cues in Location and Choice of Prey Foraging on natural prey often takes place infrequently at unpredictable locations, making it difficult to document how social wasps locate and choose prey in the field. Some authors have concentrated on cue use in one (58) or a few (24, 50) species, some have made generalizations across all of the vespid wasps (57), and some have pointed out the differences among different social wasps that have been studied (1).

Suggested mechanisms for how various social wasps locate and distinguish prey run a broad gamut. In early studies by Peckham and Peckham, foraging vespine wasps accumulated upon gauze bags full of chicken bones and tried to gain access to the bones; wasps did not accumulate upon control bags filled with gauze (101). This suggests that these wasps could locate food using solely olfactory cues. Iwata (57) states that olfactory cues are most important in detecting and locating prey from a distance.

Duncan (25), Frost (35), Heinrich (50), and Jeanne (58) observed wasps for which visual cues appear to determine what foragers choose to attack. Heinrich's (50) observations on the yellowjacket *Dolichovespula maculata* suggest that the wasps pounce first and then discriminate between prey and non-prey; this is also suggested by the observations of Duncan (25) and Frost (35), who report the seeming lack of skill of *D. maculata* foragers as they repeatedly pounce upon and release nailheads and other wasps, and miss when attempting to capture flies. Jeanne (58) observed a similar method of hunting by *Mischocyttarus drewseni:* The wasps pounced upon nodules on plants and the barbs of barbed wire fences. Jeanne concluded that the wasps pounced in response to irregular silhouettes.

Evans & West Eberhard (30) and Raveret Richter & Jeanne (108) proposed that visual cues determine the area in which certain social wasps search intensively for prey and olfactory cues determine where a forager lands. Raveret Richter & Jeanne (108) demonstrated experimentally that *Polybia sericea* foragers in Brazil used visual and olfactory prey cues to relocate a foraging area and that at close range olfactory prey cues were more likely to elicit landing on stationary prey than were visual cues.

Social wasps, like bees (132), can learn to associate color (82, 116) or odor (82) with food rewards. McPheron (82) found that *Mischocyttarus flavitarsis* foragers learned to associate odors with rewards more quickly than they learned to associate colors with rewards. When trained to a compound (presented simultaneously) color and odor stimulus and subsequently tested for responses to individual stimulus components, *M. flavitarsis* foragers showed a preference for odor over color.

Reflecting these learning abilities, the behavior of foraging wasps is influenced by cues that are associated with prey but do not emanate directly from the prey. Raveret Richter (105) observed that both *Polybia occidentalis* and *Polybia diguetana* foragers landed preferentially on rolled or damaged leaves (more than five percent of the leaf was chewed) when foraging in a clump of low, herbaceous plants upon which leafroller larvae were feeding, suggesting that these cues might be associated with prey by foraging wasps. Cornelius (21) observed that, in choice tests conducted in a greenhouse, naive *M. flavitarsis* foragers captured more *Manduca sexta* (Lepidoptera: Sphingidae) and *Trichoplusia ni* (Lepidoptera: Noctuidae) larvae from tobacco plants with leaves previously damaged by caterpillars than from those without leaf damage. Interestingly, the wasps showed no such preference when presented with a choice of prey on damaged or undamaged tomato leaves. The presence of both visual and olfactory plant cues was necessary for the wasps to exhibit a foraging preference; neither visual nor olfactory cues by themselves elicited preferential foraging.

Additional greenhouse studies on experienced (given two days of experience in finding caterpillars on test plants prior to testing) *M. flavitarsis* (81) foraging for *T. ni* caterpillars on tomato, broccoli and bean plants demonstrated that, in contrast to Jeanne's (58) field observations on *Mischocyttarus drewseni* in which foragers pounced on irregular silhouettes, visual prey cues (heat-cured Fimo clay caterpillar models) alone did not elicit higher frequency of approaches to or higher mean time spent hovering next to plants with green or brown caterpillar models. However, when caterpillar regurgitate was applied to brown clay models, wasps spent more time hovering near and landed more frequently upon the plants bearing these models. In the absence of visual prey cues, more time was spent hovering next to broccoli plants treated with caterpillar regurgitate relative to untreated plants. In contrast with Cornelius's (21) work, visual cues associated with artificially induced plant damage (3 to 4 mm diameter compared to the 8 mm diameter in Cornelius's experiment) were not associated with changes in the wasps' foraging behavior, but leaves damaged by caterpillars were landed upon preferentially by foraging wasps. In her test of *M. flavitarsis* foragers, McPheron concluded that olfactory cues emanating from both prey and from plants damaged by caterpillars were more important than visual cues (either visual prey cues or visible plant damage) in influencing both the hovering and landing behavior of experienced foragers (81).

Hunting wasps may cue in upon chemical messages sent by other insect species and use them to locate food. Hendrichs et al (52) demonstrated that the yellowjacket *Vespula germanica* responds to the odor of the pheromone used by Mediterranean fruit fly males, *Ceratitis capitata* (Diptera: Tephritidae), in their lek-based courtship displays. In dense foliage, where medfly leks are typically located (51), foraging wasps approached this odor from downwind and were effectively directed to aggregations of calling male flies; visual and acoustical signals were important prey location cues only at close range. However, in more open situations wasps capturing female medflies ovipositing in fruit switched to visually-based patrolling behavior, indicating that the hunting tactics of these wasps are situation-dependent (52) and that, like honey bees (115) and bumble bees (49), the wasps can assess and respond to changes in their environment.

Patterns of variability in hunting behavior within and among social wasp species remain to be explored. In different situations or for different wasp species, response to a particular cue may vary. The variability observed thus far suggests that the hunting tactics of social wasps are likely to be facultative and situation-dependent, with the cues used for prey location and choice most likely dependent upon the context in which the hunt takes place, the type of prey being hunted (108), and the past experience of the forager (82).

CARBOHYDRATE FORAGING

Social wasps collect carbohydrates from a variety of natural and anthropogenic sources, seeming to opportunistically exploit any available source of concentrated sugar (30). These carbohydrates serve as an energy source for both adults and developing brood (2). In addition to the collection of nectar from flowers, social wasps imbibe plant sap (117) and sweet liquid from fruits (30). The yellowjacket *Vespula vulgaris* collects seeds from *Trillium ovatum* Pursh. (Liliaceae), removing and consuming the carbohydrate-rich eliaosome and dispersing the seeds (68). Social wasp foragers consume honeydew excreted from plant-feeding insects such as aphids, psyllids, and coccids (71, 84, 117) and rob bees of their stored honey (30). Some social wasps collect sweet beverages and foods and can become pests at recreational areas, outdoor concession stands, and bakeries (2), where large numbers of wasps will gather and feed.

Carbohydrates are at times stored as honey in the nests of some social polistine wasps; wasp honey also serves as a source of stored amino acids (reviewed in 54). Supplementation of colonies of the paper wasp *Polistes metricus* with honey enhances brood development. This finding suggests that carbohydrate availability can be a limiting factor in wasp development (113).

Ecological data also support the idea that carbohydrates may be a critical resource for social wasp colony development. Keyel observed that natural sugar sources are often diffuse in nature (71). However, wasps able to take advantage of concentrated carbohydrate sources may build up large populations. *V. germanica,* introduced from Europe into the United States, has become abundant, replacing the native *Vespula maculifrons* in many urban and suburban habitats (88, 2). The ascendancy of this species may be in large part because of its ability to dominate rich human food sources (71, 88). Similarly, introduced vespine wasps in New Zealand have invaded the carbohydrate-rich honeydew beech forests, in which beech scale insects *Ultracoelostoma assimile* (Homoptera: Margarodidae) cover the trunks of trees and exude sweet liquid. *V. vulgaris* nest densities in the beech forests have been recorded at 45 nests/ha, and wasp densities on scale-infested tree trunks can reach 360 insects/m^2, at which point tree trunks are yellow with wasps (124). These examples illustrate the potential importance of carbohydrate availability as a determinant of wasp population size.

RECRUITMENT

Recruitment in the context of foraging is communication that serves to bring nestmates to a food source (137). In nest-based recruitment by social insects, a solitary forager communicates the presence of a resource to other foragers at the nest (65). Simple nest-based recruitment might involve behavior such as a returning forager stimulating a group of foragers to depart from the nest merely as a result of its arrival on the nest (32). In addition, a "departure dance" (1) in which a forager runs around rapidly in the nest and is licked and antennated by nestmates prior to its return to a food source (1, 90) could further stimulate other foragers to leave the nest. Nestmates might thus be incited to forage and predisposed to visit food bearing the odor carried on the body of the recently returned or active forager. Clumped departures of wasp foragers from the nest after such a "departure dance" would provide evidence for this mechanism of nest-based social facilitation of foraging.

Although clumped departures from the nest may seem apparent to observers in the field, such clumping is difficult to demonstrate rigorously (32). Pallett & Plowright (97) contend that, despite a subjective perception that arrivals and departures from *Vespula* colonies appear to be clumped, quantitative analyses of contagion in departures from wasp nests (12, 27, 97) have failed to provide irrefutable evidence of this clumping. Kasuya (70) similarly found no evidence that the return of foragers to nests of *Polistes chinensis antennalis* stimulated nestmates to leave the nest and forage. Such clumping of departures would be consistent with social facilitation of the passage of foragers through the nest entrance or of departure from the nest (128). Existing evidence, based upon patterns of departure from the nest, is instead consistent with the interpretation that foragers arrive and depart from the nest independently. However, tests for randomness in events are sensitive to the time interval chosen, and thus the issue of independence in arrivals and departures remains unresolved (RL Jeanne, personal communication).

Social wasps in the swarm-founding Polistinae are able to use sternal gland secretions, deposited upon landmarks between the old and new nest site, to direct swarming nestmates to new nest sites (60, 63, 65). Secretions of mandibular glands are used in a similar manner by stingless bees (Meliponinae) to direct nestmates to food sources (76, 83). Swarm-founding wasps could potentially use trail pheromones, with which they direct the movement of swarms, in the context of foraging, but they are not known to do so (65). In fact, social wasps are the only group of eusocial insects (bees, wasps, ants, and termites) in which communication of distance and direction to food have not been reported (63). Despite the lack of recruitment mechanisms providing distance and direction information to nestmates, evidence exists for some simple mechanisms of nest- and field-based recruitment in social wasps.

Nest-Based Mechanisms of Recruitment

In studies of foraging *Metapolybia azteca, M. docilis, M. cingulata, M. suffusa, Polybia occidentalis, P. diguetana, Parachartergus fraternus, Synoeca surinama, Brachygastra lecheguana,* and *Agelaia pallipes*, Forsyth found no evidence of recruitment to a specific foraging site (32). However, he stated that activity of returning and departing foragers may stimulate other wasps to leave the nest, and that workers grooming foragers that had just returned to the nest might obtain odor cues that would alert them to the availability of a particular type of food. These odor cues could serve as an important aid in food location. Naumann (90) suggested this function for the rapid running on the nest envelope and trophallactic exchanges with nestmates by returning *Protopolybia pumila* foragers, although Akre (1) observes that there is no direct evidence for this interpretation. Lindauer (76) proposed that *Polybia scutellaris* foragers may similarly alert nestmates to the presence of food. These authors suggest functional interpretations consistent with observed behavior. However, as discussed by Jeanne et al (65), experimental details and quantitative results are scant.

Recent work by Overmyer & Jeanne (95) provides the clearest experimental evidence to date of a nest-based mechanism of recruitment in a social wasp. In their study, yellowjacket (*V. germanica*) foragers were presented with a choice between a dilute corn syrup solution scented the same as that being carried to the nest by trained foragers, and a corn syrup solution bearing a different scent. Two series of tests were conducted. In the first, vanilla was the test scent and strawberry the control; in the second series, the odors of the training and control solutions were reversed. In both series, foragers were significantly more likely to visit feeders that provided the same olfactory cues as the food brought back to the nest by the trained wasps. Unlike past studies (for example, 77), Overmyer & Jeanne (95) controlled for the possibility of local enhancement of foraging (128), in which foragers might be attracted to a particular resource by the actions or presence of conspecifics (discussed in 112). In Overmyer and Jeanne's experiment (95), naive foragers did not encounter experienced foragers at the feeders as they approached the experimental setup. In addition, by frequently replacing the feeders used in the experiments, Overmyer & Jeanne (95) controlled for the effects of a food site marking substance deposited by feeding *V. germanica* (SL Overmyer & RL Jeanne, unpublished data). Their work clearly demonstrates the existence of nest-based, food-odor mediated recruitment of *V. germanica* foragers.

In contrast, experimental tests by Jeanne et al (65) on the neotropical wasp *Agelaia multipicta* did not provide evidence for recruitment of foragers to carrion in this species, either by communication of information at the nest, laying scent trails, or by flying as a group to the foraging site, as observed by Ishay et al (56) in *Vespa orientalis*. Although their experimental protocol did not directly test for local enhancement of foraging, there was no indication that such enhancement might be occurring at the feeding site.

Field-based Mechanisms of Recruitment

Several studies have reported a tendency for social wasp foragers to distribute themselves non-randomly on carbohydrate (34, 95, 100, 110) and protein (32, 33, 106, 112) resources. Forsyth (32) suggested that the tendency of *Agelaia pallipes* foragers to accumulate on one of two identical meat baits must be due in part to visual signals, and that aggregations of foragers may be attractive because they advertise the quality of resources. SL Overmyer & RL Jeanne (unpublished data) demonstrate that a food site marking substance is attractive to conspecific *Vespula germanica* foragers approaching unoccupied resources; odor cues emanating from or deposited by feeding wasps, in conjunction with visual cues provided by their presence, may also be an important component of forager attraction.

Parrish & Fowler observed that *Vespula germanica* and *Vespula maculifrons* foragers responded differently to carbohydrate feeders surrounded by varying numbers of a mixture of freshly killed *V. germanica* and *V. maculifrons* foragers or by hand-painted pushpins serving as odor-free wasp models (100). *Vespula germanica* foragers tended to land and feed preferentially on feeders having the highest number of pinned wasps, while *V. maculifrons* foragers preferentially avoided pinned wasps, landing on unoccupied feeders or those having the fewest pinned wasps. Similar patterns of response to the painted pushpins suggest that visual cues are sufficient to explain the observed distribution of foragers.

Parrish & Fowler (100) apply Wilson's definition of social facilitation, "behavior initiated or increased by the action of another individual," to the foraging behavior of the *V. germanica* foragers (137). More recent studies (96, 110, 112) suggest that local enhancement (128), a variant of social facilitation in which an animal's attention is directed to a particular location or object by the actions or presence of conspecifics, is a more accurate descriptor of this aggregation tendency (110, 112).

In further investigations of the mechanisms involved in the discovery of food and the attraction of foragers to foraging conspecifics, Reid et al (112) observed *Vespula germanica* and *V. maculifrons* foragers scavenging on traps baited with meat. Contrary to the observations of Parrish & Fowler (100), baits upon which confined groups of conspecifics were visible were more attractive to conspecifics not only of *V. germanica* foragers, but also to *V. maculifrons* foragers, who had selectively avoided carbohydrate feeders occupied with freshly killed wasps or painted pushpins in Parrish & Fowler's trials (100).

Reid et al (112) hypothesized that local enhancement of foraging is characteristic of wasps in the *V. vulgaris* species group. Wasps in the *V. vulgaris* species group characteristically have large, well-populated nests, and foragers have a broad diet that includes both natural and anthropogenic resources (43); these wasps are often pests on human food. In contrast, wasps in the *V. rufa* species group have shorter colony duration and smaller colony size and are more inclined to forage on live arthropod prey and natural carbohydrate resources (43), although they will also forage at caterpillar baits (107) and honey feeders (110).

Using hexane-extracted, dried, posed wasps to provide visual cues at feeders, Raveret Richter & Tisch studied the possible role of local enhancement on carbohydrate foraging in the paper wasp *Polistes fuscatus* and in the yellowjackets *V. germanica, V. maculifrons, V. flavopilosa, V. vidua,* and *V. consobrina* (110). They compared the foraging behavior of the *Polistes* foragers with that of yellowjackets in the *V. vulgaris* (including *V. germanica, V. maculifrons,* and *V. flavopilosa* foragers) and *V. rufa* (including *V. vidua* and *V. consobrina*) species groups. Both *P. fuscatus* and *V. germanica* preferentially fed on feeders and flowers with posed wasps.

Reid et al (112) hypothesized that the ecological success of species in the *V. vulgaris* group relative to species in the *V. rufa* group is due both to their tendency to scavenge and their ability to recruit workers to scavenging sites by means of local enhancement. The preference of *V. germanica* foragers (*V. vulgaris* species group) for occupied feeders in the studies of Parish & Fowler (100) and Raveret Richter & Tisch (110) is consistent with these predictions. In the experiments of Raveret Richter & Tisch (110), where wasps visited a feeder array having both unoccupied feeders and occupied feeders on which there were unoccupied flowers, as predicted *V. germanica* visited occupied flowers on occupied feeders and fed next to the largest wasp on the feeder. Consistent with the predictions of Reid et al (112) for foragers of the *V. rufa* species group, *V. consobrina* foragers preferentially visit unoccupied feeders, and if either *V. consobrina* or *V. vidua* foragers land on occupied feeders, they preferentially visit unoccupied flowers at those feeders. However, *V. maculifrons* (*V. vulgaris* species group) foragers selectively avoided occupied feeders and flowers, consistent with the earlier results of Parrish & Fowler (100) and contrary to the predictions of Reid et al (112).

Differences between the findings of Reid et al (112) and those of other authors on the foraging behavior of wasps in the *V. vulgaris* species group may be a result of Reid et al's (112) use of live wasps, which can move and provide odor cues, on the baits used in their tests. Other studies (100, 110) used freshly killed or hexane-extracted pinned wasps or models in which all visual cues were stationary, and odor was not always present. Stationary visual cues may be sufficient for local enhancement in *V. germanica* foragers, but may elicit avoidance by *V. maculifrons* foragers. Scale may also be important: *V. maculifrons* foragers avoided other foragers when foraging on a large array, but showed local enhancement of foraging when presented with a choice among only two or three closely spaced feeders (discussed in 110). Geographic or population level differences may underlie these behavioral differences (RL Jeanne, personal communication). The studies of Reid et al (112) were conducted in the Midwest, while those of Parrish & Fowler (100) and Raveret Richter & Tisch (110) were conducted on the East

Coast of the United States. Finally, Reid et al (112) used protein baits in their studies, whereas all of the work demonstrating local enhancement of *V. maculifrons* foragers used carbohydrate baits—foragers may behave differently on different types of resources.

Keyel (71) suggested that the ability of the introduced *V. germanica* to monopolize resources, despite the fact that these wasps are displaced from resources in one-on-one encounters with native *V. maculifrons* and *V. flavopilosa* foragers, is due to their tendency for local enhancement. *V. germanica* foragers are recruited to a resource more rapidly than are *V. maculifrons* or *V. flavopilosa* foragers. Once present, *V. germanica* foragers are able to maintain control of the site because of their increased tendency to grapple with intruders at resources.

Why is Recruitment Poorly Developed in Social Wasps?

Compared to the behavior of other social insects, the recruitment of social wasps to resources is poorly developed. Despite possession of mechanisms, such as trail pheromones (60), that could enable communication of distance and direction information, and despite the potential advantages that could accrue from efficient recruitment of nestmates directly to specific locations, social wasps do not recruit. They remain the only group of eusocial insects in which communication of distance and direction to food have not been reported (63). Even less sophisticated forms of recruitment, such as local enhancement, are not necessarily the norm. Many wasp species (106, 110) preferentially avoid occupied resources. The question, then, is why aren't social wasps more inclined to recruit? Jeanne et al (65) reviewed some possible explanations, including social constraints on recruitment (is it necessary to pass a threshold colony size in order to make recruitment profitable?), lack of ability to store proteinaceous food (might social wasp species that store nectar recruit to carbohydrate sources?), and the influence of ecological factors.

Ecological factors including the distribution and abundance of resources, competitors, and predators are likely to be important determinants of foraging strategies. For example, recruitment activities, because of the time involved, could leave resources undefended for extended periods of time. If resources are generally of a size that can be carried back to the nest by an individual in a few trips, those resources might be most efficiently harvested—and left unattended and vulnerable to theft at the capture site for the least amount of time—if a forager were to gather a load, carry it to the nest, quickly hand off her load to a nestmate, and immediately return to the resource in the field (106, 109). A forager's ability to learn, during an orientation flight, the landmarks associated with the location of the resource (18) enables her to return to a resource efficiently.

Very large resources present the wasps with stiff competition from able recruiters such as ants and stingless bees, and ants may also pose a predation risk for the wasps (65). Mobilizing groups of wasps large enough to hold off these foes would make the wasps, clumped offensively or defensively on resources, less able to disperse themselves widely over the landscape, and thus less likely to encounter new resources. In discussing the foraging strategies of *Trigona* bees, Johnson (67) reasoned that opportunistic species with large forager forces that search independently are more likely to find new resources first. She suggests that environments presenting many small resources and rare, large, high-quality resources of transient availability would select for an independent, opportunistic foraging strategy, rather than recruitment to and defense of resources. This type of resource distribution may well be consistent with that encountered by many social wasp species.

INTERACTIONS AND ACCESS TO RESOURCES

Interactions among foragers are important determinants of whether wasps gain and/or retain control of resources in the field (71, 106). The previous section documents the use of visual and olfactory cues in recruitment of foragers. In order to ascertain the roles of particular cues in recruitment, many of these experiments used odors, stationary visual cues, or restrained wasps that could not freely interact with incoming foragers (96, 100, 110, 112). Unrestrained wasps that encounter one another on resources may behave agonistically or tolerantly. The outcomes of these interactions determine which foragers are permitted or denied access to which resources.

Interactions on Carbohydrate Resources

Keyel (71), Parrish & Fowler (100), and Parrish (99) observed agonistic interactions among foraging yellowjackets on carbohydrate resources, suggesting that wasps compete for these resources (71). In agonistic interactions, an approaching forager may fly close to a feeding wasp or hover and contact a feeding wasp with her legs, two wasps may hover face to face and fly upward together, or wasps may grapple, which is sometimes accompanied by sting attempts (71, 99). Keyel (71) observed yellowjackets foraging on trees with scale insects producing honeydew. He artificially enriched these resources by adding basswood honey to randomly selected leaves. Wasps tended to have the lowest percentage of aggressive encounters on natural or highly enriched leaves and had their highest percentage of agonistic interaction on slightly enriched leaves.

Parrish (99), in observations of *Vespula maculifrons* and *V. germanica* foragers of known nest origin, found that interspecific aggression was greater than intraspecific aggression. Within a species, nestmates were not treated preferentially if encountered on a resource. Levels of aggression were related to bait size and persistence—wasps accumulated more quickly on large baits. While the number of wasps increased on a feeder over time, the number of aggressive acts per sample period increased and then decreased at high densities, where the cost of fighting was likely to outweigh any possible benefits. Parrish (99) suggested that the probability of encountering a nestmate on a resource when it is first available and when wasps are most likely to fight to gain access to the resource is low, and thus foraging yellowjackets should always defend food aggressively in this situation, leading to the unexpected aggression among nestmates at resources.

Interactions on Prey

Raveret Richter (106) experimentally determined the effects of wasp species, nest affiliation, arrival order and prey size on the agonistic behavior of *Polybia occidentalis* and *Polybia diguetana,* two species of Neotropical, swarm-founding social wasps. Foragers of both species hovered near pieces of *Bagisara repanda* (Lepidoptera: Noctuidae) prey occupied by another *Polybia* forager in preference to unoccupied prey, but their subsequent responses differed. Foragers of *P. occidentalis,* the larger of the two species, tended to land with other *Polybia* foragers and gained access to prey by either displacing the resident wasp or splitting the prey with her. In contrast, *Polybia diguetana* foragers did not land preferentially on occupied prey, and in only 2 out of 50 trials, both when landing with a conspecific, did they successfully displace a resident wasp from her prey.

In intraspecific and interspecific tests, tolerance among wasps varied with prey size. The greatest percentage of highly agonistic activity occurred when a resident wasp was on the largest piece of prey that she had the potential of carrying away; beyond that size, levels of agonistic activity decreased. The overall percentage of agonistic behavior in trials was lower in cases where the larger wasp, *P. occidentalis,* arrived first at the prey. Nest affiliation did not influence agonistic interactions between foragers; there was no preferential treatment of nestmates by conspecific foragers.

Wasps appeared to behave according to the following rules: If a forager was able to fly off with a piece of prey and thus avoid conflict, she departed with the prey; she fought if she had a chance of excluding others and monopolizing the prey; two foragers would split large prey. The defensibility of the prey and the wasps' ability to opportunistically take advantage of this defensibility structured the interactions between wasps.

When tracking hunting wasps in the field, Raveret Richter (105, 106) often observed foragers flying toward other wasps standing stationary on vegetation. Large *Polistes* foragers approaching smaller *Polybia* would often pounce upon the smaller wasp, and if the smaller *Polybia* had prey the larger *Polistes* would take it from her. Smaller wasps approaching larger wasps that had prey often hovered, backed away, and repeatedly re-approached the site until the resident wasp had flown. These observations, combined with the experiments described here and in the previous section, suggest that the presence of other wasps on prey may increase the chance of a forager perceiving or locating the prey. However, one might consider this preferential attentiveness and, when possible, landing and feeding on the prey, as opportunistic theft of prey, rather than facilitation or enhancement of foraging. Theft might even be a foraging tactic employed facultatively by individual foragers. LJ McPheron (unpublished observation) reports that, in greenhouse studies of *Mischocyttarus flavitarsis,* she observed wasps chasing one another at sites where food was presented, and one individual appeared to specialize on taking prey from other wasps rather than hunting for her own prey.

Competition for prey need not require direct interactions among foragers. Raveret Richter (105) observed a *Mischocyttarus immarginatus* forager kill a large caterpillar and fly to the nest with a portion of the flesh. While the *M. immarginatus* was at the nest, a *Polistes instabilis* forager landed and took a portion of the prey. The *P. instabilis* left, and *M. immarginatus* returned within three minutes, took a portion of the prey, and flew. While the prey was unoccupied, it was discovered by a *Polybia diguetana* forager who bit off a piece of flesh and flew with it as a *P. occidentalis* forager approached the carcass. This example clearly illustrates how, while prey is left unattended as a forager transports loads to the nest, it can be exploited by other wasps.

INFLUENCES OF WASP FORAGING ON COMMUNITIES

Selective Pressures on Lepidopteran Prey

Lepidopteran larvae, most of which are specialist herbivores that feed on plants in one or a few genera or in a single family or subfamily (9), are an important prey source for many social wasps (1). In choice tests presenting pairs of caterpillars, one a generalist and one a specialist, to foraging *Mischocyttarus flavitarsus,* Bernays (9) found that the generalist caterpillars were preyed upon more readily than were the specialists. Subsequent tests with other generalist predators such as the predatory ant *Paraponera clavata* (26), the Argentine ant *Linepithema humile* Mayr (=*Iridomyrmex humilis*, 116a) (11), and the coccinellid beetle *Hippodamia convergens* (10) also demonstrate preferential selection of generalist caterpillars by these predators. This preference might be partially explained by the association of host-plant specialization by caterpillars with either the sequestration of host plant phytochemicals or the presence of toxic leaf material in the gut of these specialist herbivores (13, 85). Preference by the above-mentioned predators for generalist prey supports the hypothesis that natural enemies such as social wasps provide selective pressures that influence the host-plant specificity of their herbivore prey (9, 26, 118).

Generalist caterpillars tend to have cryptic coloration and behavior (121). Specialist caterpillars, on the other hand, are sometimes warningly colored and unpalatable, and thus aposematic and conspicuous (13). Aposematic caterpillars may be less likely to be attacked by most predators (13), but this is not always the case. Stamp (118) demonstrated that caterpillars of the non-cryptic, specialist buck moth *Junonia coenia* (Lepidoptera: Nymphalidae), which has passive, chemical defenses, were actually more susceptible to predation by wasps (*Polistes fuscatus*) and predatory stinkbugs (*Podisus maculiventris*) than were caterpillars of the cryptic generalist tiger moth *Spilosoma congrua* (Lepidoptera: Arctiidae), which employs crypsis and rapid fleeing as protection from predators. These invertebrate predators were undeterred by the buck moths' passive, chemical defenses. As they gained experience, however, wasp foragers began to reject the specialist forager, although their rate of predation on the cryptic generalist remained constant (85, 118).

One might expect gregarious, unpalatable caterpillars to be protected from vertebrate predation (22). However, defenses effective against vertebrate predators are not always effective against invertebrates. In field observations on social wasp (*Polistes dominulus* and *P. fuscatus*) foragers and gregarious, conspicuous, black and spiny *Hemileuca lucina* (Saturniidae) larvae, the wasps were undeterred by the urticating spines of the caterpillars, and killed 77% to 99% of the caterpillars at the study site (13, 119). Upon killing the caterpillars by biting, *Polistes* foragers proceeded to process the caterpillar flesh by biting off the spines, and then transported loads of caterpillar flesh back to their nests (105). Because these wasps tend to return to sites of past hunting success, aggregated caterpillars are particularly vulnerable to depredation by returning foragers.

In the case of *Hemileuca lucina* caterpillars, attacks by *Polistes* wasps and other natural enemies not only result directly in caterpillar deaths by predation (119), but indirectly reduce the survival of the caterpillars by inducing alterations in the foraging behavior of the caterpillars. Aggregations of the caterpillars are disrupted (22) and the caterpillars are forced to forage in cooler microhabitats with poorer-quality leaves. This alteration results in poorer larval growth (119) and reduces survivorship. Of the reduction in *H. lucina* survivorship attributed to the presence of *Polistes* foragers, one-third was due to these indirect effects (120), which may have their greatest impact upon easy-to-locate, warningly colored caterpillars that are gregarious (13).

Social Wasps as Introduced Species

Humans have inadvertently introduced non-native social wasps into many areas: for example, *Vespula germanica* is thought to have been accidentally brought to New Zealand during World War II in a crate of airplane parts (124). Vespinae in the *Vespula vulgaris* species group (wasps having large, long-lived colonies) are the most apparent of these introductions, both because of the sheer abundance of the wasps, and because of their pestiferous interactions with humans (2). Introductions of social wasps with smaller colony size, such as the European paper wasp *Polistes dominulus,* recently introduced into the Boston vicinity (48), receive lesser notice. However, the ecological impacts of all introduced wasps, including competition with (45) and predation upon (37) native species, also warrant attention.

V. germanica, native to Europe, has become established in many locations outside of its native range (2, 71). Most of these areas had no native yellowjackets

(71). However, in the northeastern United States where it was introduced in the late 1960s (88), *V. germanica* has become common despite the presence of nine native yellowjackets, three of which are close relatives (71). The establishment of *V. germanica* under these circumstances may have occurred in part because of its tendency to nest within the walls of houses (71, 88) and in part due to its numerical dominance on rich, human-related food sources (71) to which it recruits by both nest-based, odor-mediated (95) mechanisms, and by field-based mechanisms that include marking of feeding sites with a substance that is attractive to conspecific foragers (Sl Overmyer & RL Jeanne, unpublished data) and local enhancement of foraging (71, 100, 110, 112).

Vespula colonies typically are annual and are initiated by a single queen (2). However, in the warmer portions of the wasps' natural range and in some mild climates where they have been introduced, *Vespula* may attain enormous colony size by overwintering, enabling the nests and their resident populations to grow much larger than those of the typical annual colonies and leading to an extended foraging season (43). A high incidence of overwintering colonies occurs in Maui, where *Vespula pensylvanica* became established in 1978. These colonies exert a much greater level of predation pressure on local fauna than the level of pressure expected from smaller annual colonies with shorter seasonal windows of activity (38). In addition, unlike most other invasive species that have colonized the low-elevation coastal regions of the Hawaiian archipelago, *V. pensylvanica* is best established at high elevations, at sites corresponding with refuges of high endemicity, and the wasps include endemic arthropods as a substantial portion of their diet (37, 38). Small populations of endemic arthropods, having evolved in the absence of vespine predation, may be unable to recover from such perturbations (37).

In New Zealand, where *V. germanica* was introduced in approximately 1945 (47), these wasps may build huge perennial colonies attached to tree trunks (117). *Vespula vulgaris,* introduced to New Zealand in the late 1970s, is replacing *V. germanica* in honeydew beech forest habitats (46, 47). *V. vulgaris* foragers have great dietary overlap with native insectivorous birds (7, 45). In the honeydew beech forests of the Nelson region on New Zealand's South Island, the estimated current biomass of *V. vulgaris* in the community, 5,200 g/ha at the peak of colony growth, is greater than the maximum estimate of the combined biomass of birds, rodents, and stoats, approximately 1150g/ha (127). In the northern portion of South Island, Harris estimates prey harvest by these wasps at the rate of 8.1 kg/ha during the portion of the year when they are active, an amount similar in magnitude to the approximate yearly consumption by the entire bird fauna in an equivalent area (45). Carbohydrate collection by the wasps, primarily from secretions of the abundant beech scale insect *Ultracoelostoma assimile* (Homoptera: Margarodidae), is estimated to be at the rate of 343 liters/ha (45), which greatly reduces both the abundance and quality of honeydew available for native wildlife and makes the resource virtually unavailable to birds for three to four months of each year (84). If invertebrate prey or carbohydrate honeydew availability are limiting resources for native birds, there is clear potential for the wasps to have a negative impact on bird populations (7, 45, 84).

Social Wasps in Biological Control

Virtually all examples of successful pest control using social wasps involve the control of caterpillars. The presence of *Polistes* wasps has been associated with decreased damage from lepidopterous pests of cotton (6, 55, 72), tobacco (75, 104), and cabbage (40, 103). Studies in Brazil on coffee plantations showed that *Polybia occidentalis* and *Brachygastra lecheguana* predation on larvae can be an important factor in checking outbreaks of the coffee leaf miner [*Perileucoptera coffeella* (Lepidoptera: Lyonetiidae; 42, 98)]. Fall webworms [*Hyphantria cunea* (Lepidoptera: Arctiidae; 87, 92)] and gypsy moth larvae [*Lymantria dispar* (Lepidoptera: Lymantriidae; 36)] are also preyed upon by *Polistes* wasps, demonstrating that the wasps are not necessarily thwarted by spines or silk webbing. *Polistes* wasps have also been used in conjunction with pesticides to control crop pests (75). In plots the size of a home garden, *Polistes* foragers have a significant impact on caterpillar populations (40; MA Raveret Richter, personal observation).

Certain features of the feeding biology of social wasps seem especially well-suited to dealing with outbreaks of pest species. The wasps return repeatedly to sites where they have had success in feeding. This behavior would tend to concentrate them disproportionately in areas where there are high caterpillar densities (89). Group-feeding caterpillars may disperse in response to harassment or attacks by wasps (22) and thereby escape wasp predators, suggesting that wasps might be of limited effectiveness in controlling caterpillars with this escape response. However, short-term avoidance of direct predation may come at a cost: If caterpillars escape danger by leaving their host plants or moving to less favorable foraging sites, their growth rate may decrease and larval development time may increase, broadening their window of vulnerability to parasites, predators, diseases, and inclement weather. Stamp & Bowers suggested that such indirect effects on caterpillars may, in addition to the effects of direct predation, contribute to decreases in the populations of pest caterpillars (119).

Wasps have clearly demonstrated promise in providing some degree of biological control of pests under circumstances described here. Unfortunately, the life cycles of the wasps are not always well synchronized with those of sympatric pest Lepidoptera. For example, gypsy moth outbreaks in northeastern North America occur long before social wasps in the area have experienced their seasonal population buildup. In addition, unlike species specific parasitoids, these generalist predators may switch to alternate, more abundant prey when they have decreased the pest population, leaving the target species at higher densities than those required for effective pest control (40).

Methods have been developed for managing colonies of wasps for purposes of biological control (39). However, social wasps vary greatly in their life histories, behavior, and ecology. Many different situation-specific management techniques will need to be developed to fully exploit the biological control potential of these insects to the same degree that, for example, we exploit the behavior of managed colonies of honey bees in crop pollination.

CONCLUSIONS

Water, prey, and carbohydrate availability can limit the productivity of social wasp colonies (53, 113). The studies on which this review focuses demonstrated enormous variability both within and among social wasp species in the mechanisms of foraging for these resources. Although as species social wasps are generalist foragers, at any given time individual foragers may specialize on particular foraging tasks (91), foraging locations (106), and similar to many solitary wasps, on specific prey types (106). Recruitment to resources is poorly developed in these social insects, despite the existence of behavioral mechanisms such as trail pheromones that could be used by swarm-founding social wasps for this purpose. Social wasps associate cues with particular resources, respond to the presence of other foragers on resources, and are influenced by their past foraging experiences. Better understanding of the relationship between environmental conditions and wasp foraging behavior is key to understanding the foraging strategies and tactics of these wasps, including how they use cues to locate and choose resources, and how they respond to changes in resource distribution and abundance. Greenhouse or screen enclosure-based studies of wasps would provide opportunities to follow individual foraging histories and to manipulate resource distribution and abundance under controlled conditions. Such studies could be useful in clarifying behavioral mechanisms of foraging.

Social wasps, by virtue of their behavior and their numbers, can have an enormous impact upon other organisms. Predation by social wasps has both direct and indirect effects on prey (37, 119). A clearer understanding of these effects will prove useful in evaluation of the potential impacts of generalist predators on the evolution of their prey. In addition, an improved understanding of the role of social wasps in communities will have both predictive and possibly ameliorative potential in areas where non-native wasps have been introduced and are having impacts upon the native fauna. Such knowledge may also prove useful in employing the wasps as biocontrol agents in integrative pest management.

Social wasps possess a wide variety of life histories and social systems, which, combined with the varying ecological circumstances in which the wasps occur, present a rich opportunity for comparative studies to illustrate the relationships among social systems, foraging behavior, and the distribution and abundance of resources. Work so far has illustrated many of the possibilities. General patterns, such as the apparent relationship between life history and foraging behavior in the *Vespula vulgaris* and *V. rufa* species groups of yellowjackets (2, 43), have begun to emerge. Even among these vespines, however, additional observation (107, 110) reveals exceptions to past generalizations (2) and predictions (112). Additional data on the mechanisms and patterns of foraging, the variability within and among species, and the relationship between environmental parameters and foraging behavior, particularly in less well-known groups such as the Stenogastrinae and the diverse, swarm-founding Polistinae, is necessary in order to distill general patterns and enable a synthetic view of the foraging behavior and ecology of social wasps.

ACKNOWLEDGMENTS

Thanks to Mary Jane West Eberhard for encouraging me to embark on this project, and to George Eickwort, Robert Jeanne, and Mary Jane West Eberhard for insights, ideas, inspiration, and support as my wasp research has progressed. The Hagnauer, Hagmann, West Eberhard and Joyce families have shared much appreciated hospitality, logistical support and local savvy during my tropical wasp research. I would like to thank Linda McPheron, Robert Jeanne, Wayne Richter, and anonymous reviewers for their helpful comments on this manuscript. Eric Olson, Robert Jeanne, Linda McPheron, Stephanie Overmyer, Linda Rayor, and Karin Warkentin have generously shared pre-publication manuscripts and observations with me. Thanks to my students and colleagues at Skidmore College and my family for their patience while I have been immersed in this project, and to Mino's and Mrs. London's for fulfilling my family's foraging needs when it counted the most.

LITERATURE CITED

1. Akre RD. 1982. Social wasps. In *Social Insects,* ed. H Hermann, 4:1–105. New York: Academic. 385 pp.
2. Akre RD, Green A, MacDonald JF, Landolt PJ, Davis HG. 1980. *The Yellowjackets of America North of Mexico. US Dep. of Agric. Handb.* No. 552. 102 pp.
3. Akre RD, Myhre EA. 1992. Nesting biology and behavior of the baldfaced hornet, *Dolichovespula maculata* (L.) (Hymenoptera: Vespidae) in the Pacific Northwest. *Melanderia* 48:1–33
4. Alexander B. 1985. Predator-prey interactions between the digger wasp *Clypeadon laticinctus* and the harvester ant *Pogonomyrmex occidentalis. J. Nat. Hist.* 19:1139–54
5. Alexander RD. 1974. The evolution of social behavior. *Annu. Rev. Ecol. Syst.* 5:325–83
6. Ballou HA. 1915. West Indian wasps. *Agric. News* 14:298
7. Barr K, Moller H, Christmas E, Lyver P, Beggs J. 1996. Impacts of introduced common wasps (*Vespula vulgaris*) on experimentally placed mealworms in a New Zealand beech forest. *Oecologia* 105:266–70
8. Barth FG. 1985. *Insects and Flowers: The Biology of a Partnership.* Princeton, NJ: Princeton Univ. Press. 297 pp.
9. Bernays EA. 1988. Host specificity in phytophagous insects: selection pressure from generalist predators. *Entomol. Exp. Appl.* 49:131–40
10. Bernays EA. 1989. Host range in phytophagous insects: the potential role of generalist predators. *Evol. Ecol.* 3:299–311
11. Bernays EA, Cornelius ML. 1989. Gen-

eralist caterpillar prey are more palatable than specialists for the generalist predator *Iridomyrmex humilis. Oecologia* 79:427–30
12. Blackith RE. 1957. Social facilitation at the nest entrance of some Hymenoptera. *Physiol. Comp. Oecol.* 4:388–402
13. Bowers DM. 1993. Aposematic caterpillars: life-styles of the warningly colored and unpalatable. In *Caterpillars: Ecological and Evolutionary Constraints on Foraging,* eds. NE Stamp, TM Casey, pp. 331–71. New York: Chapman & Hall. 587 pp.
14. Braverman Y. 1998. Increasing parasitism by the German yellow jacket wasp, *Paravespula germanica,* on dairy cattle in Israel. *Med. Vet. Entomol.* 12:192–95
15. Caron DM, Schaefer PW. 1986. Social wasps as bee pests. *Am. Bee J.* 126:269–71
16. Carpenter JM. 1982. The phylogenetic relationships and natural classification of the Vespoidea (Hymenoptera). *Syst. Entomol.* 7:11–38
17. Carpenter JM. 1991. Phylogenetic relationships and the origin of social behavior in the Vespidae. See Ref. 112a, pp. 7–32
18. Collett TS. 1995. Making learning easy: the acquisition of visual information during the orientation flights of social wasps. *J. Comp. Physiol. A* 177:737–47
19. Collett TS. 1996. Insect navigation *en route* to the goal: multiple strategies for the use of landmarks. *J. Exp. Biol.* 199:227–35
20. Collett TS, Lehrer M. 1993. Looking and learning: a spatial pattern in the orientation flight of the wasp *Vespula vulgaris. Proc. R. Soc. London Ser. B* 252:129–34
21. Cornelius ML. 1993. Influence of caterpillar-feeding damage on the foraging behavior of the paper wasp *Mischocyttarus flavitarsis* (Hymenoptera: Vespidae). *J. Insect Behav.* 6:771–81
22. Cornell JC, Stamp NE, Bowers MD. 1987. Developmental change in aggregation, defense and escape behavior of buckmoth caterpillars, *Hemileuca lucina* (Saturniidae). *Behav. Ecol. Sociobiol.* 20:383–88
23. Cowan DP. 1991. The solitary and presocial Vespidae. See Ref. 112a, p. 33–73
24. Dew HE, Michener CD. 1978. Foraging flights of two species of *Polistes* wasps (Hymenoptera: Vespidae). *J. Kans. Entomol. Soc.* 51:380–85
25. Duncan C. 1939. A contribution to the biology of North American vespine wasps. *Stanford Univ. Publ. Biol. Sci.* 8:1–272
26. Dyer LA. 1995. Tasty generalists and nasty specialists? Antipredator mechanisms in tropical lepidopteran larvae. *Ecology* 76:1483–96
27. Edgeworth FY. 1907. Statistical observations on wasps and bees. *Biometrika* 5:365–86
28. Edwards R. 1980. *Social Wasps: Their Biology and Control.* Sussex, UK: Rentokil. 398 pp.
29. Eickwort GC. 1981. Presocial insects. In *Social Insects,* ed. H Hermann, 2:199–280. New York: Academic. 491 pp.
30. Evans HE, West Eberhard MJ. 1970. *The Wasps.* Ann Arbor: Univ. Mich. Press. 265 pp.
31. Faegri K, van der Pijl L. 1979. *The Principles of Pollination Ecology.* Oxford/New York/Toronto/Sydney/Braunschweig: Pergamon. 248 pp.
32. Forsyth AB. 1978. *Studies on the behavioral ecology of polygynous social wasps.* PhD diss. Harvard Univ., Cambridge, MA. 226 pp.
33. Fowler HG. 1992. Social facilitation during foraging in *Agelaia* (Hymenoptera: Vespidae). *Naturwissenschaften* 79:424
34. Free JB. 1970. The behavior of wasps (*Vespula germanica* L. and *V. vulgaris* L.) when foraging. *Insectes. Soc.* 17:11–19

35. Frost R. 1995. The white-tailed hornet or, the revision of theories. In *Collected Poems, Prose and Plays*. New York: Libr. Am., 1036 pp.
36. Furuta K. 1983. Behavioral response of the Japanese paper wasp (*Polistes jadwigae* Dalla Torre; Hymenoptera: Vespidae) to the gypsy moth (*Lymantria dispar* L.; Lepidoptera: Lymantriidae). *Appl. Entomol. Zool.* 18:464–74
37. Gambino P. 1992. Yellowjacket (*Vespula pensylvanica*) predation at Hawaii Volcanoes and Haleakala National Parks: identity of prey items. *Proc. Hawaii Entomol. Soc.* 31:157–64
38. Gambino P, Medeiros AC, Loope LL. 1987. Introduced vespids *Paravespula pensylvanica* prey on Maui's endemic arthropod fauna. *J. Trop. Ecol.* 3:169–70
39. Gillaspy JE. 1979. Management of *Polistes* wasps for caterpillar predation. *Southwest. Entomol.* 4:334–52
40. Gould WP, Jeanne RL. 1984. *Polistes* wasps (Hymenoptera: Vespidae) as control agents for lepidopterous cabbage pests. *Environ. Entomol.* 13:150–56
41. Grant J. 1959. Hummingbirds attacked by wasps. *Can. Field Nat.* 73:174
42. Gravena S. 1983. Táticas de manejo integrado do bicho mineiro do cafeeiro *Perileucoptera coffeella* (Geurin-Meneville, 1842): I-Dinâmica populacional e inimigos naturais. *An. Soc. Entomol. Bras.* 12:61–71 (In Portuguese)
43. Greene A. 1991. *Dolichovespula* and *Vespula.* See Ref. 112a, pp. 263–305
44. Hamilton WD. 1972. Altruism and related phenomena, mainly in social insects. *Annu. Rev. Ecol. Syst.* 3:193–232
45. Harris RJ. 1991. Diet of the wasps *Vespula vulgaris* and *V. germanica* in honeydew beech forest of the South Island New Zealand. *NZ J. Zool.* 18:159–69
46. Harris RJ, Moller H, Winterbourn MJ. 1994. Competition for honeydew between two social wasps in South Island beech forests, New Zealand. *Insectes Soc.* 41:379–94
47. Harris RJ, Thomas CD, Moller H. 1991. The influence of habitat use and foraging on the replacement of one introduced wasp species by another in New Zealand. *Ecol. Entomol.* 16:441–48
48. Hathaway MA. 1981. *Polistes gallicus* in Massachusetts (Hymenoptera: Vespidae). *Psyche* 88:169–73
49. Heinrich B. 1979. "Majoring" and "minoring" by foraging bumblebees, *Bombus vagans:* an experimental analysis. *Ecology* 60:245–55
50. Heinrich B. 1984. Strategies of thermoregulation and foraging in two vespid wasps *Dolichovespula maculata* and *Vespula vulgaris. J. Comp. Physiol. B* 154:175–80
51. Hendrichs J, Katsoyannos BI, Papaj DR, Prokopy RJ. 1991. Sex differences in movement between natural feeding and mating sites and trade-offs between food consumption, mating success and predator evasion in Mediterranean fruit flies (Diptera: Tephritidae). *Oecologia* 86: 223–31
52. Hendrichs J, Katsoyannos BI, Wornoayporn V, Hendrichs MA. 1994. Odour-mediated foraging by yellowjacket wasps (Hymenoptera: Vespidae): predation on leks of pheromone-calling Mediterranean fruit fly males (Diptera: Tephritidae). *Oecologia* 99:88–94
53. Horwood MA, Toffolon RB, Brown GR. 1993. Establishment and spread of *Vespula germanica* (F.) (Hymenoptera: Vespidae) in New South Wales and the influence of rainfall on its abundance. *J. Aust. Entomol. Soc.* 32:241–48
54. Hunt JH, Rossi AM, Holmberg NJ, Smith SR, Sherman WR. 1998. Nutrients in social wasp (Hymenoptera: Vespidae, Polistinae) Honey. *Ann. Entomol. Soc. Am.* 91:466–72
55. Inst. Agric. For. Sci., China. 1976. A preliminary study on the bionomics of hunting wasps and their utilization in cotton

insect control. *Acta Entomol. Sin.* 19:303–8

56. Ishay J, Bytinski-Salz H, Shulov A. 1967. Contributions to the bionomics of the Oriental hornet (*Vespa orientalis* Fab.). *Isr. J. Entomol.* 2:45–106
57. Iwata K. 1976. *Evolution of Instinct: Comparative Ethology of Hymenoptera.* New Delhi: Amerind. 535 pp.
58. Jeanne RL. 1972. Social biology of the Neotropical wasp *Mischocyttarus drewseni. Bull. Mus. Comp. Zool. Harvard* 144:63–150
59. Jeanne RL. 1980. Evolution of social behavior in the Vespidae. *Annu. Rev. Entomol.* 25:371–96
60. Jeanne RL. 1981. Chemical communication during swarm emigration in the social wasp *Polybia sericea* (Olivier). *Anim. Behav.* 29:102–13
61. Jeanne RL. 1986. The organization of work in *Polybia occidentalis:* costs and benefits of specialization in a social wasp. *Behav. Ecol. Sociobiol.* 19:333–41
62. Jeanne RL. 1987. Do water foragers pace nest construction activity in *Polybia occidentalis? Experientia* 54(Suppl.):241–51
63. Jeanne RL. 1991. The swarm-founding Polistinae. See Ref. 112a, pp. 191–231
64. Jeanne RL. 1996. Regulation of nest construction behavior in *Polybia occidentalis. Anim. Behav.* 52:473–88

64a. Jeanne RL, Davidson DL. 1984. Population regulation in social insects. In *Ecological Entomology*, eds. CB Huffaker, RL Rabb, pp. 559–590. New York: Wiley Interscience. 844 pp.

65. Jeanne RL, Hunt JH, Keeping MG. 1995. Foraging in social wasps: *Agelaia* lacks recruitment to food (Hymenoptera: Vespidae). *J. Kans. Entomol. Soc.* 68:279–89
66. Jennings DT, Houseweart MW. 1984. Predation by eumenid wasps (Hymenoptra: Vespidae) on spruce budworm (Lepidoptera: Tortricidae) and other lepidopterous larvae in spruce-fir forests of Maine. *Ann. Entomol. Soc. Am.* 77:39–45
67. Johnson LK. 1983. Foraging strategies and the structure of stingless bee communities in Costa Rica. In *Social Insects in the Tropics,* ed. P Jaisson, 2:31–58. Paris: Univ. Paris-Nord. 252 pp.
68. Jules ES. 1996. Yellow jackets (*Vespula vulgaris*) as a second seed disperser for the myrmecochorous plant, *Trillium ovatum. Am. Midl. Nat.* 135:367–69
69. Kasuya E. 1982. Central place water collection in the Japanese paper wasp, *Polistes chinensis antennalis. Anim. Behav.* 30:1010–14
70. Kasuya E. 1984. Absence of social facilitation in foraging of workers of the Japanese paper wasp, *Polistes chinensis antennalis. J. Ethol.* 2:139–40
71. Keyel RE. 1983. *Some aspects of niche relationships among yellowjackets (Hymenoptera: Vespidae) of the northeastern United States.* PhD diss. Cornell Univ., Ithaca, NY. 161 pp.
72. Kirkton RM. 1970. Habitat management and its effects on populations of *Polistes* and *Iridomyrmex. Proc. Tall Timbers Conf.,* 2:243–46
73. Koeniger N, Koeniger G, Gries M, Tingek S, Kelitu A. 1996. Observations on colony defense of *Apis nuluensis* Tingek, Koeniger and Koeniger, 1996 and predatory behavior of the hornet *Vespa multimaculata* Pérez, 1910. *Apidologia* 27:341–52
74. Lacey LA. 1979. Predação em girinos por uma vespa e outras associações de insetos com ninhos de duas espécias de rãs de Amazônia. *Acta Amaz.* 9:755–62 (In Portuguese)
75. Lawson FR, Rabb RL, Guthrie FE, Bowery TG. 1961. Studies of an integrated control system for hornworms on tobacco. *J. Econ. Entomol.* 54:93–97
76. Lindauer M. 1971. *Communication Among Social Bees.* Cambridge, MA: Harvard Univ. Press. 161 pp.
77. Maschwitz U, Beier W, Deitrich I, Keidel W. 1974. Futterverständigung bie Wes-

pen der Gattung *Paravespula. Naturwissenschaften* 61:506
78. Matsuura M. 1968. Ecology of *Vespa* in Japan. *Jpn. Bee J.* 21:319–22 (In Japanese)
79. Matsuura M, Sakagami S. 1973. A bionomic sketch of the Giant Hornet *Vespa mandarinia,* a serious pest for Japanese apiculture. *J. Fac. Sci. Hokkaido Univ. VI (Zool.)* 19:125–62
79a. Matsuura M. Sakagami S. 1984. *Biology of the Vespine Wasps.* Berlin: Springer-Verlag. 323 pp.
80. Matthews RW, Matthews JR. 1978. *Insect Behavior.* New York: Wiley & Sons. 507 pp.
81. McPheron LJ. 1996. *The role of learning in the foraging behavior of the paper wasp Mischocyttarus flavitarsis.* PhD diss. Univ. Calif., Berkeley.
82. McPheron LJ. 1999. Learning colour and odour in a paper wasp, *Mischocyttarus flavitarsis. Anim. Behav.* In press.
83. Michener CD. 1974. *The Social Behavior of the Bees.* Cambridge, MA: Belknap Press Harvard. 404 pp.
84. Moller H, Tilley JAV, Thomas BW, Gaze PD. 1991. Effect of introduced social wasps on the standing crop of honeydew in New Zealand beech forests. *NZ J. Zool.* 18:171–79
85. Montllor CB, Bernays EA. 1993. Invertebrate predators and caterpillar foraging. In *Caterpillars: Ecological and Evolutionary Constraints on Foraging,* ed. NE Stamp, TM Casey, pp. 170–202. New York: Chapman & Hall. 587 pp.
86. Montllor CB, Bernays EA, Cornelius ML. 1991. Responses of two hymenopteran predators to surface chemistry of their prey: significance for an alkaloid-sequestering caterpillar. *J. Chem. Ecol.* 17:391–99
87. Morris RF. 1972. Predation by wasps, birds and mammals on *Hyphantria cunea. Can. Entomol.* 104:1581–91
88. Morse RA, Eickwort GC, Jacobson RS. 1977. The economic status of an immigrant yellowjacket, *Vespula germanica* (Hymenoptera: Vespidae), in northeastern United States. *Environ. Entomol.* 6:109–10
89. Nakasuji F, Yamanaka H, Kiritani K. 1976. Predation of larvae of the tobacco cutworm *Spodoptera litura* (Lepidoptera, Noctuidae) by *Polistes* wasps. *Kontyû* 44:205–13
90. Naumann MG. 1970. *The nesting behavior of Protopolybia pumila in Panama (Hymenoptera: Vespidae).* PhD diss. Univ. Kans., Lawrence. 182 pp.
91. O'Donnell S, Jeanne RL. 1990. Forager specialization and the control of nest repair in *Polybia occidentalis* Olivier (Hymenoptera: Vespidae). *Behav. Ecol. Sociobiol.* 27:359–64
92. Oliver AD. 1964. Studies on the biological control of the fall webworm, *Hyphantria cunea* in Lousiana. *J. Econ. Entomol.* 57:314–18
93. Olson E. 1999. *Parachartergus fraternus* (Gribodo) (Hymenoptera: Vespidae: Polistinae) uses venom when taking caterpillar prey. *Psyche.* In press
94. Ono M, Igarashi T, Ohno E, Sasaki M. 1995. Unusual thermal defense by a honeybee against mass attack by hornets. *Nature* 377:334–36
95. Overmyer SL, Jeanne RL. 1998. Recruitment to food by the German yellowjacket, *Vespula germanica. Behav. Ecol. Sociobiol.* 42:17–21
96. Deleted in proof.
97. Pallett MJ, Plowright RC. 1979. Traffic through the nest entrance of a colony of *Vespula arenaria* (Hymenoptera: Vespidae). *Can. Entomol.* 111:385–90
98. Parra JRP, Gonçalves W, Gravena S, Marconato AR. 1977. Parasitos e predadores do bicho-mineiro do caffeeiro *Perileucoptera coffeella* (Guérin-Méneville, 1842) en São Paulo. *An. Soc. Entomol. Bras.* 6:138–43 (In Portuguese)
99. Parrish MD. 1984. Factors influencing aggression between foraging yellowjacket wasps, *Vespula* spp. (Hymenop-

tera: Vespidae). *Ann. Entomol. Soc. Am.* 77:306–11
100. Parrish MD, Fowler HG. 1983. Contrasting foraging related behaviours in two sympatric wasps (*Vespula maculifrons* and *V. germanica*). *Ecol. Entomol.* 8:185–90
101. Peckham GW, Peckham EC. 1905. *Wasps Social and Solitary.* Westminster: Archibald Constable. 311 pp.
102. Phipps J. 1974. The vampire wasps of British Columbia. *Bull. Entomol. Soc Can.* 6:134
103. Pimental D. 1961. Natural control of caterpillars on cole crops. *J. Econ. Entomol.* 54:889–92
104. Rabb RL, Lawson FR. 1957. Some factors influencing the predation of *Polistes* wasps on the tobacco hornworm. *J. Econ. Entomol.* 50:778–84
105. Raveret Richter MA. 1988. *Prey hunting and interactions among social wasp (Hymenoptera: Vespidae) foragers and responses of caterpillars to hunting wasps.* PhD diss. Cornell Univ., Ithaca, NY. 190 pp.
106. Raveret Richter MA. 1990. Hunting wasp interactions: influence of prey size, arrival order, and wasp species. *Ecology* 71:1018–30
107. Raveret Richter MA, Colvin CL. 1994. *Vespula vidua* wasps scavenge caterpillar baits. *J. Kans. Entomol. Soc.* 67:426–28
108. Raveret Richter MA, Jeanne RL. 1985. Predatory behavior of *Polybia sericea* (Olivier), a tropical social wasp (Hymenoptera: Vespidae). *Behav. Ecol. Sociobiol.* 16:165–70
109. Raveret Richter MA, Jeanne RL. 1991. Hunting behavior, prey capture and ant avoidance in the tropical social wasp *Polybia sericea* (Hymenoptera: Vespidae). *Insectes Soc.* 38:139–47
110. Raveret Richter MA, Tisch VL. 1999. Resource choice of social wasps: influence of presence, size, and species of resident wasps. *Insectes Soc.* 46:131–36
111. Rayor LS, Mooney LJ, Renwick JA. 1999. The effects of cardenolides and glucosinolates in *Pieris napi* caterpillars on the predatory behavior of *Polistes* wasps. *J. Insect Behav.* In press
112. Reid BL, MacDonald JF, Ross DR. 1995. Foraging and spatial dispersion in protein-scavenging workers of *Vespula germanica* and *V. maculifrons* (Hymenoptera: Vespidae). *J. Insect Behav.* 8:315–30
112a. Ross KG, Matthews RW, eds. 1991. *The Social Biology of Wasps.* Ithaca, NY: Cornell Univ. Press. 678 pp.
113. Rossi AM, Hunt JH. 1988. Honey supplementation and its developmental consequences: evidence for food limitation in a paper wasp, *Polistes metricus. Ecol. Entomol.* 13:437–42
114. Seeley TD. 1985. *Honeybee Ecology: A Study in Adaptation in Social Life.* Princeton, NJ: Princeton Univ. Press. 201 pp.
115. Seeley TD. 1986. Social foraging by honeybees: how colonies allocate foragers among patches of flowers. *Behav. Ecol. Sociobiol.* 19:343–54
116. Shafir S. 1996. Color discrimination conditioning of a wasp, *Polybia occidentalis* (Hymenoptera: Vespidae). *Biotropica* 28:243–51
116a. Shattuck SO. 1992. Review of the dolichoderine ant genus *Iridomyrmex* Mayr with descriptions of three new genera (Hymenoptera: Formicidae). *J. Aust. Entomol. Soc.* 31:13–18
117. Spradbery JP. 1973. *Wasps. An Account of the Biology and Natural History of Solitary and Social Wasps.* Seattle: Univ. Wash. Press. 408 pp.
118. Stamp NE. 1992. Relative susceptibility to predation of two species of caterpillar on plantain. *Oecologia* 92:124–29
119. Stamp NE, Bowers MD. 1988. Direct and indirect effects of predatory wasps (*Polistes* sp.: Vespidae) on gregarious caterpillars (*Hemileuca lucina*: Saturniidae). *Oecologia* 75:619–24

120. Stamp NE, Bowers MD. 1991. Indirect effect on survivorship of caterpillars due to presence of invertebrate predators. *Oecologia* 88:325–30
121. Stamp NE, Wilkens RT. 1993. On the cryptic side of life: being unapparent to enemies and the consequences for foraging and growth of caterpillars. In *Caterpillars: Ecological and Evolutionary Constraints on Foraging,* eds. NE Stamp, TM Casey, pp. 283–330. New York: Chapman & Hall. 587 pp.
122. Steiner AL. 1976. Digger wasp predatory behavior (Hymenoptera, Sphecidae) II. Comparative study of closely related wasps (Larrinae: *Liris nigra,* Palearctic; *L. argentata* and *L. aequalis,* Nearctic) that all paralyze crickets (Orthoptera, Gryllidae). *Z. Tierpsychol.* 42:343–80
123. Steiner AL. 1979. Digger wasp predatory behavior (Hymenoptera: Sphecidae): fly hunting and capture by *Oxybelus uniglumis* (Crabroninae: Oxybelini); a case of extremely concentrated stinging pattern and prey nervous system. *Can. J. Zool.* 57:953–62
124. Stringer BA. 1989. Wasps—the honeydew thieves of New Zealand. *Am. Bee J.* 126:465–67
125. Suzuki T. 1978. Area, efficiency and time of foraging in *Polistes chinensis antennalis* Pérez (Hymenoptera, Vespidae). *Jpn. J. Ecol.* 28:179–89
126. Takagi M, Hirose Y, Yamasaki M. 1980. Prey-location learning in *Polistes jadwigae* Dalla Torre (Hymenoptera, Vespidae). *Kontyû* 48:53–58
127. Thomas CD, Moller H, Plunkett GM, Harris RJ. 1990. The prevalence of introduced *Vespula vulgaris* wasps in a New Zealand beech forest community. *NZ J. Ecol.* 13:63–72
128. Thorpe WH. 1963. *Learning and Instinct in Animals.* London: Methuen. 558 pp.
129. Tinbergen N. 1972 (1935). On the orientation of the digger wasp *Philanthus triangulum* Fabr. II. The hunting behavior. In *The Animal in Its World: Vol. I. Field Studies,* ed. N Tinbergen, p. 128–45. Cambridge, MA: Harvard Univ. Press. 343 pp.
130. Turillazzi S. 1983. Extranidal behavior of *Parischnogaster nigricans serrei* (DuBuysson) (Hymenoptera, Stenogastrinae). *Z. Tierpsychol.* 63:27–36
131. Turillazzi S, West Eberhard MJ, eds. 1996. *Natural History and Evolution of Paper-Wasps.* New York: Oxford Univ. Press. 400 pp.
132. von Frisch K. 1967. *The Dance Language and Orientation of Bees.* Cambridge, MA: Belknap Press. Harvard Univ. Press. 566 pp.
133. Wenzel JW. 1991. Evolution of nest architecture. See Ref. 112a, pp. 480–519
134. West Eberhard MJ. 1969. The social biology of Polistine wasps. *Misc. Publ. Mus. Zool. Univ. Mich.* 140:1–101
135. West Eberhard MJ. 1975. The evolution of social behavior by kin selection. *Q. Rev. Biol.* 50:1–33
136. Williams FX. 1927. *Euparagia scutellaris* Cresson, a masarid wasp that stores its cells with young of a curculionid beetle. *Pan-Pac. Entomol.* 4:38–39
137. Wilson EO. 1971. *The Insect Societies.* Cambridge, MA: Belknap Press, Harvard Univ. Press. 548 pp.

Annu. Rev. Entomol. 2000. 45:151–174

Blood Barriers of the Insect

Stanley D. Carlson[1], Jyh-Lyh Juang[2], Susan L. Hilgers[3], and Martin B. Garment[4]

[1]*Department of Entomology, Neuroscience Training Program, University of Wisconsin-Madison, Madison, WI 53706; e-mail: carlson@entomology.wisc.edu*
[2]*Division of Molecular and Genomic Medicine, National Health Research Institutes, 128 Yen-Chiu-Yuan Road, Sec. 2, Taipei, Taiwan 11529; e-mail: juang@nhri.org.tw*
[3]*Department of Entomology, University of Wisconsin-Madison, Madison, WI 53706*
[4]*Department of Entomology, University of Wisconsin-Madison, Madison, WI 53706; e-mail: mgarment@facstaff.wisc.edu*

Key Words blood barriers, glia, perineurium, insects, membrane specializations

■ **Abstract** The blood-brain barrier (BBB) ensures brain function in vertebrates and insects by maintaining ionic integrity of the neuronal bathing fluid. Without this barrier, paralysis and death ensue. The structural analogs of the BBB are occlusive (pleated-sheet) septate and tight junctions between perineurial cells, glia and perineurial cells, and possibly between glia. Immature Diptera have such septate junctions (without tight junctions) while both junctional types are found in the imago. Genetic and molecular biology of these junctions are discussed, namely tight (occludin) and pleated-sheet septate (neurexin IV). A temporal succession of blood barriers form in immature Diptera. The first barrier forms in the peripheral nervous system where pleated-sheet septate junctions bond cells of the nascent (embryonic) chordotonal organs in early neurogenesis. At the end of embryonic life, the central nervous system is fully vested with a blood-brain barrier. A blood-eye barrier arises in early pupal life. Future prospects in blood-barrier research are discussed.

PERSPECTIVES AND OVERVIEW

Insects have avascular nervous systems in which hemolymph bathes all outer surfaces of ganglia and nerves. This situation is unlike the vertebrate condition where arteries and veins ramify throughout all neural masses. In insects, a thin layer of perineurial and glial cells overlies nerves, forming the first barriers to disadvantageous ions, molecules, and small solutes of a polar nature. These covering cells have occluding junctions (tight and pleated-sheet septate junctions) that block paracellular passage of blood-borne elements from contact with the underlying neuronal surfaces.

Blood barriers in insects arise sequentially during their immature stages to partition cells (peripheral nervous system and central nervous system neurons,

0066-4170/00/0107-0151/$14.00

visual system neurons and testis spermatogonia) from direct access to circulating hemolymph. It is probable that these developing cells require blood-borne components in their early growth phase but neurons can only become electrophysiologically competent if they are protected from bathing hemolymph. Hemolymph is deleterious to neurons because of its excessive ionic constitution (especially K^+, which is unsuitable in a bathing solution for electrically active neurons. A selectively permeable partition between nerve and blood establishes the appropriate ionic microenvironment of the neurons. A compromised barrier system in either vertebrates or invertebrates (especially insects) would probably lead to swift death by "natural causes." The blood barrier concept is long-standing and began with workers using vertebrate models for this work.

More than a century ago Paul Ehrlich (25) discovered certain pharmacologically active compounds and dyes that, when introduced into the mammalian blood system, never entered the central nervous system (CNS). In the first half of this century, these and related findings formed the concept of a blood-brain barrier (BBB), a barrier that blocked blood-borne hydrophilic polar materials from entering the neuronal sector. This barrier also prevents permeability in the outward direction so that neurotransmitters, neuromodulators, neuropeptides, and the like released from nerve cells into the bathing fluids do not mix with blood. Studies to find the anatomical correlates of this permeability barrier in vertebrates did not begin with ultrastructural studies (e.g. 5), and thus "the concept of a blood-brain barrier was viewed with skepticism by many" (6). However, Reese & Karnovsky (62), using transmission electron microscopy, discovered tight junctions between endothelial cells of mammalian brain capillaries that prevented circulating tracer (horseradish peroxidase) from entering the nervous system.

Insect neurobiologists were slower to confirm that such a barrier existed in insects. Since Scharrer's report in 1939 (68) it was known that cockroaches had a perineurial cell layer (overlaid by a neural lamella), which wraps the glial-enshrouded nerve cells within. By 1953, Roeder (63) was at least partly right when he stated that the neural lamella played "an important regulatory function as a potassium barrier between neurons and hemolymph." That pronouncement was based on Hoyle's finding (34) that intact locust nerves could tolerate 140 mM potassium for two to three hours before conduction block ensued, but the block was immediate when these nerves were desheathed. What Roeder did not know, however, was that Hoyle's desheathing operation removed the neural lamella and underlying perineurial cells, thus surgically eliminating the BBB. Almost twenty years passed after Hoyle's and Roeder's work before Lane & Treherne (49, 50) revealed that desheathed cockroach ganglia were also open to extracellular tracers. Further, tight junctions between perineurial cells and between perineurial cells and sheath glia in adult cockroaches were shown to prevent inward tracer diffusion onto axonal surfaces. In those and later studies, the septate junction was rarely mentioned other than to acknowledge its presence as a membrane specialization. However, our intensive search for tight junctions in *Delia* and *Drosophila* larva revealed none (40), as also reported by Tepass &

Hartenstein (79). The dipteran larva, *Calliphora*, was examined by Lane & Swales (48) whose work spurred our present efforts. Those workers asserted that *Calliphora* lacks a fully formed BBB, but that the barrier forms over the five-day period of larval development. We believe that a less-than-complete BBB, even for a short period of post-embryonic time, would have critical consequences to survival. In 1981, Chi & Carlson (16) demonstrated that the P face particle ridges, the "tight junctions" seen by Lane & Swales (48) in the dipteran larva, were actually a continuous junction, a leaky variant of the pleated-sheet septate junction. From our earliest sampling point, less than a day into larval life, *Delia* (seed corn maggot) was shown to possess a functioning BBB. In *Drosophila*, the barrier forms in late (stage 17) embryonic life, during neurogenesis. This developmental tableau caused us to reexamine the issues of genesis, junction type, location and time of barrier formation in *Delia*. (Taxonomically, *Delia*, *Drosophila,* and *Calliphora,* and other true flies are physiologically similar but we speak for only *Delia* and *Drosophila.*)

What follows are synopses about the two kinds of non-neural cells (perineurial and glial) making up the main barriers, the barrier systems per se, and the molecular biology of the barrier. The latter topic will likely be resolved in the new millennium (as proposed in the section, Future Prospects).

CELLULAR COMPONENTS OF THE BARRIERS

Perineurial Cells

Barrier properties are presented by perineurial (P) cells through the septate (homophilic) junctions that bind P cells together. These four- to six-sided polygonal cells (Figures 1, 10) have expansive surfaces that far exceed their thickness, and form a monolayer of flattened cells over nerve and neuropile. This thin skin of cells overlies the core of each nerve, as well as the CNS ganglia, from late (S-17) stage embryo to imago. On their outer (distal) surface P cells have an adhering fibrous (acellular) neural lamella; the proximal surface of the P cell apposes the sheath glia of nerves or ganglia and those glia enshroud the axons. Polygonal (en face) cell borders contain extensive pleated-sheet septate junctions (Figures 2, 3, and 4). P cells never contact neurons, although they abut glial cells, which in turn, wrap or otherwise adhere to neurons. This disposition of P cells to neurons brings up the argument about whether P cells are indeed glial cells. Edwards et al (23) point out that these cells are derived from mesoderm, while true glia arise from ectoderm. Other opinions on nomenclature were given by Lane (44) and Ito et al (35). Ito and coworkers stipulated that the " . . . perineurial glia [are] in the glial population regardless of their origin and . . . probably function with other glial cells to support neurons."

Perineurial cells are ultrastructurally complicated cells with meandering lateral cell borders (seen in longitudinal sections: Figure 3) that form internal (extracel-

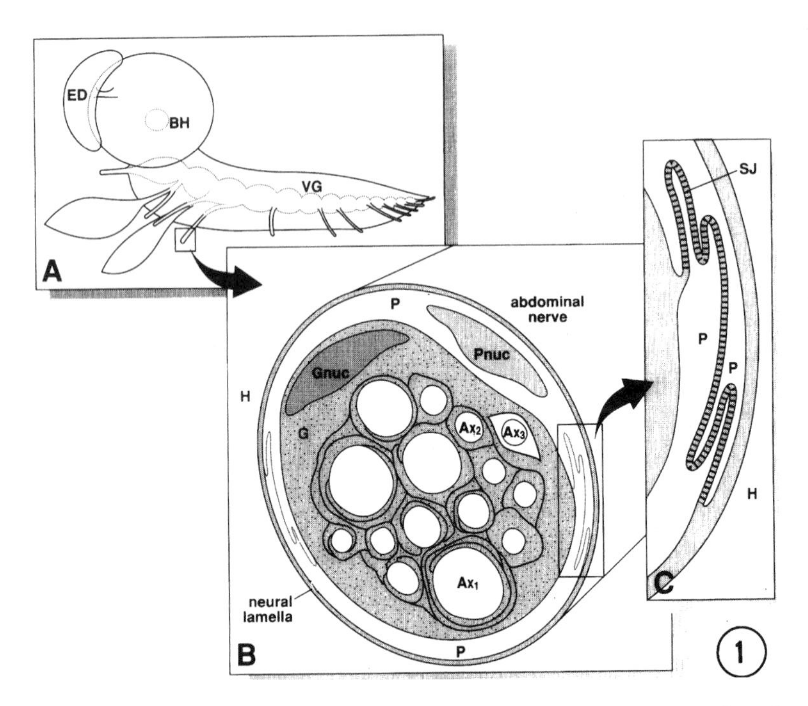

Figure 1 Artist's rendition of the larval *Drosophila* nervous system at three levels of resolution. After Hertweck (33b), (a) shows a CNS dissected out. (b) A cross-sectioned larval, abdominal nerve with more than a dozen axons (*ax*), several encompassing sheath glial cells (*G*) and the overlying perineurial cells (*P*) covered with the neural lamella. (c) A sector of abutting perineurial cells magnified to show the folded nature of adjacent perineurial cells, along with extensive pleated-sheet septate junctions (*SJ*) in the intercellular space between perineurial cells. Magnifications are about 80, 10,000 and 30,000 times for a, b, and c, respectively. (Reprinted from 40a, Elsevier Science Ltd.)

lular) channels and clefts (44). At least four kinds of membrane specializations are present between these cells: continuous junctions, septate junctions (pleated-sheet: Figure 2), spot desmosomes, and gap junctions (40). Perineurial cells also bond to the underlying glia by septate junctions (Figures 8, 9).

Perineurial cells are first formed in the early portion of the last stage (stage 17) *Drosophila* embryo and their precursors are presently unknown. Carlson & Hilgers (10a) revealed that at stage 17 P cells are sealing and their septate junctions block tracer. Thus, the CNS barrier is both formed and made intact during the last six hours of embryonic development. At about the same time, chemical synapses become functional, a coincidence that rather suddenly produces a working nervous system.

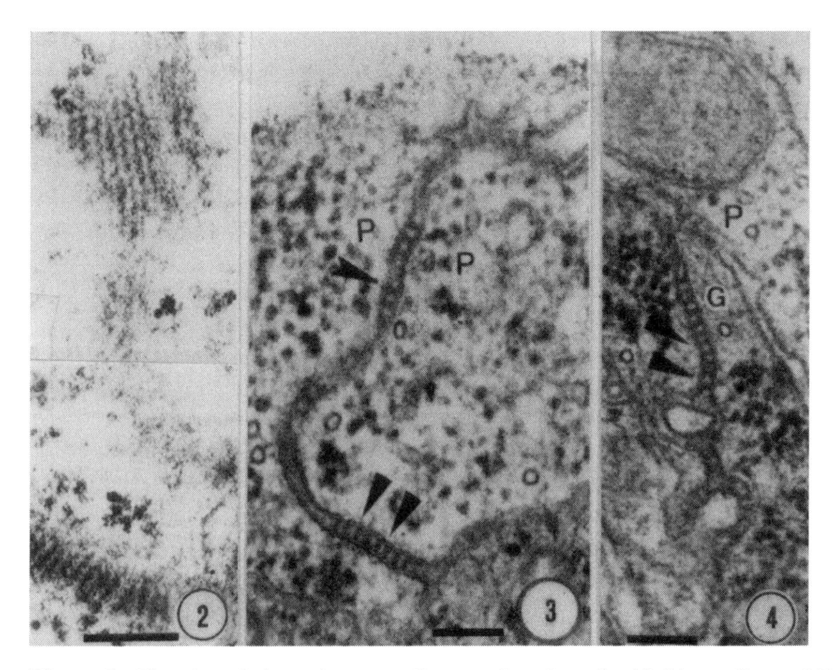

Figure 2 The pleated-sheet character of septate junctions (in third instar *Drosophila*) is revealed in oblique sections of the perineurial (homocellular) cell associations. The tracer lanthanum penetrated some distance but stopped short of total paracellular passage. The different orientations of the junction reflect the undulating contours of the cell. Bar = 0.5 μm. (From 40, reprinted courtesy of Springer Verlag.)

Figure 3 Longitudinal section similar to Figure 1c. Perineurial (*P*) lateral cell borders are bonded with pleated-sheet septate junctions (especially at *arrowheads*). *Drosophila* stage 17 embryo. Bar = 0.1 μm. (From 11a, reprinted by permission from Elsevier Science Ltd.)

Figure 4 Stage 17 *Drosophila* embryo. Portion of an abdominal nerve showing a perineurial (*P*) cell enclosing a glial (*G*) cell whose processes adhere to each other through septate junctions. Bar = 0.1 μm. (From 11a, reprinted by permission from Elsevier Science Ltd.)

Glial Cells

Glial cells are the almost indispensable companions of nerve cells (although some insect nerves are naked). Glia are as ubiquitous and as numerous (or more so) as nerve cells. Glial cell geometry is highly variable, depending on the spaces to fill between adjacent neurons. Glia are a heterogeneous lot and given-type glial cells can usually be distinguished from other glia by ultrastructural appearance, by using Golgi methods for the light microscope, or by identifying anatomical location. An example of fine structural differentiation of glial cell types relates to the adult blood-eye barrier, with the stacking of five different glial cell types over and throughout the first optic neuropils of *Musca*—a distance of only a few score microns (13, 65, 66, 67).

Functional interactions of glia with neurons and neurons' role in blood barriers have been posited recently by many authors from the topical focus of biochemistry [Tsacopoulos & Poitry (82)]; phagocytosis [Cantera & Technau (8)] electrophysiology [Walz (83)]; migration [Choi & Benzer (18)]; embryonic characterization [Ito et al (35)]; axon guidance [Tessier-Lavigne & Goodman (80)]; and broad-based reviews [Pfrieger & Barnes (61), Abbott (1), Carlson & Saint Marie (13), Treherne (81), and Edwards & Tolbert (24)].

For some time there has been an intuitive belief in a blood barrier (in insects) effected by glial cells because of their insulating, isolating, and trophic nature toward neurons. About 30 years ago, Lane & Treherne desheathed (removed neural lamella and perineurium from) cockroach ganglia and determined that the underlying neurons were then laid open to tracers (49). The Treherne Lab with NJ Lane published extensively (44, 45, 81) on the adult barrier based on tight junctions between perineurial cells, perineuria and glia, and glial-glial associations. Conversely, the Shaw Lab (73) at that time conceptualized a blood-brain barrier in which neither tight junctions nor perineurial cells affect diffusion paths of extracellular tracer. Tracer does not penetrate beyond the sheath glia after passing through a number of septa in the septate junction. Thus two camps of thought existed in the 1980s. The majority view was that the BBB originated in larval life and was caused by tight junctions. We now know that the blood-brain and blood-nerve barriers assemble in embryonic time and that one anatomical correlate of this interdiction is the pleated-sheet septate junction (38, 39, 40). Tight junctions are not found in immature dipteran forms (79).

While the neural lamella is a fibrous, durable acellular mat overlying the perineurium, it has no sealing qualities. Using cationic colloidal gold, we found that this tunic possesses abundant anionic domains wherein blood-borne cations are trapped (40). That effect would add little to overall barrier properties.

MEMBRANE SPECIALIZATIONS RESPONSIBLE FOR BARRIERS

Pleated-Sheet Septate Junctions

The blood barriers, for the most part, arise in the pre-adult life stages. The anatomical correlate of nascent blood barriers in insect embryos is the pleated-sheet septate junction. There should be no confusion between pleated-sheet and smooth or honeycomb septate junctions. The consensus, emerging over the course of the past 20 years, is that pleated-sheet septate junctions are an effective intercellular barrier. Evidence for this claim is summarized here (also see 27): (*a*) Septate junctions bind epithelial cells, encircling the apical portion of the cell similar to that noted for tight junctions in vertebrates (27, 58, 59); (*b*) compartmentalization of septa increases the cross-sectional area of the intercellular cleft (e.g. 300 fold in Hydra) (26, 27, 28, 32); (*c*) increasing impermeability to tracer seems to coincide with increasing development of septate junctions (29, 77); (*d*) septate junctions are associated with belt desmosomes, as the latter are with tight junctions in vertebrates (28); (*e*) many invertebrate epithelial cells require sealing properties to maintain bipolar physiological functions, yet tight junctions are lacking and only septate junctions are found (3, 77); (*f*) La^{3+} cannot penetrate completely through septate junctions (3, 43, 56, 65, 77, 85), but can infiltrate many tight junctions (17, 19, 31, 53, 69, 74); (*g*) both invertebrate septate junctions and vertebrate tight junctions exhibit a similar response to hypertonic solutions (52, 75).

Tight Junctions

Such membrane-to-membrane adhesions form a sealing barrier to many blood-borne compounds, macromolecules and ions that would otherwise gain paracellular passage through an epithelial layer. Tight junctions (TJs) form selective permeability barriers and, as such, (in late pupa and adult) are principal occluding junctions, although they also effect cell cohesion. Several forms exist. Two neighboring plasma membranes may fuse to each other at discrete ("punctate") loci. In other cases the surface area of the fusion is extensive. TJs are identified by transmission electron microscopy (TEM) in longitudinal sections of the opposing membranes as showing a pentalaminar construct in which the two opposing leaflets merge into one (monomolecular) resolved layer. We see both punctate and extended TJs in adult insects. Not only are opposing membranes essentially fused, but cytoskeletal elements at these consolidated areas may span intercellular space (fusing) adjoining cells or adhering to the extracellular matrix.

Tight junctions are not found in embryonic and larval Diptera (40, 79) but they arise in late pupal life and exist throughout the imago. Such junctions are an anatomical correlate of the adult blood-barrier system. Tight junctions and pleated-sheet septate junctions (in adult *Musca*) may exist side by side (66). The "blood-worthiness" of tight junctions is especially well expressed in the adult dipteran (compound) eye. Here 17 different examples are given of tight junctions forming between specific glia and laminar neurons (66).

Credit is given to Dr. Che Chi (14, 15) for first visualizing tight junctions in freeze fracture replicas in an insect (*Musca*) compound eye whose characteristics are comparable to those of vertebrate tight junctions as to both thin sections and freeze-fracture criteria. This claim is also made by Lane (45). Lane & Treherne (50) had earlier recognized TJs from ultrathin sections of cockroach CNS. Lane is also to be greatly praised for her comprehensive review of tight junctions in insects and other invertebrates (47). Her review is expansive and covers the ZO-1 and tight junction kinship, and the actin-like microfilaments at tight junctions that appear to be cytoskeletal reinforcing elements. Furuse et al (30a) also discovered in 1994 a (or the) protein that "plays a key role in the formation of tight junctions," that is, occludin, whose essentials are discussed here in the section on molecular biology.

SPECIFIC BARRIERS

The vignettes of four different barrier systems presented in this section share a skein of information about these barriers: i.e. the commonality of the junctions involved (tight and septate). The protection of these cells from hemolymph by such barriers permits a benign, ionically balanced bathing solution for the newly formed and developing cells, which could not exist if they were drenched in hemolymph.

Blood-Nerve Barrier

The blood-nerve barrier (BNB) is the earliest formed and blocks hemolymph from the afferent nerves of sensilla of the peripheral nervous system (PNS). The concept of a BNB is rather new, and everything we convey about the barrier protecting chordotonal organs (COs) of the *Drosophila* embryo may not be appropriate in regard to other neural components of the PNS. The experimental questions were: when in embryonic life does the BNB arise; in what anatomical form; and where on the sensillum is this barrier zone. The experimental PNS element was the *Drosophila* chordotonal organ and from scores of reports published by others, the molecular genetics and cell biology, placement and number of chordotonal organs were well known (7, 20, 33). These organs are internal proprioceptors (stretch receptors) that report on visceral/muscle motions in embryonic and larval life. In wild-type *Drosophila* a series of dorsally descending, parasegmental nerves convey the axons of chordotonal organs to the ventral nerve cord (CNS). The COs (also called scolopidial organs) are in serial clusters of one to five. Axons of COs branch off at intervals along the main nerve of each parasegment. CO placement, innervation and numbers are essentially invariant (in wild type).

The data presented here are from Carlson et al (11, 12). Because COs form in embryonic stage 13 to 14 (9.5–11.5 hours) we began the ultrastructural/tracer search for a barrier at the earlier stage (12). An intact BNB was present nearly as soon as the CO was constructed. How the four-celled organ is cellularly synthesized is in Carlson et al (12). Pleated-sheet septate junctions underlie the barrier. The barrier is located at the outer surface of the scolopale cell and its articulation with the cap cell (another glial cell). COs exist from early embryonic time through larval life. Interestingly, in stages 13 through 16 there is no blood-brain barrier (CNS) because the sealing perineurial cells form in the ultimate stage, i.e. stage 17 (16.5 to 22 hours). If CO electrical activity could reach the

CNS, the latter's incompetence (no barrier) would negate any positional sense the COs would be reporting.

The chronology of this story is intriguing. Nascent COs are assembled over a three-hour timeframe and then they must wait another six hours before a truly functioning CNS is available to integrate and process the collective electrical activity of the COs. Thus, barrier (BNB) readiness of the COs was not contingent on an intact BBB. We can only speculate on the dynamics of a CO that was "all dressed up with no place to go;" that is, until there was a sheltered, competent CNS. In the haste of *Drosophila* neurogenesis, no aspect of CNS development seems slothful.

Blood-Brain Barrier

This particular barrier system was the first studied, most basic, enduring, pervasive, and extensive. The blood-brain barrier (BBB) encompasses the entire avascular CNS, which is by far the largest mass of neural tissue in the insect body. The barrier system is based on tight and pleated-sheet septate junctions between neurons, glia, and perineurial cells, all of which form in embryonic time and persist to and through the imago.

We cite the historical studies concerning this barrier in the introduction. Our objective in this section is to chronologically view the evolution of the BBB concept based on data largely from the imago (Lane group), and from the immature life forms (Carlson group). The Lane group's work began in 1969 with a page-and-a-half *Nature* note (49) on horseradish peroxidase uptake by adult cockroach glial cells in a desheathed nerve preparation. This experiment was the most apt one to do considering the precursor work done in the 1950s on electrophysiology of desheathed elements of the CNS. Lane & Treherne (49) found that desheathing was a dire violation that permitted peroxidase tracer to gain access to axonal surfaces past two strata of glial cells, breaching the BBB. Minor problems with the content of that paper are that potassium is not considered and that trans-glial ion commerce occurs through "transversely permeable tight junctions." Nevertheless the Lane & Treherne work was seminal in regard to how to assay BBB systems at given CNS sites, using a range of species and developmental times with varying degrees of tissue invasiveness. Thus, before the 1970s Lane proffered a rudimentary understanding of the neuropathology of desheathed (BBB-less) neurons. Subsquently, for more than 20 years Lane pursued BBB questions, mostly in adult invertebrates (largely insects), laying the basis of our present-day concept of the adult insect BBB. This persuasive work is based on thin-section TEM work, freeze-fracture studies and biochemical analysis of glial, neuronal, and cytoskeletal elements. Lane (47) has made a definitive statement about the adult barrier properties:

> The entry of exogenous substances (e.g. tracers) was restricted to the perineurial clefts and never invaded the underlying nervous tissue . . . lanthanum penetrated both septate and gap junctions so the remaining punctate

appositions or TJ's (tight junctions) found between adjacent perineurial cell were presumed to be the morphological basis for the observed physiological barrier.

What is the barrier in immature insects? It is in these early stages (at least in dipteran flies) that all of the known blood barriers form. Before determining the embryonic condition (9), our group had studied the BBB in each of the three larval (fly) instars (38, 39, 40). The embryonic stage was first examined by Edwards et al (22, 23) who found a mesodermal origin for perineurial cells, and that glial, neuronal, and even tracheae stemmed from ectoderm. *Drosophila* mutants (*twist* and *Delta*) without mesoderm failed to develop a perineurium. Nevertheless, their BBBs were intact, presumably because the junctions between underlying glia were sealing (based on exclusion of lanthanum). Edwards (21) contended that, in wild-type *Drosophila* embryos, the ion permeability barrier resides in the sheath (or barrier) glia. The *Delta* embryo, possessing flawed or missing perineurial cells and barrier glial cells " . . . fails to form a blood-brain barrier." Edwards and coworkers maintain that the neural lamella acts as a barrier itself, and glial cells compose a secondary barrier. No membrane specializations were sought in the Edwards work so the anatomical correlates of the barriers are not known for *Delia*.

Our focus was directed initially at the last instar (wild type) *Drosophila* and *Delia* larvae to see if an intact barrier was present, then what the sealing junctions were, where and when these junctions were situated, and whether they could be breached and later rescued. Lanthanum studies (38) revealed the BBB was based on occluding septate (pleated-sheet) junctions located between perineurial cells. We retreated in developmental time to the early postembryonic larva and similar findings occurred (39). In addition to sealing junctions, the complex borders and topography of the perineurial cells were cited as additional factors in barrier formation. In *Drosophila* larva essentially no barrier qualities exist in the neural lamella, although the latter structure is a "charge-selective barrier for cations." With those findings, we could form an analogy between the vertebrate and insect BBBs: Endothelial cells plus basal lamina equal neural lamella; pericytes equal perineurial cells, and astrocytes equal sheath glia. No tight junctions were found (see also 79). We advocate that pleated-sheet septate junctions are the equivalent of tight junctions of adult forms. Adults, however, have both junctional types and each is restrictive in its own way. It cannot be contested that septate junctions are initially permeable to tracers, but the numerous septae attenuate tracer ingress to the point of exclusion. True tight junctions abruptly stop tracer diffusion. At least in flies (*Delia, Drosophila*) the BBB has its genesis in last stage (stage 17) embryos and the barrier persists into pupal time. We found that the barrier mainly resides in the lateral borders of perineurial cells that are bonded to each other with pleated-sheet septate junctions. In our hands, such junctions binding underlying sheath glial cells together seldom, if ever, accumulated tracer. Perineurial cells rather than glial cells are the key elements in the BBB of immature forms.

If blood-brain barriers assemble and function, can these barriers then be disassembled, breached, and made to become dysfunctional? In seeking additional analogs between vertebrate and invertebrate blood-brain barriers, Juang (37) found that the *Drosophila* (third instar) blood-brain barrier could be opened by two mechanisms (used singly): hyperosmotic (mannitol) and a chelating agent (EGTA). Both compounds disrupted the barrier, and tracer infiltrated into neuronal areas. These experimental results mimicked those of analogous vertebrate experiments using the same compounds (42, 55).

Blood-Eye Barrier

The exceedingly complex cellular structure of the compound eye with its myriad neurons might lead one to suppose that a barrier system might have evolved here. Insects, for the most part, are visual creatures. Flies, in particular, while flying need to make millisecond reflexive decisions based on fleeting optical cues. Other sight-based insect sensory modalities are pattern vision, color vision, and e-vector detection, among others.

The neuronal cell mass of the insect retina is also correspondingly large. For example, the tiny *Drosophila* adult packs well over 2500 photoreceptor neurons into one peripheral retina, plus 20 times that number in the neuronal inventory of the first optic neuropile (as well as seven kinds of glial cells per ommatidium).

Shaw (70, 71, 72, 73) defined the first parameters of the blood-eye barrier (BEB) when he showed that resistance barriers existed throughout the adult locust eye. Specifically, there was a barrier to extracellular current flow in the peripheral portion of this compound eye. In addition, dyes injected above the basement membrane never penetrated that vestment, leaving the first optic neuropile immaculate. Overall, Shaw (72) concluded that extracellular barriers existed in the eye, and the barrier " . . . was not a discrete structure at a single level." When dyes were introduced into circulating hemolymph, they never entered the optic lobes (73).

In various sectors of the adult house fly peripheral retina and first optic neuropile, tight and pleated-sheet septate junctions were discovered by Chi & Carlson (16, 17), Lane (45, 46), and Saint Marie & Carlson (65, 66). It became clear that tight and septate junctions associated with glial, neuronal and perineurial cells formed a complex system of blood barriers whose sites could be charted by electrophysiological methods (73, 87).

Beyond the existence of a BEB, other questions arose regarding the period in pre-imago life in which the BEB arises; the life stage involved; and the anatomical correlates, and conceptually, how the development of the BEB affects electrical activity of developing photoreceptor axons and visual interneurons.

Using *Drosophila*, Carlson et al (10b) revealed that the BEB arises in the first part (0–60 hours) of pupal development (Figures 5, 6, 7). At that time no tight

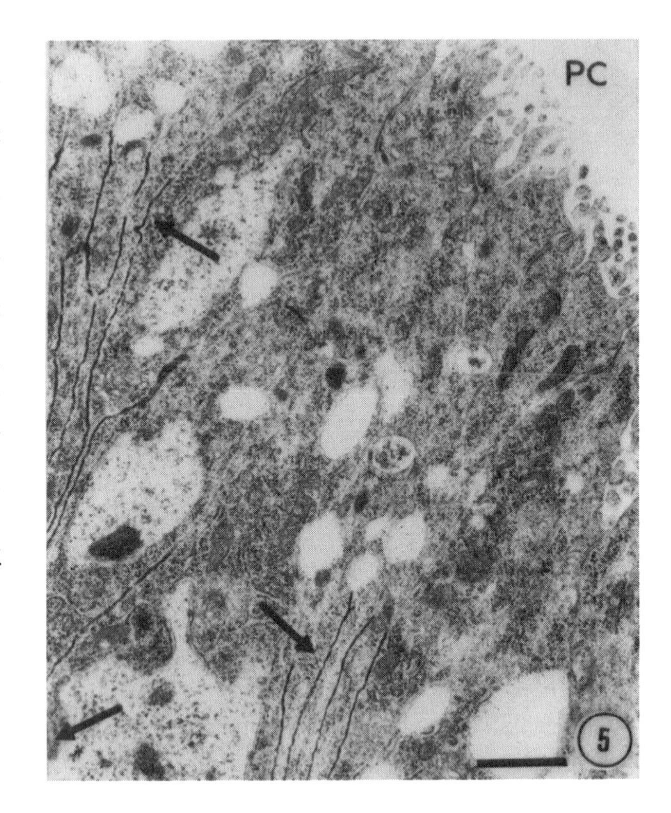

Figure 5 Longitudinal section of eye disk in last instar larva (*Drosophila*). Lanthanum is present between adjacent ommatidial cells. *PC:* peripodial cavity. Microvilli are distally located atop of each cell. Also, adherence junctions are found in that distal region. Septate junctions are more proximally placed and are not sealing the incoming tracer from paracellular entrance. No barrier is present in this stage. *Arrows:* tracer ingress. Bar = 1 μm. (From 11b, reprinted courtesy of Elsevier Science Ltd.)

junctions are found, and barrier properties are (we think) the sole responsibility of pleated-sheet septate junctions. Later in pupal life, tight junctions form and both specializations create the system of extracellular barriers that ramifies throughout the visual system of the compound eye. In our opinion an early lack of a BEB in larval life allows trophic factors, metabolites, and ions access to developing neurons; but soon electrical activity is initiated and the barrier is required. Electrophysiological competence of visual-system neurons is possible only when the ionic microenvironment of these nerve cells is set apart from the ionic constitution of hemolymph.

Blood-Testis Barrier

While the title of this section readily infers a function, one is misled because, as Szöllösi (76) pointed out, the barrier is not between hemolymph and the testis. The blood-testis barrier (BTB) is situated between the germ cells themselves (as they reside in the permeable testis). Indeed, "blood-germ cell barrier" is in prior literature (36). Model systems in use include moths (78), locusts (54), and the bug *Triatoma* (56). We summarize the barriers in moths, and migratory locusts and note that these two types are very different.

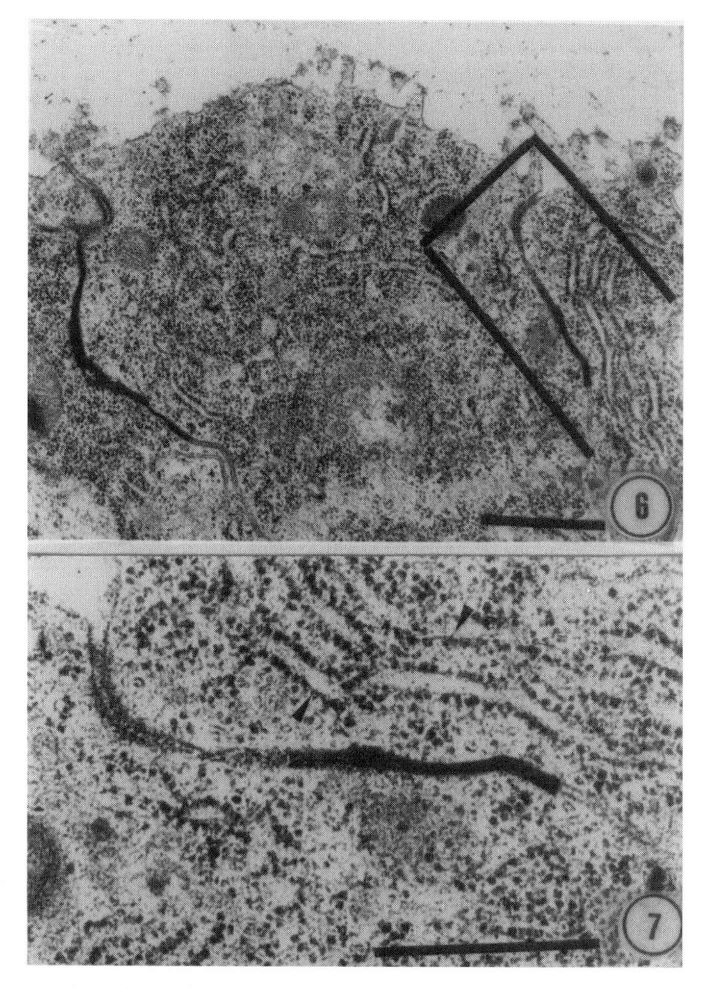

Figure 6 Longitudinal section of three ommatidial cells of the retinal epithelium at 40 hours post pupariation. Distal (upper) surface of each cell is covered with microvilli which is washed by hemolymph circulating around the eye disc. Tracer has penetrated paracellularly to a limited degree between the three cells. Two tracer-filled septate junctions are present indicating the incomplete extent of tracer penetration. Tracer tracks appear to stop after about 2 μm of penetration, a level just above the distal surface of the cell's nucleus. *Boxed area* is enlarged in subsequent micrograph. Bar = 0.5 μm. (From 11b, reprinted courtesy of Elsevier Science Ltd.)

Figure 7 Enlargement of *boxed area* in Figure 6. The septae in the extracellular cleft are better seen, as well as the abrupt cessation of tracer penetration (which is not a plane of section event). Rough endoplasmic reticulum is particularly abundant *(arrowhead)*. Bar = 0.5 μm.

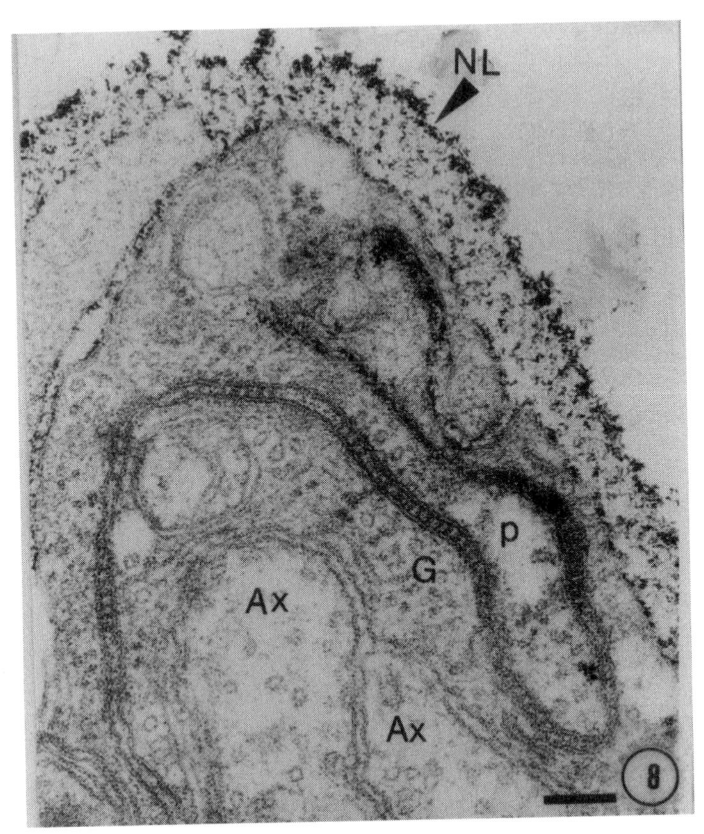

Figure 8 Flocculent nature of the neural lamella *(NL)* in third instar wild-type *Drosophila* larva. Lanthanum cations complex with myriads of NL anionic domains. A convoluted perineurial (*P*) cell border is in this field. Perineurial cells bond to each other via septate junctions. Lanthanum tracer enters this P-P intercellular space but soon fails to appreciably penetrate onto neuronal (*Ax*) surfaces. The result of this sealing effect is that a blood-nerve barrier is present. Bar = 0.1μm.

Depending on the lepidopteran species, the testis is distinctly paired, or the two lobes are fused. Szöllösi (76) showed a four-part eliptical testis in a moth. In each lobe, an apical cell extensively divides forming spherical gonia which, in turn, develop into cysts (surrounded by a cyst envelope). Spermatids mature within cysts that are immersed in a copious fluid filling the testis follicle. While smooth septate junctions along with gap junctions and desmosomes are present between cyst cells, none of these are occlusive to exogenous macromolecules. Most interestingly, extracellular tracers in hemolymph readily penetrate the two walls of the testis and are found in the testicular fluid. These compounds (horse radish peroxidase or lanthanum) are barred from entry into the cyst (78), and therein (cyst wall) lies the barrier. Sealing properties may be caused by septate

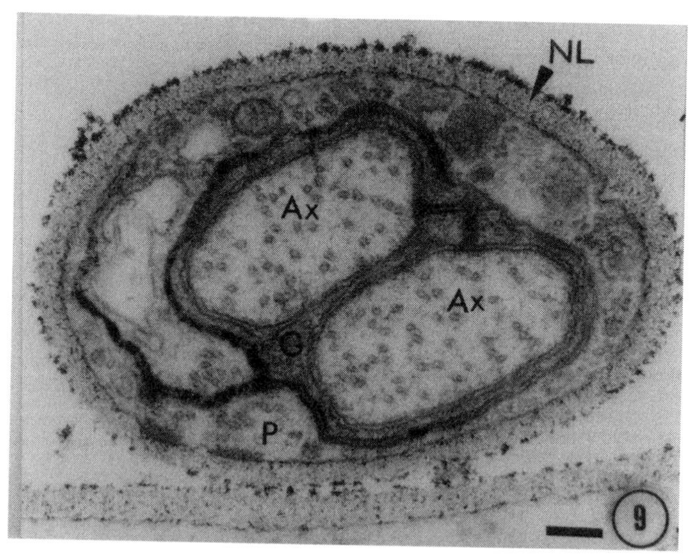

Figure 9 Last larval instar *Delia*. Even the smallest (*cross-sectioned*) branches of an abdominal nerve possess a blood-nerve barrier. Only two axons (*Ax*) are present. Lanthanum has entered through the neural lamella (*NL*) and partly penetrated through septate junctions between the embracing perineurial cells, but tracer was stopped at the outer perineurium-glia (*G*) interface. The two axons at the core are tracer free as a result of this barrier. Bar = 0.1μm.

junctions, but that point is not strongly made by the authors, who acknowledge they cannot find tight junctions and propose that freeze-fracture (FF) studies should be employed to see extended vistas of FF planes. In 1982, Szöllösi discussed the barrier from FF data (76). Replicas show aspects of local tight junctions, but she points to septate junctions with their occlusive powers as the putative bonding junction.

Locust testes have spatulate follicles surmounting a sperm duct (54, 76). Each club-shaped follicle has an apical cell that generates cyst envelopes containing germ cells. The follicle walls are thin relative to that of the moth, although the former are double layered. On the inner layers (inner parietal layer) extensive septate junctions are present, but true tight junctions are not. As we have seen in the eye, pleated-sheet septate junctions are initially permeable to lanthanum tracer. But the numerous septal ribbons quickly attenuate tracer diffusion to ultimately form a complete blockade based on Szöllösi's La^{3+} preparations.

MOLECULAR BIOLOGY OF THE BARRIER

The exciting news about the sealing junctions of insects is that these structures are involved in numerous dynamic intracellular mechanisms, many of which are far removed from the mundane task of barrier formation. The past decade has seen many additions to the proteinaceous *dramatis personnae* of sealing junctions; we discuss the underlying proteins of septate, then tight junctions.

An important class of multi-domain proteins are the membrane associated guanylate kinase homologs (MAGUKs), which reside at these junctions. MAGUKs contain peptide domains, among which are PDZs (ZO 1 and 2). That group contains the discs large gene (*dlg*), which is required for the expression of insect septate junctions, and is also found at insect synapses (19a). Without *dlg*, an uncontrolled proliferation of epithelial cells causes a neoplasm due to an overgrown disc arising from the lack of restraining junctions. The protein ZO-3 (the Z is from PDZ) is a member of the MAGUK family (33a) and is present at tight junctions. MAGUKs may be key mediators in the coupling of extracellular signals to intracellular signal transduction pathways via guanine nucleotides. Cytoskeletal elements are also associated with these proteins (30a).

We now know that PDZ domains are critical for G-protein coupled signalling (82a). Certain calcium (TRP) channels in insect photoreceptor cells are formed in a signalling complex by a PDZ protein (34a) and one of the PDZ domains binds to actin filaments (67a). These bare mentions are but a tiny fraction of the burgeoning literature about G-proteins, which coordinate a spectrum of submembrane responses to the many extracellular signals that constantly bombard the cell.

Other integral and membrane-spanning proteins present at tight junctions deserve mention. Chick liver first furnished a tight junction-associated (with four transmembrane domains) protein termed ZO-1 (35a). Some months later, Furuse et al (in Itoh's lab; 30) isolated, identified, sequenced, cloned, and made antibodies of a new integral (220kD) protein (called occludin), which is identical to ZO-1. That protein underlies both adherens and tight junctions. Furuse et al (30b) have now introduced chick occludin to insect (S-19) cells. The resultant buildup of intracellular multi-lamellar bodies seen in such transfected insect cells meets freeze-fracture and thin section criteria for TJs and advances " . . . the idea that occludin plays a key role in the formation of tight junctions." Furuse and coworkers note that occludin is exclusively at tight junctions of a variety of mammals and insects; is phosphorylated in TJ assembly; and has five known mAbs. In addition, neural TJs are more occlusive than non-neural ones, with ZO-1 forming an association with cytoskeletal elements at the TJ locus. Overall, McCarthy et al (with Furuse; 54a) asserted that " . . . occludin contributes to the electrical barrier function of the tight junction and possibly to the formation of aqueous pores within tight junction strands."

The work of Willott et al (84) revealed ZO-1 is homologous to the *dlg* tumor suppressor gene of *Drosophila*, and that gene product is located in septate junctions of *Drosophila*. While neurexins I-III are of vertebrate origin, neurexin IV is exclusively insectan and a denizen of septate junctions between glial cells. ZO-1 and the *dlg* gene product are known to be involved in intracellular signal transduction in *Drosophila*.

That signal transduction pathway involving the *Drosophila dlg* gene product also was proffered by Woods & Bryant (86). The supposition is that a guanylate kinase (GK) is part of the tumor suppressive gene and is based on one of its domains being homologous to yeast GK. Willott et al (84) believed:

> [ZO-1] is very unlikely to have GK activity [but this possibility] is better in *dlg* . . . . We suggest the N-terminal domain of ZO-1 which is homologous to *dlg* may be involved in signalling that is predicted to occur during regulation of paracellular permeability . . .

Those last two words mean blood barriers. This statement may mark the beginning of understanding the explicit function of these proteins.

An array of novel glycoproteins (e.g. glutactin, neurotactin, neuromusculin, neurexin IV, gliotactin, etc.) have recently been discovered in *Drosophila* (see 41 for a review). At least two (gliotactin, neurexin IV) of these molecules have a role in septate junctions per se or blood-barrier properties in general. As to the latter, Auld et al (2) cloned, characterized, and made a genetic analysis of gliotactin, a transmembrane protein in the *Drosophila* CNS. The phenotype of the mutant gliotactin fly is loss of perineurial cells so that axons and glia (formerly wrapped by the perineurium) are directly exposed to hemolymph. As a result, neural function is lost and the flies are essentially paralyzed. In these mutant flies, tracer (ruthenium red) enters perineurial-less nerves freely and contacts axon surfaces. No specific cell-to-cell associations are mentioned in that report, so that one wonders if gliotactin is a necessary element in septate junction formation and/or perineurial cell development.

A more likely candidate for effecting barrier genesis is neurexin IV (4). This too is a transmembrane protein and the only neurexin to be found so far in Insecta. Pleated-sheet septate junctions of mutant embryos have either degenerated or changed to smooth septate junctions whose ionic permeability is well known. Perineurial cells were not mentioned or depicted in this account, so we don't know if they are as affected as ordinary (sheath) glial cells. Baumgartner et al (4) asserted that neurexin IV is a fundamental portion of pleated-sheet septate junctions, and instrumental in the formation of this membrane specialization.

Some mysteries remain. To date, no one knows when and where the first expression of this gene product arises in the embryo. Might neurexin IV be involved in the pleated-sheet septate junctions of both perineurial cells and neighboring (sheath) glial cells even though germ line derivation of perineurial (mesoderm) and underlying glia (ectoderm) are different? The answer is yes (JT Littleton, personal communication), as pleated-sheet septate junctions between perineurial cells in late stage (stage 17) *Drosophila* embryos show presence of neurexin (Figure 10). Neurexin IV may also have an extended role in synapse modification, vesicle docking, and synapse formation (51). Localization of neurexin IV at septate junctions and synaptic organelles is of fundamental import. Parenthetically, Carlson & Hilgers (9), show that stage 15 *Drosophila* embryos exhibit first ultrastructural signs of docking vesicles at axo-axonal synapses in neuropil. It would be useful to know of neurexin's presence (or absence) at the synaptic locus at stage 15. This time (stage 15) is about three to four hours before the existence of the blood-brain barrier of the CNS. It now seems that neurexin

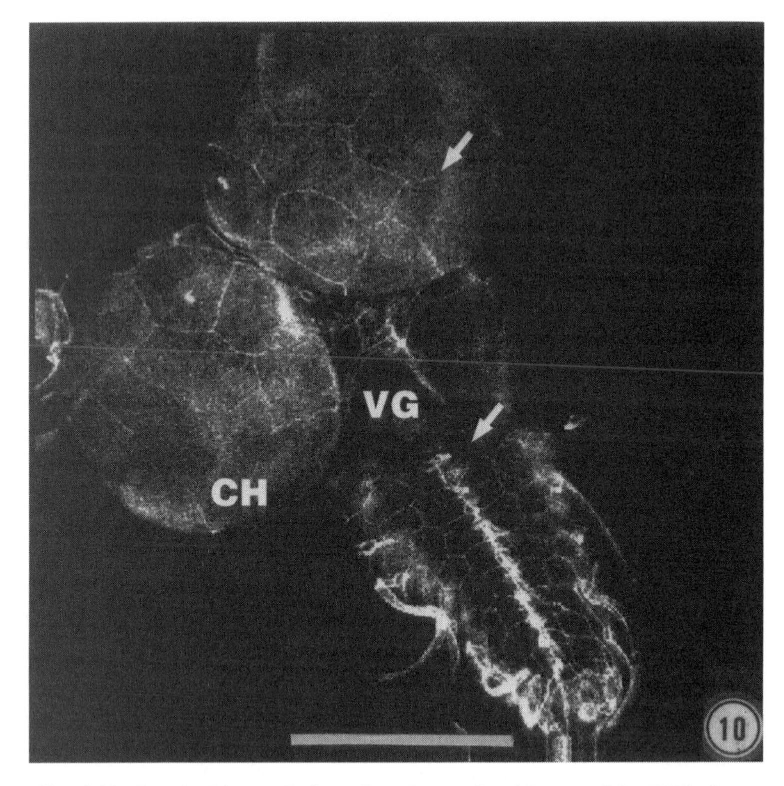

Figure 10 A black and white rendering of a color confocal image of the CNS of a mature *Drosophila* larva. The CNS was immunostained with anti-nrx IV antisera that marks pleated-sheet septate junctions. Two cerebral hemispheres (*CH*) are present surmounting the anterior portion of the ventral ganglion (*VG*). Lateral borders of perineurial (*P*) cells are noted as fine white lines (*arrows*) showing presence of that transmembrane protein associated with septate junctions between perineurial cells. P cells exhibit an *en face* polygonal form and packing pattern as they cover these hemispheres and ventral ganglion. Bar = 100 μm. (This micrograph was kindly supplied by members of the Hugo Bellen Lab, Howard Hughes Medical Institute, Baylor College of Medicine, Houston.)

IV might function in both neural and non-neural cells in several distinct roles. The complexity of reactions instituted by a host of proteins underlying blood-barrier junctions are a tale to tell.

FUTURE PROSPECTS

Future prospects are always exciting to discuss, and are particularly so now that elegant techniques in genetics and cell biology can be so acutely applied to the barriers. A basic step would be to examine epithelial sealing properties in some very early insect orders. There may be appreciable differences in the ionic composition of insect hemolymph in those less highly evolved species, necessitating more or less sophisticated barrier systems. In discussing impermeability to blood-borne tracers we are overlooking a spectrum of plant products that are ingested and would be neurocytotoxic if these gained access to the CNS, e.g. the metabolic BBB of the tobacco hornworm (*Manduca sexta*) (57). *Manduca* has a nicotine pump with its β-glycoprotein homolog (if only humans possessed such immunity to nicotine!). Basic barrier studies should and will be integrated into more endocrinological experiments to see how chronological correlates in barrier formation and deconstruction relate to hormone titers. *Drosophila* geneticists will reward us with mutants such as the *dlg* tumor-suppressant gene whose gene product is located in pleated-sheet septate junctions (84). The same could essentially be said for neurexin IV, which is the first identified transmembrane molecule of pleated-sheet septate junctions in insects (4). It is likely that definitive genetic dissection can be carried out for both tight and septate junctions. Such research would better delineate the roles of fasciculin III (a close relative of neurexin IV), as well as the *coracle* protein, which also lurks around septate junctions. These offshoots of BBB studies will focus on the barriers' role in interdicting paracellular pathways between glial cells and glia and perineurial cells. Glial cells account for at least 25% of the cells in the insect brain (61), and that is a stout minority. Until recently, septate and other junctions from the glia could not be gleaned in quantities sufficient for an immunocytochemical analysis. Ryerse (64) showed how gap junction protein in *Heliothis* sp. is brought to immunoblot status or into NaOH and NaOH/OG extracted fractions. Possibly, similar methods could be used to harvest insect septate and tight junctions in quantity.

There is so much, literally and conceptually, to accomplish with the insect blood-brain barrier. Our colleagues investigating the human BBB are spending millions to learn to transiently diminish the human BBB and so instill therapeutic proteinaceous compounds through the barrier. It is not too visionary to suppose that humans will learn to breach the insect barrier and so suppress insect populations with less or no neurotoxic pesticides. Then again, humans are highly concerned about their own BBB, if the 50-chapter tome edited by Partridge (60) is any measure.

ACKNOWLEDGMENTS

We gratefully acknowledge financial support from a Hatch Grant (No. 3857), a State Transitional Grant, and N. S. F. (BNS 8908081). Betsy True, University of Wisconsin medical illustrator, produced Figure 10 from authors' data. Professor Walter G. Goodman, Entomology Department, University of Wisconsin-Madison, critiqued an earlier draft of this manuscript.

LITERATURE CITED

1. Abbott AJ. 1991. Permeability and transport of glial blood-brain barriers. In *Glial and Neuronal Interaction*, ed. AJ Abbott, pp. 378–94. New York: NY Acad. Sci.
2. Auld VJ, Fetter RD, Broadie K, Goodman CS. 1995. Gliotactin, a novel transmembrane protein on peripheral glia is required to form the blood-nerve barrier in *Drosophila*. *Cell* 81:757–67
3. Baldwin KM, Loeb MJ, Riemann JG. 1987. A novel occluding junction which lacks membrane fusion in insect testis. *Tissue Cell* 19(3):413–21
4. Baumgartner S, Littleton JT, Broadie K, Bhat MA, Harbecke R, et al. 1996. A *Drosophila* neurexin is required for septate junction and blood-nerve barrier formation and function. *Cell* 87:1059–68
5. Bennett HS, Luff JH, Hampton JC. 1959. Morphological classification of vertebrate blood capillaries. *Am. J. Physiol.* 196:381–90
6. Bradbury M. 1979. *The Concept of a Blood-Brain Barrier*. New York: Wiley & Son. 465 pp.
7. Campos-Ortega JA, Hartenstein V, eds. 1985. *The Development of Drosophila melanogaster*. Berlin: Springer-Verlag
8. Cantera R, Technau GM. 1996. Glial cells phagocytose neuronal debris during the metamorphosis of the central nervous system in *Drosophila melanogaster*. *Dev. Genes Evol.* 206:277–80
9. Carlson SD, Hilgers SL. 1997. Ultrastructural elements of CNS development in the stage 15 *Drosophila* embryo. *Int. J. Insect Morphol. Embryol.* 26:43–47

10a. Carlson SD, Hilgers SL. 1998. Perineurium in the *Drosophila* (Diptera: *Drosophilidae*) embryo and its role in the blood-brain/nerve barrier. *Int. J. Insect Morphol. Embryol.* 27:61–66

10b. Carlson SD, Hilgers SL, Garment MB. 1998. Blood eye barrier of the developing *Drosophila melanogaster*. *Int. J. Insect Morphol. Embryol.* 27:241–47

11. Carlson SD, Hilgers SL, Juang JL. 1997. Ultrastructure and blood-nerve barrier of chordotonal organs in the *Drosophila* embryo. *J. Neurocytol.* 26:377–88
12. Carlson SD, Hilgers SL, Juang JL. 1997. First developmental signs of the scolopale (glial) cell and neuron comprising the chordotonal organ in the *Drosophila* embryo. *Glia* 19:269–74
13. Carlson SD, Saint Marie RL. 1990. Structure and function of insect glia. *Annu. Rev. Entomol.* 35:597–621
14. Chi C, Carlson SD. 1980. Membrane specializations in the first optic neuropil of the housefly (*Musca domestica* L.) I. Junctions between neurons. *J. Neurocytol.* 9:429–49
15. Chi C, Carlson SD. 1980. Membrane specializations in the first optic neuropil of the housefly (*Musca domestica* L.) II. Junctions between glial cells. *J. Neurocytol.* 9:451–69
16. Chi C, Carlson SD. 1981. Lanthanum and freeze fracture studies on the retinular cell junctions in the compound eye of the housefly. *Cell Tissue Res.* 214:541–52
17. Chi C, Carlson SD. 1981. The perineurium of the adult housefly: ultrastructure and permeability to lanthanum. *Cell Tissue Res.* 217:373–86
18. Choi K-W, Benzer S. 1994. Migration of glia along photoreceptor axons in the developing *Drosophila* eye. *Neuron* 12:423–31
19. Claude P, Goodenough DA. 1973. Fracture faces of *zonulae occludens* from "tight" and "leaky" epithelia. *J. Cell Biol.* 58:390–400

19a. Craven SE, Bredt DS. 1998. PDZ pro-

teins organize synaptic signaling pathways. *Cell* 93:495–98
20. Dambly-Chaudiere C, Ghysen A. 1986. The sense organs in the *Drosophila* larva and their relation to the embryonic pattern of sensory neurons. *Roux's Arch. Dev. Biol.* 195:222–28
21. Edwards JS. 1996. Status of the superficial cellular components of the insect nervous system. In *Proc. 20th Int. Congr. Entomol.*, p. 110 (Abstr.)
22. Edwards JS, Swales LS, Bate M. 1991. Development of non neural elements in the central nervous system of *Drosophila. Ann. NY Acad. Sci.* 633:617–18
23. Edwards JS, Swales LS, Bate M. 1993. The differentiation between neuroglia and connective tissue sheath in insect ganglia revisited: the nerural lamella and perineurial sheath cells are absent in a mesodermless mutant of *Drosophila. J. Comp. Neurol.* 333:301–8
24. Edwards JS, Tolbert LP. 1998. Insect neuroglia. In *Microscopic Anatomy of Invertebrates,* Vol. 11B. *Insects*, pp. 449–66. New York: Wiley-Liss
25. Ehrlich P. 1885. *Das Sauerstoff-Bedurfnis des Organismus. Eine Farbenanalytische Studie.* Berlin: Hirschwald
26. Filshie BK, Flower NJ. 1977. Junctional structures in *Hydra. J. Cell Sci.* 23:151–72
27. Flower NE. 1971. Septate and gap junctions between the epithelial cells of an invertebrate, the mollusc *Cominella moculosa. J. Ultrastruct. Res.* 37:259–68
28. Flower NE. 1986. Sealing junctions in number of arachnid tissues. *Tissue Cell* 18(6):899–913
29. Fristrom DK. 1982. Septate junctions in imaginal disks of *Drosophila*: a model for the redistribution of septate during cell rearrangement. *J. Cell Biol.* 94(1):77–87
30. Furuse M, Hirase T, Itoh M, Nagafuchi A, Yonemura S, et al. 1993. Occludin: a novel integral membrane protein localizing at tight junctions. *J. Cell Biol.* 123(6):1777–88
30a. Furuse M, Itoh M, Hirase T, Nagafuchi A, Yonemura S, et al. 1994. Direct association of occludin with ZO-1 and its possible involvement in the localization of occludin at tight junctions. *J. Cell Biol.* 127:1617–26
30b. Furuse M, Fujimoto K, Sato N, Hirase H, Tsukita S. 1996. Overexpression of occludin, a tight junction-associated integral membrane protein, induces the formation of intracellular multilamellar bodies bearing tight junction-like structures. *J. Cell Sci.* 109:429–35
31. Gotow T, Hashimoto PH. 1981. Graded differences in tightness of ependymal intercellular junctions within and in the vicinity of the rat median eminence. *J. Ultrastruct. Res.* 76:293–311
32. Hand AR, Gobel S. 1972. The structural organization of the septate and gap junction of *Hydra. J. Cell Biol.* 52:397–98
33. Hartenstein V, Campos-Ortega JA. 1986. The peripheral nervous system of mutants of early neurogenesis in *Drosophila melanogaster. Roux's Arch. Dev. Biol.* 195:210–21
33a. Haskins J, Gu LJ, Wittchen ES, Hibbard J, Stevenson BR. 1998. ZO-3, a novel member of the MAGUK protein family found at the tight junction, interacts with ZO-1 and occludin. *J. Cell Biol.* 141:199–208
33b. Hertwig H. 1931. Anatomie und variabilitat des nervensystems und der simmesorgane von *Drosophila melanogaster* (Meigen). *Z. Wiss. Zool.* 139:559–663
34. Hoyle G. 1952. High blood potassium in insects in relation to nerve conduction. *Nature* 169:281–82
34a. Huber A, Sander P, Bahner M, Paulsen R. 1998. The TRP $Ca2^+$ channel assembled in a signaling complex by the PDZ domain protein INAD is phosphorylated through the interaction with protein kinase C (ePKC). *FEBS Lett.* 425:317–22
35. Ito K, Urban J, Technau GM. 1995. Dis-

tribution, classification, and development of *Drosophila* glial cells in the late embryonic and early larval ventral nerve cord. *Roux's Arch. Dev. Biol.* 204:284–307

35a. Itoh M, Nagafuchi A, Yonemura S, Kitaniyasuda T, Tsukita S, et al. 1993. The 220 kd protein colocalizing with cadherins in non-epithelial cells is identical to ZO-1, a tight junction-associated protein in epithelial cells, cDNA cloning and immunoelectron microscopy. *J. Cell Biol.* 121:491–502

36. Jones RT. 1978. The blood/germ cell barrier in male *Schistocerca gregaria*: the time of its establishment and factors affecting its formation. *J. Cell. Sci.* 31:145–64

37. Juang JL. 1993. Discovery and characterization of the blood-brain barrier of *Drosophila* and *Delia*. PhD diss. Univ. Wis., Madison, 153 pp.

38. Juang JL, Carlson SD. 1992. Fine structure and blood-brain properties of the central nervous system of a dipteran maggot. *J. Comp. Neurol.* 324:343–52

39. Juang JL, Carlson SD. 1992. A blood-brain barrier without tight junctions in the fly central nervous system in the early postembryonic stage. *Cell Tissue Res.* 270:95–103

40. Juang JL, Carlson SD. 1994. Analog of vertebrate anionic sites in blood-brain interface of larval *Drosophila*. *Cell Tissue Res.* 277:87–95

40a. Juang JL, Carlson SD. 1995. X-ray microanalysis with transmission electron microscopy determined presence and movement of tracer (lanthanum chloride) at blood-neuron barrier of *Drosophila melanogaster* (Diptera: *Drosophilidae*) larva. *Int. J. Insect Morphol. Embryol.* 24:435–41

41. Keynes R, Cook GMW. 1995. Axon guidance molecules. *Cell* 83:161–69

42. Koenig H, Gladstone AD, Lu CY. 1989. Polyamines mediate the reversible opening of the blood-brain barrier by the intracarotid infusion of hyperosmolal mannitol. *Brain Res.* 483:110–16

43. Kukulies J, Komnick H. 1983. Plasma membranes, cell junctions and cuticle of the rectal chloride epithelia of the larval dragonfly *Aeshna cyanea*. *J. Cell Sci.* 59:159–82

44. Lane NJ. 1974. The organization of insect nervous systems. In *Insect Neurobiology*, ed. JE Treherne, p. 1–71. Amsterdam & Oxford: North Holland

45. Lane NJ. 1981. Tight junctions in arthropod tissue. *Int. Rev. Cytol.* 73:243–318

46. Lane NJ. 1981. Vertebrate-like tight junctions in the insect eye. *Exp. Cell Res.* 132:482–88

47. Lane NJ. 1993. Anatomy of the tight junction: invertebrates. In *Tight Junctions*, ed. M. Cereijido, pp. 23–48. Boca Raton, FL: CRC

48. Lane NJ, Swales LS. 1978. Changes in the blood-brain barrier of the central nervous system in the blowfly during development with special reference to the formation and disaggregation of gap and tight junctions, I. Larval development. *Dev. Biol.* 62:389–414

49. Lane NJ, Treherne JE. 1969. Peroxidase uptake by glia in desheathed ganglia of the cockroach. *Nature* 223:861–62

50. Lane NJ, Treherne JE. 1972. Studies on perineurial junctional complexes and the sites of uptake of microperoxidase and lanthanum in the cockroach central nervous system. *Tissue Cell* 4:427–36

51. Littleton JT, Bhat MA, Baumgartner S, Broadie K, Belin HJ. 1996. A *Drosophila* neurexin is required for septate junction formation in embryonic development, blood-brain barrier formation, and imaginal patterning. (Abstr.) *Soc. Neurosci.* 22(1–3):1501

52. Lord BA, DiBona DR. 1976. Role of the septate junction in the regulation of paracellular transepithelial flow. *J. Cell Biol.* 71(3):967–72

53. Machen TE, Erlij D, Wooding FBP.

1972. Permeable junctional complexes: the movement of lanthanum across rabbit gallbladder and intestine. *J. Cell Biol.* 54:302–12
54. Marcaillou C, Lauverjat S. 1986. Blood-testis barrier and variation in permeability of the gonial compartment of the testicular follicles of Locus migratoria implanted in a culture without ecdysteroids. *Can. J. Zool.* 64:2053–61
54a. McCarthy KM, Skare IB, Stankewich MC, Furuse M, Tsukita S, et al. 1996. Occludin is a functional component of the tight junction. *J. Cell Sci.* 109:2287–98
55. Meldolesi J, Castiglioni G, Parma R, Nassivera N, De Camilli P. 1978. Ca2+-dependent disassembly and reassembly of occluding junctions in guinea pig pancreatic acinar cells. *J. Cell Biol.* 79:156–72
56. Miranda JC, Cavicchia JC. 1986. A permeability barrier in the testis of an insect *Triatoma*: a freeze-fracture and lanthanum tracer study. *Tissue Cell* 18(3):461–68
57. Murray CL, Quaglia M, Arnason J, Morris CE. 1994. A putative nicotine pump at the metabolic blood-brain barrier of the tobacco hornworm. *J. Neurobiol.* 25:23–34
58. Noirot-Timothée C, Smith DS, Cayer ML, Noirot C. 1978. Septate junctions in insects: comparison between intercellular and intramembranous structures. *Tissue Cell* 10(1):125–36
59. Noirot-Timothée C, Noirot C. 1980. Septate and scalariform junctions in arthropods. *Int. Rev. Cytol.* 63:94–140
60. Partridge WM, ed. 1998. *Introduction to the Blood-Brain Barrier: Methodology, Biology and Pathology*. New York: Cambridge Univ. Press. 550 pp.
61. Pfrieger FW, Barnes BA. 1995. What the fly's glia tell the fly's brain. *Cell* 83:671–74
62. Reese TS, Karnovsky MJ. 1967. Fine structure localization of a blood-brain barrier to exogenous peroxidase. *J. Cell Biol.* 34:207–17
63. Roeder KD. 1953. Electric activity in nerves and ganglia. In *Insect Physiology*, ed. KD Roeder, pp. 423–62. New York: Wiley & Sons
64. Ryerse JS. 1993. Structural, immumocytochemical and initial biochemical characterization of NaOH-extracted gap junctions from an insect, *Heliothis virescens. Cell Tissue Res.* 274:393–403
65. Saint Marie RL, Carlson SD. 1983. Glial membrane specializations and the compartmentalization of the lamina ganglionaris of the housefly compound eye. *J. Neurocytol.* 12(2):243–75
66. Saint Marie RL, Carlson SD. 1983. The fine structure of neuroglia, in the lamina ganglionaris of the housefly, *Musca domestica. J. Neurocytol.* 12:213–41
67. Saint Marie RL, Carlson SD, Chi C. 1984. The glial cells of insects. In *Insect Ultrastucture*, ed. RC King, H Akai, 2:437–75. New York: Plenum
67a. Satoh A, Nakanishi H, Obaishi H, Wada M, Takahashi K, et al. 1998. Neurabin-II/spinophilin—an actin filament-binding protein with one PDZ domain localized at cadherin-based cell-cell adhesion sites. *J. Biol. Chem.* 273:3470–75
68. Scharrer BCJ. 1939. The differentiation between neuroglia and connective tissue sheath in the cockroach (*Periplaneta americana*). *J. Comp. Neurol.* 70:77–88
69. Schiller A, Forsmann WG, Taugner R. 1980. The tight junctions of renal tubules in the cortex and outer medulla. *Cell Tissue Res.* 212:395–413
70. Shaw SR. 1975. Retinal resistance barriers and electrical lateral inhibition. *Nature* 255:480–83
71. Shaw SR. 1977. Restricted diffusion and extracellular space in the insect retina. *J. Comp. Physiol.* 113:257–82
72. Shaw SR. 1978. The extracellular space and blood-eye barrier in an insect retina.

An ultrastructural study. *Cell Tissue Res.* 188:35–61

73. Shaw SR. 1984. Early visual processing in insects. *J. Exp. Biol.* 112:225–51
74. Shivers R. 1979. Occluding-like junctions at mesaxons of central myelin in *Anolis carolinensis* are not "tight." A freeze-fracture-protein tracer analysis. *Tissue Cell* 11:353–58
75. Skaer HB, Maddrell SH, Harrison JB. 1987. The permeability properties of septate junctions in Malpighian tubules of *Rhodnius. J. Cell Sci.* 88(2):251–65
76. Szöllösi A. 1982. Relationships between germ and somatic cells in the testes of locusts and moths. In *Insect Ultrastructure*, ed. RC King, H Akai, 1:32–60. New York: Plenum
77. Szöllösi A, Marcaillou C. 1977. Electron microscope study of the blood-brain barrier in an insect: *Locusta migratoria. J. Ultrastruct. Res.* 59:158–72
78. Szöllösi A, Riemann J, Marcaillou C. 1980. Localization of the blood-testis barrier in the testis of the moth *Anagasta kuehniella, J. Ultrastruct. Res.* 72:189–99
79. Tepass V, Hartenstein V. 1994. The development of cellular junctions in the *Drosophila* embryo. *Dev. Biol.* 161:563–96
80. Tessier-Lavigne M, Goodman CS. 1996. The molecular biology of axon guidance. *Science* 274:1123–35
81. Treherne JE. 1985. Blood brain barrier. In *Comprehensive Insect Physiology, Biochemistry, and Pharmacology*, ed. GA Kerkut, LI Gilbert, 5:115–37. London: Pergamon
82. Tsacopoulos M, Poitry S. 1995. Metabolite exchanges and signal trafficking between glial cells and neurons in the insect retina. In *Neuron-Glia Interrelations During Phylogeny. II. Plasticity and Regeneration,* ed. A Vernadakis, B Roots, pp. 79–94. Totowa, NJ: Humana

82a. van Huizen R, Miller K, Chen DM, Li Y, Lai ZC, et al. 1998. Two distantly positioned PDZ domains mediate multivalent INAD-phospholipase C interactions essential for G protein-coupled signaling. *EMBO J.* 17:2285–97

83. Walz W. 1982. Do neuronal signals regulate potassium flow in glial cells? Evidence from an invertebrate central nervous system. *J. Neurosci. Res.* 7:71–79
84. Willott E, Balda MS, Fanning AS, Jameson B, VanItallie C, Anderson JM. 1993. The tight junction protein Z0-1 is homologous to the *Drosophila* discs large tumor supressor protein of septate junctions. *Proc. Natl. Acad. Sci.* USA 90:7834–38
85. Wood RL. 1990. The septate junction limits mobility of lipophilic marker in plasma membranes of *Hydra vulgaris* (*attenuata*). *Cell Tissue Res.* 259(1):61–66
86. Woods DF, Bryant PJ. 1991. The discs-large tumor suppressor gene of *Drosophila* encodes a guanylate kinase homolog localized at septate junctions. *Cell* 66:451–64
87. Zimmerman RP. 1978. Field potential analysis and the physiology of second order neurons in the visual system of the fly. *J. Comp. Physiol.* 126:297–316

Annu. Rev. Entomol. 2000. 45:175–201

HABITAT MANAGEMENT TO CONSERVE NATURAL ENEMIES OF ARTHROPOD PESTS IN AGRICULTURE

Douglas A. Landis[1], Stephen D. Wratten[2], and Geoff M. Gurr[3]

[1]*Department of Entomology and Center for Integrated Plant Systems, Michigan State University, E. Lansing, Michigan 48824; e-mail: landisd@pilot.msu.edu*
[2]*Division of Soil, Plant and Ecological Sciences, PO Box 84, Lincoln University, Canterbury, New Zealand; e-mail: wrattens@lincoln.ac.nz*
[3]*Pest Management Group, Orange Agricultural College, The University of Sydney, Orange, NSW, 2800, Australia; e-mail: ggurr@oac.usyd.edu.au*

Key Words conservation biological control, agricultural landscapes, predators, parasitoids, disturbance

■ **Abstract** Many agroecosystems are unfavorable environments for natural enemies due to high levels of disturbance. Habitat management, a form of conservation biological control, is an ecologically based approach aimed at favoring natural enemies and enhancing biological control in agricultural systems. The goal of habitat management is to create a suitable ecological infrastructure within the agricultural landscape to provide resources such as food for adult natural enemies, alternative prey or hosts, and shelter from adverse conditions. These resources must be integrated into the landscape in a way that is spatially and temporally favorable to natural enemies and practical for producers to implement. The rapidly expanding literature on habitat management is reviewed with attention to practices for favoring predators and parasitoids, implementation of habitat management, and the contributions of modeling and ecological theory to this developing area of conservation biological control. The potential to integrate the goals of habitat management for natural enemies and nature conservation is discussed.

INTRODUCTION

Conservation biological control involves manipulation of the environment to enhance the survival, fecundity, longevity, and behavior of natural enemies to increase their effectiveness. Such conservation efforts may be directed at mitigating harmful conditions or enhancing favorable ones. Conservation practices can be further categorized as those that focus on reducing mortality, providing supplementary resources, controlling secondary enemies, or manipulating host

0066–4170/00/0107–0175/$14.00

plant attributes to the benefit of natural enemies (129, 160). Because of its importance in enhancing natural enemy performance, conservation biological control should be a keystone of all biological control efforts (61). Of the major forms of biological control, conservation biological control has received the least amount of attention (53). However, this trend has begun to reverse (12, 125).

Habitat Management and Conservation Biological Control

Habitat management can be considered a subset of conservation biological control methods that alters habitats to improve availability of the resources required by natural enemies for optimal performance. Habitat management may occur at the within-crop, within-farm, or landscape levels. Underlying these practices is the understanding that agricultural landscapes often do not provide resources for natural enemies at the optimal time or place. The need for habitat management is directly linked to the biology of specific pests and natural enemies, and the qualities of the environment in which they occur. As a result of frequent and intense disturbance regimes, many agricultural systems are recognized as particularly difficult environments for natural enemies (91, 96, 127, 156). This is especially true for annual monocultural cropping systems where the rates of establishment of imported natural enemies and their success in controlling the target pest are lower than in more stable cropping systems (14, 63, 64, 147). Many of the proximate factors identified as limiting the effectiveness of natural enemies in agricultural systems (pesticides, lack of adult food, lack of alternative hosts) (48, 127, 129) can be viewed as direct results of the disturbance regimes imposed on these systems (93). In particular, the ubiquity of pesticide use in crop production systems has posed a limitation to the successful implementation of biological control. A focus of many past conservation efforts has been to seek more selective pesticides, or to time the use of pesticides to minimize their negative impacts on natural enemies (135). Recently, increasing attention has been paid to conservation practices that seek to alter the quality of the natural enemies' habitat (93a).

Provision of Resources: Short Versus Long Term

Conservation efforts can take various routes to mitigate undesirable conditions. For example, well-timed food sprays have been used to supplement in agricultural systems lacking pollen and nectar resources (62, 113). Alternatively, the establishment of perennial flowering plant habitats may provide similar resources in a more stable fashion over the entire season and for years to come (98). While food sprays address the immediate symptom (i.e. lack of food resources), they fail to address the ultimate problem (i.e. herbicide and land use intensity that reduces the diversity and spatial availability of flowering plants). Similarly, removal of broad-spectrum pesticides from crops may be of limited benefit if habitats supporting populations of natural enemies are not spatially and temporally available in the landscape.

Scope of this Review

Interest in the role of habitat alteration on pests and natural enemies can be traced to observations of insect outbreaks in relation to increasing crop monoculture (3), early empirical work (162) and the subsequent study of the role of plant diversity on insect stability in agricultural ecosystems (126, 165).

The role of non-crop plants on beneficial insects was first summarized by van Emden (163) and later reviewed by Altieri & Whitcomb (4) and Altieri & Letourneau (3). Subsequent studies focused on examining the role of uncultivated corridors (83) or adjacent habitats on influencing the natural enemy communities of fields (13, 31, 44, 47, 80, 86, 168, 169). In this contribution, we will focus on the practice of habitat management to enhance the impact of arthropod predators and parasites of arthropod pests, with an emphasis on the recent literature. For information on enhancement of pathogens, nematode natural enemies of arthropods, conservation of natural enemies of plant pathogens, and weeds, see Barbosa (12).

Just as habitat management can reduce pest attack by top-down effects operating via an enhancement of the third trophic level (natural enemies), pests may also be suppressed by bottom-up effects operating via the first trophic level (flora) of diverse habitats. These resource concentration effects (134), though an important complementary area, are beyond the scope of this review. The relative contribution of top-down and bottom-up effects and the interactions between the resource concentration hypothesis effects and natural enemy-mediated pest regulation have been explored recently (59, 177).

MECHANISMS FOR HABITAT MANAGEMENT OF ARTHROPOD PREDATORS AND PARASITOIDS

Providing the "Right" Diversity

Diversity in agroecosystems may favor reduced pest pressure and enhanced activity of natural enemies (2, 5, 136, 144). However, several authors have noted that to selectively enhance natural enemies, the important elements of diversity should be identified and provided rather than encouraging diversity per se (59, 143, 164, 165, 170). Indeed, it has been shown that simply increasing diversity can exacerbate certain pest problems (7, 10, 29, 59). Identifying the key elements of diversity may be a difficult process, but the process can be guided by an understanding of the resources needed by natural enemies (177, 178). Potential mechanisms include improving the availability of alternative foods such as nectar, pollen, and honeydew; providing shelter or a moderated microclimate in which natural enemies may overwinter or seek refuge from factors such as environmental extremes or pesticides; and providing habitat in which alternative hosts or prey are present. In addition, the temporal availability of such resources may be manipulated to encourage early season activity of natural enemies. Finally, the spatial arrangement of such resources to enhance natural enemy activity within the crop must be considered.

Alternative Food Sources

While some parasitoids are able to obtain needed resources from hosts (76), others require access to non-host foods. Floral nectar is taken by many species (78), and can result in increased rates of parasitism (127). Extrafloral nectar is produced by various plants such as faba bean (*Vicia faba* L.) and cotton (*Gossypium hirsutum* L.), and is an important food source for adult parasitoids (21, 157). Pollen may also be consumed directly (68, 75) or as a contaminant within nectar. The presence of honeydew-producing insects has been suggested as desirable for some parasitoids (54). While most habitat management attempts with alternative food sources have involved hymenopteran parasitoids, Diptera may benefit as well (155).

Quantifying the impact of different nectar sources on parasitoid survival and fecundity has yielded important information on which plant species to retain or introduce into an agroecosystem. Studies have examined a range of wildflowers as nectar sources (74) and identified significant differences in the accessibility of nectaries as a result of floral architecture (119). Work with *Pimpla turionellae* L. showed that rapid weight loss occurred for insects caged on flowers with inaccessible nectaries, illustrating an important advantage of nectar over pollen in providing water as well as nutrition (167). Laboratory experimentation has been used to identify which plants to use in the field. Patt et al (120) screened a range of flowers to determine which were most favorable to *Edovum puttleri* Grissell and *Pediobius foveolatus* Crawford, parasitoids of the Colorado potato beetle (*Leptinotarsa decemlineata* Say). They found that floral architecture influenced the selection of nectar plants, with *E. puttleri* feeding effectively only upon flowers with exposed nectaries, while *P. foveolatus* could also utilize flowers with partially concealed nectaries. Provision of nectar resources may provide increased benefits to herbivores as well as parasitoids (183), but careful selection of plants can reduce this possibility (11, 11a).

Substantial work has been conducted on the enhancement of aphidophagous syrphids by provision of flowering plants. Several of these studies (69, 70, 71, 100) have used the North American annual plant *Phacelia tanacetifolia* Bentham, since it produces large quantities of pollen and nectar, though some work has examined other food plants (39, 104). Borders of *P. tanacetifolia* have also been explored in cabbage, *Brassica oleracea* L. (172), where syrphid numbers increased, and aphid populations declined. Patt et al (121) screened a range of flowers to determine which were suitable for Colorado potato beetle (*Leptinotarsa decemlineata* Say) predators. They found that that dill (*Anethum graveolens* L.) and coriander (*Coriandrum sativum* L.) had flowers compatible with the head morphology of *Coleomegilla maculata* (Degeer) and *Chrysoperla carnea* Stephens. Field observations of foraging behavior supported the utility of these plants and their subsequent use in eggplant (*Solanum melongena* L.) led to enhanced predator numbers, increased consumption rates of *L. decemlineata* egg masses and decreased larval survivorship.

Other studies have focused on the management of existing plants rather than the use of artificially introduced plants. Smith & Papacek (139) showed that the mite *Amblyseius victoriensis* (Womersley), a predator of the phytophagous eriophyid *Tegolophus australis* Keifer, was favored where growers of orange *Citrus aurantium* L. retained and allowed flowering of the widely established ground cover plant, Rhodes grass (*Chloris gayana* Kunth). The wind-blown pollen of the Rhodes grass provided a supplementary food source, and growers are now advised to maximize the benefit by mowing alternate grass strips between tree rows. Crop plants can also be important in providing pollen and nectar resources. Bowie et al (16) studied the effect of a crop of flowering canola (*Brassica napus* L.) on arthropod distribution in an adjacent crop of wheat (*Triticum aestivum* L.). Dissection of adult syrphids from within the canola and at varying distances into the wheat showed that these natural enemies made extensive use of canola pollen, resulting in lower densities of the aphid, *Rhopalosiphum padi* (L.) from sample positions close to the crop interface.

Food sprays can also benefit natural enemies, but this approach may be economically viable only in relatively high-value crops. In cotton, field testing of the product "Envirofeast" showed treated areas to be attractive to various natural enemies including Coccinellidae and Melyridae (Coleoptera); Lygaeidae and Nabidae (Hemiptera); and Chrysopidae (Neuroptera) (113). Though no effects of these predators on prey were reported, separate mesh-house and field testing indicated that treated plants were relatively unattractive to ovipositing *Helicoverpa* spp. (112).

Shelter and Microclimate

Perennial crop systems are potentially more amenable to conservation biological control than are ephemeral annual systems because they are subject to lower levels of disturbance. Thus, resident populations of natural enemies may persist from year to year in perennial crops. However, in some perennial crops such as alfalfa (*Medicago sativa* L.), the normal practice of harvesting entire fields causes disruption to the resident arthropod fauna (159). Among the earliest examples of habitat management were attempts to provide a refuge for natural enemies of alfalfa pests displaced by cutting (148, 158). This practice has also been explored in more recent work where strip harvesting was associated with lower pest densities and lower aphid-to-predator ratios, except in spring (24). Altered harvesting patterns also offer refuge to the brown lacewing *Micromus tasmaniae* (Walker) (95) and a range of coccinellid and hemipteran predators (72a). In these approaches, the provision of a moderated microclimate of uncut alfalfa is likely to be important (158). Overwintering of natural enemies has also been investigated in a number of temperate perennial systems. Shelter has been provided by augmenting leaf debris on the orchard floor with peppermint (*Mentha* x *piperita* L.) (115), wrapping the bases of apple (*Malus domestica* Borth.) trees in vegetable debris held in place with plastic, placing similar debris around the base of smaller trees (43), or providing on-tree refugia of burlap and aluminum in peach, *Punus persica* (L.) Batsch (149).

Perennial crop systems may be disrupted by unavoidable pesticide applications, so refugia outside the treated area can be critical. The predatory mite *A. victoriensis* was able to re-colonize citrus orchards more readily after pesticide-induced disruption where the nearby windbreaks were of *Eucalyptus torelliana* F. Muell. (140). This tree species has also been tested in citrus-growing areas of South Africa, where high populations of the predatory mite *Amblyseius tutsi* Pritchard & Baker occurred on trees nine months after release of mites at the density of one per leaf. Use of this tree is believed to aid in the management of phytophagous mites and thrips (58). An equivalent refuge within tea (*Thea sinestris* L.) production systems exists because the tea bush is stratified into the plucking surface and the interior of the perennial bush. The latter constitutes a refuge for natural enemies that are able to contribute to pest suppression even where pesticides are used (81).

In annual crop systems, maximizing the overwinter survival of natural enemies may be critical in ensuring adequate biological control in the following growing season. This consideration has led to efforts to identify optimal overwintering habitat for arthropod predators of British cereal fields. Overwinter survival of these natural enemies is low because field interiors contain very little vegetation. For this reason, the vast majority of predators overwinter in the field margins and disperse into the crop in the spring (152, 154). However, such colonization of the crop is too slow for optimal pest regulation. Attempts were made to provide suitable overwintering habitat within fields by creating a raised earth bank, termed beetle banks, sown with perennial grasses (174). Comparisons of several grass species led to a recommendation to use cocksfoot (*Dactylis glomerata* L.) and Yorkshire fog (*Holcus lanatus* L.), perennials that have a dense tussock-forming growth habit and harbor the greatest numbers of predators (153, 154).

During the growing season, high temperature and low humidity may constrain natural enemy populations or the activity of individuals. Orr et al interplanted ryegrass (*Lolium multiflorum* Lambert) in seed maize (*Zea mays* L.) fields to reduce the temperature of the soil surface, increasing survival of augmentatively released *Trichogramma brassicae* Bezdenko (118). Alderweireldt reported dramatic increases in spider densities after holes 10 to 12 cm deep were made in the soil surface (1). The microclimate within such depressions is certain to be different from that of untreated soil and is likely to be part of the explanation for the observed effect. However, the observation that web-building spiders (Linyphiidae) were favored more than non-web-building species, combined with the fact that holes with diameters of 5 or 9.5 cm were preferred over smaller holes, suggests that the structural heterogeneity of the manipulated environment influenced web construction.

Non-crop vegetation may be favored by natural enemies as oviposition sites. It has been observed that *Coleomegilla maculata* (Coleoptera: Coccinellidae) lays more eggs on a native weed, *Acalypha ostryaefolia* Ridell than the sweet corn (*Zea mays* L.) crop, even though the plant supported few prey. Larvae then disperse from the weed and climb maize plants. Maize plots bordered by *A. ostryaefolia* contained significantly more *C. maculata* than did plots without a border (38).

Other examples of ground covers or intercrops influencing natural enemy density, activity or impact include carabids in maize (18), parasitoids in cabbage (*Brassica oleracea capitata* L.) (151), mite predators in citrus (97), and various natural enemies in pecan (*Carya illinoensis* Koch) (123) and cotton (180). Although microclimate factors are likely to be involved, in at least some cases the availability of alternative foods in the form of floral resources and/or prey or hosts in the non-crop vegetation may have contributed to the observed effects (142). It is important to note that other studies indicate inconsistent or limited benefits of ground covers or intercrops on natural enemies (22, 28, 37, 40, 72, 139) or even negative effects (35, 36).

Alternative Prey or Hosts

Given sufficient alternative prey, populations of generalist predators may establish within a crop before the arrival and seasonal increase of pests (164). Kozar et al showed that for homopteran pests of apple the distribution of predators was determined by the presence of alternative prey on weeds or in surrounding vegetation (85). Management of crop residues or organic matter may also be effective in enhancing natural enemy populations. A classic demonstration in rice, *Oryza sativa* L., involved increasing the amount of organic matter in test plots leading to increases in detritivores, plankton-feeders, and generalist predators (138). Further, early-season insecticide applications that killed predators led to pest resurgence later in the season. Thus ensuring that natural enemies are protected from the adverse effects of pesticides is a prerequisite for successful habitat management.

Organic matter can also be beneficial when applied to the soil surface of field crops. Manure and straw increased numbers of the carabid *Bembidion lampros* (Herbst), an egg predator of the cabbage root fly *Delia radicum* (L.), and increased total carabid populations in cabbage. This increase was apparent even into the year following the last application and was attributed to increased reproduction by the predators, which in turn may have been the result of the observed increase in alternative prey availability (73). No-tillage production systems that leave crop residue on the soil surface increased the populations and impacts of predatory carabids (18, 26, 27).

Providing alternate hosts may be more difficult for parasitoids, which are often host- or habitat-specific (127). The best known example is of *Anagrus* (Hymenoptera: Mymaridae) parasitoids of the grape leafhopper (*Erythroneura elegantula* Osborne) (46). These egg parasitoids must overwinter on alternative hosts outside the grape (*Vitis vinifera* L.) vineyard because the grape leafhopper does not overwinter in the egg stage. Wild plants in the genus *Rubus* support overwintering eggs of alternative leafhopper hosts, as do cultivated French prune (*Prunus domestica* L.) trees. Prune refuges adjacent to vineyards as well as nearby riparian areas containing *Rubus* spp. increase *Anagrus* parasitism and contribute to control of grape leafhopper (34). As observed by van Emden (164), there may be many unknown examples of parasitoids with requirements for alternate hosts. This lack of alternative hosts may be a contributing factor in the observation that, although many biological control introductions result in establishment, most are unsuccessful in reducing pest densities (61).

Using the ability of some parasitoids to parasitize more than one host species has been explored by Stary (145) in glasshouse crop production. Beans were selected as a model crop and *Myzus persicae* (Sulzer) as the target pest. In this approach, the aphid parasitoids, *Aphidius colemani* Vierck and *Lysiphlebus testaceipes* Cresson were introduced into a pest-free greenhouse along with wheat plants infested with *Schizaphis graminum* (Rond.) as the alternative host. Any subsequent infestation of *M. persicae* within the glasshouse could result in parasitoids moving from the alternative host reservoirs to the target pest population.

Timing the dispersal of natural enemies by manipulating alternative prey has also been explored. The fact that faba beans are commonly infested with aphids was used as a rationale for their selection as a groundcover in hops (*Humulus lupulus* L.) to encourage natural enemies of the hop aphid, *Phorodon humuli* Schrank (57). In this habitat management approach, cutting the bean plants when aphid infestation was detected within the hops was planned as a means of encouraging dispersal of natural enemies to the crop. Though no hop-aphid infestation was observed in this study, dispersal of natural enemies was elicited by cutting the ground cover plants. Cutting is also likely to be important in other cases. Bugg et al showed that the weed *Polygonum aviculare* L. provided both floral resources and alternative prey and was attractive to many natural enemies, especially *Geocoris* spp. (Hemiptera) (20). The authors suggested the plant should be tolerated at field edges as a breeding site for natural enemies, and they speculated that timely removal of the weeds would be useful in encouraging natural enemy dispersal.

Multiple Mechanisms

Habitat management may benefit natural enemies by the simultaneous operation of more than one mechanism. Work with *Eriborus terebrans* (Gravenhorst), the primary parasitoid of *Ostrinia nubilalis* (Hübner) in maize in Michigan showed that females were most frequently captured in maize fields close to wooded field edges and that parasitism of *O. nubilalis* was higher in these areas (90). Laboratory, greenhouse, and field cage studies showed that adults lived longer when provided with sugar and that longevity increased at 25°C versus 35°C (49). The wasps were also shown to be more active on hotter days, which could result in their leaving crop habitats to find shelter in adjacent wooded areas (50). It was suggested that wooded edges benefited *E. terebrans* by providing both a source of adult food (nectar and honeydew) and access to a moderated microclimate (51). Intentionally established "insectary hedgerows" containing a diversity of shrubs and herbaceous plants to provide continuous sources of pollen, nectar, and shelter have been studied (98). Marking studies showed that natural enemies utilize these habitats and disperse into adjacent crops. Up to 47% of the marked *Hyposoter* wasps and 23% of lady beetles were found up to 75 m into the crop.

Because of the potential for natural enemies to be simultaneously favored by more than one mechanism, the small number of studies that have explicitly sought to elucidate which mechanism is most important are noteworthy. Irvin et al used ground covers of either buckwheat (*Fagopyrum esculentum* Moench) or faba bean to increase the impact of natural enemies of leafrollers (Tortricidae) in apple orchards (75). Sticky trap catches of the leafroller parasite *Dolichogenidea tasmanica* Cameron were significantly greater in plots of flowering buckwheat than in plots of buckwheat where flower buds were removed. This showed that floral rewards, rather than either availability of shelter or presence of alternative hosts on the buckwheat foliage was responsible for enhancement of parasitoid activity (60).

In a second study, buckwheat had the greatest effect on *Anacharis* sp., a parasitoid of the brown lacewing (*Micromus tasmaniae* Walker), itself an important natural enemy (146). In these trials, sticky traps in the buckwheat plots caught more parasitoids that those in the control except in the period prior to buckwheat flowering. This indicated that feeding upon flowers, rather than the presence of aphids on the buckwheat upon which host lacewings were feeding, influenced parasitoid catches. Vacuum sampling the buckwheat confirmed the very low densities of lacewings and *Anacharis* suggesting they were not using plants for shelter.

IMPLEMENTING HABITAT MANAGEMENT

Here we address five key issues in the implementation of habitat management: (*a*) the selection of the most appropriate plant species; (*b*) the predator/parasitoid behavioral mechanisms which are influenced by the manipulation; (*c*) the spatial scale over which the habitat enhancement operates, with implications for the area, shape and spacing of resources and refugia for predators and parasitoids; (*d*) the negative aspects associated with adding new plants to an agroecosystem, such as the use of the plant resources by the pest being targeted; (*e*) the degree of uptake by the agricultural/horticultural community of the proposed habitat changes.

Plant Species Selection

Several publications have attempted to rank candidate plants based on their use by the targeted natural enemy (25, 100, 103). These rankings are often carried out on replicated blocks of candidate plants and the only criterion used for the ranking is the number of predators found on, or in traps near, the candidate plants (25). Annual flowers may require repeated planting or cutting of the plants to promote regrowth of flower buds (75). These practices can be labor intensive and may interfere with growers' practices. The advantage of annuals, however, is that there is no persistence of the vegetation into the next growing season, although they may contribute to the seed bank. For perennial crops, the ideal plant may be a species that spreads within or between years and is a good competitor with weeds. One such candidate is *Lobularia maritima* (*L.*) Desv. (alyssum) (Brassicaceae) (25). Though initially slow-growing and susceptible to weed competition, it persists through most "Mediterranean" climates, behaves as a perennial, and spreads well (75). While an excellent source of pollen and nectar, it, like most other potential companion plants, may also be used by certain pests (25).

Phacelia tanacetifolia has been used extensively in Europe and Australia (10, 11, 69, 172). The annual nature of *P. tanacetifolia* however, may be a disadvantage, as discussed, and possibly more importantly, its deep corolla makes the nectar unavailable to short-tongued insects such as hover flies (69). Despite these limitations, significant reductions in aphid populations were achieved by the use of *P. tanacetifolia* around cereal fields in the United Kingdom (69), and around cabbages in New Zealand (172).

Selection of plants for habitat management should consider their suitability in the agricultural or horticultural regime in which they will be placed. On beetle banks, two tussock-forming grasses, *D. glomerata* and *H. lanatus* (174), offered the best overwinter habitat for aphid predators (152, 153, 154, 178). These European species were formerly important components of many pasture seed mixes. They are not particularly invasive, and to date, have not become problems in adjacent fields. However, where arable systems include livestock in the rotation, the palatability of such refuge grasses and other plants should be considered. Evidence from New Zealand and elsewhere indicates that, although livestock such as sheep may graze these grasses, they are not preferred over more common pasture grasses such as *Lolium* spp. (52). However, if grazed the grasses usually recover from their tussock base. In non-European countries, resistance to using these grasses on such banks because of a desire to use a native species is an important consideration (82). Because of the multitude of factors which may influence the choice of plant species for habitat management, the use of a semi-quantitative decision-making tool, the graded weighted checklist, (122), has been suggested (59).

Behavioral Mechanisms

Understanding the behaviors of natural enemies that may be altered by habitat management is a key to success. For example, parasitoids also make extensive use of semiochemicals in host location, some of which emanate from plants (41).

Recently it has been observed that use of semiochemicals may hold potential to manipulate an agroecosystem in a "push-pull" or stimulodeterrent diversionary strategy. Kahn et al studied lepidopteran stem borers and the parasitoid *Cotesia sesamiae* (Cameron) in Africa (84). In this study, the grass *Melinis minutiflora* Beauv. produced volatiles that repel female stem borers and attract the foraging female parasitoids. Intercropping maize with this grass led to reduced infestation by the stem borer and increased rates of parasitism compared with a maize monoculture.

Various techniques may be used to investigate how habitat management alters natural enemy behavior (77). Rubidium has been employed as a marker with notable success. Rubidium levels in plants can be artificially enhanced and this can result in marking of the host and subsequent transfer to natural enemies (34). Alternatively, marking of plant nectar has been used to trace movement of natural enemies as they disperse from refuge habitats (98). Pollen can also be a useful marker, particularly if the plant is not native to the area (e.g. *Phacelia* in Europe) or otherwise easy to identify (68, 75). For *Phacelia,* saffranin staining of predator or parasitoid gut contents can be used to answer questions about the proportion of individuals containing pollen over time or in relation to the distance from the pollen source (178a). This information can help in decision making regarding the size, shape and location of pollen sources. Once the value of a habitat has been established, a crucial but very difficult question to answer is the extent of pest suppression resulting from the emigration of the predators and parasitoids into the adjacent crop. Given the sporadic spatial and temporal distribution of many pests, and the experimental difficulties of evaluating predation and parasitism rates in field populations (77, 175), some sort of experimental crop or refuge manipulation may be needed to evaluate predation rates associated with the refuge (110).

Scale and Spatial Arrangement

Habitat structure may influence arthropod natural enemies over a variety of spatial scales (19, 88, 92, 93, 94, 133, 176) and impact the structure of natural enemy communities (93, 106). Increasing habitat fragmentation at a local scale can result in the loss of parasitoid species and the release of herbivores from parasitism (87). Corbett & Rosenheim showed that arrival of *Anagrus* parasitoids into a vineyard was affected by both the presence of adjacent prune refuges (via a windbreak effect) as well as the distance from presumed overwintering sites in riparian areas (34). The importance of landscape heterogeneity was also shown for predatory coccinellids that were favored by the presence of uncultivated habitats in the landscape (30, 104). A similar association was evident for hymenopteran parasitoids of *Pseudaletia unipuncta* Haworth, where increased parasitism was associated with complex landscapes characterized by smaller field sizes and higher proportions of edges with woodland or wide hedgerow (105). A similar result was shown for oilseed rape (canola) in Germany (151a). Subsequent work has shown this association to be related to the specific landscape composition rather than landscape complexity per se. (111). Structural diversity in the landscape may sometimes impede natural enemy movement between fields as in the case of certain Carabidae (55, 107), although, for others, it may also facilitate natural enemy movement (23).

The question of the optimal shape and distribution of habitats to enhance natural enemies is not well understood. In the spring carabid beetles emigrate from beetle banks to a distance of at least 60 m on either side (152, 153, 154). This finding led to a recommendation in the United Kingdom that these banks should be spaced at least 100 m apart (174). In that work there was a 5 to 10 m gap between the end of the bank and the existing field margin to allow for farm machinery to pass without crossing the bank. Concepts of landscape connectivity (55) indicate that connecting such refugia with existing boundaries may facilitate their initial colonization and interchange within the field margin predatory community. Given that field margins can also act as impediments to predators, connecting such refugia to existing boundaries is even more important to maximize colonization from the boundary to the refuge (175) and to facilitate movement along such features.

Negative Aspects of Added Habitat Diversity

Probably the most obvious potential disadvantage of increasing habitat diversity is that some land may be taken out of production. This potential disadvantage may be a major consideration for high-value crops and, for example, mitigate against the use of alfalfa strips within cotton as proposed by Mensah & Khan (114). In the case of beetle banks in wheat (*Triticum aestivum* L.) fields, such losses have been shown to be more than offset by savings from reduced need for pesticides to control aphids (141). Where within-crop botanical diversification is used, an additional concern is that any advantage from increased natural enemy activity and pest suppression may be more than offset by a reduction in yield resulting from competition (42).

Once a habitat is established, little can be done to control the community of organisms by which it is colonized. Ideally, basic experimental work should quantify these effects (175), but surprises may still occur (25, 146). An example of such a surprise is the use of pollen and nectar sources to enhance parasitism of the potato tuber moth *Phthorimaea operculella* (Zell.) by *Copidosoma koehleri.* In this work the first sources of pollen and nectar selected, although enhancing parasitism rates, were also exploited extensively by tuber moth adults. Subsequent laboratory work exposed moths and parasitoids to a range of candidate pollen and nectar sources, which led to the decision to use borage (*Borago officinalis* L.) in subsequent fieldwork. Borage enhanced parasitism of the moth but was not used by the pest itself, providing a rare example of the careful selection of a "selective food plant" as a pollen and nectar source, taking into account pest and parasitoid use (10, 11).

Another hurdle is grower reluctance to establish plants that could become invasive in the future. Selected plants should have minimal weed status, and/or knowledge should be gained concerning herbicides that control them. Sometimes selective herbicides that do not affect the companion plant will also be necessary, especially during the establishment phase. Selective herbicides have been used effectively to manage the dicotyledenous weed species in crop edges (146). Agronomic factors such as plant phenology, flowering periods, and performance in different seasons of the year, assume increasing importance as the research approaches the technology-transfer stage. Simple repeat-planting experiments can be carried out to measure agronomic characteristics independent of the insects involved (17).

Producer Acceptance

Relatively few cases of widespread grower adoption of habitat management to enhance natural enemies have occurred. These include beetle banks (153), weed strip-management in agriculture (116, 117), cotton-wheat intercropping systems in China, and use of pollen-producing ground covers in citrus orchards in China (97) and Australia (140).

Adoption of beetle banks is widespread in Europe with hundreds occurring in Britain, the Netherlands, Scandinavia, and elsewhere (N.W. Sotherton, personal communication). The term "beetle bank" is about to be included in the Oxford English Dictionary (D Pudley, personal communication). Recommendations on how to manage these habitats continue to evolve and include periodic mowing every few years (8). Such banks are also being turned into a multiple function habitat manipulation sites as the needs of other predators and parasitoids are incorporated. For example, Boatman (15) has added the concept of "conservation headlands" (146) to the banks as well as drilling winter brassicas alongside the headland. The strip of winter brassicas provides a refuge and some food for gray partridge (*Perdix perdix* L.) in winter. Current work in New Zealand is evaluating alyssum for beneficial insects (25), as well as adapting beetle banks for southern hemisphere conditions. One strong option for the future will be some form of beetle bank with alyssum in strips alongside it.

"Weed strip management" has been researched in Europe for several years (117). The practice involves establishing diverse mixtures of native flowering plants in strips in and around fields. These strips have achieved a degree of acceptance in Swiss agriculture where they contribute to increased activity density of Carabidae (Coleoptera) (101, 102), spiders (Araneida), Nabidae (Hemiptera), Dolichopodidae (Diptera) and Syrphidae (Diptera) (67). The results for spiders have indicated only a minor benefit (79). Weed strip management appears to increase the availability of food for carabids and result in enhanced reproduction (181, 182).

Cotton-wheat relay intercropping is practiced on 2.3 million hectares in northern China (179). The primary benefits are reduced damage by cotton aphid, *Aphis gossypii* Glover on seedling cotton and increased productivity. Natural enemies are maintained in the field because they feed on prey in wheat and then easily disperse to emerging cotton seedlings where they can prevent population increase by *A. gossyppi.* In the absence of wheat, predators arrive in cotton too late for effective control (180).

Citrus growers in China are reported to have planted or conserved "weeds," principally *Ageratum conyzoides* L., on 135,000 ha of orchards. These plants are beneficial to natural enemies of the citrus red mite, *Panonychus citri* (Mc. G.), primarily *Amblyseius* spp., through the provision of alternative food in the form of pollen (97). A similar case is evident for citrus growers in Queensland, Australia where *A. victoriensis* is important in the management of eriophyid mites (140). Of growers in the major districts of Central Burnett, Bundaberg, and Emerald, occupying some 3,000 ha, it is estimated that between 80% and 95% actively encourage the flowering of Rhodes grass during the fruit-growing season to provide pollen as an alternative food for the predatory mite. Doing so typically involves mowing alternate inter-rows every three weeks, allowing time for the grass to produce pollen while maintaining a neat orchard. Further, between 30% and 50% of growers used *Eucalyptus torelliana* (F.v. Muell) trees in windbreaks, making use of its hairy leaves to intercept grass pollen and provide an alternative refuge for the predator (D Papacek, personal communication).

CONTRIBUTIONS OF THEORY AND MODELING TO HABITAT MANAGEMENT

While individual researchers have been guided by theoretical considerations, habitat management as a whole has proceeded largely on an empirical basis (6). Increasingly, however, modeling and ecological theory are beginning to inform habitat management decisions.

Modeling

Models can be used to investigate the interaction of refuge habitat arrangement and natural enemy dispersal on colonization of crop habitats (32). Corbett & Plant (33) showed that a simple model could explain a wide variety of observed natural enemy responses to refuge habitats based on the dispersal capability of the enemy, refuge placement, and timing. Others used simple models to investigate critical parameters for biological control of aphids by predators and compared these models to the results of field experiments (161). Under the assumptions that best simulated reality, this model indicated that initial predator density and immigration rate were critical parameters determining if prey suppression would occur. Van der Werf concluded that the landscape features that influenced these rates should be investigated to allow advances in model precision (161). The role of landscape heterogeneity on linyphiid spiders was modeled by Halley et al (65), who found that inclusion of small amounts of grassland in a cereal landscape greatly increased the population size of spiders in cereal fields, while pesticide use and crop rotation decreased population size, and field size had no effect up to a size of 4 km^2, revealing the importance of the dispersal capabilities of spiders.

Ecological Theory

Several authors have summarized the contributions of ecological theory to the practice of habitat management. Gurr et al explored the ecological principles guiding habitat management, including diversity and stability arguments, the natural enemies hypothesis, and life history strategies (59). They integrated these principles with practical considerations to suggest crop system effects, habitat management strategies, and spatial factors that should guide habitat management practices. Landis & Menalled reviewed the ecological and entomological literature on the effect of disturbance on landscape structure and its effects on limiting parasitoid communities within agricultural landscapes (93). They suggested that moderating disturbance regimes through habitat management is a key to conserving parasitoids in agricultural landscapes. Letourneau examined the conservation biology, island biogeography, and metapopulation ecology literature for lessons related to conservation of natural enemies (96). She pointed to the need for landscape-level management if conservation biological control is to succeed on a large scale and concluded that policy and economic strategies are at least as important as the ecological strategies if success is to be achieved. Andow examined the use of vegetational diversity to augment natural enemies (6) and Schellhorn et al reviewed the role of cultural practices to enhance natural enemies from an ecological perspective (137). Theoretical and applied considerations of the role of generalist predators in regulating prey populations have been examined (128, 131, 132, 175).

Finally, a novel contribution has been the synthesis by Wissinger, who contrasted the life history traits of many insects that cyclically colonize ephemeral habitats (173). He argued that annual crops are ephemeral, but predictable, habitats. He outlined a series of life history traits for cyclic colonizers that include (*a*) dispersal from overwintering sites prior to reproduction, followed by (*b*) reproductive onset and loss of flight ability following migration, (*c*) one or more generations of highly fecund but sedentary individuals, and finally (*d*) emigration with delayed reproduction. Wissinger suggested that effective biological control strategies in annual crops must include provision of permanent habitats to act as reservoirs for cyclic colonizing natural enemies.

SYNTHESIS

Implementation

We have reviewed the growing literature on habitat management for natural enemies in agricultural systems. Applying these concepts is a challenge because of the complexity and frequently case-specific nature of the interactions. Landis & Menalled outlined several principles to guide habitat management for parasitoids (93) that are expanded below to include other natural enemies.

Many of the factors that limit the effectiveness of natural enemies in agricultural systems (pesticides, lack of adult food, lack of alternative hosts) are direct results of the disturbance regimes imposed on these systems. Subsequently, conservation of natural enemies by amelioration of these conditions ultimately must be achieved by managing disturbance, not just the symptoms it produces. Habitat management to reduce disturbance may need to occur at various spatial scales. For example, while eliminating a pesticide treatment within a field may permit the establishment or persistence of a natural enemy population, if viable metapopulations (66) do not exist at the landscape-level to provide immigrants, the within-field effort may be ineffective.

Successful implementation of natural enemy conservation involves assessing levels of disturbance in agricultural systems. Practices such as cover cropping, intercropping and reduced tillage relax the overall disturbance regime, although they may require some new disturbances (i.e. herbicides) in order to manage weeds. Alternatively, some new technologies such as transgenic maize expressing *Bacillus thuringiensis* Berliner toxins may appear to reduce disturbance by eliminating pesticide treatments, but may in fact represent a more pervasive disturbance through the potential for cascading multitrophic level impacts.

Habitat management may not always demand a radical change in farming practices as illustrated by the relative ease with which beetle banks and border plantings can be introduced into annual row crop systems. The success of these tactics within such economically important systems indicates that habitat management can be packaged in an agronomically acceptable form. Moreover, the introduction of perennial vegetation does not always require that land be taken out of productive use. Habitat management may be as simple as taking into consideration adjacent crops when planning rotations. Observations such as those by Bowie et al could lead to very simple recommendations such as "wherever possible plant canola and wheat in neighboring fields" (16). However, improved methods for extending conservation methods to producers are needed (166). Synergistic interactions of habitat management for natural enemies with agroforestry (45, 124), weed management (24a, 104a, 111a), and soil and water conservation (89) may help in this process.

Finally, habitat management will normally be complemented by other methods and should not be promoted as a standalone method. Commonly these will employ biological control agents that have been released in classical or augmentative manners. In such instances habitat management holds considerable potential for enhancing the success rates of classical agents, and to maximize the persistence and impact on pest population of augmentative agents. In the future, these formerly separate branches of biological control will be merged to synergistic effect in "integrated biological control" (sensu 61).

BENEFITS TO NATURE CONSERVATION

Because of the difficulties of nature conservation within reserves alone, it has been argued that the agricultural ecosystems, which occupy large areas of land, are critical in maintaining biodiversity (108). Thus the encouragement of natural enemies by strategic increases in habitat diversity offers potential to align the goals of agriculture with those of nature conservation (56). The benefits arise partly from the lessened need for synthetic pesticides and the attendant direct and indirect off-target impact on organisms such as butterflies (99), birds (130), and small mammals (15a, 150); and partly from the introduction or maintenance of structural heterogeneity, for example, as observed for skylarks *Alauda arvensis* L. (141). Flora, too, may be conserved. One approach is to establish and preserve a matrix of native vegetation in which agriculture is nested. This process would lead to the creation of a "variegated landscape" (109) which, in the case of rangelands of New South Wales, Australia, has been suggested as an approach to conserve native grassland flora. Initiatives such as these may be made more palatable to economic rationalists if the potential value of conserved species is stressed. Such species might include direct biocontrol effect of previously unrecognized predators such as the big brown bat, *Eptesicus fuscus* Palisotde Beauvois, which was shown only recently to feed extensively on pest insects (171).

CONCLUSION

The science of habitat management is still in its infancy. Publications on the topic date from the first half of the century, but close to 80% of the literature reviewed herein was published after 1990. While this is in part attributable to our intention to focus on recent literature, within the current decade a marked trend toward increasing activity is evident. The international community of scientists engaged in this field appear well poised to meet the challenge of making agricultural pest management more effective, and production systems more sustainable, as well as being increasingly compatible with nature conservation.

ACKNOWLEDGMENTS

We thank FD Menalled, A Raman and JN Landis who provided helpful comments and discussion on earlier versions and A Kenady who assisted in typing the manuscript. This work was supported in part by the Michigan Agricultural Experiment Station (DAL), The University of Sydney (GMG) and by The Foundation for Research, Science, and Technology (New Zealand); The Agricultural Marketing Research and Development Trust (New Zealand); and The Royal Society of London (SDW).

Visit the Annual Reviews home page at www.AnnualReviews.org.

LITERATURE CITED

1. Alderweireldt M. 1994. Habitat manipulations increasing spider densities in agroecosystems: possibilities for biological control. *J. Appl. Entomol.* 118:10–16
2. Altieri MA. 1991. Increasing biodiversity to improve insect pest management in agro-ecosystems. In *Biodiversity of Microorganisms and Invertebrates: Its Role in Sustainable Agriculture,* ed. DL Hawksworth, pp. 165–82. Wallingford, UK: CAB Int. 302 pp.
3. Altieri MA, Letourneau DK. 1982. Vegetation management and biological control in agroecosystems. *Crop Prot.* 1:405–30
4. Altieri MA, Whitcomb WH. 1979. The potential use of weeds in manipulation of beneficial insects. *Hortic. Sci.* 14:12–18
5. Andow D. 1991. Vegetational diversity and arthropod population response. *Annu. Rev. Entomol.* 36:561–86
6. Andow D. 1996. Augmenting natural enemies in maize using vegetational diversity. In *Biological Pest Control in Systems of Integrated Pest Management.* Food Fertil. Technol. Cent. Book Ser. No. 47, pp. 137–53
7. Andow DA, Risch SJ. 1985. Predation in diversified agroecosystems: relations between a coccinellid predator *Coleomegilla maculata* and its food. *J. Appl. Ecol.* 22:357–72
8. Anon., not dated. *Helping Nature To Control Pests.* Research carried out by the Univ. Southampton and the Game Conserv. Trust under Contract from MAFF. 6 pp.
9. Deleted in proof
10. Baggen LR, Gurr GM. 1998. The influence of food on *Copidosoma koehleri* (Hymenoptera: Encyrtidae), and the use of flowering plants as a habitat management tool to enhance biological control of potato moth, *Phthorimaea operculella* (Lepidoptera: Gelechiidae). *Biol. Control* 11:9–17
11. Baggen LR, Gurr GM, Meats A. 1999. Flowers in tri-trophic systems: mechanisms allowing selective exploitation by insect natural enemies for conservation biological control. *Entomol. Exp. Appl.* 91:155–61

11a. Baggen LR, Gurr GM, Meats A. 1999. Field observations on seletive food plants in habitat manipulation for biological control of potato moth, *Phthorimaea operculella* Zeller (Lepidoptera: Gelechiidae) by *Copidosoma koehleri* Blanchard (Hymenoptera: Encyrtidae). *Proc. 4th Int. Hymenoptera Conf., January 6–11, Aust. Nat. Univ., Canberra.* CSIRO: Canberra. In press

12. Barbosa P, ed. 1998. *Conservation Biological Control.* San Diego, CA: Academic. 396 pp.
13. Bedford SE, Usher MB. 1994. Distribution of arthropod species across the margins of farm woodlands. *Agric. Ecosyst. Environ.* 48:295–305
14. Beirne BP. 1975. Biological control attempts by introductions against pest insects in the field in Canada. *Can. Entomol.* 107:225–36
15. Boatman ND. 1989. Selective weed control in field margins. *Brighton Crop Prot. Conf.—Weeds,* pp. 785–94. Brighton, UK: Brighton Br. Crop Prot. Counc.

15a. Boatman ND. 1998. The value of buffer zones for the conservation of biodiversity. 1998 *Brighton Crop Prot. Conf.—Pests and Diseases,* pp. 939–50. Brighton, UK: Brighton Br. Crop Prot. Counc.

16. Bowie MH, Gurr GM, Hossain Z, Baggen LR, Frampton CM. 1999. Effects of

distance from field edge on aphidophagous insects in a wheat crop and observations on trap design and placement. *Int. J. Pest Manage.* 45:69–73
17. Bowie MH, Wratten SD, White AJ. 1995. Agronomy and phenology of "companion plants" of potential for enhancement of biological control. *NZ J. Crop Hortic. Sci.* 23:423–27
18. Brust GE, Stinner BR, McCartney DA. 1985. Tillage and soil insecticide effects on predator-black cutworm (Lepidoptera: Noctuidae) interactions in corn agroecosystems. *J. Econ. Entomol.* 78:1389–92
19. Bugg RL. 1993. Habitat manipulation to enhance the effectiveness of aphidophagous hover flies (Diptera: Syrphidae). *Sust. Agric.* 5:12–15
20. Bugg RL, Ehler LE, Wilson LT. 1987. Effect of common knotweed (*Polygonum aviculare*) on abundance and efficiency of insect predators of crop pests. *Hilgardia* 55:1–53
21. Bugg RL, Ellis RT, Carlson RW. 1989. Ichneumonidae (Hymenoptera) using extra floral nectar of faba bean (*Vicia faba* L.: Fabaceae) in Massachusetts. *Biol. Agric. Hortic.* 6:107–14
22. Bugg RL, Waddington C. 1994. Using cover crops to manage arthropod pests of orchards: a review. *Agric. Ecosyst. Environ.* 50:11–28
23. Burel F, Baudry J. 1990. Hedgerow networks as habitats for colonization of abandoned agricultural land. In *Species Dispersal in Agricultural Landscapes*, ed. RHG Bunce, DC Howard, pp. 238–55. Lymington, UK: Belhaven
24. Cameron PJ, Allan DJ, Walker GP, Wightman JA. 1983. Management experiments on aphids (*Acyrthosiphon* spp.) and beneficial insects in lucerne. *NZ J. Exp. Agric.* 11:343–49
24a. Carmona DM, Menalled FD, Landis DA. 1999. *Gryllus pensylvanicus* (Orthoptera: Gryllidae): laboratory weed seed predation and within-field activity-density. *J. Econ. Entomol.* 92:825–29
25. Chaney WE. 1998. Biological control of aphids in lettuce using in-field insectaries. See Ref. 125, pp. 73–84
26. Clark MS, Gage SH, Spence JR. 1997. Habitats and management associated with common ground beetles (Coleoptera: Carabidae) in a Michigan agricultural landscape. *Environ. Entomol.* 26:519–27
27. Clark MS, Luna JM, Stone ND, Youngman RR. 1994. Generalist predator consumption of armyworm (Lepidoptera: Noctuidae) and effect of predator removal on damage in no-till corn. *Environ. Entomol.* 23:617–22
28. Coll M, Bottrell DG. 1996. Movement of an insect parasitoid in simple and diverse plant assemblages. *Ecol. Entomol.* 21:141–49
29. Collins FL, Johnson SJ. 1985. Reproductive response of caged adult velvetbean caterpillar and soybean looper to the presence of weeds. *Agric. Ecosyst. Environ.* 14:139–49
30. Colunga-Garcia M, Gage SH, Landis DA. 1997. Response of an assemblage of Coccinellidae (Coleoptera) to a diverse agricultural landscape. *Environ. Entomol.* 26:797–804
31. Coombes DS, Sotherton NW. 1986. The dispersal and distribution of polyphagous predatory Coleoptera in cereals. *Ann. Appl. Biol.* 108:461–74
32. Corbett A. 1998. The importance of movement in response of natural enemies to habitat manipulation. See Ref. 125, pp. 25–48
33. Corbett A, Plant RE. 1993. Role of movement in the response of natural enemies to agroecosystem diversification: a theoretical evaluation. *Environ. Entomol.* 22:519–31
34. Corbett A, Rosenheim JA. 1996. Impact of a natural enemy overwintering refuge and its interaction with the surrounding landscape. *Ecol. Entomol.* 21:155–64
35. Costello MJ, Altieri MA. 1994. Living

mulches suppress aphids in broccoli. *Calif. Agric.* 48:24–28

36. Costello MJ, Altieri MA. 1995. Abundance, growth rate and parasitism of *Brevicoryne brassicae* and *Myzus persicae* (Homoptera: Aphididae) on broccoli grown in living mulches. *Agric. Ecosyst. Environ.* 52:187–96
37. Costello MJ, Daane KM. 1998. Influence of groundcover on spider populations in a table grape vineyard. *Ecol. Entomol.* 23:33–40
38. Cottrell TE, Yeargan KV. 1999. Factors influencing dispersal of larval *Coleomegilla maculata* from the weed *Acalypha ostryaefolia* to sweet corn. *Entomol. Exp. Appl.* 90:313–22
39. Cowgill SE, Wratten SD, Sotherton NW. 1993. The effect of weeds on the numbers of hoverfly (Diptera: Syrphidae) adults and the distribution and composition of their eggs in winter wheat. *Ann. Appl. Biol.* 123:499–14
40. Daane KM, Costello MJ. 1998. Can cover crops reduce leafhopper abundance in vineyards? *Calif. Agric.* 52:27–33
41. De Moraes CM, Lewis WJ, Pare PW, Alborn HT, Tumlinson JH. 1998. Herbivore-infested plants selectively attract parasitoids. *Nature* 393:570–73
42. Dempster JP. 1969. Some effects of weed control on the numbers of the small cabbage white (*Pieris rapae* L.) on Brussels sprouts. *J. Appl. Ecol.* 6:339–45
43. Deng X, Zheng ZQ, Zhang NX, Jia XF. 1988. Methods of increasing winter-survival of *Metaseiulus occidentalis* (Acari: Phytoseiidae) in north west China. *Chin. J. Biol. Contr.* 4:97–101 (Chinese with Engl. summ)
44. Dennis P, Fry GLA. 1992. Field margins: can they enhance natural enemy population densities and general arthropod diversity on farms? *Agric. Ecosyst. Environ.* 40:95–115
45. Dix ME, Johnson RJ, Harrell MO, Case RM, Wright RJ, et al. 1995. Influence of trees on abundance of natural enemies of insect pests: a review. *Agrofor. Syst.* 29:303–11
46. Doutt RL, Nakata J. 1973. The Rubus leafhopper and its egg parasitoid: an endemic biotic system useful in grape-pest management. *Environ. Entomol.* 2:381–86
47. Duelli P, Studer M, Marchand I, Jakob S. 1990. Population movements of arthropods between natural and cultivated areas. *Biol. Conserv.* 54:193–207
48. Dutcher JD. 1993. Recent examples of conservation of arthropod natural enemies in agriculture. In *Pest Management: Biologically Based Technologies,* ed. RD Lumsden, JL Vaughn, pp. 101–8. Washington, DC: Am. Chem. Soc.
49. Dyer LE, Landis DA. 1996. Effects of habitats, temperature, and sugar availability on longevity of *Eriborus terebrans* (Hymenoptera: Ichneumonidae). *Environ. Entomol.* 25:1192–201
50. Dyer LE, Landis DA. 1997. Diurnal behaviour of *Eriborus terebrans* (Hymenoptera: Ichneumonidae). *Environ. Entomol.* 26:1385–92
51. Dyer LE, Landis DA. 1997. Influence of noncrop habitats on the distribution of *Eriborus terebrans* (Hymenoptera: Ichneumonidae) in cornfields. *Environ. Entomol.* 26:924–32
52. Edwards GR, Lucas RJ, Johnson MR. 1993. Grazing preference for pasture species by sheep is affected by endophyte and nitrogen fertility. *Proc. NZ Grassland Assoc.* 55:137–41
53. Ehler LE. 1998. Conservation biological control: past, present, and future. See Ref. 12, pp. 1–8
54. England S, Evans EW. 1997. Effects of pea aphid (Homoptera: Aphididae) honeydew on longevity and fecundity of the alfalfa weevil (Coleoptera: Curculionidae) parasitoid *Bathyplectes curculionis* (Hymenoptera: Ichneumonidae). *Environ. Entomol.* 26:1437–41
55. Frampton GK, Cilgi T, Fry GLA, Wrat-

ten SD. 1995. Effects of grassy banks on the dispersal of some carabid beetles (Coleoptera: Carabidae) on farmland. *Biol. Conserv.* 71:347–55

56. Gillespie RG, New TR. 1998. Compatibility of pest management and conservation strategies. In *Pest Management—Future Challenges: Proc. 6th Aust. Appl. Entomol. Res. Conf. 29 Sept.–2 Oct.. Vol. 2*, ed. MP Zaluki, RAI Drew, GG White, pp. 195–208. Brisbane: Univ. Queensland. 356 pp.
57. Goller E, Nunnenmacher L, Goldbach HE. 1997. Faba beans as a cover crop in organically grown hops: influence on aphids and aphid antagonists. *Biol. Agric. Hortic.* 15:279–84
58. Grout TG, Stephen PR. 1995. New windbreak tree contributes to integrated pest management of citrus. *Citrus J.* 5:26–27
59. Gurr GM, van Emden HF, Wratten SD. 1998. Habitat manipulation and natural enemy efficiency: implications for the control of pests. See Ref. 12, pp. 155–83
60. Gurr GM, Wratten SD, Irvin NA, Hossain Z, Baggen LR, et al. 1998. Habitat manipulation in Australasia: recent biological control progress and prospects for adoption. In *Pest Management—Future Challenges: Proc. 6th Aust. App. Entomol. Res. Conf. 29 Sept.–2 Oct. Vol. 2*, ed. MP Zaluki, RAI Drew, GG White, pp. 225–35. Brisbane: Univ. Queensland. 356 pp.
61. Gurr GM, Wratten SD. 1999. "Integrated biological control": a proposal for enhancing success in biological control. *Int. J. Pest Manage.* 45:81–84
62. Hagen KS, Sawall EF Jr, Tassan RL. 1971. The use of food sprays to increase effectiveness of entomophagous insects. *Proc. Tall Timber Conf. Ecol. Anim. Control Habitat Manage.*, pp. 59–81. Tallahassee, FL: Tall Timbers Res. Stn.
63. Hall RW, Ehler LE. 1979. Rate of establishment of natural enemies in classical biological control. *Bull. Entomol. Soc. Am.* 25:280–82
64. Hall RW, Ehler LE, Bisabri-Ershadi B. 1980. Rates of success in classical biological control of arthropods. *Bull. Entomol. Soc. Am.* 26:111–14
65. Halley JM, Thomas CFG, Jepson PC. 1996. A model for the spatial dynamics of linyphiid spiders in farmland. *J. Appl. Ecol.* 33:471–92
66. Hanski I. 1989. Metapopulation dynamics: does it help to have more of the same? *Trends Ecol. Evol.* 4:113–14
67. Hausmmann A. 1996. The effects of weed strip-management on pests and beneficial arthropods in winter wheat fields. *Z. Pflanzenkr. Pflanzenschutz* 103:70–81
68. Hickman JM, Lövei GL, Wratten SD. 1995. Pollen feeding by adults of the hoverfly Melanostoma fasciatum (Diptera: Syrphidae). *NZ J. Zool.* 22:387–92
69. Hickman JM, Wratten SD. 1996. Use of *Phacelia tanacetifolia* strips to enhance biological control of aphids by hoverfly larvae in cereal fields. *J. Econ. Entomol.* 89:832–40
70. Holland JM, Thomas SR, Courts S. 1994. *Phacelia tanacetifolia* flower strips as a component of integrated farming. In *Field Margins: Integrating Agriculture and Conservation,* ed. N Boatman, pp. 215–20. Farnham: Br. Crop Prot. Counc. 404 pp.
71. Holland JM, Thomas SR. 1996. *Phacelia tanacetifolia* flower strips: their effect on beneficial invertebrates and gamebird chick food in an integrated farming system. *Acta Jutl.* 71:171–82
72. Hooks CRR, Valenzula HR, Defrank J. 1998. Incidence of pests and arthropod natural enemies in zucchini grown with living mulches. *Agric. Ecosyst. Environ.* 69:217–31

72a. Hossain Z, Gurr GM, Wratten SD. 1999. On the potential to manipulate lucerne insects by strip cutting. *Austr. J. Entomol.* In press

73. Humphreys IC, Mowat DJ. 1994. Effects of some organic treatments on predators

(Coleoptera: Carabidae) of cabbage root fly, *Delia radicum* (L.) (Diptera: Anthomyiidae), and on alternative prey species. *Pedobiologia* 38:513–18
74. Idris AB, Grafius E. 1995. Wildflowers as nectar sources for *Diadegma insulare* (Hymenoptera: Ichneumonidae), a parasitoid of diamondback moth (Lepidoptera: Yponomeutidae). *Environ. Entomol.* 24:1726–35
75. Irvin N, Wratten SD, Frampton CM. 1999. Understory management for the enhancement of the leafroller parasitoid *Dolichogenidea tasmanica* Cameron in Canterbury, New Zealand orchards. In *The Hymenoptera: Evolution, Biodiversity and Biological Control.* Melbourne, Austr.: CSIRO. In press
76. Jervis MA, Kidd NAC. 1986. Host-feeding strategies in hymenopteran parasitoids. *Biol. Rev.* 61:395–434
77. Jervis MA, Kidd NAC. 1996. *Insect Natural Enemies.* London: Chapman & Hall. 691 pp.
78. Jervis MA, Kidd NAC, Fitton MG, Huddleston T, Dawah HA. 1993. Flower-visiting by hymenopteran parasitoids, *J. Nat. Hist.* 27:67–105
79. Jmhasly P, Netwig W. 1995. Habitat management in winter wheat and evaluation of subsequent spider predation on insect pests. *Acta Oecol.* 16:389–403
80. Kajak A, Lukasiewicz J. 1994. Do semi-natural patches enrich crop fields with predatory epigean arthropods? *Agric. Ecosyst. Environ.* 49:149–61
81. Kawai A. 1997. Prospect for integrated pest management in tea cultivation in Japan. *Jpn. Agric. Res. Q.* 31:213–17
82. Keesing VF, Wratten SD. 1997. Integrating plant and insect conservation. In *Plant Genetic Conservation: The In Situ Approach,* ed. N Maxted, L Ford, JG Hawkes, pp. 220–35. London: Chapman & Hall
83. Kemp JC, Barrett GW. 1989. Spatial patterning: impact of uncultivated corridors on arthropod populations within soybean agroecosystems. *Ecology* 70:114–28
84. Khan ZR, Ampong-Nyarko K, Chiliswa P, Hassanali A, Kimani S, et al. 1997. Intercropping increases parasitism of pests. *Nature* 388:631–32
85. Kozar F, Brown MW, Lightner G. 1994. Spatial distribution of homopteran pests and beneficial insects in an orchard and its connection with ecological plant protection. *J. Appl. Entomol.* 117:519–29
86. Kromp B, Steinberger KH. 1992. Grassy field margins and arthropod diversity: a case study of ground beetles and spiders in eastern Austria (Coleoptera: Carabidae; Arachnida: Opiliones) *Agric. Ecosyst. Environ.* 40:71–93
87. Kruess A, Tscharntke T. 1994. Habitat fragmentation, species loss, and biological control. *Science* 264:1581–84
88. Landis DA. 1994. Arthropod sampling in agricultural landscapes: ecological considerations. In *Handbook of Sampling Methods for Arthropod Pests in Agriculture,* ed. LP Pedigo, GD Buntin, pp. 15–31. Boca Raton, FL: CRC Press
89. Landis DA, Dyer LE. 1998. *Conservation Buffers and Beneficial Insects, Mites and Spiders.* (Conserv. Inf. Sheet, Agron. Ser.) Lansing, MI: USDA Nat. Res. Conserv. Serv. 4 pp.
90. Landis DA, Haas MJ. 1992. Influence of landscape structure on abundance and within-field distribution of European corn borer (Lepidoptera: Pyralidae) larval parasitoids in Michigan. *Environ. Entomol.* 21:409–16
91. Landis DA, Marino PC. 1999. Landscape structure and extra-field processes: impact on management of pests and beneficials. In *Handbook of Pest Management,* ed. J Ruberson, pp. 79–104. New York: Marcel Dekker
92. Landis DA, Marino PC. 1999. Conserving parasitoid communities of native pests: implications for agricultural landscape structure. In *Biological Control of Native or Indigenous Pests,* ed. L Char-

let, G Brewer, pp. 38–51 Thomas Say Publ. In Entomol., *Entomol. Soc. Am. Proc.*

93. Landis DA, Menalled FD. 1998. Ecological considerations in the conservation of effective parasitoid communities in agricultural systems. See Ref. 12, pp. 101–21

93a. Landis DA, Menalled FD, Lee JC, Carmona DM, Perez-Valdez A. 1999. Habitat modification to enhance biological conrol in IPM, In Emerging Technologies for Integrated Pest Management: Concepts, Research, and Implementation. ed. GG Kennedy, TB Sutton. St. Paul: APS Press In press

94. Landis DA, van der Werf W. 1977. Early-season aphid predation impacts establishment and spread of sugar beet yellows virus in the Netherlands. *Entomophaga* 42:499–516

95. Leathwick DM. 1989. *Applied ecology of the Tasmanian Lacewing* Micromus tasmaniae *Walker (Neuroptera: Hemerobiidae).* PhD diss. Univ. Canterbury, NZ. 120 pp.

96. Letourneau DK. 1998. Conservation biology: lessons for conserving natural enemies. See Ref. 12, pp. 9–38

97. Liang W, Huang M. 1994. Influence of citrus orchard ground cover plants on arthropod communities in China: a review. *Agric. Ecosyst. Environ.* 50:29–37

98. Long RF, Corbett A, Lamb C, Reberg-Horton C, Chandler J, Stimmann M. 1998. Beneficial insects move from flowering plants to nearby crops. *Calif. Agric.* 52:23–26

99. Longley M, Sotherton NW. 1997. Factors determining the effects of pesticides upon butterflies inhabiting arable farmland. *Agric. Ecosyst. Environ.* 61:1–12

100. Lövei GL, McDougall D, Bramley G, Hodgson DJ, Wratten SD. 1992. Floral resources for natural enemies: the effect of *Phacelia tanacetifolia* (Hydrophyllaceae) on within-field distribution of hoverflies (Diptera: Syrphidae). *Proc. 45th NZ Plant Prot. Conf.* pp. 60–61

101. Lys JA.1994. The positive influence of strip-management on ground beetles in a cereal field: increase, migration and overwintering. In *Carabid Beetles: Ecology and Evolution,* ed. K Desender, M Dufrene, M Loreau, ML Luff, JP Maelfait. pp. 451–55. Dortdrecht/Boston/London: Kluwer. 474 pp.

102. Lys JA, Zimmermann M, Netwig W. 1994. Increase in activity density and species number of carabid beetles in cereals as a result of strip-management. *Entomol. Exp. Appl.* 73:1–9

103. MacLeod A. 1992. Alternative crops as floral resources for beneficial hoverflies (Diptera: Syphidae). *Proc. Brighton Crop Prot. Conf.,* pp. 997–1002. Brighton: Br. Crop Prot. Counc.

104. Maredia KM, Gage SH, Landis DA, Scriber JM. 1992. Habitat use patterns by the seven-spotted lady beetle (Coleoptera: Coccinellidae) in a diverse agricultural landscape. *Biol. Control.* 2:159–65

104a. Marino PC, Gross KL, Landis DA. 1997. Post-dispersal weed seed loss in Michigan maize fields. *Agric. Ecosyst. Environ.* 66:189–96

105. Marino PC, Landis DL. 1996. Effect of landscape structure on parasitoid diversity and parasitism in agroecosystems. *Ecol. Appl.* 6:276–84

106. Marino PC, Landis DA. 1999. Parasitoid community structure: implications for biological control in agricultural landscapes. In *Interchanges of Insects Between Agricultural and Surrounding Habitats,* ed. B Ekbom. pp. 181–91. Dordrecht: Kluwer.

107. Mauremootoo JR, Wratten SD, Worner SP, Fry GL. 1995. Permeability of hedgerows to predatory carabid beetles. *Agric. Ecosyst. Environ.* 52:141–48

108. McIntyre S. 1994. Integrating agricultural land-use and management for the conservation of a native grassland flora

in a variegated landscape. *Pac. Conserv. Biol.* 1:236–44

109. McIntyre S, Barrett GW, Kitching RL, Recher HF. 1992. Species triage—seeing beyond wounded rhinos. *Conserv. Biol.* 6:604–6
110. Menalled F, Lee J, Landis D. 1999. Manipulating carabid beetle abundance alters prey removal rates in corn fields. *Biocont.* 43:451–66
111. Menalled FD, Marino PC, Gage SH, Landis DA. 1999. Does agricultural landscape structure affect parasitism and parasitoid diversity? *Ecol. Appl.* 9:634–41

111a. Menalled FD, Marino PC, Renner KA, Landis DA. 1999. Post-dispersal weed seed predation in crop fields as a function of agricultural landscape structure. *Agric. Ecosys. Environ.* In press

112. Mensah RK. 1996. Suppression of *Helicoverpa* spp. (Lepidoptera: Noctuidae) oviposition by use of the natural enemy food supplement Envirofeast(R). *Aust. J. Entomol.* 35:323–29
113. Mensah RK. 1997. Local density responses of predatory insects of *Helicoverpa* spp. to a newly developed food supplement "Envirofeast" in commercial cotton in Australia. *Int. J. Pest Manage.* 43:221–25
114. Mensah RK, Khan M. 1997. Use of *Medicago sativa* (L.) interplantings/trapcrops in the management of the green mirid *Creontiades dilutus* (Stal) in commercial cotton in Australia. *Int. J. Pest Manage.* 43:197–202
115. Morris MA, Croft BA, Berry RE. 1996. Overwintering and the effects of autumn habitat manipulation and carbofuran on *Neoseiulus fallacis* and *Tetranychus urticae* in peppermint. *Exp. Appl. Acarol.* 20:249–58
116. Nentwig W. 1988. Augmentation of beneficial arthropods by strip management. 1. Succession of predaceous arthropods and long-term change in the ratio of phytophagous and predaceous species in a meadow. *Oecologia* 76:597–606
117. Nentwig W. 1998. Weedy plant species and their beneficial arthropods: potential for manipulation in field crops. See Ref. 125, pp. 49–72
118. Orr DB, Landis DA, Mutch DR, Manley GV, Stuby SA, King RL. 1997. Ground cover influence on the microclimate and *Trichogramma* (Hymenoptera: Trichogrammatidae) augmentation in seed corn production. *Environ. Entomol.* 26:433–38
119. Orr DB, Pleasants JM. 1996. The potential of native prairie plant species to enhance the effectiveness of the *Ostrinia nubilalis* parasitoid *Macrocentrus grandii. J. Kans. Entomol. Soc.* 69:133–43
120. Patt JM, Hamilton GC, Lashomb JH. 1997. Foraging success of parasitoid wasps on flowers: interplay of insect morphology, floral architecture and searching behavior. *Entomol. Exp. Appl.* 83:21–30
121. Patt JM, Hamilton GC, Lashomb JH. 1997. Impact of strip-insectary intercropping with flowers on conservation biological control of the Colorado potato beetle. *Adv. Hortic. Sci.* 11:175–81
122. Pearson AW. 1990. The management of research and development. In o-*Technology and Management,* ed. R Wild, pp. 28–41. New York: Nichols
123. Peng RK, Christian K, Gibb K. 1998. The effect of non-crop vegetation on the insect pests and their natural enemies in cashew (*Anacardium occidentale* L.) plantations. *Plant Prot. Q.* 13:16–20
124. Peng RK, Incoll LD, Sutton SL, Wright C, Chadwick A. 1993. Diversity of airborne arthropods in a silvoarable agroforestry system. *J. Appl. Ecol.* 30:551–62
125. Pickett CH, Bugg RL, eds. 1998. *Enhancing Biological Control: Habitat Management to Promote Natural Enemies of Agricultural Pests.* Berkeley: Univ. Calif. Press. 422 pp.
126. Pimentel D. 1961. Species diversity and

insect population outbreaks. *Ann. Entomol. Soc. Am.* 54:76–86
127. Powell W. 1986. Enhancing parasite activity within crops. In *Insect Parasitoids,* ed. JK Waage, D Greathead, pp. 314–40. London: Academic. 389 pp.
128. Provencher L, Reichert SE. 1994. Model and field test of prey control effects by spider assemblages. *Environ. Entomol.* 23:1–17
129. Rabb RL, Stinner RE, van den Bosch R. 1976. Conservation and augmentation of natural enemies. In *Theory and Practice of Biological Control,* ed. CB Huffaker, PS Messenger, pp. 23–254. New York: Academic
130. Rands MRW. 1985. Pesticide use on cereals and survival of grey partridge chicks: a field experiment. *J. Appl. Ecol.* 22:49–54
131. Reichert SE. 1998. The role of spiders and their conservation in the agroecosystem. See Ref. 125, pp. 211–37
132. Reichert SE, Bishop L. 1990. Prey control by an assemblage of generalist predators: spiders in garden test systems. *Ecology* 71:1441–50
133. Roland J, Taylor PD. 1997. Insect parasitoid species respond to forest structure at different spatial scales. *Nature* 386:710–13
134. Root RB. 1973. Organization of a plant-arthropod association in simple and diverse habitats: the fauna of collards (*Brassica oleracea*). *Ecol. Monogr.* 43:95–124
135. Ruberson JR, Nemoto H, Hirose Y. 1998. Pesticides and conservation of natural enemies. See Ref. 12, pp. 207–20
136. Ryszkowski L, Karg J, Margalit G, Paoletti MG, Zlotin R. 1993. Above-ground insect biomass in agricultural landscapes of Europe. In *Landscape Ecology and Agroecosystems,* ed. RG Bunce, HL Ryszkowski, MG Paoletti, pp. 71–82. Boca Raton, FL: Lewis
137. Schellhorn NA, Harmon JP, Andow DA. 1999. Using cultural practices to enhance pest control by natural enemies. In *Environmentally Sound Approaches to Insect Pest Management,* ed. J Rechcigl, pp. 147–70. Chelsea, MI: Ann Arbor Press
138. Settle WH, Ariawan H, Astuti ET, Cahyana W, Hakim AL, et al. 1996. Managing tropical rice pests through conservation of generalist natural enemies and alternative prey. *Ecology* 77:1975–88
139. Smith D, Papacek DF. 1991. Studies of the predatory mite *Amblyseius victoriensis* (Acarina: Phytoseiidae) in citrus orchards in south-east Queensland: control of *Tegolophus australis* and *Phyllocoptruta oleivora* (Acarina: Eriophyidae), effects of pesticides, alternative host plants and augmentative release. *Exp. Appl. Acarol.* 12:195–217
140. Smith MW, Arnold DC, Eikenbary RD, Rice NR, Shiferaw A, et al. 1996. Influence of ground cover on beneficial arthropods in pecan. *Biol. Control.* 6:164–76
141. Sotherton NW. 1995. Beetle banks—helping nature to control pests. *Pest. Outlook* 6:13–17
142. Sotherton NW. 1996. Conservation headlands: a practical combination of intensive cereal farming and conservation. In *The Ecology of Temperate Cereal Fields,* ed. LG Firbank, N Carter, JF Darbyshire, GR Potts, pp. 373–98. London: Blackwell Sci.
143. Speight MR. 1983. The potential of ecosystem management for pest control. *Agric. Ecosyst. Environ.* 10:183–99
144. Stamps WT, Linit MJ. 1998. Plant diversity and arthropod communities: implications for temperate agroforestry. *Agrofor. Syst.* 39:73–89
145. Stary P. 1993. Alternative host and parasitoid in first method in aphid pest management in glasshouses. *J. Appl. Entomol.* 116:187–91
146. Stephens MJ, France CM, Wratten SD, Frampton C. 1998. Enhancing biological control of leafrollers (Lepidoptera: Tor-

tricidae) by sowing buckwheat (*Fagopyron esculentum*) in an orchard. *Biocont. Sci. Technol.* 8:547–58

147. Stiling P. 1990. Calculating the establishment rates of parasitoids in classical biological control. *Am. Entomol.* 36:225–30
148. Summers CG. 1976. Population fluctuations of selected arthropods in alfalfa: influence of two harvesting practices. *Environ. Entomol.* 5:103–10
149. Tamaki G, Halfhill JE. 1968. Bands on peach trees as shelters for predators of the green peach aphid. *J. Econ. Entomol.* 61:707–11
150. Tew TE, MacDonald DW, Rands MRW. 1992. Herbicide application affects microhabitat use by arable wood mice (*Apodemus sylvaticus*). *J. Appl. Ecol.* 29:532–39
151. Theunissen J, Booji CJH, Lotz LAP. 1995. Effects of intercropping white cabbage with clovers on pest infestation and yield. *Entomol. Exp. Appl.* 74:7–16

151a. Thies C, Tscharntke T. 1999. Landscape structure and biological control in agroecosystems. *Science.* 285:893–95

152. Thomas MB, Sotherton NW, Coombes DS, Wratten SD. 1992. Habitat factors influencing the distribution of polyphagous predatory insects between field boundaries. *Ann. Appl. Biol.* 120:197–202
153. Thomas MB, Wratten SD, Sotherton NW. 1991. Creation of "island" habitats in farmland to manipulate populations of beneficial arthropods: predator densities and emigration. *J. Appl. Ecol.* 28:906–17
154. Thomas MB, Wratten SD, Sotherton NW. 1992. Creation of "island" habitats in farmland to manipulate populations of beneficial arthropods: predator densities and species composition. *J. Appl. Ecol.* 29:524–31
155. Topham M, Beardsley JW Jr. 1975. Influence of nectar source plants on the New Guinea sugar cane weevil parasite, *Lixophaga sphenophori* (Villeneuve). *Proc. Hawaii. Entomol. Soc.* 22:145–55
156. Townes H. 1972. Ichneumonidae as biological control agents. In *Proc. Tall Timbers Conf. Ecol. Anim. Control Habitat Manage. 3rd,* pp. 235–48. Tallahassee, FL: Tall Timbers Res. Stn.
157. Treacy MF, Benedict JH, Walmsley MH, Lopez JD, Morrison RK. 1987. Parasitism of bollworm (Lepidoptera: Noctuidae) eggs on nectaried and nectariless cotton. *Environ. Entomol.* 16:420–23
158. van den Bosch R, Lagace CF, Stern VM. 1967. The interrelationship of the aphid, *Acyrthosiphon pisum,* and its parasite, *Aphidius smithi,* in a stable environment. *Ecology* 48:993–1000
159. van den Bosch R, Messneger PS, Gutierrez AP. 1982. *An Introduction to Biological Control.* New York: Plenum. 247 pp.
160. van den Bosch R, Telford AD. 1964. Environmental modification and biological control. In *Biological Control of Pests and Weeds,* ed. P DeBac, pp. 459–88. New York: Reinhold
161. van der Werf W. 1995. How do immigration rates affect predator/prey interactions in field crops? Predictions from simple models and an example involving the spread of aphid-borne viruses in sugar beet. In *Arthropod Natural Enemies in Arable Land. I. Density, Spatial Heterogeneity and Dispersal. Nat. Sci. Ser. 9.* ed. S Toft, W Riedel, pp. 295–312. Aarhus, Den: Aarhus Univ. Press. 314 pp.
162. van Emden HF. 1962. Observations of the effects of flowers on the activity of parasitic Hymenoptera. *Entomol. Mon. Mag.* 98:265–70
163. van Emden HF. 1965. The role of uncultivated land in the biology of crop pests and beneficial insects. *Sci. Hortic.* 17:121–36
164. van Emden HF. 1990. Plant diversity and natural enemy efficiency in agroecosystems. In *Critical Issues in Biological Control,* ed. M Mackauer, LE Ehler, J Roland, p. 63–80. Andover, MA: Intercept. 330 pp.

165. van Emden HF, Williams GF. 1974. Insect stability and diversity in agroecosystems. *Annu. Rev. Entomol.* 19:455–75
166. Waage JK. 1996. "Yes, but does it work in the field?" The challenge of technology transfer in biological control. *Entomophaga* 41:315–32
167. Wackers FL, Bjornsen A, Dorn S. 1996. A comparison of flowering herbs with respect to their nectar accessibility for the parasitoid *Pimpla turionellae. Exp. Appl. Entomol.* 7:177–82
168. Wallin H. 1985. Spatial and temporal distribution of some abundant carabid beetles (Coleoptera: Carabidae) in cereal and adjacent habitats. *Pedobiologia* 28:19–34
169. Wallin H. 1986. Habitat choice of some field-inhabiting carabid beetles (Coleoptera: Carabidae) studied by recapture of marked individuals. *Ecol. Entomol.* 11:457–66
170. Way MJ. 1966. The natural environment and integrated methods of pest control. *J. Appl. Ecol.* 3:29–32
171. Whitaker JO Jr. 1995. Food of the big brown bat *Eptesicus fuscus* from maternity colonies in Indiana and Illinois. *Am. Midl. Nat.* 134:346–60
172. White AJ, Wratten SD, Berry NA, Weigmann U. 1995. Habitat manipulation to enhance biological control of *Brassica* pests by hover flies (Diptera: Syrphidae). *J. Econ. Entomol.* 88:1171–76
173. Wissinger SA. 1997. Cyclic colonization in predictably ephemeral habitats: a template for biological control in annual crop systems. *Biol. Control.* 10:4–15
174. Wratten SD. 1992. Weeding out the cereal killers. *New Sci.* 1835:31–35
175. Wratten SD, Powell W. 1990. Cereal aphids and their natural enemies. In *The Ecology of Temperate Cereal Fields,* ed. LG Firbank, N Carter, JF Darbyshire, GR Potts, pp. 233–58. London: Blackwell Sci.
176. Wratten SD, Thomas CFG. 1990. Farm-scale dynamics of predators and parasitoids in agricultural landscapes. In *Landscape Ecology and Agroecosystems,* ed. RGH Bunce, L Ryszkowski, MG Paoletti, pp. 219–37. Boca Raton, FL: Lewis Publishers
177. Wratten SD, van Emden HF. 1995. Habitat management for enhanced activity of natural enemies of insect pests. In *Ecology and Integrated Farming Systems,* ed. DM Glen, MP Greaves, HM Anderson, pp. 117–45. Chichester, UK: John Wiley. 329 pp.
178. Wratten SD, van Emden HF, Thomas MB. 1998. Within field and border refugia for the enhancement of natural enemies. See Ref. 125, pp. 375–403
178a. Wratten SD, White AJ, Bowie MH, Berry NA, Weigmann U. 1995. Phenology and ecology of hover flies (Diptera: Syrphidae) in New Zealand. *Environ. Entomol.* 24:595–600
179. Xia J. 1994. An integrated cotton insect pest management system for cotton-wheat intercropping in North China. *Proc. World Cotton Res. Conf.—I: Challenging the Future,* ed. GA Constable, NW Forrester. Brisbane, Aust., 617 pp.
180. Xia J. 1997. *Biological control of cotton aphid (Aphis gossypii Glover) in cotton (inter) cropping systems in China; a simulation study.* PhD diss. Landouwuniversiteit, Wageningen, 173 pp.
181. Zangger A. 1994. The positive influence of strip-management on carabid beetles in a cereal field: accessibility of food and reproduction in *Poecilus cupreus.* In *Carabid Beetles: Ecology and Evolution,* ed. K Desender, M Dufrene, M Loreau, ML Luff, JP Maelfait. pp. 469–72 Dortdrecht/Boston/London: Kluwer. 474 pp.
182. Zangger A, Lys JA, Nentwig W. 1994. Increasing the availability of food and the reproduction of *Poecilus cupreus* in a cereal field by strip-management. *Entomol. Expl. App.* 71:111–20
183. Zhao JZ, Ayers GS, Grafius EJ, Stehr FW. 1992. Effects of neighboring nectar-producing plants on populations of pest Lepidoptera and their parasitoids in broccoli plantings. *Great Lakes Entomol.* 25:253–58

Annu. Rev. Entomol. 2000. 45:203–231

FUNCTION AND MORPHOLOGY OF THE ANTENNAL LOBE: New Developments

B. S. Hansson and S. Anton
Department of Ecology, Lund University, SE-223 62 Lund, Sweden; e-mail: bill.hansson@ekol.lu.se

Key Words olfaction, neuroethology, integration, time coding, pharmacology

■ **Abstract** The antennal lobe of insects has emerged as an excellent model for olfactory processing in the CNS. In the present review we compile data from areas where substantial progress has been made during recent years: structure-function relationships within the glomerular array, integration and blend specificity, time coding and the effects of neuroactive substances and hormones on antennal lobe processing.

INTRODUCTION

The olfactory system plays a very important role for survival and reproduction in the large majority of insects. Sexual partners are located via sex pheromones, food plants are found via kairomones, conspecifics can be gathered using aggregation pheromones, oviposition can be deterred or induced by oviposition pheromones, and nectar-rich flowers are found using synomones. In all of these interactions and many more the olfactory system is indispensable.

Here we endeavor to review progress in studies of the primary olfactory center of the insect central nervous system, the antennal lobe (AL). The morphology and physiology of the insect AL and its neuronal elements has been extensively reviewed in the past (5, 25, 56, 63, 68, 103, 122). Recent years have, however, seen substantial progress in certain areas. The coding of odor blends has now been studied in a number of species and some general conclusions can be reached. The encoding of fluctuating odor concentrations is another area where new results from different organisms provide a more comprehensive view of the neural mechanisms underlying time coding. The functional significance of olfactory glomeruli has been investigated in several species, and also here general patterns have been established. Finally we will discuss the influence of some neuroactive substances and of hormones on AL information processing. In the introduction the stage will be set for the more detailed discussions in the following paragraphs.

0066–4170/00/0107–0203/$14.00

The Antennal Lobe and its Glomeruli

The AL of insects is a sphere-shaped part of the deutocerebrum which receives sensory input from olfactory receptor neurons (ORN) on the antennae and mouth parts (Figure 1*a*) The anatomical and physiological organization of the lobe and its neuronal elements have similar features as primary olfactory centers in other organisms throughout the animal kingdom. The insect AL, as all other primary olfactory centers, consists of so called glomeruli, spheroidal neuropilar structures, housing synaptic contacts between receptor axons and AL interneurons. The arrangement and number of glomeruli within the AL are largely species specific. The number of glomeruli varies from about 32 in the mosquito *A. aegypti* (12) to more than 1000 in locusts and social wasps (41, 61). In most insect species the AL contains 40 to 160 individually identifiable glomeruli arranged in one or two layers around a central fibrous core (8, 12, 23, 121–123, 132). The small glomeruli of locusts and social wasps, which are arranged in a multiglomerular layer around a central fiber core are, however, not individually identifiable (41, 61). Glomeruli are to different extents separated from each other by glial processes (48a, 139).

In a number of well-studied insect species using sex-pheromone communication, a sexual dimorphism in glomerular structure has been observed. The AL of male Lepidoptera, Dictyoptera, and Hymenoptera contains one or several enlarged glomeruli in addition to the sexually isomorphic glomeruli, which are also present in females (for review see 122). The enlarged glomeruli form the macroglomerular complex (MGC) in moths and bees and the macroglomerulus (MG) in cockroaches. The MGC or MG exclusively receive input from sex pheromone–sensitive ORNs (8, 19, 26, 52, 64, 121–123). These findings led to the hypothesis of a functional identity of individual glomeruli, which will be discussed further in this paper, concerning structure-function relationships in ORNs and AL interneurons.

Olfactory Receptor Neuron Anatomy and Physiology

ORN axons originating from cell bodies in the antennae enter the AL through the antennal nerve (Figure 1*a*). In most insects the ORNs arborize in a peripheral layer within one single ipsilateral glomerulus (Figure 1*b*) (for review see 68, 103). Some glomeruli in the honey bee, in some cockroaches, and in the moths *Manduca sexta* and *Mamestra brassicae* do, however, receive ORN terminals penetrating the entire glomeruli (8, 18, 69, 121). In flies ORNs arborize bilaterally and terminate in single corresponding glomeruli in each AL (132). In some Orthoptera species single ORNs arborize in several small glomeruli (Figure 1*c*) (41, 59; Ignell, Anton, and Hansson, personal observation).

ORNs respond very specifically to odors, e.g. to single sex pheromone components (51). Traditionally, ORNs have been characterized as specialists or generalists, where the specialists were examplified by the sex pheromone-specific

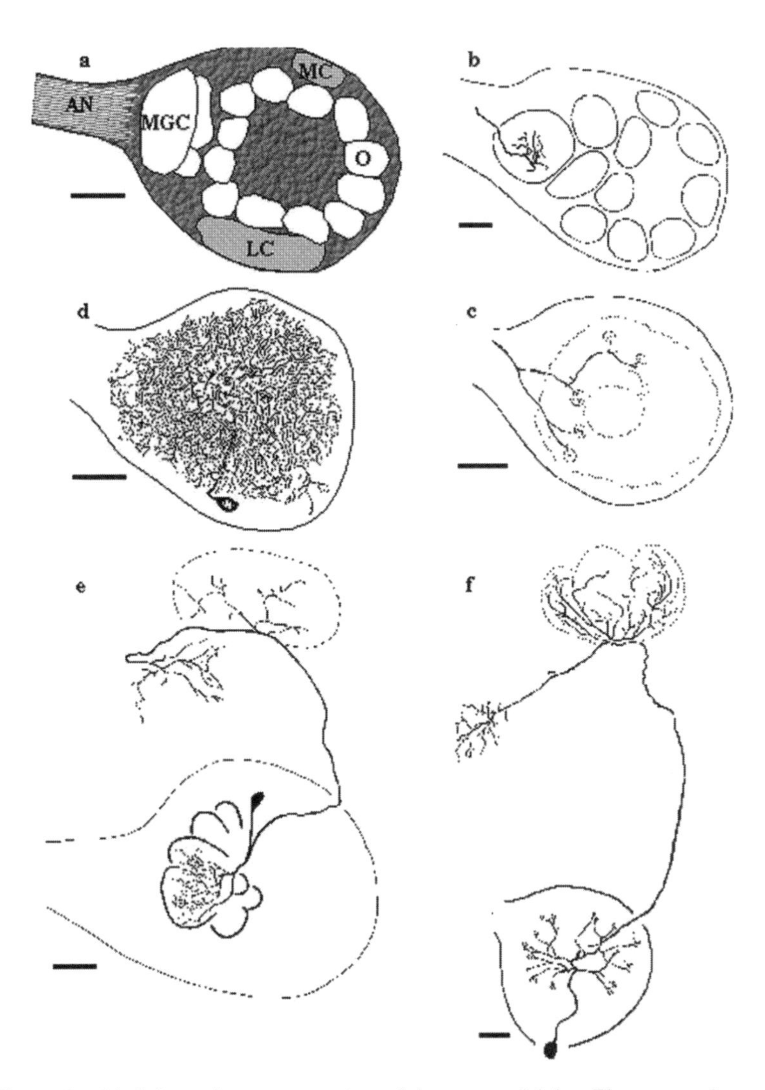

Figure 1 (*a*) Schematic representation of the antennal lobe. The antennal nerve (*AN*) enters from *top, left.* In the antennal lobe the glomerular array is represented by ordinary glomeruli (*O*) and the macroglomerular complex (*MGC*). The MGC is found only in the receiving sex (typically the male) in species using long-distance pheromones. In the periphery of the antennal lobe the two main cell body clusters are situated laterally (*LC*) and medially (*MC*). (*b*) Branching patterns of a pheromone-sensitive olfactory receptor neuron in the antennal lobe of a *Spodoptera littoralis* male. Reconstruction from frontal sections. Scale bar 100 μm. (*c*) Multiglomerular projection of an ORN in *Schistocerca gregaria.* Reconstruction from frontal sections. Scale bar 100 μm. (*d*) Branching pattern of a local interneuron in a *Spodoptera littoralis* female. Reconstruction from frontal sections. Scale bar 100 μm. (*Continued on page 206.*)

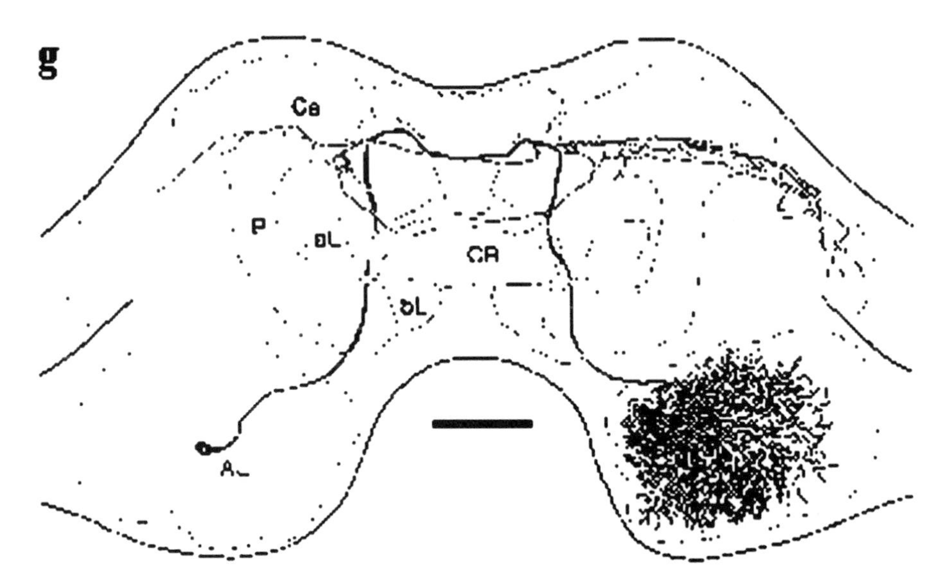

Figure 1 (*e*) Projection neuron arborizing within MGC glomeruli a and b in the antennal lobe of a *Trichoplusia ni* male. The axon runs through the median antenno-cerebral tract, arborizes in the calyces of the ipsilateral mushroom body (dashed outline) and in the inferior lateral protocerebrum. Reconstruction from frontal sections. Scale bar 50 μm. (*f*) Projection neuron with multiglomerular arborizations in *Schistocerca gregaria.* The axon runs through the median antenno-cerebral tract, arborizes in the calyces of the ipsilateral mushroom body (dotted outline) and in the inferior lateral protocerebrum. Reconstruction from frontal sections. Scale bar 100 μm. (*g*) Serotonin immunoreactive centrifugal neuron in the antennal lobe of the sphinx moth, *Manduca sexta.* The neuron arborizes in the superior protocerebrum from where an axon extends into the contralateral antennal lobe, where it ramifies in all glomeruli. Calyces (*Ca*), peduncle (*P*), alpha lobe (*aL*) and beta lobe (*bL*) of the mushroom body, central body (*CB*) (81).

ORNs in moths, and generalists were found among food odor detectors. Today, more examples of extremely specific food- and oviposition-site-detecting ORNs are reported. Most likely, very few generalistic ORNs exist.

Antennal Lobe Interneurons and Their Synaptic Interactions

Three types of AL interneurons have branches within the glomeruli: intrinsic AL neurons [local neurons (LN)], output neurons [projection neurons (PN)], and centrifugal neurons. Most AL neurons have their cell bodies in a number of cell clusters at the periphery of the AL. The location of the cell clusters varies between species (for review see 5). Some cell bodies of PNs and centrifugal neurons are situated in the protocerebrum or in the ventral nerve cord. LNs and PNs both receive direct synaptic input from ORNs. PNs also receive indirect input from

ORNs via LNs (20, 35–37, 99, 100, 136). Output and input synapses have been identified between LNs and PNs and between neurons of the same type. ORNs receive synaptic input from LNs, but not from PNs (20, 35, 37, 100, 136). Synaptic connections of centrifugal neurons have not been studied so far.

Local Neuron Anatomy and Physiology

Three types of LNs have been described in different insect species. Multiglomerular LNs with homogenous arborizations throughout the AL (Figure 1*d*) can be distinguished from multiglomerular LNs with heterogenous arborizations, asymmetrically distributed within the glomeruli of the AL, and from oligoglomerular LNs with branches in only few glomeruli (2, 32, 40, 42, 55, 90, 104, 131, 134).

LNs display different spontaneous activity and different response patterns to odors in different insect species. In the sphinx moth, *M. sexta,* LNs generally seem to have a bursting spontaneous activity, while only some LNs in noctuid moth species show such an activity (2, 3, 32, 55). In bees no bursting activity was found in LNs (42). Intracellular recordings from LNs often show different spike sizes, which might result from several spike initiating zones (32). In *M. sexta* LNs odor stimulation can elicit three different response types. Stimulation of most LNs results in a short-latency excitatory response, while a few neurons show a delayed excitatory or inhibitory response to odor stimulation. Both short-latency and delayed excitatory responses are followed by an inhibitory period (32). In locusts only non-spiking LNs have been described so far. These LNs respond to odors with graded potentials and membrane potential oscillations, oscillations forming a base for synchronization of PNs and protocerebral neurons (91, 96).

A large proportion of LNs in different insect species studied so far show GABA-like immunoreactivity, supporting the physiological findings that output from LNs is inhibitory (13, 32, 34, 68, 72, 93, 125, 129, 145).

Projection Neuron Anatomy and Physiology

PNs usually have their cell bodies in cell clusters in the periphery of the AL. Dendritic arborizations of PNs are either uniglomerular or multiglomerular within the AL, and the axons leave the AL via a number of antenno-cerebral tracts (82), connecting the AL with different areas of the protocerebrum, most prominently the calyces of the mushroom bodies and the lateral protocerebrum (Figures 1*e*-*f*). The most common type of PN in different insect species has uniglomerular arborizations (multiglomerular in the locust) and the axon projects through the inner antenno-cerebral tract (IACT) to the calyces of the mushroom bodies and to the lateral protocerebrum (for review see 5). PNs leaving the AL through the middle antenno-cerebral tract (MACT) have multiglomerular arborizations within the AL and have been found in moths, flies, and bees (43, 70, 132). Their axons project to the lateral protocerebrum and to areas adjacent to the pedunculus of the mushroom body. Uni- and multiglomerular PNs leave the AL through the outer antenno-cerebral tract (OACT) in different insect species, including moths, bees, and flies (43, 55, 69, 70, 101, 132). Their axons terminate unilaterally or bilaterally in different areas of the lateral protocerebrum and in some cases also in the calyces of the mushroom bodies. PNs leaving the AL through the dorsal antenno-cerebral tract (DACT) have only been described in the sphinx moth so far (70, 78). These PNs have their cell body in the protocerebrum, have multi-glomerular arborizations in the contralateral AL, and project their axon to the lateral horn of the protocerebrum. Very few PNs leaving the AL through the dorso-medial antenno-cerebral tract (DMACT) have been described (78, 132). These neurons have their soma in the subesophageal ganglion and innervate single glomeruli in both ALs. In *M. sexta,* the axon projects to the calyces of the mushroom bodies and to the lateral horn of the protocerebrum.

The response characteristics of PNs have mainly been studied in moths, cockroaches, bees, and locusts. In all species studied, PNs with excitatory responses to certain odors were described, but the exact time pattern of the response varies. In locusts and in the noctuid moths *Helicoverpa zea* and *Heliothis virescens,* inhibitory responses to odor stimulation occurred and the same odors elicited either excitatory or inhibitory responses in different neurons (4, 75, 142). In the sphinx moth, one of the two behaviorally active sex pheromone components elicits an excitatory response in certain PNs, whereas the second component elicits an inhibitory response (25, 26, 29). The spontaneous activity and response of PNs to odors exhibit equally spaced action potentials in *M. sexta,* in contrast to burst-like activity in LNs (32). In noctuid moths, this difference in physiological characteristics has not been consistently found (3; S Anton & BS Hansson, in preparation).

Centrifugal Neuron Anatomy and Physiology

Centrifugal neurons have been studied in a number of insect species including bees, cockroaches, moths and locusts, primarily with immunocytochemical staining methods (Figure 1*g*). The cell body of centrifugal neurons is usually situated outside the AL, in the protocerebrum, in the SOG or in the ventral nerve cord. Exceptionally, serotonin-immunoreactive centrifugal neurons with their cell bodies in the AL cell clusters were found in *M. sexta* and *P. americana* (81, 124). While the dendritic branches of centrifugal neurons can receive input in different areas of the brain or the ventral nerve cord, they exhibit multiglomerular varicose branching within the AL, but can have other output areas in addition (see 5). Both descending neurons from the protocerebrum and ascending neurons from the ventral nerve cord send axonal branches into the AL, but also centrifugal neurons with more restricted arborization areas have been found (21, 49, 65, 66, 118, 132). Neuroanatomical findings suggest a modulatory function of centrifugal neurons, which will be discussed below.

FUNCTIONAL CORRELATES IN THE GLOMERULAR ARRAY

Antennal Lobe Projections of Olfactory Receptor Neurons

Ever since the discovery of glomeruli in different olfactory systems, a general question has been whether these structures have a functional meaning. In insects we have reached far in answering this question. In 1987 Koontz & Schneider (88) could show that the MGC is targeted exclusively by pheromone-detecting receptor neurons in a male moth. More or less simultaneously, Christensen & Hildebrand (26) showed that the same structure was innervated by PNs responding to antennal stimulation with pheromone. A strong case for a functional separation between the MGC and sexually isomorphic glomeruli was thus present.

In the beginning of the 1990s, a method to stain single physiologically defined neurons was developed (58). Using this method, investigators could determine how the axonal arborizations of ORNs tuned to different pheromone components distributed themselves among the glomeruli of the MGC. A number of studies based on this method have now been published and have also been reviewed in detail elsewhere (14, 52, 54, 58, 120, 137). The main result from all of these studies is that a clear odotopic projection pattern of pheromone-specific ORNs is present in more or less all species studied to date. Each glomerulus of the MGC receives input from one type of ORN. Information regarding behavioral antagonists in the pheromone communication system is also received in MGC glomeruli, and ORNs tuned to these compounds subsequently target other specific glomeruli of the MGC.

A very good correspondence between the number of pheromone components and antagonists used by a species and the number of glomeruli included in its MGC has also been shown. We can thus state that there is high probability that olfactory glomeruli are specific projection stations for receptor neurons displaying the same olfactory specificity. These results coincide very well with what was later shown in transgenic mice (109). Olfactory receptor neurons expressing a specific receptor molecule all target the same glomeruli.

Glomerulus Innervation by Antennal Lobe Neurons

What happens at the next neural levels of the antennal lobe? The incoming message is transferred synaptically from ORNs to LNs and to PNs (35–37). LNs often target a majority of the AL glomeruli (104), so from these neurons very little is to be gained in the discussion of specific innervation patterns, except that they can definitely distribute incoming signals all over the AL. PNs, on the other hand, often show uniglomerular dendritic innervation (26, 79) and can thus be of great interest in a comparison with the identity of ORNs innervating the same glomerulus. These innervation patterns have been compared in a number of species, and the results are not as clear as for the ORNs. The first detailed study of PN innervation patterns in the MGC was performed in *M.sexta* (57). A clear pattern was observed, showing a functional projection pattern also in the PNs. PNs responding to the same pheromone component always targeted the same MGC glomerulus. A drawback was, and still is, the lack of information regarding the exact projection patterns of pheromone-specific ORNs.

A number of investigators have now attempted to compare PN dendritic innervation patterns with well-established ORN branching patterns. The results from different studies do not agree unanimously, and totally consistent results should not be expected in biological systems. In *Heliothis virescens,* results from a comparison of twelve physiologically defined and morphologically reconstructed PNs with well-established branching patterns of ORNs show that the PNs always innervate the glomerulus targeted by ORNs expressing the same physiological specificity as the PN (14, 54, 142). Some of the PNs also innervated MGC glomeruli that were not expected from their physiology (142).

In another noctuid moth, *Trichoplusia ni,* a different pattern has been established (4a). This moth uses a very complicated communication system involving seven different types of ORNs. When the innervation patterns of ORNs and PNs were compared, only part of the PNs branched in the glomerulus expected from their physiological characteristics. For PNs responding to the main pheromone component, about two-thirds branched as expected, but for PNs responding to minor components or to a behavioral antagonist none to one-third of them branched in a glomerulus innervated by ORNs of the same specificity. Many neurons also innervated several glomeruli and vice versa.

Optical Imaging of Antennal Lobe Activity

A dream for researchers working on the functional morphology of the AL has been to be able to map olfactory responses over the AL glomeruli in real time. This dream has now come true thanks to the development of optical imaging techniques and their adaptation to the insect system. Optical imaging utilizes changes in, for instance, calcium concentration or voltage and converts these changes to changes in light intensity. By advanced image capture and processing techniques, the minute changes in light emission can be observed and quantified. The pioneers in this research (45, 46, 76) have shown how very distinct activity patterns can be registered in the honey bee AL after stimulation with different odors. A single odor is normally not represented by activity in a single glomerulus, but rather by activity in a number of glomeruli. Obviously a single type of molecule interacts with a number of different types of ORNs resulting in a spatial representation of the odor among the AL glomeruli (76). The response to an odor is not static over time, but can develop as a changing pattern, moving from one pattern to another as time progresses after stimulation. By use of the optical imaging technique it has also been possible to show that the coding of odors is bilaterally symmetric in the two ALs (46).

Functional Correlates in the Glomerular Array: Conclusions

An odor is represented as a spatial map, usually over a number of AL glomeruli. These maps are most likely formed as a consequence of the ability of a single odor molecule type to interact with different types of ORNs expressing different receptor sites in their dendritic membranes. Single glomeruli are clearly the target site for specific ORN types. One glomerulus—one ORN type. As mentioned, this fact now has support from other olfactory systems. What happens after this level? Here we see differences between species. Some species seem to preserve the odotopic patterns established by the ORNs also in the PN innervation patterns; others seem to have less odotopically influenced dendritic innervation patterns among their PNs. Maybe none of these solutions should be any surprise. We know that incoming signals turn glomeruli into "information packages," with each glomerulus holding information regarding a single or a few odors. We also know that this message can be widely distributed over the entire AL by LNs that are often the first postsynaptic element to ORNs. Obviously, through evolution some species have maintained a strictly chemotopically organized system also at the PN level, while others have for some reason let this pattern dissolve into a morphologically less organized system. The neural basis to achieve both these solutions is clearly present.

CODING OF ODOR BLENDS

Rarely do insects rely on single odors as key stimuli releasing a specific behavioral sequence. Whether the odor cues are involved in food, partner or oviposition site search, they usually come as a bouquet composed of a number of chemical constituents. The importance of odor blends has been studied in detail, especially in the sexual communication systems of moths. Many moth species have had their female-produced, long distance pheromones identified in detail (7), and in the great majority the pheromone consists of a number of components, each adding to the attractivity of the blend. The importance of the ratios of components within the blend varies from species to species, where some species have a very narrow response window and others accept a wide range of component ratios.

Also in other behavioral contexts blends of single odors have been shown to play a very important role. In female cotton leafworm moths, *Spodoptera littoralis,* oviposition is deterred by an oviposition deterring pheromone (ODP) released from the feces of conspecific larvae. The ODP consists of six components, and they all must be present to conserve the activity of the pheromone (1). The Colorado potato beetle, *Leptinotarsa decemlineata,* identifies its host by the composition of a number of six-carbon compounds, so-called green leaf volatiles. These compounds are emitted by almost every green plant, but the ratio between the components constitutes a "fingerprint" of different plant species. The potato odor blend *and* component ratio is thus of crucial importance for the food search in *L. decemlineata* (143).

All of these systems, relying on blends of odors, raise a certain demand on the olfactory system. Not only must single odors be identified, but the presence of a combination of odors, sometimes in specific ratios, must be signaled. Such a task can be solved at different levels of the olfactory system, from peripheral receptor neurons to nerves innervating muscles of executing organs such as wings, legs, or ovipositor. At the receptor neuron level, very few studies have shown interactions between single components. Generally, receptor neurons are tuned to single compounds or a range of compounds, but the response of the neuron is not affected by the presence of several different molecules (51). Exceptions to this rule have been presented in the food odor detecting system of some insects (47), but so far never in pheromone detecting ORNs.

The identification of blend configuration thus resides at higher neural levels, from the antennal lobe and onward. To achieve blend specifity, i.e. that the response of a neuron to the blend is different from the summed response of its components, cross-fiber patterns must be present in the nervous system. From many studies we know that labeled lines, carrying information regarding single components, remain present also at very high levels in the CNS (25, 31, 55, 56). The response to a blend can either be stronger than the summed responses to the single components (synergism, Figure 2), or weaker (suppression). In a number of moth species, neurons that show no or a very low response to single pheromone components but that are strongly excited by the complete pheromone blend, have been found. In both the heliothine moths *H. zea* and *H. virescens,* blend-specific PNs have been identified that respond weakly to one or both components of the pheromone. These responses are often short bursts of action potentials. When the blend of the two components is used to stimulate the antenna, however, the same neurons show strongly synergistic responses with long-lasting excitation that outlasts the stimulus presentation period by several seconds (24, 30–31).

Blend-Detection in Polymorphic Systems

The blend specificity described in heliothine moths prompted studies in other species exhibiting different traits in the pheromone communication system. One trait that makes blend specificity especially interesting is when a species displays polymorphism in its pheromone communication system, i.e. different populations utilize different compounds or different proportions of the same compound to ensure sexual communication. Such polymorphism has been described in a number of species. One of the more well studied is the European corn borer, *Ostrinia nubilalis.* In this species the female produces a mixture of the two isomers of 11-tetradecenyl acetate (11–14:OAc). One strain produces and is attracted to the Z and the E isomer in a 97:3 ratio (Z-strain), while the other uses a 1:99 ratio (E-strain) (86). In areas of co-occurrence, the two strains interbreed, producing hybrids emitting and responding to intermediate isomer ratios (87, 149, 150).

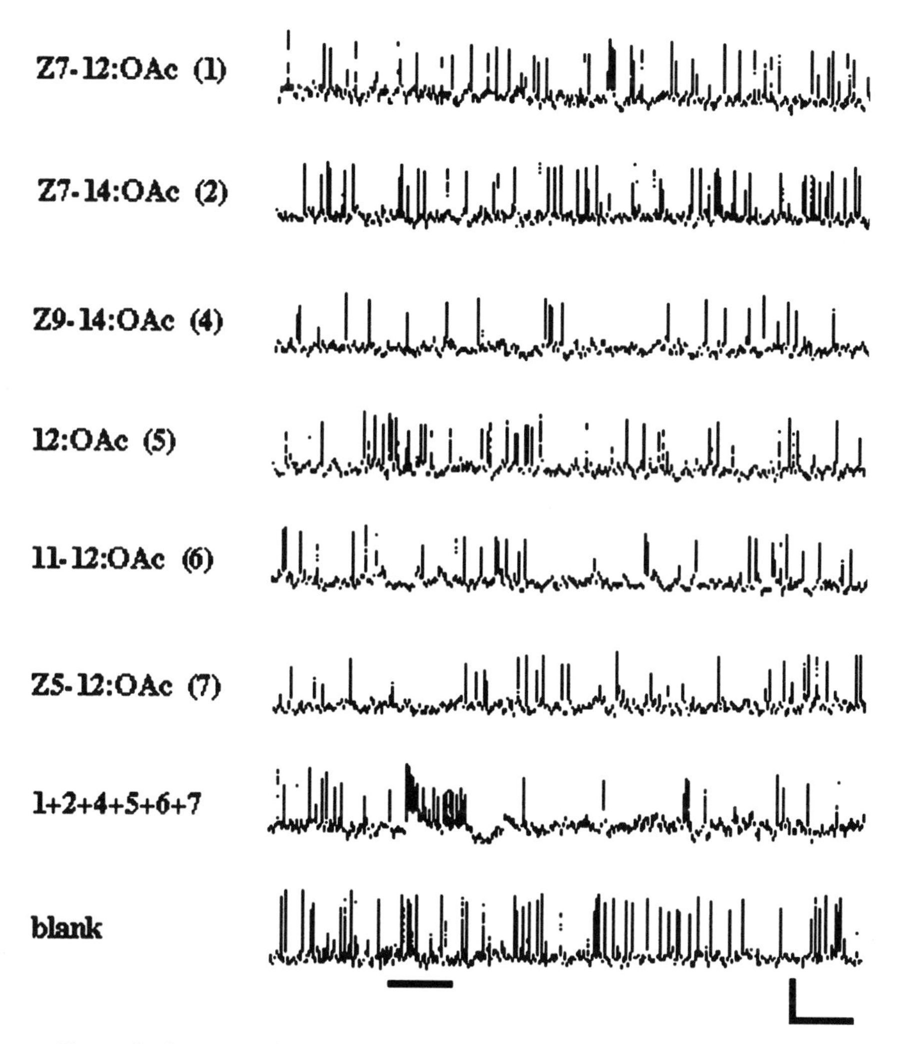

Figure 2 Response characteristics of a blend-specific neuron in *Trichoplusia ni.* The neuron responded to the six-component mixture at 10 ng but not to the single components (1, 2, 4–7). The bar underneath the registration marks the stimulus duration. Vertical scale bar 10 mV, horizontal scale bar 500 ms.

In order to elucidate differences in AL blend specificity between the different strains and the hybrids formed between them, AL neurons were penetrated with intracellular electrodes, and the responses to pure pheromone components, to the parental blends and to some intermediate, hybrid blends were recorded. Synergistic blend specificity was found in 11 out of the 100 investigated neurons. Blend specificity expressed as suppression was found in 40 neurons. Among the blend-specific neurons were neurons specific to both the parental blends and to the hybrid blends. In the parental strains, neurons responding to "their own" blend were more common than those responding to the blend of the other strain. Neurons specific to the hybrid blends were found in only the AL of hybrid males. Among the 49 percent non-blend–specific neurons, a clear correlate was also found, where the male AL of each parental strain contained more neurons tuned to the main component of "their own" blend. The arsenal of pheromone-specific AL neurons thus correlated well with the blend produced by females belonging to the same strain (6).

Another polymorphic pheromone communication system has been demonstrated in the turnip moth, *A. segetum.* In this species a large variation both in female pheromone production and male sensory setup was initially shown between different parts of Europe and western Asia (60, 95). However, the largest differences were found when two races, occurring north and south of the Sahara, were compared (140). North of the Sahara the pheromone blend consists of four components in a 1:5:2.5:0.1 ratio. South of the Sahara the blend is radically different, a 1:0.25:0.03:0.1 ratio. The turnip moth is thus an excellent candidate species for testing blend specificity.

When responses of AL neurons were investigated in the two races, neurons displaying blend interactions were found in both (148). In the European population 58% of the neurons showed interactions. Of these, 14% showed synergism, while 44% showed suppression. In the African population the percentages were 9% and 19%, respectively. Again, more than half of the neurons of the European population show blend interactions. This result is most likely the product of a very careful analysis and the use of a large number of blends; in the European population two four-component blends, three three-component blends and six two-component blends were tested, while only four four-component blends were tested in the African population. The more complete the stimulus set, the more interactions will most likely be revealed.

When the responses of the neurons were studied in detail, a large number of interactions were revealed. In 14 of the neurons, the response spectrum allowed a European male to separate the four-component blends of the two populations. Thirteen of these neurons responded with an excitation to the European blend while remaining unaffected after stimulation with the African blend. One neuron displayed the opposite response pattern. In the African population, four blend-specific neurons responded to the African blend, while six were excited exclusively by the European blend (148). These differences are interesting in an ecological and evolutionary perspective. Does the dominance of "European" neurons in European-population moths and the presence of both types in equal number in African-population moths imply that the African population stems from the European one, and is now diverging? Could the "simplification" of the African blend, as compared to the Swedish, be a product of ecological character release, so that fewer competitors in the pheromone communication channel are present in Africa? These questions remain to be answered.

The results from stimulation with binary blends revealed the full extent of interactions between single pheromone component input to the AL. Response patterns that would not be deduced from the responses to single components were registered in more than half of the neurons after binary blend stimulation. In some neurons only suppressions were present, in others some binary blends resulted in suppression while others resulted in synergistic responses. Still, in others, only synergistic responses were found. This enormous diversity in response characteristics adds another proof of the extensive integrative properties of AL neurons.

Blend Specificity in a Pheromone Communication System Showing Redundancy

In the noctuid moth *Trichoplusia ni* another speciality of the pheromone communication system makes it a good model to study integration of different combinations of pheromone components. The females of this species produce a very complex pheromone blend (94). The complete, behaviorally active sex pheromone of *T. ni* consists of six components (15, 17). Wind-tunnel experiments have shown, however, that several four-component blends elicit full expression of pheromone-mediated behavior in *T. ni* males (94). The pheromone communication system of *T. ni* thus displays what has been termed "redundancy in the pheromone communication channel." Additionally, (*Z*)-7-dodecenol (Z7–12:OH) has been shown to act as a behavioral antagonist (105, 141).

In an investigation of male antennal lobe neurons 112 pheromone-responding neurons were stimulated with single components, with binary blends, with four-component blends and with the full six-component blend. Of the tested neurons 60% exhibited blend interactions. As described previously, two types of blend-specific neurons were present: neurons expressing suppression and neurons expressing synergism. Forty-one neurons (36%) displayed suppression of some kind. Out of these, seven did not respond to the six-component blend, 15 did not respond to four-component blends, and 19 did not respond to binary blends containing single components to which the respective neuron responded. Twenty-seven neurons (24%) displayed synergism. Out of these, 24 did not respond to any single pheromone component or to the behavioral antagonist, whereas three neurons responded to some single components that were not present in the binary blends analysed. Four synergistic neurons responded to binary blends only, three responded to four-component blends only, and 20 neurons responded only to the complete six-component blend (Figure 2). No synergistic neurons were found that responded to the six-component blend and also to four-component blends (S Anton & BS Hansson, personal observation).

Again a very high proportion of the neurons investigated displayed different kinds of blend interactions. However, the redundancy in the pheromone communication system, where several four-component mixtures can substitute for the full blend, was not mirrored in AL blend-specific neurons. Most synergistic blend-specific neurons responded to the full blend and not to the four-component blends.

Neither among synergistic nor among suppressive neurons was a clear neural correlate to the redundancy found.

Concentration Effects in Blend-Specific Neurons

In several of the species investigated and where blend specificity has been demonstrated, the specificity was shown to be highly dose dependent (6, 62). Highly blend-specific neurons could change their characteristics over a single decadic step in stimulus concentration, from responding strictly to the complete pheromone blend to displaying a more generalistic response to some or all of the single components involved. The response threshold in synergistic blend-specific neurons, however, was generally higher than in component-specific neurons, indicating that blend information might be used within a closer range of the emitter. These results point out the importance of using behaviorally relevant stimulus concentrations in olfactory experiments. If not, much information regarding specificity can be lost.

Blend Interactions: Conclusions

Interactions between different components of a behaviorally relevant blend is a common phenomenon in AL neurons. In all three lepidopteran species where detailed analyses have been performed, more than half the neurons investigated displayed some kind of blend interactions. These high proportions demonstrate the importance of the integrative features of the AL. Early speculations that the AL could be a mere relay center for information on its way to higher, integrative centers of the brain are clearly wrong. In the AL, inputs from many different types of ORNs are received and neural comparisons of many of these inputs form the different output signals of the lobe.

TIME CODING

In a natural situation, due to air turbulence, an odor plume is not homogeneous. Instead, it has a filamentous structure. Packages of odor-laden air are intermixed with odor-free air (113). Temporal fluctuations in odor concentration can have a dramatic impact on searching behavior. Many animals are unable to locate a source of odor unless the stimulus presentation is intermittent. Neural circuits in the AL must thus be able to code events in time along with information concerning quality and quantity.

Wind-tunnel observations have demonstrated that the filamentous structure of odor plumes is of crucial importance for a male moth to proceed toward a pheromone source. If a homogeneous pheromone cloud is presented, the male engages in casting, i.e. across-wind flight without upwind progress, similar to when he loses contact with a normal plume (10, 11, 80, 98). For each filament hitting the antenna, the male performs an upwind surge followed by casting. If the frequency is high enough, in heliothine moths about 5 Hz (10), the flight becomes more or less a straight line towards the pheromone source.

While flying in a pheromone plume, male antennal ORNs receive intermittent pheromone stimulation with changing intervals. A few meters away from a point source, the filaments are about 100 ms long and separated by about 500 ms clean air (114). This intermittency can be translated into a frequency below 2 Hz. Under such discontinuous stimulation, male moth ORNs in several moth species have been shown to be able to follow stimuli mimicking the temporal patterns in a pheromone plume (9, 77, 102).

Antennal-lobe neuron processing of temporally dynamic information supplied by ORNs has been investigated in only a few species. In *Manduca sexta,* PNs have been observed to fire discrete bursts of action potentials following each stimulus pulse, sometimes up to 10 Hz (27). This encoding of time is strongly dependent on interactions between input from two discrete ORN types, each detecting one of the two pheromone-components (28). Typically, input from ORNs detecting the major pheromone component elicits an excitatory response, while input from ORNs specific to a minor component elicits an inhibitory response. When single components were used as stimuli, pulses could in the majority of neurons not be resolved (28), while stimulation with the two-component-blend resulted in a very good time resolution. The response to the blend stimulation had three clear phases: a short, inhibitory postsynaptic potential (IPSP) followed by a depolarization associated with action potentials. The depolarization was subsequently followed by a second, longer inhibiton. A correlation was found between the amplitude of the IPSP, and the maximum rate of pulsing that the neuron could resolve. These results again confirm the notion that many olfactory PNs respond optimally to a specific odor blend rather than to the individual odorants that comprise the blend. The characteristics of the system are also an example of how lateral interactions between neurons residing in different glomeruli can increase molecular differences between odor signals. A type of blend detection is thus at play also in the time-coding system. Integration of two odor pathways synchronises PN activity to the intermittent input.

In a noctuid moth, *A. segetum,* a similar investigation gave a somewhat different picture of how time is encoded in AL neurons. Instead of the triphasic response pattern encountered in *Manduca,* neurons that were able to encode fast transitions in stimulus intensity were characterized by a biphasic pattern (Figure 3). An initial depolarization was followed by a hyperpolarization. An initial IPSP, as found in *Manduca,* was not observed. A second difference was the fact that time resolution was as good for single components as for the full pheromone blend in *A. segetum* (92).

Time Coding: Conclusions

From the two investigations described here it is clear that different mechanisms are in action in the two olfactory systems. The main difference is the reliance on interaction between different pheromone components for time resolution in one

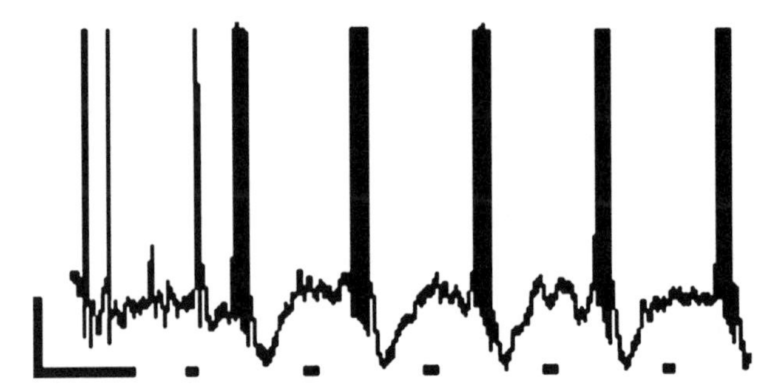

Figure 3 Physiological response of an antennal lobe neuron of the male turnip moth, *Agrotis segetum* when challenged with 2 Hz pulses of the female-produced pheromone blend. The neuron follows the pulses very well. Note the strong inhibitory period following each excitation. Black dashes indicate stimulus pulses (200 ms delay time in the delivery system causes the mismatch between delivery and response). Horizontal scale bar 500 ms. Vertical scale bar 10 mV.

species but not in the other. *M. sexta* uses two main components in its pheromone, while the *A. segetum* female emits a pheromone blend composed of four components. Could the more complex blend of *A. segetum* be the reason why a different strategy is used for time coding? The complexity of a four component blend might preclude the interactions observed in *Manduca*.

NEUROACTIVE SUBSTANCES IN THE ANTENNAL LOBE

A large number of neuroactive substances have been localized within the antennal lobe. They range from neurotransmitters such as acetylcholine (ACh) and γ-amino-butyric acid (GABA) to biogenic amines, neuropeptides and nitric oxide. The majority of the literature on neuroactive substances in the insect brain deals with their immunocytochemical localization within neuropil and identifiable neurons. During the last years, however, the function of some neuroactive substances within the primary olfactory neuropil have been revealed. Here, we summarize the major anatomical and functional findings.

The Neurotransmitters Acetylcholine and GABA

In Lepidoptera, Hymenoptera, and Diptera, acetylcholine or enzymes synthesizing or degrading ACh have been localized in axons of ORNs and in some subpopulations of PNs (for review see 71). The anatomical data suggest that insect ORNs are cholinergic, although functional studies are sparse. In the sphinx moth *M. sexta,* Waldrop & Hildebrand (146) could show in pharmacological experiments that acetylcholine can elicit excitatory or inhibitory responses in AL interneurons.

In most insect species studied so far, γ-amino-butyric acid (GABA) has been localized in LNs (13, 34, 72, 93, 125, 129). Therefore GABA seems to be the main transmitter for LNs, and elicits inhibition in the AL of *M. sexta* (145). Also, some subpopulations of PNs are GABA-ergic in the sphinx moth, the honeybee, and the cockroach (68, 72, 101,125). The effects of GABA, GABA-agonists, and GABA-antagonists on PNs have been studied in some detail. In *P. americana,* GABA injection into the MGC reduced the excitatory responses in PNs to sex-pheromone stimuli, whereas injection of the GABA-antagonist picrotoxin elicited enhanced excitatory responses in the same neurons (20, 74). In *M. sexta,* PNs show responses to odors that are not only dependent on the quality of the stimulus, but also on the temporal pattern (see time coding, above). Pharmacological experiments revealed that GABA plays an important role in shaping the response pattern of PNs to pulsed and to long-lasting odor stimuli through its inhibitory effect. The GABA antagonist bicuculline, which blocks $GABA_A$ receptors/chloride channels in PNs, changed the phase-locked bursting response to pulsed stimuli and the slowly oscillating reponse to long-lasting stimuli into tonic responses, which coded only for the beginning and the end of the whole stimulus (33). In the locust, however, the GABA-antagonist picrotoxin did not change the temporal response pattern in individual PNs, but abolished synchronization of PNs, which is found in non-treated preparations in the form of oscillating field potentials in the calyces of the mushroom bodies (96). Desynchronization of neural activity in the AL of the honey bee through picrotoxin injections impaired the discrimination of similar odorants in learning experiments, which indicates that refined odor discrimination is dependent on GABA-ergic inhibition in the AL (133).

Biogenic Amines

Centrifugal neurons innervating the AL are thought to have feedback and modulatory function. Accordingly, most biogenic amines with potential modulatory function were shown to be present in different types of centrifugal neurons.

Serotonin was localized in centrifugal neurons in moths and cockroaches, where ultrastructural studies revealed output synapses in the AL (124, 135). In *M. sexta,* serotonin was shown to have a modulatory effect on K^+ channels (107, 108). Application of serotonin in the AL led to an increase in excitability connected with an increased input resistance of AL neurons (85).

Dopamine staining revealed processes of centrifugal neurons in bees and crickets (71, 83, 84, 128, 126), whereas no dopaminergic neurons were identified in the AL of flies and locusts (for review see 71). Dopaminergic neurons in the olfactory pathway of honey bees seem to be involved in the retrieval of olfactory memory (106) and in regulating the response threshold to olfactory stimuli in AL neurons (97).

Immunocytochemical staining revealed the presence of octopamine in centrifugal neurons with blebbed branches in the AL in the locust (67), the sphinx moth (67), and the honey bee (89). These neurons have their cell bodies close to the midline of the subesophageal ganglion and send fibers not only into the AL but also in different regions of the protocerebrum. A uniquely identifiable octopaminergic centrifugal neuron (VUMmx1) was shown to mediate the unconditioned sucrose stimulus in olfactory conditioning experiments in the honey bee (49). Depolarization of the VUMmx1 as well as injection of octopamine into the AL or the calyces of the mushroom bodies substituted for the reinforcing stimulus in olfactory conditioning of the proboscis extension reflex (49, 50). Biogenic amines with unknown function have also occasionally been localized in other neuron types in the AL in different insect species (for review see 71).

Neuropeptides

Neuropeptides, such as allatostatins, allatotropin, FMRFamide, and tachykinins were localized in LNs in different insect species with immunocytochemical methods. These peptides are often co-localized with GABA in LNs, but nothing is known about their function in olfactory processing (for review see 71, 117). Neuropeptide-immunoreactivity in ORNs and PNs seems to be rare, whereas neuropeptides with potential modulatory function were shown to be present in different types of centrifugal neurons. Some allatostatin-immunoreactive centrifugal neurons with fibers in the AMMC send processes into the AL in locusts and crickets (126, 144). Centrifugal neurons that descend from the protocerebrum to the ventral cord and send varicose branches into the AL show leucokinin immunoreactivity in locusts and cockroaches (117, 118). Nothing is known, however, about the function of these neuropeptides in the olfactory pathway.

Nitric Oxide

Nitric oxide (NO), a highly reactive free radical gas, has been found to be a mediator for signals in the nervous system throughout the animal kingdom (for review see 112). NO has been found in ORNs in bees, fruit flies, and crickets (110, 114), but seems to be absent in ORNs of locusts (16, 39, 113). The function of NO has been studied in behavioral contexts, such as adaptation processes and learning in bees. NO seems to mediate olfactory information locally (within the ipsilateral AL): Habituation is impaired only ipsilaterally by blocking NO (114). Also, the formation of long-term memory is influenced by blocking NO, whereas short-term memory is not affected (111). In *M. sexta,* the finding that soluble guanylyl cyclase, the target of NO, is expressed especially in PN dendrites, leads to the hypothesis that ORNs might signal directly to PNs via NO within glomeruli, while the main synaptic pathway connects ORNs with PNs via LNs (118).

In different insect species LNs were shown to contain NO (16, 39, 110, 113, 114). In the locust, NO is co-localized with GABA in some LNs (129). However,

NO was not shown to have any effect on odor-evoked synchronization of neuronal assemblies, whereas GABA-antagonists impair synchronization (91, 96).

Hormonal Control in the AL

Recent studies have shown that juvenile hormone (JH) could be involved in plasticity in the AL. In honey bee workers, the change of tasks in the hive is correlated with changes in the JH level and with an increase of the size of an identifiable glomerulus. There is evidence, however, that the plasticity in the size of the glomerulus is not directly dependent on the JH level, but that JH induces foraging activity and this activity leads in turn to an increase in size (130, 147).

Recent electrophysiological studies indicate that JH can affect the sensitivity of AL neurons. In the noctuid moth *A. ipsilon*, newly emerged males are not sexually mature and do not respond behaviorally to the female-produced sex pheromone. The behavioral responsiveness increases with age coinciding with an increasing JH biosynthesis activity (38, 44). Intracellular recordings showed that the sensitivity of AL neurons to the sex pheromone blend increased with the age of the male moths in parallel with the increase in behavioral responsiveness. Males surgically deprived of JH showed a significant decrease in AL neuron sensitivity and high sensitivity could be restored by JH injection. High sensitivity of AL neurons could even be induced in young males by injecting JH. The results indicate that JH plays an important role in synchronizing the development of the reproductive apparatus with the sensitivity of the olfactory system (1a).

In the locust, *Schistocerca gregaria,* JH seems to have the reversed effect on the olfactory system, compared to *A. ipsilon* (Figure 4). The sensitivity of AL neurons to aggregation pheromones in freshly emerged adult individuals is high, whereas AL neurons in four-week-old locusts are often not responding to aggregation pheromones at all. The responses of AL neurons are correlated with behavioral responsiveness to aggregation pheromones, which is high before and during the reproductive time (occuring at an age of two weeks) and clearly decreases thereafter to reach indifferent behavior in four-week-old locusts (Ignell, Couilland, and Anton, personal observation).

In conclusion, JH seems to coordinate general physiological processes with the regulation of sensory input, but it is still unknown if these effects are directly elicited by JH or if they are mediated by other neuroactive substances such as biogenic amines or neuropeptides.

Neuroactive Substances in the Antennal Lobe: Conclusions

In spite of an increasing rate of identification and localization of potentially neuroactive substances in the insect olfatory pathway, including the AL, the substances with unknown function far outnumber neuroactive substances with a known role in central olfactory processing. Biogenic amines seem to play a similar modulatory role in many different insect species, while the occurrence of neuropeptides varies widely between species and might be responsible for specific

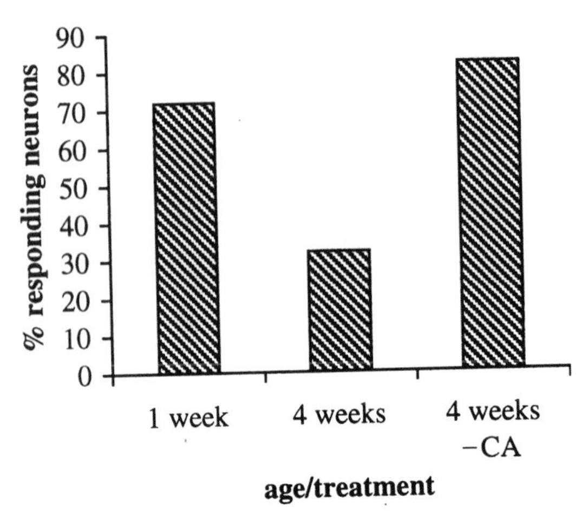

Figure 4 Responsiveness of antennal lobe neurons in male *Schistocerca gregaria* to aggregation pheromone. In young adults, which have a low juvenile hormone level, a large proportion of the neurons respond. In four-week-old adults, having a high juvenile hormone level, only 30% of the neurons respond. In four-week-old adults surgically deprived of juvenile hormone (allatectomised, –CA) the proportion of responding neurons in the antennal lobe was as high as in young adults. Responsiveness of antennal lobe neurons in locusts seems therefore to be controlled by juvenile hormone.

modulatory functions. The different effects of JH in moths and locusts on the sensitivity and responsiveness of AL neurons could be due to the presence of different neuropeptides or other neuroactive substances acting as mediators between the hormone and neuron membrane events. Modulatory substances, such as NO, are also involved in higher brain functions such as learning processes and the formation of long-term memory.

GENERAL CONCLUSIONS

The structural framework underlying AL olfactory processing is the glomerular array. The importance of single glomeruli as target stations for ORNs of identical specificity is becoming an accepted fact, while the innervation patterns of second- and third-level neurons into different glomeruli show a more divergent pattern between insect species and are still under investigation in other organisms. Are different design principles reflecting different odor coding mechanisms? The elucidation of how function is mirrored in the morphology of higher-order neurons of the AL is an important future research direction.

The general function of AL neurons as potent integrators of odor information is clearly very similar between insect species. Input regarding different types of molecules are compared within the AL. As the investigations get more detailed regarding the nature of the stimuli used, the true integrative powers are revealed. In most species investigated, more than half of the neurons show blend interactions. Fast temporal patterns of odor stimulation are followed by AL neurons, a capacity sometimes depending on blend input as well.

As the physiological characteristics of neurons are revealed, the underlying mechanisms of signal transfer and modulation within the AL get increasingly interesting. Our knowledge regarding the function of all the neuroactive substances occurring among AL neurons is still very scarce, and is one of the areas that should be prioritized in future investigations. We already know that quality, quantity, and temporal coding of odors are closely linked, but further studies must also show how the physiological state of insects interferes with these coding mechanisms. By combining the knowledge from morphological, physiological, and pharmacological studies we can reach considerably farther in our understanding of olfactory processing in the antennal lobe.

LITERATURE CITED

1. Anderson P, Hilker M, Hansson BS, Bombosch S, Klein B, Schildknecht H. 1993. Oviposition deterring components in larval frass of *Spodoptera littoralis* (Boisd.) (Lepidoptera: Noctuidae): a behavioural and electrophysiological evaluation. *J. Insect Physiol.* 39:129–37

1a. Anton S, Gadenne C. 1999. Effect of juvenile hormone on the central nervous processing of sex pheromone in an insect. *Proc. Natl. Acad. Sci.* USA 96:5764–67

2. Anton S, Hansson BS. 1994. Central processing of sex pheromone, host odour, and oviposition deterrent information by interneurons in the antennal lobe of female *Spodoptera littoralis* (Lepidoptera: Noctuidae). *J. Comp. Neurol.* USA 350:199–214

3. Anton S, Hansson BS. 1995. Sex pheromone and plant-associated odour processing in the antennal lobe interneurons of male *Spodoptera littoralis* (Lepidoptera: Noctuidae). *J. Comp. Physiol. A* 176:773–89

4. Anton S, Hansson BS. 1996. Antennal lobe interneurons in the desert locust *Schistocerca gregaria* (Forskal): Processing of aggregation pheromones in adult males and females. *J. Comp. Neurol.* 370:85–96

4a. Anton S, Hansson BS. 1999. Functional significance of olfactory glomeruli in a moth. *Proc. R. Soc. London Ser. B* 266:1813–20

5. Anton S, Homberg U. 1999. Antennal lobe structure. See Ref. 53, pp. 97–124

6. Anton S, Löfstedt C, Hansson BS. 1997. Central nervous processing of sex pheromones in two strains of the European corn borer *Ostrinia nubilalis* (Lepidoptera: Pyralidae). *J. Exp. Biol.* 200:1073–87

7. Arn H, Tóth M, Priesner E. 1997. List of sex pheromones of Lepidoptera and related attractants. *Technol. Transf. Mat-*

ing Disruption IOBC wprs Bull. 20:257–93
8. Arnold G, Masson C, Budharugsa S. 1985. Comparative study of the antennal lobes and their afferent pathway in the workerbee and the drone *Apis mellifera* L. *Cell Tissue. Res.* 242:593–605
9. Baker TC, Hansson BS, Löfstedt C, Löfqvist J. 1989. Adaptation of male moth antennal neurons in a pheromone plume is associated with cessation of pheromone-mediated flight. *Chem. Senses* 14:439–48
10. Baker TC, Vickers NJ. 1996. Pheromone-mediated flight in moths. See Ref. 22, pp. 248–64
11. Baker TC, Willis MA, Haynes KF, Phelan PL. 1985. A pulsed cloud of sex pheromone elicits upwind flight in male moths. *Physiol. Entomol.* 10:257–65
12. Bausenwein B, Nick P. 1998. Three dimensional reconstruction of the antennal lobe in the mosquito *Aedes aegypti.* In *New Neuroethology on the Move,* ed. R Wehner, N Elsner, p. 386. Stuttgart: Thieme. 614 pp.
13. Becker M, Breidbach O. 1993. Distribution of GABA-like immunoreactivity throughout metamorphosis of the supraoesophageal ganglion of the beetle *Tenebrio molitor* L. (Coleoptera, Tenebrionidae). In *Gene-Brain-Behaviour,* ed. N Elsner, M Heisenberg, p. 738. Stuttgart: Thieme. 1079 pp.
14. Berg BG, Almaas TJ, Bjaalie JG, Mustaparta H. 1998. The macroglomerular complex of the antennal lobe in the tobacco budworm moth *Heliothis virescens:* specified subdivision in four compartments according to information about biologically significant compounds. *J. Comp. Physiol. A* 183:669–82
15. Berger RS. 1966. Isolation, identification, and synthesis of the sex attractant of the cabbage looper, *Trichoplusia ni. Ann. Entomol. Soc. Am.* 59:767–71
16. Bicker G, Schmachtenberg O, De Vente J. 1997. Geometric considerations of nitric oxide cyclic GMP signalling in the glomerular neuropil of the locust antennal lobe. *Proc. R. Soc. London Ser. B* 264:1177–81
17. Bjostad LB, Linn CE, Du J-W, Roelofs WL. 1984. Identification of new sex pheromone components in *Trichoplusia ni* predicted from biosynthetic precursors. *J. Chem. Ecol.* 10:1309–23
18. Boeckh J, Ernst K-D, Sass H, Waldow U. 1976. On the nervous organization of antennal sensory pathways in insects with special reference to the olfactory system. *Verh. Dtsch. Zool. Ges.* 123–39
19. Boeckh J, Sandri C, Akert K. 1970. Sensorische Eingänge und synaptische Verbindungen im zentralen Nervensystem von Insekten. *Z. Zellforsch. Mikrosk. Anat.* 103:429–46
20. Boeckh J, Tolbert LP. 1993. Synaptic organisation and development of the antennal lobe in insects. *Microsc. Res. Tech.* 24:260–80
21. Bräunig P. 1991. Suboesophageal DUM neurons innervate the principal neuropils of the locust brain. *Philos. Trans. R. Soc. London Ser. B* 332:221–40
22. Cardé RT, Minks AK. 1996. *Insect Pheromone Research: New Directions.* New York: Chapman & Hall. 684 pp.
23. Chambille I, Rospars JP. 1985. Neurons and identified glomeruli of antennal lobes during postembryonic development in the cockroach *Blaberus craniifer* Burm. (Dictyoptera: Blaberidae). *Int. J. Insect Morphol. Embryol.* 14:203–26
24. Christensen TA, Harrow ID, Cuzzocrea C, Randolph PW, Hildebrand JG. 1995. Distinct projections of two populations of olfactory receptor axons in the antennal lobe of the sphinx moth *Manduca sexta. Chem. Senses* 20:313–23
25. Christensen TA, Hildebrand JG. 1987. Functions, organisation, and physiology of the olfactory pathways in the lepidopteran brain. In *Arthropod Brain: Its Evolution, Development, Structure, and*

Functions, ed. AP Gupta, pp. 457–84. Chichester, UK: Wiley. 588 pp.
26. Christensen TA, Hildebrand JG. 1987. Male-specific, sex pheromone-selective projection neurons in the antennal lobes of the moth *Manduca sexta. J. Comp. Physiol. A* 160:553–69
27. Christensen TA, Hildebrand JG. 1988. Frequency coding by central olfactory neurons in the sphinx moth *Manduca sexta. Chem. Senses* 13:123–30
28. Christensen TA, Hildebrand JG. 1997. Coincident stimulation with pheromone components improves temporal pattern resolution in central olfactory neurons. *J. Neurophysiol.* 177:775–81
29. Christensen TA, Hildebrand JG, Tumlinson JH, Doolittle RE. 1989. Sex pheromone blend of *Manduca sexta* responses of central olfactory interneurons to antennal stimulation in male moths. *Arch. Insect Biochem. Physiol.* 10:281–91
30. Christensen TA, Mustaparta H, Hildebrand JG. 1989. Discrimination of sex pheromone blends in the olfactory system of the moth. *Chem. Senses* 14:463–77
31. Christensen TA, Mustaparta H, Hildebrand JG. 1991. Chemical communication in heliothine moths. II. Central processing of intra- and interspecific olfactory messages in the male corn earworm moth *Helicoverpa zea. J. Comp. Physiol. A* 169:259–74
32. Christensen TA, Waldrop BR, Harrow ID, Hildebrand JG. 1993. Local interneurons and information processing in the olfactory glomeruli of the moth *Manduca sexta. J. Comp. Physiol. A* 173:385–99
33. Christensen TA, Waldrop BR, Hildebrand JG. 1998. Multitasking in the olfactory system: context-dependent responses to odors reveal dual GABA-regulated coding mechanisms in single olfactory projection neurons. *J. Neurosci.* 18:5999–6008
34. Distler PG. 1989. Histochemical demonstration of GABA-like immunoreactivity in cobalt labeled neuron individuals in the insect olfactory pathway. *Histochemistry* 91:245–49
35. Distler PG, Boeckh J. 1996. Synaptic connection between olfactory receptor cells and uniglomerular projection neurons in the antennal lobe of the American cockroach, *Periplaneta americana. J. Comp. Neurol.* 370:35–46
36. Distler PG, Boeckh J. 1997. Synaptic connections between identified neuron types in the antennal lobe glomeruli of the cockroach, *Periplaneta americana:* I. Uniglomerular projection neurons. *J. Comp. Neurol.* 378:307–19
37. Distler PG, Boeckh J. 1997. Synaptic connections between identified neuron types in the antennal lobe glomeruli of the cockroach, *Periplaneta americana:* II. Local multiglomerular interneurons. *J. Comp. Neurol.* 383:529–40
38. Duportets L, Dufour MC, Couillaud F, Gadenne C. 1998. Biosynthetic activity of corpora allata, growth of sex accessory glands and mating in the male moth *Agrotis ipsilon* (Hufnagel). *J. Exp. Biol.* 201: 2425–32
39. Elphick MR, Rayne RC, Riveros-Moreno V, Moncada S, O'Shea M. 1995. Nitric oxide synthesis in locust olfactory interneurones. *J. Exp. Biol.* 198:821–29
40. Ernst K-D, Boeckh J. 1983. A neuroanatomical study on the organisation of the central antennal pathways in insects. III. Neuroanatomical characterisation of physiologically defined response types of deutocerebral neurons in *Periplaneta americana. Cell Tissue Res.* 229:1–22
41. Ernst K-D, Boeckh J, Boeckh V. 1977. A neuroanatomical study on the organisation of the central antennal pathways in insects. II. Deutocerebral connections in *Locusta migratoria* and *Periplaneta americana. Cell Tissue Res.* 176:285–308
42. Flanagan D, Mercer AR. 1989. Morphology and response characteristics of

neurons in the deutocerebrum of the brain in the honeybee *Apis mellifera*. *J. Comp. Physiol. A* 164:483–94
43. Fonta C, Sun XJ, Masson C. 1993. Morphology and spatial distribution of bee antennal lobe interneurons responsive to odours. *Chem. Senses* 18:101–19
44. Gadenne C, Renou M, Sreng L. 1993. Hormonal control of pheromone responsiveness in the male black cutworm *Agrotis ipsilon*. *Experientia* 49:721–24
45. Galizia CG, Joerges J, Kuttner A, Faber T, Menzel R. 1997. A semi-in-vivo preparation for optical recording of the insect brain. *J. Neurosci. Methods* 76:61–69
46. Galizia CG, Nägler K, Hölldobler B, Menzel R. 1998. Odour coding is bilaterally symmetrical in the antennal lobes of honeybees (*Apis mellifera*). *Eur. J. Neurosci.* 10:2964–74
47. Getz WM, Akers RP. 1994. Honeybee olfactory sensilla behave as integrated processing units. *Behav. Neural Biol.* 61:191–95
48. Grant AJ, O'Connell RJ. 1986. Neurophysiological and morphological investigations of pheromone-sensitive sensilla on the antennae of *Trichoplusia ni*. *J. Insect Physiol.* 32:503–15
48a. Hähnlein I, Bicker G. 1996. Morphology of neuroglia in the antennal lobes and mushroom bodies of the brain of the honeybee. *J. Comp. Neurol.* 367:235–45
49. Hammer M. 1993. An identified neuron mediates the unconditioned stimulus in associative olfactory learning in honeybees. *Nature* 366:59–63
50. Hammer M, Menzel R. 1995. Learning and memory in the honeybee. *J. Neurosci.* 15:1617–30
51. Hansson BS. 1995. Olfaction in Lepidoptera. *Experientia* 51:1003–27
52. Hansson BS. 1996. Antennal lobe projection patterns of pheromone-specific olfactory receptor neurons in moths. See Ref. 22, pp. 164–83
53. Hansson BS, ed. 1999. *Insect Olfaction*. Berlin: Springer-Verlag. 457 pp.
54. Hansson BS, Almaas TJ, Anton S. 1995. Chemical communication in heliothine moths. V. Antennal lobe projection patterns of pheromone-detecting olfactory receptor neurons in the male *Heliothis virescens* (Lepidoptera: Noctuidae). *J. Comp. Physiol. A* 177:535–43
55. Hansson BS, Anton S, Christensen TA. 1994. Structure and function of antennal lobe neurons in the male turnip moth, *Agrotis segetum* (Lepidoptera: Noctuidae). *J. Comp. Physiol. A* 175:547–62
56. Hansson BS, Christensen TA. 1999. Functional characteristics of the antennal lobe. See Ref. 53, pp. 125–61
57. Hansson BS, Christensen TA, Hildebrand JG. 1991. Functionally distinct subdivisions of the macroglomerular complex in the antennal lobe of the male sphinx moth *Manduca sexta*. *J. Comp. Neurol.* 312:264–78
58. Hansson BS, Ljungberg H, Hallberg E, Löfstedt C. 1992. Functional specialisation of olfactory glomeruli in a moth. *Science* 256:1313–15
59. Hansson BS, Ochieng' SA, Grosmaitre X, Anton S, Njagi PGN. 1996. Physiological responses and central nervous projections of antennal olfactory receptor neurons in the adult desert locust, *Schistocerca gregaria* (Orthoptera: Acrididae). *J. Comp. Physiol. A* 179:157–67
60. Hansson BS, Tóth M, Löfstedt C, Szöcs G, Subchev M, Löfqvist J. 1990. Pheromone variation among eastern European and a western Asian population of the turnip moth *Agrotis segetum*. *J. Chem. Ecol.* 16:1611–22
61. Hanström B, ed. 1928. In *Vergleichende Anatomie des Nervensystems der wirbellosen Tiere*. Berlin: Springer-Verlag
62. Hartlieb E, Anton S, Hansson BS. 1997. Dose-dependent response characteristics of antennal lobe neurons in the male moth *Agrotis segetum* (Lepidoptera: Noctuidae). *J. Comp. Physiol. A* 181:469–76
63. Hildebrand JG. 1996. Olfactory control

of behavior in moths: central processing of odor information and the functional significance of olfactory glomeruli. *J. Comp. Physiol. A* 178:5–19

64. Hildebrand JG, Matsumoto SG, Camazine SM, Tolbert LP, Blank S, et al. 1980. Organisation and physiology of antennal centres in the brain of the moth *Manduca sexta.* In *Insect Neurobiology and Pesticide Action,* pp. 375–82. London: Soc. Chem. Ind. 517 pp.
65. Homberg U. 1990. Immunocytochemical demonstration of transmitter candidates in the central olfactory pathways in the sphinx moth, *Manduca sexta.* In *Olfaction and Taste X,* ed. K Døving, pp. 151–58. Oslo: Oslo Univ. Press. 402 pp.
66. Homberg U. 1994. Distribution of neurotransmitters in the insect brain. In *Progress Zoology,* Vol 40. Stuttgart: Fischer
67. Homberg U, Binkle U, Lehman HK, Vullings HGB, Eckert M, et al. 1992. Octopamine-immunoreactive neurons in the brain of two insect species. In *Rhythmogenesis in Neurons and Networks,* ed. N Elsner, DW Richter, p. 477. Stuttgart, New York: Thieme. 965 pp.
68. Homberg U, Christensen TA, Hildebrand JG. 1989. Structure and function of the deutocerebrum in insects. *Annu. Rev. Entomol.* 34:477–501
69. Homberg U, Hoskins SG, Hildebrand JG. 1995. Distribution of acetycholinesterase activity in the deutocerebrum of the sphinx moth *Manduca sexta. Cell Tissue Res.* 279:249–59
70. Homberg U, Montague RA, Hildebrand JG. 1988. Anatomy of antenno-cerebral pathways in the brain of the sphinx moth *Manduca sexta. Cell Tissue Res.* 254:255–81
71. Homberg U, Müller U. 1999. Neuroactive substances in the antennal lobe. See Ref. 53, pp. 181–206
72. Hoskins SG, Homberg U, Kingan TG, Christensen TA, Hildebrand JG. 1986. Immunocytochemistry of GABA in the antennal lobes of the sphinx moth *Manduca sexta. Cell Tissue Res.* 244:243–52
73. Deleted in proof
74. Hösl M. 1991. Pheromonsensitive Neurone im Deutocerebrum von *Periplaneta americana:* Exzitation und Inhibition nach Stimulation antennaler Rezeptorzellen. PhD diss. Univ. Regensburg, Germany
75. Ignell R, Anton S, Hansson BS. 1998. Central nervous processing of behaviourally relevant odours in solitary and gregarious fifth instar locusts, *Schistocerca gregaria. J. Comp. Physiol. A* 183:453–65
76. Joerges J, Küttner A, Galizia CG, Menzel R. 1997. Representations of odours and odour mixtures visualized in the honeybee brain. *Nature* 387:285–88
77. Kaissling K-E. 1986. Temporal characteristics of pheromone receptor cell responses in relation to orientation behaviour of moths. In *Mechanisms in Insect Olfaction,* ed. TL Payne, MC Birch, CEJ Kennedy, pp. 193–200. Oxford: Clarendon. 364 pp.
78. Kanzaki R, Arbas EA, Strausfeld NJ, Hildebrand JG. 1989. Physiology and morphology of projection neurons in the antennal lobe of the male moth, *Manduca sexta. J. Comp. Physiol. A* 165:427–53
79. Kanzaki R, Shibuya T. 1986. Identification of the deutocerebral neurons responding to the sexual pheromone in male silkworm moth brain. *Zool. Sci.* 3:409–18
80. Kennedy JS, Ludlow AR, Sanders CJ. 1980. Guidance system used in moth sex attraction. *Nature* 295:475–77
81. Kent KS, Hoskins SG, Hildebrand JG. 1987. A novel serotonin-immunoreactive neuron in the antennal lobe of the sphinx moth *Manduca sexta* persists throughout postembryonic life. *J. Neurobiol.* 18: 451–65
82. Kenyon FC. 1896. The brain of the bee. A preliminary contribution to the mor-

phology of the nervous system of the arthropoda. *J. Comp. Neurol.* 6:133–210
83. Klemm N. 1974. Vergleichend-histochemische Untersuchungen über die Veteilung monoamin-haltiger Strukturen im Oberschlundganglion von Angehörigen verschiedener Insekten-Ordnungen. *Entomol. Ger.* 1:24–49
84. Klemm N. 1976. Histochemistry of putative transmitter substances in the insect brain. *Progr. Neurobiol.* 7:99–169
85. Kloppenburg P, Hildebrand JG. 1995. Neuromodulation by 5-hydroxytryptamine in the antennal lobe of the sphinx moth *Manduca sexta. J. Exp. Biol.* 198:603–11
86. Klun JA, Chapman O, Mattes JC, Wojtkowski PW, Beroza M, Sonnett PE. 1973. Insect sex pheromones: minor amount of opposite geometrical isomer critical to attraction. *Science* 181:661–63
87. Klun JA, Maini S. 1979. Genetic basis of an insect chemical communication system: the European cornborer. *Environ. Entomol.* 8:423–26
88. Koontz MA, Schneider D. 1987. Sexual dimorphism in neuronal projections from the antennae of silk moths (*Bombyx mori, Antheraea polyphemus*) and the gypsy moth (*Lymantria dispar*). *Cell Tissue Res.* 249:39–50
89. Kreissl S, Eichmüller S, Bicker G, Rapus J, Eckert M. 1994. Octopamine-like immunoreactivity in the brain and suboesophageal ganglion of the honeybee. *J. Comp. Neurol.* 348:583–95
90. Laurent G. 1996. Dynamical representation of odours by oscillating and evolving neural assemblies. *Trends Neurosci.* 19:489–96
91. Laurent G, Davidowitz H. 1994. Encoding of olfactory information with oscillating neuronal assemblies. *Science* 265:1872–75
92. Lei H, Hansson BS. 1999. Central processing of pulsed pheromone signals by antennal lobe neurons in the male moth *Agrotis segetum. J. Neurophysiol.* 81:1113–22
93. Leitch B, Laurent G. 1996. GABAergic synapses in the antennal lobe and mushroom body of the locust olfactory system. *J. Comp. Neurol.* 372:487–514
94. Linn CE, Bjostad LB, Du JW, Roelofs WL. 1984. Redundancy in a chemical signal: behavioural responses of male *Trichoplusia ni* to a 6-component sex pheromone blend. *J. Chem. Ecol.* 11:1635–58
95. Löfstedt C, Löfqvist J, Lanne BS, van der Pers JNC, Hansson BS. 1986. Pheromone dialects in European turnip moths *Agrotis segetum. Oikos* 46:250–57
96. MacLeod K, Laurent G. 1996. Distinct mechanisms for synchronization and temporal patterning of odor-encoding neural assemblies. *Science* 274:976–79
97. Macmillan CS, Mercer AR. 1987. An investigation of the role of dopamine in the antennal lobes of the honeybee, *Apis mellifera. J. Comp. Physiol. A* 160:359–66
98. Mafra-Neto A, Cardé RT. 1994. Fine-scale structure of pheromone plumes modulates upwind orientation of flying moths. *Nature* 369:142–44
99. Malun D. 1991. Inventory and distribution of synapses of identified uniglomerular projection neurons in the antennal lobe of *Periplaneta americana. J. Comp. Neurol.* 305:348–60
100. Malun D. 1991. Synaptic relationships between GABA-immunoreactive neurons and an identified uniglomerular projection neuron in the antennal lobe of *Periplaneta americana:* a double-labeling electron microscopic study. *Histochemistry* 96:197–207
101. Malun D, Waldow U, Kraus D, Boeckh J. 1993. Connections between the deutocerebrum and the protocerebrum, and neuroanatomy of several classes of deutocerebral projection neurons in the brain of male *Periplaneta americana. J. Comp. Neurol.* 329:143–62
102. Marion-Poll F, Tobin TR. 1992. Temporal coding of pheromone pulses and

trains in *Manduca sexta. J. Comp. Physiol. A* 171:505–12
103. Masson C, Mustaparta H. 1990. Chemical information processing in the olfactory system of insects. *Physiol. Rev.* 70:199–245
104. Matsumoto SG, Hildebrand JG. 1981. Olfactory mechanisms in the moth *Manduca sexta:* response characteristics and morphology of central neurons in the antennal lobes. *Proc. R. Soc. London Ser. B* 213:249–77
105. McLaughlin JR, Mitchell ER, Chambers DL, Tumlinson JH. 1974. Perception of (Z)-7-dodecenol and modification of the sex pheromone response of male cabbage loopers. *Environ. Entomol.* 3:677–80
106. Menzel R, Hammer M, Braun G, Mauelshagen J, Sugawa M. 1991. Neurobiology of learning and memory in honeybees. In *The Behaviour and Physiology of Bees,* ed. LJ Goodman, RC Fisher, pp. 323–53. Wallingford, UK: CAB Int. 362 pp.
107. Mercer AR, Hayashi JH, Hildebrand JG. 1995. Modulatory effects of 5-hydroxytryptamine on voltage-activated currents in cultured antennal lobe neurones of the sphinx moth *Manduca sexta. J. Exp. Biol.* 198:613–27
108. Mercer AR, Kloppenburg P, Hildebrand JG. 1996. Serotonin-induced changes in the excitability of cultured antennal-lobe neurons of the sphinx moth *Manduca sexta. J. Comp. Physiol. A* 178:21–31
109. Mombaerts P. 1996. Targeting olfaction. *Curr. Opin. Neurobiol.* 6:481–86
110. Müller U. 1994. Ca2+/calmodulin-dependent nitric oxide synthase in *Apis mellifera* and *Drosophila melanogaster. Eur. J. Neurosci.* 6:1362–70
111. Müller U. 1996. Inhibition of nitric oxide synthase impairs a distinct form of long-term memory in the honeybee, *Apis mellifera. Neuron* 16:541–49
112. Müller U. 1997. The nitric-oxide system in insects. *Prog. Neurobiol.* 51:363–81
113. Müller U, Bicker G. 1994. Calcium-activated release of nitric oxide and cellular distribution of nitric oxide-synthesizing neurons in the nervous system of the locust. *J. Neurosci.* 14:7521–28
114. Müller U, Hildebrandt H. 1995. The nitric oxide/cGMP system in the antennal lobe of *Apis mellifera* is implicated in integrative processing of chemosensory stimuli. *Eur. J. Neurosci.* 7:2240–48
115. Murlis J, Elikton JS, Cardé RT. 1992. Odour plumes and how insects use them. *Annu. Rev. Entomol.* 37:505–32
116. Murlis J, Jones CD. 1981. Fine scale structure of odour plumes in relation to insect orientation to distant pheromone and other attractant sources. *Physiol. Entomol.* 6:1–86
117. Nässel DR. 1993. Neuropeptides in the insect brain: a review. *Cell Tissue Res.* 273:1–29
118. Nässel DR, Cantera R, Karlsson A. 1992. Neurons in the cockroach nervous system reacting with antisera to the neuropeptide leucokinin I. *J. Comp. Neurol.* 322:45–67
119. Nighorn A, Gibson NJ, Rivers DM, Hildebrand JG, Morton DB. 1998. The nitric oxide-cGMP pathway may mediate communication between sensory afferents and projection neurons in the antennal lobe of *Manduca sexta. J. Neurosci.* 18:7244–55
120. Ochieng' SA, Anderson P, Hansson BS. 1995. Antennal lobe projection patterns of olfactory receptor neurons involved in sex pheromone detection in *Spodoptera littoralis* (Lepidoptera: Noctuidae). *Tissue Cell* 27:221–32
121. Rospars JP. 1983. Invariance and sex-specific variations of the glomerular organisation in the antennal lobes of a moth, *Mamestra brassicae,* and a butterfly, *Pieris brassicae. J. Comp. Neurol.* 220:80–96
122. Rospars JP. 1988. Structure and development of the insect antennodeutocerebral system. *Int. J. Insect Morphol. Embryol.* 17:243–94

123. Rospars JP, Hildebrand JG. 1992. Anatomical identification of glomeruli in the antennal lobes of the male sphinx moth *Manduca sexta. Cell Tissue Res.* 270:205–27
124. Salecker I, Distler P. 1990. Serotonin-immunoreactive neurons in the antennal lobes of the American cockroach *Periplaneta americana:* light- and electron microscopic observations. *Histochemistry* 94:463–73
125. Schäfer S, Bicker G. 1986. Distribution of GABA-like immunoreactivity in the brain of the honeybee. *J. Comp. Neurol.* 246:287–300
126. Schäfer S, Rehder V. 1989. Dopamine-like immunoreactivity in the brain and suboesophageal ganglion of the honeybee. *J. Comp. Neurol.* 280:43–58
127. Schildberger K, Agricola H. 1992. Allatostatin-like immunoreactivity in the brains of crickets and cockroaches. In *Rhythmogenesis in Neurons and Networks,* ed. N Elsner, DW Richter, p. 489. Stuttgart: Thieme. 965 pp.
128. Schürmann FW, Elekes K, Geffard M. 1989. Dopamine-like immunoreactivity in the bee brain. *Cell Tissue Res.* 256:399–410
129. Seidel C, Bicker G. 1997. Colocalization of NADPH-diaphorase and GABA-immunoreactivity in the olfactory and visual system of the locust. *Brain Res.* 769:273–80
130. Sigg D, Thompson CM, Mercer AR. 1997. Activity-dependent changes to the brain and behavior of the honey bee, *Apis mellifera* (L.). *J. Neurosci.* 17:7148–56
131. Stocker RF. 1994. The organization of the chemosensory system in *Drosophila melanogaster:* A review. *Cell Tissue Res.* 275:3–26
132. Stocker RF, Lienhard MC, Borst A, Fischbach K-F. 1990. Neuronal architecture of the antennal lobe in *Drosophila melanogaster. Cell Tissue Res.* 262:9–34
133. Stopfer M, Bhagavan S, Smith BH, Laurent G. 1997. Impaired odour discrimination on desynchronization of odour-encoding neural assemblies. *Nature* 390:70–74
134. Sun XJ, Fonta C, Masson C. 1993. Odour quality processing by bee antennal lobe interneurons. *Chem. Senses* 18:355–77
135. Sun XJ, Tolbert LP, Hildebrand JG. 1993. Ramification pattern and ultrastructural characteristics of the serotonin-immunoreactive neuron in the antennal lobe of the moth *Manduca sexta:* a laser scanning confocal and electron microscopic study. *J. Comp. Neurol.* 338:5–16
136. Sun XJ, Tolbert LP, Hildebrand JG. 1997. Synaptic organization of the uniglomerular projection neurons of the antennal lobe of the moth *Manduca sexta:* a laser scanning confocal and electron microscopic study. *J. Comp. Neurol.* 379:2–20
137. Todd JL, Anton S, Hansson BS, Baker TC. 1995. Functional organization of the macroglomerular complex related to behaviorally expressed olfactory redundancy in male cabbage looper moth. *Physiol. Entomol.* 20:349–61
138. Todd JL, Haynes KF, Baker TC. 1992. Antennal neurones specific for redundant pheromone components in normal and mutant *Trichoplusia ni* males. *Physiol. Entomol.* 17:183–92
139. Tolbert LP, Hildebrand JG. 1981. Organization and synaptic ultrastructure of glomeruli in the antennal lobes of the moth *Manduca sexta:* a study using thin sections and freeze-structure. *Phil. Trans. R. Soc. London Ser. B* 213:279–301
140. Tóth M, Löfstedt C, Blair BW, Cabello T, Farag AI, et al. 1992. Attraction of male turnip moths *Agrotis segetum* (Lepidoptera: Noctuidae) to sex pheromone components and their mixtures at 11 sites in Europe, Asia, and Africa. *J. Chem. Ecol.* 18(8):1337–47
141. Tumlinson JH, Mitchell ER, Bromer SM, Lindquist DA. 1972. Cis-7-dodecen-1-ol, a potent inhibitor of the cabbage

looper sex pheromone. *Environ. Entomol.* 1:466–68

142. Vickers NJ, Christensen TA, Hildebrand JG. 1998. Combinatorial odor discrimination in small arrays of uniquely identifiable glomeruli. *J. Comp. Neurol.* 400:35–56
143. Visser JH, Avé DA. 1978. General green leaf volatiles in the olfactory orientation of the Colorado beetle, *Leptinotarsa decemlineata*. *Entomol. Exp. Appl.* 24:538–49
144. Vitzthum H, Homberg U, Agricola H. 1996. Distribution of Dip-allatostatin I-like immunoreactivity in the brain of the locust *Schistocerca gregaria* with detailed analysis of immunostaining in the central complex. *J. Comp. Neurol.* 369:419–37
145. Waldrop B, Christensen TA, Hildebrand JG. 1987. GABA-mediated synaptic inhibition of projection neurons in the antennal lobes of the sphinx moth, *Manduca sexta*. *J. Comp. Physiol. A* 161:23–32
146. Waldrop B, Hildebrand JG. 1989. Physiology and pharmacology of acetylcholinergic responses of interneurons in the antennal lobes of the moth *Manduca sexta*. *J. Comp. Physiol. A* 164:433–41
147. Winnington A, Napper RM, Mercer AR. 1996. Structural plasticity of the antennal lobes of the brain of the adult worker honey bee. *J. Comp. Neurol.* 365:479–90
148. Wu W-Q, Anton S, Löfstedt C, Hansson BS. 1996. Discrimination among pheromone component blends by interneurons in male antennal lobes of two populations of the turnip moth, *Agrotis segetum*. *Proc. Natl. Acad. Sci. USA* 93:8022–27
149. Zhu J-W, Löfstedt C, Bengtsson BO. 1996. Genetic variation in the strongly canalised sex pheromone communication system of the European corn borer, *Ostrinia nubilalis* Hübner (Lepidoptera; Pyralidae). *Genetics* 144:757–66
150. Zhu J-W, Zhao CH, Lu F, Bengtsson M, Löfstedt C. 1996. Reductase specificity and the ratio regulation of *E/Z* isomers in pheromone biosynthesis of the European corn borer, *Ostrinia nubilalis* (Lepidoptera: Pyralidae). *Insect Biochem. Mol. Biol.* 26:171–76

Annu. Rev. Entomol. 2000. 45:233–260

LIPID TRANSPORT BIOCHEMISTRY AND ITS ROLE IN ENERGY PRODUCTION

Robert O. Ryan[1] and Dick J. van der Horst[2]

[1]Lipid and Lipoprotein Research Group, Department of Biochemistry, University of Alberta, Edmonton, Alberta, Canada T6G 2S2; e-mail: robert.ryan@ualberta.ca

[2]Biochemical Physiology Research Group, Faculty of Biology and Institute of Biomembranes, Utrecht University, Padualaan 8, 3584 CH Utrecht, The Netherlands; e-mail: d.j.vanderHorst@bio.uu.nl

Key Words lipophorin, adipokinetic hormone, apolipophorin III, diacylglycerol, receptor

■ **Abstract** Recent advances on the biochemistry of flight-related lipid mobilization, transport, and metabolism are reviewed. The synthesis and release of adipokinetic hormones and their function in activation of fat body triacylglycerol lipase to produce diacylglycerol is discussed. The dynamics of reversible lipoprotein conversions and the structural properties and role of the exchangeable apolipoprotein, apolipophorin III, in this process is presented. The nature and structure of hemolymph lipid transfer particle and the potential role of a recently discovered lipoprotein receptor of the low-density lipoprotein receptor family, in lipophorin metabolism and lipid transport is reviewed.

PERSPECTIVES AND OVERVIEW

For many biochemical processes and their regulation, insects provide a fascinating yet relatively simple model system. In the last few years, considerable progress has been made in a number of areas pertinent to lipid mobilization and transport in insects. The scope of this research encompasses regulation of the release of neurohormones (adipokinetic hormones, AKHs), the mechanism of action of fat body triacylglycerol (TAG) lipase, molecular details of hemolymph lipid transport, and cell surface receptors involved in endocytosis of lipoproteins.

Advances made in the area of the biosynthesis and structure of the multifunctional insect lipoprotein, high-density lipophorin (HDLp) are of particular interest. The functioning of this lipoprotein differs from those found in well studied, but complex, vertebrate systems. Indeed, it is the relative simplicity of the invertebrate system that offers possibilities for insight into processes pertinent to both systems. Flight-induced lipid loading onto HDLp is accompanied by association of multiple copies of the exchangeable apolipoprotein, apolipophorin III (apoLp-III).

Because insect apoLp-III is the only apolipoprotein for which a complete three-dimensional structure is known, this system represents a unique model for investigating lipid-protein interactions.

In contrast to the functioning of lipophorin as a lipid shuttle in adult insects, during larval development endocytic uptake of HDLp is apparent. An endocytic receptor belonging to the low-density lipoprotein (LDL) receptor gene superfamily has recently been cloned, offering an exciting new avenue of research opportunity.

Long-distance flight of insects is a complex process that offers an attractive model for prolonged physical exercise. For instance, a pest insect such as the migratory locust is capable of uninterrupted flights of 10 h, covering more than 200 km. Although carbohydrate provides most of the energy during the initial period of flight, after 30 min, principally stored TAG is mobilized to power sustained flight. Consequently, the pivotal enzyme on which long-term flight of insects depends is fat body TAG lipase. The activity of this lipase seems to be controlled by AKH, which is released from the corpus cardiacum in response to flight activity. The enzyme catalyses the hydrolysis of the stored TAG to diacylglycerol (DAG), which is subsequently released into the insect blood and transported by lipophorin. Specific lipophorin subspecies involved in transporting DAG in hemolymph are believed to have receptor-mediated roles in flight, vitellogenesis and general lipid metabolism. Consequently, their primary sequence and properties are of great interest.

ADIPOKINETIC HORMONE-INDUCED LIPID MOBILIZATION

Release of Adipokinetic Hormones

Adipokinetic hormones (AKHs) comprise a family of structurally related N- and C-blocked small peptides. Despite their strong hydrophobicity, AKHs appear to be transported in hemolymph in their free form and not associated with a carrier protein (65). Structures of AKHs have been reported for many insects (33). Although many insects possess a single AKH species, three distinct AKHs have been identified in *Locusta migratoria*: a decapeptide AKH I and two octapeptides (AKH II and III), of which AKH I is by far the most abundant (62).

Details on AKH biosynthesis are particularly confined to a few locust species. The three AKHs of *L. migratoria* are synthesised as preprohormones (prepro-AKH), the amino acid sequences of which have been deduced from their cDNA sequences (14). All three preprohormones contain a 22-amino-acid signal peptide, one single copy of AKH, and a peptide portion termed adipokinetic hormone-associated peptide. The number and sequence of amino acid residues in the preprohormone of AKH III is surprisingly different from those of AKH I and II.

The processing of the preprohormones of AKH I and II has been elucidated in the closely related locust species *Schistocerca gregaria* (59); this species lacks AKH III (62). The signal peptide is co-translationally cleaved from the prepro-AKH, generating pro-AKH. Subsequent proteolytic processing is preceded by dimerization of two pro-AKHs (I/I, I/II, or II/II). Data on the precise processing and possible dimerization of pro-AKH III are lacking to date. In *L. migratoria*, the synthesis of the adipokinetic prohormones, their packaging into secretory granules, and their processing to the bioactive hormones is completed after approximately 75 min (63).

In situ hybridization showed that mRNA encoding *L. migratoria* prepro-AKHs are co-localized in the adipokinetic cells (14). Moreover, immuno-electron microscopic studies demonstrated that these three AKHs co-localize to the secretory granules (27; LF Harthoorn and DJ van der Horst, unpublished data). Consequently, in response to flight activity, the only known natural stimulus for the release of AKHs known, all three AKHs are released simultaneously.

Expression of the distinct AKH precursor genes is increased by flight activity. Northern blot analysis of the AKH precursor mRNAs in corpora cardiaca of locusts at rest and after a 1 h flight indicates that steady-state levels of AKH mRNAs are elevated; AKH I and II mRNAs increase approximately twofold, and AKH III mRNA increases approximately four times (14). There seems to be no acute need, however, for a marked increase in the production of AKH. Only a fraction of the total store of AKH is released during flight activity whereas AKHs are synthesized continuously, resulting in an increase in the amount of hormone stored in the adipokinetic cells with age (64), which is reflected by an increase in the number of secretory granules (28). On the other hand, Sharp-Baker et al (88, 89) discovered that, despite the huge stores of AKH, newly synthesised AKH molecules are preferentially released over older ones, suggesting that a major portion of the stored hormones belong to a non-releasable pool of older hormone. A coupling between the release of AKH and stimulation of hormone gene expression has recently been proposed at the level of signal transduction processes (61).

The adipokinetic cells in the corpus cardiacum appear to be subject to a multitude of regulatory, stimulating, inhibiting, and modulating substances as recently reviewed by Vullings et al (128) and Van der Horst et al (112). Neural influences come from secretomotor cells in the lateral part of the protocerebrum, via the nervus corporis cardiaci II. Up to now, only peptidergic factors have been established to be present in the neural fibers that make contact with the adipokinetic cells. Locustatachykinins initiate the release of AKHs (56, 57), whereas FMRFamide-related peptides inhibit the release of AKHs induced by release-initiating substances (127). Humoral factors that act on adipokinetic cells via the hemolymph are peptidergic and aminergic in nature. Crustacean cardioactive peptide initiates the release of AKHs (123), whereas the amines octopamine, dopamine, and serotonin only potentiate the effect of stimulation of the adipokinetic cells by release-initiating stimulatory substances (66). In addition, high concentrations of trehalose inhibit both the spontaneous release of AKHs and the release induced by release-initiating substances (67). While these several substances may act in concert to regulate release of AKHs, their relative contributions during flight activity (or other conditions in which AKH may be implicated) remains to be established.

Activation and Specificity of Fat Body Lipases

In both vertebrates and invertebrates, sustained physical exercise is fueled largely by the oxidation of long chain fatty acids (FA), which are derived from stored TAG reserves. In vertebrates, FA are mobilized from TAG stores in adipose tissue, and a crucial role in this process is played by hormone-sensitive lipase (HSL). This enzyme controls the rate of lipolysis, catalyzing the first and rate-limiting step in the hydrolysis of the stored TAG, and also the subsequent hydrolysis of DAG and monoacylglycerols (MAG) (43). The FA liberated in adipose tissue are released into the blood and transported bound to serum albumin for uptake and oxidation in muscle.

Unlike FA in vertebrates, lipid in insects is released as DAG and transported to the flight muscles by lipophorin. At the flight muscles, DAG is hydrolyzed and the liberated FA is taken up and oxidized to provide energy. A clear functional similarity exists between vertebrate adipose tissue HSL and insect fat body TAG lipase, as both enzymes catalyze the hydrolysis of TAG stores to meet energy demands. However, there is an essential difference in the mode of action of both lipolytic enzymes. The vertebrate HSL catalyzes hydrolysis of TAG as well as DAG and MAG, resulting in the release of FA, whereas the action of the insect TAG lipase eventually causes the formation and release of DAG. How these differences in the mode of action are brought about is a physiologically relevant question that remains to be answered.

Regulation of Hormone-Sensitive Lipase in Vertebrate Adipocytes

HSL cDNA has been cloned from rat, human and mouse adipose tissue and the sequences obtained show a high degree of homology, whereas no significant relation to other mammalian lipases is discernable (60). Domain structure analyses suggest that HSL is composed of two domains: an N-terminal, presumably lipid-binding, domain and a C-terminal catalytic domain, containing the catalytic triad and a regulatory module containing potential phosphorylation sites. Specific residues have been identified in rat HSL as regulatory (Ser563) and basal (Ser565) sites. The regulatory site is phosphorylated in vitro by cAMP-dependent protein kinase (PKA), resulting in activation of HSL. AMP-activated protein kinase is the most likely kinase responsible for phosphorylation of the basal site in unstimulated adipocytes in vivo. Phosphorylation of the two sites seems to be mutually exclusive, suggesting that phosphorylation of the basal site has an antilipolytic role in vivo, and that phosphorylation of the regulatory site by PKA is preceded by dephosphorylation of the basal site (43). Recently, two other PKA phosphorylation sites (Ser659 and Ser660) have been identified, and these novel sites were shown to be critical activity-controlling sites in rat HSL in vitro, while Ser563 plays a minor role in direct activation (1).

The mechanism behind activation of HSL upon phosphorylation by PKA is poorly understood. The mechanism seems to involve both translocation of HSL from the cytosol to the lipid droplet (30) and conformational changes in the HSL molecule. Different phosphorylation sites in HSL may play different roles in the process of translocation and increase the specific activity of the enzyme. In this connection it is worth noting that perilipins have been suggested to be involved in lipolysis (35, 36). Perilipins are a family of unique proteins intimately associated with the limiting surface of neutral lipid storage droplets. These proteins are acutely polyphosphorylated by PKA on lipolytic stimulation, hinting at a role in this process. Phosphorylated perilipin may serve as a docking protein for HSL, allowing lipase association only when cells are hormonally stimulated. Alternatively, conformational changes of phosphorylated perilipins may expose the neutral lipid cores of the lipid droplets, facilitating hydrolysis. Results of a recent study provide evidence in favor of the latter possibility, demonstrating that, in stimulated adipocytes, translocation of perilipin away from the lipid droplet occurs in concert with movement of HSL toward the droplet (17). Thus, perilipin might act as a barrier to deny HSL access to its lipid substrate in unstimulated adipocytes.

Regulation of Substrate Mobilization in Insects

Substrate mobilization for insect flight is controlled by AKHs, which are involved in both lipid and carbohydrate mobilization. Studies on signal transduction in flight-directed carbohydrate mobilization in *L. migratoria* have shown the involvement of G_s and G_q proteins (124, 126), cAMP and PKA (118, 120), inositol phosphates (121, 125) and Ca^{2+} ions (119) in AKH-induced activation of glycogen phosphorylase, the key enzyme in the conversion of fat body glycogen. Evidence was obtained for the presence of a store-operated Ca^{2+} entry mechanism in the fat body (122). In spite of the importance of lipid mobilization for sustained flight, our knowledge of the regulation of lipolysis in insect fat body is much more restricted, mainly due to technical problems in isolating or activating the lipase. The involvement of AKHs in lipolysis was demonstrated in vivo by an enhanced level of DAG in hemolymph of insects which had been injected with the hormones (reviewed in 10) and in vitro by the accumulation of DAG in isolated locust fat body tissue that was incubated with AKH (53, 134). In in vitro experiments, the effect of AKH on lipolysis was shown to be mediated at least in part by cAMP, suggesting a role for PKA in phosphorylation and activation of TAG lipase. Recently an approximate twofold increase in TAG lipase activity was reported in fat body of locusts injected with AKH-I (58).

In two insect species that rely on lipid mobilization during flight activity, it has been shown that the DAG in the hemolymph is stereospecific. While older data had demonstrated that the DAG released from the fat body of *L. migratoria* by the stimulatory action of AKH has the *sn*-1,2-configuration (52, 104), this finding was recently confirmed for *Manduca sexta* (5). The pathway for stereospecific synthesis of this *sn*-1,2-DAG is largely unknown. Proposed pathways involve stereospecific hydrolysis of TAG into *sn*-1,2-DAG by a stereospecific lipase acting at the *sn*-3 position of the TAG or hydrolysis of TAG into *sn*-2-MAG, followed by stereospecific reacylation of *sn*-2-MAG (reviewed in 111). A third suggested possibility is de novo synthesis of *sn*-1,2-DAG from *sn*-glycerol-3-phosphate via phosphatidic acid using FA produced by TAG hydrolysis (4).

Other experiments have provided more information about this important metabolic pathway. From young adult *M. sexta* fat body, a TAG lipase was purified and showed highest activity for TAG and DAG solubilized in Triton X-100 (4). Although the enzyme exhibited a preference for the primary ester bonds of acylglycerols, it did not show stereoselectivity toward either the *sn*-1 or *sn*-3 position of trioleoylglycerol. Phosphorylation of the enzyme by bovine heart PKA did not affect the activity of the enzyme toward TAG. It remains possible that, in the same manner as HSL (99), phosphorylation of fat body lipase in vivo results in translocation of the enzyme to the fat droplet, which may account for activation of the enzyme.

Because the main end products of TAG hydrolysis by the purified lipase were FA and *sn*-2-MAG and the lipase lacked stereospecificity, direct conversion of stored TAG to *sn*-1,2-DAG in *M. sexta* fat body stimulated by the action of AKH was concluded to be unlikely (4). However, the two alternative pathways for synthesis of *sn*-1,2-DAG appear to be unlikely as well. Studies on stereospecific acylation of *sn*-2-MAG catalyzed by a MAG-acyltransferase revealed that the activity of the transferase, which was found to be primarily a microsomal enzyme with only moderate stereospecificity, was not stimulated by AKH, nor did the size of the microsomal *sn*-2-MAG pool change during AKH-stimulated synthesis of *sn*-1,2-DAG (2). When decapitated adult *M. sexta* were used to measure lipid mobilization (3), AKH treatment showed that FA from TAG were converted to DAG without increasing the fat body content of either FA or phosphatidic acid, suggesting that the *sn*-glycerol-3-phosphate pathway is not involved either (5). The accumulation of *sn*-1,2-DAG led subsequent authors to conclude that the pathway for AKH-stimulated synthesis is stereospecific hydrolysis of TAG.

On the other hand, because this conclusion is in contrast with data obtained on both the stereospecificity and the end products of TAG hydrolysis by the purified fat body TAG lipase, one must speculate that this discrepancy could be an artifact of the experimental conditions under which the purified enzyme was tested in vitro (5). In addition, stereospecific hydrolysis of TAG into *sn*-1,2-DAG involves hydrolysis of the ester bond of the *sn*-3 fatty acid. The fate of this *sn*-3 fatty acid, which did not accumulate in the fat body or in the hemolymph, presently remains unclear.

Unlike the functional similarity between HSL in vertebrate adipose tissue and TAG lipase in insect fat body, a dicrepancy exists in the mode of action of both lipolytic enzymes. The vertebrate HSL catalyzes nonstereospecific hydrolysis of TAG as well as of DAG and MAG resulting in the release of FA, whereas the most likely stereospecific action of the insect TAG lipase causes the formation and release of *sn*-1,2-DAG. So far, the reason for this discrepancy is unknown, but both structural and regulatory aspects may be involved. Elucidation of the mechanism of TAG lipase action and the regulation of its activity is essential for a better insight into the poorly understood mechanism of flight-directed lipolysis in insects. In addition, data obtained may provide clues for a better understanding of the basic processes involved in activation and inactivation of HSL in vertebrates.

STRUCTURE OF LIPOPHORIN

Lipid transport via the circulatory system of animals constitutes a vital function that generally requires lipoprotein complexes, the apolipoprotein components of which serve to stabilize the lipids and modulate metabolism of the lipoprotein particle. For studies of plasma lipid transport, insects offer an attractive and relatively simple system. Because insects are physiologically highly active they require efficient lipid transport for a variety of specialised purposes. Hemolymph generally contains a single major lipoprotein particle, lipophorin, which is found in relatively large quantities. This multifunctional transport vehicle falls into the high-density lipoprotein class (high-density lipophorin, HDLp; d ~1.12 g/ml) and transports a variety of lipophilic biomolecules in the hemolymph. A characteristic feature of lipophorin is an ability to function as a reusable lipid shuttle by the selective loading and unloading of lipids at different target tissue sites (reviewed in 48, 72, 111). As a function of physiological (or developmental) needs for lipid distribution, lipophorin may exist in several forms with respect to relative lipid content and apolipoprotein composition, leading to differences in size and density of the particle. Thus, in species that rely on lipids during flight, lipophorins are loaded with additional DAG and exchangeable apolipoprotein, converting them to low-density lipophorin (LDLp), a particle with considerably more capacity to transport DAG fuel molecules between fat body and flight muscle. Lipophorin structures and that of flight-induced subspecies have been the subject of several recent reviews (11, 73, 74, 97, 113). Briefly, HDLp in insect hemolymph is a spherical particle in the range of Mr = 450,000–600,000. Its lipid cargo contains DAG as a major component, in addition to phospholipids, sterols, and hydrocarbons. In addition, lipophorin has been implicated in the transport of other hydrophobic ligands, such as juvenile hormone, pheromones, and carotenes, from internal biosynthetic sites to sites of utilization (37, 84, 87, 105, 106). Recent data on the lipid composition of HDLp of the yellow fever mosquito, *Aedes aegypti*, showed differences with other insect lipophorins studied to date.

In contrast to the usual prevalence of DAG, the most abundant neutral lipid in this lipophorin is TAG (32, 68).

The protein moiety of the HDLp particle is typically comprised of two integral, nontransferable glycosylated apolipoproteins, apolipophorin I (apoLp-I, ~240 kDa) and apolipophorin II (apoLp-II, ~80 kDa), which are invariably present in a 1:1 molar ratio. Recently, however, exceptions to this paradigm have been reported. Very recently, a different structure of HDLp apolipophorins was found in the cochineal scale insect, *Dactylopius confusus*: Although molecular mass and density of this lipophorin are in the normal range, two small glycosylated apolipophorins of approximately 25 and 22 kDa were isolated, the smaller one being the more abundant (149). Likewise, from the hemolymph of larval *Musca domestica*, in addition to HDLp, a second lipoprotein has been isolated which contains at least four small apolipoproteins of 20 to 26 kDa (25). It was suggested that the latter lipoprotein might have a role in pupal or adult cuticle formation.

Lipophorin Biosynthesis

In addition to its function in storage of lipids and carbohydrates, the insect fat body is the primary site of synthesis of proteins, including lipophorin. Depending on the species, lipophorin appears to be secreted into the hemolymph either as a nascent neutral lipid-deficient particle that loads specific lipids from other sites (reviewed in 49) or as a mature particle with a full complement of lipid (113). With respect to the TAG-rich *A. aegypti* lipophorin, wherein expression of apolipophorins is induced by a blood meal, the latter scenario appears to be the most plausible (115).

Studies of early events in lipophorin biosynthesis revealed that apoLp-I and -II arise from a common precursor (137). Using pulse-chase experiments and specific immunoprecipitation, a proapolipophorin was isolated from locust fat body homogenates, which was converted to apoLp-I and apoLp-II through post-translational cleavage. These results explain the 1:1 stoichiometry of these apolipoproteins in all lipophorin subspecies.

Important data on the cDNA and amino acid sequence of the apolipophorin precursor have recently become available for a few species. When a cDNA fragment (2.4 kb) coding for part of *L. migratoria* apoLp-I was challenged with fat body mRNA in a Northern hybridization, the length of the complete apolipophorin precursor was determined to be approximately 10.3 kb (113). A retinoid- and fatty acid-binding glycoprotein has been purified from *Drosophila melanogaster* heads, and its cDNA cloned (47). The 10.05 kb cDNA open reading frame encoded a 3351 amino acid protein that was identified as the *D. melanogaster* apolipophorin precursor. The N-terminal amino acid sequences determined for apoLp-I and -II of larval sheep blowfly *Lucilia cuprina* lipophorin (105), showed 54% and 70% identity to those of *D. melanogaster* apoLp-I and -II, respectively (47). At the same time, a 10.14 kb cDNA encoding the complete apolipophorin precursor of *M. sexta* was cloned and sequenced (100). The latter two studies demonstrated that the precursor protein is arranged with apoLp-II at the N-terminus of proapolipophorin with apoLp-I at the C-terminus. In *D. melanogaster* proapolipophorin, a single consensus cleavage site, RXRR for dibasic endoprotease processing, was reported (47). It may be assumed that such a protease processes the precursor protein at this site, generating apoLp-II and apoLp-I. In another study, a cDNA fragment (1.2 kb) coding for part of *A. aegypti* apoLp-II was cloned and sequenced. Its deduced amino acid sequence showed approximately 40% identity with the corresponding regions of the *D. melanogaster* and *M. sexta* apoLp-II sequences (115). Northern blot analysis revealed that the size of *A. aegypti* apolipophorin precursor mRNA (approximately 10 kb) is similar to that of the other apolipophorin precursor mRNAs.

Recent studies, which included molecular characterization of the complete apolipophorin precursor cDNA from *L. migratoria* (J Bogerd & DJ van der Hosrt, unpublished data), revealed significant sequence homology among apolipophorin precursor proteins. The apolipophorin precursor protein of *L. migratoria* was 27% identical and 45% similar to that of *M. sexta* (100) and 25% identical and 43% similar to the retinoid- and fatty acid-binding glycoprotein of *D. melanogaster* (47). In addition, 35% identity and 52% similarity was observed with the partial, deduced apoLp-II amino acid sequence of *A. aegypti* (115). Moreover, statistically significant similarities were found between two distinct domains of locust apolipophorin precursor protein and two groups of extracellular proteins. The region between residues 31 and 1016 is similar to human apolipoprotein B-100 (21% identical and 38% similar; 46) and showed some similarity to various vertebrate and nematode vitellogenin precursors. Another region, located between residues 2816 and 3027, is related to the D domain of various von Willebrand factor precursors (23% to 26% identity, 39% to 42% similarity) and various mucins. The apolipophorin precursor proteins of *L. migratoria, D. melanogaster,* and *M. sexta* were used to identify contiguous conserved sequence motifs in alignments of large, nonexchangeable lipid transport proteins, demonstrating that the genes encoding the apolipophorin precursor—human apolipoprotein B, invertebrate and vertebrate vitellogenins, and the large subunit of the mammalian microsomal triaglyceride transfer protein—are members of the same multigene superfamily and are derived from a common ancestor gene (6).

Lipophorin Subspecies Interconversions

In insects capable of migratory flights, such as *L. migratoria* and *M. sexta*, AKH-induced mobilization of fat body TAG stores requires that this lipid be converted to DAG. DAG is released from the cell and loaded onto circulating HDLp particles, ultimately transforming them into LDLp (d < 1.06 g/ml). This loading process is facilitated by lipid transfer particles (see below) and requires that, concomitant with DAG uptake, several copies of the amphipathic exchangeable apolipoprotein, apolipophorin III (apoLp-III; ~18–20 kDa), associate with the particle (140). Whereas HDLp is abundant during all developmental stages of the locust, apoLp-III expression is developmentally regulated and its hemolymph level is high only in adults (26). At the flight muscle, LDLp-associated DAG is hydrolyzed by a lipophorin lipase and the resulting free fatty acids are taken up and oxidized to provide energy. As the lipid content of the particle diminishes, apoLp-III dissociates with both constituents (HDLp and apoLp-III) recovered in the hemolymph. As such, they are free to return to the fat body for further DAG uptake, LDLp formation and transport. This process constitutes a vivid example of the lipophorin reusable shuttle hypothesis first proposed by Chino and coworkers.

APOLIPOPHORIN III

An important feature of all exchangeable apolipoproteins is an inherent structural adaptability, providing them with the ability to reversibly associate with circulating lipoproteins. It is implied, therefore, that these proteins can exist both in lipid-poor and lipid-associated states. Indeed, it is likely that exchangeable apolipoproteins can adopt more than one conformation in the lipid-bound state, as exemplified by human apolipoprotein A-I in disk structures versus spherical lipoproteins (31). Also, it is noteworthy that human apolipoprotein E (apoE) is not found on nascent chylomicron particles, yet acquisition of this apolipoprotein is essential for clearance of chylomicron remnant particles via receptor-mediated endocytosis (139). Importantly, the receptor recognition properties of apoE are manifest only upon lipid association, suggesting that a lipid-induced conformational adaptation may be a critical event in terms of this function. In another example, human apolipoprotein A-IV is known to partition between lipid-free and lipoprotein-associated states, perhaps serving as a reservoir of lipoprotein surface material (29).

Structural Properties

One of the best examples of the reversible existence of exchangeable apolipoproteins in lipid-free and lipoprotein-associated states is apoLp-III (74). In resting animals apoLp-III is recovered as a lipid-free hemolymph protein. However, during flight activity, or under the influence of AKH, apoLp-III associates with the surface of lipophorin, where it stabilizes DAG-enriched LDLp. Characterization of apoLp-III has revealed that it is rich in α-helical secondary structure. Sequence analysis reveals that the α-helices are amphipathic in nature, further suggesting a close similarity in structural properties with mammalian apolipoproteins (18, 93). Support for the concept that apoLp-III is structurally related to mammalian apolipoproteins also comes from comparison of the three-dimensional structures of *L. migratoria* apoLp-III and the 22 kDa N-terminal domain of human apoE (15, 144). These structures, which were determined for the proteins in the lipid-free state, reveal globular proteins comprised of a series of amphipathic α-helices, organized in a bundle. ApoLp-III is a five-helix bundle while the N-terminal domain of human apoE is a four-helix bundle. In each case the helices are organized such that their hydrophobic faces orient toward the center of the bundle and their hydrophilic faces are directed at the aqueous media. This molecular architecture provides a rational explanation for the water solubility of these proteins in the absence of lipid and provides a framework for postulating conformational changes that may accompany lipid association.

Study of apoLp-III has been improved by the availablility of recombinant proteins (79, 95, 138). In the case of *L. migratoria* apoLp-III, in contrast to the natural protein (38), bacterially expressed protein is nonglycosylated. This feature allowed Soulages et al (95) to investigate the effect of apoLp-III carbohydrate on the lipid binding properties of this protein. Interestingly, these authors found that recombinant apoLp-III interacts with phospholipid vesicles more efficiently than natural apoLp-III and suggested that the carbohydrate structures on the protein may function to mask hydrophobic domains, which are important for lipid recognition. When it was discovered that *M. sexta* apoLp-III could be produced in bacteria at more than 100 mg/L culture (79), it became evident that this protein may be amenable to structure determination by heteronuclear multidimensional NMR techniques. Following confirmation that apoLp-III meets essential criteria for protein NMR including existence as a monomer at concentrations up to 1 mM and retention of a stable, native conformation for one week at 37° C in an appropriate buffer, a variety of three-dimensional NMR experiments were employed to obtain a complete assignment of apoLp-III NMR spectra (129). Recently, this information has been used to calculate a high-resolution solution structure (J Wang, BD Sykes, RO Ryan, unpublished data). This NMR structure, which represents the first solution structure determination of a full-length apolipoprotein, reveals a five-helix bundle architecture, which is similar to the X-ray crystal structures of *L. migratoria* apoLp-III and human apoE N-terminal domain. In addition, a short connecting helix was identified that links helix 3 and helix 4. It is noteworthy that a similar short helix is also present in human apoE N-terminal domain (144). We have proposed that this short helix may funtion in lipid recognition and/or initiation of binding to lipoprotein surfaces (129).

An important goal of apoLp-III NMR studies is to develop techniques that may be generally applicable to exchangeable apolipoproteins in their biologically active, lipid-associated state. Recently, a measure of success has been achieved in this endeavor through complex formation with the micelle-forming lipid, dodecylphosphocholine (131, 132, 138). Results obtained reveal significant spectral differences between lipid-free and lipid-bound conformations of apoLp-III, indicative of a major conformational change upon binding to lipid.

Open Conformation Model

We, and others, have obtained indirect experimental evidence in support of the conformational opening hypothesis. Early studies, using a monolayer balance, revealed that apoLp-III occupies more area at the air/water interface than can be explained if the globular state was retained (44). In studies of the properties of apoLp-III in association with model phospholipids, Wientzek et al (143) and Weers et al (135) showed it is capable of transforming bilayer vesicles of dimyristoylphosphatidylcholine (DMPC) into uniform disk-like particles. Characterization of these complexes revealed that five or six apoLp-III molecules associate with each disk and, based on compositional analysis and the molecular dimensions of the protein, it was hypothesized that apoLp-III adopts an open conformation and aligns around the perimeter of the bilayer structure with its α-helices oriented perpendicular to the fatty acyl chains of the phospholipids (74, 143). Subsequently, spectroscopic evidence in support of this interpretation has been reported (69). In another study, Raussens et al (70) found that association with lipid has a significant effect on the amide proton hydrogen/deuterium exchange rate in apoLp-III, further suggesting that lipid association induces a change in protein structure that alters the solvent accessibility of amino acid residues in the protein.

Using an alternate approach to investigate conformational changes associated with lipid interaction, site-directed mutagenesis has been employed to introduce two cysteine residues into apoLp-III, which were designed to permit formation of a disulfide bond (54). Characterization of the mutant protein revealed that it contained the desired mutations and that it folds properly in solution. Furthermore, under oxidizing conditions, the cysteine residues (which are located in the loop segments that connect helices 1 and 2 and 3 and 4) spontaneously form a disulfide bond. Subsequent study of reduced and oxidized mutant apoLp-III revealed that formation of the disulfide bond abolishes the ability of the protein to interact with lipoprotein surfaces. Upon reduction of the disulfide bond, however, full lipoprotein binding activity is restored. The data suggests that disulfide bond formation tethers the helix bundle, preventing exposure of its hydrophobic interior through conformational opening. As such, these data support the hypothesis that apoLp-III opens about "hinged" loop regions to create an elongated amphipathic structure that represents the lipid binding conformation of the protein.

Lipid Binding Properties

Early studies (140) showed a correlation between the DAG content of lipophorins and the amount of apoLp-III associated. Since that time several strategies have been employed to characterize this relationship. Enrichment of isolated lipoproteins with DAG via facilitated lipid transfer (92) or direct addition of exogenous short chain di C_8 DAG (98) both induce apoLp-III association with lipoprotein surfaces. In another approach, treatment of isolated lipoproteins with phospholipase C to create DAG (51, 90) causes exchangeable apolipoprotein association as a function of the extent of conversion of surface phospholipid into DAG. Taken together, these data support a model wherein apoLp-III binding to the surface monolayer of lipoproteins involves an interaction between the amphipathic α-helices of the protein with phospholipid and DAG (74, 130). It is reasonable to consider that positively charged amino acids, located at the edge of the helices, form salt bridges either with adjacent helices or with charged groups on the lipid (148).

In terms of the initiation of lipoprotein association, studies have been performed using apoLp-III as a model system. It has been suggested that electrostatic interactions between amino acid side chains and phospholipids may localize the apoprotein at the particle surface (148). In cases where the lipoprotein has available hydrophobic binding sites or surface defects, a conformational change ensues and a stable binding interaction occurs. This process can be reversed by metabolic events that reduce the available hydrophobic surface, wherein the apolipoprotein would be released from the particle and adopt its solution conformation.

An alternate hypothesis (93, 96) proposes that apolipoprotein binding is regulated by a small region of hydrophobic amino acids, a "hydrophobic sensor," located at one end of the apolipoprotein. These workers noted that apoLp-III molecules from several insect species contain a region of exposed hydrophobic amino acids in the loop regions connecting helices 1 and 2 and helices 3 and 4. It has been proposed that these amino acids play a critical role in initiation of binding, associating with the lipoprotein with the long axis of the protein perpendicular to the particle surface. In the presence of sufficient surface-exposed hydrophobic material, the protein opens to form a stable interaction (54). Evidence in favor of the "hydrophobic sensor" mechanism has been obtained through experiments employing surface plasmon resonance spectroscopy (96), and indirectly from studies with disulfide crosslinked apoLp-III (54). More recently, Soulages & Bendavid (94) have extended this model, providing evidence that *M. sexta* apoLp-III interaction with phospholipid vesicles is greatly affected by solution pH. These authors propose that below pH 6.5, apoLp-III adopts a partially folded "molten globule" conformation with an increased affinity for binding to phosphatidylcholine vesicles. This hypothesis has been examined further in both *M. sexta* and *L. migratoria* apoLp-III (55, 136). In both studies, site-directed mutagenesis was employed to replace specific residues in apoLp-III and the effect of such changes on lipid binding evaluated. In *M. sexta* apoLp-III, it was demonstrated that valine 97, located in the approximate center of the short connecting helix (129) plays a critical role in the lipoprotein binding ability of the protein (55). In the case of *L. migratoria* apoLp-III, Weers et al (136) mutagenized key leucine residues in the loops connecting helices 1 and 2 and helices 3 and 4 (93). Upon conversion of these residues to arginine it was noted that, although the lipoprotein binding ability of the protein was decreased, the mutant protein displayed an increased ability to disrupt phospholipid bilayer vesicles. On the basis of these results, it was suggested that electrostatic attraction plays an important role in phospholipid binding while hydrophobic interactions are an important force in lipoprotein binding. Although both mutagenesis studies suggest apoLp-III interacts with lipid surfaces specifically via one of its ends, the precise mechanism of reversible lipid binding remains to be determined.

LIPID TRANSFER PARTICLE

Since the discovery by Zilversmit et al (150) of specialized proteins that function in the redistribution of hydrophobic lipid molecules, a great deal of interest in this class of protein has been generated. A wide variety of distinct lipid transfer proteins have been characterized and their metabolic roles investigated. Among these are the human plasma cholesteryl ester and phospholipid transfer proteins (103), microsomal triglyceride transfer protein (MTP; 141) and the intracelleular phospholipid transfer proteins (145).

The lipophorin shuttle hypothesis, with repeated cycles of lipid loading and depletion, implies the existence of additional factors to facilitate these interconversions. A heat-labile factor (78, 81), now termed lipid transfer particle (LTP), was isolated in 1986 and shown to facilitate vectorial redistribution of lipids among plasma lipophorin subspecies. In addition, a flight muscle lipase has been described which is hypothesized to lipolyze lipophorin-associated DAG at the flight muscle tissue (116, 142). In subsequent studies LTP was strongly implicated to function in vivo in formation of LDLp from HDLp in response to adipokinetic hormone (114) as well as in the conversion of HDLp into egg very high density lipophorin upon receptor mediated endocytosis into oocytes (50). The apparent obligatory role of LTP in hemolymph lipoprotein conversions is analagous to the recent postulate that mammalian MTP plays an important role in the biogenesis of triacylglycerol-rich lipoproteins (141). The concept that LTP functions in flight-related lipophorin conversions is congruent with an increased hemolymph concentration of LTP in adults versus other developmental stages (106, 117).

Physical and Structural Properties

Compared to other types of lipid transfer catalyst, *M. sexta* LTP exhibits novel structural characteristics. Data from early studies indicate it is a high molecular weight complex of three apoproteins (apoLTP-I, Mr $\sim$320,000; apoLTP-II, Mr $=$85,000 and apoLTP-III, Mr $=$55,000) and 14% noncovalently associated lipid (80). Interestingly, catalysts with similar structural properties have been isolated from *Locusta migratoria, Periplaneta americana* and *Bombyx mori* hemolymph (42, 102, 107). The lipid component resembles that of lipophorin in that it contains predominantly phospholipid and DAG (72). Its apparent large size, together with the presence of lipid, have been employed in the purification scheme to isolate LTP (81). An important question arising from these physical characteristics relates to the requirement of the lipid component as a structural entity and/or its involvement in catalysis of lipid transfer. Although its precise function is not known, studies employing lipoproteins containing radiolabeled lipids in incubations with LTP revealed that the lipid component of the particle is in dynamic equilibrium with that of lipoprotein substrates (80). Thus, it was concluded that the lipid moiety of LTP is not merely a static structural entity of the particle, but instead may play a key role in facilitation of lipid transfer.

The morphology of LTP has been investigated by electron microscopy (77, 102). LTP is a highly asymmetric structure with two major structural features: a roughly spherical head and an elongated, cylindrical tail, which has a central hinge. Sedimentation equilibrium experiments conducted in the analytical ultracentrifuge revealed a native particle molecular weight of approximately 900,000 (76). This data agrees with results obtained from native pore-limiting gradient gel electrophoresis (80) and calculations based on the dimensions of LTP determined by electron microscopy. In sedimentation velocity experiments evidence of a reversible, ionic-strength–dependent aggregation of LTP was obtained. Both aggregated and monomeric LTP are catalytically active, however, and have similar amounts of secondary structure conformers, indicating that large changes in structure do not accompany ionic strength-induced changes in sedimentation coefficient. The ability of LTP to self associate may be important in its physiological function because ionic strength conditions in hemolymph would favor formation of aggregated LTP (76). The presence of a 150-fold excess of lipophorin in hemolymph, however, may dictate that LTP interact with these potential substrates rather than self associate, thereby facilitating its biological function. It is possible that a differential affinity of LTP for lipid particle substrates or biomembranes, on the basis of their lipid content, net charge, or otherwise, may be a factor in modulating activity of LTP.

Catalytic Activity and Substrate Specificity

A number of studies have investigated the question of whether LTP is capable of facilitating lipid transfer to or from substrate lipid particles other than lipophorins. Ryan et al (75) showed that LTP can facilitate lipid transfer between lipophorin and an unrelated hemolymph chromolipoprotein from *Heliothis zea*. The observation that LTP can mediate DAG exchange between these substrate particles suggests it may function in vivo to redistribute lipid among lipophorin and unrelated very-high-density lipoproteins present in hemolymph.

In related experiments examining the ability of LTP to utilize other substrates, the ability to transfer lipid between lipophorin and human LDL has been examined (82). When lipophorin and LDL were incubated together in the absence of LTP and re-isolated by density gradient ultracentrifugation, no changes in their respective density distribution were detected. When catalytic amounts of LTP were added, however, a dramatic shift in the lipoprotein density profile was seen with LDL floating to a lower-density position and lipophorin recovered at a higher-density position. Neither apoprotein transfer nor particle fusion were responsible for these changes in lipoprotein density. When the lipophorin substrate was labeled with [^{3}H]DAG, it was shown that LTP-mediated vectorial net transfer of DAG from lipophorin to LDL was responsible for the observed changes in lipoprotein density profile. These data provide strong support for the concept that transfer activity inherent in LTP is sufficent to induce major changes in lipophorin lipid content and composition that normally occur in vivo. Furthermore, the data suggest that the direction and extent of LTP-mediated lipid redistribution is dependent upon the structural properties and hydrophobic environment provided by potential lipid donors and acceptors.

In other studies of the lipid substrate specificity of LTP, [1-^{14}C]acetate has been used to label the DAG and hydrocarbon moiety of lipophorin in vivo (91). Subsequent transfer experiments revealed that LTP is capable of facilitating transfer of hydrocarbon from lipophorin to LDL, suggesting LTP may play a role in movement of these extremely hydrophobic, specialized lipids from their site of synthesis to their site of deposition at the cuticle (see also 102). When acceptor LDL particles were analyzed prior to complete transfer of lipophorin-associated lipid it was observed that DAG was transferred preferentially during the initial stage of the reaction after which hydrocarbon transfer increased. Recently, *B. mori* LTP was demonstrated to facilitate transfer of carotenes among lipophorin particles (106). Again, when compared to DAG transfer, the rate of LTP-mediated carotene redistribution was much slower. Taken together, these results suggest that LTP may have a preference for DAG rather than hydrocarbon or carotenes. Alternatively, the observed preference for DAG may be a function of the relative accessibility of the substrates within the donor lipoprotein.

Mechanism of LTP-Mediated Lipid Transfer

A major question regarding the ability of LTP to facilitate net vectorial lipid transfer relates to the mechanism of this process. Transfer catalysts may act as carriers of lipid between donor and acceptor lipoproteins or, alternatively, transfer may require formation of a ternary complex between donor, acceptor and LTP. Based on the well characterized LTP-mediated vectorial transfer of DAG from HDLp to human LDL (82) a strategy was developed to address this question experimentally (13). It had been previously shown that, in solution, LTP catalyzes an extensive vectorial net transfer of DAG from lipophorin to LDL. LDL particles become enriched in DAG while lipophorin particles are depleted of neutral core lipid (82). This reaction is not accompanied by apoprotein exchange or transfer and DAG is not metabolized during this reaction. [^{3}H]DAG-HDLp and unlabeled LDL were covalently bound to Sepharose matrices and packed into separate columns connected in series, whereafter LTP or buffer was circulated. Control experiments revealed that the lipoproteins remained fixed to the solid phase and that DAG did not transfer to LDL in the absence of LTP. In the presence of LTP, however, there was a concentration-dependent increase in the amount of labeled DAG recovered in the LDL fraction, demonstrating that LTP can facilitate net lipid transfer via a carrier-mediated mechanism.

In another approach, Blacklock & Ryan (12) employed apolipoprotein-specific antibodies to probe the structure and catalytic properties of *M. sexta* LTP. In antibody inhibition studies, evidence was obtained that apoLTP-II is a catalytically important apoprotein. In a separate study, Van Heusden and coworkers (117) provided evidence from antibody inhibition studies that all three LTP apoproteins are important for lipid transfer activity. Furthermore, evidence was obtained that, unlike apoLp-III, apoLTP-III is not found as a free protein in hemolymph (117) although it can dissociate from the complex upon treatment with a nonionic detergent, while apoLTP-I and -II remain associated (12).

LIPOPHORIN RECEPTORS

A major difference between insect lipophorins and vertebrate lipoproteins is the selective mechanism by which insect lipoproteins transport their hydrophobic cargo. For example, at the fat body cell, DAG can be loaded onto circulating HDLp particles for lipid mobilization during flight activity, an adult-specific process. DAG may also be extracted from HDLp for lipid storage, a prominent process in larval and younger adult insects. It is likely that for both lipid removal and accretion, interaction between lipophorins and fat body cells takes place. In *L. migratoria*, a high-affinity HDLp binding site ($Kd \sim 10^{-7}$ M) has been characterized in intact fat body tissue as well as in fat body membranes of larval and adult locusts (23, 86, 110). The interaction between HDLp and this binding site did not require Ca^{2+}-ions (23), in contrast to other insect lipoprotein receptors (71, 109, 133). The number of HDLp binding sites at the locust fat body cell surface increased between day seven and day 11 after imaginal ecdysis (23), suggesting a role in an adult-specific process such as flight activity—this increase coincides with changes in flight muscle metabolism required for flight activity (9, 39). In adult fat body tissue, no differences were observed in the binding characteristics of AKH-injected and control locusts (23), suggesting that if this HDLp binding site is involved in LDLp formation, the abundance of these sites is sufficient to cope with the increased lipid mobilization during sustained fligh activity. As yet, however, no conclusive evidence has been presented for the involvement of HDLp binding sites in LDLp formation, although both HDLp binding and LDLp formation are inhibited by monoclonal antibodies specific for apoLp-II and not by those directed against apoLp-I (41, 86).

Interaction of Lipophorin with Fat Body Cells

Recently, the HDLp binding site in larval and young adult locusts was identified as an endocytic receptor involved in uptake of HDLp (19). Endocytic internalization of HDLp appeared to be developmentally down-regulated in the adult stage. When insects were starved, however, internalization of HDLp remained evident. Also, in larvae of the dragon fly, *Aeshna cyanea,* it had been shown that HDLp is internalized by the fat body (7), although this was not observed in the midgut epithelium (8), implying that endocytosis of HDLp does not occur in all target tissues. The specific HDLp binding sites identified in fat body and gut of larval *M. sexta* appear not to be involved in endocytotic uptake of this ligand (108, 109). Although endocytic uptake of HDLp seems to conflict with the selective process of lipid transport between HDLp and fat body cells without degradation of the lipophorin matrix, the pathway followed by the internalized HDLp may be different from the classical endosomal/lysosomal pathway. In studies with radiolabelled HDLp neither substantial degradation nor accumulation of HDLp in fat body cells were apparent (19). On the basis of these results it was proposed that lipid transport may occur by a retroendocytic pathway, as previously postulated for larval dragon fly fat body by Baurfeind & Komnick (7). Internalization of HDLp by locust fat body cells has been studied at the electron-microscopic level using ultrasmall gold-labeled HDLp and fluorescent dye (DiI)-labeled HDLp (20). In the latter experiments, visualization of the DiI-labeled lipophorin was achieved by diaminobenzidine photoconversion (21). Internalized labeled HDLp was observed in the endosomal/lysosomal compartment of fat body cells of both young and older adults, although labeling in older adults was much less abundant. Still, in view of the accumulation of HDLp in lysosomes, whether part of the internalized HDLp is re-secreted after intracellular trafficking and possible unloading of lipid cargo must be evaluated. The function of receptor-mediated endocytosis remains unclear because inhibition of endocytosis did not reduce the exchange of DAG or cholesterol between HDLp and the fat body cell (19).

Human Low-Density Lipoprotein Receptor Homologue

Very recently, the putuative receptor involved in endocytic uptake of HDLp in the locust fat body has been identified (22). Receptor-mediated endocytosis of lipoproteins has been studied in detail in vertebrate cells (34). Cloning of different receptors that function in lipoprotein uptake has resulted in the identification of the LDL receptor family, the members of which share structural and functional features (16, 24, 40, 45, 101, 146). These receptors appear to originate from an ancient receptor in view of the identification of a similarly composed cell surface molecule in *Caenorhabditis elegans* (147) and two insect vitellogenin receptors belonging to this gene superfamily (83, 85). The endocytic uptake of HDLp by locust fat body cells was therefore hypothesized to be mediated by a member of the LDL receptor family (22). A novel member of the LDL receptor family was cloned and sequenced. Northern blot analysis revealed expression in locust fat body as well as in oocytes, brain, and midgut. This receptor appeared to be a homologue of the mammalian very-low-density lipoprotein receptor: it contains eight cysteine-rich repeats in its putative ligand-binding domain. This receptor represents the first identification of an invertebrate LDL receptor family member with an extracellular domain composed of a single ligand-binding domain and EGF-precursor domain, a receptor organization that has been found in many vertebrates. When expressed in COS-7 cells, the receptor mediates endocytic uptake of HDLp. Expression of the receptor mRNA in fat body cells is down-regulated during adult development, which is consistent with the previously reported down-regulation of receptor-mediated endocytosis of lipophorins in fat body tissue (19).

DIRECTIONS FOR FUTURE RESEARCH

In assessing the current status of this research field, we note that studies of the regulation of insect lipid mobilization and transport during flight activity has led to a valuable research model. Knowledge of the key events of this process have provided a coherent understanding as well as an integrated view. In terms of future exploration, the multifactorial process of AKH release represents an important ongoing research effort, whereas neuropeptide-receptor binding interactions will require considerable research effort. The mechanisms whereby neuropeptides regulate and modulate a sequence of metabolic processes, including the mobilization of DAG through activation of fat body TAG lipase is another important future research goal. Transport of lipid released in the hemolymph requires the hormone-stimulated transformation of lipophorins, which are capable of alternating between a relatively lipid-poor and lipid-enriched forms. In these reversible conversions the exchangeable apolipoprotein, apoLp-III, which exists in alternate lipid-free and a lipid-bound states, plays an essential role. On all aspects of lipophorin (structure, biosynthesis, interconversion of subspecies and evolutionary relationships) considerable information has been gained, but much additional data is required for a full understanding of this unique system and to fully apply insight into corresponding processes in higher organisms.

The lack of sequence information on apoprotein components is an impediment to a fuller understanding of LTP structure and activity. This information is critical to addressing questions regarding the specific roles played by various subunits of the holoparticle. Since the discovery of LTP, considerable information has been gained about its in vivo function, lipid and lipid particle substrate specificity, mechanism of action, and morphology. In the future it will be important to extend this knowledge by employing molecular biology techniques to clone and sequence LTP apoproteins and to further investigate the structure and interactions of this unique catalyst. Important progress in the area of insect lipoprotein receptors has opened the door to a potential understanding of the molecular basis of the lipophorin shuttle hypothesis and its integration with emerging endocytic pathways for lipoprotein uptake from hemolymph. Through continued exploitation of the remarkable physiology surrounding flight related lipid transport, this model system will continue to provide fundamental discoveries of profound biological significance.

ACKNOWLEDGMENTS

This project was supported by a NATO collaborative research grant to DJ Van der Horst and RO Ryan. Research support from the Alberta Heritage Foundation for Medical Research and the Medical Research Council of Canada to RO Ryan is gratefully acknowledged. Support was provided by a grant from the Life Science Foundation (SLW), subsidized by the Netherlands Organization for Scientific Research (NWO).

Visit the Annual Reviews home page at www.AnnualReviews.org.

LITERATURE CITED

1. Anthonsen MW, Rönnstrand L, Wernstedt C, Degerman E, Holm C. 1998. Identification of novel phosphorylation sites in hormone-sensitive lipase that are phosphorylated in response to isoproterenol and govern activation properties in vitro. *J. Biol. Chem.* 273:215–21
2. Arrese EL, Rojas-Rivas BI, Wells MA. 1996. Synthesis of sn-1,2-diacylglycerols by monoacylglycerol acyltransferase from *Manduca sexta* fat body. *Arch. Insect Biochem. Physiol.* 31:325–35
3. Arrese EL, Rojas-Rivas BI, Wells MA. 1996. The use of decapitated insects to study lipid mobilization in adult *Manduca sexta*: effects of adipokinetic hormone and trehalose on fat body lipase activity. *Insect Biochem. Mol. Biol.* 26:775–82
4. Arrese EL, Wells MA. 1994. Purification and properties of a phosphorylatable triacylglycerol lipase from the fat body of an insect, *Manduca sexta*. *J. Lipid Res.* 35:1652–60
5. Arrese EL, Wells MA. 1997. Adipokinetic hormone-induced lipolysis in the fat body of an insect, *Manduca sexta*: synthesis of *sn*-1,2-diacylglycerols. *J. Lipid Res.* 38:68–76
6. Babin PJ, Bogerd J, Kooiman FP, Van Marrewijk WJA, Van der Horst DJ. 1999. Apolipophorin II/I, apolipoprotein B, vitellogenin, and microsomal triglyceride transfer protein genes are derived from a common ancestor. *J. Mol. Evol.* 49:150–60
7. Bauernfeind R, Komnick H. 1992. Immunocytochemical localization of lipophorin in the fat body of dragonfly larvae (*Aeshna cyanea*). *J. Insect Physiol.* 38:185–98
8. Bauernfeind R, Komnick H. 1992. Lipid loading and unloading of lipophorin in the midgut epithelium of dragonfly larvae (*Aeshna cyanea*). A biochemical and immunocytochemical study. *J. Insect Physiol.* 38:147–60
9. Beenakkers AMT, Van den Broek ATM, De Ronde TJA. 1975. Development of catabolic pathways in insect flight muscles. A comparative study. *J. Insect Physiol.* 21:849–59
10. Beenakkers AMT, Van der Horst DJ, Van Marrewijk WJA. 1985. Insect lipids and their role in physiological processes. *Prog. Lipid Res.* 24:19–67
11. Blacklock BJ, Ryan RO. 1994. Hemolymph lipid transport. *Insect Biochem. Mol. Biol.* 24:855–73
12. Blacklock BJ, Ryan RO. 1995. Structural studies of *Manduca sexta* lipid transfer particle with apolipoprotein-specific antibodies. *J. Lipid Res.* 36:108–16
13. Blacklock BJ, Smillie M, Ryan RO. 1992. *Manduca sexta* lipid transfer particle can facilitate lipid transfer via a carrier-mediated mechanism. *J. Biol. Chem.* 267:14033–37
14. Bogerd J, Kooiman FP, Pijnenburg MAP, Hekking LHP, Oudejans RCHM, et al. 1995. Molecular cloning of three distinct cDNAs, each encoding a different adipokinetic hormone precursor, of the migratory locust, *Locusta migratoria*. *J. Biol. Chem.* 39:23038–43
15. Breiter DB, Kanost MR, Benning MM, Wesenberg G, Law JH, et al. 1991. Molecular structure of an apolipoprotein

determined at 2.5-Å resolution. *Biochemistry* 30:603–8
16. Bujo H, Hermann M, Kaderli MO, Jacobsen L, Sugawara S, et al. 1994. Chicken oocyte growth is mediated by an eight ligand binding repeat member of the LDL receptor family. *EMBO J.* 13:5165–75
17. Clifford GM, McCormick DKT, Vernon RG, Yeaman SJ. 1997. Translocation of perilipin and hormone-sensitive lipase in response to lipolytic hormones. *Biochem. Soc. Trans.* 25:S672
18. Cole KD, Fernando-Warnakulasuriya GJP, Boguski MS, Freeman M, Gordon JI, et al. 1987. Primary structure and comparative sequence analysis of an insect apolipoprotein: apolipophorin III from *Manduca sexta*. *J. Biol. Chem.* 262:11794–800
19. Dantuma NP, Pijnenburg MAP, Diederen JHB, Van der Horst DJ. 1997. Developmental down-regulation of receptor-mediated endocytosis of an insect lipoprotein. *J. Lipid Res.* 38:254–65
20. Dantuma NP, Pijnenburg MAP, Diederen JHB, Van der Horst DJ. 1998. Multiple interactions between insect lipoproteins and fat body cells: extracellular trapping and endocytic trafficking. *J. Lipid Res.* 39:1877–88
21. Dantuma NP, Pijnenburg MAP, Diederen JHB, Van der Horst DJ. 1998. Electron microscopic visualization of receptor-mediated endocytosis of DiI-labeled lipoproteins by diaminobenzidine photoconversion. *J. Histochem. Cytochem.* 46:1085–89
22. Dantuma NP, Potters M, De Winther MPJ, Tensen CP, Kooiman FP, et al. 1999. An insect homolog of the vertebrate low density lipoprotein receptor mediates endocytosis of lipophorins *J. Lipid Res.* 40:973–78
23. Dantuma NP, Van Marrewijk WJA, Wynne HJ, Van der Horst DJ. 1996. Interaction of an insect lipoprotein with its binding site at the fat body. *J. Lipid Res.* 37:1345–55
24. Davail B, Pakdel F, Bujo H, Perazzolo LH, Waclawek M, et al. 1998. Evolution of oogenesis: the receptor for vitellogenin from the rainbow trout. *J. Lipid Res.* 39:1929–37
25. De Bianchi AG, Capurro M de L. 1991. *Musca domestica* larval lipids. *Arch. Insect Biochem. Physiol.* 17:15–27
26. De Winther MPJ, Weers PMM, Bogerd J, Van der Horst DJ. 1996. Apolipophorin III levels in *Locusta migratoria*. Developmental regulation of gene expression and hemolymph protein concentration. *J. Insect Physiol.* 42:1047–52
27. Diederen JHB, Maas HA, Pel HJ, Schooneveld H, Jansen WF, et al. 1987. Co-localization of the adipokinetic hormones I and II in the same glandular cells and in the same secretory granules of the corpus cardiacum of *Locusta migratoria* and *Schistocerca gregaria*. *Cell Tissue Res.* 249:379–89
28. Diederen JHB, Peppelenbosch MP, Vullings HGB. 1992. Ageing adipokinetic cells in *Locusta migratoria*: an ultrastructural morphometric study. *Cell Tissue Res.* 268:117–21
29. Duverger N, Tremp G, Caillaud J-M, Emmanuel F, Castro G, et al. 1996. Protection against atherogenesis in mice mediated by human apolipoprotein A-IV. *Science* 273:966–68
30. Egan JJ, Greenberg AS, Chang M-K, Wek SA, Moos MC Jr, et al. 1992. Mechanism of hormone-stimulated lipolysis in adipocytes: translocation of hormone-sensitive lipase to the lipid storage droplet. *Proc. Natl. Acad. Sci.USA* 89:8537–41
31. Fielding CJ, Fielding PE. 1995. Molecular physiology of reverse cholesterol transport. *J. Lipid Res.* 36:211–28
32. Ford PS, Van Heusden MC. 1994. Triglyceride-rich lipophorin in *Aedes aegypti* (Diptera: Culicudae). *J. Med. Entomol.* 31:435–41
33. Gäde G. 1997. The explosion of struc-

tural information on insect neuropeptides. *Prog. Chem. Org. Nat. Prod.* 71:1–128

34. Goldstein JL, Brown MS, Anderson RGW, Russell DW, Schneider WJ. 1985. Receptor-mediated endocytosis: concepts emerging from the LDL receptor system. *Annu. Rev. Cell Biol.* 1:1–39
35. Greenberg AS, Egan JJ, Wek SA, Garty NB, Blanchette-Mackie EJ, et al. 1991. Perilipin, a major hormonally regulated adipocyte-specific phosphoprotein associated with the periphery of lipid storage droplets. *J. Biol. Chem.* 266:11341–46
36. Greenberg AS, Egan JJ, Wek SA, Moos MC Jr, Londos C, et al. 1993. Isolation of cDNAs for perilipins A and B: sequence and expression of lipid droplet-associated proteins of adipocytes. *Proc. Natl. Acad. Sci.USA* 90:12035–39
37. Gu X, Quilici D, Juarez P, Blomquist GJ, Schal C. 1995. Biosynthesis of hydrocarbons and contact sex pheromone and their transport by lipophorin in females of the German cockroach (*Blattella germanica*). *J. Insect Physiol.* 41:257–67
38. Hård K, Van Doorn JM, Thomas-Oates JE, Kamerling JP, Van der Horst DJ. 1993. Structure of the Asn-linked oligosaccharides of apolipophorin III from the insect *Locusta migratoria*. Carbohydrate-linked 2-aminoethylphosphonate as a constituent of a glycoprotein. *Biochemistry* 32:766–75
39. Haunerland NH, Andolfatto P, Chisholm JM, Wang Z, Chen X. 1992. Fatty-acid-binding protein in locust flight muscle. Developmental changes of expression, concentration and intracellular distribution. *Eur. J. Biochem.* 210:1045–51
40. Herz J, Hamann U, Rogne S, Myklebost O, Gausepohl H, et al. 1988. Surface location and high affinity for calcium of a 500 kD liver membrane protein closely related to the LDL-receptor suggests a physiological role as a lipoprotein receptor. *EMBO J.* 7:4119–27
41. Hiraoka T, Hayakawa Y. 1990. Inhibition of diacylglycerol uptake from the fat body by a monoclonal antibody against apolipophorin II in *Locusta migratoria. Insect Biochem.* 20:793–99
42. Hirayama Y, Chino H. 1990. Lipid transfer particle in locust hemolymph: purification and characterization. *J. Lipid Res.* 31:793–99
43. Holm C, Langin D, Manganiello V, Belfrage P, Degerman E. 1997. Regulation of hormone-sensitive lipase activity in adipose tissue. *Method. Enzymol.* 286:45–67
44. Kawooya JK, Meredith SC, Wells MA, Kézdy FJ, Law JH. 1986. Physical and surface properties of insect apolipophorin III. *J. Biol. Chem.* 261:13588–91
45. Kim DH, Iijima H, Goto K, Sakai J, Ishii H, et al. 1996. Human apolipoprotein E receptor 2. A novel lipoprotein receptor of the low density lipoprotein receptor family predominantly expressed in brain. *J. Biol. Chem.* 271:8373–80
46. Knott TJ, Pease RJ, Powell LM, Wallis SC, Rall SC Jr, et al. 1986. Complete protein sequence and identification of structural domains of human apolipoprotein B. *Nature* 323:734–38
47. Kutty RK, Kutty G, Kambadur R, Duncan T, Koonin EV, et al. 1996. Molecular characterization and developmental expression of a retinoid- and fatty acid-binding glycoprotein from *Drosophila.* A putative lipophorin. *J. Biol. Chem.* 272:20641–49
48. Law JH, Ribeiro JMC, Wells MA. 1992. Biochemical insights derived from insect diversity. *Annu. Rev. Biochem.* 61:87–111
49. Law JH, Wells MA. 1989. Insects as biochemical models. *J. Biol. Chem.* 264:16335–38
50. Liu H, Ryan RO. 1991. Role of lipid transfer particle in transformation of lipophorin in *Manduca sexta* oocytes. *Biochim. Biophys. Acta* 1085:112–18
51. Liu H, Scraba DG, Ryan RO. 1993. Pre-

vention of phospholipase-C induced aggregation of low density lipoprotein by amphipathic apolipoproteins. *FEBS Lett.* 316:27–33
52. Lok CM, Van der Horst DJ. 1980. Chiral 1,2-diacylglycerols in the haemolymph of the locust, *Locusta migratoria*. *Biochim. Biophys. Acta* 618:80–87
53. Lum PY, Chino H. 1990. Primary role of adipokinetic hormone in the formation of low density lipophorin in locusts. *J. Lipid Res.* 31:2039–44
54. Narayanaswami V, Wang J, Kay CM, Scraba DG, Ryan RO. 1996. Disulfide bond engineering to monitor conformational opening of apolipophorin III during lipid binding. *J. Biol. Chem.* 271:26855–62
55. Narayanaswami V, Wang J, Schieve D, Kay CM, Ryan RO. 1999. A molecular trigger of lipid-binding induced opening of a helix bundle exchangeable apolipoprotein. *Proc. Natl. Acad. Sci. USA* 96:4366–71
56. Nässel DR, Passier PCCM, Elekes K, Dircksen H, Vullings HGB, et al. 1995. Evidence that locustatachykinin I is involved in release of adipokinetic hormone from locust corpora cardiaca. *Regul. Pept.* 57:297–310
57. Nässel DR, Vullings HGB, Passier PCCM, Lundquist CT, Schoofs L, et al. 1999. Several isoforms of locustatachykinins may be involved in cyclic AMP-mediated release of adipokinetic hormones from the locust corpora cardiaca. *Gen. Comp. Endocrinol.* 113:401–12
58. Ogoyi DO, Osir EO, Olembo NK. 1998. Fat body triacylglycerol lipase in solitary and gregarious phases of *Schistocerca gregaria* (Forskal) (Orthoptera: Acrididae). *Comp. Biochem. Physiol.* B119: 163–69
59. O'Shea M, Rayne RC. 1992. Adipokinetic hormones: cell and molecular biology. *Experientia* 48:430–38
60. Østerlund T, Danielsson B, Degerman E, Contreras JA, Edgren G, et al. 1996. Domain-structure analysis of recombinant rat hormone-sensitive lipase. *Biochem. J.* 319:411–20
61. Oudejans RCHM, Harthoorn LF, Diederen JHB, Van der Horst DJ. 1999. Adipokinetic hormones: coupling between biosynthesis and release. *Ann. NY Acad. Sci.* 897:In press
62. Oudejans RCHM, Kooiman FP, Heerma W, Versluis C, Slotboom AJ, et al. 1991. Isolation and structure elucidation of a novel adipokinetic hormone (Lom-AKH-III) from the glandular lobes of the corpus cardiacum of the migratory locust, *Locusta migratoria*. *Eur. J. Biochem.* 195:351–59
63. Oudejans RCHM, Kooiman FP, Schulz TKF, Beenakkers AMT. 1990. In vitro biosynthesis of locust adipokinetic hormones: isolation and identification of the bioactive peptides and their prohormones. In *Chromatography and Isolation of Insect Hormones and Pheromones*, ed. AR McCaffery, ID Wilson, pp. 183–94. New York: Plenum
64. Oudejans RCHM, Mes THM, Kooiman FP, Van der Horst DJ. 1993. Adipokinetic peptide hormone content and biosynthesis during locust development. *Peptides* 14:877–81
65. Oudejans RCHM, Vroemen SF, Jansen RFR, Van der Horst DJ. 1996. Locust adipokinetic hormones: carrier-independent transport and differential inactivation at physiological concentrations during rest and flight. *Proc. Natl. Acad. Sci. USA* 93:8654–59
66. Passier PCCM, Vullings HGB, Diederen JHB, Van der Horst DJ. 1995. Modulatory effects of biogenic amines on adipokinetic hormone secretion from locust corpora cardiaca in vitro. *Gen. Comp. Endocrinol.* 97:231–38
67. Passier PCCM, Vullings HGB, Diederen JHB, Van der Horst DJ. 1997. Trehalose inhibits the release of adipokinetic hormones from the corpus cardiacum in the

African migratory locust, *Locusta migratoria*, at the level of the adipokinetic cells. *J. Endocrinol.* 153:299–305
68. Pennington JE, Nussenzveig RH, Van Heusden MC. 1996. Lipid transfer from insect fat body to lipophorin: comparison between a mosquito triacylglycerol-rich lipophorin and a sphinx moth diacylglycerol-rich lipophorin. *J. Lipid Res.* 37:1144–52
69. Raussens V, Goormaghtigh E, Narayanaswami V, Ryan RO, Ruysschaert JM. 1995. Alignment of apolipophorin III α-helices in complex with dimyristoylphosphatidylcholine: a unique spatial orientation. *J. Biol. Chem.* 270:12542–47
70. Raussens V, Narayanaswami V, Goormaghtigh E, Ryan RO, Ruysschaert JM. 1996. Hydrogen/deuterium exchange kinetics of apolipophorin III in lipid free and phospholipid bound states: an analysis by Fourier transform infrared spectroscopy. *J. Biol. Chem.* 271:23089–95
71. Röhrkasten A, Ferenz H-J. 1986. Properties of vitellogenin receptor of isolated locust oocyte membranes. *Int. J. Invertebr. Reprod. Dev.* 10:133–42
72. Ryan RO. 1990. Dynamics in insect lipophorin metabolism. *J. Lipid Res.* 31:1725–39
73. Ryan RO. 1994. The structures of insect lipoproteins. *Curr. Opin. Struct. Biol.* 4:499–506
74. Ryan RO. 1996. Structural studies of lipoproteins and their apolipoprotein components. *Biochem. Cell Biol.* 74:155–74
75. Ryan RO, Haunerland NH, Bowers WS, Law JH. 1988. Insect lipid transfer particle catalyzes diacylglycerol exchange between high density and very high density lipoproteins. *Biochim. Biophys. Acta* 962:143–48
76. Ryan RO, Hicks LD, Kay CM. 1990. Biophysical studies on the lipid transfer particle from the hemolymph of the tobacco hornworm, *Manduca sexta*. *FEBS Lett.* 267:305–10
77. Ryan RO, Howe A, Scraba DG. 1990. Studies of the morphology and structure of the plasma lipid transfer particle from the tobacco hornworm, *Manduca sexta*. *J. Lipid Res.* 31:871–79
78. Ryan RO, Prasad, SV, Henriksen, EJ, Wells MA, Law JH. 1986. Lipoprotein interconversions in an insect, *Manduca sexta*. Evidence for a lipid transfer factor in the hemolymph. *J. Biol. Chem.* 261:563–68
79. Ryan RO, Schieve DS, Wientzek M, Narayanaswami V, Oikawa K, et al. 1995. Bacterial expression and site directed mutagenesis of a functional recombinant apolipoprotein. *J. Lipid Res.* 36:1066–72
80. Ryan RO, Senthilathipan R, Wells MA, Law JH. 1988. Facilitated diacylglycerol exchange between insect hemolymph lipophorins. Properties of *Manduca sexta* lipid transfer particle. *J. Biol. Chem.* 263:14140–45
81. Ryan RO, Wells MA, Law JH. 1986. Lipid transfer protein from *Manduca sexta* hemolymph. *Biochem. Biophys. Res. Commun.* 136:260–65
82. Ryan RO, Wessler A, Ando S, Price HM, Yokoyama S. 1990. Insect lipid transfer particle catalyzes bidirectional vectorial transfer of diacylglycerol from lipophorin to human low density lipoprotein. *J. Biol. Chem.* 265:10551–55
83. Sappington TW, Kokoza VA, Cho WL, Raikhel AS. 1996. Molecular characterization of the mosquito vitellogenin receptor reveals unexpected high homology to the *Drosophila* yolk protein receptor. *Proc. Natl. Acad. Sci. USA* 93:8934–39
84. Schal C, Sevala V. 1998. Novel and highly specific transport of a volatile sex pheromone by hemolymph lipophorin in moths. *Naturwissenschaften* 85:339–42
85. Schonbaum CP, Lee S, Mahowald AP. 1995. The *Drosophila* yolkless gene encodes a vitellogenin receptor belonging to the low density lipoprotein super-

family. *Proc. Natl. Acad. Sci. USA* 92:1485–89
86. Schulz TKF, Van der Horst DJ, Beenakkers AMT. 1991. Binding of locust high-density lipophorin to fat body proteins monitored by an enzyme-linked immunosorbant assay. *Biol. Chem. Hoppe-Seyler* 372:5–12
87. Sevala VL, Bachmann JAS, Schal C. 1997. Lipophorin: a hemolymph juvenile hormone binding protein in the German cockroach, *Blattella germanica. Insect Biochem. Mol. Biol.* 27:663–70
88. Sharp-Baker HE, Diederen JHB, Mäkel KM, Peute J, Van der Horst DJ. 1995. The adipokinetic cells in the corpus cardiacum of *Locusta migratoria* preferentially release young secretory granules. *Eur. J. Cell Biol.* 68:268–74
89. Sharp-Baker HE, Oudejans RCHM, Kooiman FP, Diederen JHB, Peute J, et al. 1996. Preferential release of newly synthesized, exportable neuropeptides by insect neuroendocrine cells and the effect of ageing of secretory granules. *Eur. J. Cell Biol.* 71:72–78
90. Singh TKA, Liu H, Bradley R, Scraba DG, Ryan RO. 1994. The effect of phospholipase C and apolipophorin III on the structure and stability of lipophorin subspecies. *J. Lipid Res.* 35:1561–69
91. Singh TKA, Ryan RO. 1991. Lipid transfer particle-catalyzed transfer of lipoprotein associated diacylglycerol and long chain aliphatic hydrocarbons. *Arch. Biochem. Biophys.* 286:376–82
92. Singh TKA, Scraba DG, Ryan RO. 1992. Conversion of human low density lipoprotein into a very low density lipoprotein-like particle in vitro. *J. Biol. Chem.* 267:9275–80
93. Smith AF, Owen LM, Strobel LM, Chen H, Kanost MR, et al. 1994. Exchangeable apolipoproteins of insects share a common structural motif. *J. Lipid Res.* 35:1976–84
94. Soulages JL, Bendavid OJ. 1998. The lipid binding activity of the exchangeable apolipoprotein apolipophorin III correlates with the formation of a partially folded conformation. *Biochemistry* 37:10203–10
95. Soulages JL, Pennington J, Bendavid O, Wells MA. 1998. Role of glycosylation in the lipid-binding activity of the exchangeable apolipoprotein, apolipophorin-III. *Biochem. Biophys. Res. Commun.* 243:372–76
96. Soulages JL, Salamon Z, Wells MA, Tollin G. 1995. Low concentrations of diacylglycerol promote the binding of apolipophorin III to a phospholipid bilayer: a surface plasmon resonance spectroscopy study. *Proc. Natl. Acad. Sci. USA* 92:5650–54
97. Soulages JL, Wells MA. 1994. Lipophorin: the structure of an insect lipoprotein and its role in lipid transport in insects. *Adv. Protein. Chem.* 45:371–415
98. Soulages JL, Wells MA. 1994. Effect of diacylglycerol content on some physicochemical properties of the insect lipoprotein, lipophorin. Correlation with the binding of apolipophorin III. *Biochemistry* 33:2356–62
99. Strålfors P, Belfrage P. 1983. Phosphorylation of hormone-sensitive lipase by cyclic AMP-dependent protein kinase. *J. Biol. Chem.* 258:15146–52
100. Sundermeyer K, Hendricks JK, Prasad SV, Wells MA. 1996. The precursor protein of the structural apolipoproteins of lipophorin: cDNA and deduced amino acid sequence. *Insect Biochem. Mol. Biol.* 26:735–38
101. Takahashi S, Kawarabayasi Y, Nakai T, Sakai J, Yamamoto T. 1992. Rabbit very low density lipoprotein receptor: a low density lipoprotein receptor-like protein with distinct ligand specifity. *Proc. Natl. Acad. Sci. USA* 89:9252–56
102. Takeuchi N, Chino H. 1993. Lipid transfer particle in the hemolymph of the American cockroach: evidence for its capacity to transfer hydrocarbons

between lipophorin particles. *J. Lipid Res.* 34:543–51
103. Tall AR. 1995. Plasma lipid transfer proteins. *Annu. Rev. Biochem.* 64:235–57
104. Tietz A, Weintraub H. 1980. The stereospecific structure of haemolymph and fat-body 1,2-diacylglycerol from *Locusta migratoria. Insect Biochem.* 10:61–63
105. Trowell SC, Hines ER, Herlt AJ, Rickards RW. 1994. Characterization of a juvenile hormone binding lipophorin from the blowfly *Lucilia cuprina. Comp. Biochem. Physiol.* B109:339–57
106. Tsuchida K, Arai M, Tanaka Y, Ishihara R, Ryan RO, et al. 1998. Lipid transfer particle catalyzes transfer of carotenoids between lipophorins of *Bombyx mori. Insect Biochem. Mol. Biol.* 28:927–34
107. Tsuchida K, Soulages JL, Moribayashi A, Suzuki K, Maekawa H, et al. 1997. Purification and properties of a lipid transfer particle from *Bombyx mori*: comparison to the lipid transfer particle from *Manduca sexta. Biochim. Biophys. Acta* 1337:57–65
108. Tsuchida K, Wells MA. 1988. Digestion, absorption, transport and storage of fat during the last larval stadium of *Manduca sexta.* Changes in the role of lipophorin in the delivery of dietary lipid to the fat body. *Insect Biochem.* 18:263–68
109. Tsuchida K, Wells MA. 1990. Isolation and characterization of a lipoprotein receptor from the fat body of an insect, *Manduca sexta. J. Biol. Chem.* 265:5761–67
110. Van Antwerpen R, Wynne HJA, Van der Horst DJ, Beenakkers AMT. 1989. Binding of lipophorin to the fat body of the migratory locust. *Insect Biochem.* 19:809–14
111. Van der Horst DJ. 1990. Lipid transport function in lipoproteins in flying insects. *Biochim. Biophys. Acta* 1047:195–211
112. Van der Horst DJ, Van Marrewijk WJA, Vullings HGB, Diederen JHB. 1999. Metabolic neurohormones: release, signal transduction and physiological responses of adipokinetic hormones in insects. *Eur. J. Entomol.* 96:299–308
113. Van der Horst DJ, Weers PMM, Van Marrewijk WJA. 1993. Lipoproteins and lipid transport. In *Insect Lipids: Chemistry, Biochemistry, and Biology*, ed. DW Stanley-Samuelson, DR Nelson, pp. 1–24. Lincoln, NE: Univ. Nebr. Press
114. Van Heusden MC, Law JH. 1989. An insect lipid transfer particle promotes lipid loading from fat body to lipoprotein. *J. Biol. Chem.* 264:17287–92
115. Van Heusden MC, Thompson F, Dennis J. 1998. Biosynthesis of *Aedes aegypti* lipophorin and gene expression of its apolipoproteins. *Insect Biochem. Mol. Biol.* 28:733–38
116. Van Heusden MC, Van der Horst DJ, Van Doorn JM, Wes J, Beenakkers AMT. 1986. Lipoprotein lipase activity in the flight muscle of *Locusta migratoria* and its specificity for hemolymph lipoproteins. *Insect Biochem.* 16:517–23
117. Van Heusden MC, Yepiz-Plascencia GM, Walker AM, Law JH. 1996. *Manduca sexta* lipid transfer particle: synthesis by fat body and occurrence in hemolymph. *Arch. Insect Biochem. Physiol.* 31:39–51
118. Van Marrewijk WJA, Van den Broek ATM, Beenakkers AMT. 1980. Regulation of glycogenolysis in the locust fat body during flight. *Insect Biochem.* 10:675–79
119. Van Marrewijk WJA, Van den Broek ATM, Beenakkers AMT. 1991. Adipokinetic hormone is dependent on extracellular Ca^{2+} for its stimulatory action on the glycogenolytic pathway in locust fat body in vitro. *Insect Biochem.* 21:375–80
120. Van Marrewijk WJA, Van den Broek ATM, De Graan PNE, Beenakkers AMT. 1988. Formation of partially phosphorylated glycogen phosphorylase in the fat body of the migratory locust. *Insect Biochem.* 18:821–27
121. Van Marrewijk WJA, Van den Broek

ATM, Gielbert M-L, Van der Horst DJ. 1996. Insect adipokinetic hormone stimulates inositol phosphate metabolism: roles for both Ins(1,4,5)P_3 and Ins(1,3,4,5)P_4 in signal transduction? *Mol. Cell. Endocrinol.* 122:141–50

122. Van Marrewijk WJA, Van den Broek ATM, Van der Horst DJ. 1993. Adipokinetic hormone-induced influx of extracellular calcium into insect fat body cells is mediated through depletion of intracellular calcium stores. *Cell. Signal.* 5:753–61
123. Veelaert D, Passier P, Devreese B, Vanden Broeck J, Van Beeumen J, et al. 1997. Isolation and characterization of an adipokinetic hormone release-inducing factor in locusts: the crustacean cardioactive peptide. *Endocrinology* 138:138–42
124. Vroemen SF, De Jonge H, Van Marrewijk WJA, Van der Horst DJ. 1998. The phospholipase C signaling pathway in locust fat body is activated via G_q and not affected by cAMP. *Insect Biochem. Mol. Biol.* 28:483–90
125. Vroemen SF, Van Marrewijk WJA, De Meijer J, Van den Broek ATM, Van der Horst DJ. 1997. Differential induction of inositol phosphate metabolism by three adipokinetic hormones. *Mol. Cell. Endocrinol.* 130:131–39
126. Vroemen SF, Van Marrewijk WJA, Schepers CCJ, Van der Horst DJ. 1995. Signal transduction of adipokinetic hormones involves Ca_{2+} fluxes and depends on extracellular Ca^{2+} to potentiate cAMP-induced activation of glycogen phosphorylase. *Cell Calcium* 17:459–67
127. Vullings HGB, Ten Voorde SECG, Passier PCCM, Diederen JHB, Van der Horst DJ, et al. 1998. A possible role of SchistoFLRFamide in inhibition of adipokinetic hormone release from locust corpora cardiaca. *J. Neurocytol.* 27:901–13
128. Vullings HGB, Diederen HGB, Veelaert D, Van der Horst DJ. 1999. The multifactorial control of the release of hormones from the locust retrocerebral complex. *Microsc. Res. Tech.* 45:142–53
129. Wang J, Gagné SM, Sykes BD, Ryan RO. 1997. Insight into lipid surface recognition and reversible conformational adaptations of an exchangeable apolipoprotein by multidimensional heteronuclear NMR techniques. *J. Biol. Chem.* 272:17912–20
130. Wang J, Liu H, Sykes BD, Ryan RO. 1992. ^{31}P-NMR study of the phospholipid moiety of lipophorin subspecies. *Biochemistry* 31:8706–12
131. Wang J, Sahoo D, Schieve D, Gagne S, Sykes BD, Ryan RO. 1997. Multidimensional NMR studies of an exchangeable apolipoprotein and its interaction with lipids. *Tech. Protein Chem.* VIII:427–38
132. Wang J, Sahoo D, Sykes BD, Ryan RO. 1998. NMR evidence for a conformational adaptation of apolipophorin III upon lipid association. *Biochem. Cell Biol.* 76:276–83
133. Wang Z, Haunerland NH. 1994. Storage protein uptake in *Helicoverpa zea*: arylphorin and VHDL share a single receptor. *Arch. Insect Biochem. Physiol.* 26:15–26
134. Wang Z, Hayakawa Y, Downer RGH. 1990. Factors influencing cyclic AMP and diacylglycerol levels in fat body of *Locusta migratoria*. *Insect Biochem.* 20:325–30
135. Weers PMM, Kay CM, Oikawa O, Wientzek M, Van der Horst DJ, et al. 1994. Factors affecting the stability and conformation of *Locusta migratoria* apolipophorin III. *Biochemistry* 33:3617–24
136. Weers PMM, Narayanaswami V, Kay CM, Ryan RO. 1999. Interaction of an exchangeable apolipoprotein with phospholipid vesicles and lipoprotein particles. Role of leucines 32, 34 and 95 in *Locusta migratoria* apolipophorin III. *J. Biol. Chem.* 274:21804–10
137. Weers PMM, Van Marrewijk WJA, Beenakkers AMT, Van der Horst DJ.

1993. Biosynthesis of locust lipophorin: apolipophorin I and II originate from a common precursor. *J. Biol. Chem.* 268:4300–3

138. Weers PMM, Wang J, Van der Horst DJ, Kay CM, Sykes BD, et al. 1998. Recombinant locust apolipophorin III: characterization and NMR spectroscopy. *Biochim. Biophys. Acta* 1393:99–107
139. Weisgraber KH. 1994. Apolipoprotein E: structure-function relationships. *Adv. Protein Chem.* 45:249–302
140. Wells MA, Ryan RO, Kawooya JK, Law JH. 1987. The role of apolipophorin III in in vivo lipoprotein interconversions in adult *Manduca sexta. J. Biol. Chem.* 262:4172–76
141. Wetterau JR, Lin MC, Jamil H. 1997. Microsomal triglyceride transfer protein. *Biochim. Biophys. Acta* 1345:136–50
142. Wheeler CH, Van der Horst DJ, Beenakkers AMT. 1984. Lipolytic activity in the flight muscles of *Locusta migratoria* measured with haemolymph lipoproteins as substrates. *Insect Biochem.* 14:261–66
143. Wientzek M, Kay CM, Oikawa K, Ryan RO. 1994. Binding of insect apolipophorin III to dimyristoylphosphatidylcholine vesicles: evidence for a conformational change. *J. Biol. Chem.* 269:4605–12
144. Wilson C, Wardell MR, Weisgraber KH, Mahley RW, Agard DA. 1991. Three dimensional structure of the LDL-receptor binding domain of human apolipoprotein E. *Science* 252:1817–22
145. Wirtz KWA. 1991. Phospholipid transfer proteins. *Annu. Rev. Biochem.* 60:73–99
146. Yamamoto T, Davis CG, Brown MS, Schneider WJ, Casey ML, et al. 1984. The human LDL receptor: a cysteine-rich protein with multiple *Alu* sequences. *Cell* 39:27–38
147. Yochem J, Greenwald I. 1993. A gene for a low density receptor-related protein in the nematode *Caenorhabditis elegans. Proc. Natl. Acad. Sci. USA* 90:4572–76
148. Zhang Y, Lewis RNAH, McElhaney RN, Ryan RO. 1993. Calorimetric and spectroscopic studies of the interaction of *Manduca sexta* apolipophorin III with zwitterionic, anionic and nonionic lipids. *Biochemistry* 32:3942–52
149. Ziegler R, Willingham LA, Engler DL, Tolman KJ, Bellows D, et al. 1999. A novel lipoprotein from the hemolymph of the cochineal insect, *Dactylopius confusus. Eur. J. Biochem.* 26:285-90
150. Zilversmit DB, Hughes LB, Balmer J. 1975. Stimulation of cholesterol ester exchange by lipoprotein free rabbit plasma. *Biochim. Biophys. Acta* 409:393–98

Annu. Rev. Entomol. 2000. 45:261–285

ENTOMOLOGY IN THE TWENTIETH CENTURY

R. F. Chapman
ARL Division of Neurobiology, University of Arizona, Tucson, Arizona 85721; e-mail: chapman@neurobio.arizona.edu

Key Words history, technology, communication, interdisciplinary impacts

■ **Abstract** A number of landmark events in applied entomology are listed together with some insect-related studies that have had a major impact on biology in general. In large part, however, advances in our understanding of insects have depended on technological advances, especially in the second half of the century. The exponential increase in the ease and extent of communication has been critical. Sometimes, as in the field of insect/plant relations, the ideas of a few individuals have been critical with technological advances having a facilitating role. Elsewhere, as in the study of olfaction, major changes in understanding have been directly dependent on new technology. Very brief accounts of the impacts on insect-related science of developments in the fields of radio, radioactivity, immunology, imaging techniques, and chemical analysis are given. Despite the importance of technology, the lovers of their insects continue to have a key role.

INTRODUCTION

When the *History of Entomology* was published by Annual Reviews in 1973 (90) the technological revolution of the twentieth century was already well under way, but the changes since that time have been astonishing. In this review, I will discuss developments that, to me, seem to have most influenced the ways in which we now understand insects. Inevitably, this is a personal view, but it has been my good fortune to have a career that has encompassed several widely different approaches to the study of insects.

The major discovery that has revolutionized biology, including entomology, is, of course, the discovery of the structure of DNA and the manner in which it provides the code for the manufacture of proteins. Medawar, writing in 1977, wrote:

> the greatest scientific discovery of the twentieth century [is] without qualification . . . that the chemical makeup of the compound deoxyribonucleic acid (DNA) . . . encodes genetic information and is the material vehicle of instructions by which one generation of organisms governs the development of the next.

0066-4170/00/0107-0261/$14.00

This single discovery has led to a revolution in our understanding of the mechanisms of insect development and evolution (15).

Computers did not begin to impact us in a general way until the 1970s. Then, they very quickly made it possible for all of us to utilize increasingly sophisticated statistical procedures. Subsequently, computers came to be used for tasks that previously could not possibly have been undertaken, extending even to the rapid collection and transfer of anatomical information (94) and to modeling at many levels from population to neural function.

None of the people involved in these discoveries was an entomologist, and this is the striking thing about much of biology in this century: The unbelievable strides that have been made have been driven largely by technological advances and, especially, through vastly improved communication across disciplines. As a result, many of the questions we ask today could not even have been conceived 100 or even 50 years ago. Of course, that is not to say that entomologists have not contributed in great measure but they have been using tools, and often ideas, that originated outside entomology.

Much of this technological development has occurred in the last 50 years, following World War II. This becomes obvious as changes in different fields of research are considered.

LANDMARK EVENTS

Figure 1 lists the landmark events of the century. I consider "landmark events" to be discoveries or developments involving insects, the effects of which have been far-reaching and affecting the biological sciences in general, such as Keilin's discovery of the changes in cytochrome in relation to oxidation (47); or affecting large numbers of people, such as the biological control operations (19, 114); or simply astonishing, such as the locust crossing of the Atlantic in 1988 (75). I will not dwell further on most of these events, but I will use selected examples to try to show how discoveries and developments in physics and chemistry have been used to extend our knowledge of insects. First, however, I will consider communication.

COMMUNICATION

Stemming from technological changes that allowed for ease of travel and exchange of information, together with the post–World War II blossoming of scientific research, communication has come to play an increasingly significant role in entomology, as it has in all walks of life. Scientific meetings abound, the numbers of journals have multiplied exponentially and many are more specialized, while more people than ever before study insects, although many of them do not regard themselves as entomologists.

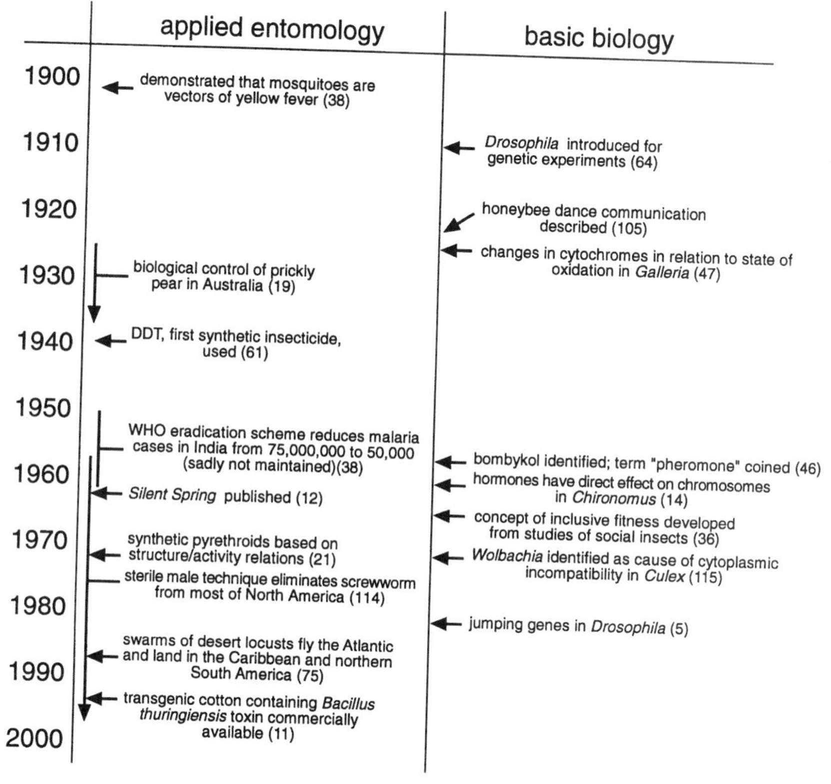

Figure 1 Landmark events in insect-related biology

The first International Congress of Entomology was held in Brussels in 1910 and at this meeting, as in the next two, fewer than 100 papers were presented. The number had climbed to nearly 300 by 1938 (Berlin) and the meetings after World War II show a progressive increase to around 1500 in Kyoto (1980). From that time, the introduction of posters permitted a greater number of individuals to present their work. National meetings have shown a similar increase, while meetings in every subject area and for many individual orders of insects have come into existence.

At the same time, the very nature of meetings has changed. As a student, I went to meetings expecting to hear the limited number of recognized authorities talking about their work. Who would ever think of inviting a graduate student to give a talk? One was lucky to be at the meeting at all. We kept abreast of new developments by regular visits to the library, where it was fairly easy to glance at all the new acquisitions of biological journals in an afternoon each week. We now commonly have a situation where most people at a meeting make presentations. This has been driven partly by economics: Financial support for attending a meeting commonly requires the justification of a presentation. Meetings have also become marketplaces where you go to impress your peers. For many younger people this is an important, sometimes perhaps primary, function of meetings. Perhaps the scientific community was once small enough that one could know all the people in the field, but, from 1950 onwards, the rapid expansion of all branches of science, especially in North America, made this familiarity impossible. Perhaps, too, the 1960s movement calling for greater openness in all walks of life, was an important factor.

The numbers of entomological journals reflects what has occurred throughout science. Gilbert & Hamilton (32) list 149 journals one might regard as professional sources dealing largely with entomology. Eighteen of these are continuations of journals already in existence at the end of the nineteenth century, although sometimes with a changed title; the rest were started in the present century. A few new journals appeared each year throughout the first half of the twentieth century, but then the information explosion began. Ninety-eight new primary entomological journals have originated in the last 40 years, with 44 others dealing specifically with the Lepidoptera, Coleoptera, Diptera, and Odonata.

Not only did the numbers of journals increase, so did the numbers of papers in those journals. The number of papers in the *Annals of the Entomological Society of America*, for example, did not rise above 30 per year before 1920, but in the 1930s the numbers hovered around 50, and through the late 1960s and 1970s more than 200 papers were published each year.

Before 1950, most entomological journals were produced by societies. Subsequently, many publishing houses saw the scientific journal as a source of revenue (see, for example, 100) although often, in fairness, this view came about with prompting from scientists who felt a need for new outlets for their work. The *Journal of Insect Physiology* was initiated in 1957 by Pergamon Press at the behest of Howard Hinton, who edited the journal for many years thereafter. The approach of some society publications came to appear somewhat antiquated and the titles sometimes uninformative, resulting in a revamping of some journals in order to apply more scientific rigor in the editing. The Royal Entomological Society of London, for example, went from producing *Proceedings* A, B, and C which dealt with general entomology, taxonomy, and proceedings of meetings, respectively, to the *Journal of Entomology* A and B which, along with the *Transactions*, soon became *Physiological Entomology*, *Ecological Entomology* and *Systematic Entomology*. These changes were not just changes of name. The Society was anxious to establish itself in a thriving scientific world and the new editors established these publications as high-quality journals. This is not to imply that, previously, the journals did not contain much excellent work, but there was a need to change the perception. Subsequently, new journals were added as new areas of study emerged and emphasis changed: *Environmental Entomology* by the Entomological Society of America in 1972 and *Molecular Entomology* by the Royal Entomological Society in 1993 are examples.

Accompanying these quantitative changes have also been important qualitative changes. Publication of the increased volume of work has necessitated stylistic changes in the way people write. FD Morice, describing the sixth abdominal segment in male *Osmia*, wrote in 1901 (65):

> Unfortunately the segment cannot be viewed without dissection of the specimen. But when extracted its beautiful forms and most interesting structure amply repay the trouble of bringing it to light; and the characters presented by it in the various species are so clear and constant, that I think they well deserve an attention which has not yet been paid to them by the framers of specific diagnoses.

It is beautifully written and conveys his love for his insects, but it does not really tell us very much. What editor would allow that today with space in journals at a premium and some journals restricting the lengths of papers? I must confess to feeling that this is not all bad. While I enjoy nice prose, my main aim in reading a scientific paper is usually to find out what the paper is about as quickly as possible. To cater for those of us who are in such a hurry (and there must be a large number of us), since about 1975 most journals have included at the start of each paper an abstract written by the author.

Another important change is that many relevant biological journals have anglicized their titles and publish all their papers in English. This is not so apparent in entomological journals, but the Springer journals are a good example where *Zeitschrift für vergleichende Physiologie* is now the *Journal of Comparative Physiology* and *Zeitschrift für Zellforschung und Mikroskopische Anatomie* is now *Cell and Tissue Research*. These journals, and many others, are of primary importance and their emphasis on production in English gives enormous privilege to those of us for whom English is the native tongue. Gone are the days when we struggled, dictionary in hand, just to get the sense " . . . is he saying it is like that, or it's not like that?" and the subtleties passed most of us by completely. Many journals have also become more international with editors from countries other than the country of publication.

The greatly increased volume of papers coupled with escalating costs, increasing demand for accountability, and citation indices being used to measure quality, have had a serious negative effect on the quality of editing and reviewing. This is not intended as criticism of editors and reviewers, but even those who are most competent and well-intentioned have less time to give per paper at a time when papers are often more hurriedly produced and in need of careful attention.

The great volume of published research created a need for authoritative reviews making information available to a wider audience. A small number of general biological journals, such as *Biological Reviews of the Cambridge Philosophical Society*, already existed but were not sufficient to cope with the growing needs, and their specialist reviews were never intended to cater to the needs of a wider, often less knowledgeable audience. This led Annual Reviews to initiate the *Annual Review of Entomology* in 1956, and in 1963 Pergamon Press followed with *Advances in Insect Physiology*. These two periodicals have maintained their high standards and the former, in particular, does provide the general entomologist with a valuable introduction to the literature across a very wide field.

From 1913 on (long before papers normally included an abstract written by the authors) the Imperial Bureau of Entomology, later to become the Commonwealth Institute of Entomology, produced the *Review of Applied Entomology,* which contained abstracts written by entomologists on the editorial staff. This was only possible because the Commonwealth Institute was government funded, but it meant that people studying agricultural or medical pest insects were well catered for from an early date. The section on Insecta in the *Zoological Record* has also provided a valuable source of information since the 1860s, but its coverage tended to be restricted and publication rate far too slow to meet the rapidly expanding information base of a century later. In an attempt to meet this need, Information Retrieval Ltd. established *Entomology Abstracts* in 1969, and *Abstracts of Entomology*, derived from *Biological Abstracts*, first appeared in 1970. These publications made it possible for the average entomologist to keep abreast of the expanding literature most effectively. It was not too long, however, before these periodicals largely gave place to the computerized data bases that we all now take for granted. Distressingly, though, this is not the total blessing that it should be because there is a strong tendency for people to think that anything published before computers existed is too hard to find, and probably not worth knowing about anyway. However, as more societies make back issues available on the World-Wide Web, as the Ecological Society of America has already done, even this will cease to be a problem.

What does not seem to have changed very much is the quality and abundance of upper-level entomological textbooks. During the first years of the century, the emphasis was taxonomic and anatomical (with relatively little attention to function), reflecting the general field of knowledge. Gillott's newer book (33) takes the opposite approach: More than half the book is devoted to physiology and ecology. This tendency to reduce the emphasis on classification is also seen in numerous introductory entomology texts, reflecting the general change in emphasis toward animal biology as opposed to taxonomy, a change that became prevalent some 30 years ago.

With increasing knowledge of insect function, it became clear that the standard entomological texts could not encompass general entomology as most earlier books had done. The first major departure from the taxonomic approach was Wigglesworth's *Principles of Insect Physiology*, which, even in its original 1939 edition, remains an amazing book (110). Other physiology books followed: Roeder's *Insect Physiology* in 1953 (78) and then the multi-author *Physiology of Insecta* edited by Rockstein that appeared first in three volumes and then in six (76, 77). When first produced, in 1964–1965, this work was authoritative, but in the 1960s the volume of information entomology students were required to know was overwhelming, and I felt there was a need to produce a book that presented structure and function in an integrated way and in a form that was accessible to the non-specialist. Taxonomy was already adequately covered by existing texts and so, in producing *The Insects: Structure and Function,* I set out to produce a book that would complement the standard taxonomic texts (13).

The abundance of new information about even the most basic aspects of insect physiology led to the publication in 1985 of the 13 volumes of *Comprehensive Insect Physiology, Biochemistry and Pharmacology,* edited by Kerkut & Gilbert (50). At the same time, advances in anatomy have appeared in a number of books, culminating in the three volumes, edited by Harrison & Locke, on *Insecta* in *Microscopic Anatomy of Invertebrates* (37). Both series are wonderful compendia of knowledge, but where do we go from here? Expense makes such treatises beyond the grasp of most individuals and even some libraries, and without continual updating many parts must soon be overtaken by new developments. My own view is that these are probably the last major multivolume works that we shall see and that World-Wide Web–based materials will take their place.

In addition to the small core of basic entomology texts, there has been an increasing flood of more specialist books. Since 1950, for example, there have been 11 new books on insect acoustics and 10 on insect flight. These are a reflection of the increasing power of the tools available to those studying insects, but the 21 on insect conservation since 1970 are the product of more global, forward-looking thinking that was rare in earlier days, perhaps partly because it seemed unnecessary.

But where does this mass of literature leave the student, or the person coming to insects for the first time? I believe that for some time to come, most individuals will continue to need a handy source of reference, probably in book form, but it becomes increasingly desirable to have the hard copy linked to a computerized base where it will be possible to pursue references and update the text with relative ease.

An increasing number of journals is already available on the World-Wide Web, with a few publishers, such as Springer-Verlag, and societies, such as the Society for Neuroscience and, in entomology, the Florida Entomological Society, taking the lead. To date, much of this has depended on the efforts of a few forward-looking, enthusiastic, and persuasive people. If one dares to prophesy, this surely has to be a major trend, with less emphasis on paper versions. The issues are well set out by Walker (106). The potential for multidisciplinary approaches to entomology that this trend offers seems almost unbounded.

Computers, linked via satellites, now enable individual scientists to communicate almost instantaneously across the world. We can send complex data files and illustrations that facilitate collaboration between scientists in remote parts of the globe. And it should be remembered that the first rockets were only launched in 1944 during World War II and the first satellite went into orbit in 1956.

THE RATE OF CHANGE

During the first decades of the century, research in many fields was not qualitatively different from what had gone before. Advances rested on the accumulation of more data or on the insights of a few individuals. But further development was often limited until technological advances made it possible to obtain information that differed qualitatively from what had gone before. I will illustrate this by first describing developments in the field of olfaction, where new technology had a major impact, and then discussing the field of insect/plant relations, in which technology was not the key issue.

The Insect Olfactory System

In the 19th century a few experiments had been carried out suggesting that the antennae were olfactory organs, and this view was reinforced by more behavioral work at the start of the 20th century, including some very convincing experiments on honey bees by von Frisch in the 1920s (104). Despite this, Eltringham, writing *The Senses of Insects* in 1933 (22), still felt it necessary to present the evidence that insects did have an olfactory sense, saying " . . . one cannot attribute the 'assembling' [the attraction of males to a female], as it is called, to any other than a kind of olfactory sense . . . " but considered "It is in every way probable that the male olfactory sense is specialized to perceive this odour [of the female] only, and is quite insensible to other non-tactile chemical stimuli." He was convinced by the experiments of von Frisch and others that the antennae were the principal olfactory organs (104), and Wigglesworth stated categorically that the antennae were the principal olfactory organs in the first edition of *The Principles of Insect Physiology* (110). But the issue was clouded by the inability to identify the olfactory sensilla with any certainty. Snodgrass (91) had already described a variety of sensilla, often innervated by a number of sensory neurons, that he considered to have an olfactory function. He " . . . supposed [the thin membranous walls of these sensilla] to be pervious to chemical stimuli . . . " (92). But McIndoo, who published several major papers on the olfactory behavior and sensory systems of insects, remained skeptical (58, 59). In 1929, he concluded, "It seems doubtful whether these hairs [on the antennae] . . . can serve as olfactory organs owing to their hair covering of chitin" (59). He believed that campaniform sensilla were olfactory pores. A principal, and logical, reason for his belief was that he could see, in sections viewed with the light microscope, that the dendrite of a campaniform sensillum ended immediately below a pore and so he thought that it " . . . comes into direct contact with the external air," an essential prerequisite for an olfactory receptor, whereas he could not see pores on the antennal hair sensilla. Consequently, he believed that the halteres of Diptera were olfactory organs, although their possible role in steering had been suggested long before (107).

Thus, the first 40 years of this century passed without any significant change in the view of insect olfaction. The first breakthrough came in 1938, when Pringle demonstrated electrophysiologically that campaniform sensilla were cuticular stress receptors (70), removing much of the doubt cast on the role of the antennae by McIndoo's beliefs. Further study was brought to an abrupt halt by World War II, and it was not until the 1950s that the story was resumed.

In the mid-1950s, Boistel's group in France (6) and Schneider in Germany (81) extended Pringle's electrophysiological technique to the antennae. Schneider developed the electroantennogram, which recorded the combined electrical activity of the neurons in the antenna responding to a particular stimulus. For the first time it was possible to demonstrate that olfactory receptors were present on the antennae. These developments built on advances in other fields. It had been known since 1850 that nerves conducted electrical activity, and the development of the cathode ray tube before the end of the nineteenth century made it possible to visualize, and hence to preserve in photographs, images of this electrical activity.

Slifer and her associates in the United States were the first to use a transmission electron microscope to study the receptors on the antennae (88). These early studies showed that there were pores in the cuticle of basiconic sensilla on the antennae, but these were less than 50 nm in diameter and so were below the limit of resolution of the light microscope. At last the question of how odor molecules could reach the dendrites through "impermeable" cuticle, the question that had dominated McIndoo's thoughts for so long, was answered. As the quality of images produced by the electron microscope improved, largely through the development of improved methods for producing thin sections with ultramicrotomes, more detail of structure was revealed. Steinbrecht, in Germany, showed that the cuticular pores were not simple tunnels through the cuticle, but were associated with bundles of pore tubules which in some cases extended across the space between the cuticle and the dendrites (95). Could this be the path along which odor molecules were carried through the lymph to the dendrites? If so, what about those frequent examples in which the pore tubules did not contact the dendrites or were not visible at all? Was that due to just bad fixation? So discussion developed about a problem that had not even been conceived of only 10 years earlier. Since the 1970s, the study of insect olfaction has exploded, driven largely by technological advances in a variety of fields (69a). As a result, we are asking about the structure of receptor molecules and carrier proteins and integration within the central nervous system (13a, 103a).

Insect/Plant Relations

That most phytophagous insects feed on a more or less restricted range of host plants was already recognized in the 19th century by Kirby & Spence (51) and Fabre (25). The basis of the insects' "botanical instinct" was not clear, however. A significant step was made in 1891 when Kossel proposed that the chemicals in plants comprised two functionally distinct groups: "primary metabolites" concerned with normal life processes, and "secondary metabolites" with separate functions; they were not just the end products of primary metabolism as was previously believed (52). The possible ecological roles of the secondary metabolites were not yet appreciated, although by 1890 Errera, a Belgian botanist who compiled a list of chemicals in plants, and Abbott both considered that "many of the chemicals may serve the plant as a means of defense against animals," including insects (1, 23, 84). This concept of secondary metabolites has subsequently proved to be a cornerstone in our understanding of insect/plant relationships. Experimental support for an antiherbivore role for secondary compounds soon came in studies on snails, but critical studies on insects lagged far behind.

Such was the state of things at the beginning of the twentieth century, and the first direct evidence that secondary compounds influenced the feeding behavior of insects came from a botanist, Verschaffelt, at the University of Amsterdam. In a paper read to the Royal Netherlands Academy in 1910, he showed that the secondary chemicals found in some plants could induce feeding by some insects that fed on those plants. Most notably, he showed that feeding by larval *Pieris* was promoted by glucosinolates, and the distribution of glucosinolates could account for the diet breadth of the insects (85, 103).

Entomologists seem to have taken little interest in this work, and Brues does not mention it in his 1920 paper. Discussing the accumulated literature, he comments " . . . any idea of parallel evolution [between insect and host plant] must be restricted to a development of undesirable attributes on the part of the plants and adaptations on the part of insects to overcome such barriers to feeding" (10). It was almost 40 years before a few figures emerged who were responsible for the blossoming of the study of insect/plant relations into its current active enterprise.

Gottfried Fraenkel had left Germany in the 1930s to work in Britain. After the war he moved to the United States, to the University of Illinois. From this base, he published in *Science* in 1959 a paper entitled "The raison d'être of secondary plant substances" (29). It was still not generally accepted that secondary compounds played a significant role in determining the host plant ranges of insects, and in this paper he surveyed the evidence. He came down strongly in favor of their importance, but ten years later, in the opening paper of the 2nd International Insect/Hostplant Symposium in Wageningen, he felt the need to reinforce his thesis in the face of "criticism, almost bordering on animosity" (30). This criticism came from no less than Beck, Kennedy, and Thorsteinson, all of them leaders in the field and respected scientists (2, 49, 102).

To some extent, this was a storm in a teacup, but it nevertheless delayed progress. At issue was whether or not all insects have essentially similar nutritional requirements, as Fraenkel had claimed, or whether the nutritional requirements of insect species differed to an extent that affiliation with a particular host plant provided a species with its specific needs. The crux of the argument was the possible nutritional value of secondary chemicals. Writing in an early volume of the *Annual Review of Entomology*, Thorsteinson argued that it was not possible to know whether or not "odd" (secondary) chemicals had nutritional value for particular insects (102). Beck, also writing in the *Annual Review of Entomology*, went even further. He considered Fraenkel's "token stimuli" theory to be "obsolete" because the "token stimuli are . . . by definition secondary chemicals of no nutritional significance . . . ." (2).

Reading Fraenkel's 1959 paper today, it seems inconceivable (at least to me) that his intent should be so much misunderstood. He clearly did not intend to imply that nutritional chemicals were unimportant in host-selection behavior, but he argued that they could not account for discrimination between plant species. He attempted to clarify these issues in his 1969 paper (30). In the interim, between 1959 and 1969, a number of important studies had established, without any possibility of doubt, the roles of secondary compounds as sign stimuli having either a positive or negative effect on the feeding and oviposition behavior of a variety of insects.

Notice that Fraenkel's view was not dependent on technological developments, but on a clarification of ideas based on an increasing mass of experimental evidence on insect development and survival. At the same time however, corroborative evidence from a new technique became available. In 1955, Hodgson and his colleagues had developed the tip recording method for recording the activity of neurons in the presumed taste receptors of insects (40). This simple technique was first applied to phytophagous insects by Stürckow, working on the Colorado potato beetle, *Leptinotarsa decemlineata* (96). Ishikawa showed that caterpillars of the silkmoth, *Bombyx mori*, had a neuron responding specifically to deterrent compounds and extracts of non-host plants (43). This work seemed to establish that these insects had the equivalent of the human bitter taste and that such a receptor could be responsible for the inhibitory effects on feeding of many plant secondary compounds. At about the same time, Schoonhoven, working with larvae of *Pieris brassicae*, showed that one of the neurons in the galeal sensilla responded specifically to sinigrin, thus providing a neural basis for Verschaffelt's much earlier observation that glucosinolates were positive sign stimuli for these insects (82, 83, 103). Thus, the arguments against Fraenkel's thesis were largely removed.

Perhaps partly as a result of the controversy, Fraenkel lost some of the credit for appreciating the real importance of plant secondary compounds in insect/plant interactions and their evolution. I believe he felt this to be the case because I have a copy of his 1969 paper on which he has written, "This paper has been rather overlooked." By the time he published this paper, Ehrlich and Raven had entered the arena, and on the basis of the literature on the host plant ranges of butterflies they reached a similar conclusion that "secondary plant substances play the leading role in determining patterns of utilization [of plants by the larvae of some groups of butterflies]." Based on this conclusion, Ehrlich and Raven developed their concept of coevolution, bringing home the realization that interacting biological entities, in this case the insect and its host plant, could not evolve independently of each other: Changes in one would affect the other (20). Suddenly evolution of insect/plant interactions was seen in a much more dynamic light. Whereas we might readily have agreed that host plant selection by phytophagous insects was strongly influenced by secondary compounds, now we were faced with the idea that "the evolution of secondary plant substances and the stepwise evolutionary responses to these by phytophagous organisms have clearly been the dominant factors in the evolution of butterflies and other phytophagous groups."

This concept gave rise to a great flurry of activity, especially in the United States, where evolutionary thought was perhaps less constrained than it was in Europe. Coevolution was "in" and its impact on the field of insect/plant relations has been enormous.

Why did Ehrlich and Raven have such an impact, while Fraenkel, with a much less unusual (for the time) interpretation had to struggle to make himself heard? Perhaps because, reflecting their backgrounds, they were addressing different audiences. Fraenkel, although he published in *Science*, was best known amongst a group of biologists whose interests were primarily in mechanisms. For them evolution was a thing that happened, or had happened; it was not something that they thought about very much and probably believed that it was not possible fully to understand the selection pressures that might have been responsible for particular traits. Ehrlich and Raven, on the other hand, published in the journal *Evolution* and addressed an audience that regularly thought and talked about evolution. For them, it became obvious that insects must have exerted significant pressure on the evolution of plant secondary compounds and vice versa. Interestingly, though, Feeny credits the "coevolutionary interpretation of phytochemical diversity" to Fraenkel (27).

Not everyone shared in the euphoria. Few argued with the likelihood that secondary compounds played "a leading role in determining patterns of utilization [of plants]" (20), although there was still not very much direct experimental evidence. The issue was the extent to which plants and insects coevolved, for the evidence for that was solely correlative. Tibor Jermy, who had contributed significantly to understanding the importance of secondary compounds as determinants of host plant range because of their deterrent effects on most insects, was an outspoken critic. In 1976, he countered with the theory of sequential evolution. He argued that there was ample evidence to show that plants exerted considerable selection pressure on insects, but (at that time) very little evidence for strong selection by insects on plants (44). He believed, and still does believe, that "the evolution of herbivorous insects follows the evolution of plants without, however, significantly affecting plant evolution" (86).

With time, the ideas of Ehrlich and Raven have been tested more rigorously, sometimes including critical phylogenetic studies. This was facilitated, but not driven, by molecular techniques which now became available. For example, Mitter & Farrell, reviewing the fast-accumulating literature, conclude that although some of the evidence supported the general thesis of coevolution, most of it did not (62). [See Feeny (27a) for a much more complete account.]

Prior to 1950, most biological research tended to be investigative; after 1950 there was a move toward more hypothesis-driven research. In the field of insect/plant relations, and building on the general concept of coevolution of insects and plants, Feeny suggested that the extent to which plants were defended might reflect the extent to which they typically escaped from herbivores, the apparency hypothesis (26). McKey suggested that the extent to which particular tissues within a plant contained secondary compounds should reflect the value of those tissues to the plant, the tissue value hypothesis (60). Coley et al put forward the resource availability hypothesis (16). These studies arose from and gave rise to many further studies in chemical ecology.

Others, like Bernays, while accepting the importance of secondary compounds in host plant recognition, were not convinced that the compounds were necessarily the driving force behind the restriction of insect host range. She demonstrated that predation could, in some cases at least, potentially drive the need for specialization for which the secondary compounds might largely only be only indicators of plant species (4). This work led to further studies on behavioral attentiveness, seemingly a far cry from coevolution, but contributing to the general understanding of the evolution of insect/plant relations (3).

Thus a rather simple idea produced a great mushrooming of research, most of which has not depended on advanced technology but on the interplay of different ideas. As with the more technology-driven work on olfaction, however, most progress has occurred in the latter part of the century.

THE NEW TECHNOLOGIES

So much has changed over the last century that it is impossible to attempt any comprehensive account of the impact of technology, and I have selected a small number of fields where the impact has been enormous. It is apparent, even from this very limited sample, that a number of developments in the 1940s and 1950s were largely responsible for the striking advances of the twentieth century. It will also be clear from the following brief accounts that, in many cases, the new developments involved integrating a number of divergent technologies. The current state of the confocal microscope, for example, involves laser illumination, video monitoring and computer logging often coupled with use of antibodies linked to fluorescent dyes in the object. See Magner (56) for an account of the earlier background to these developments.

Radio

From the time the first radio signal was transmitted across the Atlantic by Marconi in 1901, there was a steady development in the use of radio, but until transistors were developed in 1948, size precluded the use of radio as a means of tracking animals. Transistors made miniaturization possible, and by the 1960s miniature transmitters implanted into or attached to animals were used successfully to follow the migrations of a wide range of birds, mammals and marine animals (87). In addition to being used simply to obtain positional information, coupled with sensors implanted into different tissues, radiotelemetry made it possible to monitor physiological activities remotely. However, it was not until the late 1990s that transmitters small enough to attach to insects without impeding their flight ability were produced (111). The potential for this method is clearly enormous.

Radar The idea that radio waves were reflected from solid objects, the principle of radar, was already known at the turn of the century, but the practical application of this phenomenon to detect moving objects began in many countries in the 1930s as preparations were made for the impending war. After the war, radar came to be used routinely on ships and at airports. Echoes from small unidentified targets, known as "angels," were periodically reported by these radars, but following some success in observing high-flying birds, GW Schaefer, of Loughborough University in England, modified a radar system specifically to observe large insects. With support from the Anti-Locust Research Centre in London he made the first attempt to study insect flight using the new tool in Niger in 1968 (80). Knowing the fauna of the area in which he was working, he was able to make an intelligent guess at the identity of the flying objects from the modulation of the intensity of their image, which enabled him to measure the wingbeat frequency. He even claimed to be able to distinguish males from females on this basis. This work was subsequently developed by JR Riley, who brought much more rigorous standards to the work. Subsequently, the Anti-Locust Research Centre established a permanent radar entomology group headed by Riley, and in subsequent years insect-oriented radar groups were established in Australia, the United States, Canada, and China. Because of the costs, these groups were government funded, but, sadly, much of that support has now declined markedly (71).

Radar enables us to visualize individual insects from great distances and to scan huge volumes of air very rapidly. Echoes from individual locusts can be perceived at ranges of up to 5 km, and the radar scans 10^7 m^3 in about 3 seconds. It thus allows us to observe movements of insects behaving normally at high altitudes and in low densities. Initially, it was only possible to observe larger insects such as locusts and nocuid moths, but the development of a system using shorter wavelengths made it possible to detect planthopper-sized insects at distances of 1 km. Taking radar into the air further enhanced our capacity to study wide-scale movements (73, 112).

Because of this, the impact of weather systems on insects' dispersal, often previously surmised, could now be observed, and the magnitude and extent of nocturnal insect movements are beyond belief. Some species of acridids, moths, and planthoppers deliberately climb to high altitudes so that, outside their boundary layers, they are swept along by the winds, often covering hundreds of kilometers in one night. There is also evidence that some insects choose not to fly on nights when the wind is blowing in the "wrong" direction (72).

Despite its extraordinary power, radar cannot be used to observe low-flying insects because of the presence of ground clutter such as buildings, trees, and other objects. The first attempts to overcome this problem were made by Mascanzoni & Wallin based on a system developed to locate avalanche victims and also tried in relation to automobile collision avoidance systems (57). This "harmonic" radar depends on the target being tagged with a transponder that picks up energy from an illuminating radar at a wavelength to which the transponder is tuned and reradiates some of the energy at some harmonic of the incoming signal. Thus it can be distinguished even in the presence of strong signals from background clutter. As originally used, for detecting carabid beetles, this technology was basically a direction finder, and the best results were obtained when the insect carried two diodes with a radio antenna about 10 cm long. Subsequently, however, the method has been refined and the apparatus made small enough that it can be carried by honey bees or tsetse flies in flight. With such refinement the method offers the potential for learning about insect movements close to, or even on, the ground (74).

Radioactivity

After the discovery of radioactivity and of α, β, and γ rays at the turn of the century, the concept of radioactive isotopes of normal elements was proposed in 1912. After that, new radioactive elements were reported in a steady stream, but it was not until the late 1940s that radioactive isotopes became widely used for marking insects in studies of dispersal, population assessment, and predation. A range of isotopes was used, either as external markers or incorporated into the tissues (93). One valuable aspect of a radioactive mark is that, depending on its half-life, it is retained when the insect molts, and this can be important in assessing mortality as in the studies on larvae and pupae of *Panaxia dominula* (17). Radioactive marking in field studies was used fairly extensively until the increasing awareness of problems with environmental contamination in the 1970s resulted in a marked reduction.

Autoradiography Becquerel had, almost by accident, used autoradiography in his discovery of radioactivity in 1896: It was the blackening of a photographic plate when in contact with a piece of rock in darkness that alerted him to the fact that he had discovered something new. It was not, however, until after World War II that the method came to be widely used in biology. By that time radioisotopes were being produced artificially and could be incorporated into compounds. Thus by using autoradiography it was possible to localize the positions of compounds in whole animals or, more usually, in sections viewed with the light microscope and, later, the electron microscope. It is in this field that radioactivity has had the greatest impact on insect studies, most recently combined with electrophoresis in the Southern and Northern blot techniques for analysis of DNA and RNA.

Sterile Insect Technique The potential of X-rays and γ-rays to cause sterilization was explored as a means of insect control in the 1950s and was of major importance in what surely must be the most spectacular insect control event of the century, the eradication of screwworm, *Cochliomyia hominivorax,* from the greater part of North America (31, 114).

This insect inhabits the tropical and subtropical regions of the Americas and prior to 1962 it regularly spread northward in the summers, commonly infesting more than a million cattle in the United States alone. Winters in the north were generally too cold for it to survive, but populations persisted in the states along the border with Mexico. The idea of control by the release of sterile males originated with EF Knipling and, in 1947, USDA scientists initiated experiments with chemosterilization. These experiments were not successful, and in 1950, the team turned to radiation-induced sterilization. Following successful trials on a small island off the coast of Florida and then on the island of Curacao, trial releases of sterile males were made in the United States in 1957. By 1970, the fly was virtually eliminated from the United States, only to exhibit a sudden resurgence in 1972. Following efforts to increase the competitive ability of the sterile flies compared with their wild counterparts, the species was finally eradicated from the United States. The onslaught continued so that by 1991 it was eliminated from Mexico, and in subsequent years it was eliminated successively from Belize, Guatemala, El Salvador, and Honduras.

This program involved the greatest mass rearing of insects that has ever been undertaken. During 1974, the USDA plant at Mission, Texas was producing 200×10^6 flies each week. Later, a rearing facility in Mexico maintained a level of 500×10^6 flies per week. This level of production made it possible to eradicate the fly in Libya where it had been accidentally introduced around 1988. By 1990, before eradication procedures were initiated, the infested area in Libya extended over 26,000 km^2. Shipments of sterile flies from Mexico resulted in the complete elimination of the flies by the end of 1991. Of course, this success was not achieved without cost, and most other eradication programs have not met with equal success (66).

Immunology

Although the concept of immunity to diseases had been known for hundreds, perhaps thousands of years, it was not until 1890 that Behring discovered that immunity depended on molecules in the blood that were produced in response to a particular disease organism. This gave rise in the early part of the 20th century to the development of vaccines against a variety of human diseases. The finding soon afterwards that antibodies were also produced in response to other foreign proteins not associated with disease was the start of immunochemistry.

The Precipitin Test It was soon found that mixing an antigen with its antiserum in vitro produced a precipitate, and the reaction became known as the precipitin test. It was recognized that this provided a potential means of identifying the sources of unknown proteins. This method was soon used by zoologists to examine taxonomic relationships. The earlier entomological studies seem to have never extended beyond the point of supporting existing classifications (54), but with the development of monoclonal antibodies it became possible to obtain information that could be used in phylogenetic analysis (35).

From 1927, the method was also used with some success in identifying the hosts of some bloodsucking insects, notably tsetse flies (45, 101). These early workers had to prepare their own antisera in the field, starting by shooting the potential host animals in order to obtain blood samples because this was long before tranquilizing darts were available. The animals included zebras, giraffes, hippopotamus, elephants, and crocodiles. The antisera produced were relatively non-specific and it was not until the 1950s that the method was significantly improved (108).

The idea that this method could also be used to determine the predators of particular insects did not develop until later when Brook & Proske investigated the predators of mosquito larvae (9). It was only in the late 1950s that Dempster attempted a quantitative analysis of the predators of the beetle, *Phytodecta olivaceae* (18), and in 1993 Greenstone & Hunt regarded "Serological gut analysis [as] our most powerful tool for determining the impact of arthropod predators on prey populations" (34).

Immunocytochemistry The principal impact of immunology on insect studies has, however, been in immunocytochemistry, which enables us to identify and locate precisely the positions of chemicals within organisms. The impact of this technique has been enormous in studies of the nervous system, where our knowledge of the complexity of chemical signaling has developed out of all recognition from the simple synaptic connections of 20 years ago. Nässel (67) lists 20 neuropeptides, probably representing distinct peptide families, detected in insects by this technique. For the first time, it is possible to ascribe chemical phenotypes to the neurons and even to attempt to homologize them across taxa. The astonishing power of this method is, however, most evident in the field of molecular biology, where it is an essential tool in the labeling of gene products. Now, for the first time, we are starting to learn about the genetic control of segmentation, and that segments are not what they seem to be (53).

Imaging techniques

Light Microscopy Changes in the early part of the century were principally concerned with alterations in the light path in the compound microscope. Phase contrast microscopy, which was first used in the 1930s, became commonplace in the 1950s, and interference microscopy was introduced commercially in the late 1940s. These techniques were especially valuable for improving contrast in unstained material (8). The subsequent development of video systems had a marked impact on light microscopy because they made it possible to visualize wavelengths outside the human visible spectrum. More importantly, video systems made it possible to extract information from very low-contrast signals (41). Video is a key component of the most recent tool in the imaging armory, the confocal laser-scanning microscope, which greatly increases the resolution that can be obtained with fluorescent signals. Although originally conceived in 1957, it is only in the last 10 years that the instrument has been marketed with its current degree of refinement (42). The principle of this technique is that the object is scanned by a small spot of light and the reflected or, more commonly, emitted light from the spot is recorded electronically. In this way, loss of resolution due to light scatter is reduced to a minimum. The incident light can be set to scan at a certain level in the specimen to produce an image of a very thin optical section, and by scanning at different levels a three-dimensional image of the object can be built up electronically. Hence the problem of limited depth of field for fluorescence microscopy can be overcome and structures can be viewed in situ, provided they reflect or emit light differentially from the surrounding tissues. This is usually achieved by marking (tagging) the structures of interest with a fluorescent dye, for instance by attaching the dye molecule to a specific antibody or by filling the cells with fluorescent dye. Thus it is now possible to follow the paths of axons within whole brains of insects, and by differential labeling of neurons to observe the points at which synapses are likely to occur. In this way, the confocal microscope can provide an important link between light and electron microscopy (98). Examples of the confocal microscope applied to insect microanatomy can be seen in Sun et al (99).

Electron Microscopy Electrons have shorter wavelengths than visible light, and so the electron microscope provides a mechanism by which resolution can be greatly increased. The first commercial electron microscopes were produced in the United States in the early 1940s, but it was not until a decade later that cutting thin sections for the study of biological material with the transmission electron microscope became routine (69). This made possible the study of animal cells at a level of detail that was previously not possible, and the understanding of insect cells benefited from the start (89).

With transmission electron microscopy we also learned the details of cuticular structure. Rudall first showed that the chitin was present as microfibrils (79), and Bouligand, working on Crustacea, showed that the orientation of the fibrils varied with time to give the helicoidal arrangement that contributes to the strength of the cuticle (7). From Neville we learned, too, that in many species the orientation varies diurnally so it is possible to determine the age of insects from the banding patterns in sections of the cuticle (68). In most insects the fibrils are parallel with the surface of the cuticle, but Wolfgang & Riddiford found that vertical fibrils, produced by extensions of the epidermal cells, allow the cuticle of caterpillars to grow between molts—an unexpected finding (113).

The scanning electron microscope, which detects electrons that bounce off specimens (usually coated with a thin layer of metal), enabled Hinton, in particular, to explore the distribution of plastrons throughout the insects and to discover that they were of widespread occurrence in the eggs and pupae of many terrestrial insects living in habitats subject to flooding (39). By now, of course, we have many beautiful examples of the power of the scanning microscope that have revealed unimagined intricacies, not only of cuticular structures, but also of the structures on the surfaces of soft, internal tissues. See, for example, the stretch receptor on the bursa copulatrix of *Pieris* pictured by Sugawara (97).

A variation of electron microscopy, the electron probe has made it possible to detect heavy metals in insect cuticle, so that we now know that there is usually zinc or manganese in the cusps of the mandibles and maxillae and even in the tarsal claws (28).

Chemical Analysis

Advances in chemistry have equaled those in the physical sciences but can only be given the briefest consideration.

The ability to separate and identify chemicals has dramatically altered our understanding of the chemistry of insects. Chromatography was used by dye chemists before the 19th century, but it was not until the 20th century that the physico-chemical basis of the process became understood. Although chromatography was used to separate plant pigments as early as 1910, it was not until the 1930s that partition chromatography was developed, leading in the 1940s to paper chromatography. Thin-layer chromatography was developed at the same time.

Gas chromatography was first exploited in the 1950s and has played a major role in the development of our knowledge of pheromones, defensive chemicals and the cuticular lipids of insects. An important entomological development was to pass the output from the chromatograph continuously over an antenna to record electroantennograms. This method made it possible to determine which of the compounds in an odor mix produced a physiological response and so might be behaviorally important (63). Subsequently, the method has been further refined by recording from single olfactory cells rather than the whole antenna. By further linking the effluent to a mass spectrometer it is also possible to identify the active compounds (109). Considering that Butenandt required the pheromone glands from more than 300,000 silkmoths to isolate and identify bombykol in the 1950s, when the only available bioassay was behavioral, the advances can only be considered astounding.

A FEELING FOR THE ORGANISM

Despite all the advances resulting from amazing new technologies and the facility of communication, there is not, and cannot be, any substitute for the people who love and know their insects, who, like Barbara McClintock, have a feeling for their organism (48). As we inevitably become more specialized, we must become more and more dependent on those few who remain "real entomologists" if our work is to be put into the context of the organism.

And there are still those new discoveries that can be made only by people with a proper understanding of the insect. A beautiful example of this is Locke's recent description of bunches of tracheae and tracheoles in the hemocoel of the caterpillar of *Calpodes ethlius* which, he believes, may serve as stations at which hemocytes collect and obtain oxygen (55). Who knows where this extraordinary observation will lead? Are there comparable structures in other insects that have been overlooked because no one asked the right questions?

Undoubtedly such novel observations will continue to expand our knowledge and wonderment of the insects in the next century and long after that. But they can only be made if people have time to think and enjoy "their" insect.

ACKNOWLEDGMENTS

Many people have helped at various stages in the preparation of this review. In particular I should like to thank Peter Credland, John Hildebrand, Linda Restifo, and Leslie Tolbert for their suggestions and help in various ways. Liz Bernays has been, as always, constant in her support.

Visit the Annual Reviews home page at www.AnnualReviews.org.

LITERATURE CITED

1. Abbott HC de S. 1887. Comparative chemistry of higher and lower plants. *Am. Nat.* 21:800–10
2. Beck SD. 1965. Resistance of plants to insects. *Annu. Rev. Entomol.* 10:207–32
3. Bernays EA. 1998. The value of being a resource specialist: behavioral support for a neural hypothesis. *Am. Nat.* 151:451–64
4. Bernays EA, Graham M. 1988. On the evolution of host specificity in arthropods. *Ecology* 69:886–92
5. Bingham PM, Kidwell MG, Rubin GM. 1982. The molecular basis of P-M hybrid dysgenesis: the role of the P element, a P-strain-specific transposon family. *Cell* 29:995–1004
6. Boistel J, Coraboeuf E. 1953. L'activité électrique dans l'antenne isolée de Lépidoptère au cours de l'étude de l'olfaction. *C. R. Soc. Biol.* 147:1172–75
7. Bouligand Y. 1965. Sur un architecture toradée répandue dans de nombreuses cuticles d'arthropodes. *C. R. Hebd. Séanc. Acad. Sci., Paris* 261:3665–68
8. Bradbury S. 1967. *The Evolution of the Microscope*. Oxford: Pergamon. 357 pp.
9. Brook MM, Proske HO. 1946. Precipitin test for determining natural insect predators of immature mosquitoes. *J. Natl. Malaria Soc.* 5:45–56
10. Brues CT. 1920. The selection of food-plants by insects, with special reference to lepidopterous larvae. *Am. Nat.* 54:313–32
11. Carozzi N, Koziel M, eds. 1997. *Advances in Insect Control: Transgenic Plants*. London: Taylor & Francis. 301 pp.
12. Carson R. 1962. *Silent Spring*. Boston: Houghton Mifflin. 368 pp.
13. Chapman RF. 1969. *The Insects: Structure and Function*. London: Engl. Univ. Press

13a. Christensen TA, Mustaparta H, Hilde-

brand JG. 1995. Chemical communication in heliothine moths VI. Parallel pathways for information processing in the macroglomerular complex of the male tobacco budworm moth *Heliothis virescens*. *J. Comp. Physiol. A* 177:545–57
14. Clever U, Karlson P. 1960. Induktion von Puff-Veränderung in den Speicheldrüsen Chromosomen von *Chironomus tentans* durch Ecdyson. *Exp. Cell Res.* 20: 623–26
15. Cohen IB. 1985. *Revolution in Science.* Cambridge, MA: Belknap. 711 pp.
16. Coley PD, Bryant JP, Chapin T. 1985. Resource availability and plant antiherbivore defense. *Science* 22:895–99
17. Cook LM, Kettlewell HBD. 1960. Radioactive labelling of lepidopterous larvae: a method of estimating larval and pupal mortality in the wild. *Nature* 187:301–2
18. Dempster JP. 1960. A quantitative study of the predators on the eggs and larvae of the broom beetle, *Phytodecta olivacea* Forster, using the precipitin test. *J. Anim. Ecol.* 29:149–67
19. Dodd AP. 1940. *The Biological Campaign Against Prickly-Pear.* Brisbane: Commonw. Prickly Pear Board. 177 pp.
20. Ehrlich PR, Raven PH. 1965. Butterflies and plants: a study in coevolution. *Evolution* 18:586–608
21. Elliott M, Janes NF, Potter C. 1978. The future of pyrethroids in insect control. *Annu. Rev. Entomol.* 23:443–69
22. Eltringham H. 1933. *The Senses of Insects*. London: Methuen. 126 pp.
23. Errera L. 1886. L'efficacité des structures défensives des plantes. *Bull. R. Soc. Bot. Belg.* 25:80–99
24. Essig EO. 1931. *A History of Entomology.* New York: Macmillan. 1029 pp.
25. Fabre JH. 1886. *Souvenirs Entomologiques,* Vol. 3. Paris: Delagrave
26. Feeny P. 1975. Biochemical coevolution between plants and their insect herbivores. In *Coevolution of Animals and Plants*, ed. LE Gilbert, PH Raven, pp. 3–19. Austin: Univ. Tex. Press
27. Feeny P. 1991. Theories of plant chemical defense: a brief historical survey. *Symp. Biol. Hung.* 39:163–75
27a. Feeny P. 1992. The evolution of chemical ecology: contributions from the study of herbivorous insects. In *Herbivores. Their Interactions with Secondary Plant Metabolites,* Vol. 2, ed GA Rosenthal, MR Berenbaum, pp. 1–44. San Diego: Academic
28. Fontaine AR, Olsen N, Ring RA, Singla CL. 1991. Cuticular metal hardening of mouthparts and claws of some forest insects of British Columbia. *J. Entomol. Soc. B. C.* 88:45–55
29. Fraenkel GS. 1959. The raison d'être of secondary plant substances. *Science* 129:1466–70
30. Fraenkel GS. 1969. Evaluation of our thoughts on secondary plant compounds. *Entomol. Exp. Appl.* 12:473–86
31. Galvin TJ, Wyss JH. 1996. Screwworm eradication program in Central America. *Ann. NY Acad. Sci.* 791:233–40
32. Gilbert P, Hamilton CJ. 1990. *Entomology. A Guide to Information Sources.* London: Mansell. 259 pp.
33. Gillott C. 1980. *Entomology*. New York: Plenum. 729 pp.
34. Greenstone MH, Hunt JH. 1993. Determination of prey antigen half-life in *Polistes metricus* using a monoclonal antibody-based immunodot assay. *Entomol. Exp. Appl.* 68:1–7
35. Greenstone MH, Stuart MK, Haunerland NH. 1991. Using monoclonal antibodies for phylogenetic analysis: an example from the Heliothinae (Lepidoptera: Noctuidae). *Ann. Entomol. Soc. Am.* 84:457–64
36. Hamilton WD. 1964. The genetical evolution of social behaviour. I. *J. Theor. Biol.* 7:1–16
37. Harrison FW, Locke M, eds. 1998. *Microscopic Anatomy of Invertebrates.*

Vol. 11. *Insecta.* New York: Wiley-Liss. Vol. 11A, 455 pp.; Vol. 11B, 533 pp.; Vol. 11C, 456 pp.
38. Harrison G. 1978. *Mosquitoes, Malaria and Man: A History of the Hostilities Since 1880.* New York: Dutton. 314 pp.
39. Hinton HE. 1976. Plastron respiration in bugs and beetles. *J. Insect Physiol.* 22:1529–50
40. Hodgson ES, Lettvin JY, Roeder KD. 1955. Physiology of a primary chemoreceptor unit. *Science* 122:417–18
41. Inoué S. 1986. *Video Microscopy.* New York: Plenum. 584 pp.
42. Inoué S. 1995. Foundations of confocal scanned imaging in light microscopy. In *Handbook of Biological Confocal Microscopy*, ed. JB Pawley, pp. 1–17. New York: Plenum
43. Ishikawa S. 1966. Electrical response and function of a bitter receptor associated with the maxillary sensilla of the silkworm, *Bombyx mori* L. *J. Cell. Physiol.* 67:1–11
44. Jermy T. 1976. Insect/host-plant relationship: co-evolution or sequential evolution? *Symp. Biol. Hung.* 16:109–13
45. Johnson WB, Rawson PH. 1928. Use of the precipitin test to determine the food supply of tsetse flies. *Trans. R. Soc. Trop. Med. Hyg.* 21:135–49
46. Karlson P, Butenandt A. 1959. Pheromones (ectohormones) in insects. *Annu. Rev. Entomol.* 4:39–58
47. Keilin D. 1925. On cytochrome, a respiratory pigment, common in animals, yeast and higher plants. *Proc. R. Soc. London Ser. B* 98:312–39
48. Keller EF. 1983. *A Feeling for the Organism. The Life and Work of Barbara McClintock.* New York: Freeman. 235 pp.
49. Kennedy JS. 1965. Mechanisms of host plant selection. *Ann. Appl. Biol.* 56:317–22
50. Kerkut GA, Gilbert LI, eds. 1985. *Comprehensive Insect Physiology, Biochemistry and Pharmacology*, 13 Vols. Oxford: Pergamon
51. Kirby W, Spence W. 1856. *An Introduction to Entomology.* London: Longman. 607 pp.
52. Kossel A. 1891. Über die chemische Zusammensetzung der Zelle. *Arch. Anat. Physiol. Physiol. Abt.* 1891:181–86
53. Lawrence PA. 1992. *The Making of a Fly: The Genetics of Animal Design.* Oxford: Blackwell. 228 pp.
54. Leone CA. 1947. A serological study of some Orthoptera. *Ann. Entomol. Soc. Am.* 40:417–33
55. Locke M. 1998. Caterpillars have evolved lungs for hemocyte gas exchange. *J. Insect Physiol.* 44:1–20
56. Magner LN. 1994. *A History of the Life Sciences.* New York: Dekker. 496 pp.
57. Mascanzoni D, Wallin H. 1986. The harmonic radar: a new method of tracing insects in the field. *Ecol. Entomol.* 11:387–90
58. McIndoo NE. 1918. The olfactory organs of Diptera. *J. Comp. Neurol.* 29:457–84
59. McIndoo NE. 1929. Tropisms and sense organs of Lepidoptera. *Smithson. Misc. Collect. 81.* 59 pp.
60. McKey D. 1974. Adaptive patterns in alkaloid physiology. *Am. Nat.* 108:305–20
61. Mellanby K. 1992. *The DDT Story.* Farnham: Br. Crop Prot. Counc. 113 pp.
62. Mitter C, Farrell C. 1991. Macroevolutionary aspects of insect-plant relations. In *Insect-Plant Interactions,* ed. EA Bernays, 3:35–78. Boca Raton, FL: CRC
63. Moorhouse JE, Yeadon R, Beevor PS, Nesbitt BF. 1969. Method for use in studies of insect chemical communication. *Nature* 223:1174–75
64. Morgan TH. 1910. Sex-linked inheritance in *Drosophila. Science* 32:120–22
65. Morice FD. 1901. Illustrations of the 6th male segment in 17 *Osmia*-species of the *adunca*-group, with a note on the synonymy of four species, and descriptions

of four which seem new. *Trans. Entomol. Soc. London* 1901:161–80
66. Myers JH, Savoie A, van Randen E. 1998. Eradication and pest management. *Annu. Rev. Entomol.* 43:471–91
67. Nässel DR. 1996. Advances in the immunocytochemical localization of neuroactive substances in the insect nervous system. *J. Neurosci. Methods* 69:3–23
68. Neville AC. 1967. Chitin orientation in cuticle and its control. *Adv. Insect Physiol.* 4:213–86
69. Newberry SP. 1992. *EMSA and Its People. The First Fifty Years.* Magnolia, NJ: Electron Microscopy Soc. Am.
69a. Payne TL, Birch MC, Kennedy CEJ. eds. 1986. *Mechanisms in Insect Olfaction.* Oxford: Clarendon. 364 pp.
70. Pringle JWS. 1938. Proprioception in insects. I. A new type of mechanical receptor from the palps of the cockroach. *J. Exp. Biol.* 15:101–13
71. Reynolds DR. 1988. Twenty years of radar entomology. *Antenna* 12:44–49
72. Reynolds DR, Riley JR. 1997. Flight behaviour and migration of insect pests. *NRI Bull.* 71. Chatham, UK: Nat. Resourc. Inst. 114 pp.
73. Riley JR. 1989. Remote sensing in entomology. *Annu. Rev. Entomol.* 34:247–71
74. Riley JR, Valeur P, Smith AD, Reynolds DR, Poppy GM, Löfstedt C. 1998. Harmonic radar as a means of tracking the pheromone-finding and pheromone-following flight of male moths. *J. Insect Behav.* 11:287–96
75. Ritchie M, Pedgley D. 1989. Desert locusts cross the Atlantic. *Antenna* 13:10–12
76. Rockstein M, ed. 1964–1965. *The Physiology of Insecta,* 3 Vols. New York: Academic
77. Rockstein M, ed. 1973–1974. *The Physiology of Insecta,* 6 Vols. New York: Academic
78. Roeder KD. 1953. *Insect Physiology.* New York: Wiley. 1100 pp.
79. Rudall KM. 1965. Skeletal structure in insects. *Biochem. Soc. Symp.* 25:83–92
80. Schaefer GW. 1969. Radar studies of locust, moth and butterfly migration in the Sahara. *Proc. R. Entomol. Soc. London Ser. C* 34:39–40
81. Schneider D. 1957. Electrophysiological investigation on the antennal receptors of the silk moth during chemical and mechanical stimulation. *Experientia* 13:89–91
82. Schoonhoven LM. 1967. Chemoreception of mustard oil glucosides in larvae of *Pieris brassicae. Proc. Kon. Ned. Akad. Wet. C* 70:556–58
83. Schoonhoven LM. 1968. Chemosensory bases of host plant selection. *Annu. Rev. Entomol.* 13:115–36
84. Schoonhoven LM. 1991. Insects and hostplants: 100 years of botanical instinct. *Symp. Biol. Hung.* 39:3–14
85. Schoonhoven LM. 1997. Verschaffelt 1910: a founding paper in the field of insect-plant relationships. *Proc. Kon. Ned. Akad. Wet.* 100:355–68
86. Schoonhoven LM, Jermy T, van Loon JJA. 1998. *Insect-Plant Biology. From Physiology to Evolution.* London: Chapman & Hall. 409 pp.
87. Slater LE, ed. 1965. Biotelemetry. *BioScience* 15:79–121
88. Slifer EH, Prestage JJ, Beams HW. 1957. The fine structure of the long basiconic pegs of the grasshopper (Orthoptera, Acrididae) with special reference to those on the antenna. *J. Morphol.* 101:359–97
89. Smith DS. 1968. *Insect Cells: Their Structure and Function.* Edinburgh: Oliver & Boyd. 327 pp.
90. Smith RF, Mittler TE, Smith CM, eds. 1973. *History of Entomology*. Palo Alto, CA: Annu. Rev. 517 pp.
91. Snodgrass RE. 1926. The morphology of insect sense organs and the sensory nervous system. *Smithson. Misc. Collect.* 77. 80 pp.
92. Snodgrass RE. 1935. *Principles of Insect*

Morphology. New York: McGraw Hill. 667 pp.

93. Southwood TRE. 1978. *Ecological Methods.* London: Chapman & Hall. 524 pp.
94. Speck PT, Strausfeld NJ. 1983. Portraying the third dimension in neuroanatomy. In *Functional Neuroanatomy*, ed. NJ Strausfeld, pp. 156–82. Berlin: Springer-Verlag
95. Steinbrecht RA. 1973. Der Feinbau olfaktorischer Sensillen der Seidenspinners (Insecta, Lepidoptera). *Z. Zellforsch.* 139:533–65
96. Stürkow B. 1959. Über den Geschmackssin und den Tastsinn von *Leptinotarsa decemlineata* Say (Chrysomelidae). *Z. Vergl. Physiol.* 42:255–302
97. Sugawara T. 1981. Fine structure of the stretch receptor in the bursa copulatrix of the butterfly, *Pieris rapae crucivora. Cell Tissue Res.* 217:23–26
98. Sun XY, Tolbert LP, Hildebrand JG. 1995. Using laser scanning confocal microscopy as a guide for electron microscope study: a simple method for correlation of light and electron microscopy. *J. Histochem. Cytochem.* 43:329–35
99. Sun XY, Tolbert LP, Hildebrand JG. 1997. Synaptic organization of the uniglomerular projection neurons of the antennal lobe of the moth *Manduca sexta*: a laser scanning confocal and electron microscopic study. *J. Comp. Neurol.* 379:2–20
100. Sutherland J. 1999. Who owns John Sutherland? *London Rev. Books* 21:3,6
101. Symes CB, MacMahan JP. 1937. The food of tsetse-flies (*Glossina swynnertoni* and *G. palpalis*) as determined by the precipitin test. *Bull. Entomol. Res.* 28:31–42
102. Thorsteinson AJ. 1960. Host selection in phytophagous insects. *Annu. Rev. Entomol.* 5:193–218
103. Verschaffelt E. 1910. The cause determining the selection of food in some herbivous insects. *Proc. Kon. Ned. Akad. Wet.* 13:536–42

103a. Vogt RG. 1995. Molecular genetics of moth olfaction: a model for cellular identity and temporal assembly of the nervous system. In *Molecular Model Systems in the Lepidoptera,* ed. MR Goldsmith, AS Wilkins, pp. 341–67. Cambridge: Cambridge Univ. Press

104. von Frisch K. 1921. Über den Sitz des Geruchssines bei Insekten. *Zool. Jahrb. Allg. Zool. Physiol.* 38:1–68
105. von Frisch K. 1923. Über die 'Sprache' der Bienen. *Zool. Jahrb. Allg. Zool. Physiol.* 40:1–186
106. Walker TJ. 1998. The future of scientific journals: free access or pay per view. *Am. Entomol.* 44:135–38
107. Weinland E. 1890. Uber die Schwinger der Dipteren. *Z. Wiss. Zool.* 51:55–166
108. Weitz B. 1952. The antigenicity of sera of man and animals in relation to the preparation of specific precipitating antisera. *J. Hyg.* 50:275–94
109. Wibe A, Borg-Karlson AK, Norm T, Mustaparta H. 1997. Identification of plant volatiles activating single receptor neurons in the pine weevil (*Hylobius abietis*). *J. Comp. Physiol. A* 180:585–95
110. Wigglesworth VB. 1939. *Principles of Insect Physiology.* London: Methuen. 434 pp.
111. Willis MA, Yamada M, Takasaki T, Kuwana Y, Shimoyama I, Kanzaki R. 1998. Pheromone-modulated flight in *Manduca sexta*: variability in individual performances revealed by radio telemetry in flight muscle activity. *5th Int. Congr. Neuroethol.* 141 (Abstr.)
112. Wolf WW, Westbrook JK, Raulston J, Pair SD, Hobbs SE. 1990. Recent airborne radar observations of migrant pests in the United States. *Philos. Trans. R. Soc. London Ser. B* 328:619–30
113. Wolfgang WJ, Riddiford LM. 1981. Cuticular morphogenesis during continuous growth of the final instar larva of a moth. *Tissue Cell* 13:757–72

114. Wyss JH, Galvin TJ. 1996. Central America regional screwworm eradication program (benefit/cost study). *Ann. NY Acad. Sci.* 791:241–47

115. Yen JH, Barr AR. 1971. New hypothesis of the cause of cytoplasmic incompatibility in Culex pipiens. *Nature* 232:657–58

Annu. Rev. Entomol. 2000. 45:287–306

CONTROL OF INSECT PESTS WITH ENTOMOPATHOGENIC NEMATODES: The Impact of Molecular Biology and Phylogenetic Reconstruction

J. Liu, G. O. Poinar, Jr., and R. E. Berry
Department of Entomology, Oregon State University, Corvallis, Oregon 97331-2907; e-mail: jiel@bcc.orst.edu, poinarg@bcc.orst.edu, berryr@bcc.orst.edu

Key Words *Heterorhabditis*, *Steinernema*, genetic diversity, systematics, taxonomy

■ **Abstract** Entomopathogenic nematodes are excellent biological control agents. Utilization of these nematodes is developing rapidly with almost a doubling of newly described species in the past five years. Advances in molecular biology and phylogenetic reconstruction have revolutionized understanding of population structure, identification, genetic improvement, systematics, and the symbiosis between entomopathogenic nematodes and their bacteria. Population structure provides the most fundamental information for reliable identification of species and unique genetic variants. Such information could be further assessed for nematode potential as biological control agents. Phylogenetic reconstruction is an important approach for understanding multitrophic interactions among entomopathogenic nematodes, symbiotic bacteria, and their insect hosts. Phylogenetic reconstruction is also important for the development of a natural and stable type of systematics, which can provide guidelines for selecting appropriate entomopathogenic nematode species for particular biological control programs.

INTRODUCTION

Entomopathogenic nematodes are lethal obligatory parasites of insects (12, 42, 64, 92, 93, 103). They are ubiquitously distributed and comprise the families Heterorhabditidae (91) and Steinernematidae (26). The families are not closely related phylogenetically (9, 73) but share similar life histories through convergent evolution (92, 94). These nematodes are characterized by their ability to carry specific pathogenic bacteria, *Photorhabdus* (19) with Heterorhabditidae and *Xenorhabdus* (108) with Steinernematidae, which are released into the insect hemocoel after penetration of the insect hosts has been achieved by the infective stage of the nematode. The Heterorhabditidae is monotypic, represented by the

0066-4170/00/0107-0287/$14.00

genus *Heterorhabditis* (91). The Steinernematidae comprises two genera: *Steinernema* (110) and *Neosteinernema* (84).

Entomopathogenic nematodes possess many attributes of an excellent biological control agent: They are environmentally safe and acceptable (42, 64, 103), can be produced in large quantities with artificial media (10, 39), and are easily applied with standard spraying equipment or irrigation systems. Most species have a broad host range and are highly virulent, killing their host rapidly. This combination of attributes has generated intense interest in the development of these nematodes for use against insect pests (13, 42, 64, 65). More than 30 entomopathogenic nematode species exist (54, 86, 92), and hundreds of isolates have been collected from every continent. Many of these isolates have been tested for biological control of insect pests. Fully developing biological control methods using entomopathogenic nematodes will improve their virulence, host range, and environmental stability (41, 42).

Two comprehensive books on entomopathogenic nematodes and their symbiotic bacteria have been published since 1990 (12, 42). Several reviews have been written on these nematodes since 1993, including general aspects of biology and biological control (64), biodiversity and biosystematics (54, 56), symbiotic bacteria (17, 37, 38), and genetic improvement (41, 93). In the past several years, molecular biology and phylogenetic reconstruction have begun to play a more important role in the application of entomopathogenic nematodes to insect pest control. The initial stages of this trend were summarized in books describing molecular techniques in taxonomy (27, 28), biotechnology, and genetics (36). The advent of new techniques in molecular biology has created opportunities for examining the genetic diversity of entomopathogenic nematodes. Molecular methods have been used to assess the relationships of the major species and to examine intraspecific variations. Recombinant DNA technology shows early promise for the rational improvement of entomopathogenic nematodes (41, 93). Therefore, this review focuses on recent developments in molecular biology and phylogenetic reconstruction relating to entomopathogenic nematodes, and the impact of these developments on using nematodes for biological control of insect pests.

GENETIC DIVERSITY

Surveys using Galleria traps (11) have shown that entomopathogenic nematodes are distributed worldwide (6, 7, 8, 21, 52–54, 57, 68, 76, 78, 79, 102, 114). High degrees of natural genetic variability have been observed among both species and strains or isolates.

The concept of an "isolate" is generally used to describe a nematode population recovered from a particular environmental habitat or a host. Nematode isolates also have been referred to as strains. The concept of a "strain" is used to refer to a group of organisms having characteristic properties within a species.

The individuals of a strain have certain characteristics in common but different from other strains, such as virulence, persistence, and host-finding ability. Genetic differences among strains may correlate with differences that are biologically important. The logic behind this argument is that populations within species may be isolated from each other. Therefore, they should gradually diverge through genetic drift and natural selection by exposure to particular hosts and local environment (55).

In modern nematology, numerous species concepts have been proposed and advocated for general use in species delimitation. Of these, the Linnean, biological, evolutionary, and phylogenetic species concepts are prominent in the substantial literature on the subject (2). A review of taxonomic literature over the last two decades suggests that most nematode taxonomists are operating within the Linnean system: The traditional phenetic species concept combining morphology and morphometric still holds sway, regardless of discipline (2, 59). However, Linnean and biological species concepts are more prone to serious systematic error than are evolutionary or phylogenetic species concepts (2). Estimating that at least 100 species exist throughout the world is reasonable. Defining and measuring genetic diversity is fundamental to understanding identification, geographical distribution, and habitat specificity. Molecular approaches show particular potential where traditional techniques lack the necessary precision to discriminate closely related taxa. The seminal value of molecular techniques is, in the absence of fossil nematodes of any geologically significant age, likely to prove particularly potent in elucidating phylogenetic relationships (59). As an alternative to the current paradigm, an amalgamation of evolutionary and phylogenetic species concepts has been advocated with a set of discovery operations designed to minimize the risk of making systematic errors. These operations have been applied to species and isolates of *Heterorhabditis* (2).

Genome Size and Complexity

The genome of eucaryotes consists of a single-copy DNA sequence interspersed with arrays of repeated sequences. The genome size and the repetitive DNA diversity are fundamental information for molecular genetic investigations (45). Genome sizes are estimated at 2.3×10^8 bp in *S. carpocapsae* and 3.9×10^7 bp in *H. bacteriophora.* Repetitive DNA content represents 39% and 51% of these respective genomes (45). The highly repetitive components are similar in proportion for both entomopathogenic nematodes. Sequences such as satellite DNA have been described in several *Heterorhabditis* and *Steinernema* species (44, 46). The satellite DNA has been found to represent between 5% and 10% of their genomes, respectively. These results could account for the genome differences, both in size and complexity, between *H. bacteriophora* and *S. carpocapsae.* The ability of entomopathogenic nematodes to produce enormous populations in a short period is likely reinforced by the small size of their genome.

Measures of Genetic Diversity

Advances in the polymerase chain reaction (PCR), DNA sequencing, and data analysis have resulted in powerful techniques offering the promise of nearly unlimited loci for measuring and analyzing genetic diversity in entomopathogenic nematodes. Randomly amplified polymorphic DNA (RAPD) involves the use of a single arbitrary primer in a PCR reaction and results in the amplification of several discrete DNA products (112, 113). Each product is derived from a region of the genome that contains two short segments in inverted orientation, on opposite strands, that are complementary to the primer and sufficiently close together for the amplification to work. The amplified products are separated on agarose gels in the presence of ethidium bromide and visualized under ultraviolet light.

Several arbitrary primers have been used to generate RAPDs for 11 *Steinernema* isolates (66). Reproducible RAPDs from the 11 isolates showed clear inter- and intraspecific polymorphism, and different species and isolates could be separated easily. Combining RAPDs with morphological characters, a new species was characterized (66) and described (70). RAPDs were also used to examine genetic relatedness between the genera *Heterorhabditis* and *Steinernema* (49). These studies demonstrated the feasibility of RAPDs for the assessment of genetic diversity in entomopathogenic nematodes. However, it is now widely recognized that maintaining consistent reaction conditions for reproducible RAPDs is absolutely essential. In assessing the influence of PCR buffer compositions, adjunct additions, temperature cycling programs, and template DNA concentrations on RAPDs, Liu & Berry (67) found that these conditions profoundly affected the number of RAPDs.

Applying RAPD to measure genetic diversity in entomopathogenic nematodes has the advantage because no prior sequence information is required. RAPDs can be used as a first screen to determine genetic variation within new unidentified nematode isolates, and then other molecular techniques, as well as morphological characters, should be used subsequently to confirm the genetic diversity in entomopathogenic nematodes.

Restriction fragment length polymorphisms (RFLPs) involves the digestion of genomic DNA with restriction endonucleases, which generates a unique set of different-size DNA restriction fragments depending upon the DNA sequences. RFLPs are the most reliable polymorphisms that can be used for accurate scoring of genotypes. Also, RFLPs are co-dominant and can identify a unique locus. RFLPs can be used to distinguish a number of *Heterorhabditis* and *Steinernema* species as well as different populations (26, 27, 98). Furthermore, the entire ribosomal DNA repeat unit of a *Steinernema* species has been cloned (99, 100). These clones have been used to identify recognized species and to determine whether length heterogeneities exist within the rDNA repeat unit for any of the five RFLP types from a UK survey (52).

Although RFLPs are valuable in preliminary studies, the relatively large quantities of DNA required and RFLP analysis is labor intensive and time consuming

(30). Amplified fragment length polymorphism (AFLP) is based on PCR amplification of restriction fragments generated by specific restriction enzymes and oligonucleotide primers. Restricted digestions of amplified DNA from the mitochondrial genome, the nuclear ribosomal internal transcribed spacer (ITS) region, and 26S ribosomal DNA (rDNA) regions have been evaluated as genetic markers for entomopathogenic nematode species and isolates (62, 83, 101). The RFLPs of PCR-amplified ITS and 26S regions provided specific banding patterns for all nematode isolates examined except *S. feltiae* and *S. glaseri* (83). The PCR-based RFLP method allows the rapid categorization of different *Heterorhabditis* species based on an analysis of amplified DNA from individual infective juveniles. Using this method, six RFLP profiles have been identified (63). The ITS region of 19 *Steinernema* isolates belonging to 17 species was amplified with PCR and the PCR products were digested with 17 different restriction endonucleases (101).

Satellite DNA sequences are usually species specific, perhaps because this DNA evolves at a very high rate (58). An *Hae* III satellite DNA family has been cloned from *S. carpocapsae* and the DNA sequences of 13 monomers have been determined (44). The *Hae* III satellite DNA sequence was found to be specific to *S. carpocapsae* because there was no cross-hybridization to the closely related species. Three satellite DNAs isolated from *S. carpocapsae*, *H. bacteriophora*, and *H. indicus* have hybridization signals with only their own populations. A species-specific satellite DNA from *H. indicus* has been further described (1). Hybridization of genomic DNA from other *Heterorhabditis* species showed that this satellite DNA sequence is specific to the *H. indicus* genome and could provide a rapid and powerful tool for identifying *H. indicus* strains. However, satellite DNA analysis does not guarantee solutions to all identification problems of closely related species (44, 45).

DNA sequences have been used to describe genetic diversity in entomopathogenic nematodes (3, 72–74). One major advantage of DNA sequence data is the large number of potential characters available for inferring relationships (80, 81). Partial 18S rRNA gene sequences were obtained from 17 species of *Heterorhabditis* and *Steinernema*, revealing more sequence divergence in *Steinernema* species than in *Heterorhabditis* species (73). DNA sequences of the ITS region of the ribosomal tandem repeating unit showed a high degree of variability among six described *Heterorhabditis* species (3). Thus DNA sequences of the ITS region may be a reliable source of homologous characters for resolving genetic diversity among closely related species and isolates. Partial *ND4* gene sequences of mtDNA from 15 *Heterorhabditis* isolates, representing five species collected from different regions of the world, have been determined using polymerase chain reaction (PCR) and direct sequencing of PCR products. Aligned nucleotides as well as amino acid sequences were used to differentiate nematode species by comparing sequence divergence (72). Seven distinct haplotypes were detected among 15 isolates examined. The distribution of these haplotypes within the species revealed that *H. indicus*, *H. megidis*, and *H. marelatus* were polymorphic. The DNA sequences of the five *H. bacteriophora* isolates were identical. The species

H. bacteriophora, H. indicus, H. megidis, and H. marelatus can be clearly distinguished. The *ND4* gene sequences of the species *H. hepialius* were identical to that of *H. marelatus* OH10 isolate (72).

Morphological characters have long been used to identify families, genera, and species in entomopathogenic nematodes. Moreover, morphological as well as biological traits have been studied in genetics and genetic improvements (39, 87). Levels of variability can be estimated for morphological characters, and their response to environmental conditions and their genetic background can be determined. However, the identification of entomopathogenic nematodes by standard morphological criteria alone is rarely straightforward (54). Most newly described species include molecular characters in their descriptions (24, 34, 61, 69, 70, 103–106, 109, 111).

Genotype Distribution

It is now clear that entomopathogenic nematodes can exist as strains that differ biologically (55, 56) and as genotypes that differ genetically (15, 16, 29, 31, 66, 72). A number of molecular techniques have been used to identify different genotypes and to describe their distribution and population structure.

Europe is the most intensively surveyed region for entomopathogenic nematodes (56). Five distinct RFLP types were determined from 89 *Steinernema* isolates collected throughout the United Kingdom (98). Prevalence of the five types was not equal, and no association with geographical location, soil type, or habitat was detected. The most prevalent group, identified as *S. feltiae*, contained two of the RFLP types.

The few genotypes found in the United Kingdom may be the result of dispersal of nematodes by infected adult insects before the insects die, thus allowing the introduction of nematodes to a new location. In a separate study, RFLP analysis identified all 76 *Heterorhabditis* isolates recovered from Ireland and Britain as belonging to the Irish group of *Heterorhabditis* (48). PCR products of the ITS region, upon restriction enzyme digestion, identified nine *S. feltiae* isolates as one RFLP type from West Flanders, Belgium (79). RFLP types have been used to describe entomopathogenic nematodes isolated from other regions: Two *Heterorhabditis* and three *Steinernema* RFLP types were discovered in Sri Lanka (8).

Population Structure and Species Delimitation

Population structure provides the most fundamental information for reliable identification of species and unique genetic variants that could be further assessed as potential biological control agents. To describe the genotype distribution of *H. marelatus* along the coast of Oregon and California, the *ND4* gene of mtDNA from a total of 58 nematode isolates was sequenced (14, 15). Only four distinct genotypes were found, and no association with geographical location, soil type, or habitat was detected. Predicted amino acid sequences of the partial *ND4* gene revealed that two haplotypes of the species *H. marelatus* had identical amino acid sequences (72). Movement of the nematodes from one location to another may be one reason why only four genotypes were found. The movement of these nematodes may be more extensive than supposed (98). Ocean currents may play a role in dispersing entomopathogenic nematodes. As predicted, *H marelatus* shows much lower mtDNA diversity within populations and over the species as a whole, and has a much more strongly subdivided population structure. These data suggest that it may be fruitful to search for useful new nematode strains over a small geographic area (15, 16).

Morphologically, *S. oregonense* (70, 76) and *S. feltiae* are very similar, and they fall into the same group in phylogenetic reconstruction (71, 73). Hybridization tests revealed that the males and the females of the two species were mutually attracted and capable of mating. The eggs were fertilized and embryonization was initiated. However, in all cases, approximately 12 hours after development initiated, DNA was rejected from the embryos and the developing eggs eventually disintegrated (GO Poinar, unpublished observation). DNA sequencing analysis of the ITS region revealed only two nucleotide differences (J Liu, unpublished observation). While these results confirmed the distinctness of the two species, they also showed a high degree of relatedness.

PHYLOGENETIC RECONSTRUCTION

Phylogenetic reconstruction is a necessary component of the comparative method in evolutionary biology and provides a critical ingredient for studies of gene flow, population structure, biogeography, coevolution, coadaptation, cospeciation, and historical ecology (2, 3, 22, 80, 82). Different nematode species and strains exhibit differences in survival, infectivity, and efficacy against particular insect pests. Phylogenetic information is critical to rational implementation and monitoring programs when entomopathogenic nematodes are used as biological control agents. Recent advances in molecular systematics facilitate rigorous analysis of the history of the association by providing mutually independent phylogenies for each of them.

Data Types

Several data sets have been used to infer phylogenetic relationships among entomopathogenic nematodes, including morphological characters (71, 94), the RAPDs and RFLPs (71, 101), and DNA sequencing data (3, 72, 73). Using 53 small subunit ribosomal DNA sequences from a wide range of nematodes, researchers have revealed phylogenetic relationships of entomopathogenic nematodes with other nematodes (9).

A range of morphological characters has been used for the characterization and separation of entomopathogenic nematode species (54, 85, 86, 92). Generally, estimating phylogenetic relationships among entomopathogenic nematode species based on morphological characters is difficult because most of the characters are quantitative and continuous (71). The divergence of RAPDs among species has been great, and thus may provide little information about phylogeny. DNA sequencing data from nuclear ribosomal RNA genes are very useful for resolving genus- and family-level relationships in nematodes, but often fail to adequately resolve relationships among closely related congeners (9, 14, 73). *ND4* gene sequences of mtDNA may provide better information on *Heterorhabditis* phylogeny than the ITS region provides (72).

Interpretive Tools

Morphological characters as well as RAPDs and RFLPs are coded to establish data matrices. After coding, phylogenetic analysis is performed (71, 101). Discrete coding of continuous characters and ratios for entomopathogenic nematodes is difficult (71). For DNA sequencing data, multiple sequence alignment can be performed with a number of computer programs based on pairwise genetic distance and parsimony criteria with a variety of sequence alignment parameters (3, 109). An alignment is a statement of homology between particular nucleotides in a set of sequence data and explicitly dictates the tree topology derived from it (22, 23). Thus, using a repeatable alignment method is critical. When gaps are introduced and extension penalties are manipulated with the alignment of the ITS1 region, even small perturbations of the data set can result in different topological arrangement (3). No gaps have been introduced for the alignment of the *ND4* gene. Thus, more reliable sequence alignment can be obtained from the *ND4* gene than from the ITS1 region (72).

Maximum parsimony methods and maximum likelihood methods (35, 106) are extensively used for phylogenetic reconstruction in entomopathogenic nematodes (3, 71–73). Both maximum parsimony and maximum likelihood methods are used in phylogenetic analyses for a same data set because the degree of agreement between different methods in phylogenetic analyses may be an indicator of a reliable phylogenetic reconstruction. However, this is an issue to debate (23, 72). Other methods, such as distance matrix methods, are also used in phylogenetic analysis (3, 101).

Phylogenies and Their Congruency

The available morphological, distributional, physiological, and biological evidence show that the genera *Heterorhabditis* and *Steinernema* are widely divergent and have completely separate origins (i.e. they are polyphyletic) (94). This conclusion has been confirmed by two independent phylogenetic reconstructions with DNA sequencing data (9, 73).

Phylogenetic relationships among *Heterorhabditis* species have been examined with DNA sequencing data from the ITS region of the ribosomal DNA (3) and *ND4* gene of mtDNA (72). Both phylogenetic analyses provide a clear assignment of well-described species defined by morphological characters into different clades (3, 72). However, the phylogenetic relationships among some species are less well supported. Phylogenetic relationships among *Steinernema* species appear less resolved because extensive DNA sequencing work has yet to be done. However, a certain degree of congruency has been observed between the phylogeny based on morphological characters (71) and the phylogeny based on RFLPs molecular data (101). Use of *C. elegans* as an outgroup species in our phylogenetic analysis seems acceptable because of its phylogenetic relationship to the species *Heterorhabditis bacteriophora* (9). Applying phylogenetics to the taxonomy of the entomopathogenic nematodes will lead to more accurate identification and provide testable means of assessing relationships of pathogenic and nonpathogenic forms.

MULTITROPHIC INTERACTIONS

Entomopathogenic nematodes are characterized by mutualistic relationships with symbiotic bacteria. The nematodes provide protection and transportation for their bacterial symbionts. The bacterial symbionts contribute to the mutualistic relationship by killing insect hosts, by establishing and maintaining suitable conditions for nematode reproduction, and by providing nutrients and antimicrobial substances that inhibit growth of a wide range of microorganisms. Interaction among entomopathogenic nematodes, symbiotic bacteria, and insect hosts may be one of the principal forces driving diversification and specialization in these organisms. Understanding of these multitrophic interactions among the nematodes, their symbiotic bacteria, and the insect hosts is of fundamental importance for nematode infectivity, mass reproduction, and registration as biological control agents.

Mutualistic Specificity

Taxonomic studies of entomopathogenic nematodes and their symbiotic bacteria have demonstrated that almost every nematode species possesses a specific symbiotic bacteria species (17–19, 33, 75). However, the nature of the nematode-bacterium symbiosis has not been precisely defined. Generally, bacteria in the genera *Photorhabdus* and *Xenorhabdus* are considered the main symbionts of the nematodes and primarily responsible for the death of the insect hosts and the development of the nematodes within the insect cadaver. However, a number of other bacterial species have been isolated from the nematodes (4, 60, 77), and entomopathogenic nematodes can develop and reproduce on a number of other bacterial species (5, 60, 97). These findings appear to be laboratory artifacts because the other bacterial species have not been linked to field-collected nematodes (18).

Partial 16S rRNA gene sequences of the bacteria from different stages of the individual nematodes *H. marelatus* and *S. oregonense* have been determined, and only symbiotic bacteria in the genera *Photorhabdus* and *Xenorhabdus* were detected. DNA sequencing data also indicates that the symbiotic bacteria associated with *H. marelatus* and *S. oregonense* are unique species (75; J Liu, unpublished information). A new mechanism for the maintenance of the specificity of the association between the entomopathogenic nematode *S. scapterisci* and its symbiotic bacteria has been proposed (47). It was found that the noninfective stages of *S. scapterisci* reproduced in monoxenic cultures with different symbiotic bacteria species. The infective juveniles produced in such cultures were also highly pathogenic to *Galleria mellonella* larvae, indicating the retention of other symbiotic bacteria. However, a strong symbiotic-specific "dauer" recovery demonstrates a remarkably fine level of recognition of the closely related bacteria by the infective juveniles (47).

Mutualistic History

Mutualistic symbioses between distantly related organisms have generated major innovations in the evolution of biological complexity (32, 88–90). Molecular data provide a common yardstick for comparing phylogenies in taxonomically disparate groups, such as entomopathogenic nematodes and their symbiotic bacteria (74). Comparison of the topology of the symbiotic bacteria phylogeny with the phylogeny of the nematodes revealed a congruent branching pattern, suggesting a long history of symbioses (74). However, incongruence between the phylogenies was also found. Shifts among the association are one cause of the incongruence. The incongruence can also arise from a number of other causes, such as the presence of multiple lineages of parasites coupled with bacteria extinction, failure of the symbiotic bacteria to colonize the nematodes, or collection failure (88, 89). Co-speciation between the nematodes and their bacteria is an excellent example of how symbionts, often unseen or overlooked microorganisms, can underlie the preservation of more obvious and much-studied macrobiota (25, 32).

Host Preference

It is now apparent from molecular and phylogenetic analyses that entomopathogenic nematode species fall into various physiomorphs, defined as one or more species sharing certain morphological and physiological characters expressed by a degree of genetic relatedness. Such physiological characters are climatic preference and preference for another host. Another could be edaphic preference.

As an example, the tropical–warm temperate physiomorphs of *Steinernema,* including *S. glaseri, S. cubanum, S. puertoricense*, and *S. arenaria,* have infective-stage juveniles averaging slightly more than one mm in length. It is apparent that this physiomorph is preferentially adapted to locate and develop on larvae of various Coleoptera, especially members of the family Scarabaeidae. For the biological control of specific insects—especially scarabaeid larvae—in a warm to hot climate, nematodes from this physiomorph would be the choice. If one species was ineffective due to unknown causes, another from this physiomorph could be selected and tested. As more information becomes available about edaphic and micro-climatic preferences, the selection process could become more precise.

GENETIC IMPROVEMENT

In the past decade entomopathogenic nematodes have been established as an important biological control agent among the commercially available products. However, the markets for entomopathogenic nematodes are relatively small and the target pests are of limited importance (40). Genetic improvement of the nematodes has been based on selective breeding, mutation, or genetic engineering (43). Molecular biology has provided fundamental knowledge and technical protocols for the improvement of entomopathogenic nematodes by genetic engineering (41). Molecular markers to study gene expression have been found. Several genes that encode for useful traits have been identified, and efficient transformation methods have been developed. Although many countries have regulations on the release of genetically engineered organisms, the strategy to receiving approval of the release may include inducing a commercial rather than ecological gene into the nematodes and overexpressing an exiting gene rather than introducing a foreign one.

Evolutionary Aspects

To improve entomopathogenic nematodes, their genetics, including linkage maps, should be elaborated. In principle, any kind of DNA fragments of *Caenorhabditis elegans* that can be hybridized with entomopathogenic nematode DNA can be used as a molecular probe for mapping chromosomes (36). However, it is better to use a well-conserved gene of known function and location in the genome of *C. elegans*.

Identification of Genes

Most genes currently available for insertion into entomopathogenic nematodes originate from *C. elegans* or *D. melanogaster* (41). Two heat-shock protein genes from *C. elegans* have been transferred to entomopathogenic nematodes (41, 50). PCR and RFLP analyses of heat-inducible *hsp70* genes in entomopathogenic nematodes demonstrated a putative homology with the *C. elegans hsp70* A gene, thus indicating its evolutionary conservation among different nematode species (50).

Molecular Markers

Several molecular markers are available for monitoring gene expression and protein localization within cells (41). The *E. coli lacZ*-encoded enzyme, β-

galactosidease has been used as a reporter molecule for studies of gene expression in entomopathogenic nematodes (41). However, this reporter system requires exogenously added substrates and cofactors that kill the cells or organism. The jellyfish green fluorescent protein *gfp* gene has been used as a marker for gene expression in entomopathogenic nematodes (51).

Despite various attempts to utilize the powerful molecular tools for the genetic improvement of entomopathogenic nematodes, recombinant DNA technology offers no panacea (41, 93). None of the genetically improved nematode strain has demonstrated a continued improvement, and none is being used in biological control programs today. However, there are no substantial technological, regulating, and implementing barriers to genetic engineering of the nematodes. Further progress in the areas of molecular biology and biochemistry will probably identify genes and develop traits such as shortening life cycle, enhancing progeny production, extending temperature tolerances, and altering host or habitat preferences (41). Highly efficient techniques will also be developed for the nematode transformation.

CONCLUSIONS

An important aspect of using entomopathogenic nematodes concerns identification of strains and species of the nematodes and their symbiotic bacterial associates. Identification of species will continue to be based on morphological criteria, supplemented by molecular characters. At the level of sibling species and intraspecific groups, molecular methods will come into their own (55). Intraspecific differences in molecular analyses may reflect different biological characters, and these may provide foundation populations for selecting worthwhile characters. Molecular characterization may provide an initial screen to identify useful nematode strains. It is more efficient to bioassay populations that differ genetically than to randomly test a number of different isolates, many of which are identical (55). Molecular biology contributed significantly to elucidating population genetic structure and identifying isolates and species of entomopathogenic nematodes. Several genotypic differences among *Heterorhabditis* isolates, as determined by RFLP or *ND4* gene sequence analyses, were associated with biological differences in their ability to control strawberry root weevils or Colorado potato beetles (13, 31, 72).

It is apparent that entomopathogenic nematodes can be separated into several phylogenetic groups corresponding to the phylogenetic groups of the hosts. Both genera contain species that have preferred climatic and geographical ranges, many of which still have to be realized. Of greatest significance is the apparent finding that identification and phylogenetic relationships inferred from molecular data are basically congruent with those based on morphological and distributional data.

Furthermore, a congruent phylogenetic pattern between entomopathogenic nematodes and their symbiotic bacteria suggests a long history of symbioses, thus providing new information for understanding multitrophic interactions among insect hosts, entomopathogenic nematodes, and the symbiotic bacteria. Different symbiotic bacteria may produce different toxins that act on the insect immune system (13). Several toxins from *P. luminescens* have been identified (20). These toxins represent potential alternatives to *Bacillus thuringiensis* toxins for transgenic plants. Molecular biology and phylogenetic reconstruction approaches will assist in the development of future studies concerning species concepts and population structure, as well as better strategies for their use as biological control agents against insect pests.

Of the molecular approaches available, DNA sequencing provides the most prolific and dependable data for studying population structure, identification, and phylogenetic reconstruction. Because the phylogenetic information contained in different genes can vary, analysis methods must be carefully matched to the particular taxonomic level or question under investigation (23, 80–82). The *ND4* gene of mtDNA and ITS1 region of ribosomal DNA should be useful for determining relationships among closely related species. Nuclear ribosomal RNA gene sequences are useful for resolving generic level relationships in nematodes. It is obvious that the rate of speciation and morphological differentiation differs between the members of the two genera. The degree of DNA sequence divergence of the same gene can be very different between *Heterorhabditis* and *Steinernema*. For *Heterorhabditis* species, both *ND4* gene and ITS region DNA sequences provided a clear species delimitation and supported the following species recognition: *H. bacteriophora* (91), *H. megidis* (95), *H. zealandica* (92), *H. indicus* (96), and *H. marelatus* (69).

Entomopathogenic nematodes are one of the most important biological agents for insect pests. However, control of insect pests with entomopathogenic nematodes has not achieved its mission of reducing the use of chemical pesticides to a significant degree. Molecular biology contributed significantly in elucidating population genetic structure and identification of isolates and species and of entomopathogenic nematodes. Phylogenetic reconstruction played an important role in illuminating systematics, cospeciation, and coadaptation of entomopathogenic nematodes and their symbiotic bacteria. Accurate identification and systematic information are fundamental to rational implementation when entomopathogenic nematodes are used as biological control agents. Genetic engineering based on the techniques from molecular biology and genetics has shown promise for the rational improvement of entomopathogenic nematodes to overcome limitations related to inadequate efficacy, stability, and economics to realize their full pest control potential. There is every reason to believe that the use of entomopathogenic nematodes for insect pest control will increase substantially in the near future.

ACKNOWLEDGMENTS

We thank Frank Rodovsky, Michael Blouin, Noel Boemare, Alison Moldenke, and Hans Luh for valuable comments and discussions.

Visit the Annual Reviews home page at www.AnnualReviews.org.

LITERATURE CITED

1. Abadon M, Grenier E, Laumond C, Abad P. 1998. A species specific satellite DNA from the entomopathogenic nematode *Heterorhabditis indicus*. *Genome* 41:148–53
2. Adams BJ. 1998. Species concepts and the evolutionary paradigm in modern nematology. *J. Nematol.* 30:1–21
3. Adams BJ, Burnell AM, Powers TO. 1998. A phylogenetic analysis of *Heterorhabditis* (Nemata: Rhabditidae) based on internal transcribed spacer 1 DNA sequence data. *J. Nematol.* 30:22–29
4. Aguillera MM, Hodge NC, Stall RE, Smart GC Jr. 1993. Bacterial symbionts of *Steinernema scapterisci*. *J. Invertebr. Pathol.* 62:68–72
5. Aguillera MM, Smart GC Jr. 1993. Development, reproduction, and pathogenicity of *Steinernema scapterisci* in monoxenic culture with different species of bacteria. *J. Invertebr. Pathol.* 62:289–94
6. Akhurst RJ, Bedding RA. 1986. Natural occurrence of insect pathogenic nematodes (Steinernematidae and Heterorhabditidae) in soil in Australia. *J. Aust. Entomol. Soc.* 25:241–44
7. Akhurst RJ, Brooks WM. 1984. The distribution of entomophilic nematodes (Heterorhabditidae and Steinernematidae) in North Carolina. *J. Invertebr. Pathol.* 44:140–45
8. Amarasinghe LD, Hominick WM, Briscoe BR, Reid AP. 1994. Occurrence and distribution of entomopathogenic nematodes (Rhabditida: Heterorhabditidae and Steinernematidae) in Sri Lanka. *J. Helminthol.* 68:277–86
9. Baxter ML, DeLey P, Garey JR, Liu LX, Scheldeman P, et al. 1998. A molecular evolutionary framework for the phylum Nematoda. *Nature* 392:71–75
10. Bedding RA. 1984. Large scale production, storage, and transport of the insect-parasitic nematodes *Neoaplectana* spp. and *Heterorhabditis* spp. *Ann. Appl. Biol.* 104:117–20
11. Bedding RA, Akhurst RJ. 1975. A simple technique for the determination of insect parasitic rhabditid nematodes in soil. *Nematologica* 21:109–10
12. Bedding RA, Akhurst RJ, Kaya HK. 1993. *Nematodes and the Biological Control of Insect Pests*. East Melbourne, Victoria, Aust.: CSIRO
13. Berry RE, Liu J, Reed G. 1997. Comparison of endemic and exotic entomopathogenic nematode species for control of Colorado potato beetle (Coleoptera: Chrysomelidae). *J. Econ. Entomol.* 90:1528–33
14. Blouin MS. 1998. Mitochondrial DNA diversity in nematodes. *J. Helminthol.* 72:285–89
15. Blouin MS, Liu J, Berry RE. 1999. Life cycle variation and the genetic structure of nematode populations. *Heredity* 83:253–59
16. Blouin MS, Yowell CA, Courtney CH, Dame JB. 1998. Substitution bias, rapid saturation, and the use of mtDNA for nematode systematics. *Mol. Biol. Evol.* 15:1719–27
17. Boemare N, Givaudan A, Brehelen M,

Laumond C. 1997. Symbioses and pathogenicity of nematode-bacterium complexes. *Symbiosis* 22:21–45

18. Boemare NE. 1995. Controversial interpretations on the symbiotic relationships in *Steinernema* and *Heterorhabditis* nematode bacterium complexes. In *Abstr. 2nd Int. Symp. Entomopathogenic Nematodes Their Symbiotic Bacteria*, Honolulu, HI: Univ. Hawaii Manon Campus
19. Boemare NE, Akhurst RJ, Mourant RG. 1993. DNA relatedness between *Xenorhabdus* spp. (Enterobacteriaceae), symbiotic bacteria of entomopathogenic nematodes, and a proposal to transfer *Xenorhabdus luminescens* to a new genus, *Photorhabdus* gen. nov. *Int. J. Syst. Bacteriol.* 43:249–55
20. Bowen D, Rocheleau TA, Blackburn M, Andreev O, Golubeva E, et al. 1998. Insecticidal toxins from the bacterium *Photorhabdus luminescens. Science* 280:2129–32
21. Briscoe BR, Hominick WM. 1990. Occurrence of entomopathogenic nematodes (Rhabditida: Steinernematidae and Heterorhabditidae) in British soils. *Parasitology* 100:295–302
22. Brower AVZ. 1994. Phylogeny of *Heliconius* butterflies inferred from mitochondrial DNA sequences (Lepidoptera: Nymphalidae). *Mol. Phylogenet. Evol.* 3:159–74
23. Brower AVZ, DeSalle R. 1994. Practical and theoretical considerations for choice of a DNA sequence region in insect molecular systematics, with a short review of published studies using nuclear gene regions. *Ann. Entomol. Soc. Am.* 87:702–16
24. Cabanillas HE, Poinar GO Jr, Raulston RJ. 1993. *Steinernema riobravis* n. sp. (Rhabditida: Steinernematidae) from Texas. *Fundam. Appl. Nematol.* 17:123–31
25. Chapela IH, Rehner SA, Schulz TR, Mueller UG. 1994. Evolutionary history of the symbiosis between fungus-growing ants and their fungi. *Science* 266:1691–97
26. Chitwood BG, Chitwood MB. 1937. *An Introduction to Nematology.* Baltimore, MD: Monumental Printing
27. Curran J. 1990. Molecular techniques in taxonomy. See Ref. 42, pp. 63–74
28. Curran J. 1991. Application of DNA analysis to nematode taxonomy. In *Manual of Agricultural Nematology*, ed. WR. Nickle, pp. 125–43. New York: Marcel Dekker
29. Curran J, Baillie DL, Webster JM. 1985. Use of genomic DNA restriction fragment length differences to identify nematode species. *Parasitology* 90:137–44
30. Curran J, Robinson MP. 1993. Molecular aid to nematode diagnosis. In *Plant Parasitic Nematodes in Temperate Agriculture*, ed. K Evans, DL Trudgill, JM Webster, pp. 545–64. Wallingford, UK: CAB Int.
31. Curran J, Webster JM. 1989. Genotypic analysis of *Heterorhabditis* isolates from North Carolina. *J. Nematol.* 21:140–45
32. Douglas AE. 1997. Parallels and contrasts between symbiotic bacteria and bacterial-derived organelles: evidence from *Buchnera*, the bacterial symbiont of aphids. *FEMS Microbiol. Ecol.* 24:1–9
33. Ehlers RU, Nieman I. 1998. Molecular identification of *Photorhabdus luminescens* strains by amplification of specific fragments of the 16S ribosomal DNA. *Syst. Appl. Microbiol.* 21:509–19
34. Elawad S, Ahmad W, Reid AP. 1997. *Steinernema abbasi* sp. n. (Nematoda: Steinernematidae) from the Sultanate of Oman. *Fundam. Appl. Nematol.* 20:435–42
35. Felsenstein J. 1995. PHYLIP: Phylogenetic inference package, version 3.572. Seattle, WA: Univ. Washington
36. Fodor A. 1997. Isolation of EPN homologues of *C. elegans* genes. In *COST 819 Biotechnology: Genetic and molecular biology of entomopathogenic nematodes, Proc. Symp. Workshop,* ed. P Abad, AM

Burnell, C Laumond, pp. 59–74. Luxembourg: Eur. Comm.

37. Forst S, Dowds B, Boemare N, Stackebrandt E. 1997. *Xenorhabdus* and *Photorhabdus* spp: Bugs that kill bugs. *Annu. Rev. Microbiol.* 60:47–72
38. Forst S, Nealson K. 1996. Molecular biology of the symbiotic-pathogenic bacteria *Xenorhabdus* spp. and *Photorhabdus* spp. *Microbiol. Rev.* 60:21–43
39. Friedman MJ. 1990. Commercial production and development. See Ref. 42, pp. 153–72
40. Gaugler R, Glazer I, Campbell JF, Liran N. 1993. Laboratory and field evaluation of an entomopathogenic nematode genetically selected for improved host-finding. *J. Invertebr. Pathol.* 63:68–73
41. Gaugler R, Hashmi S. 1996. Genetic engineering of an insect parasite. *Genet. Eng.* 18:135–55
42. Gaugler R, Kaya HK, eds. 1990. *Entomopathogenic Nematodes in Biological Control.* Boca Raton, FL: CRC
43. Gaugler R, McGuire T, Campbell J. 1989. Genetic variability among strains of the entomopathogenic nematode *Steinernema feltiae. J. Nematol.* 21:247–53
44. Grenier E, Bonifassi E, Abad P, Laumond C. 1996. Use of species-specific satellite DNAs as diagnostic probes in the identification of Steinernematidae and Heterorhabditidae entomopathogenic nematodes. *Parasitology* 113:483–89
45. Grenier E, Catzeflis FM, Abad P. 1997. Genome sizes of the entomopathogenic nematodes *Steinernema carpocapsae* and *Heterorhabditis bacteriophora* (Nematoda: Rhabditida). *Parasitology* 114:497–501
46. Grenier E, Laumond C, Abad P. 1995. Characterization of a species-specific satellite DNA from the entomopathogenic nematode *Steinernema carpocapsae. Mol. Biochem. Parasitol.* 69:93–100
47. Grewal PS, Matsuura M, Converse VP. 1997. Mechanisms of specificity of association between the nematode *Steinernema scarpterisci* and its symbiotic bacterium. *Parasitology* 114:483–88
48. Griffin CT, Joyce SA, Dix I, Burnell AM, Downes MJ. 1994. Characterization of the entomopathogenic nematode *Heterorhabditis* (Nematoda: Heterorhabditidae) from Ireland and Britain by molecular and cross-breeding techniques, and the occurrence of the genus in these islands. *Fundam. Appl. Nematol.* 17:245–54
49. Hashmi G, Glazer I, Gaugler R. 1996. Molecular comparisons of entomopathogenic nematodes using randomly amplified polymorphic DNA (RAPD) markers. *Fundam. Appl. Nematol.* 19:399–406
50. Hashmi G, Hashmi S, Selvan S, Grewal P, Gaugler R. 1997. Polymorphism in heat shock protein gene (*hsp*70) in entomopathogenic nematodes (Rhabditida). *J. Therm. Biol.* 22:143–49
51. Hashmi S, AbuHatab MA, Gaugler RR. 1997. GFP: green fluorescent protein as a versatile gene marker for entomopathogenic nematodes. *Fundam. Appl. Nematol.* 20:323–27
52. Hominick WM, Briscoe BR. 1990. Occurrence of entomopathogenic nematodes (Rhabditida: Steinernematidae and Heterorhabditidae) in British soils. *Parasitology* 100:295–302
53. Hominick WM, Briscoe BR. 1990. Survey of 15 sites over 28 months for entomopathogenic nematodes (Rhabditida: Steinernematidae). *Parasitology* 100:289–94
54. Hominick WM, Briscoe BR, Del Pino FG, Heng JA, Hunt DJ, et al. 1997. Biosystematics of entomopathogenic nematodes: current status, protocols and definitions. *J. Helminthol.* 71:271–98
55. Hominick WM, Reid AP. 1990. Perspectives on entomopathogenic nematology. See Ref. 42, pp. 327–45
56. Hominick WM, Reid AP, Bohan DA, Briscoe BR. 1996. Entomopathogenic nematodes: biodiversity, geographical

distribution and the convention on biological diversity. *Biocontr. Sci. Tech.* 6:317–31

57. Hominick WM, Reid AP, Briscoe BR. 1995. Prevalence and habitat specificity of steinernematid and heterorhabditid nematodes isolated during soil surveys of the UK and the Netherlands. *J. Helminthol.* 69:27–32
58. Hoy MA. 1994. *Insect Molecular Genetics, An Introduction to Principles and Applications*, p. 363. San Diego, CA: Academic
59. Hunt DJ. 1997. Nematode species: concepts and identification strategies exemplified by the Longidoridae, Steinernematidae and Heterorhabditidae. In *Species: The Units of Biodiversity*, ed. MF Claridge, HA Dawah, MR Wilson, pp. 221–45. London, UK: Chapman & Hall
60. Jackson TJ, Wang HY, Nugent M, Griffin CT, Burnell AM, Dowds MJ. 1995. Isolation of insect pathogenic bacteria, *Providencia rettgeri*, from *Heterorhabditis* spp. *J. Appl. Bacteriol.* 78:237–44
61. Jian H, Reid AP, Hunt DJ. 1997. *Steinernema ceratophorum* n. sp. (Nematoda: Steinernematidae), a new entomopathogenic nematode from north-east China. *Syst. Parasitol.* 37:115–25
62. Joyce SA, Burnell AM, Powers TO. 1994. Characterization of *Heterorhabditis* isolates by PCR amplification of segments of mtDNA and rDNA genes. *J. Nematol.* 26:260–70
63. Joyce SA, Reid AP, Driver F, Curran J. 1994. Application of polymerase chain reaction (PCR) methods to the identification of entomopathogenic nematodes. In *COST 812 Biotechnology: Genetics of entomopathogenic nematode-bacterium complexes, Proc. Symp. Workshop,* ed. AM Burnell, RU Ehlers, JP Masson, pp. 178–87. Luxembourg: Eur. Comm.
64. Kaya HK, Gaugler R. 1993. Entomopathogenic nematodes. *Annu. Rev. Entomol.* 38:181–206
65. Klein MG. 1990. Efficacy against soil-inhabiting insect pests. See Ref. 42, pp. 195–214
66. Liu J, Berry RE. 1995. Differentiation of isolates in the genus *Steinernema* (Nematoda: Steinernematidae) by random amplified polymorphic DNA fragments and morphological characters. *Parasitology* 111:119–25
67. Liu J, Berry RE. 1995. Determination of PCR conditions for RAPD analysis in entomopathogenic nematodes (Rhabditida: Heterorhabditidae and Steinernematidae). *J. Invertebr. Pathol.* 65:79–81
68. Liu J, Berry RE. 1995. Natural distribution of entomopathogenic nematodes (Rhabditida: Heterorhabditidae and Steinernematidae) in Oregon soils. *Environ. Entomol.* 24:159–63
69. Liu J, Berry RE. 1996. *Heterorhabditis marelatus* n. sp. (Rhabditida: Heterorhabditidae) from Oregon. *J. Invertebr. Pathol.* 67:48–54
70. Liu J, Berry RE. 1996. *Steinernema oregonensis* n. sp. (Rhabditida: Steinernematidae) from Oregon, U. S. A. *Fundam. Appl. Nematol.* 19:375–80
71. Liu J, Berry RE. 1996. Phylogenetic analysis of the genus *Steinernema* by morphological characters and randomly amplified polymorphic DNA fragments. *Fundam. Appl. Nematol.* 19:463–69
72. Liu J, Berry RE, Blouin MS. 1999. Molecular differentiation and phylogeny of entomopathogenic nematodes (Rhabditida: Heterorhabditidae) based on ND4 gene sequences of mitochondrial DNA. *J. Parasitol.* 85:709–15
73. Liu J, Berry RE, Moldenke AF. 1997. Phylogenetic relationships of entomopathogenic nematodes (Heterorhabditidae and Steinernematidae) inferred from partial 18s rRNA gene sequences. *J. Invertebr. Pathol.* 69:246–52
74. Liu J, Berry RE, Poinar GO Jr. 1998. Phylogenetic analysis of symbiosis between entomopathogenic nematodes and their bacteria. In *Proc. Abstr. COST*

819 Biotechnology: Bacterial symbionts and survival of entomopathogenic nematodes. Wellesbourne, UK: Horticulture Res. Int.

75. Liu J, Berry RE, Poinar GO Jr, Moldenke AF. 1997. Phylogeny of *Photorhabdus* and *Xenorhabdus* species and strains as determined by comparison of partial 16S rRNA gene sequences. *Int. J. Syst. Bacteriol.* 47:948–51
76. Liu J, Poinar GO Jr, Berry RE. 1998. Taxonomic comments on the genus *Steinernema* (Nematoda: Steinernematidae): Specific epithets and distribution record. *Nematologica* 44:321–22
77. Lysenko O. 1974. Bacteria associated with the nematode *Neoaplectana carpocapsae* and the pathogenicity of this complex for *Galleria mellonella* larvae. *J. Invertebr. Pathol.* 24:332–36
78. Mason JM, Razak AR, Wright DJ. 1996. The recovery of entomopathogenic nematodes from selected areas within Peninsular Malaysia. *J. Helminthol.* 70:303–7
79. Miduturi JS, Moens M, Hominick WM, Briscoe BR, Reid AP. 1996. Naturally occurring entomopathogenic nematodes in the province of West-Flanders, Belgium. *J. Helminthol* 70:319–27
80. Nadler SA. 1992. Phylogeny of some ascaridoid nematodes, inferred from comparison of 18S and 28S rRNA sequences. *Mol. Biol. Evol.* 9:932–44
81. Nadler SA. 1995. Advantages and disadvantages of molecular phylogenetics: a case study of ascaridoid nematodes. *J. Nematol.* 27:423–32
82. Nadler SA, Near T. 1995. Genetic structure of Midwestern *Ascaris suum* populations: a comparison of isoenzyme and RAPD markers. *J. Parasitol.* 81:385–94
83. Nasmith D, Speranzini RJ, Hubbes M. 1996. RFLP analysis of PCR amplified ITS and 26S ribosomal RNA genes of selected entomopathogenic nematodes (Steinernematidae, Heterorhabditidae). *J. Nematol.* 28:15–25
84. Nguyen KB, Smart GC Jr. 1994. *Noesteinernema longicurvicauda* n. gen., n. sp. (Rhabditida: Steinernematidae), a parasite of the termite *Reticulitermes flavipes* (Koller). *J. Nematol.* 26:162–74
85. Nguyen KB, Smart GC Jr. 1995. Morphometrics of infective juveniles of *Steinernema* spp, and *Heterorhabditis bacteriophora* (Nemata: Rhabditida). *J. Nematol.* 27:206–12
86. Nguyen KB, Smart GC Jr. 1996. Identification of entomopathogenic nematodes in the Steinernematidae and Heterorhabditidae (Nemata: Rhabditida). *J. Nematol.* 28:286–300
87. O'Leary SA, Burnell AM. 1997. The isolation of mutants of *Heterorhabditis megidis* (Strain UK211) with increased desiccation tolerance. *Fundam. Appl. Nematol.* 20:197–205
88. Page RDM. 1993. Parasites, phylogeny and cospeciation. *Int. J. Parasitol.* 23:499–506
89. Page RDM. 1994. Maps between trees and cladistic analysis of historical associates among genes, organisms, and areas. *Syst. Biol.* 43:58–77
90. Page RDM, Lee PLM, Becher A, Griffiths R, Clayton DH. 1998. A different tempo of mitochondrial DNA evolution in birds and their parasitic lice. *Mol. Phylogenet. Evol.* 9:276–93
91. Poinar GO Jr. 1976. Description and biology of a new insect parasitic rhabditoid, *Heterorhabditis bacteriophora* n. gen. n. sp. (Rhabditida; Heterorhabditidae n. fam.). *Nematologica* 21:463–70
92. Poinar GO Jr. 1990. Taxonomy and biology of Steinernematidae and Heterorhabditidae. See Ref. 42, pp. 23–61
93. Poinar GO Jr. 1991. Genetic engineering of nematodes for pest control. In *Biotechnology for Biological Control of Pests and Vectors*, ed. K Maramorosch, pp. 77–93. Boca Raton, FL: CRC
94. Poinar GO Jr. 1993. Origins and phylogenetic relationships of the entomophilic rhabditids, *Heterorhabditis* and *Steiner-*

nema. Fundam. Appl. Nematol. 16:333–38

95. Poinar GO Jr, Jackson T, Klein M. 1987. *Heterorhabditis megidis* sp. n. (Heterorhabditidae: Rhabditida), parasitic in the Japanese beetle, *Popillia japonica* (Scarabidae: Coleoptera), in Ohio. *Proc. Helminthol. Soc. Wash.* 53:53–59
96. Poinar GO Jr, Karunakar GK, David H. 1992. *Heterorhabditis indicus* n. sp. (Rhabditida, Nematoda) from India: separation of *Heterorhabditis* spp. by infective juveniles. *Fundam. Appl. Nematol.* 15:467–72
97. Poinar GO Jr, Thomas GM. 1966. Significance of *Achromobacter nematophilus* Poinar and Thomas (Achromobacteraceae: Eubacteriales) in the development of the nematode, DD-136 (*Neoaplectana* sp. Steinernematidae). *Parasitology* 56:385–90
98. Reid AP, Hominick WM. 1992. Restriction fragment length polymorphisms within the ribosomal DNA repeat unit of British entomopathogenic nematodes (Rhabditida: Steinernematidae). *Parasitology* 105:317–23
99. Reid AP, Hominick WM. 1993. Cloning of the rDNA repeat unit from a British entomopathogenic nematode (Steinernematidae) and its potential for species identification. *Parasitology* 107:529–36
100. Reid AP, Hominick WM. 1993. Isolation and use of a species specific clone for the identification of the rhabditid entomopathogenic nematode *Steinernema feltiae* (Filipjev, 1934). *Fundam. Appl. Nematol.* 16:115–20
101. Reid AP, Hominick WM, Briscoe BR. 1997. Molecular taxonomy and phylogeny of entomopathogenic nematode species (Rhabditida: Steinernematidae) by RFLP analysis of the ITS region of the ribosomal DNA repeat unit. *Syst. Parasitol.* 37:187–93
102. Rueda LM, Osawaru SO, Georgi LL, Harrison RE. 1993. Natural occurrence of entomogenous nematodes in Tennessee nursery soils. *J. Nematol.* 25:181–88
103. Smart GC Jr. 1995. Entomopathogenic nematodes for the biological control of insects. *J. Nematol.* 27:529–34
104. Stock SP. 1997. *Heterorhabditis hepialius* Stock, Strong & Gardner, 1996, a junior synonym of *H. marelatus* Liu & Berry, 1996 (Rhabditida: Heterorhabditidae) with a redescription of the species. *Nematologica* 43:455–63
105. Stock SP, Choo HY, Kaya HK. 1997. An entomopathogenic nematode, *Steinernema monticolum* sp. n. (Rhabditida: Steinernematidae) from Korea with a key to other species. *Nematologica* 43:15–29
106. Swofford DL. 1998. PAUP*: Phylogenetic analysis using parsimony (and other methods), version 4.0. Sunderland, MA: Sinauer
107. Tallosi B, Peters A, Ehlers RU. 1995. *Steinernema bicornutum* sp. n. (Rhabditida: Steinernematidae) from Vojvodina, Yugoslavia. *Russian J. Nematol.* 3:71–80
108. Thomas GM, Poinar GO Jr. 1979. *Xenorhabdus* gen. nov., a genus of entomopathogenic, nematophilic bacteria of the family *Enterobacteriaceae. Int. J. Syst. Bacteriol.* 29:352–60
109. Thompson JD, Higgins DG, Gibson TJ. 1994. CLUSTAL W: improving the sensitivity of progressive multiple sequence alignment through sequence weighting, position-specification gap penalties and weight matrix choice. *Nucleic Acids Res.* 22:4673–80
110. Travassos L. 1927 Sobro o genero Oxysomatium. *Boletim Biologico, Sao Paulo* 5:20–21
111. Waturu CN, Hunt DJ, Reid AP. 1997. *Steinernema karii* sp. n. (Nematoda: Steinernematidae), a new entomopathogenic nematode from Kenya. *Int. J. Nematol.* 7:68–75
112. Welsh J, McClelland M. 1990. Fingerprinting genomes using PCR with arbitrary primers. *Nucleic Acids Res.* 18:7213–18

113. Williams AR, Kubelik KJ, Livak J, Rafalski A, Tingey SV. 1990. DNA polymorphisms amplified by arbitrary primer are useful as genetic markers. *Nucleic Acids Res.* 18:6531–35

114. Zhang GY, Yang HW, Zhang SG, Jian H. 1992. Survey on the natural occurrence of entomophilic nematodes (Steinernematidae and Heterorhabditidae) in Beijing area. *Chin. J. Biol. Contr.* 8:157–59

Annu. Rev. Entomol. 2000. 45:307–340

Culicoides Biting Midges: Their Role as Arbovirus Vectors

P. S. Mellor, J. Boorman, and M. Baylis
Institute for Animal Health, Ash Rd., Pirbright, Woking, Surrey, GU24 0NF, UK; e-mail: philip.mellor@bbsrc.ac.uk, matthew.baylis@bbsrc.ac.uk

Key Words *Culicoides*, arboviruses, vectors, climate, satellite imagery

■ **Abstract** *Culicoides* biting midges are among the most abundant of haematophagous insects, and occur throughout most of the inhabited world. Across this broad range they transmit a great number of assorted pathogens of human, and domestic and wild animals, but it is as vectors of arboviruses, and particularly arboviruses of domestic livestock, that they achieve their prime importance. To date, more than 50 such viruses have been isolated from *Culicoides* spp. and some of these cause diseases of such international significance that they have been allocated Office International des Épizooties (OIE) List A status. *Culicoides* are world players in the epidemiology of many important arboviral diseases. In this context this paper deals with those aspects of midge biology facilitating disease transmission, describes the factors controlling insect-virus interactions at the individual insect and population level, and illustrates the far-reaching effects that certain components of climate have upon the midges and, hence, transmission potential.

PERSPECTIVES AND OVERVIEW

Culicoides biting midges (Diptera: Ceratopogonidae) are among the world's smallest haematophagous flies measuring from 1 to 3 mm in size. More than 1400 species have been identified, and they occur on virtually all large land masses with the exception of Antarctica and New Zealand, ranging from the tropics to the tundra and from sea level to 4000 m.

Species within the genus are a biting nuisance to humans, cause an acute allergic dermatitis in horses, and also transmit protozoa and filarial worms affecting birds, humans, and other animals. However, it is as vectors of viruses of both man and animals that they achieve their prime importance. Worldwide, more than 50 viruses have been isolated from *Culicoides* species. In terms of disease caused to humans and/or other animals, several of these viruses are of major international significance and as such there is a wealth of literature dealing with all aspects of their transmission, epidemiology, and pathology they cause in their definitive hosts. In relation to the *Culicoides*-transmitted human pathogen, Oropouche virus,

0066-4170/00/0107-0307/$14.00

the interested reader is directed to the publication edited by Travassos da Rosa et al (186), which contains several excellent papers on the subject. Similarly, the book edited by Walton & Osburn (200) is devoted entirely to the major *Culicoides*-transmitted viral pathogens of domestic and wild animals. Several other excellent and authoritative texts based wholly or partly on this subject also exist (6, 34, 104, 154, 169).

The success of *Culicoides* midges as arbovirus vectors is related to the vast population sizes that can be reached under appropriate climatic conditions and to their means of dispersal. Consequently, the epidemiologies of *Culicoides*-borne viral diseases are strongly linked to climate and weather. Recently, progress in our understanding of how climate affects midges, and the newer technologies of remote sensing and GIS, have allowed climate-driven risk models for *Culicoides* to be produced for the first time.

CULICOIDES BIOLOGY AND TAXONOMY

Culicoides are small biting flies belonging to the family Ceratopogonidae. In different parts of the English speaking world they are variously known as gnats, midges, punkies and no-see-ums. On occasion they are also called sandflies, though this term is more properly applied to phlebotomids. The first description of *Culicoides* in the literature has been attributed to Rev. W Derham, the then Rector of Upminster, Essex, who wrote of their life history and biting habits as long ago as 1713 (41).

The family Ceratopogonidae contains some 125 genera with about 5500 species. Of these genera, four are known to contain species that suck the blood of vertebrates: *Austroconops, Culicoides, Forcipomyia* subgenus *Lasiohelea,* and *Leptoconops*.

Culicoides are easily differentiated from these others by wing characters. Pictures of the wings of many of the medically and veterinary important species are given by Boorman (23).

Species of *Culicoides* have been found in almost all parts of the world except the extreme polar regions, New Zealand, Patagonia, and the Hawaiian Islands, and in general they seem as prevalent in temperate areas as they are in the tropics. More than 1400 species of *Culicoides* are known worldwide of which about 96% are obligate blood suckers attacking mammals (including humans) and birds (98).

The life cycle of *Culicoides* includes egg, four larval stages, pupa, and imago. Immatures require a certain amount of free water or moisture and are found in an astonishingly wide range of habitats that meet that criterion. Breeding sites include pools, streams, marshes, bogs, beaches, swamps, tree holes, irrigation pipe leaks, saturated soil, animal dung, and rotting fruit and other vegetation (19, 210).

Eggs are usually laid in batches adhering to the substrate, and are banana-shaped and about 400 μm long by 50 μm wide. They are white when laid but darken rapidly. They are not resistant to drying and usually hatch within two to seven days (19, 98). Larvae are vermiform and swim with a characteristic serpentine or eel-like motion. The duration of the four larval stages varies with the species and ambient temperature, from as little as four to five days to several weeks (94, 98). In temperate countries these periods may be considerably extended because most species overwinter as fourth-instar larvae in diapause (81). Larval food for many species consists of particles of vegetable matter, but other species are predaceous, feeding upon nematodes, rotifers, protozoans, and small arthropods (19, 98).

Pupae may be free floating or loosely attached to debris. The pupal stage is brief, usually lasting for only two to three days but occasionally three to four weeks, depending on species and temperature.

Most adult *Culicoides* are crepuscular, and therefore peak activity is around sunset and sunrise and to a lesser extent through the night, though a few species bite during the day. Females undertake flight activity to seek a mate, a blood meal, or an oviposition site. Males do not blood-feed. The flight range of *Culicoides* usually is short and most species disperse only a few hundred meters from their breeding sites (81) or at most 2 to 3 km (88). However, *Culicoides* are capable of being dispersed passively as aerial plankton over much greater distances (62).

In general, adult *Culicoides* are short-lived and most individuals probably survive for fewer than 10 to 20 days but exceptionally they may live for much longer periods (44 to 90 days) and during this time may take multiple blood meals.

VIRUSES ASSOCIATED WITH *CULICOIDES*

Worldwide, more than 50 arboviruses have been isolated from *Culicoides,* most from within the families Bunyaviridae (20 viruses), Reoviridae (19 viruses), and Rhabdoviridae (11 viruses) (98). Many of these viruses have been isolated more frequently from other arthropod groups and their association with *Culicoides* species is probably incidental. The human pathogens Rift Valley fever, vesicular stomatitis, Mitchell River and eastern equine encephalomyelitis viruses, which are primarily mosquito-borne, all fall within this category. Nevertheless, 45% of the viruses isolated from *Culicoides* have not been isolated from other arthropods (98). The only significant viral pathogen of humans transmitted biologically by *Culicoides* is Oropouche virus (OROV). It will be apparent, therefore, that it is as transmitters of viral pathogens of non-human animals that *Culicoides* spp. are chiefly known. The more important of these are African horse sickness virus (AHSV), bluetongue virus (BTV), epizootic hemorrhagic disease virus (EHDV), equine encephalosis virus (EEV), Akabane virus (AKAV), bovine ephemeral fever virus (BEFV) and the Palyam viruses. Two of these, AHSV and BTV, cause diseases of such international significance that they have been allocated OIE list A status.

The eight most important *Culicoides*-borne viruses are dealt with individually to provide a realistic indicator of the global significance of this genus of tiny insects in terms of the diseases caused, the financial losses incurred, and the geographical areas at risk.

Oropouche Virus

OROV is a member of the Simbu group of bunyaviruses and is the cause of one of the most important arboviral diseases (ORO) in the Americas. It was first isolated in 1955 from a febrile forest worker, a resident of Vega de Oropouche in Trinidad (5). Since that time the virus has caused at least 27 epidemics and many thousands of clinical cases in Brazil, Panama and Peru (147, 205). Serological surveys suggest that up to half a million people may have been infected since the beginning of the 1960s in Brazil alone (148). The more common symptoms include fever, chills, headache, arthralgia, myalgia, anorexia, vomiting, photophobia, dizziness and meningitis (145, 147).

In the field, OROV has been isolated occasionally from mosquitoes and frequently, from the midge *C. paraensis* (5, 144, 153). This species is typically found at high density during epidemics of ORO and bites humans both inside houses and outside (64, 153). Biological transmission has been demonstrated via *C. paraensis,* from infected to susceptible hamsters and from infected humans to susceptible hamsters with transmission rates as high as 83% (146, 147). Transmission of OROV between hamsters has also been demonstrated using the mosquito *Culex quinquefasciatus,* although this is much less efficient. Concentrations of virus higher than those usually occurring in humans were necessary to initiate an infection in the mosquito and even then transmission rates never exceeded 5% (147).

These findings, taken as a whole, strongly suggest that *C. paraensis* is the major biological vector of OROV between humans during urban epidemics of the disease. However, the vector(s) of the virus in its "silent" sylvatic cycle remain unknown.

African Horse Sickness Virus

AHSV is a ds RNA virus, within the genus *Orbivirus* of the family Reoviridae, which causes an infectious, non-contagious, disease of equids. The disease, African horse sickness (AHS), is characterized by clinical signs that develop from impaired function of the circulatory and respiratory systems and give rise to serous effusions and hemorrhage in various organs and tissues (68). In susceptible populations of equids AHS can be devastating and in horses mortality rates frequently exceed 90%. AHS one of the most lethal of infectious horse diseases.

The virus exists as nine distinct serotypes all of which are enzootic in sub-Saharan Africa. In the field AHSV is transmitted between its equid hosts almost exclusively by the bites of certain species of *Culicoides* (102). Because of this transmission mode the distribution of the virus is limited to geographical areas where vector species of *Culicoides* are present and to those seasons when conditions favor vector activity.

Periodically, AHSV expands out of sub-Saharan Africa and has caused major epizootics extending as far as Pakistan and India in the east, where more than 300,000 equids died during the great epizootic of 1959–61 (68), and as far as Morocco, Spain, and Portugal in the west (100). However, until recently the virus has not survived for more than two years in any of these epizootic areas. This short duration has been put down to the absence of a long-term vertebrate reservoir from these regions and to an absence or to a seasonal incidence of efficient vector species of *Culicoides* (103). The recent outbreak of AHSV in the western Mediterranean basin, which lasted for five years (1987–91), has forced a reassessment of the situation (100, 101, 102).

The only confirmed field vector of AHSV is *C. imicola*. This Afro-Asiatic species is common in much of Africa and SE Asia (94). However, the species is also found in Spain and Portugal in western Europe, and on several of the Greek Islands in eastern Europe, ranging as far north as 41° 17' N (21, 108, 152). Furthermore, in many areas of Portugal and Spain (and also on some Greek Islands) *C. imicola* is present throughout the year in the adult phase (152; O Papadopoulos, 1998, personal communication). As a direct result of the year-round presence of the vector, these regions are potential enzootic zones for AHS. Rawlings et al (152) have suggested that the northward expansion of *C. imicola* may be continuing in the Iberian peninsula aided by global climate change. If this is indeed the case, with a projected temperature increase during the early part of the twenty-first century in the order of 1 to 3.5° C and bearing in mind that a 1° C rise in temperature corresponds to 90 km of latitude and 150 m of altitude, this could extend the distribution of *C. imicola* and hence the risk of disease, into central Europe.

It is now known that in Africa *C. imicola sl* is a complex of at least 10 sibling species (97). The adults are morphologically similar but exhibit widely differing biologies and distributions. Recently one of these species, *C. bolitinos,* was tested for vector competence and found to be susceptible to oral infection with AHSV, just as *C. imicola* can be infected, and supports replication to transmissive levels. *Culicoides bolitinos* should therefore be considered as a second field vector of AHSV.

Culicoides imicola sl is absent from the New World. However, the North American midge, *C. variipennis sonorensis,* is a highly efficient vector of AHSV in the laboratory (24, 206). There can be little doubt that, should AHS viremic equids gain entry into those parts of North America where *C. v. sonorensis* occurs (most of the southern and western United States), transmission of the virus would be likely.

Although *C. imicola* has long been considered a proven vector of AHSV, the most widely quoted references to support this claim merely show that the inoculation of an emulsion of unidentified wild-caught *Culicoides* caused AHS in a susceptible horse (46), and that AHSV was transmitted by the bite of unidentified *Culicoides* from an infected to a susceptible horse (207). Superficially, therefore, there is nothing in these papers to connect *C. imicola* with the transmission of AHSV. Furthermore, there is no published work proving that *C. imicola* can transmit AHSV by bite. Accordingly, *C. imicola* does not fulfill one of the four criteria regarded by a World Health Organization (WHO) study group as essential for the recognition of a vector (209). However, this is a theoretical omission only, brought about by the extreme difficulty in persuading *C. imicola* to blood feed not only once (the infecting feed) but also, after 8 to 10 days incubation, for a second time (the transmitting feed). In reality, the numerous isolations of AHSV from blood-free, wild-caught *C. imicola*, the extensive published works describing the oral infection, replication, and persistence of AHSV in this species, the invariable presence of populations of *C. imicola* in countries where AHS has been reported, and the seasonal and geographical coincidences of virus transmission and *C. imicola* abundance leave no doubt that this species is the major field vector of AHSV (9, 16, 95, 96, 106, 192, 193).

Bluetongue Virus

Like AHSV, BTV is an orbivirus and exists as a number of serotypes; 24 have been identified to date. BTV is thought to infect all known species of ruminant, but severe disease usually occurs only in certain breeds of sheep and some species of deer (92, 184). Clinical signs may include fever, depression, nasal discharge, excessive salivation, facial oedema, hyperanemia, and ulceration of the oral mucosa, coronitis, muscle weakness, secondary pneumonia, and death.

The global distribution of BTV lies approximately between latitudes 35° S and 40° N, although in parts of western North America it may extend up to almost 50° N (45). Within these areas the virus has a virtually worldwide distribution, being found in North, Central, and South America, Africa, the Middle East, the Indian sub-continent, China, Southeast Asia, and Australia (99, 169). BTV has also at times made incursions into Europe, although it has not been able to establish itself permanently in that continent (107). Nevertheless, the 1956–1960 epizootic in Spain and Portugal resulted in the deaths of almost 180,000 sheep and is the most severe outbreak of BT on record (55).

BTV is transmitted between its ruminant hosts almost entirely by the bites of vector species of *Culicoides*. In consequence, its world distribution is restricted to areas where these vector species occur and transmission is limited to those times of the year when adult insects are active. In epizootic zones this usually means during the late summer and autumn, and this is therefore the time when BT is most commonly seen (107).

The Americas Throughout almost the whole of the BTV zone in North America the major vector has long been considered to be *C. variipennis* (99). However, compelling evidence has recently been put forward to suggest that *C. variipennis* is, in fact, a complex of at least three genetically defined subspecies (*C. v. occidentalis, C. v. sonorensis, C. v. variipennis*) that are sufficiently different from each other to warrant consideration as separate species (181, 183). Of these, *C. v. sonorensis* is believed to be the primary vector of BTV on the basis of field isolations, a close relationship between the distribution of BTV antibodies and the range of *sonorensis*, and vector competence studies (181, 183). Furthermore, populations of *C. v. variipennis* have been found to be significantly more resistant than *C. v. sonorensis* is to oral infection with BTV (91, 182). On the basis of these findings it has been suggested that the low oral susceptibility rates of *C. v. variipennis* are a major factor in the apparent absence of BTV transmission in the northeastern United States, a region where *C. v. variipennis* is the sole representative of the species complex, and it has been proposed that such areas should be considered BTV-free zones (183).

In the southeastern United States BTV may be co-vectored by *C. variipennis* and other species of *Culicoides*. In east-central Alabama, an area where *C. variipennis* is scarce and where *C. stellifer* is very common, BTV monthly seroconversion rates of up to 87% in cattle and white-tailed deer were recorded and BTV RNA products have been detected in both *C. variipennis* and *C. stellifer* (116).

Further south, C. *variipennis* is absent from southern Florida, the Caribbean region, most of Central America, and all of South America, yet these are regions where BTV also occurs. In these areas *C. insignis* and *C. pusillus* are thought to be the major vectors, and BTV 2, 3, and 6 have been isolated from *C. insignis*, and BTV 3 and 4 from *C. pusillus* (57, 83, 113).

Africa *Culicoides imicola* is the major vector of BTV throughout this region and numerous isolations of the virus have been made in South Africa (98), Zimbabwe (16), Kenya (199), and the Sudan (109). The species extends throughout Africa and virtually all countries within and adjacent to the African region (e.g. Spain, Portugal, Greece, Cyprus, Israel, Turkey, Yemen, and Oman) that have been shown to support populations of *C. imicola* have reported serological or clinical evidence of BTV at some time (99). Consequently, in this region the presence of *C. imicola* invariably indicates the presence of a BT enzootic or epizootic zone.

In Kenya, in addition to *C. imicola*, BTV has been isolated from *C. tororoensis* and *C. milnei* (199), while in Cyprus the virus has been isolated from *C. obsoletus* (110). However, further evidence linking these species to BTV transmission has not been forthcoming and they are probably of only minor significance in the epidemiology of BT.

In southern Africa, *C. imicola* is the major vector of BTV (98, 136). It cannot be the only vector, however, because it is rare or absent from some of the cooler or more arid areas where BTV also occurs (98). In this context, recent work has shown that *C. bolitinos*, a sibling species of *C. imicola,* is able to support replication of BTV to high titer after ingestion and is highly likely to be a competent vector of this virus (197). Furthermore, because this species is more common than *C. imicola* in many of the cooler BT enzootic areas and breeds in the dung of cattle, the major host species for the virus, *C. bolitinos,* is likely to be the primary BTV vector in such restricted locales (194).

Asia Many countries in Asia have reported the presence of BT and/or positive BTV serology and it is likely that the virus is present, at times, in most countries in the tropical, subtropical and temperate parts of the region (169). The northernmost affected areas are in northern China at 40° to 48° N (59, 150). Information on the *Culicoides* vectors of BTV across this huge region is patchy, however.

Culicoides imicola has been recorded from many of the countries of the Near and Middle East (22, 27, 73). It also occurs in India (= *C. minutus*), Sri Lanka, China, Thailand, Laos, and Vietnam (121) but no virus isolations have been made from *C. imicola* across this part of the region. BTV has been isolated from *Culicoides* in India and Yunnan, China, but the species were not identified (71, 212). The main vectors in Yunnan are considered to be *C. schultzei, C. gemellus, C. peregrinus, C. arakawae,* and *C. circumscriptus*; however, no data are published to justify this selection (14). In Anhui, *C. actoni* is reportedly confirmed as being able to transmit BTV in China but again no data are presented to support this assertion (213). It is apparent that the identity of the major BTV vectors in China is still uncertain and further work is urgently required to resolve this situation. Because *C. imicola, C. fulvus*, and *C. actoni*, have all been recorded in China and are confirmed vectors elsewhere, it seems prudent to allocate a high priority to a study of these species.

In Indonesia, more work has been carried out on *Culicoides* than in any other country in the east Asian region. Almost 50 species of *Culicoides* have been identified including four proven BTV vectors from other regions (*C. actoni, C. brevitarsis, C. fulvus, C. wadai*) (176, 177). Several other potential vectors (*C. brevipalpis, C. peregrinus, C. oxystoma, C. nudipalpis, C. orientalis*) have also been identified and BTV 21 has been isolated from mixed pools of *C. fulvus* and *C. orientalis* (165, 177).

Australasia At least eight serotypes of BTV occur in Australia (53), three in Papua New Guinea and one in the Solomon Islands (169). Several *Culicoides* species are thought to transmit the virus in the region and of these, C. *fulvus, C. wadai, C. actoni,* and C. *brevitarsis* are considered most important. The most efficient vector is *C. fulvus* (174) but this species is restricted to areas with high summer rainfall and does not occur in the drier sheep-rearing areas of Australia, and so is unable to transmit BTV to this highly susceptible host. *Culicoides wadai,* also an efficient BTV vector, was first recorded in Australia in 1971, probably having been introduced as aerial plankton from Indonesia where it is common

(103). This species initially was restricted to the Darwin area but from 1978 to 1988 it extended its range southward into the Kimberleys, eastward to the Queensland coast and from there southward again into New South Wales. This extension covered some 2000 miles in little more than 10 years. The known distribution of C. *wadai* now puts this efficient vector in close proximity to some of the major sheep-rearing areas, causing justifiable concern among the Australian authorities (44).

Culicoides brevitarsis is an inefficient vector (174), but because it is more widespread and more abundant than either *C. fulvus* or *C. wadai,* and has a distribution similar to that of BTV-specific antibodies in cattle, it is considered to be of major importance to BT epidemiology in the region.

Europe BTV is not enzootic in Europe but incursions occur at intervals, the most recent being into Spain and Portugal between 1956 and 1960, and to the Greek Islands of Lesbos and Rhodes in 1979 (107). The major vector in the region is *C. imicola,* which is present in these areas (21, 108). Indeed, in the Mediterranean basin the known distribution of *C. imicola* is virtually identical to that of BTV incursions. Since all of Europe has been free from BTV for at least 18 years the overall picture is of a continent with a serologically naive, highly susceptible ruminant population, completely vulnerable to any new incursion. The recently reported (October 1998 to January 1999) outbreaks of BTV 9 on the Greek Islands of Rhodes, Kos, Samos, and Leros, where populations of *C. imicola* are suspected or known to be present, is not, therefore, unexpected (138). In this context, the further extension of the BTV 9 outbreak, at the time of writing in late 1999, into European Turkey, northern Greece, and Bulgaria is an especially worrying development. These are areas where *C. imicola* has not been reported but where *C. obsoletus* is common. The implication of this new outbreak will require careful analysis.

Epizootic Hemorrhagic Disease Virus

Like BTV and AHSV, EHDV is an orbivirus. Eight serotypes have been identified. The virus occurs in North, Central, and South America, Africa, Southeast Asia, Japan, and Australia, and its vertebrate hosts include a wide range of domestic and wild ruminant species (45, 58). In most ruminants, infection is inapparent, but in certain species of deer the disease can be severe and is indistinguishable from BT (65, 167). Cattle are usually subclinically affected, but in 1959 Ibaraki virus (EHDV 2) caused BT-like clinical signs in 39,000 Japanese cattle, killing approximately 400 of them (139). More recently suspicions have been growing that EHDV may also be a cause of disease in cattle in other parts of the world, including the United States (67) and South Africa (7).

In North America the major vector is *C. variipennis* (51). However, other species of *Culicoides*, such as *C. lahillei*, may also be involved in locations where

C. variipennis is scarce or absent (166). In Africa, EHDV has been isolated from *C. schultzei* group midges (109), and in Australia numerous isolations have been made from *C. brevitarsis* (142). The vectors of EHDV in Central America, South America, Japan, and Southeast Asia are unknown.

Palyam Viruses

Palyam viruses are typical orbiviruses. At present the serogroup has 16 members. Antibodies have been identified in cattle, and with decreasing frequency in sheep, goats, and humans. More than 100 isolations of Palyam serotypes have been made from cattle in Australia alone (54). Infection is frequently inapparent but Palyam viruses have been linked to congenital abnormalities and abortion (179, 208), and in Japan a member of the serogroup (Chuzan virus) has been incriminated as a cause of hydranencephaly and cerebella hypoplasia in 2463 calves (56). Palyam viruses occur in Africa, Asia, and Australia (178).

Although several Palyam viruses have been isolated from ticks and mosquitoes, the vast majority of isolations from arthropods have been from *Culicoides* (16, 43, 85, 87). Transmission of these viruses by arthropods has not yet been demonstrated but a Palyam virus (Nyabira) replicates in orally infected *C. imicola* and *C. zuluensis*, and can survive in infected midges for at least 14 days (30). The general indications are, therefore, that *Culicoides* are the major vectors of these viruses.

Equine Encephalosis Virus

EEV is the fifth group of orbiviruses that is transmitted by *Culicoides.* Six serotypes have been identified, all from horses in South Africa (33) and there is evidence that the virus also occurs in Botswana (40). Ninety percent of horses infected with EEV exhibit few or no clinical signs but in 5% to 10% of animals the virus causes abortion, cardiac failure and an African horse sickness-like disease (33).

EEV has been isolated from a mixed pool of *Culicoides* (more than 95% *C. imicola*) in South Africa (186) and the virus replicates to high titer in orally infected *C. imicola* (196).

Bovine Ephemeral Fever Virus

BEFV is a rhabdovirus which infects a range of ruminant species but seems to cause disease only in cattle and water buffalo. The disease (BEF) is characterized by a short period of fever, stiffness, and disinclination to move; hence the common name of three-day sickness (168). The major obvious losses are due to mortality (less than 3%), abortion (5%), loss of draft power and, most importantly, a severe decline in milk production (80%) (168, 186). Insidious losses result from weight loss, lowered fertility in bulls and lost export markets (131, 168).

BEFV occurs across Africa, the Middle East, India, southern China, Southeast Asia, Japan, Indonesia, and northern and eastern Australia (170). Epidemiological studies indicate that insects transmit the virus but their identity across much of its range remains unclear (170). This is probably because the virus is difficult to isolate by standard cell culture techniques, and the situation is further complicated by the fact that the BEFV group contains at least three and possibly seven other viruses that are not known to cause disease (37, 77).

In Kenya BEFV has been isolated from a pool of five species of *Culicoides* (*C. kingi, C. nivosus, C. bedfordi, C. imicola, C. cornutus*) (38) and in Zimbabwe from *C. imicola* and *C. coarctatus* (16). In Australia, the virus has been isolated both from mosquitoes and *C. brevitarsis* (36, 175), and has been shown to replicate in two species of *Culicoides* (*C. brevitarsis, C. marksi*) and one species of mosquito (*Culex annulirostris*) (172). Furthermore, studies focusing on the distribution of BEFV and its likely vectors in Australia suggest that the virus may be transmitted by several species of insect: two mosquito species (*Cx. annulirostris, An. bancrofti*) and one *Culicoides* species (*C. brevitarsis*) are the most likely candidates (122).

Clearly, much work remains to identify the primary vectors of BEFV and to determine the relative importance of *Culicoides* and mosquitoes in the epidemiology of BEF.

Akabane Virus

AKAV is a member of the Simbu group of the family Bunyaviridae and infects a wide range of animals including buffalo, cattle, camel, sheep, goats, and horses (75). Infection of non-pregnant animals is usually inapparent. However, the virus is able to cross the ruminant placenta, and should this happen early in pregnancy a variety of congenital abnormalities, including arthrogryposis and hydranencephaly, are seen at parturition (185).

Serological or clinical evidence of AKAV has been reported from Africa, the Middle East, Southeast Asia, and Australia (2, 171, 185).

The virus has been isolated from *C. brevitarsis* in Australia (43, 173), *C. oxystoma* in Japan (84), *C. imicola* and *C. milnei* in Zimbabwe (16), *C. imicola* in Oman (3), and a mixed pool consisting mainly of *C. imicola* in South Africa (187). AKAV has also been isolated from mosquitoes in Japan and Kenya (112, 141). AKAV replicates in *C. brevitarsis* and reaches the salivary glands after 10 days of incubation (171). The virus also replicates in orally infected *C. variipennis* by up to 1000-fold and transmission can occur after 7 to 10 days of incubation at 25° C (75). Replication and transmission of the virus have not been demonstrated in mosquitoes. These studies suggest that *Culicoides* species are likely to be the major vectors of AKAV and that mosquitoes are probably of only minor importance.

INFECTION OF *CULICOIDES* WITH VIRUSES AND FACTORS INFLUENCING TRANSMISSION

Barriers to Virus Dissemination Within Individual *Culicoides*

In the wild, vector *Culicoides* become infected with arboviruses only by imbibing viremic blood from vertebrate hosts. Subsequent transmission of the virus occurs only through biting. There is no published evidence to suggest that any *Culicoides*-transmitted arbovirus can be passed vertically (transovarially) or venereally through its vectors, unlike many mosquito, phlebotomid, and tick-transmitted arboviruses (199).

Female *Culicoides* ingest a range of liquid foods including blood, sugars, water, and nectar. Most of these are deposited in an acellular sac, the mid-gut diverticulum. If the food is blood, contraction of a sphincter muscle at the mouth of the diverticulum ensures that the meal is directed into the hind part of the mid-gut (93). Virus ingested in a blood meal follows the same route. In a vector species of *Culicoides,* virus particles attach to the luminal surface of the gut cells, enter them probably by direct plasma membrane penetration or receptor-mediated endocytosis, and replicate (47). Progeny virus particles escape through the basolateral membrane of the gut cells into the hemocoel and infect a range of secondary target organs, including the salivary glands. Replication in salivary gland cells is followed by release of virus particles into the salivary ducts where they accumulate and are available for transmission during subsequent biting activity (32, 52).

However, even within a vector species of *Culicoides* only a proportion of individuals is likely to be susceptible to oral infection with a particular arbovirus (74, 180). A series of barriers may act to limit or constrain virus dissemination through the insect and so prevent transmission.

Figure 1 summarizes the hypothesized barriers to arbovirus infection of hematophagous insects from the stage of ingestion until the virus is transmitted orally or transovarially. The figure also shows those barriers that have been identified in the BTV/AHSV-*C. variipennis* system, which is by far the best-defined *Culicoides*-arbovirus system. Expression of these barriers seem to be hereditary traits, and populations of *C. variipennis* with high or low incidences of one or another of the barriers have been derived by selective breeding (52, 76, 206). So far salivary-gland infection and escape barriers have not been identified in any *Culicoides* species.

Early work on the genetic control of *C. variipennis* susceptibility to BTV infection demonstrated the presence of a single controlling gene, and crosses between susceptible and resistant lines of the midge provided evidence for a major locus and a modifier, controlling susceptibility (76). In relation to the major controlling locus the maternal genotype determines the progeny phenotype and the paternal gene is always dominant in the offspring (180). This inheritance pattern

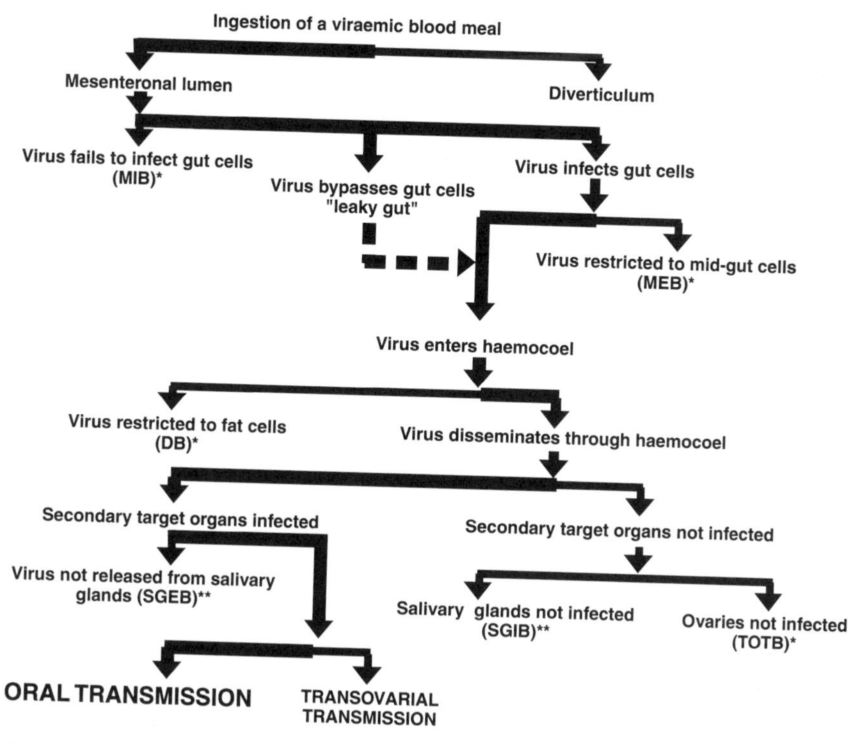

Figure 1 Hypothesized barriers to the arbovirus infection of hematophagous insects. * Barriers shown to be present in the AHSV/BTV-*Culicoides variipennis* system. ** Barriers not yet shown to be present in the AHSV/BTV-*C. variipennis* system. MIB, mesenteron infection barrier; MEB, mesenteron escape barrier; DB, dissemination barrier. TOTB, transovarial transmission barrier; SGIB, salivary gland infection barrier; SGEB, salivary gland escape barrier. (Adapted from 60.)

has allowed the construction of isogenic pools of midges and identification of a candidate controlling protein that has been used to isolate a cDNA clone for sequencing to determine function (182). These and many related factors are discussed in detail in the authoritative review of Tabachnick (182).

Effect of Temperature on Virus Infection of *Culicoides* and Transmission

Vector Species of Culicoides AHSV and BTV infection rates, and rates of virogenesis within vector *Culicoides* have been shown to be temperature dependent (120, 206). At elevated temperatures, infection rates were higher, rates of virogenesis were faster, and transmission occurred earlier. However, individual midges survived for a relatively shorter period of time. As temperature was reduced, infection and virogenesis rates fell, the time to earliest transmission was extended, but midge survival rates were enhanced. With both viruses, replication was not detected in midges maintained at or below15° C, transmission was never recorded at these temperatures, and the apparent infection rate rapidly fell to zero. However, when midges that had been maintained for extended periods at such temperatures were transferred to temperatures within the virus permissive range (more than 20° C) "latent" virus replicated to high titers (120, 206). Other work with AHSV has shown that under variable temperature regimes chosen to simulate nocturnal-diurnal variations, the infection rates and rates of virogenesis in vector midges were proportional to the time spent at permissive temperatures, and the total time spent at these temperatures was the major factor controlling transmission potential (105). On the basis of these and related studies (161), an overwintering mechanism for AHSV in temperate areas has been suggested (103, 206).

"Non-Vector" Species of Culicoides The Palaearctic midge *C. nubeculosus* has an oral susceptibility rate for AHSV of less than 1% when reared at 25° C, and when the adults are fed upon virus at a standard titer. Unlike *C. variipennis*, this rate of oral susceptibility cannot be enhanced by selective breeding from susceptible parents (111). However, if the rearing temperature is raised to 30° to 35° C, the oral susceptibility rate increases to more than 10% and virus replicates to levels suggesting transmission is possible. Similar results have also been obtained when using *C. nubeculosus* and BTV (211).

Because selective breeding had no effect upon the *C. nubeculosus* oral susceptibility rate, in this instance, oral susceptibility is clearly not a genetically controlled, heritable trait. Instead, it may be that the increase in developmental temperature not only gives rise to smaller adults but to adults with an increased incidence of the so-called "leaky gut" phenomenon (20) (see Figure 1). A similar increase in infection rates in small adults, brought about by poor larval nutrition and crowding, has been described by Tabachnick (182) when working with BTV and *C. variipennis,* though the underlying mechanism was not speculated upon.

In female *C. nubeculosus* with a "leaky gut," virus may be able to pass directly from the ingested blood meal in the gut lumen into the hemocoel without first infecting and replicating in the gut cells (111). Once in the hemocoel it is well documented that most arboviruses will replicate and then may be transmitted, even by normally non-vector insects. Such a sequence of events may be envisaged as a hybrid mechanism whereby infection is initiated by a mechanical event (i.e. passage of virus from the lumen of the mid-gut through the gut wall and into the hemocoel, without replication in the gut cells) but transmission is the result of a series of subsequent biological events (i.e. virus infection of the salivary gland cells, replication in these cells, and release of progeny virus particles into the salivary ducts) (103). This work with *C. nubeculosus* suggests that under the appropriate environmental conditions (i.e. increasing temperature) species of *Culicoides,* not normally considered vectors, may be able to transmit AHSV and possibly other viruses. Indeed, the isolations of AHSV 4 from mixed pools of *C. obsoletus* and *C. pulicaris* in Spain during 1988 (106), two *Culicoides* species that are not considered AHSV vectors may be an example of the leaky gut phenomenon in operation, potentiated by the warm conditions prevailing at the time. Increasing temperature due to climate change could increase the likelihood of such unusual isolations by "creating" new vector species.

CLIMATE, WEATHER, AND *CULICOIDES*

Climate and weather have dramatic effects on *Culicoides* populations and, consequently, the epidemiologies of midge-borne viral diseases are similarly affected. Armed with an understanding of these meteorological effects, together with the modern techniques of remote sensing and GIS, we can develop models capable of predicting both when and where disease outbreaks are likely to occur and we can also ask how things might alter with climate change. However, efforts to define a "climatic envelope" outside which a species cannot occur have been criticized for ignoring the biotic environment (39). An equally significant problem is the sheer range of patterns observable in a single species. Consider seasonal distributions of *C. imicola* (Figure 2). In Morocco, Israel, and the winter rainfall region of South Africa a pronounced peak in abundance occurs at the end of summer, coincident with or shortly after the hottest time of year but several months after the last significant rainfall. Adult *C. imicola* are absent or rare during the coldest time of year, when most rainfall occurs.

In direct contrast, the annual peak in abundance in Nigeria occurs shortly after the rainy season and during the coldest part of the year and the insect is rare or absent during the hot, dry season. In the Sudan and the summer rainfall region of South Africa, where the hot season and rains coincide, the peak occurs towards the end of these seasons, while the pattern in Kenya is different again.

Explaining these and other patterns is a challenge still to be fully met, but many underlying principles have been elucidated. We shall consider the effects of weather and climate on the activity or biting rates of *Culicoides*, their dispersal, larval development, adult survival, seasonality, and abundance. These different aspects overlap, but for clarity they are considered separately.

Activity Rates

Adult midges may not be active on certain nights because of adverse meteorological conditions (9). Identification of these conditions is hindered by certain complexities. Different trapping methods can suggest different activity patterns (133) and some meteorological conditions can affect the efficiency of trapping as well as the activity of the midge. Light traps operate less efficiently when alternative sources of light are present and catches may be reduced by moonlight. Both light traps and suction traps may operate less efficiently at higher wind speeds because their ability to draw in air is lessened. Better sampling methods

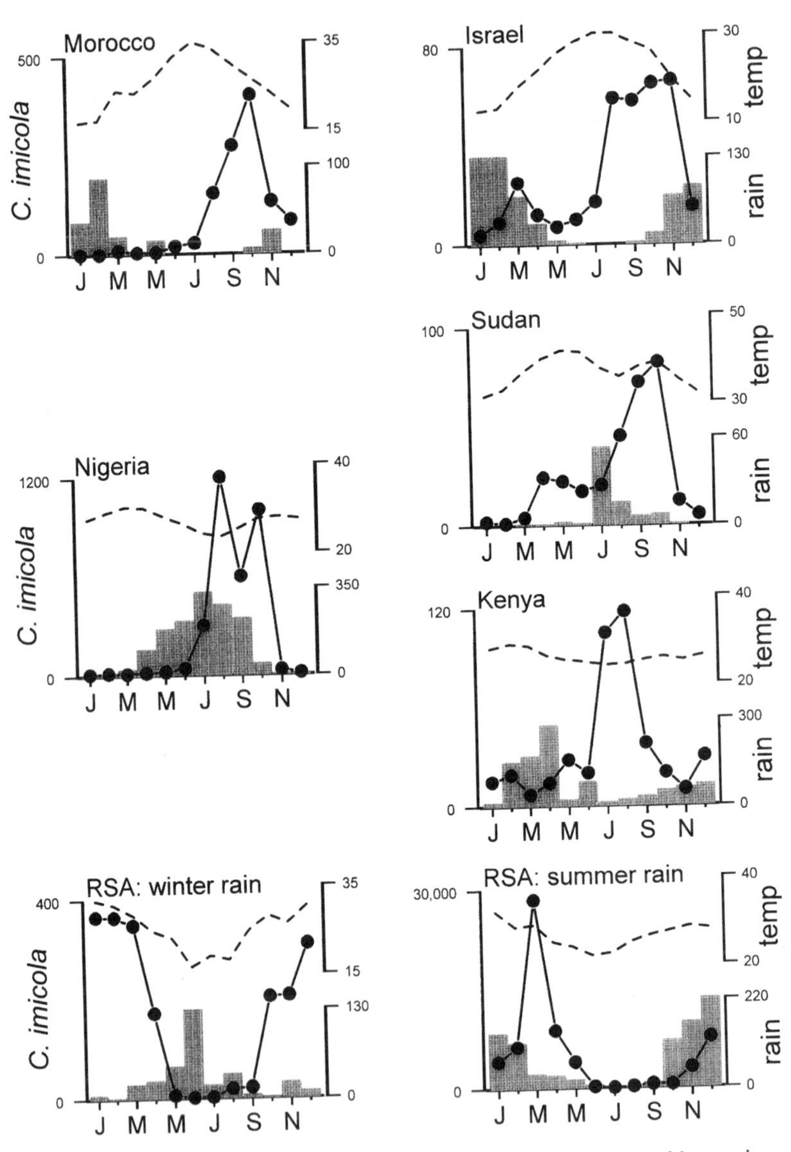

Figure 2 Seasonal abundance of *Culicoides imicola (solid circles)*, monthly maximum temperature *(dashed line)* and rainfall *(bars)* in seven locations. Sites are Sidi Moussa (Morocco), Bet Dagan (Israel), Khartoum (Sudan), Vom (Nigeria), Naivasha (Kenya), and Porterville and Onderstepoort (South Africa, winter and summer rainfall, respectively). (Data from 8, 9, 29, 63, 115, 195, 199; R Meiswinkel & M Baylis, unpublished data, and HMSO meteorological tables.)

may be the use of vehicle-mounted traps or human and animal baits for biting rate studies.

Most species of *Culicoides* are crepuscular or nocturnal. The onset of activity is triggered usually by falling light intensity (26, 78, 80, 133). Complete darkness may suppress activity, however. For example, more *C. variipennis* are active during moonlight than during darker periods (133). Meteorological variables other than light intensity appear to trigger activity in *C. impunctatus* (17) and *C. sanguisuga* (72).

With suitable light intensity, both temperature and wind speed strongly affect activity levels. Negative correlations between activity and wind speed have been reported for species as diverse as *C. impunctatus* in Scotland (17, 78) and *C. brevitarsis* in Australia (82). Almost all activity is suppressed at wind speeds greater than three m/s for *C. imicola* in Kenya (198), 2.7 m/s for *C. furens* and *C. barbosai* in the Caribbean (79), and 2.2 m/s for *C. brevitarsis* in Australia (128). Limits are lower for mating swarm formation by certain species (31). Activity is positively correlated with temperature (17, 80, 82, 198), with activity suppressed at temperatures lower than 10° C for *C. variipennis* (133) and 18° C for *C. brevitarsis* (128). Even species living in the cold environment of Alaska face lower limits to activity of about 12° C (189). Upper temperature limits to activity exist as well: *C. variipennis* is inactive at temperatures greater than 32° C (133) and *C. furens* and *C. barbosai* are less active at temperatures greater than 24° C (79).

At high temperatures air can hold considerable amounts of moisture and midges are at risk of desiccation. Activity of both *C. imicola* (198) and *C. impunctatus* (17) decreases at low moisture levels. The nocturnal activity of many *Culicoides* may, in fact, be an adaptation to exploit the lower risk of desiccation that results from the combination of low temperature and high relative humidity at night. Where moisture levels are particularly important (in arid environments, for example), peak activity levels may occur at dawn when the saturation deficit is minimized (125). Rainfall, even when light, inhibits the activity of some and maybe all species (125).

Dispersal

To a greater extent than with other insect vectors, the long-range dispersal of *Culicoides* is associated with transport on prevailing winds (156). Before it was known even that BEFV is transmitted by *Culicoides*, Seddon (155) demonstrated that the spread of BEF in Australia was explainable by wind patterns. Subsequently, numerous studies have related the spread of *Culicoides*-borne viral diseases to wind movements (25, 28, 50, 66, 115, 124, 127, 137, 158, 159, 160, 162, 163, 164). In all cases the evidence is circumstantial, and alternative modes of introduction can rarely be ruled out. Also, allowing for an approximate two-week leeway in the time of the introduction and for the fact that winds at different heights may travel in different directions, it is not clear how difficult it is to find a putative prevailing wind for any hypothesized wind-borne introduction. Nevertheless, taking the literature as a whole, the concept of disease introduction through wind-borne dispersal of midges is convincing.

In general, suitable winds tend to be at heights of 0.5 to 2 km, with velocities of 10 to 40 km/h and temperatures of 12° to 35° C. On such winds, it is suggested that midges may be blown as far as 700 km (157). It is unclear why colder winds should be considered unfavorable because in the laboratory adult midges are quite capable of surviving exposure to cold for fairly long periods.

The prevailing wind carries infected and uninfected midges from one site to another. Upon landing, these midges bite local animals and virus is transmitted. Local midges then feed upon the infected animals and become infected themselves. In this way, only one infected midge is needed to initiate an outbreak. It has also been argued, however, that in parts of Australia infected *Culicoides* are transported wholesale to sites where livestock—but no other vector midges—occur and, therefore, all transmission is attributable to those midges carried on the wind (127). This idea is derived from outbreaks of Akabane occurring south of the supposed southernmost limits of the distribution of *C. brevitarsis*. Alternatively, the supposed distribution of *C. brevitarsis* may be wrong (202). Whatever the truth, it is notable that no other *Culicoides*-borne disease outbreaks occurring outside of the known distribution of a suitable vector have been reported.

Larval Development

Field studies show that *C. furens*, *C. melleus*, *C. insignis*, and *C. variipennis* that emerge as adults during colder times of year are larger and of higher fecundity than those emerging in warmer seasons (83, 89, 90, 117). Laboratory studies confirm that rearing larvae of *C. variipennis* and *C. brevitarsis* at higher temperatures results in shorter times to emergence and smaller adults (1, 15, 119, 191).

Larval survivorship is optimized at intermediate temperatures. Emergence rates of *C. brevitarsis* are greatest at 25° to 36° C and are reduced at higher and lower temperatures (4, 15). Similar patterns, but with lower optima, are apparent for *C. variipennis* and *C. nubeculosus* (211). Although protracted exposure to cold will kill *Culicoides* larvae, under field conditions some can survive freezing temperatures by moving to the liquid interface between ice and the adjacent frozen substrate (190).

The effects of rainfall and soil moisture on *Culicoides* larval development have not received significant experimental attention. It has been noted that larval *C. variipennis* die when their habitat dries out, unless they are able to move to moist conditions (118). Also, water content is one of the most important factors determining suitability of a habitat for larval *C. belkini* (86), *C. impunctatus* (18), and *C. imicola* (95). At the other extreme, the pupae of some species, including *C. imicola*, do not float on water and will drown if breeding sites are flooded (134).

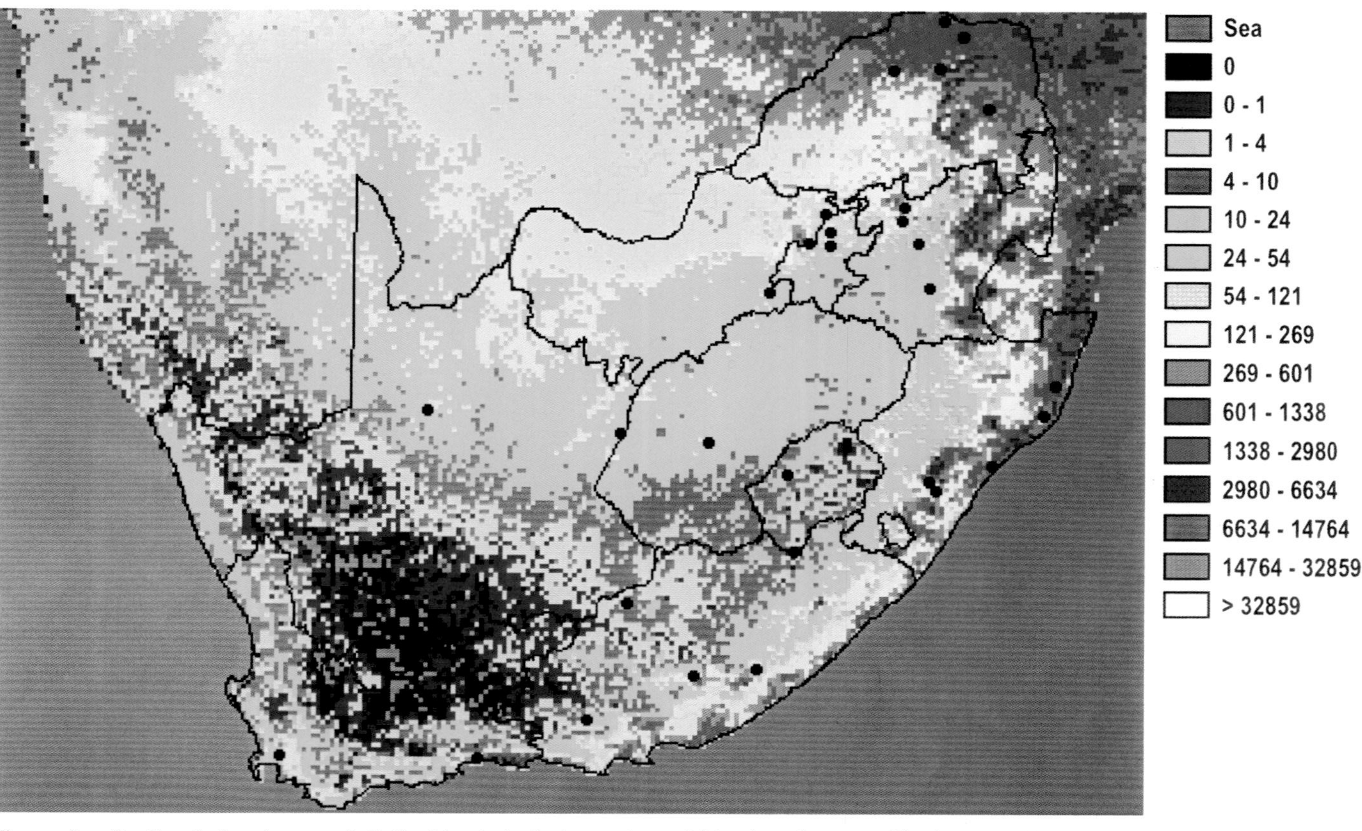

Figure 3 Predicted abundances of *Culicoides imicola* in southern Africa based on satellite images of the annual minimum Land Surface Temperature and the annual minimum Normalized Difference Vegetation Index. Values are the predicted annual mean light-trap catches of *C. imicola*. (From 10).

Adult Survival

Laboratory studies show that, when adult *Culicoides* are maintained at constant temperatures, their survival rate is lower at higher temperatures (70, 206, 211). If the survival rates of all age classes are affected equally, these findings imply that at higher temperatures a smaller proportion of adult *Culicoides* will live long enough to mature and then transmit a viral infection, although this effect will be tempered by a faster rate of virogenesis. As saturation deficit increases with temperature, in the absence of controlled experiments these findings may be attributable to effects of moisture rather than temperature per se. Murray (129) kept adult *C. brevitarsis* for short periods at different temperatures while maintaining constant humidity. Only at low relative humidity was there a noticeable effect of temperature on survival. At all temperatures, however, there were clear effects of humidity on survival.

Because most species of *Culicoides* are crepuscular or nocturnal, adults are active when temperatures are relatively low. It is not clear to what extent adult midges can escape high temperatures during daytime. Apparently, *C. brevitarsis* survival is adversely affected by several days with maxima over 30° C (130), although this report is based on a decrease in the parous rate that would also be observed if such high temperatures increase rates of emergence.

After rainfall *C. brevitarsis* survival appears to increase (129). On the coastal plains of Australia many midges feed twice and a few three times but after rainfall the age structure changes to one in which most midges have fed twice and many three times. In this way "Rainfall can thus transform a population from one maintaining a viral cycle to one capable of explosive transmission within a few days" (126).

In Morocco the mortality rates of *C. imicola,* recorded as averages over several weeks, were compared with concurrently recorded night-time meteorological variables (13). In a comparison across sites, there were no correlations between mortality rate and night-time temperatures or moisture levels, but there was a significant correlation with wind speed. At windier sites fewer *C. imicola* appear to survive. It is not clear whether more midges are killed at such sites or simply dispersed, although for the majority of dispersed midges dispersal may amount to being killed. The findings suggest that the extensive wind dispersal of midges noted earlier may give rise to detectable effects on the age structure of *Culicoides* populations.

Seasonality

Determining the factors that underlie the seasonal distributions of *Culicoides*, particularly those of arbovirus vectors, is important for two reasons. First, the seasonality of disease outbreaks is associated with the timing of the annual peak in vector numbers (9, 63, 69, 114, 115). Second, the occurrence of adults in all winter months, thereby allowing midge-to-host-to-midge cycling of viruses year-round, appears to underpin the ability of many viral diseases to occur for more than one season, and hence their endemic/epidemic status (102).

For a single species of *Culicoides*, the relationship of seasonality to certain climatic variables (temperature and rainfall) appears to vary from place to place (Figure 2). Other factors and meteorological variables that cycle annually, such as day length or wind speed, could influence seasonality. In addition, the number of generations per year will affect whether populations peak earlier (univoltine) or later (multivoltine) in the season. Nevertheless, for *C. imicola* and probably most species of *Culicoides,* the effects of rainfall and/or temperature are paramount. In areas where winters are colder, populations of *C. imicola* tend to disappear or be much reduced in the winter months. In spring the population builds up, with multiple generations, and peaks a few weeks after the hottest time of year. In these areas (Morocco, Israel, South Africa) the timing of the peak appears to be related to the annual temperature cycle and is largely independent of rainfall. The seasonal cycle is unaffected by whether the rainy season is in winter (South Africa, winter rainfall) or summer (South Africa, summer rainfall).

In more tropical areas, a temperature-induced annual loss of the adult population does not occur if the winter temperatures remain sufficiently high. In these areas temperature may still play a role (i.e. higher temperatures will still favor more rapid larval development and hence faster population increase) but the annual cycle appears to be more related to the timing of the rainy season. In both Nigeria (63) and the Sudan (114, 115), which have marked dry seasons, the seasonality of *C. imicola* is closely related to the timing of the rainy season. In the Sudan this is in turn governed largely by the movements of the Intertropical Convergence Zone (115).

Early studies in Kenya suggested that there, too, the seasonality of *C. imicola* is determined by the timing of the rainy season (199), but a subsequent, long-term study failed to find a clear association (198). The factors underlying the population cycles of *C. imicola* in Kenya remain to be determined.

Similar variation in the relationship between seasonality and climate has been reported for *C. brevitarsis* in Australia. In the south, with colder winters, *C. brevitarsis* adults largely or completely disappear in winter and the population builds to a peak in summer or autumn. Further north, where temperature is not limiting, the peaks in population size are related to the timing of the rainy season (123, 125).

A key aspect of the epidemiology of *Culicoides*-borne arboviral diseases is whether adult midges persist all year or disappear for a period when conditions are unsuitable. If the latter then, in the absence of reservoir hosts, virus must be introduced from elsewhere for disease outbreaks to occur, either by the carriage of infected midges on winds, or by the movement of infected hosts. Such an annual disappearance of adults may occur if part of the year is too arid for breeding of midges [e.g. the Sudan (115)], but it is related more often to cold winters, such as those in Mediterranean climates.

The degree of coldness required to prevent viral overwintering is difficult to define. In South Africa it has been long known that frost stops the progress of AHS epizootics (143), and the degree and frequency of frost affects the seasonal cycle of *C. imicola* (195). Freezing temperatures kill adult midges so the significance of frosts is not unexpected. Similarly, in a seven-year survey, midges were caught all winter in only one year: that with the warmest winter (135). In contrast, it has been suggested that overwintering of several *C. imicola*-borne arboviruses requires an average daily maximum temperature of at least 12.5° C in the coldest month of the year (102, 161). Interpretation of this suggestion is difficult because this temperature is likely to affect activity but not survival of the midges and, also, daily maxima occur long before the vectors are active. Perhaps this temperature should be taken simply as an indicator or correlate of the lower temperatures that occur later and to which the midges are exposed.

While temperature and rainfall probably drive the seasonality of the majority of *Culicoides* species, some may be influenced by other factors. The seasonal distribution of *C. barbosai* is dependent on mean sea level with a superimposed lunar cycle, while that of *C. furens* fluctuates with the moon phase (80). These species breed in mangrove swamps that are subject to tidal influence.

Abundance

Abundance and the special case where abundance equals zero (i.e. the distribution limit) is an integration of the factors already discussed. High rates of activity, low rates of dispersal, rapid larval development, low adult mortality, and year-round breeding will all favor high abundance at a given site. As with seasonality, however, in different places meteorological variables are likely to have different strengths of effect.

At the latitudinal limits of a species' distribution, temperature is often correlated strongly with abundance. Across a number of sites in Spain, for example, *C. imicola* abundance is positively correlated with temperature (140, 151). Similarly, the abundance of *C. imicola* and the incidence of disease spread by it are low where cold temperatures are imposed by altitude (48, 194).

In areas where temperatures are more suitable, abundances tend to be more strongly related to rainfall. In a seven-year study of *C. imicola* in South Africa the greatest annual abundance occurred in the year of the greatest rainfall (135). The role of rainfall (either summer or winter) in the winter rainfall region of South Africa is not clear, but an association between winter rainfall and summer abundance has been reported in Israel (29). In the Caribbean, where *C. insignis* is the principal vector of BTV, the incidence of BT also parallels levels of rainfall (66). In Australia the distributional limits of *C. brevitarsis* are related to temperature, while rainfall has a stronger effect on summer abundance (130). Several studies in Queensland have successfully related both temperature and rainfall to rates of infection or seroconversion of cattle to BTV (201, 203, 204).

In all cases the significance of rainfall for the abundance of *Culicoides* is likely to be through the provision and improvement of breeding sites. As already mentioned, the water content of soil is an important factor determining its suitability for breeding, and it is likely that more rain provides more or better breeding sites. In this way, winter rain in Israel may affect the summer abundance of *C. imicola* (29). However, the situation may be more complex. During the past 200 years in South Africa, epizootics of AHS have been particularly severe during events of the El Niño Southern Oscillation (ENSO) (11). In South Africa the ENSO disrupts the usual patterns of precipitation and may cause heavy rain or severe drought or both. The AHS epizootics have occurred in the subset of ENSO events that bring a particular combination of drought followed by heavy rainfall. For unknown reasons, equally rainy years that are not preceded by drought are less likely to experience epizootics. Interestingly, studies in Kenya in 1968 through 1969 concluded that the abundance of *C. imicola* in that exceptionally rainy year was strongly related to the amount of rainfall (199), but subsequent studies failed to identify clear relationships between rainfall and either *C. imicola* abundance (198) or rates of seroconversion of cattle to BTV (37). The 1968–1969 study occurred during an ENSO event, and the rainfall pattern clearly shows a period of drought preceding the heavy rains and an ensuing surge in *C. imicola* abundance (199). Effects on the vector population probably mediate the relationship between the ENSO and AHS in South Africa.

Factors other than temperature and rainfall may also affect abundance. A study in Morocco found strong negative correlations between wind speeds and the relative abundances of *C. imicola* across 22 sites (8). Although activity lessens during higher-speed winds, in this study abundance was calculated in a manner designed to minimize negative effects on activity. Subsequently, similar patterns have been found in the Iberian region (12). As mentioned earlier, at higher wind speeds the mortality rate of *C. imicola* increases, and this appears to affect the abundance of the midge. Catches of *Culicoides* in Nigeria have also been related to wind speed (42).

The significance of temperature and soil water content for determining the abundance of *C. imicola* allows us to apply the techniques of satellite imagery to modeling distributions. Predicting distributions based on meteorological variables poses several problems, including the difficulty or cost of obtaining data in many countries and the sparse distribution of recording stations. Satellite images, which are readily available at low cost and have global coverage at high resolution, offer solutions to these problems. Several earth-viewing satellite sensors record images that may act as surrogates for climatic variables (61) and two are of particular significance. Land Surface Temperature (LST) estimates correlate with temperature (149) and the Normalized Difference Vegetation Index (NDVI), which is a measure of photosynthetic activity, correlates with soil moisture (49, 132). The annual minimum NDVI is correlated with the abundance of *C. imicola* in Morocco (8) and the Iberian region (12), and the combination of annual minimum LST and annual minimum NDVI accounts for two-thirds of the variance in the abundance of the midge across sites in southern Africa (10). Another advantage of satellite images is the ease with which they can be manipulated to generate meaningful maps. For southern Africa, LST and NDVI have been combined to show predicted abundances of *C. imicola* across the region (Figure 3; see color insert). This is the first such map for any *Culicoides* (10). Challenges for the future include incorporating wind speed in predictive models, identifying upper and lower temperature limits to distributions, and extending predictions beyond the limits of southern Africa.

CONCLUSION: *CULICOIDES* RESEARCH INTO THE TWENTY-FIRST CENTURY

For several reasons research on *Culicoides* has lagged behind that on many other insect vectors. The huge number of species and their minute size have made field study laborious or uncertain while most species have proved intractable to laboratory colonization. They transmit few pathogens of human health significance while the livestock pathogens that they vector were for many years not considered a significant threat to developed countries. Recent severe epizootics of *Culicoides*-borne arboviral diseases, together with the spread of these diseases into new areas of the world has, however, raised their international profile, and rapid progress is now being made in our understanding of the epidemiology of these diseases and of vector *Culicoides* biology.

The sheer abundance and prevalence of *Culicoides* makes the prospect of effective vector control a more distant dream than is the case with many other insect vectors. Research emphasis is likely, therefore, to concentrate on the development of methods to predict when and where disease outbreaks can or will occur. A key issue is the need to identify areas of the world with or without competent vector species. At present we lack the means to identify many species reliably from among species complexes and, for almost all species, we do not have the ability to orally infect individuals under laboratory conditions or to keep them alive for long enough to demonstrate virus replication and transmission. Progress on these fronts awaits developments in molecular biology and, in particular, molecular methods of species differentiation and identification of markers to distinguish between vector and non-vector species. Following the identification of vector species, we need to define the theoretical limits to their distributions and to investigate how these will change under different scenarios of climate change. These limits are set, almost certainly, by climatic factors, and climatic models must be refined through the use of satellite-derived correlates of climatic variables combined with GIS. Equally, we need to identify causes of annual variability in *Culicoides* abundance so that we can predict, climatically, years of greater or lesser risk of disease outbreaks. Ideally, this information could be integrated with a greater understanding of how temperature itself can change the levels of vector competence of *Culicoides* spp. through the agency of mechanisms such as the "leaky gut" phenomenon, as described in this paper.

Visit the Annual Reviews home page at www.AnnualReviews.org.

LITERATURE CITED

1. Akey DH, Potter HW, Jones RH. 1978. Effects of rearing temperature and larval density on longevity, size, and fecundity in the biting gnat *Culicoides variipennis*. *Ann. Entomol. Soc. Am.* 71:411–18
2. Al Busaidy SM, Hamblin C, Taylor WP. 1987. Neutralising antibodies to Akabane virus in free-living wild animals in Africa. *Trop. Anim. Health Prod.* 19:197–202
3. Al Busaidy SM, Mellor PS. 1991. Isolation and identification of arboviruses from the Sultanate of Oman. *Epidemiol. Infect.* 106:403–13
4. Allingham PG. 1991. Effect of temperature on late immature stages of *Culicoides brevitarsis* (Diptera: Ceratopogonidae). *J. Med. Entomol.* 28:878–81
5. Anderson CR, Spence L, Downs WG, Aitken THG. 1961. Oropouche virus: a new human disease agent from Trinidad, West Indies. *Am. J. Trop. Med. Hyg.* 10:574–78
6. Barber TL, Jochim MM, eds. 1985. *Bluetongue and Related Orbiviruses*. New York: Liss. 746 pp.
7. Barnard BJH, Gerdes GH, Meiswinkel R. 1998. Some epidemiological and economic aspects of a bluetongue-like disease in cattle in South Africa—1995/96 and 1997. *Onderstepoort J. Vet. Res.* 65:145–51
8. Baylis M, Bouayoune H, Touti J, El Hasnaoui H. 1998. Use of climatic data and satellite imagery to model the abundance of *Culicoides imicola*, the vector of African horse sickness virus, in Morocco. *Med. Vet. Entomol.* 12:255–66
9. Baylis M, El Hasnaoui H, Bouayoune H, Touti J, Mellor PS. 1997. The spatial and seasonal distribution of African horse sickness and its potential *Culicoides* vectors in Morocco. *Med. Vet. Entomol.* 11:203–12
10. Baylis M, Meiswinkel R, Venter GJ. 1999. A preliminary attempt to use climate data and satellite imagery to model the abundance and distribution of *Culicoides imicola* (Diptera: Ceratopogonidae) in southern Africa. *J. S. Afr. Vet. Assoc.* 70:80–89
11. Baylis M, Mellor PS, Meiswinkel R. 1999. Horse sickness and ENSO in South Africa. *Nature* 397:574
12. Baylis M, Rawlings P. 1998. Modelling the distribution and abundance of *Culicoides imicola* in Morocco and Iberia using climatic data and satellite imagery. See Ref. 104, pp. 137–53
13. Baylis M, Touti J, Bouayoune H, Moudni L, Taoufiq B, El Hasnaoui H. 1998. Studies of the mortality rate of *Culicoides imicola* in Morocco. See Ref. 104, pp. 127–36
14. Bi Y, Li C, Li S, Qing B, Zhong N, et al. 1996. An epidemiological survey of bluetongue disease in Yunnan Province, China. See Ref. 169, pp. 51–56
15. Bishop AL, McKenzie HJ, Barchia IM, Harris AM. 1996. Effect of temperature regimes on the development, survival and emergence of *Culicoides brevitarsis* Kieffer (Diptera: Ceratopogonidae) in bovine dung. *Aust. J. Entomol.* 35:361–68
16. Blackburn NK, Searle L, Phelps RJ. 1985. Viruses isolated from *Culicoides* (Dipt.: Cerat.) caught at the Veterinary Research Farm, Mazowe, Zimbabwe. *J. Entomol. Soc. S. Afr.* 48:331–36
17. Blackwell A. 1997. Diel flight periodic-

ity of the biting midge *Culicoides impunctatus* and the effects of meteorological conditions. *Med. Vet. Entomol.* 11:361–67

18. Blackwell A, Young MR, Mordue W. 1994. The microhabitat of *Culicoides impunctatus* (Diptera: Ceratopogonidae) larvae in Scotland. *Bull. Entomol. Res.* 84:295–301
19. Blanton FS, Wirth WW. 1979. The sandflies (*Culicoides*) of Florida (Diptera: Ceratopogonidae). *Arthropods Fla. Neighb. Land Areas* 10:1–204
20. Boorman J. 1960. Observations on the amount of virus present in the haemolymph of *Aedes aegypti* infected with Uganda S, yellow fever and Semliki Forest viruses. *Trans. R. Soc. Trop. Med. Hyg.* 54:362–65
21. Boorman J. 1986. Presence of bluetongue virus vectors on Rhodes. *Vet. Rec.* 118:21
22. Boorman J. 1989. *Culicoides* of the Arabian Peninsula with notes on their medical and veterinary importance. *Fauna Saudi Arabia* 10:160–224
23. Boorman J. 1993. Biting midges (Ceratopogonidae). In *Contributions to a Manual of Palaearctic Diptera*, ed. L Papp, B Darvas, 2:349–68. Budapest: Sci. Herald
24. Boorman J, Mellor PS, Penn M, Jennings DM. 1975. The growth of African horse sickness virus in embryonating hens eggs and the transmission of virus by *Culicoides variipennis*. *Arch. Virol.* 47:343–49
25. Braverman Y. 1992. The possible introduction to Israel of *Culicoides* (Diptera, Ceratopogonidae) borne animal diseases by wind. See Ref. 200, pp. 291–96
26. Braverman Y. 1992. Host detection, hourly activity, and the preferred biting sites of *Culicoides imicola* (Diptera, Ceratopogonidae) on a calf in Israel. See Ref. 200, pp. 327–32
27. Braverman Y, Boorman J, Kremer M. 1976. Faunistic list of *Culicoides* (Diptera: Ceratopogonidae) from Israel. *Cah. ORSTOM* 14:179–85
28. Braverman Y, Chechik F. 1996. Air streams and the introduction of animal diseases borne on *Culicoides* (Diptera, Ceratopogonidae) into Israel. *Rev. Sci. Tech. OIE* 15:1037–52
29. Braverman Y, Galun R. 1973. The occurrence of *Culicoides* in Israel with reference to the incidence of bluetongue. *Ref. Vet.* 30:121–27
30. Braverman Y, Swanepoel R. 1981. Infection and transmission trials with Nyabira virus in *Aedes aegypti* (Diptera: Culicidae) and two species of *Culicoides* (Diptera: Ceratopogonidae). *Zimb. Vet. J.* 12:13–17
31. Campbell MM, Kettle DS. 1979. Swarming of *Culicoides brevitarsis* Kieffer (Diptera: Ceratopogonidae) with reference to markers, swarm size, proximity of cattle, and weather. *Aust. J. Zool.* 27:17–30
32. Chandler LJ, Ballinger ME, Jones RH, Beaty BJ. 1985. The virogenesis of bluetongue virus in *Culicoides variipennis*. See Ref. 6, pp. 245–53
33. Coetzer JAW, Erasmus BJ. 1994. Equine encephalosis. See Ref. 34, pp. 476–79
34. Coetzer JAW, Thomson GR, Tustin RC, eds. 1994. *Infectious Diseases of Livestock with Special Reference to Southern Africa*, Vol. 1. Cape Town: Oxford Univ. Press. 729 pp.
35. Cybinski DH, Gard GP. 1986. Ephemeral fever group viruses. See Ref. 168a, pp. 289–92
36. Cybinski DH, Muller MJ. 1990. Isolation of arboviruses from cattle and insects at two sentinel sites in Queensland, Australia, 1979–1985. *Aust. J. Zool.* 38:25–32
37. Davies FG. 1978. Bluetongue studies with sentinel cattle in Kenya. *J. Hyg.* 80:197–204
38. Davies FG, Walker AR. 1974. The isolation of ephemeral fever virus from cattle and *Culicoides* midges in Kenya. *Vet. Rec.* 95:63–64
39. Davis AJ, Jenkinson LS, Lawton JH,

Shorrocks B, Wood S. 1998. Making mistakes when predicting shifts in species range in response to global warming. *Nature* 391:783–86

40. *Dep. Anim. Health Mon. Rep.* 1990. Vet. Diagn. Lab., Gaborone, Botswana
41. Derham W. 1713. *Physico-Theology: or a demonstration of the being and attributes of God, from His works of creation. Being the substance of XVI sermons preached in St. Mary Le Bow-Church, London, at the Hon. Mr. Boyle's lectures in the years 1711 and 1712. With large notes and many curious observations never before published.* London: Innys. 483 pp.
42. Dipeolu OO. 1976. Studies on the *Culicoides* species of Nigeria. II. Species collected around wild animals at Ibadan. *Vet. Parasitol.* 1:257–63
43. Doherty RL, Carley JG, Standfast HA, Dyce AL, Snowdon WA. 1972. Virus strains isolated from arthropods during an epizootic of bovine ephemeral fever in Queensland. *Aust. Vet. J.* 48:81–86
44. Doyle KA. 1992. An overview and perspective on Orbivirus disease prevalence and occurrence of vectors in Australia and Oceania. See Ref. 200, pp. 44–57
45. Dulac GC, Dubuc C, Myers DJ, Afshar A, Taylor EA. 1989. Incursion of bluetongue virus type 11 and epizootic haemorrhagic disease of deer type 2 for two consecutive years in the Okanagan Valley. *Can. Vet. J.* 30:351
46. Du Toit RM. 1944. The transmission of bluetongue and horsesickness by *Culicoides*. *Onderstepoort J. Vet. Sci. Anim. Ind.* 19:7–16
47. Eaton BT, Hyatt AD, Brookes SM. 1990. The replication of bluetongue virus. *Curr. Top. Microbiol. Immunol.* 162:89–118
48. Eisa M, Karrar AE, Abu Elrahim AH. 1979. Incidence of bluetongue virus precipitating antibodies in sera of some domestic animals in the Sudan. *J. Hyg.* 83:539–45
49. Farrar TJ, Nicholson SE, Lare AR. 1994. The influence of soil type on the relationships between NDVI, rainfall, and soil moisture in semiarid Botswana. II. NDVI response to soil moisture. *Remote Sens. Environ.* 50:121–33
50. Fassi-Fehri MM. 1993. Hypothèses sur le mode d'introduction de la peste équine au Maroc en 1989. *Revue Méd. Vét.* 144:615–19
51. Foster NM, Breckon RD, Luedke AJ, Jones RH, Metcalf HE. 1977. Transmission of two strains of epizootic haemorrhagic disease virus in deer by *Culicoides variipennis*. *J. Wildl. Dis.* 13:9–16
52. Fu H, Leake CJ, Mertens PPC, Mellor PS. 1996. The barriers to bluetongue virus infection, dissemination and transmission in the vector, *Culicoides variipennis* (Diptera: Ceratopogonidae). *Arch. Virol.* 144:747–61
53. Gard GP. 1987. Studies of bluetongue virulence and pathogenesis in sheep. *Tech. Bull.* 103, Gov. Printer, p. 58. Dep. Ind. Dev. North. Territ., Aust.
54. Gard GP, Shorthose JE, Weir RP, Walsh SJ, Melville LF. 1988. Arboviruses recovered from sentinel livestock in northern Australia. *Vet. Microbiol.* 18:109–18
55. Gorman BM. 1990. The bluetongue viruses. *Curr. Top. Microbiol. Immunol.* 162:1–20
56. Goto Y, Miura Y, Kono Y. 1988. An epidemic of congenital abnormalities with hydranencephaly-cerebellar syndrome of calves in Japan: serologic evidence for the etiological role of Chuzan virus, a new Orbivirus. *Am. J. Vet. Res.* 49:2026–29
57. Greiner EC, Barber TC, Pearson JE, Kramer WL, Gibbs EPJ. 1985. Orbiviruses from *Culicoides* in Florida. See Ref. 6, p. 195–200
58. Gumm ID, Taylor WP, Roach CJ, Alexander FCM, Greiner EC, Gibbs EPJ. 1984. Serological survey of ruminants in

some Caribbean and South American countries for type-specific antibody to bluetongue and epizootic haemorrhagic disease viruses. *Vet. Rec.* 114:635–38
59. Guo Z, Hao J, Chen J, Li Z, Zhang K, Hu Y, et al. 1996. Investigation of bluetongue disease in the Bayannur Meng of Inner Mongolia. See Ref. 169, pp. 80–83
60. Hardy JL, Houk GJ, Kramer LD, Reeves WC. 1983. Intrinsic factors affecting vector competence of mosquitoes for arboviruses. *Annu. Rev. Entomol.* 28:229–62
61. Hay SI, Tucker CJ, Rogers DJ, Packer MJ. 1996. Remotely sensed surrogates of meteorological data for the study of the distribution and abundance of arthropod vectors of disease. *Ann. Trop. Med. Parasitol.* 90:1–19
62. Hayashi K, Suzuki H, Makino Y. 1979. Notes on the transoceanic insects captured on the East China Sea in 1976–1978. *Trop. Med.* 21:1–10
63. Herniman K, Boorman J, Taylor W. 1983. Bluetongue virus in a Nigerian dairy cattle herd. 1. Serological studies and correlation of virus activity to vector population. *J. Hyg.* 90:177–93
64. Hoch AL, Roberts DR, Pinheiro FP. 1990. Host-seeking behaviour and seasonal abundance of *Culicoides paraensis* (Diptera: Ceratopogonidae) in Brazil. *J. Am. Mosq. Control Assoc.* 6:110–14
65. Hoff GL, Trainer DO. 1978. Bluetongue and epizootic haemorrhagic disease viruses: their relationship to wildlife species. *Adv. Vet. Sci. Comp. Med.* 22:111–32
66. Homan EJ, Mo CL, Thompson LH, Barreto CH, Oviedo M-T, et al. 1990. Epidemiologic study of bluetongue viruses in Central America and the Caribbean: 1986–1988. *Am. J. Vet. Res.* 51:1089–94
67. House C, Shipman LD, Weybridge G. 1998. Serological diagnosis of epizootic haemorrhagic disease in cattle in the USA with lesions suggestive of vesicular disease. *Ann. NY. Acad. Sci.* 849:497–500
68. Howell PG. 1960. The 1960 epizootic of African horse sickness in the Middle East and SW Asia. *J. S. Afr. Vet. Med. Assoc.* 31:329–34
69. Howell PG. 1979. The epidemiology of bluetongue in South Africa. In *Arbovirus Research in Australia, 2nd, Brisbane*, ed. TD St George, EL French, pp. 3–13. Brisbane: CSIRO/QIMR
70. Hunt GJ, Tabachnick WJ, McKinnon CN. 1989. Environmental factors affecting mortality of adult *Culicoides variipennis* (Diptera: Ceratopogonidae) in the laboratory. *J. Am. Mosq. Control Assoc.* 5:387–91
71. Jain NC, Prasad G, Gupta Y, Mahajan NK. 1988. Isolation of bluetongue virus from *Culicoides* sp. in India. *Rev. Sci. Tech. OIE* 7:375–78
72. Jamnback H, Watthews T. 1963. Studies of populations of adult and immature *Culicoides sanguisuga* (Diptera: Ceratopogonidae). *Ann. Entomol. Soc. Am.* 56:728–32
73. Jennings DM, Boorman JPT, Ergun H. 1983. *Culicoides* from western Turkey in relation to bluetongue disease of sheep and cattle. *Rev. Elev. Med. Vet. Pays Trop.* 36:67–70
74. Jennings DM, Mellor PS. 1987. Variation in the response of *Culicoides variipennis* to oral infection with bluetongue virus. *Arch. Virol.* 95:177–82
75. Jennings DM, Mellor PS. 1989. *Culicoides*: biological vectors of Akabane virus. *Vet. Microbiol.* 21:125–31
76. Jones RH, Foster NM. 1974. Oral infection of *Culicoides variipennis* with bluetongue virus: development of susceptible and resistant lines from a colony population. *J. Med. Entomol.* 11:316–23
77. Kaneko N, Inaba Y, Akashi H, Miura Y, Shorthose J, Kurashige K. 1986. Isolation of a new bovine ephemeral fever group virus. *Aust. Vet. J.* 63:29
78. Kettle DS. 1957. Preliminary observations on weather conditions and the

activity of biting flies. *Proc. R. Entomol. Soc. London Ser. A* 32:13–20
79. Kettle DS. 1968. The biting habits of *Culicoides furens* (Poey) and *C. barbosai* Wirth & Blanton. I. The 24-h cycle, with a note on differences between collectors. *Bull. Entomol. Res.* 59:21–31
80. Kettle DS. 1969. The biting habits of *Culicoides furens* (Poey) and *C. barbosai* Wirth & Blanton. II. Effect of meteorological conditions. *Bull. Entomol. Res.* 59:241–58
81. Kettle DS. 1984. Ceratopogonidae (Biting Midges). In *Medical and Veterinary Entomology*, pp. 137–58. London: Croom Helm
82. Kettle DS, Edwards PB, Barnes A. 1998. Factors affecting numbers of *Culicoides* in truck traps in coastal Queensland. *Med. Vet. Entomol.* 12:367–77
83. Kramer WL, Greiner EC, Gibbs EPJ. 1985. Seasonal variations in population size, fecundity and parity rates of *Culicoides insignis* (Diptera: Ceratopogonidae) in Florida, USA. *J. Med. Entomol.* 22:163–69
84. Kurogi H, Akiba K, Inaba Y, Matumoto M. 1987. Isolation of Akabane virus from the biting midge *Culicoides oxystoma* in Japan. *Vet. Microbiol.* 15:243–48
85. Kurogi H, Suzuki T, Akahashi H, Ito T, Inaba Y, Matumoto M. 1989. Isolation and preliminary characterization of an orbivirus of the Palyam serogroup from biting midge *Culicoides oxystoma* in Japan. *Vet. Microbiol.* 19:1–11
86. Lardeux FJR, Ottenwaelder T. 1997. Density of larval *Culicoides belkini* (Diptera: Ceratopogonidae) in relation to physicochemical variables in different habitats. *J. Med. Entomol.* 34:387–95
87. Lee VH, Causey OR, Moore DL. 1974. Bluetongue and related viruses in Ibadan, Nigeria: isolation and preliminary identification of viruses. *Am. J. Vet. Res.* 35:1105–8
88. Lillie TH, Marquard WC, Jones RH. 1981. The flight range of *Culicoides variipennis* (Diptera: Ceratopogonidae). *Can. Entomol.* 133:419–26
89. Linley JR, Evans FDS, Evans HT. 1970. Seasonal emergence of *Culicoides furens* (Diptera: Ceratopogonidae) at Vero Beach, Florida. *Ann. Entomol. Soc. Am.* 63:1332–39
90. Linley JR, Hinds MJ. 1976. Seasonal changes in size, female fecundity and male potency in *Culicoides melleus* (Diptera: Ceratopogonidae). *J. Med. Entomol.* 13:151–56
91. Lopez JW, Dubovi EJ, Cupp EW, Lein DH. 1992. An examination of the bluetongue virus status of New York State. See Ref. 200, pp. 140–46
92. MacLachlan NJ. 1994. The pathogenesis and immunology of bluetongue virus infections of ruminants. *Comp. Immunol. Microbiol. Infect. Dis.* 17:197–206
93. Megahed MM. 1956. Anatomy and histology of the alimentary tract of the female of the biting midge *Culicoides nubeculosus* Meigen. (Diptera: Heleididae = Ceratopogonidae). *Parasitology* 46:22–47
94. Meiswinkel R. 1989. Afrotropical *Culicoides*: a redescription of *C. (Avaritia) imicola* Kieffer, 1913 (Diptera: Ceratopogonidae) with description of the closely allied *C. (A.) bolitinos* sp. nov. reared from the dung of the African buffalo, blue wildebeest and cattle in South Africa. *Onderstepoort J. Vet. Res.* 56:23–39
95. Meiswinkel R. 1997. Discovery of a *Culicoides imicola*-free zone in South Africa: preliminary notes and potential significance. *Onderstepoort J. Vet. Res.* 64:81–86
96. Meiswinkel R. 1998. The 1996 outbreak of African horse sickness in South Africa —the entomological perspective. See Ref. 104, pp. 69–83
97. Meiswinkel R. 1998. The ten species in the *Culicoides imicola* Kieffer complex: an update (Ceratopogonidae). *Int. Congr.*

Dipterol., *4th*, pp. 114–15. Oxford, UK. (Abstr.)

98. Meiswinkel R, Nevill EM, Venter GJ. 1994. Vectors: *Culicoides* spp. See Ref. 36, pp. 68–89
99. Mellor PS. 1990. The replication of bluetongue virus in *Culicoides* vectors. *Curr. Top. Microbiol. Immunol.* 162:143–61
100. Mellor PS. 1993. African horse sickness: transmission and epidemiology. *Vet. Res.* 24:199–212
101. Mellor PS. 1994. Epizootiology and vectors of African horse sickness virus. *Comp. Immunol. Microbiol. Infect. Dis.* 17:287–96
102. Mellor PS. 1996. *Culicoides*: vectors, climate change and disease risk. *Vet. Bull.* 66:301–6
103. Mellor PS. 1998. Climate-change and the distribution of vector-borne diseases with special reference to African horse sickness virus. *Proc. Int. Symp. Diagn. Control. Livestock Dis. Nucl. Related Tech.* Vienna, Austria, pp. 439–54. Vienna: IAEA
104. Mellor PS, Baylis M, Hamblin C, Calisher CH, Mertens PPC, eds. 1998. *African Horse Sickness. Arch. Virol. Suppl.* 14. Wein: Springer-Verlag. 342 pp.
105. Mellor PS, Baylis M, Rawlings P, Wellby MP. 1999. Effect of temperature on African horse sickness virus serotype 9 infection of vector species of *Culicoides* (Diptera: Ceratopogonidae). *Equine Infect. Dis.* 8:246–51
106. Mellor PS, Boned J, Hamblin C, Graham S. 1990. Isolations of African horse sickness virus from vector insects made during the 1988 epizootic in Spain. *Epidemiol. Infect.* 105:447–54
107. Mellor PS, Boorman J. 1995. The transmission and geographical spread of African horse sickness and bluetongue viruses. *Ann. Trop. Med. Parasitol.* 89:1–15
108. Mellor PS, Jennings DM, Wilkinson PJ, Boorman JPT. 1985. *Culicoides imicola*; a bluetongue virus vector in Spain and Portugal. *Vet. Rec.* 116:589–90
109. Mellor PS, Osborne R, Jennings DM. 1984. Isolation of bluetongue and related viruses from *Culicoides* species in the Sudan. *J. Hyg.* 93:621–28
110. Mellor PS, Pitzolis G. 1979. Observations on breeding sites and light-trap collections of *Culicoides* during an outbreak of bluetongue in Cyprus. *Bull. Entomol. Res.* 69:229–34
111. Mellor PS, Rawlings P, Baylis M. Wellby MP. 1998. Effect of temperature on African horse sickness virus infection in *Culicoides*. See Ref. 104, pp. 155–63
112. Metselaar D, Robin V. 1976. Akabane virus isolated in Kenya. *Vet. Rec.* 99:86
113. Mo CL, Thompson LH, Homan EJ, Oviedo MT, Greiner EC, Gonzalez J, et al. 1994. Bluetongue virus isolation from vectors and ruminants in Central America and the Caribbean. *Am. J. Vet. Res.* 55:211–15
114. Mohammed MEH, Mellor PS. 1990. Further studies on bluetongue and bluetongue-related Orbiviruses in the Sudan. *Epidemiol. Infect.* 105:619–32
115. Mohammed MEH, Taylor WP. 1987. Infection with bluetongue and related orbiviruses in the Sudan detected by the study of sentinel calf herds. *Epidemiol. Infect.* 99:533–45
116. Mullen GR, Anderson RR. 1998. Transmission of Orbiviruses by *Culicoides* Latreille species (Ceratopogonidae) among cattle and white-tailed deer in the southeastern United States. *Int. Congr. Dipterol.*, 4th, pp. 155–56. Oxford, UK. (Abstr.)
117. Mullens BA. 1987. Seasonal size variability in *Culicoides variipennis* (Diptera: Ceratopogonidae) in southern California. *J. Am. Mosq. Control Assoc.* 3:512–13
118. Mullens BA, Rodriguez JL. 1992. Survival and vertical distribution of larvae of *Culicoides variipennis* (Diptera: Ceratopogonidae) in drying mud habitats. *J. Med. Entomol.* 29:745–49

119. Mullens BA, Rutz DA. 1983. Development of immature *Culicoides variipennis* (Diptera: Ceratopogonidae) at constant laboratory temperatures. *Ann. Entomol. Soc. Am.* 76:747–51
120. Mullens BA, Tabachnick WJ, Holbrook FR, Thompson LH. 1995. Effects of temperature on virogenesis of bluetongue virus serotype 11 in *Culicoides variipennis sonorensis. Med. Vet. Entomol.* 9:71–76
121. Muller MJ, Li H. 1996. Preliminary results of trapping for *Culicoides* in south China: future bluetongue vector studies. See Ref. 169, pp. 129–35
122. Muller MJ, Standfast HA. 1986. Vectors of ephemeral fever group viruses. See Ref. 168a, pp. 295–98
123. Muller MJ, Standfast HA, St George TD, Cybinski DH. 1982. *Culicoides brevitarsis* (Diptera: Ceratopogonidae) as a vector of arboviruses in Australia. In *Arbovirus Research in Australia, 3rd, Brisbane,* ed. TD St George, BH Kay, pp. 43–49. Brisbane: CSIRO/QIMR
124. Murray MD. 1970. The spread of ephemeral fever of cattle during the 1967–68 epizootic in Australia. *Aust. Vet. J.* 46:77–82
125. Murray MD. 1975. Potential vectors of bluetongue in Australia. *Aust. Vet. J.* 51:216–20
126. Murray MD. 1986. The influence of abundance and dispersal of *Culicoides brevitarsis* on the epidemiology of arboviruses of livestock in New South Wales. See Ref. 168a, pp. 232–34
127. Murray MD. 1987. Akabane epizootics in New South Wales: evidence for long-distance dispersal of the biting midge *Culicoides brevitarsis. Aust. Vet. J.* 64:305–8
128. Murray MD. 1987. Local dispersal of the biting midge *Culicoides brevitarsis* Kieffer (Diptera: Ceratopogonidae) in south-eastern Australia. *Aust. J. Zool.* 35:559–73
129. Murray MD. 1991. The seasonal abundance of female biting-midges, *Culicoides brevitarsis* Kieffer (Diptera: Ceratopogonidae), in coastal south-eastern Australia. *Aust. J. Zool.* 39:333–42
130. Murray MD. 1995. Influences of vector biology on transmission of arboviruses and outbreaks of disease: the *Culicoides brevitarsis* model. *Vet. Microbiol.* 46:91–99
131. Nandi S, Negi BS. 1999. Bovine ephemeral fever: a review. *Comp. Immunol. Microbiol. Infect. Dis.* 22:81–91
132. Narasimha Rao PV, Venkataratnam L, Krishna Rao PV, Ramana KV. 1993. Relation between root zone soil moisture and normalized difference vegetation index of vegetated fields. *Int. J. Remote Sens.* 14:441–49
133. Nelson RL, Bellamy RE. 1971. Patterns of flight activity of *Culicoides variipennis* (Coquillett) (Diptera: Ceratopogonidae). *J. Med. Entomol.* 8:283–91
134. Nevill EM. 1967. *Biological studies on some South African* Culicoides *species (Diptera: Ceratopogonidae) and the morphology of their immature stages*. MSc thesis. Univ. Pretoria, Pretoria
135. Nevill EM. 1971. Cattle and *Culicoides* biting midges as possible overwintering hosts of bluetongue virus. *Onderstepoort J. Vet. Res.* 38:65–72
136. Nevill EM, Erasmus BJ, Venter GJ. 1992. A six-year survey of viruses associated with *Culicoides* biting midges throughout South Africa (Diptera: Ceratopogonidae). See Ref. 200, pp. 314–19
137. Newton LG, Wheatley CH. 1970. The occurrence and spread of ephemeral fever of cattle in Queensland. *Aust. Vet. J.* 46:561–68
138. OIE, Dis. Inf. 1998. Bluetongue in Greece, confirmation of diagnosis. *OIE, Dis. Inf.* 11:166–67
139. Omori T, Inaba Y, Morimoto T, Tanaka Y, Ishitani R, et al. 1969. Ibaraki virus, an agent of epizootic disease of cattle

resembling bluetongue. I. Epidemiologic, clinical and pathologic observations and experimental transmission to calves. *Jpn. J. Microbiol.* 13:139–57
140. Ortega MD, Lloyd JE, Holbrook FR. 1997. Seasonal and geographical distribution of *Culicoides imicola* Kieffer (Diptera: Ceratopogonidae) in southwestern Spain. *J. Am. Mosq. Control Assoc.* 13:227–32
141. Oya A, Okuno T, Ogata T, Kobayashi I, Makuyana T. 1961. Akabane virus, a new arbovirus isolated in Japan. *Jpn. J. Med. Sci. Biol.* 14:101–8
142. Parsonson IM, Snowdon WA. 1985. Bluetongue, epizootic haemorrhagic disease of deer and related viruses: current situation in Australia. See Ref. 6, pp. 27–36
143. Paton T. 1863. The "horse sickness" of the Cape of Good Hope. *The Vet.* 36:489–94
144. Pinheiro FP, Hoch AL, Gomes MLC, Roberts DR. 1981. Oropouche virus. IV. Laboratory transmission by *Culicoides paraensis*. *Am. J. Trop. Med. Hyg.* 30:172–76
145. Pinheiro FP, Rocha AG, Freitas RB, Ohana BA, Travassos, et al. 1982. Meningite associada as infeccoes por virus Oropouche. *Rev. Inst. Med. Trop. São Paulo*, 24:246–51
146. Pinheiro FP, Travassos Da Rosa APA, Gomes MLC, Leduc JW, Hoch AL. 1982. Transmission of Oropouche virus from man to hamster by the midge *Culicoides paraensis*. *Science* 215:1251–53
147. Pinheiro FP, Travassos Da Rosa APA, Travassos Da Rosa JF, Ishak R, Freitas RB, et al. 1981. Oropouche virus. I. A review of clinical, epidemiological and ecological findings. *Am. J. Trop. Med. Hyg.* 30:149–60
148. Pinheiro FP, Travassos Da Rosa APA, Vasconcelos PFC. 1998. An overview of Oropouche fever epidemics in Brazil and neighbouring countries. See Ref. 188, pp. 186–92
149. Price JC. 1984. Land surface temperature measurements from the split window channels of the NOAA 7 Advanced Very High Resolution Radiometer. *J. Geophys. Res.* 89:7231–37
150. Qin Q, Tai Z, Wang L, Luo Z, Hu J, Lin H. 1996. Bluetongue epidemiological survey and virus isolation in Xinjiang, China. See Ref. 169, pp. 67–71
151. Rawlings P, Capela R, Pro MJ, Ortega MD, Pena I, et al. 1998. The relationship between climate and the distribution of *Culicoides imicola* in Iberia. See Ref. 104, pp. 93–102
152. Rawlings P, Pro MJ, Pena I, Ortega MD, Capela R. 1997. Spatial and seasonal distribution of *Culicoides imicola* in Iberia in relation to the transmission of African horse sickness virus. *Med. Vet. Entomol.* 11:49–57
153. Roberts DR, Hoch AL, Dixon KE, Llewellyn CH. 1981. Oropouche virus. III. Entomological observations from three epidemics in Para, Brazil, 1975. *Am. J. Trop. Med. Hyg.* 30:165–71
154. Roy P, Gorman BM, eds. 1990. *Bluetongue Viruses*. Berlin/Heidelberg: Springer-Verlag. 200 pp.
155. Seddon HR. 1938. The spread of ephemeral fever (three-day sickness) in Australia. *Aust. Vet. J.* 14:90–101
156. Sellers RF. 1980. Weather, host and vector—their interplay in the spread of insect-borne animal virus diseases. *J. Hyg.* 85:65–102
157. Sellers RF. 1992. Weather, *Culicoides*, and the distribution and spread of bluetongue and African horse sickness viruses. See Ref. 200, pp. 284–90
158. Sellers RF, Gibbs EPJ, Herniman KAJ, Pedgley DE, Tucker MR. 1979. Possible origin of the bluetongue epidemic in Cyprus, August 1977. *J. Hyg.* 83:547–55
159. Sellers RF, Maarouf AR. 1989. Trajectory analysis and bluetongue virus sero-

type 2 in Florida 1982. *Can. J. Vet. Res.* 53:100–2

160. Sellers RF, Maarouf AR. 1991. Possible introduction of epizootic hemorrhagic disease of deer virus (serotype 20) and bluetongue virus (serotype 11) into British Columbia in 1987 and 1988 by infected *Culicoides* carried on the wind. *Can. J. Vet. Res.* 55:367–70
161. Sellers RF, Mellor PS. 1993. Temperature and the persistence of viruses in *Culicoides* spp. during adverse conditions. *Rev. Sci. Tech. OIE* 12:733–55
162. Sellers RF, Pedgley DE. 1985. Possible windborne spread to Western Turkey of bluetongue virus in 1977 and of Akabane virus in 1979. *J. Hyg.* 95:149–58
163. Sellers RF, Pedgley DE, Tucker MR. 1977. Possible spread of African horse sickness on the wind. *J. Hyg.* 79:279–98
164. Sellers RF, Pedgley DE, Tucker MR. 1978. Possible windborne spread of bluetongue to Portugal, June–July 1956. *J. Hyg.* 81:189–96
165. Sendow I, Sukarsih, Soleha E, Pearce M, Bahri S, Daniels PW. 1996. Bluetongue virus research in Indonesia. See Ref. 169, pp. 28–32
166. Smith KE, Stallknecht DE, Nettles VF. 1996. Experimental infection of *Culicoides lahillei* (Diptera: Ceratopogonidae) with epizootic haemorrhagic disease virus serotype 2 (Orbivirus: Reoviridae). *J. Med. Entomol.* 33:117–21
167. Sohn R, Yuill TM. 1991. Bluetongue and epizootic haemorrhagic disease in wild ruminants. *Bull. Soc. Vector Biol.* 16:17–24
168. St George TD. 1994. Bovine ephemeral fever. See Ref. 34, pp. 553–62

168a. St George TD, Kay BH, Blok J, eds. 1986. *Arbovirus Research in Australia, 4th, Brisbane.* Brisbane: QIMR/CSIRO

169a. St George TD, Peng K, eds. 1996. *Bluetongue Disease in Southeast Asia and the Pacific.* Canberra: ACIAR. 264 pp.

170. St George TD, Standfast HA. 1988. Bovine ephemeral fever. In *The Arboviruses: Epidemiology and Ecology*, ed. TP Monath, 2:71–86. Boca Raton, FL: CRC
171. St George TD, Standfast HA. 1994. Diseases caused by Akabane and related Simbu-group viruses. See Ref. 34, pp. 681–87
172. St George TD, Standfast HA, Christie DG, Knott SG, Morgan IR. 1977. The epizootiology of bovine ephemeral fever in Australia and Papua-New Guinea. *Aust. Vet. J.* 53:17–28
173. St George TD, Standfast HA, Cybinski DH. 1978. Isolations of Akabane virus from sentinel cattle and *Culicoides brevitarsis*. *Aust. Vet. J.* 54:558–61
174. Standfast HA, Dyce AL, Muller MJ. 1985. Vectors of bluetongue virus in Australia. See Ref. 6, pp. 177–86
175. Standfast HA, St George TD, Dyce AL. 1976. The isolation of ephemeral fever from mosquitoes in Australia. *Aust. Vet. J.* 52:242
176. Sukarsih, Daniels PW, Sendow I, Soleha E. 1993. Longitudinal studies of *Culicoides* spp. associated with livestock in Indonesia. In *Arbovirus Research in Australia, 6th, Brisbane,* ed. Uren MF, Kay BH, pp. 203–9. Brisbane: CSIRO/QIMR
177. Sukarsih, Sendow I, Bahri S, Pearce M, Daniels PW. 1996. *Culicoides* survey in Indonesia. See Ref. 169, pp. 123–28
178. Swanepoel R. 1994. Palyam serogroup Orbivirus infections. See Ref. 34, pp. 480–83
179. Swanepoel R, Blackburn NK, Lander KP, Vickers DB, Lewis AR. 1975. An investigation of infectious infertility and abortion of cattle. *Rhod. Vet. J.* 6:42–55
180. Tabachnick WJ. 1991. Genetic control of oral susceptibility to infection of *Culicoides variipennis* for bluetongue virus. *Am. J. Trop. Med. Hyg.* 45:666–71
181. Tabachnick WJ. 1992. Genetics, population genetics and evolution of *Culicoides variipennis*: implications for bluetongue virus transmission in the USA and its international impact. See Ref. 200, pp. 262–70
182. Tabachnick WJ. 1996. *Culicoides vari-*

ipennis and bluetongue-virus epidemiology in the United States. *Annu. Rev. Entomol.* 41:23–43
183. Tabachnick WJ, Holbrook FR. 1992. The *Culicoides variipennis* complex and the distribution of the bluetongue viruses in the United States. *Proc. US Anim. Health Assoc.* 96:207–12
184. Taylor WP. 1986. The epidemiology of bluetongue. *Rev. Sci. Tech. OIE* 5:351–56
185. Taylor WP, Mellor PS. 1994. The distribution of Akabane virus in the Middle East. *Epidemiol. Infect.* 113:175–86
186. Theodoridis A, Giesecke WH, Du Toit IJ. 1973. Effect of ephemeral fever on milk production and reproduction of dairy cattle. *Onderstepoort J. Vet. Res.* 40:83–91
187. Theodoridis A, Nevill EM, Els HJ, Boshoff ST. 1979. Viruses isolated from *Culicoides* midges in South Africa during unsuccessful attempts to isolate bovine ephemeral fever virus. *Onderstepoort J. Vet. Res.* 46:191–98
188. Travassos da Rosa APA, Vasconcelos PFC, Travassos da Rosa JFC, eds. 1998. *An Overview of Arbovirology in Brazil and Neighbouring Countries.* Belem, Brazil: ECI. 296 pp.
189. Travis BV. 1949. Studies on mosquito and other biting-insect problems in Alaska. *J. Econ. Entomol.* 42:451–57
190. Vaughan JA, Turner EC Jr. 1985. Spatial distribution of immature *Culicoides variipennis* (Coq.). See Ref. 6, pp. 213–19
191. Vaughan JA, Turner EC Jr. 1987. Development of immature *Culicoides variipennis* (Diptera: Ceratopogonidae) from Saltville, Virginia, at constant laboratory temperatures. *J. Med. Entomol.* 24:390–95
192. Venter GJ. 1999. *Arboviral diseases in southern Africa—identification of the vectors and development of a climate-driven risk-assessment model.* Ann. Rep. Eur. Comm., 3rd, pp. 35–49
193. Venter GJ, Hill E, Pajor ITP, Nevill EM. 1991. The use of a membrane feeding technique to determine the infection rate of *Culicoides imicola* (Diptera: Ceratopogonidae) for 2 bluetongue virus serotypes in South Africa. *Onderstepoort J. Vet. Res.* 58:5–9
194. Venter GJ, Meiswinkel R. 1994. The virtual absence of *Culicoides imicola* (Diptera: Ceratopogonidae) in a light-trap survey of the colder, high-lying area of the Orange Free State, South Africa, and implications for the transmission of arboviruses. *Onderstepoort J. Vet. Res.* 61:327–40
195. Venter GJ, Nevill EM, Van der Linde TC de K. 1997. Seasonal abundance and parity of stock-associated *Culicoides* species (Diptera: Ceratopogonidae) in different climatic regions in southern Africa in relation to their viral vector potential. *Onderstepoort J. Vet. Res.* 64:259–71
196. Venter GJ, Paweska JT, Groenewald D, Venter E, Howell PG. 1999. Vector competence of selected South African *Culicoides* species (Diptera: Ceratopogonidae) for the Bryanston serotype of equine encephalosis virus (EEV). *Med. Vet. Entomol.* 13:393–400
197. Venter GJ, Paweska JT, Van Dijk AA, Mellor PS, Tabachnick WJ. 1998. Vector competence of *Culicoides bolitinos* and *C. imicola* for South African bluetongue virus serotypes 1, 3 and 4. *Med. Vet. Entomol.* 12:378–85
198. Walker AR. 1977. Seasonal fluctuations of *Culicoides* species (Diptera: Ceratopogonidae) in Kenya. *Bull. Entomol. Res.* 67:217–33
199. Walker AR, Davies FG. 1971. A preliminary survey of the epidemiology of bluetongue in Kenya. *J. Hyg.* 69:47–60
200. Walton TE, Osburn BI, eds. 1992. *Bluetongue, African Horse Sickness, and Related Orbiviruses.* Boca Raton, FL: CRC. 1042 pp.
201. Ward MP. 1994. Climatic factors associated with the prevalence of bluetongue virus infection of cattle herds in Queensland, Australia. *Vet. Rec.* 134:407–10

202. Ward MP. 1995. Bluetongue and Douglas virus activity in New South Wales in 1989: further evidence for long-distance dispersal of the biting midge *Culicoides brevitarsis*. *Aust. Vet. J.* 72:197
203. Ward MP. 1996. Climatic factors associated with the infection of herds of cattle with bluetongue viruses. *Vet. Res. Commun.* 20:273–83
204. Ward MP, Thurmond MC. 1995. Climatic factors associated with risk of seroconversion of cattle to bluetongue viruses in Queensland. *Prev. Vet. Med.* 24:129–36
205. Watts DM, Phillips I, Callahan JD, Griebenow W, Hyams KC, Hayes CG. 1997. Oropouche virus transmission in the Amazon River basin of Peru. *Am. J. Trop. Med. Hyg.* 56:148–52
206. Wellby M, Baylis M, Rawlings P, Mellor PS. 1996. Effect of temperature on survival and rate of virogenesis of African horse sickness virus in *Culicoides variipennis sonorensis* (Diptera: Ceratopogonidae) and its significance in relation to the epidemiology of the disease. *Bull. Entomol. Res.* 86:715–20
207. Wetzel H, Nevill EM, Erasmus BJ. 1970. Studies on the transmission of African horse sickness. *Onderstepoort J. Vet. Res.* 37:165–68
208. Whistler T, Swanepoel R. 1988. Characterization of potentially foetotropic Palyam serogroup Orbiviruses isolated in Zimbabwe. *J. Gen. Virol.* 69:2221–27
209. WHO. 1967. Arboviruses and human disease. *WHO Tech. Rep. Ser.* No. 369, p. 22. Geneva: WHO
210. Wirth WW, Hubert AA. 1989. *The* Culicoides *of Southeast Asia (Diptera: Ceratopogonidae)*. 44. *Mem. Am. Entomol. Inst.* Gainesville, FL: Am. Entomol. Inst. 508 pp.
211. Wittmann EJ. 1999. *Temperature and the transmission of arboviruses by* Culicoides. PhD thesis. Univ. Bristol, Bristol.
212. Zhang N, Li Z, Zhang K, Hu Y, Li G, et al. 1996. Bluetongue history, serology and virus isolations in China. See Ref. 169, pp. 43–50
213. Zhang N, MacLachlan NJ, Bonneau KR, Zhu J, Li Z, et al. 1999. Identification of seven serotypes of bluetongue virus from the People's Republic of China. *Vet. Rec.* In press

Annu. Rev. Entomol. 2000. 45:341–369

EVOLUTIONARY ECOLOGY OF PROGENY SIZE IN ARTHROPODS

Charles W. Fox and Mary Ellen Czesak
Department of Entomology, S-225 Agricultural Science Center North, University of Kentucky, Lexington, Kentucky 40546-0091; e-mail: cfox@ca.uky.edu

Key Words egg size, geographic variation, life history, natural selection, parental investment

■ **Abstract** Most models of optimal progeny size assume that there is a trade-off between progeny size and number, and that progeny fitness increases with increasing investment per young. We find that both assumptions are supported by empirical studies but that the trade-off is less apparent when organisms are iteroparous, use adult-acquired resources for reproduction, or provide parental care. We then review patterns of variation in progeny size among species, among populations within species, among individuals within populations, and among progeny produced by a single female. We argue that much of the variation in progeny size among species, and among populations within species, is likely due to variation in natural selection. However, few studies have manipulated progeny environments and demonstrated that the relationship between progeny size and fitness actually differs among environments, and fewer still have demonstrated why selection favors different sized progeny in different environments. We argue that much of the variation in progeny size among females within populations, and among progeny produced by a single female, is probably nonadaptive. However, some species of arthropods exhibit plasticity in progeny size in response to several environmental factors, and much of this plasticity is likely adaptive. We conclude that advances in theory have substantially outpaced empirical data. We hope that this review will stimulate researchers to examine the specific factors that result in variation in selection on progeny size within and among populations, and how this variation in selection influences the evolution of the patterns we observe.

INTRODUCTION

Progeny size is an especially interesting life history trait because it is simultaneously a maternal and progeny character—mothers make eggs and determine egg size, but egg size can have substantial fitness effects for progeny. Thus, progeny size is subject to selection in both the parental and progeny generations. This selection often varies in direction and/or magnitude among generations (parental versus offspring), among environments, and even among siblings within a family, such that understanding the factors that influence the evolution of progeny size can become quite a challenge. In this review, we focus on understanding the causes and consequences of egg and progeny size variation in arthropods. Although most arthropods lay eggs, many crustaceans brood their eggs and studies of progeny size in crustaceans measure progeny after eggs hatch and are released by the parent. We thus use the phrases "egg size" or "progeny size" interchangeably. We also acknowledge that eggs and progeny often vary in ways other than size (e.g. egg composition) and that this variation may be ecologically and evolutionarily as important as variation in size (17). However, due to space constraints we limit our discussion to progeny size.

We begin with a brief discussion of the optimality model developed by Smith & Fretwell that laid the foundation for how we think about the evolution of progeny size (197). We focus on empirical studies that examine the two primary assumptions of this model. Next, we explore patterns of variation in progeny size among species and among populations within species. Lastly, we review the sources of variation in progeny size within populations and discuss proposed explanations for this variation.

CONCEPTUAL FRAMEWORK

Christopher Smith and Steven Fretwell (197) offered the first mathematical analysis of optimal progeny size. They asked "what size progeny should a female produce to maximize her total number of grandprogeny?" The number of grandprogeny a female will produce depends on both the number of progeny she produces and the fitness of those progeny. To model this, Smith & Fretwell started with two assumptions: *(a)* progeny fitness (W_{Young}) increases with increasing parental investment per offspring (I_{Young}) (i.e. larger progeny will have higher fitness), and *(b)* for any fixed amount of parental investment into reproduction (I_{Total}), a female can produce $N = I_{Total} / I_{Young}$ progeny. In other words, there is a trade-off between the number of progeny a female can make and the amount of resources allocated to each of those progeny. If a female makes larger progeny, I_{Young} increases and N decreases. To increase N, a female must either decrease I_{Young} or increase I_{Total}. Smith & Fretwell assumed that I_{Total} is a constant. Maternal fitness, $W_{Parent} = N \times (W_{Young}) = (I_{Total} / I_{Young}) \times (W_{Young})$, the product of the number of progeny that she produces times the fitness of each of those progeny. The value of I_{Young} that results in the highest parental fitness is the value that maximizes $(I_{Total} / I_{Young}) \times (W_{Young})$. Maternal fitness thus increases as W_{Young} increases, but also increases as (I_{Total} / I_{Young}) increases (i.e. fecundity increases). The constraint here is that for any fixed amount of resources (I_{Total}), females can increase W_{Young} only by increasing I_{Young}, which necessarily results in a decrease in fecundity (I_{Total} / I_{Young}).

This model illustrates three points that have become the subject of much empirical and theoretical exploration. First, for any fixed parental allocation to reproduction, progeny size is under balancing selection; large progeny are favored because W_{Young} increases as I_{Young} increases, and small progeny are favored because N increases as I_{Young} decreases. Second, there is a conflict of interest between parents and their progeny. Because progeny fitness (W_{Young}) increases with increasing investment per progeny (I_{Young}), the value of I_{Young} that maximizes progeny fitness is larger than the value that maximizes parental fitness. Third, any environmental variable that affects the relationship between investment per progeny and progeny fitness (i.e. between I_{Young} and W_{Young}) can result in a change in the optimal progeny size and thus a change in the size of progeny that should evolve in a population. The first and third of these points will be discussed in this paper. The consequences of conflicts of interest between parents and their offspring has been reviewed extensively elsewhere and will not be discussed here.

Since the original development of the Smith-Fretwell model, more complex models have been developed to examine optimal progeny size under more specific conditions (46, 182). It is not the objective of this paper to review the various models and their specific assumptions (see 17). However, most of these models start with the same basic assumptions that Smith & Fretwell started with, that *(a)* there is a trade-off between progeny size and number, and *(b)* progeny fitness (W_{Young}) increases with increasing parental investment per offspring. We thus focus first on these two assumptions.

Trade-Offs Between Progeny Size and Number

The concept of trade-offs is an integral part of life history theory (182). If an individual has a fixed amount of resources available, those resources can be divided into three basic functions—growth, somatic maintenance, or reproduction. Resources directed to reproduction can subsequently be divided into either many small progeny or a few larger progeny. Thus, for a fixed amount of resources allocated to reproduction it necessarily follows that there is a trade-off between the number and size of progeny

Phenotypic correlations between egg size and number *(a)* among species (18, 19, 33, 42, 65, 84, 86, 127, 129, 139, 143, 175, 177, 193, 204, 217), *(b)* among populations within species (2, 56, 127, 230) and *(c)* among individuals within populations (Table 1) generally indicate a trade-off between egg size and number. Most of these studies examine only phenotypic correlations between egg size and number, but a genetically based trade-off has been demonstrated for *Daphnia* (57, 140).

In general, trade-offs have been detected in most studies of relatively semelparous arthropods that use larval-acquired resources for egg production and exhibit no parental care (Table 1). In studies of more complex systems (especially vertebrates), in which females are iteroparous, use adult-acquired resources for reproduction (e.g. shrimp, mosquitoes), or exhibit parental care (e.g. birds), a trade-off has been more difficult to demonstrate (87), leading some authors to suggest that such a trade-off is not universal (e.g. 17). Failure to demonstrate

TABLE 1 Evidence for (or against) a trade-off between egg/progeny size and number, based on variation among females within a population (number of species)

Taxon	Trade-off	No Trade-off	Reference (trade-off)	Reference (no trade-off)
Crustacea (cladoceran)	7	1	15[2], 22, 45, 87[1], 140, 153, 213	15[2]
Crustacea (copepod)	2	0	2, 93	
Crustacea (shrimp)	0	4		42
Crustacea (isopod)	2	1	135, 230	50
Orthoptera	5	0	39, 68[3]	
Heteroptera	3	0	132, 154, 198	
Lepidoptera	3	3	66[2], 136, 180	16, 26, 146[4]
Coleoptera	3	0	81, 94, 219	
Diptera	3	2	4[2,5], 138	4[2], 29

[1]Varied with age of the female (iteroparous organism)
[2]Based on variation in egg size through the season
[3]Confounded with maternal age
[4]Did not control for female size
[5]Varied among studies

trade-offs probably has less to do with their absence than with the complexity of the system. For a trade-off between egg size and number to be evident, we must assume that the quantity of resources allocated to reproduction (I_{Total}) is constant. Yet I_{Total} is often not constant. For example, variation in larval growth can produce substantial variation in body size at maturation, which generally corresponds closely to total reproductive effort (within a population). Thus, larger individuals generally lay both more and larger eggs, leading to a positive correlation between egg size and number. In this case, the relationship between egg size and number will be negative only when body size is controlled (e.g. 16, 40, 81, 154).

Other sources of variation in reproductive effort are less easily quantified and controlled, including variation in adult feeding rates, the proportion of adult-acquired resources allocated to reproduction, degree of parental care, etc. Our conclusion is that the assumption of a trade-off between egg size and number is generally supported by empirical studies in arthropods; studies that have failed to detect such a trade-off have generally been on animals in which there may be substantial variation in reproductive effort obscuring the patterns.

We suggest that rather than testing for the presence or absence of trade-offs between progeny size and number, future research should focus on two general issues. First, we know of little empirical data on the shape of the relationship between progeny size and number, although theoretical predictions often depend on an assumed shape (but see 40a). Smith & Fretwell (197) originally proposed that the number of offspring produced by a female is a simple function of I_{Total} and I_{Young}; the female can produce I_{Total}/I_{Young} progeny. However, it is likely that, due to inefficiencies in resource allocation, allocating I_{Total} resources to reproduction does not allow for the production of I_{Total} / I_{Young} progeny of size I_{Young}; dividing resources among progeny may not be as simple as dividing a pie into pieces. Second, we have little understanding of how changes in reproductive effort affect the relationship between progeny size, progeny number, and maternal fitness (230a). Reproductive effort may evolve as a result of changes in female survival probabilities, changes in resource availability, or due to selection on progeny size or fecundity (e.g. 182). Smith & Fretwell (197) and most models since have assumed that total reproductive effort is constant (but see 230a). More theoretical and empirical exploration of these two issues is needed.

Fitness Consequences of Progeny Size

Many studies have examined the relationship between egg size and fitness components of progeny. They often demonstrate that smaller eggs hatch more quickly (7, 72) or are brooded for a shorter time (231), but are less likely to hatch (7, 48, 69, 70, 154; but see 93, 146, 214). Progeny hatching from smaller eggs tend to be smaller hatchlings (7, 9, 27, 31, 38, 46a, 92–94, 130, 132, 161, 178, 187, 225) that grow into smaller-than-normal later instars (118, 132, 135, 219) and have lower juvenile survivorship (27, 36, 38, 40, 70, 78, 81, 110, 119, 168, 214, 219; but see 49, 72, 202).

Smaller-than-average young have three developmental options: *(a)* mature at a smaller-than-average size (27, 36, 38, 79, 90, 109, 118, 130, 154, 202; but see 9, 49, 179, 180), *(b)* extend development to fully or partially compensate for their small starting size (7, 27, 56, 70, 72, 73, 76, 94, 101, 137, 154, 179, 184, 190, 202, 214; but see 9, 118, 219, 229), or *(c)* increase their rate of growth to mature at a normal size. Most arthropods exhibit some degree of developmental plasticity by which progeny partially compensate for their small hatchling/birth size by extending development time (72, 73). Few studies have examined the influence of juvenile size on growth rates in arthropods (but see 7). Progeny hatching from larger eggs can often better withstand environmental stresses such as larval competition (7), starvation (38, 89, 145, 199, 212), desiccation (201), oxygen stress (97), cold stress (36, 105), nutritional stress (27, 74, 78, 219), and environmental toxins (62). Some studies have failed to detect fitness advantages of hatching from large eggs; most of these studies have raised progeny in high-quality environments (e.g. 118, 228, 229), suggesting that selection is generally weak in high-quality environments but favors larger eggs in lower-quality environments (74, 189).

Most of the studies cited here are correlational studies that confound relationships between egg size and progeny fitness with genetic correlations between morphological and life history characters (194). For example, larger females generally lay larger eggs and produce progeny that mature at a larger size (because body size is generally heritable) such that there is a positive correlation between egg size and progeny size at maturation (71). Experimental approaches were thus developed to study the consequences of, and selection on, egg size variation (194).

By manipulating egg size physically or physiologically we can quantify effects of egg size variation on progeny fitness. A few studies have manipulated egg size in invertebrates (61, 100, 195) including one insect species (72). They have generally demonstrated that progeny hatching from larger eggs do indeed have higher fitness or improved performance (but see 99).

Time Limitation, Parental Care, Clutch Size, and Constraints on Progeny Size

Smith & Fretwell assumed that all eggs of size I_{young} have the same influence on a female's fitness such that maternal fitness is the product of the average fitness of her offspring times the number of progeny produced (197). However, this model assumes that females can actually lay all of their matured eggs. In many parasitic insects (e.g. herbivores and parasitoids) females may be incapable of finding enough hosts to lay all their eggs, relaxing selection for increased fecundity (177, 228) and potentially shifting the optimal egg size to a larger value than predicted by the Smith-Fretwell model (86). Thus, shifts in the abundance of hosts may result in a change in optimal egg size, even without changes in the relationship between egg size and progeny fitness (183).

The Smith-Fretwell model also assumes that maternal fecundity influences progeny fitness only by affecting progeny size. However, for organisms that exhibit parental care, large clutches may be less easily tended/defended than smaller clutches, such that progeny survivorship decreases with increasing maternal fecundity even if progeny size is constant. Similarly, progeny within larger clutches may experience increased competition or conflict that decreases progeny fitness (169). Thus, both parental care and sibling competition can select against large clutches (but see 193), resulting in a change in optimal progeny size without a change in the relationship between progeny size and progeny fitness.

Finally, there may be morphological and physiological constraints on the ability of females to make especially large or small eggs. For example, the necessity for progeny to fit into the brood pouch of a female may constrain the evolution of large progeny in *Daphnia* (181), even when large progeny are favored by environmental conditions. Unfortunately, although some physical and physiological constraints on progeny size have been studied in vertebrates (17, 46, 182), constraints have been little examined in arthropods.

VARIATION IN PROGENY SIZE AMONG SPECIES AND AMONG POPULATIONS WITHIN SPECIES

Selection on Progeny Size Varies Across Space and Time

When environmental conditions vary, the relationship between progeny size and progeny fitness is likely to vary, resulting in different optimal progeny sizes in different environments. However, few studies have manipulated progeny environments and quantified the relationship between egg size and progeny growth or survival in each environment. These studies have demonstrated that selection on egg size varies across environments (27, 40, 74, 78, 154, 180). In general, it appears that the fitness difference between progeny hatching from large vs. small eggs is greatest in lower quality or more stressful environments (27, 74, 78).

Climatic conditions vary substantially across space and time and may result in substantial variation in selection on progeny size (6, 180). In some insects, selection on egg size may depend on whether progeny need to overwinter before hatching (39, 66, 105, 125, 131, 178). Variation in season length or the amount of time left before winter may impose variable selection on development time, in which selection for rapid development of progeny produced late in the season (or progeny living in areas with short or cool summers) favors progeny hatching from large eggs (8, 169).

Selection on egg size can vary with the depth at which eggs are laid in the soil (crickets, 39) and the host species upon which eggs are laid (herbivores, 27, 74, 78, 161). Population density can affect the amount of competition for food that progeny will encounter, which may affect selection on egg size (169). At low population densities, sperm limitation becomes important for free-spawning arthropods, and selection may favor the evolution of large eggs that are more likely to be fertilized (206). Egg size may affect the ability and tendency of larvae to disperse (9, 16), such that variation in the need to disperse will influence selection on egg size.

Size-specific predation may represent an important source of selection on progeny size (128) either by influencing the demographic environment or because smaller progeny may be less susceptible to visual predators (30, 123, 139). For terrestrial insects size-selective egg predators and parasites impose selection on egg size that will vary with predation intensity. When predation on immature stages is high, selection may favor progeny that spend less time as juveniles, thus favoring progeny that start life larger (188). Egg size of predators may be constrained by the minimum size at which hatchlings can capture prey (204) such that selection intensity varies with prey size (1, 129; but see 204). Also, larger eggs may be favored at low prey densities to protect against periods of starvation encountered after egg hatch (129, 209).

Selection may also vary among progeny produced by a single female due to small-scale environmental variation. For example, selection on egg size varies among trees in the seed beetle *Stator limbatus*. Theoretical models predict that increased variability in selection on progeny size within populations will result in selection for larger progeny than predicted by the Smith-Fretwell model (67). Alternatively, variation in selection can result in the evolution of increased variance in progeny size (115, 174) or plasticity in progeny size (46, 182).

Variation in Progeny Size Among Species

Within genera or families, females of larger species generally lay larger eggs than females of smaller species (3, 18, 19, 33, 84–86, 90, 108, 139, 176, 204, 223), suggesting morphological constraints on egg size. However, in many taxa variation in female body size does not explain among-species variation in egg size (28, 64, 65, 129, 175, 176, 211, 228) and, even when female body size is correlated with egg size, there is generally substantial variation around the regression line (85, 124, 181, 193). Although females of larger-bodied species generally lay larger eggs, they often allocate a smaller proportion of their resources to each egg (3, 18, 85, 139, 147, 181; but see 223).

Few studies have examined the causes of variation in egg size among species. In many crustaceans, marine- and brackish-water species differ from inland species in both the size and number of eggs laid (98, 144). Higher-latitude shrimp (42) and satyrid (84) species generally lay larger eggs than lower-latitude species, while higher-latitude cladoceran species on average lay smaller eggs (175; but see 176), suggesting climate-mediated adaptive differentiation. Mode of parasitism explains some of the interspecific variation in egg size among parasitic cladocerans (175). Variation in relative egg size among species of cladocerans (in which smaller species produce proportionately larger eggs; 181) may be due in part to size-specific predation on progeny (139); small cladoceran species may minimize juvenile mortality by producing relatively larger progeny that quickly attain adult body size and reproduce before they are subject to predation. However, this pattern of negative allometry is observed in many other arthropods for which size-selective predation is not likely a source of selection (see above).

Marine arthropods with planktotrophic larvae produce smaller eggs than species with direct developing larvae (41, 100). Host plant toughness may influence the evolution of skipper (hesperiid) egg sizes; species that oviposit on hosts with tougher leaves lay larger eggs (160; see also 178; but see 85). In stored-products insects, the relationship between body size and egg size is different for semelparous versus iteroparous species; semelparous insects produce smaller eggs (relative to body size) and the slope of the relationship between egg size and body size (among species) is less steep, but the explanation for this pattern is unclear (108).

In ponerine ants, selection for large colony size appears to explain among-species variation in egg size (small eggs in species that produce large colonies; 217). Egg size of carabids varies among species according to prey type (219), and in some herbivore taxa specialist feeders lay larger eggs than generalist feeders (64, 86; but see 178), possibly as a result of relaxed selection on fecundity due to difficulty finding enough hosts (183). In Lepidoptera, species that overwinter as eggs tend to lay larger eggs than species that overwinter in other stages (178; but see 85).

Variation in Progeny Size Among Populations Within Species

Within species, females from larger-bodied populations tend to lay larger eggs (13, 228). However, variation in body size alone cannot account for the substantial geographic variation observed in many arthropods (43, 55, 83, 199, 230). Egg size often follows a cline in latitude (6, 13, 20, 43, 44, 91, 103), altitude (13, 98,

158), or, for crustaceans, habitat predictability (permanent versus temporary pools; 13, 158), from coastal to inland waterways (149, 150, 163, 165, 220) or from deep-sea benthic to shallow coastal waters (165). Some of these clines are known to be genetically based (6, 103, 151). Most cannot be explained entirely by clines in female body size (6, 13, 44, 103).

Most intraspecific latitudinal clines go from smaller eggs produced at lower latitudes to larger eggs at higher latitudes (6, 13, 43, 44, 91, 99, 103; see also 2, 21, 199), although some insects exhibit the opposite pattern (3, 20, 83). The commonness of these latitudinal clines is often interpreted as evidence that large eggs are selectively favored at low temperatures. However, environmental effects of temperature on egg size often mimic the geographic clines observed in nature (larger eggs at lower temperatures; see below).

Only one study (by Azevedo et al.) has experimentally demonstrated that eggs evolve to be larger when populations are reared at low temperatures; *Drosophila* maintained for nine years at 16.5° C evolved larger egg sizes than flies maintained at 25° C (6). However, it is unclear why larger eggs are favored at lower temperatures (6, 63, 180; see below). For some insects, short growing seasons may constrain fecundity of females in northern latitudes, relaxing selection for small eggs (228), but this hypothesis does not explain the results of Azevedo et al (6). Egg size clines in aquatic crustaceans have been argued to be due to variation in water temperature (165) or salinity (98). However, clines vary substantially in form and direction among species (148, 150, 151, 163, 165), suggesting alternative explanations.

Other explanations have been proposed to account for latitudinal clines. For example, food availability in polar environments may select for relatively K-selected life history strategies, including the production of a small number of highly competitive progeny, a pattern typical of polar benthic organisms (42). Variation in food availability has also been proposed to explain coastal-to-inland clines (149) and altitudinal clines (98) in crustacean egg size.

For many arthropods, variation in egg size among populations does not appear to be clinal. In many crustaceans, egg size varies among lakes or bays (12, 25, 56, 127, 142, 155, 205). In some herbivores, egg size varies among populations using different host plants (27, 78). Variation in egg size among populations may be due to variation in the need to resist desiccation (201; but see 200) or compete with conspecifics for food (169). Each of these studies suggests adaptive differentiation of egg size among populations, but in most cases the explanation for the differentiation is unclear or untested.

VARIATION IN PROGENY SIZE AMONG FEMALES WITHIN POPULATIONS

Female Size

Within populations, larger females tend to lay larger eggs (Table 2), suggesting some morphological constraints on egg size. However, there are many exceptions to this pattern: In some butterflies larger females lay smaller eggs, and in most

TABLE 2 Phenotypic correlations between maternal size and egg/progeny size, within populations (number of species)

Taxon	Positive	Negative	Variable or No Relationship	Reference
Crustacea (cladoceran)	9	0	0	Positive (22, 87, 88, 90, 97, 123, 130, 153, 172, 181)
Crustacea (copepod)	1	0	0	Positive (101, 155)
Crustacea (shrimp)	1	0	6[1], 1[2]	Positive (220); Variable (12, 13); None (158, 165)
Crustacea (isopod)	2	0	5[1], 1[2], 1[5]	Positive (211, 230); Variable (43); None (49, 50, 135, 211)
Crustacea (lobster)	0	0	1[1]	None (203)
Ephemeroptera	0	0	1[1]	None (46a)
Orthoptera	4	0	6[1]	Positive (38, 131); None (35 40, 68)
Heteroptera	5	0	2[1], 1[4]	Positive (52, 53, 132, 154, 162); Variable (208); None (110, 199)
Lepidoptera	6	2	3[1]	Positive (16, 26, 107, 116, 117, 146, 164); Negative (102, 113); None (21, 180, 207)
Coleoptera	5	0	1[1], 1[5]	Positive (69, 71, 94, 112, 122, 134, 168); Variable (81); None (114)
Diptera	2	0	5[1]	Positive (202, 224); None (29, 65, 99, 138)
Hymenoptera	7	0	0	Positive (133, 166)

[1]No relationship
[2]Varied among populations
[3]Varied among clones
[4]Varied among wing morphs
[5]Varied among studies

isopods and orthopterans there is no relationship between progeny size and female size (Table 2). Even when the relationship between female size and progeny size is positive, it is generally weak (e.g. 81), and larger females generally allocate a smaller proportion of their resources to each egg. Numerous authors have advanced adaptive explanations for why egg size should increase with female size within populations (24, 46, 152). We suggest that an equally interesting question is why (physiologically and evolutionarily) the proportion of a female's resources allocated to each egg generally decreases with increasing body size. It is likely that the degree to which egg size varies with body size is in part influenced by where resources come from during egg maturation. For insects that obtain most resources for egg production from adult feeding, the size of eggs laid by females may be more dependent on female diet than female size (202), while the reverse may be true if mostly larval-derived resources are used.

Maternal Diet/Food Availability

Maternal diet influences egg size in many arthropods. Generally, unfed or food-stressed females lay smaller eggs than well-fed females (28, 69, 104, 117, 124, 141, 159, 202, 219). However, there are many examples in which maternal diet does not affect egg size (63, 96, 99, 107), has only a small effect on egg size (106), or affects egg size only when females are extremely food stressed (104).

Theoretical models generally predict that, as food availability decreases, and thus progeny mortality increases, females should shift to laying larger eggs (46, 189). In some crustaceans females produce larger progeny at low food concentrations (Daphnia: 23, 24, 34, 58, 87–89, 92, 153, 173, 179; *Euterpina:* 93; and one isopod: 32), although progeny size may decrease at very low food levels (22, 213, 215). This increased progeny size often results in higher survivorship under food stress (89; references in 22). In some *Daphnia* the response to food concentration varies among clones (60, 87, 88, 213), indicating the potential for adaptive evolution of egg size plasticity. The environmental cues to which females respond, and the physiological mechanisms by which they respond, are still unknown (88).

Oviposition Host

Some insects modify egg size in response to the host plant upon which they mature eggs (81, 136). For the seed beetle *Stator limbatus,* hosts vary in the degree to which their seeds are defended against larvae. On well-defended hosts larval mortality is high and selection favors females that lay large eggs (74, 78, 81, 82). On undefended hosts larval mortality is low and selection favors females that lay small eggs (and thus have high fecundity; 78, 81). Apparently in response to this variation in selection, females have evolved egg size plasticity—they lay large eggs on seeds of the well-defended host and small eggs on seeds of undefended hosts (81, 82). The degree of plasticity exhibited by females is genetically variable within populations (75). Interestingly, this plasticity appears to mediate a diet shift by *S. limbatus* onto an exotic legume (80).

Maternal Density

Females reared at high densities often lay eggs that are smaller than those of females reared at low density (73, 76, 79, 154; but see 65, 185), likely due to effects of competition on female size or nutritional status. In some cladocerans, females respond to increased population density by producing larger progeny (173), which can better tolerate periods of starvation (45, 89) and may compete better for food. This plasticity may be mediated by sensitivity to the chemical (e.g. waste products) or physical cues emitted by other individuals (34), or by effects of density on food availability. Similar egg size plasticity in response to perceived larval competition has been reported for a seed beetle (122).

Paternal Effects

In most arthropods, nutrients and other substances are transferred to females during mating and provide a pool of resources for females to use during egg maturation (216). These contributions may affect female egg size by being incorporated directly into eggs or by changing female energy budgets. Many studies have examined how male contributions affect female survivorship or fecundity (216), but few have examined whether they affect the size or composition of eggs. Female insects sometimes lay larger eggs when they receive more (95, 96, 164, 192) or larger spermatophores (96), although the effect is sometimes seen only late in a female's life (69, 222). Some insects lay larger eggs when they mate with larger males (154, 225), possibly as a result of paternal investment or because males manipulate female allocation to the eggs they have fertilized (167).

Other studies have failed to find effects of female mating frequency (35, 207), spermatophore size, or male size (76, 79, 186) on egg size. Some authors have suggested that effects of male-derived nutrients on female reproduction may be detectable only when females are food stressed (but see 69). The relative influence of male-derived nutrients on egg size is still unclear.

Rearing and Oviposition Temperature

Many studies show that females lay larger eggs when reared (104) or ovipositing (4, 5, 63, 101) at lower temperatures (232), although some arthropods lay larger eggs when reared at intermediate temperatures (10), lay larger eggs at high temperatures (110), exhibit variable responses to temperature depending on other environmental conditions (e.g. food availability, 153), or show no response to temperature (221). Unfortunately, many studies do not distinguish between the effects of rearing versus oviposition temperature (10, 30, 47, 172, 181, 187). Interestingly, the temperature at which *D. melanogaster* males are reared affects the size of eggs laid by their daughters (47), but the mechanism and adaptive significance for this environmentally-based paternal effect is unknown.

The rate of oocyte production relative to the rate of oocyte growth (vitellogenesis) may change with temperature, affecting both the size and number of eggs (63). If so, the temperature at which vitellogenesis occurs should affect egg size (218), and an increase in egg size should be accompanied by a decrease in fecundity, as generally observed. The size of a female's fat body may be affected by temperature (but see 63) and may in turn affect the rate of vitellogenin uptake. This hypothesis predicts that only temperatures experienced during fat deposition (prior to oviposition) should affect egg size, and that both egg size and fecundity should be affected similarly by temperature (both increase or decrease), neither of which is generally observed.

Temperatures experienced by adults may affect the metabolic rate of females (4): If low temperature reduces the cost of somatic maintenance, a greater proportion of the female's resources may be shunted to vitellogenesis. Some arthropods mature at a larger body size when reared at lower temperature (e.g. most cladocerans; 156), potentially resulting in an increase in egg size (e.g. Table 2). However, females of many species respond to oviposition temperature independent of rearing temperature (4, 5, 63, 101), indicating that a change in body size is not a general explanation (see also 172).

Other arthropods delay oviposition at lower temperatures (e.g. *Drosophila;* 111), potentially resulting in increased vitellogenesis. This hypothesis predicts that egg size should vary with manipulations of oviposition rate independent of temperature. In some insects, delaying oviposition or changing oviposition rate affects egg size (219; references in 4), but in other insects, females forced to delay oviposition do not lay larger eggs (e.g. 81). Also, in some insects the production of larger eggs does not result in an increase in the period of oogenesis or delayed oviposition (54).

Increasing egg size at low temperatures may represent an adaptive response to temperature (232). For example, at lower temperatures growth is slower, so selection may favor the production of larger progeny that mature sooner, reducing their exposure to sources of mortality (232) or simply decreasing generation time (91, 172). Few other adaptive hypotheses have been proposed (see 181, 232) and none have been tested.

Seasonal Variation

In many arthropods, progeny size varies throughout the year (4, 5, 28, 32, 46a, 50, 63, 66, 110, 125, 161). In some cases this is due to aging of females in the population and corresponding changes in egg size (Table 3). However, seasonal variation sometimes reflects variation among generations (27, 28, 161), and maternal age cannot explain some of the patterns observed within generations (46a, 63).

Most species of crustaceans that exhibit seasonal variation in progeny size produce larger progeny in winter (15, 30, 33, 93, 123, 130, 155, 187, 196, 231), although some species produce larger progeny in summer (15), and others show some other seasonal pattern (15). In some species, females may be responding primarily to temperature, but it is unclear whether the responses are adaptations to temperature itself, non-adaptive physiological responses to temperature, or whether temperature is used as a cue to predict some other environmental condition. However, temperature cannot explain the seasonal pattern observed in some other species (93, 123, 155).

Seasonal variation in progeny size often reflects variation in female size (30, 123, 155) although it is unclear whether this reflects a cause-and-effect relationship or whether body size and egg size are influenced by the same external factors. In some crustaceans seasonal differences in body size cannot explain all of the variation in egg size (230). Seasonal changes in cladoceran progeny size often correspond to changes in predator abundance, suggesting an adaptive response to variation in size-specific predation (22, 30, 123, 130). Alternatively, seasonal variation may reflect a plastic response to variation in food availability (91; but

TABLE 3 Change in egg or progeny size as females age (number of species)

Taxon	Increase	Decrease	Varied or no change	Reference
Crustacea (cladoceran)	2	0	0	Increase (14, 24, 62, 87, 130, 153)
Crustacea (isopod)	1	0	0	Increase (32)
Orthoptera	2	4	1[1], 3[2]	Increase (35, 131); Decrease (40, 68, 131); No change (131); Varied (37, 40)
Heteroptera	6	2	2[2,3]	Increase (51, 52, 120, 154, 162); Decrease (110, 132); Varied (155a, 198)
Lepidoptera	0	20	2[1], 1[4]; 1[5], 1[6]	Decrease (26, 28, 36, 83, 102, 107, 113, 116, 118, 119, 126, 136, 137, 164, 180, 185, 207, 210, 226, 227, 229); No change (16, 146); Varied (28, 106, 157, 159)
Coleoptera	0	2	1[4]	Decrease (69, 77, 219, 222); Varied (81)
Diptera	0	0	1[1]	No change (221)

[1]No change
[2]Varied among females
[3]Varied among morphs
[4]Varied among treatments
[5]Varied among seasons
[6]Varied among studies

see 30) or clonal replacement, in which natural selection results in the replacement of large-egg clones with small-egg clones in warmer seasons (130).

In terrestrial arthropods, no consistent seasonal patterns are apparent, so temperature is not a general explanation for seasonal variation. In some insects, seasonal changes in body size correspond to changes in egg size (27). Eggs laid by second-generation females of a tortricid moth (which enter diapause) may be larger to ensure overwinter survivorship of diapausing eggs (66; see also 125, 131). In some herbivores selection on egg size varies among host plants and a seasonal change in egg size may be an adaptation to changes in host plants availability (27, 161). In isopods, seasonal variation in egg size may reflect a response to food availability (32).

Responses to Predation Risk

Some cladocerans exhibit plasticity in progeny size in response to predator-associated chemical cues (97, 179). Populations of *Daphnia magna*, and clones within populations, vary in their responses to fish kairomones, with clones from lakes with fish generally more sensitive (25). Other arthropods may also respond to predation risk by varying egg size. For example, females of the shield bug

Elasmucha ferrugata lay smaller eggs at the periphery of their clutch where the eggs are most susceptible to predation (145) and thus have lower reproductive value.

Other Environmental Sources of Variation

Many insects exhibit complex polymorphisms in suites of morphological and life history characters, and egg size often differs substantially among morphs. For example, macropterous individuals of both *Lygaeus equestris* and *Orgyia thyellina* lay smaller (and more) eggs than brachypterous individuals, possibly as a result of selection for rapid population increase (and thus high fecundity) on females that colonize new habitats, and selection for producing large, competitive progeny on females that stay in established populations (125, 198). However, the opposite pattern is observed in *Jadera aeola* (208). Alatae of polymorphic aphids generally produce smaller offspring than apterae (52, 162). This pattern has been attributed to competition between gonads and flight muscles for limited resources (162). Obligate asexual clones of *Daphnia* produce larger progeny than sexual clones (232), but parthenogenetic eggs do not differ in size from fertilized eggs in a stick insect (37).

Genetic Variation

There are surprisingly few data available on genetic variation in egg size within populations of arthropods. Estimates of the heritability of egg size, and its genetic correlations with other life history traits, are even fewer. In two seed beetles, egg size is highly heritable (range of $h^2 = 0.22$ to 0.91; h^2 varies among hosts and populations; 71, 75). Likewise, variation in egg size is heritable in spruce budworm (104). Comparisons of clones of *Daphnia* indicate substantial genetic variation in both progeny size (25, 56, 59, 87, 140) and egg size plasticity (25, 60, 87) within populations. Laboratory selection experiments have also demonstrated that egg size is heritable (7, 38, 170, 186a, 221) and is genetically correlated with body size. Selection on other life history characters, such as development rate (8) and resistance to desiccation (200) have also resulted in the evolution of egg size, indicating genetic correlations between egg size and these traits.

Little is understood about the genetic basis of among-species or among population differences in egg size. Crosses among strains of *D. melanogaster* (221) suggest that at least one autosomal and one sex-linked gene affect the variation in egg size (among strains). Egg size variation among species of *Choristoneura* (36) and among strains of silkworms (121) is also partially sex linked.

VARIATION AMONG EGGS AND PROGENY PRODUCED BY A SINGLE FEMALE

For many arthropods, the variation in size among progeny produced by a single female may be as large as or larger than the variation among females within a population. Much of this variation is an effect of maternal age (below), but in many arthropods there is substantial variation in egg size within individual clutches of eggs. Variation in egg size within and among clutches may be selected for as a diversified bet-hedging strategy to minimize variation in fitness (174; see 67a for a related adaptive explanation). Alternatively, physiological limitations in the ability to make identically sized eggs may explain much of the egg size variation within clutches.

Female Age/Egg Order

Most life-history models posit the production of uniform-sized progeny throughout a female's life (e.g. 197). In most arthropods, however, progeny size decreases with maternal age (Table 3), although an increase is commonly observed in orthopterans and heteropterans (Table 3). Only a few insects exhibit no change in egg size (16, 40, 146). An increase in progeny size is commonly observed in cladocerans (24; Table 3), but this is because females continue to grow after beginning reproduction; the ratio of progeny size to maternal size actually decreases with age (153). In some insects, maternal age effects are not observable until females near their last clutch (157). The variance in egg size sometimes also increases as mothers age (171), but too few studies present estimates of variance to allow generalization.

The effect of age often varies substantially among females (37, 102, 131, 119, 228, 229) but the degree to which variation among females reflects genetic differences is unclear. At least some of the variation is environmentally based. For example, the direction or magnitude of the maternal age effect can differ among host plants (81), between macropterous and brachypterous bugs (198), and between alatae and apterae of aphids (51). The maternal effect may also vary with maternal diet (28; but see 107), with the decrease generally steepest for food-stressed mothers (69, 159) or for females that had been food stressed as larvae (102).

A decrease in progeny size with increasing age is often attributed to a depletion of the female's resources (36, 180, 210, 228). Studies in which maternal diet is manipulated (69, 159) support this hypothesis. Alternatively, decreasing progeny size with increasing maternal age may be adaptive when female clutch size is constrained (11); young females should allocate a larger proportion of their resources to reproduction when their chances of surviving to lay the next clutch are lower. However, this hypothesis assumes that age-specific fecundity does not evolve, an assumption that is unlikely to be realistic.

Maternal age effects on progeny size may reflect a bet-hedging strategy (174); selection favors variation among progeny to ensure that at least some progeny are well suited for future environmental conditions. However, this raises the question of why females do not simply produce the full range of offspring sizes within each clutch or age class. That changes in egg size are sometimes non-adaptive is suggested by the observation that female *Daphnia* produce the size offspring that maximized maternal fitness when they were youngest (24).

In general, eggs laid by older females are less likely to hatch (68, 69, 77, 229) and progeny hatching from these eggs have higher mortality (69, 75, 110, 222; but see 102), produce smaller nymphs/larvae (132), and take longer to reach maturity (69, 77, 137, 222; but see 180) or longer to pupate (102). Sometimes progeny produced by older mothers mature smaller (110, 185) but more often they mature at normal size (69, 77, 102, 180, 222), generally by increasing development time. The sex ratio of progeny may also change as females age (102). These effects on progeny are probably in part mediated by the changes in egg size. However, egg composition (e.g. proportion yolk) also often changes with maternal age (210), such that maternal age effects on progeny cannot be attributed to a decrease in egg size without more careful and creative experimentation.

CONCLUSIONS

Arthropods exhibit substantial variation in progeny size among species, among individuals within species, and sometimes even among progeny produced by a single female. Many theoretical models have been developed to explain some of this variation, but most start with the same two assumptions as Smith & Fretwell (197)—they assume that progeny fitness increases with increasing progeny size, and that there is a trade-off between progeny size and number. We find that these two assumptions are generally supported by data but that the trade-off between progeny size and number is less apparent when organisms are iteroparous, use adult-acquired resources for reproduction, or provide parental care. This is because variation in total reproductive effort is difficult to quantify for these species. Most models solve for optimal progeny size by assuming that total reproductive effort is constant. However, reproductive effort may vary substantially among individuals and may evolve in response to natural selection. Thus, selection for increased progeny size may lead to increased reproductive effort rather than a decrease in fecundity. This possibility has been examined theoretically but needs to be explored empirically.

Much of the variation in progeny size among species, and among populations within species, appears to have evolved in response to differences in natural selection among environments. Many environmental factors covary with variation in progeny size, and these factors may be the cause of the species or population differences. However, few studies have manipulated progeny environments and demonstrated that the relationship between progeny size and fitness actually differs among environments, and fewer still have demonstrated why selection favors different sized progeny in different environments (e.g. why does selection favor larger eggs at lower temperatures?). Understanding the evolution of intra- and interspecific variation in progeny size will require more empirical studies that identify sources of natural selection within environments and that demonstrate how selection varies among environments.

Much of the variation in progeny size within populations appears to be nonadaptive. For example, smaller females generally lay smaller eggs as an inevitable consequence of phenotypic and genetic correlations between body size and egg size (due to morphological or physiological constraints). However, maternal body size explains a surprisingly small amount of the variation in progeny size within and among populations of many species. Much of the remaining variance is probably also nonadaptive, due to variation in factors such as maternal diet (e.g. food-stressed females generally produce smaller progeny). Yet some species of arthropods, especially crustaceans but also a few insects, exhibit plasticity in progeny size in response to several environmental factors, and much of this plasticity is likely adaptive. Unfortunately, few studies have examined the fitness consequences of plasticity in progeny size, and results of these studies are not always consistent (e.g. comparisons among *Daphnia* studies). The evolution of life history plasticity, including adaptive plasticity in progeny size, is one of the most exciting topics in the study of life histories. We thus suggest that substantially more research effort should be dedicated to understanding the evolution of reaction norms for progeny size.

Variation in size of progeny produced by a single female has been more difficult to explain than variation among females or among populations. Most theoretical models predict that females should produce progeny of a single size. Yet progeny size sometimes varies greatly within families (e.g. changes with female age). We suggest that much of the variation within families is probably nonadaptive. However, some authors have suggested that at least some of the variation within families is an adaptive response to living in a variable environment. At this time, however, there are few experimental studies and too little theoretical work to generalize.

The evolution of progeny size has been extensively modeled by theoretical evolutionary ecologists. However, advances in theory have substantially outpaced empirical data—few empirical studies have progressed much beyond documenting patterns of variation in progeny size within or among population, measuring phenotypic correlations between progeny size and maternal fecundity, or quantifying the relationship between progeny size and a few components of progeny fitness in one environment. We hope that this review will stimulate researchers to examine the specific factors that result in variation in selection on progeny size within and among populations, and how this variation in selection influences the evolution of the patterns that we observe.

ACKNOWLEDGMENTS

We thank J Rosenheim, U Savalli, P Spinelli, D Wise, and F Messina for comments. The project was funded by NSF DEB-98-07315 (to C Fox).

Visit the Annual Reviews home page at www.AnnualReviews.org.

LITERATURE CITED

1. Albuquerque GS, Tauber MJ, Tauber CA. 1997. Life history adaptations and reproductive costs associated with specialization in predacious insects. *J. Anim. Ecol.* 66:307–17
2. Allan JD. 1984. Life history variation in a freshwater copepod: evidence from population crosses. *Evolution* 38:280–91
3. Anderson J. 1990. The size of spider eggs and estimates of their energy content. *J. Arachnol.* 18:73–78
4. Avelar T. 1993. Egg size in *Drosophila* —standard unit of investment or variable response to environment?—the effect of temperature. *J. Insect Physiol.* 39:283–89
5. Avelar T, Rocha Pité MT. 1989. Egg size and number in *Drosophila subobscura* under semi-natural conditions. *Evol. Biol.* 3:37–48
6. Azevedo RBR, French V, Partridge L. 1996. Thermal evolution of egg size in *Drosophila melanogaster. Evolution* 50:2338–45
7. Azevedo RBR, French V, Partridge L. 1997. Life-history consequences of egg size in *Drosophila melanogaster. Am. Nat.* 150:250–82
8. Bakker K. 1969. Selection for rate of growth and its influence on competitive ability of larvae of *Drosophila melanogaster. Neth. J. Zool.* 19:541–95
9. Barbosa P, Capinera JL. 1978. Population quality, dispersal and numerical change in the gypsy moth, *Lymantria dispar* (L.). *Oecologia* 36:203–9
10. Beckwith RC. 1982. Effects of constant laboratory temperatures on the Douglas-fir tussuck moth (Lepidoptera: Lymantriidae). *Environ. Entomol.* 11:1159–63
11. Begon M, Parker GA. 1986. Should egg size and clutch size decrease with age? *Oikos* 47:293–302
12. Belk D. 1977. Evolution of egg size strategies in fairy shrimps. *Southwest. Nat.* 22:99–105
13. Belk D, Anderson G, Hsu SY. 1990. Additional observations on variations in egg size among populations of *Streptocephalus seali* (Anostraca). *J. Crust. Biol.* 10:128–33
14. Bell G. 1983. Measuring the cost of reproduction III. The correlation structure of the early life history of *Daphnia pulex. Oecologia* 60:378–83
15. Berberovic R, Bikar K, Geller W. 1990. Seasonal variability of the embryonic development time of three planktonic crustaceans—dependence on temperature, adult size, and egg weight. *Hydrobiology* 203:127–36
16. Berger A. 1989. Egg weight, batch size and fecundity of the spotted stalk borer, *Chilo partellus* in relation to weight of females and time of oviposition. *Entomol. Exp. Appl.* 50:199–207
17. Bernardo J. 1996. The particular maternal effect of propagule size, especially egg size: patterns, models, quality of evidence and interpretations. *Am. Zool.* 36:216–36
18. Berrigan D. 1991. The allometry of egg size and number in insects. *Oikos* 60:313–21
19. Blackburn TM. 1991. Evidence for a "fast-slow" continuum of life-history traits among parasitoid Hymenoptera. *Funct. Ecol.* 5:65–74
20. Blackenhorn WU, Fairbairn DJ. 1995. Life history adaptation along a latitudinal cline in the water strider *Aquaris remigis* (Heteroptera: Gerridae). *J. Evol. Biol.* 8: 21–41
21. Blau WS. 1981. Life history variation in the black swallowtail butterfly. *Oecologia* 48:116–22

22. Boersma M. 1995. The allocation of resources to reproduction in *Daphnia galeata*: against the odds? *Ecology* 76:1251–61
23. Boersma M. 1997. Offspring size in *Daphnia*: does it pay to be overweight? *Hydrobiology* 360:79–88
24. Boersma M. 1997. Offspring size and parental fitness in *Daphnia magna*. *Evol. Ecol.* 11:439–50
25. Boersma M, Spaak P, de Meester L. 1998. Predator-mediated plasticity in morphology, life history, and behavior of *Daphnia*: the uncoupling of responses. *Am. Nat.* 152:237–48
26. Boggs CL. 1986. Reproductive strategies of female butterflies: variation in and constraints on fecundity. *Ecol. Entomol.* 11:7–15
27. Braby MF. 1994. The significance of egg size variation in butterflies in relation to host plant quality. *Oikos* 71:119–29
28. Braby MF, Jones RE. 1995. Reproductive patterns and resource allocation in tropical butterflies: influence of adult diet and seasonal phenotype on fecundity, longevity and egg size. *Oikos* 72:189–204
29. Bradshaw WE, Holzapfel CM, O'Neill T. 1993. Egg size and reproductive allocation in the pitcherplant mosquito *Wyeomyia smithii* (Diptera, Culicidae). *J. Med. Entomol.* 30:384–90
30. Brambilla DJ. 1982. Seasonal variation of egg size and number in a *Daphnia pulex* population. *Hydrobiologia* 97: 233–48
31. Brittain JE, Lillehammer A, Saltveit SJ. 1984. The effect of temperature on intraspecific variation in egg biology and nymphal size in the stonefly, *Capnia atra* (Plecoptera). *J. Anim. Ecol.* 53:161–69
32. Brody MS, Lawlor LR. 1984. Adaptive variation in offspring size in a terrestrial isopod, *Armadillidium vulgare*. *Oecologia* 61:55–59
33. Burgis MJ. 1967. A quantitative study of reproduction in some species of *Ceriodaphnia* (Crustacea: Cladocera). *J. Anim. Ecol.* 36:61–75
34. Burns CW. 1995. Effects of crowding and different food levels on growth and reproductive investment of *Daphnia*. *Oecologia* 101:234–44
35. Butlin RK, Woodhatch CW, Hewitt GW. 1987. Male spermatophore investment increases female fecundity in a grasshopper. *Evolution* 41:221–25
36. Campbell IM. 1962. Reproductive capacity in the genus *Choristoneura* Led. (Lepidoptera: Tortricidae). I. Quantitative inheritance and genes as controllers of rates. *Can. J. Gen. Cytol.* 4:272–88
37. Carlberg U. 1984. Variation in the egg-size of *Extatosoma tiaratum* (MacLeay) (Insecta: Phasmida). *Zool. Anz.* 212:61–67
38. Carlberg U. 1991. Egg-size variation in *Extatosoma tiaratum* (MacLeay) and its effect on survival and fitness of newly hatched nymphs (Insecta, Phasmida). *Biol. Zent.* 110:163–73
39. Carriere Y, Masaki S, Roff DA. 1997. The coadaptation of female morphology and offspring size: a comparative analysis in crickets. *Oecologia* 110:197–204
40. Carriere Y, Roff DA. 1995. The evolution of offspring size and number: a test of the Smith-Fretwell model in three species of crickets. *Oecologia* 102:389–96
40a. Charnov EL, Downhower JF. 1995. A trade-off-invariant life-history rule for optimal offspring size. *Nature* 376:418–19
41. Christiansen FB, Fenchel TM. 1979. Evolution of marine invertebrate reproductive patterns. *Theor. Popul. Biol.* 16:267–82
42. Clarke A. 1979. On living in cold water: K-strategies in Antarctic benthos. *Mar. Biol.* 55:111–19
43. Clarke A, Gore DJ. 1992. Egg size and composition in *Ceratoserolis* (Crustacea, Isopoda) from the Weddell sea. *Polar Biol.* 12:129–34

44. Clarke A, Hopkins CCE, Nilssen EM. 1991. Egg size and reproductive output in the deep-water prawn *Pandalus borealis* Kroyer, 1838. *Funct. Ecol.* 5:724–30
45. Cleuvers M, Goser B, Rattle H-T. 1997. Life-strategy shift by intraspecific interaction in *Daphnia magna*: change in reproduction from quantity to quality. *Oecologia* 110:337–45
46. Clutton-Brock TH. 1991. *The Evolution of Parental Care.* Princeton, NJ: Princeton Univ. Press
46a. Corkum LD, Ciborowski JJH, Poulin RG. 1997. Effects of emergence date and maternal size on egg development and sizes of eggs and first-instar nymphs of a semelparous aquatic insect. *Oecologia* 111:69–75
47. Crill WD, Huey RB, Gilchrist GW. 1996. Within- and between-generation effects of temperature on the morphology and physiology of *Drosophila melanogaster. Evolution* 50:1205–18
48. Curtsinger JW. 1976. Stabilizing selection in *Drosophila melanogaster. J. Hered.* 67:59–60
49. Dangerfield JM. 1997. Growth and survivorship in juvenile woodlice: is birth mass important? *Ecography* 20:132–36
50. Dangerfield JM, Telford SR. 1990. Breeding phenology, variation in reproductive effort and offspring size in a tropical population of the woodlouse *Porcellionides pruinosus. Oecologia* 82:251–58
51. Dixon AFG, Kundo R, Kindlmann P. 1993. Reproductive effort and maternal age in iteroparous insects using aphids as a model group. *Funct. Ecol.* 7:267–72
52. Dixon AFG, Wratten SD. 1971. Laboratory studies on aggregation, size and fecundity in the black bean aphid, *Aphis fabae* Scop. *Bull. Entomol. Res.* 61:97–111
53. Dodson GN, Marshall LD. 1984. Male aggression and female egg size in a mate-guarding ambush bug: are they related? *Psyche* 91:193–99
54. Dunlap-Pianka HL. 1979. Ovarian dynamics in *Heliconius* butterflies: correlations among daily oviposition rates, egg weights, and quantitative aspects of oogenesis. *J. Insect Physiol.* 25:741–49
55. Eberhard W. 1979. Rate of egg production by tropical spiders in the field. *Biotropica* 11:292–300
56. Ebert D. 1991. The effect of size at birth, maturation threshold and genetic differences on the life-history of *Daphnia magna. Oecologia* 86:243–50
57. Ebert D. 1993. The trade-off between offspring size and number in *Daphnia magna*: The influence of genetic, environmental, and maternal effects. *Arch. Hydrobiol.* 1993:453–73
58. Ebert D. 1994. Fractional resource allocation into few eggs: *Daphnia* as an example. *Ecology* 75:568–71
59. Ebert D, Yampolsky L, Stearns SC. 1993. Genetics of life history in *Daphnia magna.* I. Heritabilities at two food levels. *Heredity* 70:335–43
60. Ebert D, Yampolsky L, van Noordwijk AJ. 1993. Genetics of life history in *Daphnia magna.* II. Phenotypic plasticity. *Heredity* 70:344–52
61. Emlet RB, Hoegh-Guldberg O. 1997. Effects of egg size on postlarval performance: experimental evidence from a sea urchin. *Evolution*: 51:141–52
62. Enserink L, Luttmer W, Mass-Diepeveen H. 1990. Reproductive strategy of *Daphnia magna* affects the sensitivity of its progeny in acute toxicity tests. *Aquat. Toxicol.* 17:15–25
63. Ernsting G, Isaaks JA. 1997. Effects of temperature and season on egg size, hatchling size and adult size in *Notiophilus biguttatus. Ecol. Entomol.* 22:32–40
64. Fitt GP. 1990. Variation in ovariole number and egg size of species of *Dacus* (Diptera, Tephritidae) and their relation to host specialization. *Ecol. Entomol.* 15:255–64

65. Fitt GP. 1990. Comparative fecundity, clutch size, ovariole number and egg size of *Dacus tryoni* and *D. jarvisi*, and their relationship to body size. *Entomol. Exp. Appl.* 55:11–21
66. Fitzpatrick SM, Troubridge JT. 1993. Fecundity, number of diapause eggs, and egg size of successive generations of the blackheaded fireworm (Lepidoptera, Tortricidae) on cranberries. *Environ. Entomol.* 22:818–23
67. Forbes LS. 1991. Optimal size and number of offspring in a variable environment. *J. Theor. Biol.* 150:299–304

67a. Forbes LS. 1999. Within-clutch variation in propagule size: the double-default model. *Oikos* 85:146–50

68. Forrest TG. 1986. Oviposition and maternal investment in mole crickets (Orthoptera: Gryllotalpidae): effects of season, size, and senescence. *Ann. Entomol. Soc. Am.* 79:918–24
69. Fox CW. 1993. The influence of maternal age and mating frequency on egg size and offspring performance in *Callosobruchus maculatus* (Coleoptera: Bruchidae). *Oecologia* 96:139–46
70. Fox CW. 1994. The influence of egg size on offspring performance in the seed beetle, *Callosobruchus maculatus*. *Oikos* 71:321–25
71. Fox CW. 1994. Maternal and genetic influences on egg size and larval performance in a seed beetle: multigenerational transmission of a maternal effect? *Heredity* 73:509–17
72. Fox CW. 1997. Egg size manipulations in the seed beetle, *Stator limbatus*: consequences for progeny growth. *Can. J. Zool.* 75:1465–73
73. Fox CW. 1997. The ecology of body size in a seed beetle, *Stator limbatus*: persistence of environmental variation across generations? *Evolution* 51:1005–10
74. Fox CW. 2000. Natural selection on seed beetle egg size in the field and the lab: variation among environments. *Ecology*. In press
75. Fox CW, Czesak ME, Mousseau TA, Roff DA. 1999. The evolutionary genetics of an adaptive maternal effect: egg size plasticity in a seed beetle. *Evolution*. In press
76. Fox CW, Czesak ME, Savalli UM. 1999. Environmentally-based maternal effects on development time in the seed beetle, *Stator pruininus* (Coleoptera: Bruchidae): consequences of larval density. *Environ. Entomol.* 28:217–23
77. Fox CW, Dingle H. 1994. Dietary mediation of maternal age effects on offspring performance in a seed beetle (Coleoptera: Bruchidae). *Funct. Ecol.* 8:600–6
78. Fox CW, Mousseau TA. 1996. Larval host plant affects the fitness consequences of egg size in the seed beetle *Stator limbatus*. *Oecologia* 107:541–48
79. Fox CW, Savalli UM. 1998. Inheritance of environmental variation in body size: Superparasitism of seeds affects progeny and grandprogeny body size via a nongenetic maternal effect. *Evolution* 52:172–82
80. Fox CW, Savalli UM. 1999. Maternal effects mediate diet expansion in a seed-feeding beetle. *Ecology*. In press
81. Fox CW, Thakar MS, Mousseau TA. 1997. Egg size plasticity in a seed beetle: an adaptive maternal effect. *Am. Nat.* 149:149–63
82. Fox CW, Waddell KJ, des Lauriers J, Mousseau TA. 1997. Seed beetle survivorship, growth and egg size plasticity in a paloverde hybrid zone. *Ecol. Entomol.* 22:416–24
83. García-Barros E. 1992. Evidence for geographic variation of egg size and fecundity in a satyrine butterfly, *Hipparchia semele* (L.) (Lepidoptera, Nymphalidae-Satyrinae). *Graellsia* 48:45–52
84. García-Barros E. 1994. Egg size variation in European satyrine butterflies (Nymphalidae, Satyrinae). *Biol. J. Linn. Soc.* 51:309–24
85. García-Barros E, Munguira ML. 1997. Uncertain branch lengths, taxonomic sampling error, and the egg to body size allometry in temperate butterflies (Lepidoptera). *Biol. J. Linn. Soc.* 61:201–21
86. Gilbert F. 1990. Size, phylogeny and life

history evolution of feeding specialization in insect predators. In *Insect Life Cycles, Genetics, Evolution and Co-ordination*, ed. F Gilbert, pp. 101–24. Berlin: Springer-Verlag

87. Glazier DS. 1992. Effects of food, genotype, and maternal size and age on offspring investment in *Daphnia magna*. *Ecology* 73:910–26
88. Glazier DS. 1998. Does body storage act as a food-availability cue for adaptive adjustment of egg size and number in *Daphnia magna*? *Freshw. Biol.* 40:87–92
89. Gliwicz ZM, Guisande C. 1992. Family planning in *Daphnia*: resistance to starvation in offspring born to mothers at different food levels. *Oecologia* 91:463–67
90. Green J. 1956. Growth, size and reproduction in *Daphnia* (Crustacea: Cladocera). *Proc. Zool. Soc. London* 126:173–204
91. Green J. 1966. Seasonal variation in egg production by Cladocera. *J. Anim. Ecol.* 35:77–104
92. Guisande C, Gliwicz ZM. 1992. Egg size and clutch size in two *Daphnia* species grown at different food levels. *J. Plankton Res.* 14:997–1007
93. Guisande C, Sanchez J, Maneiro I, Miranda A. 1996. Trade-off between offspring number and offspring size in the marine copepod *Euterpina acutifrons* at different food concentrations. *Mar. Ecol. Prog. Ser.* 143:37–44
94. Guntrip J, Sibly RM, Smith RH. 1997. Controlling resource acquisition to reveal a life history trade-off: egg mass and clutch size in an iteroparous seed predator, *Prostephanus truncatus*. *Ecol. Entomol.* 22:264–70
95. Gwynne DT. 1984. Courtship feeding increases female reproductive success in bushcrickets. *Nature* 307:361–63
96. Gwynne DT. 1988. Courtship feeding and the fitness of female katydids (Orthoptera: Tettigoniidae). *Evolution* 42:545–55
97. Hanazato T, Dodson SI. 1995. Synergistic effects of low oxygen concentration, predator kairomone, and a pesticide on the cladoceran *Daphnia pulex*. Limnol. *Oceanogr.* 40:571–77
98. Hancock MA. 1998. The relationship between egg size and embryonic and larval development in the freshwater shrimp *Paratya australiensis* Kemp (Decapoda: Atyidae). *Freshw. Biol.* 39:715–23
99. Hard JJ, Bradshaw WE. 1993. Reproductive allocation in the western tree-hole mosquito, *Aedes sierrensis*. *Oikos* 66:55–65
100. Hart MW. 1995. What are the costs of small egg size for a marine invertebrate with feeding planktonic larvae? *Am. Nat.* 146:415–26
101. Hart RC, McLaren IA. 1978. Temperature acclimation and other influences on embryonic duration in the copepod, *Pseudocalanus* sp. *Mar. Biol.* 45:23–30
102. Harvey GT. 1977. Mean weight and rearing performance of successive egg clusters of eastern spruce budworm (Lepidoptera: Tortricidae). *Can. Entomol.* 109:487–96
103. Harvey GT. 1983. A geographical cline in egg weights in *Choristoneura fumiferana* (Lepidoptera: Tortricidae) and its significance in population dynamics. *Can. Entomol.* 115:1103–8
104. Harvey GT. 1983. Environmental and genetic effects on mean egg weight in spruce budworm (Lepidoptera: Tortricidae). *Can. Entomol.* 115:1109–17
105. Harvey GT. 1985. Egg weight as a factor in the overwintering survival of spruce budworm (Lepidoptera: Tortricidae) larvae. *Can. Entomol.* 117:1451–61
106. Hill CJ. 1989. The effect of adult diet on the biology of butterflies 2. The common crow butterfly, *Euploea core corinna*. *Oecologia* 81:258–66
107. Hill CJ, Pierce NE. 1989. The effect of adult diet on the biology of butterflies. I. The common imperial blue, *Jalmenus evagoras*. *Oecologia* 81: 249–57
108. Holloway GJ, Smith RH, Wrelton AE,

King PE, Li LL, Menendez GT. 1987. Egg size and reproductive strategies in insects infesting stored-products. *Funct. Ecol.* 1:229–35
109. Honěk A. 1987. Regulation of body size in a heteropteran bug, *Pyrrhocoris apterus*. *Entomol. Exp. Appl.* 44:257–62
110. Honěk A. 1992. Female size, reproduction and progeny size in *Pyrrhocoris apterus* (Heteroptera, Pyrrhocoridae). *Acta Entomol. Bohemoslov* 89:169–78
111. Huey RB. 1995. Within- and between-generation effects of temperature on early fecundity of *Drosophila melanogaster*. *Heredity* 74:216–23
112. Johnson LK. 1982. Sexual selection in brentid weevils. *Evolution* 36:251–62
113. Jones RE, Hart JR, Bull GD. 1982. Temperature, size and egg production in the cabbage butterfly, *Pieris rapae* L. *Aust. J. Zool.* 30:223–32
114. Juliano SA. 1985. The effects of body size on mating and reproduction in *Brachinus lateralis* (Coleoptera: Carabidae). *Ecol. Entomol.* 10:271–80
115. Kaplan RH, Cooper WS. 1988. On the evolution of coin-flipping plasticity: a response to McGinley, Temme, and Geber. *Am. Nat.* 132:753–55
116. Karlsson B. 1987. Variation in egg weight, oviposition rate and reproductive reserves with female age in a natural population of the speckled wood butterfly, *Pararge aegeria*. *Ecol. Entomol.* 12:473–76
117. Karlsson B, Wickman P-O. 1990. Increase in reproductive effort as explained by body size and resource allocation in the speckled wood butterfly, *Pararge aegeria* (L.). *Funct. Ecol.* 4:609–17
118. Karlsson B, Wiklund C. 1984. Egg weight variation and lack of correlation between egg weight and offspring fitness in the wall brown butterfly *Lasiommata megera*. *Oikos* 43:376–85
119. Karlsson B, Wiklund C. 1985. Egg weight variation in relation to egg mortality and starvation endurance of newly hatched larvae in some satyrid butterflies. *Ecol. Entomol.* 10:205–11
120. Kasule FK. 1991. Egg size increases with maternal age in the cotton stainer bugs *Dysdercus fasciatus* and *D. cardinalis* (Hemiptera: Pyrrhocoridae). *Ecol. Entomol.* 16:345–49
121. Kawamura N. 1990. Is the egg size determining gene, ESD, on the W-chromosome identical with the sex-linked giant egg gene, GE, in the silkworm? *Genetica* 81:205–10
122. Kawecki TJ. 1995. Adaptive plasticity of egg size in response to competition in the cowpea weevil, *Callosobruchus maculatus* (Coleoptera: Bruchidae). *Oecologia* 102: 81–85
123. Kerfoot WC. 1974. Egg size cycle of a cladoceran. *Ecology* 55:1259–70
124. Kessler A. 1971. Relation between egg production and food consumption in species of the genus *Pardosa* (Lycosidae, Araneae) under experimental conditions of food abundance and shortage. *Oecologia* 8:93–109
125. Kimura K, Masaki S. 1977. Brachypterism and seasonal adaptation in *Orgyia thyellina* Butler (Lepidoptera: Lymantriidae). *Kontyû* 45:97–106
126. Kimura K, Tsubaki Y. 1985. Egg weight variation associated with female age in *Pieris rapae crucivora* Boisduval (Lepidoptera: Pieridae). *Appl. Entomol. Zool.* 20:500–1
127. Kolding S, Fenchel TM. 1981. Patterns of reproduction in different populations of five species of the amphipod genus *Gammarus*. *Oikos* 37:167–72
128. Kozłowski J. 1996. Optimal initial size and adult size of animals: consequences for macroevolution and community structure. *Am. Nat.* 147:101–14
129. Lamb RJ, Smith SM. 1980. Comparisons of egg size and related life-history characteristics for two predaceous tree-hole mosquitos (*Toxorhynchites*). *Can. J. Zool.* 58:2065–70

130. Lampert W. 1993. Phenotypic plasticity of the size at first reproduction in *Daphnia*: the importance of maternal size. *Ecology* 74:1455–66
131. Landa K. 1992. Adaptive seasonal variation in grasshopper offspring size. *Evolution* 46:1553–58
132. Larsson FK. 1989. Female longevity and body size as predictors of fecundity and egg length in *Graphosoma lineatum* L. *Dtsch. Entomol. Z.* 36:329–34
133. Larsson FK. 1990. Female body size relationships with fecundity and egg size in two solitary species of fossorial Hymenoptera (Colletidae and Sphecidae). *Entomol. Gen.* 15:167–71
134. Larsson FK, Kustvall V. 1990. Temperature reverses size-dependent male mating success of a cerambycid beetle. *Funct. Ecol.* 4:85–90
135. Lawlor LR. 1976. Parental investment and offspring fitness in the terrestrial isopod *Armadillidium vulgare* (Latr.), (Crustaceae: Oniscoidea). *Evolution* 30:775–85
136. Leather SR, Burnand AC. 1987. Factors affecting life-history parameters of the pine beauty moth, *Panolis flammea* (D&S): the hidden costs of reproduction. *Funct. Ecol.* 1:331–38
137. Leonard DE. 1970. Intrinsic factors causing qualitative changes in populations of *Porthetria dispar* (Lepidoptera: Lymantriidae). *Can. Entomol.* 102:239–49
138. Leprince DJ, Foil LD. 1993. Relationships among body size, blood meal size, egg volume, and egg production of *Tabanus fuscicostatus* (Diptera: Tabanidae). *J. Med. Entomol.* 30:865–71
139. Lynch M. 1980. The evolution of cladoceran life histories. *Q. Rev. Biol.* 55:23–42
140. Lynch M. 1984. The limits to life history evolution in *Daphnia*. *Evolution* 38:465–82
141. Lynch M. 1989. The life history consequences of resource depression in *Daphnia pulex*. *Ecology* 70:246–56
142. Lynch M, Pfrender M, Spitze K, Lehman N, Hicks J, et al. 1999. The quantitative and molecular genetic architecture of a subdivided species. *Evolution* 53:100–10
143. Maeta Y, Takahashi K, Shimada N. 1998. Host body size as a factor determining the egg complement of Strepsiptera, an insect parasite. *J. Insect Morphol. Embryol.* 27:27–37
144. Magalhães C, Walker I. 1988. Larval development and ecological distribution of Central Amazonian palaemonid shrimp (Decapoda, Caridea). *Crustaceana* 55:279–92
145. Mappes J, Mappes T, Lappalainen T. 1997. Unequal maternal investment in offspring quality in relation to predation risk. *Evol. Ecol.* 11:237–43
146. Marshall LD. 1990. Intraspecific variation in reproductive effort by female *Parapediasia teterella* (Lepidoptera: Pyralidae) and its relation to body size. *Can. J. Zool.* 68:44–48
147. Marshall SD, Gittleman JL. 1994. Clutch size in spiders: is more better? *Funct. Ecol.* 8:118–24
148. Mashiko K. 1982. Differences in both the egg size and the clutch size of the freshwater prawn *Palaemon paucidens* de Haan in the Sagami river. *Jpn. J. Ecol.* 32:445–51
149. Mashiko K. 1983. Differences in egg and clutch sizes of the prawn *Macrobrachium nipponense* (de Haan) between brackish and freshwaters of a river. *Zool. Mag.* 92:1–9
150. Mashiko K. 1990. Diversified egg and clutch sizes among local populations of the fresh-water prawn *Macrobrachium nipponense* (de Haan). *J. Crust. Biol.* 10:306–14
151. Mashiko K. 1992. Genetic egg and clutch size variations in freshwater prawn populations. *Oikos* 63:454–58
152. McGinley MA. 1989. The influence of a positive correlation between clutch size and offspring fitness on the optimal offspring size. *Evol. Ecol.* 3:150–56

153. McKee D, Ebert D. 1996. The interactive effects of temperature, food level and maternal phenotype on offspring size in *Daphnia magna*. *Oecologia* 107:189–96
154. McLain DK, Mallard SD. 1991. Sources and adaptive consequences of egg size variation in *Nezara viridula* (Hemiptera: Pentatomidae). *Psyche* 98:135–64
155. McLaren IA. 1965. Some relationships between temperature and egg size, body size, development rate, and fecundity, of the copepod *Pseudocalanus*. *Limnol. Oceanogr.* 10:528–38
155a. Mohaghegh J, De Clerco P, Tirry L. 1998. Effects of maternal age and egg weight on development time and body weight of offspring of *Podisus maculiventris* (Heteroptera: Pentatomidae). *Ann. Entomol. Am.* 91:315–322.
156. Moore M, Folt C. 1993. Zooplankton body size and community structure: effects of thermal and toxicant stress. *TREE* 8:178–83
157. Moore RA, Singer MC. 1987. Effects of maternal age and adult diet on egg weight in the butterfly *Euphydryas editha*. *Ecol. Entomol.* 12:401–8
158. Mura G. 1991. Additional remarks on cyst morphometrics in anostracans and its significance. 1. Egg size. *Crustaceana* 61:241–52
159. Murphy DD, Launer AE, Ehrlich PR. 1983. The role of adult feeding in egg production and population dynamics of the checkerspot butterfly *Euphydrya editha*. *Oecologia* 56:257–63
160. Nakasuji F. 1987. Egg size of skippers (Lepidoptera: Hesperiidae) in relation to their host specificity and to leaf toughness of host plants. *Ecol. Res.* 2:175–83
161. Nakasuji F, Kimura M. 1984. Seasonal polymorphism of egg size in a migrant skipper, *Parnara guttata guttata* (Lepidoptera, Hesperiidae). *Kontyû* 52:253–59
162. Newton C, Dixon AFG. 1990. Embryonic growth and birth rate of the offspring of apterous and alate aphids: a cost of dispersal. *Entomol. Exp. Appl.* 55:223–29
163. Nishino M. 1990. Geographic variation of body size, brood size and egg size of a freshwater shrimp, *Palaemon paucidens* de Haan with some discussion on brood habitat. *Jpn. J. Limnol.* 41:185–202
164. Oberhauser KS. 1997. Fecundity, lifespan and egg mass in butterflies: Effects of male-derived nutrients and female size. *Funct. Ecol.* 11:166–75
165. Odinetz-Collart O, Rabelo H. 1996. Variation in egg size of the fresh-water prawn *Macrobrachium amazonicum* (Decapoda, Palaemonidae). *J. Crust. Biol.* 16:684–88
166. O'Neill KM, Skinner SW. 1990. Ovarian egg size and number in relation to female size in five species of parasitoid wasps. *J. Zool.* 220:115–22
167. Pagel M. 1999. Mother and father in surprise genetic agreement. *Nature* 397:19–20
168. Palmer JO. 1985. Life-history consequences of body-size variation in the milkweeed leaf beetle, *Labidomera clivicollis* (Coleoptera: Chrysomelidae). *Ann. Entomol. Soc. Am.* 78:603–8
169. Parker GA, Begon M. 1986. Optimal egg size and clutch size: effects of environment and maternal phenotype. *Am. Nat.* 128:573–92
170. Parsons PA. 1964. Egg lengths in *Drosophila melanogaster* and correlated responses to selection. *Genetica* 35:175–81
171. Parsons PA. 1964. Parental age and the offspring. *Q. Rev. Biol.* 39:258–75
172. Perrin N. 1988. Why are offspring born larger when it is colder? Phenotypic plasticity for offspring size in the cladoceran *Simocephalus vetulaus* (Müller). *Funct. Ecol.* 2:283–88
173. Perrin N. 1989. Population density and offspring size in the cladoceran *Simocephalus vetulus* (Müller.). *Funct. Ecol.* 3:29–36
174. Philippi T, Seger J. 1989. Hedging one's evolutionary bets, revisited. *TREE* 4:41–44

175. Poulin R. 1995. Clutch size and egg size in free-living and parasitic copepods: a comparative analysis. *Evolution* 49:325–36
176. Poulin R, Hamilton WJ. 1997. Ecological correlates of body size and egg size in parasitic *Ascothoracida* and *Rhizocephala* (Crustacea). *Acta Oecol.* 18:621–35
177. Price PW. 1973. Reproductive strategies in parasitoid wasps. *Am. Nat.* 107:684–93
178. Reavey D. 1992. Egg size, first instar behaviour and the ecology of Lepidoptera. *J. Zool. London* 227:277–97
179. Reede T. 1997. Effects of neonate size and food concentration on the life history responses of a clone of the hybrid *Daphnia hyalina X galeata* to fish kairomones. *Freshw. Biol.* 37:389–96
180. Richards LJ, Myers JH. 1980. Maternal influences on size and emergence time of the cinnabar moth. *Can. J. Zool.* 58:1452–57
181. Robertson AL. 1988. Life histories of some species of Chydoridae (Cladocera: Crustacea). *Freshw. Biol.* 20:75–84
182. Roff DA. 1992. *The Evolution of Life Histories: Theory and Analysis.* New York: Chapman & Hall
183. Rosenheim JA. 1996. An evolutionary argument for egg limitation. *Evolution* 50:2089–94
184. Rossiter MC. 1991. Maternal effects generate variation in life history: consequences of egg weight plasticity in the gypsy moth. *Funct. Ecol.* 5:386–93
185. Ruohomäki K, Hanhimäki S, Haukioja E. 1993. Effects of egg size, laying order and larval density on performance of *Epirrita autumnata* (Lep, Geometridae). *Oikos* 68:61–66
186. Savalli UM, Fox CW. 1998. Sexual selection and the fitness consequences of male body size in the seed beetle *Stator limbatus*. *Anim. Behav.* 55:473–83
186a. Schwarzkopf L, Blows MW, Caley MJ. 1999. Life history consequences of divergent selection on egg size in *Drosophila melanogaster*. *Am. Nat.* 29:333–40
187. Sheader M. 1996. Factors influencing egg size in the gammarid amphipod *Gammarus insensibilis*. *Mar. Biol.* 124:519–26
188. Shine R. 1989. Alternative models for the evolution of offspring size. *Am. Nat.* 134:311–17
189. Sibly R, Calow P. 1986. *Physiological Ecology of Animals: An Evolutionary Approach.* Oxford: Blackwell Sci.
190. Sibly R, Monk K. 1987. A theory of grasshopper life cycles. *Oikos* 48:186–94
191. Deleted in proof
192. Simmons LW. 1990. Nuptial feedings in tettigoniids: male costs and the rates of fecundity increase. *Behav. Ecol. Sociobiol.* 27:43–47
193. Simpson MR. 1995. Covariation of spider egg and clutch size—the influence of foraging and parental care. *Ecology* 76:795–800
194. Sinervo B. 1993. The effect of offspring size on physiology and life history. *BioScience* 43:210–18
195. Sinervo B, McEdward LR. 1988. Developmental consequences of an evolutionary change in egg size: an experimental test. *Evolution* 42:885–89
196. Skadsheim A. 1984. Coexistence and reproductive adaptations of amphipods: the role of environmental heterogeneity. *Oikos* 43:94–103
197. Smith CC, Fretwell SD. 1974. The optimal balance between size and number of offspring. *Am. Nat.* 108:499–506
198. Solbreck C. 1986. Wing and flight muscle polymorphism in a lygaeid bug, *Horvathiolus gibbicollis*: determinants and life history consequences. *Ecol. Entomol.* 11:435–44
199. Solbreck C, Olsson R, Anderson DB, Forare J. 1989. Size, life history and responses to food shortage in two geographical strains of a seed bug *Lygaeus equestris*. *Oikos* 55:387–96

200. Sota T. 1993. Response to selection for desiccation resistance in *Aedes albopictus* eggs (Diptera: Culicidae). *Appl. Entomol. Zool.* 28:161–68
201. Sota T, Mogi M. 1992. Interspecific variation in desiccation survival time of *Aedes* (*Stegomyia*) mosquito eggs is correlated with habitat and egg size. *Oecologia* 90:353–58
202. Steinwascher K. 1984. Egg size variation in *Aedes aegypti*: relationship to body size and other variables. *Am. Midl. Nat.* 112:76–84
203. Deleted in proof
204. Stewart LA, Hemptinne J-L, Dixon AFG. 1991. Reproductive tactics of ladybird beetle: relationships between egg size, ovariole number and developmental time. *Funct. Ecol.* 5:380–85
205. Strong DR. 1972. Life history variation among populations of an amphipod (*Hyalella azteca*). *Ecology* 53:1103–11
206. Styan CA. 1998. Polyspermy, egg size, and the fertilization kinetics of free-spawning marine invertebrates. *Am. Nat.* 152:290–97
207. Svard L, Wiklund C. 1988. Fecundity, egg weight and longevity in relation to multiple matings in females of the monarch butterfly. *Behav. Ecol. Sociobiol.* 23:39–43
208. Tanaka S, Wolda H. 1987. Seasonal wing length dimorphism in a tropical seed bug: ecological significance of the short-winged form. *Oecologia* 75:559–65
209. Tauber CA, Tauber MJ, Tauber MJ. 1991. Egg size and taxon: their influence on survival and development of chrysopid hatchlings after food and water deprivation. *Can. J. Zool.* 69:2644–50
210. Telfer WH, Rutberg LD. 1960. Effects of blood protein depletion on the growth of the oocytes in the cecropia moth. *Biol. Bull.* 118:352–66
211. Telford SR, Dangerfield JM. 1995. Offspring size variation in some southern African woodlice. *Afr. J. Ecol.* 33:236–41
212. Tessier AJ, Consolatti NL. 1989. Variation in offspring size in *Daphnia* and consequences for individual fitness. *Oikos* 56:269–76
213. Tessier AJ, Consolatti NL. 1991. Resource quality and offspring quality in *Daphnia*. *Ecology* 72:468–78
214. Toda S, Fujisaki K, Nakasuji F. 1995. The influence of egg size on development of the bean bug, *Riptortus clavatus* Thunberg (Heteroptera: Coreidae). *Appl. Entomol. Zool.* 30:485–87
215. Trubetskova I, Lampert W. 1995. Egg size and egg mass of *Daphnia magna*—response to food availability. *Hydrobiology* 307:139–45
216. Vahed K. 1998. The function of nuptial feeding in insects: a review of empirical studies. *Biol. Rev.* 73:43–78
217. Villet M. 1990. Qualitative relations of egg size, egg production and colony size in some ponerine ants (Hymenoptera, Formicidae). *J. Nat. Hist.* 24:1321–31
218. Wall R. 1990. Ovarian aging in tsetse flies (Diptera: Glossinidae)—interspecific differences. *Bull. Entomol.* 12:109–14
219. Wallin H, Chiverton PA, Ekbom BS, Borg A. 1992. Diet, fecundity and egg size in some polyphagous predatory carabid beetles. *Entomol. Exp. Appl.* 65:129–40
220. Walsh CJ. 1993. Larval development of *Paratya australiensis* Kemp, 1917 (Decapoda: Caridea: Atyidae), reared in the laboratory, with comparisons of fecundity and egg and larval size between estuarine and riverine environments. *J. Crust. Biol.* 13:456–80
221. Warren DC. 1924. Inheritance of egg size in *Drosophila melanogaster*. *Genetics* 9:41–69
222. Wasserman SS, Asami T. 1985. The effect of maternal age upon fitness of progeny in the southern cowpea weevil, *Callosobruchus maculatus*. *Oikos* 45:191–96
223. Wasserman SS, Mitter C. 1978. The rela-

tionship of body size to breadth of diet in some Lepidoptera. *Ecol. Entomol.* 3:155–60

224. Webber LG. 1955. The relationship between larval and adult size of the Australian sheep blowfly, *Lucilia cuprina. Aust. J. Zool.* 3:346–53
225. Weigensberg I, Carriere Y, Roff DA. 1998. Effects of male genetic contribution and paternal investment to egg and hatchling size in the cricket, *Gryllus firmus. J. Evol. Biol.* 11:135–46
226. Wickman P-O, Karlsson B. 1987. Changes in egg color, egg weight and oviposition rate with the number of eggs laid by wild females of the small heath butterfly, *Coenonympha pamphilus. Ecol. Entomol.* 12:109–14
227. Wiklund C, Karlsson B. 1984. Egg size variation in satyrid butterflies: adaptive vs. historical, "Bauplan," and mechanistic explanations. *Oikos* 43:391–400
228. Wiklund C, Karlsson B, Forsberg J. 1987. Adaptive versus constraint explanations for egg-to-body size relationships in two butterfly families. *Am. Nat.* 130:828–38
229. Wiklund C, Persson A. 1983. Fecundity, and the relation of egg weight to offspring fitness in the speckled wood butterfly *Pararge aegeria*, or why don't butterfly females lay more eggs? *Oikos* 40:53–63
230. Willows RI. 1987. Intrapopulation variation in the reproductive characteristics of two populations of *Ligia oceanica* (Crustacea: Oniscidea). *J. Anim. Ecol.* 56:331–40

230a. Winkler DW, Wallin K. 1987. Offspring size and number: a life history model linking effort per offspring and total effort. *Am. Nat.* 129:708–20

231. Wittmann KJ. 1981. On the breeding biology and physiology of marsupial development in Mediterranean *Leptomysis* (Mysidacea: Crustacea) with special reference to the effects of temperature and egg size. *J. Exp. Mar. Biol. Ecol.* 53:261–79
232. Yamplosky LY, Scheiner SM. 1996. Why larger offspring at lower temperatures? A demographic approach. *Am. Nat.* 147:86–100

Annu. Rev. Entomol. 2000. 45:371–391

INSECTICIDE RESISTANCE IN INSECT VECTORS OF HUMAN DISEASE

Janet Hemingway and Hilary Ranson
School of Biosciences, University of Wales Cardiff, P.O. Box 915, Cardiff, Wales CF1 3TL; e-mail: sabjh@cardiff.ac.uk

Key Words insecticide, mosquito, esterases, monooxygenases, glutathione S-transferases

■ **Abstract** Insecticide resistance is an increasing problem in many insect vectors of disease. Our knowledge of the basic mechanisms underlying resistance to commonly used insecticides is well established. Molecular techniques have recently allowed us to start and dissect most of these mechanisms at the DNA level. The next major challenge will be to use this molecular understanding of resistance to develop novel strategies with which we can truly manage resistance. State-of-the-art information on resistance in insect vectors of disease is reviewed in this context.

INTRODUCTION

Insecticides play a central role in controlling major vectors of diseases such as mosquitoes, sandflies, fleas, lice, tsetse flies, and triatomid bugs. In 1955 the World Health Organization (WHO) assembly proposed the global eradication of the most prevalent vector-borne human disease, malaria, by the use of residual house-spraying of DDT. However, the insecticide euphoria soon ended and in 1976 WHO officially reverted from malaria eradication to malaria control. This marked shift from malaria eradication to primary health care was an emotive issue, eliciting a rapid and complete change of rhetoric from WHO (12). Several issues had prompted this switch, but a major cause of the change in policy was the appearance of DDT resistance in a broad range of the mosquito vectors. In 1975 WHO reported that 256 million people were living in areas where DDT and/or BHC resistance was undermining malaria control efforts. (This did not include the African region, where 90% of malaria occurs and where DDT resistance had already been noted in *Anopheles gambiae,* the major malaria vector.)

The resistance problems continued with the switch to newer insecticides such as the organophosphates, carbamates and pyrethroids. Operationally, many control programs have switched from blanket spraying of house interiors to focal use of insecticides on bednets. Focal spraying limits the insecticides of choice largely to pyrethroids due to the speed of kill required to protect the occupant of the bednet and the safety margin needed for insecticides used in such close contact with people. Today the major emphasis in resistance research is on the molecular mechanisms of resistance and rational resistance management, with a view to controlling the development and spread of resistant vector populations. In Africa, WHO and the World Bank have instigated major new initiatives with other major donors and the scientific community internationally to "roll back" malaria. One major problem these initiatives are tackling is the presence of two developing foci of pyrethroid resistance in the most important African malaria vector, *An. gambiae*.

SCALE OF THE PROBLEM

The amount of resistance in insect vector populations is dependent both on the volume and frequency of applications of insecticides used against them and the inherent characteristics of the insect species involved. Tsetse flies, for example, were controlled by wide-scale spraying of DDT for many years, but DDT resistance has never developed in this species. Another example of an insect vector exhibiting little or no resistance to insecticides is the triatomid bug. In both cases the major factor influencing insecticide resistance development is the life cycle of the insect pest, in particular the long life cycles for the bugs, and the production of very small numbers of young by the tsetses. In contrast, mosquitoes have all the characteristics suited to rapid resistance development, including short life cycles with abundant progeny.

Mosquito Resistance

The major mosquito vectors span the *Culex, Aedes,* and *Anopheles* genera. *Culex* are the major vectors of filariasis and Japanese encephalitis, *Aedes* of dengue and dengue hemorrhagic fever, and *Anopheles* of malaria. The range of many of these species is not static. For example, several *Aedes* species recently extended their range in Asia and Latin America, leading to an increased risk of dengue in these areas.

DDT was first introduced for mosquito control in 1946. In 1947 the first cases of DDT resistance occurred in *Aedes tritaeniorhynchus* and *Ae. solicitans* (15). Since then more than 100 mosquito species are reported as resistant to one or more insecticide, and more than 50 of these are anophelines (113). Insecticides used for malaria control have included -BHC, organophosphorus, carbamate, and pyrethroid insecticides, with the latter now taking increasing market share for both indoor residual spraying and large-scale insecticide-impregnated bednet programs. Other insecticide groups, such as the benzylphenyl ureas and Bti, have had limited use against mosquitoes. Resistance has tended to follow the switches of insecticides. Resistance to -BHC/dieldrin is widespread despite the lack of use of these insecticides for many years. Organophosphate (OP) resistance, either in the form of broad-spectrum OP resistance or malathion-specific resistance, occurs in the major vectors *An. culicifacies* (59), *An. stephensi* (30, 44), *An. albimanus* (3, 51), *An. arabiensis* (45) and *An. sacharovi* (54). *An. culicifacies* is recognized as a species complex (101): In Sri Lanka malathion resistance occurs in *An. culicifacies* species B, while in India resistance is in species B and C (59, 88). Species B in Sri Lanka is resistant to fenitrothion, which is independent of the malathion-specific resistance (58, 60), and is developing pyrethroid resistance (SHPP Karunaratne, personal communication). Organophosphorus insecticide resistance is widespread in all the major *Culex* vectors (53), and pyrethroid resistance occurs in *C. quinquefasciatus* (1, 7, 19). Pyrethroid resistance has been noted in *An. albimanus* (13), *An. stephensi* (107) and *An. gambiae* (18, 20, 112) among others, while carbamate resistance is present in *An. sacharovi* and *An. albimanus* (57). Pyrethroid resistance is widespread in *Ae. aegypti* (6, 48, 70) and cases of OP and carbamate resistance have also been recorded in this species (72, 76).

The development of pyrethroid resistance in *An. gambiae* is particularly important given the recent emphasis by the WHO and other organizations on the use of pyrethroid-impregnated bednets for malaria control. Two cases of pyrethroid resistance in *An. gambiae,* from the Ivory Coast and Kenya, are well documented (18, 112). The west African focus appears to be larger and has higher levels of resistance than that in east Africa.

Sandfly Resistance

The peridomestic vectors of *Leishmania, Plebotomus papatasi, Lutzomyia longipalpis,* and *L. intermedia* are controlled primarily by insecticides throughout their range. The control of these sandflies is often a by-product of anti-malarial house-spraying. The only insecticide resistance reported to date in sandflies is to DDT in Indian *P. papatasi* (29).

Head and Body Louse Resistance

The body louse *Pediculus humanus* has developed widespread resistance to organochlorines (16), is malathion resistant in parts of Africa (113), and has low-level resistance to pyrethroids in several regions (35). Resistance to organochlorine insecticides, such as DDT and lindane, has been recorded in the human head louse *Pediculus capitis* in Israel, Canada, Denmark, and Malaysia (16, 73, 113). Permethrin has been extensively used for head louse control since the early 1980s (77, 103). The first reports of control failure with this insecticide were in the early 1990s in Israel (77), the Czech Republic (97), and France (21).

Simulium Resistance

Some cytospecies of the *Simulium damnosum* complex are vectors of onchocerciasis. In 1974 the Onchocerciasis Control Programme in west Africa established a long-term insecticide-based control program for this vector. Temephos resistance occurred initially, prompting a switch to chlorphoxim, but resistance to this insecticide occurred within a year (49). Resistance in this species is currently being managed by a rotation of temephos, Bti, and permethrin, the insecticide usage being determined by the rate at which water is flowing in rivers forming the major breeding sites of these vectors.

THE BIOCHEMISTRY OF RESISTANCE

Insecticide Metabolism

Three major enzyme groups are responsible for metabolically based resistance to organochlorines, organophosphates, carbamates, and pyrethroids. DDT-dehydrochlorinase was first recognized as a glutathione S-transferase in the house fly, *Musca domestica* (23). It has been shown to have this role commonly in anopheline and *Aedes* mosquitoes (40, 85). Esterases are often involved in organophosphate, carbamate, and to a lesser extent, pyrethroid resistance. Monooxygenases are involved in the metabolism of pyrethroids, the activation and/or detoxication of organophosphorus insecticides and, to a lesser extent, carbamate resistance.

Esterase-Based Resistance

The esterase-based resistance mechanisms have been studied most extensively at the biochemical and molecular level in *Culex* mosquitoes and the aphid *Myzus persicae*. Work is in progress on related and distinct esterase resistance mechanisms in a range of *Anopheles* and *Aedes* species. Broad-spectrum organophosphate resistance is conferred by the elevated esterases of *Culex*. All these esterases act by rapidly binding and slowly turning over the insecticide: They sequester rather than rapidly metabolize the pesticide (62).

Two common esterase loci, *estα* and *estβ,* are involved alone or in combination in this type of resistance in *Culex* (109). In *C. quinquefasciatus* the most common elevated esterase phenotype involves two enzymes, est$\alpha 2^1$ and est$\beta 2^1$ (A_2 and B_2 on an earlier classification) (110). The classification of these esterases is based on their preferences for α- or β-naphthyl acetate, their mobility on native polyacrylamide gels, and their nucleotide sequence (53). Smaller numbers of *C. quinquefasciatus* populations have elevated estβ1 alone, elevated estα1 alone or co-elevated estβ1 and estα3 (27, 53). When purified estα and estβ from the insecticide-susceptible PelSS strain were compared to various enzymes purified from resistant strains, up to 1000-fold differences among the inhibition-kinetic constants occurred for the oxon analogues of various OPs (63).

The superiority of insecticide binding in enzymes from the resistant strains suggests that there has been positive insecticide selection pressure to maintain elevation of favorable alleles of the esterases in insecticide-resistant insects. Although there are minor variations between the inhibition kinetics of the different elevated alleles, the reason why the est$\alpha 2^1$/est$\beta 2^1$ phenotype is so common (in more than 90% of resistant populations) compared to the other elevated esterase phenotypes is not obvious. This advantage may be linked to a third gene, which is co-elevated with esterases estα2/estβ2 but not with the other esterase phenotypes (50).

Metabolic studies on *Culex* homogenates suggests that increased rates of esterase-mediated metabolism plays little or no role in resistance. One exception to this is *C. tarsalis,* where two resistance mechanisms co-exist: one involving elevated sequestering esterases, the other involving non-elevated metabolically active esterases (120). In contrast to the situation in *Culex,* a number of *Anopheles* species have a non-elevated esterase mechanism that confers resistance specifically to malathion through increased rates of metabolism (44–46) (11, 68). In *An. stephensi* three esterases with malathion carboxylesterase activity have been isolated and characterized (47, 52).

Glutathione S-Transferase-Based Resistance

Many studies have shown that insecticide-resistant insects have elevated levels of glutathione S-transferase activity in crude homogenates, which suggests a role for GSTs in resistance (37, 38). GSTs are dimeric multifunctional enzymes that play a role in detoxification of a large range of xenobiotics (86). The enzymes catalyze the nucleophilic attack of reduced glutathione (GSH) on the electrophilic centers of lipophilic compounds. Multiple forms of these enzymes have been reported for mosquitoes, house fly, *Drosophila*, sheep blow fly, and grass grub (22, 24, 105).

Two families of insect GST are recognized, and both appear to have a role in insecticide resistance in insects. In *Ae. aegypti* at least two GSTs are elevated in DDT-resistant insects (39, 41), while in *An. gambiae* a large number of different GSTs are elevated, some of which are class I GSTs (84, 85). The *Ae. aegypti* and *An. gambiae* GSTs in resistant insects are constitutively over-expressed. The GST-2 of *Ae. aegypti* is over-expressed in all tissues except the ovaries of resistant insects (39).

Monooxygenase-Based Resistance

The monooxygenases are a complex family of enzymes found in most organisms, including insects. These enzymes are involved in the metabolism of xenobiotics and have a role in endogenous metabolism. The P450 monooxygenase are generally the rate-limiting enzyme step in the chain. These enzymes are important in adaptation of insects to toxic chemicals in their host plants. P450 monooxygenases are involved in the metabolism of virtually all insecticides, leading to activation of the molecule in the case of organophosphorus insecticides, or more generally to detoxification. P450 enzymes bind molecular oxygen and receive electrons from NADPH to introduce an oxygen molecule into the substrate.

Elevated monooxygenase activity is associated with pyrethroid resistance in *An. stephensi*, *An. subpictus, An. gambiae* (14, 55, 112), and *C. quinquefasciatus*

(65). Currently this enzyme system is poorly studied in insect vectors of disease. The nomenclature of the P450 superfamily is based on amino acid sequence homologies, with all families having the CYP prefix followed by a numeral for the family, a letter for the subfamily, and a numeral for the individual gene. To date insect P450s have been assigned to six families: five are insect-specific and one, CYP4, has sequence homologies with families in other organisms (8).

Target-Site Resistance

The organophosphorus, carbamates, organochlorine, and pyrethroid insecticides all target the nervous system. Newer classes of insecticides are available for vector control, but the high cost of developing and registering new insecticides inevitably means that insecticides are developed initially for the agricultural market and then utilized for public health vector control, where their activities and safety profile are appropriate and where the market is sufficiently large to warrant the registration costs for public health use. Compounds targeting the nicotinic acetylcholine receptor have recently made this transition from agriculture into public health.

Acetylcholinesterase

The organophosphates and carbamates target acetylcholinesterase (AChE). AChE hydrolyzes the excitatory neurotransmitter acetylcholine on the post-synaptic nerve membrane. Insect AChE has a substrate specificity intermediate between vertebrate AChE and butyrylcholinesterase. The predominant molecular form in insects is a globular amphiphilic dimer which is membrane-bound via a glycolipid anchor. Alterations in AChE in organophosphate-and carbamate-resistant insects result in a decreased sensitivity to inhibition of the enzyme by these insecticides (5, 51). The organophosphorus insecticides are converted to their oxon analogues via the action of monooxygenases before acting as AChE inhibitors. In *C. pipiens,* AChE1 and AChE2 differ in their substrate specificity, inhibitor sensitivity, and electrophoretic migration pattern (69). Only AChE1 appears to be involved in conferring insecticide resistance.

GABA Receptors

Resistance to dieldrin was recorded in the 1950s, but the involvement of the GABA receptors in this resistance was not elucidated until the 1990s. The GABA receptor in insects is a heteromultimeric gated chloride-ion channel, a widespread inhibitory neurotransmission channel in the insect's central nervous system and in neuromuscular junctions (9). The insect GABA receptor is implicated as a site of action for pyrethroids and avermectins as well as cyclodienes. Studies showing that cyclodiene-resistant insects are resistant to picrotoxin and phenylpyrazole insecticides, and that the effect of ivermectin on cultured neurons can be reversed by picrotoxin pretreatment, suggest that these insecticides exert their effect by interacting with the chloride ionophore associated with the insect GABA receptor (10, 62).

Sodium Channels

The pharmacological effect of DDT and pyrethroids is to cause persistent activation of the sodium channels by delaying the normal voltage-dependent mechanism of inactivation (100). Insensitivity of the sodium channels to insecticide inhibition was first recorded in *Musca domestica* (32). In mosquitoes there have been many reports of suspected "kdr"-like resistance inferred from cross resistance between DDT and pyrethroids, which act on the same site within the sodium channel. These reports have been validated by electrophysiological measurements in *Ae.* aegypti and *An. stephensi* (48, 107).

THE MOLECULAR BIOLOGY OF RESISTANCE

Metabolic Mechanisms

Over-expression of enzymes capable of detoxifying insecticides or amino acid substitutions within these enzymes, which alter the affinity of the enzyme for the insecticide, can result in high levels of insecticide resistance. Increased expression of the genes encoding the major xenobiotic metabolizing enzymes are the most common cause of insecticide resistance in mosquitoes. These large enzyme families may contain multiple enzymes with broad overlapping substrate specificities, and there is a high probability that at least one member of the family will be capable of metabolizing one or more insecticides. Increased production of these enzymes may have a lower associated fitness cost than those associated with alterations in the structural genes because the primary function of the enzyme is not disrupted.

Mutations in Structural Genes

In many cases of resistance caused by increased metabolism of the insecticide the exact genetic mechanism is not known. As yet no validated reports exist of mutations within detoxifying enzymes leading to resistance to insecticides in disease vectors. Two examples have been reported in non-vector species: because both of these mechanisms may be present in disease vectors, we describe them here.

Resistance to the organophosphate insecticide malathion is caused by a single amino acid substitution (Trp^{251}-Leu) within the E3 esterase of the sheep blow fly, *Lucilia cuprina* (17). Malathion-resistant strains of *L. cuprina* have very low levels of activity with aliphatic esters that are conventionally used as stains for esterase activity (80). A similar phenotype has been observed in malathion-resistant strains of *An. stephensi*, *An. arabiensis*, and *An. culicifacies*. At least three enzymes are able to metabolize malathion in *An. stephensi* but it is not yet known whether a point mutation similar to that described in *Lucilia* is responsible for malathion resistance in *Anopheles* (51).

A second distinct amino acid substitution (Gly^{137}-Asp) within the active site of the *Lucilia* E3 isozyme confers broad cross resistance to many organophosphorus insecticides but not to malathion (79). This same mutation is present in OP-resistant strains of house fly.

Gene Amplification

The metabolic resistance mechanism studied in most detail in insect disease vectors is the elevated esterase-based system in *Culex* (110, 111). In the three *Culex* species studied to date at the molecular level the homologous *estβ* gene is amplified in resistant insects (64, 111, 114, 115). Insecticide resistance via amplification of genes involved in their detoxification is common in several insects. The most common amplified esterase-based mechanism in *Culex* involves the co-amplification of two esterases, *estα2*^{1} and *estβ2*^{1} in *C. quinquefasciatus* and other members of the *C. pipiens* complex worldwide (109). Other strains of *C. quinquefasciatus* have *estα3* and *estβ1* co-amplified (27), while the TEM-R strain has amplified *estβ1* alone (75). Similarly, *C. tritaeniorhynchus* has a single amplified *est*β, *Ctrest*β*1* (64).

The *estα* and *estβ* genes have arisen as the result of a gene duplication event, which must have occurred prior to speciation within the *Culex* genus. The genes occur in a head-to-head arrangement approximately 1.7 Kb apart in susceptible insects (109). In resistant insects with *estα2*^{1}/*est*β2^1 the amplified genes are 2.7 Kb apart, the difference being accounted for by expansion with three indels in the intergenic spacer (109). The indels may have introduced further regulatory elements to the amplicon intergenic spacer (52).

The identical RFLP patterns of the *estα2*^{1} and *estβ2*^{1} loci in resistant *C. pipiens* complex populations worldwide suggest that the amplification of these alleles occurred once and has since spread by migration (92). Other alleles of the esterase A and B loci are amplified in some *Culex* strains. For example, a strain of *C. pipiens* from Cyprus has an estimated 40 to 60 copies of the *estα5*^{1} and *estβ5*^{1} genes (42) whereas in the TEM-R strain from California amplification is found only at the B locus. The chromosomal region containing these esterase genes presumably represents an amplification hot spot, a theory supported by the amplification of the homologous esterase B gene in a distinct *Culex* species, *C. tritaeniorhynchus* (64).

An extensive study examining the spread and fitness of insects containing different esterase amplicons has been undertaken in southern France (93).

Transcriptional Regulation Esterases

Elevated levels of esterases may not always be the result of gene amplification. The expression of estα1 in the Barriol strain of *C. pipiens* from southern France is thought to be increased due to changes in an unidentified regulatory element rather than underlying amplification of the estα gene (42). Amplified esterases can also be expressed at different levels. For example, there is fourfold more estβ than estα in resistant *C. quinquefasciatus,* although the genes are present in a 1:1 ratio. This difference in expression is reflected at the protein and mRNA level (63, 81a).

Glutathione S-Transferases

The primary role of GSTs in mosquito insecticide resistance is in the metabolism of DDT to nontoxic products, although they also have a secondary role in organophosphate resistance (54). GST-based DDT resistance is common in a number of anopheline species, reflecting the heavy use of this insecticide for malaria control over several decades. Molecular characterization of GSTs is most developed in *An. gambiae*, although work on *An. dirus* from Thailand suggests that a similar arrangement of GSTs occurs. However, there is evidence for more limited allelic diversity in this species (52, 86, 87).

Two classes of insect GSTs have been recognized and members of both classes are important in the metabolism of insecticides in mosquitoes and other insects. The sequence of a single class II *An. gambiae* GST has been published (94), and we have cloned and sequenced the major class II GST from *Ae. aegypti* (D Grant, M Wajidi, H Ranson, and J Hemingway, unpublished data). This *Aedes* GST, GST-2, is over-expressed in the DDT resistant GG strain of *Ae. aegypti.* In this species the resistance mutation is thought to lead to disruption of a transacting repressor. The mutation prevents the normal function of the repressor leading to elevated levels of GST-2 enzyme in resistant mosquitoes (39).

The insect class I GSTs are encoded by a large gene family in *An. gambiae, M. domestica,* and *D. melanogaster.* The genomic organization of this GST class in these three insect species is strikingly different (89). In *D. melanogaster* eight divergent intronless genes are found within a 14 Kb DNA segment (106). In *An. gambiae* multiple class I GST genes are also clustered in a single location. At least one member of this family is intronless (91) but other class I GSTs in *An. gambiae* contain one or more introns (Figure 1). One of these genes, aggst1α, is alternatively spliced to produce four distinct mRNA transcripts, each of which shares a common 5′ exon with a different 3′ exon (89). The products of these spliced genes differ in their ability to metabolize DDT (91) and some of these metabolically active GSTs are upregulated in resistant mosquitoes (90).

The organization of this class I GST gene family in insecticide-resistant and insecticide-susceptible *An. gambiae* is very similar. Hence the actual GST-based resistance mechanism is probably caused by a *trans*-acting regulator. The development of a fine-scale microsatellite map (119) and bacterial artificial chromosome (BAC) library for *An. gambiae* have now made this species amenable to a positional cloning approach. Such an approach is being used to define the regulator responsible for this GST resistance.

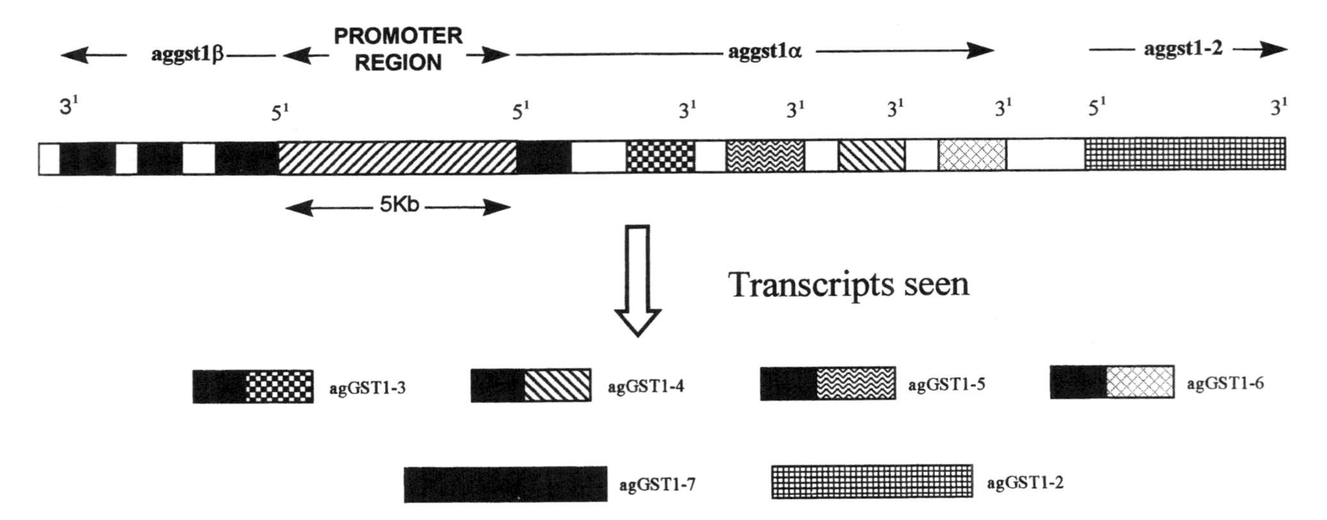

Figure 1 The class I glutathione S-transferase family of *Anopheles gambiae*, which includes an intronless gene *aggst1-2*, an alternatively spliced gene *aggst1α* with four distinct viable transcripts, and the *aggst1β* gene with two introns.

Monooxygenases

Insect microsomal P450 monooxygenases belong to six families. Increased transcription of genes belonging to the CYP4, CYP6, and CYP9 families has been observed in insecticide-resistant strains in different insect species. As yet, it is not known which enzymes are responsible for insecticide metabolism in mosquitoes. Seventeen partial cDNAs encoding CYP4 P450s have been identified in *An. albimanus* (98) and a similar level of diversity is present in *An. gambiae* (H Ranson & F Collins, unpublished) but the role, if any, that this family plays in resistance in the mosquito is not known. Studies on the Australian cotton bollworm, *Helicoverpa armigera,* show that the resistance-associated P450s can vary between different strains; CYP6B2 is over-expressed in one pyrethroid-resistant strain (118) whereas in another, CYP4G8 is over-expressed (83). Increased expression of CYP9A1, a member of a third family, is found in the related species *Heliothis virescens* (95).

There is strong evidence to suggest that P450 monooxygenase-based resistance in *M. domestica* and *D. melanogaster* is mediated by mutations in *trans*-acting regulatory genes. CYP6A8 is highly expressed in the DDT-resistant 91-R strain of *D. melanogaster* but not detectable in the uninduced 91-C susceptible strain (28, 67). Hybrids between the two strains show low levels of expression, suggesting that the 91-C strain carries a repressor that suppresses transcription of CYP6A8. A mutation in this repressor is thought to be responsible for the high level of expression of CYP6A8 in 91-R. In the house fly, CYP6D1-mediated metabolic resistance to pyrethroids is controlled by a nearly completely dominant *cis*-factor on autosome 1 and an incompletely recessive *trans*-factor on autosome 2 (66).

Target-Site Resistance

Non-silent point mutations within structural genes are the most common cause of target-site resistance. For selection of the mutations to occur, the resultant amino acid change must reduce the binding of the insecticide without causing a loss of primary function of the target site. Therefore the number of possible amino acid substitutions is very limited. Hence, identical resistance-associated mutations are commonly found across highly diverged taxa. The degree to which function is impaired by the resistance mutation is reflected in the fitness of resistant individuals in the absence of insecticide selection. This fitness cost has important implications for the persistence of resistance in the field.

The main sodium and GABA channel genes in insects have been cloned and their sequences compared in resistant and susceptible insects. Acetylcholinesterase-based resistance has been well characterized in *Drosophila* (36), but the elucidation of this mechanism at the molecular level in mosquitoes has proved more difficult.

GABA Receptor Changes

The GABA receptors belong to a superfamily of neurotransmitter receptors that also includes the nicotinic acetylcholine receptors. These receptors are formed by the oligomerization of five subunits around a central transmitter-gated ion channel. Five different subunits have been cloned from vertebrates. To date only three subunits have been cloned from *Drosophila melanogaster*, but these do not fit readily into the vertebrate GABA subunit classification (61).

An alanine-to-serine substitution in the putative channel-lining domain of the GABA receptor confers resistance to cyclodienes such as dieldrin (γ- HCH) (34). The mutation was first identified in *Drosophila* but has since been shown to occur in a broad range of dieldrin-resistant insects, including *Ae. aegypti* (104). The only variation in resistant insects is that glycine rather than serine can sometimes be the substituted amino acid residue. Despite the widespread switch away from the use of cyclodiene insecticides for agricultural and public health use the resistance allele is still found at relatively high frequencies in insect field populations (4).

Sodium Channels

A reduction in the sensitivity of the insect's voltage-gated sodium channels to the binding of insecticides causes the resistance phenotype known as "kdr." Changes associated with pyrethroid/DDT resistance in the sodium channels of insects are more variable than those seen in the GABA receptors but still appear to be limited to a small number of regions on this large channel protein.

The para sodium channel of houseflies contains 2108 amino acids, which fold into 4 hydrophobic repeat domains (I-IV) separated by hydrophilic linkers. The first mutation to be characterized in *kdr* insects was a leucine to phenylalanine point mutation in the S6 transmembrane segment of domain II in the sodium channel sequence of *M. domestica* (116, 117) which produces 10- to 20-fold resistance to DDT and pyrethroids. In "*super-kdr*" houseflies, this mutation also occurs with a second methionine to threonine substitution further upstream in the same domain, resulting in more than 500-fold resistance (117). Analysis of the domain region of the *para*-sodium channel gene in pyrethroid-resistant *An. gambiae* from the Ivory Coast showed an identical Leu to Phe mutation in this species (71).

A PCR-based diagnostic test discriminates between homozygous-susceptible, homozygous-resistant, and heterozygous individuals with the Leu to Phe mutation (71). Because *kdr* is semi-recessive or fully recessive (81), the ability to detect heterozygotes is of paramount importance in the early detection and management of resistance in the field.

The limited number of changes associated with kdr-type resistance may be constrained by the number of modifications that can influence pyrethroid/DDT binding to the sodium channels. However, a note of caution should be added:

There is already a tendency to investigate pyrethroid-resistant insects with a PCR approach confined to regions where a *kdr* mutation has already been seen. Hence consistent resistance-associated changes in other parts of the sodium-channel gene could be missed. A different approach to isolating kdr-type mutants has been used in *Drosophila,* utilizing the relative ease with which large numbers of mutants in the para sodium channel gene can be isolated based on their temperature sensitivity. Two classes of these mutations are in positions equivalent to the *kdr* and *super-kdr* mutations in different domains. The third class is in a novel position (Figure 2) (34).

Acetylcholinesterase

Vertebrates have two cholinesterases: acetylcholinesterase and butyrylcholinesterase. In *D. melanogaster* only a single cholinesterase gene, *Ace,* coding for acetycholinesterase has been cloned, based on a knowledge of its location via

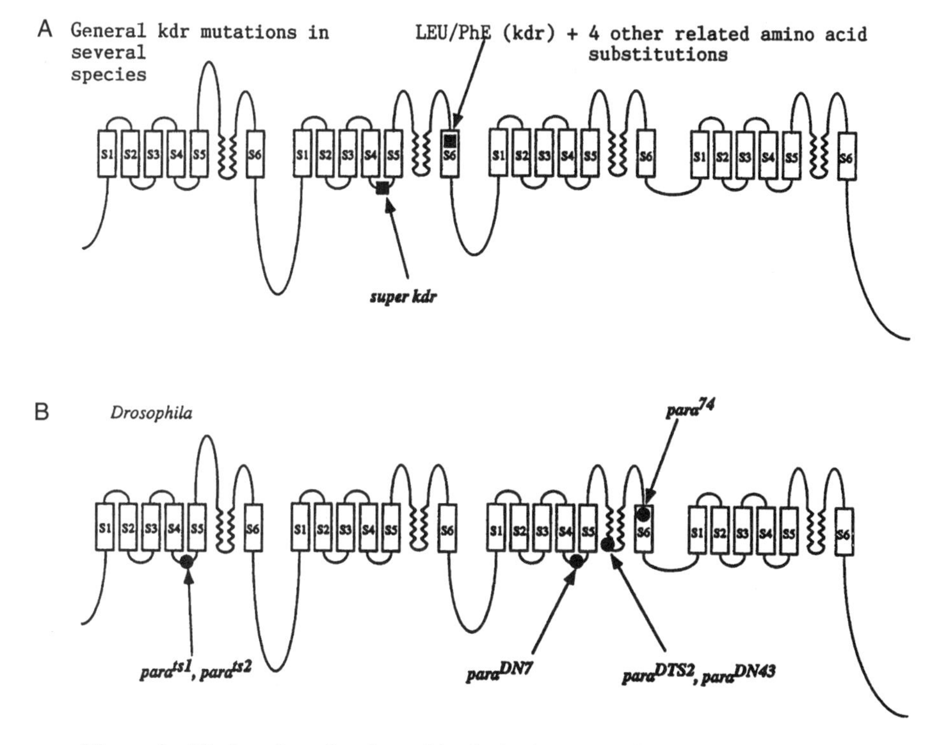

Figure 2 The location of amino acid substitutions occurring in the para-sodium channel gene of resistant strains of various insects. Mutations that have been documented in field strains of medical vectors are shown in *A. B* shows replacements in *Drosophila* isolated on the basis of temperature sensitivity paralytic phenotypes which confer DDT/pyrethroid resistance (adapted from 34).

isolation of a range of different mutants (43). A range of different amino acid substitutions in the *Ace* genes of *Drosophila* and the house fly *M. domestica* putatively cause resistance (reviewed in 33). Many of these resistance-associated residues are predicted to lie close to or within the acetylcholinesterase active site gorge (78). In *C. pipiens* at least two cholinesterase genes occur, both with acetylcholinesterase-like activity (69). A similar organization occurs in *Amphioxus*. To date the only acetylcholinesterase to have been cloned in *C. pipiens* is AChE2, which is not involved in insecticide resistance (69). The AChE2 gene is sex linked, unlike the resistance-conferring AChE1, which is autosomal. AChE genes have been cloned from the mosquitoes *Ae. aegypti* and *An. stephensi*, but both of these genes are also sex linked (2, 99).

As yet, there is no recorded AChE-based resistance mechanism in *An. stephensi*, but to date none of the acetylcholinesterase resistances recorded in insects have been sex linked, suggesting that these genes do not represent the insecticidal target site. However, detailed analysis of inhibition profiles for acetylcholinesterase from *Ae. aegypti* suggests that there is only one AChE locus in this species. If such is the case, altered acetylcholinesterase-based resistance in this species should be sex linked. At least one case of altered AChE has been reported in *Ae. aegypti* from Trinidad (108).

Five point mutations associated with resistance to organophosphorus and carbamate insecticides have been identified within the *D. melanogaster* acetylcholinesterase gene (78) and site-directed mutagenesis of the sex-linked AChE from *Ae. aegypti* has demonstrated that these same mutations also confer resistance in the mosquito enzyme (74). However, none of these mutations have been identified in field-collected or laboratory-selected strains of mosquitoes.

Insensitive AChE has a severe fitness cost in *C. pipiens* populations in southern France (31), which is probably caused by a reduction in the AChE activity of the mutated enzyme compared to the wild type.

MANAGEMENT OF INSECTICIDE-RESISTANT VECTOR POPULATIONS

The practice of using an insecticide until resistance becomes a limiting factor is rapidly eroding the number of suitable insecticides for vector control. Rotations, mosaics, and mixtures have all been proposed as resistance management tools (25, 26, 96). Numerous mathematical models have been produced to estimate how these tools should be optimally used (102). However, these models have rarely been tested under field conditions for insect vectors, due to the practical difficulties in estimating changes in resistance gene frequencies in large samples of insects (56). With the advent of different biochemical and molecular techniques for resistance-gene frequency estimation, field trials of resistance management strategies have become more feasible.

A large-scale trial of the use of rotations or mosaics of insecticides compared to single use of DDT or a pyrethroid is currently underway in Mexico (56, 82). Changes in resistance gene frequencies in *An. albimanus* are being monitored over a four-year period (82). Information resulting from such large-scale trials may allow us to establish rational strategies for long-term insecticide use in disease control programs.

As our ability to manipulate the insect genome improves and our understanding of the regulation of insecticide resistance mechanisms increase, new strategies should be devised for incorporating this new knowledge into these control programs.

Visit the Annual Reviews home page at www.AnnualReviews.org.

LITERATURE CITED

1. Amin AM, Hemingway J. 1989. Preliminary investigation of the mechanisms of DDT and pyrethroid resistance in *Culex quinquefasciatus* Say (Diptera: Culicidae) from Saudi Arabia. *Bull. Entomol. Res.* 79:361–66
2. Anthony NM, Rocheleau T, Mocelin G, Lee HJ, ffrench-Constant RH. 1995. Cloning, sequencing and functional expression of an acetylcholinesterase gene from the yellow fever mosquito *Aedes aegypti*. *FEBS Lett.* 368:461–65
3. Ariaratnam V, Georghiou GP. 1974. Carbamate resistance in *Anopheles albimanus*. Cross-resistance spectrum and stability of resistance. *Bull. WHO* 51:655–59
4. Aronstein K, Ode P, ffrench-Constant RH. 1995. PCR based monitoring of specific *Drosophila* (Diptera: Drosophilidae) cyclodiene resistance alleles in the presence and absence of selection. *Bull. Entomol. Res.* 85:5–9
5. Ayad H, Georghiou GP. 1975. Resistance to organophosphates and carbamates in *Anopheles albimanus* based on reduced sensitivity to acetylcholinesterase. *J. Econ. Entomol.* 68:295–97
6. Bang YH, Tonn RJ, Punurai P. 1969. Insecticide susceptibility and resistance found in 14 strains of *Aedes aegypti* collected from Bangkok-Thonburi, Thailand. *WHO/Vector Biol. Control/* 69:117
7. Bencheikh H, BenAliHaouas Z, Marquine M, Pasteur N. 1998. Resistance to organophosphorus and pyrethroid insecticides in *Culex pipiens* (Diptera: Culicidae) from Tunisia. *J. Med. Entomol.* 35:251–60
8. Berge JB, Feyereisen R, Amichot M. 1998. Cytochrome P450 monooxygenases and insecticide resistance in insects. *Philos. Trans. R. Soc. London Ser. B* 353:1701–5
9. Bermudez I, Hawkins CA, Taylor AM, Beadle DJ. 1991. Actions of insecticides on the insect GABA receptor complex. *J. Recept. Res.* 11:221–32
10. Bloomquist JR. 1994. Cyclodiene resistance at the insect GABA receptor chloride channel complex confers broad cross-resistance to convulsants and experimental phenylpyrazole insecticides. *Arch. Insect Biochem. Physiol.* 26:69–79
11. Boddington RG. 1992. Characterization of malathion carboxylesterases and non-specific esterases in the malaria mosquito *Anopheles stephensi*. PhD diss. Univ. London
12. Bradley DJ. 1998. The particular and general. Issues of specificity and verti-

cality in the history of malaria control. *Parassitologia* 40:5–10
13. Brogdon WG, Barber AM. 1990. Fenitrothion-deltamethrin cross-resistance conferred by esterases in Guatemalan *Anopheles albimanus. Pest. Biochem. Physiol.* 37:130–39
14. Brogdon WG, McAllister JC, Vulule J. 1997. Heme peroxidase activity measured in single mosquitoes identifies individuals expressing the elevated oxidase mechanism for insecticide resistance. *J. Am. Mosq. Control Assoc.* 13:233–37
15. Brown AWA. 1986. Insecticide resistance in mosquitoes: a pragmatic review. *J. Am. Mosq. Control Assoc.* 2:123–40
16. Brown AWA, Pal R. 1971. Insecticide resistance in arthropods. *WHO Monogr. Ser.* 38
17. Campbell PM, Newcomb RD, Russell RJ, Oakeshott JG. 1998. Two different amino acid substitutions in the ali-esterase, E3, confer alternative types of organophosphorus insecticide resistance in the sheep blowfly, *Lucilia cuprina. Insect Biochem. Mol. Biol.* 28:139–50
18. Chandre F, Darriet F, Brengues C, Manguin S, Carnevale P, Guillet P. 1999. Pyrethroid cross-resistance spectrum among populations of *Anopheles gambiae* from Cote D'Ivoire. *J. Am. Mosq. Control Assoc.* 15:53–59
19. Chandre F, Darriet F, Darder M, Cuany A, Doannio JMC, et al. 1998. Pyrethroid resistance in *Culex quinquefasciatus* from West Africa. *Med. Vet. Entomol.* 12:359–66
20. Chandre F, Darriet F, Manga L, Akogbeto M, Faye O, et al. 1999. Status of pyrethroid resistance in *Anopheles gambiae* s.l. *Bull. WHO* 77:230–34
21. Chosidow O, Chastang C, Brue C, Bouvet E, Izri M, et al. 1995. Controlled study of malathion and *d*-phenothrin lotions for *Pediculus humanus* var *capitis*-infested schoolchildren. *Lancet* 345:731–32
22. Clark AG, Dick GL, Martindale SM, Smith JN. 1985. Glutathione S-transferases from the New Zealand grass grub, *Costelytra zealandica. Insect Biochem.* 15:35–44
23. Clark AG, Shamaan NA. 1984. Evidence that DDT-dehydrochlorinase from the house fly is a glutathione S-transferase. *Pestic. Biochem. Physiol.* 22:249–61
24. Clark AG, Shamaan NA, Dauterman WC, Hayaoka T. 1984. Characterization of multiple glutathione transferases from the housefly, *Musca domestica (L). Pestic. Biochem. Physiol.* 22:51–59
25. Curtis CF. 1985. Theoretical models of the use of insecticide mixtures for the management of resistance. *Bull. Entomol. Res.* 75:259–65
26. Curtis CF, Hill N, Kasim SH. 1993. Are there effective resistance management strategies for vectors of human disease? *Biol. J. Linn. Soc.* 48:3–18
27. DeSilva D, Hemingway J, Ranson H, Vaughan A. 1997. Resistance to insecticides in insect vectors of disease: Estα3, a novel amplified esterase associated with estα1s from insecticide resistant strains of the mosquito *Culex quinquefasciatus. Exp. Parasitol.* 87:253–59
28. Dombrowski SM, Krishnan R, Witte M, Maitra S, Diesing C, et al. 1998. Constitutive and barbital-induced expression of the CYP6A2 allele of a high producer strain of CYP6A2 in the genetic background of a low producer strain. *Gene* 221:69–77
29. El-Sayed S, Hemingway J, Lane RP. 1989. Susceptibility baselines for DDT metabolism and related enzyme systems in the sandfly *Phlebotomus papatasi (Scopoli) (Diptera:Psychodidae). Bull. Entomol. Res.* 79:679–84
30. Eshgy N. 1978. Tolerance of *Anopheles stephensi* to malathion in the province of Fars, Southern Iran 1977. *Mosq. News* 38:580–83
31. Evgenev MB, Corces VG, Lankenau DH. 1992. Ulysses transposable element

of *Drosophila* shows high structural similarities to functional domains of retroviruses. *J. Mol. Biol.* 225:917–24

32. Farnham AW, Sawicki RM. 1976. Development of resistance to pyrethroids in insects resistant to other insecticides. *Pestic. Sci.* 7:278–82
33. Feyereisen R. 1995. Molecular biology of insecticide resistance. *Toxicol. Lett.* 82:83–90
34. ffrench-Constant RH, Pittendrigh B, Vaughan A, Anthony N. 1998. Why are there so few resistance-associated mutations in insecticide target genes? *Philos. Trans. R. Soc. London Ser. B* 353:1685–93
35. Fine BC. 1963. The present status of resistance to pyrethroid insecticides. *Pyrethrum Post* 7:18–21
36. Fournier D, Mutero A, Pralavorio M, Bride J-M. 1993. *Drosophila* acetylcholinesterase: mechanisms of resistance to organophosphates. *Chem.-Biol. Interact.* 87:233–38
37. Grant DF. 1991. Evolution of glutathione S-transferase subunits in Culicidae and related Nematocera: Electrophoretic and immunological evidence for conserved enzyme structure and expression. *Insect Biochem.* 21:435–45
38. Grant DF, Dietze EC, Hammock BD. 1991. Glutathione S-transferase isozymes in *Aedes aegypti*: purification, characterization, and isozyme specific regulation. *Insect. Biochem.* 4:421–33
39. Grant DF, Hammock BD. 1992. Genetic and molecular evidence for a trans-acting regulatory locus controlling glutathione-s transferase-2 expression in *Aedes aegypti. Mol. Gen. Genet.* 234:169–76
40. Grant DF, Matsumura F. 1988. Glutathione S-transferase-1 in *Aedes aegypti* larvae. Purification and properties. *Insect Biochem.* 18:615–22
41. Grant DF, Matsumura F. 1989. Glutathione S-transferase 1 and 2 in susceptible and insecticide resistant *Aedes aegypti. Pestic. Biochem. Physiol.* 33:132–43
42. Guillemaud T, Makate N, Raymond M, Hirst B, Callaghan A. 1997. Esterase gene amplification in *Culex pipiens. Insect Mol. Biol.* 6:319–27
43. Hall LMC, Spierer P. 1986. The Ace locus of *Drosophila melanogaster*: structural gene for acetylcholinesterase with an unusual 5' leader. *EMBO J.* 5:2949–54
44. Hemingway J. 1982. The biochemical nature of malathion resistance in *Anopheles stephensi* from Pakistan. *Pestic. Biochem. Physiol.* 17:149–55
45. Hemingway J. 1983. Biochemical studies on malathion resistance in *Anopheles arabiensis* from Sudan. *Trans. R. Soc. Trop. Med. Hyg.* 77:477–80
46. Hemingway J. 1985. Malathion carboxylesterase enzymes in *Anopheles arabiensis* from Sudan. *Pest. Biochem. Physiol.* 23:309–13
47. Hemingway J. 1999. Insecticide resistance in malaria vectors: A new approach to an old subject. *Parassitologia.* In press
48. Hemingway J, Boddington RG, Harris J, Dunbar SJ. 1989. Mechanisms of insecticide resistance in *Aedes aegypti* (L.) (Diptera:Culicidae) from Puerto Rico. *Bull. Entomol. Res.* 79:123–30
49. Hemingway J, Callaghan A, Kurtak DC. 1991. Biochemical characterization of chlorphoxim resistance in adults and larvae of the *Simulium damnosum* complex (Diptera: Simulidae). *Bull. Entomol. Res.* 81:401–6
50. Hemingway J, Coleman M, Vaughan A, Patton M, DeSilva D. 1999. Aldehyde oxidase is co-amplified with the worlds most common *Culex* mosquito insecticide resistance-associated esterases. *Insect Mol. Biol.* In press
51. Hemingway J, Georghiou GP. 1983. Studies on the acetylcholinesterase of *Anopheles albimanus* resistant and susceptible to organophosphate and carbamate insecticides. *Pestic. Biochem. Physiol.* 19:167–71
52. Hemingway J, Hawkes N, Prapanthadara

L, Jayawardena KGI, Ranson H. 1998. The role of gene splicing, gene amplification and regulation in mosquito insecticide resistance. *Philos. Trans. R. Soc. London Ser. B* 353:1695–99
53. Hemingway J, Karunaratne SHPP. 1998. Mosquito carboxylesterases: a review of the molecular biology and biochemistry of a major insecticide resistance mechanism. *Med. Vet. Entomol.* 12:1–12
54. Hemingway J, Malcolm CA, Kissoon KE, Boddington RG, Curtis CF, Hill N. 1985. The biochemistry of insecticide resistance in *Anopheles sacharovi*: comparative studies with a range of insecticide susceptible and resistant *Anopheles* and *Culex* species. *Pestic. Biochem. Physiol.* 24:68–76
55. Hemingway J, Miyamoto J, Herath PRJ. 1991. A possible novel link between organo phosphorus and DDT insecticide resistance genes in *Anopheles*: supporting evidence from fenitrothion metabolism studies. *Pestic. Biochem. Physiol.* 39:49–56
56. Hemingway J, Penilla RP, Rodriguez AD, James BM, Edge W, et al. 1997. Resistance management strategies in malaria vector mosquito control. A large-scale trial in Southern Mexico. *Pestic. Sci.* 51:375–82
57. Hemingway J, Small GJ, Monro A, Sawyer BV, Kasap H. 1992. Insecticide resistance gene frequencies in *Anopheles sacharovi* populations of the Cukurova plain, Adana province, Turkey. *Med. Vet. Entomol.* 6:342–48
58. Herath PRJ, Davidson G. 1981. Multiple resistance in *Anopheles culicifacies* Giles. *Mosq. News* 41:325–27
59. Herath PRJ, Hemingway J, Weerasinghe IS, Jayawardena KGI. 1987. The detection and characterization of malathion resistance in field populations of *Anopheles culicifacies* B in Sri Lanka. *Pestic. Biochem. Physiol.* 29:157–62
60. Herath PRJ, Miles SJ, Davidson G. 1981. Fenitrothion (OMS 43) resistance in the taxon *Anopheles culicifacies* Giles. *J. Trop. Med. Hyg.* 84:87–88
61. Hosie AM, Aronstein K, Sattelle DB, ffrench-Constant RH. 1997. Molecular biology of insect neuronal GABA receptors. *Trends Neurosci.* 20:578–83
62. Kadous AA, Ghiasuddin SM, Matsumura F, Scott JG, Tanaka K. 1983. Difference in the picrotoxinin receptor between the cyclodiene-resistant and susceptible strains of the German-cockroach. *Pest. Biochem. Physiol.* 19:157–66
63. Karunaratne SHPP, Hemingway J, Jayawardena KGI, Dassanayaka V, Vaughan A. 1995. Kinetic and molecular differences in the amplified and non-amplified esterases from insecticide-resistant and susceptible *Culex quinquefasciatus* mosquitoes. *J. Biol. Chem.* 270:31124–28
64. Karunaratne SHPP, Vaughan A, Paton MG, Hemingway J. 1998. Amplification of a serine esterase gene is involved in insecticide resistance in Sri Lankan *Culex tritaeniorhynchus*. *Insect Mol. Biol.* 7:307–15
65. Kasai S, Weerasinghe IS, Shono T. 1998. P^{450} Monooxygenases are an important mechanism of permethrin resistance in *Culex quinquefasciatus* say larvae. *Arch. Insect Biochem. Physiol.* 37:47–56
66. Liu N, Scott JG. 1997. Inheritance of CYP6D1-mediated pyrethroid resistance in house fly (Diptera: Muscidae). *J. Econ. Entomol.* 90:1478–81
67. Maitra S, Dombrowski SM, Waters LC, Gunguly R. 1996. Three second chromosome-linked clustered CYP6 genes show differential constitutive and barbital-induced expression in DDT-resistant and susceptible strains of *Drosophila melanogaster*. *Gene* 180:165–71
68. Malcolm CA, Boddington RG. 1989. Malathion resistance conferred by a carboxylesterase in *Anopheles culicifacies* Giles (Species B) (Diptera: Culicidae). *Bull. Entomol. Res.* 79:193–99
69. Malcolm CA, Bourguet D, Ascolillo A,

Rooker SJ, Garvey CF, et al. 1998. A sex-linked *Ace* gene, not linked to insensitive acetylcholinesterase-mediated insecticide resistance in *Culex pipiens*. *Insect Mol. Biol.* 7:107–20
70. Malcolm CA, Wood RJ. 1982. Location of a gene conferring resistance to knockdown by permethrin and bioresmethrin in adults of the BKPM3 strain of *Aedes aegypti*. *Genetica* 59:233–37
71. Martinez-Torres D, Chandre F, Williamson MS, Darriet F, Berge JB, et al. 1998. Molecular characterization of pyrethroid knockdown resistance (*kdr*) in the major malaria vector *Anopheles gambiae s.s.* *Insect Mol. Biol.* 7:179–84
72. Matsumura F, Brown AWA. 1961. Biochemical study of malathion tolerant *Aedes aegypti*. *Mosq. News* 21:192–94
73. Maunder JW. 1971. Resistance to organochlorine insecticides in head lice and trials using alternative compounds. *Med. Off.* 258:27–31
74. Milatovic D, Moretto A, Osman KA, Lotti M. 1997. Phenyl valerate esterases other than neuropathy target esterase and the promotion of organophosphate polyneuropathy. *Chem. Res. Toxicol.* 10: 1045–48
75. Mouches C, Pasteur N, Berge JB, Hyrien O, Raymond M, et al. 1986. Amplification of an esterase gene is responsible for insecticide resistance in a Californian *Culex* mosquito. *Science* 233:778–80
76. Mourya DT, Hemingway J, Leake CJ. 1993. Changes in enzyme titres with age in four geographical strains of *Aedes aegypti* and their association with insecticide resistance. *Med. Vet. Entomol.* 7:11–16
77. Mumcuoglu KY, Miller J, Uspensky I, Hemingway J, Klaus S, et al. 1995. Pyrethroid resistance in the head louse *Pediculus humanus capitis* from Israel. *Med. Vet. Entomol.* 9:427–32
78. Mutero A, Pralavorio M, Bride JM, Fournier D. 1994. Resistance-associated point mutations in insecticide-insensitive acetylcholinesterase. *Proc. Natl. Acad. Sci. USA* 91:5922–26
79. Newcomb RD, Campbell PM, Russell RJ, Oakeshott JG. 1997. cDNA cloning, baculovirus-expression and kinetic properties of the esterase, E3, involved in organophosphorus resistance in *Lucilia cuprina*. *Insect Biochem. Mol. Biol.* 27:15–25
80. Newcomb RD, East PD, Russell RJ, Oakeshott JG. 1996. Isolation of α cluster esterase genes associated with organophosphate resistance in *Lucilia cuprina*. *Insect Mol. Biol.* 5:211–16
81. Ozers MS, Friesen PD. 1996. The *Env*-like open reading frame of the baculovirus-integrated retrotransposon ted encodes a retrovirus-like envelope protein. *Virology* 226:252–59
81a. Paton MG, Karunaratne SHPP, Giakoumaki E, Roberts N, Hemingway J. 2000. Quantitative analysis of gene amplification in insecticide resistant *culex* mosquitoes. *Biochem. J.* In press
82. Penilla RP, Rodrigues AD, Hemingway J, Torres JL, Arredondo-Jimenez JI, Rodriguez MH. 1998. Resistance management strategies in malaria vector mosquito control. Baseline data for a large-scale field trial against *Anopheles albimanus* in Mexico. *Med. Vet. Entomol.* 12:217–33
83. Pittendrigh B, Aronstein K, Zinkovsky E, Andreev O, Campbell BC, et al. 1997. Cytochrome P450 genes from *Helicoverpa armigera*: expression in a pyrethroid-susceptible and -resistant strain. *Insect Biochem. Mol. Biol.* 27:507–12
84. Prapanthadara L, Hemingway J, Ketterman AJ. 1993. Partial purification and characterization of glutathione S-transferase involved in DDT resistance from the mosquito *Anopheles gambiae*. *Pest. Biochem. Physiol.* 47:119–33
85. Prapanthadara L, Hemingway J, Ketterman AJ. 1995. DDT-resistance in *Anopheles gambiae* Giles from Zanzibar Tanzania, based on increased DDT-dehy-

drochlorinase activity of glutathione S-transferases. *Bull. Entomol. Res.* 85:267–74

86. Prapanthadara L, Koottathep S, Promtet N, Hemingway J, Ketterman AJ. 1996. Purification and characterization of a major glutathione S-transferase from the mosquito *Anopheles dirus* (Species B). *Insect Biochem. Mol. Biol.* 26:277–85
87. Prapanthadara LA, Ranson H, Somboon P, Hemingway J. 1998. Cloning, expression and characterization of an insect class I glutathione S-transferase from *Anopheles dirus* species B. *Insect Biochem. Mol. Biol.* 28:321–29
88. Raghavendra K, Vasantha K, Subbarao SK, Pillai MKK, Sharma VP. 1991. Resistance in *Anopheles culicifacies* sibling species B and C to malathion in Andhra Pradesh and Gujarat states, India. *J. Am. Mosq. Control Assoc.* 7:255–59
89. Ranson H, Collins FH, Hemingway J. 1998. The role of alternative mRNA splicing in generating heterogeneity within the *Anopheles gambiae* class I glutathione S-transferase family. *Proc. Natl. Acad. Sci. USA* 95:14284–89
90. Ranson H, Cornel AJ, Fournier D, Vaughan A, Collins FH, Hemingway J. 1997. Cloning and localization of a glutathione *S*-transferase class I gene from *Anopheles gambiae*. *J. Biol. Chem.* 272:5464–68
91. Ranson H, Prapanthadara L, Hemingway J. 1997. Cloning and characterisation of two glutathione S-transferases from a DDT resistant strain of *Anopheles gambiae*. *Biochem. J.* 324:97–102
92. Raymond M, Callaghan A, Fort P, Pasteur N. 1991. Worldwide migration of amplified insecticide resistance genes in mosquitoes. *Nature* 350:151–53
93. Raymond M, Chevillon C, Guillemaud T, Lenormand T, Pasteur N. 1998. An overview of the evolution of overproduced esterases in the mosquito *Culex pipiens*. *Philos. Trans. R. Soc. London Ser. B* 353:1707–11
94. Reiss RA, James AA. 1993. A glutathione S-transferase gene of the vector mosquito, *Anopheles gambiae*. *Insect Mol. Biol.* 2:25–32
95. Rose RL, Goh D, Thompson DM, Verma KD, Heckel DG, et al. 1997. Cytochrome P450 (CYP)9A1 in *Heliothis virescens*: the first member of a new CYP family. *Insect Biochem. Mol. Biol.* 27:605–15
96. Roush RT. 1989. Designing resistance management programmes: how can you choose? *Pestic. Sci.* 26:423–42
97. Rupes V, Moravec J, Chmela J, Ledvidka J, Zelenkova J. 1994. A resistance of head lice (*Pediculus capitis*) to permethrin in Czech republic. *Centr. Eur. J. Public Health* 3:30–32
98. Scott JA, Collins FH, Feyereisen R. 1994. Diversity of cytochrome P^{450} genes in the mosquito, *Anopheles albimanus*. *Biochem. Biophys. Res. Commun.* 205:1452–59
99. Severson DW, Anthony NM, Andreen O, ffrench-Constant RH. 1997. Molecular mapping of insecticide resistance genes in the yellow fever mosquito (*Aedes aegypti*). *J. Hered.* 88:520–24
100. Soderlund DM, Bloomquist JR. 1989. Neurotoxic action of pyrethroid insecticides. *Annu. Rev. Entomol.* 34:77–96
101. Subbarao SK, Vasantha K, Joshi H, Raghavendra K, Devi CU, et al. 1992. Role of *Anopheles culicifacies* sibling species in malaria transmission in Madhya Pradesh state, India. *Trans. R. Soc. Trop. Med. Hyg.* 86:613–14
102. Tabashnik BE. 1989. Managing resistance with multiple pesticide tactics: theory, evidence and recommendations. *J. Econ. Entomol.* 82:1263–69
103. Taplin D, Meinking TL. 1987. Pyrethrins and pyrethroids for the treatment of scabies and pediculosis. *Semin. Dermatol.* 6:125–35
104. Thompson M, Shotkoski F, ffrench-Constant RH. 1993. Cloning and sequencing of the cyclodiene insecticide resistance gene from the yellow fever

mosquito *Aedes aegypti*. *FEBS Lett.* 325:187–90

105. Toung YS, Hsieh T, Tu CD. 1990. *Drosophila* glutathione S-transferase 1-1 shares a region of sequence homology with maize glutathione S-transferase III. *Proc. Natl. Acad. Sci. USA* 87:31–35
106. Toung YS, Hsieh T, Tu CD. 1993. The glutathione S-transferase *D* genes: a divergently organized, intronless gene family in *Drosophila melanogaster*. *J. Biol. Chem.* 268:9737–46
107. Vatandoost H, McCaffery AR, Townson H. 1996. An electrophysiological investigation of target site insensitivity mechanisms in permethrin-resistant and susceptible strains of *Anopheles stephensi*. *Trans. R. Soc. Trop. Med. Hyg.* 90:216
108. Vaughan A, Chadee DD, ffrench-Constant RH. 1998. Biochemical monitoring of organophosphorus and carbamate insecticide resistance in *Aedes aegypti* mosquitoes from Trinidad. *Med. Vet. Entomol.* 12:318–21
109. Vaughan A, Hawkes N, Hemingway J. 1997. Co-amplification explains linkage disequilibrium of two mosquito esterase genes in insecticide-resistant *Culex quinquefasciatus*. *Biochem. J.* 325:359–65
110. Vaughan A, Hemingway J. 1995. Mosquito carboxylesterase Est$\alpha 2^1$ (A2). Cloning and sequence of the full length cDNA for a major insecticide resistance gene worldwide in the mosquito *Culex quinquefasciatus*. *J. Biol. Chem.* 270:17044–49
111. Vaughan A, Rodriguez M, Hemingway J. 1995. The independent gene amplification of indistinguishable esterase B electromorphs from the insecticide resistant mosquito *Culex quinquefasciatus*. *Biochem. J.* 305:651–58
112. Vulule JM, Beach RF, Atieli FK, Roberts JM, Mount DL, Mwangi RW. 1994. Reduced susceptibility of *Anopheles gambiae* to permethrin associated with the use of permethrin-impregnated bednets and curtains in Kenya. *Med. Vet. Entomol.* 8:71–75
113. WHO. 1992. Vector resistance to pesticides. Fifteenth report of the expert committee on vector biology and control. In *WHO Tech. Rep. Ser.* 818:1–55
114. Whyard S, Downe AER, Walker VK. 1994. Isolation of an esterase conferring insecticide resistance in the mosquito *Culex tarsalis*. *Insect Biochem. Mol. Biol.* 24:819–27
115. Whyard S, Downe AER, Walker VK. 1995. Characterization of a novel esterase conferring insecticide resistance in the mosquito *Culex tarsalis*. *Arch. Insect Biochem. Physiol.* 29:329–42
116. Williamson MS, Denholm I, Bell CA, Devonshire AL. 1993. Knockdown resistance (kdr) to DDT and pyrethroid insecticides maps to a sodium channel gene locus in the housefly (*Musca domestica*). *Mol. Gen. Genet.* 240:17–22
117. Williamson MS, Martinez-Torres D, Hick CA, Devonshire AL. 1996. Identification of mutations in the housefly para-type sodium channel gene associated with knockdown resistance (kdr) to pyrethroid insecticides. *Mol. Gen. Genet.* 252:51–60
118. Xiao-Ping W, Hobbs AA. 1995. Isolation and sequence analysis of a cDNA clone for a pyrethroid inducible P450 from *Helicoverpa amrigera*. *Insect Biochem. Mol. Biol.* 25:1001–9
119. Zheng L, Benedict MQ, Cornel AJ, Collins FH, Kafatos FC. 1996. An integrated genetic map of the African human malaria vector mosquito *Anopheles gambiae*. *Genetics* 143:941–45
120. Ziegler R, Whyard S, Downe AER, Wyatt GR, Walker VK. 1987. General esterase, malathion carboxylesterase, and malathion resistance in *Culex tarsalis*. *Pestic. Biochem. Physiol.* 8:279–85

Annu. Rev. Entomol. 2000. 45:393–422

APPLICATIONS OF TAGGING AND MAPPING INSECT RESISTANCE LOCI IN PLANTS

G.C. Yencho[1], M.B. Cohen[2], and P.F. Byrne[3]

[1]*Department of Horticultural Science, Vernon G. James Research and Extension Center, North Carolina State University, 207 Research Station Road, Plymouth, North Carolina 27962; e-mail: Craig_Yencho@NCSU.edu*

[2]*Entomology and Plant Pathology Division, International Rice Research Institute, MCPO Box 3127, Makati City 1271, Philippines*

[3]*Department of Soil & Crop Sciences, Colorado State University, Fort Collins, Colorado 80523*

Key Words insect-resistant crops, host-plant resistance, plant breeding, molecular markers, QTL

■ **Abstract** This review examines how molecular markers can be used to increase our understanding of the mechanisms of plant resistance to insects and develop insect resistant crops. We provide a brief description of the types of molecular markers currently being employed, and describe how they can be applied to identify and track genes of interest in a marker-assisted breeding program. A summary of the work reported in this field of study, with examples in which molecular markers have been applied to increase understanding of the mechanistic and biochemical bases of resistance in potato and maize plant/pest systems, is provided. We also describe how molecular markers can be applied to develop more durable insect-resistant crops. Finally, we identify key areas in molecular genetics that we believe will provide exciting and productive research opportunities for those working to develop insect-resistant crops.

INTRODUCTION

During the last decade, rapid advances have been made in developing powerful molecular genetic tools for use in the life sciences. These techniques are increasingly being employed to improve the yield, abiotic and biotic stress resistance, and quality traits of crop species. The development of these biotechnological tools coincides with the recognition of the utility of the traditional varieties, landraces, and wild relatives of many crop species as sources of valuable genes for resistance to insects and pathogens, and countless traits of agronomic and/or horticultural value. Introgressing these resources into the cultivated crop germplasm base is critical because, along with providing economically beneficial traits, related germ-

0066-4170/00/0107-0393/$14.00

plasm can be used to increase the genetic diversity of the cultivated crop germplasm pool, which is already precariously narrow (16, 93).

Entomologists and plant breeders have long been aware that wild species and/or non-domesticated crop relatives, possess many valuable genes for resistance to insects (16, 54, 68, 69, 82, 87). However, while fertile crosses can often be made between donor and recipient genetic materials, introgressing desirable genes for insect resistance into an adapted background suitable for commercial release is often a slow and difficult task. When the donor is a wild species or non-domesticated relative of the crop species, the process of introgression into an elite cultivar can take as long as 15 years (73).

This review examines the potential utility and benefits that molecular markers provide entomologists, breeders, biochemists, and molecular biologists working collaboratively to develop insect-resistant crops. We first provide a brief review of the molecular genetic tools currently employed for this work and describe how they can be used to overcome some longstanding problems in breeding for insect resistance. Next, we summarize the work previously reported in this field of study with special reference to our own efforts in potato and maize. We then speculate how these techniques, and the information derived from their application, might be applied to develop more durable insect-resistant crops. We conclude by describing key areas that are now becoming active and exciting research directions.

MOLECULAR MARKERS AND GENE MAPPING

One of the most significant advances to occur in the last decade for the development of improved crops is the use of molecular markers to identify and track genes of interest (93, 94). Tanksley (92), Paterson (70), Liu (51), and Hoisington & Ribaut (35) provide detailed reviews of the theory and practical use of molecular markers in plant breeding. Beckman & Osborn (4) and Liu (51) review the development of the computational procedures employed to associate molecular markers with single and/or polygenic traits, and Powell et al (74) and Staub et al (90) compare the positive and negative aspects of various marker systems. For readers unfamiliar with molecular marker terminology commonly employed in the plant breeding literature, we provide a brief overview of the use of molecular markers for gene mapping.

Molecular markers are based on differences in the DNA nucleotide sequences of chromosomes of different individuals. These differences are referred to as DNA polymorphisms, and they arise as a result of insertions, deletions, duplications, and substitutions of nucleotides (51). DNA polymorphisms can be visualized using a wide variety of molecular marker methodologies (90). The most common molecular markers currently in use are restriction fragment length polymorphisms (RFLPs), random amplified polymorphic DNA (RAPDs), amplified fragment length polymorphisms (AFLPs), and simple sequence repeats (SSRs) or microsatellites. The utility of these markers in a breeding program can be appreciated by recognizing that breeders have traditionally developed improved varieties by selecting on the basis of phenotype. However, a plant's phenotype is determined by its genetic composition and the environment in which it is grown. Often, the effect of the environment masks genotypic effects. Therefore, breeders obtain an imprecise estimate of a plant's true genetic potential, and often they select plants that do not contain the target genes at all (94). Molecular markers enable breeders to exercise selection on genotypic or DNA-based differences rather than phenotypic differences, and they therefore have the potential to greatly increase selection efficiency. For most crop species, high- to medium-density molecular genetic maps covering the entire genome are already available (65, 70).

Molecular markers are phenotypically neutral (i.e. they do not have an effect on a plant's phenotype) and can be used to map both simply inherited dominant or recessive traits controlled by segregation at a single locus (i.e. qualitative traits), and traits that are controlled by multiple loci, the so-called quantitatively inherited traits. Quantitatively inherited traits, which include such common characters as plant yield, size, nutritional quality, stress tolerance, disease resistance, and insect resistance, exhibit continuous variation in phenotypic expression, display gene x gene interactions, are often strongly influenced by differences in environment, and, compared to simply inherited traits, are much more difficult to work with in a breeding program.

Until recently, quantitative traits were studied using statistical techniques based on quantitative genetic theory using appropriate experimental populations (22). However, the models used to study these traits are complex and it is difficult to precisely interpret the genetic effects of individual loci. Molecular markers have the advantage that they can be used to dissect quantitative traits into discrete genetic loci allowing the effects of individual loci to be studied (107).

The individual loci controlling the expression of a quantitative trait are referred to as polygenes or quantitative trait loci (QTL) (70, 92). If a given phenotypic trait is tightly linked to a molecular marker, the genetic segregation of the gene of interest can be determined by the presence or absence of the molecular marker instead of the phenotype, using a process referred to as marker-aided selection (MAS) (90). Molecular markers can thus be used to reduce environmental variation resulting in increased selection efficiency. They can also be used to select for genes of interest, while simultaneously exercising selection against unwanted genomic segments, thus reducing linkage drag and speeding cultivar development (108). MAS therefore promises to increase not only the types of traits incorporated into cultivated germplasm, but the speed with which they are introgressed into adapted cultivars (48, 94).

Constraints to the Development of Insect-Resistant Plants and the Advantages of Molecular Markers

Although there have been many notable successes in conventional breeding for improved plant resistance to insects, the breeding process is often slow and laborious, and sufficient levels of resistance have not been achieved for some pests.

One factor impeding the development of resistant cultivars is the fact that plant resistance to insects is most often a quantitatively inherited trait (44, 54, 68, 87). While quantitative resistance is often more durable than major gene resistance, major genes are relatively easy to identify in germplasm and to incorporate into a commercial cultivar using a backcross breeding program. As summarized in Table 1, with few exceptions major genes in plants have been identified for resistance to only two groups of insects, the order Homoptera and the dipteran family Cecidomyiidae. This is in contrast to plant pathogens (viruses, bacteria, fungi, and nematodes), where major genes for resistance to many more species have been identified and used in plant breeding. The more abundant occurrence in plants of major genes for resistance to microbial pathogens, homopterans, and cecidomyiids would appear in part to be a consequence of the intimate physiological relationships of these parasites with their hosts.

Identifying and selecting for quantitative resistance is complicated by the often strong effect of the environment (light, wind, temperature, plant and insect nutritional status, etc.) and the genetic variability of insects on the results of bioassays. These factors result in a high degree of genotype-by-environment error (87, 88). In addition, the biochemical and/or physical mechanisms mediating resistance, which can be used to screen and select insect-resistant individuals in segregating progenies, are often not known or only partially explain the observed variation in resistance in families segregating for resistance (54, 87, 104). Identifying individual components of resistance with the intent to develop physical and/or biochemical screens for insect resistance increases understanding of the mechanisms of plant resistance to insects and can facilitate the breeding process. However, this approach is time consuming and biochemical and/or physical traits are often only moderately correlated with resistance, or result from multiple complex interactions among a number of factors that are often insect specific (45, 87).

Molecular genetic markers and QTL analysis offer entomologists and plant breeders a more efficient approach for working with quantitative traits, whether or not specific genes or gene products are known. Detection of QTLs is a demanding process, requiring that a mapping population of several hundred plants or lines be produced from a cross between two parents differing in insect resistance, and that each member of the mapping population be "genotyped" (scored for a series of molecular markers) and "phenotyped" (scored for the resistance traits of interest). However, if the QTL mapping is successful, and markers within approximately 15 to 20 cM of useful resistance loci can be identified (49), the subsequent need for scoring the resistance traits directly can be greatly reduced during the remainder of the breeding process. In practice, the attractiveness of MAS versus phenotypic selection for insect resistance will depend on a comparison of economic and logistical advantages offered by the different systems.

For those studying plant resistance to insects, in addition to assisting in the identification of genes associated with resistance traits, molecular markers can facilitate understanding of the genetic basis of resistance and the influence of individual QTLs on its expression. Also, because molecular markers can be used

TABLE 1 Major genes for insect resistance that have been tagged or mapped.

Crop	Pest	Gene	Chromosome	Closest Marker[b] Distance (cM)	Closest Marker[b] Class[d]	Reference[e]
Wheat	*Mayetiola destructor*	*H3*	5A	—	RAPD	20
	M. destructor	*H5*	1A	—	RAPD	20
	M. destructor	*H6*	5A	—	RAPD	20
	M. destructor	*H9*	5A	—	RAPD	20
	M. destructor	*H10*	5A	—	RAPD	20
	M. destructor	*H11*	1A	—	RAPD	20
	M. destructor	*H12*	5A	—	RAPD	20
	M. destructor	*H13*	—	—	RAPD	20
	M. destructor	*H14*	—	—	RAPD	20
	M. destructor	*H15*	5A	—	RAPD	66
	M. destructor	*H16*	5A	—	RAPD	20
	M. destructor	*H17*	5A	—	RAPD	20
[a]	*M. destructor*	*H21*	2R	—	RAPD	85
[a]	*M. destructor*	*H25*	6R	—	RFLP	17
	Schizaphis graminum	*Gb2*	1A	—	[f]	36
	S. graminum	*Gb4*	7D	—	[f]	36
	Diuraphis noxia	*Dn2*	7D	3.3	SCAR	52, **61**
	D. noxia	*Dn4*	1D	11.6	RFLP	52
	D. noxia	*Dn5*	7D	—	[f]	19
Rice	*Nilaparvata lugens*	*Bph1*	12	10, 11.7	RFLP, RAPD	**34,** 37
	N. lugens	*bph2*	12	3.5	RFLP	59
	N. lugens	*Bph10*	12	3.7	RFLP	38
	Nephotettix virescens	—	4	5.5, 5.1	RFLP, RFLP	84
	Orseolia oryzae	*Gm2*	4	1.3, 3.4	RFLP, RFLP	56
	O. oryzae	*Gm4*	8	—	RAPD	57
	O. oryzae	*Gm6*	4	1.2	RFLP	43
Maize	*Spodoptera frugiperda*	*mir1*	6	0	RFLP	71
Apple	*Dysaphis divecta*	Sd_1	—	1, 13	SCAR, RAPD	78, 79

(continued)

to analyze any quantifiable phenotypic trait, biochemical and/or physical mechanisms and direct measures of insect resistance can be simultaneously mapped. Comparisons of the resulting QTLs identified for each trait can then be used to increase understanding of the genetic and physiochemical mechanisms of plant defense (104). Molecular markers therefore represent valuable tools for entomologists, ecologists, and plant breeders interested in understanding or devel-

TABLE 1 *(continued)* Major genes for insect resistance that have been tagged or mapped.

Crop	Pest	Gene	Chromosome	Closest Marker[b] Distance (cM)	Closest Marker[b] Class[d]	Reference[e]
Mungbean	*Callosobruchus* spp., *Riptortus clavatus*	*Br*	Linkage group 9	0.2, 0.5	RFLP, RFLP	**40,** 107
Tomato	*Macrosiphum euphorbiae*	*Mi*	6	[c]		80

[a]Gene from rye, mapped in wheat-rye translocation lines
[b]Where flanking markers have been identified, both are listed
[c]The *Mi* gene has been cloned
[d]RAPD, randomly amplified polymorphic DNA; RFLP, restriction fragment length polymorphism; SCAR, sequence characterized amplified region
[e]Where more than one paper has reported tagging or mapping of the gene, the reference reporting the closest marker(s) is indicated in bold type
[f]This gene has been mapped to the indicated chromosome, but linked molecular markers have not been reported.

oping insect-resistant germplasm because they can be used to provide understanding of the mechanistic bases of insect resistance, and facilitate the identification and introgression of genes for insect resistance even though the mechanistic basis of resistance is unknown (104).

Progress in Tagging and Mapping Major Genes and QTLs for Plant Resistance to Insects

Molecular markers have been used to map genes for insect resistance in most major crop species (Tables 1 and 2). Thirty major or single genes for insect resistance have been tagged or mapped in six crop species, conferring resistance to species from five orders: Homoptera, Hemiptera, Diptera, Lepidoptera, and Coleoptera. (Table 1). Each gene is known to confer resistance to only one insect species or to closely related species within the same genus, with the exception of the *Br* gene from mung bean. This gene is effective against several species of bruchids and the bean bug, *Riptortus clavatus* (40). The *Mi* gene from tomato provides an interesting example because it was originally identified as a dominant gene for resistance to a root-knot nematode, *Meloidogyne incognita.* After the gene was cloned (41), further studies demonstrated that it was located at the same locus as that previously known as *Meu-1*, which confers resistance to some biotypes of the potato aphid, *Macrosiphum euphorbiae* (80). Currently, *Mi* is the only gene for insect resistance to be cloned from a plant. Interestingly, it is a member of the nucleotide-binding, leucine-rich repeat family of resistance genes, many members of which have been found to confer isolate-specific resistance to viruses, bacteria, fungi, and nematodes (31).

The distinction between a major gene and a QTL with a strong effect (i.e. a QTL that explains a significant proportion of the additive phenotypic variation of a given trait) is not always clear. For example, we have included the loci identified for resistance to the Russian wheat aphid, *Diuraphis noxia*, in the list of QTLs (Table 2), although these loci could also be considered major genes (58, 64). In both studies, only two loci for resistance were found in each resistant parent, and each locus explained a high proportion of phenotypic variance (17% to 49 %). It is generally the case that major genes confer a higher level of resistance than QTLs and confer resistance to only some pest populations or biotypes. Resistance-breaking biotypes are known for many of the major genes listed in Table 1 (87).

QTLs for resistance to 11 species of insects from 3 orders (Homoptera, Lepidoptera and Coleoptera) have been mapped in 6 plant species (Table 2). Within a given plant species, it is difficult to make direct comparisons among QTL studies because of the substantial variations in the statistical analyses employed, criteria used to declare a QTL "significant," density of linkage maps, numbers of lines or plants phenotyped, and traits quantified. Traits evaluated include direct measures of insect fitness or behavior (e.g. larval weight, population growth, ovipositional preference); plant damage (e.g. scores on scales of one-to-nine, tunnel length, leaf area defoliated); plant morphology (e.g. trichomes, leaf toughness); and plant chemical content or enzyme activity (e.g. acyl sugars, maysin, polyphenol oxidase). For any single trait scored, the number of QTLs identified range from 1 to 10, and the percentage of variation explained by any single QTL ranges from 1.3 to 58. In studies where statistical models were developed to estimate the total variation of a trait that could be explained by the QTLs identified, this value has ranged from 21% to 98%.

Using Molecular Markers to Determine Mechanisms of Insect Resistance: Examples from Potato and Maize

As mentioned, in addition to their utility as selectable markers to facilitate breeding efforts, molecular markers can be employed to increase our understanding of the mechanisms of plant resistance to insects. By mapping QTLs coding for specific plant physical and/or biochemical attributes associated with insect resistance, and comparing the locations of these QTLs with those identified for the phenotypic expression of resistance to a pest species, valuable insights into the nature of resistance can be obtained. Often, these insights have both basic and applied implications that can be employed to develop insect-resistant crops more efficiently. The advantages of these techniques are well illustrated by plant resistance to insect research focused on developing new potato and maize varieties with resistance to the Colorado potato beetle *Leptinotarsa decemlineata,* and the corn earworm *Helicoverpa zea*, respectively.

In potato, RFLP markers have recently been used to elucidate the genetic and mechanistic attributes of resistance to Colorado potato beetle (CPB) derived from the wild Bolivian potato, *Solanum berthaultii* (8, 104). The CPB is considered to

TABLE 2 Summary of QTL studies for insect resistance or associated traits

Crop	Pest	Traits scored	Parents[a]	Mapping population[i]	No. of QTL	Chromosomal location of QTL	Variation explained per locus (%)	Total variation explained (%)	Reference
Maize	*Ostrinia nubilalis*	2nd generation tunnel length	B73 (S), B52 (R)	300 F_3 lines	7	1, 2, 3, 7, 10	3.4-15.7	38	
	O. nubilalis	2nd generation damage rating	B73, Mo17	112 $F_{2:3}$ lines topcrossed to V78	3	7, 8, 9	8-20	36	3
	Helicoverpa zea	Silk maysin concentration	GT119 (S), GT114 (R)	285 F_2 plants	6	1, 3, 5, 9, 10	2.3-58	75.9	13
	H. zea	Larval weight	GT119 (S), GT114 (R)	76 $F_{2:3}$ lines	4	1, 9, 10		97.8	14
	H. zea	Silk maysin concentration	FF8 (S), GE37 (R)	250 $F_{2:3}$ lines	5	1, 2, 6, 8, 9	4.5-13.6	40.1	15
	H. zea	Larval weight	FF8 (S), GE37 (R)	250 $F_{2:3}$ lines	3	1, 2, 6	3.7-8.9	46.9	15
	H. zea	Larval weight	NC7A(S), GT114 (R)	90 $F_{2:3}$ lines	3	5, 6, 9	14.2-15.2		50
	H. zea	Silk maysin concentration	NC7A(S), GT114 (R)	316 F_2 plants	1	9	55.3	55.3	50
	H. zea	Silk apimaysin concentration	NC7A(S), GT114 (R)	316 F_2 plants	1	5	64.7	64.7	50
	Diatraea spp.	1st generation *D. saccharalis* leaf damage	CML131 (S), CML67 (R)	171 F_3 lines	10	1, 2, 5, 7, 8, 9, 10	7.2-15.4	65	6
	Diatraea spp.	1st generation *D. saccharalis* leaf damage	CML131 (S), CML67 (R)	171 F_3 lines	10	1, 2, 3, 5, 7, 8, 9, 10	3.8-30.8	60.2	7
	Diatraea spp.	1st generation *D. grandiosella* leaf damage	CML131 (S), CML67 (R)	171 F_3 lines	6	1, 5, 7, 9	1.6-14.9	32.4	7

	Diatraea spp.	1st generation *D. grandiosella* damage	Ki3 (S), CML139 (R)	472 F_3 lines	7	3, 5, 6, 8, 9		11.3-29.9	46
	Diatraea spp	1st generation *D. grandiosella* leaf damage	CML131 (S), CML67 (R)	166 RIL topcrossed to CML216 (S)	4	1, 3, 7	2.1-6.5	25.3	28
	Diatraea spp.	1st generation *D. grandiosella* leaf damage	CML131 (S), CML67 (R)	170 RIL	9	1, 5, 7, 8, 9	3.2-14.0	52.4	29
	Diatraea spp.	1st generation *D. saccharalis* leaf damage	CML131 (S), CML67 (R)	170 RIL	8	1, 5, 7, 8, 9	2.1-25.8	52.8	29
	Diatraea spp.	Leaf protein content	CML131 (S), CML67 (R)	170 RIL	5	1, 5, 8, 9	6.6-17.6	38.5	29
	Diatraea spp.	Leaf toughness	CML131 (S), CML67 (R)	145 RIL	5	1, 4, 7, 8	5.3-13.2	39.0	29
	Diatraea spp.	1st generation *D. grandiosella* leaf damage	Ki3 (S), CML139 (R)	135 RIL	5	1, 6, 8, 9	2.6-14.6	35.5	29
	Diatraea spp.	Leaf toughness	Ki3 (S), CML139 (R)	135 RIL	2	5, 8	13.8-14.7	23.9	29
Barley	*Rhopalosiphum* spp.	Aphid density	Harrington (S), TR306 (R)	150 DHL	2	1, 5	?-31	—	58
	Diuraphis noxia	Leaf chlorosis	Stark (S), PI366444(R)	50 F_2 plants	1	5	49		64
	D. noxia	Leaf rolling	Stark (S), PI366444(R)	50 F_2 plants	2	2, 5	17-34	50	64
	D. noxia	Leaf chlorosis	Bearpaw (S), PI366444 (R)	50 F_2 plants	1	5	42		64
	D. noxia	Leaf rolling	Bearpaw (S), PI366444 (R)	50 F_2 plants	1	5	45		64
	D. noxia	Leaf chlorosis	Bearpaw (S), PI366453 (R)	50 F_2 plants	1	5	29		64
	D. noxia	Leaf rolling	Bearpaw (S), PI366453 (R)	50 F_2 plants	1	5	35		64

(*continued*)

TABLE 2 (*continued*) Summary of QTL studies for insect resistance or associated traits

Crop	Pest	Traits scored	Parents[a]	Mapping population[i]	No. of QTL	Chromosomal location of QTL	Variation explained per locus (%)	Total variation explained (%)	Reference
Rice	*Nilaparvata lugens*	Damage score (seedbox)	Azucena (S), IR64 (R)	131 DHL	5	1, 2, 4, 6, 8	5.6-10.1	—	1
	N. lugens	Damage score (field)	Azucena (S), IR64 (R)	131 DHL	3	2, 4, 6	5.6-16.6	—	1
	N. lugens	Antixenosis (settling)	Azucena (S), IR64 (R)	131 DHL	3	3, 6, 8	5.6-8.1	—	1
	N. lugens	Antixenosis (oviposition)	Azucena (S), IR64 (R)	131 DHL	3	1, 6, 8	6.0-7.4	—	1
	N. lugens	Tolerance	Azucena (S), IR64 (R)	131 DHL	1	1	7.1		1
	N. lugens	Feeding rate	Azucena (S), IR64 (R)	131 DHL	1	3	13.0		1
	Sogatella furcifera	Watery lesions	IR24 (S), Asominori (R)	71 RIL	7	1, 2, 3, 6, 8, 10, 12	10.1-69.9	—	103
	S. furcifera	Egg mortality	IR24 (S), Asominori (R)	71 RIL	7	1, 2, 2, 6, 6, 10, 12	11.6-46.0	—	103
Tomato	a	2-tridecanone content	c	96 F_2 plants	3	Linkage groups C, D, I	—	38	63
	a	Type VI trichome density	c	96 F_2 plants	1	Linkage group D	—		63
	Trialeurodes vaporariorum	Oviposition	d	278 F_2 plants	2	1, 12	6.4-8.0	13	53
	a	Type IV trichome density	d	278 F_2 plants	2	5, 9	—	—	53
	a	Type VI trichome density	d	278 F_2 plants	1	1	—		53

	a	Acylsucroses	e	196 F_2 plants	4	2, 3, 11	8.1-15.6	—	60
	a	Acylglucoses	e	196 F_2 plants	1	4	7.4		60
	a	Total acylsugars	e	196 F_2 plants	3	2, 3	9.8-12.9	—	60
	a	Mole % acylglucoses	e	196 F_2 plants	3	2, 4, 11	7.8-22.2	—	60
	a	Type IV trichome density	f	231 F_2 plants	7	2, 4, 5, 6, 7, 10, 11	2.6-8.1	19.1	5
	a	Total acylsugars	f	231 F_2 plants	6	2, 3, 4, 5, 7, 10	0-13	38.3	5
	a	Mole % acylglucoses	f	231 F_2 plants	4	3, 7, 9, 10	1.3-50	51.1	5
Potato	b	Enzymatic browning	g	150 BC_1 clones	2	6, 10	20.2-52.0	63.4	8
	b	Type A trichome density	g	150 BC_1 clones	2	6, 10	27.4-40.0	57.6	8
	b	Polyphenol oxidase	g	150 BC_1 clones	2	2, 5	13.2-23.0	27.0	8
	b	Fatty acids of sucrose esters	g	150 BC_1 clones	5	1, 2, 4, 5	6.1-49.4	67.6	8
	b	Type B trichome density	g	150 BC_1 clones	2	5, 11	8.6-35.4	38.1	8
	b	Enzymatic browning	h	150 BC_1 clones	1	6	34.0		8
	b	Type A trichome density	h	150 BC_1 clones	1	6	32.0		8
	b	Polyphenol oxidase	h	150 BC_1 clones	1	8	11.1		8
	Leptinotarsa decemlineata	Eggs/female	g	112 BC_1 clones	4	1, 5, 6, 10	4.2-6.9	15	104

(*continued*)

TABLE 2 (*continued*) Summary of QTL studies for insect resistance or associated traits

Crop	Pest	Traits scored	Parents[a]	Mapping population[i]	No. of QTL	Chromosomal location of QTL	Variation explained per locus (%)	Total variation explained (%)	Reference
	L. decemlineata	Eggs/egg mass	g	112 BC_1 clones	3	1, 2, 8	3.6-4.0	—	104
	L. decemlineata	Egg mass/female	g	112 BC_1 clones	4	1, 5, 6, 10	3.4-7.5	—	104
	L. decemlineata	Leaf consumption	g	75 BC_1 clones	3	5, 7, 10	5.8-11.4	25	104
	L. decemlineata	Defoliation	g	123 BC_1 clones	5	1, 5, 6, 8, 11	4.4-6.5	21	104
	L. decemlineata	Eggs/female	h	121 BC_1 clones	4	1, 2, 4, 8	4.4-10.5	23	104
	L. decemlineata	Eggs/egg mass	h	121 BC_1 clones	3	4, 6, 8	3.5-6.5	—	104
	L. decemlineata	Egg mass/female	h	121 BC_1 clones	3	1, 3, 4	3.7-12.6	—	104
	L. decemlineata	Defoliation	h	105 BC_1 clones	5	1, 6, 7, 8, 12	5.0-9.5	27	104
Soybean	*H. zea*	Antixenosis (defoliation)	Cobb (S), PI229358 (R)	103 $F_{2:3}$ lines	3	Linkage groups D1, H, M	10-37	66	76

[a]The trait scored confers resistance to numerous arthropod pests (95).
[b]The trait scored confers resistance to numerous arthropod pests (5).
[a](R), resistant parent; (S), susceptible parent
[c]*Lycopersicon esculentum* cv. Manapal (S), *L. hirsutum* f. *glabratum* PI134417 (R)
[d]*L. esculentum* cv. Moneymaker (S), *L. hirsutum* f. *glabratum* CGN1.1561 (R)
[e]*L. esculentum* cv. New Yorker ^{Lp4} (S), *L. pennellii* LA716 PI246502 (R)
[f]*L. pennellii* LA1912 (S), *L. pennellii* LA716 (R)
[g][*S. berthaultii* PI473331-a (R) x *S. tuberosum* USW2230 (S)] x *S. berthaultii* PI473331-b (R)
[h][*S. berthaultii* PI473331-a (R) x *S. tuberosum* USW2230 (S)] x *S. tuberosum* HH1-9 (S)
[i]DHL, doubled-haploid line; RIL, recombinant inbred line

be one of the most destructive pests of potato in North America and Europe, and is notorious for its ability to rapidly develop resistance to insecticides used for its control (23). All cultivated potatoes are considered susceptible to CPB damage. Several groups have screened potato germplasm to identify in wild potatoes high levels of resistance to CPB, which can then be introgressed into cultivated potato (24, 25). Numerous studies have demonstrated that the wild tuber-bearing diploid potato *S. berthaultii* ($2n = 2x = 24$), is highly resistant to CPB as well as to a wide range of arthropod pests, including aphids, leafhoppers, leaf miner flies and potato tuber moths (95).

Studies to determine the mechanistic basis of resistance to CPB have shown that resistance is associated with the presence of the type A and B glandular trichomes present on the foliage of *S. berthaultii* (95). Tingey & Yencho (96) described the physiochemical properties of the type A and B trichomes and their effects upon small-bodied insects. Neal et al (62) postulated that the type A trichomes, which contain the enzyme polyphenol oxidase, an as yet unidentified substrate, and a mixture of known and unknown sesquiterpenes, were required for the expression of resistance to CPB, while the sucrose esters produced by the type B trichomes enhanced resistance in the presence of the type A trichomes. Pelletier & Smilowitz (72) studied two sibling clones of *S. berthaultii* PI 473340, which had type A trichomes but differed in the presence or absence of secretory type B trichomes. Each of these clones was resistant to CPB feeding, but differences in the level of antifeedant activity present in leaf rinses obtained from each clone led them to hypothesize that *S. berthaultii* possessed more than one mechanism of resistance.

In order to study the complex relationship observed between the glandular trichomes of *S. berthaultii* and resistance to CPB, Bonierbale et al (8) and Yencho et al (104) conducted simultaneous QTL mapping studies using two reciprocal backcross, *S. tuberosum* × *S. berthaultii*, potato progenies, termed BCB (backcross to *S. berthaultii*) and BCT (backcross to *S. tuberosum*). These genetic materials, which segregated for a wide range of traits (97, 98, 105), were mapped with RFLP markers spaced an average of 10 cM apart (8). Bonierbale et al (8) and Yencho et al (104) identified weak (i.e. each QTL explained only 4% to 12% of the phenotypic variation observed in a given measure of resistance), but consistent QTLs for resistance to CPB consumption, oviposition, and defoliation. The locations of the QTLs for insect resistance were then compared to QTLs associated with the physiochemical traits of the glandular trichomes (type A and B trichome density, modified enzymic browning activity, polyphenol oxidase activity, and acyl sucrose production), which have been implicated in resistance.

Figure 1 presents the combined results of the separate QTL analyses for the trichome traits and resistance to CPB. In BCB, QTLs for CPB resistance were located on chromosomes 1, 5, and 10, while QTLs for the trichome traits were located on chromosomes 1, 2, 4, 5, 6, 10, and 11. Comparing the QTLs for insect resistance with the trichome traits in BCB revealed that the QTLs located on chromosomes 5 and 10 overlapped. Likewise, in BCT, QTLs for insect resistance

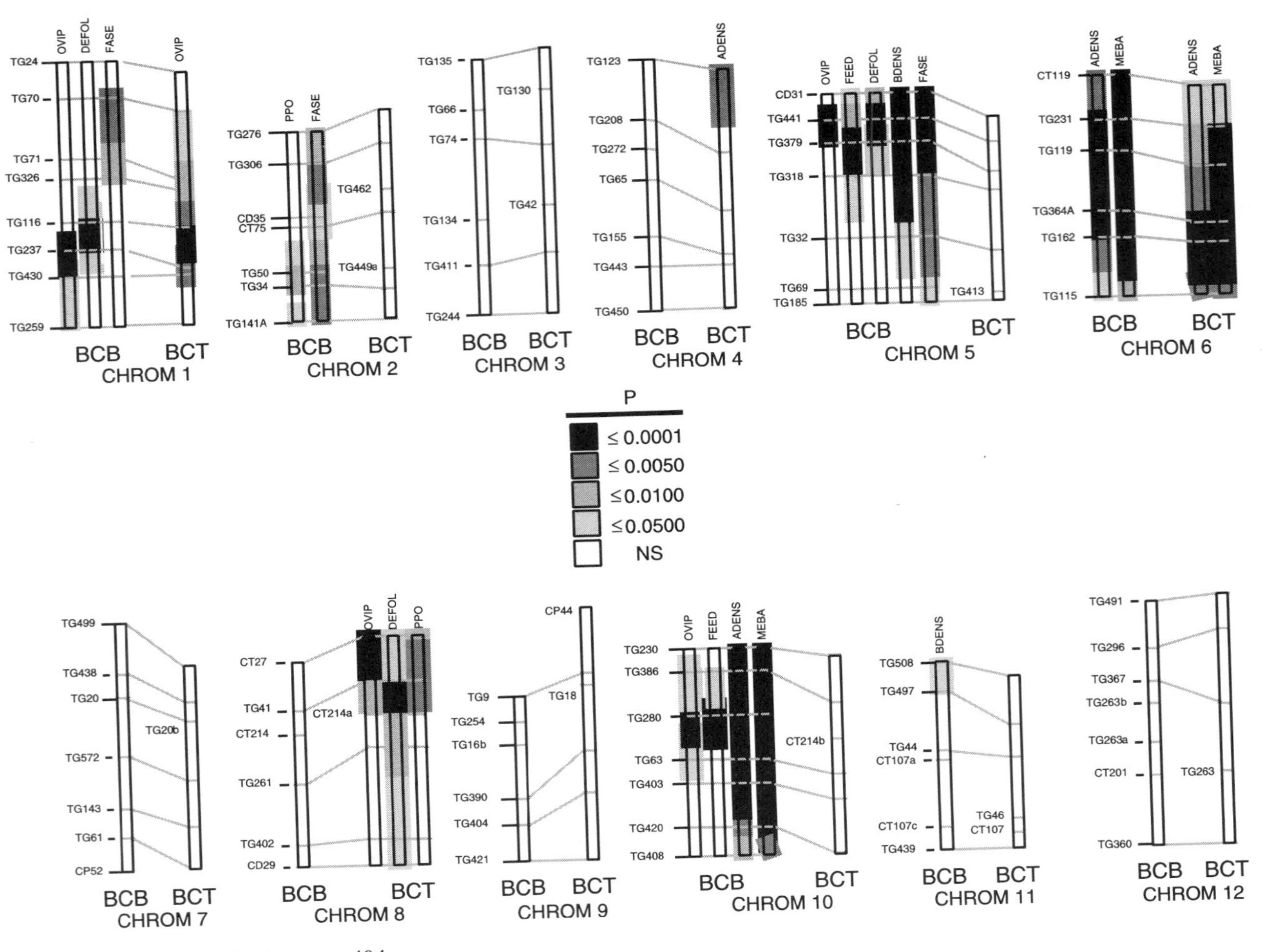

Figure 1 See legend on facing page 404.

were detected on chromosomes 1 and 8, and QTLs for the trichome traits were located on chromosomes 4, 6, and 8, with the QTLs for insect resistance and the trichomes on chromosome 8 coinciding. A closer inspection of the coincidental QTLs revealed that the QTL detected for CPB resistance in BCT on chromosome 8 near TG41 coincided with a QTL detected for polyphenol oxidase, a polymerizing enzyme located in the glandular head of the type A trichome. Polyphenol oxidase has been implicated in resistance to CPB as well as aphids and leafhoppers (47, 96). Similarly, in BCB the QTLs for insect resistance located on chromosome 5 mapped to a locus that exerted significant effects on type B trichome density and sugar ester production, while the QTL on chromosome 10 influenced type A trichome density and modified enzymic browning activity, respectively. Each of these traits have been implicated as causal agents of resistance to CPB (95, 96).

The comparative CPB/trichome QTL studies confirmed that the trichomes were indeed associated with resistance to CPB. However, the real power that comparative studies of this nature provide is best illustrated by the fact that this work also indicated that, while the trichomes may be an important component of resistance, they may not account for all of the resistance to CPB observed in these progenies. Evidence for this originates from the QTLs for insect resistance located along the interval TG71-TG237 on chromosome 1. The QTLs for oviposition and field defoliation located along this interval were among the strongest and most consistent detected in the study. But this interval was not associated with any of the known trichome traits. Part of the importance of this finding stems from the fact that the breeding program, in addition to conducting yearly field evaluations of insect resistance, has relied on the trichome traits as indirect measures of CPB resistance. The nature of the resistance factor(s) localized in this region remains unknown. Our initial speculation was that foliar glycoalkaloids may be responsible for this unaccounted resistance. However, the clones present in these progenies do not synthesize high levels of glycoalkaloids. Further, Yencho et al (105) recently completed a study of the inheritance of glycoalkaloids in these same progenies and found that the CPB resistance and glycoalkaloid QTLs did not

Figure 1 Location of QTLs for resistance to Colorado potato beetle and the trichome traits in backcross to *berthaultii* (BCB) and backcross to *tuberosum* (BCT). The vertical box plots represent the chromosomes, and the locations of the RFLP markers are indicated by horizontal lines. The traits compared were: CPB oviposition (OVIP); CPB defoliation (DEFOL); CPB leaf disk feeding (FEED); sugar ester concentration (FASE); type A and B trichome density (ADENS and BDENS); modified enzymic browning activity (MEBA); and polyphenol oxidase concentration (PPO). The location of QTLs and single-point ANOVA significance level (P) in for each trait in each population are indicated by the shading within the box plots. Markers located to the left of the QTL box plots of BCT indicate markers that were not used in BCB. For a complete description of the crosses, construction of the genetic maps of BCB and BCT, and the QTL analyses, see Bonierbale et al (8) and Yencho et al (104).

coincide. Additional factors that may account for these QTLs are the unidentified feeding deterrents recently extracted from the foliage of *S. berthaultii* PI 473334 (106) or plant phenotypic factors such as increased plant vigor or early plant emergence, which would enable the plants to tolerate or partially escape early season defoliation. These questions are yet to be addressed. These findings were significant from a basic research perspective because they demonstrated that molecular markers are useful tools for dissecting complex insect resistance traits into specific component traits.

Byrne et al (13–15) and Lee et al (50) have also used molecular markers and QTL mapping techniques to unravel the genetic mechanisms of resistance in maize to the corn earworm (CEW), *Helicoverpa zea*. This work has been demonstrated in a series of collaborative studies involving geneticists, entomologists, and biochemists, their ultimate goal being to engineer a plant metabolic pathway for improved CEW resistance. Corn earworm larvae initially feed on maize silks and later on developing kernels, causing considerable direct yield loss as well as introducing kernel-rotting fungi. Waiss et al (99) and Elliger et al (21) isolated from maize silks a group of related C-glycosyl flavones that retarded CEW growth. The major component of this group was named maysin and its analogs were designated apimaysin and 3′-methoxymaysin (Figure 2). Highly significant negative correlations were reported between maysin concentrations of fresh silks and growth of CEW larvae in dried-silk bioassays (100, 101).

Because C-glycosyl flavones are synthesized via a branch of the well-characterized flavonoid biosynthetic pathway (Figure 3) (33), Byrne et al (13) hypothesized that loci of that pathway would explain a large portion of the quantitative variation in maysin concentration, and by extension, resistance to CEW. These loci were proposed as "candidate genes" in a series of QTL analyses. (A candidate gene is one that is hypothesized to affect expression of the trait of interest, either *a priori* based on knowledge of trait biology, or *a posteriori*, guided by similar locations of QTLs and genes of known function.) The *p1* locus, a

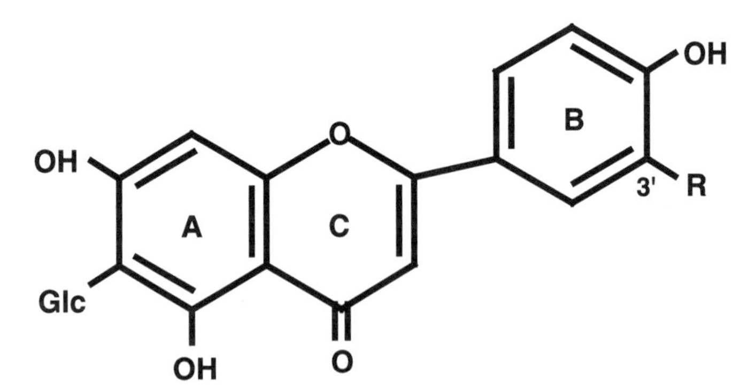

Figure 2 Structure of maysin and its analogs. For maysin, R = OH; for apimaysin, R = H, and for 3′-methoxymaysin, R = OCH_3. Glc = glycosyl group.

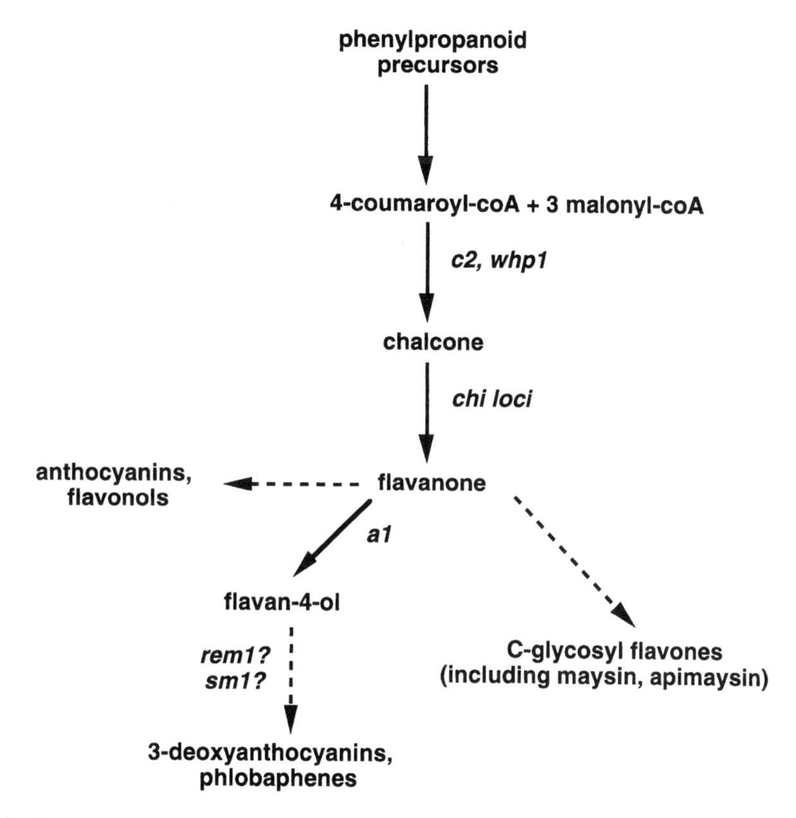

Figure 3 Portion of the flavonoid pathway of maize controlled by *p1*, with loci in italics. *Dashed lines* represent conversions that require more than one enzyme or are incompletely described.

transcription activator controlling expression of several flavonoid pathway genes and potentially a key regulatory element for silk maysin concentration, was a prime candidate gene based on previous work (12, 13, 30, 91). Byrne et al (13) examined RFLP genotypes at 39 flavonoid pathway loci or linked markers in the population (GT114 × GT119)F_2. Their results indicated a major effect on maysin concentration (r^2 = 58%) by *p1*, and a significant effect (r^2 = 11%) by a locus on chromosome 9S that enhanced maysin when homozygous recessive (subsequently designated *rem1* for recessive enhancer of maysin). Interestingly, the *rem1* allele for higher maysin concentration was from GT119, the low-maysin parent. A strong epistatic interaction between *p1* and *rem1* was also observed. Based on a number of inferences and previous reports, *rem1* was hypothesized to act on the pathway branch leading to the 3-dexyoanthocyanins (Figure 3), either as structural (enzyme-encoding) gene or a regulatory locus. According to this hypothesis, a *rem1*-induced reduction in flux through that branch would shunt common intermediates to the intersecting maysin-synthesis pathway, thereby increasing maysin levels and resistance to CEW.

When a subset of the population (GT114 × GT119)F_2 was evaluated for CEW antibiosis, *p1* was highly significant ($p < 0.001$) in explaining variation in larval weights. However, the *rem1* region was not significant, although a nearby region on chromosome 9S was significant at $p < 0.01$ (14). QTL analysis of a second population, (GE37 × FF8)$F_{2:3}$, which did not segregate at *p1*, again revealed major effects on maysin concentration by the *rem1* region and suggested additional effects by the flavonoid pathway loci *whp1* and *sm1* (Figure 3). Other effects on maysin levels were observed on chromosomes 1S, 6S, and 8L. The most striking result of this study was that neither the *rem1* region nor any other locus on chromosome 9 affected CEW larval weight, despite large *rem1* effects on maysin concentration (r^2 = 12.3% and 13.6%, respectively, in Georgia and Missouri environments). In contrast, all other chromosome regions of this population with large effects on maysin were also significant for larval weight. To explain this unpredicted result, the authors speculated that perhaps the *rem1* region effected changes in concentrations of other silk compounds that interfered with maysin's antibiotic properties (15).

The placement of *rem1* on the pathway branch leading to the 3-deoxyanthocyanins was supported by results from a population segregating for functional and nonfunctional alleles at the *a1* locus (MD McMullen, USDA-ARS, Columbia, MO, personal communication). Because the *a1* gene product (dihydroflavonol reductase) catalyzes a step on the 3-deoxyanthocyanin branch of the pathway (Figure 3), a block at *a1* should affect maysin concentration similarly to the *rem1* block. In fact, the investigators observed precisely that: a major effect by *a1* on maysin concentration (r^2 = 22%), with the nonfunctional *a1* allele associated with increased maysin.

Another population in this series, (GT114 × NC7A)F_2, was designed to detect QTLs responsible for accumulation of silk maysin and apimaysin, compounds that differ solely by the presence or absence of a 3'-hydroxyl group (Figure 2) (50). This change has a significant impact on CEW antibiosis, because the growth-retardant effect of apimaysin is only about half that of maysin (89). Because *pr1* is responsible for 3′-hydroxylation of anthocyanins, the authors hypothesized that it also controlled the conversion of apimaysin to maysin, compounds that differ solely by the presence or absence of a 3′-hydroxyl group (Figure 2). Contrary to expectations, the results showed that while the *pr1* region had a major effect on apimaysin concentration (r^2 = 64.7%), it had no effect on maysin level. Conversely, the largest QTL for maysin (in the *rem1* region, *R2* = 55.3%) did not influence apimaysin concentrations. These surprising results suggested independent synthesis of two very closely related compounds. When a subset of the population was analyzed for QTLs associated with CEW larval weight, both the *pr1* and *rem1* regions had significant effects on larval weight, and in each case the high-flavone genotypic class was associated with increased larval growth. These results suggested that the loci in question had pleiotropic effects on other compounds that influenced larval growth.

Several conclusions can be drawn from these studies that may be relevant to other attempts at manipulating metabolic pathways for enhanced insect resistance. First is the importance of regulatory loci such as *p1* in controlling trait expression. Although biochemical pathways require both enzyme-encoding and regulatory loci, the regulators provide strategic leverage points for influencing end-product concentrations. Second, this work leads us to believe that predicting the outcome of genetic changes in even well-documented biosynthetic pathways can be difficult. The apparently independent synthesis of maysin and apimaysin was contrary to expectations of a single locus difference. Third, these studies point out the limitations of relying on a single biochemical assay for evaluation of insect resistance, which is generally affected by multiple plant factors. Increased maysin concentrations associated with *rem1* did not raise resistance levels, possibly because the additional maysin was synthesized at the expense of other antibiotic compounds.

Using Molecular Markers to Increase the Durability of Insect-Resistant Crops

Once a major gene or QTL has been identified and mapped, MAS and/or map-based gene cloning can be initiated. (For the sake of simplicity in this section, when referring to plants we use "gene" both in the sense of "allele" at a resistance locus and in the more general sense of a polypeptide-encoding DNA sequence.) Both of these activities open new opportunities to use the locus in a breeding program to achieve more durable host-plant resistance. The lines produced will also be useful in testing long-standing hypotheses about the nature of durable resistance. Two possible approaches to greater durability are pyramiding of multiple resistance genes and selective breeding for particular modalities of resistance. MAS facilitates gene pyramiding because it is easier to determine when a plant carries multiple resistance genes by detecting molecular markers than by using insect bioassays, where the presence of one major gene generally masks the presence of other genes. Prior to the tagging of pest-resistance genes and the development of MAS and gene cloning, it was in principle possible to develop pyramided cultivars by screening crosses in a breeding program with multiple insect biotypes, each adapted to a different resistance gene. However, this approach has not been widely applied, in part because appropriate biotypes are often not available for new resistance genes. The use of MAS to pyramid major genes for insect resistance is in progress in the Russian wheat aphid, *Diuraphis noxia* (52, 61), and Hessian fly, *Mayetiola destructor* (20), in wheat; and Asian rice gall midge, *Orseolia oryzae*, in rice (43, 57).

The principle underlying the potentially increased durability of cultivars possessing multiple major genes is straightforward: If insect alleles conferring adaptation to a major gene occur at low frequencies in pest populations (at the time when a new resistance gene is released), then individuals that carry alleles conferring adaptation to two (or more) plant-resistance genes will be extremely rare. An associated assumption is that no insect alleles should confer cross-adaptation to multiple resistance genes in the pyramid. Simulation modeling indicates that the durability of pyramided varieties will also be enhanced if the insect adaptive alleles are highly recessive and if there is strong epistasis between the plant-resistance genes (such that an insect with alleles for adaptation to only one of the resistance genes will still have very low fitness on the pyramided cultivar) (26, 27). Use of simulation models has also indicated that, under most conditions, a variety with two pyramided major genes for resistance will provide more total years of plant protection than will sequential release of the two genes in two separate varieties, or a seed mixture of the two varieties. However, pyramided cultivars will not provide substantially increased durability under all conditions, and thus maintaining refuges (e.g. susceptible plants within fields, or fields of susceptible plants interspersed among fields of resistant plants) is still a valuable component of a resistance management strategy when using such varieties (26, 27, 81).

In addition to its application in pyramiding major genes, MAS can also be used to pyramid major genes with QTLs, and to facilitate combining multiple QTLs from different sources. We are not aware of any studies that have tested whether pyramiding a major gene with QTLs will result in increased durability of the major gene. Increased durability of the major gene would be expected if the QTLs decrease the fitness of genotypes adapted to the major gene relative to unadapted genotypes. Using MAS to increase the genetic complexity of quantitative resistance is a promising approach to increasing the durability of resistant cultivars, but there are several caveats to bear in mind. Polygenic resistance is often considered potentially more durable than monogenic resistance because, among other reasons, adaptive alleles at multiple pest loci might be required to overcome multiple, unrelated plant resistance factors. However, as noted by Kennedy et al (45) and described earlier for the maysin/CEW system, resistance that is polygenic may nonetheless result in the synthesis of only a single antibiotic chemical factor, and conversely a single gene [e.g. a regulatory locus in a defense signalling pathway (102)] may influence the production of a diverse set of resistance factors. In addition, while the plant pathology literature contains many cases in which "horizontal" resistance (which is often moderate and polygenic) has been found to be relatively durable in comparison to monogenic resistance, far fewer studies document the relative durability of polygenic quantitative resistance to insects (10, 86). Finally, while a minor gene may confer resistance in a particular genetic background, or a set of minor resistance genes may work well when in combination, these genes may not perform as well when crossed into a new background or when combined with other resistance genes (50).

Given these caveats, before embarking on a MAS program with QTLs for insect resistance, it would be wise to determine if the resistance genes targeted for introgression are indeed potentially durable. In choosing the resistant parent(s) for a mapping population, or choosing among existing mapping populations for a study of insect resistance, knowledge of the mechanisms of resistance involved or prior observation of the durability of a resistant cultivar in the field or in selection experiments can identify cultivars that may be sources of promising QTLs (2, 45). The results of a QTL analysis itself will indicate whether the insect resistance in the resistant parent of the mapping population indeed has a polygenic basis. Insight into whether the QTLs influence multiple resistance factors acting on multiple targets within the pest can be gained by analyzing the mapping population for a series of carefully chosen traits.

Selective breeding for QTLs conferring a particular modality of insect resistance [antibiosis, antixenosis, or tolerance (68, 87)] is another approach to achieving more durable varietal resistance. Tolerance, the ability of a plant to produce a high yield of good quality in spite of insect infestation, is in principle of unlimited durability because it does not exert selection on pest populations. Kennedy et al (45) described scenarios under which antibiosis or antixenosis might also be durable because of limited effects on pest fitness, e.g. the case of an antixenotic cultivar in a landscape where equally preferred host species are readily available to the pest population. A QTL analysis of six traits associated with rice resistance to the brown planthopper found a total of seven QTLs, one of which was predominantly associated with antixenosis and a second with tolerance (1). Most of the other QTL analyses of insect resistance conducted to date (Table 2) have scored chemical or morphological antibiotic resistance factors, or plant damage ratings under free-choice conditions, the results of which can be influenced by all three resistance modalities.

FUTURE PROSPECTS

Clearly, the beginning of the twenty-first century is an exciting time to be working in the plant sciences. Although major genes and QTLs for resistance to numerous insect pests have been mapped (Tables 1 and 2), demonstrations of the usefulness of this information for MAS breeding programs have not yet been reported. However, it is only a matter of time before these technologies are put to routine use. In the coming years, breeders and entomologists working to develop insect-resistant crops will be provided an assemblage of new tools to develop insect-resistant crops. This work will be increasingly facilitated by rapid advances in the genomic sciences. Indeed, many universities, private companies, and international agricultural research centers are in the process of establishing genomics laboratories capable of automated, high-throughput DNA sequencing and microarray analysis to facilitate the mapping and cloning of major genes and QTLs. The conserved gene content and order (synteny) observed among the genomes of related plant taxa will serve to multiply the effects of discoveries made in each crop species, while analysis of signaling and metabolic pathways will be harnessed to increase the power of MAS and genetic engineering for crop improvement.

The rapid development of these new technologies presents scientists studying the mechanisms and genetics of plant resistance to insects with numerous challenges. One such challenge is to develop effective collaborations to utilize these new technologies. Another challenge is to maintain research interest and funding for the areas in which progress in plant resistance to insects is founded: understanding the physiological basis of resistance and developing meaningful bioassays or other quantitative measurements to assess resistance to insects in the laboratory, greenhouse, and field. It is difficult to predict exactly how advances in molecular genetics will be applied in the future to develop crops resistant to insects. However, below we describe a few areas that will likely be capitalized upon in the coming years.

Cloning Major Genes and QTLs for Insect Resistance

Once a gene has been cloned, it is no longer necessary to rely on linked markers to detect it in segregating progenies. Instead, the gene itself can be scored directly, removing the risk of segregation between a marker and the gene of interest. Determining the DNA sequence of a cloned resistance gene can also greatly facilitate the discovery of additional genes, as demonstrated by the increase of cloned disease-resistance genes that followed the discovery of the first few genes containing the nucleotide binding site and leucine-rich repeat (NBS/LRR) motifs (31). Now that the *Mi* gene conferring resistance to the potato aphid has been determined to be a member of the NBS/LRR family (80), we can anticipate the isolation of additional major genes for insect resistance. Subsequent studies of cloned genes are likely to lead to an understanding of how they function to confer resistance. Ultimately, the manipulation of DNA sequences of major genes and QTLs, followed by their reinsertion into plants, may become a productive approach to developing more durably resistant cultivars.

Manipulation of Defense-Related Pathways in Plants

Over the past decade rapid growth has occurred in the understanding of systemic acquired resistance and associated signal pathways (42). The jasmonic acid (JA) pathway is of particular importance to plant responses to herbivory and other stresses. The first gene encoding a component of the JA pathway, *COI1*, was recently cloned (102). Further advances in gene cloning and analysis of gene function should enable these pathways to be manipulated for improved plant resistance.

Linking QTL Mapping with Genomics and Proteomics

Large-scale projects are underway or planned in many crops to develop an inventory of expressed sequence tags (ESTs) corresponding to transcribed genes. ESTs are partial sequences (300 to 500 bp) of cDNA clones, derived from expressed mRNA, that can be obtained rapidly and relatively inexpensively (11). Two uses of ESTs for gene mapping have been suggested. First, by comparing EST sequences to gene sequences of known function in public databases (e.g. GenBank, http://www.ncbi.nlm.nih.gov/), the biochemical functions of genes corresponding to ESTs can often be deduced. Because ESTs can be mapped to chromosomal locations, they will serve as a potentially rich source of candidate genes to explain the effects of QTLs detected in specific chromosome regions. Second, it has been suggested that EST analysis can be combined with bulked segregant analysis (55) and differential display techniques to rapidly identify expressed genes for a wide variety of phenotypic traits (32). This technique could conceivably be used to rapidly identify expressed genes for insect resistance using near-isolines or insect-resistant and -susceptible sibs. Map-based cloning of these genes could then be attempted.

Capitalizing on Synteny Among the Genomes of Related Plant Taxa

The maysin pathway is an example of how discoveries in one cereal can be used to stimulate progress in other members of the grass family Poaceae. The maize *p1* locus has been identified as an important regulator of maysin synthesis, a C-glycosylflavone that is antibiotic to the corn earworm (see previous section). Related compounds and some flavonoid pathway genes have been identified in other grasses, including barley, rice, and wheat (39, 77), but *p1* homologs have not yet been detected in those species. Based on known synteny relationships (18), likely locations of genes homologous to *p1* can be hypothesized, helping to focus mapping or cloning efforts, and possibly leading to candidate gene hypotheses for explaining the observed insect resistance. Alternatively, if *p1* homologs cannot be found in related species, then maize *p1* might be inserted through transgenic methods into one of the other grass family genomes to determine its effects on accumulation of C-glycosylflavones toxic to insects. Extensive synteny has also be reported in the Brassicaceae (67), and Solanaceae (9, 75), and can be likewise exploited in those families.

Exploiting Genetic Resources for Insect Resistance

Wild relatives, traditional varieties, and landraces of crops have long been recognized as a bountiful resource for crop improvement, and extensive national and international systems to preserve these genetic resources are well established. Despite numerous examples of insect-resistance genes or mechanisms first discovered in wild or traditional materials (16, 38, 52, 95, 99), these accomplishments are dwarfed by the number of genetic resources that have not yet been screened for resistance. Although traditional phenotypic screening for resistance will remain a key part of identifying useful variation, the molecular genetic techniques described in this review will have increasing importance as tools to access these valuable resources. QTL mapping and MAS of crosses involving wild or unadapted resistance sources will continue to reveal locations of resistance loci. Once an insect resistance gene is cloned, a large number of accessions can be screened to search for additional alleles at that locus, some of which may confer a greater level of resistance than the initial cloned version. In cases where resistance is attributed to a known metabolic pathway, another strategy would be to search for variation in genes of that pathway to identify loci and alleles with the best prospects for maximizing resistance levels while minimizing negative consequences.

ACKNOWLEDGMENTS

The potato QTL mapping work reported was supported by USDA/NRI Competitive Grants Program, Plant Genome Awards 9101420 and 9400667, and a contract from the International Potato Center (CIP), Lima, Peru. Research on maysin was supported by USDA/NRI, Competitive Grants Program, Plant Genome Awards 940074, 9500636, and 9701357. Mark Clough, NCSU assisted with the preparation of the figures.

LITERATURE CITED

1. Alam SN, Cohen MB. 1998. Detection and analysis of QTLs for resistance to the brown planthopper, *Nilaparvata lugens*, in a doubled-haploid rice population. *Theor. Appl. Genet.* 97:1370–79
2. Alam SN, Cohen MB. 1998. Durability of brown planthopper, *Nilaparvata lugens*, resistance in rice variety IR64 in greenhouse selection studies. *Entomol. Exp. Appl.* 89:71–78
3. Beavis WD, Smith OS, Grant D, Fincher R. 1994. Identification of quantitative trait loci using a small sample of topcrossed and F_4 progeny from maize. *Crop Sci.* 34:882–96
4. Beckman JS, Osborn TS, eds. 1992. *Plant Genomes: Methods for Genetic and Physical Mapping*. Dordrecht: Kluwer
5. Blauth SL, Churchill GA, Mutschler MA. 1998. Identification of quantitative trait loci associated with acylsugar accumulation using intraspecific populations of the wild tomato, *Lycopersicon pennellii*. *Theor. Appl. Genet.* 96:458–67
6. Bohn M, Khairallah MM, Gonzalez-de-Leon D, Hoisington DA, Utz HF, et al. 1996. QTL mapping in tropical maize. I. Genomic regions affecting leaf feeding resistance to sugarcane borer and other traits. *Crop Sci.* 36:1352–61
7. Bohn M, Khairallah MM, Jiang C, Gonzalez-de-Leon D, Hoisington DA, et al. 1997. QTL mapping in tropical maize. II.

Comparison of genomic regions for resistance to *Diatraea* spp. *Crop Sci.* 3:1892–902
8. Bonierbale MW, Plaisted RL, Pineda O, Tanksley SD. 1994. QTL analysis of trichome mediated insect resistance in potato. *Theor. Appl. Genet.* 87:973–87
9. Bonierbale MW, Plaisted RL, Tanksley SD. 1988. RFLP maps based on a common set of clones reveal modes of chromosomal evolution in potato and tomato. *Genetics* 120:1095–1103
10. Bosque-Perez NA, Buddenhagen IW. 1992. The development of host-plant resistance to insect pests: outlook for the tropics. In *Proc. 8th Int. Symp. Insect-Plant Relationships. March 9–13, Wageningen, The Netherlands.* pp. 235–49. Dordrecht: Kluwer
11. Bouchez, D, Hofte H. 1998. Functional genomics in plants. *Plant Physiol.* 118:725–32
12. Byrne, PF, Darrah LL, Snook ME, Wiseman BR, Widstrom NW, et al. 1996. Maize silk-browning, maysin content, and antibiosis to the corn earworm, *Helicoverpa zea* (Boddie). *Maydica* 41:13–18
13. Byrne PF, McMullen MD, Snook ME, Musket TA, Theuri JM, et al. 1996. Quantitative trait loci and metabolic pathways: genetic control of the concentration of maysin, a corn earworm resistance factor, in maize silks. *Proc. Natl. Acad. Sci. USA* 93:8820–25
14. Byrne PF, McMullen MD Wiseman BR, Snook ME, Musket TA, et al. 1997. Identification of maize chromosome regions associated with antibiosis to corn earworm (Lepidoptera: Noctuidae) larvae. *J. Econ. Entomol.* 90:1039–45
15. Byrne PF, McMullen MD, Wiseman BR, Snook ME, Musket TA, et al. 1998. Maize silk maysin concentration and corn earworm antibiosis: QTLs and genetic mechanisms. *Crop Sci.* 38:461–71
16. Clement SL, Quisenberry SS, eds. 1999. *Global Plant Genetic Resources for Insect-Resistant Crops.* Boca Raton, FL: CRC. 295 pp.
17. Delaney DE, Friebe BR, Hatchett JH, Gill BS, Hulbert SH. 1995. Targeted mapping of rye chromatin in wheat by representational difference analysis. *Genome* 38:458–66
18. Devos KM, Gale MD. 1997. Comparative genetics in the grasses. *Plant Mol. Biol.* 35:3–15
19. Du Toit F, Wessels WG, Marais GF. 1995. The chromosomal arm location of the Russian wheat aphid resistance gene, *Dn5. Cereal Res. Commun.* 23:15–17
20. Dweikat I, Ohm H, Patterson F, Cambron S. 1997. Identification of RAPD markers of 11 Hessian fly resistance genes in wheat. *Theor. Appl. Genet.* 94:419–23
21. Elliger CA, Chan BG, Waiss AC, Lundin RE, Haddon WF. 1980. C-glycosylflavones from *Zea mays* that inhibit insect development. *Phytochemistry* 19:293–97
22. Falconer DS, Mackay TF. 1996. *Introduction to Quantitative Genetics.* Essex, UK: Longman. 4th ed. 464 pp.
23. Ferro DN. 1985. Pest status and control strategies of the Colorado potato beetle. In *Proc. Symp. Colorado Potato Beetle, 17th Int. Congr. Entomol.*, ed. DN Ferro, RH Voss. Amherst, MA: Mass. Agric. Exp. Stn., Res. Bull. No. 704. 144 pp.
24. Flanders KL, Hawkes JG, Radcliffe EB, Lauer F. 1992. Insect resistance in potatoes: sources, evolutionary relationships, morphological and chemical defenses, and agroecological associations. *Euphytica* 61:83–111
25. Flanders KL, Radcliffe EB. 1992. Host plant resistance in *Solanum* germplasm, an appraisal of resistance to Colorado potato beetle, potato leafhopper, and potato flea beetle. *Bull. 599.* St. Paul, MN: Minn. Agric. Exp. Stn.
26. Gould F. 1986. Simulation models for predicting durability of insect-resistant germplasm: a deterministic diploid, two-locus model. *Environ. Entomol.* 15:1–10
27. Gould F. 1998. Sustainability of trans-

genic insecticidal cultivars: integrating pest genetics and ecology. *Annu. Rev. Entomol.* 43:701–26

28. Groh S, Gonzalez-de-Leon D, Khairallah MM, Jiang CZ, Bergvinson D, et al. 1998. QTL mapping in tropical maize: III. Genomic regions for resistance to *Diatraea* spp. and associated traits in two RIL populations. *Crop Sci.* 38:1062–72
29. Groh S, Khairallah MM, Gonzalez-de-Leon D, Willcox M, Jiang CZ, et al. 1998. Comparison of QTLs mapped in RILs and their test-cross progenies of tropical maize for insect resistance and agronomic traits. *Plant Breed.* 117:193–202
30. Grotewold E, Drummond BJ, Bowen B, Peterson T. 1994. The myb-homologous P gene controls phlobaphene pigmentation in maize floral organs by directly activating a flavonoid biosynthetic gene subset. *Cell* 76:543–53
31. Hammond-Kosack KE, Jones JDG. 1997. Plant disease resistance genes. *Annu. Rev. Plant Physiol. Plant Mol. Biol.* 48:575–607
32. Hannappel U, Balzer HJ, Ganal MW. 1995. Direct isolation of cDNA sequences from specific chromosomal regions of the tomato genome by the differential display technique. *Mol. Gen. Genet.* 249:19–24
33. Heller W, Forkmann G. 1994. Biosynthesis of flavonoids. In *The Flavonoids: Advances In Research Since 1986*, ed. JB Harborne, pp. 499–535. New York: Chapman & Hall. 676 pp.
34. Hirabayashi H, Ogawa T. 1995. RFLP mapping of *Bph-1* (brown planthopper resistance gene) in rice. *Breed. Sci.* 45:369–71
35. Hoisington D, Ribaut JM. 1998. Marker-assisted selection: new tools and strategies. *Trends Plant Sci.* 3:236–38
36. Hollenhorst MM, Joppa LR. 1983. Chromosomal location of genes for resistance to greenbug in 'Largo' and 'Amigo' wheats. *Crop Sci.* 23:91–93
37. Huang N, Parco A, Mew T, Magpantay G, McCouch S, et al. 1997. RFLP mapping of isozymes, RAPD and QTLs for grain shape, brown planthopper resistance in a doubled haploid rice population. *Mol. Breed.* 3:105–13
38. Ishii T, Brar DS, Multani DS, Khush GS. 1994. Molecular tagging of genes for brown planthopper resistance and earliness introgressed from *Oryza australiensis* into cultivated rice, *O. sativa*. *Genome* 37:217–21
39. Jende-Strid B. 1993. Genetic control of flavonoid biosynthesis in barley. *Hereditas* 119:187–204
40. Kaga A, Ishimoto M. 1998. Genetic localization of a bruchid resistance gene and its relationship to insecticidal cyclopeptide alkaloids, the vignatic acids, in mungbean (*Vigna radiata* L. Wilczek). *Mol. Gen. Genet.* 258:378–84
41. Kaloshian I, Yaghoobi J, Liharska T, Hontelez J, Hanson D, et al. 1998. Genetic and physical localization of the root-knot nematode resistance locus *Mi* in tomato. *Mol. Gen. Genet.* 257:376–85
42. Karban R, Baldwin IT. 1997. *Induced Responses to Herbivory*. Chicago: Univ. Chicago Press
43. Katiyar SK, Tan Y, Zhang Y, Huang B, Xu Y, et al. 1995. Molecular tagging of gall midge resistance genes in rice. In *Fragile Lives in Fragile Ecosystems. Proc. Int. Rice Res. Conf., 13–17 Feb.*, pp. 935–48. Manila: Int. Rice Res. Inst.
44. Kennedy GG, Barbour JD. 1992. Resistance in natural and managed systems. In *Plant Resistance to Herbivores and Pathogens Ecology, Evolution, and Genetics*, ed. RS Fritz, EL Simms, pp. 13–41. Chicago: Univ. Chicago Press
45. Kennedy GG, Gould F, Deponti OMB, Stinner RE. 1987. Ecological, agricultural, genetic, and commercial considerations in the deployment of insect-resistant germplasm. *Environ. Entomol.* 16:327–38
46. Khairallah MM, Bohn M, Jiang C,

Deutsch JA, Jewell DC, et al. 1998. Molecular mapping of QTL for southwestern corn borer resistance, plant height and flowering in tropical maize. *Plant Breed.* 117:309–18
47. Kowalski SP, Plaisted RL, Steffens JC. 1993. Immunodetection of polyphenol oxidase in glandular trichomes of *S. berthaultii, S. tuberosum* and their hybrids. *Am. Potato J.* 70:185–99
48. Lande R. 1992. Marker-assisted selection in relation to traditional methods of plant breeding. In *Plant Breeding in the 1990s,* ed. HT Stalker, JP Murphy, pp. 437–51. Wallingford, UK: CAB Int. 539 pp.
49. Lee EA, Byrne PF, McMullen MD, Snook ME, Wiseman BR, et al. 1998. Genetic mechanisms underlying apimaysin and maysin synthesis and corn earworm antibiosis in maize (*Zea mays* L.). *Genetics* 149:1997–2006
50. Lee M. 1995. DNA markers and plant breeding programs. *Adv. Agron.* 55:265-344
51. Liu BH. 1997. *Statistical Genomics: Linkage, Mapping and QTL Analysis.* Boca Raton, FL: CRC Press. 611 pp.
52. Ma ZQ, Saidi A, Quick JS, Lapitan NLV. 1998. Genetic mapping of Russian wheat aphid resistance genes *Dn2* and *Dn4* in wheat. *Genome* 41:303–6
53. Maliepaard C, Bas N, van Heusden S, Kos J, Pet G, et al. 1995. Mapping of QTLs for glandular trichome densities and *Trialeurodes vaporariorum* (greenhouse whitefly) resistance in an F2 from *Lycopersicon esculentum* X *L. hirsutum* f. *glabratum. Heredity* 75:425–33
54. Maxwell FG, Jennings PR, eds. 1980. *Breeding Plants Resistant to Insects.* New York: Wiley. 683 pp.
55. Michelmore RW, Paran I, Kesseli RV. 1991. Identification of markers linked to disease resistance by bulked segregant analysis: a rapid method to detect markers in specific genomic regions by using segregating populations. *Proc. Natl. Acad. Sci. USA* 88:9828–32
56. Mohan M, Nair S, Bentur JS, Rao UP, Bennett J. 1994. RFLP and RAPD mapping of the rice *Gm2* gene that confers resistance to biotype 1 of gall midge (*Orseolia oryzae*). *Theor. Appl. Genet.* 87:782–88
57. Mohan M, Sathyanarayanan PV, Kumar A, Srivastava MN, Nair S. 1997. Molecular mapping of a resistance-specific PCR-based marker linked to a gall midge resistance gene (*Gm4t*) in rice. *Theor. Appl. Genet.* 95:777–82
58. Moharramipour S, Tsumuki H, Sato K, Yoshida H. 1997. Mapping resistance to cereal aphids in barley. *Theor. Appl. Genet.* 94:592–96
59. Murata M, Fujiwara M, Kaneda C, Takumi S, Mori N, et al. 1998. RFLP mapping of a brown planthopper (*Nilaparvata lugens* STAL.) resistance gene *bph2* of Indica rice introgressed into a Japonica breeding line 'Norin-PL4'. *Genes Genet. Syst.* 73:359–64
60. Mutschler MA, Doerge RW, Liu SC, Kuai JP, Liedl BE, et al. 1996. QTL analysis of pest resistance in the wild tomato *Lycopersicon pennellii*: QTLs controlling acylsugar level and composition. *Theor. Appl. Genet.* 92:709–18
61. Myburg AA, Cawood M, Wingfield BD, Botha AM. 1998. Development of RAPD and SCAR markers linked to the Russian wheat aphid resistance gene *dn2* in wheat. *Theor. Appl. Genet.* 96:1162–69
62. Neal JJ, Steffens JC, Tingey WM. 1989. Glandular trichomes of *Solanum berthaultii* and resistance to the Colorado potato beetle. *Entomol. Exp. Appl.* 51:133–40
63. Nienhuis J, Helentjaris T, Slocum M, Ruggero B, Schaefer A. 1987. Restriction fragment length polymorphism analysis of loci associated with insect resistance in tomato. *Crop Sci.* 27:797–803
64. Nieto-Lopez RM, Blake TK. 1994. Russian wheat aphid resistance in barley:

inheritance and linked molecular markers. *Crop Sci.* 34:655–59
65. O'Brien SJ. 1993. *Genetic Maps.* Cold Spring Harbor, NY: Cold Spring Harbor Press. 6th ed. 261 pp.
66. Ohm HS, Sharma HC, Patterson FL, Ratcliffe RH, Obanni M. 1995. Linkage relationships among genes on wheat chromosome 5A that condition resistance to Hessian fly. *Crop Sci.* 35:1603–7
67. Osborn TC, Kole C, Parkin IAP, Sharpe AG, Kuiper M, et al. 1997. Comparison of flowering time genes in *Brassica rapa, B. napus,* and *Arabidopsis thaliana. Genetics* 146:1123–29
68. Painter RH. 1958. *Insect Resistance in Crop Plants.* Lawrence: Univ. Kans. Press
69. Panda N, Khush GS. 1995. *Host Plant Resistance to Insects.* Wallingford, UK: CAB Int. 520 pp.
70. Paterson AH. 1996. *Genome Mapping in Plants.* Austin, TX: Landes. 330 pp.
71. Pechan TB, Jiang D, Steckler L, Ye L, Lin L, et al. 1999. Characterization of three distinct cDNA clones encoding cysteine proteinases from maize (*Zea mays* L.) callus. *Plant Mol. Biol.* 40:111–19
72. Pelletier Y, Smilowitz Z. 1990. Effect of type B exudate of *Solanum berthaultii* Hawkes on consumption by the Colorado potato beetle, *Leptinotarsa decemlineata* (Say). *J. Chem. Ecol.* 16:1547–55
73. Plaisted RL, Tingey WM, Steffens JC. 1992. The germplasm release of NYL 235-4, a clone with resistance to the Colorado potato beetle. *Am. Potato J.* 69:843-46
74. Powell W, Morgante M, Andre C, Hanafey M, Vogel J, et al. 1996. The comparison of RFLP, RAPD, AFLP, and SSR (microsatellite) markers for germplasm analysis. *Mol. Breed.* 2:225–38
75. Prince JP, Pochard E, Tanksley SD. 1993. Construction of a molecular linkage map of pepper and a comparison of synteny with tomato. *Genome* 36:404–17
76. Rector BG, All JN, Parrott WA, Boerma HR. 1998. Identification of molecular markers linked to quantitative trait loci for soybean resistance to corn earworm. *Theor. Appl. Genet.* 96:786–90
77. Reddy VS, Dash S, Reddy AR. 1995. Anthocyanin pathway in rice (*Oryza sativa* L.): identification of a mutant showing dominant inhibition of anthocyanins in leaf accumulation of proanthocyanidins in pericarp. *Theor. Appl. Genet.* 91:301–12
78. Roche P, Alston FH, Maliepaard C, Evans KM, Vrielink R, et al. 1997. RFLP and RAPD markers linked to the rosy leaf curling aphid resistance gene (*Sd1*) in apple. *Theor. Appl. Genet.* 94 :528–33
79. Roche P, Arkel G, Heusden AW. 1997. A specific PCR assay for resistance to biotypes 1 and 2 of the rosy leaf curling aphid in apple based on an RFLP marker closely linked to the *Sd1* gene. *Plant Breed.* 116:567–72
80. Rossi M, Goggin FL, Milligan SB, Kaloshian I, Ullman DE, et al. 1998. The nematode resistance gene *Mi* of tomato confers resistance against the potato aphid. *Proc. Natl. Acad. Sci. USA* 95:950–54
81. Roush RT. 1997. Managing resistance to transgenic crops. In *Advances in Insect Control: The Role of Transgenic Plants,* ed. N Carozzi, M Koziel, pp. 271–94. London: Taylor & Francis. 301 pp.
82. Russell GE. 1978. *Plant Breeding for Pest and Disease Resistance.* Boston: Butterworth. 485 pp.
83. Schon CC, Lee M, Melchinger AE, Guthrie WD, Woodman WL. 1993. Mapping and characterization of quantitative trait loci affecting resistance against second generation European corn borer in maize with the aid of RFLPs. *Heredity* 70:648–59
84. Sebastin LS, Ikeda R, Huang N, Imbe T, Coffman WR, McCouch SR. 1996.

Molecular mapping of resistance to rice tungro spherical virus and green leafhopper. *Phytopathology* 86:25–30
85. Seo YW, Johnson JW, Jarret RL. 1997. A molecular marker associated with the *H21* Hessian fly resistance gene in wheat. *Mol. Breed.* 3:177–81
86. Simmonds NW. 1991. Genetics of horizontal resistance to diseases of crops. *Biol. Rev.* 66:189–241
87. Smith CM. 1989. *Plant Resistance to Insects: A Fundamental Approach.* New York: Wiley. 286 pp.
88. Smith CM, Khan ZR, Pathak MD. 1994. *Techniques for Evaluating Insect Resistance in Crop Plants.* Boca Raton, FL: Lewis/CRC Press
89. Snook ME, Gueldner RC, Widstrom NW, Wiseman BR, Himmelsbach DS, et al. 1993. Levels of maysin and maysin analogs in silks of maize germplasm. *J. Agric. Food Chem.* 41:1481–85
90. Staub JE, Serquen FC, Gupta M. 1996. Genetic markers, map construction, and their application in plant breeding. *HortScience* 31:729–41
91. Styles ED, Ceska O. 1989. Pericarp flavonoids in genetic strains of *Zea mays. Maydica* 34:227–37
92. Tanksley SD. 1993. Mapping polygenes. *Annu. Rev. Genet.* 27:205–33
93. Tanksley SD, McCouch SR. 1997. Seed banks and molecular maps: unlocking genetic potential from the wild. *Science* 277:1063–66
94. Tanksley SD, Young ND, Patterson AH, Bonierbale MW. 1989. RFLP mapping in plant breeding: new tools for an old science. *BioTechnology* 7:257–64
95. Tingey WM. 1991. Potato glandular trichomes: defensive activity against insect attack. In *Naturally Occurring Pest Bioregulators. Am. Chem. Soc. Symp. Ser. 449,* ed. PA Hedin, pp. 126–35 Washington, DC: ACS Press. 456 pp.
96. Tingey WM, Yencho GC. 1994. Insect resistance in potato: a decade of progress. In *Advances in Potato Pest Biology and Management*, ed. GW Zehnder, ML Powelson, RK Jansson, KV Raman, pp. 405–25. St. Paul, MN: APS Press. 655 pp.
97. Van den Berg JH, Ewing EE, Plaisted RL, McMurry S, Bonierbale MW. 1996. QTL analysis of potato tuberization. *Theor. Appl. Genet.* 93:307–16
98. Van den Berg JH, Ewing EE, Plaisted RL, McMurry S, Bonierbale MW. 1996. QTL analysis of potato tuber dormancy. *Theor. Appl. Genet.* 93:317–24
99. Waiss AC, Chan BG, Elliger CA, Wiseman BR, McMillian WW, et al. 1979. Maysin, a flavone glycoside from corn silks with antibiotic activity toward corn earworm. *J. Econ. Entomol.* 72:256–58
100. Wiseman BR, Snook ME, Isenhour DJ, Mihm JA, Widstrom NW. 1992. Relationship between growth of corn earworm and fall armyworm larvae (Lepidoptera: Noctuidae) and maysin concentration in corn silks. *J. Econ. Entomol.* 85:2473–77
101. Wiseman BR, Snook ME, Widstrom NW. 1996. Feeding responses of corn earworm larvae (Lepidoptera: Noctuidae) on corn silks of varying flavone content. *J. Econ. Entomol.* 89:1040–44
102. Xie DX, Feys BF, James S, Nieto-Rostro M, Turner JG. 1998. *COI1*: an *Arabidopsis* gene required for jasmonate-regulated defense and fertility. *Science* 280:1091–94
103. Yamasaki M, Tsunematsu H, Yoshimura A, Iwata N, Yasui H. 1999. Quantitative trait locus mapping of ovicidal response in rice (*Oryza sativa* L.) against whitebacked planthopper (*Sogatella furcifera* Horvath). *Crop Sci.* 39:1178–83
104. Yencho GC, Bonierbale MW, Tingey WM, Plaisted RL, Tanksley SD. 1996. Molecular markers locate genes for resistance to the Colorado potato beetle, *Leptinotarsa decemlineata*, in hybrid *Solanum tuberosum* x *S. berthaultii* potato progenies. *Entomol. Exp. Appl.* 81:141–54

105. Yencho GC, Kowalski SP, Kobayashi RS, Sinden SL, Bonierbale MW, Deahl KL. 1998. QTL mapping of foliar glycoalkaloid aglycones in *Solanum tuberosum* x *S. berthaultii* potato progenies: quantitative variation and plant secondary metabolism. *Theor. Appl. Genet.* 97:563–74
106. Yencho GC, Renwick JAA, Steffens JC, Tingey WM. 1994. Leaf surface extracts of *Solanum berthaultii* deter Colorado potato beetle feeding. *J. Chem. Ecol.* 20:991–1007
107. Young ND. 1996. QTL mapping and quantitative disease resistance in plants. *Annu. Rev. Phytopathol.* 34: 479–501
108. Young ND, Kumar L, Menancio-Hautea D, Danesh D, Talekar NS, et al. 1992. RFPL mapping of a major bruchid resistance gene in mungbean (*Vigna radiata*, L. Wilczek). *Theor. Appl. Genet.* 84:839–44
109. Young ND, Tanksley SD. 1989. Restriction fragment length polymorphism maps and the concept of graphical genotypes. *Theor. Appl. Genet.* 77:95–101

Annu. Rev. Entomol. 2000. 45:423–448

OVARIAN DYNAMICS AND HOST USE

Daniel R. Papaj

Department of Ecology and Evolutionary Biology, Tucson, Arizona 85721; e-mail: papaj@u.arizona.edu

Key Words ovarian development, egg load, oviposition behavior, parasitoids, oogenesis

■ **Abstract** Oviposition behavior in herbivorous and frugivorous insects and parasitoids is dynamic at the level of the individual, responding to variation in host quality and availability. Patterns of variation in egg load in response to host presence and quality suggest that ovarian development also responds to variation in the host environment. Ovarian dynamics are mediated by feedback from oviposition, by host feeding, and by sensory input from the host. The last of these mechanisms, host sensory cuing, is known to occur in three major orders and provides strong evidence that ovarian dynamics are adaptive by design. Conditions favoring host effects on ovarian development include trade-offs between egg production and either survival or dispersal, uncertainty in the host environment, and a correlation in host conditions between the time that oogenesis is initiated and the time that eggs are laid. Some host defenses block ovarian development, suggesting that ovarian dynamics in host-specific insects should be viewed from a coevolutionary perspective.

INTRODUCTION

A substantial body of theoretical and empirical literature on phytophagous, frugivorous, and parasitic insects has been devoted to understanding how variation in the quality and abundance of host resources influences oviposition behavior (9, 37, 71, 106). An equally substantial body of literature has sought to describe how aspects of an insect's internal state, such as level of experience and motivation, influence oviposition behavior (9, 37, 54, 69, 71, 87, 88, 94, 102). Taken together, this work indicates that oviposition behavior in host-specific insects is highly dynamic at the level of the individual, responding to variation in host quality and abundance in adaptive ways that depend on an individual's physiological state, including its stage of ovarian development.

One attribute of ovarian development of particular interest to students of insect-host interactions has been egg load. Egg load (defined here as the total number of mature oocytes in the ovaries) exerts a variety of effects on oviposition behavior. Females with high egg load typically expend more effort in foraging for hosts, are less selective regarding the quality of hosts used, lay larger clutches, allocate sex differently, and invest more in contests with other females over host resources than do females with low egg load (12, 33, 48, 73, 88, 103, 119).

Issues relating to egg load and, hence, ovarian development have taken center stage in a recent debate on the relative importance of egg and time limitation in host use (40, 42, 68, 93, 99). Egg limitation occurs when females deplete their egg supply before opportunities to oviposit are exhausted. Time limitation occurs when females die or otherwise lose reproductive competence before all mature eggs have been laid. A high risk of egg limitation should cause females to become choosier with respect to the quality of hosts used in oviposition, even if such choosiness reduces the rate at which eggs are laid (53, 67). A high risk of time limitation should, in contrast, cause females to adopt strategies that increase the rate at which hosts are found, even if such strategies reduce the quality of hosts on which eggs are laid (105).

Egg and time limitation are a consequence of trade-offs, most plausibly trade-offs between survival and reproduction in an unpredictable environment (93, 99). An animal has a finite amount of resources to allocate to survival and reproduction; investment in one trades off against investment in the other. To the extent that opportunities for reproduction cannot be predicted with certainty in advance of egg maturation, natural selection cannot generate a perfect match between reproductive opportunity and allocation to reproduction (93). However, selection is expected to provide the best possible match. Adjustment of selectivity, clutch size, and other behavior in response to egg load can be regarded as adaptations for reducing an insect's risk of egg limitation (68). Egg maturation processes that contribute to egg load have been cited as central to our understanding of how behavior balances the opposing risks of time and egg limitation (41, 42).

Notwithstanding its significance for the dynamics of oviposition behavior, ovarian development is itself under selection (18, 24, 25, 27, 57, 86, 89) and is conceivably dynamic in ways that permit insects to cope with variation in the host resource. How ovarian development responds to variation in host quality and availability is the subject of this review. Attention is devoted to patterns of egg maturation in relation to host quality and availability, mechanisms underlying those patterns, and trade-offs thought to favor those mechanisms.

BASIC DESIGN AND CONTROL OF OVARIAN DEVELOPMENT

Insect ovaries appear to be well suited for adjusting egg production to environmental conditions. First, once initiated, egg production is rapid. Commenting on the diversity of mechanisms associated with oogenesis in animals, Eckelbarger (25) noted:

> A major reason for [the arthropods'] success has been the evolution of specialized ovaries and vitellogenic mechanisms that enable them to manufac-

ture, sometimes in a single day, an egg mass exceeding half their body weight.

Key to rapid production of ovaries of large mass are heterosynthetic mechanisms of yolk synthesis, wherein yolk material is produced not only within the oocyte but in other tissues as well. Vitellogenesis in insects, for example, is marked by manufacture in the fat bodies of a class of yolk proteins known as vitellogenins that are transported through the hemolymph to the developing ovarioles. In the meroistic-type ovaries characteristic of more derived orders, abortive germ cells known as nurse cells, or trophocytes, furnish oocytes with metabolites and organelles via cytoplasmic bridges. As a consequence of such nourishment, vitellogenesis in meroistic ovaries is often completed in just one to two days.

Second, the quantity of eggs available to lay can be adjusted at the level of the individual. Females of many species (sometimes labeled "synovigenic;" 29) mature eggs throughout adult life, sometimes eclosing with few or no mature eggs. Egg production in such species is regulated through two jointly acting processes: oogenesis, in which follicles form in ovarioles, yolk is deposited in oocytes within those follicles and a chorion laid down (25); and oosorption, in which vitellogenesis ceases and yolked oocytes degenerate (7). Less commonly, females eclose with their entire complement of eggs ready to lay. In such species (sometimes labeled "pro-ovigenic;" 29), the quantity of eggs matured is probably regulated little or not at all in the adult stage, but may nonetheless be modified in the juvenile stage.

Third, timing of egg maturation can be adjusted at the level of the individual. An elaborate system of neural and hormonal control regulates the timing of ovarian development (58, 97, 114, 126, 127). Via this system, a variety of extrinsic factors influence the timing and degree of ovarian development, including temperature (6, 32, 117), humidity (64), photoperiod (1, 6, 70, 108, 117), mating (8, 34, 126), social context (11, 97, 114) and diet (126).

PATTERNS OF EGG MATURATION IN RELATION TO THE HOST RESOURCE

In principle, the ability to mature eggs rapidly, in variable quantity and at the appropriate time, should permit insects to adjust rates of egg maturation to variation in host quality and availability. Yet discussions of egg load as a factor in oviposition behavior have routinely treated ovarian development as blind to such variation. In one scenario, eggs accumulate in an insect's abdomen as time since last oviposition elapses (17, 54, 80, 81). The increase in egg load affects oviposition behavior, for example, by reducing selectivity. When eggs are next laid, egg load declines and selectivity is recovered. Eggs begin again to accumulate, completing the cycle. In some models, eggs mature at a linear rate until either egg load reaches some capacity set by the physical limits of a female's abdominal cavity or nutrient reserves are depleted (17, 67, 71). The host has no effect on ovarian development except through egg deposition or, in species that feed on hosts, nutrient input.

Direct evidence in support of this paradigm is lacking. No one has directly tracked changes in an individual's egg load and correlated these with changes in behavior, owing to the difficulty of counting mature oocytes without killing the individual whose behavior is being followed. Support has instead been marshaled from two independent lines of evidence, one showing that variation in egg load among individuals affects selectivity (12, 48, 73, 88, 103), and one showing that an individual's selectivity increases with time since last oviposition (102).

Resorting to indirect evidence might merit little concern if such evidence uniformly supported the existing paradigm. As this review will illustrate, it does not. While variation in egg load is a powerful predictor of variation in behavior in many host-specific insects, it has little or no significance in others. At the same time, patterns of ovarian maturation in relation to host use appear to be more complex than sometimes assumed.

Responses to Variation in Host Availability

Effects of host availability on egg load and behavior are typically assessed by depriving females of hosts. In some insects, host deprivation has no detectable effect on egg load. In the pollen beetle, *Meligethes aeneus*, egg load of individuals deprived of oviposition sites did not differ significantly from that of individuals exposed continuously to such sites (45). Neither a beetle's propensity to oviposit nor its clutch size showed detectable changes over time since last oviposition. Mediterranean fruit fly (*Ceratitis capitata*) females deprived of hosts for three weeks similarly retained no more eggs in their ovaries than females deprived for two weeks (79). Nevertheless, the difference in deprivation period was accompanied by a difference in behavior. More deprived flies foraged more persistently for fruit and were more aggressive in defense of fruit than less deprived flies.

In other insects, eggs accumulate under host deprivation. In *Bemisia tabaci* whiteflies, total vitellin and vitellogenin content increased steadily over a 24-hour period during which females were denied access to host melon leaves (121). Similarly, in pear psylla (*Cacopsylla pyricola*), eggs accumulated over a 16-hour period of host deprivation. Increased egg load in pear psylla was associated with increased oviposition activity when host leaves were subsequently made available (47).

In yet another group of insects, host deprivation is associated with smaller egg loads, not larger ones as expected. Females of an aphidiid parasitoid, *Monoctonus paulensis*, provided with host aphids during day one post-eclosion had higher egg loads on day three than females deprived of hosts continuously from emergence (72). Egg loads for females of an encyrtid parasitoid, *Leptomastix dactylopii*, declined slightly as host deprivation progressed over ten days (90).

Variability in results may partly reflect the timescale over which responses to host deprivation were evaluated. In synovigenic parasitoids, egg load increases initially under deprivation, but declines subsequently as eggs are resorbed (23, 29). Waxing and waning of egg load is paralleled by changes in host selectivity. The early literature on parasitoids described a "faculty of restraint" in oviposition that broke down as eggs accumulated (29, 65). As deprivation continued, females were progressively less likely to attack hosts. Generally, the appearance of hosts restores egg maturation (23). If the period of deprivation continues for too long, however, ovarian development may not revive and females may not lay eggs again, a phenomenon termed "ecological castration" (30).

Timescale probably does not account entirely for variation in ovarian responses to host deprivation. Some variation may reflect species differences in synchrony of ovariolar maturation. In order for egg load to change gradually in response to host deprivation, development within different ovarioles within the ovaries must necessarily be asynchronous. Some insects, such as sheep blow flies (*Lucilia cuprina*), which lay just a few large clutches in their lifetime, appear to be constrained to mature oocytes synchronously across ovarioles (13). Oviposition site deprivation in *L. cuprina* affects neither egg load nor oviposition behavior.

Variability in results may also reflect variation in strategies for coping with host deprivation. An intriguing alternative response to host deprivation has been reported in females of a tachinid fly species, *Chetogena edwardsii* (111). Host-deprived females retain no more eggs in their ovaries than females given unlimited access to their nymphalid caterpillar hosts. However, host-deprived females differ from undeprived counterparts in one key respect: They retain a fertilized egg in a uterus where it undergoes embryogenesis. Maggots hatching from embryonated eggs reach pupation sooner than maggots hatching from eggs fertilized shortly before oviposition. By incubating eggs, females conceivably broaden host range to include late-instar caterpillars that offer a relatively narrow window of time for parasitoid development. In short, *C. edwardsii* females may effectively increase host availability by generating embryonated eggs that fare better in low-quality hosts, and expanding host range to include those hosts. Ovoviviparity in females denied access to preferred breeding sites has also been reported in some Hawaiian *Drosophila* and *Scaptomyza* flies (57).

While insects surely experience periods of complete host deprivation in nature, host abundance often varies in a more graded fashion. Data, though sparse, indicate that ovarian maturation responds to patterns of graded variation. The egg parasitoid *Trichogramma minutum* is highly opportunistic, adjusting its fecundity schedule to current host density (5). Females offered host flour-moth eggs at low density and just once every three days spread out egg production evenly over a period of several weeks. In contrast, females offered eggs daily in essentially unlimited supply realized 50% of their lifetime fecundity in the first few days after emergence, after which egg production plummeted below levels achieved by females offered hosts at low density.

In a study exceptional for its strong focus on ovarian dynamics (90), female *Leptomastix dactylopii* parasitoids provided with mealybug hosts at high density had egg loads resembling those of host-deprived females and females provided with hosts at low density. Because wasps foraging at high density were depositing eggs at relatively high rates and retaining no fewer eggs on average, it follows that wasps in that treatment were producing eggs at higher rates than wasps in other treatments.

In encyrtid parasitoids, oosorption leaves visible chitinized remnants. By counting remnants, the boost in egg production at high host density was determined to be due to an increase in oogenesis and not a decrease in oosorption. Finally, effects of host regime on egg production vanished immediately upon deprivation of hosts; in terms of ovarian development, there was no "memory" of the host environment experienced prior to deprivation.

Responses to Variation in Host Quality

Ovarian maturation is also sensitive to variation in host quality. In *M. aeneus* pollen beetles, egg load in individuals offered highly acceptable cruciferous host plant species was greater than egg load in individuals offered species of low acceptability—not less, as might be expected if individuals provided with less acceptable plants were accumulating eggs that those provided with acceptable plants were depositing (45). Egg maturation was linked to oviposition. Given free access to highly acceptable hosts, more eggs were laid, but proportionately more were matured. In beetles given free access to less acceptable hosts, in contrast, egg maturation was virtually shut down. Finally, individuals proved capable of adjusting egg production in response to changes in host quality. Being switched from low-quality to a high-quality host boosted egg production; switches in the opposite direction depressed egg production (46).

In seedcorn flies (*Delia platura*), rates of egg maturation were likewise matched to rates of oviposition (125). Seedcorn flies develop as larvae on both living and dead plant material, and females oviposit in moist sand even in the absence of germinating host lima beans. However, females provided with lima bean produced two to three times more eggs and produced them sooner than host-deprived females. Bean-exposed females retained no fewer eggs in their ovaries at death than bean-deprived females, suggesting that rate of egg maturation was matched to rate of egg deposition. When flies were switched from lima bean and sand to sand only, egg production and deposition initially slowed but soon matched the rate at which flies maintained continuously on lima bean and sand produced and laid eggs.

In the bethylid parasitoid, *Goniozus nigrifemur*, rates of egg maturation depended on the size of host pink bollworm larvae recently encountered (66). Females permitted to oviposit in a large host (20–28 mg) and dissected five days later retained nearly twice as many eggs in their ovaries as females permitted to oviposit in a small host (2–6 mg) (means of 12.6 versus 6.8 respectively), despite the fact that large-host females laid larger clutches than small-host females. Same-aged, host-deprived females retained just two eggs on average. Large-host females also oviposited over a day sooner in the next host than did small-host females.

Adaptiveness of Patterns

Many studies cited here were not intended to evaluate ovarian dynamics in relation to the host resource. Nevertheless, data suggest that ovarian development responds to variability in host quality and availability in adaptive ways. For example, boosting egg production when hosts are common, as observed in *Trichogramma* wasps (5), reduces risk of egg limitation, at least in the short term. Slowing egg production when the quality of the host is low, as observed in *Meligethes* pollen beetles, may allow females to wait out periods of poor host quality, at least if egg production trades off against survival (as discussed below: a common trade-off). In an insect such as a pollen beetle, which has a relatively long reproductive season, the benefits of delaying a commitment to reproduce may be particularly high (45). If host quality remains low, restoring rates of egg maturation to high levels, as observed in *Delia* flies (125) makes adaptive sense too. Eggs that are made and laid on poor-quality hosts have some chance of survival; eggs that are never made and never laid do not. Hence, if poor conditions persist, an insect should "make the best of a bad job" and resume laying eggs. Finally, oosorption under host deprivation has long been regarded as adaptive; by reallocating nutrients from eggs to somatic tissues in times when oviposition resources are scarce, an insect may survive periods of host shortage (7, 29).

MECHANISMS UNDERLYING OVARIAN DYNAMICS

Ovarian development responds to variation in the host environment by three mechanisms: *(a)* Egg maturation is triggered by oviposition, and oviposition varies in frequency with host quality and abundance; *(b)* hosts serve as a source of nourishment for oogenesis to a degree related to their quality and abundance; *(c)* host stimuli directly promote egg maturation, again to a degree related to host quality and abundance.

Oviposition-Mediated Effects of the Host

Many patterns in egg maturation cited here can be explained by a simple mechanism: Oviposition begets oogenesis; failure to oviposit begets oosorption. Because oviposition behavior is sensitive to host quality and abundance, a cause-and-effect relationship between oviposition and ovarian development would generate a de facto match between ovarian development and variation in the host resource.

That failure to oviposit begets oosorption is well known (7, 29, 30, 32). Likewise, oviposition is known to facilitate oogenesis. In cockroaches, oogenesis is stimulated by oothecal deposition (97); oothecal release delivers input (possibly via stretch receptors) along the ventral nerve cord to the brain. Stimulation of the corpora allata by the brain results in production of juvenile hormone (JH); high JH titres transmitted through the hemolymph stimulate oogenesis. In cyclorrhaphous flies, a different mechanism serves the same end. An oostatic hormone produced by the primary follicles in the ovaries inhibits development of more distal follicles. Once eggs are laid, inhibition is released and a subsequent gonotrophic cycle begins (2, 15, 128); involvement of neural input has not been demonstrated in this case.

Does the effect of an oviposition on egg maturation depend on the quality of the last host on which eggs were laid? Does rate of host encounter influence the degree to which oviposition promotes egg maturation? To date, no study has manipulated oviposition history systematically enough to answer these questions [although a step in that direction, with negative results, has been made; (90)].

Nutritional Effects of the Host

Adult nourishment is profoundly important for oogenesis in many insects (126). Of interest here are instances in which insects feed as adults on the same kind of host used for juvenile development. If hosts nourish oogenesis to a degree related to their quality and abundance, ovarian development might, as a consequence, respond adaptively to variation in the host resource.

Adult host feeding in relation to oviposition has been considered in depth for hymenopterous parasitoids (40, 56, 118). The relationship between host feeding and oviposition constitutes a classic example of a trade-off between current and future reproduction, particularly when host feeding renders individual hosts unsuitable for oviposition. Dynamic programming models predicated on this trade-off routinely predict that parasitoids should host feed when egg loads are low (and risk of egg limitation correspondingly high), a prediction generally supported by laboratory data (40, but see 95). Models also predict that parasitoids will restrict feeding to hosts of lower quality and use higher-quality hosts for oviposition, a prediction also supported by data (40).

Developmental realism has been added to models in the form of egg maturation delays and costs of oosorption (16). Nevertheless, in most discussions, ovarian maturation simply tracks oviposition, albeit through an indirect path: Oviposition reduces egg load to levels low enough to stimulate feeding, which in turn facilitates egg production. Does the effect of host-nutrient input on egg maturation depend on the quality of the host used in feeding? Does host abundance influence the degree to which nutrition promotes egg maturation? Can parasitoids regulate allocation of nutrients to reproductive versus somatic processes according to the quality and quantity of host nutrients imbibed? At this point, too little is known about the physiological details of nutrient allocation to reproduction to answer these questions.

DIRECT SENSORY EFFECTS OF HOSTS ON OVARIAN DEVELOPMENT

The discussion thus far provides little evidence of adaptive design in ovarian dynamics. Based on current understanding, both oviposition- and nutrient-mediated mechanisms simply link ovarian development to oviposition, which responds in turn to variation in the host environment. Conceivably, it is oviposition behavior that responds by design to variation in host quality and availability; ovarian dynamics follow only as an incidental consequence. Evidence that host stimuli act as sensory promoters of ovarian development, independent of oviposition or host nourishment, would constitute stronger evidence of design in ovarian dynamics.

Many studies on egg production in host-specific insects were not intended to evaluate direct sensory effects of the host on ovarian development and their experimental design precludes that evaluation. Investigators counted eggs laid rather than dissecting ovaries or, if ovaries were dissected, performed dissections only after eggs were laid. Such methodology cannot distinguish between a direct effect of host stimuli on ovarian development and an effect on egg deposition, which then stimulates production of more eggs (60). In cabbage root flies, for example, exposure to cabbage stimulates oogenesis, but only after some eggs are laid (59). Only by using periods of host exposure prior to dissection that are too brief to permit oviposition, or using surrogate hosts that somehow prevent oviposition, can oviposition-mediated effects be ruled out.

Likewise, in cases where females feed on hosts, it can be difficult to distinguish between nutritional and sensory influences (58). These issues notwithstanding, solid evidence for sensory effects of hosts on ovarian development spans three major orders (Lepidoptera, Coleoptera, and Diptera) in work dating back to the 1960s.

Order Lepidoptera

Host plant stimuli commonly have direct effects on ovarian development in Lepidoptera (4, 8, 21, 44, 82, 91, 112, 113). Hillyer and Thorsteinson's classic work on the diamondback moth, *Plutella xylostella*, showed direct effects of host plants on ovarian development and demonstrated that a single volatile component of the insects' cruciferous host plants, allyl isothiocyanate, induced effects of a magnitude similar to that of whole plants (44). This work is noteworthy for the completeness with which it surveyed the effect of host stimuli and the effects of other factors, such as non-host plants and mating on ovarian development. Exposure to host cabbage for three days following emergence of an adult female hastened onset and degree of ovarian maturation. Two non-host species, pea and wheat, had no effect. An experiment in which screening prevented contact with a cabbage leaf failed to find evidence of an effect of host odor. Nevertheless, females exposed for three days to paper treated with allyl isothiocyanate, a volatile compound known to be an oviposition stimulant for *P. xylostella* (38), retained almost as many eggs per ovariole as females provided access to cabbage, and many more than control females.

Implicit in some of Hillyer and Thorsteinson's experiments is the assumption that male behavior is not influenced by host stimuli. In fact, exposure to host plants accelerates reproductive maturity in males (82). Because mating promotes egg maturation in *P. xylostella* (44), an effect of the host on egg maturation was likely due in part to an effect of the host on mating.

Order Coleoptera

Bruchid beetles have been the subject of a large literature on the role of host stimuli on ovarian development. Egg production in *Acanthoscelides obtectus* has long been known to be greater in the presence of its leguminous host seeds than in their absence (123). This result does not depend on oviposition: After just two days of exposure to host beans and before any eggs were laid, egg load exceeded that of control females not exposed to host beans (62).

Because females do not feed on the beans (83), input from the host must be non-nutritional in nature. Facilitation of egg production required females to make physical contact with beans: Bean odor alone had no effect (83). Assays with ether extracts of beans further suggested that host stimuli were gustatory in nature (75, 83). Amputation experiments indicated that the maxillary palps were required for detection of active components (83).

While *A. obtectus* has been studied most, effects of host beans or bean pods on ovarian development have been demonstrated in at least four other bruchid species (*Brucidius atrolineatus, Callosobruchus maculatus, Zabrotes subfasciatus,* and *Caryedon serratus*) (20, 84). Ovarian development in all four species is facilitated only when the host plants are at, or nearly at, the phenological stage suitable for feeding by first-instar larvae. Lower-quality hosts do not promote oogenesis. In *B. atrolineatus*, a number of factors, including short photophase, high humidity and host presence, trigger an end to reproductive diapause in both adult males and females. These factors stimulate the previtellogenic phase of ovarian development, during which nurse cells transfer metabolites to oocytes; however, the presence of host inflorescences and bean pods alone or in combination appears to be required for vitellogenesis (36, 51, 64). Inflorescences furnish nutrients in the form of pollen; however, volatile, non-nutritive compounds play a role as well (51). While a nutritional contribution of the pods has not been ruled out, the effect of pods is specific to host cowpeas; presence of non-host legume pods has little or no effect on maturation (64).

The only other beetle family subject to detailed analysis of host resource effects on ovarian development is the carrion-feeding Silphidae. Both the behavioral context and underlying physiological mechanism by which the oviposition resource enhances ovarian development have been described for the burying beetle, *Nicrophorous tomentosus*. During a 15- to 20-day period of post-emergence feeding on carrion, ovaries increase in mass to a "resting stage" characterized by low titres of juvenile hormone (JH) (114). In this stage, females forage for an oviposition host (a small vertebrate carcass). Upon discovery of a carcass, beetles make dorso-ventral circuits around the carcass. During transits, beetles remove hair or feathers from the carcass, deposit anal secretions that delay decomposition, and shape the carcass into a ball. The prepared carcass is eventually buried in a chamber around which eggs are laid.

Within 24 hours of carcass discovery, ovarian mass increases two- or threefold (98, 114, 115). To determine whether feeding on the carcass was sufficient to stimulate oogenesis, some females were provided with a mouse carcass and some with pieces of a mouse large enough to feed upon but not large enough to prepare and bury. Proportionately fewer females in the mouse-parts treatment developed ovaries, relative to the whole-mouse treatment, suggesting that feeding alone does not facilitate oogenesis. Evidently, carcass preparation behavior is a prerequisite for ovarian maturation.

The precise stage of host use contributing to stimulation of oogenesis was resolved by assays of JH titres (115). Within 10 minutes of discovery, JH titre approximately doubled and rose to a peak over the subsequent 50 minutes. During that first hour, females palpated, lifted, and circumambulated the carcass, and occasionally plowed through the soil around the carcass. No feeding on the carcass was observed during that period, excluding nutritional input from the host as a factor in stimulation of oogenesis (114, 115).

The primary significance of the burying beetle work lies in its demonstration that the process of host examination is associated with a rapid rise in JH titre that initiates vitellogenesis and consequently egg production (114). No other study of host-specific insects has clearly demonstrated a role for this so-called "master regulator" (sensu 97) in mediating a sensory effect of the host resource on ovarian development. In other respects, this work demonstrates the difficulty of studying host sensory effects on egg production. Precisely what stimuli facilitate oogenesis is not known and could be difficult to determine. Is it host stimuli detected during carcass assessment, proprioceptive input associated with assessment, or something else, perhaps even input from anal secretions deposited on the carcass? If host stimuli are directly involved, during execution of which behavior patterns (palpation, lifting, circumambulation, or plowing) are those stimuli detected? Palpation seems an obvious candidate and ablation studies might furnish answers.

Order Diptera

Both preoviposition period length and number of ovariole cycles in the frit fly, *Oscinella frit*, were influenced by exposure to hosts (43). The number of cycles, computed as egg load divided by the number of ovarioles, for females exposed to hosts was higher than the number of cycles in females held with non-hosts or no plants at all. Among host species, oats stimulated oogenesis more than other graminaceous plants.

Most work relating to host sensory effects on ovarian development involves fruit flies of the family Tephritidae. The difficulty of distinguishing sensory effects of the host resource on ovarian development from nutritional effects is illustrated in work on the olive fly, *Bactrocera oleae*. In field cage assays (31, 32), olives enhanced development of the first clutch, but the effect was attributed to adult feeding on olive juice. It was proposed that flies imbibe bacteria on the olive surface that liberate nutrients in the gut and thereby promote ovarian development (32, 61).

Evidence of direct sensory effects of host olives on oogenesis was put forward in two studies relating presence of wax domes (a kind of surrogate fruit) to high rates of egg maturation (35, 60). However, one study (35) used oviposition counts exclusively, possibly confounding direct sensory effects on ovarian development with indirect effects mediated through oviposition (60). Experimental design helped eliminate this problem in the second study, but results were mixed (60). In three assays, wax domes had a significant effect (though substantially weaker than that of real fruit) but in four assays there was no effect at all. In all assays, olives strongly promoted ovarian development.

Work on a tephritid fly, the walnut-infesting *Rhagoletis juglandis*, demonstrated that host fruit influenced egg maturation and ruled out both oviposition and host feeding as factors (3). In initial experiments, host fruit enhanced egg load; host foliage, by contrast, had no effect. A small number of eggs were laid in fruit during these manipulations, leaving open the possibility that egg maturation was facilitated by oviposition. This possibility was ruled out with the use of yellow plastic spheres that resembled fruit in shape and size. Females vigorously attempted to oviposit into these surrogate fruit, but because the plastic could not be penetrated with their ovipositors, eggs were never laid. Mean egg load of flies held with surrogate fruit was significantly higher than that of control flies and similar to that of flies held with real fruit. Because surrogate fruit provide nothing in the way of nutrition, oogenesis enhancement evidently does not require nutrients from host fruit. The possibility remains that host stimuli promote non-host–based nutrient uptake. In the absence of a protein source, no eggs are produced, regardless of the presence of surrogates (D Papaj, H Alonso-Pimentel, and C Nufio, unpublished data).

In *R. juglandis*, both shape and color play a role in the enhancement of oogenesis by host fruit (3). In one experiment, mean egg load of flies held with a yellow sphere was significantly greater than that of control flies or flies held with a yellow cube of like surface area. Either tactile or visual stimuli could account for the shape effect. In another experiment, mean egg load of flies held with green and yellow spheres (colors most similar to yellow-green walnut husks) was significantly higher than that of flies held without spheres. By contrast, mean egg load of flies held with blue, black, or red spheres was not significantly higher than that of flies held without spheres. Whether ovarian development is responding to hue, saturation, intensity, or some combination of these factors is unclear, but visual stimuli are almost certainly involved.

Finally, exposure to host stimuli both hastens onset of vitellogenesis and increases the proportion of oocytes undergoing vitellogenesis. A detailed morphological analysis (A Lachmann and D Papaj, unpublished data) revealed that host experience during the first gonotrophic cycle influenced maturation of each of the first two follicles in a given ovariole. Maturation of the second follicle in a given ovariole does not ordinarily begin until the first follicles in all ovarioles contain mature, chorionated eggs. Females provided with surrogate fruit commonly matured eggs in all first and many second follicles; in contrast, females deprived of access to a surrogate fruit almost never matured the full complement of first-follicle eggs and, evidently for this reason, rarely initiated vitellogenesis in second follicles.

Host Sensory Cuing in Relation to Host Quality and Availability

Whether efficacy of sensory cuing of ovarian development depends on host availability (beyond mere presence versus absence) is an open question. Manipulations of host availability have simply not been carried out in this context. However, data cited here indicate that efficacy of sensory cuing of ovarian development depends on host quality. First, effects of plant stimuli vary according to host status. Non-host stimuli have little or no effect. Second, different host species vary in their effect on ovarian development. Whether such effects correlate with either oviposition preference or juvenile performance is often unclear. Only in bruchid beetles is there clear evidence that sensory cuing is sensitive to variation in the suitability of the host for juvenile growth and survival.

The overall impact of host sensory input on egg maturation is difficult to gauge. Evidence for direct host sensory effects on ovarian maturation to date is restricted to the first gonotrophic cycle; effects in subsequent gonotrophic cycles (which may comprise the greater part of a female's lifetime fecundity) have not yet been distinguished from facilitative effects of oviposition.

CONDITIONS FAVORING OVARIAN RESPONSES TO HOST VARIATION

Sensory cuing of ovarian development by host resources constitutes reasonably strong evidence that ovarian development is matched, by design, to variation in the host environment. But is it good design? Why should ovarian development be forestalled until host stimuli are detected? At least three conditions favor adjustments in ovarian development in response to variation in the host resource. First, the host environment must be unpredictable over time and/or space (although, as later discussed, not too unpredictable). Second, there must be a trade-off, such that costs and benefits of ovarian development trade off against those of some other process contributing to fitness. Finally, host stimuli must be more useful indicators of the status of the host environment than other stimuli. Each condition is addressed in turn below.

Uncertainty of the Host Environment

For sensory cuing of ovarian development to be adaptive, the host environment must be unpredictable in time and/or space. To take an obvious example, if hosts were continuously available, timing of development could be regulated internally (or according to another extrinsic factor). Illustrative of the role of uncertainty in host cuing of ovarian development is a pattern of inter-population variation in the bruchid beetle *A. obtectus* (49, 50). Females of a strain originating from higher elevations (2000 m) develop more slowly and depend more on the presence of host pods and seeds for onset of vitellogenesis than females originating from lower elevations (1000 m). The greater reliance of high-elevation females on host stimuli is evidently a consequence of a higher level of uncertainty in host conditions. At low elevations, hosts are continuously and reliably available; at high elevations, there is considerable year-to-year variability in host availability. An analogous pattern of population variation was reported in the parasitoid *A. tabida* (27).

Trade-Offs Associated with Ovarian Development

In order for host sensory cuing or other host effects on ovarian development to be adaptive, egg maturation must trade off against another process or set of processes important to fitness. Otherwise, all insects would have all eggs ready to lay at the time that reproductive competence is first achieved. The body of life history theory equipped to deal with these trade-offs is only now being brought to bear upon the details and diversity of ovarian development. At present, two kinds of trade-offs, survival-reproduction trade-offs and dispersal-reproduction trade-offs, appear to have special significance for ovarian dynamics in relation to the host resource.

Survival-Reproduction Trade-Offs

Pouzat et al (84) suggested that forestalling oogenesis until suitable host beans were available improved bruchid beetle survival during periods unfavorable for oviposition and so accounted for an observed effect of host stimuli on oogenesis. This proposition implies that investment in reproduction trades off against survival. Survival-reproduction trade-offs are common in insects (26, 27, 92). In *Asobara tabida* parasitoids, for example, females allowed to deposit eggs die sooner on average, indicative of a survival-reproduction trade-off. Egg production appears to involve a "borrowing" against lipid reserves that could otherwise be used to prolong longevity (26).

In a lady beetle, *Epilachna niponica*, a specialist herbivore on thistle, females turn egg production on and off according to the availability of larval food plant (77, 78). Oosorption increased seasonally as host availability declined, a change coincident with increased female survival. Oosorption also increased in response to habitat perturbation such as flooding, which reduced host availability. Increased female survival associated with oosorption demonstrably improved chances of future oviposition in a subsequent season, thereby increasing a female's lifetime reproductive success.

Precisely how reproduction affects survival is rarely known and may be complex. In the bruchid beetle *Callosobruchus maculatus*, two distinct survival trade-offs are associated with egg production (110). An early trade-off was conditional upon current diet and possibly related to depletion of energy reserves. A late trade-off was independent of diet and may reflect senescence. Both trade-offs depended on an interaction between early reproduction and nutritional state at the time of reproduction. High food availability at a time of intensive egg production abolished the early trade-off and mitigated the late trade-off.

Sensitivity of survival-reproduction trade-offs to diet can profoundly affect patterns of senescence in host-specific insects. Mediterranean fruit flies postpone senescence when deprived of protein, a nutrient required for egg production (14). The remaining life expectancy for a protein-deprived individual that had already lived 60 days, for example, was not detectably shorter than the remaining life expectancy for a deprived fly that had lived just 30 days. In effect, protein-deprived females entered a "waiting mode" during which they evidently minimized reproductive costs that would reduce survival. Once protein was furnished in the diet, females began producing eggs and remaining life expectancy rapidly declined. Curiously, the decline depended little on the length of time over which protein was deprived. This dual mode of aging presumably results in higher reproductive success than a single determinate mode that is independent of protein intake. It would be interesting to know if, in addition to protein, host stimuli facilitate transfer to a reproductive mode.

Finally, while survival-reproduction trade-offs must often reflect competition between egg maturation and survival-related processes for a limited pool of nutrients, there may be other causes. Reproduction may, for example, inflict some somatic insult that directly reduces survival (110). Only by manipulating both egg production and nutrient availability in assessments of trade-offs can these alternatives can be distinguished. This is rarely done and so little can be said regarding the extent to which nutrient allocation underlies reproduction-survival trade-offs.

Dispersal-Reproduction Trade-Offs

In an environment that is patchy with respect to host quality or availability, females may need to disperse to find hosts. Host sensory cuing of ovarian development or, more generally, any matching between ovarian development and host conditions, might be of particular benefit if producing eggs trades off against effective dispersal (3, 44). Dispersal-reproduction trade-offs are common in host-specific insects (52, 100, 101) and are central to the so-called oogenesis-flight syndrome, wherein oogenesis and dispersal tend to be separated in time over an insect's life (22).

In some cases, dispersal-reproduction trade-offs are a special case of survival-reproduction trade-offs. Long-winged forms of the cricket, *Modicogryllus confirmatus*, for example, have lower fecundity during early adult life than short-winged forms have (107). Forms did not differ in fat content at emergence, implying that the difference in fecundity is related to nutrient allocation in the adult (109). Moreover, longevity was strongly negatively correlated with egg production, suggesting that the apparent dispersal–egg production trade-off is actually a survival–egg production trade-off, in which long-winged females mainly allocated energy from food to flight muscle development and somatic maintenance, whereas short-winged females allocated it to egg production and longevity.

In other cases, dispersal-reproduction trade-offs are better framed as egg production–reproduction trade-offs. If egg maturation reduces an insect's ability to find a resource required for oviposition, for example, realized fecundity may paradoxically be diminished by egg maturation, even if survival is unaffected. Under such conditions, an insect may profit by forestalling egg maturation until hosts are found. How maturation of eggs hinders dispersal to host habitats is not well understood. Possibly, nutritional investment in eggs trades off against investment in flight muscle, as in crickets (74; see also 126). There may be a basic metabolic cost of maintaining an egg load (3, 7, 16, 90). Alternatively, dispersal effectiveness may be affected by the eggs themselves. In one of few measurements made of this type, a batch of mature eggs in a sarcophagid fly's ovaries reduced lift production during flight by 40% (10).

Host Stimuli As Useful Indicators Of Host Conditions

Phenotypic plasticity theory predicts that the adaptive value of tracking environmental change will depend on the occurrence of cues that are reliably correlated with changes in the environment (76). It is not a given that host stimuli, whether through ovipositional, nutritional, or sensory routes, are predictive of host conditions. Because making eggs takes time (16, 120), host stimuli will be predictive only if conditions at the time that oogenesis is stimulated are correlated with conditions at the time of egg laying readiness. Thus, while uncertainty in the host environment is a prerequisite for sensory cuing of ovarian development, the host environment must not be too uncertain. To paraphrase a remark made by Stephens in regard to learning, today's host environment must tell the insect something about tomorrow's host environment, or host-mediated cuing of ovarian development will have little value (104).

Even if host stimuli are the most reliable signals of changes in host conditions, they are not necessarily the most useful. For example, where host availability is highly seasonal, photoperiod or temperature changes may be nearly as reliable as host stimuli in predicting the appearance of hosts. Moreover, such cues may provide an earlier indication of host availability than stimuli from the host itself.

Where intraspecific competition places a premium on having eggs ready as soon as hosts are available, the early warning provided by photoperiod or temperature cues may make these cues more valuable than host cues themselves. In fact, photoperiod (1, 6, 55, 70, 108, 117) and temperature (6, 32, 117) commonly influence both reproductive diapause and termination of diapause in host-specific insects. Similarly, where mating occurs in close association with the host, mating rather than host input may regulate ovarian development. In the seedcorn fly, *D. platura*, host priming of egg production may reflect in part a facilitatory effect of the host on mating, which in turn stimulates egg production (124, 125).

Host stimuli must commonly interact with a variety of factors, including temperature, humidity, photoperiod, mating status, and social context (11, 19, 64, 82, 125). In *Anastrepha* flies, extensive systematic manipulations of host stimuli, adult diet, male pheromone, and social context revealed the existence of complex, higher-order interactions among factors in effects on egg production (63). Presumably, such interactions permit insects to cope better with uncertainty in an assortment of ecological factors than in the absence of such interactions.

Finally, certain host stimuli might be more useful than other host stimuli as predictors of host conditions. Long-range cues, such as host odor, may be less useful than short-range cues, especially if egg production trades off against the ability to travel to far-away hosts. Some host stimuli such as nutrients may act more directly on oogenesis than other stimuli and may be less costly in terms of sensory reception and neural processing. In this regard, JH itself, or a JH mimic that directly regulates onset of vitellogenesis, might be an ideal signal. In *Varroa* mites, the host honey bee's JH, ingested during host feeding, promoted oogenesis (39). Conceivably, specialist herbivores might exploit JH analogs found in host plants for the same purpose.

COEVOLUTIONARY ASPECTS

Plant Defenses Aimed at Ovarian Development

In examining the adaptive value of ovarian dynamics in relation to the host resource, the evolutionary interests of the host have so far been neglected. Yet hosts and their insect enemies are involved in a coevolutionary arms race, and hosts might conceivably evolve defenses that suppress oogenesis or facilitate oosorption in insects that attempt to consume the hosts (cf. 85). The evolutionary potential for such defenses is illustrated in a recently discovered pattern of resistance in Sitka spruce against white pine weevils (*Pissodes strobi*) (96 and references within). Feeding on resistant spruce genotypes causes ovarian regression in female weevils, an effect attributable neither to decreased feeding nor to reduced oviposition. Resistant and susceptible trees do not differ in content of phagostimulants and deterrents; moreover, JH treatment of weevils feeding on resistant leaders restores ovarian function. The possibility that feeding on resistant leaders blocks JH function suggests that *P. strobi* may be affected in diverse ways. Because JH regulates oogenesis and dispersal in a complementary manner (22), blocking of JH function may not only inhibit oogenesis but also increase insect dispersal from the resistant tree. Given the central role of JH in insect development (97), circumventing such a defense may pose an evolutionary challenge to the weevil.

Ovarian Dynamics and Host Specialization

Ovarian development has received surprisingly scant attention in discussions of the evolution of host specialization. Yet surely host specificity (sensu 102) could be as much a function of specificity in sensory cuing of oogenesis as it is a function of specificity in host acceptance behavior, for example. Estimates of host preference that do not take into account the effects of the host on ovarian development may be misleading (121).

Level of specialization is associated with the pattern of ovarian maturation in some tephritid fly species. In the specialist olive fly, *Bactrocera oleae*, the proportion of gravid females in a population declined as an early-season olive crop disappeared (32). This decline was due in part to suppression of oogenesis in newly emerging females and in part to ovarian regression (including both cessation of oogenesis and onset of oosorption) in older females. Ovarian maturation resumed as a late-summer crop of olives became available. Similarly, egg maturation in three Australian dacine specialist species, *Bactrocera cacuminatus*, *B. cucumis* and *B. jarvisi*, was retarded when females were deprived of hosts (28). The first two Australian species exhibited no decline in host selectivity in response to deprivation, whereas the third did.

Egg maturation in the highly polyphagous *B. tryoni*, in contrast, continued unabated and egg load increased steadily over a period of host deprivation, a pattern accompanied by a decline in selectivity (28). These species differences in ovarian development make sense intuitively. A specialist faced with a shortage of hosts can either disperse to find hosts or wait out the period of scarcity. In the face of a dispersal-reproduction and/or survival-reproduction trade-off, reducing egg production might improve dispersal and/or ensure survival. A generalist faced with a shortage of a host species, by contrast, can resort to use of alternative species. At emergence, the generalist would use the most preferred host. However, if that species was not available, eggs would accumulate, selectivity would decline, and less preferred species would be used.

Tritrophic Patterns in Sensory Cuing

Evidence of host sensory effects on ovarian development in entomophagous parasitoids is conspicuously lacking. Possibly, the absence of evidence on sensory cuing of ovarian development in parasitoids reflects a lack of attention to the appropriate cues. Signals that influence ovarian development might derive not from the host directly but, as is the case for signals involved in host finding (116,

122), from the host's resource (for example, a plant exploited by a parasitoid's host). The rationale for involvement of a third trophic level is derived from co-evolutionary theory. The host insect is expected to evolve ways to conceal itself from a parasitoid and thus its cues, though reliable, may not be highly detectable. The host's food plant is under no such selection and may, in fact, be selected to signal the herbivore's presence to natural enemies. Such signals, while highly detectable, are indirect indicators of host presence and may be less reliable than host cues; in terms of host finding, this reliability problem is mitigated by parasitoids through learning (116, 122). It is conceivable that a third trophic level plays a role in sensory cuing of ovarian development in parasitoids, and even that something analogous to learning is involved.

Alternatively, conditions favoring sensory cuing may be uncommon in parasitoids. Synovigenic parasitoids frequently host feed (29, 40, 56), perhaps precluding and at least obscuring sensory cuing. Moreover, use of yolk-deficient, hydropic eggs by some hymenopterous parasitoids (29) may significantly lower costs of egg maturation and thereby relax selection for its timing. Not surprisingly, parasitoids with hydropic eggs are often pro-ovigenic, oogenesis being completed by eclosion and, thus, well before females encounter hosts (29, 30).

CLOSING REMARKS

Biologists tend to put physiology in one box and behavior in another. Evidence of sensory cuing of oogenesis suggests that, in terms of mechanisms, insects blur the boundary between ovarian physiology and oviposition behavior. In this regard, one future direction for research is an in-depth comparison of host stimuli cuing ovarian development and host stimuli used in oviposition. Evidence to date suggests that stimuli affecting oogenesis overlap broadly with those eliciting oviposition. However, if each set of stimuli is tuned to the particular ecological context (for example, the phenological context) in which ovarian development or oviposition behavior is initiated, differences in stimuli might be subtle and require close examination to detect.

Another issue worthy of attention is the extent to which our understanding of dynamics in oviposition behavior (an area of intensive research in host-specific insects) depends on an understanding of ovarian dynamics. We need to intensify our study of mechanisms underlying ovarian dynamics. Such knowledge is worthwhile for its own sake but will also improve the sophistication with which models of oviposition behavior depict ovarian development.

We also need to give thought to the separate contributions that ovipositional and ovarian dynamics make to effective host use. One obvious difference between the two sets of dynamics is the scale of variation over time and space to which each is best suited to respond. Ovarian responses to environmental change, though dynamic and adaptive in form, are comparatively slow (16). As such, oviposition behavior can respond to a finer scale of variation in the host resource than can ovarian maturation. Other differences may be found in the details of trade-offs affecting oviposition on one hand and ovarian development on the other. The diversity in patterns of host use is likely to reflect in part a diversity of interactions between ovarian development and oviposition behavior.

ACKNOWLEDGMENTS

I am grateful to Paul Feeny and members of his lab group for comments made on an earlier draft. This review was completed while on sabbatical leave from the University of Arizona. I thank Ron Hoy for providing a quiet place to write. This review is dedicated to the late William Bell whose jointly authored review on oosorption (7) demonstrated an appreciation both of the mechanisms of ovarian development and of the ecological factors shaping those mechanisms.

Visit the Annual Reviews home page at www.AnnualReviews.org.

LITERATURE CITED

1. Adams AJ. 1985. The photoperiodic induction of ovarian diapause in the cabbage whitefly *Aleyrodes proletella* (Homoptera: Aleyrodidae). *J. Insect Physiol.* 31:693–700
2. Adams TS, Hintz AM, Pomonis JG. 1968. Oöstatic hormone production in houseflies *Musca domestica*, with developing ovaries. *J. Insect Physiol.* 14: 983–93
3. Alonso-Pimentel H, Korer JB, Nufio C, Papaj DR. 1998. Role of colour and shape stimuli in host-enhanced oogenesis in the walnut fly, *Rhagoletis juglandis*. *Physiol. Entomol.* 23:97–104
4. Arnault C, Loevenbruck C. 1986. Influence of host plant and larval diet on ovarian productivity in *Acrolepiopsis assectella* (Lepidoptera: Acrolepiidae). *Experientia* 42:448–50
5. Bai B, Smith SM. 1993. Effect of host availability on reproduction and survival of the parasitoid wasp *Trichogramma minutum*. *Ecol. Entomol.* 18:279–86
6. Barbosa P, Frongillo EAJ. 1979. Photoperiod and temperature influences on egg number in *Brachymeria intermedia* (Hymenoptera: Chalcididae), a pupal parasitoid of *Lymantria dispar* (Lepidoptera: Lymantriidae). *J. NY Entomol. Soc.* 87:175–80
7. Bell WJ, Bohm MK. 1975. Oosorption in insects. *Biol. Rev.* 50:373–96
8. Benz G. 1969. Influence of mating, insemination, and other factors on the oogenesis and oviposition in the moth *Zeiraphera diniana*. *J. Insect Physiol.* *15*:55–71
9. Bernays EA, Chapman RF. 1994. *Host-Plant Selection by Phytophagous Insects*. New York: Chapman & Hall. 312 pp.
10. Berrigan D. 1991. Lift production in the flesh fly, *Neobellieria (= Sarcophaga) bullata* Parker. *Funct. Ecol.* 5:448–56
11. Biemont JC, Jarry M. 1983. Inhibitory effects of the presence of a congener on the maturation of ovocytes in *Acanthoscelides obtectus* (Coleoptera: Bruchidae). *Can. J. Zool.* 61:2329–37
12. Bjorksten TA, Hoffmann AA. 1998. Separating the effects of experience, size, egg load, and genotype on host response in *Trichogramma* (Hymenoptera: Tricho-

grammatidae). *J. Insect Behav.* 11:129–48
13. Browne LB, van Gerwen ACM, Vogt WG. 1990. Readiness to lay in the blowfly *Lucilia cuprina* is unaffected by oviposition site-deprivation or egg-load. *Entomol. Exp. Appl.* 55:33–40
14. Carey JR, Liedo P, Muller HG, Wang JL, Vaupel JW. 1998. Dual modes of aging in Mediterranean fruit fly females. *Science* 281:996–98
15. Clift AD. 1971. Control of germarial activity and yolk deposition in non-terminal oocytes of the *Lucilia cuprina. J. Insect. Physiol.* 17:601–6
16. Collier TR. 1995. Adding physiological realism to dynamic state variable models of parasitoid host feeding. *Evol. Ecol.* 9:217–35
17. Courtney SP, Chen GK, Gardner A. 1989. A general model for individual host selection. *Oikos* 55:55–65
18. Craddock EM, Kambysellis MP. 1997. Adaptive radiation in the Hawaiian *Drosophila* (Diptera: Drosophilidae): Ecological and reproductive character analyses. *Pac. Sci.* 51:475–89
19. DeClercq P, Degheele D. 1997. Effects of mating status on body weight, oviposition, egg load, and predation in the predatory stinkbug *Podisus maculiventris* (Heteroptera: Pentatomidae). *Ann. Entomol. Soc. Am.* 90:121–27
20. Delobel A. 1989. Effect of groundnut pods (*Arachis hypogaea)* and imaginal feeding on oogenesis mating and oviposition in the seed beetle *Caryedon serratus. Entomol. Exp. Appl.* 52:281–90
21. Deseo KV. 1976. The oviposition of the Indian meal moth (*Plodia interpunctella* Hbn., Lep., Phyticidae) influenced by olfactory stimuli and antennectomy. In *The Host-Plant in Relation to Insect Behaviour and Reproduction*, ed. T Jermy, pp. 61–65. Budapest: Akadémiai KiadÓ
22. Dingle H. 1996. *Migration: The Biology of Life on the Move*. New York: Oxford Univ. Press. 474 pp.
23. Droste YC, Carde RT. 1992. Influence of host deprivation on egg load and oviposition behaviour of *Brachymeria intermedia* a parasitoid of gypsy moth. *Physiol. Entomol.* 17:230–34
24. Dunlap Pianka HL. 1979. Ovarian dynamics in *Heliconius* butterflies: correlations among daily oviposition rates, egg weights, and quantitative aspects of oogenesis. *J. Insect Physiol.* 25:741–50
25. Eckelbarger KJ. 1994. Diversity of metazoan ovaries and vitellogenic mechanisms: implications for life history theory. *Proc. Biol. Soc. Wash.* 107:193–218
26. Ellers J. 1996. Fat and eggs: an alternative method to measure the trade-off between survival and reproduction in insect parasitoids. *Neth. J. Zool.* 46:227–35
27. Ellers J, van Alphen JJM. 1997. Life history evolution in *Asobara tabida*: plasticity in allocation of fat reserves to survival and reproduction. *J. Evol. Biol.* 10:771–85
28. Fitt GP. 1986. The influence of a shortage of hosts on the specificity of oviposition behavior in species of *Dacus* (Diptera: Tephritidae). *Physiol. Entomol.* 11:133–44
29. Flanders SE. 1942. Oosorption and ovulation in relation to oviposition in the parasitic Hymenoptera. *Ann. Entomol. Soc. Am.* 35:251–66
30. Flanders SE. 1950. Regulation of ovulation and egg disposal in the parasitic Hymenoptera. *Can. Entomol.* 82:134–40
31. Fletcher BS, Kapatos ET. 1983. The influence of temperature, diet and olive fruits on the maturation rates of female olive flies at different times of the year. *Entomol. Exp. Appl.* 33:244–52
32. Fletcher BS, Pappas S, Kapatos ET. 1978. Changes in the ovaries of olive flies (*Dacus oleae* (Gmelin)) during the summer, and their relationship to tem-

perature, humidity and fruit availability. *Ecol. Entomol.* 3:99–107
33. Fletcher JP, Hughes JP, Harvey IF. 1994. Life expectancy and egg load affect oviposition decisions of a solitary parasitoid. *Proc. R. Soc. London Ser. B* 258:163–67
34. Fox CW. 1993. The influence of maternal age and mating frequency on egg size and offspring performance in *Callosobruchus maculatus* (Coleoptera: Bruchidae). *Oecologia* 96:139–46
35. Girolami V, Strapazzon A, Brian E. 1987. Host plant stimulation of oogenesis in *Dacus oleae* Gmel. *Int. Symp. Fruit Flies Econ. Importance, Rome, Italy, 7–10 April,* pp. 159–67
36. Glitho IA, Lenga APD, Huignard J. 1996. Changes in the responsiveness during two phases of diapause termination in *Bruchidius atrolineatus* Pic (Coleoptera: Bruchidae). *J. Insect Physiol.* 42:953–60
37. Godfray HCJ. 1994. *Parasitoids: Behavioral and Evolutionary Ecology.* Princeton, NJ: Princeton Univ. Press. 473 pp.
38. Gupta PD, Thorsteinson AJ. 1960. Food plant relationships of the diamondback moth (*Plutella maculipennis* (Curt.)). II. Sensory regulation of oviposition of the adult female. *Entomol. Exp. Appl.* 3:305–14
39. Haenel H, Koeniger N. 1986. Possible regulation of the reproduction of the honey bee mite *Varroa jacobsoni* (Mesostigmata: Acari) by a host's hormone, Juvenile Hormone III. *J. Insect Physiol.* 32:791–98
40. Heimpel GE, Collier TR. 1996. The evolution of host-feeding behaviour in insect parasitoids. *Biol. Rev.* 71:373–400
41. Heimpel GE, Rosenheim JA. 1998. Egg limitation in parasitoids: a review of the evidence and a case study. *Biol. Control* 11:160–68
42. Heimpel GE, Rosenheim JA, Mangel M. 1996. Egg limitation, host quality, and dynamic behavior by a parasitoid in the field. *Ecology* 77:2410–20
43. Hillyer RJ. 1965. Individual variation in ovary development and reproductive behaviour of *Oscinella frit* L. (Diptera). *Int. Congr. Entomol. Proc. 12th, 1964,* p. 390
44. Hillyer RJ, Thorsteinson AJ. 1969. The influence of the host plant or males on ovarian development or oviposition in the diamondback moth *Plutella maculipennis* (Curt.).*Can. J. Zool.* 47:805–16
45. Hopkins RJ, Ekbom B. 1996. Low oviposition stimuli reduce egg production in the pollen beetle *Meligethes aeneus. Physiol. Entomol.* 21:118–22
46. Hopkins RJ, Ekbom B. 1999. The pollen beetle, *Meligethes aeneus*, changes egg production rate to match host quality. *Oecologia* 120:274–78
47. Horton DR, Krysan JL. 1991. Host acceptance behavior of pear psylla (Homoptera: Psyllidae) affected by plant species, host deprivation, habituation, and egg load. *Ann. Entomol. Soc. Am.* 84:612–27
48. Hughes JP, Harvey IF, Hubbard SF. 1994. Host-searching behavior of *Venturia canescens* (Grav.) (Hymenoptera: Ichneumonidae): interference: the effect of mature egg load and prior behavior. *J. Insect Behav.* 7:433–54
49. Huignard J, Biemont JC. 1978. Comparison of four populations of *Acanthoscelides obtectus* (Coleoptera Bruchidae) from different Colombian ecosystems: assay of interpretation. *Oecologia* 35:307–18
50. Huignard J, Biemont JC. 1979. Vitellogenesis in *Acanthoscelides obtectus* (Coleoptera: Bruchidae). 2. The conditions of vitellogenesis in a strain from Colombia: comparative study and adaptive significance. *Int. J. Invertebr. Reprod.* 1:233–44
51. Huignard J, Germain JF, Monge JP. 1987. Influence of the inflorescences and pods of *Vigna unguiculata* on the termi-

nation of the reproductive diapause of *Bruchidius atrolineatus* Pic. In *Insects and Plants*, ed. V Labeyrie, G Fabres, D Lachaise, pp. 183–88. Dordrecht: Junk. 459 pp.
52. Isaacs R, Byrne DN. 1998. Aerial distribution, flight behaviour and eggload: their inter-relationship during dispersal by the sweetpotato whitefly. *J. Anim. Ecol.* 67:741–50
53. Iwasa Y, Suzuki Y, Matsuda H. 1984. Theory of oviposition strategy of parasitoids. *Theor. Popul. Biol.* 26:205–27
54. Jaenike J, Papaj DR. 1992. Behavioral plasticity and patterns of host use by insects. In *Insect Chemical Ecology*, ed. BD Roitberg, MB Isman, pp. 245–64. New York: Chapman & Hall. 359 pp.
55. James DG. 1983. Induction of reproductive dormancy in Australian monarch butterflies *Danaus plexippus*. *Aust. J. Zool.* 31:491–98
56. Jervis MA, Kidd NAC. 1986. Host-feeding strategies in Hymenopteran parasitoids. *Biol. Rev.* 61:395–434
57. Kambysellis MP, Heed WB. 1971. Studies of oogenesis in natural populations of Drosophilidae. I. Relation of ovarian development and ecological habitats of the Hawaiian species. *Am. Nat.* 105:31–49
58. Klowden MJ. 1997. Endocrine aspects of mosquito reproduction. *Arch. Insect Biochem. Physiol.* 35:491–512
59. Kostal V. 1993. Oogenesis and oviposition in the cabbage root fly, *Delia radicum* (Diptera: Anthomyiidae), influenced by food quality, mating, and host plant availability. *Eur. J. Entomol.* 90:137–47
60. Koveos DS, Tzanakakis ME. 1990. Effect of the presence of olive fruit on ovarian maturation in the olive fruit fly, *Dacus oleae*, under laboratory conditions. *Entomol. Exp. Appl.* 55:161–68
61. Koveos DS, Tzanakakis ME. 1993. Diapause aversion in the adult olive fruit fly through effects of the host fruit, bacteria, and adult diet. *Ann. Entomol. Soc. Am.* 86:668–73
62. Labeyrie V. 1960. Action de la presence de grains de haricot sur l'ovogénese d'*Acanthoscelides obtectus* Say. *C. R. Acad. Sci.* 250:2626–28
63. Lagunes Hernandez G. 1998. Efecto del contexto social, dieta previa, y la presencia de hospedero artificial en el desarrollo de los ovariaos de dos especies de *Anastrepha* (Diptera: Tephritidae) con diferentes estrategias de oviposicion. Thesis. Univ. Xalapa, Veracruz, Mex. 97 pp.
64. Lenga A, Glitho I, Huignard J. 1993. Interactions between photoperiod, relative humidity and host-plant cues on the reproductive diapause termination in *Bruchidius atrolineatus* Pic (Coleoptera Bruchidae). *Invertebr. Reprod. Dev.* 24:87–96
65. Lloyd DC. 1938. A study of some factors governing the choice of hosts and distribution of progeny by the chalcid *Ooencyrtus kuvanae* Howard. *Philos. Trans. R. Soc. London Ser. B* 229:275–322
66. Luft PA. 1993. Experience affects oviposition in *Goniozus nigrifemur* (Hymenoptera: Bethylidae). *Ann. Entomol. Soc. Am.* 86:497–505
67. Mangel M. 1989. An evolutionary interpretation of the "motivation to oviposit." *J. Evol. Biol.* 2:157–72
68. Mangel M, Heimpel GE. 1998. Reproductive senescence and dynamic oviposition behaviour in insects. *Evol. Ecol.* 12:871–79
69. Mangel M, Roitberg BD. 1989. Dynamic information and host acceptance by a tephritid fruit fly. *Ecol. Entomol.* 14:181–89
70. Martinson TE, Dennehy TJ. 1995. Influence of temperature-driven phenology and photoperiodic induction of reproductive diapause on population dynamics of *Erythroneura comes* (Homoptera: Cicadellidae). *Environ. Entomol.* 24:1504–14
71. Mayhew PJ. 1997. Adaptive patterns of

host-plant selection by phytophagous insects. *Oikos* 79:417–28
72. Michaud JP, Mackauer M. 1995. Oviposition behavior of *Monoctonus paulensis* (Hymenoptera: Aphidiidae): factors influencing reproductive allocation to hosts and host patches. *Ann. Entomol. Soc. Am.* 88:220–26
73. Minkenberg OPJM, Tatar M, Rosenheim JA. 1992. Egg load as a major source of variability in insect foraging and oviposition behavior. *Oikos* 65:134–42
74. Mole S, Zera AJ. 1994. Differential resource consumption obviates a potential flight-fecundity trade-off in the sand cricket (*Gryllus firmus*). *Funct. Ecol.* 8:573–80
75. Monge JP. 1983. Oviposition behavior of the bean weevil *Acanthoscelides obtectus* on artificial substrate impregnated with host plant extract *Phaseolus vulgaris*. *Biol. Behav.* 8:205–13
76. Moran NA. 1992. The evolutionary maintenance of alternative phenotypes. *Am. Nat.* 131:971–89
77. Ohgushi T. 1996. A reproductive tradeoff in an herbivorous lady beetle: egg resorption and female survival. *Oecologia* 106:345–51
78. Ohgushi T, Sawata H. 1985. Population equilibrium with respect to available food resource and its behavioral basis in an herbivorous lady beetle *Henosepilachna niponica*. *J. Anim. Ecol.* 54:781–96
78a. Papaj DR, Lewis AC, eds. 1993. *Insect Learning: Ecological and Evolutionary Perspectives.* New York: Chapman & Hall 398 pp.
79. Papaj DR, Messing R. 1998. Asymmetries in dynamical state as an explanation for resident advantage in contests. *Behaviour* 135:1013–30
80. Papaj DR, Rausher MD. 1983. Individual variation in host location by phytophagous insects. In *Herbivorous Insects: Host-Seeking Behavior and Mechanisms*, ed. S Ahmad, pp. 77–124. New York: Academic. 257 pp.
81. Pilson D, Rausher MD. 1988. Clutch size adjustment by a swallowtail butterfly. *Nature* 333:361–63
82. Pivnick KA, Jarvis BJ, Slater GP, Gillott C, Underhill EW. 1990. Attraction of the diamondback moth (Lepidoptera: Plutellidae) to volatiles of oriental mustard: the influence of age, sex and prior exposure to mates and host plants. *Environ. Entomol.* 19:704–9
83. Pouzat J. 1978. Host plant chemosensory influence on oogenesis in the bean weevil, *Acanthoscelides obtectus,* (Coleoptera: Bruchidae). *Entomol. Exp. Appl.* 24:401–8
84. Pouzat J, Bilal H, Nammour D, Pimbert M. 1989. A comparative study of the host plant's influence on the sex pheromone dynamics of three bruchid species. *Acta Oecol. Oecol. Gen.* 10:401–10
85. Prabhu VKK, John M. 1975. Ovarian development in juvenilized adult *Dysdercus cingulatus* affected by some plant extracts. *Entomol. Exp. Appl.* 18:87–95
86. Price PW. 1973. Reproductive strategies in parasitoid wasps. *Am. Nat.* 107:684–93
87. Prokopy RJ, Luna S, Cooley SV, Duan JJ. 1995. Combined influence of protein hunger and egg load on the resource foraging behavior of *Rhagoletis pomonella* flies (Diptera: Tephritidae). *Eur. J. Entomol.* 92:655–66
88. Prokopy RJ, Roitberg BD, Vargas RI. 1994. Effects of egg load on finding and acceptance of host fruit in *Ceratitis capitata* flies. *Physiol. Entomol.* 19:124–32
89. R'Kha S, Moreteau B, Coyne JA, David JR. 1997. Evolution of a lesser fitness trait: egg production in the specialist *Drosophila sechellia*. *Genet. Res.* 69:17–23
90. Rivero-Lynch A, Godfray HCJ. 1997. The dynamics of egg production, oviposition and resorption in a parasitoid wasp. *Funct. Ecol.* 11:184–88
91. Robert PC. 1971. Influence of host plant

and nonhost plants on oogenesis and egg laying of the sugar beet moth *Scrobipalpa ocellatella* (Lepidoptera: Gelechiidae). *Acta Phytopathol. Acad. Sci. Hung.* 6:235–41
92. Roitberg BD. 1989. The cost of reproduction in rosehip flies, *Rhagoletis basiola*: eggs are time. *Evol. Ecol.* 3:183–88
93. Rosenheim JA. 1996. An evolutionary argument for egg limitation. *Evolution* 50:2089–94
94. Rosenheim JA, Rosen D. 1991. Foraging and oviposition decisions in the parasitoid *Aphytis lingnanensis* distinguishing the influences of egg load and experience. *J. Anim. Ecol.* 60:873–94
95. Rosenheim JA, Rosen D. 1992. Influence of egg load and host size on host-feeding behaviour of the parasitoid *Aphytis lingnanensis*. *Ecol. Entomol.* 17:263–72
96. Sahota TS, Manville JF, Peet FG, White EE, Ibaraki AI, Nault JR. 1998. Resistance against white pine weevil: effects on weevil reproduction and host finding. *Can. Entomol.* 130:337–47
97. Schal C, Holbrook GL, Bachmann JAS, Sevala VL. 1997. Reproductive biology of the German cockroach, *Blattella germanica*: juvenile hormone as a pleiotropic master regulator. *Arch. Insect Biochem. Physiol.* 35:405–26
98. Scott MP, Traniello JFA. 1987. Behavioral cues trigger ovarian development in the burying beetle, *Nicrophorus tomentosus*. *J. Insect Physiol.* 33:693–96
99. Sevenster JG, Ellers J, Driessen G. 1998. An evolutionary argument for time limitation. *Evolution* 52:1241–44
100. Shirai Y. 1995. Longevity, flight ability and reproductive performance of the diamondback moth, *Plutella xylostella* (L.) (Lepidoptera: Yponomeutidae), related to adult body size. *Res. Popul. Ecol.* 37:269–77
101. Shirai Y. 1998. Laboratory evaluation of flight ability of the Oriental corn borer, *Ostrinia furnacalis* (Lepidoptera: Pyralidae). *Bull. Entomol. Res.* 88:327–33
102. Singer MC. 1982. Quantification of host preference by manipulation of oviposition behaviour in the butterfly *Euphydryas editha*. *Oecologia* 52:224–29
103. Sirot E, Ploye H, Bernstein C. 1997. State dependent superparasitism in a solitary parasitoid: egg load and survival. *Behav. Ecol.* 8:226–32
104. Stephens DW. 1993. Learning and behavioral ecology: incomplete information and environmental unpredictability. See Ref. 78a, pp. 195–218
105. Stephens DW, Krebs J. 1986. *Foraging Theory*. Princeton, NJ: Princeton Univ. Press. 274 pp.
106. Strong DR, Lawton JH, Southwood TRE. 1984. *Insects on Plants. Community Patterns and Mechanisms*. Oxford: Blackwell. 313 pp.
107. Tanaka S. 1993. Allocation of resources to egg production and flight muscle development in a wing dimorphic cricket, *Modicogryllus confirmatus*. *J. Insect Physiol.* 39:493–98
108. Tanaka S, Okuda T. 1996. Effects of photoperiod on sexual maturation, fat content and respiration rate in adult *Locusta migratoria*. *Jpn. J. Entomol.* 64:420–28
109. Tanaka S, Suzuki Y. 1998. Physiological trade-offs between reproduction flight capability and longevity in a wing-dimorphic cricket, *Modicogryllus confirmatus*. *J. Insect Physiol.* 44:121–29
110. Tatar M, Carey JR. 1995. Nutrition mediates reproductive trade-offs with age-specific mortality in the beetle *Callosobruchus maculatus*. *Ecology* 76:2066–73
111. Terkanian B. 1993. Effect of host deprivation on egg quality, egg load, and oviposition in a solitary parasitoid, *Chetogena edwardsii* (Diptera: Tachinidae). *J. Insect Behav.* 6:699–713
112. Thibout E. 1974. Influences respectives de la plante-hote et de la copulation sur la longevite, la ponte, la production ovar-

ienne et la fertilite des femelles d'*Acrolepia assectella* Zell. (Lepidoptera: Plutellidae). *Ann. Zool. Ecol. Anim.* 6:81–96

113. Traynier RMM. 1984. Influence of plants and adult food on the fecundity of the potato moth, *Phthorimaea operculella. Entomol. Exp. Appl.* 33:145–54
114. Trumbo ST. 1997. Juvenile hormone-mediated reproduction in burying beetles: From behavior to physiology. *Arch. Insect Biochem. Physiol.* 35:479–90
115. Trumbo ST, Borst DW, Robinson GE. 1995. Rapid elevation of juvenile hormone titer during behavioral assessment of the breeding resource by the burying beetle, *Nicophorus orbicollis. J. Insect Physiol.* 41:535–43
116. Turlings TCJ, Wackers FL, Vet LEM, Lewis WJ, Tumlinson JH. 1993. Learning of host-finding cues by hymenopterous parasitoids. See Ref. 78a, pp. 51–78
117. Tzanakakis ME, Koveos DS. 1986. Inhibition of ovarian maturation in the olive fruit fly, *Dacus oleae* (Diptera: Tephritidae), under long photophase and increase of temperature *Ann. Entomol. Soc. Am.* 79:15–18
118. van Lenteren JC, van Vianen A, Gast HF, Kortenhoff A. 1987. The parasite-host relationship between *Encarsia formosa* Gahan (Hymenoptera: Aphelinidae) and *Trialeurodes vaporariorum* Westwood (Homoptera: Aleyrodidae) XVI. Food effects on oogenesis oviposition life-span and fecundity of *Encarsia formosa* and other hymenopterous parasites. *J. Appl. Entomol.* 103:69–84
119. van Randen EJ, Roitberg BD. 1996. The effect of egg load on superparasitism by the snowberry fly. *Entomol. Exp. Appl.* 79:241–45
120. van Vianen A, van Lenteren JC. 1986. The parasite-host relationship between *Encarsia formosa* (Hymenoptera, Aphelinidae) and *Trialeurodes vaporariorum* (Homoptera, Alyerodidae). *J. Appl. Entomol.* 102:130–39
121. Veenstra KH, Byrne DN. 1998. Effects of starvation and oviposition activity on the reproductive physiology of the sweet potato whitefly, *Bemisia tabaci. Physiol. Entomol.* 23:62–68
122. Vet LEM, Wackers FL, Dicke M. 1991. How to hunt for hiding hosts: the reliability-detectability problem in foraging parasitoids. *Neth. J. Zool.* 41:202–13
123. Vourkassovitch P. 1949. Facteurs conditionnels de la ponte chez *Acanthoscelides obtectus* Say. *Bull. Mus. Hist. Natl. Belgrade,* B(1–2):224–34
124. Weston PA, Keller JE, Miller JR. 1992. Ovipositional stimulus deprivation and its effect on lifetime fecundity of *Delia antiqua* (Meigen) (Diptera: Anthomyiidae). *Environ. Entomol.* 21:560–65
125. Weston PA, Miller JR. 1987. Influence of ovipositional resources quality on fecundity of the seedcorn fly (Diptera: Anthomyiidae). *Environ. Entomol.* 16:400–4
126. Wheeler D. 1996. The role of nourishment in oogenesis. *Annu. Rev. Entomol.* 41:407–31
127. Wigglesworth VB. 1970. *Insect Hormones.* San Francisco: Freeman. 159 pp.
128. Yin C-M, Stoffolano JG Jr. 1997. Juvenile hormone regulation of reproduction in the cyclorrhaphous diptera with emphasis on oogenesis. *Arch. Insect Biochem. Physiol.* 35:513–37

Annu. Rev. Entomol. 2000. 48:449–466

CYCLODIENE INSECTICIDE RESISTANCE: From Molecular to Population Genetics

Richard H. ffrench-Constant[1], Nicola Anthony[2], Kate Aronstein[2], Thomas Rocheleau[2], and Geoff Stilwell[2]

[1]Department of Biology and Biochemistry, University of Bath, Bath BA2 7AY, United Kingdom: e-mail: bssrfc@bath.ac.uk

[2]Department of Entomology, University of Wisconsin-Madison, Wisconsin 53706

Key Words insecticide resistance, cyclodienes, Resistance to dieldrin, GABA receptor

■ **Abstract** This review follows progress in the analysis of cyclodiene insecticide resistance from the initial isolation of the mutant, through cloning of the resistance gene, to an examination of the distribution of resistance alleles in natural populations. Emphasis is given to the use of a resistant *Drosophila* mutant as an entry point to cloning the associated γ-aminobutyric acid (GABA) receptor subunit gene, *Resistance to dieldrin*. Resistance is associated with replacements of a single amino acid (alanine302) in the chloride ion channel pore of the protein. Replacements of alanine302 not only directly affect the drug binding site but also allosterically destabilize the drug preferred conformation of the receptor. Resistance is thus conferred by a unique dual mechanism associated with alanine302, which is the *only* residue replaced in a wide range of different resistant insects. The underlying mutations appear either to have arisen once, or multiply, depending on the population biology of the pest insect. Although resistance frequencies decline in the absence of selection, resistance alleles can persist at relatively high frequency and may cause problems for compounds to which cross-resistance is observed, such as the novel fipronils.

INTRODUCTION

The molecular basis of insecticide resistance has recently been reviewed elsewhere (1). The purpose of the current review is therefore to focus on our progress in cyclodiene resistance over the past decade and highlight its importance as a model system for studying the genetics of pesticide resistance. In particular we emphasize the progression of work from the initial isolation of the field-collected *Drosophila* mutant, through gene cloning, analysis of the mutation, and conclude with a molecular analysis of pest insect populations. Although cyclodiene resistance is historically very widespread and in the past accounted for over 60% of reported cases of resistance (2), cyclodienes themselves have been largely with-

0066-4170/00/0107-0449/$14.00

drawn from use, and therefore, in relative terms, the overall frequency of resistance cases is declining. However, cyclodiene resistance still remains a model of target site–mediated resistance to an insecticide. Further, cyclodiene type insecticides, such as endosulfan, are still used to control multiply resistant pests including the whitefly. Perhaps more importantly, the cyclodiene target site itself represents a proven and attractive site for novel insecticides such as the fipronils. We must therefore look to previous examples of resistance at this site to avoid or overcome future *Resistance to dieldrin* (*Rdl*)–mediated crossresistance to novel compounds.

THE MUTANT

γ-Aminobutyric acid (GABA) is the major inhibitory neurotransmitter in both insects and vertebrates (3–5). Despite the prevalence of cyclodiene resistance and the extensive biochemical work suggesting that cyclodienes interact with GABA-gated chloride channels or GABA receptors (6), resistant pest insects provided little genetic access to the cloning of the resistance gene. To this end, resistant mutants of the genetic model *Drosophila melanogaster* were searched for in field populations in order to provide an entry point into the cloning of the associated gene (7–9). One advantage of using *Drosophila* is that the gene is cloned solely on positional localization of the resistance phenotype. Thus, no prior assumptions are made regarding the nature of the gene product or its homology to other cloned receptors.

Insecticide Resistance

Dieldrin resistant mutants of *Drosophila melanogaster* were found at relatively high frequencies (1%–10%) in orchard populations in upstate New York (10). Dieldrin resistance was semidominant (10), as it is in other pest insects, meaning that the dose–response curve of heterozygous flies (Rdl^R/Rdl^S or *R/S*) is clearly intermediate between that of homozygous resistant (Rdl^R/Rdl^R or *R/R*) or susceptible (Rdl^S/Rdl^S or *S/S*) flies. These dieldrin-resistant mutants displayed similar cross-resistance patterns to those found in other cyclodiene resistant pests (11), namely, high levels of resistance to dieldrin, with progressively lower levels of resistance to aldrin and endrin. Importantly, the resistant *Drosophila* also showed cross resistance to picrotoxinin (PTX). PTX is an important antagonist and probe (12) of the GABA subtype A ($GABA_A$) receptors of vertebrates, and cross resistance to this compound is typical of target-site mediated resistance to cyclodienes (13, 14). Finally, suction electrode recordings from the central nervous system of resistant third instar *Drosophila* larvae clearly revealed that this insensitivity to dieldrin and PTX was associated with the nervous system itself (15, 16).

Other Phenotypes

Further analysis of the resistant *Drosophila* mutants revealed several other phenotypes associated with the locus. First, resistant insects show temperature sensitive paralysis (17). That is to say, when placed in a water bath at 38°C, flies show a subsequent inability to take off from a substrate when returned to room temperature, whereas wildtype flies fly away immediately. This is similar to the temperature sensitivity of other nervous system mutants such as those of the voltage-gated sodium channel gene, *para*^{ts} (18), the site of action of DDT and pyrethroids. Interestingly, as is the case for insecticide resistance, the temperature-sensitive phenotype of *Rdl* is also semidominant. Second, following genetic deletion, loss of the *Rdl* locus was also shown to be lethal. Thus, flies that are homozygous for a deletion removing *Rdl* die as late embryos and do not appear to be able to hatch from the egg (19). Again, this is consistent with *Rdl* coding for an essential component of the insect nervous system.

THE GENE

Gene Mapping

Repeated backcrossing of the resistance gene into a susceptible background showed that resistance was associated with a single major locus mapping to the left arm of chromosome 3, at 26.5 cM (11). Subsequent deficiency mapping showed that resistance was uncovered by *Df(3L)29A6*, but not by the overlapping *Df(3L)AC1* (11). This immediately localized the gene to the 66F subregion of the polytene chromosome, which is the only region unique to *Df(3L)29A6*. The ability of the resistant allele to be uncovered by a deficiency is important, not only because it allowed for fine-scale mapping of the gene, but also because resistance over a deficiency (*R/-*, where - is the deficiency) is equivalent to homozygous resistance (*R/R*); thus, flies carrying only insensitive receptors are fully resistant, as they carry only one copy of the resistance allele.

Gene Cloning

The *Rdl* locus was cloned via a chromosomal walk across the 66F subregion in a cosmid vector (average insert size ~40 kb) (20). To identify the position of the locus within the 200 kb of DNA in this region, we made several new mutants disrupting *Rdl* using gamma-irradiation. These mutants were characterized by cytology and found to be either deficiencies deleting the locus or a number of different inversions with one breakpoint in the 66F region. By taking each of the cosmids and probing these against Southern blots of the new mutants we were able to localize most of the inversion breakpoints to a 10 kb *Eco*RI restriction fragment in one of the cosmids, number 6 (20). As these inversions all have one breakpoint that disrupts the *Rdl* gene, this cluster of breakpoints must mark the location of the resistance locus.

Gene Structure and Alternative Splicing

Subsequent screening of an embryonic cDNA library produced numerous *Rdl* cDNAs, many of which were either truncated, contained improperly processed introns, or appeared to contain different exons (20). Analysis of these cDNAs and their relative locations in genomic DNA revealed that the *Rdl* locus was large and complex. Of the nine exons within the open reading frame, exons 3 and 6 showed two alternative exon choices of equal size but different predicted amino acid sequence. These were termed alternative exons 3A and 3B, and exons 6C and 6D. Subsequent analysis of the RNA revealed that all four possible splice form combinations are found in the fly, i.e. 3A/6C, 3A/6D, 3B/6C, and 3B/6D (21). Beyond the complexities of alternative splicing, the *Rdl* locus is also large. The more recent definition of the transcription unit, by identifying the position of transcripts immediately upstream and downstream, has shown it to be over 50 kb in size. Further, the mature mRNA is again large: 8.8 kb long despite the presence of only 2 kb of open reading frame. The additional length of mRNA corresponds to 1.8 kb of 5′, and an exceptionally long 5 kb of 3′, flanking untranslated region.

Rescue of Phenotypes Associated with the Locus

To prove that the *Rdl* locus encodes a susceptible copy of the resistance gene, the susceptible phenotype was rescued by genetic transformation. This involves nesting the gene in a transposable element (P-element) vector and then inserting it at random into a different location in the genome. Because resistance over a deficiency (*R/-*) is fully resistant, it is possible to test whether inserting a copy of susceptibility (*S′*) can convert these flies to effective heterozygotes again (i.e. testing whether the transformant bearing *S′; R/-* flies are as susceptible as the normal heterozygous *R/S* flies). Transformation of cosmid 6 rescued insecticide susceptibility and also allowed larvae carrying the transgene to avoid the lethal phenotype and emerge from the egg.

Initially these experiments were carried out with the full 40 kb of cosmid 6. This construct gave convincing but only partial rescue, i.e. the dose response curves of the transformants *S′; R/-* were not as susceptible as normal *R/S* heterozygotes (20), suggesting that some essential elements of the gene were missing. Interestingly the same level of rescue could be achieved by fusing the 5′ flanking DNA from cosmid 6 onto an intronless cDNA (19). Subsequent *in situ* hybridization of the transgenic mRNA also showed that the correct pattern of expression was seen in the embryonic nervous system and the larval brain (22). Combination of different transgenes with two different alternative splice forms, however, did not appear to improve the rescue of insecticide susceptibility (19). Increasing the number of transgenes in an insect also has little effect; rather it is the overall ratio of resistant to susceptible receptors that determines the effective genotype (23). Therefore, it appears that the absence of much of the long 3′ untranslated region of the message from the transformation constructs is responsible for the incomplete rescue achieved. This implies that this long 3′ region may be necessary for mRNA stability or for correct trafficking of the receptor in the cell (22).

THE GABA RECEPTOR

Classification

As predicted from previous biochemical studies of the receptor for cyclodiene insecticides, the *Rdl* gene sequence predicts a protein with high similarity to GABA receptor subunits. However, it is important to stress that both the predicted RDL sequence (24), and the genomic organization (25) are equally divergent from any of the large array of vertebrate $GABA_A$ receptor subunit subtypes, and the amino acid sequence is in fact somewhat more similar to glycine receptor subunits. Therefore, although the pharmacology of the expressed receptor closely resembles that of the vertebrate ρ subtype of GABA receptor subunit (see below), based on sequence alone *Rdl* appears to belong to a novel class of GABA receptor subunits.

Distribution

With the cloned gene and antibodies against the protein, detailed analyses of the distribution of the *Rdl* message and protein in the nervous system were carried out. Both the message and protein are highly expressed in the central nervous system of the developing embryo (26). More recent studies have also detected expression of *Rdl* message in the peripheral nervous system (22). The RDL receptor subunit is present throughout much of the brain, with the optic lobes, ellipsoid body, fan-shaped body, the ventrolateral protocerebrum, and the glomeruli of the antennal lobes showing strong antibody staining (26). Interestingly, strong anti-RDL immunoreactivity is also seen in the pedunculus and lobes of the mushroom bodies of the blowfly, *Calliphora erythrocephala* (27). These structures consist mainly of Kenyon fibers, and the pattern of staining suggests that these are in fact a heterogeneous class of neurons, only part of which receive inhibitory GABAergic input from extrinsic elements. Overall the pattern of anti-RDL staining is consistent with the receptor subunit being expressed at many synapses in the nervous system (26) and shows a staining pattern similar to the synaptic vesicle protein synaptotagmin (28).

Pharmacology

Unlike many other GABA receptor subunits, *Rdl*-encoded subunits from either *D. melanogaster* or the yellow fever mosquito *Aedes aegypti* can readily form functional homomultimeric receptors in a variety of expression systems (29–31).

Stable expression of these receptors in insect cells therefore represents an attractive tool for novel insecticide screening (32). The pharmacology of these receptors is unlike that of either class of ionotropic vertebrate GABA receptors, but bears a striking similarity to many of the GABA receptors found in the insect nervous system itself. Thus, RDL homomultimers are unaffected by high concentrations of bicuculline (29, 33–35) and can be distinguished from both $GABA_A$ and $GABA_C$ receptors by the relative potency and efficacy of GABA analogs (36). Again, as with native receptors, the potency of barbiturate and steroid modulators on RDL (37) is less than that observed on $GABA_A$ receptors.

RDL homomultimers and native insect receptors do, however, differ in their response to benzodiazepines. Thus, the GABA response of RDL homomers (37) is less enhanced by 4′-chlorodiazepam than it is in native receptors. Further flunitrazepam, which potentiates the GABA response in many native receptors, has no effect on RDL homomultimers (34, 37). Vertebrate $GABA_A$ receptors require coexpression of several different subunits (α, β, and γ) for benzodiazepine sensitivity (38), suggesting that RDL in fact assembles with another as yet unidentified subunit(s) in the insect nervous system. This hypothesis is backed up by our observations that the conductance of the GABA-gated chloride channels generated by RDL homomultimers differs from the conductance of RDL containing channels in the nervous system (35), also suggesting that other subunits are coassembling with RDL.

Although there is substantial evidence that RDL must be coassembling with other receptor subunits, the role of the alternative splice forms of RDL itself remains relatively obscure. Interestingly, the regions of the N-terminal part of the protein they encode have been previously implicated in the agonist potency of other GABA receptors (24), and recent work has shown that the GABA EC50 for RDL splice forms A/C and B/D differs by as much as three-fold (24). Although this difference seems small, it is similar to the differences seen between different vertebrate $GABA_A$ receptor subunit isoforms, and may suggest a functional role for these splice forms in dictating agonist sensitivity or kinetics (24).

Expression with Other Subunits

The evidence that RDL coassembles with other unidentified subunits in the insect nervous system led to testing whether it can coassemble with other GABA receptor-like genes subsequently cloned from *Drosophila*. One of these, *ligand gated chloride ion channel homolog 3*, encodes a subunit LCCH3 (39) that is capable of coassembling with RDL in baculovirus-infected cells (35). Interestingly LCCH3 confers bicuculline sensitivity on the resulting heteromultimer (35), suggesting that this subunit either carries, or combines with RDL to form, a bicuculline binding site. Despite this heterologous coexpression, *in situ* hybridization and immunocytochemistry with an anti-LCCH3 antibody suggests that LCCH3 is largely expressed in the cell bodies of the central nervous system (40) and may therefore not coassemble with RDL *in vivo*. In conclusion, although

LCCH3 may contribute to the bicuculline sensitivity reported for some insect GABA receptors, it seems most likely that the RDL and LCCH3 subunits are not present in the same GABA receptors in the nervous system and that accompanying subunits remain to be identified.

THE MUTATION

Functional Expression and Conservation

Sequencing of the open reading frames of susceptible and resistant *D. melanogaster* identified a point mutation that predicted the replacement of alanine302 with a serine (29, 41). Site-directed mutagenesis of the susceptible cDNA and expression in *Xenopus* oocytes confirmed that this replacement conferred resistance to both dieldrin and PTX (29). This amino acid is predicted to lie in the second membrane-spanning region of the protein, a region thought to constitute the ion channel pore itself by analogy with the closely related nicotinic acetylcholine receptor (42). Following this observation, the same region was amplified by the polymerase chain reaction from a range of pest insects for which cyclodiene resistant strains were available. These included another fruitfly, *D. simulans*, the housefly *Musca domestica*, the cockroach *Blatella germanica*, the mosquito *Aedes aegypti*, the aphid *Myzus persicae*, the whiteflies *Bemisia tabaci* and *B. argentifolii* and the beetles *Hypothenemus hampei* and *Tribolium castaneum* (41, 43–49). In all of these species alanine302 (*Drosophila* numbering) is replaced either with a serine or, in the case of some *D. simulans* strains and some *M. persicae* clones, a glycine residue. This remarkable conservation of the resistance associated replacements across several insect orders led us to ask the question: What is so unique about alanine302?

Biophysical Analysis

To assess the role of alanine302 and its replacements in GABA receptor function, we carried out a detailed analysis of the biophysics of the RDL-containing receptor in neurons cultured from resistant and susceptible *Drosophila* larvae (50). Under patch clamp analysis, resistant and susceptible RDL-containing receptors show little difference in the shape of their GABA dose–response curves. They also showed different levels of resistance to the three different GABA receptor antagonists, [^{35}S]*t*-butylbicyclophosphorothionate or TBPS (10-fold), PTX (100-fold) and lindane (1000-fold). Inhibition increased with the concentration of each antagonist in a manner consistent with saturation of a single binding site. This suggested that the mutation was affecting the ability of these three different compounds to interact with the same binding site in the ion channel pore. A more detailed comparison of the channel kinetics of the resistant and susceptible receptors, however, revealed small differences in both their inward and outward channel conductances. Further, resistance was associated with longer open and shorter closed times of the channel, reflecting a net stabilization of the open state. More importantly, the mutation was associated with a marked reduction in the rate of GABA-induced desensitization and a net destabilization of the desensitized conformation of the receptor by a factor of 29 (50). At first these differences are difficult to reconcile with resistance; however, other workers have proposed that PTX, in fact, binds more tightly to the desensitized state of the GABAA receptor (51). We therefore proposed the model that alanine302 is unique in that it both occupies the drug-binding site directly and *also* allosterically destabilizes the drug-preferred desensitized state of the receptor (50, 52). This one amino acid, therefore, has *dual* effects on drug binding and only through this combination can sufficient resistance levels be reached. This hypothesis is supported by analysis of the second resistance-associated replacement (alanine302> glycine), which shows a correspondingly greater change in the rate of receptor desensitization (53).

POPULATION GENETICS

Although biophysical constraints on the GABA receptor appear to confine resistance to replacements of alanine302, this does not tell us how *often* the underlying mutations arise in natural populations. Are equivalent mutations occurring repeatedly at the same sites? Or are the mutations occurring once and then being spread through populations by migration and selection? These questions were originally brought to a focus following the publication of a report suggesting that single alleles of amplified esterase genes arose once and then spread globally in *Culex* mosquito populations (54). To address the question of the number of origins of insecticide resistance we examined the population genetics of *Rdl* in many different insects. These included insects with a range of different life history traits, including inbreeding and rapid biotype formation, and also widely differing dispersal capabilities. One of the central aims of this section is therefore to review how an insect's life history traits and dispersal capabilities can dictate the pattern of extant resistance alleles in its different populations (55).

PCR-Based Diagnostics for Resistance Detection

A range of detection techniques based on the polymerase chain reaction (56) have been employed to look for different resistance alleles in a number of different animals. These are briefly reviewed here so that the different population level studies in the following sections may be placed in a wider context. The first of these, PCR amplification of specific alleles, or PASA, relies on primers that selectively amplify either resistant or susceptible alleles (by matching the 3′ ends of the allele specific primer to each allele). This technique has been used to document the presence of one particular allele in a number of different insects (57) or insect populations (58) or to assess allele frequencies (59). While it has the advantage of being fast, it has the disadvantage that any undocumented novel alleles will not be detected.

The second technique is single stranded conformational polymorphism analysis or SSCP. In this technique the region containing the resistance-associated mutation is PCR amplified, denatured and run on a neutral polyacrylamide gel. This technique detects electrophoretic mobility shifts arising from differences in the nucleotide sequence of a given stretch of DNA. These mobility shifts are the result of conformational differences between different sequences, which in the case of different *Rdl* alleles, can differ by as little as a single nucleotide. Thus, individuals that are heterozygous for a specific resistance mutation display banding patterns for complementary strands of DNA from both susceptible and resistant alleles and can be distinguished from both homozygous-susceptible or resistant individuals (60). This technique is more laborious to run but has the advantage that novel mutations can be detected as different banding patterns. The final technique, PCR REN, relies again on PCR amplification of the region containing the resistance-associated mutation, but this time the PCR product is subsequently cut with a diagnostic restriction enzyme. This technique can be used only if the resistance–associated mutations disrupt or add a restriction site and is therefore of more limited use.

All three of these techniques have facilitated the analysis of large numbers of individual insects and, along with nucleotide sequencing of the underlying alleles, have allowed us to extensively document the nature and frequency of different *Rdl* alleles in different insect populations. In the case of the green peach aphid *Myzus persicae,* they have also facilitated the detection of an apparent duplication of the *Rdl* locus. In this species we found both alanine302> serine and alanine302> glycine replacements present at the first *Rdl* locus (49). However, we also found evidence for a second *Rdl*-like locus which *always* appears to carry a serine in position 302. Endosulfan resistance levels are only correlated with variation at the first locus, and the relevance of this second locus in resistance is therefore open to question (49). Two intriguing possibilities are that it either represents the result of a gene duplication event unique to aphids, or that it may correspond to one of the putative subunits presumed to coassemble with RDL from the pharmacological evidence discussed above. Whatever the role of the second locus, its presence was only detected by extensive SSCP and PASA analysis, which showed the apparent presence of more than two *Rdl*-like alleles in single aphid clones.

Drosophila Species

We examined 58 resistant and 122 susceptible lines of *Drosophila* collected from all over the globe (41). We used PASA to show that all the resistant *D. melanogaster* carried the same mutation predicting the alanine302> serine replacement, again raising the question as to whether this mutation had arisen once and spread worldwide, or had arisen independently in each different resistant population. This question was addressed by assessing linkage disequilibrium in restriction enzyme sites flanking the resistance-associated mutation. We identified a novel *Eco*RI site in the intron flanking exon 7, which contains the resistance-associated mutation. This linked site is only 700 bp away, and we used PCR REN to show that it was present in all except 3 of the 48 resistant strains examined for this site (41). This tight linkage of the *Eco*RI site with the resistance-associated mutation suggests that the mutation arose once in this particular genetic background and then spread globally. In contrast to *D. melanogaster*, where a single origin of the resistance allele is a formal possibility, resistance must have arisen at least twice in the closely related *D. simulans,* as we find both alanine302> serine and alanine302> glycine replacements in the same species (41).

The Coffee Berry Borer

If the highly mobile lifestyle of *D. melanogaster*, and its commensalism with man, has facilitated a spread of a limited number of resistance alleles, then how do the dramatically different life cycles of other insects affect the spread of resistance? The coffee berry borer, the major insect pest of coffee, provides an unusual example in which cyclodiene (endosulfan) resistance is found in the presence of inbreeding polygyny. In this species, mated females enter a coffee berry and lay a large single brood. This brood has a distorted sex ratio, with ten females to every one male. The males are dwarf and flightless and must mate with their sisters, never leaving the natal coffee berry. As there is usually one brood per bean, inbreeding is almost complete. In many species this type of lifestyle is associated with haplodiploidy (61), where the males are haploid and the females diploid. In *Hypothenemus hampei*, there is "functional haplo-diploidy" where the males are in fact diploid but the paternally derived set of chromosomes are condensed in males and are then eliminated during meiosis, leaving the males effectively haploid (62, 63). Paternally derived resistance alleles, present in the condensed chromosomes, also cannot be detected in males by either PASA or SSCP; males therefore also appear genotypically haploid with these diagnostics. In this system, males are phenotypically hemizygous for resistance (*R*/-) and are therefore fully resistant (equivalent to resistance over a deficiency in *Drosophila*, see above) allowing for a more rapid selection of resistance. More important, if resistance arises in one inbreeding line it could be trapped in this line by females who are effectively continuously backcrossing to their maternal genotypes (63).

To test this hypothesis, we examined the pattern of mitochondrial DNA sequences in global collections of *H. hampei* (64). The resulting phylogeny is consistent with the spread of only two different lines (mitochondrial clades) of coffee berry borer from their presumed source in East Africa, one westward to Central and Southern America and the other eastward toward East Asia. To date, resistance has been documented only from the South Pacific island of New Caledonia and appears to be present within a single inbreeding sub-line that is restricted to the island (46, 64). This is therefore a powerful example of how an insect life cycle, in this case through inbreeding, can constrain the spread of resistant alleles despite the obvious potential for resistant *H. hampei* to be spread around the world with the trade of coffee berries in the same way that they have colonized the coffee-growing countries of the world.

Whiteflies

Many other insect species show great potential for the evolution of new races or "biotypes." These include the *Bemisia* group of whiteflies and a variety of aphids. Following the recent appearance and rapid spread of the "B-biotype" of *Bemisia tabaci*, (also considered a new species, *B. argentifolii*) (65) this species was also considered. The hypothesis tested was to ascertain whether insecticide resistance was a unique driving force in the evolution and spread of this novel biotype. To examine this question, cyclodiene resistance–associated mutations in a range of B-biotype strains using PASA and SSCP were identified and were compared with those found in other non B-biotype strains (45). Interestingly, the B-biotypes were not uniquely cyclodiene resistant, suggesting that the widespread use of endosulfan to control this multiply resistant pest is not likely to be a major force accelerating the spread of this novel biotype. However, SSCP and nucleotide sequence analysis of the different *Rdl* alleles did show far more variation in the indigenous *Bemisia* populations than in those conforming to the B-biotype, supporting the concept of the recent origin and global spread of the B-biotype itself. In fact, cyclodiene resistance was present in most biotypes that are found in crops and was notably absent from those monophagous biotypes that feed exclusively on noncrop plants. Further, we found evidence for at least two different resistant *Rdl* alleles in the non B-biotypes that carried the same predicted resistance-associated replacement but differed in the underlying codon use (45). This lack of a correspondence between resistance and the success of the novel B-biotype has also been supported by data on the distribution of resistant alleles of another insecticide target site, insensitive acetylcholinesterase (66, 67).

The Red Flour Beetle

In view of how a range of life cycles can influence the behavior of resistance alleles in insect populations, the question arises: What are the expectations for a normal outbreeding population? To address this question, global collections of the red flour beetle *Tribolium castaneum*, which we screened for dieldrin resistance (68), were examined. A limited section of the *Rdl* gene from 23 different homozygous resistant strains of *T. castaneum* was sequenced, and the exon containing the resistance-associated mutation and part of the flanking intron were examined. Phylogenetic analysis of resistant and susceptible *Rdl* sequences suggests at least six different origins of the different resistance alleles. Detailed examination of the pattern of flanking variation in these alleles showed clear evidence for the susceptible precursors to extant resistance alleles. Thus, resistant and susceptible alleles could be found differing only by the presence or absence of the resistance-associated mutation itself (68), suggesting that resistance arose independently in a number of different susceptible genetic backgrounds. Further, the relationship between these different alleles could not be readily explained by recombination, as some of the flanking variation examined was as close as one or two nucleotides to the resistance mutation (68), too small a distance for a recent recombination event to have occurred within. This evidence for multiple origins of resistance alleles in an outbreeding species provides a striking example of parallel or convergent evolution. It also suggests that multiple origins of resistance-associated point mutations (as opposed to more infrequent events, such as gene duplications) may be the rule rather than the exception.

FITNESS IN THE ABSENCE OF PESTICIDE

Finally, after an examination of the number of origins of resistance alleles it is important to know how long these can be expected to persist in insect populations. This question centers on the widely assumed cost of resistance in the absence of insecticide selection. In the case of *Rdl,* we know a considerable amount about the different phenotypes associated with resistance. Thus, the difference in GABA receptor desensitization may not only help explain the resistance phenotype associated with the resistance-associated mutation but may also explain the observed temperature-sensitive paralysis of adult flies at 38°C (17). Therefore, although replacements of alanine302 are repeatedly selected for in the field, we can begin to correlate them with physiological changes in the RDL receptor that might be expected to lead to fitness differences in the field. Interestingly, although no fitness costs are apparent in population cage experiments with different *Drosophila* strains at fixed temperatures in the laboratory (69), fitness costs have been documented for *Rdl*-mediated resistance in the Australian sheep blowfly, *Lucilia cuprina*, both in the laboratory and the field (70). In the latter case, there is significant selection against resistance in the field when *L. cuprina* enters arrested development associated with overwintering diapause in the field (70). However, the precise link between this observation and any putative temperature sensitivity of *Rdl* in *L. cuprina* is currently not established.

Despite the evidence for fitness costs, resistance *can* persist in the absence of strong pesticide selection. Thus, resistance frequencies in *Drosophila* may be as high as 1%–10% (59) in the apparent absence of cyclodiene selection (although continued use of endosulfan may cause continued low levels of selection). In the absence of any clear evidence that *Rdl*-mediated resistance is likely to disappear from a population, there is considerable concern over the introduction of novel classes of insecticide to which cross resistance may be conferred. One such class is the recently developed fipronil insecticides (71). Cyclodiene-resistant insects and also functionally expressed alanine302 > serine RDL receptors display cross resistance to these compounds (72, 73). Further binding studies with the radioligand 4'-Ethynyl-4-*n*[2,3-^{3}H2]propylbicycloorthobenzoate ([H^3]EBOB) suggest that the fipronils occupy an overlapping binding site to the cyclodienes (74). Thus, the cyclodiene resistance–associated mutation completely abolishes [H^3]EBOB binding in resistant *Drosophila* head membranes (75, 76), showing that the radioligand occupies the cyclodiene binding site. Further, in competition assays a variety of fipronils can then displace [H^3]EBOB from its binding site, suggesting that the two sites must overlap. Given the rapid ability to select *Rdl* from low frequencies in the field (69), it will be interesting to see if the widespread use of fipronil selects for *Rdl*-mediated resistance again.

CONCLUSIONS

Although what has been learned cannot be readily extrapolated to all reported cases of cyclodiene resistance from the limited number of cyclodiene insects that have been examined, all have carried replacements of alanine302, with the exception of *Helicoverpa armigera,* which probably carries metabolic resistance to endosulfan (77). This suggests that these mutations are very widespread in insect populations. The associated GABA-receptor subunit is also widespread in insect nervous system and possesses many of the pharmacological properties of insect GABA receptors. Analysis of the biophysical basis of resistance reveals that a unique dual mechanism may confine resistance to replacements of a single amino acid, alanine302. However, examination of different insect populations shows that the mutations replacing this residue can occur repeatedly at the same site, providing a striking example of convergent evolution. Interestingly, resistance seems to be able to persist in the absence of extensive insecticide selection, representing a threat for novel insecticides interacting with the cyclodiene binding site, such as the fipronils.

Despite the extent of knowledge on *Rdl,* there are still several fundamental questions relating to the nature and behavior of target-site resistance that remain unanswered (78). Many of these relate to the inability to correlate the biophysical changes observed in the RDL receptor with any possible fitness costs associated with resistance. Such an understanding is central to the ability to predict how long resistance will persist in the absence of insecticide selection, and future work should therefore focus on correlating the biophysical phenotypes of the resistant receptor with fitness correlates in the whole animal.

ACKNOWLEDGMENTS

We would like to thank RT Roush, in whose laboratory the initial cloning of *Rdl* was carried out, all in the laboratory at the University of Wisconsin, Madison who have worked on *Rdl* over the last ten years, and C Nice for comments on the manuscript. Work was supported by grants from NIH, USDA, and Rhone-Poulenc.

Visit the Annual Reviews home page at www.AnnualReviews.org.

LITERATURE CITED

1. ffrench-Constant RH, Park Y, Feyereisen R. 1998. Molecular biology of insecticide resistance. In *Molecular Biology of the Toxic Response,* ed. A Puga, KB Wallace, pp. 533–51. Philadelphia: Taylor & Francis
2. Georghiou GP. 1969. The magnitude of the resistance problem. In *Pesticide Resistance: Strategies and Tactics for Management,* ed. National Academy of Sciences, pp. 14–43. Washington, DC: Natl. Acad. Press
3. Kuffler SW, Edwards C. 1965. Mechanisms of gamma aminobutyric acid (GABA) action and its relation to synaptic inhibition. *J. Neurophysiol.* 21:589–610
4. Usherwood PNR, Grundfest H. 1965. Peripheral inhibition in skeletal muscle of insects. *J. Neurophysiol.* 28:497–518
5. Otsuka M, Iversen LL, Hall ZW, Kravitz EA. 1966. Release of gamma-aminobutyric acid from inhibitory nerves of lobster. *Proc. Natl. Acad. Sci. USA* 56:1110–15
6. Eldefrawi ME, Abalis IM, Sherby SM, Eldefrawi AT. 1986. Neurotransmitter receptors of vertebrates and insects as targets for insecticides. In *Neuropharmacology and Pesticide Action,* ed. MG Ford, GG Lunt, RC Reay, PNR Usherwood, pp. 154–73, 319–91. Southampton, UK: Camelot
7. Wilson TG. 1988. *Drosophila melanogaster* (Diptera: Drosophilidae): a model insect for insecticide resistance studies. *J. Econ. Entomol.* 81:22–27
8. ffrench-Constant RH, Roush RT. 1992. Cloning of a locus associated with cyclodiene resistance in *Drosophila*: a model system in a model insect. In *Molecular Mechanisms of Insecticide Resistance: Diversity Among Insects,* ed. CA Mullin, JG Scott, pp. 90–98. Washington, DC: Am. Chem. Soc.
9. ffrench-Constant RH, Roush RT, Carino F. 1992. *Drosophila* as a tool for investigating the molecular genetics of insecticide resistance. In *Molecular Approaches to Pure and Applied Entomology,* ed. MJ Whitten, JG Oakeshott, pp. 1–37. Berlin: Springer-Verlag
10. ffrench-Constant RH, Roush RT, Mortlock D, Dively GP. 1990. Isolation of dieldrin resistance from field populations of *Drosophila melanogaster* (Diptera: Drosophilidae). *J. Econ. Entomol.* 83: 1733–37
11. ffrench-Constant RH, Roush RT. 1991. Gene mapping and cross-resistance in cyclodiene insecticide-resistant *Drosophila melanogaster* (Mg.). *Genet. Res.* 57:17–21
12. Leeb-Lundberg F, Olsen RW. 1980. Picrotoxin binding as a probe of the GABA postsynaptic membrane receptor-ionophore complex. In *Psychopharmacology and Biochemistry of Neurotransmitter Receptors,* ed. HI Yamamura, RW Olsen, E Usdin, pp. 593–606. New York: Elsevier
13. Kadous AA, Ghiasuddin SM, Matsumura F, Scott JG, Tanaka K. 1983. Difference in the picrotoxinin receptor between cyclodiene-resistant and susceptible strains of the German cockroach. *Pestic. Biochem. Physiol.* 19:157–66
14. Tanaka K. 1987. Mode of action of insecticidal compounds acting at inhibitory synapse. *J. Pestic. Sci.* 12:549–60
15. Bloomquist JR, ffrench-Constant RH, Roush RT. 1991. Excitation of central neurons by dieldrin and picrotoxinin in susceptible and resistant *Drosophila*

melanogaster (Meigen). *Pestic. Sci.* 32:463–69

16. Bloomquist JR, Roush RT, ffrench-Constant RH. 1992. Reduced neuronal sensitivity to dieldrin and picrotoxinin in a cyclodiene-resistant strain of *Drosophila melanogaster* (Meigen). *Arch. Insect Biochem. Physiol.* 19:17–25
17. ffrench-Constant RH, Steichen JC, Ode P. 1993. Cyclodiene insecticide resistance in *Drosophila melanogaster* (Meigen) is associated with a temperature sensitive phenotype. *Pestic. Biochem. Physiol.* 46:73–77
18. Wu CF, Ganetzky B. 1980. Genetic alteration of nerve membrane excitability in temperature-sensitive paralytic mutants of *Drosophila melanogaster. Nature* 286:814–16
19. Stilwell GE, Rocheleau T, ffrench-Constant RH. 1995. GABA receptor minigene rescues insecticide resistance phenotypes in *Drosophila. J. Mol. Biol.* 253:223–27
20. ffrench-Constant RH, Mortlock DP, Shaffer CD, MacIntyre RJ, Roush RT. 1991. Molecular cloning and transformation of cyclodiene resistance in *Drosophila*: an invertebrate GABAA receptor locus. *Proc. Natl. Acad. Sci. USA* 88:7209–13
21. ffrench-Constant RH, Rocheleau T. 1993. *Drosophila* γ-aminobutyric acid receptor gene *Rdl* shows extensive alternative splicing. *J. Neurochem.* 60:2323–26
22. Stilwell GE, ffrench-Constant RH. 1998. Transcriptional analysis of the *Drosophila* GABA receptor gene *Resistance to dieldrin. J. Neurobiol.* 468–84
23. ffrench-Constant RH, Aronstein K, Roush RT. 1992. Use of a P-element mediated germline transformant to study the effect of gene dosage in cyclodiene insecticide-resistant *Drosophila melanogaster* (Meigen). *Pestic. Biochem. Physiol.* 43:78–84
24. Hosie AM, Aronstein K, Sattelle DB, ffrench-Constant R. 1997. Molecular biology of insect neuronal GABA receptors. *Trends Neurosci.* 20:578–83
25. ffrench-Constant RH, Rocheleau T. 1992. Drosophila cyclodiene resistance gene shows conserved genomic organization with vertebrate γ-aminobutyric acid A receptors. *J. Neurochem.* 59:1562–65
26. Aronstein K, ffrench-Constant R. 1995. Immunocytochemistry of a novel GABA receptor subunit *Rdl* in *Drosophila melanogaster. Invertebr. Neurosci.* 1:25–31
27. Brotz TM, Bochenek B, Aronstein K, ffrench-Constant RH, Borst A. 1997. γ-Aminobutyric acid receptor distribution in the mushroom bodies of a fly (*Calliphora erythrocephala*): a functional subdivision of Kenyon cells. *J. Comp. Neurol.* 383:42–48
28. DiAntonio A, Burgess RW, Chin AC, Deitcher DL, Scheller RH, Schwarz TL. 1993. Identification and characterization of *Drosophila* genes for synaptic vesicle proteins. *J. Neurosci.* 13:4924–35
29. ffrench-Constant RH, Rocheleau TA, Steichen JC, Chalmers AE. 1993. A point mutation in a *Drosophila* GABA receptor confers insecticide resistance. *Nature* 363:449–51
30. Lee H-J, Rocheleau T, Zhang H-G, Jackson MB, ffrench-Constant RH. 1993. Expression of a *Drosophila* GABA receptor in a baculovirus insect cell system: functional expression of insecticide susceptible and resistant GABA receptors from the cyclodiene resistance gene *Rdl. FEBS Lett.* 335:315–18
31. Shotkoski F, Lee H-J, Zhang H-G, Jackson MB, ffrench-Constant RH. 1994. Functional expression of insecticide-resistant GABA receptors from the mosquito *Aedes aegypti. Insect Mol. Biol.* 3:283–87
32. Shotkoski F, Zhang H-G, Jackson MB, ffrench-Constant RH. 1996. Stable expression of insect GABA receptors in

insect cell lines: promoters for efficient expression of *Drosophila* and mosquito *Rdl* GABA receptors in stably transformed cell lines. *FEBS Lett.* 380:257–62
33. Buckingham SD, Hue B, Sattelle DB. 1994. Actions of bicuculline on cell body and neuropilar membranes of identified insect neurones. *J. Exp. Biol.* 186:235–44
34. Chen R, Belelli D, Lambert JL, Peters JA, Reyes A, Lan NC. 1994. Cloning and functional expression of a *Drosophila* γ-aminobutyric acid receptor. *Proc. Natl. Acad. Sci. USA* 91:6069–73
35. Zhang H-G, Lee H-J, Rocheleau T, ffrench-Constant RH, Jackson MB. 1995. Subunit composition determines picrotoxin and bicuculline sensitivity of *Drosophila* GABA receptors. *Mol. Pharmacol.* 48:835–40
36. Hosie AM, Sattelle DB. 1996. Agonist pharmacology of two recombinant *Drosophila* GABA receptors. *Br. J. Pharmacol.* 119:1577–85
37. Hosie AM, Sattelle DB. 1996. Allosteric modulation of an expressed homooligomeric GABA-gated chloride channel of *Drosophila melanogaster*. *Br. J. Pharmacol.* 117:1229–37
38. Sieghart W. 1995. Structure and pharmacology of γ-aminobutyric acidA receptor subtypes. *Pharmacol. Rev.* 47:181–234
39. Henderson JE, Soderlund DM, Knipple DC. 1993. Characterization of a putative g-aminobutyric acid (GABA) receptor β subunit gene from *Drosophila melanogaster*. *Biochem. Biophys. Res. Commun.* 193:474–82
40. Aronstein K, Rocheleau T, ffrench-Constant RH. 1996. Distribution of two GABA receptor-like subunits in the *Drosophila* CNS. *Invertebr. Neurosci.* 2:115–20
41. ffrench-Constant RH, Steichen J, Rocheleau TA, Aronstein K, Roush RT. 1993. A single-amino acid substitution in a γ-aminobutyric acid subtype A receptor locus associated with cyclodiene insecticide resistance in *Drosophila* populations. *Proc. Natl. Acad. Sci. USA* 90:1957–61
42. Leonard RJ, Labarca CG, Charnet P, Davidson N, Lester HA. 1988. Evidence that the M2 membrane-spanning region lines the ion channel pore of the nicotinic receptor. *Science* 242:1578–81
43. Thompson M, Steichen JC, ffrench-Constant RH. 1993. Conservation of cyclodiene insecticide resistance associated mutations in insects. *Insect Mol. Biol.* 2:149–54
44. Thompson M, Shotkoski F, ffrench-Constant R. 1993. Cloning and sequencing of the cyclodiene insecticide resistance gene from the yellow fever mosquito *Aedes aegypti*. *FEBS Lett.* 325:187–90
45. Anthony NM, Brown JK, Markham PG, ffrench-Constant RH. 1995. Molecular analysis of cyclodiene resistance–associated mutations among populations of the sweetpotato whitefly *Bemisia tabaci*. *Pestic. Biochem. Physiol.* 51:220–28
46. ffrench-Constant RH, Steichen JC, Brun LO. 1994. A molecular diagnostic for endosulfan insecticide resistance in the coffee berry borer *Hypothenemus hampei* (Coleoptera: Scolytidae). *Bull. Entomol. Res.* 84:11–16
47. Andreev D, Rocheleau T, Phillips TW, Beeman R, ffrench-Constant RH. 1994. A PCR diagnostic for cyclodiene insecticide resistance in the red flour beetle *Tribolium castaneum*. *Pestic. Sci.* 41:345–49
48. Kaku K, Matsumura F. 1994. Identification of the site of mutation within the M2 region of the GABA receptor of the cyclodiene-resistant German cockroach. *Comp. Biochem. Physiol. C.* 108:367–76
49. Anthony N, Unruh T, Ganser D, ffrench-Constant RH. 1998. Duplication of the *Rdl* GABA receptor in an insecticide-resistant aphid, *Myzus persicae*. *Mol. Gen. Genet.* 260:165–77

50. Zhang H-G, ffrench-Constant RH, Jackson MB. 1994. A unique amino acid of the *Drosophila* GABA receptor with influence on drug sensitivity by two mechanisms. *J. Physiol.* 479:65–75
51. Newland CF, Cull-Candy SG. 1992. On the mechanism of action of picrotoxin on GABA receptor channels in dissociated sympathetic neurones of the rat. *J. Physiol.* 447:191–213
52. ffrench-Constant RH, Zhang H-G, Jackson MB. 1995. Biophysical analysis of a single amino acid replacement in the resistance to dieldrin g-aminobutyric acid receptor: novel dual mechanism for cyclodiene insecticide resistance. In *Molecular Action of Insecticides on Ion Channels,* ed. JM Clark, pp. 192–204. San Diego, CA: Am. Chem. Soc.
53. ffrench-Constant RH, Pittendrigh B, Vaughan A. 1998. Why are there so few resistance-associated mutations in insecticide target genes? *Philos. Trans. R. Soc. London Ser. B* 353:1685–93
54. Raymond M, Callaghan A, Fort P, Pasteur N. 1991. Worldwide migration of amplified insecticide resistance genes in mosquitoes. *Nature* 350:151–53
55. ffrench-Constant RH, Anthony NM, Andreev D, Aronstein K. 1995. Single versus multiple origins of insecticide resistance: inferences from the cyclodiene resistance gene *Rdl.* In *Molecular Genetics and Evolution of Pesticide Resistance,* ed. TM Brown, pp. 106–16. Big Sky, MT: Am. Chem. Soc.
56. ffrench-Constant RH, Aronstein K, Anthony N, Coustau C. 1995. Polymerase chain reaction-based monitoring techniques for the detection of insecticide resistance-associated point mutations and their potential applications. *Pestic. Sci.* 43:195–200
57. ffrench-Constant RH, Steichen JC, Shotkoski F. 1994. Polymerase chain reaction diagnostic for cyclodiene insecticide resistance in the mosquito *Aedes aegypti. Med. Vet. Entomol.* 8:99–100
58. Steichen JC, ffrench-Constant RH. 1994. Amplification of specific cyclodiene insecticide resistance alleles by the polymerase chain reaction. *Pestic. Biochem. Physiol.* 48:1–7
59. Aronstein K, Ode P, ffrench-Constant RH. 1994. Direct comparison of PCR-based monitoring for cyclodiene resistance in *Drosophila* populations with insecticide bioassay. *Pestic. Biochem. Physiol.* 48:229–33
60. Coustau C, ffrench-Constant RH. 1995. Detection of cyclodiene insecticide resistance-associated mutations by single-stranded conformational polymorphism analysis. *Pestic. Sci.* 43:267–71
61. Kirkendall LR 1993. Ecology and evolution of biased sex ratios in bark and ambrosia beetles. In *Evolution and Diversity of Sex Ratio in Insects and Mites,* ed. DL Wrensch, MA Ebbert, pp. 235–345. New York: Chapman & Hall
62. Brun LO, Borsa P, Gaudichon V, Stuart JJ, Aronstein K, Coustau C, ffrench-Constant RH. 1995. 'Functional' haplodiploidy. *Nature* 374:506
63. Brun LO, Stuart J, Gaudichon V, Aronstein K, ffrench-Constant RH. 1995. Functional haplodiploidy: a novel mechanism for the spread of insecticide resistance in an important international insect pest. *Proc. Natl. Acad. Sci. USA* 92:9861–65
64. Andreev D, Breilid H, Kirkendall L, Brun LO, ffrench-Constant RH. 1998. Lack of nucleotide variability in a beetle pest. *Insect Mol. Biol.* 7:197–200
65. Perring TM, Cooper AD, Rodriguez RJ, Farrar CA, Bellows TS. 1993. Identification of a whitefly species by genomic and behavioral studies. *Science* 259:74–77
66. Byrne FJ, Devonshire AL. 1993. Insensitive acetylcholinesterase and esterase polymorphism in susceptible and resistant populations of the tobacco whitefly *Bemisia tabaci* (Genn.). *Pestic. Biochem. Physiol.* 45:34–42

67. Anthony NM, Brown JK, Feyereisen R, ffrench-Constant RH. 1998. Diagnosis and characterization of insecticide-insensitive acetylcholinesterase in three populations of the sweetpotato whitefly *Bemisia tabaci*. *Pestic. Sci.* 52:39–46
68. Andreev D, Kreitman M, Phillips TH, Beeman R, ffrench-Constant RH. 1999. Multiple origins of cyclodiene insecticide resistance in *Tribolium castaneum* (Coleoptera: tenebrionidae). *J. Mol. Evol.* 48:615–24
69. Aronstein A, Ode P, ffrench-Constant RH. 1995. PCR based monitoring of specific *Drosophila* (Diptera: Drosophilidae) cyclodiene resistance alleles in the presence and absence of selection. *Bull. Entomol. Res.* 85:5–9
70. McKenzie JA. 1996. *Ecological and Evolutionary Aspects of Insecticide Resistance*. Austin, TX: Landes. 185 pp.
71. Colliot F, Kukorowski KA, Hawkins DW, Roberts DA. 1992. Fipronil: a new soil and foliar broad spectrum insecticide. In *Brighton Crop Protection Conference: Pests and Diseases*, pp. 29–34. Farnham, UK: Br. Crop Prot. Counc.
72. Cole LM, Nicholson RA, Cassida JE. 1993. Action of phenylpyrazole insecticides at the GABA-gated chloride channel. *Pestic. Biochem. Physiol.* 46:47–54
73. Hosie AM, Baylis HA, Buckingham SD, Sattelle DB. 1995. Actions of the insecticide fipronil on dieldrin-sensitive and -resistant GABA receptors of *Drosophila melanogaster*. *Br. J. Pharmacol.* 115:909–12
74. Deng Y, Palmer CJ, Casida JE. 1993. House fly head GABA-gated chloride channel: four putative insecticide binding sites differentiated by [3H]EBOB and [35S]TPBS. *Pestic. Biochem. Physiol.* 47:98–112
75. Lee H-J, Zhang H-G, Jackson MB, ffrench-Constant RH. 1995. Binding and physiology of 4'-ethynyl-4-n-[2,3-3H2] propylbicycloorthobenzoate ([3H] EBOB) in cyclodiene resistant *Drosophila*. *Pestic. Biochem. Physiol.* 51:30–37
76. Cole LM, Roush RT, Casida JE. 1995. *Drosophila* GABA-gated chloride channel: modified [3H]EBOB binding site assoiated with Ala > Ser or Gly mutants of *Rdl* subunit. *Life Sci.* 56:757–65
77. Trowell S, Zinkovsky E, Daly J, Russell R, ffrench-Constant RH. 1994. DNA probes for key insecticide resistance genes-1. Endosulfan resistance in Australian *H. armigera*. In *Proc. 7th Aust. Cotton Conf.*, ed. D Swallow. Broadbeach: Aust. Cotton Growers Res. Dev. Assoc.
78. ffrench-Constant RH. 1999. Target site mediated insecticide resistance: What questions remain? *Insect Biochem. Mol. Biol.* 29:397–403

Annu. Rev. Entomol. 2000. 45:467–493

LIFE SYSTEMS OF POLYPHAGOUS ARTHROPOD PESTS IN TEMPORALLY UNSTABLE CROPPING SYSTEMS

George G. Kennedy and Nicholas P. Storer
Department of Entomology, North Carolina State University, Raleigh, North Carolina 27695-7630; e-mail: george_kennedy@ncsu.edu

Key Words host plant preference, annual cropping systems, pest management, shifting mosaic, insect mobility

■ **Abstract** Annual cropping systems consist of a shifting mosaic of habitats that vary through time in their availability and suitability to insect pests. Agroecosystem instability results from changes that occur within a season with crop planting, development, and harvest. Further instability results from continuous alterations in biotic and abiotic insect life system components and from agricultural inputs. Changes to agroecosystems occur across seasons with changing agricultural practices, changing cropping patterns, and technological innovations. Much of this instability is a result of events unconnected with pest management.

The abilities of polyphagous pest species to move among and utilize different habitat patches in response to changes in suitability enable the pests to exploit unstable cropping systems. These pest characteristics determine the location and timing of damaging populations. Habitat suitability is influenced by plant species and cultivar, crop phenology, and agricultural inputs. Pest movement is affected by a suite of intrinsic factors, such as population age structure and mobility, and extrinsic factors, including weather systems and habitat distribution.

The life systems of three selected polyphagous pests are presented to demonstrate how an understanding of such systems in agricultural ecosystems improves our ability to predict and hence manage these populations.

INTRODUCTION

There is a general recognition among ecologists that environmental heterogeneity, in both space and time, is a major factor influencing the evolution and population dynamics of animal species (12, 93, 189, 211, 216). Similarly, agricultural entomologists have long recognized that patterns of crop placement as well as crop phenology and crop management practices constitute a shifting mosaic through time that is a major determinant of the population dynamics of many important pest species (125, 157, 158, 199). In this review, we examine the prominent features of the life systems of polyphagous pest species in temporally unstable cropping systems and the ways in which host preference and environmental heterogeneity act to influence the dynamics of polyphagous pests in individual crop fields.

Although pest status is determined in large part by factors other than an organism's life history attributes, as a group, pests are more apt to have life history strategies that fall toward the r-end of the continuum from r to K strategists (187, 200, 226). They are also more likely to be polyphagous (226–228). In this review article, we consider a pest species to be polyphagous if its host range includes multiple crop and weed species within the agroecosystem, even if those species fall within the same plant family. Throughout our discussion, we treat crops as both host plants and habitats that vary in both suitability and predictability.

Clark et al (51) defined the life system of a population as "that part of the ecosystem that determines the existence, abundance and evolution of a particular population." It thus includes both intrinsic and extrinsic elements. The life system concept is useful because it breaks down the myriad of forces shaping a population's abundance and evolution into its individual components, each of which can be observed and measured. These components can be grouped into four broad categories: the properties of the subject population and its individual members, resources available to the subject population, inimical agents, and the abiotic environment. Because of space limitations, we restrict our discussion to selected components that appear to be primary determinants of the life systems of most major polyphagous pests in annual cropping systems. Thus, our emphasis is on the population consequences of pest mobility and the ability to discriminate among hosts and habitats varying in availability and suitability.

The life system of an organism is overlaid across a landscape comprising a dynamic mosaic of interacting elements (216). Principal among these elements are patches of habitat and potential habitat that change in relative and absolute suitability over time and that are separated by patches of non-habitat. The life systems of insects are heavily influenced by interactions among populations within this structure. Recent development of metapopulation concepts has promoted an understanding of the role of these interactions in determining the persistence of a metapopulation over a larger area (86). Populations in less favorable habitats produce fewer surviving offspring than populations in more favorable habitats, while least favorable habitats may act as sinks. Dispersal enables emigrants from source habitats to periodically or continuously repopulate sink habitats to enhance stability in the number of occupied patches and in the size of the total metapopulation (75). In agricultural systems, metapopulation stability is disrupted by changes in relative suitability of individual patches within and across seasons, and by changes in the patterns of insect movement among habitats as dictated by landscape structure. Recognition that metapopulation processes play a key role in pest life systems enhances our ability to predict the timing and extent of insect movement between crop fields and, hence, to manage these pests (105. 173).

The life system of a mobile, polyphagous pest can be understood only when viewed from a spatial scale sufficiently large to encompass the source of immigrants into the site of interest and the destination of emigrants from that site. Thus, the scale is influenced by the ability of the pest to move among habitats (133, 190). Polyphagous pest species vary greatly in this respect. Populations of some species regularly move rapidly over great distances on weather systems, whereas others may move great distances over the course of a season by tracking the seasonal and latitudinal progression in suitability of hosts (57, 59, 81, 84, 100, 137, 161, 177, 193, 200, 210). Still others may move more locally by tracking a sequence of temporarily suitable host plants (30, 34, 37, 139, 146, 163, 174).

Fields of crops serve as habitats for the pests that attack them, providing both a source of food and a place to live. Regions vary in both the total hectares planted and the proportion of the total landscape occupied by specific crops. Within a region, changes over time in the availability, suitability, and accessibility of these specific crops to individual pest species result in a constantly changing mosaic of habitat patches across the agricultural landscape that can have profound effects on the population dynamics of polyphagous pests (27, 176). These changes, which can be dramatic, occur both within seasons and over years.

HABITAT CHANGES WITHIN SEASONS

Crop Hosts

In regions with rigorous climates and short growing seasons, the number of pest generations per year and the range of crop planting dates are limited. For polyphagous pest species in such settings, the sequential exploitation of a series of crop and weed hosts during the growing season may not be a key component in the life systems. In contrast, where the climate is more mild and the growing season longer, crops are often planted over a period of several months, and sequential cropping is common. In such settings, the landscape is in a continuous state of change, and the rate of change is rapid relative to the generation time of many pests. Consequently, many polyphagous pests exploit a sequence of crop and weed hosts over the course of several generations during a single growing season.

Most polyphagous insect pests display distinct preferences for particular plant species, cultivars and plant growth stages. These preferences can lead to the concentration of pest populations in fields that represent the most preferred hosts and habitats.The potential of a crop to serve as a source of insects that subsequently invade other crops is influenced by seasonal changes in availability and suitability of the source and recipient crops in relation to the seasonal occurrence of migratory (or dispersing) life stages of the pest. Planting and harvest dates and cultivar maturity characteristics determine not only when a particular field is available to a pest, but also when the field is most attractive for colonization and most suitable for population growth. The suitability of individual habitat patches or crop fields for the development and survival of the offspring of invading adults and for subsequent population growth therefore determines the spatial population dynamics of the pest (e.g. 1, 36, 199, 203). Within a production region, for a given crop, there is a range of planting and harvest dates that may span several weeks, and a range of cultivar maturity rates. These factors create a mosaic of suitable habitat that changes through time and has a profound effect on the extent to which an individual crop field is colonized by a particular pest species. They also influence the size of the pest population that develops in the crop and subsequently migrates to other hosts (3, 6, 112, 116, 146, 147, 170, 201, 202, 205, 213, 217). Knowledge of this relationship has formed the basis for cultural controls that involve manipulating planting and harvest dates and selection of early maturing varieties to minimize damage.

The following examples illustrate how the host preferences of mobile polyphagous species interact with planting dates, harvest dates, and crop maturation to dramatically influence the distribution of pest populations within an agroecosystem.

The southern green stinkbug, *Nezara viridula*, has a broad host range that includes a number of crops, and the adults are strong fliers. It prefers plants in the fruiting stage and seems to display a strong preference for legumes (214). Stinkbug populations in soybean increase significantly during the pod development (R3 and R4) stages, with the greatest increases occurring during pod fill (R5 and R6) (171). The preference for reproductive stage soybean is so great that nearly 70% of the stinkbug (primarily *N. viridula* and *Euchistis servis*) population in a 260-ha soybean production area was found in 19.4 ha of early planted soybeans that reached the more attractive reproductive stage before the remainder of the crop. However, as the season progressed, the early plantings became less attractive and the adult stinkbugs dispersed to later planted soybean fields (138). The size of the stinkbug population in an individual soybean field is influenced by the degree of synchrony between the dispersing populations of adult stinkbugs and the preferred phenological stages of the crop. The timing of the latter is influenced by both planting date and cultivar (171).

Lygus spp. in the San Jaoquin Valley of California prefer alfalfa and safflower over cotton. In areas where alfalfa and cotton are both produced, large populations of *Lygus* develop in alfalfa, but the immature stages are killed and the adults forced to disperse when alfalfa is harvested for hay. These dispersing adults move into cotton, where they can cause significant economic losses, but move back into the alfalfa fields when new growth develops (172). Similarly, in areas where safflower and cotton are important crops, large *Lygus* populations develop in safflower in the spring, but as the crop matures and becomes less suitable as a host, the adults disperse and concentrate in nearby cotton that is preferred over the maturing safflower (140).

The area-wide dynamics of silverleaf whitefly, *Bemisia argentifolii,* populations in diverse agroecosystems are intimately tied to changes in host availability and suitability throughout the year (36). In the lower Rio Grande valley of Texas, *B. argentifolii* overwinters in low numbers on cruciferous crops. When these are harvested in spring, the adults migrate to melons, where the populations increase dramatically as temperatures increase. The populations migrate en masse from melons into cotton when the melon crop is harvested. When the cotton is defoliated in preparation for harvest, the populations migrate into fall vegetables (cucumbers, melons, and peppers) from which they migrate to and overwinter on cruciferous crops (162). Similar patterns are seen in the Imperial Valley of California (34) and in Arizona (62). In the Imperial Valley, proximity of cotton to melons and the area of melon production within 2.5 km of a cotton field were found to be important in determining the likelihood that a high *B. argentifolii* population would develop in the cotton. There was also a tendency for populations to be lower in early- than in late-planted cotton (34). To deal with the region-wide nature of the problem, a community-wide action program for *B. argentifolli* management has been developed in Arizona (62).

The suitability of potential crop hosts as oviposition sites, food sources, and habitats is influenced by many factors that change throughout the season. The nature and timing of these changes relative to those occurring in other fields of the same crop and of alternative crop hosts can profoundly influence the area-wide population dynamics of polyphagous pests.

In addition to planting and harvest dates, habitat suitability is dramatically affected by a large number of other biotic and abiotic factors, many of which are related to or influenced by agricultural inputs (102, 114, 116). Included among these factors are microclimate as affected by cultural practices such as planting density; tillage, fertilization, and irrigation (3, 70, 85, 90); host plant variety, phenological stage, and quality (71, 95, 115, 181); parasitoids, predators, and pathogens of the pest (15, 94); pathogens of the plant (16); pesticides and plant growth regulators (96); and weather (58). For those pests that generally pass only one of several generations per year in a particular crop, there is limited opportunity for a population to recover from the effects of a period of intense unsuitability such as that associated with the application of an insecticide that takes place after most immigration into the field has ended. Consequently, the affected field ceases to be a source of insects in the spatial dynamics of the pest population. In contrast, populations of pests that pass several generations within a crop field (e.g. spider mites, whiteflies, and aphids) have the opportunity to rebound when the habitat again becomes suitable (88). The affected field can return to acting as a source of insects. Ultimately, for pest species that can be controlled by remedial agents such as insecticides, the carrying capacity of a well-managed crop for each pest species is effectively determined by the economic threshold used by the farmers to trigger the application of remedial control measures (155). Whether the population reaches this carrying capacity depends on the level of immigration into the crop and the array of pest management, crop production, and naturally occurring factors that operate to influence the suitability of the crop as a host and habitat for the pest.

Uncultivated Hosts

Uncultivated hosts are important reservoirs of arthropod pests and their natural enemies (4, 221). An abundance of suitable uncultivated hosts during critical periods of the year enables populations of many polyphagous pests to persist and, in some instances, to increase when suitable crop hosts are unavailable. Weeds are important in the overall population dynamics of the major heliothine pests (66). In Tennessee, weed hosts are abundant throughout the cropping season and serve as a trap, protecting crop hosts from high infestations of *Heliothis virescens* and *Helicoverpa zea* (208). By contrast, in the Mississippi Delta region, a limited number of weed species growing primarily along field margins and road sides provide a nursery for first generation *H. virescens* and *H. zea* larvae in the spring before the major cultivated hosts are available (141, 142, 185, 198). The importance of these local populations, relative to that of immigrants from more distant sources, in the dynamics of *H. virescens* and *H. zea* populations developing on soybean and cotton later in the season has not been well quantified (66, 91). In Australia, a greater dependence on native host plants by *H. punctigera* than *H. armigera* is an important factor in their different seasonal dynamics and in the greater intensity of insecticide resistance in the latter (66, 67, 79, 82). Other polyphagous pests for which weed hosts are important include green peach aphid (13, 209, 212, 225), *Lygus* spp. (140, 222), green stinkbug and other pentatomid pests (152, 214, 224), thrips (5), two-spotted spider mite (23, 29, 33, 139), agromyzid leafminers (233), potato leafhopper (69), and the beet leafhopper (55). The contribution of pest populations developing on noncrop hosts is likely to vary with location and year in response to changes in the abundance and suitability of non-crop hosts, which may be strongly affected by weather patterns (e.g. 79) or changes in land use patterns.

PEST MOBILITY

The response of a pest to declining habitat suitability can be to diapause or to move (188, 212). Given the dynamic nature of habitat suitability and the range of different habitat types a polyphagous insect can exploit, the manner by which insects move among habitat patches is a crucial element in their life systems (103, 200). The extent to which any suitable crop field is colonized by migrants of a particular pest is determined by its accessibility to potential immigrants. Accessibility is a function of the abundance and distribution of the crop in the environment, as well as the ability of the pest to locate the crop (133). Accessibility is influenced by the nature of the landscape and the distance separating the crop field from the source of migrants (30, 34, 36).

Patterns of pest movement are determined by processes occurring at the source habitats (e.g. changes in host suitability, and in number and density of insects in dispersal stage), the distance and direction of the field of interest from the source habitats, and the suitability of the field relative to that of surrounding potential habitats that compete for the same pool of migrants (191). The spatial layout of habitats and the presence of connecting corridors can be important for insects with short dispersal distances (e.g. 216). For longer-range movements, dispersal relies on aerial movement, and direction is random with respect to the spatial arrangement of habitats. In these situations, connecting corridors may be of limited importance to local population dynamics (65). The spatial scale of dispersal of any particular population at any particular time depends on properties of the population itself (age structure, body size, motility, and mobility) and on the abiotic (strength and pattern of weather systems, photoperiod) and biotic (e.g. habitat quality) environments (159). These in turn define the absolute number of migrants, the time spent in transit, and the distance and direction moved (e.g. 37–39, 56, 153).

Immigration of large numbers of a polyphagous pest species into a crop field of transient suitability can override the population processes operating within the field. For example, natural enemies of *B. argentifolii* tend to be ineffective at reducing pest populations in diverse agroecosystems, at least in part because they are overwhelmed by massive numbers of immigrants from other crops (99). Consequently, predicting damaging populations of highly mobile polyphagous pest species requires knowledge of the pest's movement patterns (17, 122, 158).

The recent commercialization of transgenic, insecticidal crops has heightened awareness of the role of mobility in the evolution and spread of resistant insects (45, 132, 154, 167) and has stimulated field studies of insect movement patterns (44). Numerous advances have been made in our ability to measure insect movement patterns (20, 68, 78, 80, 92, 126, 130, 151, 160). Mathematical tools are being developed (215, 229) to investigate the consequences of movement on population processes, and to model spatiotemporal dynamics (19, 35, 73, 86, 154, 166). Recognition of the temporally dynamic nature of patch quality and the development of models in which habitat suitability fluctuates represent important contributions to this effort (83). Recently, spatiotemporal dynamics have been modeled for *Bemisia* spp. (36), *H. armigera* and *H. punctigera* (68), and *H. zea* (231).

AGROECOSYSTEM CHANGES ACROSS YEARS

In the preceding discussion, we have attempted to illustrate how the interrelationships among crops and among crop and weed hosts constitute key elements in the life systems of highly mobile polyphagous pests. Because of these interrelationships, changes in the agricultural landscape can profoundly alter the life systems of polyphagous pests.

Changes in Host Availability

Agricultural landscapes change dramatically over a period of years. Table 1 illustrates the magnitude of the largest year-to-year changes in hectares of wheat, maize, soybean, and cotton in three southern and three midwestern states in the United States during the 15-year period from 1980 through 1995. The minimum and maximum number of hectares of these crops in each state during that period, presented in Table 2, indicate dramatic changes in the agricultural landscape. These changes, which can differ greatly in magnitude and direction from one area to the next, represent major shifts in the availability of hosts and habitats for polyphagous pests. While such changes can have major effects on pest populations (27), pest and pest management–related factors rarely motivate changes to the agricultural landscape, although there are notable exceptions. In the case of cotton, the elimination of the boll weevil from large areas of the southeastern

TABLE 1 Maximum year-to-year changes in production area of four agronomic crops in each of six states during the period 1980 through 1995.

	Percent change in production area			
State	**Wheat**	**Maize**	**Soybean**	**Cotton**[a]
N. Carolina	65	20	−19	128
Georgia	74	44	−33	70
Mississippi	73	84	−18	52
Illinois	+53/−53[b]	36	−3	
Iowa	47	47	9	
Indiana	−34	27	−11	

[a]Cotton not planted in Illinois, Iowa or Indiana.
[b]Production area decreased 53% from 1984 to 1985 and increased 53% from 1985 to 1986.

TABLE 2 Minimum and maximum number of hectares planted to four agronomic crops in each of six states during the period 1980 through 1995.

	Minimum / Maximum number of hectares (×1000)			
State	**Wheat**	**Maize**	**Soybean**	**Cotton**[a]
N. Carolina	130 / 325	325 / 810	465 / 870	20 / 325
Georgia	140 / 595	160 / 650	130 / 850	49 / 610
Mississippi	73 / 445	40 / 250	730 / 1340	278 / 565
Illinois	345 / 830	3320 / 4750	3560 / 4010	
Iowa	20 / 53	3700 / 5830	3220 / 3850	
Indiana	275 / 565	1950 / 2610	1620 / 2190	

Source: U.S. Dep. Agr., National Agricultural Statistics Service 1980–1995
[a]Cotton not planted in Illinois, Iowa or Indiana.

United States contributed significantly to making cotton a more profitable alternative to other crops (14, 47).

Differences among production areas, changes within a production area over time in the relative abundance of different host crops, and widespread changes in planting and harvest dates of important host crops can alter the source/sink relationships between crops with respect to a pest population. This is illustrated by *H. zea* in several different agroecosystems. During the 1950s, maize production in the Carolinas involved the use of late-maturing, open-pollinated varieties that remained attractive and suitable as hosts for *H. zea* well into August. During the 1960s, the widespread adoption of early maturing maize hybrids that ceased to be suitable as a host in late July rendered most of the maize unavailable as host habitat for the third generation of *H. zea* and was associated with a dramatic increase in the severity of this pest on cotton (27) and on soybean (116), which remained suitable much later in the season. Similarly, in the Texas High Plains, *H. zea* was not a major pest of cotton until maize production expanded dramatically during the 1970s. Maize provided an excellent early season host for *H. zea*. The large populations of moths that developed on maize emigrated to cotton when the maize matured in late July and early August. The use of cotton varieties and production systems that result in early maturity of the cotton crop is now a key element in reducing *H. zea* populations and damage in cotton because such plants are less attractive for oviposition by *H. zea* moths and less suitable for larval establishment than the longer season varieties (27, 168).

Changes in Host Suitability

Changes in production or pest management practices in a crop that occupies a key role in a pest's life system have the potential to profoundly alter the pest's population dynamics (87, 224). Kennedy et al (111) used a modeling approach to illustrate how widespread planting of maize varieties having different modalities of resistance (antixenosis, tolerance, or antibiosis) to *H. zea* in North Carolina would affect the pest populations occurring in other crops late in the season.

Transgenic insecticidal crops represent a new dimension in host plant resistance. The first transgenic crops producing *Bacillus thuringiensis* delta endotoxins were commercialized in the United States in 1996. Bt cotton, toxic to *H. virescens* larvae that feed on it (21), and Bt maize, toxic to European corn borer larvae (117), have the potential to render a substantial proportion of these crops unsuitable as hosts for these pests. Crops that were once a nursery for these pests may soon become trap crops. Where the crop represents the major host plant for the target pest, as maize does for the European corn borer in the midwest and southeast, and cotton does for *H. virescens* in the mid-south, the entire life systems of the populations may be altered. Because these pests are polyphagous, populations on other crops and non-crop hosts and immigrants from sources outside of the agroecosystem will assume relatively greater importance in the pests' life systems. Further, because resistance management efforts require that farmers plant a portion of their corn and cotton crops to non-transgenic varieties (77), the pest production in these refugia may come to dominate the regional population dynamics. The spatial arrangement of refuge fields in relation to fields planted to transgenic crops, the extent of pest movement between these fields, and the agronomic and pest management practices used in managing the refugia will profoundly affect both the population dynamics of the pest and the evolution of resistance in the pest to the transgenic plants (9). The scarcity of data on the size of populations in crop and non-crop hosts, on the movement of moths within and among regions, and on the subdivision of populations makes the effects of the widespread deployment of insecticidal transgenic crops difficult to predict. Spatial models incorporating population dynamics and population genetics of resistance evolution are being developed (e.g. 2, 43, 154, 207) and are helping to identify key processes and interactions, as well as gaps in information on the life systems of targeted polyphagous pests.

LIFE SYSTEMS OF SELECTED POLYPHAGOUS PESTS

The following summary descriptions of the life systems of selected polyphagous pests are presented to illustrate the complex array of biological, ecological, and agricultural factors that influence the dynamics of polyphagous pest populations in diverse agroecosystems.

Two-Spotted Spider Mite

The highly polyphagous two-spotted spider mite, *Tetranychus urticae,* is a pest on a variety of annual and perennial crops. Its life history is characterized by a short developmental time, a high reproductive rate, and the ability to overwinter in a state of diapause in harsh environments or as actively reproducing populations in more mild climates. It is capable of dispersing by crawling and by wind. Wind dispersal involves a specific adaptive behavior (124, 133, 182). Most aerially dispersing mites appear to traverse relatively short distances (100 m or less), although long distance aerial dispersal certainly occurs (25, 30, 101, 139, 183). Natural enemies, weather, host plant quality, and pesticides are important in the population dynamics of *T. urticae* in agricultural systems (33).

In Chowan County, NC, a diversity of crops is grown commercially. Damaging spider mite populations frequently develop on most of the crops, but the timing and magnitude of infestations in specific crops are heavily influenced by events that occur in maize during the early part of the season. Mite populations overwinter on weed hosts growing along field margins and, in the spring, disperse into adjacent crops (29, 135). For a variety of reasons, including the use of soil-applied systemic insecticides in peanut and cotton for thrips control, significant spring populations of *T. urticae* do not develop in crops other than maize (23, 134, 135). The development of mite populations on maize is accelerated by the use of turbufos insecticide applied to the soil at planting for control of soil insect pests (136). Because the value of maize does not justify the application of additional pesticides, the mite populations are not controlled and they grow at a modest rate when the maize is in the whorl stage. The population growth rate increases significantly when the maize tassels (134), and is enhanced by hot, dry weather (33). The growing population begins to disperse on air currents, and dispersal continues until the maize can no longer support mite populations (143, 180, 183). The size of the mite population that develops in maize and subsequently disperses to other hosts is dramatically influenced by *Neozygites floridana*, a fungal pathogen of *T. urticae*. Epizootics of this pathogen, which typically occur following several days of cool, wet weather, can decimate the mite population (31, 184). If this occurs prior to dispersal from maize, few mites disperse; if it occurs during dispersal, the number of dispersing mites is reduced and many dispersers are infected (184).

Mites dispersing from maize infest peanut and a number of horticultural crops produced in the area (30, 135). Mite populations are aggressively controlled in the horticultural crops, and significant populations are not permitted to develop. The mite population thus becomes concentrated in peanut, with the highest infestations occurring in peanut fields adjacent to mite-infested maize (30, 135). Population growth in peanut is enhanced by hot dry conditions and by suppression of *N. floridana* by fungicides used to control foliar diseases. Insecticides applied to control thrips and lepidopterous pests may also enhance spider mite populations (24, 26, 32, 42). If populations reach damaging levels, acaricides are applied and the population is suppressed to very low levels. In late summer and early fall, mites disperse and become established on the weed hosts growing in the field margins where they overwinter (135). Because maize is frequently planted in rotation with peanut, maize is readily accessible to the overwintered mite populations the following spring (112).

Although maize occupies a central role in this life system, damaging mite populations still occur in areas where maize has been replaced by cotton. As the dynamics of the system have been altered, without maize to serve as a nursery, massive invasions of mites into peanut no longer occur. The damaging populations that develop in peanut originate primarily from weed hosts and generally occur later in the season (RL Brandenburg & GG Kennnedy, unpublished). In this situation, factors operating within the peanut crop assume greater importance in the overall life system.

European Corn Borer

The highly polyphagous European corn borer, *Ostrinia nubilalis*, is an important pest on a number of crops. In the United States, there are three known ecotypes that differ in voltinism (uni-, bi-, and multivoltine) and two strains that differ in the sex pheromone blend they use (175, 176). In some areas, more than one ecotype and sex pheromone strain occur sympatrically, complicating our ability to understand the life system(s) of local corn borer populations (46, 60, 61, 165, 186). The corn borer overwinters as a diapausing 5th instar in host stems (40). Overwinter survival is influenced by tillage practices (22, 40, 50, 218). However, there is little relationship between overwinter survival and population size the following season (50, 178). The timing of diapause and of spring emergence relative to the availability of suitable crop hosts influences the extent to which a crop or a planting of a crop is used as a host and, consequently, the size of the population that develops within it (63, 176, 196).

Regardless of the agroecosystem, the life systems of corn borer populations exhibit a number of common elements. Host and habitat preferences of ovipositing moths largely determine the distribution of the population within an agroecosystem. Host preference is related, in part, to plant size, phenological stage, and development of the crop canopy, all qualities that are influenced by crop, variety, planting date, and weather (2, 6, 48, 149, 170, 217). Suitability of the host and habitat are important as well (2, 120, 148).

A number of parasitoids and predators attack eggs and larvae but their impact varies depending on year, location, crop stage, and availability of alternative prey (7, 8, 52, 53, 72, 194, 230). The microsporidium *Nosema pyrausta* is a complex but important factor in the population dynamics of the corn borer (10, 98, 118, 123, 129, 179, 220, 232). Density-independent factors, including weather (49, 50, 64, 106, 178, 195) and host suitability, are important sources of larval mortality (18, 120). Several population and pest management models have been developed for the corn borer (2, 41, 119, 120, 148–150).

In areas where a diversity of crops is grown and the corn borer is multivoltine, the distribution of the population changes from one generation to the next as ovipositing moths abandon less preferred habitats in favor of those that are more preferred. The result is that the corn borer population utilizes a sequence of crop and weed hosts during the growing season (112, 196).

For example, in eastern North Carolina, the corn borer population is partitioned among a mosaic of suitable hosts at any point in time. That mosaic changes through the season (135). At the time moths of the overwintered generation oviposit in April and early May, potato and wheat are more attractive than maize; consequently, egg laying is concentrated in potato and wheat (6, 108, 217): Because this preference is related, at least in part, to plant size and development of the crop canopy, infestations of first generation larvae occur earlier and reach higher levels in early-planted than in late-planted potatoes (BA Nault & GG Kennedy, unpublished). During the 1970s and early 1980s, the European corn borer was routinely controlled in the potato crop with insecticides (110). Consequently, much of the potato production served as a trap crop for the first generation. However, as evidence mounted that corn borer infestations did not suppress potato yields (109, 143), control efforts on potato were reduced and the importance of potato as a suitable host for the first generation increased dramatically (217). Wheat also has become increasingly important as a host for the first generation as its production area has increased and production practices have changed. Both potato and wheat have become nursery crops for corn borer (108). When first generation moths emerge from potato and wheat in early June, maize is the most attractive crop for oviposition, and most of the population moves into maize fields (16, 217). The attractiveness of any given maize field for oviposition and its suitability for subsequent larval development is influenced by both variety and crop phenology and by local weather conditions, as described earlier.

By the time this second generation completes development in maize, the silks are dry and cotton is the only other abundant crop host in the system, although there is also a limited amount of late-planted maize. Moths abandon the maturing maize and partition among cotton, the limited number of fields of late-planted maize that are silking, and a variety of other crop and weed hosts (169, 170). Larval infestations in cotton are controlled by insecticides. Thus, the third generation larval population is concentrated in late-planted maize and in weed hosts. A variable portion of these larvae complete development to adult and give rise to a fourth generation; the remaining larvae die or enter diapause.

The relative size of the fourth generation is determined largely by spring temperatures that influence the time when moths of the overwintered generation oviposit and the rate at which first generation larvae develop. The critical photoperiod for induction of diapause in North Carolina occurs on August 15 and the incidence of diapause declines with larval age at the time of exposure to diapause-inducing conditions. By accelerating or delaying the timing of the first and subsequent generations, spring temperatures influence the proportion of third generation larvae that are exposed to diapause-inducing conditions and hence the proportion that complete development and produce a fourth generation (63). All of the resulting fourth generation larvae enter diapause. Cotton again acts as a trap crop in which the larvae are killed either by insecticides or by harvest. The majority of the population overwinters in late-planted maize and to a lesser extent weeds.

In contrast, in the less diverse maize/soybean agroecosystems of the Midwest, there are only one or two generations of corn borer each year, depending on the ecotype present (176), and maize is by far the predominant host throughout the season. Under those conditions the planting date, genotype, and phenological stage of maize in relation to the timing of moth flights likely contribute significantly to the distribution of the corn borer population among maize fields (2, 11, 48, 149). In both production areas maize occupies a central role in the life system of the corn borer. Consequently, major changes in the amount and distribution of maize within an agroecosystem, or in production or pest management practices that significantly affect the corn borer in maize (e.g. Bt transgenic maize), could significantly alter the life system and the region-wide population dynamics of the corn borer population within the agroecosystem.

Corn Earworm/Cotton Bollworm

H. zea is a highly mobile, highly polyphagous, multivoltine pest. In North America, it overwinters as diapausing pupae in the soil south of 40–45°N (89), but in summer can be found as far north as Ontario (145). It feeds on a variety of crops and common weeds (144, 197, 208), although maize is the preferred host (89, 104). Across its host range, *H. zea* prefers flowering and fruiting structures (107, 202, 208). The generation time is long compared with the period of peak host suitability, so moths often move away from their emergence sites to reproduce. This movement can cover a few miles or less (54), or many hundreds of miles (92, 151, 192). Across its range, the spatial and temporal dynamics of *H. zea* populations show significant differences that relate to physiographic and climatic factors, as well as to differences in agricultural practices and natural vegetation. Likewise, the dynamics within a region can change through time as a result of changes in land use and agricultural practices (27).

In eastern North Carolina, numerous host crops are grown and *H. zea* can complete three or four generations per year. It emerges from winter diapause in May (127) and infests flowering wild hosts and whorl stage maize (144), which act as a nursery and create a temporal bridge between spring emergence and the appearance of the most favored host, ear stage maize. First generation adults emerge coincident with silking in the majority of the maize crop. Silking maize acts as a trap crop, diverting ovipositing moths away from cotton and soybean. However, ear stage maize also acts as a nursery producing large numbers of second generation moths at a time when other hosts are more suitable for oviposition. In Virginia, this relationship is exploited by surveying infestation of maize ears in July to predict the timing and extent of subsequent local infestations of soybean fields (97). The mix of crops and wild hosts and the timings of events vary regionally. *H. zea* spring emergence, planting dates, and maturity times of crops, and temperature-mediated development rates of insects and plants all affect the succession of host usage and the source/sink relationships among crops (28, 116, 213).

Changes in production practices over time also affect the dynamics of this pest. (27, 164). The widespread planting of transgenic Bt maize and Bt cotton will dramatically affect the life systems of *H. zea* in the southern United States. Mortality of *H. zea* larvae in whorl stage Bt maize is virtually 100%, and mortality in ear stage Bt maize and in Bt cotton ranges between 70% and 90% (121, 131, 156, 208) In eastern North Carolina, where maize is used almost exclusively by the early generations of the insect, the widespread planting of Bt maize could potentially suppress populations subsequently invading cotton and soybean (e.g. 111). Not enough is yet known about the number of insects produced on hosts other than maize during July, nor the extent of immigration into the region from elsewhere, to be able to determine the extent to which late season populations in maize and soybean will be reduced. Where Bt maize and Bt cotton are planted in the same area, there is potential for a high proportion of the population to be exposed to Bt endotoxin for three or four generations each year. Changing deployment levels of these transgenic crops, variability in susceptibility to the Bt toxins within and among *H. zea* populations (128, 204), and the complexity of *H. zea*'s life systems indicate that the consequences of widespread production of Bt transgenic crops will be manifold and complex.

CONCLUSIONS

In this review, we have described how annual cropping systems are especially variable in time, both within a season and across seasons, and in space, on local and on regional scales. The within-season variation encompasses spatial and temporal components of habitat availability and suitability as affected by host phenology and agricultural production practices. The timing and sequence of habitat availability and suitability relative to the timing of population processes and the ability of the insects to move among habitat patches determine the spatial dynamics of the pest population. The high risks and uncertainties associated with living in a diverse annual cropping system are reduced by the traits of polyphagy and mobility: Polyphagy enables an insect to utilize more habitat patches in space and time; mobility allows the insect to view its landscape on a larger scale, making widely distributed habitat patches accessible.

The among-season variation encompasses changes in the crops and varieties that are planted, changes in the social, political, and economic climate that affect cropping patterns, and the adoption of new agronomic practices and technologies. Driving these changes is farmers' overriding need to obtain high yields efficiently, dependably and profitably. Understanding the integration of insect life systems with the spatial and temporal habitat template, and using this information to determine how the life systems can best be managed, are the challenges for the agricultural entomologist.

We have described the life systems of a few polyphagous pests where our knowledge approaches the level needed and encompasses the relevant spatial scales. Armed with such information, entomologists are better able to predict and manage damaging local populations, and to anticipate the pests' responses to changes in the agroecosystem.

ACKNOWLEDGMENTS

We thank Fred Gould and Amy Sheck for their valuable comments on an earlier draft of this manuscript, and Rick Brandenburg for sharing his insights on the life system of the *T. urticae*. We also thank Vann Covington for his valuable assistance in obtaining references.

LITERATURE CITED

1. Allen JC, Brewster CC, Paris JF, Riley DG, Summers CG. 1996. Spatiotemporal modeling of whitefly dynamics in a regional cropping systems using satellite data. See Ref. 74, pp. 111–24
2. Alstad DN, Andow DA. 1995. Managing

the evolution of insect resistance to transgenic plants. *Science* 268:1894–96
3. Alston DG, Bradley JR Jr, Schmitt DP, Coble HD. 1991. Response of *Helicoverpa zea* (Lepidoptera: Noctuidae) populations to canopy development as influenced by *Heterodera glycine* (Nematoda: Heteroderidae) and annual weed population densities. *J. Econ. Entomol.* 84:267–76
4. Altieri MA. 1988. The dynamics of insect populations in crop systems subject to weed interferences. See Ref. 95, pp. 433–51
5. Ananthakrishnan TN. 1993. Bionomics of thrips. *Annu. Rev. Entomol.* 38:71–92
6. Anderson TE, Kennedy GG, Stinner RE. 1984. Distribution of the European corn borer, *Ostrinia nubilalis* (Hübner) (Lepidoptera: Pyralidae), as related to oviposition preference of the spring-colonizing generation in eastern North Carolina. *Environ. Entomol.* 13:248–51
7. Andow DA. 1990. Characterization of predation on egg masses of *Ostrinia nubilalis* (Lepidoptera: Pyralidae). *Ann. Entomol. Soc. Am.* 83:482–86
8. Andow DA. 1992. Fate of eggs of first-generation *Ostrinia nubilalis* (Lepidoptera: Pyralidae) in three conservation tillage systems. *Environ. Entomol.* 21: 388–93
9. Andow DA, Hutchison WD. 1998. Bt-corn resistance management. In *Now or Never: Serious New Plans to Save a Natural Pest Control,* ed. M Mellon, J Rissler, pp. 19–66. Cambridge, MA: Union Concerned Sci. 149 pp.
10. Andreadis TG. 1984. Epizootiology of *Nosema pyrausta* in field populations of European corn borer (Lepidoptera: Pyralidae). *Environ. Entomol.* 13:882–87
11. Andrew RH, Carlson JR. 1976. Preference differences of egg laying European corn borer (*Ostrinia nubilalis*) adults among maize genotypes. *HortScience* 11:143
12. Andrewartha HG, Birch LC. 1954. *The Distribution and Abundance of Animals.* Chicago: Univ. Chicago Press. 782 pp.
13. Annis B, Tamaki G, Berry RE. 1981. Seasonal occurrence of wild secondary hosts of the green peach aphid, *Myzus persicae* (Sulzer), in agricultural systems in the Yakima Valley. *Environ. Entomol.* 10:307–13
14. Bacheler JS. 1991. Life without the boll weevil: the status of cotton insect pests in North Carolina following eradication. *Proc. Beltwide Cotton Prod. Res. Conf., San Antonio,* 2:615–17. Memphis: Natl. Cotton Counc.
15. Barbosa P, ed. 1998. *Conservation Biological Control.* San Diego: Academic. 396 pp.
16. Barbosa P, Krischik VA, Jones CG, eds. 1991. *Microbial Mediation of Plant-Herbivore Interactions.* New York: Wiley. 530 pp.
17. Barfield CS, Stimac JL. 1981. Understanding the dynamics of polyphagous, highly mobile insects. *Proc. Int. Congr. Plant Prot., 9th,* 1:43–46. New York: Burgess. 411 pp.
18. Barry D, Darrah LL. 1991. Effect of research on commercial hybrid maize resistance to European corn borer (Lepidoptera: Pyralidae). *J. Econ. Entomol.* 84:1053–59
19. Bascompte J, Sole RV. 1995. Rethinking complexity: modelling spatiotemporal dynamics in ecology. *Trends Ecol. Evol.* 10:361–66
20. Beerwinkle KR, Lopez JD Jr, Witz JA, Schleider PG, Eyster RS, Lingren PD. 1994. Seasonal radar and meteorological observations associated with nocturnal flights at altitudes to 900 meters. *Environ. Entomol.* 23:676–83
21. Benedict JH, Sachs ES, Altman DW, Deaton WR, Kohel RJ, et al. 1996. Field performance of cottons expressing transgenic CryIA insecticidal proteins for resistance to *Heliothis virescens* and *Helicoverpa zea* (Lepidoptera: Noctuidae). *J. Econ. Entomol.* 89:230–38

22. Bigger JH, Petty HB. 1953. Reduction of corn borer numbers from October to June—a ten year study. *Ill. Agric. Exp. Stn. Bull.* 566
23. Boykin LS. 1983. *Ecology of the two-spotted spider mite,* Tetranychus urticae *Koch, on peanut in North Carolina.* PhD thesis. NC State Univ., Raleigh. 97 pp.
24. Boykin LS, Campbell WV. 1982. Rate of population increase of the twospotted spider mite (Acari: Tetranychidae) on peanut leaves treated with pesticides. *J. Econ. Entomol.* 75:966–71
25. Boykin LS, Campbell WV. 1984. Wind dispersal of the two-spotted spider mite (Acari: Tetranychidae) in North Carolina peanut fields. *Environ. Entomol.* 13:221–27
26. Boykin LS, Campbell WV, Beute MK. 1984. Effect of pesticides on Neozygites floridana (Entomophthorales: Entomophthoraceae) and arthropod predators attacking the twospotted spider mite (Acari: Tetranychidae) in North Carolina peanut fields. *J. Econ. Entomol.* 77:969–75
27. Bradley JR Jr. 1993. Influence of habitat on the pest status and management of *Heliothis* species on cotton in the southern United States. See Ref. 113, pp. 375–91
28. Bradley JR Jr, Van Duyn JW. 1980. Insect pest management in North Carolina soybeans. *World Soybean Res. Conf. II: Proc.*, ed. FT Corbin, pp. 343–54. Boulder, CO: Westview
29. Brandenburg RL, Kennedy GG. 1981. Overwintering of the pathogen *Entomophthora floridana* and its host, the twospotted spider mite. *J. Econ. Entomol.* 74:428–31
30. Brandenburg RL, Kennedy GG. 1982. Intercrop relationships and spider mite dispersal in a corn/peanut agro-ecosystem. *Entomol. Exp. Appl.* 32:269–76
31. Brandenburg RL, Kennedy GG. 1982. Relationship of *Neozygites floridana* (Entomophthorales: Entomophthoraceae) to twospotted spider mite (Acari: Tetranychidae) populations in field corn. *J. Econ. Entomol.* 75:691–94
32. Brandenburg RL, Kennedy GG. 1983. Interactive effects of selected pesticides on the two-spotted spider mite and its fungal pathogen *Neozygites floridana. Entomol. Exp. Appl.* 34:240–44
33. Brandenburg RL, Kennedy GG. 1987. Ecological and agricultural considerations in the management of twospotted spider mite (*Tetranychus urticae* Koch). *Agric. Zool. Rev.* 2:185–236
34. Brazzle JR, Heinz KM, Parella MP. 1997. Multivariate approach to identifying patterns of *Bemisia argentifolii* (Homoptera: Aleyrodidae) infesting cotton. *Environ. Entomol.* 26:995–1003
35. Brewster CC, Allen JC. 1997. Spatiotemporal model for studying insect dynamics in large-scale cropping systems. *Environ. Entomol.* 26:473–82
36. Brewster CC, Allen JC, Schuster DJ, Stansly PA. 1997. Simulating the dynamics of *Bemisia argentifolii* (Homoptera: Aleyrodidae) in an organic cropping system with a spatiotemporal model. *Environ. Entomol.* 26:603–16
37. Byrne DN, Blackmer JL. 1996. Examination of short-range migration by *Bemisia.* See Ref. 74, pp. 17–28
38. Byrne DN, Houck MA. 1990. Morphometric identification of wing polymorphism in *Bemisia tabaci* (Homoptera, Aleyrodidae). *Ann. Entomol. Soc. Am.* 83:487–93
39. Byrne DN, Rathman RJ, Orum TV, Palumbo JC. 1996. Localized migration and dispersal by the sweet potato whitefly *Bemisia tabaci. Oecologia* 105:320–28
40. Caffrey DJ, Worthley LH. 1927. A progress report on the investigations of the European corn borer. *USDA Bull.* 1476
41. Calvin DD, Welch SM, Poston FI. 1988. Evaluation of a management model for second generation European corn borer (Lepidoptera: Pyralidae) for use in Kansas. *J. Econ. Entomol.* 81:335–43

42. Campbell WV. 1978. Effect of pesticide interactions on the twospotted spider mite on peanuts. *Peanut Sci.* 5:83–86
43. Caprio MA. 1998. Evaluating resistance management strategies for multiple toxins in the presence of external refuges. *J. Econ. Entomol.* 91:1021–31
44. Caprio MA, Tabashnik BE. 1992. Allozymes used to estimate gene flow among populations of diamondback moth (Lepidoptera: Plutellidae) in Hawaii. *Environ. Entomol.* 21:808–16
45. Caprio MA, Tabashnik BE. 1992. Gene flow accelerates local adaptation among finite populations: simulating the evolution of insecticide resistance. *J. Econ. Entomol.* 85:611–20
46. Carde RT, Kochansky J, Stimmel JF, Wheeler AG Jr, Roelofs WL. 1975. Sex pheromone of the European corn borer (Ostrinia nubilalis): cis- and trans-responding males in Pennsylvania. *Environ. Entomol.* 4:413–14
47. Carlson GA, Sappie GP, Hammig MD. 1989. Economic returns to boll weevil eradication. *USDA Agric. Econ. Rep.* No. 621, Resourc. Technol. Div. Washington, DC: USDA Econ. Res. Serv.
48. Chiang HC, Hodson AC. 1959. Distribution of the first-generation egg masses of the European corn borer in corn fields. *J. Econ. Entomol.* 52:295–99
49. Chiang HC, Hodson AC. 1972. Population fluctuation of the European corn borer, *Ostrinia nubilalis*, at Waseca, Minnesota, 1948–70. *Environ. Entomol.* 1:7–16
50. Chiang HC, Jarvis JL, Bunkhardt CL, Fairchild ML, Weekman GT, Triplehorn CA. 1961. Populations of European corn borer, *Ostrinia nubilalis* (Hbn.) in field corn, *Zea mays* (L.) *Mo. Agric. Exp. Stn. Res. Bull.* 776
51. Clark LR, Geier PW, Hughes RD, Morris RF. 1967. *The Ecology of Insect Populations in Theory and Practice.* London: Methuen. 232 pp.
52. Coll M, Bottrell DG. 1988. Mortality by natural enemies in the European corn borer, *Ostrinia nubilalis* (Hübner) (Lepidoptera: Pyralidae) in field corn. *Proc. Int. Congr. Entomol., 18th, Vancouver,* p. 363. Vancouver: Intl. Congr. Entomol.
53. Coll M, Bottrell DG. 1992. Mortality of European corn borer larvae by natural enemies in different corn microhabitats. *Biol. Control* 2:95–103
54. Culin JD. 1995. Local dispersal of male *Helicoverpa zea*. *Entomol. Exp. Appl.* 74:165–76
55. Davis RM, Wang H, Falk BW, Nunez JJ. 1998. Curly top virus found in perennial shrubs in foothills. *Calif. Agric.* 52(5):38–40
56. Dixon AFG. 1998. *Aphid Ecology.* London: Chapman & Hall. 302 pp. 2nd ed.
57. Drake DC, Chapman RK. 1965. Evidence for long distance migration of the six-spotted leafhopper in Wisconsin. In *Migration of the Six-Spotted Leafhopper* Macrosteles fascifrons *(Stal). Part 1.* Madison: Univ. Wis. Agric. Exp. Stn. 45 pp.
58. Drake VA. 1994. The influence of weather and climate on agriculturally important insects: an Australian perspective. *Aust. J. Agric. Res.* 45:487–509
59. Drake VA, Gatehouse AG. 1996. Population trajectories through space and time: a holistic approach to insect migration. In *Frontiers in Population Ecology,* pp. 399–408, ed. RB Floyd, AW Sheppard, PJ De Barro. Melbourne, Aust.: CSIRO. 639 pp.
60. DuRant JA, Manley DG. 1987. Relative efficiencies of two European corn borer sex pheromone blends in South Carolina. *J. Agric. Entomol.* 4:82–86
61. Eckenrode CJ, Robbins PS, Andaloro JT. 1983. Variations in flight patterns of European corn borer (Lepidoptera: Pyralidae) in New York. *Environ. Entomol.* 12:393–96
62. Ellsworth PC, Diehl JW, Husman SH. 1996. Establishment of integrated pest management infrastructure: a commu-

nity-based action program for *Bemisia* management. See Ref. 74, pp. 681–93
63. Ellsworth PC, Umeozor OC, Kennedy GG, Bradley JR Jr, Van Duyn JW. 1989. Population consequences of diapause in a model system: the European corn borer. *Entomol. Exp. Appl.* 53:45–55
64. Everett TR, Chiang HC, Hibbs ET. 1958. Some factors influencing populations of European corn borer (*Pyrausta nubilalis* [Hbn.]) in the north central states. *Minn. Agric. Exp. Stn. Tech. Bull.* 229
65. Fahrig L, Paloheimo J. 1988. Effect of spatial arrangement of habitat patches on local population size. *Ecology* 69:468–75
66. Fitt GP. 1989. The ecology of *Heliothis* species in relation to agroecosystems. *Annu. Rev. Entomol.* 34:17–52
67. Fitt GP. 1994. Cotton pest management: Part 3. An Australian perspective. *Annu. Rev. Entomol.* 39:543–62
68. Fitt GP, Dillon ML, Hamilton JG. 1995. Spatial dynamics of *Helicoverpa* populations in Australia: simulation modelling and empirical studies of adult movement. *Comput. Electron. Agric.* 13:177–92
69. Flanders KL, Radcliffe EB. 1989. Origins of potato leafhoppers (Homoptera: Cicadellidae) invading potato and snap bean in Minnesota. *Environ. Entomol.* 18:1015–24
70. Flint HM, Naranjo SE, Leggett JE, Henneberry TJ. 1996. Cotton water stress, arthropod dynamics and management of *Bemisia tabaci* (Homoptera: Aleyrodidae). *J. Econ. Entomol.* 89:1288–300
71. Fritz RL, Simms EL, eds. 1992. *Plant Resistance to Herbivores and Pathogens: Ecology, Evolution, and Genetics.* Chicago: Univ. Chicago Press. 590 pp.
72. Frye RD. 1972. Evaluation of insect predation on European corn borer in North Dakota. *Environ. Entomol.* 1:535–36
73. Gathman FO, Williams DD. 1998. Intersite: a new tool for the simulation of spatially-realistic population dynamics. *Ecol. Model.* 113:125–39
74. Gerling D, Mayer RT, eds. 1996. *Bemisia 1995: Taxonomy, Biology, Damage, Control and Management.* Andover, UK: Intercept. 702 pp.
75. Gilpin ME, Hanski I, eds. 1991. *Metapopulation Dynamics: Empirical and Theoretical Investigations.* London: Academic
76. Gould F. 1994. Potential and problems with high-dose strategies for pesticidal engineered crops. *Biocontrol Sci. Technol.* 4:451–61
77. Gould F. 1998. Sustainability of transgenic insecticidal cultivars: integrating pest genetics and ecology. *Annu. Rev. Entomol.* 43:701–26
78. Graham HA, Wolfenbarger DA, Nosky JR, Hernandez NS Jr, Llanes JR, Tamago JA. 1978. Use of rubidium to label corn earworm and fall armyworm for dispersal studies. *Environ. Entomol.* 7:435–38
79. Gregg PC. 1991. Ecology of *Heliothis* spp. in non-cultivated areas. In *A Review of Heliothis Research in Australia,* ed. PH Twine, MP Zalucki, pp. 63–73. Brisbane, Aust.: Queensland Dept. Primary Ind.
80. Gregg PC. 1993. Pollen as a marker for migration of *Helicoverpa armigera* and *H. punctigera* (Lepidoptera: Noctuidae) from western Queensland. *Aust. J. Ecol.* 18:209–19
81. Gregg PC. 1995. Migration of cotton pests: patterns and implications for management. In *Challenging the Future: Proceedings of the World Cotton Conference I,* ed. GA Constable, NW Forrester, pp. 423–33. Melbourne, Aust.: CSIRO
82. Gregg PC, Fitt GP, Zalucki MP, Murray DAH. 1995. Insect migration in an arid continent II. *Helicoverpa* spp in eastern Australia. In *Insect Migration: Tracking Resources Through Time and Space,* ed. VA Drake, AG Gatehouse, pp. 151–72. Cambridge/New York: Cambridge Univ. Press. 478 pp.
83. Gyllenberg M, Hanski I. 1997. Habitat deterioration, habitat destruction, and

metapopulation persistence in a heterogeneous landscape. *Theor. Popul. Biol.* 52:198–215
84. Hamilton JG, Rochester WA, Gregg PC. 1994. Predicting long-distance migration of insect pests in eastern Australia. *Proc. Conf. Biometeorol. Aerobiol., 11th, San Diego,* Boston: Am. Meteorol. Soc. 468 pp.
85. Hammond RB, Stinner BR. 1987. Soybean foliage insects in conservation tillage systems: effects of tillage, previous cropping history, and soil insecticide application. *Environ. Entomol.* 16:524–31
86. Hanski I. 1998. Metapopulation dynamics. *Nature* 396:41–49
87. Hardee DD, Bryan WW. 1997. Influence of *Bacillus thuringiensis*-transgenic plants and nectariless cotton on insect populations with emphasis on the tarnished plant bug (Heteroptera: Miridae). *J. Econ. Entomol.* 90:663–68
88. Hardin MR, Benrey B, Coll M, Lamp WO, Roderick GK, Barbosa P. 1995. Arthropod pest resurgence: an overview of potential mechanisms. *Crop Prot.* 14:3–18
89. Hardwick DF. 1965. The corn earworm complex. *Mem. Entomol. Soc. Can.* No. 40. 250 pp.
90. Harper LA, Giddens JE, Langdale GW, Sharpe RR. 1989. Environmental effects on nitrogen dynamics in soybean under conservation and clean tillage systems. *Agron. J.* 81:623–31
91. Harris VE, Phillips JR. 1986. Mowing spring host plants as a population management technique for *Heliothis* spp. *J. Agric. Entomol.* 3:125–34
92. Harstack AW, Lopez JD, Muller RA, Sterling WL, King EG, et al. 1982. Evidence of long-range migration in *Heliothis zea* into Texas and Arkansas. *Southwest. Entomol.* 7:188–201
93. Hassel MP, Comins HN, May RM. 1991. Spatial structure and chaos in insect population dynamics. *Nature* 353:255–58
94. Hawkins BA, Cornell HV, Hochberg ME. 1997. Predators, parasitoids, and pathogens as mortality agents in phytophagous insect populations. *Ecology* 78:2145–52
95. Heinrichs EA, ed. 1988. *Plant Stress—Insect Interactions.* New York: Wiley & Sons. 492 pp.
96. Henneberry TJ, Bariola LA, Chu CC, Deeter B, Meng T. 1990. Plant growth regulators in pink bollworm management systems. *Proc. Beltwide Cotton Prod. Res. Conf., Las Vegas,* pp. 187–89. Memphis: Natl. Cotton Counc.
97. Herbert DA Jr, Zehnder GW, Day ER. 1991. Evaluation of a pest advisory for corn earworm (Lepidoptera: Noctuidae) infestations in soybean. *J. Econ. Entomol.* 84:515–19
98. Hill RE, Gary WJ. 1979. Effects of the microsporidium, *Nosema pyrausta,* on field populations of European corn borers *Ostrinia nubilalis* in Nebraska. *Environ. Entomol.* 8:91–95
99. Hoelmer KA. 1996. Whitefly parasitoids: Can they control field populations of *Bemisia?* See Ref. 74, pp. 451–76
100. Hoy CW, Heady SE, Koch TA. 1992. Species composition, phenology, and possible origins of leafhoppers (Cicadellidae) in Ohio vegetable crops. *J. Econ. Entomol.* 85:2336–43
101. Hoy MA, Groot JJR, van de Baan HE. 1985. Influence of aerial dispersal on persistence and spread of pesticide-resistant *Metaseiulus occidentalis* in California almond orchards. *Entomol. Exp. Appl.* 37:17–31
102. Huffaker CB, Gutierrez AP. 1999. *Ecological Entomology.* New York: Wiley. 756 pp. 2nd ed.
103. Hughes RD. 1979. Movement in population dynamics. See Ref. 160, pp. 14–32
104. Isely D. 1935. Relation of host to abun-

dance of cotton bollworm. *Ark. Exp. Stn. Bull.* 320. 30 pp.

105. Ives AR, Settle WH. 1997. Metapopulation dynamics and pest control in agricultural systems. *Am. Nat.* 149:220–46
106. Jarvis JL, Guthrie WD. 1987. Ecological studies of the European corn borer (Lepidoptera: Pyralidae) in Boone County, Iowa. *Environ. Entomol.* 16:50–58
107. Johnson MW, Stinner RE, Rabb RL. 1975. Ovipositional response of *Heliothis zea* (Boddie) to its major hosts in North Carolina. *Environ. Entomol.* 4:291–97
108. Jones KD. 1994. Aspects of the biology and biological control of the European corn borer in North Carolina. PhD thesis. NC State Univ., Raleigh. 127 pp.
109. Kennedy GG. 1983. Effects of European corn borer (Lepidoptera: Pyralidae) damage on yields of spring-grown potatoes. *J. Econ. Entomol.* 76:316–22
110. Kennedy GG, Anderson TE. 1981. Considerations for the management of the European corn borer on potatoes. In *Advances in Potato Pest Management,* ed. JH Lashomb, R Casagrande, pp. 204–22. Stroudsburg, PA: Hutchinson Ross. 288 pp.
111. Kennedy GG, Gould F, Deponti OMB, Stinner RE. 1987. Ecological, agricultural, genetic, and commercial considerations in the deployment of insect-resistant germplasm. *Environ. Entomol.* 16:327–38
112. Kennedy GG, Margolies DC. 1985. Mobile arthropod pests: management in diversified agroecosystems. *Bull. Entomol. Soc. Am.* 31:21–27
113. Kim KC, McPheron BA, eds. 1993. *Evolution of Insect Pests: Patterns of Variation.* New York: Wiley & Sons. 479 pp.
114. Kogan M, ed. 1986. *Ecological Theory and Integrated Pest Management Practice.* New York: Wiley. 362 pp.
115. Kogan M, Fischer DC. 1991. Inducible defenses in soybean against herbivorous insects. In *Phytochemical Induction by Herbivores,* ed. DW Tallamy, MJ Raupp, pp. 347–78. New York: Wiley. 431 pp.
116. Kogan M, Turnipseed S. 1987. Ecology and management of soybean arthropods. *Annu. Rev. Entomol.* 32:507–38
117. Koziel MG, Beland GL, Bowman C, Carozzi NB, Crenshaw L, et al. 1993. Field performance of elite transgenic maize plants expressing an insecticidal protein derived from *Bacillus thuringiensis. Bio-Technology* 11:194–200
118. Kramer JP. 1959. Observations on the seasonal incidence of microsporidiosis in European corn borer populations in Illinois. *Entomophaga* 4:37–42
119. Labatte J-M, Meusnier S, Migeon A, Chaufaux J, Couteaudier Y, et al. 1996. Field evaluation of and modeling the impact of three control methods on the larval dynamics of *Ostrinia nubilalis* (Lepidoptera: Pyralidae). *J. Econ. Entomol.* 89:852–62
120. Labatte J-M, Meusnier S, Migeon A, Piry S, Got B. 1997. Natural mortality of European corn borer (Lepidoptera: Pyralidae) larvae: field study and modeling. *J. Econ. Entomol.* 90:773–83
121. Lambert AL, Bradley JR Jr, Van Duyn JW. 1997. Interactions of *Helicoverpa zea* and Bt cotton in North Carolina. *Proc. Beltwide Cotton Prod. Res. Conf., New Orleans,* pp. 870–73. Memphis: Natl. Cotton Counc.
122. Lamp WO, Zhao L. 1993. Prediction and manipulation of movement by polyphagous, highly mobile pests. *J. Agric. Entomol.* 10:267–81
123. Lewis LC, Lynch RE. 1976. Influence on the European corn borer of *Nosema pyrausta* and resistance in maize to leaf feeding. *Environ. Entomol.* 5:139–42
124. Li J, Margolies DC. 1993. Quantitative genetics of aerial dispersal behavior and life-history traits in *Tetranychus urticae. Heredity* 70:544–52
125. Lincoln CJ, Isely D. 1947. Corn as a trap crop for the cotton bollworm. *J. Econ. Entomol.* 40:437–38

126. Lingren PD, Bryant VM Jr, Raulston JR, Pendleton M, Westbrook J, Jones GD. 1993. Adult feeding host range and migratory activities of corn earworm, cabbage looper and celery looper (Lepidoptera: Noctuidae) moths as evidenced by attached pollen. *J. Econ. Entomol.* 86:1429–39
127. Logan JA, Stinner RE, Rabb RL, Bacheler JS. 1979. A descriptive model for predicting spring emergence of *Heliothis zea* populations in North Carolina. *Environ. Entomol.* 8:141–46
128. Luttrell RG, Wan L, Knighten K. 1999. Variation in susceptibility of noctuid (Lepidoptera) larvae attacking cotton and soybean to purified endotoxin proteins and commercial formulations of *Bacillus thuringiensis*. *J. Econ. Entomol.* 92:21–32
129. Lynch RE, Lewis LC. 1976. Influence on the European corn borer of *Nosema pyrausta* and resistance in maize to sheath-collar feeding. *Environ. Entomol.* 5:143–46
130. Mackenzie DR, Barfield CS, Kennedy GG, Berger RD, Taranto DJ, eds. 1985. *The Movement and Dispersal of Agriculturally Important Biotic Agents.* Baton Rouge, LA: Claitors. 611 pp.
131. Mahaffey JS, Bradley JR Jr, Van Duyn JW. 1996. B.T. cotton: field performance in North Carolina under conditions of unusually high bollworm populations. *Proc. Beltwide Cotton Prod. Res. Conf. Nashville,* pp. 795–98. Memphis: Natl. Cotton Counc.
132. Mallet J, Porter P. 1992. Preventing insect adaptation to insect-resistant crops: Are seed mixtures or refugia the best strategy? *Proc. R. Soc. London Ser. B* 250:165–69
133. Margolies DC. 1993. Adaptation to spatial variation in habitat: spatial effects in agroecosystems. See Ref. 113, pp. 129–41
134. Margolies DC, Kennedy GG. 1984. Population response of the twospotted spider mite, *Tetranychus urticae*, to host phenology in corn and peanut. *Entomol. Exp. Appl.* 36:193–96
135. Margolies DC, Kennedy GG. 1985. Movement of the twospotted spider mite, *Tetranychus urticae*, among hosts in a corn-peanut agroecosystem. *Entomol. Exp. Appl.* 37:55–61
136. Margolies DC, Kennedy GG, Van Duyn JW. 1985. Effect of three soil-applied insecticides in field corn on spider mite (Acari: Tetranychidae) pest potential. *J. Econ. Entomol.* 78:117–20
137. McNeil JN. 1987. The true armyworm, *Pseudaletia unipuncta*: a victim of the 'Pied Piper' or a seasonal migrant? *Insect Sci. Appl.* 8:591–97
138. McPherson RM, Newsom LD. 1984. Trap crops for control of stink bugs in soybean. *J. Ga. Entomol. Soc.* 19:470–80
139. Miller RW, Croft BA, Nelson RD. 1985. Effects of early season immigration on cyhexatin and formetanate resistance of *Tetranychus urticae* (Acari: Tetranychidae) on strawberry in central California. *J. Econ. Entomol.* 78:1379–86
140. Mueller AJ, Stern VM. 1974. Timing of pesticide treatments on safflower to prevent *Lygus* from dispersing to cotton. *J. Econ. Entomol.* 67:77–80
141. Mueller TF, Harris VE, Phillips JR. 1984. Theory of *Heliothis* (Lepidoptera: Noctuidae) management through reduction of the first spring generation: a critique. *Environ. Entomol.* 13:625–34
142. Mueller TF, Phillips JR. 1988. Population dynamics of *Heliothis* spp. in spring weed hosts in southeastern Arkansas: survivorship and stage-specific parasitism. *Environ. Entomol.* 12:1846–50
143. Nault BA, Kennedy GG. 1996. Evaluation of Colorado potato beetle (Coleoptera: Chrysomelidae) defoliation with concomitant European corn borer (Lepidoptera: Pyralidae) damage on potato yield. *J. Econ Entomol.* 89:475–80
144. Neunzig HH. 1963. Wild host plants of

the corn earworm and the tobacco budworm in eastern North Carolina. *J. Econ. Entomol.* 56:135–39

145. Neunzig HH. 1969. The biology of the tobacco budworm and the corn earworm in North Carolina with particular reference to tobacco as a host. *NC Agric. Exp. Stn. Tech. Bull. 196*. 63 pp.
146. North RC, Shelton AM. 1986. Colonization and intraplant distribution of *Thrips tabaci* (Thysanoptera: Thripidae) on cabbage. *J. Econ. Entomol.* 79:219–23
147. North RC, Shelton AM. 1986. Ecology of Thysanoptera within cabbage fields. *Environ. Entomol.* 15:520–26
148. Onstad DW. 1988. Simulation model of the population dynamics of *Ostrinia nubilalis* (Lepidoptera: Pyralidae) in maize. *Environ. Entomol.* 17:969–76
149. Onstad DW, Gould F. 1998. Modeling the dynamics of adaptation to transgenic maize by European corn borer (Lepidoptera: Pyralidae). *J. Econ. Entomol.* 91:585–93
150. Onstad DW, Maddox JV. 1989. Modeling the effects of the microsporidium, *Nosema pyrausta*, on the population dynamics of the insect, *Ostrinia nubilalis*. *J. Invertebr. Pathol.* 53:410–21
151. Pair SD, Raulston JR, Sparks AN, Westbrook JK, Wolf WW, Goodenough JL. 1995. A prospectus—impact of *Helicoverpa zea* (Boddie) production from corn in the Lower Rio-Grande Valley on regional cropping systems. *Southwest. Entomol. Suppl.* 18:155–67
152. Panizzi AR. 1997. Wild hosts of pentatomids: ecological significance and role in their pest status on crops. *Annu. Rev. Entomol.* 42:99–122
153. Parella MP. 1987. Biology of *Liriomyza*. *Annu. Rev. Entomol.* 32:201–24
154. Peck SL, Gould F, Ellner S. 1999. The spread of resistance in spatially extended regions of transgenic cotton: implications for the management of *Heliothis virescens* (Lepidoptera: Noctuidae). *J. Econ. Entomol.* 92:1–16
155. Pedigo LP, Hutchins SH, Higley LG. 1986. Economic injury levels in theory and practice. *Annu. Rev. Entomol.* 31:314–68
156. Pilcher CD, Rice ME, Obrycki JJ, Lewis LC. 1997. Field and laboratory evaluations of trangenic *Bacillus thuringiensis* corn on secondary lepidopteran pests (Lepidoptera: Noctuidae). *J. Econ. Entomol.* 90:669–78
157. Quaintance AL, Brues CT. 1905. The cotton bollworm. *USDA Bull.* 50:1–155
158. Rabb RL. 1978. A sharp focus on insect populations and pest management from a wide-area view. *Bull. Entomol. Soc. Am.* 24:55–61
159. Rabb RL. 1985. Conceptual bases to develop and use information on the movement and dispersal of biotic agents in agriculture. See Ref. 130, pp. 5–34
160. Rabb RL, Kennedy GG, eds. 1979. *Movement of Highly Mobile Insects: Concepts and Methodology in Research.* Raleigh: NC State Univ. 456 pp.
161. Rabb RL, Stinner RE. 1978. The role of insect dispersal and migration in population processes. See Ref. 223, pp. 3–16
162. Riley DG, Ciomperlik MA. 1997. Regional population dynamics of whitefly (Homoptera: Aleyrodidae) and associated parasitoids (Hymenoptera: Aphelinidae). *Environ. Entomol.* 26:1049–55
163. Riley DG, Nava-Camberos U, Allen J. 1996. Population dynamics of *Bemisia* in agricultural systems. See Ref. 74, pp. 93–109
164. Roach SH. 1981. Emergence of overwintered *Heliothis* spp. moths from three different tillage systems. *Environ. Entomol.* 10:817–18
165. Roelofs WL, Du J-W, Tang X-H, Robbins PS, Eckenrode CJ. 1985. Three European corn borer populations in New York based on sex pheromones and voltinism. *J. Chem. Ecol.* 11:829–36

166. Rossi RE, Borth PW, Tollefson RV. 1993. Stochastic simulation for characterizing ecological spatial patterns and appraising risk. *Ecol. Appl.* 3:719–35
167. Roush R. 1997. Managing resistance to transgenic crops. In *Advances in Insect Control: The Role of Transgenic Crops,* ed. N Carozzi, M Koziel, pp. 271–94. London: Taylor & Francis. 301 pp.
168. Rummel DR, Leser JF, Slosser JE, Puterka GJ, Neeb CW, et al. 1986. Cultural control of *Heliothis* spp. in southwestern U.S. cropping systems. *South. Coop. Ser. Bull.* 316:38–53
169. Savinelli CE, Bacheler JS, Bradley JR Jr. 1986. Nature and distribution of European corn borer (Lepidoptera: Pyralidae) larval feeding damage to cotton in North Carolina. *Environ. Entomol.* 15:399–402
170. Savinelli CE, Bacheler JS, Bradley JR Jr. 1988. Ovipositional preferences of the European corn borer (Lepidoptera: Pyralidae) for field corn and cotton under field cage conditions in North Carolina. *Environ. Entomol.* 17:688–90
171. Schumann FW, Todd JW. 1982. Population dynamics of southern green stink bug (Heteroptera: Pentatomidae) in relation to soybean phenology. *J. Econ. Entomol.* 75:748–53
172. Sevacherian V, Stern VM. 1975. Movements of lygus bugs between alfalfa and cotton. *Environ. Entomol.* 4:163–65
173. Shea K. 1998. Management of populations in conservation, harvesting and control. *Trends Ecol. Evol.* 13:371–75
174. Shelton AM, North RC. 1986. Species composition and phenology of Thysanoptera within field crops adjacent to cabbage fields. *Environ. Entomol.* 15:512–19
175. Showers WB. 1979. Effect of diapause on the migration of the European corn borer *Ostrinia nubilalis* into the southeastern United States. See Ref. 160, pp. 420–30
176. Showers WB. 1993. Diversity and variation of European corn borer populations. See Ref. 113, pp. 287–309
177. Showers WB. 1997. Migratory ecology of the black cutworm. *Annu. Rev. Entomol.* 42:393–425
178. Showers WB, De Rozari MB, Reed GL, Shaw RH. 1978. Temperature-related climatic effects on survivorship of the European corn borer. *Environ. Entomol.* 7:717–23
179. Siegel JP, Maddox JV, Ruesink WG. 1986. Impact of *Nosema pyrausta* on a braconid, *Macrocentrus grandii* in central Illinois. *J. Invert. Pathol.* 47:271–76
180. Silberman L. 1983. *Aerial dispersal of the twospotted spider mite,* Tetranychus urticae *(Koch), from its host plants.* MS thesis. NC State Univ., Raleigh. 68 pp.
181. Smith CM. 1989. *Plant Resistance to Insects: A Fundamental Approach.* New York: Wiley. 286 pp.
182. Smitley DR, Kennedy GG. 1985. Photo-oriented aerial-dispersal behavior of *Tetranychus urticae* (Acari: Tetranychidae) enhances escape from the leaf surface. *Ann. Entomol. Soc. Am.* 78:609–14
183. Smitley DR, Kennedy GG. 1988. Aerial dispersal of the two-spotted spider mite (*Tetranychus urticae*) from field corn. *Exp. Appl. Acarol.* 5: 33–46
184. Smitley DR, Kennedy GG, Brooks WM. 1986. Role of the entomogenous fungus *Neozygites floridana,* in population declines of the twospotted spider mite, *Tetranychus urticae,* on field corn. *Entomol. Exp. Appl.* 41:255–64
185. Snow JW, Hamm JH, Brazzel JR. 1966. *Geranium carolinianum,* as an early host for *Heliothis zea* and *H. virescens* (Lepidoptera: Noctuidae) in the southeastern United States, with notes on associated parasites. *Ann. Entomol. Soc. Am.* 59:506–9
186. Sorenson CE, Kennedy GG, Van Duyn JW, Bradley JR Jr, Walgenbach JF. 1992. Geographic variation in pheromone response of the European corn borer,

Ostrinia nubilalis, in North Carolina. *Entomol. Exp. Appl.* 64:177–85
187. Southwood TRE. 1962. Migration of terrestrial arthropods in relationship to habitat. *Biol. Rev.* 37: 171–214
188. Southwood TRE. 1975. The dynamics of insect populations. In *Insects, Science and Society,* ed. D Pimentel, pp. 151-99. New York: Academic. 284 pp.
189. Southwood TRE. 1977. Habitat, the template for ecological strategies? *J. Anim. Ecol.* 46: 337–65
190. Southwood TRE. 1981. Ecological aspects of insect migration. In *Animal Migration,* ed. DJ Aidley, pp. 196–208. Cambridge: Cambridge Univ. Press. 264 pp.
191. Southwood TRE, Way MJ. 1970. Ecological background to pest management. In *Concepts of Pest Management,* ed. RL Rabb, FE Guthrie, pp. 6–29. Raleigh: NC State Univ. 242 pp.
192. Sparks AN. 1979. An introduction to the status, current knowledge and research on movement of selected Lepidoptera in south-eastern United States. See Ref. 160, pp. 382–85
193. Sparks AN. 1986. Fall armyworm (Lepidoptera: Noctuidae): potential for area-wide management. *Fla. Entomol.* 69:603–14
194. Sparks AN, Chiang HC, Burkhardt CC, Fairchild ML, Weekman GT. 1966. Evaluation of predation on corn borer populations. *J. Econ. Entomol.* 59:104–7
195. Sparks AN, Chiang HC, Triplehorn CA, Guthrie WD, Brindley TA. 1967. Some factors influencing populations of the European corn borer, *Ostrinia nubilalis* (Hübner), in the north central states: resistance of corn, time of planting and weather conditions, part 2, 1958–1962. *Iowa Agric. Home Econ. Exp. Stn. Res. Bull.* 559:66–103
196. Sparks AN, Showers WB. 1975. Current status of European corn borer in south Georgia. *J. Ga. Entomol. Soc.* 10:342–52
197. Stadelbacher EA. 1979. *Geranium dissectum:* an unreported host of the tobacco budworm and bollworm and its role in their seasonal and long term population dynamics in the delta of Mississippi. *Environ. Entomol.* 8:1153–56
198. Stadelbacher EA. 1981. Role of early season wild and naturalized host plants in the build-up of the F1 generation of *Heliothis zea* and *H. virescens* in the delta of Mississippi. *Environ. Entomol.* 10:776–70
199. Stinner RE. 1979. Biological monitoring essentials in studying wide-area moth movement. See Ref. 160, pp. 199–211
200. Stinner RE, Barfield CS, Stimac JL, Dohse L. 1983. Dispersal and movement of insect pests. *Annu. Rev. Entomol.* 28:319–35
201. Stinner RE, Rabb RL, Bradley JR Jr. 1974. Population dynamics of *Heliothis zea* (Boddie) and *H. virescens* (F.) in North Carolina: a simulation model. *Environ. Entomol.* 3:163–68
202. Stinner RE, Rabb RL, Bradley JR Jr. 1977. Natural factors operating in the population dynamics of *Heliothis zea* in North Carolina. *Proc. Int. Congr. Entomol. 15th, Washington, DC, 1976,* pp. 622–42. College Park, MD: Entomol. Soc. Am.
203. Stinner RE, Regniere J, Wilson K. 1982. Differential effects of agroecosystem structure on dynamics of three soybean herbivores. *Environ. Entomol.* 11:538–43
204. Stone TB, Sims SR. 1993. Geographic susceptibility of *Heliothis virescens* and *Helicoverpa zea* (Lepidoptera: Noctuidae) to *Bacillus thuringiensis. J. Econ. Entomol.* 86:989–94
205. Stoner KA, Shelton AM. 1988. Effect of planting date and timing of growth stages on damage to cabbage by onion thrips (Thysanoptera: Thripidae). *J. Econ. Entomol.* 81:1186–89
206. Storer NP. 1999. The corn earworm, Bt transgenic corn and Bt resistance evolution in a mixed cropping system. PhD thesis. NC State Univ., Raleigh. 319 pp.

207. Storer NP, Van Duyn JW, Gould F, Kennedy GG. 1999. Ecology and biology of cotton bollworm in reference to modeling Bt resistance development in a Bt cotton/Bt corn system. *Proc. Beltwide Cotton Prod. Res. Conf., Orlando,* Natl. Cotton Counc. 2:949–52.
208. Sudbrink DL Jr, Grant JF. 1995. Wild host plants of *Helicoverpa zea* and *Heliothis virescens* (Lepidoptera: Noctuidae) in eastern Tennessee. *Environ. Entomol.* 24:1080–85
209. Tamaki G. 1975. Weeds in orchards as important alternate sources of green peach aphids in late spring. *Environ. Entomol.* 4:958–60
210. Taylor LR. 1985. An international standard for synoptic monitoring and dynamic mapping of migrant pest populations. See Ref. 130, pp. 337–419
211. Taylor LR, Taylor RAJ. 1977. Aggregation, migration and population mechanics. *Nature* 265:415–21
212. Taylor RAJ. 1985. Migratory behavior in the Auchenorrhynca. In *The Leafhoppers and Planthoppers,* ed. LR Nault, JG Rodriguez, pp. 259–88. New York: Wiley & Sons. 500 pp.
213. Terry I, Bradley JR Jr, Van Duyn JW. 1987. Survival and development of *Heliothis zea* (Lepidoptera: Noctuidae) larvae on selected soybean growth stages. *Environ. Entomol.* 16:441–45
214. Todd JW. 1989. Ecology and behavior of *Nezara viridula. Annu. Rev. Entomol.* 34:273–92
215. Turchin P, Omland KS. 1999. Quantitative analysis of insect movement. See Ref. 102, pp. 463–98
216. Turner MG. 1989. Landscape ecology: the effect of pattern on process. *Annu. Rev. Ecol. Syst.* 20:171–97
217. Umeozor OC, Bradley JR Jr, Van Duyn JW, Kennedy GG. 1986. Intercrop effects on the distribution of populations of the European corn borer, *Ostrinia nubilalis,* in maize. *Entomol. Exp. Appl.* 40:293–96
218. Umeozor OC, Van Duyn JW, Bradley JR Jr, Kennedy GG. 1985. Comparison of the effect of minimum tillage treatments on the overwintering emergence of European corn borer (Lepidoptera: Noctuidae) in cornfields. *J. Econ. Entomol.* 78: 937–39
219. Deleted in proof
220. Van Denburgh RS, Burbutis PP. 1962. The host-parasite relationship of the European corn borer, *Ostrinia nubilalis,* and the protozoan *Perezia pyraustae* in Delaware. *J. Econ. Entomol.* 55:65–67
221. Van Emden HF. 1965. The role of uncultivated land in the biology of crop pests and beneficial insects. *Sci. Hortic.* 17:121–36
222. Van Steenwyk RA. 1978. Short-range movement of major agricultural pests. See Ref. 223, pp. 17–21
223. Vaughn CR, Wolf W, Klassen W, eds. 1978. *Radar, Insect Population Ecology, and Pest Management.* Wallops Island, VA: NASA
224. Velasco LRI, Walter GH. 1992. Availability of different host plant species and changing abundance of the polyphagous bug *Nezara viridula* (Hemiptera: Pentatomidae). *Environ. Entomol.* 21:751–59
225. Wallis RL. 1967. Green peach aphids and the spread of beet western yellow virus on the Northwest. *J. Econ. Entomol.* 60:513–15
226. Wallner WE. 1987. Factors affecting insect population dynamics: differences between outbreak and non-outbreak species. *Annu. Rev. Entomol.* 32:317–40
227. Ward LK, Spalding DF. 1993. Phytophagous British insects and mites and their food-plant families: total numbers and polyphagy. *Biol. J. Linn. Soc.* 49:257–76
228. Way MJ. 1977. Pest and disease status in mixed stands vs. monocultures: the relevance of ecosystem stability. In *Origins of Pest, Parasite, Disease and Weed Problem,* ed. JM Cherrett, GR Sagar, pp. 127–38. Oxford: Blackwell. 413 pp.
229. Wiens JA, Crist TO, Milne BT. 1993. On

quantifying insect movements. *Environ. Entomol.* 22:709–15

230. Wilson JA Jr, DuRant JA. 1992. Parasites of the European corn borer (Lepidoptera: Pyralidae) in South Carolina. *J. Agric. Entomol.* 8:109–16
231. Yu Y, Gold HJ, Stinner RE, Wilkerson GG. 1992. Leslie model for the population dynamics of corn earworm in soybean. *Environ. Entomol.* 21:253–63
232. Zimmack HL, Brindley TA. 1957. The effect of the protozoan parasite *Perezia pyraustae* Paillot on the European corn borer. *J. Econ. Entomol.* 50:637–40
233. Zoebisch TG, Schuster DJ. 1987. Suitability of foliage of tomatoes and three weed hosts for oviposition and development of *Liriomyza trifolii* (Diptera: Agromyzidae). *J. Econ. Entomol.* 80:758–62

Annu. Rev. Entomol. 2000. 45:495–518

ACCESSORY PULSATILE ORGANS: Evolutionary Innovations in Insects

Günther Pass
Institut für Zoologie, Universität Wien, Althanstrasse 14, A-1090 Vienna, Austria; e-mail: guenther.pass@univie.ac.at

Key Words circulation, heart, hemolymph, body appendage, phylogeny

■ **Abstract** In addition to the dorsal vessel ("heart"), insects have accessory pulsatile organs ("auxiliary hearts") that supply body appendages with hemolymph. They are indispensable in the open circulatory system for hemolymph exchange in antennae, long mouthparts, legs, wings, and abdominal appendages. This review deals with the great diversity in the functional morphology and the evolution of these accessory pulsatile organs. In primitive insects, hemolymph is supplied to antennae and cerci by arteries connected to the dorsal vessel. In higher insects, however, these arteries were decoupled and associated with autonomous pumps that entered their body plan as evolutionary innovations. To ensure hemolymph supply to legs, wings, and some other appendages, completely new accessory pulsatile organs evolved. The muscular components of these pulsatile organs and their elastic antagonists were recruited from various organ systems and assembled to new functional units. In general, it seems that the evolution of accessory pulsatile organs has been determined by developmental and spatial constraints imposed by other organ systems rather than by changes in circulatory demands.

INTRODUCTION

In the open circulatory system of insects, the pumping organs of the central body cavity cannot circulate hemolymph in long body appendages. Diffusion must also be ruled out as a feasible exchange mechanism because low velocity restricts it to cellular dimensions. Therefore, auxiliary circulatory structures are indispensable for hemolymph circulation. In some noninsect arthropods and primitive insects (hexapods[1]), appendages are supplied by arteries originating from major vessels (52, 58). In most modern insects, however, this task is accomplished by means of so-called accessory pulsatile organs (39, 68).

[1]Modern taxonomic nomenclature utilizes "Hexapoda" for "Insecta s.l." (47). Various conceptions prevail concerning the term "Insecta s. str."(21). For the sake of simplicity, "insects" is used synonymously for "hexapods" in this review.

0066-4170/00/0107-0495/$14.00

As a rule, these auxiliary hearts are separate from the dorsal vessel and function autonomously. An insect can possess a considerable number, so we may think of insects as the animals richest in hearts (Figure 1). Most accessory pulsatile organs are evolutionary innovations that emerged in primitive pterygotes and have since become integral elements of their body plan. Auxiliary hearts have long failed to receive due recognition, and even their great diversity has only recently been uncovered (45, 46, 67, 68). This review examines (*a*) the functional morphology

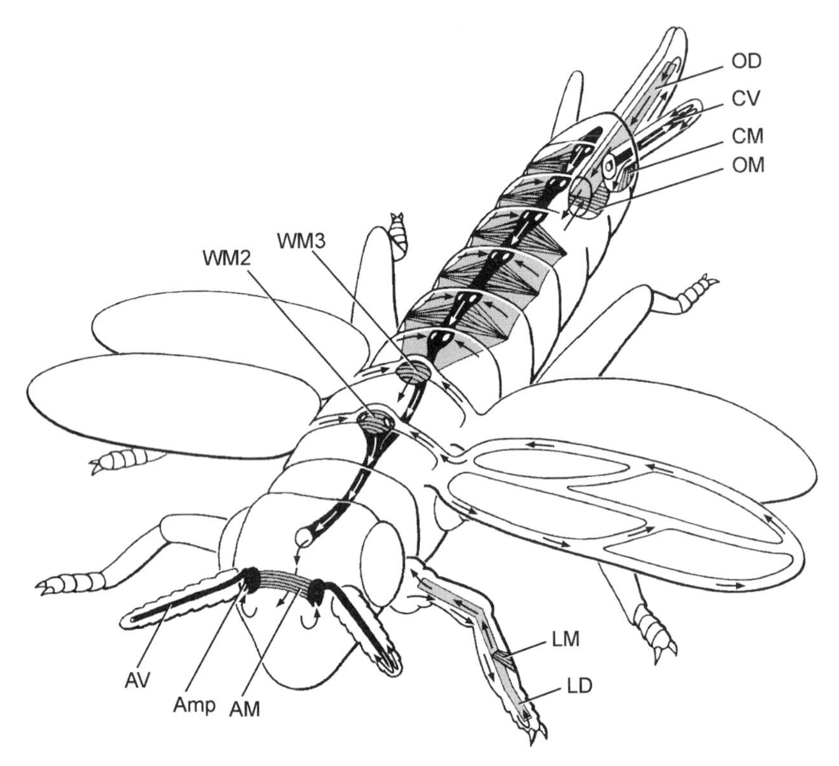

Figure 1 Diagram of an idealized insect with the maximum possible set of circulatory organs. Vessels are in *solid black*, diaphragms and pumping muscles are in *gray; arrows* indicate hemolymph flow directions. Central body cavity: dorsal vessel composed of anterior aorta and posterior heart region (with paired ostia, dorsal diaphragm, and alary muscles); ventral diaphragm concealed. Antennae: ampullae *(Amp)* with ostia connected to antennal vessels *(AV);* pumping muscle *(AM)* associated with ampullae. Legs: diaphragm *(LD)* pulsatile owing to associated pumping muscle *(LM)*. Wings: dorsal vessel muscle plate with ostia pair in mesothorax *(WM2);* separate pumping muscle in metathorax *(WM3)*. Cerci: cercal vessel *(CV)* with basal suction pump *(CM)*. Ovipositor: each valvula with nonpulsatile diaphragm *(OD)* and basal forcing pump *(OM)*. Among insect species, the functional morphology of the accessory pulsatile organ in a given body appendage may be heterogeneous.

and physiology, (*b*) the phylogenetic pathways, and (*c*) the evolutionary innovations in insect circulatory organs.

STRUCTURE AND FUNCTION OF CIRCULATORY ORGANS

Methodological intricacies related to small body size and open circulatory systems contribute to the gaps in the understanding of hemolymph circulation and routing through an insect's body. From a hemodynamic point of view, it is crucial to know that hemolymph is not exclusively transported by pumps. Shifting of hemolymph bulk by certain muscle contractions and volume changes of the abdomen may also play an essential role (85, 86, 91, 95–99). Moreover, the tracheal system of some endopterygotes has turned out to be a powerful antagonist to the action of the circulatory organs (97–101).

Central Body Cavity

Although the accessory pulsatile organs of various insect body appendages constitute nearly autonomous systems, they may also be linked to the circulatory organs of the central body cavity. Because the structure and function of dorsal vessels have been treated in several other reviews (39, 57, 107), a detailed analysis is not required here. The basic type of dorsal vessel is a rather uniform muscular tube bearing valved ostia in most thoracic and abdominal segments (45). In many species, the dorsal vessel is partitioned into a posterior pumping region, the heart, and an anterior poorly contractile region, the aorta. The two regions differ not only in the strength of the muscular wall but also in the presence of ostia, dorsal diaphragm, and alary muscles (35, 46). An accessory pulsatile organ associated with the aorta was recently discovered in the head of the blow fly (102).

The dorsal vessel can be linked to other vessels that may then be regarded as arteries. Some of these serve hemolymph circulation in long body appendages (antennae, cerci, and terminal filament). Other vessels distributing hemolymph from the dorsal vessel without detour to certain regions of the central body cavity are the segmental vessels in the thorax and the abdomen of cockroaches, mantids, and orthopterans (51, 57, 62) and the circumesophageal vessel ring in the head of apterygotes[2] (8, 24).

The view generally held presumes that the dorsal vessel collects hemolymph in the abdomen via incurrent ostia and transports it toward the head by peristaltic contractions. The flow mode, however, may be substantially different. All apterygotes and the mayflies feature a bidirectional flow in their dorsal vessel: Hemo-

[2]Despite their status as paraphyletic taxa, the terms "apterygotes" and "exopterygotes" are retained for clarity in this review.

lymph is propelled toward the front except in the most posterior vessel portion, where it exclusively flows toward the rear; intracardiac valves shunt these flow directions (24, 54). Where this bidirectional flow mode applies, the dorsal vessel is posteriorly open and communicates with vessels supplying the caudal appendages. Heartbeat reversal, with its periodic change of pumping directions, is another flow mode common among endopterygotes in particular (25, 39). This mode is associated either with a posteriorly open dorsal vessel, the presence of excurrent ostia, or even two-way ostia allowing selective shunting (95, 100, 101).

Diaphragms of connective tissue and/or muscle channel the various hemolymph flows in the central body cavity and can also actively aid hemolymph circulation via undulating movements. In particular, a ventral diaphragm is widespread and can be variously developed (76).

Antennae

The functional morphology of antennal circulatory organs has been quite thoroughly investigated in many insects (6, 13, 16–18, 50a, 63, 64, 66–68, 81). This investigation yielded results on the stupendous diversity of these accessory pulsatile organs (for details on functional types and distribution among insect orders, see Figures 2 and 3). Insects usually possess vessels in their antennae, except for a few species whose short antennae lack circulatory organs altogether or have tiny diaphragms. The vessels serve the efferent hemolymph supply; the afferent current into the head capsule leads through the antennal hemocoel. Hemolymph enters the antennal vessels in different ways in the various insect taxa (67, 68). They may be directly linked with the dorsal vessel (Figure 2*a*), but in most insects they are separate and have ampullary enlargements with valved ostia at their bases. Ampullae are paired or unpaired; in a few species, they constitute large frontal sacs. Rarely, the ampullae just funnel hemolymph into the antennal vessels without acting as pumps (Figure 2*b*) or are indirectly compressed via dilation of the pharynx. As a rule, however, the ampullae are autonomous pulsatile circulatory organs and can be called antenna-hearts because they are independent from the dorsal vessel.

Muscles associated with the ampullae vary considerably in anatomy and function. In the majority of species, the muscles dilate the ampullae (Figures 2*c, d, e*); in only a few do the muscles compress them (Figure 2*f*; 63, 66). Elastic properties of the ampulla wall or of suspending structures antagonize the action of these muscles. They may attach to a number of structures: compressor muscles to the frontal cuticle or to the pharynx; dilator muscles always have one attachment site at the ampulla wall and the other at the pharynx (Figure 2*d*), the frontal cuticle, or the anterior end of the aorta (Figure 2*e*). Another type of dilator muscle spans the two ampullae and causes their simultaneous dilation upon contraction (Figure 2*d* and *e*; 64).

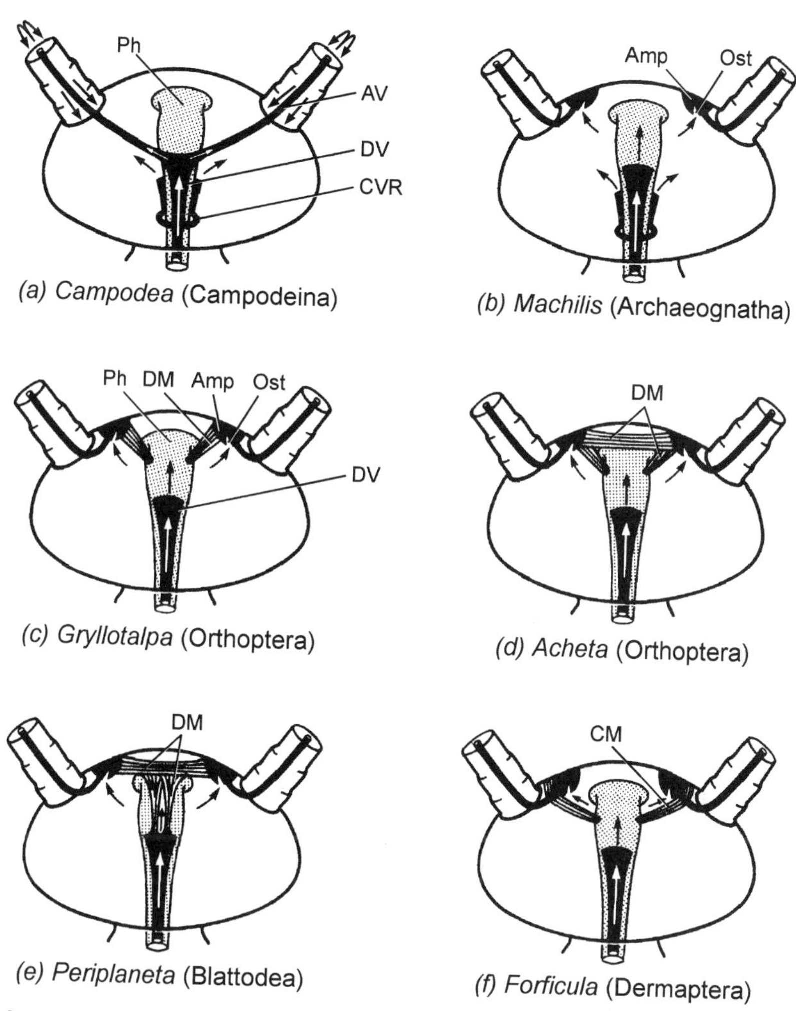

Figure 2 Antennal circulatory organs. Head diagram in dorsal view in different species. Vessels and their derivative structures in *solid black;* foregut is *stippled*; *arrows* indicate hemolymph flow. In nonpulsatile organs, antennal vessels are either *(a)* connected to the dorsal vessel or *(b)* separate from the dorsal vessel having basal nonpulsatile ampullae with ostia (Ost) communicating with the frontal sinus. In pulsatile organs, the basic layout is uniform except for the attachment sites and functions of the associated pumping muscles: *(c)* ampullo-pharyngeal dilator, *(d)* ampullo-pharyngeal and ampullo-ampullary dilators, *(e)* ampullo-ampullary dilator and accessory dilators attached to anterior end of aorta, *(f)* fronto-pharyngeal compressor. In *(a)* and *(b),* also note the circumesophageal vessel ring with ventral trumpet-shaped opening. Abbreviations: ampulla *(Amp),* antennal vessel *(AV),* compressor muscle *(CM),* circumesophageal vessel ring *(CVR),* dilator muscle *(DM),* dorsal vessel *(DV),* ostium *(Ost),* pharynx *(Ph).*

	Non-pulsatile organs			Pulsatile organs					
				ampulla compressor		ampulla dilator			
	1	2	3	4	5	6	7	8	9
Apterygotes									
COLLEMBOLA	-	-	-	-	-	-	-	-	-
PROTURA	-	-	-	-	-	-	-	-	-
CAMPODEINA	■	□	□	□	□	□	□	□	□
JAPYGINA	■	□	□	□	□	□	□	□	□
ARCHAEOGNATHA	□	■	□	□	□	□	□	□	□
ZYGENTOMA	□	■	□	□	□	□	□	□	□
Exopterygotes									
EPHEMEROPTERA	-	-	-	-	-	-	-	-	-
ODONATA	□	□	■	□	□	□	□	□	□
PLECOPTERA	□	■	□	□	□	■	□	□	□
BLATTODEA	□	□	□	□	□	□	■	■	□
ISOPTERA	□	□	□	□	□	□	■	■	□
MANTODEA	□	□	□	□	□	□	■	■	□
GRYLLOBLATTODEA	□	■	□	□	□	□	□	□	□
DERMAPTERA	□	□	□	■	□	□	□	□	□
ORTHOPTERA	□	□	□	□	□	■	■	■	□
PHASMATODEA	□	□	□	□	□	□	■	■	□
EMBIOPTERA	□	□	□	□	□	□	□	□	■
ZORAPTERA	?	?	?	?	?	?	?	?	?
PSOCOPTERA	?	?	?	?	?	?	?	?	?
PHTHIRAPTERA	-	-	-	-	-	-	-	-	-
HEMIPTERA	□	□	□	□	■	□	□	□	□
THYSANOPTERA	?	?	?	?	?	?	?	?	?
Endopterygotes									
STREPSIPTERA	?	?	?	?	?	?	?	?	?
COLEOPTERA	□	□	□	□	■	□	□	□	□
MEGALOPTERA	□	□	□	□	□	□	□	■	□
RAPHIDIOPTERA	?	?	?	?	?	?	?	?	?
NEUROPTERA	□	□	□	□	□	■	□	□	□
MECOPTERA	□	□	□	□	□	■	□	□	□
SIPHONAPTERA	-	-	-	-	-	-	-	-	-
DIPTERA	□	□	□	□	□	■	■	■	□
TRICHOPTERA	□	□	■	□	□	□	□	□	□
LEPIDOPTERA	□	□	■	□	□	□	□	□	□
HYMENOPTERA	□	□	■	□	□	□	□	□	□

Figure 3 Functional types of antennal circulatory organs and their occurrence in insect orders. Numbers 1–9 indicate different organ designs. Nonpulsatile organs: *1 antennal vessels connected to dorsal vessel; 2 antennal vessel with non-pulsatile ampulla; 3 ampullae or frontal sacs indirectly compressed by pharynx movements.* Pulsatile organs with associated muscles: *4 fronto-pharyngeal compressor; 5 fronto-frontal compressor; 6 ampullo-pharyngeal dilator; 7 ampullo-ampullary dilator; 8 ampullo-aortic dilator; 9 ampullo-frontal dilator.* ■ trait present; □ trait absent; - no organ, ? not investigated. Data compiled from references 13, 16–18, 66–68, 81.

Mouthparts

Hemolymph pumping organs have hitherto been found linked only to the lepidopteran proboscis and are involved in its hydraulic extension (4, 42). They are located near the proboscis base and consist of compressible cuticular tubes with associated muscles. Contraction presses the tubes together whereby hemolymph is forced into the proboscis. Elongate mouthparts of other insects presumably also contain special circulatory organs for hemolymph exchange. In some maxillary and labial palps, diaphragms partition the hemocoel into two sinuses, in which countercurrent hemolymph streams can be observed (G Pass, unpublished data); propagation of these streams remains unclear.

Legs

Pulsatile organs in legs are known only in Orthoptera and Hemiptera (39, 68). In these insects a delicate nonmuscular diaphragm takes over the channeling function for each extremity. It longitudinally spans the entire leg and ends just short of the tip. The leg hemocoel is thereby partitioned into two sinuses with countercurrent flows. The efferent flow in one sinus doubles back at the tip where it becomes the afferent flow returning to the thorax in the other sinus.

The pulsatile apparatuses are dissimilar in the two taxa. In the trochanter of the locust middle leg, two small muscles attach to the diaphragm, raising it upon contraction (38); muscle relaxation is followed by flattening of the diaphragm, which forces hemolymph towards the thorax. In Hemiptera, the pumping muscle is associated with the longitudinal diaphragm in the tibia (14, 26, 41; Figure 4*a, b*). Its contraction narrows one sinus and propels hemolymph towards the thorax channeled by a valve flap. Consequently, hemolymph is sucked out of the thorax into the other now dilated sinus. Attachment sites of the leg-heart muscle vary among Hemiptera (cf. Figure 4*a, b*).

Leg diaphragms without associated pumping muscles are reported from a number of other insects (68, 83); how the observed countercurrent flows are generated is not yet fully understood.

In the legs of adult Lepidoptera, an entirely different scheme regulates hemolymph circulation: A large tracheal sac replaces the diaphragm as a partitioner of the leg hemocoel. Mutual dependent fluctuations in the tracheal volume and the thoracic hemolymph bulk caused by heartbeat reversal are considered the impetus for hemolymph exchange (100, 101). The distension of the tracheal sac may be more pronounced in one sinus than in the other, resulting in a forced hemolymph propulsion through the countercurrent sinuses.

Wings

Insect wing veins contain living cells that rely on hemolymph supply. Hemolymph flow generally follows a basic circulation pattern: Anterior veins carry efferent flows and posterior veins afferent ones (1). Pulsatile organs in the thorax power

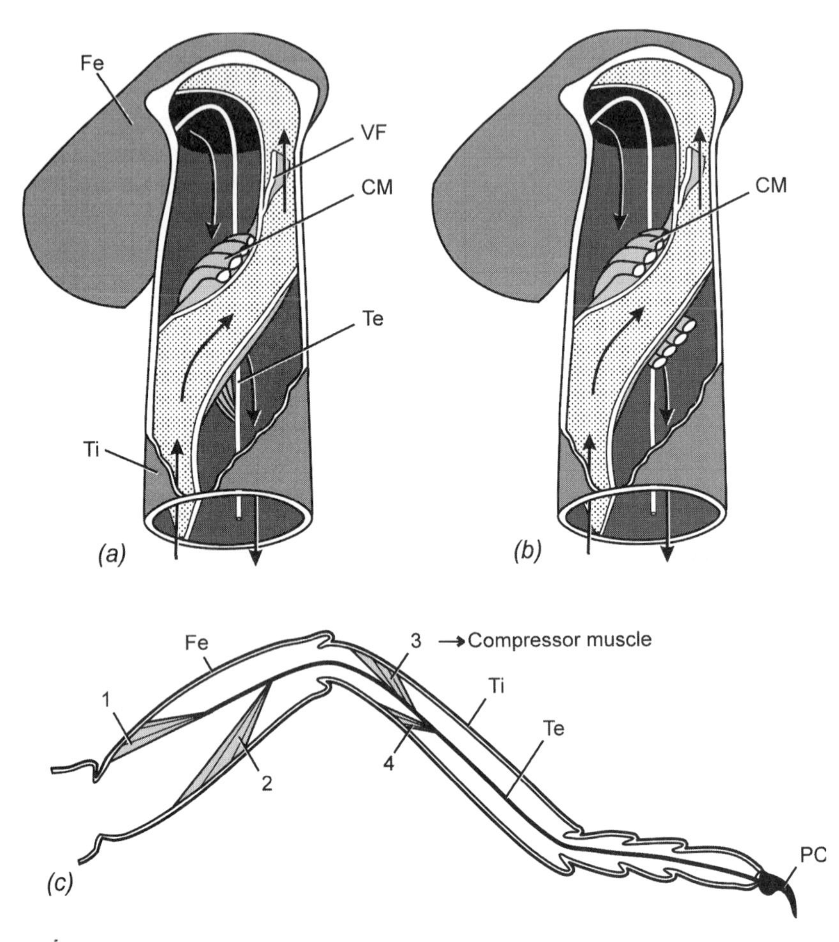

Figure 4 Leg circulatory organs. *(a, b)* Diagrams of the joint region between femur and tibia in two different hemipteran species. Proximal part of tibia opened to show leg-heart region, *arrows* indicate hemolymph flow direction. Longitudinal diaphragm *(stippled)* twisted within the leg hemocoel separate efferent from afferent sinus; valve flap in efferent sinus precludes backflow. The pumping muscle may attach to different locations: *(a)* to the tibia cuticle and tendon of the pretarsal claw flexor as in most Hemiptera; *(b)* the leg-heart muscle has both attachment sites at the tibia cuticle as in Belostomatidae, Nepidae, and Reduviidae partim. *(c)* The pretarsal claw flexor consists of four separate muscle portions (indicated by numbers) inserting at the long tendon, which runs through the entire leg. In Hemiptera, one tibial portion of the pretarsal claw flexor (labeled *3*) was obviously recruited for the leg-heart function. Abbreviations: compressor muscle *(CM)*, femur *(Fe)*, pretarsal claw *(PC)*, tendon of pretarsal claw flexor *(Te)*, tibia *(Ti)*, valve flap *(VF)*.

this circulation (for details on functional types and distribution among insect orders, see Figures 5 and 6). They are located in the winged segments directly beneath the scutellum, which forms the pumping case, and communicate with the posterior wing veins via cuticular tubes. Although these cuticular components are rather invariable in their design, the anatomy of the pulsatile apparatus is not. However, it always functions by the same principle, sucking hemolymph out of the posterior veins. In many species, the pulsatile structures are modifications of the dorsal vessel (6, 11, 43, 45, 46, 104; Figures 5*a* ,*b, c*). They represent simple enlargements or diverticles whose dorsal wall musculature is reinforced and attached to the basal ridge of the scutellum. When relaxed, they occupy part of the small sinus beneath (Figure 5*a*). Contractions flatten the dorsal portion of the pulsatile structure, thereby widening that sinus. This action results in hemolymph suction out of the wing veins (Figure 5*b*). Elastic suspending strands antagonize the muscle contraction and redilate this portion of the dorsal vessel, compressing the small sinus beneath the scutellum. Consequently, hemolymph streams into the lumen of the dorsal vessel via ostia; concurrently, a valve prevents backflow into the wing veins. In other species, the pumping apparatus is made up of a muscular plate called the pulsatile diaphragm (10, 44–46). It may be attached to or be entirely separate from the dorsal vessel (Figures 5*d* and *e*, respectively). In the latter case, hemolymph is being carried through a valved opening directly into the thoracic hemocoel instead of into the dorsal vessel lumen. Typically, one pulsatile diaphragm exists per winged segment, although paired pulsatile diaphragms, one at each wing base, may also occur (46, 92).

In some Coleoptera and Lepidoptera, the so-called tidal flow is still another mode of hemolymph exchange in the wings (95, 98–101). Hemolymph flow into and out of all wing veins occurs simultaneously. The flows are correlated with periodic heartbeat reversal, intermittent pulse activity of the wing-hearts, slow volume changes of the abdomen and the consequential fluctuations of the wing trachea volume. The withdrawal of hemolymph from the wing veins by the concerted action of the pumping organs effects widening of these elastic wing tracheae; upon their relaxation, hemolymph is again sucked back into the wing veins.

Abdominal Appendages

Information is scarce on circulation in the various abdominal body appendages. Vessels communicating with the dorsal vessel supply the long cerci and the terminal filament of apterygotes and mayflies (5, 24, 79; Figure 7*a*). This condition goes along with a bidirectional hemolymph flow within the dorsal vessel. Ephemeroptera have a caudal pulsatile ampulla with a conspicuous muscular wall (54). This structure is linked to the posterior end of the dorsal vessel but contracts with different beat frequencies. The activity of the ampulla contributes significantly to hemolymph propulsion through the abdominal appendages. In Plecoptera, the cercal vessels are separate from the dorsal vessel, which is posteriorly closed. Here, specific pulsatile organs in the anal lobes assume hemolymph circulation

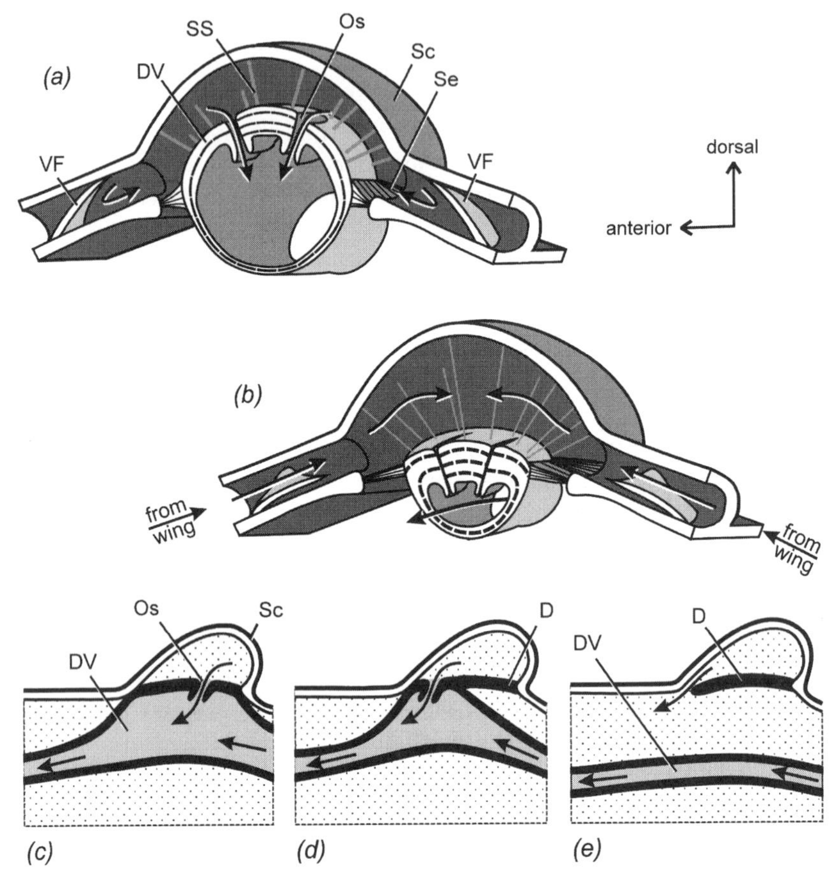

Figure 5 Wing circulatory organs. *(a, b)* Dorsal vessel modification in two distinct phases of action. Diagrams show dorsal portion of a winged thoracic segment in cross section. Scutellum and supplying tubes cut open to show pulsatile apparatus and hemolymph flow directions (*arrows*). The pulsatile apparatus consists of an enlarged vessel portion with strengthened dorsal wall bearing a pair of ostia; it is attached to cuticular structures by a connective tissue septum and numerous elastic suspending strands. See text for functional explanation. *(c, d, e)* Organization levels in different taxa. Midline views of thorax regions. The basic type is dorsal vessel modification *(c)* where hemolymph from wing veins is propelled into dorsal vessel lumen via ostia; in the attached pulsatile diaphragm *(d)*, hemolymph is propelled by action of a muscular plate linked to the dorsal vessel; in the separate pulsatile diaphragm *(e)*, no link to the dorsal vessel exists and hemolymph is forced directly into the thoracic cavity. Abbreviations: dorsal vessel *(DV)*, ostium *(Os)*, pulsatile diaphragm *(D)*, scutellum *(Sc)*, suspending septum *(Se)*, suspending strands *(SS)*, valve flap *(VF)*.

	Dorsal vessel modifications	Pulsatile diaphragms		
		attached unpaired	separate unpaired	separate paired
Exopterygotes				
EPHEMEROPTERA	■	□	□	□
ODONATA	■	□	□	□
PLECOPTERA	■	□	□	□
BLATTODEA	■	□	□	□
ISOPTERA	■	□	□	□
MANTODEA	■	□	□	□
GRYLLOBLATTODEA	-	-	-	-
DERMAPTERA	■	□	□	□
ORTHOPTERA	■	□	□	□
PHASMATODEA	■	□	□	□
EMBIOPTERA	■	□	□	□
ZORAPTERA	?	?	?	?
PSOCOPTERA	■	□	□	□
PHTHIRAPTERA	-	-	-	-
HEMIPTERA	□	□	■	□
THYSANOPTERA	□	□	□	□
Endopterygotes				
STREPSIPTERA	?	?	?	?
COLEOPTERA	■	□	□	□
MEGALOPTERA	□	□	■	□
RAPHIDIOPTERA	□	□	■	□
NEUROPTERA	□	■	■	□
MECOPTERA	□	□	■	□
SIPHONAPTERA	-	-	-	-
DIPTERA	□	□	■	■
TRICHOPTERA	□	■	■	□
LEPIDOPTERA	□	■	■	■
HYMENOPTERA	■	□	■	□

Figure 6 Functional types of wing circulatory organs and their occurrence in insect orders. ■ trait present; □ trait absent; - wingless no organ; ? not investigated. (Data compiled from 45, 46.)

in the cerci (65, Figure 7*b*). Cercal vessels have been found in no other insects investigated so far. Instead, the hemocoel is partitioned by a diaphragm channeling countercurrent flows, whose source of propulsion is unknown (60).

In many insects, the ovipositors can be of considerable length. Very recently, autonomous pulsatile organs have been discovered in the cricket ovipositor (G Pass, BA Gereben-Krenn, R Hustert, unpublished data). The hemocoels of the

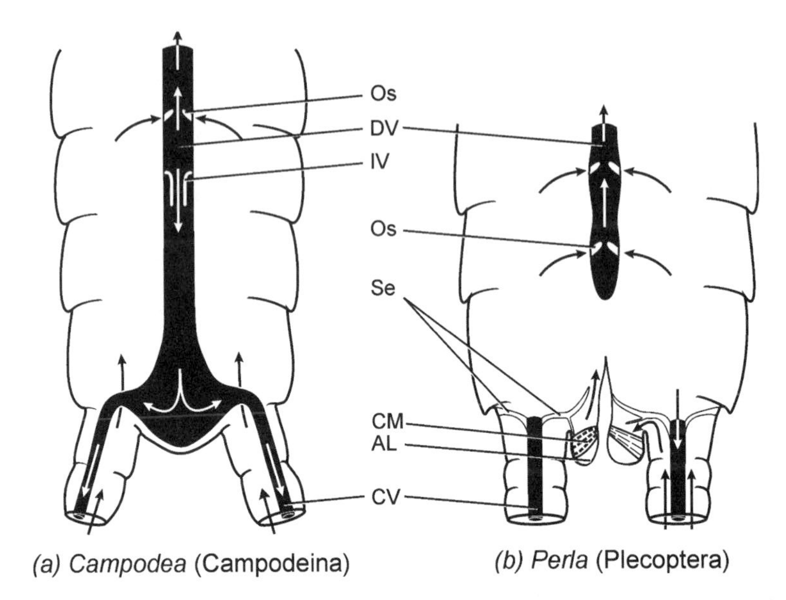

Figure 7 Cercal circulatory organs. Diagram of posterior abdominal segments with cerci. Vessels in *solid black; arrows* indicate hemolymph flow directions. *(a)* Intracardiac valve enforces posteriorly directed hemolymph flow within dorsal vessel continuing into cercal vessels. *(b)* Dorsal vessel separate from cercal vessels; cercus hemocoel isolated by septum from central body cavity. Pulsatile organ in the anal lobes depicted during different phases of action: left anal lobe indented by contraction of pumping muscle thereby forcing hemolymph through valve opening into body cavity; in right anal lobe, relaxation of muscle allows lobe cuticle to return to original shape drawing hemolymph from cercus into anal lobe and from body cavity into cercal vessel. Abbreviations: anal lobe *(AL),* cercal vessel *(CV),* compressor muscle *(CM),* dorsal vessel *(DV),* intracardiac valve *(IV),* ostium *(Os),* septum *(Se).*

ovipositor valvulae are all partitioned by delicate diaphragms. Countercurrent streams in the two sinuses are generated by the action of a pumping muscle at the base of each valvula.

Cardiac Activity and Physiology

Heart physiology is rather well known in the dorsal vessel (39, 56, 57, 107), yet investigation on accessory pulsatile organs is often limited to the registration of pumping activity. As a rule, the auxiliary hearts pulse independently, and their beat frequencies are not synchronized. The rates may be considerably faster (26) or slower (31) than those of the dorsal vessel. Where there is a series of pulsatile organs, such as the six leg-hearts, the beat frequencies are about the same, although they are not exactly in phase (26). Some accessory pulsatile organs work continuously (31, 44) and others discontinuously, with rests of up to a few minutes

(26). Lepidopteran wing-hearts are known to beat intermittently in coordination with the periodic beat reversal in the dorsal vessel (95, 100); during adult eclosion and wing expansion, however, the dorsal vessel pumps only toward the head and the wing-hearts work continuously.

A myogenic pacemaker is inherent in the dorsal vessels of all investigated insects (39, 57). Beat frequencies may be neuronally or hormonally modulated (55, 56). All this also holds true for the investigated accessory pulsatile organs. The cockroach *Periplaneta americana* has been the favorite vehicle for the bulk of studies on auxiliary hearts. Research covers a wide range of topics spanning functional morphology (64), neuroanatomy (7, 69), neurochemistry (70, 72, 106), pharmacology and electrophysiology (31–34, 77). This interest makes the cockroach antenna-heart the best understood of any insect accessory pulsatile organ. Its myogenic rhythm (31, 77) is modulated by neurons located in the subesophageal ganglion (69). Neuroactive substances involved in regulation include octopamine as an inhibitor and some neuropeptides, especially proctolin, as powerful excitors (31–34). Further electrophysiological investigation is as yet restricted to the locust leg-heart; here, the myogenic rhythm is normally controlled by neurons located in the mesothoracic ganglion and occurs synchronously with abdominal ventilatory movements (38).

PHYLOGENETIC PATHWAYS OF CIRCULATORY ORGANS

The newly discovered striking similarities in the developmental genetics of insect and vertebrate hearts point to the very ancient roots of circulatory organs (9, 27). Their common ancestor obviously already had a circulatory system with a contractile dorsal vessel (15). Hence, the notion of the arthropod circulatory organ as a newly emerged functional system (12) becomes implausible. The general tenet holds that the open circulatory system in arthropods is derived from the closed vascular system of annelidlike ancestors (87). Despite doubts about a close relationship between these two taxa (19), the traces of metamery in the vascular system of many primitive arthropods imply a derivation from an ancestor with a segmented body cavity (80). When the metameric organization of the body cavity disappeared in arthropod evolution, a complex vascular system became dispensable. Simple circulatory designs, such as those in insects, must then be derived designs.

In trying to unearth the circulatory design of the common ancestor of insects, the enigmatic and controversial arthropod relationships prove a major obstacle (21, 90, 105). The traditional notion of tight links between insects and myriapods (50, 103) contrasts with the new idea that certain crustaceans are the sister group of insects (3, 22, 75). Remarkably, primitive taxa of both groups possess complex vascular systems whose major components are well-developed dorsal and ventral longitudinal vessels (52, 58, 84). Their body appendages are supplied by arteries originating from these vessels. Accessory pumping structures have also been described: Rather well known are the frontal hearts in Malacostraca (36, 88, 89), which are widenings of the aorta associated with rhythmically contracting esophageal muscles. Separate and autonomous pulsatile organs are absent in nonhexapod arthropods.

Apterygotes

A recurrent trait in apterygotes is small body size, a feature also postulated for their common ancestor, which may have been a reason for reductions of the original arthropod vascular system. Dorsal vessels have nevertheless prevailed in all apterygotes. They appear as uniform, unchambered tubes with segmental ostia. The dorsal diaphragm and alary muscles are poorly developed, and a ventral diaphragm is generally absent. A peculiarity of the dorsal vessel in apterygotes is its inherent bidirectional flow (24). The functional significance of the posteriorly directed flow lies in the supply of the abdominal appendages whose vessels may be linked in different ways with the dorsal vessel.

In apterygotes, there are distinct differences in the hemolymph supply of the antennae. In diplurans, supply is achieved via arteries linked with the anterior end of the dorsal vessel (67). Typically in insects, however, the antennal vessels are separate from the dorsal vessel and have ampullae at their bases. Outgroup comparison suggests that the situation in diplurans is the plesiomorphic character state in insects, whereas the presence of separate antennal vessels is a synapomorphy of Ectognatha.

A circulatory structure unique to apterygotes is the circumesophageal vessel ring in their head (Figures 2*a, b*); its absence is a synapomorphy of Pterygota. Remarkably, a similar structure called the mandibular arch exists in the head of some chilopods (20); homology of these ring vessels is currently intangible.

Exopterygotes

Distinct accessory pulsatile organs first appeared in Pterygota. In most exopterygotes, autonomous pulsatile organs serve to supply the antennae. Ampulla muscles may attach at very different anatomical structures and may also vary in their modes of function (Figure 3; 67, 68). The fact that evolutionary changes in muscle attachment sites cannot be clearly recognized suggests a multiple and independent development of the pulsatile circulatory organs serving antennae.

In pterygotes, the need to supply hemolymph to wings is a new objective, and additional circulatory organs have evolved to meet this demand. The scutellum as a pumping case, with its supplying tubes, is derived from parts of the tergal cuticle. These cuticular structures are uniform in their basic design among all winged insects and can be considered a synapomorphy of Pterygota (45). In almost all orders of exopterygotes, the pulsatile structures of the wing circulatory organs are dorsal vessel modifications and can be considered the plesiomorphic type (cf. Figure 6). Hemiptera are the only exopterygotes to have pulsatile diaphragms separate from the dorsal vessel. This type of wing circulatory organ is normally associated only with endopterygotes and is presumably the result of parallel evolution.

Because data on leg circulatory organs are scarce, conclusions regarding phylogenetic pathways of these organs are not yet possible, except for the respective pulsatile apparatuses of Orthoptera and Hemiptera, which have undoubtedly evolved independently from each other.

Hemolymph supply to abdominal appendages is accomplished in different ways among exopterygotes. Vessels exist only in Ephemeroptera and Plecoptera and diaphragms elsewhere. The caudal pulsatile ampulla of mayflies is probably an autapomorphy; remarkably, these insects are the only pterygotes that share with apterygotes the bidirectional flow design in the dorsal vessel. A clear autapomorphy of stoneflies, on the other hand, is their pulsatile pump in the anal lobes serving cercal circulation.

Endopterygotes

The trend of differentiating the dorsal vessel into a cephalo-thoracic aorta and an abdominal heart is most pronounced in endopterygotes. Within the thorax, the aorta may bend in one or two long loops or follow a straight central course (35, 46). Heartbeat reversal in several endopterygote orders is reflected in various anatomical modifications of the rear end of the dorsal vessel (25, 101).

The great variety of antennal circulatory organ types does not allow for a reconstruction of major evolutionary pathways (Figure 3; 68). The plesiomorphic condition in Endopterygota also remains unresolved at this time. Potential candidates, given their basic design, could be the ampullae of Hymenoptera, which are compressed via the pharynx (50a).

Regarding circulation in wings, dorsal vessel modifications can be found in only two orders (Coleoptera and Hymenoptera; Figure 6). Because the primitive taxa in both orders share these modifications with exopterygotes, these modifications may therefore be regarded as the plesiomorphic condition of wing circulatory organs in endopterygotes (46). Elsewhere, pulsatile diaphragms are either attached to or separate from the dorsal vessel. Transformation lines in some groups suggest that the character state polarity leads from an attached to a separate mode. The distribution of separate pulsatile diaphragms along a cladogram of Endopterygota clearly shows that they must have evolved multiple times, notwithstanding their nearly identical designs (Figure 6). Paired pulsatile diaphragms are unique to endopterygotes (46).

Hemolymph exchange in wing veins and legs supported by volume changes of elastic tracheae or air sacs has so far been substantiated in only Coleoptera and Lepidoptera (98, 100, 101) and is likely to be a derived condition.

EVOLUTIONARY INNOVATIONS IN CIRCULATORY ORGANS

Accessory pulsatile organs are genuine body plan innovations of insects. In the course of their formation, existing organs have been subjected to modifications and other sets of structures have been assembled to build new functional units. Discussion of evolutionary innovation usually focuses on diversification and adaptive radiation (37, 61, 93). However, the internal workings of morphological innovations, such as the questions of from where organ components are recruited and how organ reassembly is triggered, have received less attention (59, 74). The striking diversity of these innovations in accessory pulsatile organs is another aspect that calls for an explanation. In fact, it seems implausible that the pulsatile organs should have such different designs, in one given appendage, when they all serve the same function.

The next sections attempt to give some framework for the emergence and morphological radiation of accessory pulsatile organs. Circulatory organs seem to be choice material for this kind of study given their relatively simple design and the fact that modifications are rather obvious. Starting from the hemodynamic basics, change in functional demands and spatial constraints imposed by other organ systems are discussed as possible forces for the evolution of accessory pulsatile organs.

Hemodynamic Principles

To make hemolymph circulation in body appendages possible, a structure is required separating efferent from afferent flows. This may be a vessel, a diaphragm, or even a tracheal sac. Some smaller appendages such as mouthparts, tracheal gills, and styli demonstrate that accessory pumps are not strictly needed, although countercurrent hemolymph flows occur. Conceivably, different hemolymph flow velocities at the efferent and afferent sinus bases may effectuate slight differences between the hydraulic pressure there; in turn, this condition may enforce a bulk propulsion through the two sinuses. However, almost all body appendages of greater length evolved specific pulsatile organs that enable hemolymph circulation independent from that in the central body cavity.

Still, hemodynamics in insects is an uncharted field. Because hemolymph is a suspension of hemocytes in plasma liquid, its currents in narrow appendages are subjected to a different set of variables than in the wide sinuses of the central body cavity (23, 48, 82). Viscosity of hemolymph within tubes of diameters from hemocyte cell size up to about 500 μm must be substantially lower than its bulk viscosity due to the Fahraeus-Lindquist effect (73).

The hydraulic pressure needed to force hemolymph through long body appendages may be considerable, but it is not accessible to experimental measurement or to calculation owing to a multitude of unmeasurable parameters. Remarkably, some insects manage to move body appendages by creating highly localized fluctuations in hydraulic pressures by means of accessory pulsatile organ action. Examples include the uncoiling of the lepidopteran proboscis (4, 42) and the spreading of the lamellate antennae of scarabaeid beetles (63).

Circulation in appendages is rather slow. For example the hemolymph in the antennae of *Periplaneta americana* requires 10 minutes to exchange (70). In pierid butterfly wings, major veins take 10 to 20 minutes to be dyed in vital staining experiments, whereas the entire veinal system may take up to one hour (99).

Functional Demands

Bearing in mind the huge variation in the dimensions of certain insect body appendages, it would be plausible to assume that alterations in appendage size and volume have been an essential factor in accessory pulsatile organ evolution. Antennae provide good examples of this issue. In an investigation of several orthopterans with drastically different antenna lengths, positive correlations were found between antenna length and strength of the antenna-heart. Its design, however, was basically the same in all (67). This result may be an indication against variation in appendage size as a decisive reason for morphological radiation of accessory pulsatile organs.

Deletion or acquisition of functions is thought to be another impetus for the emergence of evolutionary innovations. The shift of oxygen transport from the circulatory to the tracheal system is widely believed to have effected the profound reductions in the insect vessel system compared with that of their arthropod ancestors (53, 71, 87). In some endopterygotes, however, the circulatory and tracheal systems have again evolved close functional ties (100, 101). These mutual interactions between circulation and tracheal ventilation constitute a highly efficient mechanism for both oxygen supply and hemolymph exchange. They go along with heartbeat reversal and are also the basis for hemolymph exchange via trachea volume fluctuations in body appendages. This new functional link is prone to replace accessory pulsatile organs in legs and wings of some endopterygotes (46, 101).

Thermoregulation as an additional functional innovation has become another chore of the circulatory system in "hot-blooded" insects (28–30). Long aorta loops in the thorax are morphological adaptations enhancing heat absorption by the passing hemolymph. The accessory pulsatile organs, however, are not known to play a significant part in thermoregulation (40, 94). Preliminary investigations reveal that the hemolymph volume in the appendages would be too small and its flow too slow to be relevant for heat dissipation; this condition at least holds true for larger insects.

Hormone distribution certainly requires a high-performance circulatory system. Beyond the mere pumping function, accessory pulsatile organs can also be the sites of hormone release. For example, the antenna-heart of the cockroach *Periplaneta americana* acts as a neurohemal organ (7, 69). Hormones released there are pumped into the antennae and the complex antennal sensory system is their probable target site. Neurochemical studies have uncovered high concentrations of octopamine in the neurohemal areas of the cockroach antenna-heart (70); octopamine released there might well modulate antennal receptor sensitivity. Such a neurohemal function could be quite widespread in accessory pulsatile organs because many insect appendages bear numerous sensilla.

Principles of Organ Design

Evolutionary changes in circulatory organs may have their roots in modifications of their functional demands as well as in constraints appearing in the course of reconstruction in other organ systems. Generally, a hierarchy seems to exist determining whether a certain structure or organ can be easily subjected to alteration or whether it is invariably fixed in the body plan of the respective organism (59, 74). The degree of evolutionary plasticity of a structure may be related to its functional burden (78). In insects, alterations in other organ systems can obviously impose a number of reconstructions on circulatory organs. In several cases, the resulting spatial constraints may have decoupled the task of hemolymph supply from the dorsal vessel. Decoupling of previously linked structures or functions is seen to be one of the most common ways to induce evolutionary innovations (2, 59). In this manner, body appendage supply became independent from the dorsal vessel, and accessory pulsatile organs eventually appeared. In wing circulatory organs, for example, reconstructions of the flight apparatus may have triggered the individualization of wing-hearts from the dorsal vessel (46). Thus the appearance of accessory pulsatile organs during insect evolution may be interpreted as a result of alteration in other organ systems.

Conceivably, building blocks for new organs are recruited from various systems and assembled in new ways. Construction of a pump requires muscles as well as elastic antagonists such as connective tissue structures or flexible cuticle. Comparative investigation of the attachment sites, innervation, and ultrastructure of accessory pulsatile organ muscles has revealed their heterogeneous provenance (Table 1). Circulatory muscles may have been recruited by splitting some fibers off a muscle and shifting their attachment sites, or by displacement of a muscle portion. The former mode has probably been realized in some antenna-hearts (67),

Table 1 Survey of the accessory pulsatile organ components and their supposed provenance

	Pulsatile organ		
Appendage	**Contractile component**	**Elastic antagonist**	**Hemolymph flow conduit**
Antenna	Pharynx dilator	Connective tissue	Vessel
Mouthpart	Skeletal muscle	Flexible cutile	Diaphragm
Leg	Skeletal muscle	Connective tissue	Diaphragm
Wing	Myocardium	Connective tissue	Cuticular tube
Cercus	Rectum dilator	Flexible cuticle	Vessel
Ovipositor	Genital chamber muscle	Flexible cuticle	Diaphragm

whereas the recruitment of an entire muscle portion is evident in leg-hearts (26, 29; Figure 4*c*). The wing-hearts, on the other hand, are supposed to be individualized portions of the myocardium (46), a conclusion that is supported by developmental genetics (49). Obviously, any muscle system is liable to become a component of a circulatory pump given that its location is close to the reconstruction site. Further developmental studies on this subject would certainly be rewarding. Because myogenic autonomy is inherent to all known accessory pulsatile organs, the recruited heart muscles must have attained pacemaker rhythmicity. Analogously, nervous control of the newly assembled hearts must have evolved toward a modulation of the autonomous rhythmic muscle contraction.

CONCLUSION

Because of the limited number and clear arrangement of the involved components, accessory pulsatile organs graphically illustrate the ways in which new organs enter and prevail in the insect bodyplan. Several building blocks originating from various organ systems have been reassembled to form new functional units with their own new physiological properties. These new entities cannot be homologized with any predecessor organ and are therefore evolutionary innovations (59). The functional design of an accessory pulsatile organ can be realized in very different ways in a body appendage, depending on the respective anatomical situation and the available building blocks. Even more astonishing is the diversity of the pulsatile organs of a given appendage in different species. The huge gaps in the understanding of microcirculation notwithstanding, changes in circulatory demands do not seem to be the decisive evolutionary forces for accessory pulsatile organ modification. Instead, developmental and spatial constraints resulting from reconstructions in other organ systems may be responsible for both the appearance of the new pulsatile organs and their subsequent morphological radiation. Accessory pulsatile organs therefore well illustrate the principle that changes in single organ systems can be fully grasped only if examined along with the evolution of the entire organism.

ACKNOWLEDGMENTS

I am deeply grateful to J Edwards, BA Gereben-Krenn, H Krenn, F Ladich, N Szucsich, A Tadler, LT Wasserthal, and C Wirkner for their helpful comments in various stages of the draft. I owe the figures of this manuscript to the graphical expertise of H Grillitsch and T Gatschnegg. Special thanks to T Micholitsch for his linguistic advice in the translation process. Supported by the Austrian Science Foundation in project 10631-Bio.

Note: The figures of this article can be seen in color and animated in the Supplementary Materials of the *Annual Review of Entomology* (www.annualreviews.org).

Visit the Annual Reviews home page at www.AnnualReviews.org.

LITERATURE CITED

1. Arnold JW. 1964. Blood circulation in insect wings. *Mem. Entomol. Soc. Can.* 38:5–60
2. Arnold SJ, Alberch P, Csány V, Dawkins RC, Emerson SB, et al. 1989. Group report: How do complex organisms evolve? In *Complex Organismal Functions: Integration and Evolution in Vertebrates*, ed. DB Wake, G Roth, pp. 403–33. New York: Wiley
3. Averof M, Akam M. 1995. Insect crustacean relationships—insights from comparative developmental and molecular studies. *Philos. Trans. R. Soc. London Ser. B* 347:293–303
4. Bänziger H. 1971. Extension and coiling of the lepidopterous proboscis—a new interpretation of the blood-pressure theory. *Mitt. Schweiz. Entomol. Ges.* 43:225–39
5. Barth R. 1963. Über das Zirkulationssystem einer Machilide (Thysanura). *Mem. Inst. Oswaldo Cruz* 61:371–439
6. Bayer R. 1968. Untersuchungen am Kreislaufsystem der Wanderheuschrecke (*Locusta migratoria migratorioides* R. et F., Orthopteroidea). *Z. Vergl. Physiol.* 58:76–155
7. Beattie TM. 1976. Autolysis in axon terminals of a new neurohaemal organ in the cockroach *Periplaneta americana*. *Tissue Cell* 8:305–10
8. Bitsch J. 1963. Morphologie céphalique des Machilides (Insecta, Thysanura). *Ann. Sci. Nat. Zool. Paris* Sér. #12, 5:501–706
9. Bodmer R. 1993. The gene tinman is required for specification of the heart and visceral muscles in *Drosophila*. *Development* 118:719–29
10. Brocher F. 1919. Les organes pulsatiles méso- et métatergaux des Lépidoptères. *Arch. Zool. Exp. Gén.* 60:1–45
11. Bruserud A. 1985. The ultrastructure of the larval heart and aortic diverticula in *Coenagrion hastulatum* Charpentier (Odonata, Zygoptera). *Zool. Anz.* 214: 25–32
12. Clarke UK. 1979. Visceral anatomy and arthropod phylogeny. In *Arthropod Phylogeny*, ed. AP Gupta, pp. 467–549. New York: Van Nostrand. 762 pp.
13. Clements AN. 1956. The antennal pulsatile organs of mosquitoes and other Diptera. *Q. J. Microsc. Sci* . 97:429–35
14. Debaisieux P. 1936. Organes pulsatiles des tibias de Notonectes. *Ann. Soc. Sci. Bruxelles Ser. B* 56:77–87
15. De Robertis EM, Sasai Y. 1996. A common plan for dorsoventral patterning in Bilateria. *Nature* 380:37–40
16. Dudel H. 1977. Vergleichende funktionsanatomische Untersuchungen über die Antennen der Dipteren. I. Bibiomorpha, Homoeodactyla, Asilomorpha. *Zool. Jahrb. Abt. Anat. Ontog. Tiere* 98:203–308
17. Dudel H. 1978. Vergleichende funktionsanatomische Untersuchungen über die Antennen der Dipteren. II. Cyclorrhapha (Aschiza and Schizophora-Acalyptratae). *Zool. Jahrb. Abt. Anat. Ontog. Tiere* 99:224–98
18. Dudel H. 1978. Vergleichende funktionsanatomische Untersuchungen über die Antennen der Dipteren. III. Calyptratae (I.O.Cyclorrhapha). *Zool. Jahrb. Abt. Anat. Ontog. Tiere* 99:301–70
19. Eernisse DJ. 1998. Arthropod and annelid relationships re-examined. In *Arthropod Relationships*, ed. RA Fortey, RH Thomas, pp. 43–56. London: Chapman & Hall. 383 pp.

20. Fahlander K. 1938. Beiträge zur Anatomie und systematischen Einteilung der Chilopoden. *Zool. Beitr. Uppsala* 17:1–148
21. Fortey RA, Thomas RH, eds. 1998. *Arthropod Relationships*. London: Chapman & Hall. 383 pp.
22. Friedrich M, Tautz D. 1995. Ribosomal DNA-phylogeny of the major arthropod classes and the evolution of myriapods. *Nature* 376:165–67
23. Fung YC. 1984. *Biodynamics: Circulation*. Berlin/Heidelberg/New York: Springer-Verlag. 404 pp.
24. Gereben-Krenn BA, Pass G. 1999. Circulatory organs in Diplura: the basic design in Hexapoda? *Int. J. Insect Morphol. Embryol.* 28:71–79
25. Gerould JH. 1933. Orders of insects with heartbeat reversal. *Biol. Bull.* 64:424–31
26. Hantschk A. 1991. Functional morphology of accessory circulatory organs in the legs of Hemiptera. *Int. J. Insect Morphol. Embryol.* 20:259–73
27. Harvey RP. 1996. NK-2 homeobox genes and heart development. *Dev. Biol.* 178:203–16
28. Heinrich B. 1971. Temperature regulation of the sphinx moth *Manduca sexta*. II. Regulation of heat loss by control of blood circulation. *J. Exp. Biol.* 54:153–66
29. Heinrich B. 1976. Heat exchange in relation to blood flow between thorax and abdomen. *J. Exp. Biol.* 64:561–85
30. Heinrich B. 1993. *The Hot-Blooded Insects, Strategies and Mechanisms of Thermoregulation*. Cambridge, MA: Harvard Univ. Press. 601 pp.
31. Hertel W, Pass G, Penzlin H. 1985. Electrophysiological investigation of the antennal heart of *Periplaneta americana* and its reactions to proctolin. *J. Insect Physiol.* 31:563–72
32. Hertel W, Pass G, Penzlin H. 1988. The effects of the neuropeptide proctolin and of octopamine on the antennal heart of *Periplaneta americana*. *Symp. Biol. Hung.* 36:351–62
33. Hertel W, Penzlin H. 1992. Function and modulation of the antennal heart of *Periplaneta americana*. *Acta Biol. Hung.* 43:113–25
34. Hertel W, Rapus J, Richter M, Eckert M, Vettermann S, Penzlin H. 1997. The proctolinergic control of the antenna-heart in *Periplaneta americana* (L.). *Zoology* 100:70–79
35. Hessel JH. 1969. The comparative morphology of the dorsal vessel and accessory structures of the Lepidoptera and its phylogenetic implications. *Ann. Entomol. Soc. Am.* 62:353–70
36. Huber B. 1992. Frontal heart and arterial system in the head of Isopoda. *Crustaceana* 63:57–69
37. Hunter JP. 1998. Key innovations and the ecology of macroevolution. *TREE* 13:31–36
38. Hustert R. 1999. Locust middle legs are supplied with an accessory pump. *Int. J. Insect Morphol. Embryol.* 28:91–96
39. Jones JC. 1977. *The Circulatory System of Insects*. Springfield, IL: Thomas. 255 pp.
40. Kammer AE, Bracchi J. 1973. Role of the wings in the absorption of radiant energy by a butterfly. *Comp. Biochem. Physiol.* 45:1057–63
41. Kaufman WR, Davey KG. 1971. The pulsatile organ in the tibia of *Triatoma phyllosoma pallidipennis*. *Can. Entomol.* 103:487–96
42. Krenn HW. 1990. Functional morphology and movements of the proboscis of Lepidoptera (Insecta). *Zoomorphology* 110:105–14
43. Krenn HW. 1993. Postembryonic development of accessory wing circulatory organs in *Locusta migratoria* (Orthoptera: Acrididae). *Zool. Anz.* 230:227–36
44. Krenn HW, Pass G. 1993. Wing-hearts in Mecoptera (Insecta). *Int. J. Insect Morphol. Embryol.* 22:63–76
45. Krenn HW, Pass G. 1994. Morphological

diversity and phylogenetic analysis of wing circulatory organs in insects. Part I: Non-Holometabola. *Zoology* 98:7–22
46. Krenn HW, Pass G. 1995. Morphological diversity and phylogenetic analysis of wing circulatory organs in insects. Part II: Holometabola. *Zoology* 98:147–64
47. Kristensen NP. 1991. Phylogeny of extant hexapods. In *The Insects of Australia*, ed. Commonw. Sci. Ind. Res. Organ. (CISRO), pp. 125–40, Carlton, Victoria: Melbourne Univ. Press. 2nd ed.
48. LaBarbera M. 1990. Principles of design of fluid transport systems in zoology. *Science* 249:992–1000
49. Lawrence PA. 1982. Cell lineage of the thoracic muscles of *Drosophila*. *Cell* 29:493–503
50. Manton SA. 1977. *The Arthropoda: Habits, Functional Morphology and Evolution.* Oxford: Claredon Press. 527 pp.
50a. Matus S, Pass G. 1999. Antennal circulatory organ of *Apis Mellifera* L. (Hymenoptera: Apidae) and other Hymenoptera: functional morphology and phylogenetic aspects. *Int. J. Insect Morphol. Emtryol.* 28:97–109
51. McIndoo NE. 1939. Segmental blood vessels of the American cockroach (*Periplaneta americana* L.). *J. Morphol.* 65:323–51
52. McLaughlin PA. 1983. Internal anatomy. In *The Biology of Crustacea,* ed. LH Mantel, 5:1–53. New York: Academic. 479 pp.
53. McMahon BR, Burnett LE. 1990. The crustacean open circulatory system: a reexamination. *Physiol. Zool.* 63:35–71
54. Meyer E. 1931. Über den Blutkreislauf der Ephemeriden. *Z. Morphol. Oekol. Tiere* 22:1–51
55. Miller TA. 1979. Nervous versus neurohormonal control of insect heartbeat. *Am. Zool.* 19:77–86
56. Miller TA. 1985. Heart and diaphragms. In *Comprehensive Insect Physiology, Biochemistry and Pharmacology,* ed. GA Kerkut, LI Gilbert, 11:119–29. Oxford: Pergamon. 575 pp.
57. Miller TA. 1985. Structure and physiology of the circulatory system. In *Comprehensive Insect Physiology, Biochemistry and Pharmacology,* ed. GA Kerkut, LI Gilbert, 3:289–353. Oxford: Pergamon. 625 pp.
58. Minelli A. 1993. Chilopoda. In *Microscopic Anatomy of Invertebrates*, ed. FW Harrison, ME Rice, 12:57–114. New York: Wiley. 484 pp.
59. Müller GB, Wagner GP. 1991. Novelty in evolution: restructuring the concept. *Annu. Rev. Ecol. Syst.* 22:229–56
60. Murray JA. 1967. Morphology of the cercus in *Blattella germanica* (Blattaria: Pseudomopinae). *Ann. Entomol. Soc. Am.* 60:10–16
61. Nitecki MW, ed. 1990. *Evolutionary Innovations.* Chicago: Univ. Chicago Press. 304 pp.
62. Nutting WL. 1951. A comparative anatomical study of the heart and accessory structures of the orthopteroid insects. *J. Morphol.* 89:501–97
63. Pass G. 1980. The anatomy and ultrastructure of the antennal circulatory organs in the cockchafer beetle *Melolontha melolontha* L. (Coleoptera, Scarabaeidae). *Zoomorphology* 96:77–89
64. Pass G. 1985. Gross and fine structure of the antennal circulatory organ in cockroaches (Blattodea, Insecta). *J. Morphol.* 185:255–68
65. Pass G. 1987. "Cercus heart" in stoneflies—a new type of accessory circulatory organ in insects. *Naturwissenschaften* 74:440–41
66. Pass G. 1988. Functional morphology and evolutionary aspects of unusal antennal circulatory organs in *Labidura riparia* Pallas (Labiduridae), *Forficula auricularia* L. and *Chelidurella acanthopygia* Géné (Forficulidae) (Dermaptera: Insecta). *Int. J. Insect Morphol. Embryol.* 17:103–12
67. Pass G. 1991. Antennal circulatory

organs in Onychophora, Myriapoda and Hexapoda: functional morphology and evolutionary implications. *Zoomorphology* 110:145–64
68. Pass G. 1998. Accessory pulsatile organs. In *Microscopic Anatomy of Invertebrates*, ed. F Harrison, M Locke, 11B:621–40. New York: Wiley. 296 pp.
69. Pass G, Agricola H, Birkenbeil H, Penzlin H. 1988. Morphology of neurones associated with the antennal heart of *Periplaneta americana* (Blattodea, Insecta). *Cell Tissue Res.* 253:319–26
70. Pass G, Sperk G, Agricola H, Baumann E, Penzlin H. 1988. Octopamine in a neurohaemal area within the antennal heart of the American cockroach. *J. Exp. Biol.* 135:495–98
71. Paul RJ, Bihlmayer S, Colmorgen M, Zahler S. 1994. The open circulatory system of spiders (*Eurypelma californicum, Pholcus phalangioides*): a survey of functional morphology and physiology. *Physiol. Zool.* 67:1360–82
72. Predel R, Agricola H, Linde D, Wollweber L, Veenstra A, Penzlin H. 1994. The insect neuropeptide corazonin: physiological and immunocytochemical studies in Blattariae. *Zoology* 98:35–49
73. Pries AR, Neuhaus D, Gaehtgens P. 1992. Blood viscosity in tube flow: dependence on diameter and hematocrit. *Am. J. Physiol.* 263:1770–78
74. Raff RA. 1996. *The Shape of Life: Genes, Development and the Evolution of Animal Form.* Chicago: Univ. Chicago Press. 520 pp.
75. Regier JC, Shultz JW. 1997. Molecular phylogeny of the major arthropod groups indicates polyphyly of crustaceans and a new hypothesis for the origin of hexapods. *Mol. Biol. Evol.* 14:902–13
76. Richards AG. 1963. The ventral diaphragm of insects. *J. Morphol.* 113:17–47
77. Richter M, Hertel W. 1997. Contributions to physiology of the antenna-heart in *Periplaneta americana* (L.) (Blattodea: Blattidae). *J. Insect Physiol.* 43:1015–21
78. Riedl R. 1978. *Order in Living Organisms: A Systems Analysis of Evolution.* Chichester: Wiley. 313 pp.
79. Rousset A. 1974. Les différenciations postérieures du vaisseau dorsal de *Thermobia domestica* (Packard) (Insecta, Lepismatida). Anatomie et innervation. *C. R. Acad. Sci. Paris D* 278:2449–52
80. Ruppert EE, Carle KJ. 1983. Morphology of metazoan circulatory systems. *Zoomorphology* 103:193–208
81. Schneider D, Kaissling KE. 1959. Der Bau der Antenne des Seidenspinners *Bombyx mori* L. III. Das Bindegewebe und das Blutgefäß. *Zool. Jahrb. Abt. Anat. Ontog. Tiere* 77:111–32
82. Secomb TW. 1995. Mechanics of blood flow in the microcirculation. In *Biological Fluid Dynamics*, ed. CP Ellington, TJ Pedley, pp. 305–21. Cambridge: Co. Biol. 363 pp.
83. Selman BJ. 1965. The circulatory system of the alder fly *Sialis lutaria. Proc. Zool. Soc. London* 144:487–535
84. Siewing R. 1956. Untersuchungen zur Morphologie der Malacostraca (Crustacea). *Zool. Jahrb. Abt. Anat. Ontog. Tiere* 75:39–176
85. Sláma K. 1976. Insect haemolymph pressure and its determination. *Acta Entomol. Bohemoslov.* 74:362–74
86. Sláma K, Baudry-Partiaogolou N, Provansal-Baudez A. 1979. Control of extracardiac haemolymph pressure pulses in *Tenebrio molitor. J. Insect Physiol.* 25:825–31
87. Snodgrass RE. 1938. Evolution of the Annelida, Onychophora and Arthropoda. *Smithson. Misc. Collect.* 97:1–159
88. Steinacker A. 1978. The anatomy of the decapod auxiliary heart. *Biol. Bull. Woods Hole* 154:497–507
89. Steinacker A. 1979. Neural and neurosecretory control of the decapod crustacean auxiliary heart. *Am. Zool.* 19:67–75
90. Stys P, Zrzavý J. 1994. Phylogeny and

classification of extant Arthropoda: review of hypotheses and nomenclature. *Eur. J. Entomol.* 91:257–75

91. Tartes U, Kuusik A. 1994. Periodic muscular activity and its possible functions in pupae of *Tenebrio molitor. Physiol. Entomol.* 19:216–22
92. Thomsen E. 1938. Über den Kreislauf im Flügel der Musciden, mit besonderer Berücksichtigung der akzessorischen pulsierenden Organe. *Z. Morphol. Oekol. Tiere* 34:416–38
93. Thomson KS. 1992. Macroevolution: the morphological problem. *Am. Zool.* 32:106–112
94. Wasserthal LT. 1975. The role of butterfly wings in regulation of body temperature. *J. Insect Physiol.* 21:1921–30
95. Wasserthal LT. 1976. Heartbeat reversal and its coordination with accessory pulsatile organs and abdominal movements in Lepidoptera. *Experientia* 32:577–78
96. Wasserthal LT. 1980. Oscillating haemolymph "circulation" in the butterfly *Papilio machaon* L. revealed by contact thermography and photocell measurements. *J. Comp. Physiol.* 139:145–63
97. Wasserthal LT. 1981. Oscillating haemolymph "circulation" and discontinuous tracheal ventilation in the giant silk moth *Attacus atlas* L. *J. Comp. Physiol.* 145:1–15
98. Wasserthal LT. 1982. Antagonism between haemolymph transport and tracheal ventilation in an insect wing (*Attacus atlas* L.): a disproof of the generalized model of insect wing circulation. *J. Comp. Physiol.* 147:27–40
99. Wasserthal LT. 1983. Haemolymph flows in the wings of pierid butterflies visualized by vital staining (Insecta, Lepidoptera). *Zoomorphology* 103:177–92
100. Wasserthal LT. 1996. Interaction of circulation and tracheal ventilation in holometabolous insects. *Adv. Insect Physiol.* 26:297–351
101. Wasserthal LT. 1998. The open haemolymph system of Holometabola and its relation to the tracheal space. In *Microscopic Anatomy of Invertebrates,* ed. F Harrison, M Locke, 11B:583–620. New York: Wiley. 296 pp.
102. Wasserthal LT. 1999. Functional morphology of the heart and of a new cephalic pulsatile organ in the blowfly *Calliphora vicina* (Diptera: Calliphoridae) and their role in hemolymph transport and tracheal ventilation. *Int. J. Insect Morphol. Embryol.* 28:111–29
103. Weygoldt P. 1979. Arthropod interrelationships—the phylogenetic-systematic approach. *Z. Zool. Syst. Evolutionsforsch.* 24:19–35
104. Whedon A. 1938. The aortic diverticula of the Odonata. *J. Morphol.* 63:229–61
105. Wheeler WC, Cartwright P, Hayashi CY. 1993. Arthropod phylogeny: a combined approach. *Cladistics* 9:1–39
106. Woodhead AP, Stoltzman CA, Stay B. 1992. Allatostatins in the nerves of the antennal pulsatile organ muscle of the cockroach *Diploptera punctata. Arch. Insect Biochem. Physiol.* 20:253–63
107. Woodring JP. 1985. Circulatory systems. In *Fundamentals of Insect Physiology*, ed. MS Blum, pp. 5–183. New York: Wiley. 598 pp.

Annu. Rev. Entomol. 2000. 45:519–548

PARASITIC MITES OF HONEY BEES: Life History, Implications, and Impact

Diana Sammataro[1], Uri Gerson[2], and Glen Needham[3]
[1]*Department of Entomology, The Pennsylvania State University, 501 Agricultural Sciences and Industries Building, University Park, PA 16802; e-mail: acarapis@psu.edu*
[2]*Department of Entomology, Faculty of Agricultural, Food and Environmental Quality Sciences, Hebrew University of Jerusalem, Rehovot 76100, Israel; e-mail: gerson@agri.huji.ac.il*
[3]*Acarology Laboratory, Department of Entomology, 484 W. 12th Ave., The Ohio State University, Columbus, Ohio 43210; e-mail: Needham.1@osu.edu*

Key Words bee mites, *Acarapis*, *Varroa*, *Tropilaelaps*, *Apis mellifera*

■ **Abstract** The hive of the honey bee is a suitable habitat for diverse mites (Acari), including nonparasitic, omnivorous, and pollen-feeding species, and parasites. The biology and damage of the three main pest species *Acarapis woodi, Varroa jacobsoni*, and *Tropilaelaps clareae* is reviewed, along with detection and control methods. The hypothesis that *Acarapis woodi* is a recently evolved species is rejected. Mite-associated bee pathologies (mostly viral) also cause increasing losses to apiaries. Future studies on bee mites are beset by three main problems: *(a)* The recent discovery of several new honey bee species and new bee-parasitizing mite species (along with the probability that several species are masquerading under the name *Varroa jacobsoni*) may bring about new bee-mite associations and increase damage to beekeeping; *(b)* methods for studying bee pathologies caused by viruses are still largely lacking; *(c)* few bee- and consumer-friendly methods for controlling bee mites in large apiaries are available.

INTRODUCTION

The mites (Acari) that parasitize honey bees have become a global problem. They are threatening the survival of managed and feral honey bees, the beekeeping industry and, due to the role of bees in pollination, the future of many agricultural crops. *Acarapis woodi*, *Varroa jacobsoni*, and *Tropilaelaps clareae* are the main pests, but about 100 mostly harmless mite species are associated with honey bees (66, 167). The significance of bee mites to apiculture is emphasized by the publication of specialized books (35, 131, 135, 206) and by sessions devoted to these pests in apicultural, entomological, and acarological meetings.

Bee mites have greater dispersal potential than most other Acari, first through their hosts and then by humans who move bees primarily for commerce and pollination. Two species of parasitic mites introduced into the United States in the early 1980s changed beekeeping profoundly by causing epidemic losses, ranging from 25% to 80% of managed colonies between 1995 and 1996 (79). Also, feral bee colonies are virtually gone in regions where both mites occur (115). This void of wild bees has been especially noticed by homeowners and growers who had relied on bees to pollinate their crops. Further consequences of these introduced mites are: *(a)* beekeepers and bee breeders going out of business, *(b)* fewer managed colonies, *(c)* increased demand and cost for leasing hives, *(d)* increased incidence of pathogens in colonies, and *(e)* problems with some commercially reared queens and bees (32).

THE HIVE AS A HABITAT FOR MITES

The biology of parasitic bee mites, and hence control, can best be understood by reference to their hosts. The honey bee, *Apis mellifera*, is a social insect that lives in colonies. Each colony has two female castes: the single queen and the workers who may number 20,000 to 50,000 per colony. Colony life is regulated by an array of pheromones, emitted mainly by the queen (81). The colony also has males, called drones, who serve mainly to inseminate the queen during her mating flight.

Characteristic of social insects, a division of labor exists in the colony. Young workers nurse and groom the queen and brood. Older workers forage for pollen, nectar, propolis, and water, and defend the colony. Nest homeostasis is maintained by the workers, who keep the temperature and humidity nearly constant despite external conditions. An abundance of proteinaceous foods (pollen) and carbohydrates (honey) are usually available. Total development of the queen requires about 16 days, that of workers 21 days, and the drones 24 days. The queen can survive for several years, but workers and drones usually live only a few weeks. During cold periods the workers, who may then survive for several months, cluster around the queen to incubate eggs laid during this time.

Non-bee invaders that circumvent colony defenses will thus benefit from the favorable conditions of the hive environment for their own development and reproduction. Most of the successful invaders are mites, and they make up the largest and most diverse group of honey bee associates. Different honey bee species have varied nesting habitats, and their relationships with acarine parasites are discussed in this review.

Hive mites were placed in four groups (67), namely scavengers, predators of scavengers, phoretics, and parasites. Occasional visitors, found in all four of these groups, will not be discussed. For simplicity, we divide the bee mites into non-parasitic and parasitic groups.

NON-PARASITIC BEE MITES

Three common suborders of mites associated with bees are the Astigmata, Prostigmata, and the Mesostigmata. Many astigmatic mites live on the hive's floor, feeding on bee debris, dead insects and fungi. *Forcellinia faini* (Astigmata), a common, whitish, slow-moving scavenger initially described in Puerto Rico, was abundantly collected in hive debris in northern and southern Thailand (77). A representative of the Prostigmata is the tarsonemid *Pseudacarapis indoapis* (Lindquist), a probable pollen feeder, which is apparently restricted to *Apis cerana* (122a).

Melichares dentriticus (Mesostigmata), a cosmopolitan predator on scavenger mites, is common in stored products (105), while members of *Neocypholaelaps* and *Afrocypholaelaps* live in flowers. They feed on the pollen of subtropical and tropical trees and are phoretic on bees. Several dozen *Afrocypholaelaps africana* were found on individual bees visiting mangrove umbels in Queensland (187). Mites dispersing on *A. mellifera*, mostly as egg-bearing females, board and depart the bees via the tongue. The bees did not appear to be annoyed by these mites, nor were their foraging activities disrupted. Ramanan & Ghai (161) reported that one to 400 *A. africana* occurred on individuals of *A. cerana* in India. As bees return to the hive, mites disembark, roam on the combs and subsist on the pollen.

Melittiphis alvearius has been found in hives and on bees in various parts of the world (57). Serological procedures demonstrated that, contrary to views formerly held, this mite is not a predator, but feeds on stored pollen (90).

PARASITIC MITES

Three species of parasitic bee mites are of economic importance due to their destruction of honey bee colonies worldwide. We will focus on the tracheal, varroa, and tropilaelaps mites, as well as provide some information on lesser known mites on other *Apis* species.

Tracheal Mites

The honey bee tracheal mites (HBTM) *Acarapis woodi* (Rennie) live inside the tracheae and air sacs of adult bees. Mites in tracheal systems seem to be rare in arthropods and not well studied. Approximately 15 species are known to parasitize members of Hymenoptera, Orthoptera, Lepidoptera, and Hemiptera (176).

HBTM were first observed earlier this century, when bees on the Isle of Wight were dying from an unknown disease. The die-off reached epidemic levels between 1904 and 1919 (36). At first thought to be caused by a bacterium, the disease was identified by Zander (214) as *Nosema apis*, a protozoan parasite of the bee's alimentary tract. As the disease spread throughout Europe (157), a mite living in the bee's tracheal tubes was discovered and named *Tarsonemus woodi*

(165, 166) [later renamed *Acarapis*, from *Acarus,* mite, and *apis*, bee (102)]. The disease was then called Acarine or Isle of Wight disease. The real cause of the loss of colonies during this time is still unknown and may have been several diseases or other factors causing the symptoms (207).

The identification of *A. woodi* in Europe led the United States Congress in 1922 to prohibit importation of all bees (158) and to examine bees for evidence of mites from apiaries in many states. Other species of *Acarapis*, namely *A. externis* Morgenthaler and *A. dorsalis* Morgenthaler (but not *A. woodi*), were found as a consequence of this intensive sampling conducted by the USDA honey bee laboratories (201–203). The reasons for the HBTM's absence in North America remain a mystery: At least eight honey bee subspecies had been imported to the New World since the 1600s (188). HBTM is now thought to be worldwide (wherever European bees have been introduced) except in Sweden, Norway, Denmark, New Zealand, and Australia, and in the state of Hawaii.

South America allowed bee importation and in 1980, *A. woodi* was reported in Colombia (133). The mites moved northward and reached Texas by 1984; they were later reported in seven states that same year. This fast spread was greatly facilitated by the extensive trucking of bee colonies from southern states northward for pollination and for sale as package bees and queens.

The HBTM is responsible for significant colony losses throughout North America. Reports of losses as great as 90% were recorded just two years after initial discovery (176). A heavy HBTM load causes diminished brood area, smaller bee populations, looser winter clusters, increased honey consumption, lower honey yields (10, 68, 140, 147, 170), and, ultimately, colony demise. In temperate regions, mite populations increase during winter, when bees are confined to the hive, and decrease in summer when bee populations are highest (178). In subtropical climates, the cycle is similar (192), even though bees are not so confined. Unfortunately, the introduction of varroa mites has overshadowed the impact HBTM has on bees.

Recent Evolution of HBTM? Eickwort (66) speculated that *Acarapis* might have evolved from saprophagous or predatory mites and their appearance in hives may have been due to the nesting behavior of Apinae, which provided habitats for these mites. The evolutionary history of *Acarapis* suggests that they may have been pre-adapted for this way of life. All members of the tarsonemid subfamily Acarapinae occur and feed on insects; the related Coreitarsonemini are parasitic in the thoractic odoriferous glands of coreid bugs (122a).

Morse & Eickwort (139) hypothesized that the invasion of tracheae by *A. woodi* was a recent evolutionary event, believed to have begun in England around the year 1900. HBTM evolved from one of the closely related external *Acarapis* species, probably *A. dorsalis*. Their argument was based on the (apparent) formerly restricted distribution of HBTM and on the incipient resistance to this pest in North America. Initially, the mite was believed to occur only in England, Switzerland, and Russia, whereas *A. dorsalis* and *A. externis* are worldwide in distribution. If the pest had been present in England in former times, why was it not noted? And if it was more widely distributed, what prevented it from damaging honey bees in other regions also? Finally, the observation that honey bee populations in North America show a wide range of susceptibilities, as well as resistance, to *A. woodi* could indicate quite recent exposure to the pest (139).

Acceptance of this theory presents many difficulties, such as a rather truncated evolutionary time scale and the differences between this species and *A. dorsalis*, its postulated progenitor. Table 1 lists some of these differences (based on 59, 65, 169). Such an array of changes could hardly have evolved in the brief span of a century. The discovery of HBTM in many regions of mainland Europe, where migratory beekeeping and trade in bees were limited, soon after its initial description (10) also argues against *A. woodi*'s recent evolution.

Why then was the damage attributable to *A. woodi* not noticed earlier? Nowadays beekeeping is much more productive than in the past, and factors that reduce colony yield are more critically observed. Minor diseases and pests, which cause only marginal reductions in bee longevity and yield, were probably masked in the past by more severe mortality factors. The great advances in bee health and management, however, raised the awareness of beekeepers toward pests such as *A. woodi*, whose ravages did not decline through modern control methods. The pest might have been overlooked, even in England, had not the outbreak of the Isle of Wight disease necessitated the intensive examinations of bees. Eickwort (66) suggested a similar scenario in regard to the discovery of *A. woodi* in Europe, where it was noticed only after an epidemic of viral diseases led to bee dissections.

Another explanation for their "sudden appearance" concerns beekeeping practices, which have changed significantly in the past century. Straw skep hives were

TABLE 1 Differences between *Acarapis woodi* and *Acarapis dorsalis*.

Character	*Acarapis woodi*	*Acarapis dorsalis*
Site of reproduction	Inside body, in tracheae	Outside bee, in dorsal groove and at wing bases
Adult movement	Remains in one host bee except for questing adults	Moves to other host bees
Maximal attachment to bees	1–6 days	12 days
Season of maximum bee infestation	Autumn to spring	Spring to summer
Length of development	264 hours	216–240 hours
Female anterior median apodeme	Incompletely developed, not joining transverse apodeme	Developed, joining transverse apodeme
Female posterior margin of coxal plate IV	Shallowly notched	Deeply incised
Male solenidion on tibia I	Slightly club shaped	Not club shaped
Number of setae on male femur-genu-tibia of Leg I	4–4–7	3–3–6

used by early beekeepers, and the bees tended to swarm frequently (172) or were killed to harvest honey. Such methods resulted in smaller, more remote apiaries that kept bees apart, admixing neither bee nor mite strains. Modern beekeepers maintain colonies in wooden boxes, in greater numbers, and move hives frequently to take advantage of bee forage or for pollination. Such practices result in large numbers of perennial hives from diverse regions commingling and rapidly infecting all with the mites and diseases.

Life Cycle Like other members of the prostigmatic Heterostigmata, *A. woodi* has a foreshortened life cycle with only three apparent stages, namely egg, larva, and adult. However, the mite has an apdous nymphal instar that remains inside the larval skin (122a). Males complete their development in 11 to 12 days, females in 14 to 15 days. A generation is thus raised in two weeks, explaining the rapid population growth of the HBTM (for detailed biological information see 155 and 207). Mites feed on bee hemolymph, which they obtain by piercing the tracheae with their closed-ended, sharply pointed stylets that move by internal chitinous levers (101). Once the bee trachea is pierced, the mites' mouth, located just below the stylets, is appressed to the wound and the mites suck host hemolymph through the short tube into the pharynx.

All instars live within the tracheae, except during a brief period when adult females disperse to search for new hosts (181, 194). Dispersing female mites (mated) are attracted to air expelled from the prothoracic (first thoracic) spiracle of young bees (101), as well as to specific hydrocarbons from the cuticle of callow (less than four days old) bees (156). Because older bees might not live long enough for the HBTM to complete its cycle, mites are less attracted to older bees.

Once a suitable host is found, preferably a drone (171), the female enters its spiracle to reach the tracheae and lay eggs. Tracheal mites and their eggs also occur at a lower rate in the air sacs of the bees' abdomen and head, and externally on the wing bases (91). In these alternate locations, neither their effect on the host nor their fate is known.

The mite's small size is critical to its survival (females measure 120 to190 μm long by 77 to 80 μm wide and weigh 5.5 × 10-4 mg; males 125 to 136 μm by 60 to 77 μm and weigh 2.61 × 10-4 mg; 176). The tiny mites can hide under the flat lobe that covers the bee's first thoracic spiracle (access the main tracheal trunk), which many individuals can thus occupy.

Mites begin to disperse by questing on bee setae when the host bee is more than 13 days old, peaking at 15 to 25 days. *A. woodi* is vulnerable to desiccation and starvation during this time outside the host, and survival depends on the ambient temperature and humidity as well as on its state of nourishment (86, 194). A mite can die after a few hours unless it enters a host (101). Mites are also at risk of being dislodged during bee flight and grooming.

Diagnosis HBTM are not visible to the naked eye, making diagnosis difficult. Consequently, beekeepers often use unreliable bee stress symptoms, which include dwindling populations, weak bees crawling on the ground with disjointed hind-wings (called K-wings) and abandoned, overwintered hives full of honey.

The only certain way to identify mite infestation is to dissect the tracheae of bees and visualize the parasites. Bees are collected in winter or early spring, when HBTM populations are highest; fewer mites are found in the summer, due to the dilution effect caused by the rapid emergence of many young bees.

Nine methods for diagnosing HBTM are described (190). Most involve some form of dissection, for which a dissecting microscope (at 40X to 60X magnification) and a pair of fine jeweler's forceps are necessary. Because drones are favored by HBTM and are bigger than workers, they should be collected and dissected first. Queens, even those commercially reared, often have HBTM (27, 151). Camazine et al (32) found that infested queens weighed less.

Fresh or frozen specimens are preferred for dissection over alcohol-preserved material, as alcohol darkens the tissues, making visualization of mites difficult. Tomasko et al (198) developed a sequential sampling method to determine how many samples are needed for accurate measurements. A time-consuming method of cutting thoracic discs (190) and staining the tracheae (150) can be used, but this method requires double handling and identification is delayed. Camazine (30) used a kitchen blender to pulverize bee thoraces, allowing the air-filled tracheae to float, then sampled the surface debris for mites.

Serological diagnoses using ELISA techniques have been developed (78, 93, 159, 160). The visualization of guanine, a nitrogenous waste product not excreted by bees, was advocated (144). Removing the head, pulling off the flat lobe covering the first thoracic spiracle and extracting the tracheal tube (193) is faster and provides immediate diagnosis once mastered. If live bees are used, dead versus live mites are counted when testing acaricides (68, 194).

Controlling Tracheal Mites

Chemical The overriding constraints for chemical control are that the chemicals must be effective against the target and harmless to bees, and they must not accumulate in hive products. Because bees and mites are both arthropods, many of their basic physiological processes are similar, narrowing the possibilities for finding suitable toxicants. Bees are extremely sensitive to accidental poisoning by many of the common agrochemicals (8). To control HBTM, the material must reach the bee tracheae via a volatile compound, be inhaled by the bee, and be lethal only to the parasite. A single registered treatment in the United States is pure menthol crystals, originally extracted from the plant *Mentha arvensis*. Each two-story colony requires 1.8 ounces (50 g) or one packet for two weeks. However, in cold conditions menthol sublimation is ineffective because an insufficient amount of vapor is released from the crystals; conversely, at high temperatures the vapors may repel bees from the hive.

An effective pesticide, Miticur (sold as Amitraz) has been withdrawn from the U.S. market but still is used in Israel (40). Formic acid (104), a potentially effective agent, though caustic to humans, will soon be released in the United States; it is used in Canada.

Cultural An alternate, environmentally safe control is to apply a vegetable shortening and sugar patty at peak mite populations. A quarter-pound (113 g) patty, placed on the top bars at the center of the broodnest where it comes in contact with the most bees, will protect young bees (which are most at risk) from becoming infested. The oil appears to disrupt the questing female mite searching for a new host (181). Because young bees emerge continuously, the patty must be present for an extended period. The optimal application season is in the fall and early spring, when mite levels are rising.

Resistant Bees Several lines of bees resistant to tracheal mites have been developed, starting with Brother Adam's Buckfast bees (23). Some of this stock is commercially reared and sold to beekeepers. Resistance to HBTM seems to be accomplished by the increased grooming behavior of bees (42, 43, 122, 145).

Varroa Mites

The varroa mite, *Varroa jacobsoni (Oudemans)*, is currently considered the major pest of honey bees in most parts of the world. Only Australia, New Zealand and the state of Hawaii remain free of this pest. The pathology it causes is commonly called varroasis (also seen as varroatosis or varrosis). Initially discovered in Java (148), varroa was originally confined to Southeast Asia where it parasitizes the Asian honey bee, *Apis cerana*. This bee has probably coevolved with the parasite, which adapted to keep the mite under control (163, 182).

A post–World War II increase in international travel and commerce has facilitated the worldwide dispersal of varroa (39). Once established, the mite spreads on drifting, robbing, and feral bees, on swarms, and are even reported on wasps (87) although this may an artifact of wasps robbing infected bee colonies. De Jong et al (53) documented the history of the spread of varroa (see 95 for the first reliable map of varroa worldwide distribution). An animated map of the spread of varroa is found on the Worldwide Web at http: //www.csfnet.org/image/animap2.gif. Varroa became an economic concern in Japan and China in the 1950s and 1960s, in Europe in the late 1960s and 1970s, and in Israel and North America in the 1980s.

Life Cycle Adult females measure 1.1 mm long × 1.6 mm wide, weigh approximately 0.14 mg (176) and are a reddish-brown color. The ovoid males are much smaller, about 500 μm wide, and are light in color.

Several key morphological features help make varroa a successful ectoparasite. It can survive off the host for 18 to 70 hours, depending on the substrate (46). The female's chelicerae are structurally modified, the fixed digit is lacking, and the moveable digit is a saw-like blade capable of piercing and tearing the host's integument (56). The mite's body is dorsoventrally compressed, allowing the mite to fit beneath the bee's abdominal sclerites, thus lessening water loss from transpiration (213). Its hiding there reduces varroa's vulnerability to grooming and to dislodgment during host activity.

Female varroa are often found on adult bees, which provide for dispersal (phoresy) and serve as short-term hosts. Varroa prefers young "house" bees to older workers, probably because of the lower titer of the Nasonov gland pheromone geraniol, which strongly repels the mite (103). The mite pierces the soft intersegmental tissues of the bee's abdomen or behind the bee's head, and feeds on the hemolymph. When in an actively reproducing bee colony, the mite disembarks and seeks brood cells containing third-stage bee larvae. Varroa, which prefers drone larvae but also invades workers' cells, is attracted to fatty acid esters, which are found in higher quantities on immature drones than on workers (119). Other known attractants are the aliphatic alcohols and aldehydes from bee cocoons (64) and perhaps the larger volume of drone cells.

Varroa enters the prepupal cells one to two days prior to capping and hides from the nurse bees by submerging in the remaining liquid brood food, lying upside down. The mite's modified peritremes protrude snorkel-like out of the fluid surface, enabling them to respire (63). The female remains concealed until the brood cell is capped. To keep from becoming trapped, the female attaches herself to the bee larva as it spins its cocoon. Once the prepupa is formed, the mite begins to feed at a site located on the prepupa's fifth abdominal segment.

Varroa produces its first egg 60 hours after the cell is sealed (110). The first egg is usually a haploid male, and the subsequent female eggs are laid at 30-hour intervals. Mites go through the following instars: pharate larvae, mobile protonymph, pharate deutonymph, mobile deutonymph, pharate adult, and adult (50, 63). Mobile nymphs actively feed and grow while pharate instars are quiescent—all active postembryonic instars are eight legged. The foundress mite keeps the feeding site open to allow her offspring to eat and will even push away the prepupa's posterior legs to keep the site exposed (62).

Ideal temperatures for optimal varroa development correspond to the ideal temperatures of drone brood (118, 120). Young females mature in 6.5 to 6.9 days (50), emerge with the callow bee, and may be phoretic for a time (four to five days on average) on other bees before invading new brood to repeat the cycle. Each varroa female may undergo two to three reproductive cycles (128). If mites invade drone cells, each foundress produces on average 2.6 new female offspring. In worker cells, the mean is 1.3 (186), although the numbers may vary considerably depending on the number of foundress mites in the cell, their fertility level, and/or the race of bees.

Fecal pellets (mostly guanine) are deposited on the cell wall and act as an aggregation site for immatures and a meeting place for mating (62). Males mature in 5.5 to 6.3 days, then mate frequently with each emerging sister mite, in succession, using specially modified chelicerae to inject sperm into the pair of induction pores between the bases of legs III and IV in females (62). Unmated females produce only male offspring. The total life cycle of the male is completed in the brood cell, after which it dies.

Recent molecular techniques have demonstrated that *V. jacobsoni* most likely is a complex of species represented by at least five sibling species (6a), only one of which has spread from *A. cerana* to *A. mellifera,* causing the enormous bee losses. Morphological differences have been recorded between these species and the species appear to be reproductively isolated from one another. Hence the varroa on *A. mellifera* could soon be renamed. The name *Varroa destructor* has been proposed (D Anderson, personal communication) and if several species and strains are masquerading under the name *V. jacobsoni*, and they possess different levels of virulence, the survival of untreated bee colonies despite varroa presence in South America may be explained (6, 48, 49, 52). Martin & Ball (127) reported differences in mite virulence and virus associations on honey bee survival in the United Kingdom. Other factors contributing to bee resistance may be climatic differences (51), the behavior and life cycle of the Africanized honey bee (31) *A. m. scutellata* (e.g. shorter post-capping period of bee brood), or other as yet undiscovered factors or combinations of factors.

Symptoms

Varroasis symptoms can be confused with other disorders, and even with pesticide poisoning. Here are the most notable symptoms: *(a)* Pale or dark reddish-brown mites are seen on otherwise white pupae. *(b)* Colonies are weak with a spotty brood pattern and other brood disease symptoms are evident. *(c)* The drone or worker brood has punctured cappings. *(d)* Disfigured, stunted adults with deformed legs and wings are found crawling on the combs or on the ground outside (50). Additionally, bees are seen discarding larvae and pupae and there is a general colony malaise, with multiple disease symptoms. Because mite populations increase in proportion to the available bee larvae, varroa can quickly overrun a colony and often colonies are dead by the fall.

Interrupted brood rearing during the winter slows the population increase of varroa in temperate regions, but in warmer climates, colonies can be destroyed within months (88). Treated apiaries can still perish if the beekeeper is not diligent; reinfestation occurs due to the robbing of varroa-infested and weakly defended colonies.

Detecting Varroa

Observation of Brood Mites can be detected by pulling up capped brood cells using a cappings scratcher (with fork-like tines); varroa appears as brown or whitish spots on the white pupae. Guanine, the fecal material of varroa, can be seen as white spots on the walls of brood frames in highly infested colonies (76a).

Ether Roll About 100 to 200 bees are collected in a clear glass jar and sprayed briefly with engine starter fluid; the jar is shaken to dislodge mites and then rolled so mites adhere to the sides. Adding alcohol or soapy water to the jar and re-agitating the contents will displace the remaining mites. Pouring the liquid through coarse mesh will strain out the bees, after which the mites can be fine-filtered and counted.

Sticky Board A white paper or plastic sheet covered with petroleum jelly or another sticky agent is placed on the bottom board of a colony and the hive is smoked with pipe tobacco in a smoker. After closing the hive for 10 to 20 minutes, the board is removed and the mites counted. Alternately, a sticky board by itself can be left in place for one to three days. This last method is more efficient than sampling brood or bees and is widely used as a research tool to monitor mite levels (82). Special sticky boards (145a) have been developed to ease the task of counting mites.

Economic injury levels and economic damage have not been reliably calculated for varroa. Delaplane & Hood (54) found late-season drops on a sticky board of more than 100 mites per day sufficient to justify treatment in Georgia. However, other researchers find this number too high (D Caron, personal communication). Defining a reliable ratio of critical infestation of mites to bees is problematic. If varroa is present in high numbers early in the season, treatment should be immediate. Finding mites during a honeyflow precludes chemical applications for treatment (see next section).

Control of Varroa

Chemical While long-range, non-chemical controls are vigorously being sought, beekeepers need immediate relief from existing mite infestations. Fluvalinate (Apistan®), a pyrethroid, is currently the only U.S.-registered pesticide for varroa control. Coumaphos (Bayer Bee Strips or CheckMite), an organophosphate, received Section 18 registration in Florida in 1999; it is granted this status in many other states as well. Section 18 allows a limited, short-term, single-use application. Coumaphos is the only product known to control the small hive beetle, *Aethina tumida,* introduced from South Africa and identified in the United States in 1998. These new strips will also be used on fluvalinate-resistant varroa populations. Both chemicals are applied as pesticide-impregnated plastic strips, which are hung between frames of bees in a hive. Treatment time is in the spring and again in the fall as needed, and only when there is no honeyflow. Applied in this manner, fluvalinate is released slowly and dispersed by adult bees (26).

These chemical options for varroa pose a serious problem because repeated exposure to the same pesticides select for resistant mites (89). Recent reports of fluvalinate-resistant mites have surfaced in Italy (124, 125, 134), France (38) and Israel (R Mozes-Koch, A Dag, Y Slabezki, H Efrat, H Kaleb and BA Yakobson, unpublished data) and some U.S. states (70, 71, 74, 152, 153). Coumaphos resistance is also reported in Italy (204) and the United States and recent studies (73) found low levels of resistance to Amitraz (another chemical previously used for mite control) in one beekeeping operation that exhibited fluvalinate resistance. Cross- or multiple-resistance studies need to be done. This resistance crisis is being compounded by contamination of hive products (205). In addition, drone survival is found to be lower in both varroa-infested colonies and colonies treated with fluvalinate (166a), which may also affect their mating ability. Poor quality or low numbers of drones result in poorly mated queens.

Organic Acids Formic acid kills varroa and HBTM (104), but is temperature dependent and dangerous to humans. When used with absorbent paper over the top bars, the evaporating fumes kill HBTM and varroa (82). In 1998, the USDA-ARS licensed a U.S. bee supplier to produce formic acid gel packs, which should be available soon. Other organic acids, such as oxalic and lactic, are used in Europe (22a) and are applied in sugar syrup trickled on bees. These acids require broodless bees and may cause bee mortality (22a).

Essential Oils Another approach is the use of volatile plant essential oils to control bee mites (29, 37, 84, 111) and other bee diseases (34, 80, 132). Many beekeepers are already experimenting with such "natural" products (179), but plant oils are complex compounds that may have unwanted side effects on bees and beekeepers (183), and could contaminate hive products.

Biological/Cultural Controls

Other methods have been used to control mites, but most are too labor intensive and impractical in large apiaries. Used in combination with or in an integrated pest management (IPM) project, they may be helpful.

Smoke and Dropping Mites Partial control in lightly infested apiaries can be obtained with tobacco smoke or smoke from other plant materials that cause mite knockdown (69, 72). Smoke dislodges mites and can be used periodically to remove those that subsequently emerge from brood cells. A sticky board used in conjunction with smoke traps mites dislodged by the smoke. Pettis & Shimanuki (154) found varroa that dropped to the bottom board of a hive were more likely to remain there unless a bee passed within seven mm of it. Using a screen to separate fallen varroa from bees may help keep mite levels lower.

Traps Because varroa prefer drones, combs of drone brood can be used to attract, trap, and remove mites by cutting out drone brood (184, 185). Worker brood can also be removed (83). Drone brood can also be cut out of frames (83, 184, 185). Also, caging the queen of *A. cerana* for 35 to 40 days and separating the brood frames helped interrupt the brood/mite cycle (75) in *A. cerana*.

Heat Another method employs heat: The mite succumbs at or around 111° F (44° C), whereas sealed brood survives. Using a combination of these methods, along with a lactic acid treatment, mite numbers were reduced in Denmark (22).

Resistant Bees Attributes that enhance honey bee tolerance to varroa are reviewed (24). Some beekeepers let all susceptible colonies die and then rear queens from the survivors to head new colonies. Untreated Africanized colonies were maintained in Arizona for several years with few tracheal and varroa mites; resistant mechanisms were not discussed (76).

Hygienic Behavior and Grooming Bees will open capped brood cells and remove dead or dying brood. Such hygienic activity (19, 20, 195) reduces the mite levels in untreated colonies, which require less chemical treatment to manage varroa. Bee grooming (both autogrooming and allogrooming) has been observed in bees infested with mites (25). Defensive behaviors against varroa in races of *A. cerana* were studied (163, 182) and grooming is an important component in mite reduction. However, grooming is highly variable in *A. mellifera* (24). Bees will remove mites from each other and some even kill them using their mandibles (149). Unfortunately, this trait (mite biting) may not be heritable in some European bee stock (96, 97).

Length of Post-Capping Stage The pupal period influences the number of mites completing development. Shortening this time results in fewer varroa reaching maturity; if the capped cell stage is reduced by only six hours, fewer immature mites will become adults. Two African bee races have a heritable (worker) post-capping period of only 10 days (138), whereas European races require 11 to 12 days. Some researchers (51, 136) suggest climate plays a more important role, as this trait is difficult to maintain in *A. mellifera* colonies in northern regions.

Brood Attractiveness The larvae of European bees are highly attractive to varroa; the ARS-Y-C-1 strain (*A. m. carnica*) was less attractive than other bee stocks (47). Differences in chemical components or levels in the brood may be the reason, but these possibilities were not tested.

Low Mite Fecundity At times, varroa mites do not reproduce (31), die without producing offspring, or get caught in the cocoons of bee pupae and die. Some of these traits are genetic (96, 97). Harris & Harbo (99) have further classified mite fecundity: live mites that do not lay eggs, live mites delayed in laying eggs, and mites that die before oviposition. Mites with lower or no fertility were found to have fewer (or no) spermatozoa in their seminal receptacle.

Queens selected for suppression of mite reproduction trait (SMRT) had reduced mite fecundity even after these queens were placed in susceptible colonies (J Harbo & J Harris, unpublished data). This trait, used conjointly with hygienic behavior and other IPM methods, may help solve the bee/varroa problem in the next decade (180).

Tropilaelaps clareae

This mite belongs in the mesostigmatic family Laelapidae, which has members that are mammalian parasites. *Tropilaelaps clareae*, originally obtained from a field rat (55) from the Philippines, normally occurs on *A. dorsata*; this mite also parasitizes *A. mellifera* (117). Currently, *T. clareae* is restricted to Asia, from Iran in the northwest to Papua New Guinea in the southeast (129). A single, alarming report of this mite in Kenya (116, 130) has not been repeated.

Life Cycle The life cycle is similar to that of varroa, but is shorter. Females are medium sized (1030 μm long × 550 μm wide), elongated and light reddish-brown; males are similar but less sclerotized. The foundress mites place three to four eggs on mature bee larvae shortly before capping and the progeny, usually a (first) male and several females feed only on bee brood. Development requires about one week and the adults, including the foundress mite, emerge with the adult bee and search for new hosts. The shortened life cycle, as well as a very brief stay on adult bees, explains how *T. clareae* populations can grow faster than those of varroa. *T. clareae* also out-competes the latter when both infest the same colony of *A. mellifera* (191). Nevertheless, populations of both mites can survive in the same apiary for 12 months, probably because their niches are not completely congruent (164).

Like varroa, the female mites are dispersed by bees, but phoretic survival is of short duration because the unspecialized chelicerae of tropilaelaps cannot pierce the integument of adult bees. Gravid female mites die within two days unless they deposit their mature eggs (208, 210).

The mouthparts are stubby, with an apically bidentate fixed upper digit, and a longer, unidentate and pointed moveable digit. This piercing-grasping structure is more suitable to piercing soft brood tissue, rather than the tearing-sawing type of varroa. This implies that tropilaelaps can feed only on soft tissues, such as honey bee brood (94).

Symptoms Irregular brood pattern, dead or malformed wingless bees at the hive's entrance, and the presence of fast-running, brownish mites on the combs, are diagnostic for *T. clareae* (50). The faster development rate make *T. clareae* more dangerous to European honey bees than varroa (209) where they cohabit.

Treatment Fluvalinate controls *T. clareae* (126), as do monthly dustings with sulfur (9) and treatment with formic acid (85) or with chlorobenzilate (208). The inability of this mite to feed on adult bees, or to survive outside sealed brood for more than a few days, is being used as a nonchemical control method (209, 211).

OTHER PARASITIC BEE MITES

Most parasitic bee mites (with the exception of *Acarapis*) are in the tribe Varroini (or Group V) in the family Laelapidae (see Table 2; 34a).

Euvarroa

Euvarroa sinhai is a parasite of *A. florea*, occurring in Asia from Iran through India to Sri Lanka. The mite develops naturally on the capped drone brood (142) but has been reared in the laboratory on *A. mellifera* worker brood (141). Development requires less than one week, and each female produces four to five offspring. Drones as well as workers are used for dispersal. The female mite overwinters in the colony, probably feeding on the clustering bees. Colony infestation by *E. sinhai* is somehow hindered by the construction of queen cells (1) and its population growth is inhibited in the presence of *T. clareae* and of *V.*

TABLE 2 Mesostigmatic mites parasitizing bees, arranged according to host bee species (113). Asterisk (*) denotes mites believed to be originally associated with the particular bee species.

Apis species	Mites	Source
andreniformis	*Euvarroa sinhai*	60
	*Euvarroa wongsirii**	121
cerana	*Tropilaelaps clareae* (?)	60
	*Varroa jacobsoni**	
	*Varroa underwoodi**	60
dorsata	*Tropilaelaps clareae**	61
	*Tropilaelaps koenigerum**	60
florea	*Euvarroa sinhai**	
	Tropilaelaps clareae	60
koschevnikovi	*Varroa rindereri**	45
	Varroa jacobsoni	60
laboriosa	*Tropilaelaps clareae*	
	Tropilaelaps koenigerum	60
mellifera	*Euvarroa sinhai*	114
	Tropilaelaps clareae	60
	Varroa jacobsoni	60
nigrocincta	*Varroa underwoodi*	7
nuluensis	*Varroa jacobsoni*	60
	Varroa near *underwoodi*	44

jacobsoni (191). Transfer experiments (114) confirmed that *E. sinhai* may survive on *A. mellifera* and *A. cerana*, emphasizing their ability to cross-infest exotic hosts.

Euvarroa wongsirii parasitizes drone brood of *A. andreniformis* in Thailand and Malaysia. Its biology appears similar to *E. sinhai* and it can live for at least 50 days on worker bees outside the nest (137).

PARASITIC MITES ON OTHER SPECIES OF APIS

The systematics of *Apis* is still unsettled. Only four species, namely *cerana, dorsata, florea,* and *mellifera*, were recognized until recently (174), with other entities being treated as races or subspecies. Wu & Kuang (212) distinguished *andreniformis* from *florea*, whereas the separate status of *koschevnikovi* was confirmed in 1989 (175). Koeniger (113) added *laboriosa, nigrocincta* and *nuluensis*, the latter discovered from Sabah (northern Borneo) (197). Most *Apis* species occur in south Asia and more are likely to be distinguished there (146, 194a). As these species come to be recognized, along with their unique acarine associates, a more complete pattern of the host relationships should emerge, based on phylogenetic analyses of both bees and mites.

Another untapped source for additional parasitic Acari could be the associates of the African "races" of *A. mellifera* (100, 173), about which very little is known (18). Most parasitic mites are described from the four "classic" bees. However, several new mites were recently discovered parasitizing the classic as well as the newly-recognized honey bee species (see Table 2).

Delfinado et al (60) and Koeniger (112) noted the relationship between bee nesting habitat and the genus or family of associated parasitic mites. The small bush-nesting species with single combs (*andreniformis* and *florea*) are attacked by *Euvarroa* spp. (121). The bees that build multi-comb nests in caves and trees (*cerana, koschevnikovi,* and *mellifera*) are parasitized by *Varroa* spp. (45), and the tree bees with single combs (*dorsata* and *laboriosa*) by *Tropilaelaps* spp. (61). This pattern also seems to hold for the more recently described *V. rindereri*, but it could change due to the mites' facility for cross infestation (e.g. *T. clareae* occurring on *cerana* and *mellifera*).

These data, along with the tentative suggestion that a bee from Luzon in the Philippines may be distinct from *A. cerana* and from *A. nigrocincta* (41), indicate that more and perhaps unexpected associations between honey bees and parasitic mites remain to be discovered. Signs of things to come are the significant differences found between *E. sinhai* from India and from Thailand (137), and the collection of *Varroa* sp. near *underwoodi* on *A. nuluensis* in Borneo (44). More species of varroids and *Tropilaelaps* are likely to be found in southeast Asia; if these mites find their way to European or the Western Hemisphere beehives, significant problems could result.

MITE-ASSOCIATED BEE PATHOLOGIES

Several mite-related pathologies have been reported (177), but their etiology and precise relation to bees are obscure. Even the Isle of Wight disease, which precipitated the discovery of *A. woodi*, is now considered to have been "of unknown origin" (10).

Bee Parasitic Mite Syndrome (BPMS)

First reported when bee colonies were stressed by varroa mites, the name BPMS or PMS was given (189) to explain why colonies infested with both HBTM and varroa were not thriving. BPMS may be related to both mites vectoring a virus, such as acute paralysis virus (106, 108). The symptoms, which can be present any time of the year, include the presence of mites, the presence of various brood diseases with symptoms similar to that of the foulbroods and sacbrood but without any predominant pathogen, American foulbrood-like symptoms, spotty brood pattern, increased supersedure of queens, bees crawling on the ground, and a lowered adult bee population. Although BPMS remains an enigma, feeding colonies Terramycin (antibiotic) syrup or patties and pollen supplements and using resistant bee stock have shown promise in keeping bees alive.

Viruses

Honey bees are subject to many viruses (4, 17), five of which are associated with varroa and one with HBTM. Virus particles are probably always present in latent or inapparent form in or on bees, or in the hive environment. Some viruses may be induced or activated by puncturing healthy bees (109), similar to the wounds inflicted by mites. Such etiology could explain why these diseases appear mostly in infested colonies and when colonies are heavily parasitized by mites that the mortality is so acute (14, 16). Challenges to controlling bee viruses include early diagnosis and identification, inconsistent association with mites, or methodical problems in establishing the relationship. In addition, many viral strains from different countries may be related. Molecular and serological techniques are expected to clarify these issues. Most of these viruses were unknown prior to the introduction of varroa.

Acute Paralysis Virus (APV) This virus kills both adult bees and brood, especially in varroa-infested colonies in Europe and the U.S. (5, 15, 107, 109). The virus, activated by an unknown mechanism, multiplies when mites feed on infected bees. Nurse bees parasitized by mites infected with this activated virus can transmit it to larvae in the brood food, or to other adult bees. In this manner the activated virus can spread quickly and, once systemic, overwhelm the colony.

Kashmir Bee Virus (KBV) This disease was originally reported from *A. cerana* in Kashmir (13). Similar in size to several picornavirus-like agents, it is the most virulent and widespread disease, found everywhere bees are present. Like APV, KBV may be activated in the presence of varroa, multiplying to lethal levels, but KBV was held responsible for bee deaths in Australia (12) where varroa is not present. Strains of KBV from Canada and Spain resemble APV in serological tests (3). The pathology of KBV is still being studied and new molecular techniques may help to identify it (196).

Deformed Wing Virus (DWV) First reported from Poland (199), where young bees had malformed wings and were stunted, DWV is now found wherever varroa is found, even on *A. cerana* in China (17, 21). Originally attributed to varroa feeding on bee pupae, DWV is transmitted to healthy brood by varroa (17).

Other viral associations include slow paralysis virus (SPV), found in the United Kingdom (17), and cloudy wing virus (CWV), reported in the United States, Greece, United Kingdom, and Australasia. Pathologies are far from clear. CWV may be carried by air via the tracheal tubes. A recent report from Pennsylvania (33) verified the presence of an unknown iridovirus in varroa; the significance of this is not yet understood.

HBTM and Virus

Chronic Paralysis Virus (CPV) First reported in 1933, this is the only specific tracheal mite-associated virus (11, 28). CPV may have caused the Isle of Wight disease: The symptoms are very similar to HBTM infestation. CPV comes in two syndromes. Type I is distinguished by bees trembling, unable to fly, with K-wings and distended abdomens. Type II, called the hairless black syndrome, is recognized by hairless, black shiny bees crawling at the hive entrance. The virus is prevalent in colonies where bees are confined for long periods, and the circumstances of the disease are the same as those that aggravate mite infestations. While CPV can cause occasional outbreaks in the absence of mites, in mite-infested colonies the virus was markedly increased (15). Susceptibility to CPV may also be inherited.

Bacteria Varroa may transmit *Serratia marcescens,* a bacterium that causes septicemia in bees (92). About 20% of healthy brood become diseased if infected mites are allowed to feed on bee larvae. Varroa may also transmit other bacteria, such as *Hafnia alvei* (15). Although an increase in European foulbrood (EFB) was reported from infested colonies (200), the transmission of the bacterium by varroa was not confirmed. Because EFB is a stress-related disorder, colonies that are heavily infested with varroa are susceptible to EFB. American foulbrood

(AFB) spores have been photographed on the surface of varroa, but the mite has not been implicated in transmitting the disease (2).

Fungi Fungi are ubiquitous organisms found in all bee colonies, but varroa-stressed colonies appear to have an increased incidence of chalkbrood disease, *Ascosphera apis* (123). Such outbreaks may be attributable to inadequate brood care by mite-stressed workers and depleted bee populations. Chilled brood is more prone to succumb to the fungus than brood adequately incubated by nurse bees. Spores of *Aspergillus* spp. (the cause of stonebrood) have also been found on varroa (123).

CONCLUSIONS

The sudden global emergence of bee mites during the past decade prompted increased research on detecting, monitoring, and controlling them. Meantime, basic mite studies have been lagging. Several areas of research must be addressed if bees are to remain a viable segment of agriculture. For example, it is not clear how particular bee mites actually damage their hosts, and what the role various disease organisms play in the ensuing colony decline. The presence of viruses is a special challenge and controlling these mites would almost certainly diminish viral effects. Next, the emerging recognition of new bee species and their mite parasites makes continued research a high priority. We need to determine whether any of these "new" mites will also cross-infest the dominant domesticated honey bee, *A. mellifera,* should they ever be introduced. More parasitic mites would be disastrous for the bee industry.

That *V. jacobsoni* may actually consist of several genotypes (or species) attests to large gaps in our knowledge. Further investigation of this matter must be pursued vigorously. Then, the development of chemicals or other modes of controlling bee mites would be facilitated by rearing them apart from bees. Much more work on in vitro culture of all parasitic bee mites is needed, although laboratory methods for culturing varroa and *T. clareae* were developed by Rath (162).

Finally, the most pressing problem remains control, preferably without synthetic chemicals. Thus we conclude by advocating the vigorous testing of various active ingredients, such as "botanicals," the breeding of bees for resistance to both mites, and exploring other biological or cultural techniques. The challenges ahead include attracting mites away from bees, killing mites without contaminating hive products or injuring bees, and keeping the controls economic, effortless, and easy for both the hobbyist and commercial beekeeper. Another problem is to determine an economic injury level of mite infestation and a threshold to indicate treatment times; doing so would take the guesswork out of when to medicate. The long-term solution to parasitic bee mites is in developing an integrated pest management program to manage mites by multiple means, not relying on any one or two chemical treatments.

ACKNOWLEDGMENTS

Diana Sammataro would not have been able to finish this work without the help of Pennsylvania State University colleagues Maryann Frazier, Nancy Ostiguy, Scott Camazine, and Jennifer Finley. They made this task easier, more productive, and enjoyable. Special thanks also to Dr. Evert E. Lindquist, Ingemar Fries, and Dennis Anderson for their helpful suggestions. Uri Gerson wishes to thank his colleagues A Dag, H Efrat, Y Slabezki, and especially R Mozes-Koch for their consistent cooperation. Parts of this study were supported by Grant No. IS-2508-95 from the United States–Israel (Binational) Agricultural Research and Development Fund (BARD).

LITERATURE CITED

1. Aggarwal K, Kapil RP. 1988. Observations on the effect of queen cell construction on *Euvarroa sinhai* infestation in drone brood of *Apis florea*. In *Africanized Honey Bees and Bee Mites*, ed. GR Needham, RE Page Jr, M Delfinado-Baker, CE Bowman, pp. 404–8. Chichester, UK: Ellis Horwood
2. Alippi AM, Albo GN, Marcangeli J, Leniz D, Noriega A. 1995. The mite *Varroa jacobsoni* does not transmit American foulbrood from infected to healthy colonies. *Exp. Appl. Acarol.* 19:607–13
3. Allen MR, Ball BV. 1995. Characterisation and serological relationships of strains of Kashmir bee virus. *Ann. Appl. Biol.* 126:471–84
4. Allen MR, Ball BV. 1996. The incidence and world distribution of honey bee viruses. *Bee World* 77:141–62
5. Allen MR, Ball BV, White RF, Antoniw JF. 1986. The detection of acute paralysis virus in *Varroa jacobsoni* by the use of a simple indirect ELISA. *J. Apic. Res.* 25:100–5
6. Anderson DL. 1999. Genetic and reproductive variation in *Varroa jacobsoni*. *Proc. XIII Int. Congr. IUSSI*, Adelaide, p. 33

6a. Anderson DL. 1999. Are there different species of *Varroa jacobsoni?* In *Proc. Apimondia 99, Congr. XXXVI, Vancouver, Can.* Sept., pp. 59–62

7. Anderson DL, Halliday RB, Otis GW. 1997. The occurrence of *Varroa underwoodi* (Acarina: Varroidae) in Papua New Guinea and Indonesia. *Apidologie* 28:143–47
8. Atkins L, Kellum D, Atkins KW. 1981. *Reducing Pesticide Hazards to Honey Bees*. Leaflet 2883. Univ. Calif. Div. Agric. Sci.
9. Atwal AS, Goyal NP. 1971. Infestation of honey bees colonies with *Tropilaelaps*, and its control. *J. Apic. Res.* 10:137–42
10. Bailey L, Ball BV. 1991. *Honey Bee Pathology*. San Diego: Academic. 193 pp. 2nd ed.
11. Bailey L, Ball BV, Carpenter JM, Woods RD. 1980. Small virus-like particles in honey bees associated with chronic paralysis virus and with a previously undescribed disease. *J. Gen. Virol.* 46:149–55
12. Bailey L, Carpenter JM, Govier DA, Woods RD. 1979. Egypt bee virus and Australian isolates of Kashmir bee virus. *J. Gen. Virol.* 43:641–47
13. Bailey L, Woods RD. 1977. Two more

small RNA viruses from honey bees and further observations on sacbrood and acute bee-paralysis viruses. *J. Gen. Virol.* 37:175–82

14. Ball B. 1988. The impact of secondary infections in honey-bee colonies infested with the parasitic mite *Varroa jacobsoni*. See Ref. 1, pp. 457–61
15. Ball B. 1994. Host-parasite-pathogen interactions. See Ref. 131, pp. 5–11
16. Ball BV, Allen FM. 1988. The prevalence of pathogens in honey bee (*Apis mellifera*) colonies infested with the parasitic mite *Varroa jacobsoni*. *Ann. Appl. Biol.* 113:237–44
17. Ball BV, Bailey L. 1997. Viruses. In *Honey Bee Pests, Predators, and Diseases*, ed. RM Morse, PK Flottum, 2:13–31. Medina, OH: Root. 3rd ed.
18. Benoit PLG. 1959. The occurrence of the acarine mite, *Acarapis woodi*, in the honey-bee in the Belgian Congo. *Bee World* 40:156
19. Boecking O, Rath W, Drescher W. 1993. Grooming and removal behavior—strategies of *Apis mellifera* and *Apis cerana* bees against *Varroa jacobsoni*. *Am. Bee J.* 133:117–19
20. Boecking O, Spivak M, Drescher W. 1999. In search of tolerance mechanisms of the honey bee *Apis mellifera* to the mite *Varroa jacobsoni*. See Ref. 206. In press
21. Bowen-Walker PL, Martin SJ, Gunn A. 1999. The transmission of deformed wing virus between honeybees (*Apis mellifera*) by the ectoparasitic mite *Varroa jacobsoni* Oud. *J. Invertebr. Pathol.* 73:101–6
22. Brødsgaard CJ, Hansen H. 1994. An example of integrated biotechnical and soft chemical control of varroa in a Danish apiary. See Ref. 131, pp. 101–5

22a. Brødgaard CJ, Hansen H, Hansen CW. 1997. Effect of lactic acid as the only control method of varroa mite populations during four successive years in honeybee colonies with a brood-free period. *Apiacta: An Int. Tech. Mag. Apic. Econ. Inf.* 32:81–88

23. Brother A. 1968. "Isle of Wight" or acarine disease: its historical and practical aspects. *Bee World* 49:6–18
24. Büchler R. 1994. Varroa tolerance in honey bees occurrence, characters and breeding. *Bee World* 75:54–70
25. Büchler R, Drescher W, Tornier I.1992 (1993). Grooming behaviour of *Apis cerana, A. mellifera* and *A. dorsata* and its effect on the parasitic mites *Varroa jacobsoni* and *Tropilaelaps clareae*. *Exp. Appl. Acarol.* 16:313–19
26. Burgett DM, Kitprasert C. 1990. Evaluation of Apistan as a control for *Tropilaelaps clareae* (Acari: Laelapidae), an Asian honey bee brood mite parasite. *Am. Bee J.* 130:51–53
27. Burgett DM, Kitprasert C. 1992. Tracheal mite infestation of queen honey bees. *J. Apic. Res.* 31:110–11
28. Burnside CE. 1933. Preliminary observation on "paralysis" of honeybees. *J. Econ. Entomol.* 26:162–68
29. Calderone NW, Wilson WT, Spivak M. 1997. Plant extracts used for control of the parasitic mites *Varroa jacobsoni* (Acari: Varroidae) and *Acarapis woodi* (Acari: Tarsonemidae) in colonies of *Apis mellifera* (Hymenoptera: Apidae). *J. Econ. Entomol.* 90:1080–86
30. Camazine S. 1985. Tracheal flotation: a rapid method for the detection of honey bee acarine disease. *Am. Bee J.* 125:104–5
31. Camazine S. 1986. Differential reproduction of the mite, *Varroa jacobsoni* (Mesostigmata: Varroidae), on Africanized and European honey bees (Hymenoptera: Apidae). *Ann. Entomol. Soc. Am.* 79:801–3
32. Camazine S, Çakmak I, Cramp K, Finley J, Fisher J, Frazier M. 1998. How healthy are commercially-produced U.S. honey bee queens? *Am. Bee J.* 138:677–80
33. Camazine S, Liu TP. 1998. A putative iridovirus from the honey bee mite, *Var-*

roa jacobsoni Oudemans. *J. Invertebr. Pathol.* 71:177–78
34. Carpana E, Cremasco S, Baggio A, Capolongo F, Mutinelli F. 1996. Prophylaxis and control of honeybee American foulbrood using essential oils. *Apic. Mod.* 87:11–16 (In Italian)
34a. Casanueva ME. 1993. Phylogenetic studies of the free-living and arthropod associated Laelapidae (Acari: Mesostigmata). *Guyana Zool.* 57:21–46
35. Cavalloro R, ed. 1983. Varroa jacobsoni *Oud. Affecting Honey Bees: Present Status and Needs.* Rotterdam: Balkema. 107 pp.
36. Clark KJ. 1985. *Mites (Acari) associated with the honey bee,* Apis mellifera *L. (Hymenoptera: Apidae), with emphasis on British Columbia.* Ms. thesis. Burnby, Can.: Simon Fraser Univ.
37. Colin ME. 1990. Essential oils of Labiatae for controlling honey bee varroosis. *J. Appl. Entomol.* 110:19–25
38. Colin ME, Vandame R, Jourdan P, Di Pasquale S. 1997. Fluvalinate resistance of *Varroa jacobsoni* (Acari: Varroidae) in Mediterranean apiaries of France. *Apidologie.* 28:375–84
39. Crane E. 1988. Africanized bee, and mites parasitic on bees, in relation to world beekeeping. See Ref. 1, pp. 1–9
40. Dag A, Slabezki Y, Efrat H, Damer Y, Yakobson BA, et al. 1997. Control of honey bee tracheal mite infestations with amitraz fumigation in Israel. *Am. Bee J.* 137:599–602
41. Damus MS, Otis GW. 1997. A morphometric analysis of *Apis cerana* F. and *Apis nigrocincta* Smith populations from southeastern Asia. *Apidologie* 28:309–23
42. Danka RG, Villa JD. 1998. Evidence of autogrooming as a mechanism of honey bee resistance to tracheal mite infestation. *J. Apic. Res.* 37:39–46
43. Danka RG, Villa JD, Rinderer TE, DeLatte FT. 1995. Field test of resistance to *Acarapis woodi* (Acari: Tarsonemidae) and of colony production by four stocks of honey bees (Hymenoptera: Apidae). *J. Econ. Entomol.* 88:584–91
44. de Guzman LI, Delfinado-Baker M. 1996. A scientific note on the occurrence of Varroa mites on adult worker bees of *Apis nuluensis* in Borneo. *Apidologie* 27:329–30
45. de Guzman LI, Delfinado-Baker M. 1996. A new species of *Varroa* (Acari: Varroidae) associated with *Apis koschevnikovi* (Apidae: Hymenoptera) in Borneo. *Int. J. Acarol.* 22:23–27
46. de Guzeman LI, Rinderer TE, Beaman LD. 1993. Survival of *Varroa jacobsoni* Oud. (Acari: Varroidae) away from its living host *Apis mellifera* L. *Exp. Appl. Acarol.* 17:283–90
47. de Guzman LI, Rinderer TE, Lancaster VA. 1995. A short test evaluating larval attractiveness of honey bees to *Varroa jacobsoni. J. Apic. Res.* 34:89–92
48. de Guzman LI, Rinderer TE, Stelzer JA. 1997. DNA evidence of the origin of *Varroa jacobsoni* Oudemans in the Americas. *Biochem. Genet.* 34:327–35
49. de Guzman LI, Rinderer TE, Stelzer JA, Anderson D. 1998. Congruence of RAPD and mitochondrial DNA markers in assessing *Varroa jacobsoni* genotypes. *J. Apic. Res.* 37:49–51
50. De Jong D. 1997. Mites: varroa and other parasites of brood. See Ref. 17, pp. 281–327
51. De Jong D. 1999. The effect of climate on the development of resistance to *Varroa jacobsoni.* See Ref. 6, p. 130
52. De Jong D, Gonçalves LS. 1999. The Africanized bees of Brazil have become tolerant of varroa. See Ref. 6, p. 131
53. De Jong D, Morse RA, Eickwort GC. 1982. Mite pests of honey bees. *Annu. Rev. Entomol.* 27:229–52
54. Delaplane KS, Hood WM. 1997. Effects of delayed acaricide treatment in honey bee colonies parasitized by *Varroa jacobsoni* and a late-season treatment threshold for the south-eastern USA. *J. Apic. Res.* 36:125–32

55. Delfinado M, Baker EW. 1961. *Tropilaelaps*, a new genus of mites from the Philippines (Laelapidae s. lat.) Acarina. *Fieldiana Zool.* 44:53–56
56. Delfinado M, Baker EW. 1974. Varroidae, a new family of mites on honey bees (Mesostigmata: Acarina). *J. Wash. Acad. Sci.* 64:4–10
57. Delfinado-Baker M. 1994. A harmless mite found on honey bees—*Melittiphis alvearius*: from Italy to New Zealand. *Am. Bee J.* 134:199
58. Delfinado-Baker M, Baker EW. 1982. A new species of *Tropilaelaps* parasitic on honey bees. *Am. Bee J.* 122:416–17
59. Delfinado-Baker M, Baker EW. 1982. Notes on honey bee mites of the genus *Acarapis* Hirst (Acari: Tarsonemidae). *Int. J. Acarol.* 8:211–26
60. Delfinado-Baker M, Baker EW, Phoon ACG. 1989. Mites (Acari) associated with bees (Apidae) in Asia, with description of a new species. *Am. Bee J.* 122:416–17
61. Delfinado-Baker M, Underwood BA, Baker EW. 1985. The occurrence of *Tropilaelaps* mites in brood nests of *Apis dorsata* and *A. laboriosa* in Nepal, with description of the nymphal stages. *Am. Bee J.* 125:703–6
62. Donzé G, Guerin PM. 1994. Behavioral attributes and parental care of varroa mites parasitizing honeybee brood. *Behav. Ecol. Sociobiol.* 34:305–19
63. Donzé G, Guerin PM. 1997. Time-activity budgets and space structuring by the different life stages of *Varroa jacobsoni* in capped brood of the honey bee, *Apis mellifera*. *J. Insect Behav.* 10:371–93
64. Donzé G, Schnyder-Candrian S, Bogdanov S, Diehl P-A, Guerin PM, et al. 1998. Aliphatic alcohols and aldehydes of the honey bee cocoon induce arrestment behavior in *Varroa jacobsoni* (Acari: Mesostigmata), an ectoparasite of *Apis mellifera*. *Arch. Insect Biochem. Physiol.* 37:129–45
65. Eckert JE. 1961. Acarapis mites of the honey bee, *Apis mellifera* Linnaeus. *J. Insect Pathol.* 3:409–25
66. Eickwort GC. 1988. The origins of mites associated with honey bees. See Ref. 1, pp. 327–84
67. Eickwort GC. 1997. Mites: an overview. See Ref. 17, pp. 241–50
68. Eischen FA. 1987. Overwintering performance of honey bee colonies heavily infested with *Acarapis woodi* (Rennie). *Apidologie* 18:293–304
69. Eischen FA. 1997. Natural products, smoke and varroa. *Am. Bee J.* 137:107
70. Eischen FA. 1998. Varroa control problems: some answers. *Am. Bee J.* 138:107–08
71. Eischen FA. 1998. Varroa's response to fluvalinate in the Western U.S. *Am. Bee J.* 138:439–40
72. Eischen FA, Wilson WT. 1998. Natural products, smoke and varroa. *Am. Bee J.* 138:293
73. Elzen PJ, Eischen FA, Baxter JR, Elzen GW, Wilson WT. 1999. Detection of resistance in U.S. *Varroa jacobsoni* Oud. (Mesostigmata: Varroidae) to the acaricide fluvalinate. *Apidologie.* In press
74. Elzen PJ, Eischen FA, Baxter JR, Pettis J, Elzen GW, Wilson WT. 1998. Fluvalinate resistance in *Varroa jacobsoni* from several geographic locations. *Am. Bee J.* 138:674–76
75. Enayet Hossain ABM, Sharif M. 1991. Control of mite infestations of hives of *Apis cerana* Fabr. Hymenoptera, Apidae. *Bangladesh J. Zool.* 19:101–6
76. Erickson EH, Atmowidjojo AH, Hines L. 1998. Can we produce varroa-tolerant honey bees in the United States? *Am. Bee J.* 138:828–32
76a. Erickson EH, Cohen AC, Cameron BE. 1994. Mite excreta: a new diagnostic for varroasis. *BeeScience* 3:76–78
77. Fain A, Gerson U. 1990. Notes on two astigmatic mites (Acari) living in beehives in Thailand. *Acarologia* 31:381–84
78. Fichter BL. 1988. ELISA detection of *Acarapis woodi*. See Ref. 1, pp. 526–29

79. Finley J, Camazine S, Frazier M. 1996. The epidemic of honey bee colony losses during the 1995–1996 season. *Am. Bee J.* 136:805–8
80. Floris I, Carta C, Moretti MDL. 1996. Activity of various essential oils against *Bacillus larvae* White in vitro and in apiary trials. *Apidologie* 27:111–19 (In French)
81. Free JB. 1987. *Pheromones of Social Bees*. Chapman & Hall. 218 pp.
82. Fries I. 1989. Short-interval treatments with formic acid for control of *Varroa jacobsoni* in honey bee (*Apis mellifera*) colonies in cold climates. *Swed. J. Agric. Res.* 19(4):213–16
83. Fries I, Hansen H. 1989. Use of trapping comb to decrease the populations of *Varroa jacobsoni* in honeybees *Apis mellifera* colonies in cold climate. *Tidsskr. Planteavl* 93:193–98
84. Gal H, Slabezki Y, Lensky Y. 1992. A preliminary report on the effect of Originum oil and thymol applications in honey bee colonies (*Apis mellifera* L.) in a subtropical climate on population levels of *Varroa jacobsoni*. *BeeScience* 2:175–80
85. Garg R, Sharma OP, Dogra GS. 1984. Formic acid: an effective acaricide against *Tropilaelaps clareae* Delfinado & Baker (Laelapidae: Acarina) and its effect on the brood and longevity of honey bees. *Am. Bee J.* 124:736–38
86. Gary NE, Page RE Jr. 1989. Tracheal mite (Acari: Tarsonemidae) infestation effects on foraging and survivorship of honey bees (Hymenoptera: Apidae). *J. Econ. Entomol.* 82:734–39
87. Gerig L. 1988. Wespen als Varroatragerinnen. *Allg. Dtsch. Imkerztg.* (ADIZ) 22:274–77 (In German)
88. Gerson U, Lensky Y, Lubinevski Y, Slabezki Y, Stern Y. 1988. *Varroa jacobsoni* in Israel, 1984–1986. See Ref. 1, pp. 420–24
89. Gerson U, Mozes-Koch R, Cohen E. 1991. Enzyme levels used to monitor pesticide resistance in *Varroa jacobsoni*. *J. Apic. Res.* 30:17–20
90. Gibbins BL, van Toor RF. 1990. Investigation of the parasitic status of *Melittiphis alvearius* (Berlese) on honeybees, *Apis mellifera* L., by immunoassay. *J. Apic. Res.* 29:46–52
91. Giordani G. 1967. Laboratory research on *Acarapis woodi* Rennie, a causative agent of acarine disease of the honey bee. Note 5. *J. Apic. Res.* 6:147–57
92. Glinski Z, Jarosz J. 1992. *Varroa jacobsoni* as a carrier of bacterial infections to a recipient bee host. *Apidologie* 23:25–31
93. Grant GM, Nelson DL, Olsen PE, Rice WA. 1993. The ELISA detection of tracheal mites in whole honey bee samples. *Am. Bee J.* 133:652–55
94. Griffiths DA. 1988. Functional morphology of the mouthparts of *Varroa jacobsoni* and *Tropilaelaps clareae* as a basis for the interpretation of their life-styles. See Ref. 1, pp. 479–86
95. Griffiths DA, Bowman CE. 1981. World distribution of the mites *Varroa jacobsoni*, a parasite of honeybee. *Bee World* 62:154–63
96. Harbo JR, Harris JW. 1999. Heritability in honey bees (Hymenoptera: Apidae) of characteristics associated with resistance to *Varroa jacobsoni* (Mesostigmata: Varroidae). *J. Econ. Entomol.* 92:261–265
97. Harbo JR, Hoopingarner RA. 1997. Honey bees (Hymenoptera: Apidae) in the United States that express resistance to *Varroa jacobsoni* (Mesostigmata: Varroidae). *J. Econ. Entomol.* 90:893–98
98. Deleted in proof.
99. Harris JW, Harbo JR. 1999. Low sperm counts and reduced fecundity of mites in colonies of honey bees (Hymenoptera: Apidae) that are resistant to *Varroa jacobsoni* (Mesostigmata: Varroidae). *J. Econ. Entomol.* 92:83–90
100. Hepburn HR, Radloff SE. 1998. *Honeybees of Africa*. Berlin: Springer-Verlag. 370 pp.

101. Hirschfelder H, Sachs H. 1952. Recent research on the acarine mite. *Bee World* 33:201–9
102. Hirst S. 1921. On the mites (*Acarapis woodi* (Rennie) associated with Isle of Wight bee disease. *Ann. Mag. Nat. Hist.* 7:509–19
103. Hoppe H, Ritter W. 1989. The influence of the Nasonov pheromone on the recognition of house bees and foragers by *Varroa jacobsoni*. *Apidologie* 19:165–72
104. Hoppe H, Ritter W, Stephen EWC. 1989. The control of parasitic bee mites: *Varroa jacobsoni*, *Acarapis woodi* and *Tropilaelaps clareae* with formic acid. *Am. Bee J.* 129:739–42
105. Hughes AM. 1976. *The Mites of Stored Food and Houses*. London: HMSO. 400 pp.
106. Hung ACF, Adams JR, Shimanuki H. 1995. Bee parasitic mite syndrome (II): the role of varroa mite and viruses. *Am. Bee J.* 135:702–4
107. Hung ACF, Ball BV, Adams JR, Shimanuki H, Knox DA. 1996. A scientific note on the detection of American strains of acute paralysis virus and Kashmir bee virus in dead bees in one U.S. honey bee. *Apidologie* 27:55–56
108. Hung ACF, Shimanuki H, Knox DA. 1996. The role of viruses in bee parasitic mite syndrome. *Am. Bee J.* 136:731–32
109. Hung ACF, Shimanuki H, Knox DA. 1996. Inapparent infection of acute paralysis virus and Kashmir bee virus in the U.S. honey bees. *Am. Bee J.* 136:874–76
110. Ifantidis MD. 1983. Ontogenesis of the mite *Varroa jacobsoni* in worker and drone honeybee brood cells. *J. Apic. Res.* 22:200–6
111. Imdorf A, Charrière J-D, Maquelin C, Kilchenmann V, Bachofen B. 1995. *Alternative Varroa Control*. Fed. Dairy Res. Inst., Liebefeld, Switz. 11 pp.
112. Koeniger N. 1990. Co-evolution of the Asian bees and their parasitic mites. *Proc. 11th Int. Congr. IUSSI,* India, pp. 130–31
113. Koeniger N. 1996. The 1996 special issue of Apidologie on Asian honeybee species. *Apidologie* 27:329–30
114. Koeniger N, Koeniger G, de Guzman LI, Lekprayoon C. 1993. Survival of *Euvarroa sinhai* Delfinado and Baker (Acari, Varroidae) on workers of *Apis cerana* Fabr., *Apis florea* Fabr., and *Apis mellifera* L. in cages. *Apidologie* 24:403–10
115. Krause B, Page RE Jr. 1995. Effect of *Varroa jacobsoni* (Mesostigmata: Varroidae) on feral *Apis mellifera* (Hymenoptera: Apidae) in California. *Environ. Entomol.* 24:1473–80
116. Kumar NR, Kumar RW. 1993. *Tropilaelaps clareae* found on *Apis mellifera* in Africa. *Bee World* 74:101–2
117. Laigo FM, Morse RA. 1968. The mite *Tropilaelaps clareae* in *Apis dorsata* colonies in the Philippines. *Bee World* 49:116–18
118. Le Conte Y, Arnold G, Desenfant Ph. 1990. Influence of brood temperature and hygrometry variations on the development of the honey bee ectoparasite *Varroa jacobsoni* (Mesostigmata: Varroidae). *Environ. Entomol.* 19:1780–85
119. Le Conte Y, Arnold G, Trouiller J, Masson C, Chappe B, et al. 1989. Attraction of the parasitic mite *Varroa* to the drone larvae of honey bees by simple aliphatic esters. *Science* 245:638–39
120. Le Conte Y, Bernes YR, Salvy M, Martin C. 1999. Physical and chemical signals of importance for host recognition and development of *Varroa jacobsoni*. See Ref. 6, p. 276
121. Lekprayoon C, Tangkanasing P. 1991. *Euvarroa wongsirii*, a new species of bee mite from Thailand. *Int. J. Acarol.* 17:255–58
122. Lin H, Otis GW, Scott-Dupree CD. 1996. Comparative resistance in Buckfast and Canadian stocks of honey bees (*Apis mellifera* L) to infestation by honey bee tracheal mites (*Acarapis woodi* (Rennie)). *Exp. Appl. Acarol.* 20:87–101
122a. Lindquist EE. 1986. The world genera of

Tarsonemidae (Acari: Heterostigmata): a morphological, phylogenetic, and systematic revision, with a reclassification of family-group taxa in the Heterostigmata. *Mem. Entomol. Soc. Can.* 136:1–517

123. Liu T. 1996. Varroa mites as carriers of honey bee chalkbrood. *Am. Bee J.* 136:655
124. Lodesani M, Colombo M, Spreafico M. 1995. Ineffectiveness of Apistan treatment against the mite *Varroa jacobsoni* Oud. in several districts of Lombardy (Italy). *Apidologie* 26:67–72
125. Loglio G, Plebani G. 1992. Valutazione dellíeffecadia dellíApistan. *Apic. Mod.* 83:95–98 (In Italian)
126. Lubinevski Y, Stern Y, Slabezki Y, Lensky Y, Ben-Yossef H, Gerson U. 1988. Control of *Varroa jacobsoni* and *Tropilaelaps clareae* mites using Mavrik under subtropical and tropical climates. *Am. Bee J.* 128:48–52
127. Martin SJ, Ball B. 1999. Variations in the virulence of *Varroa* infestations. See Ref. 6, p. 303
128. Martin SJ, Kemp D. 1997. Average number of reproductive cycles performed by *Varroa jacobsoni* in honey bees (*Apis mellifera*) colonies. *J. Apic. Res.* 36:113–23
129. Mattheson A. 1993. World bee health report. *Bee World* 74:176–212
130. Mattheson A. 1997. Country records for honey bee diseases, parasites and pests. See Ref. 17, p. 587–602
131. Mattheson A, ed. 1994. *New Perspectives on Varroa*. Cardiff, UK: IBRA. 164 pp.
132. Meena MR, Sethi V. 1994. Antimicrobial activity of essential oils from spices. *J. Food Sci. Technol.* 31:68–70
133. Menapace DM, Wilson WT. 1980. *Acarapis woodi* mites found in honey bees from Colombia. *Am. Bee J.* 120:761–62
134. Milani N. 1995. The resistance of *Varroa jacobsoni* Oud. to pyrethroids: a laboratory assay. *Apidologie* 26:415–29
135. Mobus B, de Bruyn C. 1993. *The New Varroa Handbook*. Mytholmroyd, UK: North. Bee Books. 160 pp.
136. Moretto G. 1999. Heritability of some traits of *Apis mellifera* associated with resistance to the mite *Varroa jacobsoni*. See Ref. 6, p. 324
137. Morin CE, Otis GW. 1993. Observations on the morphology and biology of *Euvarroa wongsirii* (Mesostigmata: Varroidae), a parasite of *Apis andreniformis* (Hymenoptera: Apidae). *Int. J. Acarol.* 19:167–72
138. Moritz RFA. 1985. Heritability of the postcapping stage in *Apis mellifera* and its relation to varroatosis resistance. *J. Hered.* 76:267–70
139. Morse RA, Eickwort GC. 1990. *Acarapis woodi*, a recently evolved species? *Proc. Int. Symp. Recent Res. Bee Pathol., Gent, Belg.*, pp. 102–7
140. Morse RA, Nowogrodzki R. 1990. *Honey Bee Pests, Predators, and Diseases*. Ithaca, NY: Comstock. 2nd ed.
141. Mossadegh MS. 1990. Development of *Euvarroa sinhai* (Acarina: Mesostigmata), a parasitic mite of *Apis florea*, on *A. mellifera* worker brood. *Exp. Appl. Acarol.* 9:73–78
142. Mossadegh MS, Komeili BA. 1986. *Euvarroa sinhai* Delfinado & Baker (Acarina: Mesostigmata): a parasitic mite on *Apis florea* F. in Iran. *Am. Bee J.* 126:684–85
143. Deleted in proof
144. Mozes-Koch R, Gerson U. 1997. Guanine visualization, a new method for diagnosing tracheal mite infestation of honey bees. *Apidologie* 28:3–9
145. Nasr ME. 1997. Tracheal mite resistant and hygienic honey bee stocks in Ontario. *Can. Beekeep.* 20:63–34

145a. Ostiguy N, Sammataro D, Camzine S. 1999. How to count *Varroa jacobsoni* without going blind: a sane approach. *Am. Bee J.* 139:313–14

146. Otis GW. 1991. A review of the diversity of species within *Apis*. In *Diversity in the*

Genus Apis, ed. DR Smith, pp. 29–49. New Delhi: Westview

147. Otis GW, Scott-Dupree CD. 1992. Effects of *Acarapis woodi* on overwintering colonies of honey bees (Hymenoptera: Apidae) in New York. *J. Econ. Entomol.* 85:40–46
148. Oudemans AC. 1904. On a new genus and species of parasitic Acari. *Notes Leyden Mus.* 24:216–22
149. Peng CYS, Fang YZ, Xu SY, Ge LS. 1987. The resistance mechanism of the Asian honey bee *Apis cerana* Fabr. to an ectoparasitic mite, *Varroa jacobsoni* Oudemans. *J. Invertebr. Pathol.* 49:54–60
150. Peng CYS, Nasr ME. 1985. Detection of honeybee tracheal mites (*Acarapis woodi*) by simple staining techniques. *J. Invertebr. Pathol.* 46:325–31
151. Pettis JS, Dietz A, Eischen FA. 1988. Incidence rates of *Acarapis woodi* (Rennie) in queen honey bees of various ages. *Apidologie* 20:69–75
152. Pettis JS, Shimanuki H, Feldlaufer M. 1998. An assay to detect fluvalinate resistance in varroa mites. *Am. Bee J.* 138:538–41
153. Pettis JS, Shimanuki H, Feldlaufer M. 1998. Detecting fluvalinate resistance in varroa mites. *Am. Bee J.* 138:535–37
154. Pettis JS, Shimanuki H. 1999. A hive modification to reduce varroa populations. *Am. Bee J.* 139:471–73
155. Pettis JS, Wilson WT. 1996. Life history of the honey bee tracheal mite (Acari; Tarsonemidae). *Ann. Entomol. Soc. Am.* 89:368–74
156. Phelan LP, Smith AW, Needham GR. 1991. Mediation of host selection by cuticular hydrocarbons in the honey bee tracheal mite *Acarapis woodi* (Rennie). *J. Chem. Ecol.* 17:463–73
157. Phillips EF. 1922. *The Occurrence of Diseases of Adult Bees*. USDA Circ. #218
158. Phillips EF. 1923. *The Occurrence of Diseases of Adult Bees*. II. USDA Circ. #287
159. Ragsdale D, Furgala B. 1987. A serological approach to the detection of *Acarapis woodi* parasitism in honey bees using an enzyme-linked immunosorbent assay. *Apidologie* 18:1–10
160. Ragsdale D, Kjer KM. 1989. Diagnosis of tracheal mite (*Acarapis woodi* Rennie) parasitism of honey bees using a monoclonal based enzyme-linked immunosorbent assay. *Am. Bee J.* 129:550–53
161. Ramanan VR, Ghai S. 1984. Observations on the mite *Neocypholaelaps indica* Evans and its relationship with the honey bee *Apis cerana indica* Fabricius and the flowering of Eucalyptus trees. *Entomon.* 9:291–92
162. Rath W. 1995. The laboratory culture of the mites *Varroa jacobsoni* and *Tropilaelaps clareae*. *Exp. Appl. Acarol.* 10:289–93
163. Rath W. 1999. Defensive adaptations of *A. cerana* against *V. jacobsoni* and bearing for *A. mellifera*. See Ref. 6, p. 386
164. Rath W, Boeking O, Drescher W. 1995. The phenomena of simultaneous infestation of *Apis mellifera* in Asia with the parasitic mites *Varroa jacobsoni* Oud. and *Tropilaelaps clareae* Delfinado & Baker. *Am. Bee J.* 135:125–27
165. Rennie J. 1921. Acarine disease in hive bees: its cause, nature and control. *N. Scotland Coll. Agric. Bull.* 33:3–34
166. Rennie J, White PB, Harvey EJ. 1921. Isle of Wight disease in hive bees. *Trans. R. Soc. Edinburgh.* 52 (29, Part 4):737–54

166a. Rinderer TE, de Guzman LI, Lancaster VA, Delatte GT, Stelzer JA. 1999. Varroa in the mating yard: I. The effects of *Varroa jacobsoni* and Apistan on drone honey bees. *Am. Bee J.* 139:134–39

167. Robaux P. 1986. *Varroa et Varroasis*. Paris: Opida. 238 pp.
168. Deleted in proof
169. Royce LA, Krantz GW, Ibay LA, Burgett DM. 1988. Some observations on the biology and behavior of *Acarapis woodi*

and *Acarapis dorsalis* in Oregon. See Ref. 1, pp. 498–505

170. Royce LA, Rossignol PA. 1989. Honey bee mortality due to tracheal mite parasitism. *Parasitology* 100:147–51
171. Royce LA, Rossignol PA. 1991. Sex bias in tracheal mite [*Acarapis woodi* (Rennie)] infestation of honey bees (*Apis mellifera* L.). *BeeScience* 1:159–61
172. Royce LA, Rossignol PA, Burgett DM, Stringer BA. 1991. Reduction of tracheal mite parasitism of honey bees by swarming. *Philos. Trans. R. Soc. London Ser. B* 331:123–29
173. Ruttner F. 1986. Geographical variability and classification. In *Bee Genetics and Breeding*, ed. TE Rinderer, pp. 23–56. New York: Academic
174. Ruttner F. 1988. *Biogeography and Taxonomy of Honey Bees*. Berlin: Springer-Verlag. 282 pp.
175. Ruttner F, Kauhausen D, Koeniger N. 1989. Position of the red honey bee, *Apis koschevnikovi* (Buttel-Reepen 1906), within the genus *Apis*. *Apidologie* 20:395–404
176. Sammataro D. 1995. *Studies on the Control, Behavior, and Molecular Markers of the Tracheal Mite (Acarapis woodi (Rennie)) of Honey Bees (Hymenoptera: Apidae)*. PhD diss. Ohio State Univ. Columbus, OH. 125 pp.
177. Sammataro D. 1997. Report on parasitic honey bee mites and disease associations. *Am. Bee J.* 137:301–2
178. Sammataro D, Cobey S, Smith BH, Needham GR. 1994. Controlling tracheal mites (Acari: Tarsonemidae) in honey bees (Hymenoptera: Apidae) with vegetable oil. *J. Econ. Entomol.* 87:910–16
179. Sammataro D, Degrandi-Hoffman G, Needham GR, Wardell G. 1998. Some volatile plant oils as potential control agents for varroa mites (Acari: Varroidae) in honey bee colonies (Hymenoptera: Apidae). *Am. Bee J.* 138:681–85
180. Sammataro D, Needham GR. 1996. Developing an integrated pest management (IPM) scheme for managing parasite bee mites. *Am. Bee J.* 136:440–43
181. Sammataro D, Needham GR. 1996. Host-seeking behaviour of tracheal mites (Acari: Tarsonemidae) on honey bees (Hymenoptera: Apidae). *Exp. Appl. Acar.* 20:121–36
182. Sasagawa Y, Matsuyama HS, Peng CYS. 1999. Recognition of a parasite: hygienic allo-grooming behavior induced by parasitic *Varroa* mites in the Japanese honey bee, *Apis cerana japonica* RAD. See Ref. 6, p. 415
183. Schaller M, Korting HC. 1995. Allergic airborne contact dermatitis from essential oils used in aromatherapy. *Clin. Exp. Dermatol.* 20:143–45
184. Schmidt-Bailey J, Fuchs S. 1997. Experiments for the efficiency of varroa control with drone brood-trapping combs. *Apidologie* 28:184–86
185. Schmidt-Bailey J, Fuchs S, Büchler R. 1996. Effectiveness of drone brood trapping combs in broodless honey bee colonies. *Apidologie* 27:293–95
186. Schulz AE. 1984. Reproduction and population dynamics of the parasitic mite *Varroa jacobsoni* Oud. in correlation with the brood cycle of *Apis mellifera*. *Apidologie* 5:401–19
187. Seeman OD, Walter DE. 1995. Life history of *Afrocypholaelaps africana* (Evans) (Acari: Ameroseiidae), a mite inhabiting mangrove flowers and phoretic on honeybees. *J. Austral. Entomol. Soc.* 34:45–50
188. Sheppard WS. 1989. A history of the introduction of honey bee races into the United States. *Am. Bee J.* 129:617–19
189. Shimanuki H, Calderone NW, Knox DA. 1994. Parasitic mite syndrome: the symptoms. *Am. Bee J.* 134:827–28
190. Shimanuki H, Knox D. 1991. *Diagnosis of Honey Bee Diseases. USDA Agric. Handb. AH-690.* 53 pp.
191. Sihag RC. 1988. Incidence of *Varroa, Euvarroa* and *Tropilaelaps* mites in the colonies of honey bees *Apis mellifera* L.

in Haryana (India). *Am. Bee J.* 128:212–13

192. Slabezki Y, Efrat H, Dag A, Kamer Y, Yakobson BA, et al. 1999. The effect of honey bee tracheal mite infestation on colony development and honey yield of Buckfast and Italian honey bee strains in Israel. *Am. Bee J.* In press
193. Smith AW, Needham GR. 1988. A new technique for the rapid removal of tracheal mites from honey bees for biological studies and diagnosis. See Ref. 1, pp. 530–34
194. Smith AW, Page RE Jr, Needham GR. 1991. Vegetable oil disrupts the dispersal of tracheal mites, *Acarapis woodi* (Rennie), to young host bees. *Am. Bee J.* 131:44–46

194a. Smith DR. 1999. So many different honey bees! What mitochondrial DNA tells us about honey bee biogeography. *Proc. XXXVI Congr. Apimondia 99,* Vancouver, Can. p. 118

195. Spivak M. 1996. Honey bee hygienic behavior and defense against *Varroa jacobsoni*. *Apidologie* 27:245–60
196. Stoltz D, Shen X, Boggis C, Sisson G. 1995. Molecular diagnosis of Kashmir bee virus infection. *J. Apic. Res.* 34:153–60
197. Tingek S, Koeniger G, Koeniger N. 1996. Description of a new cavity nesting species of Apis (*Apis nuluensis*) from Sabah, Borneo with notes on its occurrence and reproductive biology (Hymenoptera: Apoidea: Apini). *Sencken. Bergiana Biol.* 76:115–19
198. Tomasko M, Finley J, Harkness W, Rajotte E. 1993. A sequential sampling scheme for detecting the presence of tracheal mite (*Acarapis woodi*) infestations in honey bee (*Apis mellifera* L.) colonies. *Pa. Agric. Exp. Stn. Bull.* 871
199. Topolska G, Ball B, Allen M. 1995. Identification of viruses in bees from two Warsaw apiaries. *Medycyna Weterynaryjna* 51:145–47 (In Polish)
200. Trubin AV, Chernov KS, Kuchin LA, Borzenko IE, Yalina AG. 1987. European foulbrood: transmission and sensitivity of the causal agents to antibiotics. *Veterinariya* 8:46–47 (In Russian)
201. USDA Q. Rep. 1960–1961. *Entomol. Res. Div., Bee Cult. Res. Invest.* Beltsville, MD
202. USDA Q. Rep. 1960 to 1962, 1970. *Entomol. Res. Div., Bee Cult. Res. Invest.* Laramie, WY
203. USDA Q. Rep. 1960. *Entomol. Res. Div., Bee Cult. Res. Invest.* Madison, WI
204. Vedova G, Lodesani M, Milani N. 1997. Development of resistance to organophosphates in *Varroa jacobsoni. Ape Nostra Amica* 19:6–10. (In Italian)
205. Wallner K. 1995. The use of varroacides and their influence on the quality of bee products. *Am. Bee J.* 135:817–21
206. Webster TC, Delaplane KS, eds. 1999. *Mites of the Honey Bee.* Hamilton, IL: Dadant & Sons. In press
207. Wilson WT, Pettis JS, Henderson CE, Morse RA. 1997. Tracheal Mites. See Ref. 17, pp. 255–77
208. Woyke J. 1987. Length of stay of the parasitic mite *Tropilaelaps clareae* outside sealed honeybee brood cells as a basis for its effective control. *J. Apic. Res.* 26:104–9
209. Woyke J. 1994. Repeated egg laying by females of the parasitic honeybee mite *Tropilaelaps clareae* Delfinado and Baker. *Apidologie* 25:327–30
210. Woyke J. 1994. Mating behavior of the parasitic honeybee mite *Tropilaelaps clareae. Exp. Appl. Acarol.* 18:723–33
211. Woyke J. 1994. *Tropilaelaps clareae* females can survive for four weeks when given open bee brood of *Apis mellifera. J. Apic. Res.* 33:21–25
212. Wu K-R, Kuang B. 1987. Two species of small honeybee—a study of the genus *Micrapis. Bee World* 68:153–55
213. Yoder JA, Sammataro D, Peterson JA, Needham GR, Wa B. 1999. Water

requirements of adult females of the honey bee parasitic mite, *Varroa jacobsoni* (Acari: Varroidae) and implications for control. I. *Int. J. Acarol.* In press

214. Zander E. 1909. Tierische parasiten als Krankheitserreger bei der Biene. *Leipz. Bienenz.* Jahrg. 24, 10:147–50 and 11:164–66 (In German)

Annu. Rev. Entomol. 2000. 45:549–574

INSECT PEST MANAGEMENT IN TROPICAL ASIAN IRRIGATED RICE

P. C. Matteson
FAO Programme for Community IPM in Asia, Hanoi, Vietnam; e-mail: matteson@fpt.vn

Key Words rice integrated pest management, IPM extension, farmer field school, farmer participatory research, community IPM

■ **Abstract** Abundant natural enemies in tropical Asian irrigated rice usually prevent significant insect pest problems. Integrated pest management (IPM) extension education of depth and quality is required to discourage unnecessary insecticide use that upsets this natural balance, and to empower farmers as expert managers of a healthy paddy ecosystem. Farmers' skill and collaboration will be particularly important for sustainable exploitation of the potential of new, higher-yielding and pest-resistant rices. IPM "technology transfer" through training and visit (T&V) extension systems failed, although mass media campaigns encouraging farmer participatory research can reduce insecticide use. The "farmer first" approach of participatory nonformal education in farmer field schools, followed by community IPM activities emphasizing farmer-training-farmer and research by farmers, has had greater success in achieving IPM implementation. Extension challenges are a key topic for rice IPM research, and new pest management technology must promote, rather than endanger, ecological balance in rice paddies.

INTRODUCTION

Pest management in rice, *Oryza sativa* L., was last addressed in these pages in 1979 (88). Although much has changed since then, Kiritani correctly foresaw today's priorities:

> The implementation of pest management by farmers still remains far behind. To bridge this apparent gap, it is anticipated that many obstacles should be conquered not only in technology but also in socioeconomic sectors.

The primary focus of this review is integrated pest management (IPM) implementation, customarily considered the domain of extension science, although this account opens with a summary of key advances in IPM technology. Today's rice IPM strategies can be said to have matured, in that they reflect a more sophisticated appreciation of the structure and dynamics of paddy ecosystems (90). Their implementation involves greater farmer participation, with research by scientific institutions and farmers playing a dynamic supporting role.

This review concerns irrigated rice in tropical Asia, the source of approximately one third of the world's rice (75) and the arena in which key developments have taken place. Pest management research and experimentation with different extension approaches have resulted in an IPM effort in that agroecosystem that is the largest and arguably the most innovative in the world. This review does not cover postharvest pest management, temperate rice, non-irrigated rice, or other crops and regions, except to enlarge upon specific points. Rather than attempting to review all ongoing work and cumulative literature since 1979, this review addresses important present and future directions in implementation and research. The educational principles and methodologies, as well as the training strategy, applied in Asia for rice IPM are now being used in other crops and on other continents (44a). Moreover, they are relevant to extension covering all aspects of agriculture, not just IPM in crop production.

THE MATURATION OF RICE IPM

Formerly, as described by Kiritani, control of rice insect pests was considered a central problem for Asian rice farmers. Yield losses of 15% to 25% or more were (and sometimes still are—see 108), attributed to "an abundance of pests" (88). Two or three crops a year, often overlapping, of heavily fertilized monocultures of "Green Revolution" high-yielding cultivars were considered a vulnerable pest breeding ground. Developing strongly pest-resistant rice cultivars was a high priority, but these were threatened by "resistance breakdowns," particularly to the brown planthopper (BPH), *Nilaparvata lugens* (Stål). Although the problem of insecticide-induced secondary pests (notably BPH) was recognized and attributed to the destruction of natural enemies, insecticides used according to economic thresholds were considered a valuable complement to varietal resistance and synchronized planting as the basis for integrated control (88).

Today's view of the irrigated tropical rice ecosystem and corresponding recommendations for insect pest management, summarized by Way & Heong (158), represents a radical revision of previous ideas regarding losses to insect pests and the role of insecticides. Insecticide use is considered destructive under most circumstances, and not a fundamental component of rice IPM. Instead, successful IPM depends on farmers' understanding of, and confidence in, resistance and tolerance to pests in a healthy crop protected by naturally occurring biological control. Action thresholds for insecticide use that are developed by researchers are irrelevant and should be discarded.

Crop Loss Assessment Revisited

Much of the previous attribution of high yield losses to insect pests is now considered an artifact of overestimation based on worst-case scenarios, short-term, small-plot trials at single sites frequently unrepresentative of farmers' field conditions, and misunderstanding of the effects of insecticides on paddy ecology (126). This reassessment highlights pest-related risk and the representativeness of crop loss data as researchable topics (21, 131).

Changed field conditions, including less insecticide use by farmers in some regions (e.g. 61, 119), and new rice cultivars (53), may have reduced yield losses to pests. Unstable, relatively high-level single-gene resistance has been deployed against BPH, green leafhoppers *Nephotettix* spp., and the gall midge *Orseolia oryzae* (Wood-Mason) (158). Minor resistance to BPH is also operative (18). Stem borer damage, especially by the yellow stem borer *Scirpophaga incertulas* (Walker), is ubiquitous and variable, but generally minor (31, 77). Current estimates of yield losses to stem borers have declined (158). This decline may be due to the moderate, seemingly polygenic resistance of many modern rice cultivars, and their ability to compensate for stem borer damage by increases in tillering, percentage productive tillers, and grain weight (129, 130).

Resistance breeding has had relatively little success against leaf feeders, however (85, 158). Despite that lack of varietal resistance, and although leaf-feeding insect damage is highly visible, leaf-feeding insects do not appear to cause significant yield loss under most circumstances. For instance, no yield loss was detected when up to 60% of leaves were damaged by whorl maggot (*Hydrellia* spp.) (142). Similarly, up to five larvae/hill of the leaffolder *Cnaphalocrocis medinalis* (Guenée) may damage as many as 50% of leaves (26), but Japonica rice at the tillering stage can compensate for as much as 67% of leaffolder-damaged leaves (104), and computer simulations show that leaffolder densities must reach 15/hill before any detectable yield loss results (35). Such findings indicate that farmers' common practice of early-season insecticide sprays against stem borers and defoliators is usually unnecessary (55, 123, 158).

Reliance on a Balanced Paddy Ecosystem

Food web studies (e.g. 58, 59, 72, 125, 135, 136), and investigations of predator and parasitoid biology, ecology, and impact (e.g. 26, 80, 155) have highlighted the biodiversity of rice paddy fauna, including natural enemy richness (6, 110, 140, 141, 153). Most rice pests are controlled by a complex and rich web of generalist and specialist predators and parasites that live in or on the rice plant, paddy water, or soil. Abundant early-season detritivores and plankton-feeders such as Collembola and chironomid midge larvae allow generalist predators to establish and multiply in unsprayed paddies before herbivores immigrate (139, 162). If undisturbed, these natural enemies normally prevent significant insect pest problems.

Early-season insecticide applications destroy that ecological balance. Insecticide wipes out predators along with their food supply, leaving the field open for pest buildup (5, 17, 64a, 134, 139). Insecticide suppression of natural enemies, particularly spiders, predacious water striders in the genera *Microvelia* and *Mesov-*

elia, and the mirid bug *Cyrtorhinus lividipennis* Reuter, has been confirmed as the key factor in the emergence of BPH as a secondary pest (54, 83, 112).

Instances in which natural control fails in untreated crops are not well enough understood. Insecticide use is not the only factor that may perturb the paddy ecosystem. Unusual weather patterns and/or migratory behavior are frequently associated with outbreaks of sporadic pests such as the rice hispa *Dicladispa armigera* (Olivier), the rice thrips *Stenchaetothrips biformis* (Bagnall), armyworms [*Spodoptera mauritia acronyctoides* (Guenée), *Mythimna separata* (Walker)], and black bugs (*Scotinophara* spp.) (22). Some rice insects are major pests in only one region—hispa in Bangladesh (4) and thrips in Vietnam, for example—and this raises the question of possible geographical variation in associated natural enemies (114a).

Because most insect pests of tropical rice are indigenous and have coevolved with a rich natural enemy fauna, attempts at classical and inundative biological control have generally been fruitless. Attention has turned to conserving existing natural enemies and maximizing their impact (114). In contrast to earlier advocacy of large-scale, synchronized planting with long fallow periods for controlling pests (e.g. 96, 109), some ecologists are recommending continuous, staggered planting that keeps natural enemies in mature rice within easy immigration distance of new crops (78, 132, 139, 156).

Although resistant cultivars and biological control are generally considered compatible, interactions between cultivars and natural enemies may be positive or negative (11). For example, volatile chemicals produced by certain rice genotypes can attract BPH predators (122). The influence of cropping practices and nonrice habitats—particularly bunds and field edges—on pest infestations and natural enemy biodiversity, numbers, and impact is being examined with a view to identifying better management options for farmers (e.g. 16, 94, 137, 159, 163, 164).

Aside from causing secondary pest problems, insecticide use is believed to have accelerated the adaptation of BPH to resistant varieties by favoring the survival and reproduction of virulent individuals (44, 51). In this regard, it is important to note that past extension efforts failed to instill the understanding and confidence necessary for farmers to take advantage of varietal resistance. For example, in 1992 most farmers in three villages in Central Luzon, Philippines, planted the BPH-, green leafhopper- and stem borer-resistant rice cultivar IR64, but named those pests as the main targets of their insecticide applications (115).

Economic Thresholds Reconsidered

Numerous studies indicate that in tropical Asia irrigated rice farming without insecticides is economically competitive (e.g. 31, 81, 100, 115, 144, 157). When human health and environmental costs of insecticide application are considered (89, 95, 120), "no action" appears to be the wisest pest management option (126).

Farmers who do not drop insecticide use altogether are hard pressed to identify the infrequent occasions when it will be profitable to spray.

Economic threshold levels as promulgated by researchers and used by surveillance and alert systems are not useful under most circumstances. At worst, they alarm farmers without reason and pressure them to apply insecticides unnecessarily. At best, they provide a forecast that farmers must second-guess, because general observations and thresholds are not sufficiently farm and locality specific. In practice, farmers decide for themselves whether pest control action is warranted, based on a more refined process that takes into account the condition and priority of their individual crop, as well as other factors such as the current prices of rice and insecticide, and options for more productive alternative expenditures (56, 101).

IMPLEMENTING RICE IPM

Now that many IPM specialists have arrived at a radically different message—"insecticides are usually not needed in rice"—and are focusing on a strong crop in a healthy paddy ecosystem, they have the difficult task of undoing the effect on farmers of decades of exhortations to rely on pesticides. Advocacy of pesticide use, reinforced by extensive agrichemical advertising, led both farmers and policy makers to overestimate crop losses caused by pests, and the effectiveness of insecticides (157). Unless they understand the benefits of avoiding unnecessary insecticide use, farmers tend to overreact to slight infestations and make routine preventive applications (32).

Twelve percent of pesticides sold worldwide are applied to rice crops, and no other single crop accounts for as much pesticide use (161). Rice farmers will continue to be the target of massive agrichemical industry marketing and promotion that is supported by financial resources dwarfing those of agricultural extension programs. If farmers are to have the understanding, skills, and confidence to withstand that barrage, IPM training must have depth and quality, and appropriate long-term technical support must be available afterward.

Rice IPM extension in Asia has evolved over two decades. Advances have drawn heavily on social science expertise and the incorporation of techniques developed in other fields of endeavor, notably commercial advertising, participatory nonformal education, and community organizing (101). Changes in people's roles in communications and in education during this evolution reflect a paradigm shift in agricultural development.

Scientific paradigms are universally recognized scientific achievements that, for a time, define scientists' view of nature and offer model problems and solutions to a community of practitioners. When novel experience or knowledge becomes incompatible with the prevalent paradigm, a "paradigm shift" results in a new view of the world in which scientific work is done (91). Current IPM extension approaches reflect two paradigms. The "technology transfer" paradigm proved unsuitable for farmer education but can motivate pesticide use reduction with a simple message. Under the "farmer first" paradigm, IPM extensionists seek to lay the educational foundation for farmer-led community development through farmer field schools and community IPM.

Developing IPM Training Principles

Interdisciplinary research facilitated by the International Rice Research Institute (IRRI) in the Philippines from 1978 to 1980 (46–48, 93) developed basic rice IPM training principles:

1. Group training so that farmers can learn from each other, with frequent discussions and group reinforcement of decisions.
2. A curriculum pared down to essentials and simplified, having the most important points repeated often.
3. Twenty to 40 hours of good-quality instruction in the rice paddy, distributed so that farmers can practice skills and crop protection decision-making each week during an entire growing season.
4. Class experiments and demonstrations that engage farmers' curiosity and encourage imaginative inquiry and self-reliance.
5. Periodic followup as farmers gain confidence in their independent decision-making.

Training on a pilot scale in the Philippines according to those guidelines, with highly motivated, intensely supervised field officers using conventional training methodology, made substantial long-term impact (81, 84).

The Asian rice IPM training effort has received long-term technical and financial assistance from the United Nations Food and Agriculture Organization (FAO) Intercountry Programme for Integrated Pest Control in Rice in South and Southeast Asia, recently renamed the FAO Programme for Community IPM in Asia. This program, hereinafter referred to as the FAO IPM Program, provides coordination, technical support, and training to national IPM extension and research initiatives, and currently includes 12 member countries (34, 40). In 1984, FAO increased its support to several countries in order to reach more farmers with the training approach that had been proven effective on a relatively small scale in the Philippines.

Technology Transfer

Under the longstanding "technology transfer" paradigm, agricultural research and development are carried out stepwise by a large, multipurpose hierarchy. Recommendations for farmers are defined after several stages of research, which takes place largely on experiment stations. Instructional messages about the resulting technical packages pass through a chain of extension officers to a subset of farmers, who are supposed to communicate the recommendations to their neighbors.

These systems have historically promoted the use of fixed "packages" of purchased inputs including pesticides, with or without threshold levels governing pesticide application. Insecticides are commonly subsidized, and recipients of government credit are often required to purchase a certain quantity of pesticides each season (84, 124). Decades of such programs entrenched a chemical-dependent attitude in farmers and government agriculturalists alike.

Training and Visit Extension The "technology transfer" model has been widely implemented in the form of the Training and Visit (T&V) Extension System promoted by the World Bank, long the only source of credit for major extension projects. The T&V system is a set of principles developed for increasing the effectiveness of technology transfer: improving management, fostering professionalism, providing regular training for extension agents, and preserving a strong field orientation stressing regular (biweekly) visits to farmers (8). With FAO IPM Program support in the late 1980s, Philippine master trainers presented field IPM implementation courses for selected T&V staff from Sri Lanka, Indonesia, and new areas in the Philippines. Then IPM extension responsibility was handed over to those countries' national T&V systems, supported by strategic mass media campaigns.

The strongly hierarchical T&V extension model conditioned the style and content of training at all levels. Trainer and trainee related to each other through the traditional teacher/student relationship and the corresponding conventional training methodology: The trainer as "expert" dominates, defining the curriculum and tending to lecture, trainees are expected to be interested, deferential, and accepting. Extension officers are usually trained in the classroom between cropping seasons, with little opportunity for hands-on field practice. The curriculum does not emphasize teaching skills because the accent is on periodic message delivery. As a result, extension agents emerged ill-prepared for their job of helping farmers, including their duty to pass farmers' messages "bottom-up" back to researchers. In practice, "technology transfer" is a "top-down" approach with "experts" in charge and with little feedback from below. That situation is commonly exacerbated by a lack of incentives and rewards for good work by village-level extensionists (3, 12, 45, 145, 149). Extension training under this paradigm fails to motivate field extension agents and farmers because it is unresponsive to their actual needs and ideas.

Indeed, rice IPM implementation in Asia failed under the "technology transfer" paradigm despite T&V management improvements, FAO technical support, and, in some cases, top-level willingness to be flexible and to initiate special activities to duplicate the IPM training approach that succeeded originally in the Philippines. Inertia, indifference, and extension agents' many conflicting responsibilities precluded the necessary intensive, high-quality field training effort. As a result, farmers' pest management practices did not change appreciably (101,

160; PC Matteson & H Senerath, unpublished data). Similar experiences in other crops and regions (e.g. 2, 13) support the conclusion that traditional technology transfer practices are unsuited to IPM.

Strategic Extension Campaigns Strategic extension campaigns (SECs) using mass media (1) have been more effective than T&V as an element of "technology transfer" programs. They convey research findings and recommendations in a simplified form in order to motivate attitude change. SECs can achieve rapid impact because they reach most farmers in an area all at once, including remote locations normally not visited by extension trainers.

Analysis of farmers' IPM attitudes, knowledge, and decision-making processes (105–107) has been applied to SEC planning. A number of SECs on IPM themes were carried out in Asia during the 1980s. Evaluation surveys indicate that they improved farmers' knowledge, attitudes, and practices, often strikingly, and that they can rectify misconceptions that prevent farmers from making good pest control decisions (1, 33, 66, 116). SEC materials concerning pesticide use must be planned with particular care, however. Campaign results indicate that long exposure to intensive pesticide advertising can condition farmers to react to any image of pesticide use as a recommendation, regardless of the accompanying message (32).

Motivation by Mass Media Currently, "Forty Days" SECs are being fielded in several countries in order to reduce unnecessary insecticide use in early-season rice. The objective is to rectify farmers' mistaken belief that leaf-feeding insects, particularly leaffolders, cause severe yield loss. This belief leads them to apply insecticides during the early stages of the crop, endangering applicators (126) and often triggering outbreaks of BPH and other secondary pests (63).

The Forty Days campaign conveys a heuristic, or simple rule of thumb (79), that summarizes a large volume of research findings and simplifies decision-making: "Spraying for leaffolder control in the first 40 days after planting (or 30 days after transplanting) is not necessary" (61). Such messages, which are at odds with farmer's beliefs, provoke cognitive dissonance (41), psychological conflict that can be resolved through reevaluation (25). To encourage reevaluation, the campaign includes an element of farmer participatory research. Skeptical farmers are urged to test that decision rule with a field experiment, leaving about 500 m^2 of rice field unsprayed and then comparing the yield with that of sprayed rice.

This approach to insecticide use reduction was first tried out on a pilot scale, via interpersonal contact between researchers, extensionists, and farmers rather than via mass media. In Leyte, Philippines in 1992, 101 farmers presented and discussed their experimental results in end-of-season workshops. Workshop participants concluded that early-season insecticide applications could be dropped without affecting rice yield, and this conclusion changed their beliefs, practices, and profits (62). Previously, most farmers believed that leaf-feeding insects cause severe damage (77% of farmers) and yield losses (87%), and should be sprayed for early in the season (62%). After the experiment, 28%, 9%, and 10% of farmers, respectively, held those beliefs. The proportion of farmers who applied insecticide during the first 30 days after transplanting dropped from 68% to 20% after one year, and to 11% after two years. Average insecticide applications per season dropped from 3.2 to 2 in two years. The average seasonal cost of insecticide/ha was reduced accordingly, from $17.10 to $7.60 (60). Similar results were obtained in the Mekong Delta in southern Vietnam (64).

Mass media—radio dramas and the distribution of posters and leaflets—were subsequently used in conjunction with farmers' meetings and demonstrations to implement Forty Days campaigns in the Mekong Delta. A media campaign in Long An province initiated in late 1994 with an audience of 20,000 farm families prompted 56% of farmers to perform the experiment. A year later, the typical number of seasonal insecticide applications was reduced from three or four to one or two although leaffolders and other leaf-feeding insects remained the chief targets. The proportion of farmers applying their first insecticide application within six weeks of planting had dropped from 96% to 62%. The proportion of farmers believing that early season spraying is needed for leaffolders fell from 77% to 17% (61).

Farmer First

In response to the failure of IPM technology transfer by T&V extension, the Indonesian National Programme for the Development and Training of IPM in Rice-Based Cropping Systems, supported by the FAO IPM Program, developed a more dynamic, self-replicating IPM training process. The training is carried out in Farmer Field Schools (FFSs), which retain the rice IPM training principles and season-long framework first elaborated in the Philippines, and add participatory nonformal education methodology to motivate and empower farmers (82, 117, 127, 128). This new training process is based on changed roles for farmers and trainers, reflecting a more recent agricultural development paradigm called "farmer first" (15, 138).

Proponents of "farmer first" argue that conventional top-down agricultural research and extension methods usually fail to produce appropriate innovations, and that best results are achieved when farmers are instrumental in every step of the process. Under this model, farmers, extension agents, and researchers work together as equal partners, each having specialized skills and knowledge to contribute. The latter two become collaborators, facilitators, and consultants, empowering farmers to analyze their own situation, to experiment, and to make constructive choices (29, 42, 43).

Farmer Field Schools In a typical weekly half-day Farmer Field School (FFS) session, about 25 farmers divide into groups of five to analyze the rice agroecosystem and decide what the crop needs that week. The situation in an "IPM

Practice" plot is compared to that in a "Farmer Practice" plot where customary preventive insecticide applications are made. While the farmers are observing the crop, the trainer facilitates a participatory "discovery" learning process. Farmers' questions are answered with other, leading questions that help them draw on their own knowledge and experience, or trainers help farmers design and carry out field experiments that fill information gaps. Each small group makes a drawing of the rice ecosystem that illustrates the condition of the paddy and the rice plant, along with associated pests and their natural enemies. This diagram helps group members analyze ecological interactions and draw conclusions about crop needs, which are presented to the larger group. A plenary decision is hammered out via extensive discussion that reinforces learning while allowing trainers to evaluate trainees' progress and to correct misunderstandings.

After this agroecosystem analysis process is completed, a special topic appropriate to the stage of the crop is taken up. Complementary group dynamics exercises provide fun, enhance learning, and reinforce farmer solidarity and collaboration.

The FFS curriculum focuses on community ecology and dynamics in the rice paddy, with emphasis on the natural enemies of pests. Learning more about natural enemies, often via "insect zoos" that allow farmers to observe predation and parasitism in action, is at once the most enjoyable and the most powerful FFS activity for farmers. With this new knowledge, they understand clearly why unnecessary insecticide use, so harmful to their "friends," must be avoided (111).

Farmers make crop management decisions based on their personal circumstances and the ecological balance in each paddy. The four IPM implementation principles of the Indonesian national program reflect this holism and the IPM goal of making farmers confident managers and decision-makers, eager for new ideas and information but free from dependence on a constant stream of pest control directives from outside:

1. Grow a healthy crop.
2. Observe fields weekly.
3. Conserve natural enemies.
4. Farmers are IPM experts.

Improved results in Indonesia and subsequent comparative studies of training methodologies and their impact indicate that FFSs are more effective for IPM training than conventional extension approaches, which generally still focus on economic thresholds for pesticide application (99, 150). A followup study of the first 50,000 Indonesian FFS graduates found that they reduced insecticide application from an average of 2.8 sprays per season to less than one, with most farmers not spraying at all. When farmers did apply insecticide, they could identify a specific target pest (118).

A 1993 Philippine Barangay IPM Project in Central Luzon found that both FFSs and the Forty Days approach reduced the proportion of rice farmers using insecticides from 80% to less than 20%, with no yield loss. This change in farmers' behavior lasted at least four years in each case (115).

Insecticide use reduction is only one of many agroeconomic impacts of FFSs, however. FFSs address all aspects of production, including optimal seeding rates and fertilizer application. In general, rice FFS graduates' profits rise because their insecticide and seed costs decrease while yields are as high as or higher than before, due to better crop management. For example, FFS participants in the Bangladesh Integrated Rice and Fish (INTERFISH) project of CARE, a large international nongovernmental organization, use no pesticides and are harvesting 17% to 33% more rice, which increases their gross marginal income by 33% to 54%. In their rice/fish systems, made possible by eliminating insecticides, those higher rice yields are maintained, but income increases by 96% to 387% (9). Moreover, yield variance typically decreases, reflecting lower production risk—an important consideration for farmers who rely on growing rice for their livelihoods (Vietnam National IPM Programme, unpublished data). Learning and practicing problem-solving skills for IPM enables farmers to continue learning from the field and from the consequences of each management decision from then on. The result is better cropping practices and management expertise, which are then applied to other crops and production systems (82, 97).

In the past, impact evaluations of IPM extension programs confined themselves to quantitative agroeconomic indicators. Because FFSs and post-FFS followup activities have broader objectives than IPM, however, the range of relevant indicators is much broader, and the indicators are qualitative as well as quantitative. Participation and empowerment are not just ways to improve farmers' production practices; rather, they are the goal. Therefore, indicators of increased access, leverage, status, and choices for farmers, such as farmer innovations and the degree to which farmers control the IPM agenda and act to affect policy, are important (J Pontius, unpublished data). Case studies are useful for presenting qualitative as well as quantitative evaluation findings (38, 39).

Community IPM National and local extension systems are seeking to insure the sustainability of IPM implementation. It is considered important that farmers acquire the necessary skills to establish and maintain local IPM programs. The FAO IPM Program is supporting training of Asian IPM officers in participatory planning methods. Farmer graduates of FFSs, assisted by trained IPM personnel, carry out planning exercises for followup activities at the subdistrict level, then implement the plans that they make (38, 39). This community IPM development process is most advanced in Indonesia and Vietnam. By presenting their plans to government officials up to the provincial level, farmers establish a dialogue that reinforces their role as planners, secures local funding for IPM, and is meant to lay the foundation for effective, long-term government support to farmer-led initiatives (69, 70, 82).

The broad range of resulting activities has deepened farmers' ecological understanding, strengthened FFS alumni groups, and helped them find ways to involve other farmers (38, 39). Examples include:

1. Followup FFS in rice with in-depth studies of special interest, such as rice-fish production.
2. FFS for other crops (e.g. vegetables, soybeans, tea).
3. FFS for elementary school students.
4. Farmer IPM clubs, congresses and technical meetings.
5. Marketing of "green label" pesticide-free rice.
6. IPM consultant teams.
7. Irrigation system improvement.
8. The organization of cooperatives.
9. Action Research Facilities for farmer-led experimentation (see Research Models section below).

Especially in Indonesia and Vietnam, FFS graduates who show ability and enthusiasm for involving others participate in a week-long training-of-trainers course and then conduct their own field schools. Farmer Trainers are supported by periodic visits from IPM staff and by trainer workshops conducted at least three times per season for discussing leadership and FFS issues. In addition, they have access to case studies of IPM innovations, documented with extension materials, that can enrich a community's IPM program. They also represent their local farmer groups in farmer planning meetings and farmer technical meetings that help institutionalize farmer-based extension (69, 70).

Quality Control

Farmers have many demands competing for their time. Rice IPM education and communications must be of consistent high quality at all levels in order to engage their interest and maintain attendance in FFSs. Therefore, quality control measures should be a permanent, integral part of IPM implementation programs (34). Moreover, because farmers are constantly under commercial pressure to buy and use insecticides, IPM training must be adequately followed up in order for a program to have lasting impact. To help ensure quality, the Indonesian and Vietnamese community IPM programs send teams to observe and evaluate training and followup activities, and provide feedback and guidance on the spot. In-depth case studies help interpret impact data by studying the ecological, educational, and social processes at work (39).

The Policy Environment

Farmer training programs may be insufficient for IPM implementation unless they can be conducted within a policy framework that supports them and promotes IPM. Institutional support is a particularly significant issue for FFSs. In many countries, FFS implementation and impact are hampered by national institutions working under conflicting paradigms: traditional transfer-of-technology extension systems and crop protection services that warn farmers to spray their fields when wide-scale crop surveillance and pest forecasts indicate that pest populations may exceed fixed economic threshold levels (99).

Numerous policy options can support the implementation of IPM as a way to meet social, environmental, and crop protection goals (121). Eliminating pesticide subsidies is a powerful measure for reducing unnecessary pesticide use. Indonesia saved more than $100 million/yr by phasing out an 85% pesticide subsidy between 1986 and 1989, while rice yields went up and the price effect reduced average pesticide applications/season/paddy from over 4 to about 2.5. That rate is still far more pesticide use than is justified in rice, however, and it illustrates the fact that price policies alone will not rationalize pesticide use where continued overuse is perpetuated by factors such as pesticide advertising, pesticide sales by government agricultural staff, and fear of dropping calendar spray routines that were recommended for so long. Training that motivates farmers to withstand those pressures is essential for effecting real change. High-level political will must be complemented by grassroots activism prepared to demand its implementation (119).

It is particularly important to prevent crop protection and extension staff from being able to profit from pesticide marketing, officially or as informal salespeople on commission. In some countries, the ministry of agriculture sells pesticides to farmers. Pressures associated with that conflict of interest and the opportunity to top up inadequate incomes can motivate field officers to subvert IPM training (99, 151).

Women's Participation

Women play important roles in rice production (68, 76, 148), and initial steps were taken in the Philippines to involve women in IPM development and extension (36). Nevertheless, men have been the overwhelming majority of participants in Asian FFSs organized by national extension systems, due in part to a traditional focus on male landed farmers that excludes women from extension activities for production agriculture (71). Another barrier to women's participation is the day-long workload of child-minding and housekeeping tasks, in addition to farm labor. These responsibilities leave little time for attending IPM classes, let alone workshops or planning exercises that take one or more days, or require travel (14, 102). The FAO IPM Program is developing guidelines for strengthening women's participation and leadership in national IPM programs, including post-FFS activities (87).

New Rices, New Challenges

New kinds of rice promise to change the framework within which future IPM strategies will be developed. Some new rices may be more vulnerable than present cultivars are to pests or resistance breakdown. Farmers' skill and collaboration will be important for sustainable exploitation of their potential.

Hybrid Rice Hybrid vigor can raise rice yield potential by 15% to 20%. Hybrid cultivars are under development in many tropical countries, and a few have been released for commercial production. These cultivars inherit insect resistance from their parents, depending on whether the resistance genes are dominant or recessive (154). Hybrids may differ from inbred rices, however, in pest management-related characteristics as well as agronomic ones. Some Chinese hybrids appear to exhibit superior ability to compensate for stem borer and/or defoliator damage (98, 103; PE Kenmore & X Yu, personal communications).

Transgenic Rices Insect resistance has been among the top priorities of rice biotechnology (65). Extensive research is underway in dozens of public- and private-sector laboratories around the world. New tools for improved breeding and cultivar deployment include wide hybridization, DNA markers, genetic transformation, and DNA fingerprinting of pests (7).

The incorporation into rice of protein toxins from entomopathogenic bacteria, protein inhibitors of insect digestive enzymes, and certain lectins have been prime research targets (86, 147). Genes for protease inhibitors toxic to BPH have been inserted into rice (92). "Bt rice" containing delta-endotoxins from *Bacillus thuringiensis* (Bt) is farther along. Field trials began in China in 1997, and Bt rice will probably be available to farmers within a few years (MB Cohen, personal communication). Bt toxins confer resistance to stem borers and/or leaf-feeding caterpillars (7). As discussed above in Crop Loss Assessment Revisited, infestations of those pests normally do not justify control action at the level of the individual farmer. Scientists responsible for the development and deployment of Bt rices, however, maintain that Bt cultivars could prevent the loss of 5% to 10% of overall rice production as well as much unnecessary pesticide use (74).

Farmer education and cooperation will be key to deployment plans for genetically engineered rice, which must address ecological concerns and resistance management challenges in order to ensure safety and sustainability (74). In many Asian countries, insect resistance genes could spread into wild or weedy rices, perhaps enhancing their invasiveness (19). The possibility of indirect effects of Bt rice on natural enemies must also be explored. Moreover, the useful life span of Bt as a pest management tool both in rice cultivars and as a rice insecticide will be cut short unless special efforts are made to deploy Bt toxins in a way that prevents the development of resistance (20, 49).

New Plant Type Rice breeders hope to achieve a quantum increase in yield potential through modification of the present high-yielding semidwarf plant type. Creation and subsequent hybridization of a new plant type (NPT) with increased leaf area per unit ground area and more nitrogen stored in erect leaves (143) have the potential to increase yields by up to 25%. Lower tillering and thicker stems are important for the extra stem strength required to support increased panicle weight. Those characters may be associated, however, with the heavy stem borer damage observed in preliminary field trials, particularly damage by the striped stem borer *Chilo suppressalis* (Walker) (JE Sheehy, personal communication). Research to address this problem, including the incorporation of Bt genes into NPT cultivars, is under way (24).

Meeting Challenges with Improved Technology

Research Models Successful IPM requires research at all levels of the system. Breakthroughs made by research institutions and universities enabled early programs to be built on a sound scientific foundation. Since 1990, an IPM Network coordinated by the International Rice Research Institute (IRRI) and comprised of national IPM research programs in China, India, Indonesia, Lao PDR, Malaysia, the Philippines, Thailand, and Vietnam has contributed to the current understanding of rice-field ecology and interactions with nonrice habitats, farmers' pest control practices, and insecticide misuse (73).

In addition, a new model is emerging that offers an opportunity for farmer/research/extension collaboration in village-level research: post-FFS Action Research Facilities for experimentation by expert farmers. One such farmer group in the Indramayu province, Indonesia, developed a strategy for preventing infestations of the white stem borer *Scirpophaga innotata* (Walker), a locally severe problem long considered intractable. They then organized area-wide implementation by holding village, subdistrict, and district-level seminars (113). Scientists, including IPM Network collaborators, have begun facilitating other such farmer initiatives, which can benefit all concerned. Farmers have access to technical assistance with experimental design and the interpretation of results, while researchers can ensure that their work responds to farmers' priorities, produces results adapted to local circumstances, and enhances IPM implementation through farmers' heightened awareness and participation. All are empowered by acquiring constructive, nontraditional skills (82).

Priority Research Implementation must be an integral part of IPM research and development. Extension challenges are crucial subjects for IPM research. For instance, creative ways must be found to involve women more fully in rice IPM activities and to promote and support farmer-to-farmer training. In addition, extension programs must do a better job of educating farmers about, and instilling confidence in, the special properties of resistant cultivars in order to capture their potential pesticide use reduction benefits, including the potential benefits of genetically engineered rices (74, 101). Rice scientists' responsibilities should extend to collaboration with farmers and social scientists in order to better understand farmers' perceptions and motivations, draw on their insight and ingenuity, and be more responsive to their needs (56).

Scientists must ensure that new technology promotes, rather than endangers, the ecological balance in the paddy that protects rice from insect pest problems most of the time. Stable resistance and tolerance to insect pests continue to be valuable complements to natural controls. More work is needed to identify, and draw maximum benefit from, the strengths of hybrid rices. In addition, rice breeders should evaluate the impact of varietal characteristics on important natural enemies, with a view to combining pest resistance with characteristics that favor biological controls (10). Urgent research is required in order to deploy genetically engineered rices safely and sustainably, and to solve the NPT stem borer problem in a way that keeps the rice crop compatible with natural control-based IPM (19, 158). Too little is known about the effect of agrichemicals, including fungicides and herbicides, on paddy fauna. Attention is turning to "surrogate taxa" that could function as biodiversity indicators for monitoring trends and giving early warning of detrimental environmental change (137).

The conservation and enhancement of natural controls is a key research area. There remains much scope for designing rice systems, rotation crops, and landscapes that feed and shelter natural enemies and facilitate their access to newly planted paddies (137, 139). Exceptions to the general protection afforded by natural enemies of insect pests should also receive scrutiny. A better understanding of when and why pest outbreaks are triggered is necessary for creating ecologically sound control strategies, and for planning the improvement of present levels of natural control.

SERVING ALL THE FARMERS

About two million Asian rice farmers have graduated from FFSs. Helping the other 99% of Asian rice farmers to become IPM practitioners is a formidable task by any standard (60). The relative speed, impact, and cost-effectiveness of different extension and communication approaches are being analyzed and debated (57, 133, 115; LL Price, unpublished data). That discussion is muddied by the widespread misconception that FFSs and community IPM cover only IPM, whereas these programs aim to strengthen the initiative, the crop production knowledge, and the management skills farmers now need to increase their general productivity and develop their communities (12, 82, 117, 127, 152).

It falls to those responsible for IPM programs to implement the most effective combination of approaches and activities possible, in accordance with local objectives and resources. Communications media can be important in raising awareness and creating demand for IPM education, and can improve knowledge in some cases. SECs such as the Forty Days campaign are a relatively quick and cheap means of wide-scale achievement of specific objectives that lend themselves to simple messages. Interpersonal collaboration in educational activities with a broader scope, such as FFSs and support to farmer research, provides opportunities for participation and experiential learning, which are essential for empowering farmers as expert IPM practitioners and production management decision-makers (32). FFSs and Forty Days campaigns are being implemented simultaneously in some places. For example, Forty Days campaign materials in Vietnam urge farmers to contact the Ministry of Agriculture to request further

IPM training in FFSs, and the two extension approaches appear to be complementary.

The community IPM strategy is to create a dense network of FFS alumni groups that will spread IPM implementation horizontally, supported in large part by local resources (40). That strategy is meant to reach Asian rice farmers more quickly than is possible through direct training by the relatively small number of government extension or crop protection officers. Where successful, it would also lay the groundwork for extension by farmers. Farmers own and manage the agricultural extension service in Denmark (23). The Danish model may be useful for developing countries where people are ill served or unserved by government extension agencies.

ACKNOWLEDGMENTS

Special thanks are due to PE Kenmore and KL Heong, who were instrumental in conceptualizing this review, and whose assistance made its completion possible. The author is grateful to them and to DG Bottrell, MB Cohen, R Dilts, MM Escalada, PAC Ooi, KG Schoenly, MJ Way, M Whitten, and MR Zeiss for their reviews of an earlier version of the manuscript. Thanks also to the many colleagues, too numerous to list by name here, who generously contributed information and publications.

LITERATURE CITED

1. Adhikarya R. 1994. *Strategic Extension Campaign: A Participatory-Oriented Method of Agricultural Extension. A Case-Study of FAO's Experiences.* Rome: FAO. 209 pp.
2. Agudelo LA, Kaimowitz D. 1989. *Institutional Linkages for Different Types of Agricultural Technologies: Rice in the Eastern Plains of Colombia. Linkages Discuss. Pap. No. 1,* The Hague: Int. Serv. Nat. Agric. Res.
3. Axinn GH. 1988. T & V (Tragic and Vain) extension? *Interpaks Exchange* 5(3):6–7
4. Bangladesh Rice Res. Inst. 1989. *Proc. SAARC Workshop Rice Hispa, 28–29 Dec. 1986.* Dhaka: BRRI
5. Barrion AT, Aquino GB, Heong KL. 1994. Community structures and population dynamics of rice arthropods in irrigated ricefields in the Philippines. *Philipp. J. Crop Sci.* 19(2):73–85
6. Barrion AT, Litsinger JA. 1995. *Riceland Spiders of South and Southeast Asia.* Wallingford, UK: CAB Int. Manila: IRRI. 715 pp.
7. Bennett J, Cohen MB, Katiyar SK, Ghareyazie B, Khush GS. 1997. Enhancing insect resistance in rice through biotechnology. In *Advances in Insect Control: The Role of Transgenic Plants*, ed. N Carozzi, M Koziel, 5:75–93. London: Taylor & Francis. 301 pp.
8. Benor D, Baxter M. 1984. *Training and Visit Extension.* Washington, DC: World Bank
9. Best J, McKemey K, Underwood M. 1998. *CARE-Bangladesh INTERFISH*

Project, Output-to Purpose-Review on behalf of the AID Manage. Off., Dhaka, Sept. Reading, UK: Agricult. Ext. Rural Dev. Dep. Univ. Reading

10. Bottrell DG. 1996. The research challenge for integrated pest management in developing countries: a perspective for rice in Southeast Asia. *J. Agric. Entomol.* 13(3):185–93
11. Bottrell DG, Barbosa P, Gould F. 1998. Manipulating natural enemies by plant variety selection and modification: a realistic strategy? *Annu. Rev. Entomol.* 43:347–67
12. Byerlee D. 1987. *Maintaining the Momentum in Post-Green Revolution Agriculture: A Micro-Level Perspective from Asia. Int. Dev. Pap. No. 10.* Lansing, MI: Dep. Agric. Econ., Mich. State Univ.
13. Castella J-C, Jourdain D, Trébuil G, Napompeth B. 1999. A systems approach to understanding obstacles to effective implementation of IPM in Thailand: key issues for the cotton industry. *Agric. Ecosyst. Environ.* 72:17–34
14. Cent. Family Womens' Stud. 1994. *Women and Integrated Pest Management (An Initial Study on the Participation of Women Farmers in the IPM Programme in Vietnam),* with Nat. IPM Programme, Hanoi, Vietnam. 34 pp.
15. Chambers R, Pacey A, Thrupp LA, eds. 1989. *Farmer First: Farmer Innovation and Agricultural Research.* New York: Bootstrap. 218 pp.
16. Cheng J-A, Lou Y-G. 1998. The role of non-rice habitats in predator conservation. See Ref. 50, pp. 64–68
17. Cohen JE, Schoenly K, Heong KL, Justo H, Arida G, et al. 1994. A food web approach to evaluating the effect of insecticide spraying on insect pest population dynamics in a Philippine irrigated rice ecosystem. *J. Appl. Ecol.* 31:747–63
18. Cohen MB, Alam SN, Medina EB, Bernal CC. 1997. Brown planthopper, *Nilaparvata lugens*, resistance in rice cultivar IR64: mechanism and role in successful *N. lugens* management in Central Luzon, Philippines. *Entomol. Exp. Appl.* 85: 221–29
19. Cohen MB, Jackson MT, Lu BR, Morin SR, Mortimer AM, et al. 1999. Predicting the environmental impact of transgene outcrossing to wild and weedy rices in Asia. *Proc. Br. Crop Prot. Counc. Symp. Gene Flow Agric.: Relevance for Transgenic Crops,* 72nd, Univ. Keele, pp. 151–57. Farnham, Surrey, UK: BCPC
20. Cohen MB, Romena AM, Aguda, RM, Dirie A, Gould FL. 1998. Evaluation of resistance management strategies for Bt rice. *Proc. Pac. Rim Conf. Biotechnol. Bacillus thuringiensis Impact Environ., 2nd, 4–8 Nov. 1996, Chiang Mai, Thailand,* pp. 496–505. Bangkok: Entomol. Zool. Assoc. Thailand, Kasetsart Univ., Mahidol Univ., Nat. Cent. Genet. Eng. Biotechnol., Natl. Res. Counc. Thailand, Dep. Agric. 631 pp.
21. Cohen MB, Savary S, Huang N, Azzam O, Datta SK. 1998. Importance of rice pests and challenges to their management. See Ref. 30, pp. 145–64
22. Dale D. 1994. Insect pests of the rice plant—their biology and ecology. See Ref. 52, pp. 364–85
23. Danish Agric. Advis. Cent. 1995. *Being a Farmer in Denmark. Organization, Advice, and Education. The Danish Model.* Skejby, Denmark. 28 pp.
24. Datta K, Vasquez A, Tu J, Torrizo L, Alam MF, et al. 1998. Constitutive and tissue-specific differential expression of the *cryIA(b)* gene in transgenic rice plants conferring resistance to rice insect pest. *Theor. Appl. Genet.* 97:20–30
25. De Bono E. 1977. *Lateral Thinking*. Hermondsworth, UK: Penguin
26. de Kraker J. 1996. *The potential of natural enemies to suppress rice leaffolder populations.* PhD diss. Wageningen Agric. Univ., The Netherlands. 257 pp.
27. Denning GL, Xuan V-T, eds. 1995. *Vietnam and IRRI: A Partnership in Rice*

Research. Manila: IRRI, Hanoi: Minist. Agric. Food Ind. 353 pp.

28. Denno RF, Perfect TJ, eds. 1994. *Planthoppers: Their Ecology and Management*. New York: Chapman & Hall. 799 pp.
29. Dilts R. 1984. Training: reschooling society? *Prisma* 38:78–90. Jakarta, Indonesia
30. Dowling NG, Greenfield SM, Fischer KS, eds. 1998. *Sustainability of Rice in the Global Food System*. Davis, CA: Pac. Basin Study Cent. Manila: IRRI. 404 pp.
31. Elazegui FA, Savary S, Teng PS. 1993. Management practices of rice pests in Central Luzon, Philippines, in relation to injury levels. *J. Plant Prot. Trop.* 10(2):137–50
32. Escalada MM, Heong KL. 1993. Communication and implementation of change in crop protection. In *Crop Prot. Sustainable Agric., Ciba Found. Symp. 177,* pp. 191–207. Chichester, UK: Wiley
33. Escalada MM, Kenmore PE. 1988. Communicating integrated pest control to rice farmers at the village level. See Ref. 146, pp. 221–28
34. Eveleens KG, Chisholm R, van de Fliert E, Kato M, Pham TN, Schmidt P. 1996. *FAO Intercountry Programme for the Development and Application of Integrated Pest Control in Rice in South and Southeast Asia, Mid Term Review of Phase III.* The Netherlands: Wageningen Agric. Univ.
35. Fabellar LT, Fabellar N, Heong KL. 1994. Simulating rice leaffolder feeding effects on yield using MACROS. *Int. Rice Res. Newsl.* 19:7–8
36. FAO. 1983. *Proc. FAO/NCPC Workshop Role Potential Filipina in Rice Crop Prot., 2–4 Feb., Los Baños, Philipp.* Manila: FAO Intercountry Programme Integrated Pest Control Rice South and Southeast Asia Natl. Crop Prot. Cent. Philipp.
37. Deleted in proof
38. FAO. 1997. *Community-Based IPM Case Studies.* Manila: FAO Intercountry Programme Dev. IPM Rice South and Southeast Asia. 160 pp.
39. FAO. 1998. *Community IPM: Six Cases from Indonesia.* Jakarta: FAO-Techn. Assist., Indonesian Natl. IPM Program. 260 pp. + annexes
40. FAO. 1998. *FAO/Government Cooperative Programme, Project Document, "The Intercountry Programme for the Development and Application of Integrated Pest Control in Rice in South and Southeast Asia, Phase IV," which will also be known as "The FAO Programme for Community IPM in Asia."* UN FAO, Rome
41. Festinger L. 1957. *A Theory of Cognitive Dissonance.* Stanford, CA: Stanford Univ. Press
42. Freire P. 1970. *Pedagogy of the Oppressed.* New York: Herder & Herder
43. Freire P. 1981. *Education for Critical Consciousness.* New York: Continuum
44. Gallagher KD, Kenmore PE, Sogawa K. 1994. Judicial use of insecticides deter planthopper outbreaks and extend the life of resistant varieties in southeast Asian rice. See Ref. 28, pp. 599–614

44a. Global IPM Facility. 1999. *Global IPM Facility First Prog. Rep.,* March, Rome. 39 pp.

45. Goodell GE. 1983. Improving administrators' feedback concerning extension, training, and research relevance at the local level: new approaches and findings from Southeast Asia. *Agric. Admin.* 13:39–55
46. Goodell GE. 1984. Challenges to international pest management research and extension in the Third World: Do we really want IPM to work? *Bull. Entomol. Soc. Am.* 30(3):18–26
47. Goodell GE, Kenmore PE, Litsinger JA, Bandong JP, Dela Cruz CG, Lumaban MD. 1982. Rice insect pest management technology and its transfer to small-scale farmers in the Philippines. In *Report of*

an Exploratory Workshop on the Role of Anthropologists and Other Social Scientists in Interdisciplinary Teams Developing Improved Food Production Technology, pp. 25–42. Los Baños, Philipp.: IRRI

48. Goodell GE, Litsinger JA, Kenmore PE. 1981. Evaluating integrated pest management technology through interdisciplinary research at the farmer level. In *Proc. Conf. Future Trends Integrated Pest Management, 30 May–4 June 1980, Bellagio, Italy,* pp. 72–75. London: Centre Overseas Pest Res. IOBC special issue
49. Gould F. 1998. Sustainability of transgenic insecticidal cultivars: integrating pest genetics and ecology. *Annu. Rev. Entomol.* 43:701–26
50. Hamid AA, Lum KY, Sadi T. 1998. *Integrating Science and People in Rice Pest Management. Proc. Rice IPM Conf., 18–21 Nov. 1996, Kuala Lumpur, Malaysia,* Malays. Agric. Res. Dev. Inst. (MARDI) Minist. Agric., Malaysia
51. Heinrichs EA. 1994. Impact of insecticides on the resistance and resurgence of rice planthoppers. See Ref 28, pp. 571–98
52. Heinrichs EA, ed. 1994. *Biology and Management of Rice Insects.* New Delhi: Wiley East. Ltd. Manila: IRRI. 779 pp.
53. Heinrichs EA. 1994. Host plant resistance. See Ref. 52, p. 517–47
54. Heinrichs EA, Mochida O. 1984. From secondary to major pest status: the case of insecticide-induced rice brown planthopper, *Nilaparvata lugens*, resurgence. *Prot. Ecol.* 7:201–18
55. Heong KL. 1993. Rice leaffolders: Are they serious pests? In *Research on Rice Leaffolder Management in China, Proc. Int. Workshop Economic Threshold Rice Leaffolders China, 4–6 March 1992, Beijing,* pp. 8–11. Hangzhou, PR China: China Natl. Rice Res. Inst.
56. Heong KL. 1996. Pest management in tropical rice ecosystems: new paradigms for research. See Ref. 67, pp. 139–54
57. Heong KL. 1998. *IPM in developing countries: progress and constraints in rice IPM.* Presented at Australas. Appl. Entomol. Res. Conf., 6th, 29 Sept.–2 Oct., Brisbane, Aust.
58. Heong KL, Aquino GB, Barrion AT. 1991. Arthropod community structures of rice ecosystems in the Philippines. *Bull. Entomol. Res.* 81:407–16
59. Heong KL, Aquino GB, Barrion AT. 1992. Population dynamics of plant- and leafhoppers and their natural enemies in rice ecosystems in the Philippines. *Crop Prot.* 11:371–79
60. Heong KL, Escalada MM. 1997. Perception change in rice pest management: a case study of farmers' evaluation of conflict information. *J. Appl. Commun.* 81(2):3–17
61. Heong KL, Escalada MM, Huan NH, Mai V. 1998. Use of communication media in changing rice farmers' pest management in the Mekong Delta, Vietnam. *Crop Prot.* 17(5):413–25
62. Heong KL, Escalada MM, Lazaro AA. 1995. Misuse of pesticides among rice farmers in Leyte, Philippines. See Ref. 120, pp. 97–108
63. Heong KL, Escalada MM, Mai V. 1994. An analysis of insecticide use in rice: case studies in the Philippines and Vietnam. *Int. J. Pest Manage.* 40:173-78
64. Heong KL, Nguyen TTC, Nguyen B, Fujisaka S, Bottrell DG. 1995. Reducing early-season insecticide applications through farmers' experiments in Vietnam. See Ref. 27, pp. 217–22

64a. Heong KL, Schoenly KG. 1998. Impact of insecticides on herbivore-natural enemy communities in tropical rice ecosystems. In *Ecotoxicol.: Pestic. Benef. Org.*, ed PT Haskell, P McEwen, 41:381–403. Dordrecht, Netherlands: Kluwer

65. Herdt RW. 1996. Establishing the Rockefeller Foundation's priorities for rice bio-

technology research in 1995 and beyond. See Ref. 86, pp. 17–29

66. Ho NK. 1996. Introducing integrated weed management in Malaysia. In *Herbicides in Asian Rice: Transitions in Weed Management*, ed. R Naylor, pp. 167–82. Stanford, CA: Stanford Univ. Los Baños, Philipp.: IRRI
67. Hokyo N, Norton G, eds. 1996. *Proc. Int. Workshop Pest Manage. Strateg. Asian Monsoon Agroecosyst., 15–18 Nov. 1995, Kumamoto, Jpn.* Kumamoto: Natl. Agric. Exp. Stn. Minist. Agric., For. Fish. 320 pp.
68. Hu R, Cheng J, Dong S, Sun Y. 1997. The role of women in rice pest management in Zhejiang, China. In *Pest Management Practices of Rice Farmers in Asia*, ed. KL Heong, MM Escalada, pp. 63–73. Manila: IRRI. 245 pp.
69. Indonesian Natl. IPM Program. 1994. *IPM Consolidation: Developing a Farmer-Led IPM Program.* Jakarta: FAO-IPM Secr. 66 pp.
70. Indonesian Natl. IPM Program. 1994. *Building Community-Based IPM Programs.* Jakarta: FAO-IPM Secr. 65 pp.
71. Indonesian Natl. IPM Program. 1997. IPM by Farmers. The Indonesian IPM Program. Jakarta: IPM Secr. 22 pp.
72. Inthavong S, Inthavong K, Sengsaulivong V, Schiller JM, Rapusas HR, et al. 1998. Arthropod biodiversity in Lao irrigated rice ecosystem. See Ref. 50, pp. 69–85
73. IRRI. 1995. *Rice IPM in Review. Diagnostic Workshops Summary Rep.*, IRRI, Manila, Philipp.
74. IRRI. 1996. *Bt Rice: Research and Policy Issues. IRRI Inf. Ser. No. 5*, July, Manila, Philipp.
75. IRRI. 1997. *Rice Almanac.* Los Baños, Philipp. Cali, Colombia: Int. Cent. Trop. Agric. Bouaké, Ivory Coast: West Afr. Rice Dev. Assoc. 181 pp. 2nd ed.
76. IRRI. 1998. *Filipino Women in Rice Farming Systems.* Los Baños, Philipp.: IRRI, Univ. Philipp. Philipp. Inst. Dev. Stud. 408 pp.
77. Islam Z, Karim ANMR. 1997. Whiteheads associated with stem borer infestation in modern rice varieties: an attempt to resolve the dilemma of yield losses. *Crop Prot.* 16(4):303–11
78. Ives AR, Settle WH. 1997. Metapopulation dynamics and pest control in agricultural systems. *Am. Nat.* 149(2):220–46
79. Kahneman D, Tverskey A. 1973. On the psychology of prediction. *Psychol. Rev.* 80:237–51
80. Kamal NQ, Odud A, Begum A. 1990. The spider fauna in and around the Bangladesh Rice Research Institute farm and their role as predators of rice insect pests. *Philipp. Entomol.* 8(2):771–77
81. Kenmore PE. 1987. Crop loss assessment in a practical integrated pest control program for tropical Asian rice. In *Crop Loss Assessment and Pest Management,* ed. PS Teng, pp. 225–41. St. Paul, MN: Am. Phytopathol. Soc.
82. Kenmore PE. 1996. Integrated pest management in rice. In *Biotechnology and Integrated Pest Management*, ed. GJ Persley, 4:76–97. Wallingford, UK: CAB Int.
83. Kenmore PE, Cariño FO, Perez CA, Dyck VA, Gutierrez, AP. 1984. Population regulation of the rice brown planthopper (*Nilaparvata lugens* Stål) within rice fields in the Philippines. *J. Plant Prot. Trop.* 1(1):19–37
84. Kenmore PE, Litsinger JA, Bandong JP, Santiago AC, Salac MM. 1987. Philippine rice farmers and insecticides: thirty years of growing dependency and new options for change. In *Management of Pests and Pesticides: Farmers' Perceptions and Practices*, ed. J Tait, B Napompeth, pp. 98–108. Boulder, CO: Westview
85. Khush GS. 1990. Rice breeding—accomplishments and challenges. *Plant Breed. Abstr.* 60(5):461–69

86. Khush GS, ed. 1996. *Rice Genetics III. Proc. Int. Rice Genet. Symp., 3rd, 16–20 Oct. 1995, Manila.* Los Baños, Philipp: IRRI. 1011 pp.
87. Kingsley MA, Siwi SS. 1997. *Gender Study II: Gender Analysis, Agricultural Analysis, and Strengthening Women's Leadership Development in the Natl. IPM Program.* Jakarta: Indonesian Natl. IPM Program, Minist. Agric. FAO
88. Kiritani K. 1979. Pest management in rice. *Annu. Rev. Entomol.* 24:279–312
89. Kishi M, Hirschhorn N, Djajadisastra M, Satterlee LN, Strowman S, Dilts R. 1995. Relationship of pesticide spraying to signs and symptoms in Indonesian farmers. *Scand. J. Work Environ. Health* 21:124–33
90. Kogan M. 1998. Integrated pest management: historical perspectives and contemporary developments. *Annu. Rev. Entomol.* 43:243–70
91. Kuhn TS. 1970. *The Structure of Scientific Revolutions.* Chicago, IL: Univ. Chicago Press
92. Lee SI, Lee S-H, Koo JC, Chun HJ, Lim CO, et al. 1999. Soybean Kunitz trypsin inhibitor (SKTI) confers resistance to the brown planthopper (*Nilaparvata lugens* Stål) in transgenic rice. *Mol. Breed.* 5(1):1–9
93. Litsinger JA, Bandong JP, Dela Cruz CG. 1984. Verifying and extending integrated pest control technology to small-scale farmers. In *Proc. Int. Workshop Integrated Pest Control Grain Legumes, 3–9 April 1983, Goiania, Goias, Brazil,* ed. PC Matteson, pp. 326–54. Brasilia: EMBRAPA
94. Loc NT, Nghiep HV, Luc NH, Nhan NT, Luat NV, et al. 1998. Effect of crop residue burning on rice predators: a case study in Vietnam. See Ref. 50, pp. 54–55
95. Loevinsohn ME. 1987. Insecticide use and increased mortality in rural Central Luzon, Philippines. *Lancet* 13:1359–62
96. Loevinsohn ME, Litsinger JA, Heinrichs EA. 1988. Rice insect pests and agricultural change. In *The Entomology of Indigenous and Naturalized Systems in Agriculture,* ed. MK Harris, CE Rogers, 8:161–82. Boulder, CO: Westview. 238 pp.
97. Loevinsohn M, Meijerink G. 1998. *Enhancing capacity to manage resources: assessing the Farmer Field School approach.* Presented at Meet. IPM Network for the Caribbean, 2nd, 4–6 Feb., Kingston, Jamaica
98. Luo S. 1987. Studies on the compensation of rice to the larval damage caused by the Asian rice borer (*Chilo suppressalis* (WK)). *Sci. Agric. Sin.* 20(2):67–72
99. Mangan J, Mangan MS. 1998. A comparison of two IPM training strategies in China: the importance of concepts of the rice ecosystem for sustainable insect pest management. *Agric. Hum. Values* 15: 209–21
100. Marciano VP, Mandac AML, Flynn JC. 1981. *Insect management practices of rice farmers in Laguna, Philippines. IRRI Agric. Econ. Dep. Pap. No. 81–03.* IRRI, Los Baños, Philipp.
101. Matteson PC, Gallagher KD, Kenmore PE. 1994. Extension of integrated pest management for planthoppers in Asian irrigated rice: empowering the user. See Ref. 28, pp. 656–85
102. Meenakanit L, Escalada MM, Heong KL. 1996. The changing role of women in rice pest management in central Thailand. See Ref. 67, pp. 201–12
103. Mew TW, Wang FM, Wu JT, Lin KR, Khush GS. 1988. Disease and insect resistance in hybrid rice. In *Hybrid Rice,* pp. 189–200. Manila: IRRI
104. Miyashita T. 1985. Estimation of the injury level in the rice leaffolder, *Cnaphalocrocis medinalis* Guenée (Lepidoptera: Pyralidae): 1. Relations between yield loss and injury of rice leaves at heading or in the grain filling period. *Jpn. J. Appl. Entomol. Zool.* 29:73–76
105. Norton GA. 1982. A decision analysis

approach to integrated pest control. *Crop Prot.* 1:147–64

106. Norton GA, Mumford JD. 1982. Information gaps in pest management. *Proc. Int. Conf. Plant Prot. Tropics,* pp. 589–97. Kuala Lumpur: Malays. Plant Prot. Soc.
107. Norton GA, Mumford JD, eds. 1993. *Decision Tools for Pest Management.* Oxford, UK: CAB Int.
108. Oerke EC, Dehne HW, Schonbeck F, Weber A, eds. 1994. *Crop Production and Crop Protection: Estimated Losses in Major Food and Cash Crops.* Amsterdam: Elsevier. 808 pp.
109. Oka IN. 1988. Role of cultural techniques in rice IPM systems. See Ref. 146, pp. 83–93
110. Okuma C, Kamal NQ, Hirashima Y, Alam MZ, Ogata K. 1993. *Illustrated Monograph of the Rice Field Spiders of Bangladesh.* IPSA-JICA Proj. Publ. No. 1. Salna, Gazipur. Bangladesh: Inst. Postgraduate Stud. Agric. 93 pp.
111. Ooi PAC. 1995. Village level farmer education to sustain classical biological control: a case study. *Proc. Workshop Biol. Control, 11–15 Sept., Kuala Lumpur, Malaysia,* ed. P Ferrar et al, pp. 185–92. Kuala Lumpur, Malays.: Malays. Agric. Res. Dev. Inst. (MARDI), Aust. Cent. Int. Agric. Res. (ACIAR)
112. Ooi PAC. 1986. Insecticides disrupt natural control of *Nilaparvata lugens* in Sekinchan, Malaysia. In *Biological Control in the Tropics. Proc. Reg. Symp. Biol. Control, 1st, 1985, Serdang, Malaysia,* ed. MY Hussein, AG Ibrahim, pp. 109–20. Serdang: Univ. Pertanian Malaysia
113. Ooi PAC. 1998. *Beyond the Farmer Field School: IPM and Empowerment in Indonesia. Gatekeeper Series No. 78.* London: Int. Inst. Environ. Dev.
114. Ooi PAC, Shepard BM. 1994. Predators and parasitoids of rice insect pests. See Ref. 52, pp. 585–612

114a. Ooi PAC, Waage JK. 1994. Biological control in rice: applications and research needs. In *Rice Pest Science and Management. Select. Pap. Int. Rice Res. Conf.,* ed. PS Teng, KL Heong, K Moody, pp. 209–16. Manila: IRRI. 289 pp.

115. Palis FG. 1998. Changing farmers' perceptions and practices: the case of insect pest control in Central Luzon, Philippines. *Crop Prot.* 17(7):599–607
116. Pfuhl EH. 1988. Radio-based communication campaigns: a strategy for training farmers in IPM in the Philippines. See Ref. 146, pp. 251–55
117. Pimbert MP. 1991. *Designing integrated pest management for sustainable and productive futures. Gatekeeper Ser. No. 29.* London: Int. Inst. Environ. Dev.
118. Pincus J. 1991. *Impact Study of Farmer Field Schools.* Indones. Natl. IPM Program, Jakarta
119. Pincus J. 1994. Discussion points on pesticide policy and rice production in Indonesia: combining reform from above and action from below. *Proc. Göttingen Workshop Pestic. Policies, 28 Feb.–4 March, Göttingen, Ger.,* ed. S Agne, G Fleischer, H Waibel, pp. 61–73. Göttingen: Inst. Agrarökon. Univ. Göttingen
120. Pingali PL, Roger PA, eds. 1995. *Impact of Pesticides on Farmer Health and the Rice Environment.* Norwell, MA: Kluwer. 664 pp.
121. Ramirez OA, Mumford JD. 1995. The role of public policy in implementing IPM. *Crop Prot.* 14(7):565–72
122. Rapusas HR, Bottrell DG, Coll M. 1996. Intraspecific variation in chemical attraction of rice to insect predators. *Biol. Control* 6:394–400
123. Rapusas HR, Heong KL. 1995. *Reducing early season insecticide use for leaf-folder control in rice: impact, economics, and risks. Rice IPM Network Workshop Rep.,* IRRI, Manila
124. Repetto R. 1985. *Paying the price: pesticide subsidies in developing countries. Res. Rep. 2.* Washington, DC: World Resourc. Inst. 33 pp.
125. Roger PA, Heong KL, Teng PS. 1991.

Biodiversity and sustainability of wetland rice production: role and potential of microorganisms and invertebrates. In *The Biodiversity of Microorganisms and Invertebrates: Its Role in Sustainable Agriculture*, ed. DL Hawksworth, 10:117–36. Wallingford, UK: CAB Int.

126. Rola A, Pingali P. 1993. *Pesticides, Rice Productivity, and Farmers' Health—An Economic Assessment.* Manila: IRRI Washington, DC: World Resourc. Inst. 100 pp.
127. Röling N, van de Fliert E. 1991. *Beyond the Green Revolution: Capturing Efficiencies from Integrated Agriculture. Indonesia's IPM Programme Provides a Generic Case.* Informal Rep. Natl. Programme Dev. Train. IPM Rice-based Crop. Syst., Jakarta, Indonesia
128. Röling N, van de Fliert E. 1994. Transforming extension for sustainable agriculture: the case of integrated pest management in rice in Indonesia. *Agric. Hum. Values* 11(2–3):96–108
129. Rubia EG, Heong KL, Zalucki M, Gonzales B, Norton GA. 1996. Mechanisms of compensation of rice plants to yellow stem borer *Scirpophaga incertulas* (Walker) injury. *Crop Prot.* 15:335–40
130. Rubia EG, Shepard BM, Yambao EB, Ingram KT, Arida GS, Penning de Vries F. 1989. Stem borer damage and grain yield of flooded rice. *J. Plant Prot. Trop.* 6(3):205–11
131. Savary S, Elazegui FAD, Pinnschmidt HO, Castilla NP, Teng PS. 1997. *A New Approach to Quantify Crop Losses Due to Rice Pests in Varying Production Situations, IRRI Discuss. Pap. Ser. No. 20,* IRRI, Manila, Philipp.
132. Sawada H, Gaib Subroto SW, Mustaghfirin, Wijaya ES. 1991. *Immigration, population development and outbreaks of brown planthopper, Nilaparvata lugens (Stål), under different rice cultivation patterns in Central Java, Indonesia. Tech. Bull. No. 130*, Asian Pac. Food Fertil. Technol. Cent., Taipei, Republic China, pp. 8–18
133. Schmidt P, Stiefel J, Hürlimann M. 1997. *Extension of Complex Issues: Success Factors in Integrated Pest Management.* St. Gallen, Switz.: Swiss Cent. Dev. Coop. Technol. Manage. 101 pp.
134. Schoenly KG, Cohen JE, Heong KL, Arida GS, Barrion AT, Litsinger JA. 1996. Quantifying the impact of insecticides on food web structure of rice-arthropod populations in a Philippine farmer's irrigated field: a case study. In *Food Webs: Integration of Patterns and Dynamics*, ed. GA Polis, K Wisemiller, pp. 343–51. London: Chapman & Hall
135. Schoenly KG, Cohen JE, Heong KL, Litsinger JA, Aquino GB. 1996. Food web dynamics of irrigated rice fields at five elevations in Luzon, Philippines. *Bull. Entomol. Res.* 86:451–66
136. Schoenly KG, Justo HDJ, Barrion AT, Harris MK, Bottrell DG. 1998. Analysis of invertebrate biodiversity in a Philippine farmer's irrigated rice field. *Environ. Entomol.* 27(5):1125–36
137. Schoenly K, Mew TW, Reichardt W. 1998. Biological diversity of rice landscapes. See Ref. 30, pp. 285–99
138. Scoones I, ed. 1994. *Beyond Farmer First: Rural People's Knowledge, Agricultural Research, and Extension Practice.* London: IT. 288 pp.
139. Settle WH, Ariawan H, Astuti ET, Cahyana W, Hakim AL, et al. 1996. Managing tropical rice pests through conservation of generalist natural enemies and alternative prey. *Ecology* 77(7):1975–88
140. Shepard BM, Barrion AT, Litsinger JA. 1987. *Friends of the Rice Farmer: Helpful Insects, Spiders, and Pathogens.* Los Baños, Philipp.: IRRI. 136 pp.
141. Shepard BM, Barrion AT, Litsinger JA. 1995. *Rice-Feeding Insects of Tropical Asia.* Manila: IRRI. 228 pp.
142. Shepard BM, Justo HD, Rubia EG, Estano DB. 1990. Response of the rice

plant to damage by the rice whorl maggot, *Hydriella philippina* Ferino (Diptera: Ephydridae). *J. Plant Prot. Trop.* 7:173–77

143. Sinclair TR, Sheehy JE. 1999. Erect leaves and photosynthesis in rice. *Science* 283:L1456–57
144. Smith J, Litsinger JA, Bandong JP, Lumaban MD, Dela Cruz CG. 1989. Economic thresholds for insecticide application to rice: profitability and risk analysis to Filipino farmers. *J. Plant Prot. Trop.* 6:19–24
145. Swanson BE, ed. 1990. *Rep. Global Consult. Agric. Ext., 4–8 Dec. 1989, Rome, Italy.* Rome: FAO
146. Teng PS, Heong KL, eds. 1988. *Pesticide Management and Integrated Pest Management in Southeast Asia.* College Park, MD: Consort. Int. Crop Prot. 473 pp.
147. Toenniessen GH. 1991. Potentially useful genes for rice genetic engineering. In *Rice Biotechnology*, ed. GS Khush, GH Toenniessen, 11:253–80. Wallingford, UK: CAB Int. 320 pp.
148. Truong TNC, Nguyen TK, Bui TTT, Paris TR. 1995. Gender roles in rice farming systems in the Mekong River Delta: an exploratory study. See Ref. 27, pp. 291–99
149. UNDP. 1991. *Programme Advisory Note. Agricultural Extension.* New York: UN Dev. Programme Tech. Advis. Div., Bur. Programme Policy Eval.
150. Useem M, Setti L, Pincus J. 1992. The science of Javanese management: organizational alignment in an Indonesian development programme. *Public Admin. Dev.* 12:447–71
151. van de Fliert E. 1993. *Integrated pest management: Farmer Field Schools generate sustainable practices. A case study in Central Java evaluating IPM training.* PhD diss., WAU Pap. 93–3, Agric. Univ., Wageningen, Netherlands
152. van de Fliert E, Wiyanto. 1996. A road to sustainability. *ILEIA Newsl.* 12(2):6–8
153. van Vreden G, Ahmadzabidi AL. 1986. *Pests of Rice and Their Natural Enemies in Peninsular Malaysia.* Wageningen, The Netherlands: Pudoc. 230 pp.
154. Virmani SS. 1996. Hybrid rice. *Adv. Agron.* 57:377–462
155. Vromant N, Rothuis AJ, Cuc NTT, Ollevier F. 1998. The effect of fish on the abundance of the rice caseworm *Nymphula depunctalis* (Guenée) (Lepidoptera: Pyralidae) in direct seeded, concurrent rice-fish fields. *Biocontrol. Sci. Tech.* 8:539–46
156. Wada T, Salleh NM. 1992. Population growth pattern of the rice planthoppers, *Nilaparvata lugens* and *Sogatella furcifera*, in the Muda area, West Malaysia. *Jpn. Agric. Res. Q.* 26:105–14
157. Waibel H. 1986. *The Economics of Integrated Pest Control in Irrigated Rice.* Crop Prot. Monogr. Berlin: Springer-Verlag. 196 pp.
158. Way MJ, Heong KL. 1994. The role of biodiversity in the dynamics and management of insect pests of tropical irrigated rice—a review. *Bull. Ent. Res.* 84:567–87
159. Way MJ, Islam Z, Heong KL, Joshi RC. 1998. Ants in tropical irrigated rice: distribution and abundance, especially of *Solenopsis geminata* (Hymenoptera: Formicidae). *Bull. Entomol. Res.* 88:467–76
160. Whitten MJ, Brownhall LR, Eveleens KG, Heneveld W, Khan MAR, et al. 1990. *Mid-term Review of FAO Intercountry Programme for the Dev. Appl. Integr. Pest Control Rice South and Southeast Asia, Phase II, Mission Rep., Nov.,* Jakarta, Indonesia
161. Woodburn AT. 1990. The current rice agrochemicals market. In *Proc. Conf. Pest Manage. Rice*, ed. BT Grayson, MB Green, pp. 15–30. London: Elsevier Appl. Sci.
162. Wu J, Hu GTJ, Shu ZYJ, Wan ZRZ. 1994. Studies on the regulation effect of neutral insect on the community food web in paddy field. *Acta Ecol. Sin.* 14(4):381–85

163. Yu X, Heong KL, Hu C. 1998. Effects of various non-rice hosts on the growth, reproduction and predation of mirid bug, *Cyrtorhinus lividipennis* Reuter. See Ref. 50, pp. 56–63

164. Yu X, Heong KL, Hu C, Barrion AT. 1996. Role of non-rice habitats for conserving egg parasitoids of rice planthoppers and leafhoppers. See Ref. 67, pp. 63–77

Annu. Rev. Entomol. 2000. 45:575–604

POLYENE HYDROCARBONS AND EPOXIDES: A Second Major Class of Lepidopteran Sex Attractant Pheromones

Jocelyn G. Millar
Department of Entomology, University of California, Riverside, California 92521; e-mail: email:jocelyn.millar@ucr.edu

Key Words Geometridae, Noctuidae, pheromone synthesis, enantiomeric synergist, enantiomeric antagonist

■ **Abstract** Polyene hydrocarbons and epoxides are used as pheromone components and sex attractants by four macrolepidopteran families: the Geometridae, Noctuidae, Arctiidae, and Lymantriidae. They constitute a second major class of lepidopteran pheromones, different from the C_{10}-C_{18} acetates, alcohols, and aldehydes commonly found in other species. They are biosynthesized from diet-derived linoleic or linolenic acids and are characterized by C_{17}-C_{23} straight chains, 1-3 *cis* double bonds separated by methylene groups, and 0, 1, or 2 epoxide functions. Pheromone blends are created from components with different chain lengths, numbers of double bonds, and functional groups, or from mixtures of epoxide regioisomers or enantiomers, with several examples of synergism between enantiomers. Behavioral antagonists also limit interspecific attraction, with numerous examples of antagonism by enantiomers. This review summarizes the taxonomic distribution, mechanisms used to generate unique pheromone blends, and the identification, synthesis, and biosynthesis of these compounds.

PERSPECTIVES AND OVERVIEW

Lepidopteran sex attractant pheromones are probably the most well-studied class of intraspecific semiochemicals, with sex attractant or pheromone blends identified for many hundred species. Studies of these semiochemicals have been biased (understandably) toward species of economic importance, primarily in a small subset of lepidopteran families, including the Tortricidae, Pyralidae, Gelechiidae, Sessiidae, and some noctuid subfamilies. The pheromone chemistry of these species is dominated by straight-chain 10–18 carbon acetates, aldehydes, and alcohols with 0–3 double bonds, and with the double bonds frequently being conjugated for those compounds with multiple double bonds (4). These compounds are derived de novo from saturated 16- or 18-carbon carboxylic acid intermediates, which, by chain shortening, desaturation, reduction, oxidation, and acetylation steps, are converted to the many known pheromone compounds of this general class (35).

However, in the late 1970s the first examples of a new class of lepidopteran pheromones were identified, consisting of unsaturated hydrocarbons with 1–3 all-*cis* double bonds separated by single methylene groups (a structural motif typical of polyunsaturated fatty acids), and the corresponding monoepoxides. Since then, many pheromone components based on this general structural motif have been elucidated, with chain lengths from 17–23 carbons, 1–4 double bonds, and 0, 1, or 2 epoxides (4). Unlike the acetate/alcohol/aldehyde class of pheromones, these compounds are derived from linoleic (6Z,9Z-hydrocarbons and related compounds) and linolenic acids (3Z,6Z,9Z-hydrocarbons and analogs), respectively (65). Because these two fatty acids are diet derived rather than synthesized de novo (67), the resulting pheromone components are the products of a different biosynthetic pathway than are the acetates/alcohols/aldehydes. Also unlike the latter compounds, the epoxide components have two enantiomeric forms, providing a further dimension that can be exploited to create unique pheromone signals. Indeed, examples include species that produce and respond to a single enantiomer (with the other enantiomer being inactive or antagonistic), and even a few species that require nonracemic blends of both enantiomers.

This chapter summarizes current knowledge of the occurrence, utilization, and chemistry of this class of lepidopteran pheromones. The first section focuses on the types of compounds that have been identified, and the families and subfamilies in which they have been found. The second section describes the various methods used to produce unique, species-specific pheromone signals. The final sections describe practical aspects of the identification and synthesis of these pheromone components, and their biosynthesis.

This chapter drew heavily on the Internet-based "Pherolist" (http://www.nysaes.cornell.edu/pheronet) of all known lepidopteran pheromones, which is compiled and maintained by Heinrich Arn and coworkers (4). In the interest of brevity, I will not list all citations for a particular pheromone structure but will use selected examples. For a comprehensive list of lepidopteran pheromones and their occurrence, the reader is referred to The Pherolist.

PHEROMONE STRUCTURES AND NOMENCLATURE

The structures of known pheromones of the polyene and epoxide type are shown in Figure 1. Because of the length and complexity of some of the formal Chemical Abstracts names of the pheromone structures, abbreviations will be used for simplicity and clarity. A few formal names are given here as an aid to locating these types of compounds in Chemical Abstracts or other database searches, and the abbreviations are shown with the structures in Figure 1.

The IUPAC names of the polyene hydrocarbon pheromones are straightforward, simply listing the position and conformation of each of the double bonds

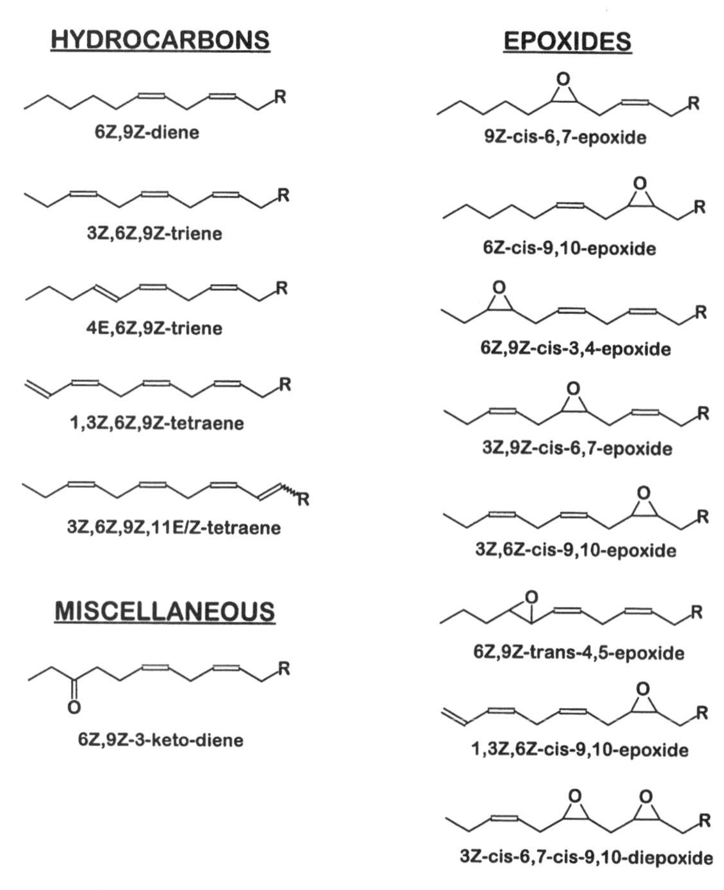

Figure 1 Structures and short-hand names for polyene and epoxide pheromone components.

[e.g. (*Z*,*Z*,*Z*)-3,6,9-heptadecatriene]. Names such as these will be abbreviated as 3*Z*,6*Z*,9*Z*-17:H. The names of the epoxides are more complicated; for example, the formal name [2*R*-[2α(2*Z*,5*Z*),3α]]-2-(2,5-octadienyl)-3-undecyloxirane, while unambiguous, is cumbersome. The name is keyed on the oxirane (epoxide) as the root structure, with the alkenyl and alkyl chains attached to C_2 and C_3 of the epoxide ring being secondary, and the descriptors providing positions and stereochemistries of functional groups. The simpler "common" name used by workers in the field for this compound is 3*Z*,6*Z*-9*R*,10*S*-epoxy-heneicosadiene, abbreviated 3*Z*,6*Z*-9*R*,10*S*-epoxy-21:H, which provides all the relevant information in an immediately understandable format. Racemic epoxides, or epoxides for which the enantiomer is unknown, are referred to as 3*Z*,6*Z*-*cis*-9,10-epoxy-21:H.

The situation with the formal names is further complicated by the fact that the precedence of the alkyl and alkenyl chains attached to the oxirane is determined alphabetically. Thus, the next highest homolog, [2*R*-[2α(2*Z*,5*Z*),3α]]-2-dodecyl-3-(2,5-octadienyl)-oxirane actually corresponds to 3*Z*,6*Z*-9*S*,10*R*-epoxy-22:H because the dodecyl group takes precedence over the octadienyl group.

For those unfamiliar with chemistry nomenclature, the term "stereoisomer" refers to one of several possible compounds in which the same atoms are linked in the same sequence, but which have different configurations in space. For our purposes, this will include *E* and *Z* geometric isomers, and each of the two mirror-image forms (enantiomers) of the chiral epoxides. "Regioisomers" are compounds that have the same molecular formulae and functional groups, but with the functional groups in different places. Thus, 3*Z*,6*Z*-9*R*,10*S*-epoxy-21:H and 3*Z*,9*Z*-6*R*,7*S*-epoxy-21:H are regioisomers.

TAXONOMIC DISTRIBUTION OF POLYENE AND EPOXIDE PHEROMONES

The occurrence of polyene and epoxide pheromone components throughout the Lepidoptera is summarized in Table 1. These pheromone components appear to be restricted to the macrolepidoptera, and in particular, to three families in the Noctuoidea (Noctuidae, Arctiidae, and Lymantridae), and to the family Geometridae (superfamily Geometroidea). However, the pheromone components used by individual species vary between and within subfamilies, tribes, and even genera, and chemotaxonomic trends remain confused for some groups, particularly in the Lymantridae and Arctiidae. Several factors have contributed to difficulties in developing chemotaxonomic relationships. First, lepidopteran classification in these superfamilies is in flux, with no clear agreement as to the placement of and relationships between the higher taxonomic levels (52, 66). Second, subfamilies with few species are underrepresented or not represented at all, so generalizations must be based on subsets of each taxon. Third, the term "pheromone" has been applied rather loosely in some cases, with many citations reporting the identification of pheromone components from female extracts in the absence of any data demonstrating a biological role for the identified compounds. This problem is particularly egregious in the Arctiidae; for the "pheromones" or "pheromone components" reported from approximately 30 species, there is no documented evidence of behavioral roles for the compounds identified from the majority of species. Fourth, many pheromones have been identified at least in part by screening trials, which are inherently biased by the limited sets of compounds screened. Consequently, the specimens obtained are not representative of all of the pheromone types within a taxon. Furthermore, in some cases it has proven difficult to obtain female moths to confirm that compounds that attracted males are indeed pheromone components. Because male moths can be attracted to homologs or mimics of their pheromone components (44, 51), the assumption that sex attractants are indeed pheromone components is risky. Incidental catches of a few

TABLE 1 Overview of identified pheromone components, sex attractants, and compounds isolated from female pheromone glands

Family and Subfamily	Genera[a]	Species[b]	Number of Species with Indicated Carbon Chain Length						
			C_{17}	C_{18}	C_{19}	C_{20}	C_{21}	C_{22}	$C_{23 \text{ or } 25}$
Geometridae									
Oenochrominae	2/2	6/6			4		3		
Geometrinae	2/3	3/4				3			
Laurentiinae	14/15	23/25		1	6	6	13	1	
Ennominae	40/42	60/65	17	5	37	4	4	1	
Noctuidae									
Herminiinae	7/7	8/8			2	3	4	1	1
Rivulinae	5/5	8/8		2	3	1	1		1
Hypeninae	3/4	3/4				2	1		
Catocalinae	8/8	14/14			1 + 1[c]	6 + 1[c]	13+2[d]	1 + 1[c]	
Arctiidae									
Amatidae	2/2	2/2			1		1		
Ctenuchinae	3/3	3/3				1	3		
Arctiinae	12/14	16/25		4[e]	1	5	16	1	4,1[f]
Lymantriidae	3/6	4/21		1	1	1	2		1[g]

[a]Number of genera attracted to or producing polyene and epoxide pheromones versus the number of genera for which sex attractants or pheromones are known or suspected.
[b]Number of species attracted to or producing polyene and epoxide pheromones versus the number of species for which sex attractants or pheromones are known or suspected.
[c]Compounds also produced by males of one species.
[d]Compounds also produced by males of two species.
[e]Eighteen carbon compounds are all 9Z,12Z,15Z-18:Ald or 9Z,12Z-18:Ald, rather than polyene hydrocarbons or epoxides.
[f]Four species produce C_{23}compounds, 1 species produces C_{25}compounds.
[g]The pheromone of *Euproctis chrysorrhoea* contains 7Z,13Z,16Z,19Z-docosatetraenyl isobutyrate; the 13Z,16Z,19Z-triene portion of this structure fits the linolenic acid structural motif.

specimens of nontarget species in screening or other trials must also be treated with caution, because those trap catches may be just incidental, and a result of factors unrelated to sex attraction (e.g. convenient resting place).

Finally, it has become obvious that the chirality of the epoxide components can be crucially important, with one enantiomer being highly attractive, whereas the other may be strongly antagonistic. Alternatively, in several cases, moths were attracted only to a nonracemic mixture of the two enantiomers, and each enantiomer individually was inactive (see below). However, because of the difficulties in synthesizing enantiomerically pure epoxides, many investigators have used only racemic or enantiomerically enriched compounds, the minor component of which may have inhibited attraction. It is anticipated that future studies with enantiomerically pure materials will increase greatly the number of species attracted.

Geometridae

Having more than 20,000 described species, the Geometridae is one of the largest lepidopteran families. Six subfamily groupings have been suggested (66), although none are monophyletic, and one (Archiearinae) is small (fewer than 20 species) with no pheromones yet described. Polyenes and epoxides comprise all reported pheromone and sex attractant components for most species in four of the remaining six subfamilies (Ennominae, Oenochrominae, Geometrinae, and Laurentiinae). There are exceptions: methyl-branched hydrocarbons are proven pheromone components for four *Lambdina* spp. in the Ennominae (4), and two *Hydriomena* spp. in the Laurentiinae (*H. furcata* Thunberg and *H. ruberata* Freyer) were attracted with *E*9-12:Ac and *Z*11-16:Ac, respectively (4). The serendipitous attraction of male *Perizoma grandis* Hulst (subfamily Ennominae) to *E*11-14:Ald comprises another possible exception to the general trend (16). In contrast, all known pheromones and sex attractants in the sixth subfamily, the Sterrhinae, are of the acetate/alcohol/aldehyde type. Thus, there appears to be a distinct chemotaxonomic division between this and the other subfamilies.

With pheromone or sex attractant blends known from about 100 geometrid species, primarily in the families Ennominae and Laurentiinae, trends are starting to appear (Table 1). C_{17} polyenes and epoxides are restricted to the Ennominae so far, and all reported examples are for 3Z,6Z,9Z-17:H or the corresponding *cis*-3,4- or 6,7-epoxides. However, the absence of examples of *cis*-9,10-epoxides of this chain length may simply reflect the fact that they have not been tested extensively in field screening. Within the Ennominae overall, compounds with odd numbers of carbons are favored (Table 1), with C_{19} compounds (dienes, trienes, and the corresponding monoepoxides) being most prevalent, followed by C_{17} compounds, for which only trienes and the analogous monoepoxydienes have been reported. $C_{20\text{-}21}$ compounds are less common, and a single C_{22} compound (6Z,9Z-22:H) of unknown function has been reported from *Theria rupicapraria* Denn. and Schiff. extracts (68).

Pheromone and sex attractant components for Laurentiine spp. tend to be of higher molecular weight, with the largest subset of species using C_{21} compounds, followed by C_{19} and C_{20} components. The majority of compounds reported to date are polyenes, with lesser numbers of epoxides. A single species (*Epirrhoe sperryi* Herbulot) produces a C_{18} compound, but it is not part of the pheromone (82, 48). Similarly, a single species (*Mesoleuca ruficillata* Hbn.) has been attracted to baits containing 3Z,6Z,9Z-22:H (47), but it is not known whether this compound is produced by females.

Within the subfamily Oenochrominae, four of the six reported species produce or are attracted to C_{19} compounds, and three to C_{21} compounds. Two *Alsophila* species, *A. pometaria* (Harris) and *A. aescularia* Den. and Schiff., produce the only known examples of 3Z,6Z,9Z,11Z/E-tetraenes (80, 69). Three examples of sex attractants have been reported from the Geometrinae, with three Asian species attracted to 3Z,6Z,9Z-20:H (2). However, no pheromones have been isolated from any Geometrine species to date.

Noctuidae

Taxonomically, the concept of the Noctuidae as a family is uncertain, and subfamily classifications remain to be resolved (66). From the approximately 20 subfamilies proposed, pheromones or sex attractants have been identified from species from 15, but polyene or epoxide pheromones have been found only in the cluster of subfamilies comprising the Herminiae, Hypeninae, Rivulinae, and Catocalinae, and almost all reported pheromone components within these four subfamilies are of this type. Two notable exceptions are the catocaline species *Alabama argillacea* Hb. and *Anomis texana* F.: The first uses (9*S*)-9-methyl-19:H as a pheromone, and the second uses a homolog, (7*S*)-7-methyl-17:H (26). However, female *A. argillacea* produce 6Z,9Z-21:H and 6Z,9Z-23:H, and female *A. texana* produce 6Z,9Z-23:H. These dienes elicited GC-EAD responses from male antennae of each species, but they do not appear to have any role in attraction of conspecifics. Although their purpose remains obscure, their presence in females is nevertheless intriguing.

The eight hermiine species primarily produce or are attracted to C_{19}-C_{21} monoene or diene *cis*-6,7-epoxides. Only one of the eight species [*Tetanolita mynesalis* (Walker)] (28) uses the corresponding diene or triene hydrocarbon as part of the pheromone blend. There is a single report of attraction to C_{22} and C_{23} compounds (1). For the rivuline species, all eight reported examples are for sex attractants only, and no obvious trend as to chain length or functional group has emerged. Similarly, the three examples of sex attractants for hypenine species (3Z,6Z-*cis*-9,10-epoxy-20- or 21:H) were all from screening trials, with no pheromone components actually identified.

Within the Catocalinae, all reported sex attractant or pheromone components have 20- or 21-carbon chains. 3Z,6Z,9Z-19:H was found in females of *Caenurgina distincta* (Cramer), but it was not part of the attractant blend (79). In all cases reported, the epoxides have been in the 9,10-position. Interestingly, males of two species produce compounds similar to the female-produced pheromone; the female *Anticarsia gemmetalis* Hbn. pheromone blend consists of 3Z,6Z,9Z-20- and 21:H, the latter of which is also produced by males (29, 30). Male moth response is bimodal, with males being attracted to 3Z,6Z,9Z-21:H as a single component, and to the female-produced blend. Males may release 3Z,6Z,9Z-21:H as an aphrodisiac during courtship, and other males also may cue in on this compound and attempt to displace the first male. Conversely, male *Mocis megas* Guenée produce 3Z,6Z,9Z-19-, 20-, 21-, and 22:H, and 6Z,9Z-21:H, whereas only 3Z,6Z,9Z-21:H and the corresponding 9,10-epoxide are produced by females (18). The behavioral roles of the *M. megas* compounds remain to be determined. In unpublished studies, the author's research group identified 3Z,6Z,9Z-21:H and 3Z,6Z-*cis*-9,10-epoxy-21:H from extracts of females from four species in the genus *Catocala*, but male moths were not attracted to any reconstructed blends. However, the epoxides used to make lures were enantiomerically enriched (approximately 88% ee) rather than enantiomerically pure, and it is possible that the small amounts of enantiomeric impurities were strongly inhibitory.

Arctiidae

The Arctiidae is a large family with approximately 11,000 described species. From the five subfamilies proposed [Arctiinae, Ctenuchinae (= Amatidae or Euchromiidae), Lithosiinae, Thyretinae, and Pericopinae], pheromone components, or putative pheromone components found in extracts of female pheromone glands, have been identified for approximately 30 species in the first two subfamilies, primarily in the Arctiinae. However, in most cases, the designation of these compounds as pheromone components remains to be proven. This may be due in part to the curious pheromone release behavior of Arctiid females, which use a rhythmic pumping action of the abdomen to release the pheromone in discrete puffs (14, 37), and in at least one case, the pheromone is actually released as liquid droplets in an aerosol, instead of as a vapor (37). The fact that the plume of vapor emitted by passive pheromone release devices such as rubber septa does not mimic the pulsed aerosol plume may have contributed to difficulties in demonstrating the role of these compounds in most species studied to date.

Disregarding these difficulties, for the 23 species within the Arctiinae for which pheromone components have been reported, 14 species produce polyene or epoxide type pheromones, whereas the remaining species (eight in the genus *Holomelina* and one in the genus *Pyrrharctia* (= *Isia*), use saturated, methyl-branched hydrocarbon pheromones derived from a completely different biosynthetic pathway involving de novo synthesis from leucine and malonic acid (12). All of the other Artiine species produce C_{21} polyene-epoxide type pheromones, and in all known cases, the epoxide is in the 9,10 position, including in the two known cases of a triene monoepoxide [1,3Z,6Z-*cis*-9,10-epoxy-21:H; *Hyphantrea cunea* Drury (74) and *Diacrisia* (= *Spilosoma*) *obliqua* (Walker)] (58). Higher and lower homologs (C_{19}-C_{25}) of the polyene components are common in pheromone gland extracts. Five species also produce the linoleic and linolenic acid–derived aldehydes, 9Z,12Z-18:Ald and 9Z,12Z,15Z-18:Ald, which have been found only in the Arctiines. These aldehydes also are the only C_{18} pheromone components found in the Arctiidae.

Attractants or pheromone gland constituents (C_{19}-C_{21} polyenes and 3Z,6Z-*cis*-9,10-epoxy-21:H) have been reported for five other species in two other subfamilies, the Ctenuchinae and the Amatidae.

Lymantriidae

There are no clear patterns of pheromone use in the lymantriids. Pheromones are known for species in six genera, three from the tribe Orgyiini (*Dasychria*, *Gynaephora*, *Orgyia*), and three from the tribe Lymantriini (*Euproctis*, *Leucoma*, and

Lymantria). In the Orgyiini, three *Dasychria* species and four *Orgyia* species were attracted to the ketone 6Z-heneicosen-11-one (Z6-11-keto-21:H), which, along with Z6,E8-11-keto-21:H, is a known pheromone component of *O. pseudotsugata* McDunnough (25). However, another species in the same tribe, *Gynaephora quinghainsis* Chou et Ying produces 3Z,6Z,9Z-20- and 21:H (C-H Zhao, personal communication). The situation is even more confused in the Lymantriini: Of the five *Lymantria* species for which pheromones or sex attractants are known, 2-methyl-*cis*-7,8-epoxy-18:H (disparlure) is the major component of the attractive blend, with or without lesser amounts of related components. However, a sixth species, *L. mathura* L., produces 3Z,6Z,9Z-19:H, and nonracemic mixtures of the enantiomers of 3Z,6Z-*cis*-9,10-epoxy-18:H, 3Z,6Z-*cis*-9,10-epoxy-19:H, and 6Z-*cis*-9,10-epoxy-19:H (23a, 57a). Although the optimal blend remains to be worked out, from preliminary tests it was clear that the enantiomeric ratios were crucial: Males were not attracted to either pure enantiomer of the major component, 3Z,6Z-*cis*-9,10-epoxy-19:H, but were attracted to a 4:1 ratio of the 9*S*:9*R* enantiomers, mimicking the female-produced ratio.

Pheromones or sex attractants have been reported for five *Euproctis* species, four of which produce and/or are attracted to butyrate, isobutyrate, or isovalerate esters of methyl-branched alcohols. The fifth species, *E. chrysorrhoea* L., produces a remarkable pheromone component, Z7,Z13,Z16,Z19-docosatetraen-1-yl isobutyrate, the largest lepidopteran pheromone known (40). This compound is a hybrid, with an isobutyrate ester similar to pheromone components of congeners, and yet the placement and stereochemistry of the Z13,Z16,Z19 unit match a chain-extended linolenic acid motif. Finally, the pheromone of a *Leucoma* species (*L. salicis* L.) has been partially identified, with females producing 3Z,6Z-*cis*-9,10-epoxy-21:H, 3Z,9Z-*cis*-9,10-epoxy-21:H, and the first example of a diepoxide, 3Z-*cis*-6,7-*cis*-9,10-diepoxy-21:H (24).

MECHANISMS TO GENERATE UNIQUE PHEROMONE SIGNALS

Sympatric species that share pheromone components develop mechanisms to avoid unproductive interspecific attraction. Species-specific pheromone blends can be generated from different sets of constituents, or by mixing the same set of constituents in different ratios. Whereas it is clear from field trials with polyene and epoxide blends that the ratios of components are important, examples of species-specific blends made up of constituents with structural variations are more clearly documented. A number of variables can be manipulated, as shown in the following examples taken from identified pheromone blends, or from sex attractant blends for which there is strong evidence of biological activity from field tests.

Use of an Unusual or Unique Pheromone Component

There are several examples of moths using unique pheromone components. For example, *Bupalus piniaria* L. uses 4*E*,6*Z*,9*Z*-19:H and the corresponding epoxide, 6*Z*,9*Z*-*trans*-4*S*,5*S*-epoxy-19:H, both of which have been described only from this moth, as components of its blend (20). Another geometrid, *Peribatodes rhomboidaria* Schiff., blends a unique compound, 6Z,9Z -3-keto-19:H, with 3Z,6Z,9Z -19:H, a much more common component (11, 75). The lymantriid *Leucoma salicis* L. creates a unique blend from the only known example of a diepoxide, 3Z -*cis*-6,7-*cis*-9,10-diepoxy-21:H, in combination with each of the corresponding monoepoxides (24).

Mixed Chain Lengths

Several species use blends of 6Z,9Z-dienes [C_{20} + C_{21}: *Theria rupicapraria* Den. and Schiff. (69)] or 3Z,6Z,9Z-trienes [C_{18} + C_{19}: *Agriopis bajaria* Den. and Schiff., (69); C_{20} + C_{21}: *Anticarsia gemmetalis* Hbn., *Caenurgina distincta* Neumüller, *C. erechtea* (Cramer), *Mocis disseverans* (Walker), and all in the noctuid subfamily Catocalinae, (30, 39, 79)]. A mixture of 3Z,9Z -6*S*,7*R*-epoxy-20- and 21:H comprises a sex attractant for the geometrid *Idia americalis* Guenée (47). It is noteworthy that almost all examples of this type of blend consist of compounds differing in chain length by one carbon. Four arctiid species whose pheromone glands contain mixtures of 9Z,12Z,15Z-18:Ald, 9Z,12Z-18:Ald and 3Z,6Z-*cis*-9,10-epoxy-21:H may represent exceptions to this generalization (31, 32, 58), but there is little or no behavioral data in support of the designation of one or more of these compounds as pheromone components. In contrast, for acetate/alcohol/aldehyde pheromone blends consisting of compounds of mixed chain length, most blends are made up of components differing in length by two carbons.

Different Numbers of Double Bonds

Blends containing compounds with the same chain length but different numbers of double bonds are common. Examples include blends containing 6Z,9Z-dienes and 3Z,6Z,9Z-trienes [C_{19}, *Alsophila quadripunctata* Esper (70); C_{21}, *Lobophora nivigerata* Wlk. (48) and *Mocis latipes* (Guenée) (38, 39)] or a 3Z,6Z,9Z-19:H with 3Z,6Z,9Z,11*E*/*Z*-19:H tetraenes [*Alsophila pometaria* (Harris)] (81). There are also several examples of blends of epoxides differing only in the degree of unsaturation [6Z- and 3Z,6Z-*cis*-9,10-epoxy-21:H, *Oraesia excavata* (Butler) (57); 6Z- and 3Z,6Z-9*S*,10*R*-epoxy-18:H, *Hemerophila atrilineata* Butler (72)].

Polyenes with the Corresponding Monoepoxides Examples of pheromone blends consisting of a polyene with the corresponding epoxide are too numerous to list, particularly within the geometrid subfamily Ennominae; a comprehensive list can be generated easily from The Pherolist (4). Examples include both blends of 6Z,9Z-dienes with the corresponding monoene monoepoxides, and 3Z,6Z,9Z-trienes with diene monoepoxides.

Mixtures of Epoxide Regioisomers Examples of these blends are less common, but this may simply reflect the fact that specific blends of epoxides have not been screened as frequently as blends of a polyene with an epoxide. Cases include the pheromone blend of *Tephrina arenacearia* Hbn. [6Z,9Z-3*S*,4*R*- and 3Z,9Z-*cis*-6,7-epoxy-17:H (76)], and sex attractants for *Anavitrinelia pampinaria* (Guenée) [6Z,9Z-*cis*-3,4- and 3Z,9Z-6*S*,7*R*-epoxy-19:H (42)] and *Probole amicaria* Herrich-Schaffer [6Z,9Z-3*S*,4*R*- and 3Z,9Z-6*R*,7*S*-epoxy-19:H (42)].

Blends of Enantiomers Synergism between enantiomers in lepidopteran pheromone blends was discovered with the geometrid species *Semiothisa signaria dispuncta* (Walker) and *Epelis truncataria* (Walker), both of which were attracted more to racemic 6Z,9Z-*cis*-3,4-epoxy-17:H than to samples enriched in one enantiomer (42). Another geometrid, *Bleptina caradrinalis* (Guenée), was most attracted to specific blends of the enantiomers of either 3Z,9Z-*cis*-6,7-epoxy-20- or 21:H (47). However, the studies described here were limited by the fact that, at the time, the technology to prove that the females actually produce blends of the enantiomers was not available. Since then, two incontrovertible cases of enantiomeric synergism have been documented, with the blend of enantiomers in female pheromone gland extracts being determined using gas chromatography with chiral stationary phases. In the first, the geometrid *Ascotis selenaria cretacea* Butler produced a 57:43 blend of the enantiomers of 6Z,9Z-*cis*-3,4-epoxy-19:H (3), and in a second study, *Lymantria mathura* L. females produced a 1:4 blend of 3Z,6Z-9*R*,10*S*- and 3Z,6Z-9*S*,10*R*-epoxy-19:H (23a). For both species, males responded optimally to blends of enantiomers mimicking the female blend.

Behavioral Antagonists

The use of behavioral antagonists as an additional method of limiting interspecific attraction is well developed in species with polyene and epoxide pheromones or sex attractants. Their patterns of use parallel those used in generation of attractant blends, but the known examples are fewer because, in working with pheromone lures, researchers are generally trying to make lures more rather than less attractive. Antagonists usually have been discovered in screening trials or in lure optimization trials when, for example, changing from using a component in racemic form to using the compound in enantiomerically enriched form has resulted in a substantial increase in attraction, providing strong circumstantial evidence that one enantiomer is inhibitory. Only a few studies have been aimed at determining the roles of antagonists in separating closely related sympatric species, and these are mentioned among the other cases cited below.

Antagonism by Chain Length Homologs The geometrid *Plagodis alcoolaria* (Guenée) was attracted to blends of 3Z,6Z,9Z-20:H and 3Z,9Z-6*R*,7*S*-epoxy-20:H. The attraction was suppressed by the enantiomers of the corresponding C_{19} epoxide, and EAG studies determined that male moth antennae were more sensitive to the C_{19} compounds than the C_{20} attractant (47).

Antagonism by Compounds with Different Functional Groups The pheromone blend of *Paleacrita vernata* (Peck) consists of 6Z,9Z-19:H, 3Z,6Z,9Z-19:H, and 3Z,6Z,9Z-20:H. Attraction of male moths was inhibited by 6Z,9Z-*cis*-3,4-epoxy-19:H (43). A *Semiothisa* spp., *S. bicolorata* F., uses 3Z,6Z,9Z-17:H as its sex pheromone, the attractiveness of which is antagonized by 6Z,9Z-*cis*-3,4-epoxy-17:H (51). Conversely, the former compound inhibits the attraction of a sympatric congener, *S. ulsterata* Pearsall, whose major pheromone component is 3Z,9Z-6*S*,7*R*-epoxy-17:H (42). Antagonism of diene monoepoxide sex attractants by the corresponding triene hydrocarbon has been reported for other species [e.g. *Metanema inatomaria* Guenée (82); *Spargoloma sexpunctata* Grote (82)] whose sex pheromones remain to be elucidated.

Antagonism by Epoxide Regioisomers A sex attractant for male *Euchlaena madusaria* Walker consists of 6Z-*cis*-9,10-epoxy-19:H plus 6Z,9Z-19:H, and the attraction is inhibited by the regioisomer, 9Z-*cis*-6,7-epoxy-19:H (46). Another geometrid species, *Sicya macularia* (Harris), was attracted to blends of 3Z,6Z,9Z-19:H with the corresponding 3*S*,4*R*-epoxide, and attraction was suppressed by either of the 3Z,9Z-*cis*-6,7-epoxy-19:H enantiomers (42). A number of other examples can be inferred from the fact that blends of monoepoxides generated by nonregioselective partial epoxidation of 3Z,6Z,9Z-trienes or 6Z,9Z-dienes frequently have not attracted moth species, but those same species were attracted in subsequent trials using lures consisting only of a single epoxide regioisomer (e.g. 82).

Antagonism By the Other Enantiomer There are numerous examples of enantiomeric antagonism, indicating that moths make full use of the structural features of epoxide pheromones. Many field trapping studies to date have used enantiomerically enriched rather than enantiomerically pure compounds, and even more cases of enantiomeric antagonism will come to light as field tests are repeated with enantiomerically pure constituents.

Some of the first examples were reported by Wong et al (82) in experiments testing single enriched enantiomers or blends of 3Z,6Z-*cis*-9,10-epoxides as lepidopteran sex attractants. Attraction of several species to a single enantiomer was suppressed upon addition of the other enantiomer. Since then, numerous examples have been reported (20, 28, 42, 47, 60, 71, 76). The most interesting cases are those involving sympatric species, in which attraction of each species is suppressed by addition of the enantiomer produced by the other species [e.g. *Itame brunneata* and *I. occiduaria*, (42); *Colotois pennaria* L. and *Erannis defoliaria*

Cl. (69, 71)]. An additional intriguing observation is worth mentioning: Haynes et al identified the pheromone of the noctuid *Tetanolita mynesalis* (Walker) as a blend of 3Z,6Z,9Z-21:H and 3Z,9Z-6*S*,7*R*-epoxy-21:H (28), and further found that the 6*R*,7*S*-enantiomer suppressed attraction of males. This species is a major prey item for the bolas spider *Mastophora hutchinsoni* Gertsch, which produces lepidopteran pheromones to lure male moths within reach of its sticky bolas. Thus, the spiders must also synthesize only one enantiomer of the epoxide in order for it to work as an attractant for *T. mynesalis*.

ANALYSIS AND IDENTIFICATION OF PHEROMONE COMPONENTS

The first identifications of triene and epoxide pheromone components were carried out with microchemical tests and multiple spectrometric methods (14, 31, 32). Ultraviolet spectrometry was used to demonstrate conjugation in 1,3Z,6Z,9Z-tetraenes (34), and the lack of conjugation in trienes (14). Fourier transform infrared (FTIR) spectra demonstrated that all of the double bonds in trienes were *cis* because of the lack of a band at approximately 960–980 cm^{-1} (32). Alternatively, the all-*Z* geometry was demonstrated by "scrambling" the double bonds in a 3Z,6Z,9Z-standard to generate all eight possible isomers, separation of isomers, and demonstration that only the 3Z,6Z,9Z-isomer matched retention times with the insect-produced isomer (14, 43). Valuable microchemical tests included hydrogenation (determination of number of double bonds, and indirectly, confirmation of an unbranched chain), ozonolysis (31, 32, 34, 41, 63) or reaction with dimethyldisulfide (41) to locate double bond positions, or oxidation with periodic acid to locate double bonds and epoxide positions (31, 57).

As further examples of these types of compounds were identified and synthetic standards became available, it was recognized that the different compounds had distinctive mass spectral features that facilitated their identification. Furthermore, the epoxide regioisomers are separable by capillary gas chromatography (GC), and by liquid chromatography (1, 27, 82), and their elution order from various columns is known (see below). The distinctive mass spectral features of each class of compounds are discussed below, and representative mass spectra are shown in Figures 2, 3, and 4. These spectra were taken on the same instrument (Hewlett-Packard 5973 mass selective detector, tuned to Autotune parameters, 70 eV ionization energy in EI mode) to allow comparisons without the added variability of the (often considerable) inter-instrument differences in spectra. Citations also are provided to examples of published spectra, particularly for the less common compounds.

The electron impact (EI) mass spectra of 3Z,6Z,9Z-trienes (Figure 2*b*) (17) are characterized by small molecular ions, a base peak at m/z 79, and a large diagnostic ion at m/z 108, $[CH_3CH_2(CH{=}CH)_3H]^+$, from a structure-specific cleav-

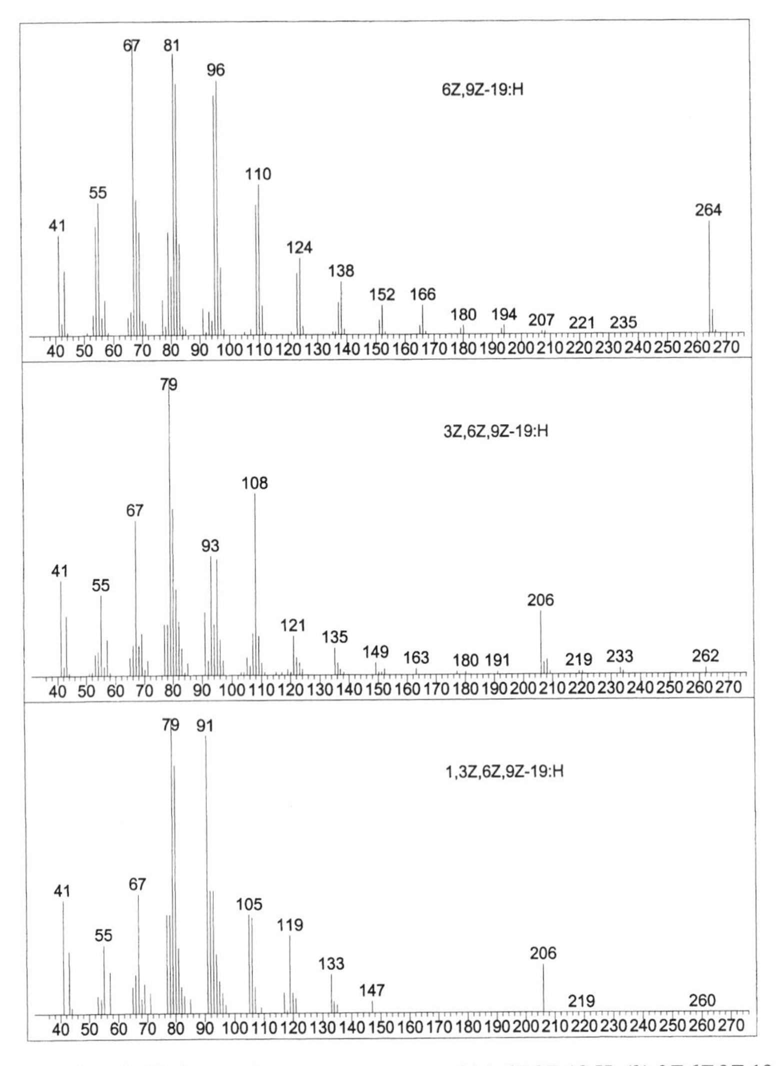

Figure 2 70 eV electron impact mass spectra of (*a*) 6Z,9Z-19:H, (*b*) 3Z,6Z,9Z-19:H, and (*c*) 1, 3Z,6Z,9Z-19:H.

age between C_8 and C_9 with a hydrogen transfer to give the even-numbered rearrangement fragment, characteristic of methylene-interrupted polyenes with one outer double bond at C_3. There is a distinct ion at M-56 (loss of C_4H_8), corresponding to $[H(CH{=}CH)_3R]^+$ from rearrangement and cleavage between C_4 and C_5, placing the other outer double bond between carbons 9 and 10 (14, 79). The molecular weights of trienes are easily confirmed by chemical ionization (CI) MS, using methane, isobutane, or nitrous oxide. The lack of conjugation is confirmed by retention times shorter than the saturated alkane of equal chain length on nonpolar GC columns [e.g. Kovats index (KI) of 3Z,6Z,9Z-19:H = 18.78 on HP5-MS, 5% phenyl methyl siloxane column, 30m x 0.25mm, 0.25 μ film thickness; 50 to 250°C at 10°/min].

The EI mass spectra of the 6Z,9Z-dienes are distinctively different (Figure 2a) (17). The base peak is now at m/z 67, and the molecular ion is much more prominent than in the corresponding triene. The spectrum is characterized by a monotonically decreasing series of even-numbered fragments (m/z 166, 152, 138, 124, 110, 96) and a distinct ion at M-98 (M-C_7H_{14}) from cleavage with a hydrogen transfer between carbons 7 and 8 (17), which suggests one of the double bonds is at C_6. However, because the m/z 110 ion, corresponding to cleavage and rearrangement between carbons 8 and 9, is part of a monotonically decreasing series of even-numbered fragments, it cannot be used reliably to place the other double bond at C_6; this must be verified by other means (e.g. 41). The retention time shorter than that of the equivalent alkane on nonpolar columns (KI of 6Z,9Z-19:H on HP5-MS = 18.71) indicates that the diene is not conjugated.

EI mass spectra of 3Z,6Z,9Z,11*Z*/*E*-hydrocarbons (80) are qualitatively similar to those of 3Z,6Z,9Z-trienes in the low mass range (< m/z 100), with a base peak at m/z 79 and significant fragments at m/z 55, 67, and 93, although these fragments are of lower intensity than in the trienes. Conversely, the m/z 108 fragment, $[CH_3CH_2(CH{=}CH)_3H]^+$, from cleavage between carbons 8 and 9, is more prominent, whereas the molecular ion is weak.

Spectra of 1,3Z,6Z,9Z-tetraenes (Figure 2*c*) (21) are readily distinguishable from those of the corresponding trienes, with a base peak at m/z 79 instead of 81, and an intense m/z 91 fragment. The diagnostic fragment at m/z 106, $[CH_2{=}CH(CH{=}CH)_3H]^+$, from cleavage between C_8 and C_9 with hydrogen transfer, places the outer double bond at C_9, and confirms that the additional double bond is between C_1 and C_8. There are also smaller fragments at m/z 119, 133, and 147, two mass units less than the equivalent fragments in the trienes. A second diagnostic fragment occurs at M-54, from loss of carbons 1 to 4 as butadiene (C_4H_6). The absence of a fragment corresponding to hexatriene (or loss of hexatriene) indicates that the 1,3-diene is not further conjugated, placing the remaining double bond unaccounted for at C_6 (21, 78). Conjugation somewhere in the molecule is indicated by retention times longer than those of equivalent alkanes on nonpolar columns (KI of 1,3Z,6Z,9Z-19:H on HP5-MS = 19.06).

The diene monoepoxide analogs of 3Z,6Z,9Z-trienes are separable on capillary GC phases of varying polarity, including DB-5 (51), SE-54 (8, 73), DB-1701 (44), and DB-210 (72). The elution order on all these columns is 6,7<3,4<9,10 epoxide. The EI mass spectra of the regioisomers are readily distinguishable (Figure 3) (2, 27, 73, 77). All three isomers give weak molecular ions, and small M-18 and M-29 ions. Spectra of 6Z,9Z-*cis*-3,4-epoxides are characterized by base peaks at m/z 79 or 80, and a distinctive fragment at M-72 from cleavage α to the epoxide, between carbons 4 and 5 with hydrogen transfer, and M-58 from rearrangement and loss of propionaldehyde (Figure 3*a*) (15, 42). A small fragment

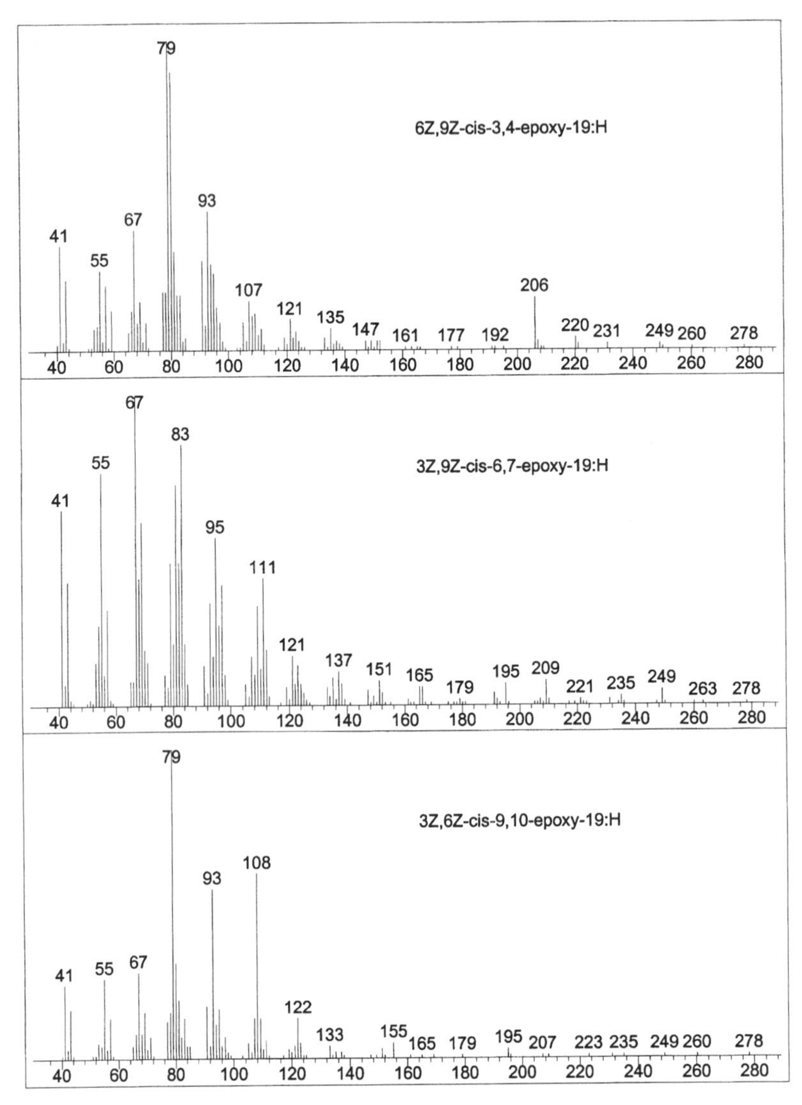

Figure 3 70 eV electron impact mass spectra of (*a*) 6Z,9Z-*cis*-3,4-epoxy-19:H, (*b*) 3Z,9Z-*cis*-6,7-epoxy-19:H, and (*c*) 3Z,6Z-*cis*-9,10-epoxy-19:H.

at m/z 59 $(C_3H_7O)^+$ is useful for distinguishing between 3,4 epoxides and the other regioisomers.

3Z,9Z-*cis*-6,7-Epoxides exhibit a base peak at 67 (Figure 3*b*) (27) or m/z 55 (73), depending on the spectrometer, and a strong diagnostic fragment at m/z 111, from cleavage α to the epoxide between C_7 and C_8, and M-69 from cleavage on the other side of the epoxide between C_5 and C_6. There is also a prominent fragment at m/z 95, which is much smaller in the other two regioisomers.

3*Z*,6*Z*-*cis*-9,10-epoxides are characterized by a base peak at m/z 79, a strong m/z 108 fragment due to $[CH_3CH_2(CH{=}CH)_3H]^+$, and a weaker but distinct ion at m/z 122 from rearrangement of the epoxide and cleavage of $[C_9H_{14}]^+$ (Figure 3*c*) (23, 57). Cleavage on the other side of the epoxide, between C_9 and C_{10}, apparently is not favored, and provides no significant fragment.

Mass spectra of monoepoxides of 6*Z*,9*Z*-dienes show some similarities to those of the diene monoepoxides (1, 46). 9*Z*-*cis*-6,7-epoxide spectra (Figure 4*a*) are not easily interpreted (1) and many of the fragments expected, while present, are relatively small. The molecular ion and M-18 ions are weak. There are small- to medium-intensity ions at m/z 113 and M-71 from simple cleavage on either side of the epoxide, and medium-intensity, even-numbered rearrangement ions at M-114, M-142, and M-156, and more intense fragments at m/z 109 and M-111. There is also a medium-intensity ion at m/z 99, $[C_5H_{11}CO]^+$, from rearrangement and cleavage between C_6 and C_7.

6*Z*-*cis* 9,10-epoxides are characterized by a base peak at m/z 55 (46, 57), 69, or 81 (Figure 4*b*) (1), a weak molecular ion and M-18 ion, and a significant M-

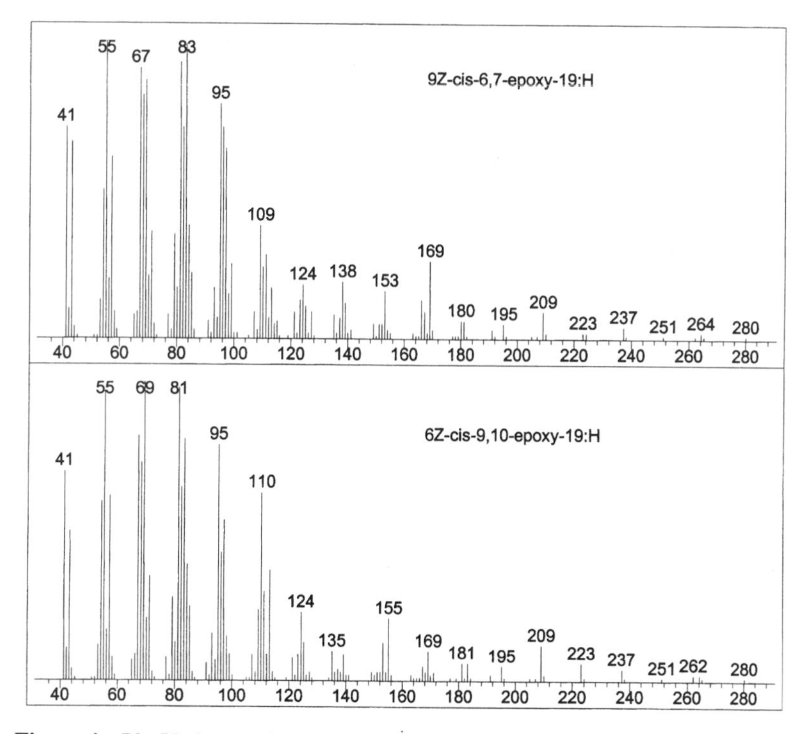

Figure 4 70 eV electron impact mass spectra of (*a*) 9*Z*-*cis*-6,7-epoxy-19:H, and (*b*) 6*Z*-*cis*-9,10-epoxy-19:H.

71 fragment. Distinct diagnostic ions are seen at m/z 110 $[C_4H_9(CH=CH)_2H]^+$ from cleavage with hydrogen transfer between C_8 and C_9, and m/z 124 $[C_5H_9(CH=CH)_2H]^+$, along with M-125 $[RCO]^+$ from rearrangement of the epoxide and cleavage between C_9 and C_{10}. A small fragment at m/z 153 is also seen from simple cleavage α to the epoxide between C_{10} and C_{11}.

DETERMINATION OF ENANTIOMERIC COMPOSITION OF EPOXIDES

All of the epoxides have two enantiomeric forms, and determination of which enantiomer or blend of enantiomers is produced by a species has been a significant problem, made more intractable by the small quantities of compounds that are usually available. The importance of the problem is clear; there are now numerous examples of antagonism of behavioral responses by the "wrong" enantiomer, and several examples of enantiomeric synergism (vide supra). During the 1980s this issue was addressed indirectly by challenging male antennae with each enantiomer to see which elicited the largest response (e.g. 23, 31, 32, 61). This method is flawed because it assumes, amongst other things, that the insect produces or responds to only one enantiomer. The development of chiral GC [and high pressure liquid chromatography (HPLC)] stationary phases has partially solved this problem. The resolutions of some 6Z,9Z-*cis*-3,4- (71, 76), 3Z,9Z-*cis*-6,7- (71), and 3Z,6Z-*cis*-9,10-epoxide enantiomers (13, 23a) and the enantiomers of 6Z-*cis*-9,10-epoxy-19:H (23a) have been achieved with custom-made chiral cyclodextrin GC stationary phases. Unfortunately, only a single resolution (6Z,9Z-*cis*-3,4-epoxy-19:H) has been reported with a commercial GC column [Chiraldex A-PH, ASTEC Inc., Whippany NJ; (2)]. It is to be hoped that this limitation will be eliminated as more selective and robust chiral GC phases are developed.

Columns and conditions for resolution of the enantiomers of each diene epoxide regioisomer by HPLC also have been developed (3,4-epoxides, Chiralpak AS; 6,7- and 9,10-epoxides Chiralpak AD, both columns from Daicel Chemical Industry Co., Tokyo) (60). Very recently, a Chiralcel OJ-R column (Daicel) was used to separate 3,4- and 6,7-epoxydienes of C_{17}-C_{23} chain lengths, and 9,10-epoxydienes of C_{17}-C_{20} chain lengths (59a).

However, none of the epoxides contain a UV chromophore and, consequently, it may not be possible to detect the small quantities found in insect extracts by HPLC using a UV detector.

SYNTHESES OF PHEROMONE COMPONENTS

Hydrocarbons

6Z,9Z-Dienes and 3Z,6Z,9Z-trienes of 18 or more carbons are synthesized most easily by reduction of linoleic or linolenic acid esters to the corresponding alcohols, followed by tosylation and reduction (18 carbon compounds) or chain extension with the appropriate dialkyllithium cuprates (Figure 5, Equation 1) (14, 30, 34, 79). An alternate route using Kolbe electrolysis is not recommended because it produces poor yields of impure material (14). Syntheses of the C_{17} homologs by chain extension is not practical because the corresponding C_{16} carboxylic acid is not readily available. Hence the C_{17} compounds must be assembled from scratch. The most flexible route (Figure 5, Equation 2), which allows synthesis of 6Z,9Z-dienes or 3Z,6Z,9Z-trienes of any length, uses the 8-carbon 1-tosyloxy-octa-2-yne or -2,5-diyne synthons, to which a chain of any desired length can be attached, followed by reduction to the diene or triene respectively (51).

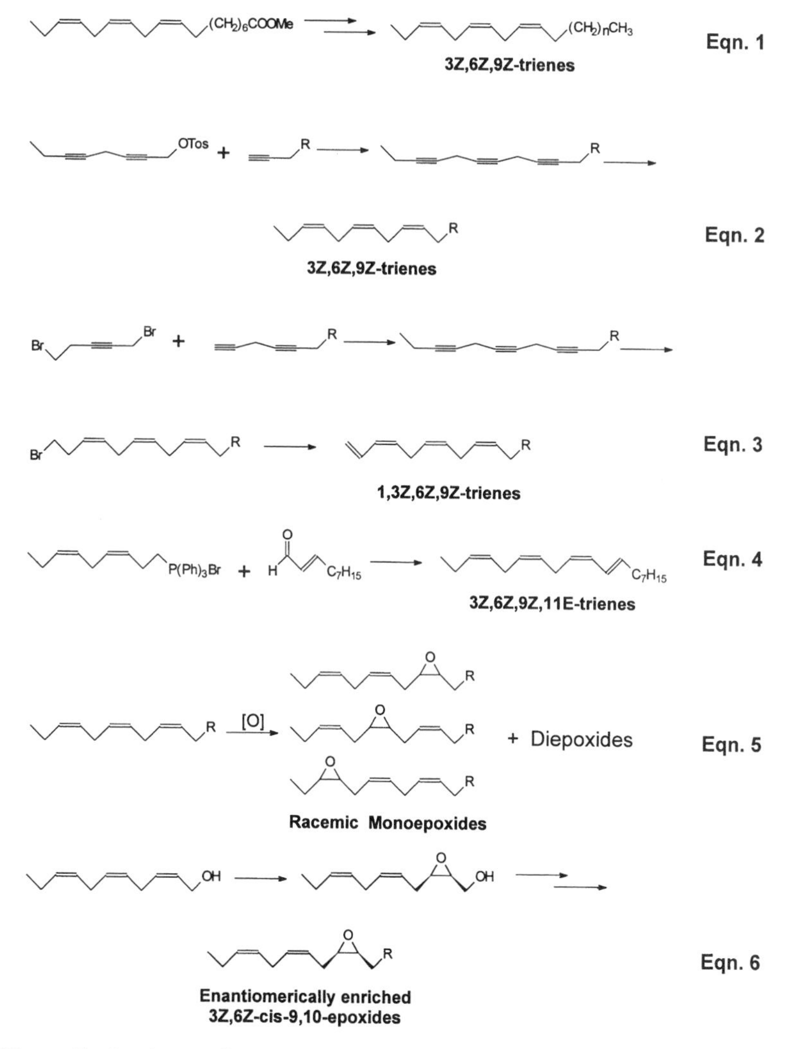

Figure 5 Syntheses of polyene hydrocarbon and epoxide pheromones.

There is no fatty acid precursor that can be converted to 1,3Z,6Z,9Z-tetraenes, so these compounds also must be synthesized. Most syntheses have assembled acetylene-containing subunits to produce triynol intermediates, which are reduced, followed by conversion of the alcohol to a leaving group and elimination of HX to put in the final double bond (e.g. 33). The shortest and most flexible synthesis is 5 steps (Figure 5, Equation 3) (49). Wittig chemistry also has been used (7, 63), but is not recommended because mixtures of *E*/*Z* isomers are produced.

3Z,6Z,9Z,11*E*/*Z*-tetraenes, pheromone components of two *Alsophila* species, were synthesized by a combination of acetylene and Wittig chemistry (Figure 5, Equation 4) (80). The key intermediate, 3Z,6Z-nonadienyltriphenylphosphonium bromide, prepared from 3,6-nonadiynol, was reacted with *E*2-decenal to produce 3Z,6Z,9Z,11*E*-19:H directly, or with 2-decynal to produce the trienyne, which was then reduced with dicyclohexylborane to 3Z,6Z,9Z,11Z-19:H. This indirect route was used because of the instability of 2Z-decenal. In both cases, the Wittig reaction produced a mixture of *Z/E* isomers at the 9 position, so purification on a silver-ion-loaded ion exchange resin was necessary (80).

Epoxides

Racemic monoepoxides of 3Z,6Z,9Z-trienes (2, 27, 82) and their 1,3Z,6Z,9Z-tetraene (74, 78) or 6Z,9Z-diene analogs (2) are prepared by nonregioselective partial epoxidation of the polyenes with metachloroperbenzoic acid (Figure 5, Equation 5), followed by liquid chromatographic separation of the regioisomers on silica gel, using three percent ether in hexane (27, 82) or one percent THF in hexane (1). They have also been separated by thin-layer chromatography on silica, eluting with hexane:benzene 1:1 (2). This simple method has recently been extended by first separating the regioisomers on silica, followed by resolution of each regioisomer into its enantiomers by HPLC on a chiral stationary phase (60). Furthermore, the method was used to generate diepoxides, used in the identification of the first pheromone component of this type (3Z-*cis*-6,7-*cis*-9,10-epoxy-21:H) (24). Overall, this method is useful for providing small quantities only. For larger quantities, regio- and stereospecific syntheses of the desired compounds are preferred.

Several strategies have been used to synthesize the epoxides in enantiomerically enriched or enantiomerically pure forms. The first and most common strategy uses Sharpless asymmetric epoxidation of an appropriate allylic alcohol precursor to generate the key epoxide function, (e.g. Figure 5, Equation 6, for the enantiomers of 3Z,6Z-*cis*-9,10-epoxides) (54, 82). Either enantiomer is available from a single precursor by changing the ligands used to form the chiral epoxidation catalyst. The key epoxyalcohol intermediates are then converted to tosylates or iodides, and alkylated with a chain of any desired length.

However, the asymmetric epoxidation reaction is stereoselective rather than stereospecific (maximum enantiomeric excess, approximately 90%), and the small amount of the other enantiomer may be inhibitory (55). The enantiomeric impurity can be removed by recrystallization of the epoxyalcohol intermediates, if they are crystalline (19), or by formation of crystalline 2,4-dinitrobenzoate derivatives and recrystallization to high enantiomeric purity (Figure 6, Equation 7) (9, 10, 53, 55, 56).

Two other synthetic routes to 3*Z*,6*Z*-*cis*-9,10-epoxides have been developed, starting from chiral natural products. The first, used to make 3*Z*,6*Z*-9*S*,10*R*-epoxy-21:H, (and 6*Z*-9*S*,10*R*-epoxy-21:H) proceeds via a chiral epoxide synthon derived from *D*-xylose (59, 64). This route yields enantiomerically pure products, but it can only be used to produce one of the two enantiomers, it is long, and it cannot be easily manipulated to produce other regioisomers. The second route used alkylative rearrangement of threo-(2*R*,3*R*)-1,2-epoxyalkan-3-ol tosylates, and the cor-

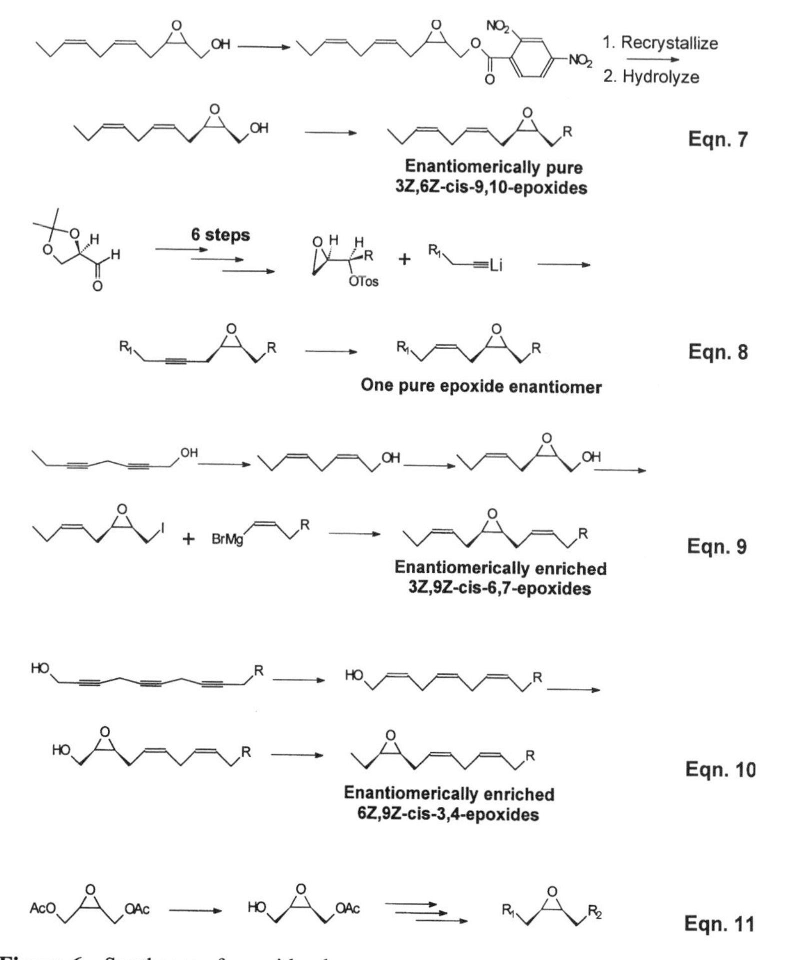

Figure 6 Syntheses of epoxide pheromones.

responding 2*S*,3*S* enantiomer, prepared in multiple steps from *R*-glyceraldehyde and *L*-(+)-diethyltartrate respectively (5, 6). The key epoxide intermediates are regioselectively alkylated with an alkyne, and the resulting alkoxide stereospecifically displaces the tosylate to form the new epoxide ring (Figure 6, Equation 8). The alkyne is then reduced by standard methods. This route has disadvantages: Each enantiomer is generated from a different starting material, the overall chain length is fixed at an early stage, and the syntheses are lengthy with moderate yields. A similar strategy was used recently to make the enantiomers of 1,3Z,6Z-*cis*-9,10-epoxy-21:H (83).

The asymmetric epoxidation strategy also was used in the syntheses of the 3*Z*,9*Z*-*cis*-6,7-epoxides (Figure 6, Equation 9). Fine-tuning of the coupling step connecting an alkenyl chain to the key epoxide intermediates proved crucial to avoid a competing rearrangement reaction. CuI-catalyzed reaction of a *Z*-alkenyl Grignard reagent with epoxyiodide intermediates (instead of tosylates) with a specific hexamethyl phosphoramide/tetrahydroforan (HMPA/THF) solvent mixture was critical in the final coupling step (50). The route was flexible because the key chiral intermediates could be alkylated with chains of any desired length. A later synthesis of the 6,7- epoxides used similar steps, but incorporated a derivatization and recrystallization step to produce material of higher enantiomeric purity (53).

In contrast to the other regioisomers, syntheses of 6Z,9*Z*-*cis*-3,4-epoxides start from the end of the chain remote from the epoxide, with the epoxide being incorporated at a late stage (10, 15, 42, 71). Consequently, homologs must be synthesized individually, rather than by simply altering the length of the chain attached to a late-stage intermediate. The key intermediates common to all routes (Figure 6, Equation 10) are 2Z,5Z,8Z-trienols, assembled from propargyllic alcohol subunits, followed by reduction to the trienols and asymmetric epoxidation. Conversion of the alcohols to iodides [with or without further purification by derivatization/recrystallization; (10)] and reaction with dimethyllithium cuprate completed the syntheses. More flexible routes, allowing the syntheses of a series of homologs from a single key intermediate, have not yet been developed.

Asymmetric epoxidation followed by chain extension of the resulting epoxides also was used to synthesize the enantiomers of 6*Z*-*cis*-9,10- and 9*Z*-*cis*-6,7-epoxides (46). A slightly different route proceeding through a crystalline epoxyalcohol intermediate provided pure enantiomers of 6*Z*-*cis*-9,10-epoxy-21:H (19). The 9*S*,10*R* enantiomer has also been made by a chemo-enzymatic synthesis (9), using the enantioselective enzymatic hydrolysis of an achiral 1,4-diacetoxy-*cis*-2,3-epoxide (Figure 6, Equation 11). Further manipulations allowed sequential coupling of alkenyl and alkyl chains, respectively. Note that the single chiral epoxide intermediate can be used to make either enantiomer of the 6*Z*-*cis*-9,10-epoxide product by simply switching the order in which the alkyl and alkenyl chains are attached. Furthermore, the same intermediate could be used to make either enantiomer of any desired epoxide pheromone by simply attaching alkyl or alkenyl chains of the correct length in the appropriate sequence.

BIOSYNTHESIS OF POLYENE AND EPOXIDE PHEROMONES

A reasonable biosynthetic pathway for polyene hydrocarbon and epoxide pheromones that accommodates all of the known pheromone components has been proposed (Figure 7) (14, 62, 65). Even-numbered carbon skeletons (18, 20, 22 carbons) are produced from 0, 1, or 2 additions of 2-carbon units to linoleic (9Z,12Z-18:COOH) or linolenic (9Z,12Z,15Z-18:COOH) acids, followed by reduction of the carboxyl group, whereas odd-numbered carbon skeletons ($C_{17, 19, \text{or } 21}$) arise from reductive decarboxylation of even-numbered acyl precursors. Subsequent regio- and stereoselective oxidation of one or two of the double bonds produces the corresponding epoxides. Biosynthetic epoxidation of alkene precursors has been demonstrated in gypsy moth (36). Circumstantial evidence suggests that epoxidation may be the final step. Pheromone glands of decapitated female *Ascotis selenaria cretacea* Butler contained 3Z,6Z,9Z-19:H, but no 3Z,6Z-*cis*-9,10-epoxy-19:H. Upon injection of subesophageal gland extract containing pheromone biosynthesis activating neuropeptide (PBAN) into the decapitated females, the epoxide appeared in the pheromone gland and, concurrently, the titer of 3Z,6Z,9Z-19:H decreased (3).

Known structural variations are easily accommodated. For example, 6Z,9Z-18:Ald and 3Z,6Z,9Z-18:Ald, from arctiid pheromone gland extracts, require only reduction of fatty acyl precursors to the alcohols and reoxidation to aldehydes, known biosynthetic steps in other moths (35, 62). 1,3Z,6Z,9Z-tetraenes and

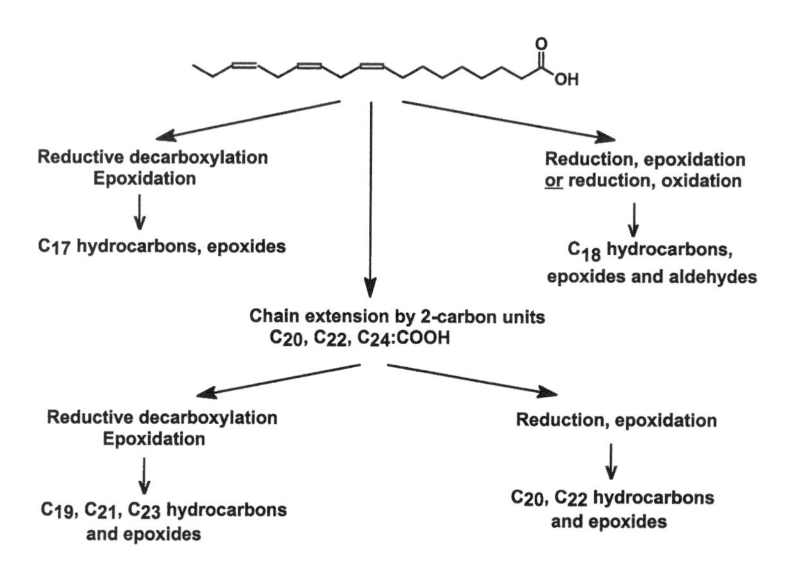

Figure 7 Biosynthetic pathways to produce polyene hydrocarbon and epoxide pheromones from linoleic or linolenic acid precursors.

3Z,6Z,9Z,11Z/E-tetraenes require an additional desaturation. As with dienes and trienes, oxidation of one of the double bonds produces epoxides, the first examples of which have been identified (58, 74). In similar fashion, 4*E*,6Z,9Z-19:H and the corresponding 6Z,9Z-*trans*-4,5-epoxy-19:H, from *Bupalus piniaria* L. (20), require an additional desaturation of a 6Z,9Z-hydrocarbon precursor, followed by oxidation. 6Z,9Z-Nonadecadien-3-one, from *Peribatodes rhomboidaria* Schiff. (11) and the only known example of a ketone with the skipped methylene structural motif, requires only rearrangement of the corresponding 6Z,9Z-*cis*-3,4-epoxy-19:H.

Two studies provide support for the proposed pathway. The first (65) used the arctiid moths *Estigmene acrea* Drury and *Phragmatobia fuliginosa* L., the pheromone glands of which contain 9Z,12Z-18:Ald, 9Z,12Z,15Z-18:Ald, and 3Z,6Z-*cis*-9,10-epoxy-21:H, and 3Z,6Z,9Z-21:H plus 3Z,6Z-*cis*-9,10-epoxy-21:H respectively. Freshly eclosed moths had a high titer of pheromone, and incorporation of labeled precursors was poor when they were applied to the pheromone gland, suggesting either that most pheromone biosynthesis occurred prior to eclosion, or that the pheromone was synthesized in other tissues and transported to the gland for release (35). With *E. acrea*, labeled 9Z,12Z,15Z-18:COOH was incorporated into the epoxide and both aldehydes, although the postulated 20, 21, and 22 carbon fatty acyl precursors to the epoxide were not found. Labeled 9Z,12Z-18:COOH was incorporated primarily into the aldehydes. Furthermore, labeled 11Z,14Z,17Z-20:COOH was incorporated into the epoxide, whereas 12Z,15Z,18Z-21:COOH was not, providing evidence for the proposed 2-carbon chain elongations followed by reductive decarboxylation. Proof of chain elongation was obtained from incorporation of labeled malonic acid into the epoxide (via elongation of endogenous 9Z,12Z,15Z-18:COOH). Analogous results were obtained with *P. fuliginosa*; labeled 9Z,12Z,15Z-18:COOH and 13Z,16Z,19Z-22:COOH were incorporated into both 3Z,6Z,9Z-21:H and 3Z,6Z-*cis*-9,10-epoxy-21:H.

In a second study, using the geometrid moth *Operophtera brumata* L., labeled 9Z,12Z,15Z-18:COOH was incorporated into the major pheromone component, 1,3Z,6Z,9Z-19:H (C-H Zhao and C Löfstedt, unpublished data). The fatty acyl intermediates 11Z,14Z,17Z-20:COOH and a tetraene acid (presumably 11Z,14Z,17Z,19-20:COOH) also were detected. These data all support the proposed biosynthetic pathway of chain elongation of 3Z,6Z,9Z-18:COOH, Δ^{19} desaturation, and reductive decarboxylation.

Indirect support for the proposed pathway can also be derived from the structures that have not been found. That is, all known examples of these types of pheromones have Z double bonds, or *cis*-epoxides, in the 3, 6, or 9 positions. Additional unsaturations in the 1, 4, or 11 positions do not disturb this motif. Furthermore, 6Z,9Z,12Z-18:COOH (gamma-linolenic acid) is much less common than 9Z,12Z,15Z-18:COOH, and no pheromones with a 6Z,9Z,12Z-motif have been found, even in sex attractant screening trials (79; JG Millar and EW Underhill, unpublished data). 6Z,9Z,12Z-trienes also elicited only weak signals from geometrid and noctuid moth antennae in EAG studies (82). Similarly, 6Z,9Z,12Z,15Z-18:COOH is not common, and no tetraene pheromones with this motif have been found; in screening trials, 3Z,6Z,9Z,12Z-18:H to 21:H attracted no moths (JG Millar and EW Underhill, unpublished data), indicating that these structures are unlikely candidates for pheromone components.

ACKNOWLEDGMENTS

I thank C Löfstedt, C-H Zhao, G Gries, and R Gries for sharing unpublished data, and H Arn and his coworkers for maintaining The Pherolist, which greatly expedited preparation of this chapter. I am also grateful for the mentorship of Dr. EW Underhill, who kindled my interest in this subject, and gave me the opportunity to conduct research on these interesting compounds from 1984–1986.

LITERATURE CITED

1. Ando T, Kishi H, Akashio N, Qin X-R, Saito N, et al. 1995. Sex attractants of geometrid and noctuid moths: chemical characterization and field test of monoepoxides of 6,9-dienes and related compounds. *J. Chem. Ecol.* 21:299–311
2. Ando T, Ohsawa H, Ueon T, Kishi H, Okamura Y, Hashimoto S. 1993. Hydrocarbons with a homoconjugated polyene system and their monoepoxy derivatives: sex attractants of geometrid and noctuid moths distributed in Japan. *J. Chem. Ecol.* 19:787–98
3. Ando T, Ohtani K, Yamamoto M, Miyamoto T, Qin X-R, Witjaksono. 1997. Sex pheromone of Japanese giant looper, *Ascotis selenaria cretacea*: identification and field tests. *J. Chem. Ecol.* 23:2413–23
4. Arn H, Tóth M, Priesner E. 1999. *The Pherolist.* http://www.nysaes.cornell.edu/pheronet/
5. Bell TW, Ciaccio JA. 1988. Alkylative epoxide rearrangement, application to stereoselective synthesis of chiral pheromone epoxides. *Tetrahedron Lett.* 29:865–68
6. Bell TW, Ciaccio JA. 1993. Alkylative epoxide rearrangement. A stereospecific approach to chiral epoxide pheromones. *J. Org. Chem.* 58:5153–62
7. Bestmann HJ, Brosche T, Koschatzky KH, Michaelis K, Platz H, et al. 1982. Pheromone-XLII. 1,3,6,9-Nonadecatetraen, das Sexualpheromon des Frostspanners *Operophtera brumata* (Geometridae). *Tetrahedron Lett.* 23:4007–10 (In German)
8. Bestmann HJ, Kern F, Mineif A, Platz H, Vostrowsky O. 1992. Pheromone 84. Der Sexualpheromonkomplex des "Braunen Baren" *Arctia caja* (Lepidoptera: Arctiidae). *Z. Naturforsch.* Teil C47:132–35 (In German)
9. Brevet J-L, Mori K. 1992. Pheromone synthesis. CXXXIX. Enzymatic preparation of (2*S*,3*R*)-4-acetoxy-2,3-epoxybutan-1-ol and its conversion to the epoxy pheromones of the gypsy moth and the ruby tiger moth. *Synthesis*, pp. 1007–12
10. Brevet J-L, Mori K. 1993. Synthesis of both the enantiomers of (6Z,9Z)-*cis*-3,4-epoxy-6,9-heptadecadiene, the pheromone component of various Geometrid

moths. *Biosci. Biotechnol. Biochem.* 57:1553–56

11. Buser HR, Guerin PM, Toth M, Szöcs G, Schmid A, et al. 1985. (*Z,Z*)-6,9-Nonadecadien-3-one and (*Z,Z,Z*)-3,6,9-nonadecatriene: identification and synthesis of sex pheromone components of *Peribatodes rhomboidaria. Tetrahedron Lett.* 26:403–6
12. Charlton RE, Roelofs WL. 1991. Biosynthesis of a volatile, methyl-branched hydrocarbon sex pheromone from leucine by arctiid moths (*Holomelina* spp.). *Arch. Insect Biochem. Physiol.* 18:81–97
13. Clarke CA, Cronin A, Francke W, Philipp P, Pickett JA, et al. 1996. Mating attempts between the Scarlet Tiger Moth, *Callimorpha dominula* L., and the Cinnabar Moth, *Tyria jacobaeae* L. (Lepidoptera: Arctiidae), involve a common sex pheromone composition. *Experientia* 52:636–37
14. Connor WE, Eisner T, Vander Meer RK, Guerrero A, Ghiringelli D, Meinwald J. 1980. Sex attractant of an arctiid moth (*Utetheisa ornatrix*): a pulsed chemical signal. *Behav. Ecol. Sociobiol.* 7:55–63
15. Cossé AA, Cyjon R, Moore I, Wysoki M, Becker D. 1992. Sex pheromone components of the giant looper, *Boarmia selenaria* Schiff. (Lepidoptera: Geometridae): identification, synthesis, electrophysiological evaluation and behavioral activity. *J. Chem. Ecol.* 18:165–81
16. Daterman GE, Robbins RG, Eichlin TD, Pierce J. 1977. Forest lepidoptera attracted by known sex attractants of western spruce budworms, *Choristoneura* spp. (Lepidoptera: Tortricidae). *Can. Entomol.* 109:875–78
17. Déscoins C, Lalanne-Cassou B, Malosse C, Milat M-L. 1986. Analysis of the sex pheromone produced by the virgin females of *Mocis latipes* (Guenee), Noctuidae, Catocalinae, from Guadeloupe (French Antilla). *C. R. Acad. Sci III* 302:509–12
18. Déscoins C, Malosse C, Renou M, Lalanne-Cassou B, Le Duchat d'Aubigny J. 1990. Chemical analysis of the pheromone blends produced by males and females of the neotropical moth, *Mocis megas* (Guenee) (Lepidoptera, Noctuidae, Catocalinae). *Experientia* 46:536–39
19. Ebata T, Mori K. 1989. Synthesis of both the enantiomers of (*Z*)-*cis*-9,10-epoxy-6-heneicosene. *Agric. Biol. Chem.* 53:801–4
20. Francke W, Brunnemann U, Bergmann J, Plass E. 1998. Semiochemistry at junctions: volatile compounds from desert locusts, caddisflies, and geometrid moths. *Int. Symp. Insect Pheromones, 2nd, Wageningen, Netherlands, 30 March–3 April,* pp. 71–73 (Abstr.)
21. Frérot B, Malosse C, Desmer DE, Chenon R, Ducrot PH, Cain A-H. 1993. Identification of the sex pheromone components in *Pareuchaetes pseudoinsulata* Rego Barros (Lepidoptera: Arctiidae). *C. R. Acad. Sci. III* 317:1045–50
22. Frérot B, Pougny J-R, Milat M-L, Rollin P, Malosse C. 1988. Study of the pheromonal secretion of the Tiger Moth, *Cymbalophora pudica* (Esper) and enantioselectivity of its pheromonal perception. *C. R. Acad. Sci. III* 306:157–60
23. Frérot B, Renou M, Malosse C, Déscoins C. 1988. Isolation and identification of pheromone compounds in female moths of *Tyria jacobaeae* (Lepidoptera, Arctiidae). Biological characterization of the absolute configuration of the main component. *Entomol. Exp. Appl.* 46:281–89

23a. Gries G, Gries R, Schaefer PW, Gotoh T, Higashiura Y. 1999. Sex pheromone components of pink gypsy moth, *Lymantria mathura. Naturwissen* 86:235–38

24. Gries G, Holden D, Gries G, Wimalaratne PDC, Slessor KN, Saunders C. 1997. 3*Z*-*cis*-6,7-*cis*-9,10-Diepoxy-heneicosene: novel class of lepidopteran pheromone. *Naturwissen* 84:219–21
25. Gries G, Slessor KN, Gries R, Khaskin G, Wimalaratne PDC, et al. 1997.

(*Z*)6,(*E*)8-heneicosadien-11-one: synergistic sex pheromone component of Douglas-Fir tussock moth, *Orgyia pseudotsugata* (McDonnough) (Lepidoptera: Lymantridae). *J. Chem. Ecol.* 23:19–34
26. Hall DR, Beevor PS, Campion DG, Chamberlain DJ, Cork A, et al. 1993. Identification and synthesis of new pheromones. *Proc. Work. Group Meet. Int. Org. Biol. Control, Chatham, UK, 11-14 May,* pp. 1–9. Montfavet, Fr: IOBC Press
27. Hansson BS, Szöcs G, Schmidt F, Francke W, Löfstedt C, Tóth M. 1990. Electrophysiological and chemical analysis of the sex pheromone communication system of the mottled umber *Erannis defoliaria* (Lepidoptera: Geometridae). *J. Chem. Ecol.* 16:1887–97
28. Haynes KF, Yeargan KV, Millar JG, Chastain BB. 1996. Identification of the sex pheromone of *Tetanolita mynesalis* (Lepidoptera: Noctuidae), a prey species of a bolas spider, *Mastophora hutchinsoni*. *J. Chem. Ecol.* 22:75–89
29. Heath RR, Landolt PJ, Leppla NC, Dueben BD. 1988. Identification of a male-produced pheromone of *Anticarsia gemmatalis* (Hübner) (Lepidoptera: Noctuidae) attractive to conspecific males. *J. Chem. Ecol.* 14:1121–30
30. Heath RR, Tumlinson JH, Leppla NC, McLaughlin JR, Dueben B, et al. 1983. Identification of a sex pheromone produced by female velvetbean caterpillar moth. *J. Chem. Ecol.* 9:645–56
31. Hill AS, Kovalev BG, Nikolaeva LN, Roelofs WL. 1982. Sex pheromone of the fall webworm moth, *Hyphantria cunea*. *J. Chem. Ecol.* 8:383–96
32. Hill AS, Roelofs WL. 1981. Sex pheromone of the saltmarsh caterpillar moth, *Estigmene acrea*. *J. Chem. Ecol.* 7:655–68
33. Huang W, Pulaski SP, Meinwald J. 1983. Synthesis of highly unsaturated insect pheromones: (*Z,Z,Z*)-1,3,6,9-heneicosatetraene and (*Z,Z,Z*)-1,3,6,9-nonadecatetraene. *J. Org. Chem.* 48:2270–74
34. Jain SC, Dussourd DE, Conner WE, Eisner T, Guerrero A, Meinwald J. 1983. Polyene pheromone components from an arctiid moth (*Utetheisa ornatrix*): characterization and synthesis. *J. Org. Chem.* 48:2266–70
35. Jurenka RA, Roelofs WL. 1993. Biosynthesis and endocrine regulation of fatty acid derived sex pheromones in moths. In *Insect Lipids: Chemistry, Biochemistry, and Biology*, ed. DW Stanley-Samuelson, DR Nelson, pp. 353–88. Lincoln, NB/London, UK: Univ. Nebr. Press. 489 pp.
36. Kasang G, Schneider D, Beroza M. 1979. Biosynthesis of the sex pheromone disparlure by olefin-epoxide conversion. *Naturwissenschaften* 61:130–31
37. Krasnoff SB, Roelofs WL. 1988. Sex pheromone released as an aerosol by the moth *Pyrrharctia isabella*. *Nature* 333:263–65
38. Landolt PJ, Heath RR. 1989. Lure composition, component ratio, and dose for trapping male *Mocis latipes* (Lepidoptera: Noctuidae) with synthetic sex pheromone. *J. Econ. Entomol.* 82:307–9
39. Landolt PJ, Heath RR, Leppla NC. 1986. (Z,Z,Z)-3,6,9-Eicosatriene and (Z,Z,Z)-3,6,9-heneicosatriene as sex pheromone components of a grass looper, *Mocis disseverans* (Lepidoptera: Noctuidae). *Environ. Entomol.* 15:1272–74
40. Leonhardt BA, Mastro VC, Schwarz M, Tang JD, Charlton RE, et al. 1991. Identification of sex pheromone of browntail moth, *Euproctis chrysorrhoea* (L.) (Lepidoptera: Lymantridae). *J. Chem. Ecol.* 17:897–910
41. McDonough LM, Bailey JB, Hoffmann MP, Leonhardt BA, Brown DF, et al. 1986. *Sabulodes caberata* Guénée (Lepidoptera: Geometridae) components of its sex pheromone gland. *J. Chem. Ecol.* 12:2107–16
42. Millar JG, Giblin M, Barton D, Morrison A, Underhill EW. 1990. Synthesis and field testing of enantiomers of 6Z,9Z-*cis*-

3,4-epoxydienes as sex attractants for geometrid moths. Interactions of enantiomers and regioisomers. *J. Chem. Ecol.* 16:2317–39

43. Millar JG, Giblin M, Barton D, Reynard DA, Neill GB, Underhill EW. 1990. Identification and field testing of female-produced sex pheromone components of the spring cankerworm, *Paleacrita vernata* Peck (Lepidoptera: Geometridae). *J. Chem. Ecol.* 16:3393–99
44. Millar JG, Giblin M, Barton D, Underhill EW. 1990. (3Z,6Z,9Z)-Nonadecatriene and enantiomers of (3Z,9Z)-*cis*-6,7-epoxy-nonadecadiene as sex attractants for two geometrid and one noctuid moth species. *J. Chem. Ecol.* 16:2153–66
45. Millar JG, Giblin M, Barton D, Underhill EW. 1990. 3Z,6Z,9Z-Trienes and unsaturated epoxides as sex attractants for geometrid moths. *J. Chem. Ecol.* 16:2307–16
46. Millar JG, Giblin M, Barton D, Underhill EW. 1991. Synthesis and field screening of chiral monounsaturated epoxides as lepidopteran sex attractants and sex pheromone components. *J. Chem. Ecol.* 17:911–29
47. Millar JG, Giblin M, Barton D, Underhill EW. 1991. Chiral lepidopteran sex attractants: blends of optically active C_{20} and C_{21} diene epoxides as sex attractants for geometrid and noctuid moths (Lepidoptera). *Environ. Entomol.* 20:450–57
48. Millar JG, Giblin M, Barton D, Underhill EW. 1992. Sex pheromone components of the geometrid moths *Lobophora nivigerata* and *Epirrhoe sperryi*. *J. Chem. Ecol.* 18:1057–68
49. Millar JG, Underhill EW. 1986. Short synthesis of 1,3Z,6Z,9Z-tetraene hydrocarbons. Lepidopteran sex attractants. *Can. J. Chem.* 64:2427–30
50. Millar JG, Underhill EW. 1986. Synthesis of chiral bis-homoallylic epoxides. A new class of lepidopteran sex attractants. *J. Org. Chem.* 51:4726–28
51. Millar JG, Underhill EW, Giblin M, Barton D. 1987. Sex pheromone components of three species of *Semiothisa* (Geometridae), (Z,Z,Z)-3,6,9-heptadecatriene and two monoepoxydiene analogs. *J. Chem. Ecol.* 13:1271–83
52. Minet J. 1991. Tentative reconstruction of the ditrysian phylogeny (Lepidoptera: Glossata). *Entomol. Scand.* 22:69–95
53. Mori K, Brevet J-L. 1991. Pheromone synthesis; CXXXIII. Synthesis of both the enantiomers of (3Z,6Z)-*cis*-6,7-epoxy-3,9-nonadecadiene, a pheromone component of *Erannis defoliaria*. *Synthesis,* pp. 1125–29
54. Mori K, Ebata T. 1981. Synthesis of optically active pheromones with an epoxy ring, (+)–disparlure and the saltmarsh caterpillar moth pheromone [(Z,Z)-3,6-*cis*-9,10-epoxyheneicosadiene]. *Tetrahedron Lett.* 22:4281–82
55. Mori K, Ebata T. 1986. Synthesis of optically active pheromones with an epoxy ring, (+)-disparlure and both enantiomers of (3Z,6Z)-*cis*-9,10-epoxy-heneicosadiene. *Tetrahedron* 42:3471–78
56. Mori K, Takeuchi T. 1989. Synthesis of the enantiomers of (3Z,6Z)-*cis*-9, 10-epoxy-heneicosatriene and (3Z,6Z)-*cis*-9,10-epoxy-eicosatriene, the new pheromone components of *Hyphantrea cunea*. *Liebigs Ann. Chem.* pp. 453–57
57. Ohmasa Y, Wakamura S, Kozai S, Sugie H, Horiike M, et al. 1991. Sex pheromone of the fruit-piercing moth, *Oraesia excavata* (Butler) (Lepidoptera: Noctuidae): isolation and identification. *Appl. Entomol. Zool.* 26:55–62

57a. Oliver JE, Dickens JC, Zlotina M, Mastro VC, Yurchenko GI. 1999. Sex attractant of the rosy Russian gypsy moth (*Lymantria mathura* Moore). *Z. NaturForsch.* 54C:387–94

58. Persoons CJ, Vos JD, Yadav JS, Prasad AR, Sighomonoy S, et al. 1993. Indo-Dutch cooperation on pheromones of Indian agricultural pest insects: sex pheromone components of *Diacrisia obliqua* (Arctiidae), *Achaea janata* (Noctuidae)

and *Amsacta albistriga* (Arctiidae). *Proc. Work. Group Meet. Int. Org. Biol. Control, Chatham, UK, 11–14 May*, pp. 136–140. Montfavet, Fr: IOBC Press

59. Pougny JR, Rollin P. 1987. Synthesis from *D*-xylose of the salt marsh caterpillar moth pheromone (3*Z*,6*Z*-9*S*,10*R*)-epoxyheneicosadiene and its (3*Z*,6*E*)-stereoisomer. *Tetrahedron Lett.* 28:2977–78

59a. Pu G, Yamamoto M. Takeuchi Y, Yamazawa H, Ando T. 1999. Resolution of epoxydienes by reversed-phase chiral HPLC and its application to stereochemistry assignment of mulberry looper sex pheromone. *J. Chem. Ecol.* 25:1151–62

60. Qin X-R, Ando T, Yamamoto M, Yamashita M, Kusano K, Abe H. 1997. Resolution of pheromonal epoxydienes by chiral HPLC, stereochemistry of separated enantiomers, and their field evaluation. *J. Chem. Ecol.* 23:1403–17

61. Renou M, Lalanne-Cassou B, Dore J-C, Milat M-L. 1988. Electroantennographic analysis of sex pheromone specificity in neotropical Catocalinae (Lepidoptera: Noctuidae): a multivariate approach. *J. Insect Physiol.* 34:481–88

62. Roelofs W, Bjostad L. 1984. Biosynthesis of lepidopteran pheromones. *Bioorg. Chem.* 12:279–98

63. Roelofs WL, Hill AS, Linn CE, Meinwald J, Jain SC, et al. 1982. Sex pheromone of the winter moth, a geometrid with unusually low temperature precopulatory responses. *Science* 217:657–59

64. Rollin P, Pougny J-R. 1986. Synthesis of (6*Z*)-*cis*-9*S*,10*R*-epoxyheneicosene, a component of the ruby tiger moth pheromone. *Tetrahedron* 42:3479–90

65. Rule GS, Roelofs WL. 1989. Biosynthesis of sex pheromone components from linoleic acid in Arctiid moths. *Arch. Insect Biochem. Physiol.* 12:89–97

66. Scroble MJ. 1992. *The Lepidoptera.* Oxford: Oxford Univ. Press. 404 pp.

67. Stanley-Samuelson DW, Jurenka RA, Cripps C, Blomquist GJ, de Renobales M. 1988. Fatty acids in insects: composition, metabolism, and biological significance. *Arch. Insect Biochem. Physiol.* 9:1–33

68. Szöcs G, Francke W, Tóth M. 1998. Winter "love story" ... of geometrids. *Work. Group Meet. Int. Org. Biol. Control, Dachau, Ger.*, p. 15. Montfavet, Fr: IOBC Press

69. Szöcs G, Plass E, Francke S, Francke W, Zhu J, et al. 1998. Homologous polyenes, or chiral epoxides: How are pheromones composed in winter geometrids (Lepidoptera)? *Int. Symp. Insect Pheromones, 2nd, Wageningen, Netherlands, 30 March–3 April*, pp. 131–32 (Abstr.)

70. Szöcs G, Tóth M, Bestmann HJ, Vostrowsky O. 1984. A two-component sex attractant for males of the geometrid moth *Alsophila quadripunctata. Entomol. Exp. Appl.* 36:287–91

71. Szöcs G, Tóth M, Francke W, Schmidt F, Philipp P, et al. 1993. Species discrimination in five species of winter-flying geometrids (Lepidoptera) based on chirality of semiochemicals and flight season. *J. Chem. Ecol.* 19:2721–35

72. Tan Z-H, Gries R, Gries G, Lin G-Q, Pu G-Q, et al. 1996. Sex pheromone components of mulberry looper, *Heremophila atrilineata* Butler (Lepidoptera: Geometridae). *J. Chem. Ecol.* 22:2263–71

73. Tóth M, Buser HR, Guerin PM, Arn H, Schmidt F, et al. 1992. *Abraxas grossulariata* L. (Lepidoptera: Geometridae): identification of (3Z,6Z,9Z)-3,6,9-heptadecatriene and (6Z,9Z)-6,9-cis-3,4-epoxyheptadecadiene in the female sex pheromone. *J. Chem. Ecol.* 18:13–25

74. Tóth M, Buser HR, Pena A, Arn H, Mori K, et al. 1989. Identification of (3Z,6Z)-1,3,6-9,10-epoxyheneicosatriene and (3Z,6Z)-1,3,6-9,10-epoxyeicosatriene in the sex pheromone of *Hyphantria cunea. Tetrahedron Lett.* 30:3405–8

75. Tóth M, Szöcs G, Francke W, Guerin PM, Arn H, Schmid A. 1987. Field activ-

ity of sex pheromone components of *Peribatodes rhomboidaria*. *Entomol. Exp. Appl.* 44:199–204

76. Tóth M, Szöcs G, Francke W, Schmidt F, Philipp P, et al. 1994. Pheromonal production of and response to optically active epoxydienes in some geometrid moths (Lepidoptera: Geometridae). *Z. Naturforsch.* Teil C49:516–21
77. Tóth M, Szöcs G, Lofstedt C, Hansson BS, Schmidt F, Francke W. 1991. Epoxyheptadecadienes identified as sex pheromone components of *Tephrina arenacearia* Hbn. (Lepidoptera: Geometridae). *Z. Naturforsch.* Teil C46:257–63
78. Underhill EW, Millar JG, Ring RA, Wong JW, Barton D, Giblin M. 1987. Use of a sex attractant and an inhibitor for monitoring winter moth and Bruce spanworm populations. *J. Chem. Ecol.* 13:1319–30
79. Underhill EW, Palaniswamy P, Abrams SR, Bailey BK, Steck WF, Chisholm MD. 1983. Triunsaturated hydrocarbons, sex pheromone components of *Caenurgina erechtea*. *J. Chem. Ecol.* 9:1413–23
80. Wong JW, Palaniswamy P, Underhill EW, Steck WF, Chisholm MD. 1984. Novel sex pheromone components from the fall cankerworm moth, *Alsophila pometaria*. *J. Chem. Ecol.* 10:463–73
81. Wong JW, Palaniswamy P, Underhill EW, Steck WF, Chisholm MD. 1984. Sex pheromone components of fall cankerworm moth, *Alsophila pometaria*. Synthesis and field trapping. *J. Chem. Ecol.* 10:1579–96
82. Wong JW, Underhill EW, MacKenzie SL, Chisholm MD. 1985. Sex attractants for geometrid and noctuid moths. Field trapping and electroantennographic responses to triene hydrocarbons and monoepoxydiene derivatives. *J. Chem. Ecol.* 11:727–56
83. Yadav JS, Valli MY, Prasad AR. 1998. Total synthesis of enantiomers of (3*Z*, 6*Z*)-*cis*-9,10-epoxy-1,3,2-heneicosatriene—the pheromonal component of *Diacrisia obliqua*. *Tetrahedron* 54:7551–62

Annu. Rev. Entomol. 2000. 48:605–630

Insect Parapheromones in Olfaction Research and Semiochemical-Based Pest Control Strategies

Michel Renou[1] and Angel Guerrero[2]

[1]*INRA, Unité de Phytopharmacie et Médiateurs Chimiques, Route de St. Cyr, 78026, Versailles Cédex, France; e-mail: renou@versailles.inra.fr*

[2]*Department of Biological Organic Chemistry, CID (CSIC), Jordi Girona, 18-29, 08034-Barcelona, Spain; e-mail: agpqob@cid.csic.es*

Key Words pheromone, pheromone analogue, isosteric replacement, synergism, inhibition

■ **Abstract** The possibility of disrupting the chemical communication of insect pests has initiated the development of new semiochemicals, parapheromones, which are anthropogenic compounds structurally related to natural pheromone components. Modification at the chain and/or at the polar group, isosteric replacements, halogenation or introduction of labeled atoms have been the most common modifications of the pheromone structure. Parapheromones have shown a large variety of effects, and accordingly have been called agonists, pheromone mimics, synergists and hyper-agonists, or else pheromone antagonists, antipheromones and inhibitors. Pheromone analogues have been used in quantitative structure-activity relationship studies of insect olfaction, and from a practical point of view they can replace pheromones when these are costly to prepare or unstable under field conditions.

INTRODUCTION

After the structural identification of the first sex pheromone of a moth (18), it was suggested that compounds structurally close to the pheromone could act as efficient communication disruptants (146). Soon thereafter, the first report on pheromone analogues disclosed that attraction to pheromone sources was inhibited by compounds present in pheromone gland extracts (10). In this regard, Roelofs & Comeau (122, 123) revealed that the attractant signal of *Argyrotaenia velutinana* can be modulated in the field by unnatural compounds emitted simultaneously with the natural attractant, resulting in an increase (synergism) or decrease (inhibition) of the number of males trapped. New synthetic analogues were prepared by Mitchell et al (91), who showed that (*Z*)-9-tetradecenyl formate, a chemical of nonbiological origin but structurally similar to (*Z*)-11-hexadecenal (in short Z11-16:Ald), the major component of the sex pheromone of *Heliothis*

0066-4170/00/0107-0605/$14.00

virescens, disrupted the communication between sexes in *Heliothis* spp. The possibility of disrupting the chemical communication system of pest species, opening therefore the prospect of a new ecologically friendly approach to pest control, induced scientists to pursue the development of new semiochemicals with potential use in pest management programs (121). In addition, pheromone analogues are invaluable tools for studying the molecular mechanisms involved in olfaction (105).

Such compounds have been called according to their effects pheromone mimics, agonists, synergists, or hyperagonists on the one hand, or pheromone antagonists, antipheromones, or inhibitors, on the other hand. The more general term, parapheromone, has been widely used without clear definition, for compounds of very different origins. In Diptera, male attractants of plant origin (61, 128) as well as potent male lures for monitoring or male annihilation (34) have been called parapheromones despite that their chemical structure is very different from that of the natural pheromone (32). Cockroach pheromone mimics of natural origin have been found after the fortuitous observation that some plant terpenoids stimulate sexual behavior in males of *P. americana* (15, 94). In Lepidoptera, many attraction-inhibiting compounds have been found by screening pheromone analogues, but such compounds may be part of the pheromone blends of other moth species, and therefore they could be regarded as kairomones. In this review we restrict the term *parapheromones* to the chemical compounds of anthropogenic origin, not known to exist in nature but structurally related to some natural pheromone components that in some way affect physiologically or behaviorally the insect pheromone communication system.

STRUCTURAL DIVERSITY OF PARAPHEROMONES

Two types of rationale have been considered in the development of parapheromones. First, synthetic compounds were screened to design a good attractant, mimicking the action of the natural pheromone. Second, new chemicals were developed through rational hypotheses according to their mode of action, either to establish structure-activity relationships or to prepare specific antagonists that might impede the olfactory communication. The great variety of parapheromone compounds that result from these researches has been classified herein according to the type and site modification of the original pheromone molecule.

Modification of the Alkyl Chain

Chain-shortened, chain-elongated and dienic analogues of (*Z*)-5-decenyl acetate (Z5-10:Ac), a pheromone component of the turnip moth, *Agrotis segetum*, have been prepared to study the effective dimensions of the acceptor site (6, 79, 80). Saturation of the double bond in the structure of (*Z*)-13-hexadecen-11-ynyl acetate (Z13yne11-16:Ac), the major component of the sex pheromone of the processionary moth, *Thaumetopoea pityocampa,* resulted in an acetylenic analogue with potent intrinsic attractivity in the field (23) and the wind tunnel (116). Other compounds derived from shortening the terminal hydrophobic part of the parent pheromone molecule resulted in a decrease or loss of activity in electroantennogram (EAG) (23). Alkyl-branched analogues required between 300 to 1000 times the amount of the corresponding unbranched analogue of the same length and double bond position to elicit equivalent EAG amplitudes in noctuid moths (112). Alternatively, removal of a methyl group in the geminal dimethyl function of the sex pheromone of *P. americana* resulted in loss of behavioral activity (87). Methyl-substituted analogues have been prepared as probes for the chirality of the receptor (11, 65). Other synthetic olefinic analogues, which have been found to be potent disruptants of male attraction to virgin females or to synthetic lures, include (*Z,Z*)-1,12,14-heptadecatriene (35), (*Z*)-1,12-heptadecadiene (30) and 7-vinyldecyl acetate (17). However, 1,12-pentadecadiene, analogue of (*E*)- and (*Z*)-tetradecenal, two pheromone components of *Choristoneura fumiferana,* displayed ambiguous results in the field (127). Cyclopropanated analogues of Z11-16:Ald showed reduced EAG activity and significantly inhibited alcohol and aldehyde oxidizing enzymes in *Heliothis virescens* male antennae and female pheromone glands (42). Cyclopropene analogues of the sex pheromones of *Musca domestica, Plutella xylostella,* and *Ephestia elutella* have been shown to cause long-term inhibition of behavior at physiological doses (2).

Modification of the Polar Group

As discussed earlier, formate analogues have been designed to substitute for aldehydes in *Heliothis virescens* (92). A number of formates, propionates, butyrates, ethers, and acids have also been prepared as substitutes for nonaldehyde pheromone components, but although the position and stereochemistry of the double bond were preserved in all compounds, the EAG activity was always lower than the natural attractant by one to several orders of magnitude (112, 113). The formate and propionate analogues of (*Z*)-11-tetradecenyl acetate (Z11-14:Ac), the main pheromone component of *Argyrotaenia velutinana* inhibited attraction when mixed with the parent compound in 1:1 ratio (123). In contrast, formate and propionate analogues of the sex pheromone of *Ostrinia nubilalis* elicited from 47% to 88% male upwind flights with regard to the natural attractant in the wind tunnel (126). Formate and propionate analogues of the pheromone of the processionary moth showed dissimilar activities in the field. The formate was intrinsically inactive by itself, was a synergist when mixed with the natural attractant in 1:10 ratio, practically ineffective in a 1:1 blend, and was an antagonist at higher proportions (23). In contrast, the propionate derivative was an agonist of the natural pheromone, the level of catches being approximately 40% of the catches with the pheromone, and a synergist in 1:10 mixture with the parent compound (23).

Liljefors et al (79) replaced the acetate group of Z5-10:Ac by other functional groups, such as formate, propionate, trifluoroacetate, ether and methyl ketone, and measured their activity on the olfactory receptor neurons of *Agrotis segetum*. The effectiveness of the compounds was 25- to 1000-fold lower than the parent natural compound. Substitution of an aldehyde group for the acetate moiety resulted in a potent antagonist of the upwind flight response of the processionary moth males when (Z)-13-hexadecen-11-ynyl acetate was admixed with the analogue in 99:1 ratio (116). In the field, the analogue significantly reduced the number of male catches by 90% when mixed with the parent compound in 1:10 ratio (116). Other aldehyde analogues were effective upwind flight inhibitors in *Ostrinia nubilalis* (126).

With respect to nonlepidopterous groups, analogues of the sex pheromone of *P. americana* were prepared by replacing the acetate group with other ester groups (86, 87). The activity was dependent on the length of the substituent; The most active analogues were those containing an ester group with enriched electron density on the oxygen atom. Thus, the methyl carbonate analogue was the most potent mimic of the *P. americana* sex pheromone ever reported. The minimum energy conformations of some periplanone structural analogues have been calculated in order to construct an atomic model capable of predicting their activities (19). The electrophysiological activity of acyl fluoride analogues of the aldehyde components of *Anthonomus grandis* was two orders of magnitude lower than that of the natural aldehydes (39).

Replacing oxygen atom(s) of the functional group with sulfur can severely decrease the activity of these analogues. Thus, males of the processionary moth (26) and the turnip moth (57) showed markedly reduced EAG or receptor neurons responses to Z13yne11-16:Ac or to Z5-10:Ac after preexposure to vapors of the sulfur analogues of their sex pheromones. The effect was, however, different according to whether the sulfur atom was substituted for the ether oxygen, the carbonyl oxygen, or both. These differences suggest that the receptor sites are differently affected by the same type of atom substitution at the polar function of the pheromone molecule (57). In the field, the thioester containing the C = S bond and the dithioester derivative behaved as very good antagonists of attraction in blends with the natural attractant (26).

Other analogues were created by replacing the acetate function with a ketone (1) or a carbamate (1, 60). Both types of substitution led to marked antagonism of response to the natural pheromone, particularly the carbamate substitution whose effect was irreversible. Since carbamates inhibit acetylcholine esterase, it is likely that an additional inhibition of the antennal esterase or aldehyde dehydrogenase by the analogue occurs in this case. In the gypsy moth, the epoxide group of the pheromone main component (disparlure) has been replaced by the cyclopropyl, difluorocyclopropyl, dichlorocyclopropyl, and aziridinyl moieties to obtain analogues with reduced EAG activity in comparison with disparlure (38).

Halogenated Esters

Halogenated esters, particularly fluorinated derivatives, have been a subject of common interest to many research groups owing to the special features of the halogen. Fluorine closely mimics the steric requirement of hydrogen at enzyme receptor sites, but its strong electronegativity significantly alters the reactivity of neighboring centers. In addition, fluorine substitution increases lipid solubility, enhancing the rates of absorption and transport of the fluorinated derivatives.

Albans et al (1) prepared the first halogenated analogues and found that the responses of *Heliothis virescens* males to their natural pheromone were inhibited after exposure to the trifluoroacetates and trichloroacetates. In addition, the trifluoroacetate analogue effectively decreased number of matings for a few days after foliar application. Among the complete series of mono-, di- and trihaloacetate analogues of (*Z*)-11-hexadecenyl acetate (Z11-16:Ac), a component of the sex pheromone blend of *Plutella xylostella,* only the fluorinated derivatives were slightly active in EAG and also displayed competitive inhibition of the antennal esterase (109). The authors proposed that the steric size of the halogenated polar group is of primary importance. Similar results were obtained in the processionary moth by Camps et al (27) with the haloacetate analogues of Z13yne11-16:Ac. Moreover, in the field, addition of the fluoroacetates to the natural attractant remarkably decreased the capture of males. Chlorinated derivatives displayed poorer inhibitory activity, confirming that a strict steric requirement is necessary to achieve an efficient competitive interaction with the antennal receptors. Fluorinated analogues of (*E,E*)-8,10-dodecadienyl acetate, the sex attractant of *Cydia medicaginis,* retained up to 73% of its EAG activity (134). In the field, the monofluorinated derivative was a good mimic of the natural pheromone: The number of males captured in traps containing the analogue was about 55% the number captured in traps containing the parent molecule.

Fluoroacetate analogues of Z11-16:Ac, the major component of the sex pheromone of *Sesamia nonagrioides*, were also EAG-active and in the field the monofluoro-, monobromo, and trifluoroacetate derivatives significantly disrupted the pheromone action (118). The monofluoroacetate analogue of (*E*)-11-tetradecenyl acetate (E11-14:Ac), the main component of the sex pheromone of the E-type *Ostrinia nubilalis,* evoked upwind flight responses in the wind tunnel, while the corresponding monofluoroacetate of (*Z*)-11-tetradecenyl acetate was only slightly attractive to males of the Z-type (126). In the wind tunnel, only the trifluoroacetate analogue in E-type and the monofluoroacetate derivative in Z-type antagonized upwind flight of males when mixed with the natural attractant of each strain in 30:1 ratio. In the same manner, trifluoroacetate analogues of Z11-16:Ac, the main component of the sex pheromone of *Mamestra brassicae*, evoked lower EAG responses than the parent pheromone, and also strongly inhibited the attraction of males in the field (93). With regard to nonlepidopterous parapheromone components, a trifluoroacetate analogue of (−)-(5*R*,6*S*)-6-acetoxy-5-hexadecanolide,

Culex pipiens oviposition pheromone, showed similar activity to that of the natural pheromone (16).

Isosteric Replacement in the Chain

A single fluorine atom has been introduced into a double bond of main components of the sex pheromones of *Diparopsis castanea, Cydia pomonella, Bombyx mori, Spodoptera littoralis* and *Thaumetopoea pityocampa* (22, 24, 25). In the two latter species, the derivatives showed EAG activities similar to that of the natural pheromone (25), but in *T. pityocampa* the analogue significantly inhibited male attraction when mixed with the natural attractant in ratios over 3:1 (24). Some fluorinated analogues of (E,E)-8,10-dodecadien-1-ol (codlemone), the major sex pheromone component of *C. pomonella,* with fluorine atom(s) on the double bond(s) or at the terminal methyl group have been synthesized (137, 138). Among them, only the difluoro derivative with the heteroatom located at positions 10,11 evoked EAG responses similar to those elicited by codlemone (84). Several fluorinated analogues of (Z)-9-dodecenyl acetate (Z9-12:Ac), the major pheromone component of *Eupoecilia ambiguella*, with the halogen atoms occupying various positions at the terminal hydrophobic chain of the molecule, have been prepared (133); only the 11,11-difluoro analogue was equipotent to the natural pheromone (8). The 12,12,12-trifluoromethyl analogue was inactive alone, and antagonized trap catches when mixed with Z9-12:Ac. On the other hand, the 11,11,12,12,12-pentafluoroethyl analogue was practically inactive as an attractant but it synergized the attractivity of the pheromone. Fluorination at the allylic position appeared to enhance or at least to preserve the pheromone activity. This was also found in (Z)-7,7-difluoro-8-dodecenyl acetate, analogue of (Z)-8-dodecenyl acetate, the main pheromone component of *Grapholita molesta*, which evoked an EAG response similar to the natural material (88). In the wind tunnel, no difference in male response was observed when the pheromone blend was compared with mixtures in which one of the components was replaced by its analogue (88). However, a similar allylic difluorinated analogue of Z13yne11-16:Ac was practically inactive in EAG tests, and in the field inhibited the pheromone attractivity to male processionary moths when mixed with the natural attractant in 1:1 ratio (47).

Other pheromone analogues containing multiple fluorine atoms in the hydrophobic terminus of the chain have been prepared and tested in *Heliothis zea* (110), *Trichoplusia ni* (81, 110), *Diatraea grandiosella* (110), and *Agrotis segetum* (145). These perfluoroalkyl analogues were more volatile than their parent compounds, but they required 100 to 10,000 times higher concentrations to elicit an equal response from pheromone receptor cells. The reduced electrophysiological activity of the analogues has been explained in terms of poor affinity of the fluorinated hydrocarbon for a lipophilic environment at the pheromone receptor (110). In the wind tunnel, however, activity of mixtures of the parent pheromone with the fluorinated materials did not significantly differ from that of the natural pheromone blend, while in the field the natural pheromone complex caught significantly more males than any mixture containing fluorinated analogues (145). In contrast, a highly fluorinated analogue of the major component of the oviposition pheromone of the mosquito *Culex pipiens* retained high biological activity both in the laboratory and in the field (36). Likewise, two fluorinated analogues of (*E*)-β-farnesene, a component of the alarm pheromone of the aphid, *Myzus persicae*, were highly active (16).

Other isosteric substitutions have been made in lepidopterous pheromone components, involving the replacement of a double bond by a thiomethylene group (26) and a methyl group by a halogen (63). Sulfur analogues of Z13yne11-16:Ac containing the heteroatom in α position to the triple bond reduced EAG responses in male processionary moths previously exposed to vapors of the chemical, while the derivative containing the sulfur at β position was practically inactive in this context (26). In the field, both analogues moderately antagonized male attraction when mixed with the natural attractant (26). Introduction of an O or S atom into the chain of Z11-14:Ac produced compounds having significant synergistic activity to *Argyrotaenia velutinana* (123). Analogues resulting from replacing the terminal methyl group of Z5-10:Ac with halogen (Cl, Br, I) were prepared to study interaction of the receptor with the terminal hydrophobic part of the pheromone molecule in *Agrotis segetum*. These elicited a significantly reduced neuronal response in comparison with the pheromone (63). A similar type of substitution was made for codlemone (136) in which a chlorinated analogue was 10 to 100 times less active than the parent molecule in EAG and single sensillum recordings (SSR) (84).

Alterations in activity have also been observed after similar substitutions in nonlepidopterous pheromone components. Replacing a methyl group with chlorine in the main trail pheromone component of the ant *Atta texana* resulted in an analogue with similar activity to that of the natural pheromone (129). ω-Fluorinated analogues of (3*Z*,6*Z*,8*Z*)-dodeca-3,6,8-trien-1-ol, a trail pheromone for *Reticulitermes* termites, induced trail following by workers as did their nonfluorinated counterparts (31). The corresponding analogue of E11-14:Ac, pheromone component of *Choristoneura occidentalis,* showed a long-range attraction of males, being potentially useful as a mating disruptant (90). One of the monofluoro analogues of eugenol methyl ether, a potent and specific attractant for the Oriental fruit fly *Bactrocera dorsalis,* showed similar attractiveness as the parent molecule in field tests (71). Introduction of a fluorine atom in α position to the alcohol group in 4-methyl-3-heptanol, one pheromone component in *Scolytus multistriatus*, retained 50% to 75% of the activity of the parent compound in a laboratory assay (102). Other analogues involving a deeper modification of the natural material were much less active or completely inactive. The EAG responses correlated well with the behavioral activity of the fluorinated analogue, but in the field, mixtures with the other pheromone components showed very little activity.

Because of the interest in bioactive organosilicon compounds (135), isosteric replacement of carbon by silicon has also been considered, and therefore, dimethylsila analogues of Z13yne11-16:Ac have been synthesized and tested on the processionary moth (4). The compound with the silicon atom in the propargylic position was very slightly active in EAG and displayed no effect on behavior when mixed with the natural material. It is likely that the required replacement of a CH_2 for the $Si(CH_3)_2$ group, to avoid the high reactivity of the isosteric SiH_2, impedes a good interaction with the receptor sites because of the bulky gem-dimethyl group.

Isotopically Labeled Pheromones

Isotopically labeled pheromones have been used in conjunction with other analogues to study pheromone detection and catabolism (105). Although different labels could in principle be used (^{125}I, ^{14}C, . . .), tritium derivatives are particularly useful because of their highly specific activity and easy monitoring. Thus, tritiated analogues of a variety of pheromones (31, 41, 49, 70, 106, 108, 142) have been prepared as probes for the characterization of catabolic proteins. Isotopic effects on biological activity have very rarely been investigated. Four-deuterated analogues showed reduced EAG response, but no significant difference in activity was observed between the natural pheromone and the analogues containing two or three deuterium atoms (12). On the other hand, the tritiated Z11-14:Ac elicited behavioral responses from *Z*-type *Ostrinia nubilalis* males as effectively as the natural pheromone (75). By contrast, the tritiated *E* isomer was significantly less effective on the *E*-type males than its protonated counterpart.

Photoactivatable analogues have been prepared to produce reactive species, such as carbenes, capable of covalently attaching to a nucleophilic residue of the acceptor site. Diazoacetate derivatives have been frequently used, but other groups, such as diazirines, diazoketones, and azides, have also been considered (105). The diazoacetate analogue of (*E,Z*)-6-11-hexadienyl acetate (E6,Z11-16:Ac), the main sex pheromone component of *Antheraea polyphemus*, was used in competitive inhibition studies of degradation of the tritiated E6,Z11-16:Ac by the sensillum esterase (111) as well as in pheromone-binding protein (PBP) mapping studies (43). The unlabeled form was an excellent stimulant for pheromone receptor neurons in *Antheraea* spp. (50, 67). A similar derivative of the *Plutella xylostella* pheromone showed, however, a weak EAG activity in comparison with the parent molecule (109). PBPs from sensory hairs and antennal branches of the processionary moth were photoaffinity labeled with a tritiated diazoacetate analogue of Z13yne11-16:Ac (48).

Miscellaneous

Because of their ability to inhibit pheromone catabolism, trifluoromethyl ketones might be considered in a rational new approach to insect control (105). A variety of these chemicals have been prepared and tested as inhibitors of antennal esterases of *Spodoptera littoralis* (44, 98, 124, 140), and *Plutella xylostella* (109) as well as behavior inhibitors (74, 97, 98, 116, 118).

Insecticide-related pheromone analogues have been examined as disruptant of pheromone-mediated behavior. In this regard, a series of dialkyl phosphorofluoridates and alkyl methylphosphonofluoridates were tested in *Grapholita molesta,* and only those chemicals containing the alkyl substituent of Z8-12:Ac disrupted communication as effectively as the natural attractant. The activity was attributed to the inactivation of the pheromone esterase by the chemicals (85). In the same context, male flight behavior of *Trichoplusia ni* was adversely affected by long exposure of the males to vapors of (*Z*)-7-dodecenyl diethyl phosphate or pyrethroid carboxylate, but none of the chemicals tested was as disruptive as the moth's own pheromone, (*Z*)-7-dodecenyl acetate (119).

PHYSIOLOGICAL AND BEHAVIORAL EFFECTS OF PARAPHEROMONES

Parapheromones as Mimics, Synergists, or Inhibitors of Natural Pheromones

The intrinsic activity of parapheromones, i. e. their direct effects on behavior or sensory system when tested alone, is highly variable. Many pheromone analogues do not show significant behavioral activity at physiological concentration (81), although some may exhibit such activity at high doses. As discussed in the previous section, the best pheromone mimics have been obtained by isosteric replacements on the parent structures. The receptor neurons are precisely tuned to pheromone components, and it now seems unlikely to find compounds more active at the receptor level than the pheromone itself. Thus, evidence of true hyperagonistic effects of parapheromones is scarce. Prestwich (104) reported that acyl fluoride analogues hyperstimulated pre-copulatory behavior in male *Heliothis virescens*. Likewise, the difluorinated analogue 11,11-difluoro-Z9-12:Ac was more effective than Z9-12:Ac, the natural pheromone of the grape berry moth, in eliciting male responses in the wind tunnel (8). In other cases, enhanced field activity as compared with the natural pheromone may result from a greater stability and/or volatility of the parapheromone (84).

The ability of parapheromones to modulate insect responses to their natural pheromone holds some promise for controlling insect behavior. An increase in attractivity as compared with the natural attractant blend has been obtained in blends of natural pheromones with brominated derivatives (123), formate and propionate analogues (23), and fluorinated pheromones (8). However, such examples of synergy appear to be exceptional.

However, a decrease in activity of an attractive blend following introduction of an analogue has been observed repeatedly. The doses of analogue needed to observe antagonism vary considerably, ranging from 0.1% for the alcohol analogue of the main pheromone component in *Mamestra brassicae* (131) up to 1000% for a double-bond configurational isomer in *Carpophilus freemani* (99).

This implies that different mechanisms are involved, and that the commonly used term "inhibition" may be misleading. Some compounds that dramatically decrease male responses to their specific pheromone are naturally emitted by other moth species as a component of their own pheromone blend. For instance, captures of male *M. brassicae* were reduced to nearly zero after addition of only 0.1% of the corresponding alcohol analogue of its main pheromone component (131). Olfactory receptor neurons specifically tuned to these interspecific inhibitors have been found in a number of species (83) and it has been postulated that they function to prevent mating mistakes by males, reinforcing reproductive isolation between sympatric species (114). On the other hand, significant decreases in insect catches have been sometimes observed in field tests when anthropogenic parapheromone components are added to the sex attractant. For instance, sulfur analogues of Z13yne11-16:Ac, the sex pheromone of the female processionary moth, reduce the number of males captured when mixed with pityolure at about 10% ratio (26). In this case, other mechanisms that are not yet fully understood may be involved.

Effects of preexposure of male receptors to parapheromone vapors before testing the male's responsiveness to pheromone have been frequently analyzed. The doses of parapheromone used have been generally very high compared with the low thresholds of response to the pheromone, and the exposure time has ranged from a few minutes (85) to several hours (27, 57, 118). Such preexposures have resulted in a reduction in behavioral or electrophysiological responses to the parent compound, although in one case, an increase in the EAG response was reported (27). For a majority of compounds, habituation and sensory adaptation are likely the main mechanisms involved (57, 69). Indeed, the level of response reduction obtained with the analogue is generally lower than that obtained with the natural pheromone itself and is well correlated with the intrinsic activity of the parapheromone. Likewise, the recovery time of olfactory neurons to parapheromone preexposure is sometimes faster than with the pheromone itself (69).

Parapheromone-Induced Modifications of Insect Behavior

Insect orientation to odor sources is a complex behavior, and a simple record of the number of individuals reaching the source is often insufficient to fully describe the effects of parapheromones. In wind tunnel studies, males are scored for key behaviors such as activation, takeoff, oriented flight, source approach, source landing, and copulation attempt that are typically exhibited in response to sex pheromones (8, 81, 116, 145). Modification of the pheromone structure diversely affects these behavioral steps. Thus, some compounds may retain the capacity to elicit the first steps of behavior—activation and takeoff—but evoke only low attraction to the source (126). However, it is not clear whether the decrease in final attraction is due to a weaker interaction with the pheromone receptor sites resulting in a lower sensory input, or to sensory mechanisms further downstream in the pathway. Kaissling et al (68) reported that (*Z,E*)-4,6-hexadecadiene, an unsaturated hydrocarbon analogue of (*Z,E*)-10,12-hexadecadiene-1-ol, the bombykol, produces intense and long-lasting impulse firing on bombykol receptor cells but little or no anemotactic behavior by *Bombyx mori* males. The authors postulated that this sustained neuronal activity represents a tonic stimulus, which is known to be insufficient to cause males to orient to the source. Thus, in this case, neuronal circuits downstream of the olfactory receptor neurons appear to be responsible for the inadequate behavioral response. Alteration of flight tracks have also been observed in *Coleophora laricella* when as little as 0.3 % of an inhibitor of natural origin was added to the pheromone (114). With regard to pre-upwind flight activity, male moths often respond to parapheromones with a longer latency than they do to the natural pheromone (8, 145). For instance, the latency of male *Trichoplusia ni* to take flight is about four times higher in the presence of blends containing the fluorinated analogue than in (*Z*)-7-dodecenyl acetate (81).

Contrasting with the number of studies focusing on attraction and orientation to a distant pheromone source, the effects of pheromone analogues on close-range behaviors have been poorly studied. In this context, the difluoro-1-norfarnesene analogue of the alarm pheromone of *Myzus persicae* caused a high level of dispersal in aphids (16). Similarly, a chloroformate and an alken-4-olide analogue of Z8-12:Ac were able to decrease mating in *Grapholita molesta* (59).

Activity on Olfactory Receptor Neurons

Electrophysiological tests have been extensively used to study detection of parapheromones by insect olfactory organs. Responses have been recorded extracellularly from the whole antenna using electroantennograms (EAG) or single sensillum recordings (SSR). The first systematic investigations of structure-activity relationships with insect pheromones were conducted by Priesner et al using electroantennography (112, 113). The test compounds included more than 130 pheromone analogues, naturally occurring or anthropogenic compounds, with different chain length, position, and configuration of double bonds, functional groups, and alkyl substituents in the chain. The natural pheromone was always found to be the most effective compound. Numerous subsequent investigations confirmed that structural changes of the pheromone molecules implied a reduction of activity. More precise quantitative EAG evaluation has resulted from dose-response curves (40), but it should be noted that a reliable comparison of activity of compounds with different chemical structures requires correction of the amplitude according to their relative volatility (60).

Differences in the temporal structure of the electrophysiological responses of the antennal receptor neurons to parapheromones have been poorly investigated, although they should provide invaluable information on their mode of action since static or temporal parameters of the EAG and receptor potentials have been shown to depend upon the stimulus compound (66). Thus, the ω-fluorinated analogue of E11-14:Ac elicited EAG responses with much shorter recovery time than the natural pheromone on male antennae of *Choristoneura occidentalis* (90). The response kinetics to parapheromones may differ drastically, as shown by post-stimulus firing triggered by a saturated parapheromone in *Bombyx mori* (68). In contrast, the shape of EAG responses to haloacetate analogues and (*Z,E*)-9,12-tetradecadien-1-yl acetate, the main pheromone component, were similar in *Ephestia kuehniella* (69), although the dose-response curves were shifted to the higher doses.

In most moth species studied to date, olfactory receptor neurons sensitive to pheromone components show a very high degree of selectivity for the compounds they are tuned for. Consequently, one might question whether effective pheromone mimics would stimulate the same or a different population of olfactory receptor neurons. In this regard, the capacity of formates to effectively substitute for aldehydes behaviorally and at receptor cell level was confirmed in several moth species (9, 52, 139) whose receptor cells showed very similar dose-response curves for the aldehyde and its formate analogue. The design of optically active parapheromones has revealed in several moth species the remarkable ability of the pheromone receptor cells to differentiate between enantiomers of parapheromones, implying that the receptor site contains elements of chirality although the natural pheromone is not chiral (11, 13, 33, 65, 73). In *Agrotis segetum* the SSR activity of the enantiomers depends on the location of the asymmetric center (65).

A large number of analogues of Z5-10:Ac have been prepared to investigate the interactions between this compound and its receptor through quantitative structure activity studies at the single receptor neuron level in *Agrotis segetum* (63, 79, 80). A model has been proposed and further quantified (95) based on the high steric complementarity between ligands and acceptor (78, 80), the conformational energy required to adopt the bioactive conformation (6, 55), and the calculated electrostatic potential at putative hydrogen-bonding sites (56). The proposed cisoid bioactive conformation for Z5-10:Ac involves three structural elements of decisive importance for activity: the terminal methyl group, the double bond, and the acetate group (56). The observed differences in single-cell activities of chain-elongated and dienic analogues of Z5-10:Ac were found to be correlated with the conformational energies required for the analogue molecule to mimic the proposed bioactive conformation of the natural pheromone component, a small conformational energy corresponding to a high electrophysiological activity and a conformational energy increase of 1.7 kcal/mol corresponding to a 10-fold decrease of the electrophysiological activity. Further studies on chain-shortened analogues indicated that the terminal alkyl chain interacts with a highly complementary hydrophobic "pocket" (7), and that introduction of bulky groups in the terminal alkyl part or between the double bond and the acetate group substantially reduced the firing response of olfactory receptor cells (62, 64).

Effects on Pheromone Catabolism

Insect olfactory tissues contain substrate-specific catabolic enzymes that convert pheromones into nonactive metabolites, thus preventing prolonged activation of the olfactory receptor neurons. In principle, inhibition of the pheromone catabolism could lead to communication disruption via enhanced sensory adaptation and in this context various parapheromones have been specifically designed. Specific catabolic enzymes have been evidenced for structurally different pheromones. Remarkable inhibition of aldehyde-oxidizing enzymes in antennal homogenates of *Heliothis virescens* has been displayed by cyclopropanated analogues (42) and by a vinyl ketone derivative of Z11-16:Ald (107). In this latter case the inhibition was irreversible. By contrast, α-fluorinated aldehydes were only modest inhibitors of alcohol and aldehyde-oxidases. In *Lymantria dispar*, the epoxide pheromone (+)-disparlure is converted at significantly different rates into the corresponding diol by an epoxide hydrolase present in the male antennae (51). Analogues substituted at the 6-position (6-hydroxy, 6-oxo and 6,6-difluorodisparlure) along with 9,9-difluorodisparlure were the most potent inhibitors, showing an IC_{50} <10 μM.

Acetate pheromones are cleared from the sensilla by hydrolysis to the corresponding alcohol by antennal esterases. The trifluoromethyl ketone 1,1,1-trifluorotetradecan-2-one is a potent inhibitor in vitro of the sensillar esterase of *Antheraea polyphemus* (IC_{50} 5 nM) (142). Hydrolysis of Z11-16:Ac, a minor component of the sex pheromone of *Plutella xylostella,* is also inhibited by haloacetate analogues, particularly by fluoroacetates (109). Catabolism of *Ostrinia nubilalis* pheromone, which involves hydrolysis of Z- or E11-14:Ac and oxidation of the alcohol to the carboxylic acid (75), is moderately inhibited in vivo by 1,1,1-trifluoro-14-heptadecen-2-one (74). Likewise, a series of trifluoromethyl ketones have been tested in vitro as inhibitors of the antennal esterase of *Spodoptera littoralis*, the most active compounds being 3-octylthio-1,1,1-trifluoropropan-2-one (OTFP, IC_{50} 0.55 μM) and 1,1,1-trifluorotetradecan-2-one (IC_{50} 1.16 μM) (44). The OTFP reversibly binds the enzyme, forming an adduct (probably a hemiacetal) of tetrahedral geometry with its active site (124). In vivo catabolism of the pheromone is also inhibited by topical application of OTFP on the male antennae (115).

Although the importance of catabolic activity in olfaction seems undisputed, the exact linkage between pheromone clearance and the transductory processes remains unclear. Data on the biological activity of parapheromones interfering with pheromone catabolism are contradictory, possibly owing to the multiple sites of action. Thus, while the pheromone analogue of *Ostrinia nubilalis* did not inhibit male behavior when it was coevaporated with the pheromone (74), the corresponding analogue of Z13yne11-16:Ac inhibited the response of male processionary moths to the natural pheromone component, particularly the number of source contacts (116). In the field, the trifluoromethyl propanone analogue of the sex pheromone of *Sesamia nonagrioides* proved to be a good synergist, as well as a notable disruptant of the attractant activity on sympatric species (118). By contrast, mixtures of the same type of analogue of the Z13yne11-16:Ac with the parent compound inhibited attraction of male processionary moths to the traps, while the trifluoromethyl ketone analogue elicited different effects depending on the pheromone-to-analogue ratio in the bait (97). In turn, good agreement was found between the capacity of a series of trifluoromethyl ketones to inhibit the

EAG responses to the pheromone and the male upwind flight in the wind tunnel (117). Most of the trifluoromethyl ketones that displayed good antiesterase activity in vitro significantly decreased the EAG amplitude and increased the EAG repolarization time after a pheromone stimulation (117). However, from electrophysiological experiments showing that trifluoromethyl ketones did not prolong the decline of the sensillar potential and even inhibited the firing of the receptor cells, it was concluded that esterase inhibition is not the sole or even the major effect of trifluoromethyl ketones on olfactory transduction (103).

Effect on Pheromone Production

Development of specific inhibitors of the enzymes responsible for pheromone biosynthesis in Lepidoptera is a potential tool in the search for new strategies in insect control. In this regard, pheromone analogues have also been used as inhibitors of the biosynthetic pathways, particularly of the two key steps, β-oxidation and desaturation. Cyclopropene fatty acids inhibited delta-9 and delta-11 desaturases, present in the pheromone glands of *Spodoptera littoralis* and *Thaumetopoea pityocampa* (45, 46). With regard to β-oxidation inhibitors, a number of fluorinated (14, 37), acetylenic, and cyclopropane fatty acids (28), structurally related to palmitic acid, have also been prepared and tested in vitro and in vivo (125). Inhibition of pheromone production in the above insects as well as in *Bombyx mori* has been accomplished by application of 2-halofatty acids in isolated glands (58). Metabolically blocked analogues of housefly pheromones have been used to investigate pheromone biosynthesis in Diptera (77).

The Toxicity of Parapheromones

Because of the introduction of reactive atoms or functions in the molecule, parapheromones might have toxic effects on target insects, interfering with their specific activity on olfaction. Parapheromone compounds possessing structural features of an insecticide and the long unsaturated chain of Z7-12:Ac showed little contact toxicity (119) to *Trichoplusia ni*. Thus, (*Z*)-7-dodecenyl diethylphosphate presented a LD_{50} 350 μg/moth, while the methyl carbamate and pyrethroid carboxylate derivatives showed a $LD_{50} > 500$ μg/moth, three orders of magnitude less toxic than permethrin (LD_{50} 0.6 μg/moth). On the contrary, some of the ω-fluorinated analogues of the trail pheromone of *Reticulitermes* termites (31) presented a relatively high toxicity, attributed to in vivo β-oxidation to fluoroacetate. In the same context, a ω-fluorinated analogue of E11-14:Ac was relatively toxic to *Choristoneura occidentalis*, the LD_{50} being 12.0 and 41.7 μg per male and female, respectively (90). Removal of the antennae reduced male mortality. However, fluoridate derivatives of the Oriental fruit moth pheromone showed low toxicity to the housefly, the LD_{50} being between 27 and 315.4 μg/g depending on the parapheromone, and no toxicity to mice at 100 mg/kg body weight (85). In the same context and although trifluomethyl ketones (TFMK) may inhibit hepatic carboxyl esterases of vertebrates, no acute toxicity was observed in mice after ingestion of up to 1000mg/kg of the TFMK analogue of the main pheromone component of *Sesamia nonagrioides* or of OTFP (G Grolleau, unpublished data).

THE PRACTICAL USES OF PARAPHEROMONES

Parapheromones have a high potentiality as alternative material in integrated pest management strategies, particularly when the natural pheromones are expensive to produce, present longevity problems, or are quickly degraded under field conditions. Three important criteria have to be taken into account to include them in synthetic formulations: The efficiency of the blend must be optimized, the specificity must be preserved, and the stability must be improved. A number of field experiments have demonstrated that parapheromones meet these prerequisites for their use in insect control or monitoring.

Monitoring and Mass Trapping

Attractants have to be sufficiently active to be used for monitoring and the release rate and composition adjusted so that the sampling rate of the traps is appropriate to the requirements of the crop (143). The chemical nature of the attractants does not necessarily need to be consistent with the stringent structural requirements of the natural pheromone. In many cases, substitution by less active analogues is not disadvantageous, and on the contrary, it may involve obvious advantages, as in *Cydia nigricana* wherein the natural pheromone, (*E,E*)-8,10-dodecadienyl acetate, attracts an impractical number of insects saturating the traps. The use of a pheromone mimic of natural origin, (*E*)-10-dodecenyl acetate, results in lower catches of males, but is more reliable for monitoring purposes (54). Moreover, the conjugated diene system of the pheromone is particularly unstable under field conditions; its activity is considerably diminished after 5 days, while lures containing *E*10-12:Ac, with or without antioxidants, are still attractive after 3-month exposure in the field. The acetylenic analogue of the sex pheromone of the processionary moth mimics the natural pheromone, attracting 65% of the number of males caught in the field with Z13yne11-16:Ac (23), which makes the chemical a good candidate for mass trapping experiments.

Parapheromones have sometimes contributed to increased specificity. Thus, the addition of 1,1,1-trifluoro-14-nonadecen-2-one to the synthetic sex attractant blend of *Sesamia nonagrioides* reduced the number of catches of *Mythimna unipuncta* and *Scotogramma trifolii* (118), increasing simultaneously the number of males of the target species. (*Z*)-2-methyl-7-octadecene, the olefinic analogue of disparlure, inhibits attraction of male gypsy moths to racemic (+)-disparlure and living-female baited traps (29, 96), and it is ineffective as mating disruptant (20). The olefin, however, synergizes the attraction of *Lymantria monacha* to (+)-disparlure-baited traps up to ninefold, depending on the ratio and dose of the components (53), and will be valuable for detecting and monitoring nun moth populations, particularly in Europe, where the insect coexists with the gypsy moth.

Structural modification of the pheromone generally reduces field activity, but a significant increase in catches has sometimes been reported. More *Cydia pomonella* males were caught in traps baited with (*E,E*)-10,11-difluoro-8,10-dodecadienol or (*E,E*)-11-chloro-8,10-undecadienol than in traps containing codlemone (84). Although the higher attractivity of haloacetate analogues has still not been rationalized, it might rely on purely physical factors, such as better stability or a modified volatility of the parapheromones.

Practical use of pheromones in pest management has often been hampered by their rapid oxidation in air or high volatility. Formates, more economical to synthesize and much more stable than aldehydes, have been successively substituted for aldehydes in the attractant blend of *Heliothis virescens* and *H. zea* (91, 92). Another strategy has been proposed by Pickett et al (82, 101) who introduced the concept of "propheromones" to designate photolabile adducts able to protect pheromone molecules under field conditions, such as acetals as precursors of the aldehyde group. Later, protection of the conjugated double bonds of the pheromones of *Cydia pomonella* and *C. nigricana* by complexation as iron diene carbonyls was reported (130).

Mating Disruption

Many disruption experiments have used analogues of the natural pheromones such as alcohols when the natural compounds are acetates, geometric isomers, monounsaturated analogues of the dienic parent pheromone, etc. Although the complete blend of the natural pheromone components is generally the best formulation for mating disruption, mixtures of blend components and attraction antagonists have been efficient for mating disruption in *Synanthedon pictipes* (100), *Eupoecilia ambiguella* (3, 141), and *Planotortrix octo* (132). Communication disruption has been studied also for *Diparopsis castanea* (89), *Spodoptera littoralis* (21) and *Chilo suppressalis* (5) by baiting traps with virgin females or synthetic pheromone surrounded with inhibitory compounds. Most of these compounds are indeed natural components of other moth species, and more complete reviews of such cases may be found elsewhere (120, 121).

On the other hand, the potential of parapheromone components sensu stricto for direct control has been more seldom tested. Disruption of male orientation in *Ostrinia nubilalis* was obtained with the acetylenic analogue 11-tetradecynyl acetate, at a rate of 66.4 g/ha (72), while the 2-fluoro analogue of the pheromone caused mating suppression in the laboratory at concentrations equivalent to that needed with the pheromone (73). Formate analogues were effective at a rate of 2 mg/hr/ha in disrupting male *Heliothis* orientation and mating (91). In large-scale field tests, pheromone communication was disrupted in the navel orangeworm *Amyelois transitella* by air permeation with (*Z,Z*)-9,11-tetradececadien-1-ol formate, the formate analogue of the main pheromone component (*Z,Z*)-11-13-hexadecadienal (76). A novel olefin analogue was later found to disrupt mating communication in the same insect (35), its effectiveness depending on the type of dispenser, release rate, and proximity to the female-baited traps. Both compounds are more stable than the natural aldehydic pheromone component. A similar type of pheromone analogue caused a significant disruption of male *Heliothis zea* behavior, comparable to the natural pheromone (30).

Registration and Marketing Aspects

Concern for the impact of pesticides on human health and the environment has resulted in increased regulatory action by the responsible authorities in almost all countries. Semiochemical registration is still under debate, each country having its own system (144). Generally, registration of pheromone is required only for mass trapping and mating disruption, but the principle of lighter registration has been advocated by pheromone experts. However, from their anthropogenic origin parapheromones will probably be subject to the general registration requirements designed for all protection agents used for direct control in agriculture. Thus, the general principle that the product should demonstrate effectiveness and will not present unacceptable hazards to users, consumers, and the environment will apply to parapheromones.

Like pheromones, parapheromones involve nontoxic modes of action and show relatively low toxicity, with the exception of some ω-fluorinated analogues, as discussed above. Moreover and from a practical point of view, the reduced risks of side effects on nontarget species owing to their species specificity, the low doses needed, and their high volatility minimize the risks of their utilization in the field. Thus, ecotoxicology should not be a major limitation to their practical use, although extensive research will be required before their registration and commercial utilization can take place. Markets for the new compounds are likely to be relatively small, but because of their character of anthropogenic compounds, parapheromones might be better protected by patents than pheromones. Finally, because of the time and cost required for registration, parapheromones will probably be developed for use on major crops only.

PROSPECTS AND CONCLUDING REMARKS

Parapheromones have been increasingly investigated in the last three decades with variable results. The variety of the effects observed after homologous modifications of pheromones of different insect species indicates that there is not a general design that could be applied to develop pheromone analogues for every species. Rather, parapheromones have proven that they may bring about specific solutions to practical problems caused by pheromones, such as stability, synthetic cost, etc. A number of experimental studies have demonstrated the capacity of parapheromones to antagonize the pheromone activity, but this result has been obtained without a precise knowledge of the mechanisms by which parapheromones interact with pheromone detection. A number of sites can be involved and more physiological work is needed to improve the efficiency of the parapheromones. Most electrophysiological studies focus on the responses to single compounds, and the importance of synergistic or inhibitory interactions between odorants at the receptor cell level is probably underestimated. New methodological approaches, like patch clamp, may setup the basis for a better knowledge of the parapheromones' mode of action. The elucidation of the structure of odorant-binding proteins and the still-to-come characterization of insect olfactory receptors should provide cues for a rational design of more active chemicals. Conversely, pheromone analogues are a model system for basic research in chemoreception. Besides other major effects, occasionally parapheromones may induce subtle modifications of behavior that deserve to be analyzed in more detail. A better knowledge of these subtle effects may also provide insights for the development of new bioactive molecules.

ACKNOWLEDGMENTS

We gratefully acknowledge CICYT (AGF95-0815 and AGF97-1217-CO2) and EC (FAIR CT966-1302) for financial support. A Picasso cooperative program enabled reciprocal visits between our laboratories. The authors are grateful to Profs. TC Baker and F Camps for helpful comments on the manuscript. We also thank Mrs. C Young for linguistic advice.

Visit the Annual Reviews home page at www.AnnualReviews.org.

LITERATURE CITED

1. Albans KR, Baker R, Jones OT, Jutsum AR, Turnbull MD. 1984. Inhibition of response of *Heliothis virescens* to its natural pheromone by antipheromones. *Crop Prot.* 3:501–6
2. Al Dulayymi JR, Baird MS, Simpson MJ, Nyman S, Port GR. 1996. Structure-based interference with insect behaviour-cyclopropene analogues of pheromones containing Z-alkenes. *Tetrahedron* 52: 12509–20
3. Arn H, Rauscher S, Buser H-R, Guerin P-M. 1986. Sex pheromone of *Eupoecilia ambiguella* female: analysis and male response to ternary blend. *J. Chem. Ecol.* 12:1417–29
4. Arsequell G, Camps F, Fabriàs G, Guerrero A. 1990. Sila-pheromones: silicon analogues of the female sex pheromone of the processionary moth *Thaumetopoea pityocampa. Tetrahedron Lett.* 31:2739–42
5. Beevor PS, Campion DG. 1979. The field use of inhibitory components of lepidopterous sex pheromones and pheromone mimics. In *Chemical Ecology: Odour Communication in Animals*, ed. FJ Ritter, pp. 313–25. Amsterdam: Elsevier/North Holland Biomed. Press
6. Bengtsson M, Liljefors T, Hansson BS. 1987. Dienic analogs of (Z)-5-decenyl acetate, a pheromone component of the turnip moth, *Agrotis segetum.* Synthesis, conformational analysis and structure-

activity relationships. *Bioorg. Chem.* 15:409–22

7. Bengtsson M, Liljefors T, Hansson BS, Löfstedt C, Copaja SV. 1990. Structure-activity relationships for chain-shortened analogs of (Z)-5-decenyl acetate, a pheromone component of the turnip moth, *Agrotis segetum*. *J. Chem. Ecol.* 16:667–84
8. Bengtsson M, Rauscher S, Arn H, Sun W-C, Prestwich GD. 1990. Fluorine-substituted pheromone components affect the behavior of the grape berry moth. *Experientia* 46:1211–13
9. Berg BG, Tumlinson JH, Mustaparta H. 1995. Chemical communication in heliothine moths. IV. Receptor neuron responses to pheromone compounds and formate analogues in the male tobacco budworm moth *Heliothis virescens*. *J. Comp. Physiol.* 177:527–34
10. Beroza M. 1967. Nonpersistent inhibitor of the gypsy moth sex attractant in extracts of the insect. *J. Econ. Entomol.* 60:875–76
11. Bestmann HJ, Hirsch HL, Platz H, Rheinwald M, Vostrowsky O. 1980. Differenzierung chiraler Pheromon-Analoga durch Chemorezeptoren. *Angew. Chem.* 92:492–93
12. Bestmann HJ, Rehefeld C. 1989. Elektrophysiologische Messungen mit deuterierten Pheromonen zum Nachweis eines Isotopieeffektes. *Naturwissenschaften* 76:422–24
13. Bestmann HJ, Wu CH, Rehefeld C, Kern F, Leinemann B. 1992. Do pheromone receptors that receive the same or similar signal molecule have the same or similar structure? *Angew. Chem. Int. Ed. Engl.* 31:330–31
14. Bosch MP, Pérez R, Lahuerta G, Hernanz D, Camps F, Guerrero A. 1996. Difluoropalmitic acids as potential inhibitors of the biosynthesis of the sex pheromone of the Egyptian armyworm *Spodoptera littoralis*. *Bioorg. Med. Chem.* 4:467–72
15. Bowers WS, Bodenstein WG. 1971. Sex pheromone mimics of the American cockroach. *Nature* 232:259–61
16. Briggs GG, Cayley GR, Dawson GW, Griffiths DC, Macaulay EDM, et al. 1986. Some fluorine-containing pheromone analogs. *Pestic. Sci.* 17:441–48
17. Burger BV, le Roux M, Mackenroth WM, Spies HSC, Hofmeyr JH. 1990. 7-Vinyldecyl acetate, novel inhibitor of pheromonal attraction in the false codling moth, *Cryptophlebia leucotreta*. *Tetrahedron Lett.* 40:5771–72
18. Butenandt A, Beckman R, Stamm D, Hecker E. 1959. Über den Sexuallockstoff des Seidenspinners *Bombyx mori*; Reindarstellung und Konstitution. *Z. Naturforsch. Teil B* 14:283–84
19. Bykhovskaia MB, Zhorov BS. 1996. Atomic model of the recognition site of the American cockroach pheromone receptor. *J. Chem. Ecol.* 22:869–83
20. Cameron EA, Schwalbe CP, Stevens LJ, Beroza M. 1975. Field tests of the olefin precursor of disparlure for suppression of mating in the gypsy moth. *J. Econ. Entomol.* 68:158–60
21. Campion DG, Bettany BW, Nesbitt BF, Beevor PS, Lester R, Poppi RG. 1974. Field studies of the female sex pheromone of the cotton leafworm *Spodoptera littoralis* (Boisd.) in Cyprus. *Bull. Entomol. Res.* 64:84–96
22. Camps F, Coll J, Fabriàs G, Guerrero A. 1984. Synthesis of dienic fluorinated analogs of insect sex pheromones. *Tetrahedron* 40:2871–78
23. Camps F, Fabriàs G, Gasol V, Guerrero A, Hernández R, Montoya R. 1988. Analogs of the sex pheromone of processionary moth *Thaumetopoea pityocampa*: synthesis and biological activity. *J. Chem. Ecol.* 14:1331–46
24. Camps F, Fabriàs G, Guerrero A. 1986. Synthesis of a fluorinated analog of the sex pheromone of the processionary moth *Thaumetopoea pityocampa* (Denis and Schiff). *Tetrahedron* 42:3623–29
25. Camps F, Fabriàs G, Guerrero A, Riba

M. 1984. Fluorinated analogs of insect sex pheromones. *Experientia* 40:933–34
26. Camps F, Gasol V, Guerrero A. 1990. Inhibitory pheromonal activity promoted by sulfur analogs of the processionary moth *Thaumetopoea pityocampa* (Denis and Schiff.). *J. Chem. Ecol.* 16:1155–72
27. Camps F, Gasol V, Guerrero A, Hernández R, Montoya R. 1990. Inhibition of the processionary moth sex pheromone by some haloacetate analogues. *Pestic. Sci.* 29:123–34
28. Camps F, Hospital S, Rosell G, Delgado A, Guerrero A. 1992. Synthesis of biosynthetic inhibitors of the sex pheromone of *Spodoptera littoralis*. Part II: Acetylenic and cyclopropane fatty acids. *Chem. Phys. Lipids* 61:157–67
29. Cardé RT, Roelofs WL, Doane CC. 1973. Natural inhibitor of the gypsy moth sex attractant. *Nature* 241:474–75
30. Carlson DA, McLaughlin JR. 1982. Diolefin analog of a sex pheromone component of *Heliothis zea* active in disrupting mating communication. *Experientia* 38:309–10
31. Carvalho JF, Prestwich GD. 1984. Synthesis of ω-tritiated and ω-fluorinated analogues of the trail pheromone of subterranean termites. *J. Org. Chem.* 49: 1251–58
32. Chambers DL. 1977. Attractants for fruit fly survey and control. In *Chemical Control of Insect Behavior. Theory and Application*, ed. HH Shorey, J McKelvey Jr, pp. 327–44. New York: Wiley Intersci.
33. Chapman OL, Mattes KC, Sheridan RS, Klun JA. 1978. Stereochemical evidence of dual chemoreceptors for an achiral sex pheromone in Lepidoptera. *J. Am. Chem. Soc.* 100:4878–84
34. Cunningham RT, Kobayashi RM, Miyashita DH. 1990. The male lures of tephritid fruit flies. In *Behavior-Modifying Chemicals for Insect Management. Applications of Pheromones and Other Attractants*, ed. RL Ridgway, RM Silverstein, MN Inscoe, pp. 255–67. New York/Basel: Marcel Dekker
35. Curtis CE, Clark JD, Carlson DA, Coffelt JA. 1987. A pheromone mimic: disruption of mating communication in the navel orangeworm, *Amyelois transitella*, with Z,Z-1,12-14-heptadecatriene. *Entomol. Exp. Appl.* 44:249–55
36. Dawson GW, Mudd A, Pickett JA, Pile MM, Wadhams LJ. 1990. Convenient synthesis of mosquito oviposition pheromone and a highly fluorinated analog retaining biological activity. *J. Chem. Ecol.* 16:1779–89
37. Delgado A, Ruiz M, Camps F, Hospital S, Guerrero A. 1991. Synthesis of potential inhibitors of the biosynthesis of the sex pheromone of *Spodoptera littoralis*. Part I: Monofluorinated fatty acids. *Chem. Phys. Lipids* 59:127–35
38. Dickens JC, Oliver JE, Mastro VC. 1997. Response and adaptation to analogs of disparlure by specialist antennal receptors of gypsy moth, *Lymantria dispar*. *J. Chem. Ecol.* 23:2197–210
39. Dickens JC, Prestwich GD, Sun W-C. 1991. Behavioral and neurosensory responses of the boll weevil, *Anthonomus grandis* Boh. (Coleoptera: Curculionidae), to fluorinated analogs of aldehyde components of its pheromone. *J. Chem. Ecol.* 17:1007–19
40. Dickens JC, Prestwich GD, Sun W-C, Mori K. 1991. Receptor site analysis using neurosensory responses of the boll weevil to analogs of the cyclohexylideneethanol of its aggregation pheromone. *Chem. Senses* 16:239–50
41. Ding Y-S, Prestwich GD. 1986. Metabolic transformation of tritium-labeled pheromone by tissues of *Heliothis virescens* moths. *J. Chem. Ecol.* 12:411–29
42. Ding Y-S, Prestwich GD. 1988. Chemical studies of proteins that degrade pheromones: cyclopropanated, fluorinated, and electrophilic analogs of unsaturated aldehyde pheromones. *J. Chem. Ecol.* 14:2033–46

43. Du J, Ng C-S, Prestwich GD. 1994. Odorant binding by a pheromone binding protein: active site mapping by photoaffinity labeling. *Biochemistry* 33:4812–19
44. Durán I, Parrilla A, Feixas J, Guerrero A. 1993. Inhibition of antennal esterases of the Egyptian armyworm *Spodoptera littoralis* by trifluoromethyl ketones. *Bioorg. Med. Chem. Lett.* 3:2593–98
45. Fabriàs G, Barrot M, Camps F. 1995. Control of the sex pheromone biosynthetic pathway in *Thaumetopoea pityocampa* by the pheromone biosynthesis activating neuropeptide. *Insect Biochem. Mol. Biol.* 25:655–60
46. Fabriàs G, Gosalbo L, Quintana J, Camps F. 1996. Direct inhibition of (*Z*)-9-desaturation of (*E*)-11-tetradecenoic acid by methylenehexadecenoic acids in the biosynthesis of *Spodoptera littoralis* sex pheromone. *J. Lipid Res.* 37:1503–9
47. Feixas J, Camps F, Guerrero A. 1992. Synthesis of (*Z*)-10,10-difluoro-13-hexadecen-11-ynyl acetate, new difluoro analogue of the sex pheromone of the processionary moth. *Bioorg. Med. Chem. Lett.* 2:467–70
48. Feixas J, Prestwich GD, Guerrero A. 1995. Ligand specificity of pheromone-binding proteins of the processionary moth. *Eur. J. Biochem.* 234:521–26
49. Ferkovich SM, Mayer MS, Rutter RR. 1973. Conversion of the sex pheromone of the cabbage looper. *Nature* 242:53–55
50. Ganjian I, Pettei M, Nakanishi K, Kaissling KE. 1978. A photoaffinity-labeled insect sex pheromone for the moth *Antheraea polyphemus*. *Nature* 271:157–58
51. Graham SM, Prestwich GD. 1994. Synthesis and inhibitory properties of pheromone analogues for the epoxide hydrolase of the gypsy moth. *J. Org. Chem.* 59:2956–66
52. Grant AJ, Mayer MS, Mankin RW. 1989. Responses from sensilla on antennae of male *Heliothis zea* to its major pheromone component and two analogs. *J. Chem. Ecol.* 15:2625–34
53. Grant GG, Langevin D, Liska J, Kapitola P, Chong JM. 1996. Olefin inhibitor of gypsy moth, *Lymantria dispar*, is a synergistic pheromone component of nun moth, *L. monacha*. *Naturwissenschaften* 83:328–30
54. Greenway AR, Wall C. 1981. Attractant lures for males of the pea moth *Cydia nigricana* (F.) containing (*E*)-10-dodecenyl acetate and (*E,E*)-8,10-dodecadienyl acetate. *J. Chem. Ecol.* 7:563–73
55. Gustavsson A-L, Liljefors T, Hansson BS. 1995. Alkyl ether and enol ether analogs of (Z)-5-decenyl acetate, a pheromone component of the turnip moth, *Agrotis segetum*: probing a proposed bioactive conformation for chain-elongated analogs. *J. Chem. Ecol.* 21:815–32
56. Gustavsson A-L, Tuvesson M, Larsson MC, Wenqi W, Hansson BS, Liljefors T. 1997. Biosteric approach to elucidation of binding of the acetate group of a moth sex pheromone component to its receptor. *J. Chem. Ecol.* 23:2755–76
57. Hansson BS, Ochieng SA, Wellmar U, Jönsson S, Liljefors T. 1996. No inhibitory effect on receptor neurone activity by sulphur analogues of the sex pheromone component (*Z*)-5-decenyl acetate in the turnip moth, *Agrotis segetum* (Lepidoptera: Noctuidae). *Physiol. Entomol.* 21:275–82
58. Hernanz D, Fabriàs G, Camps F. 1997. Inhibition of sex pheromone production in female lepidopteran moths by 2-halo-fatty acids. *J. Lipid Res.* 38:1988–94
59. Hoskovec M, Hovorka O, Kalinova B, Koutek B, Streinz L, et al. 1996. New mimics of the acetate function in pheromone-based attraction. *Bioorg. Med. Chem.* 4:479–88
60. Hoskovec M, Kalinová B, Konecny K, Koutek B, Vrkoc J. 1993. Structure-activity correlations among analogs of the currant clearwing moth pheromone. *J. Chem. Ecol.* 19:735–50

61. Jang EB, Light DM. 1996. Olfactory semiochemicals of tephritids. In *Fruit Fly Pests*, ed. BA McPheron, GJ Steck, pp. 73–90. Delray Beach, FL: St. Lucie
62. Jönsson S, Liljefors T, Hansson BS. 1991. Alkyl substitution in terminal chain of (*Z*)-5-decenyl acetate, a pheromone component of turnip moth, *Agrotis segetum*: synthesis, single sensillum recordings, and structure-activity relationships. *J. Chem. Ecol.* 17:103–22
63. Jönsson S, Liljefors T, Hansson BS. 1991. Replacement of the terminal methyl group in a moth sex pheromone component by a halogen atom. Hydrophobicity and size effects on the electrophysiological single-cell activities. *J. Chem. Ecol.* 17:1381–97
64. Jönsson S, Liljefors T, Hansson BS. 1992. Introduction of methyl groups to acetate substituted chain of (*Z*)-5-decenyl acetate, a pheromone component of turnip moth, *Agrotis segetum*. Synthesis, single-sensillum recordings, and structure-activity relationships. *J. Chem. Ecol.* 18:637–57
65. Jönsson S, Liljefors T, Hansson BS. 1993. Enantiomers of methyl substituted analogs of (*Z*)-5-decenyl acetate as probes for the chirality and complementarity of its receptor in *Agrotis segetum*: synthesis and structure-activity relationships. *J. Chem. Ecol.* 19:459–83
66. Kaissling KE. 1977. Structure of odour molecules and multiple activities of receptor cells. In *Int. Symp. Olfaction and Taste VI*, ed. J Le Magnen, P MacLeod, pp. 9–16. London: Information Retrieval
67. Kaissling KE. 1986. Chemo-electrical transduction in insect olfactory receptors. *Annu. Rev. Neurosci.* 9:121–45
68. Kaissling KE, Meng LZ, Bestmann H-J. 1989. Responses of bombykol receptor cells to *(Z,E)*-4,6-hexadecadiene and linalool. *J. Comp. Physiol. A* 165:147–54
69. Kalinova B, Svatos A, Borek V, Vrokc J. 1990. What is the basis for the antipheromone activity of some haloacetate sex pheromone analogs? *Proc. Conf. Insect Chem. Ecol.*, pp. 51–55. Tábor: Academia Prague/SPB Acad. Publ. The Hague
70. Kasang G. 1974. Uptake of the sex pheromone 3H-bombykol and related compounds by male and female *Bombyx* antennae. *J. Insect Physiol.* 20:2407–22
71. Khrimian AP, DeMilo AB, Waters RM, Liquido NJ, Nicholson JM. 1994. Monofluoro analogs of eugenol methyl ether as novel attractants for the Oriental fruit fly. *J. Org. Chem.* 59:8034–39
72. Klun J, Chapman O, Mattes K, Beroza M. 1975. European corn borer and redbanded leafroller disruption of reproduction behavior. *Environ. Entomol.* 4:871–76
73. Klun JA, Oliver JE, Khrimian AP, Dickens JC, Potts WJE. 1997. Behavioral and electrophysiological activity of the racemate and enantiomers of a monofluorinated analog of European corn borer (Lepidoptera: Pyralidae) sex pheromone. *J. Entomol. Sci.* 32:37–49
74. Klun JA, Schwarz M, Uebel EC. 1991. European corn borer: pheromonal catabolism and behavioral response to sex pheromone. *J. Chem. Ecol.* 17:317–34
75. Klun JA, Schwarz M, Uebel EC. 1992. Biological activity and in vivo degradation of tritiated female sex pheromone in the male European corn borer. *J. Chem. Ecol.* 18:283–98
76. Landolt PJ, Curtis CE, Coffelt JA, Vick KW, Doolittle RE. 1982. Field trials of potential navel orangeworm mating disruptants. *J. Econ. Entomol.* 75:547–50
77. Latli B, Prestwich GD. 1991. Metabolically blocked analogs of housefly sex pheromone. I. Synthesis of alternative substrates for the cuticular monooxygenases. *J. Chem. Ecol.* 17:1745–68
78. Liljefors T, Bengtsson M, Hansson BS. 1987. Effects of double-bond configuration on interaction between a moth sex pheromone component and its receptor:

a receptor-interaction model based on molecular mechanics. *J. Chem. Ecol.* 13:2023–40

79. Liljefors T, Thelin B, Van Der Pers JNC. 1984. Structure-activity relationships between stimulus molecule and response of a pheromone receptor cell in turnip moth, *Agrotis segetum*. Modification of the acetate group. *J. Chem. Ecol.* 10:1661–75
80. Liljefors T, Thelin B, Van der Pers JNC, Löfstedt C. 1985. Chain-elongated analogs of a pheromone component of the turnip moth, *Agrotis segetum*. A structure-activity study using molecular mechanics. *J. Chem. Soc. Perkins Trans. II* 2:1957–62
81. Linn C, Roelofs WL, Sun W-C, Prestwich GD. 1992. Activity of perfluorobutyl-containing components in pheromone blend of cabbage looper moth, *Trichoplusia ni*. *J. Chem. Ecol.* 18:737–48
82. Liu X, Macaulay EDM, Pickett JA. 1984. Propheromones that release pheromonal carbonyl compounds in light. *J. Chem. Ecol.* 10:809–22
83. Lucas P, Renou M. 1989. Responses to pheromone compounds in *Mamestra suasa* (Lepidoptera: Noctuidae) olfactory neurones. *J. Insect Physiol.* 35:837–45
84. Lucas P, Renou M, Tellier F, Hammoud A, Audemard H, Descoins C. 1994. Electrophysiological and field activity of halogenated analogs of (*E,E*)-8,10-dodecadienol, the main pheromone component, in the codling moth (*Cydia pomonella* L.). *J. Chem. Ecol.* 20:489–503
85. Malik MS, Vetter RS, Baker TC, Fukuto TR. 1991. Dialkyl phosphorofluoridates and alkyl methylphosphonofluoridates as disruptants of moth sex pheromone-mediated behavior. *Pestic. Sci.* 32:35–46
86. Manabe S, Nishino C, Matsushita K. 1985. Studies on relationship between activity and electron density on carbonyl oxygen in sex pheromone mimics of the American cockroach. *J. Chem. Ecol.* 11:1275–87
87. Manabe S, Takayanagi H, Nishino C. 1983. Structural significance of the geminal-dimethyl group of (+)-trans-verbenyl acetate, sex pheromone mimic of the American cockroach. *J. Chem. Ecol.* 9:533–49
88. Manysk M, Fried J, Roelofs WL. 1989. Synthesis and pheromonal properties of (*Z*)-7,7-difluoro-8-dodecenyl acetate, a difluoro derivative of the sex pheromone of the Oriental fruit moth. *Tetrahedron Lett.* 30:3243–46
89. Marks RJ. 1976. Field studies with the synthetic sex pheromone and inhibitor of the red bollworm *Diparopsis castanea* Hmps. (Lepidoptera, Noctuidae) in Malawi. *Bull. Entomol. Res.* 66:243–65
90. McLean JA, Morgan B, Sweeney JD, Weiler L. 1989. Behavior and survival of western spruce budworm, *Choristoneura occidentalis* Freeman, exposed to an ω-fluorinated pheromone analogue. *J. Chem. Ecol.* 15:91–103
91. Mitchell ER, Jacobson M, Baumhover AH. 1975. *Heliothis* spp.: disruption of pheromonal communication with (*Z*)-9-tetradecen-1-ol formate. *Environ. Entomol.* 4:577–79
92. Mitchell ER, Tumlinson JH, Baumhover AH. 1978. *Heliothis virescens:* attraction of males to blends of (*Z*)-9-tetradecen-1-ol formate and (*Z*)-9-tetradecenal. *J. Chem. Ecol.* 4:709–16
93. Nikonov AA, Tyazhelova TV, Nesterov YA, Rastegayeva VM, Ilyasov FE, et al. 1994. Olfactory male sensitivity and its variation in response to fluoroanalogs of the main pheromone component of female *Mamestra brassicae*. *Z. Naturforsch. Teil C* 49:508–15
94. Nishino C, Tobin TR, Bowers WS. 1977. Sex pheromone mimics of the American cockroach (Orthoptera: Blattidae) in monoterpenoids. *Appl. Entomol. Zool.* 12:287–90
95. Norinder U, Gustavsson A-L, Liljefors T.

1997. A 3D-QSAR study of analogs of (Z)-5-decenyl acetate, a pheromone component of the turnip moth, *Agrotis segetum*. *J. Chem. Ecol.* 23:2917–34
96. Odell TM, Xu CH, Schaffer PW, Leonhardt BA, Yao D-F, Wu X-D. 1992. Capture of gypsy moth, *Lymantria dispar* (L.), and *Lymantria Mathura* (L.) males in traps baited with disparlure enantiomers and olefin precursors in the People's Republic of China. *J. Chem. Ecol.* 18:2153–60
97. Parrilla A, Guerrero A. 1994. Trifluoromethyl ketones as inhibitors of the processionary moth sex pheromone. *Chem. Senses* 19:1–10
98. Parrilla A, Villuendas I, Guerrero A. 1994. Synthesis of trifluoromethyl ketones as inhibitors of antennal esterases of insects. *Bioorg. Med. Chem.* 2:243–52
99. Petroski RJ, Weisdeler D. 1997. Inhibition of *Carpophilus freemani* Dobson (Coleoptera: Nitidulidae) aggregation pheromone response by a Z-double-bond pheromone analog. *J. Agric. Food Chem.* 45:943–45
100. Pfeiffer DG, Killian JC, Rajotte EG, Hull LA, Snow JW. 1991. Mating disruption for reduction of damage by lesser peachtree borer (Lepidoptera: Sesiidae) in Virginia and Pennsylvania peach orchards. *J. Econ Entomol.* 84:218–23
101. Pickett JA, Dawson GW, Griffiths DC, Liu X, Macaulay EDM, Woodcock CM. 1984. Propheromones: an approach to the slow release of pheromones. *Pestic. Sci.* 15:261–64
102. Pignatello JJ, Grant AJ. 1983. Structure-activity correlations among analogs of 4-methyl-3-heptanol, a pheromone component of the European elm bark beetle, *Scolytus multistriatus*. *J. Chem. Ecol.* 9:615–43
103. Pophof B. 1998. Inhibitors of sensillar esterase reversibly block the responses of moth pheromone receptor cells. *J. Comp. Physiol. A* 183:153–64
104. Prestwich GD. 1986. Fluorinated sterols, hormones and pheromones: enzyme-targeted disruptants in insects. *Pestic. Sci.* 37:430–40
105. Prestwich GD. 1987. Chemical studies of pheromone reception and catabolism. In *Pheromone Biochemistry*, ed. GD Prestwich, GJ Blomquist, pp. 473–527. New York: Academic
106. Prestwich GD, Golec FAJ, Andersen NH. 1984. Synthesis of a highly tritiated photoaffinity labelled pheromone analog for the moth *Antheraea polyphemus*. *J. Label. Comp. Radiopharm.* 21:593–601
107. Prestwich GD, Graham SM, Handley M, Latli B, Streinz L, Tasayco ML Jr. 1989. Enzymatic processing of pheromones and pheromone analogs. *Experientia* 45:263–70
108. Prestwich GD, Graham SM, Kuo J-W, Vogt RG. 1989. Tritium-labeled enantiomers of disparlure. Synthesis and in vitro metabolism. *J. Am. Chem. Soc.* 111:636–42
109. Prestwich GD, Streinz L. 1988. Haloacetate analogs of pheromones: effects on catabolism and electrophysiology in *Plutella xylostella*. *J. Chem. Ecol.* 14:1003–21
110. Prestwich GD, Sun W-C, Mayer MS, Dickens JC. 1990. Perfluorinated moth pheromones. Synthesis and electrophysiological activity. *J. Chem. Ecol.* 16:1761–78
111. Prestwich GD, Vogt RG, Riddiford LM. 1986. Binding and hydrolysis of radiolabeled pheromone and several analogs by male-specific antennal proteins of the moth *Antheraea polyphemus*. *J. Chem. Ecol.* 12:323–33
112. Priesner E, Bestmann H-J, Vostrowsky O, Rösel P. 1977. Sensory efficacy of alkyl-branched pheromone analogues in noctuid and tortricid lepidoptera. *Z. Naturforsch. Teil C* 32:979–91
113. Priesner E, Jacobson M, Bestmann HJ. 1975. Structure-response relationships in

noctuid sex pheromone reception. *Z. Naturforsch. Teil C* 30:283–93

114. Priesner E, Witzgall P. 1984. Modification of pheromonal behaviour in wild *Coleophora laricella* male moths by (Z)-5-decenyl acetate, an attraction inhibitor. *J. Appl. Entomol.* 98:118–35
115. Quero C. 1996. *Estudios sobre el proceso de percepción, inhibición y catabolismo de feromonas sexuales de Lepidópteros.* PhD thesis, Univ. Barcelona. 228 pp.
116. Quero C, Camps F, Guerrero A. 1995. Behavior of processionary males (*Thaumetopoea pityocampa*) induced by sex pheromone and analogs in a wind tunnel. *J. Chem. Ecol.* 21:1957–69
117. Renou M, Lucas P, Malo E, Quero C, Guerrero A. 1997. Effects of trifluoromethyl ketones and related compounds on the EAG and behavioural responses to pheromones in male moths. *Chem. Senses* 22:407–16
118. Riba M, Eizaguirre M, Sans A, Quero C, Guerrero A. 1994. Inhibition of pheromone action in *Sesamia nonagrioides* by haloacetate analogues. *Pestic. Sci.* 41: 97–103
119. Rider DA, Berger RS. 1985. Toxicity and electrophysiological and behavioral effects of insecticide-pheromone-related chemicals on cabbage looper moths. *Environ. Entomol.* 14:427–32
120. Ridgway RL, Silverstein RM, Inscoe MN. 1990. *Behavior-Modifying Chemicals for Insect Management: Applications of Pheromones and Other Attractants.* New York: Marcel Dekker. 761 pp.
121. Roelofs WL, Cardé RT. 1977. Responses of Lepidoptera to synthetic sex pheromone chemicals and their analogues. *Annu. Rev. Entomol.* 22:377–405
122. Roelofs WL, Comeau A. 1968. Sex pheromone perception. *Nature* 220:600–1
123. Roelofs WL, Comeau A. 1971. Sex pheromone perception: synergists and inhibitors for the red-banded leaf roller attractant. *J. Insect Physiol.* 17:435–48
124. Rosell G, Herrero S, Guerrero A. 1996. New trifluoromethyl ketones as potent inhibitors of esterases: ^{19}F NMR spectroscopy of transition state analog complexes and structure-activity relationships. *Biochem. Biophys. Res. Commun.* 226:2887–92
125. Rosell G, Hospital S, Camps F, Guerrero A. 1992. Inhibition of a chain shortening step in the biosynthesis of the sex pheromone of the Egyptian armyworm *Spodoptera littoralis. Insect Biochem. Mol. Biol.* 22:679–85
126. Schwarz M, Klun JA, Uebel EC. 1990. European corn borer sex pheromone. Inhibition and elicitation of behavioral response by analogs. *J. Chem. Ecol.* 16:1591–604
127. Silk PJ, Kuenen LPS. 1986. Spruce budworm (*Choristoneura fumiferana*) pheromone chemistry and behavioral responses to pheromone components and analogs. *J. Chem. Ecol.* 12:367–83
128. Sivinski JM, Calkins C. 1986. Pheromones and parapheromones in the control of tephritids. *Fla. Entomol.* 69:157–68
129. Sonnet PE, Moser JC. 1972. Synthetic analogs of the trail pheromone of the leaf-cutting ant, *Atta texana* (Buckley). *J. Agric. Food Chem.* 20:1191–94
130. Streinz L, Horák A, Vrkoc J, Hrdy I. 1993. Propheromones derived from codlemone. *J. Chem. Ecol.* 19:1–9
131. Struble DL, Arn H, Buser HR, Städler E, Freuler J. 1980. Identification of four sex pheromone components isolated from calling females of *Mamestra brassicae. Z. Naturforsch. Teil C* 35:45–48
132. Suckling DM, Burnip GM. 1996. Orientation disruption of *Planotortrix octo* using pheromone or inhibitor blends. *Entomol. Exp. Appl.* 78:149–58
133. Sun W-C, Prestwich GD. 1990. Partially fluorinated analogs of (Z)-9-dodecenyl acetate: probes for pheromone hydrophobicity requirements. *Tetrahedron Lett.* 31:801–4

134. Svatos A, Kalinová B, Borek V, Vrock J. 1990. *Cydia medicaginis* (Lepidoptera, Tortricidae) response to halogenated analogues of (E8, E10)-dodecadien-1-yl acetate. *Acta Entomol. Bohemoslov.* 87:393–95
135. Tacke R, Linoh H. 1989. Bioorganosilicon chemistry. In *The Chemistry of Organosilicon Compounds*, ed. S Patai, Z Rappoport, 2:1144–206. New York: Wiley & Sons
136. Tellier F, Hammoud A, Ratovelomanana V, Linstrumelle G, Descoins C. 1993. Synthèse de chlorocodlémones. *Bull. Soc. Chim. Fr.* 130:281–86
137. Tellier F, Sauvêtre R. 1992. Synthesis of a new fluorinated analog of (E,E)-8,10-dodecadienol (codlemone). *Tetrahedron Lett.* 33:3643–44
138. Tellier F, Sauvêtre R, Normant J-F. 1989. Synthèse de fluorocodlémones. *J. Organomet. Chem.* 364:17–28
139. Todd JL, Millar JG, Vetter RS, Baker TC. 1992. Behavioral and electrophysiological activity of (*Z,E*)-7,9,11-dodecatrienyl formate, a mimic of the major sex pheromone component of carob moth, *Ectomyelois ceratoniae*. *J. Chem. Ecol.* 18:2331–52
140. Villuendas I, Parrilla A, Guerrero A. 1994. An efficient and expeditious synthesis of functionalized trifluoromethyl ketones through lithium-iodine exchange reaction. *Tetrahedron* 50:12673–84
141. Vogt H, Schropp A, Neumann U, Eichhorn KW. 1993. Befallsregulierung des Einbindingen Traubenwicklers, *Eupoecilia ambiguella* Hbn, durch Paarungsstörung mit synthetischen Pheromon. *J. Appl. Entomol.* 115:217–32
142. Vogt RG, Riddiford LM, Prestwich GD. 1985. Kinetic properties of a sex pheromone-degrading enzyme: the sensillar esterase of *Antheraea polyphemus*. *Proc. Natl. Acad. Sci. USA* 82:8827–31
143. Wall C. 1990. Principles of monitoring. In *Behavior-Modifying Chemicals for Insect Management*, ed. RL Ridgway, RM Silverstein, MN Inscoe, pp. 9–23. New York: Marcel Dekker
144. Weatherston I, Minks AK. 1995. Regulation of semiochemicals—global aspects. *Integr. Pest Manage. Rev.* 1:1–13
145. Wenqi W, Bengtsson M, Hansson BS, Liljefors T, Löfstedt C, et al. 1993. Electrophysiological and behavioral responses of turnip moth males, *Agrotis segetum*, to fluorinated analogs. *J. Chem. Ecol.* 19:143–57
146. Wright RH. 1965. Finding metarchons for pest control. *Nature* 207:103–4

Annu. Rev. Entomol. 2000. 45:631–659

PEST MANAGEMENT STRATEGIES IN TRADITIONAL AGRICULTURE: An African Perspective

T. Abate[1,2], A. van Huis[3], and J. K. O. Ampofo[4]
[1]*Ethiopian Agricultural Research Organization, Nazareth Research Centre, P.O. Box 436, Nazareth, Ethiopia; e-mail: tabate@y.net.ye*
[2]*c/o FAO Office, P.O. Box 1867, Sana'a, Republic of Yemen; e-mail: tabate@y.net.ye*
[3]*Laboratory of Entomology, Wageningen University, P.O. Box 8031, 6700 EH Wageningen, The Netherlands; e-mail: arnold.vanhuis@users.ento.wau.nl*
[4]*CIAT, PO Box 2704, Arusha, Tanzania; e-mail: k.ampofo@cgiar.org*

Key Words Africa, crop protection, IPM

■ **Abstract** African agriculture is largely traditional—characterized by a large number of smallholdings of no more than one ha per household. Crop production takes place under extremely variable agro-ecological conditions, with annual rainfall ranging from 250 to 750 mm in the Sahel in the northwest and in the semi-arid east and south, to 1500 to 4000 mm in the forest zones in the central west. Farmers often select well-adapted, stable crop varieties, and cropping systems are such that two or more crops are grown in the same field at the same time. These diverse traditional systems enhance natural enemy abundance and generally keep pest numbers at low levels. Pest management practice in traditional agriculture is a built-in process in the overall crop production system rather than a separate well-defined activity. Increased population pressure and the resulting demand for increased crop production in Africa have necessitated agricultural expansion with the concomitant decline in the overall biodiversity. Increases in plant material movement in turn facilitated the accidental introduction of foreign pests. At present about two dozen arthropod pests, both introduced and native, are recognized as one of the major constraints to agricultural production and productivity in Africa. Although yield losses of 0% to 100% have been observed on-station, the economic significance of the majority of pests under farmers' production conditions is not adequately understood. Economic and social constraints have kept pesticide use in Africa the lowest among all the world regions. The bulk of pesticides are applied mostly against pests of commercial crops such as cotton, vegetables, coffee, and cocoa, and to some extent for combating outbreaks of migratory pests such as the locusts. The majority of African farmers still rely on indigenous pest management approaches to manage pest problems, although many government extension programs encourage the use of pesticides. The current pest management research activities carried out by national or international agricultural research programs in Africa focus on classical biological control and host plant resistance breeding. With the exception of classical biological control of the cassava mealybug, research results have not been widely adopted. This could be due to African farmers facing heterogeneous conditions, not needing fixed prescriptions or one ideal variety but a number of options and genotypes to choose from. Indigenous pest management knowledge is site-specific and should be the basis for developing integrated pest management (IPM) techniques. Farmers often lack the biological and ecological information necessary to develop better pest management through experimentation. Formal research should be instrumental in providing the input necessary to facilitate participatory technology development such as that done by Farmer Field Schools, an approach now emerging in different parts of Africa.

INTRODUCTION

In Africa, agriculture is the most important enterprise and the key to economic development. According to FAO (49) nearly 59% of the people of Africa earn their livelihood from agriculture. At present the continent has the lowest agricultural productivity per unit area of land—a 1989–1997 average yield of nearly 1.2 tons per ha as compared with 4.5 for Europe, 4.1 for North-Central America, 3.0 for Asia, 2.5 for South America, and 1.9 for Oceania (49). The productivity of African agriculture is not only the lowest, but it has remained stagnant whereas all other regions of the world have shown substantial increases for the period mentioned. The bulk of African agriculture remains traditional—land holdings are small, crop production is labor intensive, little or no external inputs are used. Pest management practice under this condition is a built-in process in the overall crop production system rather than a separate, well-defined activity.

Arthropod pests are one of the major constraints to agricultural production in Africa. A large number of insect and mite pests attack crops during all stages of growth—seedling to storage. Of these, only about two dozen have economic significance across agro-ecological zones. The majority of these are introduced pests (200), with their inadvertent introduction time varying from several to more than 100 years. Only a few, such as the cassava mealybug (*Phenacoccus manihoti*), cassava green mite (*Mononychellus tanajoa*), and the larger grain borer (*Prostephanus truncatus*) deserve new introduction status. Other introduced species such as the spotted stemborer *Chilo partellus*, the banana weevil *Cosmopolites sordidus*, and the diamondback moth *Plutella xylostella* are already well established and harbor a good number of natural enemies. A number of indigenous arthropods rank among the highest in importance: cereal stemborers (*Busseola fusca, Coniesta ignefusalis, Heliocheilus albipunctella, Sesamia calamistis*), the African bollworm (*Helicoverpa armigera*), the bean stem maggot (*Ophiomyia spencerella*), and a number of grasshopper species. Pest problems and subsequent losses are expected to increase during the next decades as more intensive production techniques will be employed to meet the extra food demand by a growing population (124).

During the past three decades or so a significant amount of integrated pest management (IPM) or IPM-related research has been conducted by national programs and international agricultural research centers in Africa (200). Pest management research has focused on host plant resistance breeding, classical biological control, chemical control, and cultural control measures (200). With the exception of the classical biological control of the cassava mealybug (72, 73), research on other major pests has not made a lasting impact on African agriculture (118, 176). Has crop protection been considered too much in isolation from the overall crop production system of the African farmer?

In this chapter we review the traditional pest management practices and the current ongoing efforts in research and extension, and we make suggestions to improve IPM implementation in Africa.

AGRO-ECOLOGY, FARMING SYSTEMS, AND CROPS

Africa is a vast continent with a tremendous amount of physical and biological variability (54, 58, 63, 189, 191), holding a great potential for agricultural development (124). At present published information giving comprehensive coverage on agro-ecological zones (AEZs) of the African continent is inadequate, although some works cover tropical and subtropical areas in general (54, 89, 148), a specific subregion (63), country (167) or crop (191). Geddes (58) describes ten AEZs for sub-Saharan Africa (Figure 1).

These are

1. The Sahel (A2): characterized by erratic rainfall of 250 to 500 mm per annum, more than 8 months of dry season, and altitudes of less than 900 m above sea level.
2. The Sudan savannah (A3): rainfall 500 to 900 mm, dry season 8 months, altitude less than 900 m above sea level.
3. The Guinea savannah (A4): rainfall 900 to 1500 mm (mostly unimodal), dry season of 5 to 7 months, altitude less than 900 m above sea level.
4. The forest–savannah transition (A4/5): rainfall 1300 to 1800 mm (unimodal or bimodal), dry season of 4 months, altitude more than 900 m above sea level.
5. The forest (A5): rainfall 1500 to 4000 mm, virtually no dry season, altitude less than 900 m above sea level.
6. The east coast (A6): rainfall 750 to 1500 mm (bimodal in some countries), altitude less than 900 m above sea level.
7. The semi-arid east and south (A7): rainfall 250 to 750 mm, more than 8 months of dry season, altitude less than 1500 m above sea level.
8. The plateaux (B7): rainfall 750 to 1500 mm (mostly unimodal), dry season 5 to 8 months, altitude 900 to 1500 m above sea level.
9. The Uganda and Lake Victoria shore (L1): rainfall 1000 to 1500 mm (bimodal), altitude 1135 to 1300 m above sea level.
10. The mountain (B2): rainfall 750 to 1800 mm (unimodal or bimodal), altitude more than 1500 m above sea level.

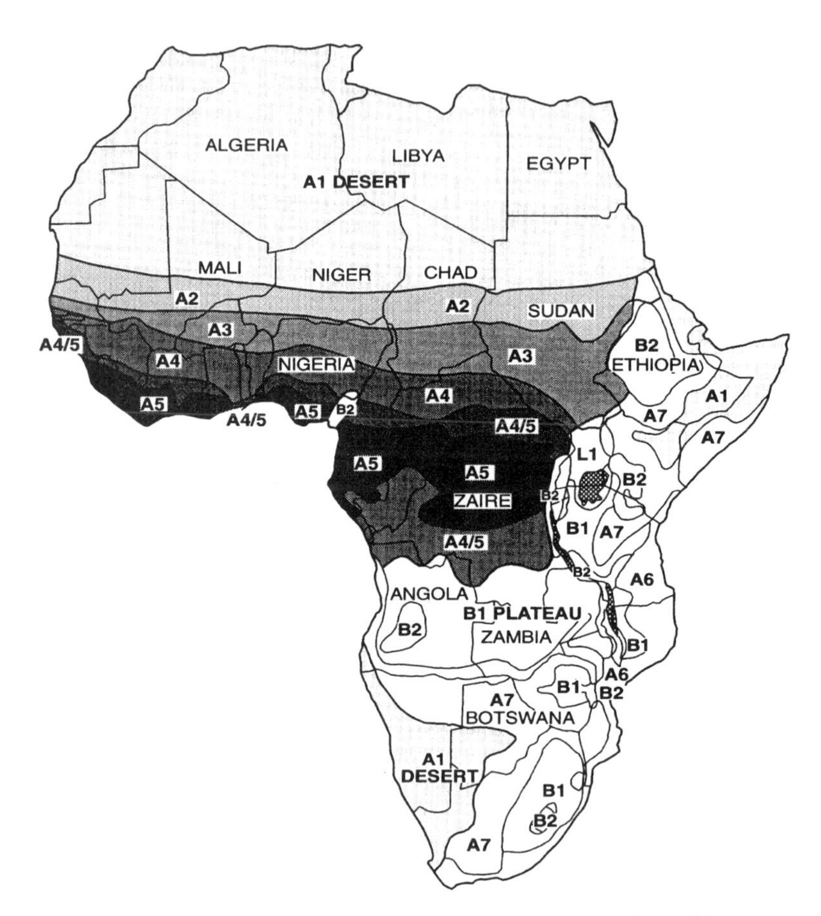

Figure 1 Major agro-ecological zones of sub-Saharan Africa (from 58)

Depending on the AEZ, crop management in African agriculture can be divided into five broad categories: grazed fallow, shifting cultivation, short fallow, permanent cultivation, and perennials (58). Grazed fallow is mostly practiced in the dry AEZ of Sahel and semi-arid east and south. Short fallow, shifting cultivation, permanent cultivation, and perennials are prevalent across the rest of the AEZs. Perennials are not common in the plateaus.

Crop production under such conditions uses mainly traditional methods: land preparation is done either manually using hand tools or with animal drawn plows; hoeing is the most practiced method of weed management; land holdings per household rarely exceed one ha (167, 200). In most instances, mixed agriculture consisting of crops and livestock is practiced. Pesticide use, at least on food crops, is almost non-existent. Traditional farmers rely on their age-old experience to tackle pest problems. Crop protection in traditional African agriculture is a subset of the overall crop production and is achieved through the use of a traditional pest management approach that consists of appropriate sowing date, optimum plant density, varietal mixtures, intercropping, good crop husbandry, and the use of home-grown available methods (11).

In recent years, however, government extension programs have encouraged farmers to look for pest solutions with pesticides, and in some African countries the indigenous knowledge is being eroded (23, 60, 169, 198) as it is elsewhere in the tropics (106). By and large, traditional farmers in Africa use one or another form of intercropping, and monoculture is rare (164, 118, 200). In such diverse crop production conditions pest outbreaks on the level known in commercial agriculture are rare (except for migratory pests such as the desert locust and African armyworm). However, this situation is rapidly changing due to increased population pressure. It has prompted many African governments to "modernize" agriculture, based mostly on the "green revolution" models. Technologies such as irrigation, fertilizers, and pesticides are used in a package to control the environment to grow high-yielding varieties. However, such fixed prescriptions under conditions of complex, diverse, and risk-prone agriculture often do not work (35).

Increased production can be achieved either through improved productivity, or through expansion into areas hitherto undisturbed. In Africa, the latter seems to have been the more prevalent scenario in recent years. For example, agricultural production showed an average increase of 2.45% between 1991 and 1996, whereas the area of cultivated land increased by 10% between 1980 and 1995 (48). In comparison, countries in North-Central America and India registered only a slight increase of about 1% in the area of cultivated land area, whereas China and Europe showed decrements of 4.4% and 4.3%, respectively, over the same period (48).

Waage (187) pointed out that the more serious pests in the developing world are the direct consequence of efforts aimed at improving agricultural production. Recent reviews (9, 42, 171) corroborate this statement. According to these authors agricultural intensification aggravates pest problems as it leads to, inter alia, a narrow genetic base of crop varieties (often grown in monoculture), reduction in natural areas around crops (depriving natural enemies of their natural habitat), and more pest-susceptible plants resulting from soil fertility decline due to shorter intervals between plantings (103). For example, soil exhaustion aggravated the problems with pests and diseases in the south of Sudan (43).

Plants that evolved outside Africa may lack resistance to insect pests from Africa (112); e.g. the indigenous sorghum grows especially thick and tall on termite mounds without suffering damage from termites (43), while non-indigenous crops such as maize (43), rice, potato, and groundnut are regularly attacked (70). Also, termite damage to native trees such as the oil palm (*Elaeis guineensis*) is not mentioned in Africa, while damage to introduced trees such as eucalyptus and the rubber plant is often severe (70). Because many crops in Africa are introduced, the crop pest–natural enemy systems have not co-evolved, indicating the potential of classical biological control (44).

Wide ranges of crops are grown in Africa, their economic importance varying from one agro-ecological zone to another. In general, root crops (such as cassava, yams, potato, and sweet potato), cereals (maize, sorghum, and millet), fruits (citrus and banana), and vegetables are the most important crops, together accounting for nearly 77% of crop production in Africa. The balance is covered by oil seeds, sugarcane, pulses, cocoa beans, cotton lint, coffee, nuts, and tea (Table 1). It can be seen that crop production in Africa showed a slow growth or even a decline in some years owing to adverse weather conditions, declines in soil fertility, and damage caused by pests and diseases.

ARTHROPOD PESTS AND THEIR ECONOMIC IMPORTANCE IN AFRICA

A wide range of insects and mites attack crops and reduce potential yields in Africa (8, 11, 16, 56, 58, 82, 83, 98, 104, 117, 123, 139, 152, 153, 155, 159, 196, 200, 201). However, only a limited number are of economic importance, depending on the agro-ecological zone, cropping system, and the crop in question (Table 2). Although yield losses of up to 100% have been reported due to major arthropod pests under on-station conditions, there is no adequate information on the economic significance of pests under farmers' conditions (9, 11, 118, 171). How-

TABLE 1 Production (million metric tons) of major crops in Africa

	Year							
Crops	1991	1992	1993	1994	1995	1996	Average	Percentage of total
Root crops	121.3	127.7	132.6	133.2	136.7	137.7	131.5	37.2
Cereals	104.1	90.7	101.8	113.1	100.1	127.6	106.2	30.1
Fruits	47.8	50.7	51.8	52.3	53.0	54.4	51.7	14.6
Vegetables + melons	32.6	33.2	33.9	34.6	34.9	35.5	34.1	9.7
Oilseeds	10.8	10.3	11.0	11.3	11.6	10.8	11.0	3.1
Sugar cane	6.0	6.9	6.4	7.3	7.4	8.4	7.1	2.0
Pulses	7.1	6.5	6.8	6.9	7.3	7.6	7.0	2.0
Cocoa beans	1.3	1.3	1.3	1.4	1.8	1.9	1.5	0.4
Cotton lint	1.4	1.3	1.4	1.3	1.4	1.5	1.4	0.4
Coffee	1.2	1.1	0.9	1.1	1.1	1.1	1.1	0.3
Nuts	0.4	0.4	0.4	0.4	0.4	0.4	0.4	0.1
Tea	0.3	0.3	0.3	0.3	0.4	0.4	0.3	<0.1
Total	334.3	330.4	348.6	363.2	356.1	387.3	353.3	100.0
% difference*	—	−1.2	5.5	4.2	−2.0	8.8	—	—

* Refers to difference between two successive years. Source: computed from FAO (48)

TABLE 2 Examples of major arthropod pests in Africa

Common name	Scientific name	Main host (crops)	AEZ where important *
Maize stemborer	*Busseola fusca*	Maize, sorghum	B1, B2, A4/5
Spotted stemborer	*Chilo partellus*	Sorghum, maize	A7, A6,B1
Pearl millet stemborer	*Coniesta ignefusalis*	Pearl millet	A2
African bollworm	*Helicoverpa armigera*	Polyphagous	Widespread
Pearl millet headminer	*Heliocheilus albipunctella*	Pearl millet	A2
African armyworm	*Spodoptera exempta*	Cereals	A6, A7, B2
Grasshoppers	Several species	Pearl millet	A2
Desert locust	*Schistocerca gregaria*	Polyphagous	A2, A3, A4, A7, A6
Pearl millet blister beetle	*Psalydolytta fusca*	Pearl millet	A2
Sugar cane aphid	*Longuinguis sacchari*	Sorghum	A7
Sorghum midge	*Contarinia sorghicola*	Sorghum	A3
Cowpea pod borer	*Maruca vitrata*	Cowpea	A7, A3, A2
Cowpea flower thrips	*Megalurothrips sjostedti*	Cowpea, beans	A2, A3, A7
Bean stem maggot	*Ophiomyia* sp.	Beans	B1, L1
Cassava green mite	*Mononychellus tanajoa*	Cassava	A4, A4/5, A5, A6, B1, L1
Cassava mealybug	*Phenacoccus manihoti*	Cassava	A4/5, A6, L1
Banana weevil	*Cosmopolites sordidus*	Banana	L1
Cowpea aphid	*Aphis craccivora*	Cowpea	A3, A7
Bean aphid	*Aphis fabae*	Beans	B1, L1
Cereal weevils	*Sitophilus* spp	Maize, sorghum	A4/5, A7, A3
Angoumois grain moth	*Sitotroga cerealella*	Maize, sorghum	B1, B2
Common bean weevil	*Acanthoscelides obtectus*	Beans	B1, L1, B2
Mexican bean weevil	*Zabrotes subfasciatus*	Beans	B1, L1, B2
Cowpea weevil	*Callosobruchus maculatus*	Cowpea	A3, A7

* Letters in the AEZ column correspond to those in Figure 1

ever, characteristically yields are highly variable. For example, yield variability can be as high as 50% for rain-fed crops in east Africa (52), mainly due to erratic rainfall and variable losses due to pests and diseases (195). Arthropod pests in Africa can be categorized into field crop, migratory, emerging, and storage pests.

Field Crop Pests

The millet head miner *Heliocheilus albipunctella* causes substantial loss in pearl millet in the Sahel zone (82, 83, 98, 119, 196). Depending on the crop season, yield losses due to this insect can range from 3% to 85% in Senegal, 16% to 85% in Burkina Faso, Gambia, and Mali, and 1% to 47% in Niger (58, 98). Several species of short-horned grasshoppers, including *Oedaleus senegalensis* and *Kraussaria angulifera* are also important on pearl millet in this region (82, 83, 155). Losses of 70% to 90% are estimated from grasshoppers in a bad year, which occurs about every 5 years, across crops in the Sahel zone (58). Other insect pests of economic importance in this zone include the cowpea flower thrips *Megalurothrips sjostedti* and the blister beetle *Psalydolytta fusca* (58, 119, 201).

M. sjostedti is a widespread species that attacks cowpea and beans throughout the continent (11, 79, 99–102, 122, 170). It is reported to be more important in the Sahel and Sudan savannah zones than in the semi-arid zone (58).

The sorghum midge *Contarinia sorghicola* may attain a major degree of economic importance in some countries in the Sudan savannah zone of western Africa. Nwanze (117) observed that its attack varied with seasons: usually low in the dry Sahelian zone, but severe south of the 13-degree-north latitude where the rainfall is higher.

The cassava green mite (CGM) *Mononychellus tanajoa* and the cassava mealybug (CMB) *Phenacoccus manihoti* are exotic arthropod pests of South American origin that were accidentally introduced into Africa with planting materials in the early 1970s. Since then CGM and CMB have become widespread in the "cassava belt" of the continent and are the most important pests of cassava in the Guinea savannah, Guinea savannah–forest transition, forest, east coast, plateaus, and Lake Victoria zones. The CGM was reported first from Uganda, and the first outbreaks of CMB were in the Kinshasa and Brazzaville areas in 1973 (73, 116). The CMB and CGM are reported to cause direct yield losses of about 80% and 60%, respectively (73, 116). Secondary losses include reduction of healthy leaves (vegetable sources in many African countries), erosion, weed invasion, and poor quality of planting material for the next planting season (116).

Stemborers are the most important pests of maize, sorghum, and millet across the continent in many agro-ecological zones (Table 2). In general, *B. fusca* favors high altitudes that are cool and wet. Others are quite important in certain areas, i.e. *Coniesta ignefusalis* in the Sahel, *Eldana saccharina* and *Sesamia* spp. in the lowland forest areas of west Africa (16, 56, 68, 69, 93, 94, 117, 152, 153, 159). *Chilo partellus* is an exotic species and occurs only in east and southern Africa, but may invade west Africa in coming years (134). Yield losses due to lepidopterous borers in Africa vary greatly (0% to 100%) among ecological zones, regions, and seasons (138). Other reports (20, 21, 133) suggest an average estimate of 15% to 40% yield loss due to stemborers on cereals in Africa.

Helicoverpa armigera, commonly known as African bollworm or Old World bollworm, is extremely polyphagous and a major pest of a large number of major crops in Africa (32, 120, 179). It is the most important pest of cotton; other crops affected include vegetables (tomato and capsicum), beans, sorghum, and maize. On beans (*Phaseolus vulgaris*), in conjunction with the cowpea pod borer *Maruca vitrata*, *H. armigera* causes estimated yield losses of 12% to 53% in Ethiopia, Kenya, and Tanzania (11). *M. vitrata* is considered to be a major pest of cowpea throughout Africa (157). Shanower et al (154) reported estimated annual yield losses of US$ 317 million and US$ 30 million due to African bollworm and *Maruca* on pigeon pea (*Cajanus cajan*), which is another important crop grown by small-scale growers in drier parts of southern Africa.

According to Singh & Jackai (157) the cowpea aphid *Aphis craccivora* is a major pest of cowpea throughout Africa. It appears that this insect is most important in the Sudan savannah and the semi-arid zones of Africa (Table 2). A related species, *A. fabae*, is a major pest of beans in Kenya, Zambia, Uganda, Burundi, and Tanzania (193) where it is reported to cause yield losses of 90%, 50%, and 37%, respectively (11). Several other species of aphids, including the maize aphid *Rhopalosiphum maidis* and the sugarcane aphid *Longuinguis sacchari* on maize and sorghum, and the cotton aphid *Aphis gossypii* on cotton are important pests in Africa across the agro-ecological zones where their respective host crops are grown (25). Although the direct damage from aphids can be significant, their main importance lies in their ability to transmit viruses, such as the groundnut rosette virus on peanut, maize streak, or maize mottle virus on maize, cassava mosaic on cassava, sweet potato virus complex, cocoa swollen shoot (58), and tristeza on citrus (3). The pod-sucking hemipterans in the genera *Anoplocnemis*, *Clavigralla*, and *Riptortus* on cowpea (157) and pigeonpea (154) are also important across the continent where these crops are grown.

The bean stem maggots (*Ophiomyia phaseoli*, *O. spencerella*, and *O. centrosematis*) are perhaps the most important pests of beans (*Phaseolus vulgaris*) in major bean-growing areas mainly in eastern and southern Africa. Yield losses ranging from 8% to 100% have been reported from these countries (11). *O. phaseoli* and *O. centrosematis* are more prevalent in warmer mid-altitude areas, whereas *O. spencerella* favors cooler and wetter high altitudes; furthermore, *O. phaseoli* is more common early in the season, while *O. spencerella* is more prevalent in late-sown crop (11, 13, 24). *O. phaseoli* and *O. spencerella* are the most important of the three bean stem maggot species; *O. centrosematis* usually occurs rarely and in small numbers (11, 13, 24). The distribution of *O. phaseoli* and *O. centrosematis* ranges throughout tropical and subtropical Africa, Asia, and Australia, but *O. spencerella* has not been recorded outside of Africa (11).

Migratory Pests

The desert locust *Schistocerca gregaria* and many other species of locusts are sporadic pests that cause substantial crop losses across Africa during plague years (110). The 1986–89 spectacular plague of the desert locust cost donors an estimated US$ 200 to 315 million for control in the Sahel region where 26 million ha of land were treated with pesticides (168, 183). Local governments in desert locust–affected countries also spend significant amounts of money for fighting the desert locust (86). The African armyworm *Spodoptera exempta* is another migratory pest that invades eastern and northeastern Africa (110) almost every other year. Research of these and other transboundary pests is handled by regional programs such as the Emergency Prevention System of the Food and Agriculture

Organization of the United Nations and the Desert Locust Control Organization for East Africa, in collaboration with the plant protection departments of affected countries.

Emerging Pests

In addition to the arthropod pests of field crops, there are new and emerging pests on horticultural crops in Africa. Notable among these are the citrus leafminer (*Phyllocnistis citrella*), the western flower thrips (*Frankliniella occidentalis*), and the vegetable leafminers (*Liriomyza trifolii* and related species) (90, 62, and 163, respectively). The citrus leafminer is native to Asia but has been a minor pest of citrus in Africa (2, 74) until recent years. It is now considered to be the major threat to the fledgling citrus industry in the Horn of Africa. Since its accidental introduction to east Africa in the early 1970s (91) the vegetable leafminer has attained major pest status on vegetables (tomato, snap beans, onions) in eastern and northeastern Africa (105; S Tibebu, personal communication).

Storage Pests

Traditionally, the grain weevils (*Sitophilus* spp.) and the Angoumois grain moth (*Sitotroga cerealella*) on cereals and three genera of bruchids (*Acanthoscelides*, *Zabrotes* and *Callosobruchus*) on legumes are the most important pests of stored grain in Africa. Crop losses due to storage pests are usually below 5% in traditional agriculture in Africa and elsewhere (37, 60, 169). However, this situation is changing due to the introduction of high-yielding crop varieties that are usually more susceptible. More damage may also be caused by new pests, such as the larger grain borer *Prostephanus truncatus* in stored maize that was accidentally introduced in east and west Africa in the 1980s (8).

PESTICIDE USE IN AFRICAN AGRICULTURE

Detailed and comprehensive data on pesticide usage in the African continent are lacking. The scanty information available mainly deals with expenditures on pesticides by a particular country or region (47–49, 108, 109, 111, 132, 151, 173, 177, 178). Detailed information on pesticide volumes as well as expenditure extending more than 10 years is available for a few countries such as Ethiopia and Madagascar (10, 64, 107).

Pesticide usage at a rate of 1.23 kg per ha has been reported for Africa, compared with 7.17 and 3.12 for Latin America and Asia, respectively (143). Data computed from pesticide import and crop production (47) and cultivated land (48) on pesticide usage in Africa are presented in Figure 2. The figure reveals that Africa's expenditure on pesticides is one of the lowest among the world regions. For example, on average, a hectare of cultivated land in Africa receives pesticides worth US$ 3.04 (Figure 2*a*), meaning that for every metric ton of crop produced

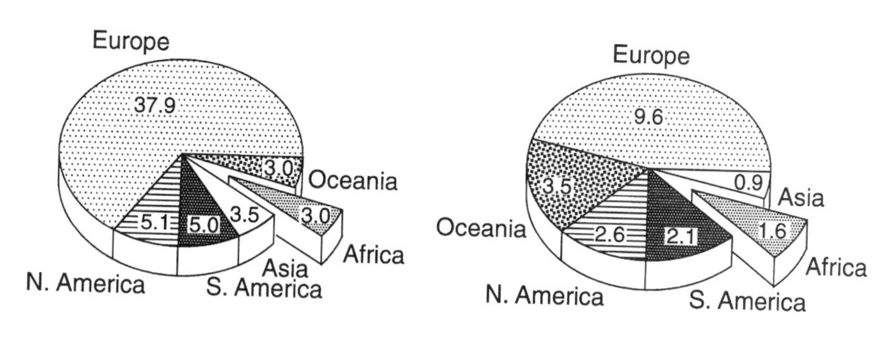

Figure 2 Expenditure (in US dollars) on pesticides (*a*) per hectare and (*b*) per metric ton of crops produced across the world regions.

per ha the investment on pesticides is about US$ 1.64 (Figure 2*b*); in Europe these figures are $ 37.86, and $ 9.55 respectively. Average figures for the period of 1991 to 1995 (48) suggest that expenditures on pesticides per metric ton of crop produced show a downward trend for Africa (-2.4%) and Asia (-8.7%), but an upward trend for Europe (13.4%), South America (19.8%), Oceania (21%), and North-Central America (34.7%). The downward trend in pesticide usage in Africa is further corroborated by an USAID report (178). Here, the annual pesticide import growth rate for 15 west-African countries was about 1.5% for the period of 1980 to 1989 compared with 19.9% for the preceding decade. According to the FAO (46) the market share of pesticide imports for Africa during 1993 to 1994 was 3% compared with 30%, 26%, and 16%, for North America, western Europe and Asia, respectively. In this respect, countries in Africa compare favorably to others of the world community in the adoption of Agenda 21 of the United Nations Conference on Environment and Development in Rio de Janeiro (175).

The bulk of pesticide usage in Africa is on commercial crops: Cocoa, coffee, and cotton receive 40%, 15% to 25%, and 13% to 15%, respectively, with the remaining 5% used against migrant pests (such as locusts and armyworm) and food crops (108, 132). In recent years, stringent regulations against pests in European countries have also made pesticides the major means of controlling pests on horticultural crops that are produced for export (78, 105). This would mean that pesticide use in agriculture in Africa is limited to fighting migratory pests such as the desert locust, and to some extent the African armyworm and pests of commercial crops.

Problems associated with pesticide usage in Africa are those relating to economic, social, and health concerns (8, 27, 81, 107, 135, 143, 149, 177, 178). One major problem, often overlooked, is the issue of disposal of obsolete and unwanted pesticides that have accumulated over the years in African countries (50, 51). These are often obtained as donations from industrialized countries to fight outbreaks of migratory pests such as the desert locust and armyworm. Recipient countries usually have inadequate storage facilities and are inadequately prepared to accurately forecast their pesticides need for a given season. FAO (46) reported an estimated 20,000 to 30,000 metric tons of obsolete pesticides in Africa in 1992. Between 1994 and 1997 FAO (50) has registered 10,099 metric tons of obsolete, unwanted and/or banned insecticides in 34 African countries, with Morocco (22.4%), Ethiopia (11.4%), Tunisia (8.7%), and Sudan (6.5%) being the most affected. Nearly four million kg of the pesticides were chlorinated hydrocarbons such as HCH, dieldrin, lindane, DDT, and aldrin. It is estimated that the cost of disposal of obsolete pesticides is between US$ 3,000 and US$ 4,500 per metric ton (50). African governments do not have the technical and financial capacities to adequately carry out disposal of these obsolete pesticides.

Another problem is associated with donated pesticides left over from emergency campaigns: these pesticides are often applied against crop pests other than those causing the emergency, and these applications are rarely economically justified. This practice is often carried out by National Plant Protection Services that are socially or politically motivated to provide chemical control free of charge to farmers (40). The end result is that farmers, who cannot afford to buy the pesticides, will eventually consider pest management the responsibility of the government, and neglect the use of home-tailored traditional pest management practices in the process (23, 185).

TRADITIONAL PEST MANAGEMENT PRACTICES

Traditional control practices are still the major means of pest management to small-scale farmers in Africa (9, 11, 200). These control practices are based on built-in features in cropping systems, such as farm plot location, crop rotation, and intercropping, or on specific responsive actions to reduce pest attack, such as timing of weeding, use of plants with repellent or insecticide action, traps, scarecrows, smoke, and digging up grasshopper egg masses. However, detailed information on traditional pest management practices widely used by African farmers is often lacking. Zethner (200) cites only two cases of such details from the Sahel (150) and from Tanzania (113). In general traditional agricultural systems are poorly understood (172), and it is often not sufficiently recognized that crop protection is a thoroughly tested and built-in process in the overall production system.

In principle, farmers have a good ecological understanding of those pests that can easily be observed (23, 36, 41). For example, farmers in Nigeria were able to develop a control method against the variegated grasshopper, *Zonocerus variegatus*: exposing egg pods to the sun after marking out and digging up egg-laying sites (144). However, farmers ignore or tend to underestimate those insect pests that are difficult to observe, such as nocturnal insects or inconspicuous pests such as the sorghum midge and thrips (29, 41).

Traditional agricultural systems are finely tuned and adapted, both biologically and socially, to counter the pressures of what are often harsh and inimical environments. Often such systems represent hundreds of years of adaptive evolution (71). Traditional pest management practices consist mainly of cultural control methods such as crop associations, planting and harvesting time, closed season, mechanical control, use of herbal products, and, sometimes, dealing with pests in a supernatural way.

Crop Associations

A longstanding practice in traditional African agriculture is growing two or more crops in the same field at the same time (129, 164). Farmers use well-adapted crop species in mixtures that are generally more stable than those in pure stands. This practice, although discouraged in favor of monocropping practices, better meets the agronomic, socio-economic, and nutritional needs of the small-scale grower. This includes better food security, optimal use of soil and space, maintenance of soil fertility (especially where intercropping involves leguminous species), better erosion control, and reduction of the need for weeding (200). There are also merits for pest control (136). Pest and disease incidence is reduced and natural enemy abundance favored (1, 6, 17–19, 99–102, 125–128, 131, 174, 186).

Farmers in west Africa also use the practice of diversionary hosts by leaving felled tree trunks and other wood remnants in rice fields to lure termites away from the crop (140). The same principle can be used in a maize/cotton intercrop when silking-stage maize diverts ovipositing females of *Helicoverpa armigera* away from the squares and bolls of cotton. However, when the timing is not correct, populations may build up in maize, increasing the levels of attack in cotton (162). The danger of pest build-up may occur when companion crops share the same pests (130). Intercropping may also affect the relative resistance of crops (59).

In plantation coffee, several species of insects, including the antestia bugs (*Antestiopsis* spp.) and leafminers (*Leucoptera* spp.) are major pests (74) when grown in pure stands. However, these pests are uneconomical in environments where coffee is produced by small-scale growers under tree canopies and in association with other crops such as *Ensete ventricosum* (15). Apparently coffee production environments under natural conditions such as those practiced by small-scale producers either enhance the performance of natural enemies, are not favorable for pest development, or both (9).

Planting and Harvesting Time

Adjusting planting or harvesting time to escape pest damage is the most important means of keeping pest damage below economic levels in Africa. For example, early planting is perhaps the most effective means of control against stemborers on sorghum and maize in many parts of Africa and is widely practiced by farmers (9, 14, 16, 57). Appropriate time of sowing and use of high plant populations are the major means of reducing damage by bean stem maggots (4, 11, 145).

In addition to influencing the level of attack and damage suffered by the crop, early planting helps to maximize yield: The emerging crop benefits from a full season's rainfall and soil nitrate fluxes and suffers less from weed competition (45, 57, 161). For some pests, populations vary between seasons as well as within a season, and time of planting is used to avoid peaks of pest populations (115, 158, 160, 161). Farmers are often aware of such pest population fluctuations and their effects on crop performance and therefore adjust their planting dates accordingly—for example, sweet potato planting is delayed to avoid sweet potato weevil (*Cylas* spp.) damage in western Kenya (161), and bean sowing is delayed to avoid *Ootheca* spp. infestation (22). In Malawi, farmers use variable planting dates to avoid damage by bean pests (147). Harvesting early reduces field infestation by post-harvest pests whose populations increase toward the crop maturity period (34, 161).

Closed Season

The use of closed season is the main means of keeping damage to cotton by pink bollworm (*Pectinophora gossypiella*) below economic levels in many parts of Africa (75, 123, 141). Recent studies reveal, however, that the economic importance of this pest is on the increase perhaps due to changes in the duration of the closed season imposed by increased demands by governments to intensify agricultural production (9).

Mechanical Control

Farmers in west Africa pick by hand important pests of pearl millet or set night fires to lure and burn nocturnal beetles (98, 197). Other conspicuous insect pests that are hand-picked and destroyed mechanically are egg hatches or first-instar gregarious larvae of the cotton leafworm (*Spodoptera littoralis*) in Egypt, those of the orange dog (*Papilio demodocus*) and the sweet potato butterfly (*Acraea* spp.) in southern Ethiopia (2), and soil insects (76). Trenches dug across the path of the moving armyworm or locust bands act as traps for locusts and armyworms. Insects fallen into it are killed by drowning or by burying (155).

Selection of Site and Clean Seeds

Farmers in the lowlands of Kenya got cuttings from high-altitude areas because cassava that is almost free of African mosaic virus was grown in these highlands. Disease-transmitting whiteflies are also scarce in these cool, high-altitude areas (166).

Use of Plant Products

Although conventional chemical pesticides are a recent introduction in African agriculture, the use of chemicals is not new. Farmers used various forms of herbal and animal products in the control of pests and diseases but the active ingredients were not known; the products were not standardized and the application was shrouded in mysticism or was culture dependent. Local communities often have their own array of insecticidal plants they use for the control of specific pests. Stoll (165) lists a number of plant species with insecticidal properties and guidelines for the extraction of the active ingredients.

Various forms of herbal products have been used to control pests in field and in storage. Some of these have been investigated and found effective. For example, *Tephrosia* spp. juice for maize stemborer control in southern Tanzania and northern Zambia (192) and *Tephrosia* plants in sweet potato fields in Uganda protect the tubers from mole rat damage (192). Seeds soaked in hot water infusions of preparations from neem seed or leaves, bark of *Balamites aegyptiaca*, leaves or roots of *Euphorbia paganorum*, *Parkia biglobosa*, or cowpea pod product are reported to protect the emerging seedlings from stemborer, termite, and bird damage, and grains from storage pests (55, 199). Wardell (188) gives account of traditional methods of termite control on forest nurseries by farmers in Tanzania, Kenya, and Malawi using locally available plant products. The Morokodo tribesmen of southern Sudan use fish poison to control termites in citrus nurseries (156).

The use of botanical products is more prevalent in storage systems. Farmers in Uganda use banana juice, pepper, Mexican marigold (*Tagetes minuta*), and eucalyptus leaves for bruchid control in storage (61). The use of vegetable oils, ash, other grains, sand, and other plant products admixed with stored grain is also widespread in traditional African farming practice (55, 60, 61, 169, 182, 199). Sometimes these products are mixed with ground chillies or other fumigants to enhance the repellent or killing potential.

Animal wastes such as goat and cow urine or dung are also used in the management of storage pests. For example, farmers in parts of Tanzania and the Sahel report that beans stored in sacks soaked and dried in goat urine give protection against storage pests (34, 55).

The introduction of modern chemicals is replacing some of these traditional strategies, especially in areas where production is market oriented.

Pest Insects Used as Food

Some insect species are not only considered a nuisance. A number of species, such as locusts, many grasshoppers, and the palm weevil *Rhynchophorus phoenicis*, are edible. Farmers may be reluctant to apply pesticides to orthopteran species because it renders them inedible. For the same reason, farmers may object to spraying campaigns by the Plant Protection Services. Harvested and marketed edible grasshoppers and locusts may yield more revenue to farmers than the sale of their millet (184). Insects may also be used for other purposes. The pentatomid *Agonoscelis pubescens*, a pest of rain-fed sorghum, is used as a source of oil to prepare food and as medicine to treat scab disease of camels (53).

Supernatural Pest Management Actions

Pests may have a religious significance. Outbreaks of locusts, grasshoppers, caterpillars, rats, and birds are often considered a punishment, after which any wrongdoing or negligence in the society must be corrected, e.g. by offerings of gifts to the poor. Farmers in western Sudan (190) and Niger (41) use charming by religious local leaders to drive away pests. In Zimbabwe farmers consulted ancestors to halt a plague of the African armyworm *Spodoptera exempta* (184).

IPM TECHNOLOGY DEVELOPMENT

Economic, social, and health concerns about pesticide use (8, 27, 81, 107, 135, 143, 149, 177, 178) have made the use of synthetic pesticides in Africa the lowest among all regions of the world. This means that most of Africa is not on the so-called pesticide treadmill and therefore an opportunity exists to learn from the negative experiences elsewhere. It should also be taken into account that the limited crop value in traditional farming systems rarely economically justifies the use of pesticides (84, 198). Therefore, the main focus of developing and implementing IPM in Africa will not be reducing pesticide usage, but rather to build IPM programs around the traditional pest management approaches that abound in small-scale agriculture in the continent (9).

A recent review of IPM activities in Africa focused on research conducted by international agricultural research centers (200). Of a total of 51 IPM or IPM-related projects between 1972 and 1992, about 34% deal with biological control using natural enemies, 27% with chemical control, 15% with scouting/monitoring, 13% with host plant resistance, and 11% with cultural practices. The most important characteristics of these projects are that all are funded by donors and they have two major components—research and training, the latter mostly for extension staff and researchers. Detailed information dealing with pest management research by national programs is not available. The majority of the research projects are run by international agricultural research centers, and the very few that are run by national governments are mostly targeted against pests of cash crops, rather than food crops grown by small-scale farmers.

Biological Control

So far classical biological control in Africa, as elsewhere, has been successful against pests that were accidentally introduced to Africa. A shining example of such success is the control of the cassava mealybug by importing the parasitic wasp *Epidinocarsis lopezi* from South America (67, 73, 116, 200). The introduction of another parasitoid *Gyranosoidea* sp. against the mango mealybug in Togo is also reported to give promising results (114, 200). Emphasis has also been given to classical biological control of cereal stemborers, but attempts have not yet resulted in lasting successes (80, 93, 94, 118, 134). However, the introduction in 1993 of the Asian parasitoid *Cotesia flavipes* is promising, causing an average of 10% parasitism of the stemborer *C. partellus* across a wide area of southeastern Kenya; in the same areas a concomitant decrease in stemborer population was observed (WA Overholt, personal communication). There are no known cases of success for other forms of biological control, such as conservation of natural enemies through habitat management. Currently in Kenya, the stimulo-deterrent diversionary strategy is investigated where stemborers are repelled from maize and simultaneously attracted to a discard or trap crop such as molasses grass *Melinis minutiflora*; when intercropped, larval parasitism by *Cotesia sesamiae* increased (95, 96).

A good number of known major pests in Africa have a large number of natural enemies associated with them (5–7, 26, 30, 38, 65, 66, 121, 137, 179–181, 194) and therefore classical biological control will not receive a high priority against those pests.

Host Plant Resistance

Research on host plant resistance has been given the major emphasis for the majority of arthropod pests that are dealt with at international agricultural centers. To date resistant varieties have been identified for all major pests with the exception of the larger grain borer (31, 171). Of the major pests shown in Table 2, resistant varieties have been released by international agricultural research centers against only four—sorghum midge, cowpea aphid, flower thrips, and cowpea weevil (171). Resistant varieties have also been identified and released against other pests (12, 76; T Abate, H Gridley, JKO Ampofo, in preparation). However, it has been argued that the availability of resistant varieties has failed to achieve a major impact on subsistence food production (75): First, because local varieties were most resistant probably because of co-evolution and selection by farmers over many years (88); second, farmers in unstable and variable environments plant mixtures of varieties (85, 113) that are more able to respond to erratic rainfall, fluctuations in soil conditions, and to pest and disease problems (77); third, breeding physical characteristics in varieties may have a detrimental effect on palatability, or cooking time. and therefore may be unacceptable to farmers. Therefore, farmers are probably best served by being able to choose from a variety of genotypes to exploit a highly variable environment.

Preventing Pests from Entering the Continent

Many governments in Africa have enacted laws governing the movement of plant materials. Plant quarantine regulations are the responsibility of plant protection departments of the Ministries of Agriculture. Very often these departments are understaffed and inadequately trained and equipped to discharge their responsibilities effectively and efficiently.

Other Control Methods

Although nonchemical control methods such as cultural and mechanical control are important means of keeping pest numbers below economic levels in small-scale agriculture (97), current research has not given the due attention these aspects deserve. Concerted research on traditional methods of pest management by international agricultural research centers is nonexistent to date. Researchers in national programs and nongovernmental organizations have shown greater interest in this area in recent years, but their efforts could benefit greatly from support by national governments and the international agricultural research and donor community as a whole.

PERSPECTIVES OF IPM

Agricultural scientists in national and internal agricultural research systems have been actively engaged in various aspects of pest management in Africa since the beginning of the 1970s. However, research results have not been able to make a lasting impact on African agriculture (118). There is consensus that IPM should be given the highest priority as a pest management strategy for Africa, but how should it be developed and implemented?

As many African farmers are challenged with less rainfall, declining soil fertility, and more marginal land to be cultivated, they must minimize risks and maximize output per unit of labor. For that reason, farmers must have at their disposal appropriate techniques that use locally available cheap procurable inputs instead of expensive external products. Another issue to be considered is that indigenous farming systems are usually grounded in the necessary detailed knowledge of the local environment because crops are grown under heterogeneous conditions (142). Fixed descriptions often do not work, and site-specific pest management solutions are needed to respond to a wide range of conditions (76, 142, 185). Thus farmers must be in the forefront of finding solutions to their specific agricultural problems. What would be the role of formal research in such an "informal research and development" process?

Farmers need information on the biology and ecology of pests. For example, farmers in southern Malawi and Uganda were unaware that sweet potato weevil (*Cylas puncticollis*) adults crawl down the cracks in the sweet potato ridges to lay their eggs on the tubers, and that tamping down the soil when the tubers set may help control the problem (145). Farmers are reluctant to tamp down the cracks in the sweet potato ridges; large cracks are a source of pride because they indicate a bumper harvest, and the time when the tamping should be done conflicts with other farming practices (145). Farmers in southern Malawi were also unaware that the swellings at the base of the bean plants are caused by pupae of the bean stem maggot (*Ophiomyia* spp.), opening the way for opportunistic infection by fungal diseases (145). These pupae are offspring of the tiny flies that they see around the bean plants in the morning. In surveys it also appeared that farmers ignore the beneficial action of natural enemies (MJ Bonhof, in preparation). When farmers realize the role of natural enemies, they would probably worry less about the herbivores in their crops and consequently be more inclined to refrain from the use of pesticides.

If farmers acquire knowledge they will be able to understand the rationale and evaluate the merits of proposed control methods (76). As a result, they will be able to make better choices, and their ability to experiment and to develop own control options will be enhanced. Farmers have proven to be not passive recipients of technology, but innovators capable of generating their own technology (87, 144). In a natural process of innovation farmers will use a cautious, low-risk experimental approach, preserving adaptive diversity (87). The formal sector could provide the inputs and interventions necessary to sustain the self reliance and empowerment of farmers in IPM research and development. Enhancement of farmer-to-farmer communication would also make their innovations more generally available, enabling integration of individual component technologies.

IPM training in the FAO Inter-Country Programme for the Development and Application of Integrated Pest control in Rice in South and Southeast Asia is organized through so-called Farmer Field Schools (FFS) (92, 146), which bring farmers together once a week during an entire growing season for intensive training. This approach has been adopted in several projects across Africa, among vegetable, wheat, and cotton growers in the Gezira Sudan (39) and among food producers in Zanzibar, Tanzania (33). Important components of these meetings are agro-ecosystem analysis, an insect zoo with pests and natural enemies, group dynamic exercises, and participatory action research. Farmers are empowered by fostering participation, self confidence, dialogue, joint decision making and self determination. Facilitators are recruited from research, extension, national plant protection services, or from the farmers' community. The approach may involve disciplines other than crop protection, such as integrated crop, nutrient, and water management and even marketing and credit. Such a holistic approach requires institutional support and change. The FFS approach seems to be emerging as the new mode of extension (28), with IPM being instrumental in making this change possible. IPM as such contributes more than just in the technical sense to a better potential of agricultural development.

ACKNOWLEDGMENTS

Funds for this research were provided in part by the Ethiopian Agricultural Research Organization and as part of the senior author's fellowship supported by the African Bureau of the United States Agency for International Development. The authors would like to thank F Meerman and WA Overholt for comments received during the preparation of this manuscript.

Visit the Annual Reviews home page at www.AnnualReviews.org.

LITERATURE CITED

1. Abate T. 1988. Experiments with trap crops against African bollworm, *Heliothis armigera*, in Ethiopia. *Entomol. Exp. Appl.* 48:135–40
2. Abate T. 1988. Insect and mite pests of horticultural and miscellaneous plants in Ethiopia. In *IAR Handbook No. 1,* pp. 42–43. Addis Ababa, Ethiop.: Inst. Agric. Res. 115 pp.
3. Abate T. 1988. The identity and bionomics of insect vectors of tristeza and greening diseases of citrus in Ethiopia. *Trop. Pest Manage.* 34(1):19–23
4. Abate T. 1990. *Studies on genetic, cultural and insecticidal controls against the bean fly,* Ophiomyia phaseoli *(Tryon) (Diptera: Agromyzidae), in Ethiopia.* PhD diss. Simon Fraser Univ., Burnaby, BC Can. 177 pp.
5. Abate T. 1991. *Entomophagous Arthropods of Ethiopia: A Catalog. Tech. Man. No. 4.* Addis Ababa, Ethiop.: Inst. Agric. Res. 50 pp.
6. Abate T. 1991. The bean fly *Ophiomyia phaseoli* (Tryon) (Dipt., Agromyzidae) and its parasitoids in Ethiopia. *J. Appl. Entomol.* 111:278–85
7. Abate T. 1991. Intercropping and weeding: effects on some natural enemies of African bollworm in bean fields. *J. Appl. Entomol.* 112:38–42
8. Abate T. 1993. Control of other pests and diseases. In *Dryland Farming in Africa*, ed. JRJ Rowland, p. 188–204. London: Macmillan
9. Abate T. 1997. Integrated pest management in Ethiopia: an overview. In *Integrating Biological Control and Host Plant Resistance: Proc. CTA/IAR/IIBC Semin.*, pp. 24–37. Wageningen, Netherlands: Tech. Coop. Rural Agric. Dev. 383 pp.
10. Abate T. 1997. Pesticides in Ethiopian agriculture: a researcher's view. *Pest Manage. J. Ethiop.* 1:49–55
11. Abate T, Ampofo JKO. 1996. Insect pests of beans in Africa: their ecology and management. *Annu. Rev. Entomol.* 41:45–73
12. Abate T, Ayalew G. 1997. Sources of resistance in tomato against fruitworms. *Pest Manage. J. Ethiop.* 1:1–8
13. Abate T, Girma A, Ayalew G. 1998. Bean stem maggots of Ethiopia: their species composition, geographical distribution, importance and population dynamics. *Pest Manage. J. Ethiop.* 2:14–25
14. Abate T, Worku E. 1998. Carbosulfan seed dressing, sowing date, and varietal effects on maize stalk borer. *Pest Manage. J. Ethiop.* 2:85–89
15. Abebe M, Mormene B. 1986. A review of coffee pest management in Ethiopia. In *A Review of Crop Protection Research in Ethiopia,* ed. T Abate, pp. 163–77. Addis Ababa, Ethiop.: Inst. Agric. Res. 685 pp.
16. Ajayi O.1989. Sorghum stemborers in West Africa. In *International Workshop on Sorghum Stem Borers*, ed. KF Nwanze, S Kearl, V Sadhana, pp. 27–31. Patancheru, India: Int. Cent. Res. Semi-Arid Trop. 188 pp.
17. Amoako-Atta B. 1983. Observations on the pest status of striped bean weevil *Alcidodes leucogrammus* Erichs under intercropped systems in Kenya. *Insect Sci. Appl.* 4:351–56
18. Amoaka-Atta B, Omolo E. 1983. Yield losses caused by the stem-/pod-borer complex within maize-cowpea-sorghum intercropping systems in Kenya. *Insect Sci. Appl.* 4(1/2):39–46
19. Amoako-Atta B, Omolo E, Kidega EK.

1983. Influence of maize, cowpea and sorghum intercropping systems on stem/ pod-borer infestation. *Insect Sci. Appl.* 4:47–57
20. Ampofo JKO. 1986. Maize stalk borer (Lepidoptera: Pyralidae) damage and plant resistance. *Environ. Entomol.* 15:1124–29
21. Ampofo JKO. 1989. The expression of maize resistance to the spotted stem borer (Lepidoptera: Pyralidae) in relation to plant age at infestation. *Acta Phytopathol. Entomol. Hung.* 24:17–24
22. Ampofo JKO, Massomo SMS. 1998. Some cultural strategies for the management of bean stem maggots (Diptera: Agromyzidae) on beans in Tanzania. *Afr. Crop Sci. J.* 6:351–56
23. Atteh OD. 1984. Nigerian farmers' perception of pests and pesticides. *Insect Sci. Appl.* 5:213–20
24. Autrique A. 1989. Bean pests in Burundi: their status and prospects for control. *Proc. First Meet. Pan-Afr. Work. Group Bean Entomo., Nairobi, Kenya, 6–9 Aug. CIAT Afr. Workshop Ser.* No. 11, pp. 1–9. Cali, Colomb.: Cent. Int. Agric. Trop. 48 pp.
25. Autrique A, Ntahimpera L. 1994. Atlas des principales espèces de pucerons rencontrées en Afrique sud-Saharienne. *Admin, Gén. Coop. Dév., Publ. Agric. No. 33.* 78 pp.
26. Autrique A, Stary P, Ntahimpera L. 1989. Biological control of pest aphids by hymenopterous parasitoids in Burundi. *FAO Plant Prot. Bull.* 37(2):71–76
27. Balk F, Koeman JH. 1984. Future hazards from pesticide use, with special reference to West Africa and Southeast Asia. *Environmentalist* 4(Suppl. 6). 99 pp.
28. Barfield CS, Swisher ME. 1994. Integrated pest management: ready for export? Historical context and internationalization of IPM. *Food Rev. Int.* 10(2):215–67
29. Bentley JW. 1992. The epistemology of plant protection: Honduran campesinos knowledge of pests and natural enemies. In *Proc. Semin. Crop Prot. Resource-Poor Farmers,* ed. RW Gibson, A Sweetmore, pp. 107–18. Chatham, Kent: Nat. Res. Inst.167 pp.
30. Bonhof MJ, Overholt WA, van Huis A, Polaszek A. 1997. Natural enemies of cereal stemborers in East Africa: a review. *Insect Sci. Appl.* 17(1):19–35
31. Bosque-Pérez NA, Buddenhagen IW. 1992. The development of host-plant resistance to insect pests: outlook for the tropics. In *Proc. 8th Int. Symp. Insect-Plant Relationships*, ed. SBJ Menken, JH Visser, P Harrewijn, pp. 235–49. London: Kluwer Acad. 424 pp.
32. Brettell JH. 1986. Some aspects of cotton pest management in Zimbabwe. *Zimbabwe J. Agric. Res.* 83:41–46
33. Bruin G, van Huis A. 1999. *Final Report Phase V (1993-1998) of the Project "Strengthening the Plant Protection Division of Zanzibar."* Wageningen, Netherlands: Dept. Entomol. Wageningen Agric. Univ.
34. Cent. Int. Agric. Trop. 1997. Project IP-2: Meeting demand for beans in sub-Saharan Africa in sustainable ways. *Annu. Rep. 1997*, pp. 80–82. Cali, Colomb.: Cent. Int. Agric. Tropic.
35. Chambers R. 1992. Scientist or resource-poor farmer—whose knowledge counts ? See Ref. 29, pp. 1–15
36. Chitere PO, Omolo BA. 1993. Farmers 'indigenous knowledge of crop pests and their damage in Western Kenya. *Int. J. Pest Manage.* 39(2):126–32
37. Compton JAF, Tyler PST, Hindmarsh PS, Golob P, Boxall RA, et al. 1993. Reducing losses in small farm grain storage in the tropics. *Trop. Sci.* 33:283–318
38. Conlong DE. 1997. Biological control of *Eldana saccharina* Walker in South African sugarcane: constraints identified from 15 years of research. *Insect Sci. Appl.* 17(1):69–78
39. Dabrowski ZT, ed. 1997. *Integrated Pest*

Management in Vegetables, Wheat and Cotton in the Sudan: a Participatory Approach. Nairobi: ICIPE Sci. Press. 244 pp.
40. de Groot AA. 1995. The functioning and sustainability of village crop protection brigades in Niger. *Int. J. Pest Manage.* 41:243–48
41. de Groot AA. 1995. *La protection des végétaux dans les cultures de subsistence: le cas du mil au Niger de l'Ouest.* Niamey: Dép. Form. Prot. Vég., Cent. Aghrymet, Niamey Niger. 90 pp.
42. Delleré R, Symoens JJ, eds. 1991. *Agricultural Intensification and Environment in Tropical Areas, Semin. Proc.* Wageningen, Netherlands: Tech. Cent. Agric. Rural Coop. 202 pp.
43. de Schlippe P. 1956. *Shifting Cultivation in Africa: the Zande System of Agriculture.* London: Routledge & Kegan Paul. 304 pp.
44. Ehler LE, Andres LA. 1983. Biological control: exotic natural enemies to control exotic pests. In *Exotic Plant Pests and North American Agriculture*, ed. CL Wilson, CL Graham, p. 395–418. New York: Academic. 522 pp.
45. Emehute JKU, Egwuatu RI. 1990. Effect of field populations of cassava mealy bug, *Phenacoccus manihoti*, on cassava yield and *Epidinocarsis lopezi* at different planting dates in Nigeria. *Trop. Pest Manage.* 36:279–81
46. Food Agric. Org. 1995. *Production Yearbook 1995, Vol. 49*. Rome: FAO. 235 pp.
47. Food Agric. Org. 1995. *Trade Yearbook 1995, Vol. 49.* Rome: FAO. 378 pp.
48. Food Agric. Org. 1996. *Production Yearbook 1996, Vol. 50.* Rome: FAO. 325 pp.
49. Food Agric. Org. 1997. *Production Yearbook 1997, Vol. 51.* Rome: FAO. 239 pp.
50. Food Agric. Org. 1998. *Inventory of Obsolete, Unwanted and/or Banned Pesticides.* Rome: FAO. 247 pp.
51. Food Agric. Org. 1998. *Obsolete Pesticides: Problems, Prevention and Disposal.* Rome: FAO. 4 pp.
52. Farrington J. 1977. Economic thresholds of insect pest infestation in peasant agriculture: a question of applicability. *PANS* 23:143–48
53. Faure JC. 1944. Pentatomid bugs as human food. *J. Entomol. Soc. South. Afr.* 7:111–12
54. Fresco LO, Westphal E. 1988. A hierarchical classification of farm systems. *Exp. Agric.* 24:399–419
55. Gahukar RT. 1988. Problems and perspectives of pest management in the Sahel: a case study of pearl millet. *Trop. Pest Manage.* 4:35–38
56. Gebre-Amlak A. 1985. Survey of lepidopterous stalk borers attacking maize and sorghum in Ethiopia. *Ethiop. J. Agric. Sci.* 7:15–26
57. Gebre-Amlak A, Sigvald R, Pettersson J. 1989. The relationship between sowing date, infestation and damage by the maize stalk borer, *Busseola fusca* (Noctuidae), on maize in Awassa, Ethiopia. *Trop. Pest Manage.* 35:143–45
58. Geddes AMW. 1990. The relative importance of crop pests in sub-Saharan Africa. *Nat. Resourc. Res. Inst. Bull. No. 36.* p. vi. Chatham, UK: Nat. Resourc. Res. Inst. 69 pp.
59. Gethi M, Omolo EO, Mueke JM. 1993. The effect of intercropping on relative resistance and susceptibility of cowpea cultivars to *Maruca testulalis* Geyer when in mono- and when intercropped with maize. *Insect Sci. Appl.* 14:305–13
60. Getu E, Gebre-Amlak A. 1998. Arthropod pests of stored maize in Sidama zone: economic importance and management practices. *Pest Manage. J. Ethiop.* 2:26–35
61. Giga DP, Ampofo JKO, Silim MN, Negasi F, Nahimana M, et al. 1992. Onfarm storage losses due to bean bruchids and farmers' control strategies: a report on a travelling workshop in eastern and southern Africa. *Occas. Publ. Ser. No. 8.* Cali, Colomb: CIAT
62. Giliomee JH. 1989. First record of

western flower thrips, *Frankliniella occidentalis* (Pergande) (Thysanoptera: Thripidae) from South Africa. *J. Entomol. Soc. South. Afr.* 52(1):179–82
63. Goldman A. 1996. Pest and disease hazards and sustainability in African agriculture. *Exp. Agric.* 32:199–211
64. Gordon H, Chiri A, Abate T. 1995. *Environmental and Economic Review of Crop Protection and Pesticide Use in Ethiopia.* Arlington, VA: Winrock Int. Environ. Alliance. 117 pp.
65. Greathead DJ, Girling DJ. 1982. Possibilities for natural enemies in *Heliothis* management and the contribution of the Commonwealth Institute of Biological Control. *Proc. Int. Workshop* Heliothis *Manage.* pp. 147–58. Patancheru, India: Int. Cent. Res. Semi-Arid Trop.
66. Greathead DJ, Girling DJ. 1989. Distribution and economic importance of *Heliothis* and of their natural enemies and host plants in southern and eastern Africa. In *Proc. Workshop Biological Control of* Heliothis*: Increasing the Effectiveness of Natural Enemies*, ed. EG King, RD Jackson, pp. 329–45. Patancheru, India: Int. Cent. Res. Semi-Arid Trop.
67. Hammond WNO, Neuenschwander P, Yaninek JS, Herren HR. 1992. Biological control in cassava: a viable crop protection package for resource-poor farmers. See Ref. 29, pp. 45–54
68. Harris KM. 1989. Recent advances in sorghum and pearl millet stem borer research. See Ref. 16, pp. 9–16
69. Harris KM. 1989. Bio-ecology of sorghum stem borers. See Ref. 16, pp. 63–71
70. Harris WV. 1969. *Termites as Pests of Crops and Trees.* London: Commonw. Inst. Entomol. 41 pp.
71. Haskell PT, Beacock T, Wortley PJ. 1981. World wide socio-economic constraints to crop protection. In *Proc. Int. Congr. Plant Prot., 9th,* ed. T Kommedahl, 1:39–41
72. Herren HR. 1991. Biological control as primary option in sustainable pest management: the cassava pest project example. See Ref. 42, p. 133–48
73. Herren HR, Neuenschwander P. 1991. Biological control of cassava pests in Africa. *Annu. Rev. Entomol.* 36:257–83
74. Hill DS. 1983. *Agricultural Insect Pests of the Tropics and Their Control*, pp. 260, 287–89. London: Cambridge Univ. Press. 746 pp.
75. Hillocks RJ. 1995. Integrated management of insect pests, diseases and weeds of cotton in Africa. *Integr. Pest Manage. Rev.* 1(Suppl. 1):31–47
76. Hillocks RJ, Logan JWM, Riches CR, Russell-Smith A, Shaxson LJ. 1996. Soil pests in traditional farming systems in sub-Saharan Africa—a review. Part II. Management strategies. *Int. J. Pest Manage.* 42(4):253–65
77. Hoekstra GJ, Kannenburg LW, Christie BR. 1985. Grain yield comparison of pure stands and equal proportion mixtures of seven hybrids of maize. *Can. J. Plant Sci.* 65:447–79
78. ICIPE. 1997. Towards improved pest management on export vegetables in Africa. *Annu. Tech. Rep. Oct. 1996–Sept. 1997.* Nairobi: Int. Cent. Insect Physiol. Ecol. 61 pp.
79. Ingram WR. 1969. Observations on the pest status of bean flower thrips in Uganda. *E. Afr. Agric. For. J.* 34:482–84
80. Ingram WR. 1983. Biological control of graminaceous stem-borers and legume pod borers. *Insect Sci. Appl.* 4(1/2):205–9
81. Jager I. 1995. The hazards of pesticide use for man and the ecosystem. In *Pesticides in Tropical Agriculture: Hazards and Alternatives*, pp. 23–62. Weikersheim, Germany: Margraf Verlag
82. Jago ND. 1992. IPM in the Sahelian zone, peasant-level farm environment of north-west Mali. See Ref. 29, pp. 25–32
83. Jago ND. 1993. *Millet Pests of the Sahel:*

Biology, Monitoring and Control. Chatham, UK: Nat. Res. Inst. 72 pp.

84. Jago ND, Kremer AR, West C. 1993. Pesticides on millet in Mali. *NRI Bull. 50*. Chatham, UK: Nat. Res. Inst.
85. Jiggins J. 1990. Crop variety mixtures in marginal environments. *Sustainable Agric. Programme Int. Inst. Environ. Dev., Gatekeeper Ser. No. SA19*. London, UK: Int. Inst. Environ. Dev. 13 pp.
86. Joffe S. 1995. Desert locust management: a time for change. *World Bank Discuss. Pap. 284*, Washington, DC: World Bank. 53 pp.
87. Johnson AW. 1972. Individuality and experimentation in traditional agriculture. *Hum. Ecol.* 1:149–59
88. Johnson RA, Lamb RW, Wood TG. 1981. Termite damage and crop loss studies in Nigeria—a survey of damage to groundnuts. *Trop. Pest Manage.* 27:325–42
89. Juo ASR. 1989. New farming systems development in the wetter tropics. *Trop. Pest Manage.* 25:145–63
90. Kamburov SS. 1986. New pests and beneficial insects on citrus in South Africa—Part I. *Citrus Subtrop. Fruit J.* 625:6–7, 9–11
91. Katundu JM. 1980. Agromyzid leafminer: a new pest to Tanzania. *Trop. Grain Legume Bull.* 20:8–10
92. Kenmore PE, Litsinger JA, Bandong JP, Santiago AC, Salac MM. 1987. Philippine rice farmers and insecticides: thirty years of growing dependency and new options for change. In *Management of Pests and Pesticides*, ed. J Tait, B Napompeth, pp. 98–109. Boulder, CO: Westview
93. Kfir R. 1997. Competitive displacement of *Busseola fusca* by *Chilo partellus*. *Ann. Entomol. Soc. Am.* 90(5):619–24
94. Kfir R. 1997. Natural control of the cereal stemborers *Busseola fusca* and *Chilo partellus* in South Africa. *Insect Sci. Appl.* 17(1):61–67
95. Khan ZR, Ampong-Nyarko K, Chiliswa P, Hassanali A, Kimani S, et al. 1997. Intercropping increases parasitism of pests. *Nature* 388: 631–32
96. Khan ZR, Chiliswa P, Ampong-Nyarko K, Smart LE, Polaszek A, et al. 1997. Utilisation of wild gramineous plants for management of cereal stemborers in Africa. *Insect Sci. Appl.* 17(1):143–50
97. Kiss A, Meerman F. 1991. Integrated pest management and African Agriculture. *World Bank Tech. Pap. No. 42*. Washington, DC: World Bank
98. Krall S, Youm O, Kogo SA. 1995. Panicle insect pest damage and yield loss in pearl millet. In *Panicle Insect Pests of Sorghum and Pearl Millet: Proceedings of an International Consultative Workshop*, ed. KF Nwanze, O Youm, p. 135–45. Patancheru, India: Int. Cent. Res. Semi-Arid Trop.
99. Kyamanywa S. 1988. *Ecological factors governing* Megalurothrips sjostedti *(Trybom) populations on cowpea/maize intercropped systems*. PhD diss. Makerere Univ. Kampala, Uganda. 253 pp.
100. Kyamanywa S, Ampofo JKO. 1988. Effect of cowpea/maize mixed cropping on the incident light at the cowpea canopy and flower thrips (Thysanoptera: Thripidae) population density. *Crop Prot.* 7(3):186–89
101. Kyamanywa S, Baliddawa CW, Ampofo JKO. 1993. Effect of maize plants on colonization of cowpea plants by bean flower thrips *Megalurothrips sjostedti*. *Entomol. Exp. Appl.* 69:61–68
102. Kyamanywa S, Tukahirwa EM. 1988. The effect of mixed cropping beans, cowpeas and maize on population densities of bean flower thrips, *Megalurothrips sjostedti* (Trybom) (Thripidae). *Insect Sci. Appl.* 9:255–59
103. Lagemann J. 1977. Traditional African farming systems in eastern Nigeria: an analysis of reaction to increasing population pressure. In *Afrika-Studien, No. 98*. Munich: Weltforum-Verlag. 269 pp.
104. Leuschner K. 1995. Insect pests of sor-

ghum panicles in eastern and southern Africa. See Ref. 98, pp. 49–56
105. Löhr B, Michalik S. 1997. GTZ-IPM horticulture project—achievements of the last three years in French bean research. In *HCDA-ICIPE-USAID Training Workshop Pest Manag. Vegetable Crops Hortic. Techn. Off., Programme Proc.* Nairobi: Int. Cent. Insect Physiol. Ecol.
106. Matteson PC, Altieri MA, Gagne WC. 1984. Modification of small farmer practices for better pest management. *Annu. Rev. Entomol.* 29:383–402
107. Matteson PC, Ferraro PJ, Knausenberger W. 1995. *Pesticide Use and Management in Madagascar: Subsector Review and Programmatic Environmental Assessment.* Washington, DC: U.S. Agency Int. Dev. 131 pp.
108. Matteson PC, Meltzer MI.1994. *Environmental Implications of Agricultural Trade and Policy Reform Programmes in Cameroon: Pest and Pesticide Management.* Washington, DC: U.S. Agency Int. Dev. 42 pp.
109. Matteson PC, Meltzer MI. 1995. *Environmental and Economic Implications of Agricultural Trade and Promotion Policies in Kenya: Pest and Pesticide Management.* Washington, DC: U.S. Agency Int. Dev. 90 pp.
110. Meinzingen WF, ed. 1993. *A Guide to Migrant Pest Management in Africa.* Rome: FAO. 184 pp.
111. Meltzer M, Matteson P, Knausenberger W. 1994. *Environmental and Economic Implications of Agricultural Trade and Promotion Policies in Uganda: Pest And Pesticide Management.* Washington, DC: USAID. 103 pp.
112. Mielke HW. 1987. Termitaria and shifting cultivation: the dynamic role of the termite in soils of tropical wet-dry Africa. *Trop. Ecol.* 19(1):117–22
113. Mohamed RA, Teri RA. 1989. Farmers' strategies of insect pest and disease management in small-scale bean productions systems in Mgeta Tanzania. *Insect Sci. Appl.* 10:821–25
114. Moore D. 1992. Biological control of mango mealybug. In *Biological Control Manual,* Vol. 2: *Case Studies of Biological Control in Africa*, ed. RH Markham, A Wodageneh, S Agboola, pp. 95–125. UNDP/IITA/FAO/CABI Int./OAU. Silwood Park, UK: Int. Inst. Biol. Control
115. Nderitu JH, Kayumbo HY, Mueke JM. 1990. Effect of date of sowing on beanfly infestation of the bean crop. *Insect Sci. Appl.* 11:97–101
116. Neuenschwander P. 1992. Biological control of cassava mealybug in Africa. See Ref. 114, pp. 3–47
117. Nwanze KF. 1985. Sorghum insect pests in West Africa. *Proc. Int. Sorghum Entomol. Workshop*, ed. K Leuschner, GL Teetes, V Kumble, p. 37–43. Patancheru, India: Int. Cent. Res. Semi-Arid Trop.
118. Nwanze KF. 1997. Integrated management of stem borers of sorghum and pearl millet. *Insect Sci. Appl.* 17(1):1–8
119. Nwanze KF, Klaij MC, Markham R. 1995. Possibilities for integrated management of millet earhead caterpillar, *Heliocheilus albipunctella.* See Ref. 98, pp. 263–71
120. Nyambo BT. 1988. Significance of host-plant phenology in the dynamics and pest incidence of the cotton bollworm, *Heliothis armigera* Hubner (Lepidoptera: Noctuidae), in western Tanzania. *Crop Prot.* 7:161–67
121. Nyambo BT. 1990. Effect of natural enemies on the cotton bollworm, *Heliothis armigera* (Hubner) (Lepidoptera: Noctuidae) in western Tanzania. *Trop. Pest Manage.* 36:50–58
122. Nyiira ZM. 1973. Pest status of thrips and lepidopterous species on vegetables in Uganda. *E. Afr. Agric. For. J.* 39:131–35
123. Nyirenda GKC. 1992. Insect pest management in cotton on small-scale farmers' fields in Malawi. See Ref. 29, p. 33–43

124. Oerke EC, Weber A, Dehne HW, Schönbeck F. 1994. Conclusions and perspectives. In *Crop Production and Crop Protection: Estimated Losses in Major Food and Cash Crops*, ed. EC Oerke, HW Dehne, F Schönbeck, A Weber, pp. 742–70. Amsterdam: Elsevier. 808 pp.
125. Ogenga-Latigo MW, Ampofo JKO, Baliddawa CW. 1992. Influence of maize row spacing on infestation and damage of inter-cropped beans by the bean aphid *(Aphis fabae* Scop.). I: incidence of aphids. *Field Crops Res.* 30:111–21
126. Ogenga-Latigo MW, Baliddawa CW, Ampofo JKO. 1992. Influence of maize row spacing on infestation and damage to intercropped beans by the bean aphid *Aphis fabae* (Scop.). II: reduction of bean yields. *Field Crops Res.* 30:123–30
127. Ogwaro K. 1983. Intensity levels of stem-borers in maize and sorghum and the effect on yield under different intercropping patterns. *Insect Sci. Appl.* 4(1/2):33–37
128. Okeyo-Owuor JB, Oloo GW, Agwaro PO. 1991. Natural enemies of the legume pod borer, *Maruca testulalis* Geyer (Lepidoptera: Pyralidae) in small scale farming systems in western Kenya. *Insect Sci. Appl.* 12:35–41
129. Okigbo BN, Greenland DJ. 1976. Intercropping systems in tropical Africa. In *Multiple Cropping: Proceedings of a Symposium of the American Society of Agronomy, ASA Spec. Publ. 27*, ed. RI Papendick, PA Sanchez, GB Triplett, pp. 63–101. Madison, WI: Am. Soc. Agronomy
130. Omolo EO.1988. Intercropping and its use in controlling stem borer complex. In *Crop Protection for Small-Scale Farms in Eastern and Central Africa—A Review*, ed. RT Prinsley, PJ Terry, pp. 44–54. London: Commonw. Secr.
131. Omolo EO, Nyambo B, Simbi COJ, Ollimo P. 1993. The role of host plant resistance and intercropping in integrated pest management (IPM) with specific reference to the Oyugis project. *Int. J. Pest Manage.* 39:265–72
132. Ondieki JJ. 1996. The current state of pesticide management in sub-Saharan Africa. In *Science of the Total Environment, Supplement 1*, ed. R Dardozzi, C Ramel, pp. S30–34. Nairobi: Int. Cent. Insect Physiol. Ecol.
133. Overholt WA. 1996. *Classical Biological Control of Cereal Stemborers in Africa: Summary of Research 1991–1996.* Nairobi: Int. Cent. Insect Physiol. Ecol.
134. Overholt WA. 1998. Biological control. See Ref. 138, pp. 349–62
135. Pelerents C. 1991. Introduction to integrated control. See Ref. 42, pp. 119–32
136. Perrin RM, Phillips ML. 1978. Some effects of mixed cropping on the population dynamics of insect pests. *Entomol. Exp. Appl.* 24:385–93
137. Polaszek A. 1997. An overview of parasitoids of African lepidopteran cereal stemborers (Hymenoptera: Chrysidoidea, Ceraphronoidea, Chalcidoidea, Ichneumonoidea, Ichneumonoidea, Plastygastroidea). *Insect Sci. Appl.* 17(1):13–18
138. Polaszek A, ed. 1998. *African Cereal Stem Borers: Economic Importance, Taxonomy, Natural Enemies and Control*, pp. 3–72. Wallingford, UK: CAB Int. 530 pp.
139. Ratnadass A, Ajayi O. 1995. Panicle insect pests of sorghum in West Africa. See Ref. 98, pp. 29–38
140. Raymundo SA. 1986. Traditional pest control practices in West Africa. *Int. Rice Res. Network* 11(1):24
141. Refera A, Demissie A. 1986. A review of research on insect pests of fibre crops in Ethiopia. See Ref. 15, pp. 215–22
142. Reij C. 1991. Indigenous soil and water conservation in Africa. In *Sustainable Agriculture Programme of the International Institute for Environment and Development Gatekeeper Ser. No. 27.* London: Int. Inst. Environ. Dev. 35 pp.
143. Repetto R, Baliga SS. 1996. *Pesticides*

and the Immune System: The Public Health Risks. Washington, DC: World Resour. Inst. 104 pp.

144. Richards P. 1985. *Indigenous Agricultural Revolution*. London: Hutchinson. 192 pp.
145. Riches CR, Shaxson LJ, Logan JWM, Munthali DC. 1993. Insect and parasitic weed problems in southern Malawi and the use of farmer knowledge in the design of control measures. In *Agricultural Administration Res. Ext. Netw., 42*. pp. 1–17. London: Overseas Dev. Inst.
146. Röling NG, van der Fliert E. 1998. Introducing integrated pest management in rice in Indonesia: a pioneering attempt to facilitate large-scale change. In *Facilitating Sustainable Agriculture*, ed. NG Röling, MAE Wagemakers, pp. 158–71. Cambridge, UK: Cambridge Univ. Press. 318 pp.
147. Ross S. 1998. Farmers' perception of bean pest problems in Malawi. In *Network on Bean Research in Africa, Occas. Publ. No. 25*. Kampala, Uganda: CIAT. 31 pp.
148. Ruthenberg H. 1976. *Farming Systems in the Tropics*. London: Oxford Univ. Press. 366 pp.
149. Sangodoyin AY. 1993. Field evaluation of the possible impact of some pesticides on the soil and water environment in Nigeria. *Exp. Agric.* 29:227–32
150. Sanou LR. 1984. *Première liste sur les méthodes traditionelles de lutte contre les ennemis des principales cultures vivières au Sahel recensées dans cinq pays membres du CILSS*. Ouagadougou: Comité Permanent Inter-États de Lutte contre Sécheresse Sahel (in French)
151. Schaefers GA. 1990. Public sector pesticide use in Africa. *J. Agric. Entomol.* 7(3):183–90
152. Seshu Reddy KV. 1989. Sorghum stem borers in eastern Africa. See Ref. 16, pp. 9–16
153. Seshu Reddy KV, Omolo FO. 1985. Sorghum insect pest situation in eastern Africa. See Ref. 116, pp. 37–43
154. Shanower TG, Romeis J, Minja EM. 1999. Insect pests of pigeonpea and their management. *Annu. Rev. Entomol.* 44:77–96
155. Sharah Uvu HA. 1992. Indigenous pest control systems and cultivation practices for resource-poor farmers. See Ref. 29, pp. 65–74
156. Sharland R. 1990. A trap, a fish poison and culturally significant pest control. *Newslett. Inf. Cent. Low-External-Input Sustainable Agric.* 6(1):12–13
157. Singh SR, Jackai LEN. 1985. Insect pests of cowpea in Africa: their life cycle, economic importance and potential for control. In *Cowpea: Research, Production and Utilization*, ed. SR Singh, KO Rachie, pp. 217–31. New York: Wiley
158. Sithanantham S. 1989. Status of bean entomology research in Zambia. See Ref. 24, pp. 17–20
159. Sithole SZ. 1989. Sorghum stem borers in southern Africa. See Ref. 16, pp. 41–47
160. Slumpa S, Kabungo D. 1989. Status of bean entomology research in Tanzania. See Ref. 24, pp. 13–16
161. Smit NEJM, Matengo LO. 1995. Farmers' cultural practices and their effects on pest control in sweet potato in south Nyanza, Kenya. *Int. J. Pest Manage.* 41:2–7
162. Southwood TRE, Way MJ. 1970. Ecological background to pest management. *Concepts of Pest Management, Proc. Conf. N.C. State Univ.*, ed. RL Rabb, FE Guthrie, pp. 6–29. N.C. State Univ., Raleigh. 242 pp.
163. Spencer KA. 1985. East African Agromyzidae (Diptera): further descriptions, revisionary notes and new records. *J. Nat. Hist.* 19(5):969–1027
164. Steiner KG. 1982. *Intercropping in tropical smallholder agriculture with special reference to West Africa*. Eschborn: Ger. Tech. Coop. 303 pp.

165. Stoll G. 1986. *Natural Crop Protection Based on Local Farm Resources in the Tropics and Subtropics.* Langen: Margraf. 188 pp. 2nd ed.
166. Storey HH. 1936. Virus diseases of African plants: VI. A progress report on the disease of cassava. *E. Afr. Agric. J.* 2:439
167. Stroud A, Mekuria M. 1992. Ethiopia's agricultural sector: an overview. In *Research with Farmers: Lessons from Ethiopia,* ed. S Franzel, H van Houten, pp. 9–27. Wallingford, UK: CAB Int.
168. Symmons P. 1992. Strategies to combat the Desert Locust. *Crop Prot.* 11:206–12
169. Tadesse A. 1997. Arthropods associated with stored maize and farmers' management practices in the Bako area, western Ethiopia. *Pest Manage. J. Ethiop.* 1:19–27
170. Taylor TA. 1969. On the population dynamics and flight activity of *Taeniothrips sjostedti* (Trybom) on cowpea. *Bull. Entomol. Soc. Nigeria* 2:60–71
171. Thomas M, Waage J. 1996. *Integration of Biological Control and Host-Plant Resistance Breeding: A Scientific and Literature Review.* Wageningen, Netherlands: Techn. Cent. Agric. Rural Cooperation. 99 pp.
172. Thurston HD. 1990. Plant disease management practices of traditional farmers. *Plant Dis.* 74(2):96–102
173. Tobin RJ. 1994. Bilateral donor agencies and the environment: pest and pesticide management. Washington, DC: U.S. Agency Int. Dev. 124 pp.
174. Tukahirwa EM, Coaker TH. 1982. Effect of mixed cropping on some insect pests of brassicas: reduced *Brevicoryne brassicae* infestations and influences on epigeal predators and the disturbance of oviposition behaviour in *Delia brassicae. Entomol. Exp. Appl.* 32:129–40
175. UNCED. 1992. *Promoting Sustainable Agriculture and Rural Development,* Agenda 21, Chapter 14, pp. 22–26. Switzerland: UN Conf. Environ. Dev., June 3–14, Rio de Janeiro
176. USAID. 1993. *The Impact of Agricultural Technology in Sub-Saharan Africa: A Synthesis of Symposium Findings.* Tech. Pap. No. 3.
177. USAID. 1994. *Opportunities for Success in Integrated Pest Management: Socioeconomic Conditions of Farmers in Mali.* Washington, DC: USAID Bur. Afr.
178. USAID. 1994. *Pesticides and the Agricultural Industry in Sub-Saharan Africa.* Proj. No. 698-0510 under 936-5555. Arlington, VA: Winrock Int. Environ. Alliance. 177 pp.
179. van den Berg H. 1993. *Natural control of* Helicoverpa armigera *in smallholder crops in East Africa.* PhD diss. Wageningen Agric. Univ., Wageningen, Netherlands. 233 pp.
180. van den Berg H, Cock MJW. 1995. Natural control of *Helicoverpa armigera* in cotton: assessment of the role of predation. *Biocontrol Sci. Technol.* 5:453–63
181. van den Berg J, van Rensburg GDJ, van der Westhuizen MC. 1994. Host plant resistance and chemical control of *Chilo partellus* (Swinhoe) and *Busseola fusca* (Fuller) in an integrated pest management system on grain sorghum. *Crop Prot.* 13(Suppl. 4):308–10
182. van Huis A. 1991. Biological methods of bruchid control in the tropics. *Insect Sci. Appl.* 12:87–102
183. van Huis A. 1992. New developments in desert locust management and control. *Proc. Exp. App. Entomol. N.E.V. Amsterdam* 3:2–18
184. van Huis A. 1996. The traditional use of arthropods in sub-Saharan Africa. *Proc. Exp. Appl. Entomol. N.E.V. Amsterdam* 7:3–20
185. van Huis A, Meerman F. 1997. Can we make IPM work for resource-poor farmers in sub-Saharan Africa? *Int. J. Pest Manage.* 43:313–20
186. van Rheenen HA, Hasselbaach OE, Muigai GS. 1981. The effect of growing beans together with maize on the inci-

dence of bean diseases and pests. *Neth. J. Plant Pathol.* 87:193–99

187. Waage JK. 1993. Making IPM work: developing country experiences and prospects. In *Agriculture and Environmental Challenges. Proc. 13th Agric. Symp.,* ed. JP Srivastava, H Alderman, pp. 119–34. Washington, DC: World Bank
188. Wardell DA. 1987. Control of termites in nurseries and young plantations in Africa: established practices and alternative courses of action. *Commonw. For. Rev.* 66(Suppl. 1):77–89
189. Westphal E. 1975. *Agricultural Systems in Ethiopia. Agric. Res. Rep. 826.* Wageningen, Netherlands: Wageningen Agric. Univ. 278 pp.
190. Widanapathirana AS. 1990. Learning from traditional dryland farmers. *ILEIA (Low External Input for Sustainable Agriculture) Newsl.* 6(1): 20–21
191. Wortmann CS, Allen DJ. 1994. *African Bean Production Environment: Their Characteristics, Constraints and Opportunities. Occas. Publ. Ser. No. 11.* Cali, Colomb: CIAT. 47 pp.
192. Wortmann CS, Fischler M, Alifugani F, Iaizzi CK. 1998. Accomplishments of participatory research for systems improvement in Iganga District, Uganda 1993–97. In *Network on Bean Research in Africa, Occas. Ser. No. 27.* Kampala, Uganda: CIAT. 40 pp.
193. Wortman CS, Kirkby RA, Eledu CA, Allen DJ. 1998. *Atlas of Common Bean (*Phaseolus vulgaris*) Production in Africa. CIAT Publ. No. 297.* Cali, Colomb: CIAT. 133 pp.
194. Yitaferu K, Walker AK. 1997. Studies on the maize stemborer, *Busseola fusca* (Lepidoptera:Noctuidae) and its major parasitoid, *Dolichogenida fuscivora* (Hymenoptera:Braconidae) in eastern Ethiopia. *Bull. Entomol. Res.* 87:319–24
195. Youdeowei A. 1989. Major arthropod pests of food and industrial crops of Africa and their economic importance. In *Biological Control: A Sustainable Solution to Crop Pest Problems in Africa,* ed. JS Yaninek, HR Herren. Cotonou, Benin: Int. Inst. Trop. Agric. 210 pp.
196. Youm O. 1995. Bioecology of scarab beetle *Rhinyptia infuscata* and millet head miner *Heliocheilus albipunctella.* See Ref. 98, pp. 115–24
197. Youm O, Baidu-Forson J. 1995. Farmers' perceptions of insect pests and control strategies, their relevance to IPM in pearl millet. See Ref. 98, pp. 291–96
198. Youm O, Gilstrap FE, Teetes GL. 1990. Pesticides in traditional farming systems in West Africa. *J. Agric. Entomol.* 7:171–81
199. Zehrer W. 1986. Traditional agriculture and integrated pest management. *ILEIA (Low External Input for Sustainable Agriculture) Newsl.*, 6(Nov.):4–6
200. Zethner O. 1995. Practice of integrated pest management in tropical and subtropical Africa: an overview of two decades (1970–1990). In *Integrated Pest Management in the Tropics: Current Status and Future Prospects,* ed. AN Mengech, KN Saxena, HNB Gopalan, pp. 1–67. Chichester, UK: Wiley
201. Zethner O, Laurense AA. 1988. The economic importance and control of the adult blister beetle *Psalydolytta fusca* Olivier (Coleoptera: Meloidae). *Trop. Pest Manage.* 34(4):407–12

Annu. Rev. Entomol. 2000. 45:661–708

THE DEVELOPMENT AND EVOLUTION OF EXAGGERATED MORPHOLOGIES IN INSECTS

Douglas J. Emlen[1] and H. Frederik Nijhout[2]
[1]*Division of Biological Sciences, The University of Montana, Missoula, Montana 59812-1002; e-mail: demlen@selway.umt.edu*
[2]*Department of Zoology, Duke University, Durham, North Carolina 27708; e-mail: hfn@acpub.duke.edu*

Key Words allometry, polyphenism, threshold traits, sexual selection, castes

■ **Abstract** We discuss a framework for studying the evolution of morphology in insects, based on the concepts of "phenotypic plasticity" and "reaction norms." We illustrate this approach with the evolution of some of the most extreme morphologies in insects: exaggerated, sexually selected male ornaments and weapons, and elaborate social insect soldier castes. Most of these traits scale with body size, and these scaling relationships are often nonlinear. We argue that scaling relationships are best viewed as reaction norms, and that the evolution of exaggerated morphological traits results from genetic changes in the slope and/or shape of these scaling relationships. After reviewing literature on sexually selected and caste-specific structures, we suggest two possible routes to the evolution of exaggerated trait dimensions: (*a*) the evolution of steeper scaling relationship slopes and (*b*) the evolution of sigmoid or discontinuous scaling relationship shapes. We discuss evolutionary implications of these two routes to exaggeration and suggest why so many of the most exaggerated insect structures scale nonlinearly with body size. Finally, we review literature on insect development to provide a comprehensive picture of how scaling relationships arise and to suggest how they may be modified through evolution.

INTRODUCTION

Insects take shape to the extreme: eyes on the ends of long stalks, forelegs longer than twice the body length, long, serrated mandibles—again, sometimes reaching lengths greater than the rest of the body, and countless knobs, spurs, and horns extending from all parts of the head and thorax (4, 41, 124, 174). In some cases, the sizes of these traits can be so extreme that they yield some of the most bizarre-looking organisms in the animal world (Figure 1).

Species with extraordinary morphologies are also characterized by extreme variation in morphology, so that not all individuals express the trait to the same extent (9, 102, 105, 124). Often, the exaggerated traits are expressed in only one sex, as, for example, in the case of the huge head and mandibles of soldier ant

0066-4170/00/0107-0661/$14.00

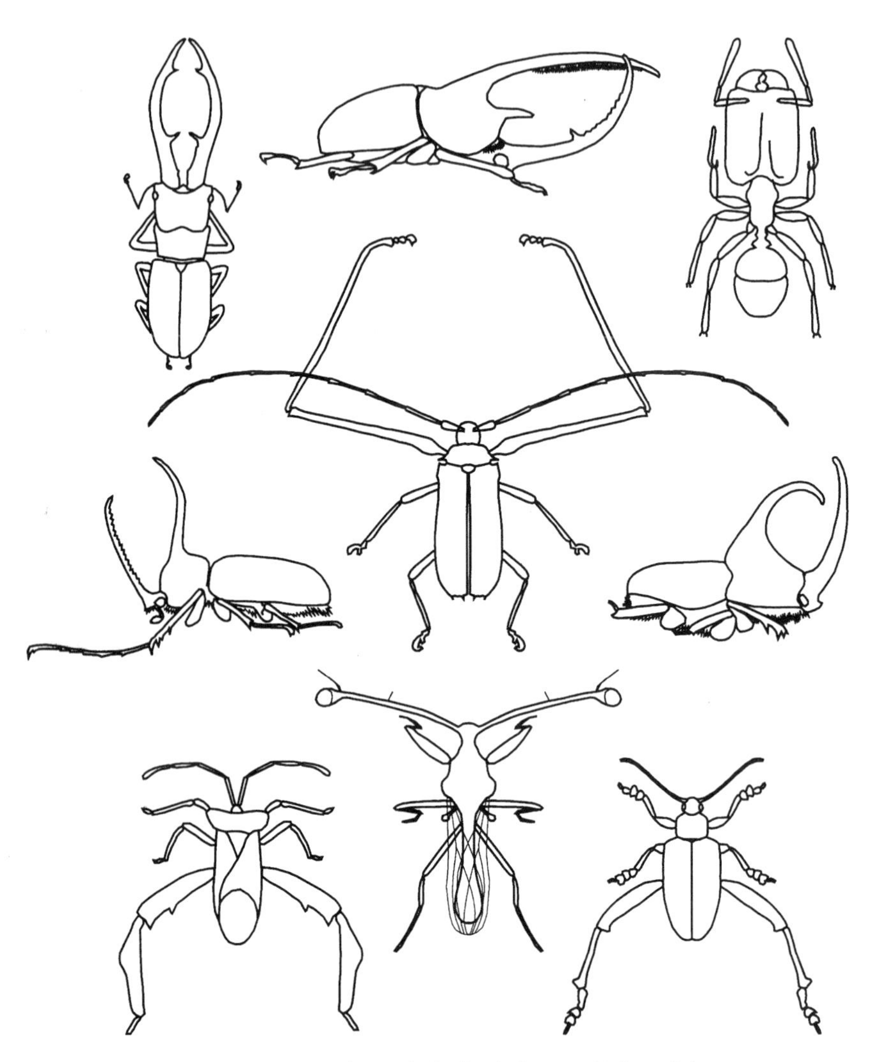

Figure 1 Examples of exaggerated morphologies in insects. *Left to right, top row:* mandibles in *Cyclommatus imperator* (Coleoptera: Lucanidae); head and thoracic horns in *Dynastes neptunus* (Coleoptera: Scarabaeidae); head width in *Pheidole tepicana* (Hymenoptera: Formicidae); *Middle row:* head and thoracic horns in *Golofa porteri* (Coleoptera: Scarabaeidae); forelegs in *Acrocinus longimanus* (Coleoptera: Cerambycidae); head and thoracic horns in *Enema pan* (Coleoptera: Scarabaeidae); *Bottom row:* hind legs in *Acanthocephala declivis* (Hemiptera: Coreidae); eyestalks in *Cyrtodiopsis whitei* (Diptera: Diopsidae); hind legs in *Sagra papuana.* (Coleoptera: Chrysomelidae).

castes (all females; e.g. 238), or the enlarged legs or horns in beetles (generally, all males; e.g. 4, 65). In addition, trait size often scales with body size, so that in a population individuals range from small, relatively normally proportioned animals, to very large animals with grossly enlarged structures.[1]

Most measurable aspects of the insect body covary with body size (e.g. large flies have larger wings than small flies). When measurements are collected for large numbers of individuals of similar age or at the same life stage, it is possible to characterize the precise relationship between the dimensions of each trait and individual variations in overall body size ("static allometry;" 28, 32, 118). The slopes of these scaling relationships vary almost as much as the shapes of the traits themselves, from no slope (size-invariant trait expression) to very steep slopes (traits become disproportionately larger with increasing body size), and even in a few cases, to negative slopes (traits become proportionately smaller with increasing body size). A large number of scaling relationships depart from linearity, with discontinuous and sigmoid patterns surprisingly widespread. What can the study of scaling relationships tell us, and what, if anything, can we learn from the variations in slope and shape of these scaling relationships?

Here we survey the range of variation that occurs in insect scaling relationships, with particular emphasis on exaggerated morphological traits. We provide a framework for viewing and studying the evolution of exaggerated traits and their scaling relationships that builds on recent developments in the fields of reaction norms and phenotypic plasticity.[2] While this view is not entirely new, it is seldom made explicit for the study of trait allometry. We make this framework explicit because it incorporates a more accurate appreciation for how exaggerated traits are inherited and because it offers new and informative avenues for future research. We summarize this review with six points: *(a)* the decades-old view of allometries as constraints to evolution is inaccurate and misleading; *(b)* scaling relationships may be considered a special type of reaction norm, whereby the expression of specific traits is influenced by growth in overall body size, and growth in body size (at least in insects) is influenced by the environment; *(c)*

[1]We use the term scaling instead of allometry as recommended by Schmidt-Nielsen (189) and LaBarbera (123). For this paper, "scaling relationships" refers to the covariation of trait magnitude with overall body size, with no assumptions as to the slope or shape of the relationship.

[2]We use the term reaction norm to refer to the range of possible morphologies that individuals with the same genotype would express were they reared across a range of different environments or growth conditions (after 15, 188). Each individual insect that a researcher captures and measures will have one morphology—one body size and one horn, foreleg, or mandible size. Yet that same individual, had it been reared in a different environment, would have matured at a different body size, with a correspondingly altered horn, foreleg, or mandible. The reaction norm encompasses the entire range of morphologies that are possible endpoints for that genetic individual. This can also be viewed as the range of morphologies that would be produced either by close relatives (e.g. siblings) or by subsequent generations of one lineage, were they reared in different growth environments.

components of the developmental mechanism producing scaling are heritable, and scaling relationships can and do themselves evolve; *(d)* rich insight may be gained from comparative studies of the shapes or slopes of scaling relationships and how these relate to physical and social selective environments; *(e)* we use this approach to suggest why many exaggerated morphological traits exhibit discontinuous, or nonlinear scaling relationships; *(f)* we draw on what is currently known about how the development of some of these traits is regulated, and suggest implications of these mechanisms for the evolution of extreme shapes in insects.

SIZE-DEPENDENT EXPRESSION OF MORPHOLOGICAL TRAITS

Scaling relationships depict the size-dependent expression of body parts (32, 102, 123, 189). Large individuals tend to have larger wings, legs, or eyes than smaller individuals have (Figure 2*a*). But what makes an individual insect large or small? In most insects growth in body size is influenced by the larval environment (e.g.

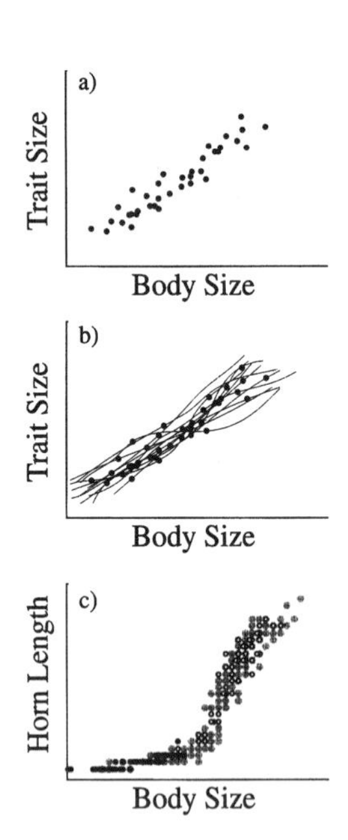

Figure 2 "Reaction-norm" view of scaling relationships. *(a)* The dimensions of most morphological traits covary with body size. The slopes or shapes of these scaling relationships are generally measured from static samples of individuals ("static allometry;" 32). *(b)* Static samples can obscure the fact that genotypes may have the potential to generate a range of different trait sizes, depending on how large each individual expressing the genotype grows to be. For example, an individual insect reared on a favorable diet will have a large body size and a large trait size. Yet that same individual, had it been reared in a poor growth environment, would have matured at a smaller body size, with a correspondingly smaller trait magnitude. The "reaction norm" view of the scaling relationship considers the breadth of possible endpoints for each genotype. In this case, the morphology actually produced by a specific individual depends on the shape of that individual's genotype-specific reaction norm (genotypes are indicated by *thin lines*), and on how large that individual grows to be (*filled circles*). *(c)* Evidence that scaling relationships can be considered reaction norms comes from controlled breeding experiments that compare the morphologies of close relatives reared across a range of nutrient conditions. For example, *Onthophagus acuminatus* males reared with large amounts of food grew to large body sizes and produced long horns (*open circles*), whereas sibling males reared on smaller food amounts remained small, and these males produced shorter horn lengths (*closed circles;* from 59).

12, 26, 103, 104, 231). Fluctuations in nutrient availability, temperature, and humidity all affect larval growth, as does the intensity of larval competition and larval population density. Through their effects on larval growth, these environmental factors all influence how large a larva grows to be, and hence partially determine the final adult size of that animal. Often sequential offspring from the same parents encounter different environmental conditions and mature at different body sizes despite the fact that they are genetically similar. This is perhaps most obvious in insects with parthenogenetic generations, such as aphids, where genetic clones vary extensively in body size. Yet even in these clones, morphological traits scale with variations in body size (e.g. 201, 202).

The scaling relationship, then, must reflect allocation to a trait across a range of possible final body sizes (53, 60, 61, 187, 223, 224). The body size actually attained, and thus the dimension of trait produced, depends on the environment that each larva encounters, as well as inherited factors. More specifically, the dimensions of each trait will depend on an interaction between the genotype of each individual, and the environmental factors influencing that larva's final adult body size (Figure 2*b*). Scaling relationships may thus be considered a special type of reaction norm (15, 187, 188), with each genotype capable of expressing a range of trait sizes (59, 60, 61, 187). Just as genotypes can produce a range of different phenotypes in response to variations in the physical environment (e.g. plant shape responding to variation in light availability; 49, 190), they can also produce different phenotypes in response to a range of final adult body sizes. In this case, the shape of the reaction norm determines the types of morphologies produced at each possible body size, and these relationships may be simple (linear), or more complex (e.g. threshold traits).

Despite this realization, the "reaction norm" view of phenotypes has seldom been applied to the study of insect allometry. One reason for this discrepancy may be that reaction norm studies and scaling relationship studies often use different types of data. The reaction norm concept arose from studies on plants where seeds of similar genotype could be planted across an array of physical environmental conditions (15, 127, 186, 187, 188). When relatives were planted in different environments it became clear that a genotype grown in one environment produced a different phenotype than that same genotype would produce in a second environment (15, 49, 186, 187, 188, 190). Furthermore, it became obvious that to understand the evolution of these characters it was necessary to consider not just the plastic phenotypes, but also the genotype-specific reaction norms that gave rise to the phenotypes (76, 77, 127, 183, 184, 186, 187, 199, 221, 223).

In contrast, studies of relative growth or scaling in insects generally do not involve rearing related individuals across a range of growth conditions. Instead, these studies often use collections of individuals sampled from wild populations or museums (Figure 2*a*). From these static samples it is less obvious that each genotype is capable of producing a range of different forms (Figure 2*b*), and it becomes easy to study the phenotypes (e.g. the exaggerated traits) and to overlook the underlying reaction norms that generated the phenotypes.

There is good evidence that the size-dependent expression of morphological structures in insects is sensitive to the growth environment (i.e. that scaling relationships can be equated with reaction norms). Body size in insects is strongly affected by the physical and social environment (e.g. 12, 26, 104, 231), and genotypes must therefore be capable of producing a range of different trait forms depending on how large any individual grows to be. The best evidence is provided by controlled laboratory experiments where relatives (or clones) are reared under a range of growth conditions. From these experiments it is possible to characterize the shapes of the reaction norms for specific genotypes or families. Several recent studies have deliberately manipulated environmental variables relevant to growth in insects, and these clearly demonstrate that single genotypes can generate the full range of possible trait dimensions (59, 100, 219). For example, manipulations of the amount of food available to growing larvae determined the final body sizes attained by male *Onthophagus acuminatus* and *O. taurus* (Coleoptera: Scarabaeidae), and these experimentally induced variations in body size were accompanied by corresponding variations in male horn length (Figure 2*c*; 59, 100). Because these diet manipulations were administered within beetle families, it was possible to illustrate that each family had the potential to generate the full range of possible horn lengths. Furthermore, in all cases where these reaction norms have been studied so far, the shapes of these relationships match closely with the general scaling relationship for the population (e.g. Figure 2*c*; 59, 100, 219).

Consequently, we suggest that a useful framework for viewing and studying scaling relationships is to recognize that the shape of the static relationship is a reflection of underlying patterns of allocation to traits by genotypes across a range of body sizes. This means that studies of morphological traits in insects must consider not just the traits themselves (e.g. the horns or mandibles) but also the underlying reaction norms that relate expression of the trait to variations in body size. It is evolutionary modifications to these scaling relationships that ultimately yield exaggerated or bizarre morphological structures in insects. It is also important to remember that these scaling relationships are not the same as growth trajectories (i.e. ontogenetic and static allometries are not the same thing; 32). It is very clear now—especially in holometabolous insects—that traits do not grow along the trajectory that we define as a static allometry (153, 200). Instead, the scaling relationship reflects the range of possible endpoints—final shapes—that would be generated by genotypes at each possible adult body size.

SCALING RELATIONSHIPS CAN AND DO EVOLVE

Scaling relationships result from developmental processes that regulate the growth of body parts (153, 200). The final dimensions of any morphological trait will be determined by patterns of gene expression, by patterns of cell growth and division, by the actions of hormones, and by the growth of other tissues (reviewed in 200). All of these processes can be influenced by the internal and external environments that growing tissues encounter, and in some cases this can result in condition sensitivity of final trait size. For example, the nutritional environment encountered by a larva will determine the rate at which that animal acquires nutrients essential for growth. Acquisition of nutrients by a larva can translate into protein and fat stores, as well as circulating levels of nutrients, and these can determine how large both the animal and the different body parts grow to be. The usual result: animals encountering favorable nutritional environments end up larger and with larger traits than individuals encountering less favorable diets. Similarly, larval exposure to environmental factors such as crowding, photoperiod, or temperature can influence levels of hormones, and these can also affect both overall growth and the growth of specific tissues.

Although these mechanisms may permit the growth of body parts to be sensitive to changes in the environment and to variations in overall body size, this characteristic does not indicate that these mechanisms are "non-genetic." Inherited differences in the expression or interaction of the various components of these mechanisms can cause different individuals to be sensitive to the environment in slightly different ways. For example, different individuals within a population may vary genetically in *how* they respond to changes in the growth environment, such that for any given final body size, some genotypes produce slightly larger or smaller traits than other genotypes (Figure 2*b*). Whenever differences among individuals result from genetic variations in components of the mechanisms that regulate trait growth, then these mechanisms may themselves evolve.

This condition is exactly analogous to the regulation of reaction norms in other animals or plants, where genotypes vary in the shape or position of their respective reaction norms: that is, genotypes differ in how they respond to the environment (e.g. 91, 161, 211). In these situations, the reaction norms may themselves evolve, and selection experiments clearly indicate that reaction norms are often capable of very rapid responses to selection (15a, 48, 97, 117, 182, 185, 214–216).

Scaling relationships for morphological traits in insects should therefore be capable of adaptive evolution (see 125, 187, 219, 242). Evidence that this is indeed the case comes from two sources: comparative studies measuring differences in the scaling relationships among related taxa, and artificial selection experiments that select directly on scaling relationships within populations.

Comparative Studies

The recognition that different taxa display scaling relationships with different slopes or shapes is not new. Huxley, Rensch, Gould, and others all used scaling relationships as a convenient way to compare populations or species (86–88, 95, 102, 173, 196, 197). Ironically, many of these same authors invoked the scaling relationship as evidence of restricted (immutable) patterns of growth, and considered these relationships to be evidence that the evolution of populations was constrained—even though they recognized that closely related taxa differed in aspects of their scaling relationships (e.g. taxa differed in the slopes of scaling relationships when measured on log-log plots; 88, 95, 102, 164, 167, 173, 196, 197). Many studies have shown that the scaling relationships of closely related taxa can differ significantly (e.g. Figure 3*a*; 23, 61, 86, 114, 116, 119, 158, 178, 189, 194, 234, 238). Although such comparative studies do not necessarily indicate that the trait allometries of *extant* populations are capable of evolving, they certainly demonstrate that such changes have occurred extensively in the past. A better way to reveal whether the size-dependent expression of morphological traits in current populations can evolve is by artificial selection experiments.

Artificial Selection Experiments

At least three recent studies have attempted to artificially select for changes in trait scaling relationships. In the first case, Weber (219) selected for several different changes in the wing morphology of flies (*Drosophila melanogaster*). In these experiments, he did not select on wing length, or wing width per se, but selected instead on the relative sizes of specific wing parts. He applied selection

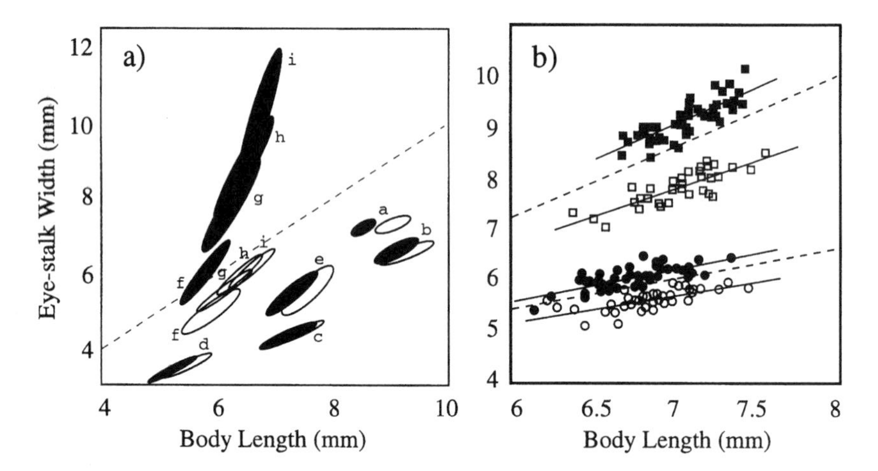

Figure 3 Evidence that scaling relationships evolve. *(a)* Comparative studies illustrating changes that have already occurred [example: bivariate distributions between eyestalk width and body length for nine Malaysian species of Diopsidae. Separate ellipses are shown for the males (black) and females (open) of each species; *a: Diopsis indica*; *b: Eurydiopsis subnotata*; *c: Cyrtodiopsis quinqueguttata*; *d: Maglabops quadriguttata*; *e: M. sexguttata*; *f: M. rubicunda*; *g: C. dalmanni*; *h: C. whitei*; *i: Teleopsis discrepans* (modified from 23); *(b)* Artificial selection experiments altering scaling relationships directly (example: scaling relationships for eyestalk width in *Cyrtodiopsis dalmanni* were subjected to artificial selection for either increased relative male eyestalk width, *closed squares,* or decreased relative male eyestalk width, *open squares.* Tenth generation individuals are shown. Shifts in the male scaling relationship were accompanied by correlated but smaller shifts in the female scaling relationship (*circles;* modified from 233).

to the scaling relationships between five different sets of landmarks on the wings. These experiments produced rapid visible changes in the dimensions of parts of the fly wings and resulted in significant shifts in the scaling relationships among these traits (219). Weber argued that all subdimensions of the wing exhibit locally acting additive genetic variation and that the heritability of the allometry was as large as the heritability of wing size itself (219).

A second study involves the stalk-eyed flies (Diptera: Diopsidae). A number of species in this family exhibit striking sexual dimorphism, with males producing eyes that are perched at the ends of long eyestalks (23, 174, 193, 234). In some individuals, the distance between the eyes can be more than twice the body length. Both males and females have eyestalks, but in dimorphic species the slope of the male eyestalk/body-size scaling relationship is much steeper than the relationship for females (23, 234). Comparisons among related taxa revealed marked differences in the slopes of the eyestalk scaling relationships, ranging from sexually monomorphic taxa, like *Cyrtodiopsis quinqueguttata*, where both males and females have the same low slope of eyestalk-to-body size relationship, to other taxa, such as *Cyrtodiopsis whitei*, where the males have eyestalk scaling relationships that are much steeper than those of females (Figure 3*a*; 23, 234). These interspecific comparisons suggest that eyestalk scaling relationships have evolved extensively in the past. The most convincing evidence for scaling relationship evolution was provided by an artificial selection experiment in one of the species. In the sexually dimorphic species *Cyrtodiopsis dalmanni*, Wilkinson artificially selected on male eyestalk allometry, by directly selecting on the ratio of eyestalk length to overall body length (233). In each of two genetically independent lines he selected males with disproportionately long eyestalks (males with high eyestalk-width/body-length ratios), and in two additional lines he selected for males with relatively short eyestalks (small eyestalk-width/body-length ratios). *Cyrtodiopsis dalmanni* populations responded rapidly and significantly to selection, so that after 10 generations the scaling relationships of the two types of treatment lines were completely nonoverlapping (Figure 3*b*; 233). The scaling relationship between eyestalk length and body size had changed dramatically in response to artificial selection (see also 235a).

A third example involves horns in male beetles. Males in many beetle taxa produce elongated extensions of the cuticle called horns (4, 41, 51–53, 55, 174). In *Onthophagus acuminatus*, males, but not females, produce a pair of cephalic horns (59, 60). Male horn length increases with overall body size, and this scaling relationship is not linear (discussed further below). The horn-length/body-size scaling relationship in this and related species is sigmoidal in shape and results in a bimodal horn-length frequency distribution: males with long horns are common, as are males with only rudimentary horns, but males with intermediate horn lengths are relatively rare (59). One of us (61) used an artificial selection design to select directly on the horn-length/body-size scaling relationship. By selecting on the relative length of male horns (i.e. selecting males with unusually long or unusually short horns for their respective body sizes), significant shifts in this scaling relationship were produced after only seven generations. Again, this indicates that populations of this species contain heritable variation for the relationship between horn length and body size.

Scaling Relationships as Traits

As noted above, scaling relationships have been considered a form of constraint on evolutionary change—a reflection of underlying immutable properties of a developmental system that restrict evolutionary changes to the direction specified by the scaling relationship. Increasingly, however, this view is being called into question as both inaccurate and misleading. The studies mentioned in this section indicate that scaling relationships vary heritably within populations, that they have evolved extensively in the past, and that they are still capable of evolving rapidly if the selective environment is changed.

These findings have important implications for the study of insect scaling relationships. If we stop viewing scaling relationships as immutable properties of development, and instead consider them to be manifestations of condition-sensitive mechanisms of trait expression that have been molded by a history of natural and sexual selection, then we can begin to explore not just *how* traits covary with body size, but *why* they covary with body size. More specifically, we can begin to examine the slopes or shapes of scaling relationships in different taxa and ask why these relationships have the shapes that they have. Again, to draw on the literature from the study of reaction norms, we can ask why the reaction norms that relate the expression of a trait to variations in the growth environment have the precise shapes that they have. Why are some slopes steeper than others? Why do some traits exhibit discontinuous or sigmoidal scaling relationships and others exhibit linear relationships? We suggest that in many cases, the shape of trait-scaling relationships will contain valuable information regarding the underlying developmental mechanisms that give rise to the scaling relationship, as well as the natural forces of selection on each trait and how those forces of selection vary with differences in body size. This is especially evident for scaling relationships of exaggerated, disproportionately large morphological traits like eyestalks and horns. With this as a framework, we explore the range of variation present in insect scaling relationships, with particular emphasis on exaggerated structures.

SURVEY OF SCALING RELATIONSHIPS OF EXAGGERATED MORPHOLOGICAL TRAITS

We reviewed literature on bivariate scaling relationships for morphological traits and body size. For exaggerated traits, we recorded the trait involved, the sex expressing the exaggerated form of the trait, and the shape of the scaling relationship (linear, curved, sigmoid, or discontinuous). Because in most cases we did not have access to the raw data, we were unable to use the quantitative methods of Happell (92) or Eberhard & Gutierrez (58) to describe scaling relationship shape. Instead, scaling relationships were characterized qualitatively by visual examination of the bivariate relationships.

To facilitate visual comparison of variations in the shapes of scaling relationships, we selected a subset of nine of the studies and digitized the points from the published figures (Figure 4). Digitization was necessary because the overall sizes of the insects and traits varied considerably, and published figures differed in the use of logarithmic and linear scales in one or both axes. Digitization of published scaling relationships allowed us to standardize the body size and trait values around the mean of each species, and to perform the same transformations on all of the data. For this comparison we selected taxa where linear measures of body size were compared with linear measures of trait dimension. All body size and trait values were mean standardized, and all data were examined on both log and untransformed scales (for studies where log values were plotted, we converted these to antilogs after digitization).

Most scaling relationships for morphological traits in insects are linear. However, when we considered the subset of morphological traits that reach unusual proportions—the exaggerated ornaments or weapons of males, or the distended heads of soldier castes in ants—the shapes of scaling relationships were quite diverse (Table 1). Scaling relationships for exaggerated traits were either linear, curved, sigmoid, or composed of completely discontinuous segments (Table 1; see Figure 4 for examples). When we considered the most extreme traits, a comparison across taxa revealed two types of scaling relationship shapes, suggesting that there are two basic ways of achieving grossly enlarged morphological structures: linear relationships with very steep slopes, or sigmoid/broken relationships incorporating a threshold (Figure 5).

Steep Allometry Slopes

Populations may produce ever larger traits by evolving ever steeper scaling relationships, so that with increases in body size, genotypes produce disproportionately larger increases in trait magnitude (Figure 5*b*). Some of the most extreme morphologies are produced this way, including the antlers and eyestalks of Tephritid and Diopsid flies, forceps in many species of earwig, and the enlarged legs of bugs, weevils, and harlequin beetles (Table 1).

With the exception of ant castes, exaggerated traits tend to be expressed primarily in males, and in all cases where behavior has been explored, the exaggerated traits play a role in competition over access to reproduction (sexual selection). Male eyestalks in Diopsid flies, for instance, are characters on which direct female choice of mates is based (23, 24, 128, 235). Enlarged male legs are used for male-male combat over access to females in bugs (*Coreidae*: 56, 75, 134–136), weevils (*Macromerus bicinctus*: 218), and harlequin beetles (*Acrocinus longimanus*: 241). Fly antlers (*Phytalmia* spp.: 45, 234) and eyestalks (160a) and

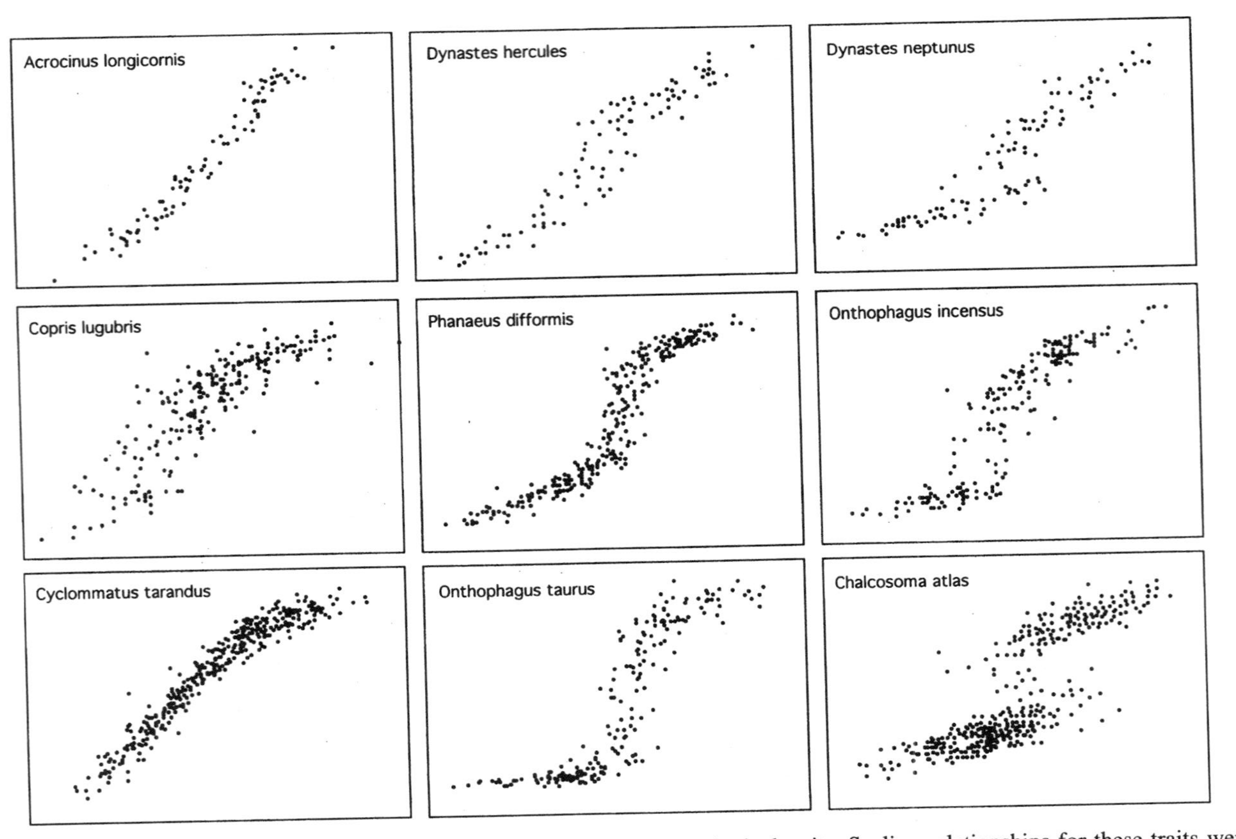

Figure 4 Examples of nine scaling relationships for exaggerated morphological traits. Scaling relationships for these traits were linear, curved, sigmoid, or completely broken. All figures included linear measures of both trait magnitude (*Y-axes*) and body size (*X-axes*). Values were digitized from published graphs and mean-standardized to facilitate shape comparisons. See Table 1 for trait descriptions and references.

TABLE 1 Scaling relationships for exaggerated traits in insects

Order	Family	Genus and Species	Exaggerated Trait	Sex-expressing trait	Scaling-relationship shape	Reference
Dermaptera	Chelisochidae	*Adiathetus tenebrator*	forceps	males	linear	J Tomkins, pers. comm.
		Chelosoches morio	forceps	males	linear	J Tomkins, pers. comm.
		Eunkrates varegatum	forceps	males	linear	J Tomkins, pers. comm.
		Proreus ludekingi	forceps	males	discontinuous	206
	Forficulidae	*Anechura harmandi*	forceps	males	linear	204
		Doru taeniatum	forceps	males	linear	58
		Eluanon bipartitus	forceps	males	discontinuous	206
		Forficula auricularia	forceps	males	sigmoid	9, 43, 170, 206, 207
		Metrasura ruficeps	forceps	males	linear	58
		Oreasiobias stolickzae	forceps	males	discontinuous	206
		Timomenus aeris	forceps	males	discontinuous	206
	Labiidae	*Chaetospania thoracia*	forceps	males	linear	J Tomkins, pers. comm.
		Paralabella dorsalis	forceps	males	linear	58
		Spongovostox assiniensis	forceps	males	discontinuous	206
		Vostox punctipennis	forceps	males	linear	J Tomkins, pers. comm.
	Labiduridae	*Forcipula gariazzi*	forceps	males	linear	J Tomkins, pers. comm.
		Forcipula quadrispinosa	forceps	males	sigmoid	J Tomkins, pers. comm.
		Labidura truncata	forceps	males	discontinuous	9, 43, 170, 206, 207
Hemiptera	Coreidae	*Acanthocephala declivis*	hind femur	males	linear	56
		Leptoglossus australis	hind femur	males	linear	136
Homoptera	Hormaphidae	*Pseudoregma alexanderi*	fore femur	females	discontinous	201
Coleoptera	Cerambycidae	*Acrocinus longimanuss*	forelegs	males	linear	241

TABLE 1 (continued) Scaling relationships for exaggerated traits in insects

Order	Family	Genus and Species	Exaggerated Trait	Sex-expressing trait	Scaling-relationship shape	Reference
Coleoptera	Cerambycidae	*Dendrobias mandibularis*	mandibles	males	sigmoid	82
	Curculionidae	*Rhinostomus barbirostris*	beak/rostra	males	linear	54, 58
		Centrinaspis sp.	ventral spine	males	linear	58
		Geraeus sp.	ventral spine	males	sigmoid	57
		Macromerus bicinctus	forelegs	males	linear?	218
	Endomychidae	*Stenotarsus rotundus*	hind trochanter	males	linear	144
	Lucanidae	*Cyclommatus bicolor*	mandibles	males	linear	116
		Cyclommatus elaphus	mandibles	males	linear	116
		Cyclommatus lunifer	mandibles	males	curved	101, 116
		Cyclommatus tarandus	mandibles	males	curved	50, 58, 101
		Hexarthris davisoni	mandibles	males	linear	158
		Lamprima alophinae	mandibles	males	linear	158
		Lucanus cervus	mandibles	males	curved	31, 58, 101
		Lucanus elephas	mandibles	males	linear	158
		Neolucanus cinglatus	mandibles	males	linear	116
		Neolucanus nitidus	mandibles	males	linear	116
		Neolucanus perarmatus	mandibles	males	linear	116
		Odontolabis cuvera	mandibles	males	sigmoid	158
		Odontolabis imperialis	mandibles	males	sigmoid	116
		Odontolabis micros	mandibles	males	discontinuous	116
		Odontolabis siva	mandibles	males	discontinuous	116, 158
		Prosopocoelus serricornis	mandibles	males	linear	158

	Psalidoremus inclinatus	mandibles	males	curved	158
	Serrognathus platymelus	mandibles	males	linear	158
	Xylotrupes gideon	mandibles	males	sigmoid	9, 101, 58
Melolonthidae	*Inca clathra*	clypeal horn	males	linear	142
Scarabaeidae	*Ageopsis nigricollis*	head and thoracic horns	males	sigmoid	55
	Allomyrina dichotoma	head horn	males	linear	103, 198
	Chalcosoma atlas	head and thoracic horns	males	discontinuous	114
	Chalcosoma caucasus	head and thoracic horns	males	discontinuous	114
	Copris lugubris	head horn	males	curved	58
	Coprophanaeus ensifer	head horn	both	linear	159
	Drepanoceros kirbyi	thoracic horn	males	sigmoid	R Knell, pers. comm.
	Dynastes centaurus	thoracic horn	males	sigmoid	58
	Dynastes hercules	head and thoracic horn	males	sigmoid	114
	Dynastes hyllus	thoracic horn	males	linear	143
	Dynastes neptunus	head and thoracic horns	males	sigmoid	114
	Megasoma elephas	head horn	males	curved	58
	Onthophagus acuminatus	head horns	males	sigmoid	59, 61
	Onthophagus australis	head horns	males	sigmoid	J Hunt, pers. comm.
	Onthophagus batesi	head horns	males	sigmoid	D Emlen, unpublished
	Onthophagus binodis	thoracic horn	males	curved	35, 195
	Onthophagus ferox	head and thoracic horns	males	linear	35
	Onthophagus fuliginosus	head horns	males	curved	J Hunt, pers. comm.
	Onthophagus gazella	head horns	males	sigmoid	J Hunt, pers. comm.
	Onthophagus haagi	head horns	males	sigmoid	J Hunt, pers. comm.
	Onthophagus hecate	thoracic horn	males	curved	D Emlen, unpublished
	Onthophagus incensus	head horns	males	sigmoid	58

TABLE 1 (continued) Scaling relationships for exaggerated traits in insects

Order	Family	Genus and Species	Exaggerated Trait	Sex-expressing trait	Scaling-relationship shape	Reference
Coleoptera	Scarabaeidae	*Onthophagus marginicollis*	head horns	males	sigmoid	D Emlen, unpublished
		Onthophagus striatulus	head horns	males	sigmoid	D Emlen, unpublished
		Onthophagus taurus	head horns	males	sigmoid	63, 100, 137, 138, 195
		Onthophagus vermiculatus	head horns	males	sigmoid	J Hunt, pers. comm.
		Phanaeus difformis	head horn	males	sigmoid	171
		Podischnus agenor	head horn	males	sigmoid	53
		Xylorectes lobicollis	head horn	males	linear	58
	Staphylinidae	*Leistotrophus versicolor*	mandibles	males	linear	70
	Tenebrionidae	*Bolitotherus cornutus*	thoracic horns	males	linear	20, 21
Diptera	Diopsidae	*Cyrtodiopsis dalmanni*	eyestalk	males	linear	233
		Cyrtodiopsis whitei	eyestalk	males	linear	23, 234
		Diasemopsis dubia	eyestalk	males	linear	234
		Diasemopsis fasciata	eyestalk	males	linear	234
		Diasemopsis sylvatica	eyestalk	males	linear	234
		Teleopsis boettcheri	eyestalk	males	linear	193
		Teleopsis breviscopium	eyestalk	males	linear	234
		Teleopsis rubicunda	eyestalk	males	linear	234
	Tephritidae	*Phytalmia alcicornis*	antler	males	linear	234
		Phytalmia mouldsi	antler	males	linear	234
Hymenoptera	Andrenidae	*Perdita portalis*	large head	males	discontinuous	38, 39
		Perdita texana	large head	males	linear	39, 40
	Formicidae	*Anomma nigricanus*	wide head	females	curved	99, 238

Atta colombica	wide head	females	curved	69
Atta texana	wide head	females	curved	238
Camponotus castaneus	wide head	females	linear	237
Camponotus floridanus	wide head	females	sigmoid	238
Camponotus maculatus	wide head	females	discontinuous	7
Camponotus novaeboracensis	wide head	females	linear	78
Camponotus rufipes	wide head	females	curved	44
Cephalotes atratus	wide head	females	linear	37
Daceton armigerum	wide head	females	linear	140
Dorylus spp.	wide head	females	curved	84, 85, 98
Eciton hamatum	wide head	females	discontinuous	69
Formica exsecta	wide head	males	discontinuous	71
Formica obscuripes	wide head	females	linear	238
Lasius flavus	wide head	females	linear	72
Lasius fulignosus	wide head	females	linear	238
Lepthothorax longispinosus	wide head	females	linear	94, 98
Megaponera foetens	wide head	females	linear	212
Messor capensis	wide head	females	linear	213
Messor spp.	wide head	females	linear	80, 98
Myrmecia brevinoda	wide head	females	linear	96
Myrmecia froggatti	wide head	females	linear	106
Neivamyrmex nigrescens	wide head	females	linear	238
Oecophylla leakeyi	wide head	females	sigmoid	238, 239
Oecophylla smaragdina	wide head	females	broken	237
Paraponera clavata	wide head	females	linear	16
Pheidole bicarinata	wide head	females	discontinuous	227, 228

TABLE 1 (continued) Scaling relationships for exaggerated traits in insects

Order	Family	Genus and Species	Exaggerated Trait	Sex-expressing trait	Scaling-relationship shape	Reference
Hymenoptera	Formicidae	*Pheidole rhea*	wide head	females	sigmoid	238
		Pheidologeton diversus	wide head	females	discontinuous	139
		Ropalidia ignobilis	wide head	females	discontinuous	222
		Solenopsis germinata	wide head	females	linear	225
		Solenopsis invicta	wide head	females	linear	225
	Vespidae	*Pseudopolybia difficilis*	large head	females	discontinuous	107
		Synagris cornuta	mandibular tusks	males	discontinuous	Longair, pers. comm.

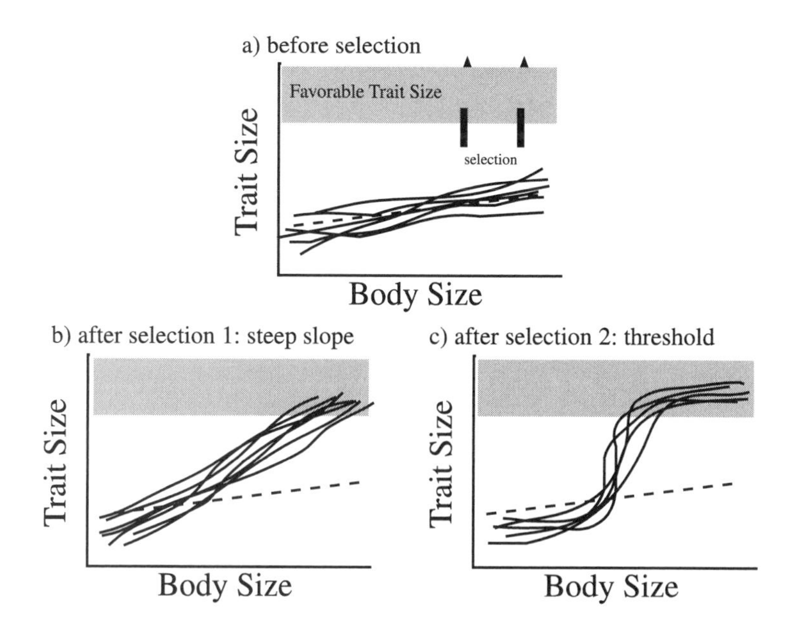

Figure 5 Alternative routes to the evolution of exaggerated traits. *(a)* Populations subjected to selection for enlarged trait dimensions (e.g. sexual selection for a large male weapon) can evolve the capacity to generate extreme forms in one of two ways: *(b)* genotypes that allocate disproportionately rapidly to the trait may be favored, so that populations evolve steep, positive linear scaling relationships, or *(c)* genotypes capable of facultatively expressing the trait may be favored, resulting in populations with sigmoid or broken scaling relationships. It is also possible that steep scaling relationships evolve first, and facultative expression evolves later (b → c). Reasons why this might occur are discussed in the text. An additional possibility not considered here would be a shift in the intercept of a linear scaling relationship.

earwig forceps (*Forficula* spp.: 18, 141, 170) are also used in aggressive encounters between males.

Given that many of these are sexually selected traits, their steep linear scaling relationships are not surprising. Theoretical models of sexual selection predict the evolution of steep positive scaling relationship slopes (89, 165, 166, 194). Sexual selection is generally manifest whenever disproportionate access to reproduction is gained by a small, nonrandom subset of males (2, 13, 30, 41). Often this means that males with the best genetic constitution, or more importantly, males whose genotypes interact the most favorably with surrounding conditions, have the highest relative fitnesses. In insects, these individuals are generally the largest individuals (examples in 30, 205), as body size can reflect both the genotype and the interaction between that genotype and the physical and social environment (e.g. competitively superior larvae gain disproportionate access to limiting nutritive resources, and emerge at larger body sizes). Sexual selection that favors large body size can operate either through females that prefer to mate with only the largest males, or through male-male competition, with only the largest males able to secure access to females or to resources utilized by females (2, 205).

In either situation (direct female choice of males, or male-male competition), individuals must assess the relative size of other individuals in the population. Morphological structures that scale with body size contain information regarding the overall body size of each individual, and traits that exhibit steep positive scaling relationships are typically the most effective indicators of body size because they amplify subtle differences in body size among individuals. In these traits, each incremental increase in overall body size is magnified into a disproportionately larger increase in trait dimension, with the result that individual variations in the size of these traits offer greater resolution to underlying variations in body size than a direct assessment of size itself would provide. For this reason, exaggerated morphological traits are predicted to be unusually effective criteria for either female choice of mates, or similarly, for male assessment of rival males (2, 89, 165, 166). If this is indeed the case, then we might expect the intensity of sexual selection present in each species to be correlated with the steepness of the slope of the scaling relationship, with stronger sexual selection leading to steeper slopes (194). Simmons & Tomkins (194) found exactly this pattern when they compared the slopes for 42 species of earwigs (Dermaptera): taxa with the most intense sexual selection showed significantly steeper allometry slopes than taxa with weaker sexual selection.

One interesting outcome of the present survey is the realization that many of the most extreme morphological characters do not show steep linear scaling relationships. In fact, many of these taxa exhibit sigmoid or discontinuous scaling relationships, suggesting the operation of threshold mechanisms during development.

Threshold Traits

Horns in most species of rhinoceros beetle (Scarabaeidae: Dynastinae), mandibles in many of the stag beetles (Scarabaeidae: Lucanidae), mandibles in Cerambycidae, tusks in wasps (Hymenoptera: Vespidae), forceps in several earwig species (Dermaptera), and head widths in ants with the most pronounced castes (Hymenoptera: Formicidae), all exhibited broken or sigmoid scaling relationships (Table 1). In fact, both the literature and our survey of taxa suggest that sigmoid and discontinuous scaling relationships have arisen repeatedly within the insects. Within the Hymenoptera, for example, sigmoid or discontinuous scaling relationships have arisen independently in at least seven ant genera (98, 157), and separately at least once each within the bees and wasps (Table 1). Likewise, sigmoid and discontinuous scaling relationships have arisen many times within the Coleoptera: at least once each in the Cerambycidae, Curculionidae, and Lucanidae, and multiple times within the Scarabaeidae (Note: These are conservative estimates based on the taxa included in Table 1, and the assumption of parsimony).

Despite the large number of independent origins, the evolution of sigmoid or discontinuous scaling relationships is consistently associated with the expression of the most exaggerated morphological structures. It is the most elaborate castes in ants, with the most extreme head morphologies, that incorporate complex scaling relationships (98, 237, 238). Similarly, it is the most elaborate of the horned beetles—the rhinoceros beetles with the largest or most dramatic horns, and the stag beetles with the most distended mandibles—that exhibit nonlinear scaling relationships. Why is the expression of enlarged or exaggerated morphological structures so often associated with sigmoid or discontinuous scaling relationships?

Trait exaggeration can arise through either sigmoid, discontinuous, or linear scaling relationships (all can produce disproportionately large structures in individuals with the largest body sizes; Figure 5*c*). But several properties distinguish sigmoid and discontinuous relationships from linear scaling relationships. First, the switch between minimal and exaggerated trait expression often occurs abruptly, over a small range of body sizes (the "critical" or threshold body size). As a result few individuals with intermediate shapes are produced. Second, because this switch occurs abruptly, the exaggerated traits are facultatively expressed: Individuals larger than a threshold body size produce one morphology, whereas individuals smaller than this size produce a different morphology. Because only large individuals express the trait, this results in co-occurrence within populations of two relatively discrete morphs (polyphenisms). Both of these factors have important implications for the evolution of exaggerated morphologies in insects, and we discuss each factor in detail in the next section.

WHY ARE SO MANY EXAGGERATED TRAITS FACULTATIVELY EXPRESSED?

The incorporation of thresholds into the development of morphological traits has several consequences that may facilitate the evolution of the most bizarre, or exaggerated morphologies; *(a)* they minimize the production of animals with intermediate forms; *(b)* they permit the morphologies of large and small individuals within a sex to evolve independently—at least with respect to the trait of interest—allowing a genotype to simultaneously specialize for more than one task or situation; *(c)* they uncouple the phenotypes of males and females so that only one sex produces the enlarged trait. These three properties of thresholds all reduce the "cost" to a genotype for producing an exaggerated morphological trait and may facilitate the evolution of such structures.

Thresholds Minimize Production of Intermediate Forms

The morphologies generated at intermediate body sizes differ for the two basic scaling relationship types (linear versus sigmoid/discontinuous; Figure 6). A linear scaling relationship produces the trait in all individuals, including those with

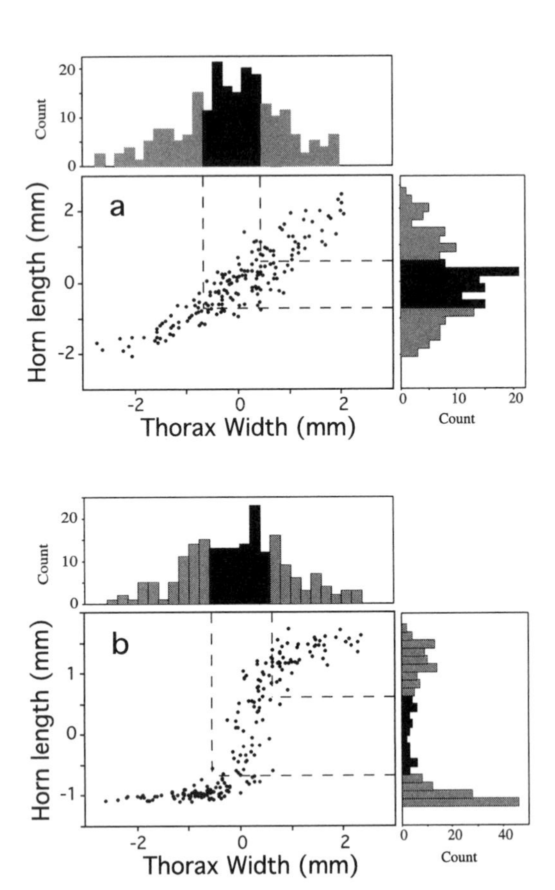

Figure 6 The production of intermediate morphologies by linear versus sigmoidal scaling relationships. Histograms show frequencies of body sizes (*top margin*), and horn lengths (*right margin*). (*a*) Genotypes with linear scaling relationships express the trait in all individuals, and consequently these genotypes generate large numbers of individuals with intermediate trait dimensions (*black bars*). (*b*) Genotypes with broken or sigmoidal scaling relationships switch abruptly from minimal to complete trait production and consequently produce fewer individuals with intermediate forms.

intermediate sizes (Figure 6*a*). Sigmoid or broken scaling relationships, in contrast, switch between minimal and complete trait expression over a narrow range of body sizes. As a result, few individuals emerge with intermediate morphologies (Figure 6*b*). Why should this matter, if both linear and sigmoid scaling relationship types result in large individuals that have the selected dramatic morphology?

To address this question it is necessary to adopt the framework we put forward in this review: that scaling relationships reflect the allocation by a genotype to a trait across the range of possible body sizes. The scaling relationship becomes the focal trait. In this context, comparing the relative reproductive success of males with different ornament sizes (e.g. horned versus hornless males) is less relevant than comparing the average success of genotypes that vary in *how* they allocate to the ornament. Do genotypes with linear patterns of allocation to the ornament do better or worse, on average, than alternative genotypes that facultatively express the ornament only in the largest individuals (e.g. sigmoid/ discontinuous scaling)? Viewed this way, the consequences of the shape of the scaling relationship become clear. In both situations, genotypes encounter a range of growth environments, and consequently both types of genotypes are expressed in individuals of all possible body sizes. Linear and sigmoid genotypes do basically the same thing at the body size extremes: in both types of scaling, large ornaments are expressed in the largest individuals, and rudimentary ornaments are expressed in the smallest individuals. But these two types of scaling differ substantially in the morphologies they generate when they are expressed in individuals with intermediate body sizes. Genotypes with linear relationships produce intermediate morphologies when expressed in individuals of average body sizes, whereas genotypes with sigmoid or broken relationships do not (Figure 6).

The relative success of each genotype will be the sum of the reproductive contributions of all individuals of that genotype. Two factors are important here: First, in most insect populations body sizes are normally distributed, so the majority of individuals that express each genotype will be of average body size (the largest individuals are actually relatively rare). Second, intermediates often do poorly. In many cases, individuals with either of the extreme morphologies perform better than individuals with intermediate morphologies. In the case of sexually selected traits, intermediate males incur the cost of producing and maintaining an ornament or weapon, but are not successful at outcompeting larger males, and so derive no reproductive benefit from this morphology. Likewise, their unwieldly shapes often make them less effective at adopting the alternative behavioral tactics often employed by smaller males (small males frequently avoid aggressive encounters and sneak access to females: reviewed in 2). Similarly, in non–sexually selected traits (e.g. soldier head morphologies in ants), intermediate shapes may perform less well at size-specific colony tasks.

We suggest that whenever intermediate morphologies perform poorly compared to either of the extremes, genotypes encoding sigmoid or discontinuous scaling relationships will outperform genotypes with linear relationships. In these situations, genotypes that generate a majority of intermediate forms will have lower average fitnesses than genotypes that switch abruptly between minimal and complete trait expression. This is what we refer to as the "cost" of producing intermediates, and we suggest that genotypes with sigmoid or broken scaling relationships may minimize the costs associated with the exaggerated production of morphological traits. Evolution of such traits may be more likely in taxa with the capacity to facultatively express the trait, perhaps explaining why so many sexually selected traits exhibit sigmoid or broken scaling relationships (61).

Thresholds Uncouple the Phenotypes of Large and Small Individuals

Genotypes with linear scaling relationships produce one basic morphology. The relative dimensions of the horn, tusk, or femur may vary with body size, but all individuals express the trait. In these genotypes, the only way to generate an extreme morphology in the largest individuals is to also express that trait in all of the intermediate and smaller individuals (Figures 5*b*, 6*a*). Thus the morphologies of large and small individuals will not be independent, and the potential for these size classes to diverge morphologically will be limited.

Threshold mechanisms permit genotypes to switch abruptly between minimal and complete trait expression (Figures 5*c*, 6*b*). Alternative morphologies regulated by thresholds often result from the expression of alternative, partially nonoverlapping sets of genes (see section on How Threshold Traits Work; and 66, 199, 223, 224), and in these animals the evolution of large and small morphologies can be at least partially independent (223, 224). This means that small individuals can dispense with the investment of producing and bearing the same traits as large individuals, "uncoupling" the phenotype produced by the two size classes.

These differences in scaling relationship shape are relevant to the evolution of exaggerated morphologies because large and small individuals often do very different things, and the enlarged trait may be beneficial in only one of these contexts. For example, large and small males often differ substantially in their competitive status, affecting their ability to garner access to critical resources or females (90, 205). In a variety of taxa, small, competitively inferior males adopt alternative behaviors that bypass direct competition with larger males (1, 6, 47, 205). These alternative reproductive tactics can cause large and small individuals to experience very different physical and social environments, and this may favor different morphologies of large and small males. For example, ornaments or weapons that contribute substantially to the reproductive success of the largest males may be useless or detrimental to smaller males. What this means is that selection is often heterogeneous: The best morphologies for large males differ from the best morphologies for smaller males. Whenever large and small individuals encounter divergent selective situations, incorporation of a threshold into trait development may be favored because it permits the shape of these size classes to evolve relatively independently.

For example, in horned beetles, large and small males often employ separate behavioral tactics to encounter and mate with females (36, 53, 60, 62, 83, 137, 137a, 171, 198). Large males generally use their horns in battles with other males over access to females (34, 51–55, 59, 60, 62, 137, 137a, 160, 171, 198), and long horns have been shown to improve male competitive ability in several of these taxa (60, 62, 137, 137a, 171). Relatively small males in many of the same species adopt nonaggressive alternative behaviors like dispersing (53) or sneaking (60, 62, 137, 137a, 171) to encounter females. Horns may directly impede performance of the alternative behavior (e.g. in *Onthophagus acuminatus* and *O. taurus* small males may be better at sneaking into tunnels containing females if they do not produce horns; 60, 62, 137, 137a), and horns may be costly to produce. Horn growth requires resources that could otherwise be used for different traits (151), and horn growth slows development time, increasing the risk of larval mortality (100). Because horns are not utilized in the sneaking or dispersing alternative behaviors, their production may be prohibitively costly for small males.

Because large and small males appear to experience disruptive selection for horns (large males do best with long horns, small males do best without horns), the mating system in these beetles may have favored genotypes capable of uncoupling the phenotypes produced by large and small individuals, so that neither size class has an inappropriate horn morphology. Although similar heterogeneous selection may be present in most or all taxa expressing a costly, exaggerated male ornament, only species with sigmoid or discontinuous scaling relationships uncouple the morphologies of large and small males. Nonaggressive, alternative reproductive behaviors of small males have been described for cerambycid beetles (*Dendrobias mandibularis*, 82, 83), rhinoceros beetles (e.g. *Podischnus agenor*, 53), dung beetles (*Onthophagus acuminatus*, 60, 62; *O. taurus*, 137, 137a; *Phanaeus difformis*, 171), and bees (*Perdita portalis*, 38), and all of these insects exhibit sigmoid or discontinuous scaling relationships.

These conditions are not exclusive to sexually selected ornaments or weapons in males. In ants, large and small females may perform very different colony tasks, and these tasks can select for divergent morphologies. Soldiers may be more effective at colony defense if they have enlarged heads with biting mandibles and extensive jaw musculature, while these same traits would be a hindrance to smaller colony workers. Many of the ants with the most specialized caste behaviors also exhibit sigmoid or discontinuous scaling relationships ("triphasic allometry" of 98, 238), with the result that exaggerated head structures are produced only in the largest females.

In all of these cases, the developmental capacity to uncouple the morphologies of large and small individuals paves the way for subsequent morphological divergence because it permits these forms to become increasingly specialized for their respective behaviors, tasks, or situations. Interestingly, this capacity to uncouple ornament expression between large and small individuals *within* a sex appears to apply *between* the sexes as well.

Thresholds Uncouple the Phenotypes of Males and Females

Species with linear scaling relationships generally show correlations between the sexes, such that the exaggerated trait is expressed to some extent in both males and females (e.g. 233, 241). Selection favoring sexual dimorphism appears to reduce this correlation so that males show steeper scaling relationship slopes than females, but these correlations persist nevertheless. For example, in harlequin beetles, males have dramatically enlarged forelegs and a steep linear scaling relationship, but females have elongated forelegs as well (241). Similar conditions apply for eyestalks in the Diopsidae. Males often have much steeper scaling relationships than females, but both sexes produce eyestalks. In this case, the genetic correlation is especially clear: when Wilkinson artificially selected on the scaling relationship for the male trait, he observed a correlated response in the scaling relationships of the females (Figure 3*b*; 233). What this means is that sexual selection favoring enlargements of a male trait will also affect the morphology of the females. Because females generally do not use these ornaments or weapons (1, 2, 30), and because these traits are often costly to produce and to bear (reviewed in 2), this viability selection on female morphology may hinder extreme enlargement of male traits.

In species with sigmoid or discontinuous scaling relationships, in contrast, the ability to facultatively produce the trait within a sex appears to transfer to the other sex as well. In most of the horned beetles, for example, large males produce elongated horns, whereas both small males and females do not. In these cases, the horns are truly absent from females. This suggests that the capacity to facultatively express an exaggerated trait permits a genotype to produce the trait only in a subset of situations (e.g. when the individual expressing the genotype is male, and grows large), and to completely dispense with the trait in other circumstances. Facultative trait expression has important consequences for the evolution of sexually selected ornaments or weapons (which includes all of the taxa in Table 1 except for the ants), because these traits are generally favored only in males, and only in the largest males. Genotypes that express the trait in either small males or females incur fitness costs because these individuals produce and bear the ornament but derive no benefits in return. Consequently, taxa with sigmoid or broken scaling relationships may be more likely to evolve exaggerated secondary sexual weapons or ornaments in response to sexual selection.

In summary, scaling relationships dictate the size-dependent expression of secondary sexual and other traits. The shapes of these relationships vary, and in many cases, differences in the shape of the scaling relationships have important consequences for the evolution of exaggerated or extreme morphologies. Linear and sigmoid scaling relationships differ substantially in how and when they express enlarged morphological structures, and this difference can influence the "cost" incurred by a genotype for producing the structure. Enlarged ornaments or weapons may be unusually expensive to generate and maintain, and often only the largest individuals profit by utilizing the structure. In these situations, smaller individuals may benefit by not producing the exaggerated trait, but this is impossible with linear scaling relationships. We suggest that the developmental capacity to minimize the production of intermediate shapes, the capacity to uncouple the phenotypes of large and small individuals, and the capacity to uncouple the phenotypes of males and females all may predispose traits to evolutionary enlargement and may help to explain why so many of the most bizarre and exaggerated structures exhibit sigmoid or discontinuous scaling relationships.

CONNECTING GENOTYPE TO PHENOTYPE: HOW MIGHT SCALING RELATIONSHIPS CHANGE IN SHAPE?

This review focuses on the evolution of scaling relationships. By considering exaggerated traits like eyestalks or horns to be the result of evolutionary changes in scaling relationship slope or shape, we hope to provide a realistic view of the evolution of insect morphology. Doing so entails re-aligning our perspective of trait evolution to consider the scaling relationship itself. One potential problem with this view is that it explicitly incorporates aspects of mechanism: it is the developmental mechanisms regulating the expression of traits that ultimately generate the shape of the scaling relationship, and evolutionary modifications to the scaling relationship must therefore result from genetic changes to the underlying regulatory mechanisms. While this view may be attractive to some (because it attempts to span the gulf between genotype and phenotype), it can also constitute a limitation. By placing emphasis on the scaling relationship, we reveal our ignorance of the underlying mechanisms. After all, what does a change in the shape of a scaling relationship really mean? How can such changes arise, and can we assume that all such changes are possible?

Insects are perhaps unique in that their development has been so well characterized that we can begin to appreciate how scaling relationships are generated, and from this information we can glimpse how they may evolve. Here we briefly describe the developmental processes regulating the expression of morphological traits in insects. We start by describing how the general scaling of body parts occurs (i.e. how a linear scaling relationship is generated). We then consider the special situation with threshold traits and discuss how polyphenisms in insects are regulated. Finally, we use this information to suggest how linear scaling relationships may have been modified during the course of evolution to generate sigmoid or discontinuous scaling relationships.

Background: Separation of Larval and Adult Tissues

We focus our discussion on the development of holometabolous (completely metamorphic) insects because the regulation of growth is best understood in these animals (primarily from work on *Drosophila melanogaster* and *Manduca sexta*) and because most of the taxa known to exhibit extreme or exaggerated morphologies are holometabolous (Table 1). Postembryonic development in insects has been recently reviewed (8, 79, 149), so we briefly describe only those aspects relevant to this review.

In holometabolous insects, larvae bear little physical resemblance to the adults. Animals proceed through several larval stages before molting into a pupa and then subsequently into the adult insect. Although the pupal stage is typically credited with the metamorphic transformation from larva to adult, many of the adult morphological structures are produced *before* the pupal stage, while the animals are still larvae. Cells that will form the adult structures (e.g. eyes, wings, legs, genitalia) are set aside very early in development, during embryogenesis (33, 73, 122, 155, 191). These "imaginal" cells (imago means adult) remain distinct from the larval cells throughout development. In many insects, as these clusters of imaginal cells divide, they fold into the body cavity of the larva, forming isolated pockets of adult cells called "imaginal disks" (33, 73). Once they have invaginated away from the body wall, the imaginal disks can grow without affecting the exterior shape or structure of the larva. At the very end of the larval period, when animals are ready to molt into pupae, these imaginal disks evert and join to form the morphological structures of the pupal cuticle. When animals shed their larval cuticle and expand to fill this new pupal cuticle, they now have all the morphological structures characteristic to the adults (pupae have visible legs, compound eyes, wing buds, genitalia, etc).

Two features of this mechanism are relevant to the generation of trait-scaling relationships. First, the adult structures form as discrete pockets of imaginal cells. For example, there are separate imaginal disks for left and right wings and for forewings and hindwings. Likewise, there are imaginal disks for each of the legs, for the genitalia, for the eyes and antennae, and there are distinct pockets of imaginal cells that form cuticular protrusions such as horns. This means that the adult structures develop from independent clusters of cells that may be regulated at least partially autonomously (22, 33, 236).

Second, these imaginal structures (i.e. the adult traits) do not grow at the same time and rate as the larvae. Imaginal disks undergo much of their growth very late in the larval period—after the animals have stopped feeding, and after all growth in overall body size has ceased (122, 133, 151, 153). In hemimetabolous insects, too, growth of appendages is independent of body growth and can be adjusted late in development (11, 68, 131, 132). Consequently, the relative growth of adult traits is not simply a result of tissues growing gradually at rates proportional to overall growth in body size. This also means that scaling relationships are not reflections of the underlying growth trajectories of traits (i.e. small animals with small traits do not simply stop the growth process earlier than larger animals). Instead, the scaling of body parts must result from some centralized system of coordination, where growth in the imaginal traits is modified depending on the body size each animal attains (200). Hormones are one way this "size" information may be communicated to the growing traits.

Background: Hormones and the Regulation of Tissue Growth

Many developmental events in insects are regulated by hormones. Circulating hormones coordinate the timing of physiological events, such as molting or metamorphosis, so that all of the tissues undergo these changes in synchrony (reviewed in 149). Hormones also regulate the fates of specific tissues, by signalling which of several possible phenotypic outcomes are expressed (as in the polyphenisms discussed below). Finally, hormones couple developmental events with the outside world. The secretion of the primary hormones (ecdysteroids and juvenile hormones) is controlled by the central nervous system (149), meaning that it is possible for these hormonal signals to become part of a transduction pathway between an environmental stimulus (perceived and integrated by the nervous system) and a developmental response (controlled by the hormone). In other words, the endocrine control of development makes it possible for certain aspects of development to become sensitive to specific environmental variables.

Cells respond to hormones only during brief periods of the life cycle. Developing tissues become sensitive to the presence or levels of specific hormones during discrete "critical" or sensitive periods, and these periods vary for different hormones, and for different stages in larval development (149). Perhaps the most important feature of hormone-sensitive periods is that they appear to be tissue- and character-specific. Thus in *Manduca* there are distinct juvenile hormone (JH) sensitive periods for pupal commitment of the imaginal disks and the larval epidermis (149, 175). Moreover, in the epidermis there are separate JH-sensitive periods for pupal commitment and for pigmentation (149, 209). In *Onthophagus*, the JH-sensitive period for horn induction affects only a small portion of the head epidermis (63), and in the alate/apterous polyphenism of aphids, JH can suppress the development of wings without affecting the normal development and metamorphosis of other body parts (93).

The simplest explanation for the tissue specificity and relatively narrow time windows of hormone sensitivity is that not all tissues express receptors for a given hormone, and those that express receptors do so for only brief periods of time. Recent advances in the molecular studies of hormone receptor expression support this view. Riddiford and her colleagues have shown that different tissues in *Manduca* have very different patterns of expression for various ecdysteroid receptors (74, 108). Ecdysteroid receptors in the imaginal disks fluctuate with a different temporal pattern from those in the epidermis. In addition, *within* tissues the temporal pattern of ecdysteroid receptor expression is complex, with many distinct peaks of expression occurring throughout larval and pupal development. Furthermore, different types of ecdysteroid receptor have different patterns of fluctuation (74, 108), and these different isoforms of the ecdysteroid receptor appear to control different downstream response pathways (27, 208). This arrangement suggests that the response of a tissue to a hormone signal may be altered by altering the expression of receptors.

It appears, then, that tissue responses to hormones show great temporal and spatial precision. Superimposed on these fluctuations in hormone receptor expression is the pattern of hormone secretion. Both ecdysteroid and JH secretion patterns exhibit great fluctuations in the course of larval and pupal development (175–177), and at least some of these peaks of hormone secretion coincide with times at which certain developmental events are most sensitive to the hormone (149, 150). The overall picture that emerges is that of a dialogue between the endocrine system and the responsive tissues: The amount and type of hormone receptors expressed in each tissue controls when, where, and how that tissue responds to a hormone.

The hormone, in turn, is a means of providing a centrally controlled synchronizing signal. At some points in development a given hormone may only affect one tissue, while at another time it may control the synchronized development of many. Because the ability to respond to a hormone can be regulated and varied at the tissue level, hormone-mediated developmental control is essentially modular. Therefore, the hormone-responsiveness of a given tissue (e.g. an exaggerated trait) could, in principle, evolve independently of that of other tissues, and this leads us to the mechanisms regulating the scaling of body parts.

Mechanisms of Scaling in Insects: The Linear Allometry

The question of how body parts scale with variations in body size can now be examined at a mechanistic level. Larvae feed and gain weight during much of the larval period (the feeding period). At the end of the feeding period, they purge their guts and begin the behavioral and physiological processes associated with the onset of metamorphosis, and it is at this time, after increases in overall body size have ceased, that the imaginal structures undergo their most prolific growth (151, 153). Some of these larvae will have encountered favorable growth conditions, and these animals will terminate growth at very large body sizes. Others will have encountered less favorable conditions, and these animals enter metamorphosis at much smaller body sizes. The mechanisms of scaling concern how the final sizes of each of the various adult traits becomes matched with the final body size of each individual larva (i.e. how a genotype generates a linear allometry; these mechanisms are reviewed in 200). How do large animals end up with larger wings and legs and eyes than smaller animals?

Somehow, the growth of the adult traits must be modified to scale with the actual size of each developing animal. This suggests that information pertaining to the actual body size of an individual (or some close correlate, e.g. growth conditions) is communicated to all of the growing tissues. Although few researchers have looked for such a signal, there is accumulating evidence that these signals exist (reviewed in 200). For example, several studies indicate that the final sizes of imaginal disks can be modified by growth factors and hormones (14, 19, 25, 42, 67, 112, 130, 168). However, for a growth factor or a hormone to modify the growth of an imaginal structure *relative to body size*, the levels or period of activity of that factor or hormone must contain information about the final body size of that animal—thus developing insects must assess their own body size. In fact, many developmental events in insects are triggered by internal assessments of body size (110, 145, 146, 154). Even insects that develop in isolation from other larvae are able to assess whether they are large or small, and their development is regulated accordingly. In some cases, these size-detection mechanisms are known, as in Hemiptera, where stretch receptor neurons respond to distensions in the abdominal wall (3, 10, 29, 147, 148, 230). In other cases the precise mechanisms remain elusive (149).

In summary, it appears that sometime late in the larval period, after all intake of resources has ceased, imaginal tissues become sensitive to the presence and level of a hormone or growth factor that modifies their growth (200). It is likely that quantitative variation in the levels of this factor communicate size information to the growing tissues, and by responding to this signal, tissues grow to dimensions appropriate for the actual body size of each animal. Although the specific factors that regulate scaling relationships have not yet been identified, and several additional aspects of these mechanisms remain to be explored, the existing picture provides an adequate framework to consider how scaling relationships arise, and, more importantly, to consider how they may change over time. The relative size of body parts will result from intrinsic properties of each developing trait (i.e. the number of cells and their rate of proliferation), as well as how those cells respond to the body size information (determined by numbers or densities of receptors expressed and/or the timing of receptor expression). Modifications to any of these components could lead to changes in how large tissues become relative to body size. Because these changes in mechanism have predictable consequences for the shape of scaling relationships, they reveal possible avenues by which exaggerated morphologies may have evolved. Here we consider two of the many possible routes to scaling relationship evolution: variations in the number of starting cells, and variations in the sensitivity of those cells to the hormone.

Changing the Slope or Intercept of a Linear Scaling Relationship

All of the imaginal disks undergo some proliferation during the earlier larval stages, but the extent of this early growth depends on the identity of the imaginal disk. Different disks grow at different rates (e.g. forelegs may grow faster than hind legs), and the same disks may grow at different rates in different genotypes (forelegs may grow faster in one genotype than in another). One possible way this variation may arise concerns the starting conditions of the imaginal disk. Variation in the number of cells present at the onset of exponential growth may lead to large differences in the absolute rates of growth of those traits at the end of larval life, because if all cells are dividing, initial differences in cell numbers will be magnified with subsequent cell proliferation (153). This suggests one mechanism for scaling-relationship evolution: Changes in the slope of a scaling relationship could arise through genetic changes in these starting conditions (e.g. changes in cell number within an imaginal disk). Large numbers of starting cells would produce fast trait growth and a steep scaling relationship and vice versa (Figure 7*a*). Consequently, differences in the number of dividing cells provide one possible explanation for how different parts of an organism could scale with body size in different ways. This mechanism could also explain how the same trait scales differently in males and females, as well as how the unusually steep scaling relationships of exaggerated insect ornaments or weapons arise.

Evolutionary changes in the relative growth of imaginal disks may also occur by modifying the expression of hormone receptors. Changing the number or density of receptors may change the way that imaginal structures "read" body size information from the hormone signal (200). Increased receptor density may cause genotypes to be more sensitive to the hormone signal and thus to begin or cease

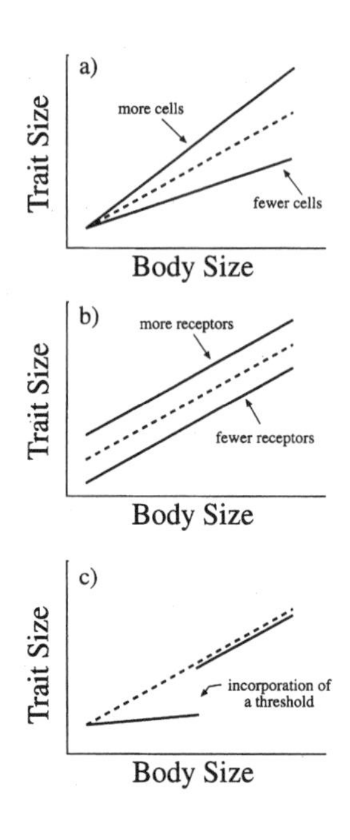

Figure 7 Suggested mechanisms for scaling relationship evolution. Adult structures in holometabolous insects derive from isolated, semiautonomous pockets of cells (imaginal disks) which undergo most of their growth during a concentrated period at the very end of the larval stage. This growth appears to be regulated by size-dependent variations in circulating levels of a hormone or growth factor. *(a)* Changes in the starting conditions (e.g. the number of simultaneously dividing cells present at the beginning of the concentrated period of trait growth) may lead to changes in the relative rates of tissue growth. In genotypes whose traits contain large numbers of cells at the outset of exponential growth, the exaggerated traits may grow at faster rates (and have steeper scaling relationship slopes) than traits in genotypes with fewer cells. *(b)* Changes in how long each trait grows may affect the relative size of the trait. Genotypes expressing large numbers of receptors for the hormone or growth factor in the exaggerated trait may be more sensitive to this factor, and subsequently may commence or terminate trait growth at slightly different times than genotypes with fewer receptors. This would change the total length of time the trait grows and could shift the relative size of the trait (i.e. shift the intercept of the scaling relationship). *(c)* Incorporation of a threshold may "uncouple" the relative growth of tissues in large and small individuals. By bringing trait expression under the regulatory control of a new hormonal stimulus, it may be possible to express the trait only in a subset of individuals (e.g. only in individuals with sufficiently high concentrations of hormones present). This can lead to sudden changes from minimal, to extensive (but still size-dependent) trait expression, and could generate sigmoid, or discontinuous scaling relationship shapes.

trait growth at slightly different times. In this manner the length of the total period of trait growth can be affected and cause traits to grow to relatively larger or smaller sizes. Such a mechanism could explain evolutionary changes in the intercept of a scaling relationship, as genotypes now express trait dimensions that previously had been appropriate for individuals with a different body size (Figure 7*b*).

Combined, these results suggest how linear scaling relationships may arise, and we have used this information to suggest how these linear relationships may change in either slope or intercept (see also 200). We now consider how these basic, linear scaling relationships might be modified to incorporate a threshold.

How Threshold Traits Work

The facultative expression of traits (polyphenism) requires a stimulus that is experienced by some members of a population and not by others. Often this stimulus comes from the external environment in the form of temperature, photoperiod, pheromone, or nutrient, but as we show in the following section, some facultative characters are allometric consequences of body size, and the stimulus in such cases is generated entirely internally.

All polyphenisms whose development has been studied in some detail appear to be controlled by hormones, and these fall into three classes: ecdysteroids, juvenile hormones, and an as yet poorly characterized set of neurosecretory hormones (149, 150). Hormones alter the fate of threshold traits by inducing the expression of specific genes (66, 149, 224). The general picture that has emerged from these studies is that the polyphenic traits are sensitive to the presence and level of hormones, that an environmental factor somehow alters the temporal pattern of hormone secretion, and that this change in the timing of hormone secretion reprograms the developmental trajectory of the polyphenic trait. The color polyphenism of *Precis coenia,* for instance, is controlled during a critical period of sensitivity to ecdysteroids that lasts from 28 to 48 hours after pupation (181). Presumably, the ecdysteroid receptors are expressed in the presumptive wings for this brief period, causing them to be sensitive to levels of hormone during that time. The control of polyphenic development lies in a shift in the timing of ecdysteroid secretion that depends on the photoperiod experienced by the larva. Under long-day conditions (summer) secretion of ecdysteroids begins about 18 hours after pupation, whereas under short-day conditions (spring) it does not begin until about 48 hours after pupation (after the critical period has ended). Hence, long-day pupae experience elevated ecdysteroids during the sensitive period and short-day pupae do not. The result is that different pigmentation patterns develop in spring and summer animals.

A number of insect polyphenisms are regulated by levels of juvenile hormone (JH). As with the ecdysteroid example, these polyphenic traits are sensitive to JH only during specific critical periods (63, 64, 93, 93a, 121, 163, 203, 228, 240, 243–245). Where they have been studied at the molecular level, the JH-sensitive periods appear to coincide with periods of ecdysteroid secretion (149, 150, 152), and levels of JH may affect patterns of gene expression indirectly by interacting with the ecdysone (149, 150, 152). Ecdysteroids initiate a series of molecular and cellular events by altering patterns of gene expression, and the presence or absence of JH during this process can affect which genes will be expressed. For instance, during metamorphosis in *D. melanogaster* and *M. sexta*, transcripts of the *Broad-complex* genes (which code for transcription factors) appear in epidermal cells within six hours of exposure to ecdysteroids in the absence of JH but not in the presence of JH (111, 177). By contrast, ecdysteroid-induced transcription of the *E-75A* gene (which also codes for a transcription factor) is enhanced in the presence of JH (177, 192). All of these genes encode transcription factors and are therefore involved in the regulation of further downstream genes. Although many of the details remain to be elucidated, these studies indicate that the levels of hormone (e.g. JH) present during critical periods of sensitivity can alter patterns of gene expression, and this mechanism can induce changes within the developing animals (e.g. switch the fate of a developing organ).

The picture that emerges is that developmental thresholds, like linear scaling relationships, result from tissue-specific responses to hormones. In both cases, tissues become sensitive to hormones by expressing receptors to the hormone, and in both cases these receptors appear to be present only during brief periods. However, scaling (graded) and threshold responses to the hormone differ in one key respect: In graded responses the hormone is present in all individuals, and variations in the levels of the hormone translate into graded differences in the dimensions of trait produced. In threshold traits, by contrast, adequate levels of the hormone are present only in a subset of the individuals. In this case, sufficient levels of hormone "reprogram" the fate of developing tissues, with the result that some individuals produce a phenotype very different from the phenotype that other individuals produce (149, 226).

Scenario for the Evolution of Sigmoid or Discontinuous Scaling Relationships

We are now in a position to speculate how linear scaling relationships may have been modified to generate sigmoid or discontinuous scaling relationships. Because the developmental system is modular, it is possible to obtain novel patterns of trait expression by deploying old, previously existing mechanisms in new contexts. Indeed, there is now excellent evidence for such redeployment at the molecular level in the evolution of eyespots in butterflies (220).

Sigmoid or discontinuous scaling relationships can be considered polyphenisms where facultative expression of the trait depends on body size (large individuals do one thing, small individuals do another). Where studied, these size-dependent polyphenisms work in the same basic ways as the polyphenisms described in the previous section: They involve hormones, the tissues are sensitive to the hormone only during brief periods, and exposure to sufficient levels of hormones appears to result in altered patterns of gene expression and subsequent reprogramming of the growth rates of specific tissues (63, 169, 172, 180, 210, 226, 240). Thus size-dependent polyphenisms may be built from the same hormone-response pathways already utilized in other tissues and in other developmental contexts. Coupling the regulatory mechanisms for an existing exaggerated trait with a different type of hormone receptor, or with receptors expressed at different developmental periods, may have permitted the expression of that structure to become sensitive to a novel environmental stimulus: in this case, the attainment of a critical body size. But how could developmental events be coupled with growth in body size?

For most studied polyphenisms, hormones couple the expression of morphology with environmental stimuli that are external to the developing animal. Photoperiod, crowding, temperature, and diet each affect the expression of some polyphenic traits. However, as discussed already, insects can also respond to stimuli that occur within each developing animal, including the growth or body size of that animal. A number of developmental events in insects are regulated by assessments of body size, and for some of these, the responses incorporate a threshold. For example, the onset of molting and metamorphosis occurs only after larvae attain a specific critical body size in *M. sexta* (145, 154), and *Trichoplusia ni* (110). These developmental events also are regulated by hormones, except that in this case the levels of hormone are associated with individual variations in growth and body size, rather than variations in photoperiod or temperature.

Whenever the levels of a hormone are influenced by growth or by body size, the potential exists to bring the size-dependent expression of traits under hormonal control. Juvenile hormone meets these criteria: Levels of JH are known to be sensitive to both diet conditions and larval growth (5, 42a, 46, 81, 109, 126, 156, 172, 210), and levels of JH appear to communicate size information to growing tissues (63, 200, 229). Thus size-dependent polyphenism could arise simply by incorporating a threshold response to JH into the development of the trait. In fact, this is exactly what appears to have occurred in every studied example of size-dependent threshold traits. JH regulates the size-dependent expression of the horn length polyphenism in dung beetles (*Onthophagus taurus*; 63), as well as size-dependent caste polyphenism in ants (*Myrmica rubra*, *Pheidole bicarinata*, *Pheidole pallidula,* and *Solenopsis invicta*; 17, 162, 179, 228, 229), honeybees (*Apis mellifera*, 169, 217, 240), bumblebees (*Bombus terrestris* and *Bombus hypnorum*, 180), wasps (*Polistes gallicus*, 180), and termites (*Kalotermes flavicollis*, *Macrotermes michaelseni,* and *Reticulitermes santonensis,* 125a, 129, 155a).

Combined, these patterns suggest an avenue for the evolution of sigmoid and discontinuous scaling. Ancestrally, the trait would be expressed in all individuals (genotypes have linear scaling relationships). In the derived situation, trait expression becomes facultative (i.e. trait expression is brought under the control of a new hormonal stimulus). Because the hormone communicates body size information, it is now possible to couple trait expression with the attainment of a specific body size (e.g. by only expressing the trait when sufficient levels of hormone are present). Genotypes regulating trait expression in this fashion do not produce the trait in small individuals. Because this developmental reprogramming results from hormone-mediated changes in patterns of gene expression, the resulting phenotypes of large and small individuals would be at least partially uncoupled and free to evolve at least partially independently.

In conclusion, we suggest that the evolution of complex scaling relationships may entail the incorporation of a threshold into the development of the trait (Figure 7*c*). The modular nature of polyphenic development suggests that this evolutionary transformation need not be a difficult one. In fact, simple changes in the timing or levels of hormone-receptor expression may be adequate to bring the development of a trait under hormonal control.

A FINAL CAVEAT: TRAITS INTERACT WITH EACH OTHER

One outcome of the study of exaggerated traits is the realization that traits interact with one another during development (120, 151, 153). These interactions cause correlations among traits, meaning that variation in one trait is coupled to variation in another trait. For example, male beetles expressing long horns also have disproportionately small eyes (151) or wings (113, 115), and ants with the largest head widths also have proportionately smaller legs (69).

The best evidence that morphological structures interact comes from perturbation experiments, where the growth of one trait is altered experimentally during development, and resultant changes in nontarget traits are monitored. In the buckeye butterfly *Precis coenia*, surgical removal of a hindwing imaginal disk prior to the period of rapid growth results in overgrowth of the adjoining forewings (151). Removal of one hindwing disk causes forewings to be disproportionately larger than they otherwise would be, and removal of both hindwings cause the forewings to be larger still (151). These experiments also affected the symmetry of the animal: when the left hindwing disk was removed, both forewings grew larger than they should have, but the forewing on the left side of the animal grew larger than the forewing on the opposite side (120). In the beetle *O. taurus* experimental reductions in the relative length of male horns were accompanied by increases in the relative sizes of male eyes (151), and this same interaction was manifest as a negative genetic correlation between horns and eyes in an artificial selection experiment in *O. acuminatus* (151).

Interactions among growing morphological traits mean that the scaling relationships of these traits will not be independent. This is particularly relevant to taxa with nonlinear scaling relationships because reprogramming of the growth of one tissue (e.g. through attainment of a threshold body size) may lead to nonlinear scaling of other tissues (153).

We advocate exploring the evolutionary significance of scaling relationship shapes. Yet we must end with a note of caution. Sigmoid or discontinuous scaling relationships make obvious targets for this type of study, and several important consequences of such scaling relationships may affect the evolution of insect morphology. However, this approach assumes that the nonlinearity in trait expression evolved in response to selection on the trait in question. If the nonlinearity arises as an indirect consequence of selection on another trait, then the observed relationship could be unrelated to selection on the focal trait. In these situations, investigators may be misled. Consequently, we suggest a multivariate approach (e.g. 118), by which scaling relationships are simultaneously examined for a number of traits. Nonlinear scaling relationships in very small traits, or traits adjacent to much larger structures also showing nonlinear scaling, may be particularly suspect and should be treated with caution.

ACKNOWLEDGMENTS

We wish to thank A Badyaev, K Bright, K Corll, ST Emlen, J Hodin, D Stern, D Tallmon, and DE Wheeler for comments on this manuscript. Funding was provided by the National Science Foundation Grant IBN-9807932 (DJE), and IBN-9728727 (HFN).

Visit the Annual Reviews home page at www.AnnualReviews.org.

LITERATURE CITED

1. Alcock J. 1984. *Animal Behavior: An Evolutionary Approach.* Sunderland, MA: Sinauer. 596 pp. 3rd. ed.
2. Andersson M. 1994. *Sexual Selection.* Princeton: Princeton Univ. Press
3. Anwyl R. 1972. The structure and properties of an abdominal stretch receptor in *Rhodnius prolixus. J. Insect Physiol.* 18:2143–53
4. Arrow GH. 1951. *Horned Beetles.* The Hague: Dr W Junk. 181 pp.
5. Asencot M, Lensky Y. 1976. The effect of sugars and juvenile hormone on the differentiation of the female honeybee larvae (*Apis mellifera* L.) to queens. *Life Sci.* 18:693–700
6. Austad SN. 1984. A classification of alternative reproductive behaviors and methods for field-testing ESS models. *Am. Zool.* 24:309–19
7. Baroni Urbani C. 1976. Réinterprétation du polymorphisme de la caste ouvrière chez les fourmis à l'aide de la régression polynomiale. *Rev. Suisse Zool.* 83:105–10
8. Bate M, Martinez-Arias A, eds. 1993. *The Development of Drosophila melanogaster.* New York: Cold Spring Harbor Lab. Press. 1558 pp.
9. Bateson W, Brindley HH. 1892. On some cases of variation in secondary sexual characters, statistically examined. *Proc. Zool. Soc. London* 1892:585–94
10. Beckel WE, Friend WG. 1964. The relation of abdominal distension and nutrition to molting in *Rhodnius prolixus* (Stahl) (Hemiptera). *Can. J. Zool.* 42:71–78
11. Blackith RE, Davies RG, Moy EA. 1963. A biometric analysis of development in *Dysdericus fasciatus* Sign (Hemiptera: Pyrhocoridae). *Growth* 27:317–34
12. Blanckenhorn WU. 1991. Life-history differences in adjacent water strider populations: phenotypic plasticity or heritable responses to stream temperature? *Evolution* 45:1520–25
13. Blum M, Blum N. 1979. *Sexual Selection and Reproductive Competition in Insects.* New York: Academic. 463 pp.
14. Bodenstein D. 1943. Hormones and tissue competence in the development of *Drosophila. Biol. Bull.* 84:34–58
15. Bradshaw AD. 1965. Evolutionary significance of phenotypic plasticity in plants. *Adv. Genet.* 13:115–55

15a. Brakefield PM, Gates J, Keys D, Kesbeke F, Wijngaarden PJ, et al. 1996. Development, plasticity and evolution of butterfly eyespot patterns. *Nature* 384: 236–242

16. Breed MD, Harrison JM. 1988. Worker

size, ovary development and division of labor in the giant tropical ant, *Paraponera clavata* (Hymenoptera: Formicidae). *J. Kans. Entomol. Soc.* 61:285–91
17. Brian MV. 1974. Caste differentiation in *Myrmica rubra*: the role of hormones. *J. Insect Physiol.* 20:1351–65
18. Briceño RD, Eberhard WG. 1995. The functional morphology of male cerci and associated characters in 13 species of tropical earwigs (Dermaptera: Forficulidae, Labiidae, Carcinophoridae, Pygidicranidae). *Smithson. Contrib. Zool. No. 555.* Washington, DC: Smithson. Inst. Press. 63 pp.
19. Britton JS, Edgar BA. 1998. Environmental control of the cell cycle in *Drosophila*: nutrition activates mitotic and endoreplicative cells by distinct mechanisms. *Development* 125:2149–58
20. Brown L, Bartalon J. 1986. Behavioral correlates of male morphology in a horned beetle. *Am. Nat.* 127:565–70
21. Brown L, Siegfried BD. 1983. Effects of male horn size on courtship activity in the forked fungus beetle, *Bolitotherus cornutus* (Coleoptera: Tenebrionidae). *Ann. Entomol. Soc. Am.* 76:253–55
22. Bryant PJ, Levinson P. 1985. Intrinsic growth control in the imaginal primordia of *Drosophila*, and the autonomous action of a lethal mutation causing overgrowth. *Dev. Biol.* 107:355–63
23. Burkhardt D, de la Motte I. 1987. Physiological, behavioural, and morphometric data elucidate the evolutive significance of stalked eyes in Diopsidae (Diptera). *Entomol. Gen.* 12:221–33
24. Burkhardt D, de la Motte I, Lunau K. 1994. Signalling fitness: larger males sire more offspring. Studies of the stalk-eyed fly *Cyrtodiopsis whitei* (Diopsidae: Diptera). *J. Comp. Physiol.* 174:61–64
25. Champlin DT, Truman JW. 1998. Ecdysteroid control of cell proliferation during optic lobe neurogenesis in the moth *Manduca sexta*. *Development* 125:269–77
26. Chapman RF. 1982. *The Insects: Structure and Function*. London: Hodder & Stoughton
27. Cherbas P, Cherbas L. 1996. Molecular aspects of ecdysteroid hormone action. See Ref. 79, pp. 175–221
28. Cheverud JM. 1982. Relationships among ontogenetic, static and evolutionary allometry. *Am. J. Phys. Anthropol.* 59:139–49
29. Chiang GR, Davey KG. 1988. A novel receptor capable of monitoring applied pressure in the abdomen of an insect. *Science* 241:1665–67
30. Choe JC, Crespi BJ. 1997. *The Evolution of Mating Systems in Insects and Arachnids*. Cambridge, UK: Cambridge Univ. Press. 387 pp.
31. Clark JT. 1977. Aspects of variation in the stag beetle *Lucanus cervus* (L.) (Coleoptera: Lucanidae). *Syst. Entomol.* 2:9–16
32. Cock AG. 1966. Genetical aspects of metrical growth and form in animals. *Q. Rev. Biol.* 41:131–90
33. Cohen SM. 1993. Imaginal disc development. See Ref. 8, pp. 747–841
34. Conner J. 1988. Field measurements of natural and sexual selection in the fungus beetle, *Bolitotherus cornutus*. *Evolution* 42:736–49
35. Cook D. 1986. Sexual selection in dung beetles I. A multivariate study of the morphological variation in two species of *Onthophagus* (Scarabaeidae: Onthophagini). *Aust. J. Zool.* 35:123–32
36. Cook D. 1990. Differences in courtship, mating and postcopulatory behavior between male morphs of the dung beetle *Onthophagus binodis* Thunberg (Coleoptera: Scarabaeidae). *Anim. Behav.* 40:428–36
37. Corn ML. 1980. Polymorphism and polyethism in the neotropical ant *Cephalotes atratus* (L.). *Insectes Soc.* 27:29–42
38. Danforth BN. 1991. The morphology and behavior of dimorphic males in *Per-*

dita portalis (Hymenoptera: Andrenidae). *Behav. Ecol. Sociobiol.* 29:235–47
39. Danforth BN, Desjardins CA. 1999. Male dimorphism in *Perdita portalis* (Hymenoptera, Andrenidae) has arisen from preexisting allometric patterns. *Insectes Soc.* 46:18–28
40. Danforth BN, Neff JL. 1992. Male polymorphism and polyethism in *Perdita texana* (Hymenoptera: Andrenidae). *Ann. Entomol. Soc. Am.* 85:616–26
41. Darwin C. 1871. *The Descent of Man, and Selection in Relation to Sex,* Vol. 1. London: Murray. 423 pp.
42. Davis KT, Shearn A. 1977. In vitro growth of imaginal disks from *Drosophila melanogaster. Science* 196:438–40
42a. de Wilde H, Beetsma J. 1982. The physiology of caste development in social insects. *Adv. Insect Physiol.* 19:167–246
43. Diakonov DM. 1923. Experimental and biometrical investigations on dimorphic variability of *Forficula. J. Genet.* 15: 200–32
44. Diniz-Filho JAF, Von Zuben CJ, Fowler HG, Schlindwein MN, Bueno OC. 1994. Multivariate morphometrics and allometry in a polymorphic ant. *Insectes Soc.* 41:153–63
45. Dodson GN. 1997. Resource defense mating system in antlered flies, *Phytalmia* spp. (Diptera: Tephritidae). *Ann. Entomol. Soc. Am.* 90:496–504
46. Dogra GS, Ulrich GM, Rembold H. 1977. A comparative study of the endocrine system of the honeybee larva under normal and experimental conditions. *Z. Naturforsch. Teil C* 32:637–42
47. Dominey WJ. 1984. Alternative mating tactics and evolutionary stable strategies. *Am. Zool.* 24:385–96
48. Druger M. 1962. Selection and the effect of temperature on scutellar bristle number in *Drosophila. Genetics* 56:39–47
49. Dudich E. 1923. Uber die variation des *Cyclommatus tarandus* Thunberg (Coleop., Lucanidae). *Arch. Naturgesch.* 2:62–89
50. Dudley SA, Schmitt J. 1996. Testing the adaptive plasticity hypothesis: density-dependent selection on manipulated stem length in *Impatiens capensis. Am. Nat.* 147:445–65
51. Eberhard WG. 1978. Fighting behavior of male *Golofa porteri* beetles (Scarabaeidae: Dynastinae). *Psyche* 83:292–98
52. Eberhard WG. 1979. The functions of horns in *Podischnus agenor* Dynastinae and other beetles. In *Sexual Selection and Reproductive Competition in Insects*, ed. MS Blum, NA Blum, pp. 231–58. New York: Academic. 463 pp.
53. Eberhard WG. 1982. Beetle horn dimorphism: making the best of a bad lot. *Am. Nat.* 119:420–26
54. Eberhard WG. 1983. Behavior of adult bottle brush weevils (*Rhinostomus barbirostris*) (Coleoptera: Curculionidae). *Rev. Biol. Trop.* 31:233–44
55. Eberhard WG. 1987. Use of horns in fights by the dimorphic males of *Ageopsis nigricollis* (Coleoptera, Scarabeidae, Dynastinae). *J. Kans. Entomol. Soc.* 60:504–9
56. Eberhard WG. 1998. Sexual behavior of *Acanthocephala declivis guatemalana* (Hemiptera: Coreidae) and the allometric scaling of their modified hind legs. *Ann. Entomol. Soc. Am.* 91:863–71
57. Eberhard WG, Garcia-C JM. 1999. Ritual jousting by horned *Geraeus* sp. weevils (Coleoptera, Curculionidae, Barydinae). *Psyche.* In press
58. Eberhard WG, Gutierrez EE. 1991. Male dimorphisms in beetles and earwigs and the question of developmental constraints. *Evolution* 45:18–28
59. Emlen DJ. 1994. Environmental control of horn length dimorphism in the beetle *Onthophagus acuminatus* (Coleoptera: Scarabaeidae). *Proc. R. Soc. London Ser. B* 256:131–36
60. Emlen DJ. 1994. *Evolution of male horn length dimorphism in the dung beetle Onthophagus acuminatus* (Coleoptera:

Scarabaeidae). PhD diss. Princeton, NJ: Princeton Univ.

61. Emlen DJ. 1996. Artificial selection on horn length-body size allometry in the horned beetle *Onthophagus acuminatus* (Coleoptera: Scarabaeidae). *Evolution* 50:1219–30
62. Emlen DJ. 1997. Alternative reproductive tactics and male-dimorphism in the horned beetle *Onthophagus acuminatus* (Coleoptera: Scarabaeidae). *Behav. Ecol. Sociobiol.* 41:335–41
63. Emlen DJ, Nijhout HF. 1999. Hormonal control of male horn length dimorphism in the horned beetle *Onthophagus taurus*. *J. Insect Physiol.* 45:45–53
64. Endo K, Funatsu S. 1985. Hormonal control of seasonal morph determination in the swallowtail butterfly, *Danaus plexippus*. *J. Insect Physiol.* 31:669–74
65. Enrodi S. 1985. *The Dynastinae of the World.* Boston: Dr W Junk. 800 pp.
66. Evans JD, Wheeler DE. 1999. Differential gene expression between developing queens and workers in the honey bee, *Apis mellifera*. *Proc. Natl. Acad. Sci. USA* 96:5575–80
67. Fain MJ, Schneiderman HA. 1979. Wound healing and regenerative response of fragments of the *Drosophila* wing imaginal disc cultured in vitro. *J. Insect Physiol.* 25:913–24
68. Fairbairn DJ. 1990. The origins of allometry: size and shape polymorphism in the common waterstrider, *Gerris remigis* Say (Heteroptera, Gerridae). *Biol. J. Linn. Soc.* 45:167–86
69. Feener DH Jr, Lighton JRB, Bartholomew GA. 1988. Curvilinear allometry, energetics and foraging ecology: a comparison of leaf-cutting ants and army ants. *Funct. Ecol.* 2:509–20
70. Forsyth A, Alcock J. 1990. Female mimicry and resource defense polygyny by males of a tropical rove beetle, *Leistotrophus versicolor* (Coleoptera: Staphylinidae). *Behav. Ecol. Sociobiol.* 26:325–30
71. Fortelius W, Pamilo P, Rosengren R, Sundström L. 1987. Male size dimorphism and alternative reproductive tactics in *Formica exsecta* ants (Hymenoptera: Formicidae). *Ann. Zool. Fenn.* 24:45–54
72. Franks NR, Healey KJ, Byrom L. 1991. Studies on the relationship between the ant ectoparasite *Antennophorus grandis* (Acarina: Antennophoridae) and its host *Lasius flavus* (Hymenoptera: Formicidae). *J. Zool.* 225:59–70
73. Fristrom D, Fristrom JW. 1993. The metamorphic development of the adult epidermis. See Ref. 8, pp. 843–97
74. Fujiwara H, Jindra M, Newitt R, Palli SR, Hiruma K, Riddiford LM. 1995. Cloning of an ecdysone receptor homolog from Manduca sexta and the developmental profile of its mRNA in wings. *Insect Biochem. Mol. Biol.* 25:845–56
75. Fujusaki K. 1981. Studies on the mating system of the winter cherry bug, *Acanthocoris sordidus* Thunberg (Heteroptera: Coreidae) II. Harem defence polygyny. *Res. Popul. Ecol.* 23:262–79
76. Gabriel W, Lynch M. 1992. The selective advantage of reaction norms for environmental tolerance. *J. Evol. Biol.* 5:41–59
77. Gavrilets S, Scheiner SM. 1993. The genetics of phenotypic plasticity. V. Evolution of reaction norm shape. *J. Evol. Biol.* 6:31–48
78. Gibson RI. 1989. Soldier production in *Campenotus novaeboracensis* during colony growth. *Insectes Soc.* 36:28–41
79. Gilbert LI, Tata JR, Atkinson BG, eds. 1996. *Metamorphosis*. New York: Academic. 687 pp.
80. Goetsch W, Eisner H. 1930. Beiträge zur biologie körnersammelnder ameisen, II. *Z. Morphol. Ökol. Tiere* 16:371–452
81. Goewie EA. 1978. Regulation of caste differentiation in the honeybee (*Apis melifera* L.). *Meded. Landbouwhogesch. Wageningen* 78:1–76
82. Goldsmith SK. 1985. Male dimorphism in *Dendrobias mandibularis* Audinet-

Serville (Coleoptera: Cerambycidae). *J. Kans. Entomol. Soc.* 58:534–38

83. Goldsmith SK. 1987. The mating system and alternative reproductive behaviors of *Dendrobias mandibularis* (Coleoptera: Cerambycidae). *Behav. Ecol. Sociobiol.* 20:111–15
84. Gotwald WH. 1978. Trophic ecology and adaptation in tropical Old World ants of the subfamily Dorylinae (Hymenoptera: Formicidae). *Biotropica* 10:161–69
85. Gotwald WH. 1982. Army ants. In *Social Insects*, ed. HR Hermann, 4:157–254. New York: Academic. 385 pp.
86. Gould SJ. 1966. Allometry and size in ontogeny and phylogeny. *Biol. Rev.* 41:587–640
87. Gould SJ. 1974. The origin and function of "bizarre" structures: antler size and skull size in the "Irish elk," *Megaloceros giganteus. Evolution* 28:191–220
88. Gould SJ. 1982. Change in developmental timing as a mechanism of macroevolution. In *Evolution and Development*, ed. JT Bonner, pp. 333–46. New York: Springer-Verlag. 356 pp.
89. Green AJ. 1992. Positive allometry is likely with mate choice, competitive display and other functions. *Anim. Behav.* 43:170–72
90. Gross MR. 1996. Alternative reproductive strategies and tactics: diversity within sexes. *Trends Ecol. Evol.* 11:92–98
91. Gupta AP, Lewontin RC. 1982. A study of reaction norms in natural populations of *Drosophila pseudoobscura. Evolution* 36:934–48
92. Happell R. 1989. Fitting bent lines to data, with applications to allometry. *J. Theor. Biol.* 138:235–56
93. Hardie J. 1980. Juvenile hormone mimics the photoperiodic apterization of the alate gynopara of aphid, *Aphis fabae. Nature* 286:602–4

93a. Hardie J. 1987. The corpus allatum, neurosecretion and photoperiodically controlled polymorphism in an aphid. *J. Insect Physiol.* 33:201–5

94. Herbers JM, Cunningham M. 1983. Social organization in *Leptothorax longispinosus* Mayr. *Anim. Behav.* 31:759–71
95. Hersh AH. 1934. Evolutionary relative growth in the titanotheres. *Am. Nat.* 68:537–61
96. Higashi S, Peeters CP. 1990. Worker polymorphism and nest structure in *Myrmecia brevinoda* Forel (Hymenoptera: Formicidae). *J. Aust. Entomol. Soc.* 29:327–31
97. Hillesheim E, Stearns SC. 1991. The response of *Drosophila melanogaster* to artificial selection on body weight and its phenotypic plasticity in two larval food environments. *Evolution* 45:1909–23
98. Hölldobler B, Wilson EO. 1990. *The Ants*. Cambridge: Belknap Press of Harvard Univ. Press. 733 pp.
99. Hollingsworth MJ. 1960. Studies on the polymorphic workers of the army ant *Dorylus (Anomma) nigricans* Illiger. *Insectes Soc.* 7:17–37
100. Hunt J, Simmons LW. 1997. Patterns of fluctuating asymmetry in beetle horns: an experimental examination of the honest signalling hypothesis. *Behav. Ecol. Sociobiol.* 41:109–14
101. Huxley JS. 1931. Relative growth of mandibles in stag-beetles (Lucanidae). *J. Linn. Soc. London Zool.* 37:675–703
102. Huxley JS. 1932. *Problems of Relative Growth*. Baltimore: Johns Hopkins Univ. Press. 276 pp. Repr. 1993
103. Iguchi Y. 1998. Horn dimorphism of *Allomyrina dichotoma septentrionalis* (Coleoptera: Scarabaeidae) affected by larval nutrition. *Ann. Entomol. Soc. Am.* 91:845–47
104. Imasheva AG, Bosenko DV, Bubli OA. 1999. Variation in morphological traits of *Drosophila melanogaster* (fruit fly) under nutritional stress. *Heredity* 82: 187–92
105. Inukai T. 1924. Statistical studies on the

variation of stag beetles. *Trans. Sapporo Natl. Hist. Soc.* 9:77–91

106. Ito F, Sugiura N, Higashi S. 1994. Worker polymorphism in the red-head bulldog ant (Hymenoptera: Formicidae), with description of nest structure and colony composition. *Ann. Entomol. Soc. Am.* 87:337–41
107. Jeanne RL. 1996. Non-allometric Queen-worker dimorphism in *Pseudopolybia difficilis* (Hymenoptera: Vespidae). *J. Kans. Entomol. Soc.* 69 (Suppl.):370–74
108. Jindra M, Malone F, Hiruma K, Riddiford LM. 1996. Developmental profiles and ecdysteroid regulation of the mRNAs for two ecdysone receptor isoforms in the epidermis and wings of the tobacco hornworm, *Manduca sexta. Dev. Biol.* 180:258–72
109. Johansson AS. 1958. Relation of nutrition to endocrine functions in the milkweed bug *Oncopeltus fasciatus* (Dallas) (Heteroptera, Lygaeidae). *Nytt Mag. Zool. (Oslo)* 7:1–132
110. Jones D, Jones G, Hammock BD. 1981. Growth parameters associated with endocrine events in larval *Trichoplusia ni* (Hübner) and timing of these events with developmental markers. *J. Insect Physiol.* 27:779–88
111. Karim FD, Guild GM, Thummel CS. 1993. The *Drosophila* Broad-Complex plays a key role in controlling ecdysone-regulated gene expression at the onset of metamorphosis. *Development* 118:977–88
112. Kawamura K, Shibata T, Saget O, Peel D, Bryant PJ. 1999. A new family of growth factors produced by the fat body and active on *Drosophila* imaginal disc cells. *Development* 126:211–19
113. Kawano K. 1995. Horn and wing allometry and male dimorphism in giant rhinoceros beetles (Coleoptera: Scarabaeidae) of tropical Asia and America. *Ann. Entomol. Soc. Am.* 88:92–99
114. Kawano K. 1995. Habitat shift and phenotypic character displacement in sympatry of two closely related rhinoceros beetle species (Coleoptera: Scarabaeidae). *Ann. Entomol. Soc. Am.* 88:641–52
115. Kawano K. 1997. Cost of evolving exaggerated mandibles in stag beetles (Coleoptera: Lucanindae). *Ann. Entomol. Soc. Am.* 90:453–61
116. Kawano K. 1998. How far can the Neo-Darwinism be extended? A consideration from the history of higher taxa in the Coleoptera. *Riv. Biol. (Biol. Forum)* 91:31–56

116a. Kercut GA, Gilbert LI, eds. 1985. *Comprehensive Insect Physiology, Biochemistry, and Pharmacology,* Vols. 2, 8. New York: Pergamon. 505 pp.

117. Kindred B. 1965. Selection for temperature sensitivity in scute *Drosophila. Genetics* 52:723–28
118. Klingenberg CP. 1996. Multivariate allometry. In *Advances in Morphometrics,* ed. LF Marcus, pp. 23–49. New York: Plenum. 587 pp.
119. Klingenberg CP, Nijhout HF. 1998. Competition among growing organs and developmental control of morphological asymmetry. *Proc. R. Soc. London Ser. B* 265:1135–39
120. Klingenberg CP, Zimmerman M. 1992. Static, ontogenetic and evolutionary allometry: a multivariate comparison in nine species of water striders. *Am. Nat.* 140:601–20
121. Koch PB, Büchmann D. 1987. Hormonal control of seasonal morphs by the timing of ecdysteroid release in *Araschina levana* L. (Nymphalidae: Lepidoptera). *J. Insect Physiol.* 33:823–29
122. Kremen C, Nijhout HF. 1998. Control of pupal commitment in the imaginal disks of *Precis coenia* (Lepidoptera: Nymphalidae) *J. Insect Physiol.* 44:287–96
123. LaBarbera M. 1989. Analyzing body size as a factor in ecology and evolution. *Annu. Rev. Ecol. Syst.* 20:97–117
124. Lameere A. 1904. L'Evolution des ornaments sexuels. *Bull. Acad. Belg.* 1904:1327–64

125. Lande R. 1979. Quantitative genetic analysis of multivariate evolution, applied to brain: body size allometry. *Evolution* 33:402–16
125a. Lelis AT, Everaerts C. 1993. Effects of juvenile hormone analogues upon soldier differentiation in the termite *Reticulitermes santonensis* (Rhinotermitidae, Heterotermitinae). *J. Morphology* 217: 239–61
126. Lenz M. 1976. The dependence of hormone effects in termite caste determination on external factors. See Ref. 129a, pp. 73–89.
127. Lloyd DG. 1984. Variation strategies of plants in heterogeneous environments. *Biol. J. Linn. Soc.* 21:357–85
128. Lorch PD, Wilkinson GS, Reillo PR. 1993. Copulation duration and sperm precedence in the stalk-eyed fly *Cyrtodiopsis whitei* (Diptera: Diopsidae). *Behav. Ecol. Sociobiol.* 32:303–11
129. Lüscher M. 1972. Environmental control of juvenile hormone (JH) secretion and caste differentiation in termites. *Comp. Endocrinol.* 3 (Suppl.):509–14
129a. Lüscher M. 1976. *Phase and Caste Determination in Insects: Endocrine Aspects.* New York: Pergamon. 130 pp.
130. Madhavan K, Schneiderman HA. 1969. Hormonal control of imaginal disc regeneration in *Galleria mellonella* (Lepidoptera). *Biol. Bull.* 137:321–31
131. Matsuda R. 1960. Morphology, evolution and classification of the Gerridae (Hemiptera: Heteroptera). *Univ. Kans. Sci. Bull.* 41:25–632
132. Matsuda R. 1961. Studies of relative growth in Gerridae (4) (Memiptera: Heteroptera). *J. Kans. Entomol. Soc.* 34:5–17
133. Milan M, Campuzano S, Garcia-Bellido A. 1996. Cell cycling and patterned cell proliferation in the *Drosophila* wing during metamorphosis. *Proc. Natl. Acad. Sci. USA* 93:11687–92
134. Mitchell PL. 1980. Combat and territorial defense of *Acanthocephala femorata* (Hemiptera: Coreidae). *Ann. Entomol. Soc. Am.* 73:404–8
135. Miyatake T. 1995. Territorial mating aggregation in the bamboo bug, *Notobitus meleagris,* Fabricius (Heteroptera: Coreidae). *J. Ethol.* 13:185–89
136. Miyatake T. 1997. Functional morphology of the hind legs as weapons for male contests in *Leptoglossus australis* (Heteroptera: Coreidae). *J. Insect Behav.* 10:727–35
137. Moczek AP. 1996. Male dimorphism in the scarab beetle *Onthophagus taurus Schreber,* 1759 (Scarabaeidae, Onthophagini): evolution and plasticity in a variable environment. *Diplomarbeit.* Würzburg, Ger.: Julius-Maximilians-Univ.
137a. Moczek AP, Emlen DJ. 2000. Male horn dimorphism in the scarab beetle *Onthophagus taurus:* Do alternative reproductive tactics favor alternative phenotypes? *Anim. Behav.* In press
138. Moczek AP, Emlen DJ. 1999. Proximate determination of male horn dimorphism in the beetle *Onthophagus taurus* (Coleoptera: Scarabaeidae). *J. Evol. Biol.* 12:27–37
139. Moffett MW. 1987. Division of labor and diet in the extremely polymorphic ant *Pheidologeton diversus. Nat. Geogr. Res.* 3:282–304
140. Moffett MW, Tobin JE. 1991. Physical castes in ant workers: a problem for *Daceton armigerum* and other ants. *Psyche* 98:283–92
141. Moore AJ, Wilson P. 1993. The evolution of sexually dimorphic earwig forceps: social interactions among adults of the toothed earwig, *Vostox apicedentatus. Behav. Ecol.* 4:40–48
142. Moron MA. 1983. Los estados inmaduros de *Inca clathrata sommeri* Westwood (Coleoptera: Melolonthidae: Trichiinae); con observaciones sobre el crecimiento alometrico del imago. *Folia Entomol. Mex.* 56:31–51 (In Spanish)
143. Moron MA. 1987. Los estados inmadu-

ros de *Dynastes hyllus* Chevrolat (Coleoptera: Melolonthidae: Dynastinae); con observaciones sobre su biologia y el crecimiento alometico del imago. *Folia Entomol. Mex.* 72:33–74 (In Spanish)
144. Nedvéd O, Windsor D. 1994. Allometry in sexual dimorphism of *Stenotarsus rotundus* Arrow. *Coleop. Bull.* 48:51–59
145. Nijhout HF. 1975. A threshold size for metamorphosis in the tobacco hornworm, *Manduca sexta* (L.). *Biol. Bull.* 149:214–25
146. Nijhout HF. 1979. Stretch-induced moulting in *Oncopeltus fasciatus*. *J. Insect Physiol.* 25:277–81
147. Nijhout HF. 1981. Physiological control of molting in insects. *Am. Zool.* 21:631–40
148. Nijhout HF. 1984. Abdominal stretch reception in *Dipetalogaster maximus* (Hemiptera: Reduviidae). *J. Insect Physiol.* 30:629–33
149. Nijhout HF. 1994. *Insect Hormones*. Princeton: Princeton Univ. Press. 267 pp.
150. Nijhout HF. 1999. Control mechanisms of polyphenic development in insects. *Bioscience* 49:181–92
151. Nijhout HF, Emlen DJ. 1998. Competition among body parts in the development and evolution of insect morphology. *Proc. Natl. Acad. Sci. USA* 95:3685–89
152. Nijhout HF, Wheeler DE. 1982. Juvenile hormone and the physiological basis of insect polymorphisms. *Q. Rev. Biol.* 57:109–33
153. Nijhout HF, Wheeler DE. 1996. Growth models of complex allometries in holometabolous insects. *Am. Nat.* 148:40–56
154. Nijhout HF, Williams CM. 1974. Control of moulting and metamorphosis in the tobacco hornworm, *Manduca sexta* (L.): Growth of the last instar larva and the decision to pupate. *J. Exp. Biol.* 61:481–91
155. Oberlander H. 1985. The imaginal disks. See Ref. 116a, 2:151–82
155a. Okkut-Kotber BM. 1980. The influence of juvenile hormone analogue on soldier differentiation in the higher termite, *Macrotermes michailseni*. *Physiol. Entomol.* 5: 407–16
156. Ono S. 1982. Effect of juvenile hormone on the caste determination in the ant *Pheidole fervida* Smith (Hymenoptera: Formicidae). *Appl. Entomol. Zool.* 17:1–7
157. Oster GF, Wilson EO. 1978. *Caste and Ecology in the Social Insects*. Princeton: Princeton Univ. Press. 352 pp.
158. Otte D, Stayman K. 1979. Beetle horns: some patterns in functional morphology. In *Sexual Selection and Reproductive Competition in Insects*, ed. M Blum, N Blum, pp. 259–92. New York: Academic. 463 pp.
159. Otronen M. 1988. Intra- and intersexual interactions at breeding burrows in the horned beetle, *Coprophanaeus ensifer*. *Anim. Behav.* 36:741–48
160. Palmer TJ. 1978. A horned beetle which fights. *Nature* 274:583–84
160a. Panhuis T, Wilkinson GS. 1999. Exaggerated male eye span influences contest outcome in stalk-eyed flies. *Behav. Ecol. Sociobiol.* 46:221–27
161. Parejko K, Dodson SI. 1991. The evolutionary ecology of an antipredator reaction norm: Daphnia pulex and *Chaoborus americanus*. *Evolution* 45:1665–74
162. Passera L, Suzzoni J-P. 1979. Le rôle de la reine de *Pheidole pallidula* (Nyl.) l'hormone juvénile. *Insectes Soc.* 26: 343–53 (In French)
163. Pener MP. 1985. Hormonal effects on flight and migration. See Ref. 116a, 8:491–550
164. Peters RH. 1983. *The Ecological Implications of Body Size*. Cambridge, UK: Cambridge Univ. Press
165. Petrie M. 1988. Intraspecific variation in structures that display competitive ability: large animals invest relatively more. *Anim. Behav.* 36:1174–79
166. Petrie M. 1992. Are all secondary sexual

display structures positively allometric and, if so, why? *Anim. Behav.* 43:173–75
167. Phleger FB. 1940. Relative growth and vertebrate phylogeny. *Am. J. Sci.* 238:643–62
168. Postlethwait JH, Schneiderman HA. 1970. Effects of an ecdysone on growth and cuticle formation of *Drosophila* imaginal discs cultured in vivo. *Drosoph. Inf. Serv.* 45:124
169. Rachinsky A, Hartfelder K. 1990. Corpora allata activity, a prime regulating element for caste-specific juvenile hormone titre in honey bee larvae (*Apis mellifera carnica*). *J. Insect Physiol.* 36:189–94
170. Radesäter T, Halldórsdöttir H. 1993. Two male types of the common earwig: male-male competition and mating success. *Ethology* 95:89–96
171. Rasmussen J. 1994. The influence of horn and body size on the reproductive behavior of the horned rainbow scarab beetle *Phanaeus difformis* (Coleoptera: Scarabaeidae). *J. Insect Behav.* 7:67–82
172. Rembold H. 1987. Caste specific modulation of juvenile hormone titers in *Apis mellifera. Insect Biochem.* 17:1003–7
173. Rensch B. 1959. *Evolution Above the Species Level.* London: Methuen. 419 pp.
174. Richards OW. 1927. Sexual selection and allied problems in the insects. *Biol. Rev.* 2:298–364
175. Riddiford LM. 1985. Hormone action at the cellular level. See Ref. 116a, 8:37–84
176. Riddiford LM. 1996. Juvenile hormone: The status of its "status quo" action. *Arch. Insect Biochem. Physiol.* 32:271–86
177. Riddiford LM. 1996. Molecular aspects of juvenile hormone action in insect metamorphosis. See Ref. 79, pp. 223–51
178. Riska B, Atchley WR. 1985. Genetics of growth predict patterns of brain-size evolution. *Science* 229:668–71
179. Robeau RM, Vinson SB. 1976. Effects of juvenile hormone analogues on caste differentiation in the imported fire ant, *Solenopsis invicta. J. Ga. Entomol. Soc.* 11:198–203
180. Röseler PF. 1976. Juvenile hormone and queen rearing in bumblebees. See Ref. 129a, pp. 55–61
181. Rountree DB, Nijhout HF. 1995. Hormonal control of a seasonal polyphenism in *Precis coenia* (Lepidoptera: Nymphalidae). *J. Insect Physiol.* 41:987–92
182. Scharloo W, Zweep A, Schuitema KA, Wijnstra JG. 1972. Stabilizing and disruptive selection on a mutant character in *Drosophila.* IV. Selection on sensitivity to temperature. *Genetics* 71:551–66
183. Scheiner SM. 1993. Genetics and evolution of phenotypic plasticity. *Annu. Rev. Ecol. Syst.* 24:35–68
184. Scheiner SM, Gavrilets S. 1993. The genetics of phenotypic plasticity. V. Evolution of reaction norm shape. *J. Evol. Biol.* 6:31–48
185. Scheiner SM, Lyman RF. 1991. The genetics of phenotypic plasticity II. Response to selection. *J. Evol. Biol.* 3:23–50
186. Schlichting CD. 1986. The evolution of phenotypic plasticity in plants. *Annu. Rev. Ecol. Syst.* 17:667–93
187. Schlichting CD, Pigliucci M. 1998. *Phenotypic Evolution: A Reaction Norm Perspective.* Sunderland, MA: Sinauer. 387 pp.
188. Schmalhausen II. 1949. *Factors of Evolution.* Philadelphia: Blakiston. 327 pp.
189. Schmidt-Nielsen K. 1984. *Scaling: Why is Animal Size so Important?* Cambridge, UK: Cambridge Univ. Press. 241 pp.
190. Schmitt J, McCormac AC, Smith H. 1995. A test of the adaptive plasticity hypothesis using transgenic and mutant plants disabled in phytochrome-mediated elongation responses to neighbors. *Am. Nat.* 146:937–53
191. Sehnal F. 1985. Growth and life cycles. See Ref. 116a, 2:1–86
192. Segraves WA, Woldin C. 1993. The E75

gene of *Manduca sexta* and comparison with its *Drosophila* homolog. *Insect Biochem. Mol. Biol.* 23:91–97

193. Shillito JF. 1971. Dimorphism in flies with stalked eyes. *Zool. J. Linn. Soc.* 50:297–305
194. Simmons LW, Tomkins JL. 1996. Sexual selection and the allometry of earwig forceps. *Evol. Ecol.* 10:97–104
195. Simmons LW, Tomkins JL, Hunt J. 1999. Sperm competition games played by dimorphic male beetles. *Proc. R. Soc. London Ser. B* 266:145–50
196. Simpson GG. 1944. *Tempo and Mode in Evolution*. New York: Columbia Univ. Press. 237 pp.
197. Simpson GG. 1953. *The Major Features of Evolution*. New York: Columbia Univ. Press. 434 pp.
198. Siva-Jothy MT. 1987. Mate securing tactics and the cost of fighting in the Japanese horned beetle, *Allomyrina dichotoma* L. (Scarabaeidae). *J. Ethol.* 5:165–72
199. Stearns SC. 1982. The role of development in the evolution of life histories. In *Evolution and Development*, ed. JT Bonner Dahlem Konferenzen: New York: Springer-Verlag. 356 pp.
200. Stern DLS, Emlen DJ. 1999. The developmental basis for allometry in insects. *Development* 126:1091–101
201. Stern DL, Moon A, Martinez del Rio C. 1996. Caste allometries in the soldier-producing aphid *Pseudoregma alexanderi* (Hormaphididae: Aphidoidea). *Insectes Soc.* 43:137–47
202. Stern DL, Whitfield JA, Foster WA. 1997. Behavior and morphology of monomorphic soldiers from the aphid genus *Pseudoregma* (Cerataphidini: Hormaphididae): implications for the evolution of morphological castes in social aphids. *Insectes Soc.* 44:379–92
203. Tanaka S. 1994. Endocrine control of ovarian development and flight muscle histolysis in a wing dimorphic cricket, *Modicogryllus confirmatus*. *J. Insect Physiol.* 40:483–90
204. Terata K. 1991. Polymorphism of *Anechura harmandi* (Burr) (Insecta: Dermaptera: Forficulidae) in the Shikoku district, Japan. *Bull. Biogeogr. Soc. Jpn.* 46:115–20
205. Thornhill R, Alcock J. 1983 *The Evolution of Insect Mating Systems*. Cambridge, MA: Harvard Univ. Press. 547 pp.
206. Tomkins JL, Simmons LW. 1996. Dimorphisms and fluctuating asymmetry in the forceps of male earwigs. *J. Evol. Biol.* 9:753–70
207. Tomkins JL, Simmons LW. 1999. Heritability of size but not symmetry in a sexually selected trait chosen by female earwigs. *Heredity* 82:151–67
208. Truman JW. 1996. Metamorphosis of the insect nervous system. See Ref. 79, pp. 283–320
209. Truman JW, Riddiford LM, Safranek L. 1974. Temporal patterns of response to ecdysone and juvenile hormone in the epidermis of the tobacco hornworm, *Manduca sexta* (L.). *J. Insect Physiol.* 19:195–203
210. Velthius HHW. 1976. Environmental, genetic and endocrine influences in stingless bee caste determination. See Ref. 129a, pp. 35–53
211. Via S. 1984. The quantitative genetics of polyphagy in an insect herbivore. II. Genetic correlations in larval performance within and among host plants. *Evolution* 38:896–905
212. Villet MH. 1990. Division of labour in the Matabele ant *Megaponera foetens* (Fabr.) (Hymenoptera: Formicidae). *Ethol. Ecol. Evol.* 2:397–417
213. Vorster H, Hewitt PH, Van Der Westhuizen C. 1991. Polymorphism and the foraging behavior of the granivorous ant, *Messor capensis* (Mayr) (Formicidae: Myrmicinae). *J. Afr. Zool.* 105:485–92
214. Waddington CH. 1957. *The Strategy of Genes*. London: Allen & Unwin

215. Waddington CH. 1960. Experiments on canalizing selection. *Genet. Res.* 1:140–50
216. Waddington CH, Robertson E. 1966. Selection for developmental canalization. *Genet. Res.* 7:303–12
217. Wang D-I. 1965. Growth rates of young queen and worker honeybee larvae. *J. Apic. Res.* 4:3–5
218. Wcislo WT, Eberhard WG. 1989. Club fights in the weevil *Macromerus bicinctus* (Coleoptera: Curculionidae). *J. Kans. Entomol. Soc.* 62:421–29
219. Weatherbee SD, Nijhout HF, Grunert LW, Halder G, Galant R, et al. 1999. Ultrabithorax function in butterfly wings and the evolution of insect wing patterns. *Curr. Biol.* 9:109–15
220. Weber KE. 1990. Selection on wing allometry in *Drosophila melanogaster. Genetics* 126:975–89
221. Weis AE, Gorman WL. 1990. Measuring selection on reaction norms: an exploration of the Eurosta-Solidago system. *Evolution* 44:820–31
222. Wenzel JW. 1992. Extreme queen-worker dimorphism in *Ropalidia ignobilis*, a small-colony wasp (Hymenoptera: Vespidae). *Insectes Soc.* 39:31–43
223. West-Eberhard MJ. 1989. Phenotypic plasticity and the origins of diversity. *Annu. Rev. Ecol. Syst.* 20:249–78
224. West-Eberhard MJ. 1992. Behavior and evolution, In *Molds, Molecules and Metazoa: Growing Points in Evolutionary Biology*, ed. PR Grant, HS Horn. Princeton: Princeton Univ. Press. 181 pp.
225. Wheeler DE. 1990. The developmental basis of worker polymorphism in fire ants. *J. Insect Physiol.* 36:315–22
226. Wheeler DE. 1991. The developmental basis of worker caste polymorphism in ants. *Am. Nat.* 138:1218–38
227. Wheeler DE, Nijhout HF. 1981. Soldier determination in ants: new role for juvenile hormone. *Science* 213:361–63
228. Wheeler DE, Nijhout HF. 1983. Soldier determination in *Pheidole bicarinata*: effect of methoprene on caste and size within castes. *J. Insect Physiol.* 29:847–54
229. Wheeler DE, Nijhout HF. 1984. Soldier determination in the ant *Pheidole bicarinata*: Inhibition by adult soldiers. *J. Insect Physiol.* 30:127–35
230. Wigglesworth VB. 1934. The physiology of ecdysis in *Rhodnius prolixus* (Hemiptera). II. Factors controlling moulting and "metamorphosis." *Q. J. Microsc. Sci.* 77:191–222
231. Wigglesworth VB. 1965. *The Principles of Insect Physiology*. London: Methuen & Co. 741 pp.
232. Deleted in proof
233. Wilkinson GS. 1993. Artificial selection alters allometry in the stalk-eyed fly *Cyrtodiopsis dalmanni* (Diptera: Diopsidae). *Genet. Res.* 62:213–22
234. Wilkinson GS, Dodson S. 1997. Function and evolution of antlers and eye stalks in flies. See Ref. 30, pp. 310–28
235. Wilkinson GS, Reillo PR. 1994. Female choice response to artificial selection on an exaggerated male trait in a stalk-eyed fly. *Proc. R. Soc. London Ser. B* 255:1–6
235a. Wilkinson GS, Taper M. 1999. Evolution and genetic variation for condition dependent ornaments in stalk-eyed flies. *Proc. R. Soc. London Ser. B* 266:1685–90
236. Williams C. 1980. Growth in insects. In *Insect Biology in the Future*, ed. M Locke, DS Smith, pp. 369–83. New York: Academic
237. Wilson EO. 1953. The origin and evolution of polymorphism in ants. *Q. Rev. Biol.* 28:136–56
238. Wilson EO. 1971. *The Insect Societies*. Cambridge, MA: Belknap Press of Harvard Univ. Press. 548 pp.
239. Wilson EO, Taylor RW. 1964. A fossil ant colony: new evidence of social antiquity. *Psyche* 71:93–103
240. Wirtz P. 1973. Differentiation in the honeybee larva. *Meded. Landbouwhogesch. Wageningen* 73–75:1–66

241. Zeh DW, Zeh JA. 1992. Sexual selection and sexual dimorphism in the harlequin beetle *Acrocinus longimanus*. *Biotropica* 24:86–96
242. Zeng ZB. 1988. Long-term correlated response, interpopulation covariation, and interspecific allometry. *Evolution* 42:363–74
243. Zera AJ, Tiebel KC. 1988. Brachypterizing effect of group rearing, juvenile hormone III and methoprene on wing length development in the wing-dimorphic cricket, *Gryllus rubens*. *J. Insect Physiol.* 34:489–98
244. Zera AJ, Tiebel KC. 1989. Differences in juvenile hormone esterase activity between presumptive macropterous and brachypterous *Gryllus rubens*: implications for the hormonal control of wing polymorphism. *J. Insect Physiol.* 35:7–17
245. Zera AJ, Tobe S. 1990. Juvenile hormone III biosynthesis in presumptive long-winged and short-winged *Gryllus rubens*: implications for the endocrine regulation of wing dimorphism. *J. Insect Physiol.* 36:271–80

Annu. Rev. Entomol. 2000. 45:709–746

PHYLOGENETIC SYSTEM AND ZOOGEOGRAPHY OF THE PLECOPTERA

Peter Zwick
Limnologische Fluss-Station des Max-Planck-Instituts fuer Limnologie, D 36110 Schlitz, Germany; e-mail: Pzwick@mpil-schlitz.mpg.de

Key Words stoneflies, morphology, monophyly, cladogram, distribution

■ **Abstract** Information about the phylogenetic relationships of Plecoptera is summarized. The few characters supporting monophyly of the order are outlined. Several characters of possible significance for the search for the closest relatives of the stoneflies are discussed, but the sister-group of the order remains unknown. Numerous characters supporting the presently recognized phylogenetic system of Plecoptera are presented, alternative classifications are discussed, and suggestions for future studies are made. Notes on zoogeography are appended. The order as such is old (Permian fossils), but phylogenetic relationships and global distribution patterns suggest that evolution of the extant suborders started with the breakup of Pangaea. There is evidence of extensive recent speciation in all parts of the world.

INTRODUCTION

The first record of Plecoptera is an excellent illustration of an adult stonefly on a magnificent color plate depicting elements of European fauna and flora (Hoefnagel 1592)—there is no accompanying text. By its habitus and coloration, the animal is probably *Perlodes* sp., at approximately natural size. The first few validly named Plecoptera were in the genus *Phryganea*, together with a number of Trichoptera (Linnaeus 1758). Overviews of the early stonefly literature are in Pictet (1841) and Zwick (177), where all references prior to 1900 mentioned in this review but not otherwise referenced can be found. Various attempts to separate (or to reunite) Plecoptera and both Trichoptera and Megaloptera, respectively, at taxonomic ranks between genus and order were made. Uncertainties about stonefly identity ended when Pictet (1832, 1833) described nymphs of several Nemouridae and confirmed Muralto's (1683) description of a perlid nymph. Burmeister (1839) proposed the name Plecoptera, and Pictet (1841) distinguished more than 100 species, presenting the last world synopsis of the order. Today more than 2000 species are recognized, on all continents except Antarctica (58, 171).

In the following analysis of phylogenetic relationships of Plecoptera I apply concepts and terms proposed and defined by Hennig (e.g. 39, 41), addressing three main aspects:

1. Monophyly of the order.
2. Phylogenetic relations with other insects.
3. Phylogenetic relations within the order.

A few comments on zoogeography are appended.

In the text, apomorphic character states are given the same numbers (in brackets, marked with an asterisk) as those appearing in the cladogram (Figure 1). To save space some references are to major works from which original papers can be traced. A complete bibliography of Plecoptera is in preparation, with a draft available upon request.

MONOPHYLY OF PLECOPTERA

Among extant Neoptera, Plecoptera are identified by a combination of mostly primitive characters (e.g. 10, 43, 177). Ordinal autapomorphies are not obvious (75). Apomorphic reduction of tarsal segments to three is the main difference from extinct Paraplecoptera, but this reduction also occurred in other insects. Some authors believe Plecoptera are distinguished by a different origin of vein M in the two wings, but there is no universally accepted system of vein homologies. In the stonefly front wing, M originates normally from the wing base, while in the hind wing it originates from Rs (for example, 57). This interpretation is insufficiently supported, and other authors (Pictet 1841; see also 42, 43) see no difference between front and hind wings in this respect. Kukalová-Peck (78) discusses a plecopteroid assemblage whose groundplan apomorphy is the basal fusion of MA and MP in both wings.

Plecopteran monophyly is supported by very few uniquely derived character expressions (171): Gonads form loops, anterior ends of left and right ovaries and testes, respectively, medially fused [*1]. In males there is a complex arrangement of two superimposed seminal vesicles each of which forms an anterior loop (illustrations in 14, 15, 171, 177) [*2]. The presence of strong oblique intersegmental ventrolongitudinal muscles in nymphs (177) facilitates laterally undulating swimming, which is unusual among invertebrates [*3] but also occurs in Zygoptera. If an ovipositor is part of the groundplan of Pterygota, its absence in Plecoptera would be apomorphic (162) [*4]. Most Plecoptera oviposit in flight, dropping an egg mass from the air or washing it into streams. Secondarily derived ovipositors occur occasionally (see Nemouridae s.l.) An accessory circulatory organ ("cercus heart") said to be unique in insects (105) was studied only in Perlidae and Perlodidae, but frequent occurrence of gills at or near the abdominal tip suggests its presence in other families. It is thus probably a derived groundplan characteristic of Plecoptera [*5] (105).

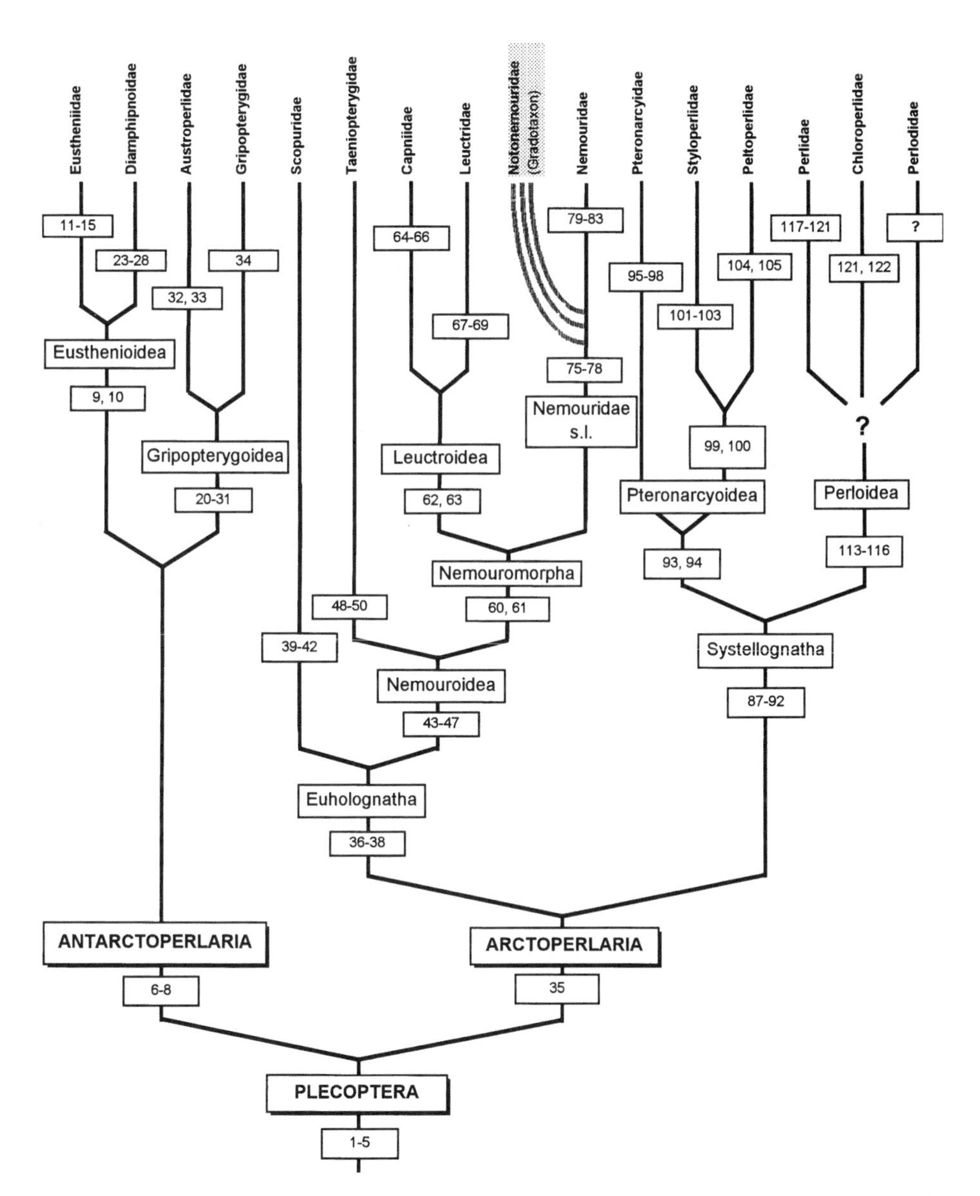

Figure 1 Cladogram of extant Plecoptera. Apomorphic characters supporting monophyly of taxa are numbered; character descriptions are in the text and are identified by the same numbers as in the diagram, in square brackets, marked by an asterisk. The text also contains characters identifying subfamilies and tribes not shown in the figure.

Nymphal osmoregulatory organs first misinterpreted as respiratory organs (171) are generally similar to other orders. Modified hooked setae along the caudal front wing margin around branches of Cu (in small species, only near the forward curvature of same margin) were proposed as a family character of Gripopterygidae (69) but these setae occur in all macropterous Antarctoperlaria and in several Arctoperlaria. These hooks may be a Plecoptera groundplan character of unknown significance. The modified setae do not seem to aid in coupling wings, which remain separate in flight (171).

Phylogenetic Relations with Other Extant Insects

Fossil Plecoptera from the early Permian are known (125), but their affinities are uncertain. Most discussions in support of the various opinions are obsolete. The dominant argument of pre-Darwinian times was whether certain morphological, anatomical, life-history, or ecological traits were essential for the establishment of methodological or natural systems, or were only auxiliary characters. Debaters arrived at conclusions arbitrarily. Even long after Darwin, well-founded, systematic, and phylogenetic concepts were still missing.

Plecoptera are Neoptera: At rest their wings fold back flat over the abdomen and the costal and subcostal areas fold down (in some families they roll slightly) to protect the sides of body (17). I accept the interpretation of neopteran wing folding as derived, despite alternative proposals (77, 78). Hennig (42, 43) considered possible relationships of Plecoptera with Polyneoptera, or alternatively with Paurometabola plus Holometabola, but provided no convincing support for either view. The main characters supporting Polyneoptera are an enlarged anal fan of the hind wing folded in front of the first anal vein and the presence of a soft, cushion-like arolium (93), which the Plecoptera indeed possess. Using Polyneoptera as an outgroup, Nelson (97) studied the plecopteran pretarsus, finding support for several subgroups of Plecoptera but no suggestion of a closest relative.

The size of the anal fan of the hindwing varies: The number of anal veins determines the size of the posterior nymphal pterotheca if wing-pads are separate from notum. In turn, size determines the position of both front and hindwing-pads, and nymphal habitus. Nymphal habitus is of no separate phylogenetic significance (P Zwick, in preparation). Mainly in Peltoperlidae, Perlidae and Chloroperlidae, wing-pads meet medially instead of leaving a separate rear margin of notum (like pannote mayflies) providing more space than required by the developing wing. Empty space may be filled by non-wing adult tissue, e.g. flanged scutum of Chloroperlini, or filamentous scutal appendages of some *Isoperla* (21) and Peltoperlidae (154, 174).

Kristensen (75) discussed possible close relationships with Polyneoptera or Embioptera, and also reached no definite conclusion. Structural details of the dorsal artery supporting a possible close relationship to Embioptera cannot be easily judged because of imperfect comparative knowledge. To me, the overall similarity of habitus of Plecoptera and Embioptera is superficial: it lies mainly in a slender body with largely independent segments (thorax forming no compact tagma), and in flat, narrow front wings.

Zwick (177) regarded Plecoptera as the possible sister group of all remaining Neoptera; Kristensen (75) admits this possibility. If Plecoptera belong to a subgroup of Neoptera, independent similar transformations must be assumed for Plecoptera characters that are more primitive than those in other Neoptera. Such characters are metameric testes and the presence of a transverse stipital muscle (171). Plecoptera sperm structure is more primitive than that of other Neoptera but also lacks distinctive traits. Although no clear distinction from orthopteroid sperm (3, 63) was eventually made, the fact remains that several structural details shared by the other Neoptera (63) would be polyphyletic or lacking secondarily in Plecoptera. A ribosomal DNA study of Holometabola used Plecoptera as one of several outgroup taxa. Results regarding the phylogenetic position of stoneflies were inconclusive (156).

Additional primitive traits (e.g. two separate penial openings) occur in other basal Neoptera. For example, terminal filament and metamerous abdominal gills are possibly primitive characters requiring further consideration. Structures called terminal filament in nymphal and adult male Austroperlidae (53) are not homologous: The mediodorsal nymphal gill filament of Austroperlidae located between epiproct and anus is homologous with the gill rosette of Gripopterygidae (178), which has a dorsal retractor muscle; a structure arising from the base of the male epiproct without muscle (171) is now called the epiproct basal process (88). The homology of nymphal structure with the terminal filament of the insect groundplan remains unproven.

Assuming a terrestrial origin of Tracheata and Insecta, aquatic insect nymphs were once regarded as derived (e.g. 107), but this status is uncertain. Gill remnants in adult Plecoptera and other hemimetabolous insects suggest (145) that in the evolution of insects tracheate pro-pterygotes became fully aquatic and only subsequently developed a terrestrial adult stage. Distribution of osmoregulatory organs across insect orders suggested to Wichard (157) that the ancestors of Pterygota invaded aquatic habitats, meaning that aquatic larvae are plesiomorphic. Kukalová-Peck believes insect wings originated from leg appendages via a gill stage (76, 77). Thus, she also assumes an aquatic life of pro-pterygotes. These ideas cannot be discussed here, but errors regarding Plecoptera require correction.

According to Kukalová-Peck (quoted in 152), moveable abdominal gills of some Plecoptera (not specified) may be homologous to wings. Insect wings are "above the spiracle and below the tergum" (76); plecopteran abdominal gills are near posterior segment borders and far ventrally from spiracular scars. Mobility of Eustheniidae gills is very limited (P Zwick, unpublished data; ID McLellan, A Neboiss, G Theischinger, personal communication); Hynes (46) states *Eusthenia venosa* can wave gills from side to side; no information is available on Diamphipnoidae, pteronarcyid gills are immobile.

According to this scenario, wings would originate from flapping gills and immobility of nymphal pterothecae would be secondary. Some (not specified) stoneflies would possess remnants of articular sclerites (78) surrounding the wing-pad base like a horseshoe, apparently resembling the suggested primary type of wing articulation (79). Plecoptera nymphs would have wing-pads "still" (my quotes) separated from terga by sutures (78). These claims are wrong. Late-instar nymphs develop wing-pads as rigid thoracic outgrowths attaining final size after (usually three) moults (7). Eventually, wing-pads may or may not be delimited by pale, hairless strips of integument, which are no sutures. Wing-pads cannot actively or passively be moved along these lines without damage to nymphs. There is no trace of articulations. Wing-pads have no connection with muscles and contain only dermal epithelium. Immediately before metamorphosis, tissue (apparently already surrounded by adult cuticle) spreads into wing-pads from the thorax and the pharate adult becomes visible through the nymphal cuticle (181).

Homology of metamerous abdominal gills (Eustheniidae, Diamphipnoidae, Pteronarcyidae and a few Perlodidae) with mayfly gills (81) is uncertain. Adult mayflies never retain abdominal gills (remnants of other gills occur) but adult Plecoptera do (145) (except Gripopterygoidea, *Scopura*, and *Taeniopteryx*). Gill remnants lack chloride cells, which abound on nymphal gills. There is no comparative study of Plecoptera and Ephemeroptera abdominal gills including embryology. Possible abdominal gill-leg homologies were considered in the past; embryology suggests abdominal gills of *Pteronarcys* are not homologous with legs (92). Most authors regard metamerous abdominal gills as homologous across families, or as part of the Plecoptera groundplan (170, 171, 177), despite differences in details of location and muscles. Ricker thought abdominal gills of Eusthenioidea were primitive and homologous with segmental gills of fossil ancestors of Plecoptera (110); abdominal gills of Pteronarcyidae would be different and homologous with substigmatic thoracic gills. To me, homology with substigmatic gills appears compatible with homology with protoperlarian abdominal gills.

There may be other variously shaped gills on different body parts in addition to metamerous gills, or by themselves; excellent micrographs are included in Wichard et al (158). Many Plecoptera have no gills. Shepard & Stewart (123) place Plecoptera among orthopteroid Neoptera and believe they originated from terrestrial gill-less (possibly grylloblattid-like) insects invading aquatic habitats. Osmoregulation was needed in the new environment, and chloride cells and eventually osmobranchiae developed. Secondarily, osmobranchiae also performed respiratory functions. Homology of chloride cells and abdominal gills with similar structures in Ephemeroptera and other insects was not discussed; the proposed topographic nomenclature of Plecoptera gills obscures recognized homologies.

PHYLOGENETIC RELATIONSHIPS AMONG PLECOPTERA

Relationships between family group taxa within Plecoptera became a focal point early in the twentieth century. Diagnostic characters distinguishing suborders recognized by Klapálek (72, 73) (Subulipalpia or Setipalpia, as opposed to Filipalpia) or Enderlein (30) (Systellognatha, as opposed to Holognatha) had already been used by Geoffroy (1762), Latreille (1796), and others to separate the genera *Perla* and *Nemoura*. Despite this historical continuity, there was long-lasting disagreement about limits and names of suborders (overview in 171). In Plecoptera systematics, derived (specialized) and primitive (archaic) character expressions were distinguished early; for example, Tillyard (151) and Illies (52, 57) calculated anagenetic indices for different taxa. However, the fundamentally different heuristic value of plesiomorphies and apomorphies was not perceived. Both derived and primitive character states were used to define taxa, or to assign taxa a position in the family tree. Also, the competing suborder concepts were treated like mutually exclusive alternatives: Klapálek's Subulipalpia (or Setipalpia) and Filipalpia against Enderlein's Systellognatha and Holognatha.

Illies (52) established a third suborder, Archiperlaria, in addition to Filipalpia and Setipalpia. Archiperlaria were completely primitive and defined by plesiomorphies only. Dendrograms by Illies were at variance with his classification, either showing members of Archiperlaria as separate branches originating from basal stem section of Setipalpia (51), or Archiperlaria as sister-group of Setipalpia plus Filipalpia (57); the abscissa indicated opposite evolutionary directions for Setipalpia and Filipalpia, respectively. The evolutionary branch of Archiperlaria was curved in order for its tip to lie in the middle, between the two other suborders. Although the terms apomorphic and plesimorphic appeared, the phylogenetic method from which they were borrowed was not applied.

Ricker (110, 111) inferred affinities from shared derived characters. Ricker's phylogenetic tree (111) showed Setipalpia families as a subgroup of Systellognatha, not as an alternative to them. His cladogram also grouped together families that were classified as Filipalpia or Holognatha, but showed southern hemisphere Plecoptera as distinct basal branches of Plecoptera. This organization is similar to the present system.

The paleontologist Sinitshenkova (125) followed lines of resemblances through geological time and derived extant stonefly families based on characters available in fossils, such as wing venation, body size, and proportions of body parts, especially tarsi. However, even she admits (125) that fossils cast no light on any of the major problems of Plecoptera phylogeny discussed here. Sinitshenkova's classification of extant Plecopera (as Paraneoptera: Perlida) partially resembles that of Illies (57) but is very different in some respects (Table 1).

A more detailed discussion is not called for here because no consistent methodology underlies that system. Some taxa are based on presumed apomorphies, others expressly on plesiomorphies (archaedictyon; in contrast, numerous crossveins are regarded as secondarily convergent development related to large body size; 19), or on nonhomologous structures (gills of Gripopterygoidea). It must be understood that group assignments of fossil stoneflies are usually based on only some diagnostic characters of extant forms and rarely on constitutive apomorphies of the taxon concerned.

Table 1

Nemourina = [{(Notonemouridae + Nemouridae) + Taeniopterygidae} + (Capniidae + Leuctridae)]

Perlina = **Perlomorpha** = [(Perlidae + Chloroperlidae) + Perlodidae + Peltoperlidae)] + **Gripopterygomorpha** [= Eusthenioidea (= Diamphiphnoidae + Eustheniidae + Pteronarcyidae) + Gripopterygoidea = (Austroperlidae + Gripopterygidae + Scopuridae)]

The following discussion is based on a suborder concept and cladogram I proposed (170, 171, 177). The main divisions of this have since been generally accepted by students of extant Plecoptera. A test employing numerical methods (95) used no additional characters, and accepted my character state interpretations; results were close but not identical. The names of some superfamilies have been corrected (principle of coordinated names; 143). The Styloperlinae were raised to family level. Results of ongoing molecular studies will be the first serious test of the present system.

There is no sound criterion for absolute category ranking of taxa. I use existing names for categories but do not propose a proliferation of categorical names for each possible split in the cladogram (Figure 1). A few unnamed taxa are given informal names here to permit unambiguous brief communication; redundant categories for monotypic taxa are not established. Sister-group relationships between taxa can be read from number codes before names; subfamilies and lower taxa are not numbered.

I. Antarctoperlaria

With no striking external distinctive characters, monophyly is evidenced by presence of a unique sternal depressor of fore trochanter [*6], absence of the normal tergal depressor of fore trochanter [*7], and presence of floriform chloride cells [*8] (which may additionally have sensory function: 64) (171, 177).

A thickened sclerotized area in the anal area of forewing suggested as synapomorphy of the suborder (177) is probably simply a toughening which, during folding, wipes over projecting vein bases of the hindwing; a similar sclerotization at the same spot occurs in other Plecoptera, for example *Pteronarcys reticulata* (my own unpublished data).

I.1. Eusthenioidea This group includes the genus containing the largest extant Plecoptera (*Diamphipnoa*), and also the most colorful ones (genera *Eusthenia*, *Thaumatoperla*); ground color of most species is green. Only two autapomorphies are known: large sensory seta on each nymphal coxa [*9]; eversible male paraproct lobes [*10] (171). Similar structures recently discovered in Perlodidae (182) are analogous. The narrow female genital opening is perhaps correlated with sperm transfer through spermatophores, a characteristic considered a superfamilial autapomorphy (177); I observed similarly narrow gonopores and similar spermatophores in living *Pteronarcys reticulata* and *P. sachalina*. Spermatophores probably occur in most Gripopterygidae and Austroperlidae (86, 88, 89), but female gonopores are not described. Spermatophores also occur in Taeniopterygidae: Brachypterainae (9, 62a). A comparative study of sperm transfer across all families to identify the Plecoptera groundplan is needed.

I.1.1. Eustheniidae The family possesses several autapomorphies (171, 176): carnivorous nymphs with slender mandibles and palpi and with shortened glossae [*11]; attachment of prosternal trochanter depressor shifted back across furcal bases to a position next to spina [*12]; abdominal nerve cord contracted, last ganglion in segment 7 [*13]; single median tibial spur [*14]. A regularly rounded posterior contour of the hindwings was regarded as plesiomorphic (e.g. 57). If this were correct, the incision delimiting the anal fan would have arisen independently in Diamphipnoidae, Gripopterygoidea, Arctoperlaria, and in many other Neoptera. A rounded hindwing contour is certainly an autapomorphy of Eustheniidae [*15].

The subfamily Eustheniinae (Australia and South America) is distinguished by an anterior extension of the female sternite 9 under the subgenital plate (segment 8) [*16] (176). Australian genera share a strongly raised, sclerotized arched anal vein in the front wing and a tough membrane surrounding each egg. Crown-shaped outgrowths of egg chorion that support the outer membrane occur in *Thaumatoperla* spp. and *Eusthenia costalis* (176) and are probably reduced in other *Eusthenia* spp. in which the membrane is detached from chorion, leaving small polar marks.

Apomorphies of subfamily Stenoperlinae (South America, New Zealand, Australia, not Tasmania) are abdominal gills reduced to 5 (6 in Eustheniinae) [*17]; elongate body [*18]; short cerci [*19]; long postocular region [*20]; extension of head capsule into prothorax [*21]; and regularly arranged spatulate spines on nymphal tibae and tarsi [*22] (87, 148, 176).

I.1.2. Diamphipnoidae Five species occur in two South American genera. Family apomorphies are abdominal gills complex, either epaulette- or brush-shaped [*23]; only four pairs of gills [*24]; condensed abdominal nerve cord (abdominal ganglion II adjoining metathorax ganglionic complex) (56) [*25]; slender mediobasal appendage on male paraprocts [*26]; serrate tarsal claws [*27] (97); and secondary absence of atypical nymphal swimming muscles [*28] (171, 177). Nymphs feed on plant material including dead wood; *Diamphipnopsis* nymphs scrape with a rasp on the galea (52).

I.2. Gripopterygoidea If paired abdominal gills are part of the Plecopteran groundplan, the gills must have been reduced several times in Plecoptera and also in the present superfamily. Monophyly is convincingly supported by modified internal genitalia: large accessory glands in males [*29], and the loss of the seminal receptacle in females [*30] (171); a small sclerite at each side of nymphal mentum is also probably apomorphic [*31] (177, 180). In nymphs, the medi-odorsal gill between epiproct and anus may be homologous to terminal filaments of ancestral insects; in other words, they may be plesiomorphic (171, 177). However, the modified gill structure is apomorphic and distinctive of families.

I.2.1. Austroperlidae Austroperlidae retain primitive traits in wing venation (57, 60). Functional gills are thick and beaded, with a parallel longitudinal pattern reflecting the arrangement of tracheoles (171). Transformation of the paraproct apex [*32] and of distal cercal segments [*33] into beaded gills is apomorphic. Two similar gills between the "terminal filament" and cercus (i.e. a total of seven beaded gills) may or may not be part of the family groundplan; dorsolateral gills are missing even in some species that have other well-developed gills. Gills of *Acruroperla atra* are poorly developed but transformation of cerci and paraprocts is recognizable; nymphs have extended pronotal corners and raised paired processes on abdominal tergites one through nine. Similar nymphal habitus ties in the aberrant *Crypturoperla paradoxa*: gills are concealed under a shield formed by tergite 10, the terminal filament is thin, bases of cerci and paraprocts have many short thin gill filaments, and the apices are unmodified.

I.2.2. Gripopterygidae Reduced wing venation in comparison with Austroperlidae may be synapomorphic or may have developed independently in subgroups. Monophyly of Gripopterygidae is confirmed by replacement of terminal filament by a retractable, often luxuriant rosette of fine gill filaments [*34]; the (secondarily) gill-less aquatic genus *Notoperla* seems to breathe via soft abdominal sternites (171), and the gill-less terrestrial taxa (e.g. *Vesicaperla, Rakiuraperla*) may have cutaneous respiration.

Several originally proposed subfamilies (see 58) based on plesiomorphies have been abandoned. McLellan (84) recognized five subfamilies in South Ameria (SAM), New Zealand (NZ), and Australia (AUS): Gripopteryginae (SAM), Leptoperlinae (SAM, AUS), Dinotoperlinae (AUS), Zelandoperlinae (NZ, SAM), and Antarctoperlinae (SAM, NZ). Antarctoperlinae has primitive topography of abdominal segmental nerves and ventral longitudinal muscles (Austroperlidae similar): Nerves lie on top of muscles; in other subfamilies the lateral part of the nerve crosses under most lateral muscle fibers (171). Details of structure suggest that the loss of tibial spurs in Zelandoperlinae and Antarctoperlinae occurred independently (86).

II. Arctoperlaria

Arctoperlaria are structurally diverse and not easily recognized as a monophyletic unit, but monophyly is strongly supported by a complex (structural and behavioral) character syndrome related to mate-finding: drumming [*35] (171, 177).

Drumming, tremulation, or behavior derived from drumming and/or related male structures (ventral lobe, hammer, vesicle, hair brush) is known in all families of Systellognatha and Nemouroidea (32, 117, 119, 137, 138, 139, 140, 142, and others), except Scopuridae. Antarctoperlaria lack similar structures and drumming and related behaviors have never been noticed (no information is available on Diamphipnoidae), despite years of conscientious observations and special efforts to record signals.

The occurrence of cervical gills, once proposed as evidence of arctoperlarian monophyly (171), is spurious and structures are variable; support of Arctoperlaria is weak (see also Nemouridae). Several types of osmoregulatory organ (unicellular caviform chloride cells and coniform and bulbiform cell complexes) are currently known only in Arctoperlaria (13, 158, 159, 160, 161): coniforms are widespread, bulbiforms are known in *Taeniopteryx* and Nemouridae (s.l.), and caviforms are known only in the latter. Inferences on family interrelationships are not possible at this time; similar kinds of chloride cells exist in other insect orders (74).

II.1. Euholognatha An unpaired corpus allatum fused to the aorta [*36] (155, 171) is good evidence of monophyly. A sclerotized hard egg chorion is almost universally present in Antarctoperlaria and Arctoperlaria: Systellognatha and suggests that this is a groundplan character of Plecoptera. The soft chorion of Euholognatha is probably apomorphic [*37]. In all euholognathan families segmental nerves cross under longitudinal abdominal muscles [*38]; in other Plecoptera, nerves lie on top of muscles (compare Gripopterygidae).

II.1.1. Scopuridae The monotypic Scopuridae are the sister-group of the large superfamily Nemouroidea. The five *Scopura* species, all completely wingless, are restricted to parts of Japan and Korea [*39]. *Scopura* was sometimes thought to be related to Austroperlidae, Gripopterygidae, and also Peltoperlidae (57, 125), because of a terminal filament and/or anal gills. The first character is plesiomorphic, and gills are not homologous. *Scopura* (except in late-instar and adult males) has a delicate terminal filament. A single gill filament originates from the soft inner paraproct base, not from the paraproct tip, which is the typical location of austroperlid gills. Nymphal gills rise from the intersegmental membrane between segments nine and 10 [*40], and not behind tergite 10 as in Gripopterygidae. *Scopura* has additional unique characters: male epiproct with an eversible membranous tube armed with sclerites [*41]; a distantly similar tube occurs only in Taeniopterygidae: Brachypterainae; a prong on male basal cercus segment is unique [*42]. Several structures are incompletely known; some of my own unpublished observations are added. The nymphal clypeus is distinct from the frons by its shape, less obviously so in adults, but there is no suture; formation of a frontoclypeus certainly occurred independently in Perloidea. The egg is round and simple, with no special structures. The chorion is described as "not very hard . . . covered with a fine membrane," and the surface is described as rasplike and scattered with many filiform processes (66). These filiform processes are not erect structures but strongly light-refracting vermicular rugosities on a very thin completely transparent membrane; no micropyles are visible. This membrane actually is the chorion and the only hull surrounding egg contents. A cytological study of chloride cells is required; I now wonder whether they are indeed coniform cells (171) because the cuticular apparatus is more similar to caviforms.

II.1.2. Nemouroidea This superfamily was long classified as a single family, Nemouridae, with several subfamilies; restricted use has been generally accepted since Illies (58). A major monophyletic group with many hundreds of species, Nemouroidea is the core of the former Filipalpia. Early authors characterized Filipalpia by a number of plesiomorphies (in comparison with Systellognatha), which made it difficult to exclude similarly primitive unrelated taxa. Nemouroidea share the following apomorphic constitutive characters: meso- and metasternal furca- and spinasterna of reduced size, modified spina transformed into a raised crest extending anteriorly and meeting medially closely adjacent furcal bases, and associated shifts of muscle attachments [*43] (171); additional tergopleural flight muscle (t13) (18) [*44]; completely reduced tenth sternite and paraprocts positioned directly behind sternite nine, laterally in contact with lateral parts of tergite 10 and the cercus bases or attached sclerites [*45]. Females of *Perlomyia* have a secondarily complete ninth sternite (100); a reduced tenth sternite occasionally occurs in other groups, such as male Perlidae. Prosternal spina is separate from prosternum, connected with the mesothoracic basisternum by a sclerite arch [*46]; male paraprocts are distinctly divided into a medial lobe (to which retractor muscles attach) and a variously shaped outer lobe with a small internal muscle [*47]; the pretarsus bears setiform basipulvilli [*48] (97).

Mode of sperm transfer and related morphological changes provide excellent characters by which the five included families can be distinguished and grouped. In the groundplan, the male sternite nine with pedunculate ventral lobe (or ventral vesicle) of characteristic fine structure is used in drumming (32, 117, 119).

II.1.2.1. Taeniopterygidae The elongate second tarsal segment is an easily seen diagnostic and constitutive character of the family [*49]. The cerci are short, and the first segment is enlarged in males [*50]; sometimes, only the first segment is well developed. The penis is short but eversible. Females have no corresponding recipient structure. The vagina is reduced. The tiny genital opening is freely exposed in the middle of sternite eight [*51]; spermatophores known for Brachypterainae (9, 177) probably also occur in *Taeniopteryx*. Male tergite 10 has lateral unsclerotized strips delimiting a shieldlike median portion of the tergite bearing the epiproct. Cu_1 in the forewing is forked, which is primitive in comparison with other Nemouroidea, but additional anteriorly directed terminal branches occurring in many Brachypterainae are derived. Two distinct subfamilies exist.

Taeniopteryginae is monotypic. *Taeniopteryx* male terminalia are simple and very similar throughout the genus: The cerci consist of a single inflated segment [*52] closely connected with the outer lobe of the paraproct. The inner paraproct lobe is simple; the epiproct is a simple erect fleshy finger with anterior sclerite strip. Female cerci are plurisegmented. The seminal receptacle is subdivided into morphologically distinct sections [*53] (171). Telescoping three-segmented gills [*54] exist on the medial face of each nymphal coxa (only scars remain in adults); the retractor muscle originating distally from gill in trochanter (171) is not homologous with structures of biramous legs. Palaearctic species form a monophyletic group distinguished by large processes on nymphal abdominal tergites. Monophyly of Nearctic species group is not evident.

Nymphs are absent from streams until autumn (e.g. 37, 38), even if eggs develop directly, as eggs of the European *T. nebulosa* (16), *T. auberti* (my own unpublished data) and *T. burksi* (Nearctic) (36) develop. At an early molt, young nymphs of *T. burksi* lose setae, cerci, and the ability to move and then spend summer in diapause (36); no information on other species is available.

Brachypterainae have amazingly complex male terminalia (171): the cercus has a large appendix to an enlarged basal segment [*55] and reduced distal segments. Sternite nine enlarged, normally covering the entire abdominal tip from below [*56]; the epiproct is sclerotized; the basal bulb contains an internal drum-like sclerite (with a narrow duct to a tiny pore on the posterior erect finger-shaped portion of the epiproct) and one to three coiled chitinous filaments [*57]; the inner paraproct lobes are complicated, of obligate asymmetry: the left usually forms a guide, the right some sort of flagellum [*58]. Eversible filaments and paraprocts together serve female stimulation during mating and perhaps open ducts permitting sperm flow from the externally attached spermatophore (9). Female sternite nine is enlarged to a postgenital plate [*59]; in all known cases, eggs lie over the gonopore and on the sternites in front of it of an ovipositing female (177), instead of being carried at the abdominal tip. Nymphs lack gills but have large soft areas on anterior abdominal sternites that are densely covered with coniform chloride cells. Nymphs are scraping sprawlers with a rasp of combed setae (158) on the apex of the galea [*60]. Stewart & Stark (141) state this for all Taeniopterygidae, but *Taeniopteryx* is at best slightly similar). Antennal sensilla are in distinct pockets of a type unknown in other taxa (158, 177) but comparative knowledge is limited. Disturbed nymphs curl up. Relationships between the genera were analyzed (113). *Zhiltzovia* is now a synonym of *Brachyptera* (168). Egg and nymphal diapause involving limited morphological change (as in Capniidae, 67) has been reported (29, 36, 68).

II.1.2.2. Nemouromorpha Taeniopterygidae are the sister-group of a previously unnamed taxon for which the name Nemouromorpha is proposed. It includes the following nemouroid families sharing two synapomorphies: reduced fork of Cu_1 in front wing [*61] and a penis that is not eversible or entirely absent [*62].

II.1.2.2.1. Leuctroidea Leuctroidea is the new superfamily name for a monophylum comprising families Capniidae and Leuctridae, sharing sperm transfer via modified inner paraproct lobes. If at all recognizable as a separate structure, the penis is tightly pressed against or firmly attached to the base of the inner paraproct lobes [*63]. Left and right ovaries are anteriorly separate, not forming an anterior arch [*64]. Male internal genitalia vary slightly between taxa (171), representing an indistinct intermediate evolutionary level. The anal fan of the hindwing is reduced and only three anal veins are present, the third of which is usually very short. Rarely, one of these veins is forked. Convergent reduction to three veins is probable; *Capnioneura* has four and *Megaleuctra* has six anal veins.

Other structures are also similar in the two families but the significance of the resemblance is uncertain. Through modifications of furca- and spinasterna typical of Nemouroidea, the strongly sclerotized portion of the sternal surface becomes small, leaving relatively large membranous areas between itself and the mesocoxae. These areas tend to be more or less raised, pilose, and pigmented, showing so-called parafurcasterna, a condition which appears to be relatively primitive in comparison with Nemouridae (s.l.) Details have been used to distinguish genera and analyse their phylogenetic relationships (34, 65), but apparent correlation between structure and variable flight ability makes this analysis questionable.

II.1.2.2.1.1. Capniidae Short-wingedness or even complete winglessness is common. Many species inhabiting intermittent streams must have some form of diapause; diapausing nymphs are slightly curled, immobile, and milky in appearance (36, 67); knowledge of this life stage is insufficient for phylogenetic inferences.

Capniidae take reduction of M-Cu-crossveins (at most two) and Cu-crossveins (not more than a single one) to a distinctive extreme [*65]. Sperm transfer normally involves medial lobes of paraprocts and epiproct. The paraproct lobes merge into an unpaired structure, the so-called fusion plate whose basic structure is uniform (34): Distal parts form a ventrally narrow open tube covered over from below by large, simple outer paraproct lobes whose medial edges almost meet [*66]. Normal intersegmental paraproct retractors attach to the anteriorly extended base of fusion plate. The basal opening of the tubular fusion plate is positioned directly over the immobile reduced penis; during copulation, sperm enters the base of the tube and exits distally. The fusion plate has also been called the specillum (8) but it is different from the leuctrid specilla: Although both originate from inner paraproct lobes (and are to that extent homologous), actual transformations differ markedly.

Primitively, the tip of the fusion plate seems to release sperm into a rearward and dorsal furrow of the epiproct. The epiproct is an anteriorly curved sclerotized structure, often of a bizarre shape, and always of a specific shape (101). In many *Capnia, Allocapnia,* and others, the epiproct apex contains infolded membranes and probably acts as an intromittent organ; Brinck (14) provides an illustration of the inflated structure with an everted membranous apex. The fusion plate must act as a functional penis if the epiproct is simplified and reduced, as in *Capnioneura*. Female Capniidae lack a spermatheca but have very wide distal parts of oviducts storing sperm [*67]; *Capnioneura* stores sperm in sac-shaped lateral extensions of the wide ducts (171). Ovoviviparity (67) is probably related to sperm storage in oviducts.

Among Nemouroidea, the multisegmented cerci of Capniidae are strikingly archaic; adult *Capnopsis* (Europe) and *Eucapnopsis* (East Asia, Nearctic Region) have only four segments; *Capnioneura* (Europe) has only one. The genus *Capnioneura* is either the most derived member of the family (i.e. its fourth anal vein has reappeared), or it is an early side branch. *Capnia* may be heterogeneous; even after removal of several new genera, numerous species in many species groups remain (102, 166) although not all faunal regions have been considered. A new synoptic study of interrelations between genera of Capniidae is needed.

Illies (54) erronously regarded Notonemourinae as a subfamily of Capniidae because they share relatively primitive long testes (as opposed to much shortened testes of Leuctridae and Nemouridae) and a primitive, relatively long abdominal nerve cord (as opposed to Nemouridae). Internal male genitalia are of symmetric structure and there is no evidence for the consistent asymmetric position that Illies claimed. Presumed similar transformations of male copulatory organs (54) are considered superficial resemblances; no structural agreement exists. Notonemouridae share no groundplan characters of Leuctridae and/or of Capniidae, but belong to Nemouridae (s.l.).

II.1.2.2.1.2. Leuctridae Leuctridae roll their wings around the body at rest and appear more slender than the closely related Capniidae, from which they differ by one-segmented cerci [*68], by loss of paired seminal vesicles in male internal genitalia (mature sperm stored in enlarged vasa deferentia) [*69], and in the way in which inner paraproct lobes are transformed for sperm transfer. In all Leuctridae, bases of both paraprocts are changed into a common bulbous structure with retractor muscles [*70]. The unpaired end of the vasa deferentia directly and firmly attached to this bulb. The penis is never developed; its role was taken by modified apices of inner paraproct lobes. Klapálek homologized them correctly and called them titillators (Klapálek 1896); the terms specillum or subanal probe were also used (1, 33). Distal parts of titillators differ between subfamilies.

Transformations of the paraproct base confirm that the genus *Megaleuctra* (several extant plus one fossil Nearctic species, one fossil from Baltic Amber) belongs to Leuctridae; it is not a notonemourid (contra 59). *Megaleuctra* is regarded as the most primitive leuctrid genus (100, 112): it represents a separate subfamily, Megaleuctrinae. The female gonopore is located basally between extensions of sternites eight and nine, forming an ovipositor [*71]; spines are present on the tarsi [*72]. Additional unique characters are incompletely known (171), and their derived or primitive conditions are in doubt. The male epiproct is large and complex, apparently guiding a long common tube (reminiscent of Capniidae) formed by paraprocts.

In Leuctrinae, each titillator contains a separate tube [*73] with an apical or subapical opening; externally, titillators remain at least partially separate, but details vary. Titillators may be slender with a narrow tube (*Leuctra*: about 10 μm) or wide with a wide canal (*Zealeuctra, Perlomyia, Rhopalopsole*) (171). Outer lobes of paraprocts vary: they are short, sclerotized and little modified, or have an unpaired hull surrounding titillators (*Paraleuctra*). In *Leuctra* outer lobes are slender apical processes alongside titillators; simple hooks without lumen bend down the female subgenital plate during mating (8). In other genera the epiproct apparently performs the same task. Internal male genitalia are modified, and the entrance of the vasa is shifted caudad to the base of the unpaired seminal vesicle [*74]. Females have a sessile seminal receptacle at the front end of vagina, between entrances of oviducts [*75]; this receptacle is lined with cuticle and often armed with sclerite rings. Its position in *Zealeuctra* is exceptional but not completely understood (171). Knowledge of these structures in several unstudied Nearctic genera might help resolve controversial views of relationships between genera; only externally visible characters have so far been considered by Nelson & Hanson (100) and Ricker & Ross (112), and also by Shepard & Baumann (122) whose cladogram resembles Nelson & Hanson's.

II.1.2.2.2. Nemouridae (s.l.) (Notonemouridae + Nemouridae) An informal designation is introduced here for a monophyletic assemblage including so-called Notonemouridae and Nemouridae. Conventionally, keys separate Nemouridae from Leuctridae by the so-called nemourid X resulting from completion of a pattern of some apical veins that are generally present in Nemouroidea by an apical crossvein distally from Sc. This crossvein is not distinctive of Nemouridae: It is a plesiomorphic remnant of former rich venation. The same crossvein is found in some Taeniopterygidae and Capniidae, which are usually keyed out before by another character. The so-called nemourid X neither supports monophyly of Nemouridae, nor speaks its absence in some Notonemouridae against membership in Nemouridae. Other primitive traits of Nemouridae (s.l.) are forked M and five anal veins in the hindwing. Monophyly of Nemouridae (s.l.) is evidenced by apomorphies of internal anatomy (171): the mesothoracic pleural arm is forked [*76]; the penis and its retractor muscles are completely absent [*77]; an unpaired strongly muscular ejaculatory duct [*78] leads directly to the gonopore at the tip of sternite nine, and is often very long. The abdominal nerve cord is shortened through posterior fusions, and no more than six ganglia are free (56) [*79].

II.1.2.2.2.1. Notonemouridae (Gradotaxon) Nemouridae (s.l.) are the only Euholognatha with southern hemisphere representatives that share the described apomorphies. However, the southern taxa do not share the additional apomorphies of northern hemisphere Nemouridae, or at best a few of these apomorphies are shared. Habitus resembles Leuctridae (wings at rest rolled around body, head narrow) or Nemouridae s.str. (wings flat on abdomen; head large); the mesosternum is close to nemouroid groundplan (e.g. *Neonemura*, *Notonemoura*), or like Nemouridae (most other genera). Ricker (110) placed Australasian genera in a new subfamily, Notonemourinae, and South American genera in Leuctrinae; later, all genera were classified as Notonemourinae (2, 54); for presumed affinities with Capniidae see Section on Capniidae. Notonemourinae and Nemouridae seem to be sister-groups and I ranked both as families; the structural diversity of Notonemouridae was noted as was their doubtful monophyly (171, 178). A tendency for separation of inner from outer lobes of male paraprocts and development of ovipositors were suggested to support monophyly.

Epiproct and paraprocts may be involved in mating, apparently without actually conveying sperm; great structural diversity is observed. Retractor muscles of apically displaced paraprocts are often employed to bend up the long subgenital plate with the apical gonopore. Information on function and placement of parts during copulation is available for *Notonemoura* whose males introduce the subgenital plate right to base of spermatheca (82), and for *Aphanicercopsis* and *Aphanicercella,* whose males introduce epiproct and paraprocts into the vulva (5). *Austronemoura,* with a microscopic gonopore (diameter 4 μm; 90) on the needle-like apex of the subgenital plate and *Udamocercia,* with complex fusion between the apex of subgenital plate and paraprocts (2; details not understood) are isolated.

Some Notonemouridae occur in marginal running-water habitats such as hygropetric rock faces, and must protect eggs against drying. Ovipositors are widespread and of several structural types (47, 54, 62, 85, 136, 171; P Zwick, unpublished data). Mostly, the female gonopore remains in a normal position, at the level of the rear margin of segment eight; either only sternite eight is extended, overlying the gonopore and somewhat hollowed sternite nine (*Spaniocerca*, New Zealand); or, sternites eight and nine are extended, together forming an egg guide (*Austrocerca, Austrocercella, Austrocercoides, Tasmanocerca,* Australia; *Madenemura*, Madagascar, 106; *Megaleuctra*, Leuctridae, similar); or extended sternite eight, or sometimes nine, and paraprocts are elongated, together forming an ovipositor (*Aphanicerca, Aphanicercopsis, Afronemoura*; Africa). Alternatively, in *Notonemoura* (New Zealand, Australia), *Halticoperla, Cristaperla, Spaniocercoides, Omanuperla* (New Zealand), *Neonemoura* and *Neofulla* (South America) sternite eight extends back as a finger, and the gonopore is terminal or subterminal (*Vesicaperla,* Gripopterygidae, similar). The ovipositor develops inside the nymphal body, with no external theca. Nymphs of several genera possess enlarged hind femora (54, 85).

This scheme differs somewhat from the one proposed by McLellan (85). The scheme's congruence with male structures requires checking; it leaves genera without ovipositors unplaced but shows that notonemourid monophyly cannot be claimed because of ovipositor presence. Instead, Notonemouridae appear to be heterogeneous, a paraphyletic assembly of early nemourid lines surviving on fragments of Gondwanaland where Nemouridae do not occur.

II.1.2.2.2.2. Nemouridae The numerous northern hemisphere Nemouridae have large transverse heads with large eyes and many (like most Notonemouridae; ID

McLellan, personal communication) are rather attentive little stoneflies noticing an approaching observer, sometimes taking to flight with a little jump. Wings rest flat on abdomen. All Nemouridae share several uncontested apomorphies (171, 177): an adult labial palpus with a large disk-shaped end segment [*80]; front coxae enlarged and transverse, with bulging median parts almost touching [*81]. The abdominal nerve cord has only five free ganglia due to fusion of posterior ones (56) [*82]. Testes are star shaped, and follicles are all attached to the efferent duct at one point [*83]. The seminal vesicle is unpaired with tube-shaped ducts entering at its anterior end [*84].

Baumann distinguished two subfamilies and discussed phylogenetic relationships between genera (6); Notonemouridae were not compared. Monophyly of subfamily Amphinemurinae is supported by a third external lobe closely connected with the cercus base separated off male paraprocts [*85]; paraprocts that are mostly heavily armed with spines [*86]. A ventral epiproct sclerite is suggested evidence for monophyly of subfamily Nemourinae [*87]. Genera with complex laterally projecting articulations of epiproct ("lateral knobs") are certainly a monophyletic taxon within Nemourinae (including *Nemoura*). The remaining nemourine genera have enlarged and expanded ventral epiproct sclerites, but seem to be structurally rather diverse.

Better knowledge of sperm transfer and function of parts is needed. An elongated subgenital plate with an apical gonopore is apparently a functional penis but in *Nemoura* the epiproct with its deep membranous pockets is also introduced into vagina and is involved in sperm transfer; short inner paraproct lobes guide the subgenital plate (14). Also in *Nemoura*, specific lock-and-key correspondences exist between epiproct and vaginal folds and sclerites (P Zwick, unpublished observations). In contrast, *Nemurella* males grasp the narrow female subgenital plate with the ventral part of trilobed epiproct, its middle and dorsal parts guiding enormously long inner paraproct lobes. Inner lobes are introduced directly into the female receptacle and sperm flows between slightly concave lobes (P Zwick, unpublished observations); this is the only known example of paraprocts as sperm-conveying organs in Nemouridae.

In some members of both subfamilies gills occur on either side of cervical sclerites (*Visoka* with exceptional submental gill); these are the only cervical gills in Euholognatha. Homology with cervical gills of Pteronarcyidae is uncertain; if confirmed, cervical gills would be a groundplan character of Arctoperlaria (171).

II.2. Systellognatha Systellognatha is the only taxon identified by derived external characters by early students of Plecoptera. Most extant species belong to a derived subgroup, Perloidea, whose additional apomorphies are not present in all Systellognatha; this condition once caused confusion. Reduction of adult mouthparts, especially weak soft mandibles and elongate, sometimes whip-shaped annulate palpi are distinctive [*88]; if adult mandibles are secondarily sclerotized, they are of a derived shape lacking a mola; see Chloroperlidae. The Systellognatha groundplan includes a complex male epiproct with modified parts of antecosta

10 (155, 171) sunk into a pouch dividing tergite 10 into hemitergites [*89]; however, homology of hemitergal lobes between groups was questioned (99). In some taxa exhibiting apomorphies typical of subgroups of Systellognatha these complex male structures may be variously modified or again reduced; reductions probably relate to insemination via an intromittent penis. Sclerotized egg chorion forms a collar surrounding a polar attachment disc, or anchor [*90]. Pterothoracic pleurae are connected to the postnota by a post-alar sclerite [*91]. Atypic intersegmental nymphal swimming muscles are remarkably developed, a character that is probably also an apomorphy of Systellognatha [*92] (171, 177). Females have accessory glands on the receptacular duct; although they occur in no other group of Plecoptera they are here for the first time proposed as apomorphy of Systellognatha [*93].

Opinions on basal branching within Systellognatha diverge. Uchida & Isobe (155) studied previously unavailable material and present discussions of many systellognathan characters, recognizing a close relationship between Peltoperlidae (s.l.) and Pteronarcyidae, as also suggested by (132). If so, several similarities between Styloperlidae, Peltoperlidae and Perloidea must have arisen independently: reduction of one pair of seminal vesicles; position of micropyles near anterior egg pole; reduced clypeofrontal suture (only Peltoperlidae and Perloidea). Additionally, mating seems to differ: Pteronarcyidae transfer spermatophores, while elaborate penis sclerites and wide vaginae of Styloperlidae and Peltoperlidae suggest actual introduction of the penis, as in most Perloidea. Classification below follows Uchida & Isobe (155) but an alternative [modified from Zwick (171)] deserves consideration: Pteronarcyidae + [Styloperlidae + (Peltoperlidae + Perloidea)].

II.2.1. Pteronarcyoidea According to Uchida & Isobe (155), former Peltoperlidae s.l. and Pteronarcyidae are sister-groups constituting a superfamily whose monophyly is supported by paired but closely adjacent corpora allata [*94] (unpaired in *Microperla*) and fused ends of tegumental head nerves [*95] (155). Tridentate nymphal laciniae [*96] and spinelike point of nymphal tergite 10 are also listed in support of this superfamily (132).

Egg polarity and chorionic structures were described in various terms; here the egg pole directed back, toward the oviduct, is termed posterior pole (44). Egg polarity of Pteronarcyoidea is not uniform: Pteronarcyidae has micropyles near the posterior egg pole carrying collar and anchor; nymphs hatch at the same pole. In Styloperlidae and Peltoperlidae, collar and anchor (where present) are located at the posterior pole but micropyles are near the anterior pole (or, in flat eggs, on the face opposite the anchor, contra 171) (127, 130, 155). Topography in Styloperlidae and Peltoperlidae agrees with Perloidea eggs whose hatching nymphs remove the anterior egg cap, i.e. they hatch also near micropyles; however, the position of developed embryo may be influenced by egg shape (71).

II.2.1.1. Pteronarcyidae Pteronarcyidae, or salmonflies, are mostly huge. Among northern hemisphere Plecoptera, adults stand out by rich wing venation. Nymphs have a distinctive rich set of branched gills. Both of these diagnostic characters are ancestral. Reduced sclerotization of thoracic ventropleurites and postfurcasterna [*97] and laterally expanded arolium [*98] are proposed family apomorphies (99). Males drum (140) but the ventral vesicle (or corresponding structure) is reduced [*99]; independent reductions in other families do not invalidate this family apomorphy (171). Disturbed nymphs curl the forebody under, a character I propose as a new apomorphy [*100]. Anteriorly directed pointed setae along the rear margin of female sternite nine (99), apparently plesiomorphic, are known in various other families (155).

Postfurcal gills are not apomorphic (contra 171, 177), occurring also in some Peltoperlinae (155), which may support Pteronarcyoidea but reduction in other families is also possible. Styloperlidae and some Peltoperlidae (secondarily) lack gills. Thoracic substigmatic gills are widespread in Systellognatha and homodynamic with abdominal gills, which are also beneath stigmatic scars. First-instar *Pteronarcys* nymphs have gill stubs on abdominal segments one and two, and two postfurcal and two basisternal (substigmatic) gills on every thorax segment. The prothoracic basisternal gills apparently correspond to median cervical gills (P Zwick, unpublished data); compare the cervical gills of Protonemurinae.

The male epiproct is always well developed and structural variations are typical of genera or species groups (99, 96); Researchers have speculated about the function of structures described as sperm cups, but no observations are offered (98, 171). Phylogenetic analyses of this small family are available to species level. However, one cladogram is based mainly on egg structures (133), whereas other proposed cladograms are based on entire adult and egg morphology (96). An allozyme analysis of eight North American species produced ambiguous results (164). Recent authors suppress *Allonarcys* and maintain integrity of the structurally diverse genus *Pteronarcys*. Monotypic tribes Pteronarcellini and Pteronarcyini (133) are redundant (96).

II.2.1.2. (Styloperlidae + Peltoperlidae) No name is proposed for this complex corresponding to former Peltoperlidae (s.l.). One of two sets of superimposed male seminal vesicles is reduced [*101]. Abdominal ganglion II coalesces with the metathoracic ganglionic complex [*102] (155), which, in all Plecoptera, includes abdominal ganglion I (56). Similar changes plus fusions of posterior abdominal ganglia occur in Perlidae; Zwick (171, 177) regarded (Styloperlidae + Peltoperlidae) as a sister-group of Perloidea simply on account of reduced absolute number of free abdominal ganglia. Recent discoveries permit a better resolution of this complex (155). Egg shapes are diverse (132).

II.2.1.2.1. Styloperlidae Both included genera (155) share a dense setal brush on male sternite nine (instead of vesicle or hammer) [*103] and a unique X-shaped sclerotization of male sternite 10 [*104]. Nymphs lack spurs but have trifurcate and other strongly modified setae on the tarsi and tibial apex [*105]. The epiproct is reduced and unsclerotized. Nymphs are slender and gill-less, the pro-spinasternum is free. The frontoclypeal suture is distinct, and mouthparts are of a generalized phytophagous type.

II.2.1.2.2. Peltoperlidae The nymph is shaped strikingly like a cockroach [*106]. The prothoracic spina- and furcasternum are firmly connected [*107] (171). A frontoclypeal suture is lacking, as in Perloidea; parallel reduction in correlation with peltoperlid head shape was assumed, but the normal nymphal head of *Microperla* also lacks a suture (155).

Monotypic subfamily Microperlinae includes two species with an unpaired corpus allatum [*108]. Nymphs obviously graze algae: modified nymphal mandibles [*109] and rake-shaped setae on galea are present [*110]. Gills are absent, the thoracic sterna are simple, and a sclerotized epiproct is absent. A vestigial anterior ocellus is retained (155).

Previously proposed family apomorphies in fact restricted to subfamily Peltoperlinae (155) are strongly shortened head lacking anterior ocellus [*111]; enlarged nymphal pro- and mesobasisterna, covering furcal pits [*112]; nymphal coxa having a flap-like lobe [*113]; and absent tibial spurs [*114]. *Microperla* has spurs, and reduction is probably independent from Styloperlidae. In several Peltoperlinae, basal male cercal segments merge into a shaft, often with apical hair brushes or spines. Styloperlidae and several Perloidea possess similar structures, the latter certainly independently derived (155). The male epiproct is variable, mostly much reduced, but all components of the systellognathan groundplan are retained in at least individual species. Lateral stylets were described in *Soliperla* and *Viehoperla* (155, 171); Stark & Stewart (132) doubt the homology (see Perloidea and Perlodidae). Nearctic genera (132) fall into distinct eastern and western groups; some Nearctic genera are apparently closely related to Asian ones (141), *Yoraperla* shared with Asia (130).

II.2.2. Perloidea This taxon was recognized early but known as Setipalpia or Subulipalpia (143). Recent studies necessitate reconsideration of previously suggested apomorphies. Details of egg structure thought to be distinctive of Perloidea (171) are actually shared with Styloperlidae and Peltoperlidae (see Pteronarcyoidea), and character polarity is uncertain. The significance of lateral stylets of epiproct is uncertain and not included in this analysis. Uchida & Isobe (155) believe in homology between styles and a sclerite band ("forklike structure") in the pouch of Pteronarcyidae (99); they do not accept styles as an apomorphy. However, development of freely projecting apices in Perloidea is certainly apomorphic. If structures in a few Peltoperlidae are true lateral stylets, they are a groundplan character of Peltoperlidae plus Perloidea (171) and secondarily absent in Chloroperlidae (perhaps except *Sweltsa;* 171) and Perlidae. Should peltoperlid structures prove not to be homologous with lateral stylets, the latter might indeed be an apomorphy of Perlodinae (134).

Reductions of some or all components of epiproctal apparatus are widespread among Perloidea. The role of the epiproct in mating is not well known. The penis is eversible and frequently highly modified; in many taxa the penis is introduced and sperm directly transferred into females. In *Arcynopteryx compacta* (Perlodidae) the penis deposits sperm directly in a receptacular duct (14). A brief note describing a cleaning action of the epiproct unfortunately spoke of spermatophore (15). Active absorption of liquid sperm by female *Hydroperla, Afroperlodes* (Perlodidae) and *Sweltsa* (Chloroperlidae) was observed (9, 104, 146).

Monophyly of Perloidea is beyond doubt, defined by carnivorous nymphs with long and slender palpi, slender mandibles without mola, slender sharply toothed laciniae, and glossae that are shorter than paraglossae [*115]. The caudal ganglionic complex is merged with abdominal ganglion eight [*116]. Adult mandibles are nonfunctional (systellognathan groundplan), in some Isoperlinae and all Chloroperlinae regaining sclerotization and used in feeding (118, 171) but a modified mandible form is retained. Currently recognized families are easily diagnosed mainly by nymphal characters; established classification is only partly supported by apomorphies and relations between families are unresolved.

Possible apomorphies supporting a sister-group Perlidae + Chloroperlidae (171, 177) are modified thoracic musculature and the rounded shape of nymphal wing-pads whose edges meet medially, without leaving a separate notal contour. If this sister-relationship were confirmed, gut caeca would be an apomorphy of Perloidea [*117] and the lack of gut caeca secondary in Chloroperlidae (171, 177). Perlodidae and Perlidae possess anterior gut caeca of variable number and relative size, no consistent differences exist between families (171).

A possible alternative monophylum Perlidae + Perlodidae is supported by diet shift to pronounced carnivory and consumption of larger prey; enlarged bases of nymphal maxillae (obscured in Perlidae by additional head capsule modifications, see below) are obviously correlated. Gut caeca in both families are possibly a response to diet shift; if so, ancestral Chloroperlidae may never have had caeca.

A close relationship between Chloroperlidae and Perlodidae has been suggested (141) because of minute cuticular spicules (acanthae; figures in 131) at bases of major setae on nymphal mandibles. If this is a significant resemblance, the lack of acanthae in Perlodidae: Isoperlinae must be secondary.

II.2.2.1. Perlodidae Many plesiomorphies (in wings, gills, epiprocts, and other) are retained, and diagnostic combinations serve as identification of Perlodidae in keys. Many adults have a striking yellow longitudinal stripe on their dark forebody and many nymphs are yellow with a contrasting dark pattern. No family apomorphies are known and monophyly is uncertain; ventral lobes used for drumming on segments seven and/or eight are widespread. Ricker (111) established most currently recognized genus group taxa and recognized three subfamilies; Isogeninae was based on plesiomorphies and was later sunk under Perlodinae (171). The latter and Isoperlinae may be monophyletic but this is not easily recognized; proposed apomorphies are never expressed in all included genera but close similarities and apparent affinities at the genus level create supporting networks of sister-group relationships.

For the subfamily Perlodinae, proposed apomorphies (134) are submental gills [*118] and lateral stylets of male epiproct [*119]; compare under Perloidea. Penes lack striking cuticular armatures. Eggs exhibit a wealth of structures and shapes, and the hatching line is often visible. Genera belong to three tribes (134): Arcynopterygini (see 135), Perlodini, and Diploperlini (including *Bulgaroperla* and *Rickera,* formerly Isoperlinae). References to numerous studies of Nearctic and east Palaearctic genera are found in (141); Palaearctic genera: 94, 129, 182, 184.

Subfamily Isoperlinae (restricted: 147) includes smaller species. Typical systellognathan epiprocts are always reduced [*120]. Tergite 10 is mostly simple, but may have distal hooks or processes; paraprocts are normally simple upcurved hooks [*121]. Sternite eight has a flat ventral lobe (often pedunculate). The penis is membranous, with specific sclerites, spicule patterns, or bizarre specific shapes. Nymphs lack gills. Several small or monotypic genera exist beside *Isoperla,* a very large genus with many different species groups; a comparative global analysis is needed.

II.2.2.2. Chloroperlidae A reduced anal lobe of hindwing, or a combination of negative characters (e.g. no gills) were used to define Chloroperlidae (31). The body is slender and cerci are distinctly shorter than the abdomen [*122]; oval pronota [*123] are common to all Chloroperlidae, and nymphs are also long and slender, with short cerci; these are not convincing synapomorphies. Modifications of palpi proposed as a family apomorphy (146, 171, 177) cannot be maintained for Paraperlinae. A U-shaped fold surrounding the scutellum seems more strongly expressed than in other Subulipalpia (P Zwick, unpublished observations), but comparative quantitative studies are needed to confirm this as chloroperlid synapomorphy.

Subfamily Paraperlinae has three genera with primitive venation, complex epiproct, well-developed hammer (*Kathroperla)*, and multisegmented cerci. Monophyly is evidenced by prolonged postocular head region [*124], although elongation in the small *Utaperla* is hardly more important than in exceptional Chloroperlinae (e.g. *Alloperla rostrata*). Paraperlinae have large flat thoracic sterna with slot-like furcal pits turned parallel to the body axis; pits are anteriorly firmly connected to the outwardly directed edge of basisternum, but furcal pits are not interconnected and the spina is independent, with a separate medial crest [*125] (177); presumably correlated shifts of muscle attachments remain to be studied.

Subfamily Chloroperlinae have a very short and narrow, nipple-like terminal maxillary palpus segment [*126] and reduced anal area of hindwing (four or fewer veins, front and hindwings almost equal) [*127]. The epiproct bar is anteriorly shortened and broadly fused with antecosta 10, and the apex is turned forward [*128]. Adults feed on conifer pollen and have sclerotized mandibles resembling nymphal tearing type (171, 177), with few exceptions (146); some Isoperlinae have similar mandibles. Cerci are short, often less than 10 segments, rarely more. The name Chloroperlinae refers to the bright green or yellow body color of most species; the abdomen usually has mid-dorsal and sometimes basilateral dark stripes, and the thoracic nota always have a U- or W-shaped black suture. Adult Chloroperlinae (171) (and many *Isoperla*: P Zwick, unpublished data; no information on other Isoperlinae is available) possess grouped microtrichiae of unknown function along laterodistal margins of segments eight or nine.

Tribes Alloperlini, Suwalliini, and Chloroperlini have been recognized and a cladogram for Nearctic genera (most of the same genera in East Asia) presented (146); it allows for addition of the west-palaearctic genera, which are all Chloroperlini.

II.2.2.3. Perlidae Within Perloidea, branched thoracic gills (distinct remnants in adults) are diagnostic of Perlidae. Gills are homologous with ancestral systellognathan gills; the substigmatic metathoracic gill of Perlidae by its tracheation is actually the gill of the largely reduced first abdominal segment (171). The male epiproct is reduced; only by exception do components remain from the systellognathan groundplan. The metamerous structure of the male testes and ducts is observed in only few genera (153) and is normally reduced. The penis is eversible, and in many genera in both subfamilies complex sclerites, teeth, or trichomes adorn a more or less sclerotized basal tube of the penis or an eversible sac in it; especially in Neoperlini examples of lock-and-key fit with female vaginal or receptacular structures is apparent. Terms for genitalia of other Neoptera are not used because of doubtful homologies.

Apomorphies supporting Perlidae (171, 177) are fusions at both ends of the abdominal nerve cord, and only six free ganglia [*129]. Nymphal head has expanded genae covering the bases of the mandibles and huge maxillae from above, inflated paraglossae, and wide mentum covering maxillae from below [*130]. Complex folds in nymphal proventriculus form an additional valve [*131]. Gill bases have hairy sclerites (gill shields) [*132] (153). Thoracic muscles show a shift of *ppm56* to postalar sclerite and the presence of *ism24* [*133] (muscle terminology of 163).

Two subfamilies are described. Hemitergites 10 of male Perlinae are modified into anteriorly curved hooks [*134]. Males mostly lack a hammer and have unmodified paraprocts; males have hair brushes on mesal areas of abdominal sternites [*135]. Nymphs have continuous transverse ridge or setal row across occiput [*136] and three (instead of two) gill shields [*137] (128, 153). Sivec et al provide a synopsis of world genera recognizing tribes Claasseniini (monotypic), Perlini and Neoperlini and analyzing phylogenetic relationships between genera (126).

In the subfamily Acroneuriinae, male paraprocts are transformed into anteriorly upcurved hooks [*138]. In most taxa, males have a hammer; the occipital setal row of nymphs is irregular, incomplete, or absent (128, 153). The tribe Acroneuriini (restricted 153) is distinguished by a transversely expanded, enlarged epiproct; the tribe Kiotinini is mainly supported by one enlarged proventricular fold forming a valve in the nymphal gut (153); in the tribe Anacroneuriini the penis has a pair of opposed sclerotized grapples, and the hammer (if present) is a raised sclerite knob or nail; eggs are often conical and the collar at the blunt egg pole is nipple shaped or absent. (see 153, p. 725).

Zoogeography

Illies repeatedly discussed phylogeny and zoogeography of Plecoptera (52, 53, 55), finally at the levels of family, genus, and species (57). Previously described distribution patterns need not be repeated in detail; Banarescu provides distribution maps of many genera (4). There have been no major discoveries or range extensions in the meantime. However, my view differs significantly from that of Illies, especially regarding the highest systematic categories.

I discuss patterns of vicariance and disjunction, in order of decreasing taxonomic rank and geographic scale. Regardless of how close relationships between taxa are and also regardless of rank of taxa concerned, disjunctions can be explained in three ways: dispersal (active and passive), relict distribution of formerly more widespread taxa, and origin of barriers subdividing formerly continuous ranges. Because of Plecoptera ecology, long-distance dispersal during single events is improbable. Provided there are hierarchic sets of similarly disjunct sister-group taxa, it is more parsimonious to assume their common origin on a once-entire but subsequently divided landmass than to postulate identical patterns of regional extinctions in several taxa. This logic was once developed for the Diptera of New Zealand (40) but is generally applicable; constructing an area cladogram would lead to similar conclusions.

Illies noticed vicariance between north and south hemisphere taxa. However, the classification, cladogram, and zoogeographic interpretation proposed were inconsistent (57). The classification recognized three suborders but the cladogram showed Archiperlaria as a sister-group of the combined Setipalpia + Filipalpia. The zoogeographic analysis presented a third variant: Archiperlaria would have originated on the Palaeozoic southern landmass incorporating present-day Antarctica, Australia, and South America, where they survive today; Filipalpia and Setipalpia would have separately originated from Archiperlaria and migrated north. Both eventually crossed the equator into the northern hemisphere, which was previously void of Plecoptera; both Filipalpia and Setipalpia radiated north. Later, some northern Perlidae (Setipalpia) and Notonemouridae (Filipalpia) returned south; subsequently, Notonemourinae became extinct in the north. Because of a systematic misconception, a Baltic amber *Megaleuctra* seemed to prove former northern hemisphere notonemourid occurrence (59). Raušer (108) doubted Illies' scenario and instead proposed a Carboniferous (or earlier) origin of Plecoptera in tropical latitudes. According to Raušer, Archiperlaria initially occurred in both hemispheres but became extinct in the north. Origin and migrations of the Filipalpia and Setipalia were not described in detail but objections were raised against notonemourid dispersal as described by Illies (57).

Suborders and Hemispheres

The area of origin of Plecoptera is unknown. The present suborder names, Antarctoperlaria and Arctoperlaria, allude to distributions on lands surrounding Antarctica, or the Arctic, respectively; Arctoperlaria occurr in the entire northern hemisphere, and also in the Oriental region to south of equator, which was always firmly connected to the palaearctic area. Two arctoperlarian families, Notonemouridae and Perlidae, are exceptions occurring in the southern hemisphere; see below. Fossils (57, 125) show the same distribution as extant suborders; however, evidence of this distribution is weak because of incomplete records and unreliable group assignments. Hennig's disjunct-hierarchy approach is not applicable because the sister-group of Plecoptera is not known; it likely inhabits both hemispheres. The second and third alternatives (extinction of Antarctoperlaria in northern and of Arctoperlaria in southern hemisphere, respectively, versus divergent evolution of a common stock following the breakup of the Palaeozoic supercontinent Pangaea into Laurasia and Gondwanaland) appear possible.

I favor the latter assumption. Although Plecoptera are an ancient insect order, their extant families may not necessarily be old, but may have evolved relatively recently; the earliest fossil specimens of extant families are presumed Jurassic Taeniopterygidae, Leuctridae and Chloroperlidae; Jurassic "Nemouridae" actually belong to the stem group Nemouridae + Notonemouridae (125). Large numbers of very closely related species in numerous genera in all parts of the world (e.g. 58, 86, 91, 171) reveal extensive recent speciation.

Antarctoperlaria and the Southern Hemisphere

Antarctoperlaria are a textbook example of a hierarchy of disjunct sister-groups whose distribution is most parsimoniously explained by assuming evolution on a common landmass that subsequently broke up. Diamphipnoidae (five species) are known only from South Andean locations but the other three families and subfamily Stenoperlinae are known on all southern lands. Subfamily Eustheniinae and five subfamilies of Gripopterygidae each inhabit only one or two areas, and patterns require no other explanation than accidental regional extinctions or relatively recent local radiation.

However, Antarctoperlaria is unrecorded from Africa (assignment of Permian *Euxenoperla* to Gripopterygidae tentative, apparently arbitrary: 114), Madagascar, and India. Phylogenetic analysis is not sufficiently advanced to determine alternative times of evolution of extant Antarctoperlaria from a Gondwanian stock and time of dispersal to current ranges. If current taxa date back to Cretaceous or earlier times (12, 120, 149) they must be extinct in Africa–India. If the only diverse family, Gripopterygidae, radiated relatively recently its members may have evolved and dispersed between South America, New Zealand, and Australia via Antarctica without having access to Africa–India. The maximum diversity at family to genus-group levels is in subantarctic to temperate latitudes, but in Australia Stenoperlinae and some genera of Gripopterygidae also have tropical species (91) (compare 20, 49, 61, and 83 for regional distribution patterns).

Arctoperlaria and the Northern Hemisphere

Taeniopterygidae, Capniidae, Perlodidae, and Chloroperlidae are strictly holarctic. Perlidae, Peltoperlidae, Nemouridae, and Leuctridae extend into tropical Asia, the latter two with few genera. Pteronarcyidae and Peltoperlidae have an Asian-Nearctic distribution; Scopuridae, Styloperlidae, and Microperlinae (Peltoperlidae) are relatively narrow Asian endemics.

The holarctic region consists of three major ancient shields that were interconnected in various ways in the past (23). Separate evolution on isolated landmasses, faunal exchange between them at different times, and recent Pleistocene displacements and extinctions took place. Plecoptera have distinct European, Asian, and North American faunas. Regionally endemic genera, range exclusions, and disjunctions make a complex picture; only a few examples can be mentioned.

Asian-American disjunctions from tribal to species levels are numerous. Disjunctions originated at various times between the Tertiary and Pleistocene (141); Kiotinini may be another example (153). There is good fit between speciation and Asian and American paleogeography in *Pteronarcys* (Pteronarcyidae) (96) and *Yoraperla* (Peltoperlidae) (130). Tribe Arcynopterygini (Perlodidae) probably includes many disjunct sister-groups, genera of either western Nearctic or east Palaearctic (a few shared) and Himalayan (*Neofilchneria*) regions. Pleistocene extinctions, displacements, and post-Pleistocene range extensions led to speciation, for example in Capniidae (101, 102, 116). Pronounced east-west American differences are related to the former Mid-Continental Seaway and to survival of deciduous forests in eastern North America (28, 48, 103, 141). Only very limited inter-American exchange with South America occurred (141).

Disjunct American-European groups are less numerous: Diploperlini and primitive Perlodini (Perlodidae with developed epiproct) may be some. Leuctridae belong here; extant Megaleuctrinae nearctic, fossil members also existed in Europe (59). Leuctrinae exhibit a hierarchic set of disjunctions (113); after Nelson & Hanson (100) this would be less pronounced. The opening Atlantic Ocean seems to have split the genera *Leuctra* and *Taeniopteryx* into distinct American and European species groups. In both genera, species in each group seem most closely interrelated among themselves; this would hardly be so if North Atlantic dispersal (141) were really involved. Transberingian dispersal is even less probable because of absence of both genera from Asia except one widespread species in each genus (*L. fusca, T. nebulosa*) extending from Europe to Mongolia and Baikal area, respectively (171). Among Brachypterainae (Taeniopterygidae), *Oemopteryx* (amphi-atlantic) is a sister-group of European *Brachyptera;* European *Rhabdiopteryx* is a sister-group of transberingian *Strophopteryx* + *Taenionema* (113). European "*Capnia*" *vidua* and American genus *Allocapnia* are probably amphi-atlantic sister-groups (171).

Pleistocene glaciations apparently displaced or restricted the original European fauna to Mediterranean areas (including Africa north of Sahara, Asia Minor, Caucasus, and Near East relic habitats). Europe north of the Alps today is inhabited by species returning from Pleistocene Mediterranean refugia of arboreal (27, 179), plus a few circumpolar species (e.g. Perlodidae: *Diura bicaudata, Arcynopteryx compacta;* Nemouridae: *Nemoura arctica*). Total species number declines rapidly toward the north (45, 80). Intra–European north-south exchange is less important than suggested (50, 150).

European genera of Perlinae are a distinct group; distribution of their closest relatives suggests dispersal along a route between Himalaya and the Caucasian–Anatolian–Balkanian highlands. Advanced Perlodini (genera lacking an epiproct) (182), *Xanthoperla* (Chloroperlidae) (167, 169), Amphinemurinae (6), *Capnia pedestris*-group (from *C. femina* on Mt. Everest to *C. arensi* in the Caucasus and Lebanon) are also examples of this European–Asian connection; compare *Neoperla*. Further north, no direct east-west route was available because the Obik Sea and Turgai Strait east of Ural Mts once separated Europe from Asia.

The Asian fauna is diverse (e.g. 35, 65, 124, 153, 165) and incompletely known. Many genera in temperate regions are shared with America or Europe, or closely related to taxa there (see above). Insular faunas have high endemicity, e.g. Japan, Taiwan (58). Several Nemouridae (6) and Leuctridae (mainly numerous undescribed species of *Rhopalopsole*; I Sivec, in a letter), Peltoperlidae, and relatives (127, 155), and Acroneuriinae endemic in subtropical and tropical Asia. Neoperlini originated in the Oriental region (126). *Neoperla,* which is very speciose, is one of most widely distributed stonefly genera. *Neoperla* entered North America from the northwest; its species abound in insular Asia, one reached New Guinea (Australian region) (171); many Afrotropical species.

Arctoperlaria on Southern Continents

Arctoperlaria, down to the level of tribe and sometimes genus, form a taxonomic hierarchy distributed over the separate northern hemisphere core areas in a coherent way, without unexplained gaps; no doubt Arctoperlaria developed in Laurasia. Both Perlidae and Notonemouridae are late apical branches in this hierarchy. Three groups of Arctoperlaria in the southern hemisphere require discussion; only the first case is easy.

1. The widespread genus *Neoperla* (Perlinae), also in Afrotropical region (not in Madagascar), is the only stonefly genus in vast parts of Africa; many endemic species of *spio*-species-group occur there (172, 173 and my own unpublished data). A Miocene or Pliocene arrival of *Neoperla* from Eurasia is probable, perhaps together with large mammals (22; Siwalik fauna, 120) through Arabia, while Asia Minor and Europe were intermittently locked off by seaways (70, 115).

2. A northern origin was assumed for southern Perlidae (57, 128), but South American Acroneuriinae appeared too diverse for late (Miocene) immigration from North America; slow acclimatization of originally cool-adapted animals to tropical conditions was assumed. Separate dispersal of Acroneuriini and Anacroneuriini from Asia or Europe through Africa to South America in the Cretaceous was suggested (128), but some problems remain. The greatest generic diversity of Anacroneuriini is found in eastern South America, in an area inhabited by many other taxa shared with Africa, for example families of river fishes (11). However, these are old Gondwanian elements; Cyprinidae arrived later in Africa from Asia and did not reach South America. While South America was still connected to Africa, other southern landmasses should also have been accessible (12, 23, 24, 120), but Acroneuriinae do not occur there. No evidence exists of past occurrence of Acroneuriinae in Europe. Acroneuriinae would be secondarily extinct in Africa. Details of egg structure now suggest that all South American Perlidae belong to Anacroneuriini (B Stark, at the *20th Int. Plec. Symp.*, and in a letter); necessary new definitions of Acroneuriinae tribes are not yet available.
3. Notonemouridae. This family is thought to be of northern origin (57); however, if the group is monophyletic, notonemourid distribution would be intriguingly reminiscent of Antarctoperlarian distribution. The African Notonemouridae is regarded as Palaeogenic element (136) in the sense of Stuckenberg (144).

Notonemouridae here is regarded as a paraphyletic gradotaxon, as independent surviving nemourid stem-group lines probably coming from the northern hemisphere at different times and along different routes. The structure of several rare and often minute animals is incompletely known, interrelationships accordingly doubtful, and actual areas occupied by various included monophyla are unknown. An indispensible prerequisite of a meaningful discussion of dispersal is therefore still lacking. Restriction to southern ends of continents may have ecological reasons; in Australia, warm temperate and tropical habitats are devoid of so-called Notonemouridae. In future investigations, the African route, immigration to Australia via recent northern connections (26), and faunal exchange with New Zealand along the Inner Melanesian Arch, as with other stream insects that survive also on New Guinea and New Caledonia (25, 175, 183), should be taken into consideration.

Whichever phylogenetic inference ultimately proves correct, the choice will forseeably be between several improbable explanations. If present distributions are old and were attained on Gondwanaland or along Transantarctic routes, the present idea of suborders originating through breakup of Gondwanaland cannot be maintained. In that case, one must also at least assume extinction of all Arctoperlarian lines everywhere on Gondwanaland, except notonemourids, which survived everywhere in the south but nowhere in the north. Alternatively, if present distributions are more recent, dispersal across seas must almost certainly have been involved, and in the case of notonemourid paraphyly, perhaps even several times.

ACKNOWLEDGMENTS

I would like to thank all the colleagues who helped me during the preparation of this manuscript. I thank Irene Tade (Bad Hersfeld) for Russian translations, and T Soldán (České Budjejovice) contributed interesting information for a discussion of gill homologies. I am deeply grateful to HB Noel Hynes (Waterloo), Ian D McLellan (Westport), Charles H Nelson (Chattanooga), C Riley Nelson (Austin), and Bill P Stark (Clinton) who read, discussed, and significantly improved a draft of this paper.

LITERATURE CITED

1. Aubert J. 1959. Plecoptera. *Ins. Helv. Fauna* 1:1–140
2. Aubert J. 1960. Contribution à l'étude des Notonemourinae (Plecoptera) de l'Amérique du Sud. *Mitt. Schweiz. Entomol. Ges.* 33:47–64
3. Bacetti BM. 1987. Spermatozoa and phylogeny in orthopteroid insects. In *Evolutionary Biology of Orthopteroid Insects*, ed. BM Bacetti, pp. 12–112. Chichester/New York: Ellis Horwood & Halsted
4. Banarescu P. 1990. *Zoogeography of Fresh Waters. General Distribution and Dispersal of Freshwater Animals,* Vol. 1. Wiesbaden: Aula-Verlag
5. Barnard KH. 1934. South African stoneflies (Perlaria), with descriptions of new species. *Ann. S. Afr. Mus.* 30(1931/1934):511–48
6. Baumann RW.1975. Revision of the stonefly family Nemouridae (Plecoptera): a study of the world fauna at the generic level. *Smithson. Contrib. Zool.* 211:3 + 74 pp.
7. Beer-Stiller A, Zwick P. 1995. Biometric studies of some stoneflies and a mayfly (Plecoptera and Ephemeroptera). *Hydrobiologia* 299:169–78
8. Berthélemy C. 1969. Contribution à la connaissance des Leuctridae. *Ann. Limnol.* 4:175–98
9. Berthélemy C. 1979. Accouplement, période d'incubation et premiers stades larvaires de *Brachyptera braueri* et de *Perlodes microcephalus* (Plecoptera). *Ann. Limnol.* 15:317–25
10. Boudreaux HB. 1979. *Arthropod Phylogeny with Special Reference to Insects.* Chichester: Wiley & Sons
11. Bowmaker AP, Jackson PBN, Jubb RA. 1978. Freshwater fishes. In *Biogeography and Ecology of Southern Africa*, ed. MJA Werger, 2:1181–230. Monogr. Biol. 31. The Hague: Junk
12. Briden JC, Drewry GE, Smith AG. 1974. Phanerozoic equal-area world maps. *J. Geol.* 82:555–74
13. Bricknell IR, Potts WTW. 1989. The distribution of the ion transporting cells of Plecoptera in relation to pH and the osmotic environment. *Entomologist* 108:176–83
14. Brinck P. 1956. Reproductive system and mating in Plecoptera. *Opusc. Entomol.* 21:57–127
15. Brinck P. 1962. Begattungsorgane und Spermaübertragung bei den isogeninen

Perlodiden. *Verh. Int. Kongr. Entomol., 11th, Wien 1960,* 3:267–69. Wien: Org.-Komm. Int. Kongr. Entomol.
16. Brittain JE. 1977. The effect of temperature on the egg incubation period of *Taeniopteryx nebulosa* (Plecoptera). *Oikos* 29:302–5
17. Brodskiy AK. 1979. Evolution of the wing apparatus in stoneflies (Plecoptera). Part 1. Functional morphology of the wings. *Entomol. Obozr.* 58:69–77. (In Russian) Engl. transl. 1980. *Entomol. Rev.* 58(1):31–36
18. Brodskiy AK. 1979. Evolution of the wing apparatus in stoneflies. Part 2. Functional morphology of the axillary apparatus, skeleton, and musculature. *Entomol. Obzr.* 58:710–15 (In Russian) Engl. transl. 1980. *Entomol. Rev.* 58(4):16–26
19. Brodskiy AK. 1982. Evolution of the wing apparatus in stoneflies (Plecoptera). Part 4. Kinematics of the wings and general conclusions. *Entomol. Obozr.* 61:491–500. (In Russian) Engl. transl. 1983. *Entomol. Rev.* 61(4):34–43
20. Campbell IC. 1981. Biogeography of some rheophilous aquatic insects in the Australian region. *Aquat. Insects* 3:33–43
21. Consiglio C. 1967. Una nuova specie italiana e considerazioni sui gruppi di specie nel genere *Isoperla* (Plecoptera, Perlodidae). *Fragm. Entomol.* 5:67–75
22. Cooke HBS. 1968. The fossil mammal fauna of Africa. *Q. Rev. Biol.* 43:234–64
23. Cox CC. 1974. Vertebrate palaeodistributional patterns and continental drift. *J. Biogeogr.* 1:75–94
24. Cracraft J. 1974. Continental drift and vertebrate distribution. *Annu. Rev. Ecol. Syst.* 6:215–61
25. Craig DA. 1969. A taxonomic revision of New Zealand Blepharoceridae and the origin and evolution of the Australasian Blepharoceridae (Diptera: Nematocera). *Trans. R. Soc. NZ Biol. Sci.* 11:101–51
26. Cranston PS, Naumann ID. 1991. Biogeography. In *The Insects of Australia,* 1:180–97. Melbourne: Melbourne Univ. Press
27. deLattin G. 1967. *Grundriss der Zoogeographie.* Jena: Gustav Fischer
28. Downes JA, Kavanaugh DH, eds. 1988. Origins of the North American insect fauna. *Mem. Entomol. Soc. Can.* 144:1–168
29. Elliott JM. 1988. Interspecific and intraspecific variations in egg hatching for British populations of *Taeniopteryx nebulosa* and *Brachyptera risi* (Plecoptera: Taeniopterygidae). *Holarctic Ecol.* 11:55–59
30. Enderlein G. 1909. Klassifikation der Plecopteren sowie Diagnosen neuer Gattungen und Arten. *Zool. Anz.* 34:385–419
31. Frison TH. 1942. Studies of North American Plecoptera, with special reference to the fauna of Illinois. *Bull. Ill. Nat. Hist. Surv.* 22:235–355
32. Gnatzy W, Rupprecht R.1972. Die Bauchblase von *Nemurella picteti* (sic!) Klapálek (Insecta, Plecoptera). *Z. Morphol. Tiere* 7:325–42
33. Hanson JF. 1941. Records and descriptions of North American Plecoptera. Part. I. Species of *Leuctra* of the Eastern United States. *Am. Midl. Nat.* 26:174–78
34. Hanson JF. 1946. Comparative morphology and taxonomy of the Capniidae (Plecoptera). *Am. Midl. Nat.* 35:193–249
35. Harper PP. 1994. Plecoptera. In *Aquatic Insects of China Useful for Monitoring Water Quality,* ed. JC Morse, L Yang, L Tian, pp. 176–209. Nanjing: Hohai Univ. Press
36. Harper PP, Hynes HBN. 1970. Diapause in the nymphs of Canadian winter stoneflies. *Ecology* 51:925–27
37. Harper PP, Lauzon M, Harper F. 1991. Life cycles of 12 species of winter stoneflies from Quebec (Plecoptera, Capniidae and Taeniopterygidae). *Can. J. Zool.* 69:787–96
38. Harper PP, Magnin E. 1969. Cycles vitaux de quelques Plécoptères des Lau-

rentides (Insectes). *Can. J. Zool.* 47:483–94
39. Hennig W. 1950. *Grundzüge einer Theorie der phylogenetischen Systematik.* Berlin: Dtsch. Zentralver.
40. Hennig W. 1960. Die Dipteren-Fauna von Neuseeland als systematisches und tiergeographisches Problem. *Beitr. Entomol.* 10:221–329
41. Hennig W. 1965. Phylogenetic systematics. *Annu. Rev. Entomol.* 10:97–116
42. Hennig W. 1968. *Die Stammesgeschichte der Insekten.* Frankfurt: Waldemar Kramer
43. Hennig W. 1981. *Insect Phylogeny.* Chichester: Wiley & Sons
44. Hinton HE. 1981. *Biology of Insect Eggs.* Oxford: Pergamon
45. Hynes HBN. 1967. A key to the adults and nymphs of the British stoneflies (Plecoptera) with notes on their ecology and distribution. *Sci. Publ. Freshw. Biol. Assoc.* 17:1–90. 2nd ed.
46. Hynes HBN. 1978. An annotated key to the nymphs of the stoneflies (Plecoptera) of the state of Victoria. *Aust. Soc. Limnol., Spec. Publ.* 2:1–63
47. Hynes HBN. 1981. Taxonomical notes on Australian Notonemouridae (Plecoptera) and a new species from Tasmania. *Aquat. Insects* 3:147–66
48. Hynes HBN. 1988. Biogeography and origins of the North American stoneflies (Plecoptera). *Mem. Entomol. Soc. Can.* 144:31–37
49. Hynes HBN, Hynes ME. 1980. The endemism of Tasmanian stoneflies (Plecoptera). *Aquat. Insects* 2:81–89
50. Illies J. 1953. Beitrag zur Verbreitungsgeschichte der europäischen Plecopteren. *Arch. Hydrobiol.* 48:35–74
51. Illies J. 1960. Phylogenie und Verbreitungsgeschichte der Ordnung Plecoptera. *Zool. Anz.* Suppl. 24:384–94
52. Illies J. 1960. Archiperlaria, eine neue Unterordnung der Plecopteren (Revision der Familien Eustheniidae und Diamphipnoidae) (Plecoptera). *Beitr. Entomol.* 10:661–97
53. Illies J. 1960. Penturoperlidae, eine neue Plecopterenfamilie. *Zool. Anz.* 164:26–41
54. Illies J. 1961. Südamerikanische Notonemourinae und die Stellung der Unterfamilie im System der Plecopteren. *Mitt. Schweiz. Entomol. Ges.* 34:97–126
55. Illies J. 1962. Verbreitungsgeschichte der Plecopteren auf der Südhemisphäre. *Verh. Int. Kongr. Entomol., 11th, Wien 1960,* 1:476–80. Wien: Org.-Komm. Int. Kongr. Entomol.
56. Illies J. 1962. Das abdominale Zentralnervensystem der Insekten und seine Bedeutung für Phylogenie und Systematik der Plecopteren. *Ber. 9. Wanderv-ers. Dtsche. Entomol.* 45:139–52
57. Illies J. 1965. Phylogeny and zoogeography of the Plecoptera. *Annu. Rev. Entomol.* 10:117–40
58. Illies J. 1966. Katalog der rezenten Plecoptera. *Das Tierreich* 81:(I-IXXX)1–632
59. Illies J. 1967. Die Gattung *Megaleuctra* (Plecopt., Ins.), Beitrag zur konsequent phylogenetischen Behandlung eines incertae-sedis-Problems. *Z. Morphol. Ökol. Tiere* 60:124–34
60. Illies J. 1969. Revision der Plecopterenfamilie Austroperlidae. *Entomol. Tidskr.* 90:19–51
61. Illies J. 1969. Biogeography and ecology of neotropical freshwater insects, especially those from running waters. In *Biogeography and Ecology in South America,* ed. EJ Fittkau, J Illies, H Klinge, GH Schwabe, H Sioli, 2:685–708. The Hague: Junk (Monogr. Biol. 19)
62. Illies J. 1975. Notonemouridae of Australia (Plecoptera, Ins.). *Int. Rev. Ges. Hydrobiol.* 60:221–49
62a. Int. Comm. on Zoological Nomenclature. 1999. BRACHYPTERINAE Zwick, 1973 (Insecta, Plecoptera): spelling amended to BRACHYPTERAINAE, so removing the homonymy with BRA-

CHYPTERINAE Erichson, [1845] (Insecta, Coleoptera); KATERETIDAE Erichson in Agassiz, [1846]: given precedence over BRACHYPTERINAE Erichson. *Bull. Zool. Nomencl.* 56:82–86
63. Jamieson BGM. 1987. *The Ultrastructure and Phylogeny of Insect Spermatozoa*. Cambridge: Cambridge Univ. Press
64. Kapoor NN, Zachariah K. 1983. Ultrastructure of the sensilla of the stonefly nymph, *Thaumatoperla alpina* Burns and Neboiss (Plecoptera: Eustheniidae). *Int. J. Insect Morphol. Embryol.* 12:157–68
65. Kawai T. 1967. Plecoptera Insecta. *Fauna Jpn.* 1:1–211. Tokyo: Biogeogr. Soc. Jpn.
66. Kawai T, Isobe Y. 1984. Notes on the egg of *Scopura longa* Ueno (Plecoptera). *Ann. Limnol.* 20:57–58
67. Khoo SG. 1964. Studies on the biology of *Capnia bifrons* (Newman) and notes on the diapause in the nymphs of this species. *Gewäss. Abwäss.* 34/35:23–30
68. Khoo SG. 1968. Experimental studies on diapause in stoneflies. III. Eggs of *Brachyptera risi* (Morton). *Proc. R. Entomol. Soc. London Ser. A* 43:141–46
69. Kimmins DE. 1951. A Revision of the Australian and Tasmanian Gripopterygidae and Nemouridae (Plecoptera). *Bull. Br. Mus. Nat. Hist. Entomol.* 2:45–93
70. Kinzelbach R. 1975. Die Skorpione der Ägäis. Beiträge zur Systematik, Phylogenie und Biogeographie. *Zool. Jahrb., Syst. Ökol. Geogr. Tiere* 102:12–50
71. Kishimoto T, Ando H. 1985. External features of the developing embryo of the stonefly, *Kamimuria tibialis* (Plecoptera, Perlidae). *J. Morphol.* 183:311–26
72. Klapálek F. 1905. Conspectus Plecopterorum Bohemiae. *Cas. Cesk. Spol. Entomol.* 2:27–32
73. Klapálek F. 1909. Plecoptera, Steinfliegen. In *Die Süsswasserfauna Deutschlands*, ed. A Brauer, 8:33–95. Jena: Gustav Fischer
74. Komnick H. 1977. Chloride cells and chloride epithelia of aquatic insects. *Int. Rev. Cytol.* 49:285–329
75. Kristensen NP. 1991. Phylogeny of extant hexapods. In *The Insects of Australia,* 1:125–40. Melbourne: Melbourne Univ. Press
76. Kukalová-Peck J. 1978. Origin and evolution of insect wings and their relation to metamorphosis, as documented by the fossil record. *J. Morphol.* 156:53–126
77. Kukalová-Peck J. 1983. Origin of the insect wing and wing articulation from the arthropodan leg. *Can. J. Zool.* 61:1618–69
78. Kukalová-Peck J. 1991. Fossil history and the evolution of hexapod structures. In *The Insects of Australia,* 1:141–79. Melbourne: Melbourne Univ. Press
79. Kukalová-Peck J, Sinichenkova ND. 1992. The wing venation and systematics of Lower Permian Diaphanopterodea from the Ural Mountains, Russia (Insecta: Paleoptera). *Can. J. Zool.* 70:229–35
80. Lillehammer A. 1988. Stoneflies (Plecoptera) of Fennoscandia and Denmark. *Fauna Entomol. Scand.* 21:1–165. Leiden: Brill & Scand. Sci.
81. Matsuda R. 1977. *Morphology and Evolution of the Insect Abdomen*. Oxford: Pergamon
82. McLellan ID. 1968. A revision of genus *Notonemoura* (Plecoptera: Notonemouridae). *Trans. R. Soc. NZ Zool.* 10:133–40
83. McLellan ID. 1975. The freshwater insects. In *Biogeography and Ecology of New Zealand,* ed. G Kuschel, pp. 537–59. The Hague: Junk (Monogr. Biol. 27)
84. McLellan ID. 1977. New alpine and southern Plecoptera from New Zealand and a new classification of the Gripopterygidae. *NZ J. Zool.* 4:119–47
85. McLellan ID. 1991. Notonemouridae (Insecta: Plecoptera). *Fauna NZ* 22:1–64
86. McLellan ID. 1993. Antarctoperlinae (Insecta: Plecoptera). *Fauna NZ* 27:1–70
87. McLellan ID. 1996. A revision of *Stenoperla* (Plecoptera: Eustheniidae) and

removal of Australian species to *Cosmioperla* new genus. *NZ J. Zool.* 23:165–82

88. McLellan ID. 1997. *Austroperla cyrene* Newman (Plecoptera: Austroperlidae). *J. R. Soc. NZ* 27:271–78
89. McLellan ID. 1998. A revision of *Acroperla* (Plecoptera: Zelandoperlinae) and removal of species to *Taraperla* new genus. *NZ J. Zool.* 25:185–203
90. McLellan ID, Zwick, P. 1998. *Austronemoura auberti* new species and other new Chilean Notonemouridae (Plecoptera). *Mitt. Schweiz. Entomol. Ges.* 69:107–15
91. Michaelis FB, Yule C. 1988. Plecoptera. In *Zoological Catalogue of Australia*, ed. Bur. Flora Fauna, Canberra, 6:133–37. Canberra: Aust. Gov. Print. Serv.
92. Miller A. 1939. The egg and early development of the stonefly *Pteronarcys proteus* Newman (Plecoptera). *J. Morphol.* 64:555–609
93. Minet J, Bourgoin T. 1986. Phylogénie et classification des hexapodes (Arthropoda). *Cah. Liaison OPIE* 20:23–28
94. Miron J, Zwick P. 1973. Un nouveau genre de Plécoptères du Haut Atlas marocain. *Bull. Soc. Sci. Nat. Phys. Maroc* 52:219–25
95. Nelson CH. 1984. Numerical cladistic analysis of phylogenetic relationships in Plecoptera. *Ann. Entomol. Soc. Am.* 77:466–73
96. Nelson CH. 1988. Note on the phylogenetic systematics of the family Pteronarcyidae (Plecoptera), with a description of the eggs and nymphs of the Asian species. *Ann. Entomol. Soc. Am.* 81:560–76
97. Nelson CH. 1991. Preliminary note on the comparative morphology of the stonefly pretarsus (Plecoptera). In *Overview and Strategies of Ephemeroptera and Plecoptera,* ed. J Alba-Tercedor, A Sanchez-Ortega. pp. 137–56. Gainesville, FL: Sandhill Crane
98. Nelson CH, Hanson JF. 1969. The external anatomy of *Pteronarcys* (*Allonarcys*) *proteus* Newman and *Pteronarcys* (*Allonarcys*) *biloba* Newman (Plecoptera: Pteronarcidae). *Trans. Am. Entomol. Soc.* 94:429–72
99. Nelson CH, Hanson JF. 1971. Contribution to the anatomy and phylogeny of the family Pteronarcidae. *Trans. Am. Entomol. Soc.* 97:123–200
100. Nelson CH, Hanson JF. 1973. The genus *Perlomyia* (Plecoptera: Leuctridae). *J. Kans. Entomol. Soc.* 46:187–99
101. Nelson CR, Baumann RW. 1987. The winter stonefly genus *Capnura* (Plecoptera: Capniidae) in North America: systematics, phylogeny, and zoogeography. *Trans. Am. Entomol. Soc.* 113:1–28
102. Nelson CR, Baumann RW. 1989. Systematics and distribution of the winter stonefly genus *Capnia* (Plecoptera: Capniidae) in North America. *Great Basin Nat.* 49:289–363 + 3 unnumbered pages
103. Noonan GR. 1988. Faunal relationships between eastern North America and Europe as shown by insects. *Mems. Entomol. Soc. Can.* 144:39–53
104. Oberndorfer RY, Stewart KW. 1977. The life cycle of *Hydroperla crosbyi* (Plecoptera: Perlodidae). *Great Basin Nat.* 37: 260–73
105. Pass G. 1987. The "Cercus Heart" in stoneflies—a new type of accessory circulatory organ in insects. *Naturwissenschaften* 74:440–41
106. Paulian R. 1959. Recherches sur les insectes d'importance biologique à Madagascar. XXXIII. Nouveaux Plécoptères malgaches. *Mém. Inst. Sci. Madagascar (E)* 11:9–16
107. Pritchard G, McKee MH, Pike EM, Scrimgeour GJ, Zloty J. 1993. Did the first insects live in water or in air? *Biol. J. Linn. Soc.* 49:31–44
108. Raušer J. 1968. Plecoptera. Ergebnisse der zoologischen Forschungen von Dr. Z. Kaszab in der Mongolei. *Entomol. Abh. Mus. Tierk.* 34:329–98
109. Raven PH, Axelrod DI. 1972. Plate tec-

tonics and Australasian palaeobiogeography. *Science* 176:1379–86

110. Ricker WE. 1950. Some evolutionary trends in Plecoptera. *Proc. Indiana Acad. Sci.* 59:197–209
111. Ricker WE. 1952. Systematic studies in Plecoptera. *Indiana Univ. Sci. Ser.* 18:1–200
112. Ricker WE, Ross HH. 1969. The genus *Zealeuctra* and its position in the family Leuctridae (Plecoptera, Insecta). *Can. J. Zool.* 47:1113–27
113. Ricker WE, Ross HH. 1975. Synopsis of the Brachypterinae (Insecta: Plecoptera: Taeniopterygidae). *Can. J. Zool.* 5:32–53
114. Riek EF. 1973. Fossil insects from the Upper Permian of Natal, South Africa. *Ann. Natal Mus.* 21:513–32
115. Rögl F, Steininger FF. 1983. Vom Zerfall der Tethys zu Mediterran und Paratethys. Die neogene Paläogeographie und Palinspastik des zirkum–mediterranen Raumes. *Ann. Naturhist. Mus. Wien* 85/A:135–63
116. Ross HH, Ricker WE. 1971. The classification, evolution and dispersal of the winter stonefly genus *Allocapnia*. *Ill. Biol. Monogr.* 45:1–166
117. Rupprecht R. 1976. Struktur und Funktion der Bauchblase und des Hammers von Plecopteren. *Zool. Jahr., Abt. Anat. Ontog. Tiere* 95:9–80
118. Rupprecht R. 1990. Can adult stoneflies utilize what they eat? In *Mayflies & Stoneflies: Life Histories and Biology*, ed. CD Campbell, pp.119–23. Dordrecht: Kluwer
119. Rupprecht R, Gnatzy W.1974. Die Feinstruktur der Sinneshaare auf der Bauchblase von *Leuctra hippopus* und *Nemoura cinerea* (Plecoptera). *Cytobiologie* 9:422–31
120. Sedlag U, Weinert E. 1987. *Biogeographie, Artbildung, Evolution*. Wörterb. Biol., UTB 1430. Stuttgart: Gustav Fischer
122. Shepard WD, Baumann RW. 1995. *Calileuctra*, a new genus, and two new species of stoneflies from California (Plecoptera: Leuctridae). *Great Basin Nat.* 55:124–34
123. Shepard WD, Stewart KW. 1983. Comparative study of nymphal gills in North American stonefly (Plecoptera) genera and a new, proposed paradigm of Plecoptera gill evolution. *Misc. Publ. Entomol. Soc. Am.* 55:1–57
124. Shimizu T. 1993. *A revision on the Japanese species of the genus Nemoura (Nemouridae: Plecoptera)*. PhD diss. Tokyo Univ. Agric. 102 pp.
125. Sinitshenkova ND. 1987. Historical development of Plecoptera. *Trudy Palaeontol. Inst.* 221:1–144. (In Russian)
126. Sivec I, Stark BP, Uchida S. 1988. Synopsis of the world genera of Perlinae (Plecoptera: Perlidae). *Scopolia* 16:1–66
127. Stark BP. 1989. Oriental Peltoperlinae (Plecoptera): a generic review and descriptions of a new genus and seven new species. *Entomol. Scand.* 19:503–25
128. Stark BP, Gaufin AR. 1976. The nearctic genera of Perlidae (Plecoptera). *Misc. Publ Entomol. Soc. Am.* 10:1–80
129. Stark BP, Gonzalez del Tanago M, Szczytko SW. 1986. Systematic studies on Western Palaeartic Perlodini (Plecoptera: Perlodidae). *Aquat. Insects* 8:91–98
130. Stark BP, Nelson CR. 1994. Systematics, phylogeny and zoogeography of genus *Yoraperla* (Plecoptera: Peltoperlidae). *Entomol. Scand.* 25:241–73
131. Stark BP, Ray DH. 1983. A revision of the genus *Helopicus* (Plecoptera: Perlodidae). *J. N. Am. Benthol. Soc.* 2:16–27
132. Stark BP, Stewart KW. 1981. The nearctic genera of Peltoperlidae (Plecoptera). *J. Kans. Entomol. Soc.* 54:285–311
133. Stark BP, Szczytko SW. 1982. Egg morphology and phylogeny in Pteronarcyidae (Plecoptera). *Ann. Entomol. Soc. Am.* 75:519–29
134. Stark BP, Szczytko SW. 1984. Egg morphology and classification of Perlodinae (Plecoptera: Perlodidae). *Ann. Limnol.* 20:99–104

135. Stark BP, Szczytko SW. 1988. Egg morphology and phylogeny in Arcynopterygini (Plecoptera: Perlodidae). *J. Kans. Entomol. Soc.* 61:143–60
136. Stevens DM, Picker MD. 1995. The Notonemouridae (Plecoptera) of southern Africa: description of a new genus, *Balinskycercella,* and a key to genera. *Afr. Entomol.* 3:77–83
137. Stewart KW, Abbott JC, Kirchner RF, Moulton SR. 1995. New descriptions of North American Euholognathan stonefly drumming (Plecoptera) and first Nemouridae ancestral call discovered in *Soyedina carolinensis* (Plecoptera: Nemouridae). *Ann. Entomol. Soc. Am.* 88:234–39
138. Stewart KW, Bottorff RL, Knight AW, Moring JB. 1991. Drumming of four North American Euholognatha stonefly species, and a new complex signal pattern in *Nemoura spiniloba* Jewett (Plecoptera: Nemouridae). *Ann. Entomol. Soc. Am.* 84:201–6
139. Stewart KW, Maketon M. 1990. Intraspecific variation and information content of drumming in three Plecoptera species. In *Mayflies and Stoneflies: Life Histories and Biology,* ed. IC Campbell, pp. 259–68. Dordrecht: Kluwer
140. Stewart KW, Maketon M. 1991. Structures used by Nearctic stoneflies (Plecoptera) for drumming and their relationship to behavioral pattern diversity. *Aquat. Insects* 13:33–53
141. Stewart KW, Stark BP. 1988. Nymphs of North American stonefly genera (Plecoptera). *Thomas Say Found. Entomol. Soc. Am.* 12:1–436
142. Stewart KW, Zeigler DD. 1984. Drumming behavior of twelve North American stonefly (Plecoptera) species: first description in Peltoperlidae, Taeniopterygidae and Chloroperlidae. *Aquat. Insects* 6:49–61
143. Steyskal GC.1976. Notes on the nomenclature and taxonomic growth of the Plecoptera. *Proc. Biol. Soc. Wash. DC* 88:408–10
144. Stuckenberg BR. 1962. The distribution of the montane palaeogenic element in the South African invertebrate fauna. *Ann. Cape Prov. Mus. Nat. Hist.* 2:190–205
145. Štys P, Soldán T. 1980. Retention of tracheal gills in adult Ephemeroptera and other insects. *Acta Univ. Carol. Biol.* 1978:409–35
146. Surdick RF. 1985. Nearctic genera of Chloroperlinae (Plecoptera: Chloroperlidae). III. *Biol. Monogr.* 54:1–146
147. Szczytko SW, Stewart KW. 1984. Descriptions of *Calliperla* Banks, *Rickera* Jewett, and two new western Nearctic *Isoperla* species (Plecoptera: Perlodidae). *Ann. Entomol. Soc. Am.* 77:251–63
148. Theischinger G. 1983. The genus *Stenoperla* in Australia (Insecta: Plecoptera: Eustheniidae). *Aust. J. Zool.* 31:541–56
149. Thenius E. 1975. Biogeographie in der Sicht der Erdwissenschaften. Die Paläogeographie als Grundlage einer historischen Biogeographie. *Verh. Dtsch. Zool. Ges.* 67:357–72
150. Thienemann A. 1950. Verbreitungsgeschichte der Süßwassertierwelt Europas. *Die Binnengewässer* 18:1–809. Stuttgart: Schweizer. Verlag
151. Tillyard RJ. 1921. A new classification of the order Perlaria. *Can. Entomol.* 53:35–44
152. Trueman JWH. 1989. Comment—Evolution of insect wings: a limb exite plus endite model. *Can. J. Zool.* 68:1333–35
153. Uchida S. 1990. *A revision of the Japanese Perlidae (Insecta: Plecoptera), with special reference to their phylogeny.* PhD diss., Tokyo Metrop. Univ. 228 pp.
154. Uchida S, Isobe Y. 1988. *Cryptoperla* and *Yoraperla* from Japan and Taiwan (Plecoptera: Peltoperlidae). *Aquat. Insects* 10:17–31
155. Uchida S, Isobe Y. 1989. Styloperlidae, stat. nov. and Microperlinae, subfam.

nov. with a revised system of the family group Systellognatha. *Spixiana* 12:145–82

156. Whiting MF, Carpenter JC, Wheeler QD, Wheeler WC. 1997. The Strepsiptera problem: phylogeny of the holometabolous insect orders inferred from 18S and 28S ribosomal DNA sequences and morphology. *Syst. Biol.* 46:1–68
157. Wichard W. 1997. Insekten erobern den aquatischen Lebensraum. *Verh. Westdtsch. Entomol. Tag.* 1966:19–28. Düsseldorf: Löbbecke Mus.
158. Wichard W, Arens W, Eisenbeis G. 1995. *Atlas zur Biologie der Wasserinsekten.* Stuttgart: Gustav Fischer
159. Wichard W, Eisenbeis G. 1979. Fine structural and histochemical evidence of chloride cells in stonefly larvae 3. The floriform chloride cells. *Aquat. Insects* 1:185–91
160. Wichard W, Komnick H.1973. Feinstruktureller und histochemischer Nachweis von Chloridzellen bei Steinfliegenlarven. 1. Die coniformen Chloridzellen. *Cytobiologie* 7:297–314
161. Wichard W, Komnick H.1974. Feinstruktureller und histochemischer Nachweis von Chloridzellen bei Steinfliegenlarven. 2. Die caviformen und bulbiformen Chloridzellen. *Cytobiologie* 8:297–311
162. Willmann R. 1997. Advances and problems in insect phylogeny. In *Arthropod Relationships,* ed. RA Fortey, RH Thomas. London: Chapman & Hall (Syst. Assoc. Spec. 55:269–79)
163. Wittig G. 1955. Untersuchungen am Thorax von *Perla abdominalis* Burm. (Larve und Imago). *Zool. Jahr. Anat. Ontog.* 74:491–570
164. Wright M, White MM. 1992. Biochemical systematics of the North American *Pteronarcys* (Pteronarcyidae: Plecoptera). *Biochem. Syst. Ecol.* 20:515–21
165. Wu CF. 1938. *Plecopterorum sinensium, A monograph of the Stoneflies of China (Order Plecoptera).* 225 pp.
166. Zhiltzova LA. 1999. The Capniidae of Russia and adjacent territories (within the limits of the former USSR). *Proc. 13th Int. Plecoptera Symp.* In press
167. Zhiltzova LA, Zwick P. 1971. Notes on Asiatic Chloroperlidae (Plecoptera), with descriptions of new species. *Entomol. Tidskr.* 92:183–97
168. Zhiltzova LA, Zwick P. 1993. On the genus *Kyphopteryx* Kimmins (Plecoptera: Taeniopterygidae). *Aquat. Insects* 15:193–98
169. Zwick P. 1967. Revision der Gattung *Chloroperla* Newman (Plecoptera). *Mitt. Schweiz. Entomol. Ges.* 40:1–26
170. Zwick P. 1969. *Das phylogenetische System der Plecopteren als Ergebnis vergleichend-anatomischer Untersuchungen.* PhD diss., Christian-Albrechts-Univ., Kiel. VI + 291 pp.
171. Zwick P.1973. Insecta Plecoptera. Phylogenetisches System und Katalog. *Tierreich* 94:XXXII + 465 pp.
172. Zwick P. 1973. Entomological explorations in Ghana by Dr. S. Endrödy-Younga 27. Notes on some species of *Neoperla* (Plecoptera). *Folia Entomol. Hung., Ser. nov. Suppl.* 26:381–98
173. Zwick P. 1976. *Neoperla* (Plecoptera) emerging from a mountain stream in Central Africa. *Int. Rev. Ges. Hydrobiol.* 61:683–97
174. Zwick P. 1977. Ergebnisse der Bhutan-Expedition 1972 des Naturhistorischen Museums in Basel. Plecoptera. *Entomol. Basil* 2:85–134
175. Zwick P. 1977. Australian Blephariceridae (Diptera). *Aust. J. Zool. Suppl.* 46:1–121
176. Zwick P. 1979. Revision of the stonefly family Eustheniidae (Plecoptera), with emphasis on the fauna of the Australian region. *Aquat. Insects* 1:17–50
177. Zwick P. 1980. Plecoptera (Steinfliegen). *Handb. Zool.* 4(2)2/7:1–115. Berlin: de Gruyter
178. Zwick P. 1981. Plecoptera. In *Ecological Biogeography of Australia,* ed. A Keast,

2:1171–82. Monogr. Biol. 41. The Hague: Junk

179. Zwick P. 1981. The Mediterranean area as glacial refuge of Plecoptera. *Acta Entomol. Jugosl.* 17:107–11. (In German)
180. Zwick P. 1990. Transantarctic relationships in the Plecoptera. In *Mayflies and Stoneflies: Life Histories and Biology,* ed. IC Campbell, pp.141–48. Dordrecht: Kluwer
181. Zwick P. 1991. Biometric studies of the growth and development of two species of *Leuctra* and of *Nemurella pictetii* (Plecoptera: Leuctridae and Nemouridae). In *Overview and Strategies of Ephemeroptera,* ed. J Alba-Tercedor, A Sanchez-Ortega pp. 515–26. Gainesville, FL: Sandhill Crane
182. Zwick P. 1997. *Rauserella,* a new genus of Plecoptera (Perlodidae), with notes on related genera. In *Ephemeroptera and Plecoptera Biology-Ecology-Systematics,* ed. P Landolt, M Sartori, pp. 489–96. Fribourg: Mauron + Tinguely & Lachat SA.
183. Zwick P. 1998. The Australian net-winged midges of tribe Apistomyiini (Diptera: Blephariceridae). *Aust. J. Entomol.* 37:289–311
184. Zwick P, Weinzierl A. 1995. Reinstatement and revision of genus *Besdolus* (Plecoptera: Perlodidae). *Entomol. Scand.* 26:1–26

Annu. Rev. Entomol. 2000. 45:747–767

IMPACT OF THE INTERNET ON ENTOMOLOGY TEACHING AND RESEARCH

J. T. Zenger and T. J. Walker
Entomology and Nematology Department, University of Florida, Gainesville, Florida 32601; e-mail: zenger@GNV.IFAS.UFL.EDU

Key Words World Wide Web, distance education, computer software, Web sites, electronic publications

■ **Abstract** The Internet is affecting entomology teaching and research. Internet tools help students communicate and easily find and access information. Entomology instructors who adopt these tools may discover they are surprisingly time consuming to implement. Requiring students to use the Internet teaches them to glean from the glut of available information and to communicate electronically, both vital skills in today's workplace. The Internet helps meet the growing need for distance education by providing a medium that allows students to conveniently access course materials and to communicate with the instructor and other students. Researchers benefit from using the Internet for one-to-one and one-to-many communication and from access to large cooperative databases, for example, in molecular biology and systematics. Perhaps the greatest impact on research will be the migration to the Web of journals and other specialized research literature. This may permit free access and will change the content and format of journal articles. In the online version of this chapter, the hyperlinks in the Literature Cited section are active.

THE IMPACT OF THE INTERNET ON ENTOMOLOGY TEACHING AND RESEARCH

The Internet is relatively new and yet is entering our lives at an extraordinary rate. Because it is new, the Internet's ultimate impact is not yet determined and cannot be foreseen. However, because the Internet's effect on the field of entomology is already great it is important to chronicle its adoption into the areas of teaching and research to *(a)* alert entomologists to the changes that have occurred, and *(b)* take advantage of the resources that are now available. We will briefly review the history and current components of the Internet, and address separately the topics of teaching and research.

A BRIEF HISTORY OF THE INTERNET

The network that became the Internet began in 1969 as a United States Department of Defense project to develop a communication network that would survive

0066-4170/00/0107-0747/$14.00

a nuclear attack. Initially four universities were connected, then a dozen or so in 1971, and eventually many sites were online. Scientists and researchers could transfer files, hold online discussions and send e-mail between universities, companies, and government agencies. The Internet as now defined was born in 1983, when a universal protocol was adopted and all research and government networks were connected (62, 79).

A major event in the evolution of the Internet was the birth in 1991 of the World Wide Web (Web). This event was accompanied by the development of the Hypertext Markup Language (HTML), which allowed links embedded within text, enabling the user to access other text or files. In 1993, the release of Mosaic, the first graphical browser, began the transition of the Web from a text-based communications system to an exciting graphic medium (62). To this point, however, the Internet was effectively restricted to the realms of higher education and government. It was not until 1995 when commercial networks such as AOL, Prodigy, and CompuServe began offering online dialup services to the public that the Internet's popularity began to accelerate (62, 66). Today more than 60 million people in the United States—[37% of the adult population (86)]—and 102 million people worldwide (89) have access to the Internet. Retailers have quickly taken advantage of the Internet's advertising and sales potential. The online sales of mainstream products such as music, clothing, automobiles, and electronic products have increased 200% between 1997 and 1998 (122), with more than $5 billion dollars in 1998 holiday sales alone (108). While not as vigorous in their adoption as sales professionals have been, entomologists are now using the Internet for a myriad of teaching and research purposes.

THE IMPACT OF THE INTERNET ON TEACHING

Though still in its infancy, the Internet's use in teaching has already had a great impact on the way instructors teach and students find and access information. We will address the topic of the Internet's impact on teaching under two major headings: *(a)* the impact on traditional on-campus teaching, and *(b)* distance education, in which the instructor and student do not regularly meet face to face and special techniques are needed to convey information and communicate (92).

On-Campus Education

World Wide Web Traditional courses are beginning to reap tangible benefits from the use of Internet tools, of which the Web is the most visible. Educators throughout higher education, including entomologists, are now using course Web sites for several purposes. What follows are descriptions of the common uses, organized in an ascending order of complexity. Citations that follow a category description refer to representative entomology course Web sites.

Course Description The most basic course Web sites are those used to simply advertise or inform students about a particular course (43, 68, 87). Such sites usually consist of single Web pages posted on a department Web site. They are most often developed by someone other than the instructor and the same format is generally used for all department courses. The information is basic and includes the instructor's name, a brief statement about course content, when it is taught, and a listing of prerequisites.

Course Syllabus The next most common application of the Web for teaching, and a logical first step for most instructors, is the placement of the course syllabus on the Web (3, 18, 24, 27, 67, 75–77, 85, 96). Some advantages of placing a syllabus online include providing additional information for potential students, updating lecture schedules, and relief from replacing syllabi students have lost. It is also a simple yet useful first step for faculty wanting to try the Web. Today it is extremely easy to use a modern word processing application to make a course syllabus available on the Web because such word processors will save documents in HTML format. In fact, some, such as Microsoft® Word 97, can be used as very capable Web page editors.

Beyond the Syllabus Once a syllabus is online, lecture notes are easily and commonly added (15, 48, 49, 53, 59, 60, 72, 78, 88, 104, 111, 120, 128, 135, 141), as are links to old exams (12, 32, 34, 39, 40, 72, 88, 112, 113, 141) and current grade information (13, 39, 40, 111–113, 134, 141). The lecture material included can be in the form of HTML text and graphics, portable document format (PDF) files of actual class handouts (2, 15, 40, 111, 112, 127, 128, 141), or Microsoft PowerPoint® presentations (Figure 1; see color insert) (20, 141). Audio (20) and video clips (15, 121) can also be added to Web sites using products such as RealAudio (Figure 1; see color insert) (107), RealVideo (107) and Quicktime (4). If desired, portions of course Web sites can be password protected to limit access to enrolled students (22, 23, 39, 70, 88, 103, 127, 137).

The Web site components discussed to this point merely provide students greater access to material and information that was likely presented in class. However, a course Web site can also act as a gateway for students to access additional information and resources. For example, links to other Web sites containing information relevant to a lecture topic are an important feature of many course Web sites (84, 103, 105, 106, 111, 113, 141). Interactive quizzes or exams can be posted for students to use for practice and review (15, 88, 111). These quizzes can be repeated as often as desired and generally give students immediate feedback on their performance and can even suggest lectures they should review. Online laboratory tutorials (5, 61, 109, 121) and simulations (56) can also be included on a course Web site and are particularly useful for courses lacking a true laboratory component.

An example of laboratory experience that can be implemented through the Web is interaction with researchers in remote field locations. Using a satellite digital phone, a laptop computer, and a digital camera, a researcher can share on a daily basis his or her field experiences with students and involve them in the decision making process (26, 93). On a less elaborate scale, a course Web site can be used to increase student involvement and interest by posting photographs of class members on field trips and in other class activities (40, 127, 128, 140). Students themselves can also be required to post projects on a course Web site to be critiqued by other students and ultimately to be published (29, 140).

Online Textbooks Textbooks are expensive, soon out of date, and often include unwanted sections. Though currently limited to larger markets, publishers are beginning to produce electronic texts that may solve some of these problems. These texts are easily updated regularly and instructors will have the ability to tailor the text to course needs by ordering specific sections (115).

Instructors can develop their own Web-based texts in a manner similar to the way study packets have been put together for years. Online texts can also be developed collaboratively by organizing chapters submitted by experts into a cohesive text such as Ratcliffe's IPM World Textbook (106).

Interaction Through E-Mail and Discussion Groups Though the Web is the most visible Internet tool used in on-campus teaching, e-mail is the most common. E-mail facilitates both teacher-student and student-student communication, and can be either in the form of direct messages to specific individuals, or mailing lists (also known as listservs). Mailing lists are a form of e-mail in which every message that is sent to the mailing list is forwarded to all subscribers. Mailing lists are particularly useful for students because students can use them to communicate with the entire class without having to know the e-mail address of every person in the course. E-mail gives students increased access to the instructor in a way that is less intrusive than office visits or telephone calls, and gives the instructor greater access to the students, allowing important announcements to be made prior to the next class time. E-mail is also a convenient and discrete way for the instructor to contact students who need extra attention. Some instructors are going a step further by requiring that assignments be submitted electronically as attachments to e-mail messages (141).

Discussion groups (also called bulletin boards) and chat rooms are two other forms of Internet communication currently used. Discussion groups are a list of posted messages and responses that all participants can view. Chat rooms are Internet sites that several participants log on to at the same time to have live (real-time) typed conversations (Figure 2; see color insert). Discussion groups have the advantage of not requiring that participants be logged on to the site at the same time. They also provide an organized record of the past and ongoing dialog. While discussion groups tend to elicit diverse topics and thoughtful responses, chat rooms seem to encourage a greater volume of discussion in number of participants and in variety of responses. Chat rooms also elicit fewer questions about course logistics than discussion groups (58).

Student Interaction with Alumni An interesting use of the Web and e-mail is to promote contact with former students through an alumni career bank. An alumni career bank is a Web page listing the alumni of a department or school along with current positions and e-mail addresses. Because students have a strong interest in career opportunities, the application and interview process, and details of the professional world, they are often eager to correspond with those working in the field they wish to enter. Alumni career banks also facilitate contact between and among teachers and alumni and allow recruiters to find qualified individuals (9).

Concerns About Internet Use in On-Campus Courses

Despite the numerous advantages of the Internet that have been discussed, some concerns and limitations exist.

New Tools The biggest concern is that instructors often adopt Internet tools for the sake of change, giving too little thought to instructional needs, benefits, and costs. Faculty purchasing Internet hardware and software tend to choose cutting-edge products hoping to extend their useful life. Unfortunately, this means the tools are generally unproven, possibly of marginal worth, or unstable. Even if the tools are effective and stable, students may avoid using them simply because they do not want to bother learning to use new tools (35). While rewards for pioneering the use of new technologies exist, a general rule of thumb is to use evolutionary rather than revolutionary tools, those that have become stable, reliable, and affordable (35).

Time Sink Instructors adding novel Internet tools to an existing course may quickly realize the new tools take up time with few tangible rewards (28). Internet time sinks include maintaining course Web sites and replying to e-mail messages, because face-to-face exchanges of information and ideas can occur much more rapidly than keyboard conversations (80). If Internet tools are simply added to an existing system, costs will increase. However, if the entire teaching approach is modified to encompass and rely on the strengths of the Internet tools, costs can be reduced (29, 35). See Ehrmann (35) for an extensive analysis of the costs of Internet implementation.

Elusive Benefits One intuitive conclusion about the use of e-mail and discussion groups is that students who would not normally participate in class or seek out an instructor to ask questions will do so through e-mail. One study, however, has shown that e-mail access may do little to encourage the involvement of less interactive students (28). Many of these students, in fact, considered required e-mail and discussion group activity as unnecessary additional pressures. Those who took advantage of the increased access to instructors were those who normally would have otherwise (28).

Benefits of Moving Ahead Despite these problems, there are reasons to adopt Internet tools. The most compelling reason is the instructor's responsibility to prepare students to enter today's dynamic work environment. In the past instructors attempted to dispense entomological knowledge while simultaneously teaching the skills needed to function successfully in more advanced courses or in the workplace. These skills included strategies for gathering relevant material from textbooks, methods for conducting research in a library, and techniques for writing a laboratory report or research paper. Today's students need a modified set of skills that will allow them to extract knowledge from the information avalanche enveloping them and to communicate through electronic media. Presently, these skills include the ability to search the Web (including online literature databases), communicate through e-mail, and publish material as Web pages. To meet these new needs, instructors must be using the technology themselves and incorporate the use of these tools into their curricula (35).

Fortunately, preliminary research indicates that the integration of the Internet, particularly the Web, has a positive impact on learning when compared to traditional courses (31, 80).

DISTANCE EDUCATION

What is Distance Education?

Distance education has a long history that most of us are at least somewhat familiar with, generally in the form of correspondence courses. From the late 1800s to the present, correspondence courses have been offered in which students receive and submit assignments and communicate with instructors through the mail (92). In most cases, the learning material provided was in the form of written material and later through audio tapes, videotapes, and TV broadcasts. Two-way video conferencing is a relatively recent advance in distance education, in which the instructor's lecture is broadcast live from a studio classroom to a remote classroom. The remote classroom contains monitors for viewing the lecturer and students in other remote classrooms, and a microphone and video camera for communicating with the instructor. The advantage of video conferencing is that students are able to see, hear, and speak to the instructor.

With the advent of the Internet, a dramatic new medium for delivering distance education has become available: cybercourses. Through the Web, instructors can now electronically distribute all material types previously available, including text, images, video, and audio, while maintaining ability to choose when and where they view them (in other words, asynchronous learning). Most importantly, the Internet facilitates efficient communication between students and instructors and between students themselves.

Today distance education is fast evolving from an obscure education novelty to one of the most exciting topics in higher education, due largely to the ability to deliver courses over the Web. As of 1997 an estimated 55% of U.S. four-year colleges and universities had distance education courses (52). In that same year approximately 1 million of the 14 million students in higher education were enrolled in cybercourses, with three million expected by the year 2000 (52). It has even been suggested that universities will become obsolete and students will soon be able to take courses from any institution (52, 138).

Why Distance Education is Growing

Some of the reasons for the current emphasis on distance learning are legitimate and should be taken seriously by entomology educators. Other reasons are less valid but need to be addressed because their influence and impacts are real and important.

Technology-Driven Need In some respects, Internet technology itself is driving the move toward distance education. The popularity of distance education among administrators and the glamor of cybercourses are creating a rush of faculty wanting to stake their claim in this new education territory (36). There may, in fact, be some basis for this concern regarding territory. The barriers of education are disappearing due to the growth in distance learning. These include not only distance barriers but also institutional barriers. For example, the Western Governors University (WGU) (136) is an alliance of over 30 universities and colleges that have agreed to contribute courses. To earn their degrees, WGU students living anywhere in the world can choose and complete courses from any of the participating schools. The Tri-State Agricultural Distance Delivery Alliance (123a, 132) is a similar organization. As institution barriers disappear, all students—not just those at prestigious schools—will be able to take courses from the best educators in a particular field (52, 138).

Another aspect of this competitive market is the development of cybercourses by professional education companies. For example, the Florida Community College system recently announced the purchase of a commercially produced introductory chemistry cybercourse that will be used by all the schools in the state (41).

Nontraditional Students Another driving force behind the distance education movement is the changing demographics of those seeking the services of higher education. In 1972, only 28% of students attending U.S. colleges and universities were over 25 years of age. By 1994 the percentage had increased to 41% and continues to grow (119). Reasons for this change include transformations in the global marketplace that are requiring many to return for more schooling in order to remain competitive (16, 35, 138). Many of these students have family and work responsibilities that prohibit them from relocating and attending school in the manner of a typical 18- to 22-year-old-student.

This growth in non-traditional student populations is exemplified in a survey of 649 agricultural science graduates in which 69% said they were interested in taking distance education courses. Of these same graduates, 82% said they were interested for professional development, 50% for advanced degrees. Most of those interested were older than 30 and married (98).

Possibly the greatest need for distance education courses and degrees is at the graduate level (16). In a survey of more than 350 current and potential graduate students, 65% said that finances and work schedules were major obstacles to attending graduate school, and 55% said that distance education courses would make attending graduate school more attractive (10).

Money The final reason for the growth of distance education is the perception that it will solve the two-headed problem of diminishing funds and increasing enrollments, which most schools face (138). The idea that a large number of students could be served in an automated fashion without additional building space is an enticing prospect to administrators. As a result most institutions of higher learning are strongly encouraging the development of distance education courses, and incorporating and relying on distance education in their long-range planning (138). Unfortunately, the current realities of distance education do not completely meet the grand expectations described.

How It Is Being Done

Distance education courses are being delivered over the Web through a wide range of methods. While the total number of higher education cybercourses is in the thousands, the number of entomology cybercourses is extremely small. On a scale of complexity of courses offered in a wide range of disciplines, the entomology courses currently offered range from very basic to moderately complex.

Simple Structure The most basic of these courses present the material as large amounts of text and simple graphics (11, 19, 99). The material in these courses is organized in a linear sequence of Web pages and links that structure the information in a manner similar to that of a textbook. More complicated are courses that deliver material, as video transmissions of actual lectures, streamed over the Internet (30, 44, 73, 118). Another moderately complicated technique is to translate Microsoft PowerPoint slides into HTML Web pages, record the audio portion of the lecture, and attach it to the corresponding slides (Figure 2; see color insert) (20). These courses of simple structure are straightforward and can be implemented relatively quickly because they maintain the linear sequence of the traditional course and much of the original course material can be recycled.

Advanced Structure Courses that utilize both advanced structure and technology have been developed in other disciplines (33, 50). These courses incorporate advanced media and present material in a manner that allows students to explore the material in a more investigative, less linear manner. This approach exploits the Internet's strengths and encourages students to be less passive and more active in the learning process (117). Such courses require the greatest amount of time to develop because much more effort must be put into producing the graphics and developing the ways students will navigate the material. In addition, little if any of the text or graphics of the traditional course can be used because the structure and visual content of the course is unlike that of a traditional course.

Cybercourse Tools Several new software tools aid in the development of cybercourses: Web CT (47), Asymetrix Toolbook II® (7), Topclass® (133), and Course in a Box® (46) are a few examples. These products help the user develop and organize the course, keep track of the time each student spends on each page, and record practice exam scores. The products also have built in e-mail and discussion groups, online glossaries, practice exams that provide immediate results, and areas where students can place and store online assignments, including student-produced Web pages (Figure 3; see color insert). These programs can also limit course access to registered students. Many institutions are adopting one or more of these programs by purchasing a site license so that all instructors have free use of the product, technical support, and server space.

Concerns and Benefits

Concerns Distance education is not for everyone. While highly motivated nontraditional students tend to do well in distance education settings, less motivated younger students often flounder without the encouragement and discipline imposed by a physically present instructor (115). This problem may be exacerbated if unmotivated students are lured to distance courses by the apparent novelty, prospect of entertainment, and freedom of cybercourses. On average, distance learning courses have a dropout rate between 30% to 50%. Contributing factors to this dropout rate include educational background, bad experiences with prior distance education courses, employment, and family responsibilities (92).

Another area of concern is distance learning's impact on those whose learning styles may be poorly suited for the method. However, with appropriate instructional design, and through the use of a variety of media and instructional techniques, a cybercourse can effectively target many different learning styles (90).

Some expectations of cybercourses may be unrealistic—specifically, that they are an inexpensive answer to increasing enrollments, that learners become interactive in the process of learning, that repeated use of a course can reduce staff, that two-way video equals face-to-face contact, and that most distance learners have access to computers and the Internet (36). Some predict that the costs of infrastructure staff and faculty time alone will keep distance education from replacing today's universities as we know them (115). While simple cybercourses are relatively easy to develop, interactive courses that use the medium to its best advantage are extremely time consuming, often requiring a team of developers and support staff (58). Though costs may be reasonable when compared with the costs of classrooms and buildings, the endeavor is larger than simply acquiring server space and some software. The main limiting factor is the time it takes to interact electronically with students, i.e. e-mailing, receiving and grading assignments, and tracking progress. Class sizes generally need to be smaller than traditional courses to allow for such time-consuming communication (80).

Most faculty members feel they lack the skill necessary to use appropriate distance learning methodologies, though they believe that using electronic technologies would enhance their teaching (95). They also perceive little incentive to take the steps needed to acquire these skills and as a result fail to adopt and support distance education technologies and methodologies (94)

One practical concern is whether accreditation agencies, other universities, or employers will consider distance education courses and degrees equivalent to traditional degrees (115). On a more philosophical note, some worry that attempts to make learning convenient, entertaining, and full of constant reinforcements are not in the long-term best interest of students. Jobs in the real world are often inconvenient, repetitive and lacking in external rewards. Distance learning may produce happy, technically savvy students that are poorly trained to be persistent and determined workers who succeed even when the work lacks fun or convenience (71).

Benefits Despite all concerns, the need for distance education is undisputed and growing. The recent advances in Internet technology and availability have produced a medium that ideally meets the needs of distance learners (90). In addition to being asynchronous, cybercourses allow the utilization of all currently available media and provide an effective means of communication.

Through distance education the potential market for a course is dramatically increased (115). Consequently, enrollment can be enlarged in courses that might otherwise be canceled, and the impact and awareness of a program can be enhanced (91). Courses can also be offered to students attending schools that lack expertise in a particular area. Online, students can access experts from other schools and even private industry (74).

In most cases the effectiveness of distance education courses has been shown to equal that of traditional courses (110), and exceeds them in some instances (80). Tools are available for evaluating distance education (116) and course Web sites (123) that can help faculty ensure that the materials produced will have the desired results.

IMPACT ON RESEARCH

The Web will affect research in diverse and unpredictable ways, but the most profound effects will probably be in the areas of communication, databases, and literature.

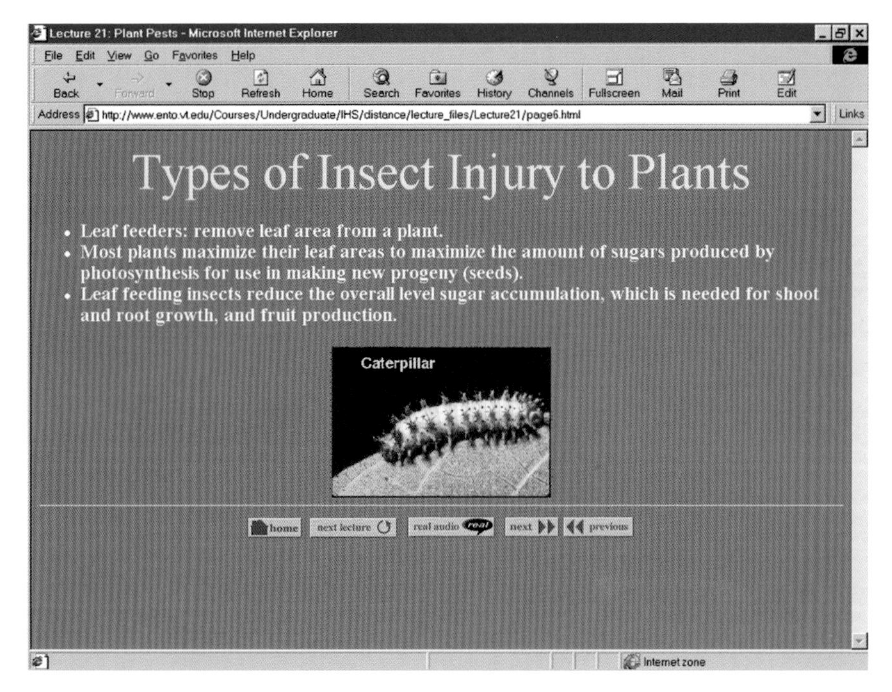

Figure 1 Example of a cybercourse utilizing transformed Microsoft PowerPoint slides and RealAudio lecture narration (see also 20).

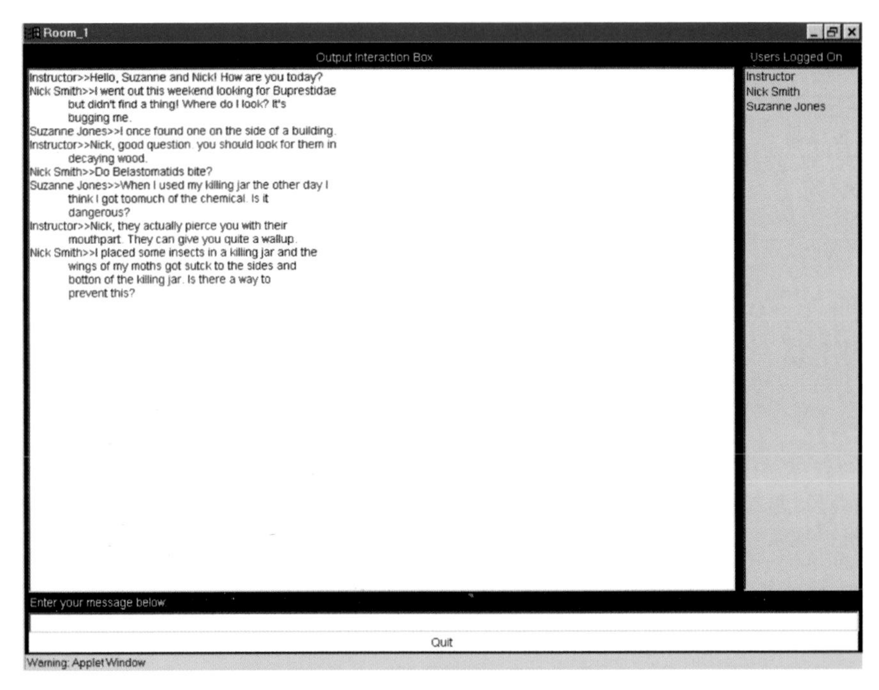

Figure 2 Example of a three-person chat room conversation.

Figure 3 Example of tools available using the Web-based course management software WebCt.

Researchers Will Increasingly Communicate Via the Web

E-Mail and FTP E-mail is quick and informal. Compared to oral communication, it encourages concise, well-thought-out statements. Because e-mail gives direct access to scientists of the highest reputation, researchers have a much broader choice of those they can informally ask for advice or information than they have had in the past, and scientists at smaller institutions and in developing countries are at less disadvantage than before. E-mail attachments permit near instantaneous transfer of manuscripts, spreadsheets, and some images. File Transfer Protocol (FTP) supplements e-mail attachments by quickly and cheaply transferring very large files, such as high-resolution images and large databases. E-mail and FTP help researchers in widely separated locales to pursue joint projects.

Discussion Groups E-mail or Web-based discussion groups enable researchers to keep up with and contribute to new developments in their areas of interest. The Entomology Index of Internet Resources (125) lists more than 60 mailing lists likely to be of special interest to entomologists. Mailing lists that do not focus on insects may also be important to insect researchers (e.g. bioacoustics, molecular biology, and museum collections). The archives of Entomo-L, the most-used general entomology mailing list, are on line from 1995 to date with the full text of 15,000 postings searchable by any word, phrase, or boolean combination thereof (124).

Internet Conferences and Meetings The viability of scientific meetings without airline tickets or registration fees is illustrated by the Fifth Internet Congress for Biomedical Sciences (63). During this 10-day, online meeting there were 580 presentations by 1555 authors, a plenary session, invited symposia, poster sessions, discussion boards for each session, and 362,000 requests for downloads from the site. As bandwidth on the Internet increases, the ease and quality of video conferencing will improve, allowing researchers everywhere to interact in real time.

Researchers Will Increasingly Establish Cooperative, Publicly Accessible Databases

Some entomological research depends on access to large databases to which many contribute. The cost of posting large databases on the Web is minimal, but the requirement of time and money to organize, establish, and continually update a cooperative database is significant. Funding such databases and distributing the work and the credit will remain a challenge. Molecular biology and systematics are two research fields that illustrate the value of such databases.

Germplasm and Genome Databases Molecular biologists, including those working with insects, are maintaining numerous Internet databases (25). The biggest of these is the International Nucleotide Sequence Database Collaboration (65), a cooperative effort of GenBank (sponsored by NIH), the European Molecular Biology Laboratory, and the DNA DataBank of Japan. These three organizations receive nucleotide sequence data, exchange data daily, and make them publicly available and searchable on their World Wide Web servers. As of December 1998, the database included 3,044,000 sequence records and 2,162,000,000 bases (45). Flybase (51) is a Web-posted genomic database of *Drosophila melanogaster,* genetically the best known of all complex organisms. In addition to chromosomal DNA base sequences, Flybase includes cytological maps, genes, structure and function of gene products, and sources of stocks and clones.

Phylogenetic and Taxonomic Databases Systematists are beginning to use the Web for their databases. The Tree of Life (81) is a distributed cooperative database for documenting evolutionary relations and biodiversity of all organisms. Some of the arthropod branches are already well leafed out (82) while others are nearly bare (e.g., Lepidoptera).

Taxonomic catalogs are especially suited to Web posting. They need continual updating and paper copies have a scant market. The Orthoptera Species File (102), the first Web-posted catalog for any major animal or plant group, contains full synonymic and taxonomic information for more than 25,000 species and genera, as well as pictures and calling songs for many species. More than 100 lesser taxonomic sites are identified by VanDyk and Bjostad under the categories of checklists, databases, and images (125).

Research Literature Will Move to the Web

The low cost of Web publishing promises free access to all research literature for which authors receive no royalties. This access will allow all the journal literature and most review articles to become a seamless Web library open all the time to almost everyone almost everywhere. Research libraries will no longer be intermediaries in the distribution of most research literature, and every researcher will have convenient access to a larger percentage of needed literature than researchers at the largest institutions have today.

Journals Access to journal articles via the Web is more convenient and certain than requesting reprints or going to a research library, finding issues, and making photocopies. Posting articles on the Web need cost, at most, a few dollars per page. However, three circumstances ensure that publication of paper issues will not soon be abandoned: *(a)* no system of archiving digital data has been certified as safe for all time (55, 57), *(b)* researchers are reluctant to abandon a familiar system that is not obviously broken, and *(c)* journal publishers fear that the end of paper will shrink or end their profits.

The first circumstance has not impeded governments, corporations, and research libraries from using digital formats for their most important data, and it will not stop journal publishers from doing the same. Making perfect copies of digital information is easy and cheap, and moving the information from an obsolescing digital format to a newer one is not difficult unless the migration is postponed too long. The second circumstance is of even less concern. Because Web access costs so little, the traditional access system need not be abandoned until an all-electronic system has near unanimous support.

The third circumstance is the most important. Publishers' profits are high and depend on researchers insisting that their libraries subscribe to journals reporting research in their areas. These same researchers sign away copyright to their articles, enabling publishers to charge whatever the market will bear for journal subscriptions. This practice has led to journal prices increasing far more rapidly than inflation and to a long-running "serials crisis," in which libraries are able to pay for fewer and fewer of the journals their researchers desire (21). Electronic distribution of articles makes it harder for publishers to control access and eliminates the need for library subscriptions—journal publishers' greatest source of revenue (129).

Parallel Publication Journal publishers realize that the end of central printing is inevitable and are preparing for it by publishing electronic versions in parallel with the traditional paper versions. This approach increases costs, and publishers are trying various means to profit from the electronic versions. For example, Blackwell Science, which publishes the six journals of the Royal Entomological Society, charges approximately 10% more for a combined online and paper subscription than for paper alone, and for the online version alone, approximately 90% of the price of a paper subscription (14). The Entomological Society of America started putting its four principal journals online with the 1999 issues (38), and now charge members a $15 surcharge for electronic access only, the same as for paper only. On the other hand, Web access to *Bulletin of Entomological Research* is free to those who subscribe to the paper version (17) and the Florida Entomological Society has made the online version of *Florida Entomologist* freely accessible to all since 1994 (42).

Two Views of Cost Recovery in the All-Electronic Future There is no consensus on how publishing costs (editing and composing) will be paid for in the all-electronic future (54, 100, 101, 130). Two competing views of the future are illustrated in Figure 4. One depends on maintaining the revenue sources of the current system, by selling online subscriptions to individuals and site licenses to research organizations. Those who lack access by either of these means would be denied access or would have to pay for individual access. As an example of the latter, users who do not subscribe to the online version of *Science* (1) can pay $10 for 24 hours of access or $5 for access to a single article. The other view is that authors or their institutions or research grants will pay the costs prior to publication, allowing unrestricted access to articles.

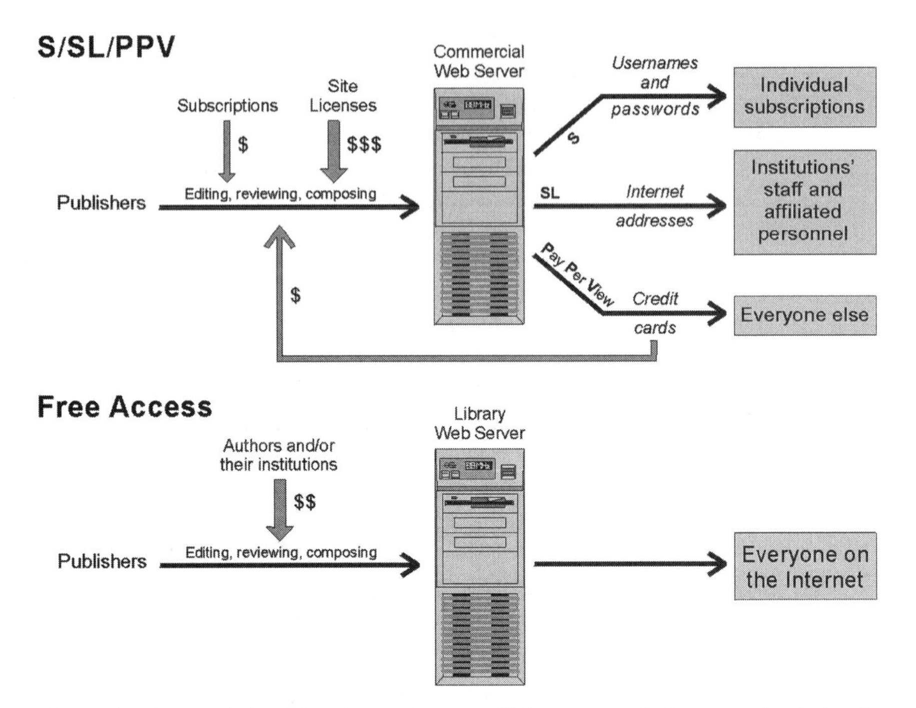

Figure 4 Two models of how journal costs will be recovered after central printing is abandoned in favor of all-electronic distribution of articles. The Subscription/Site License/ Pay-Per-View (S/SL/PPV) model requires that users pay and attempts to replace library-subscription revenues with site-license revenues. The free-access model requires that publication costs be collected prior to publication.

Restricted access will enable publishers to maintain or increase their profits (100), and libraries rather than authors to pay most of the costs of the journal system. However, the cost of free access is low and the advantages to authors and their sponsors are great.

Why Access Will Be Free At least seven forces favor eventual free access to Web-posted journal articles: *(a)* free access is less expensive than restricted access, which is complex and therefore expensive to implement (Figure 4); *(b)* administrators of research institutions will want the savings in library costs that free access yields. A small portion of the savings would be enough to pay all publication costs (100); *(c)* those who support research (mostly federal agencies and private foundations) prefer that access to the results of the research they fund be free, and they may forbid or discourage their grantees from transferring copyrights to publishers (8, 105a)—U.S. federal workers have never been allowed to do so; *(d)* researchers will seek the benefits of easy access to any current article without having to contend with the costs and delays of tollgates; *(e)* as long as traditional and e-versions of articles are co-published, members will pressure their societies to grant free access to the e-versions. For example, they will claim the right to post their manuscripts as preprints (e-prints) on their own or discipline-based Web servers (37, 83). They will ask for the option to purchase immediate, unrestricted posting of PDF files of their refereed articles (electronic-reprints), and that PDF files of all articles be posted toll free a few years after initial publication (129–131); *(f)* authors will chafe at restrictions on Web access to their articles. Some will ignore the restrictions and substitute the refereed version for the manuscript version on preprint servers (54) and most will welcome their institutions' efforts to make the terms of copyright transfers more favorable to the authors' interests; *(g)* restricted access makes publishers rather than libraries responsible for journal archives. Researchers trust libraries more than publishers to preserve their research results.

Journals Will Change in the All-Electronic Future When journals no longer need to conform to centrally printed issues, they will be free to fully exploit the advantages of the Web. Color, sound, and video will be used. Publication will no longer be delayed by queues or the wait for issues to fill. The length of an article will not matter, as long as reader friendliness is maintained. Varian (126) suggested that a typical Web-published article might consist of a one-paragraph abstract, a one-page executive summary, 20 pages of article, and a 50-page appendix. Interactivity will be great, with internal and external links to references and databases. Articles will be retrievable by searches of selected fields or the full text.

Back issues will be put on line. The cumulative costs of maintaining access to back issues of journals are so large that posting them on the Web will have fiscal as well as convenience benefits. The techniques, pioneered by Journal Storage (JSTOR) (69), include scanning the pages and reading them with optical-character-recognition software to allow automatic indexing of the full text. The scanned pages are posted on the Web and articles and pages are retrieved via online searches of the indexes. Any posted article can be found by any constituent word, phrase, or combination thereof, and a copy can be printed that is equivalent to a photocopy of the original article. The economics of this process are illustrated by the back-issue project of the Florida Entomological Society (131). Scanning and indexing the 20,000 pages of *Florida Entomologist* from 1917 to 1993 cost less than $12,000 (less than 60 cents per page). These one-time costs are no more than the estimated annual costs for 100 research libraries to maintain less convenient, less searchable access to paper versions of the same 20,000 pages. (If the 20,000 pages are divided among 40 500-page volumes and 100 libraries shelve the volumes at an operating cost of $3 per volume per year (6), then $40 \times 100 \times \$3 = \$12{,}000$.)

Other Research Literature Most theses and dissertations will soon be posted on the Web. Fifty-seven universities, including 11 land grant institutions, are in the forefront of this effort as members of the Networked Digital Library of Theses and Dissertations (NDLTD) (97). Virginia Tech, headquarters to the NDLTD, has required submission of its theses and dissertations in digital form since 1997 (139). Most are posted on the Web with free access, but authors can choose to limit access or to keep their work entirely off line.

Review articles, which often appear in expensive, late-appearing symposium volumes, will increasingly move to the Web because authors of the chapters receive no royalties, and they want their efforts published quickly and available to all. Current literature indexes will be largely replaced by Web-based search systems more powerful than those currently used by free Web services such as Alta Vista and HotBot (114).

ACKNOWLEDGMENTS

We are grateful to K. Jones (University of Florida) for her editing and research assistance, and J. VanDyk (University of Iowa) for his assistance and advice.

LITERATURE CITED

1. AAAS. 1999. *Science Online.* http://www.sciencemag.org
2. Adobe Systems, Inc. 1999. *Adobe Acrobat.* http://www.adobe.com/prodindex/acrobat/main.html
3. Appel AG. 1999. *Urban Entomology.* Auburn Univ. http://www.ag.auburn.edu/dept/ent/courses/urbanent.html
4. Apple Computer, Inc. 1999. *QuickTime.* http://www.apple.com/quicktime/
5. Ariz. Univ. 1999. *The Biology Project, An Online Interactive Resource for Learning Biology.* Univ. Arizona. http://www.biology.arizona.edu/
6. ARL. 1999. *Association of Research Libraries statistics and measurement program.* http://www.arl.org/stats
7. Asymetrix Learning Systems, Inc. 1999. *Toolbook.* http://www.asymetrix.com/
8. Bachrach S, Berry SR, Blume M, von Foerster T, Fowler A, et al. 1998. Intellectual property: who should own scientific papers? *Science* 281:1459–60. http://www.sciencemag.org/cgi/contents/full/281/5382/1459
9. Barkely AP. 1998. Bridging the distance: linking current students with alumni via the Internet. *NACTA* 42:46–49
10. Belcher MJ. 1996. *A Survey of Current and Potential Graduate Students.* Boise: Boise State Univ.
11. Bell PD. 1999. *Forest Entomology and Control.* Sir Sandford Fleming College. http://gaia.flemingc.on.ca/~pbell/entom.htm
12. Berenbaum M. 1999. *Insects and People.* Univ. Ill. http://www.life.uiuc.edu/Entomology/105
13. Bjostad LB. 1999. *Insect Behavior.* Colo. State Univ. http://www.colostate.edu/Depts/Entomology/courses/en507/outline.html
14. Blackwell Science. 1999. http://www.blacksci.co.uk/uk/journals.htm
15. Brandenburg RL. 1999. *Turf and Ornamental Entomology.* NC State Univ.

http://www.cals.ncsu.edu/course/ent063/e63-home.html
16. Brazziel WF. 1993. *Shaping graduate education's future: implications of demographic shifts for the 21st Century.* Presented at the Annu. Conf. Can. Soc. Study Higher Educ., Ottawa, Ont.
17. CABI. 1999. *Pest CABWeb journals online.* http://pest.cabweb.org/howgetin.htm
18. Caron DM. 1998. *Elements of Entomology.* Univ. Delaware. http://udel.edu/~21632/ento205.htm
19. Carrasco JV. 1999. *Morfología de Insectos.* Inst. Fitosanidad Col. Postgrad., Mex. http://www.geocities.com/CollegePark/Classroom/9932/
20. Carroll B, Mack TP. 1999. *Insects and Human Society.* Va. Tech. http://www.ento.vt.edu/Courses/Undergraduate/IHS/
21. Case M. 1998. ARL promotes competition through SPARC: the Scholarly Publishing & Academic Resources Coalition. *ARL Newsl.* Vol. 196. http://arl.cni.org/newsltr/196/sparc.html
22. Clark WE. 1999. *Entomology for Educators.* Auburn Univ. http://www.auburn.edu/~clarkwe/501.htm
23. Clark WE. 1999. *Insects.* Auburn Univ. http://www.auburn.edu/~clarkwe/204.htm
24. Coats J. 1999. *Insect Pathology.* Iowa State Univ. http://www.ent.iastate.edu/dept/courses/ent675/syllabus.html
25. Cockburn AF. 1998. Insect germplasm and genome databases. *Am. Entomol.* 44:16–99
26. Classroom Connect. 1999. *Quest Products.* http://quest.classroom.com/archive/archive.asp
27. Courtney G. 1999. *Systematic Entomology.* Iowa State Univ. http://www.ent.iastate.edu/dept/courses/ent576/syllabus.html
28. Cravener PA, Michael WB. 1998. *Students use of adjunctive CMC.* Presented at the Annu. Distance Educ. Conf.
29. Dana PH. 1999. The *Geographer's Craft.* Univ. Texas at Austin. http://www.utexas.edu/depts/grg/gcraft/contents.html
30. Danielson S. 1999. Biological Control of Pests. Univ. Nebr. http://ianrwww.unl.edu/ianr/entomol/courses/biocon.html
31. Day TM, Raven MR, Newman ME. 1998. The effects of World Wide Web instruction and traditional instruction and learning styles on achievement and changes in student attitudes in a technical writing in an agricommunication course. *J. Agric. Educ.* 39:65–75
32. Dunkel F, Wessel M, Tharp C. 1999. *Insects and Society.* Montana State Univ. http://scarab.msu.montana.edu/academic/j8102cal.htm
33. Education B. 1999. Spywatch. Br. Broadcast. Corp. http://www.bbc.co.uk/education/lookandread/lar/index.htm
34. Ehler LE, Kaya H. 1999. *Introduction to Biology Control.* Univ. Calif., Davis. http://ucdnema.ucdavis.edu/imagemap/nemmap/ENT135/135syll.htm
35. Ehrmann SC. 1996. *Adult Learning in a New Technological Era.* OECD Proc. Paris: Cent. Educ. Res. Innov.
36. Ely DP. 1996. *Distance education: by design or default?* Presented at Conf. Assoc. Educ. Commun. Technol., Tallahassee
37. Los Alamos Natl. Lab. 1999. arxiv.org e-Print archive. U.S. Dept. Energy. http://xxx.lanl.gov/
38. Entomol. Soc. Am. 1999. *Online Journals.* http://Journals.entsoc.org/
39. Fell RD. 1999. *Bees and Beekeeping.* Va. Tech. http://www.bsi.vt.edu/rfell/ent_2254/
40. Fell RD. 1999. *Insect Biology.* Va. Tech. http://www.bsi.vt.edu/rfell/InsBio3014/
41. Fla. Community Coll. 1999. *Statewide Initiatives: Archipelago.* http://www.distancelearn.org/consortnav/con_frame.htm
42. Fla. Entomologist. 1999. *An Interna-*

tional Journal for the Americas. http://www.fcla.edu/FlaEnt/

43. Foshee W. 1998. *Insecticides in the Environment.* Auburn Univ. http://www.ag.auburn.edu/dept/ent/courses/insecenvir.html
44. Foster JF. 1999. *Host Plant Resistance.* Univ. Nebr. http://ianrwww.unl.edu/ianr/entomol/courses/hpr496d.htm
45. GenBank. 1999. *National Center for Biotechnology Information.* http://www.ncbi.nlm.nih.gov/Genbank/GenbankOverview.htm
46. Godwin-Jones R, Polyson S. 1998. *Web Course In A Box (WCB).* Richmond: MadDuck Technol. http://www.madduck.com/
47. Goldberg MW. 1999. *World Wide Web Course Tools (WebCT).* Vancouver: Univ. B. C. http://www.webct.com/
48. Grant JF. 1999. *Integrated Pest Management.* Univ. Tenn. http://eppserver.ag.utk.edu/EPP530/default.html
49. Grant JF, Windham MT. 1999. *Diseases and Insects of Ornamental Plants.* Univ. Tenn. http://eppserver.ag.utk.edu/EPP410/default.html
50. Gray J, Sengupta D, Wisselink S. 1999. The living Africa. *Think Quest.* http://hyperion.advanced.org/16645
51. GSA. 1999. *Flybase.* Genet. Soc. Am. http://flybase.bio.indiana.edu/

51a. GTE Internetworking. 1998. BBN's Timeline, 1950's to 1990's. http://www.bbn.com/roles/researcher/timeline/index.htm

52. Gubernick L, Ebeling A. 1997. I got my degree through E-mail. *Forbes Mag.* http://www.forbes.com/forbes/97/0616/5912084a.htm
53. Gwinn KD. 1999. *Plant Pathology.* Univ. Tenn. http://eppserver.ag.utk.edu/EPP313/default.html
54. Harnad S. 1998. On-line journals and financial fire-walls. *Nature* 395:127–28. http://www.princeton.edu/~harnad/nature.html
55. Hayes B. 1998. Bit rot. *Am. Sci.* 86:410–15
56. Hedaya M. 1998. *Steady State and Multiple Drug Admin.* Washington State Sch. Pharm. http://virtual.phar.wsu.edu/hedaya/modules/default.htm
57. Hedstrom M, Montgomery S. 1998. *Digital Preservation Needs and Requirements in RLG Member Institutions. Study Commissioned by the Res. Libr. Group.* http://www.rlg.org/preserv/digpres.html
58. Hegngi YN. 1998. Changing roles, changing technologies: the design, development, implementation, and evaluation of a media technology and diversity online course. Presented at the Annu. Meet. Am. Educ. Res. Assoc., San Diego, CA
59. Higley LG. 1998. *Insects, Science, and Society.* Univ. Nebr. http://ianrwww.unl.edu/ianr/entomol/ent108/home108.html
60. Higley LG. 1999. *Insect Ecology.* Univ. Nebr. http://ianrwww.unl.edu/ianr/entomol/ent806/home806.html
61. Howard Hughes Med. Cent. 1998. *HHMI Virtual Lab.* http://www.hhmi.org/grants/lectures/vlab1/frames.html
62. Howe W. 1998. *A Brief History of the Internet. Delphi Internet.* http://www.delphi.com/navnet/faq/history.html
63. Internet Assoc. Biomed. Sci. 1998. 5th Internet World Congr. Biomed. Sci., 7–16 Dec. http://www.mcmaster.ca/inabis98/
64. Deleted in proof
65. INSDC. 1999. International Nucleotide Sequence Database Collaboration. http://www.ncbi.nlm.nih.gov/collab/
66. GTE Internetworking. 1998. BBN's Timeline, 1950's to 1990's. http://www.bbn.com/roles/researcher/timeline/index.htm
67. Johnsen RE. 1999. *Agric. Entomol. Lab.* Colo. State Univ. http://www.colostate.edu/Depts/Entomology/courses/en303c.html
68. Johnsen RE. 1999. Agricultural Pesti-

cides. Colorado State Univ. http://www.colostate.edu/Depts/Entomology/courses/en452.html
69. JSTOR. 1999. *Journal Storage.* http://www.jstor.org
70. Jurenka R. 1999. *Insect Physiology.* Iowa State Univ. http://www.ent.iastate.edu/dept/courses/ent555/syllabus.html
71. Karlen JM. 1994. *Technology and higher education: in debt, inept, and in loco parentis.* ERIC Doc. No. 38004
72. Kaya H, Parrella MP. 1999. *Natural History of Insects.* Univ. Calif., Davis. http://ucdnema.ucdavis.edu/imagemap/nemmap/ENT10/ent10syll.htm
73. Keith D. 1999. *Urban and Industrial Entomology.* Univ. Nebr. http://ianrwww.unl.edu/ianr/entomol/courses/ENT407.HTM
74. Kennedy D, Agnew D. 1998. Utilizing industry experts and interactive video to teach a poultry science course. *NACTA* 42:22-27
75. Knudson DL. 1995. *Molecular Entomology.* Colo. State Univ. http://www.colostate.edu/Depts/Entomology/courses/en575/en575.html
76. Kondratieff BC. 1999. *Systematic Zoology.* Colo. State Univ. http://www.colostate.edu/Depts/Entomology/courses/en424.html
77. Krafsur E. 1998. *Insect Morphology and Evolution.* Iowa State Univ. http://www.ent.iastate.edu/dept/courses/ent572/syllabus.html
78. Kuhr RJ. 1999. *Insects and People.* NC State Univ. http://www.cals.ncsu.edu/course/ent201/
79. Leiner BM, Cerf VG, Clark DD, Kahn RE, Kleinrock L, et al. 1998. *A Brief History of the Internet.* http://www.isoc.org/internet-history/brief.html
80. Lippert RM, Speziale BJ, Palmer JH, Delicio GC. 1998. Evaluating an internet distance learning course on sustainable agriculture. *J. Nat. Resour. Life Sci. Educ.* 27:75–79
81. Maddison DR. 1999. *Tree of Life.* http://ag.arizona.edu/ENTO/tree/phylogeny.html
82. Maddison W. 1999. *Tree of Life: Salticidae.* http://spiders.arizona.edu/salticidae/salticidae.html
83. Marshall E. 1998. Embargoes: good, bad, or 'necessary evil'? *Science* 282:860–67. http://www.sciencemag.org/cgi/content/full/282/5390/860
84. McMullen DW, Goldbaum H, Wolffe RJ, Sattler JL. 1998. *Using asynchronous learning technology to make the connections among faculty, students, and teachers.* Presented at the Annu. Meet. Am. Assoc. Coll. Teach. Educ, New Orleans
85. McPherson B, Camazine S, Frazier M. 1998. *Insect Connection.* Penn. State Univ. http://www.courses.psu.edu/courseweb/courses/?course = ent202_esh1
86. Mediamark R. 1998. *Fall 1998 Cyber Stats Report.* Int. Commun., Inc. http://www.headcount.com/count/datafind.htm
87. Merritt RW, Kaufman MG. 1998. *Biomonitoring of Streams and Rivers.* Mich. State Univ. http://www.ent.msu.edu/courses/ent469/index.html
88. Meyer JR. 1998. *General Entomology.* NC State Univ. http://www.cals.ncsu.edu/course/ent425/
89. MIDS. 1998. *Internet State, January 1998.* http://www.mids.org/mmq/501/pages.html
90. Miller G. 1997. Cognitive style preferences of agricultural distant learners. *NACTA* 41:23–28
91. Miller G, King J. 1994. Taking the distance out of distance education. *Agric. Educ. Mag.* 66:5–6
92. Moore MG, Kearsley G. 1996. *Distance Education.* Belmont, CA: Wadsworth. 290 pp.
93. Mountain Travel—Sobek and World-Travel Partners. 1999. TerraQuest. http://www.terraquest.com
94. Murphy T, Terry H Jr. 1998. Opportunities and obstacles for distance education in agricultural education. *J. Agric. Educ.* 39:28–36

95. Murphy T, Terry H Jr. 1998. Faculty needs associated with agricultural distance education. *J. Agric. Educ.* 39:17–27
96. NC State Univ. 1998. *An Introduction to the Honey Bee and Beekeeping.* http://www.cals.ncsu.edu/entomology/DIRECTORY/203syl.html
97. NDLTD. 1999. Networked digital library of theses and dissertations. http://www.ndltd.org
98. Nti NO. 1998. An assessment of agricultural science graduates' interest in participating in credit courses using distance education. *J. Agric. Educ.* 39:21–30
99. Obrycki JJ. 1999. *Introduction to Insects.* Iowa State Univ. http://www.ent.iastate.edu/ent201/
100. Odlyzko A. 1999. Competition and cooperation: libraries and publishers in the transition to electronic scholarly journals. http://www.research.att.com/~amo/doc/eworld.html
101. Odlyzko A. 1999. The economics of electronic journals. In *Technology and Scholarly Communication*, ed. R Ekman, RE Quandt, pp. 380–93. Berkeley: Univ. Calif. Press. http://www.research.att.com/~amo/doc/eworld.html
102. Otte D, Naskrecki P. 1999. *Orthoptera species file online.* http://viceroy.eeb.uconn.edu/Orthoptera
103. Paulson SL. 1998. *Medical and Veterinary Entomology.* Va. Tech. http://everest.ento.vt.edu/~paulson/medvet98
104. Pedigo L. 1999. *Fundamentals of Entomology and Pest Management.* Iowa State Univ. http://www.ent.iastate.edu/dept/courses/ent376/default.html
105. Pfeiffer DG. 1999. *Arthropod Management in Fruit Crops.* Va. Tech. http://www.ento.vt.edu/Fruitfiles/IPMcourse.html
105a. PubMed Central. 1999. An NIH-operated site for electronic distribution of life science research reports. http://www.nih.gov/welcome/director/pubmedcentral/pubmedcentral.htm
106. Radcliffe EB, Hutchison WD. 1999. *Radcliffe's IPM World Textbook.* Univ. Minn. http://ipmworld.umn.edu/
107. RealNetworks. 1999. *RealPlayer.* http://www.real.com/
108. Reuters Ltd. 1998. Holiday buying hits $5 billion. *PC World News.* http://www.pcworld.com/pcwtoday/article/0,1510,9184,00.html
109. Robertson D. 1999. *Virtual Frog Dissection Kit.* Lawrence Berkeley Natl. Lab. http://george.lbl.gov/ITG.hm.pg.docs/dissect/info.html
110. Russell TL. 1999. The *"No Significant Difference Phenomenon". TeleEducation NB.* http://teleeducation.nb.ca/nosignificantdifference/
111. Salmon S, Roberts A. 1997. *Insects and Human Society.* Va. Tech. http://www.ento.vt.edu/Courses/Undergraduate/IHS/oncampus/
112. Salmon SM. 1999. *Forest Protection.* Va. Tech. http://everest.ento.vt.edu/~salom/ForProt/for4514.html
113. Salmon SM, Stipes RJ. 1999. *Pet and Stress Management of Trees.* Va. Tech. http://everest.ento.vt.edu/~salom/Shade_Tree/ENT4524.html
114. Schatz BR. 1997. Information retrieval in digital libraries: bringing search to the Net. *Science* 275:327–34. http://www.sciencemag.org
115. Schurle B. 1997. What are we going to do with all of this technology stuff? *NACTA* 41:7–11
116. Shrader VE. 1997. *Designing a teacher/course assessment instrument for distance education.* Presented at Annu. Meet. Northern Rocky Mountain Educ. Res. Assoc., Jackson, WY
117. Siegel MA. 1994. Inventing the virtual textbook: changing the nature of schooling. *Educ. Technol.* 34:49–54
118. Siegfried B. 1999. *Insecticide Toxicology.* Univ. Nebr. http://ianrwww.unl.edu/ianr/entomol/courses/toxbrochure.htm
119. Snyder TD. 1997. *Digest of Education*

Statistics. Washington, DC: US Dept. Ed. Natl. Cent. Educ. Stat.

120. Stanley D. 1999. *Insect Physiology, Entomology*. Univ. Nebr. http://ianrwww.unl.edu/ianr/entomol/ent801/ent801home.html
121. Strauss R, Foss J, Kinzie M. 1994. *The Interactive Frog Dissection, An Interactive Tutorial*. http://curry.edschool.virginia.edu/go/frog/
122. Survey CsAIU. 1998. *Cyber Dialogue: Surge in Online Sales of Mainstream Products. Nua Internet Surveys*. http://www.nua.net/surveys/index.cgi?f=VS&art_id=905354394&rel=true
123. Terry RJ, Briers GE. 1996. *Case analysis of a website for an agricultural education course*. Presented at Natl. Agric. Educ. Res. Meet., Cinncinnati

123a. Tri-State Agric. Distance Delivery Alliance. 1999. http://www.aee.uidaho.edu/tadda/tadda.html

124. VanDyk JK. 1999. *Entomo-L Mailing List Archives*. http://www.ent.iastate.edu/mailinglist/archives/entomo-l/
125. VanDyk JK, Bjostad LB. 1999. *Entomology Index of Internet Resources*. http://www.ent.iastate.edu/List/
126. Varian HR. 1997. *The Future of Electronic Journals*. http://www.press.umich.edu/jep/04-01/varian.html
127. Voshell JR, 1999. *Aquatic Entomology*. Va. Tech. Univ. http://everest.ento.vt.edu/~voshell/aquaticent
128. Voshell JR. 1999. *Freshwater Biomonitoring*. Va. Tech. http://everest.ento.vt.edu/~voshell/class97.html
129. Walker TJ. 1998. Free Internet access to traditional journals. *Am. Sci.* 86:463–71. http://www.amsci.org/amsci/articles/98articles/walker.html
130. Walker TJ. 1998. The future of scientific journals: free access or pay per view? *Am. Entomol.* 44:135–38 http://csssrvr.entnem.ufl.edu/~walker/fewww/aecom3.html
131. Walker TJ. 1999. *Web Access to Traditionally Published Journals*. http://csssrvr.entnem.ufl.edu/~walker/fewww/tjwonwww.htm
132. Washington State Univ. 1999. *Extended Degree Programs, Agriculture Courses*. http://www.eus.wsu.edu/edp/home/progcrs/ag.htm
133. WBT Systems, Inc. 1998. *TopClass*. http://www.wbtsystems.com
134. Weller HG. 1996. Assessing the impact of computer-based learning in science. *J. Res. Comput. Educ.* 28:461–85
135. Westerdahl BB. 1999. *Biology of Parasitism—Nematology*. Univ. Calif., Davis. http://entomology.ucdavis.edu/courses/ent110/index.html
136. Western Governors Univ. 1999. http://www.wgu.edu/wgu/index.html
137. York A. 1999. *Insects of Field Crops, Forage, and Stored Products*. Purdue Univ. http://icdweb.cc.purdue.edu/%7Eadarnell/307b/
138. Young JR. 1997. Information Technology: rethinking the role of the professor in an age of high-tech tools. *Chron. Higher Educ.*, Oct. 3, 1997:A26–28 http://chronicle.com/data/articles.dir/eguid-44.dir/06eguide.htm
139. Young JR. 1998. *Requiring Theses in Digital Form: The First Year at Virginia Tech*. http://chronicle.com/data/articles.dir/art-44.dir/issue-23.dir/23a02901.htm
140. Zenger JT. 1998. *Insect Classification*. Univ. Fla. http://bugweb.entnem.ufl.edu/class/
141. Zenger JT. 1998. *Principles of Entomology*. Univ. of Fla. http://bugweb.entnem.ufl.edu/3005/

Annu. Rev. Entomol. 2000. 45:769–793

MOLECULAR MECHANISM AND CELLULAR DISTRIBUTION OF INSECT CIRCADIAN CLOCKS

Jadwiga M. Giebultowicz
Department of Entomology, Oregon State University, Corvallis, Oregon 97331; e-mail: giebultj@bcc.orst.edu

Key Words circadian rhythms, period, timeless, Malpighian tubules, sperm release

■ **Abstract** Circadian clocks are endogenous timing mechanisms that control molecular, cellular, physiological, and behavioral rhythms in all organisms from unicellulars to humans. Circadian rhythms influence many aspects of insect biology, fine-tuning life functions to the light and temperature cycles associated with the solar day. Genetic studies in the fruit fly *Drosophila melanogaster* have led to the cloning and characterization of several genes involved in the mechanism of the circadian clock. Periodic transcription and translation of these clock genes form the basis of a molecular feedback loop that has a "circa" 24-hour period. Rhythmic expression of clock genes in specific brain neurons appears to control behavioral rhythms in adult flies. However, clock genes are also expressed in other tissues, both within and outside of the nervous system. These observations prompted chronobiologists to investigate whether nonneural tissues possess intrinsic circadian clocks, what role they may be playing, and what the relationships are between clocks in the nervous system and those in peripheral tissues. Answers to those questions are providing important insights into the overall organization of the circadian system in insects.

PERSPECTIVE AND OVERVIEW

It does not take a biologist to realize that insects "time" their lives in a very precise way to the 24-hour daily solar cycle: one is much more likely to be bitten by a mosquito at dawn or dusk than at other times of the day, and much more likely to encounter a cockroach at night than during the day. There are many more examples of insect rhythmic behaviors. From larval hatching to adult eclosion, insects tend to limit their ecdyses to a narrow species-specific window of time (118). Female moths emit sex pheromones in a daily rhythm synchronized with the daily fluctuations in males' sensitivity to pheromones (12). Activities such as flight, foraging, and oviposition are usually restricted to certain times of day (for review, see 77, 102).

How do insects measure time? Daily changes in light intensity, temperature, and food availability, as well as interactions within and between species, have

0066-4170/00/0107-0769/$14.00

profound influences on daily rhythms. But chronobiologists have gathered ample evidence from insects, and other organisms, that daily rhythms persist in constant conditions lacking temporal cues from the environment. This is accomplished by endogenous timing mechanisms called circadian clocks. Circadian clocks not only are used to drive daily rhythms but are also involved in measuring day and night length allowing for physiological adaptations to seasonal changes in environment (102). The involvement of circadian clocks in photoperiodism has been reviewed recently (104, 122) and is not discussed here.

Circadian clocks seem to be omnipresent in living organisms, from bacteria (44) to humans (72). A set of features common to endogenously generated rhythms emerged from studies involving unicellular organisms, plants, and animals (26, 50, 78). The characteristics that distinguish endogenous circadian rhythms from other rhythms are as follows:

1. Endogenous rhythms persist (or free-run) in constant conditions maintaining close to a 24-hour circadian period (from Latin: "circa" = about, and "dies" = day). Therefore, endogenous rhythms are called circadian rhythms and the mechanisms that generate them are referred to as circadian clocks, oscillators, or pacemakers.
2. The free-running periods of endogenous rhythms are temperature compensated within a physiological range. While most biochemical processes double their rate with every 10-degree rise in temperature (Q_{10}), the periods of endogenous oscillation have a Q_{10} between 0.85 and 1.1. This temperature compensation is essential for accurate time measurement in the face of seasonal and short-term changes in ambient temperature (102).
3. Circadian rhythms are synchronized, or entrained, by environmental stimuli referred to as Zeitgebers (a German term for "timegiver"). Predictable changes of light intensity associated with the solar day constitute a powerful Zeitgeber that entrains circadian clocks. Shifting the phase of light–dark (LD) cycles will result in a similar phase shift in rhythmic process, so that the characteristic phase relationship between the rhythm and Zeitgeber is maintained. Detailed analysis of entraining factors can be obtained by constructing phase response curves (PRC), which plot changes in the phase of a given rhythm in response to entraining stimuli applied at different phase of free-running circadian cycle (57). For example, a short pulse of light, given to *Drosophila* kept in constant darkness will phase-delay their activity rhythm if applied early in subjective night, but will phase-advance the rhythm if applied in the second half of subjective night (79).

The nature of mechanisms generating circadian rhythms has intrigued scientists for many decades. First evidence for a genetic basis of circadian clocks was obtained nearly 30 years ago, when mutagenesis of *Drosophila melanogaster* (61) and the fungus *Neurospora crassa* (30) resulted in organisms with free-running circadian periods substantially different from the normal 24-hour period. These pioneering efforts were followed by the identification of essential clock genes in organisms ranging from bacteria to mammals. Timing mechanisms in all living organisms appear to involve molecular feedback loops such that transcriptions of clock genes are inhibited by their own products (22) Within the last two years it became apparent that insects and vertebrates use homologous genes in molecular loops constituting circadian clocks. These exciting new developments were recently reviewed from many different perspectives (e.g. 23, 24a, 89, 127).

Insects feature prominently in molecular chronobiology due to the powerful genetic manipulations that are possible in the fruit fly, *Drosophila melanogaster.* Genetic studies of *Drosophila* yielded extensive knowledge about the molecular basis of the fly circadian system (45). Can entomologists use this knowledge as a base to understand the rich repertoire of circadian rhythms in other insects? In this review we explore how data about clock genes and their cellular expression patterns in different organs relate to behavioral, physiological, and cellular rhythms in insects.

INTRODUCTION

The circadian system can be depicted as a cascade of processes leading from the circadian clock through molecular and cellular rhythms to the temporal organization of physiology and behavior (Figure 1). The most evident insect rhythms are those associated with behavior. They include daily rhythms of individual insects such as those affecting locomotion, flight, feeding, oviposition, and once-in-a-lifetime rhythmic events such as hatching, molting, pupation, and adult eclosion (102). Rhythms in physiological processes are not as obvious as behavioral cycles, but they also play an important role in temporal organization of body functions. Examples of insect physiological rhythms include the rhythm of pheromone synthesis (86), rhythm of spermatophore formation (65), rhythm of sperm release from testis (42, 94), and rhythmic secretion of ecdysteroids from prothoracic glands (3, 17) (for expanded references see 9, 102). Physiological rhythms can result from passive responses of tissues to a timing signal, e.g. daily fluctuations in circulating hormones. On the other hand, some physiological rhythms are tissue autonomous and thus likely to be underlined by cellular rhythms.

Rhythms in cellular physiology are well documented in a wide range of organisms. The unicellular dinoflagellate *Gonyaulax* displays a circadian rhythm of bioluminescence that results from underlying rhythms in cellular pH and the number of organelles having bioluminescent reactions (73). Basal retinal neurons of marine mollusks display cell-autonomous circadian rhythms in resting potential (70). Pinealocytes of lower vertebrates show cell-autonomous rhythms in production of melatonin (14). In insects, only a few rhythms in cellular physiology are known. One prominent example is the rhythm of cuticle deposition which arises owing to changes in the orientation of secreted chitin by the underlying epidermal cells (75, 76, 126). These rhythmic activities continue in pieces of

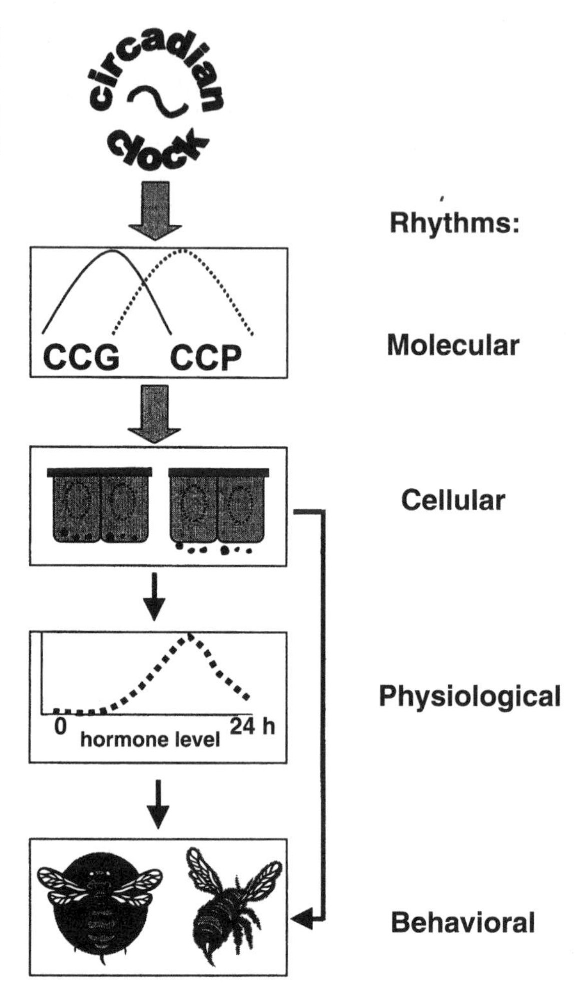

Figure 1 Scheme illlustrating levels at which the circadian clock may interact with biological processes. CCG, clock-controlled genes; CCP, clock-controlled expression of proteins

cockroach epidermis in vitro, providing indirect evidence for the occurrence of autonomous circadian rhythms in the epidermal cells (124). Circadian changes in the ultrastructure of several cell types were also reported: the neurosecretory cells in the cricket brain (24), cells forming optic cartridges in the optic lobes of the house fly (69), and photoreceptor cells of several insect species (8).

Rhythms in cellular physiology are likely the result of daily changes in the activity of certain proteins, which may be regulated via rhythmic transcription or translation. Many rhythmically expressed genes that are relevant to physiological rhythmic functions have been described in bacteria (44) , fungi (66), plants (113), and mammals (115). Among insects, only a few genes showing rhythmic fluctuations in mRNA abundance have been identified in *Drosophila* (95, 121). These rhythms are abolished in flies with nonfunctional clock genes (see below), demonstrating that they are controlled by the circadian clock. Another gene, *lark,* which encodes an RNA-binding protein, is involved in rhythmic eclosion, and seems to be controlled by a circadian clock at the protein level. *Lark* mRNA does not cycle in abundance; however, LARK protein shows circadian oscillations in a group of *Drosophila* neurons involved in eclosion (68).

The cascades of rhythmic processes, from cycling gene expression to behavior, are driven by circadian clocks. Our knowledge about insect circadian clocks is derived mainly from genetic and molecular studies in *Drosophila.* The essential part of circadian timing mechanisms is a molecular feedback loop, which involves several genes.

MOLECULAR MECHANISM OF INSECT CIRCADIAN CLOCKS

Discovery and Characterization of *per* Gene

The first significant insight into the molecular mechanism that generates behavioral rhythms was identification of the gene *period* (*per*). This gene was found in a search for mutations that would change the free-running period of the eclosion rhythm in *Drosophila* (61). Three different alleles of *per* were isolated: per^s produced a short-free-running period of 19 hours, per^l gave a long period of 29 hours, and a null mutation per^0 produced flies that had arrhythmic eclosion. Mutations in *per* not only dramatically changed the periodicity of eclosion, but similarly affected the period of another rhythmic output—locomotor activity. Therefore, *per* was considered part of the central clock mechanism affecting all rhythmic outputs. The *per* gene was cloned in 1984 (6, 87), and extensive molecular characterization of *per* has been accomplished in the last decade. Use of the *per* protein (PER) antibody revealed a circadian rhythm in PER abundance in several clusters of brain neurons, photoreceptors, and many peripheral tissues (109, 132). As a rule (with exceptions: see later), PER is detected in the cytoplasm during early night, and then in the nuclei late at night and early in the day. Levels of *per* mRNA in *Drosophila* heads also cycle, and mutations of *per* change this cycling in a way corresponding to their effects on behavioral rhythms (48). Data suggested that PER may inhibit transcription of its own gene: First, the message is low when PER protein is present in the nucleus and increases only after PER declines below detection level, and second, per^0 mutant shows no *per* mRNA cycling (48, 132). Several independent lines of evidence confirmed the existence of such negative feedback loops (49, 111, 130). The regulatory role of PER in the nucleus was also inferred from its sequence; PER contains a protein dimerization domain termed PAS that is characteristic of a family of transcription factors (55).

Homologs of PER in Other Organisms Molecular characterization of the *per* gene in species of fruit flies other than *Drosophila melanogaster* (16) revealed conserved regions within this gene, especially within the PAS domain. This knowledge facilitated isolation of structural *per* homologues from representatives of other insect orders: several species of moths, including *Manduca sexta* and *Antheraea pernyi* and a cockroach *Periplaneta americana* (90). Analysis of the gene from *A. pernyi* demonstrated that the *per* mRNA and protein are rhythmically expressed in the heads of this silkmoth (99). Further, it was demonstrated that the moth gene is a functional homologue of *Drosophila per*, since the insertion of the moth *per* into the genome of arrhythmic per^0 *Drosophila* can restore the rhythm of locomotor activity (63). In addition to species already mentioned, the *per* gene has been identified in 26 lepidopteran species for the purpose of phylogenetic studies (88) and in bees (117). Thus, *per* appears to be a conserved clock-associated gene in insects.

Recently, homologues of *per* were identified in humans and mice (114, 116), generating considerable excitement among chronobiologists. *per* was expressed in mammalian suprachiasmatic nuclei (SCN), the site of a well-characterized circadian clock, and also in a host of other tissues (1, 108, 133). There are three mouse genes related to insect *per*, and all three are expressed rhythmically, albeit with somewhat different phases (23). The discovery of functional *per* homologues in mammals suggest that molecular elements, which build circadian clocks, may have been conserved throughout animal evolution.

Isolation of Second Clock Gene, *timeless*

Two decades after discovery of the *period* gene, a second clock gene was identified in a mutagenesis screen. This gene, named *timeless (tim)*, was linked to the clock because circadian rhythms of activity and eclosion were abolished in null (tim^0) mutations (105, 123). Analysis of *tim* suggested that it may interact with the *per* gene; evidence included the fact that tim^0 flies failed to exhibit both circadian fluctuations of *per* mRNA and nuclear accumulation of PER protein. Cloning and molecular characterization of *tim* confirmed its essential role in the circadian timing; *tim* mRNA oscillates in phase with *per* mRNA (107), and *timeless* protein (TIM) is detected at the same time as PER in the nuclei of *per*-expressing brain neurons (56). Further, interactions of PER and TIM have been demonstrated using yeast two-hybrid assay (34), and in cultured *Drosophila* cells in vitro (98). Available data suggest that translated PER and TIM proteins accumulate in the cytoplasm as single species. After sufficient buildup of monomers, PER and TIM form dimers, which are then translocated into the nucleus, where they suppress transcription of their genes. Sequences responsible for physical association of PER and TIM have been mapped in both proteins (129). However, neither *tim* nor *per* coding sequences revealed known DNA-binding domains, suggesting the indirect action of this complex in the clock transcriptional feedback loop.

Positive Regulators of *per* and *tim* Transcription

As it appeared likely that other genes participate in the circadian timing mechanism, attempts continued to generate mutations that would make flies behaviorally arrhythmic. Two such mutations were uncovered recently, *jerk* (*jrk*) and *cycle* (*cyc*), both disrupting rhythms of eclosion and locomotor activity (2, 96). Homozygous *jrk* and *cyc* mutants showed decreased *per* and *tim* transcription, suggesting that products of *jrk* and *cyc* may positively regulate *per* and *tim*. This assumption was further supported by data accumulated from studies of the mouse circadian system (35, 59). Analysis of the first mammalian clock gene (named *Clock*) revealed that it encodes a transcription factor containing PAS domain and a basic helix-loop-helix (bHLH) DNA binding domain. CLOCK protein forms heterodimer with another PAS-containing protein, BMAL1 (reviewed in 23).

It turned out that two new clock mutations in *Drosopohila* are homologues of mouse clock genes; the *jrk* locus encodes *Clock* (*dClk*) and the *cyc* locus encodes *Bmal1*. Like their mammalian counterparts, dCLK and CYC proteins contain bHLH DNA binding domain. A DNA sequence (named an E-box) that binds transcription factors containing bHLH domain was identified in the promoter region of *per* and shown necessary for the activation of *per* transcription (4). The role of dCLK in the activation of *per* and *tim* transcription was investigated in *Drosophila* S2 cells in vitro, using transient transfection assays and *per* and *tim* reporter constructs (19). Those experiments demonstrated that dCLK (presumably complexed with constitutively expressed CYC) stimulate transcription of *per* and *tim* reporter constructs, and the E-box sequence on the reporter construct was essential for this stimulation. Moreover, transfection of cells with constructs producing TIM and PER proteins partially prevented stimulation of *per* and *tim* by dCLK. Further biochemical studies of interactions between the clock proteins demonstrated that PER and TIM inhibit the DNA binding activity of CLOCK/CYC complex (62a). Recent analysis of the levels of clock gene mRNAs in different clock mutants has led to the conclusion the *per-tim* negative feedback loop is interlocked with the dCLK loop, which is repressed by dCLK/CYC and derepressed by PER/TIM (4, 43a).

Role of Protein Phosphorylation Progression of a molecular circadian oscillator requires that proteins such as PER and TIM are present through only part of the cycle; therefore, their degradation is an important element in keeping time. It has been shown that PER and TIM are increasingly phosphorylated as the cycle progresses and that multiple phosphorylation promotes turnover of these proteins (25). Recently a new gene, *double-time* (*dbt*), has been isolated that encodes a homologue of mammalian kinase (60) and appears to affect the phosphorylation of PER (81). In *dbt* loss of function mutants, there is accumulation of hypophosphorylated PER, which results in slowing of the oscillator and disturbance of behavioral rhythms. Thus, *dbt* is not a part of the oscillating feedback loop but is essential in the fly timing process as a regulator of a stability of per monomers (see Figure 2).

Entrainment of the Clock Mechanism by Light

Sensitivity of TIM to Light A common feature of circadian oscillators is a shift in the oscillation phase in response to a pulse of light delivered at different times in a free-running cycle in constant dark (dark-dark or DD). Light applied during times corresponding to early night slows down the clock, while light applied toward the end of the night accelerates the clock and consequently, clock-

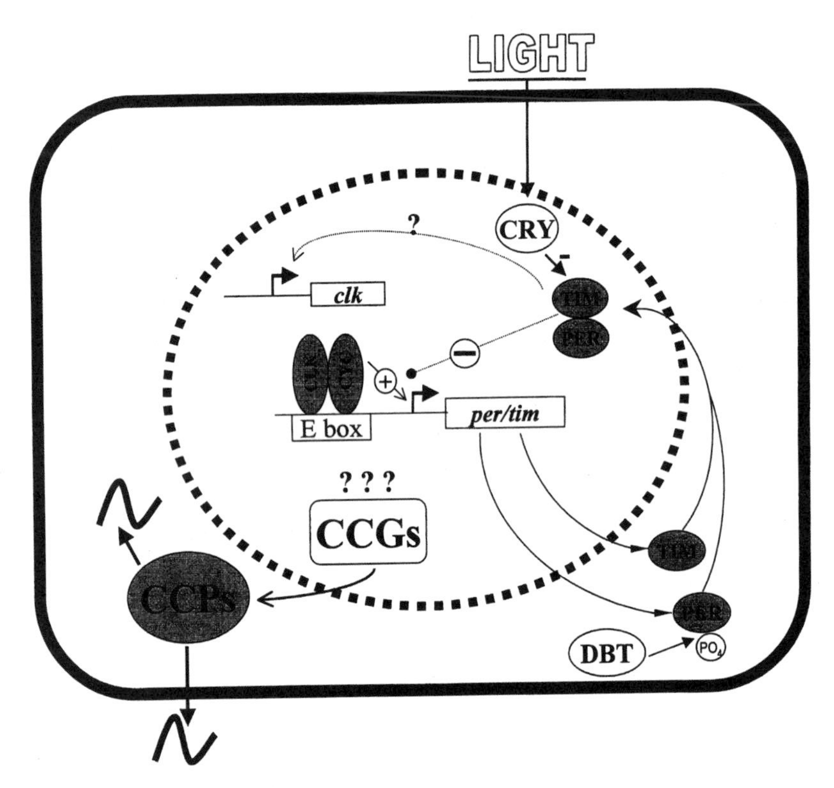

Figure 2 Scheme illustrating current knowledge about molecular feedback loop underlying circadian clock of *Drosophila*. Two positive factors, CLOCK (CLK) and CYCLE (CYC), stimulate transcription of *per* and *tim* genes via binding to E-boxes in their promoter regions. Resulting PER and TIM interact as negative elements with CLK-CYC-E box complex leading to the inhibition of *per* and *tim* transcription. PER-TIM complexes may derepress transcription of the *clock* gene. Levels of TIM are affected by light mediated by CRY. Levels of PER are affected by DBT. *Question marks* symbolize lack of understanding of how the molecular feedback loop is linked to clock-controlled genes (CCGs) and proteins (CCPs) to bring about cellular rhythms.

controlled rhythms. The mechanism of these phase adjustments, consistently observed in organisms from plants to mammals (57), has puzzled chronobiologists. Recently, this mechanism has been partially explained in fruit flies through the discovery that light affects levels of TIM protein (56, 74, 131). Pulses of light, as short as 1 minute cause a decline of TIM to undetectable levels. As a consequence, PER, which is stabilized by TIM, also disappears. If the light pulse is applied in the early night when *per* and *tim* mRNA is still abundant, degraded PER/TIM heterodimers are replaced due to continuing translation of TIM and PER. Prolonged action of these negative elements extends repression of *per* and *tim* transcriptional activity, leading to the delay of the next molecular cycle. The situation is different if the light pulse is applied in the late night, when *per* and *tim* mRNA levels have already declined. At this time, TIM lost due to the action of light cannot be replaced. Premature loss of TIM and PER proteins results in early derepression of *per* and *tim* transcriptional activity leading to the advancement of the next molecular cycle. Importantly, TIM's response to light correlates with the effects of light on behavioral rhythms (128, 129).

The sensitivity of TIM protein to light explains the phenomenon that has been observed in many insects, namely, the disrupting effects of constant light on various rhythms (40, 92; also reviewed in 102). Constant light disrupts the circadian feedback loop by preventing accumulation of TIM (82) and the abolishment of *per* mRNA cycling (85).

Role of Chryptochrome Degradation of TIM protein in response to light occurs in vivo but not in vitro, suggesting that light exerts its effects not directly on the protein, but rather via cellular mechanisms involving photoreceptors. The identity of photoreceptors that mediate circadian entrainment has remained elusive. The hypothesis that entrainment occurs through photoreceptors involved in vision was tested by examining entrainment of genetically eyeless, opsin-depleted, or blind flies. The entrainment of locomotor activity is somewhat affected (128), while the entrainment of eclosion rhythm is unaffected in flies with impaired vision (reviewed in 52). Clearly then, opsin-based photoreceptors used in vision cannot completely account for phase shifting of rhythms by light. Recently, another class of photoreceptor molecules "came to light" as candidates for mediating effects of light on circadian clocks in organisms from plants to mammals (67, 113). These are chryptochromes, blue light–sensitive, flavin-based receptors related by sequence to a family of blue light–sensitive DNA repair enzymes (13). It was an exciting development when a mutation in the gene coding for cryptochrome (cry^b) was identified in *Drosophila* and demonstrated the importance of this gene in circadian entrainment (112). Flies lacking functional cryptochrome failed to reset activity rhythms in response to brief light pulses, and light did not cause degradation of TIM protein in those flies. However, cry^b flies showed activity rhythms, despite that *cry* mutation abolishes cycling of *per* in most tissues! Careful examination revealed that per cycling was retained in a group of brain lateral neurons that are most likely involved in controlling rhythms of activity (see below). This rhythm of PER was abolished along with activity cycles in double mutants lacking both cryptochrome and rhodopsin. Thus, both photoreceptors seem to be involved in maintaining circadian clocks in brain neurons that control rhythms of locomotor activity in flies (112).

Chryptochrome is intimately involved in the proper function of other clock genes (28, 112). Both cry^b mRNA and CRY protein cycle in constant darkness, a condition not necessarily expected from the photoreceptor molecule. Moreover, cry^b mutation prevents TIM cycling in constant darkness, and conversely, clock mutants per^0, tim^0, *Clk*, and *cyc* abolish cycling of cry mRNA (28). A study on the interactions of different clock proteins in heterologous systems suggests that light-activated CRY binds to TIM and prevents PER/TIM complex from exerting negative effect on *per* and *tim* transcription (14a)

Summary of Molecular Data The data discussed above (and schematically presented in Figure 2), represent impressive progress that has been achieved in understanding the molecular basis of circadian clocks in recent years. A complex of CLK and CYC stimulate transcription of *per* and *tim* genes via binding to E-boxes in their promoter regions. As the levels of PER and TIM increase, these proteins dimerize and enter the nucleus where they interact with CLK-CYC-E-box complex, leading to the inhibition of *per* and *tim* transcription. At the same time, PER-TIM complexes derepress transcription of *cyc* gene, which has been repressed by the dCLK/CYC complex. Thus, it appears that the fly circadian clock is composed of two interlocked negative feedback loops, and the interactions of clock elements within those loops are affected by light acting via CRY photoreceptor. A recent report on effects of light and temperature on expression profiles of clock genes (67a) puts us on the road toward understanding how circadian clocks adapt insects to seasonal changes in temperature and day length.

While the time-keeping molecules revealed many of their secrets, we still do not understand (hence question marks in Figure 2) how they regulate the clock–controlled genes (CCGs) to time clock–controlled proteins (CCPs) that ultimately lead to overt rhythms. To search for links between clocks and their outputs, one has to leave the general molecular landscape and focus on specific tissues and cells. As we see in the next section, many tissues in the fly body express *per* and other clock genes. Studies of the spatial and temporal patterns of clock gene expression are important in two ways. First, they help to reveal cellular and physiological rhythms in insect body and, second, they should lead to the understanding of the overall structure of the insect circadian system.

DISTRIBUTION AND ROLES OF CLOCK GENES IN INSECT TISSUES

The "oldest" insect clock gene, *per*, has been the most thoroughly characterized in terms of its spatial expression patterns (97, 109, 132). Activity of *per* in different tissues was studied by immunocytochemistry using antibody against PER protein, by in situ hybridization, and in transgenic flies carrying *per*-reporting constructs. Flies have been transformed with constructs containing the *per* promoter region plus varying length of the *per* coding region fused to the β-galactosidase (64, 110), and with constructs containing *per* promoter fused to firefly luciferase, allowing the monitoring of *per* expression in real time (10). In the latter flies, *per*-expressing tissues produce measurable light when luciferase acts on its substrate, luciferin, which is provided in the fly diet. The sensitivity of this technique is so high that *per* activity can be monitored in individual organs cultured in mediums supplemented with luciferin (27, 80)

All available reporting techniques consistently revealed a broad distribution of *per* mRNA and protein in fruit fly tissues (64, 109). Cycling PER expression was detected in certain areas of the central nervous system (CNS), including photoreceptors, several subsets of brain neurons, and groups of glial cells (58, 110, 132), and also in many peripheral tissues outside of the CNS (38, 51, 80). This wide distribution of *per*-positive cells has puzzled chronobiologists, who expected to find *per* in the CNS, because known clock-controlled rhythms are at the behavioral level. Even more puzzling were reports that cycling of clock genes can occur in isolated peripheral organs in vitro. Below, we provide a survey of insect tissues that express clock genes and/or clock-controlled rhythms to generate a debate on how insects achieve a synchronized circadian timing system.

Expression of Clock Genes in the Central Nervous System

Central Brain Patterns of spatial expression of *per* in the brain of adult *Drosophila* have been the subject of several reports (54, 64, 109, 110, 132). Several clusters of neurons in the brain exhibit cycling of *per* mRNA and protein. Most prominent "clock-positive" groups of cell bodies include two sets of lateral neurons, ventral and dorsal, located at the border between the central brain and the optic lobes. Another distinct group of *per*-positive neurons are located dorsally; one in pars intercerebralis, near the calyces of mushroom bodies, and the other more laterally, near the optic lobes. Many glial cells in the central brain also show cycling expression of PER. Several approaches were used to locate the circadian pacemaker neurons controlling rhythms of eclosion and activity. These studies brought to focus lateral neurons (LNs) as being essential for behavioral rhythmicity. Elimination of other *per* positive cells through *per*+/per0 mosaic (29), using transgenic flies (32), or using flies with mutated brain structure (52), does not abolish locomotor rhythms, whereas genetic removal of all lateral LNs invariably renders flies arrhythmic (53). The projection patterns of the ventral group of LNs have been visualized using an antiserum against pigment-dispersing hormone (PDH) (54): These neurons have extensive arborization in the medulla portion of the optic lobes, and they also send axons into the protocerebrum. Interestingly, the latter axonal tract is enveloped by *per* positive glial cells, and there is evidence from mosaic studies that these cells also contribute to behavioral rhythmicity (29).

In addition to controlling activity, LNv's are good candidates for pacemakers controlling eclosion rhythm. The phase of both rhythms can be set in early larval stages and "remembered" so that insects subsequently kept in DD will emerge within their circadian gate or show individual rhythms of locomotor activity (11, 106). The fact that a cluster of lateral neurons is identifiable in larvae, and they are the only neurons that show cycling of PER throughout metamorphosis (58), provides correlative evidence for their involvement in both rhythms. No data are available as to the function of other *per*-positive neurons. Neurosecretory cells of the brain produce hormones regulating metabolism and behavior, and some of these hormones are secreted in a daily rhythm in various insects (86, 118, 119). It is feasible that *per*-positive neurons may be directly or indirectly involved in such rhythms.

Distribution of *per*-positive cells has been studied in the brain of a silkworm, *A. pernyi,* and a picture different from flies emerged (99). In the silkworm, four pairs of neurosecretory cells express *per* mRNA, PER, and TIM-like protein. In contrast to flies, PER is cytoplasmic at all phases of the circadian cycle and is most abundant at the same time as *per* mRNA. Interestingly, PER was also detected in 8 neurosecretory cells in the embryonic brain of *A. pernyi* but without obvious cycling. Yet, it appears that PER is necessary for the rhythm of the first behavioral event in the insect's life—larval hatching. Treatment of pharate larvae with *per* antisense oligodeoxynucleotides abolished the circadian rhythm of egg hatching. At the same time, this treatment depleted PER immunoreactivity from cells in the embryonic brain (101).

Brains of a few other insects have been studied in terms of *per* expression. Identification of cells expressing PER-like antigen (using *Drosophila* PER antibody) has been accomplished in a beetle, *Pachymorpha sexguttata* (31). The staining was more extensive than in moths, but similar in that several sets of neurosecretory neurons showed PER-like immunoreactivity in cytoplasm and axons. No obvious cycling of the antigen has been detected, except in axon terminals forming a neurohemal organ in the corpora cardiaca.

Finally, the expression of *per* has been monitored in honeybee brain . Although cell-level spatial localization of PER in the brain has not yet been determined, *per* expression has been correlated with the stages of behavioral development in adult bees. Young bees who are behaviorally arrhythmic have weak (although rhythmic) expression of *per.* Foragers, on the other hand, who have robust behavioral rhythms, show strong and cycling expression of *per* mRNA (117; G Robinson, personal communication).

Optic Lobes Optic lobes of flies do not seem to contain *per*-positive neuronal cell bodies; however, neurites from *per*-positive LNvs extensively arborize there as visualized by anti-PDH staining (52). Cycling expression of PER was detected in glial cells of the medulla and lamina (109, 132). The latter glial cells, which envelope optic cartridges in the lamina, may be involved in rhythms associated with visual processing. Optic cartridges are part of the first visual neuropile where photoreceptor terminals make synaptic contacts with large monopolar neurons (69). These neurons exhibit circadian changes in the diameter of their axons in *Musca* (83) and *Drosophila* (84a). The clock that controls changes in the diameter of monopolar neurons may be located in *per*-positive glial cells that are closely associated with these neurons. In *Musca*, cells of the glial epithelium change size in antiphase with neighboring monopolar neurons (E Pyza, personal communication). The equivalent glial cells in *Drosophila* are *per*-positive (110).

Photoreceptors Compound eyes of insects are composed of hundreds of ommatidia, each containing up to eight photoreceptor cells. Rhythmic expression and nuclear translocation of PER and TIM are consistently detected in these photoreceptors as well as in ocelli (56, 64, 109, 132). Experiments involving transgenic *Drosophila* demonstrated that the circadian oscillator in fly photoreceptors is autonomous in that it does not depend on *per* expression in other parts of the CNS (15). Patterns of *per* expression have also been studied in the photoreceptors of the silkworm, *A. pernyi* (99), and shown to match the patterns seen in flies. Just as in flies, silkworm photoreceptors exhibit a peak of *per* mRNA early in the night, followed by detection of PER in the nucleus several hours later. The consistency of *per* expression patterns in the eyes of flies and moths suggests that there may be a conserved set of clock-controlled visual processes. Several eye rhythms are known in insects. For example, circadian rhythmicity in visual sensitivity was reported in *Manduca sexta* (7). Another process that may be under clock control in insect eyes is the rhythm in screening pigment granules in photoreceptor terminals (84).

Sensory Structures Expression of PER in the fly antennae was first detected in transgenic flies carrying *per:lac Z* reporters (64). It was subsequently shown, using Green Fluorescent Protein and luciferase as reporters of *per* activity, that *per* is present in structures at the base of chemoreceptor cells in the antenna, proboscis, anterior wing margin, and legs (80). Rhythmic expression of *per:luc* persisted for several days in each of these body parts in vitro suggesting that brain-independent, autonomous clocks may be associated with chemoreceptors. Such clocks could conceivably modulate the excitability of sensory systems; this is supported by a recent demonstration of circadian rhythms in antennal responsiveness, which are abolished in clock-deprived, per^0 mutant flies (61a).

Expression of Clock Genes in Peripheral Tissues

Endocrine Glands Rhythmic expression of *per* has been reported in the *Drosophila* prothoracic gland (97), which is a portion of the ring gland producing ecdysone. Experiments using transgenes containing *per:luc* reporters demonstrated that *per* is rhythmically expressed in the prothoracic glands of pupal brain-ring gland complexes cultured in vitro in light-dark (LD) cycles or constant dark

(DD) (27). This expression continued after tetrodotoxin treatment, suggesting that brain-independent cycling of *per* may be occurring in prothoracic glands (27).

Consistent with this result are reports from "bigger" insects suggesting that release of the molting hormone ecdysone from prothoracic glands may be controlled by a circadian clock. In the silkworm, *Samia cynthia ricini*, the time when larvae quit feeding and purge their gut in preparation for metamorphosis is circadian-gated (33). The gut purge is stimulated by a small peak of ecdysone delivered from the prothoracic glands at a specific time of day. Based on experiments involving localized illuminations and the transplantation of glands from one larva to another, it appears that the photoreceptive clock that times ecdysone release is localized in the prothoracic glands themselves (71). Circadian control of ecdysone titers may also exist in the larvae of a wax moth, *Galleria melonella* (17, 18).

Substantial evidence for the circadian clock operating in prothoracic glands has been also gained from the hemipteran bug, *Rhodnius prolixus*. These insects display pronounced circadian fluctuations in circulating ecdysone levels (3). Prothoracic glands of *R. prolixus* contain their own photosensitive circadian oscillator, which is capable of timing ecdysone release in vitro (120). In intact animals, this oscillator is entrained by rhythmic release of prothoracicotropic hormone from the brain (119).

Alimentary Canal Early studies (64, 109) demonstrated strong *per* expression in segments of the alimentary canal of adult fruit flies, including salivary glands, esophagus, midgut, hindgut, and rectum. PER is detected in virtually all gut epithelial cells, showing cycling and nuclear translocation. The role of this clock gene in the digestive system is not known, but it is not hard to imagine that circadian clocks in the gut epithelium could have an adaptive value. Daily cycles of rest and activity are likely associated with feeding cycles, which in turn may be matched by daily rhythms in production of digestive enzymes, absorbtion of nutrients, and other processes associated with digestion and metabolism.

Activity of clock genes in the gut epithelium may be a more general phenomenon in insects. Cyclic PER expression and its nuclear translocation were observed in the midgut of the embryos and early larval stages of the silkworm, *A. pernyi* (101). It is thought that the cycling of PER in this tissue is not autonomous but, rather, brain dependent. This conclusion was reached after ligation of larvae behind the heads caused dysregulation of *per* movement in the nuclei of midgut epithelial cells (100).

Transformants with *per* β-galacosidase fusion gene revealed *per* expression in another segment of the fly alimentary system, the Malpighian tubules (64). More recent studies provided evidence for the existence of a brain-independent circadian mechanism in Malpighian tubules (38, 51). Both clock proteins, PER and TIM, cycle in the tubules with phases similar to those in clock-containing brain neurons. Cycling of PER and TIM persists in the tubules of decapitated flies in LD cycles and in DD. This rhythm is reset by 12 hours when decapitated flies are placed in reversed LD cycles (38). Furthermore, the adjustment of *per* expression levels in response to reversed LD cycles was identical in intact and decapitated larvae (51). There is also evidence for light-responsive cycling of luciferase *per*-reporter in cultured Malpighian tubules and hindguts dissected from transgenic flies carrying *per*-luciferase fusion gene (41a).

Even if the clock in Malpighian tubules is self-contained in vitro, it could still be affected in vivo by other circadian clocks in the body, as is the case with the clock in the *Rhodnius* prothoracic gland (119). To test this, we designed the following experiment (41a). We transplanted Malpighian tubules from the donor flies that were reared in reverse LD cycles to the host flies that were reared in the regular LD cycles, and then placed host flies in DD. At 12-hour intervals after this operation, both donor and host tubules were retrieved and stained with antibody against TIM. The results indicated that the TIM protein in the donor tubules cycled 12 hours out of phase as compared with host tubules. This suggests that the clock in the Malpighian tubules does not receive any phase-synchronizing signal from other parts of the body and therefore is completely autonomous. Experiments are underway to determine whether Malpighian tubules express circadian photoreceptor *cry*^{b} gene (M. Ivantchenko, personal communication).

Functions of the circadian clock in Malpighian tubules are not known. High expression of clock genes has been found in principal secretory cells (51). These cells are involved in producing primary urine via the action of a vacuolar ATPase that generates transepithelial voltage (TEV) (20, 21). The size of TEV has been measured at different times of day and not found to undergo circadian changes (E. Blumenthal, personal communication). Thus, a search for tubule functions that may follow a circadian rhythm has not yet produced conclusive results.

Ovaries Ovaries of *Drosophila* appear to express *per* gene (47, 64). The *per* signal in the ovaries is limited to the follicle cells associated with small previtellogenic egg chambers and appears to be cytoplasmic. In contrast to other *per*-positive cells in the fly body, no nuclear translocation of PER is observed in ovarian follicles. An interesting follow-up of this observation is the fact that *per* mRNA extracted from the ovaries does not exhibit daily cycles in abundance in contrast to *per* mRNA extracted from the rest of female abdomen or from male body (47). The role of *per* in the ovaries is not clear, but several leads indicate that it could have clock-related functions. First, rhythmic vitellogenesis and ovulation were reported in flies (reviewed in 45). Second, the critical daylength for ovarian diapause is shortened in *per*-null females (103). Finally, preliminary studies on fertility of *per*^{0}-mutants conducted in our lab suggest that flies without a functional clock develop fewer eggs than their wild-type counterparts (L Beaver, personal communication).

Testes It has been demonstrated over 10 years ago that insect testis contain autonomous circadian pacemaker (41). In moths, sperm bundles are released from the testis in a two-step circadian rhythm. The exit of sperm bundles from the testis to the vas deferens is restricted to a few hours in the evening, and subsequent transfer of sperm from the vas deferens to seminal vesicles occurs only within a few hours in the morning. Such rhythms were first observed in the flour moth, *Ephestia kuehniella* (94), and subsequently confirmed in several other moth species (36, 37, 43, 62). The testes–vas deferens complexes of the gypsy moth, *Lymantria dispar*, can be maintained in culture, which allowed demonstration that the two-step rhythm of sperm release persists in cultured complexes in DD. The rhythm of sperm release can by entrained by a shift in photoperiod applied to isolated testes–vas deferns in vitro, suggesting that these tissues contain both circadian clocks and photoreceptors involved in circadian entrainment (41). One accordingly wonders how the circadian mechanism is organized within this complex: Is there a specific group of cells endowed with timing capability, or is this capability widespread among many or even all cells comprising this complex?

We recently obtained evidence that *per* is expressed in the reproductive system of the codling moth, *Cydia pomonella* (44a). A fragment of the codling moth *per* coding region (88) has been used to perform in situ hybridizations in the male reproductive system. The results suggest that *per* mRNA is broadly distributed in the epithelial cells of the vas deferens but is not detected in the testis. The hybridization signal peaks at night and is lowest during the day, indicating that levels of *per* mRNA in the vas deferens exhibit a daily rhythm (44a).

What is the role of *per* in the moth's vas deferens epithelium? Several rhythms are associated with this epithelium. A modified segment of the vas deferens, the terminal epithelium, provides a barrier between the testis and vas deferens lumen. The terminal epithelium restricts the exit of sperm bundles and the rhythmic release of sperm from the testis appears to be regulated here (42, 93). These cells undergo dramatic daily changes in their shape associated with a rhythmic formation of exit channels that allow sperm bundles to leave the testes (42). Moreover, *per*-positive epithelial cells surrounding vas deferens lumen display daily cycles of secretory activity in moths (39, 91; P Bebas, personal communication). Thus, *per* is likely involved in the circadian system coordinating sperm release and maturation. The circadian coordination is vital for moths: Disruption of sperm release rhythms by constant light leads to male sterility (40, 92).

Expression of *per* occurs in *Drosophila* testes (17, 22); detailed studies show that *per* is active in the terminal epithelium at the base of the testis and in the vas deferens epithelium (J Giebultowicz, unpublished). The fact that *per* is found in homologous regions of the reproduction system in flies and moths suggests the existence of clock-controlled reproductive processes in *Drosophila*. Evidence in support of this would be to demonstrate that loss of *per* function interferes with reproductive rhythms. Since several mutants with disrupted clock function exist in *Drosophila*, we started to investigate whether clock-controlled reproductive rhythms may occur in this fly. Preliminary evidence suggests that a circadian clock may be associated with sperm release in fruit flies (T Vollintine & J Giebultowicz, unpublished)

Do Insects Have Cell-Autonomous Circadian Clocks?

Spatial expression patterns of clock genes in insects suggest that many tissue types, at least in flies and moths, may have independent circadian pacemaking abilities. This is supported by the following observations: (*a*) cycling of *per* and PER have very similar phases in different tissues (although with a few exceptions: see 58); (*b*) several isolated organs display rhythmic expression of clock genes in vitro; and (*c*) it has been directly demonstrated that some output rhythms are driven by local tissue-autonomous clocks (35a).

If circadian clocks are tissue autonomous, are they also cell autonomous? Does uniform expression of *per* in the epithelial cells of the moth vas deferens means that each of those cells possesses a clock? One beautiful demonstration that single cells can carry out the task of driving output rhythm was a brain-behavioral study of LNv's in *Drosophila disconnected* mutants (53). This mutation disrupts neural cell patterning and causes loss of LNv's in varying degrees. Correlative studies of brain neuroanatomy and locomotor behavior suggest that the presence of even a single LNv is sufficient to evoke a circadian rhythm of activity. This is not an unusual finding considering that circadian clocks operate in unicellular organisms. Moreover, individual cells from animal tissues can be competent circadian pacemakers; circadian rhythms were recorded in vitro from single cells derived from pacemaking tissues such as basal retinal neurons of molluscan eyes (70) and suprachiasmatic neurons of rat (125). Recently circadian rhythms were detected in a rat fibroblast cell line, and efforts are underway to monitor these rhythms in single isolated cells (5).

CONCLUSIONS AND FUTURE PROSPECTS

Data summarized in this review suggest that many cells in insect peripheral tissues have the molecular machinery necessary to "run" the circadian clock. This finding expands our knowledge of the circadian system and implies that temporal synchronization of life functions may be achieved by a set of independently working clocks rather then by a central master clock, as has been traditionally assumed. Experiments demonstrate that peripheral clocks are photoreceptive and entrainable in vitro. It is thus conceivable that solar days directly coordinate all clocks in the body, acting, in effect, as the "master oscillator." It is possible that a collection of independently entrainable clocks could fine-tune physiological processes in the organism. For example, at the time when the clock in the brain stimulates locomotor activity via its neuronal and hormonal outputs, a clock in Malpighian tubules may stimulate synthesis of enzymes that will handle waste resulting from increased metabolism associated with activity. Clocks in different cell types, such as neurons versus epithelium, may be similar in the molecular sense, but different in the scope of the outputs that they control.

The puzzle of how circadian systems are organized in insects is far from solved. We know very little about the relationships between the various clocks in insects and the nature of the rhythms they control. Particularly missing are the molecular intermediates leading from clocks to overt rhythms (Figure 1). These gaps in our knowledge may soon be patched, considering that molecular mechanisms of the *Drosophila* clocks are being unraveled at a fast pace, and homologues of *Drosophila* clock genes have been found in other insects. A better understanding of the molecular and cellular pathways downstream of circadian clocks may be achieved through cooperative efforts between *Drosophila* geneticists and insect physiologists.

ACKNOWLEDGMENTS

I thank Brian Dixon and Maria Ivantchenko for comments on the manuscript, as well as Jeff Hall for his various inputs into our circadian research. Research in the author's laboratory reported here was supported by grant IBN-972322 and REU supplement of NSF; and USDA CSREES, 98-35302-6793 and the NATO linkage grant.

Visit the Annual Reviews home page at www.AnnualReviews.org.

LITERATURE CITED

1. Albrecht U, Sun ZS, Eichele G, Lee CC. 1997. A different response of two putative mammalian circadian regulators, *mper1* and *mper2*, to light. *Cell* 91:1055–64
2. Allada R, White NE, So WV, Hall JC, Rosbash M. 1998. A mutant *Drosophila* homolog of mammalian *clock* disrupts circadian rhythms and transcription of *period* and *timeless*. *Cell* 93:791–804
3. Ampleford EJ, Steel CGH. 1985. Circadian control of a daily rhythm in hemolymph ecdysteroid titer in the insect *Rhodnius prolixus* (Hemiptera). *Gen. Comp. Endocrinol.* 59:453–59
4. Bae K, Lee C, Sidote D, Chuang KY, Edery I. 1998. Circadian regulation of a *Drosophila* homolog of the mammalian *clock* gene: PER and TIM function as positive regulators. *Mol. Cell. Biol.* 18:6142–51
5. Balsalobre A, Damiola F, Schibler U. 1998. A serum shock induces circadian gene expression in mammalian tissue culture cells. *Cell* 93:929–37
6. Bargiello TA, Young MW. 1984. Molecular genetics of a biological clock in *Drosophila*. *Proc. Natl. Acad. Sci. USA* 81:2142–46
7. Bennet RR. 1983. Circadian rhythm of visual sensitivity in *Manduca sexta* and its development from ultradian rhythm. *J. Comp. Physiol. A* 150:165–74
8. Blest AD. 1988. The turnover of the phototransductive membrane in compound eyes and ocelli. *Adv. Insect Physiol.* 20:1–53
9. Brady JN. 1974. The physiology of insect circadian rhythms. *Adv. Insect. Physiol.* 10:1–115
10. Brandes C, Plautz JD, Stanewsky R, Jamison CF, Straume M, et al. 1996.

Novel features of Drosophila *period* transcription revealed by real-time luciferase reporting. *Neuron* 16:687–92
11. Brett WJ. 1955. Persistent diurnal rhythmicity in *Drosophila* emergence. *Ann. Entomol. Soc. Am.* 48:119–31
12. Carde RT, Minks AK, eds. 1997. *Insect Pheromone Research: New Directions* London: Chapman & Hall. 684 pp.
13. Cashmore AR. 1998. The cryptochrome family of blue/UV-A photoreceptors. *J. Plant Res.* 111:267–70
14. Cassone VM, Natesan AK. 1997. Time and time again: The phylogeny of melatonin as a transducer of biological time. *J. Biol. Rhythms* 12:489–97
14a. Ceriani MF, Darlington TK, Staknis D, Mas P, Petti AA, et al. 1999. Light dependent sequestration of TIMELESS by CRYPTOCHROME. *Science* 285:553–56
15. Cheng Y, Hardin PE. 1998. *Drosophila* photoreceptors contain an autonomous circadian pacemaker that can function without *period* mRNA cycling. *J. Neurosci.* 18:741–50
16. Colot HV, Hall JC, Rosbash M. 1988. Interspecific comparison of the period gene of *Drosophila* reveals large blocks of non-conserved coding DNA. *EMBO J.* 7:3929–37
17. Cymborowski B, Muszynska-Pytel M, Porcheron P, Cassier P. 1991. Hemolymph ecdysteroid titres controlled by a circadian clock mechanism in larvae of the wax moth, *Galleria melonella. J. Insect Physiol.* 37:35–40
18. Cymborowski B, Smietanko A, Delbecque JP. 1989. Circadian modulation of ecdysteroid titer in *Galleria melonella* larvae. *Comp. Biochem. Physiol.* A94: 431–38
19. Darlington TK, Wager-Smith K, Ceriani MF, Staknis D, Gekakis N, et al. 1998. Closing the circadian loop: CLOCK-induced transcription of its own inhibitors *per* and *tim. Science* 280:1599–603
20. Dow JAT, Davies SA, Guo Y, Graham S, Finbow M, Kaiser K. 1997. Molecular genetic analysis of V-ATPase function in *Drosophila melanogaster. J. Exp. Biol.* 202:237–45
21. Dow JAT, Maddrell SH, Gortz A, Skaer NJV, Brogan S, Kaiser K. 1994. The Malpighian tubles of *Drosophila melanogaster*: a novel phenotype for studies of fluid secretion and its control. *J. Exp. Biol.* 197:421–28
22. Dunlap JC. 1996. Genetic and molecular analysis of circadian rhythms. *Annu. Rev. Genet.* 30:579–601
23. Dunlap JC. 1999. Molecular bases for circadian clocks. *Cell* 96:271–90
24. Dutkowski AB, Cymborowski B, Przelecka A. 1971. Circadian changes in the ultrastructure of the neurosecretory cells of the pars intercerebralis of the house cricket. *J. Insect Physiol.* 17:1763–22
24a. Edery I. 1999. Role of postranscriptional regulation in circadian clocks: lessons from *Drosophila. Chronobiol. Int.* 16: 377–414
25. Edery I, Zwiebel LJ, Dembinska ME, Rosbash M. 1994. Temporal phosphorylation of the *Drosophila* period protein. *Proc. Natl. Acad. Sci. USA* 91:2260–64
26. Edmunds LN. 1988. *Cellular and Molecular Bases of Biological Clocks.* New York: Springer-Verlag. 497 pp.
27. Emery IF, Noveral JM, Jamison CF, Siwicki KK. 1997. Rhythms of-*Drosophila period* gene expression in culture. *Proc. Natl. Acad. Sci. USA* 94:4092–96
28. Emery P, So V, Kaneko M, Hall JC, Rosbash M. 1998. CRY, a *Drosophila* clock and light-regulated cryptochrome, is a major contributor to circadian rhythm resetting and photosensitivity. *Cell* 95:669–79
29. Ewer J, Frish B, Hamblen-Coyle MJ, Rosbash M, Hall JC. 1992. Expression of the *period* clock gene within different cell types in the brain of *Drosophila* adults and mosaic analysis of these cells'

influence on circadian behavioral rhythms. *J. Neurosci.* 12:3321–49
30. Feldman JF, Hoyle M. 1973. Isolation of circadian clock mutants in *Neurospora crassa. Genetics* 75:605–13
31. Frisch B, Fleissner G, Fleissner G, Brandes C, Hall JC. 1996. Staining in the brain of *Pachymorpha sexguttata* mediated by antibody against a *Drosophila* clock-gene product: labeling of cells with possible importance for beetle's circadian rhythms. *Cell Tissue Res.* 286:411–92
32. Frisch B, Hardin PE, Hamblen-Coyle MJ, Rosbash M, Hall JC. 1994. A promoterless *period* gene mediates behavioral rhythmicity and cyclical *per* expression in a restricted subset of the *Drosophila* nervous system. *Neuron* 12:555–70
33. Fuishita M, Ishizaki H. 1982. Temporal organization of endocrine events in relation to the circadian clock during larval-pupal development in *Samia cynthia ricini. J. Insect Physiol.* 28:77–84
34. Gekakis N, Saez L, Sehgal A, Young M, Weitz C. 1995. Isolation of *timeless* by PER protein interaction: defective interaction between *timeless* protein and long-period mutant PER. *Science* 270:811–15
35. Gekakis N, Staknis D, Nguyen HB, Davis FC, Wilsbacher LD, et al. 1998. Role of the CLOCK protein in the mammalian circadian mechanism. *Science* 280:1564–69
35a. Giebultowicz JM. 1999. Insect circadian rhythms: Is it all in their heads? *J. Insect Physiol.* 45:791–800
36. Giebultowicz JM, Bell RA, Imberski RB. 1988. Circadian rhythm of sperm movement in the male reproductive tract of the gypsy moth, *Lymantria dispar. J. Insect Physiol.* 34:527–32
37. Giebultowicz JM, Brooks NL. 1998. The circadian rhythm of sperm release in the codling moth, *Cydia pomonella. Entomol. Exp. Appl.* 88:229–34
38. Giebultowicz JM, Hege DM. 1997. Circadian clock in Malpighian tubules. *Nature* 386:664
39. Giebultowicz JM, Joy JE, Riemann JG. 1992. Circadian rhythm of sperm release from the testis in moths: protein changes during development of the sperm release system. In *Advances in Regulation of Insect Reproduction*, ed. B Benetova, I Gelbic, T Soldan, pp. 91–95. Ćeské Budejovice: Czech. Acad. Sci.
40. Giebultowicz JM, Ridgway RL, Imberski RB. 1990. Physiological basis for sterilizing effects of constant light in *Lymantria dispar. Physiol. Entomol.* 15:149–56
41. Giebultowicz JM, Riemann JG, Raina AK, Ridgway RL. 1989. Circadian system controlling release of sperm in the insect testes. *Science* 245:1098– 100
41a. Giebultowicz JM, Stanewsky R, Hall JC, Hege DM. 2000. Autonomous circadian clocks in *Drosophila:* transplanted excretory tubules maintain out of phase cycling with the host. *Curr. Biol.* In press
42. Giebultowicz JM, Weyda F, Erbe EF, Wergin WP. 1997. Circadian rhythm of sperm release in the gypsy moth, *Lymantria dispar:* ultrastructural study of transepithelial penetration of sperm bundles. *J. Insect. Physiol.* 43:1133–47
43. Giebultowicz JM, Zdarek J. 1996. The rhythm of sperm release from testis and mating flight are not correlated in *Lymantria* moths. *J. Insect Physiol.* 42:167–70
43a. Glossop NR, Lyons LC, Hardin PE. 1999. Interlocked feedback loops within the Drosophila circadian oscillator. *Science* 286:766–68
44. Golden SS, Ishiura M, Johnson CH, Kondo T. 1997. Cyanobacterial circadian rhythms. *Annu. Rev. Plant Physiol. Plant. Mol. Biol.* 48:327—54
44a. Gvakharia BO, Kilgore JA, Bebas P, Giebultowicz JM. 2000. Temporal and spatial expression of the *period* gene in the reproductive system of the codling moth. *J. Biol. Rhythms.* 15:27–35
45. Hall JC. 1998. Genetics of biological

rhythms in *Drosophila*. *Adv. Genet.* 38:135–84

46. Hao H, Allen DL, Hardin PE. 1997. A circadian enhancer mediates PER-dependent mRNA cycling in *Drosophila melanogaster*. *Mol. Cell. Biol.* 17:3687–93
47. Hardin PE. 1994. Analysis of period mRNA cycling in *Drosophila* head and body tissues indicates that body oscillators behave differently from head oscillators. *Mol. Cell. Biol.* 14:7211–18
48. Hardin PE, Hall JC, Rosbash M. 1990. Feedback of the *Drosophila period* gene product on circadian cycling of its messenger RNA levels. *Nature* 343:536–40
49. Hardin PE, Hall JC, Rosbash M. 1992. Circadian oscillations in the *period* gene mRNA levels are transcriptionally regulated. *Proc. Natl. Acad. Sci. USA* 89:11711–15
50. Hastings JW, Rusak B, Boulos Z. 1991. Circadian rhythms: the physiology of biological timing. In *Neural and Integrative Animal Physiology*, ed. CL Prosser, pp. 435–523. New York: Wiley-Liss
51. Hege DM, Stanewsky R, Hall JC, Giebultowicz JM. 1997. Rhythmic expression of a PER-reporter in the Malpighian tubules of decapitated *Drosophila*: evidence for a brain-independent circadian clock. *J. Biol. Rhythms* 12:300–8
52. Helfrich-Förster C. 1996. *Drosophila* rhythms: From brain to behavior. *Semin. Cell Dev. Biol.* 7:791–802
53. Helfrich-Förster C. 1998. Robust circadian rhythmicity of *Drosophila melanogaster* requires the presence of lateral neurons: a brain-behavioral study of *disconnected* mutants. *J. Comp. Physiol. A* 182:435–53
54. Helfrich-Förster C. 1995. The *period* clock gene is expressed in central nervous system neurons which also produce a neuropeptide that reveals the projections of circadian pacemaker cells within the brain of *Drosophila melanogaster*. *Proc. Natl. Acad. Sci. USA* 92:612–16
55. Huang ZJ, Edery I, Rosbash M. 1993. PAS is a dimerization domain common to *Drosophila period* and several transcription factors. *Nature* 364:259–62
56. Hunter-Ensor M, Ousley A, Sehgal A. 1996. Regulation of the *Drosophila* protein *timeless* suggests a mechanism for the resetting of the circadian clock by light. *Cell* 84:677–85
57. Johnson CH. 1992. Phase response curves: What can they tell us about circadian clocks? In *Circadian Clocks from Cell to Human*, ed. T Hiroshige, K Honma, pp. 209–49. Sapporo: Hokkaido Univ. Press
58. Kaneko M, Helfrich-Förster C, Hall JC. 1997. Spatial and temporal expression of the *period* and *timeless* genes in the developing nervous system of *Drosophila*: Newly identified pacemaker candidates and novel features of clock gene product cycling. *J. Neurosci.* 17:6745–60
59. King DS, Zhao Y, Sangoram AM, Wilsbacher LD, Tanaka M, et al. 1997. Positional cloning of the mouse circadian *clock* gene. *Cell* 89:644–54
60. Kloss B, Price JL, Saez L, Blau J, Rothenfluh A, et al. 1998. The *Drosophila* clock gene *double-time* encodes a protein closely related to human casein kinase I. *Cell* 94:97–107
61. Konopka RJ, Benzer S. 1971. Clock mutants of *Drosophila melanogaster*. *Proc. Natl. Acad. Sci. USA* 68:2112–16

61a. Krishnan B, Dryer SE, Hardin PE. 1999. Circadian rhythms in olfactory responses of *Drosophila melanogaster*. *Nature* 4000:375–78

62. LaChance LE, Richard RD, Ruud RL. 1977. Movement of eupyrene sperm bundles from the testes and storage in the ductus ejaculatoris duplex of the male pink bollworm: effects of age, strain, irradiation and light. *Ann. Entomol. Soc. Am.* 70:647—52

62a. Lee C, Bae K, Edery I. 1999. PER and TIM inhibit the DNA binding activity of a *Drosophila* CLOCK-CYC/dBMAL1

heterodimer without disrupting formation of the heterodimer: a basis for circadian transcription. *Mol. Cell Biol.* 19:5316–25
63. Levine J, Sauman I, Imbalzano M, Reppert S, Jackson F. 1995. Period protein from the giant silkmoth *Antheraea pernyi* functions as a circadian clock element in *Drosophila melanogaster*. *Neuron* 15:147–57
64. Liu X, Lorenz L, Yu QN, Hall JC, Rosbash M. 1988. Spatial and temporal expression of the *period* gene in *Drosophila melanogaster*. *Genes Dev.* 2:228–38
65. Loher W. 1974. Circadian control of spermatophore formation in the cricket, *Teleogryllus commodus* Walker. *J. Insect Physiol.* 20:1155–72
66. Loros JJ, Dunlap JC. 1991. *Neurospora crassa* clock-controlled genes are regulated at the level of transcription. *Mol. Cell. Biol.* 11:558–63
67. Lucas RJ, Foster RG. 1999. Photoentrainment in mammals: A role for chryptochrome? *J. Biol. Rhythms* 14:4–10
67a. Majercak J, Sidote D, Hardin PE, Edery I. 1999. How a circadian clock adapts to seasonal decreases in temperature and day length. *Neuron* 24:219–30
68. McNeil GP, Zhang X, Genova G, Jackson FR. 1998. A molecular rhythm mediating circadian clock output in *Drosophila*. *Neuron* 20:297–303
69. Meinertzhagen IA, Pyza E. 1996. Daily rhythms in cells of the fly's optic lobe: taking time out from the circadian clock. *Trends Neurosci.* 19:285–91
70. Michel S, Geusz M, Zarisky J, Block G. 1993. Circadian rhythm in membrane conductance expressed in isolated neurons. *Science* 259:239–41
71. Mizoguchi A, Ishizaki H. 1982. Prothoracic glands of the Saturniid moth *Samia cynthia ricini* possess a circadian clock controlling gut purge timing. *Proc. Natl. Acad. Sci. USA* 79:2726–30
72. Moore-Ede MC, Sulzman FM, Fuller CA. 1982. *The Clocks That Time Us*. Cambridge: Harvard Univ. Press. 448 pp.
73. Morse DS, Fritz L, Hastings WJ. 1990. What is the clock? Translational regulation of circadian bioluminescence. *Trends Biochem. Sci.* 15:262–5
74. Myers MP, Wager-Smith K, Rothenfluh-Hilfinker A, Young MW. 1996. Light-induced degradation of TIMELESS and entrainment of *Drosophila* circadian clock. *Science* 271:1736–40
75. Neville AC. 1967. A dermal light sense influencing skeletal structure in locusts. *J. Insect Physiol.* 13:933–9
76. Neville AC. 1970. Cuticle ultrastructure in relation to the whole insect. In *Insect Ultrastructure*, ed. AC Neville, pp. 17–39. Symp. R. Entomol. Soc. London, Vol. 5. Oxford: Blackwell Sci.
77. Page TL. 1985. Clocks and circadian rhythms. In *Comprehensive Insect Physiology, Biochemistry and Pharmacology*, Vol. 6, ed. G Kerkut, L Gilbert, 6:577–652. Oxford: Pergamon
78. Pittendrigh CS. 1960. Circadian rhythms and circadian organization of the living systems. *Cold Spring Harbor Symp. Quant. Biol.* 25:159–82
79. Pittendrigh CS. 1981. Circadian systems: Entrainment. In *Handbook of Behavioral Neurobiology*, ed. J Ashoff, 4:95–124. New York: Plenum
80. Plautz JD, Kaneko M, Hall JC, Kay SA. 1997. Independent photoreceptive circadian clocks throughout *Drosophila*. *Science* 278:1632–35
81. Price JL, Blau J, Rothenfluh A, Abodeely M, Kloss B, Young MW. 1998. *double-time* is a novel *Drosophila* clock gene that regulates PERIOD protein accumulation. *Cell* 94:83–95
82. Price JL, Dembinska ME, Young MW, Rosbash M. 1995. Suppression of PERIOD protein abundance and circadian cycling by the *Drosophila* clock mutation *timeless*. *EMBO J.* 14:4044–49
83. Pyza E, Meinertzhagen IA. 1995. Monopolar cell axons in the first optic neuro-

phil of the housefly *Musca domestica* L., undergo daily fluctuations in diameter that have a circadian basis. *J. Neurosci.* 15:407–18
84. Pyza E, Meinertzhagen IA. 1997. Circadian rhythms in screening pigment and invaginating organelles in photoreceptor terminals of the housefly's first optical neuropile. *J. Neurobiol.* 32:517—29
84a. Pyza E, Meinertzhagen I. 1999. Daily rhythmic changes of cell size and shape in the first optic neuropil in *Drosophila melanogaster*. *J. Neurobiol.* 40:77–88
85. Qiu J, Hardin P. 1996. *per* mRNA cycling is locked to lights-off under photoperiodic conditions that support circadian feedback loop function. *Mol. Cell. Biol.* 16:4182–88
86. Raina AK, Menn JJ. 1987. Endocrine regulation of pheromone production in Lepidoptera. In *Pheromone Biochemistry*, ed. GD Prestwich, GJ Blomquist, pp. 159–74. Orlando, FL: Academic
87. Reddy P, Zehring WA, Wheeler DA, Pirrotta V, Hadfield C, et al. 1984. Molecular analysis of the *period* locus in *Drosophila melanogaster* and identification of a transcript involved in biological rhythms. *Cell* 38:701–10
88. Regier JC, Fang QQ, Mitter C, Peigler RS, Friedlander TP, Solis MA. 1998. Evolution and phylogenetic utility of the *period* gene in Lepidoptera. *Mol. Biol. Evol.* 15:1172–82
89. Reppert SM. 1998. A clockwork explosion! *Neuron* 21:1–4
90. Reppert SM, Tsai T, Roca AL, Sauman I. 1994. Cloning of a structural and functional homolog of the circadian clock gene *period* from the giant silk moth *Antheraea pernyi*. *Neuron* 13:1167–76
91. Riemann JG, Giebultowicz JM. 1991. Secretion in the upper vas deferens of the gypsy moth correlated with the circadian rhythm of sperm release from the testes. *J. Insect Physiol.* 37:53–62
92. Riemann JG, Ruud RL. 1974. Mediterranean flour moth: effects of continuous light on the reproductive capacity. *Ann. Entomol. Soc. Am.* 67:857–60
93. Riemann JG, Thorson BJ. 1976. Ultrastructure of the vasa deferentia of the Mediterranean flour moth. *J. Morphol.* 149:483–506
94. Riemann JG, Thorson BJ, Ruud RL. 1974. Daily cycle of release of sperm from the testes of the Mediterranean flour moth. *J. Insect Physiol.* 20:195–207
95. Rouyer F, Rachidi M, Piekielny C, Rosbah M. 1997. A new gene encoding a putative transcription factor regulated by the *Drosophila* circadian clock. *EMBO J.* 16:3944–54
96. Rutila JE, Suri V, Le M, So V, Rosbash M, Hall JC. 1998. CYCLE is a second bHLH-PAS clock protein essential for circadian rhythmicity and transcription of *Drosophila period* and *timeless*. *Cell* 93:805–14
97. Saez L, Young MW. 1988. *In situ* localization of the *per* clock protein during development of *Drosophila melanogaster*. *Mol. Cell. Biol.* 8:5378–85
98. Saez L, Young MW. 1996. Regulated nuclear entry of the *Drosophila* clock proteins PERIOD and TIMELESS. *Neuron* 17:911–12
99. Sauman I, Reppert SM. 1996. Circadian clock neurons in the silkworm, *Antheraea pernyi*: Novel mechanism of period protein regulation. *Neuron* 17:889–900
100. Sauman I, Reppert SM. 1998. Brain control of embryonic circadian rhythms in the silkworm, *Antheraea pernyi*. *Neuron* 20:741–48
101. Sauman I, Tsai T, Roca AL, Reppert SM. 1996. Period protein is necessary for circadian control of egg hatching behavior in the silkworm, *Antheraea pernyi*. *Neuron* 17:901–9
102. Saunders DS. 1982. *Insect Clocks*. Oxford: Pergamon. 279 pp. 2 ed.
103. Saunders DS. 1990. The circadian basis of ovarian diapause regulation in *Drosophila melanogaster*: Is the *period* gene casually involved in photoperiodic time

measurement? *J. Biol. Rhythms* 5:315–31
104. Saunders DS. 1997. Insect circadian rhythms and photoperiodism. *Invert. Neurosci.* 3:155–64
105. Sehgal A, Price JL, Man B, Young MW. 1994. Loss of circadian behavioral rhythms and *per* RNA oscillations in the *Drosophila* mutant *timeless*. *Science* 263:1603–6
106. Sehgal A, Price J, Young MW. 1992. Ontogeny of a biological clock in *Drosophila melanogaster*. *Proc. Natl. Acad. Sci. USA* 89:1423–27
107. Sehgal A, Rothenfluh-Hilfiker A, Hunter-Ensor M, Chen YF, Myers MP, Young MW. 1995. Rhythmic expression of *timeless*: a basis for promoting circadian cycles in *period* gene autoregulation. *Science* 270:808–10
108. Shigeyoshi Y, Taguchi K, Yamamoto S, Takekida S, Yan L, et al. 1997. Light-induced resetting of a mammalian circadian clock is associated with rapid induction of the *mPer1* transcript. *Cell* 91:1043–53
109. Siwicki KK, Eastman C, Petersen G, Rosbash M, Hall JC. 1988. Antibodies to the *period* gene product of *Drosophila* reveal diverse tissue distribution and rhythmic changes in the visual system. *Neuron* 1:141–50
110. Stanewsky R, Frisch B, Brandes C, Hamblen-Coyle MJ, Rosbash M, Hall JC. 1997. Temporal and spatial expression patterns of transgenes containing increasing amounts of the *Drosophila* clock gene *period* and a *lac*Z reporter: mapping elements of the PER protein involved in circadian cycling. *J. Neurosci.* 17:676–96
111. Stanewsky R, Jamison CF, Plautz JD, Kay SA, Hall JC. 1997. Multiple circadian regulated elements contribute to cycling *period* gene expression in *Drosophila*. *EMBO J.* 16:5006–18
112. Stanewsky R, Kaneko M, Emery P, Beretta B, Wager-Smith K, et al. 1998. The *cryb* mutation identifies cryptochrome as a circadian photoreceptor in *Drosophila*. *Cell* 95:681–92
113. Strayer C, Kay SA. 1999. The ins and outs of circadian regulated gene expression. *Curr. Opin. Plant Biol.* 2:114–20
114. Sun ZS, Albrecht U, Zhuchenko O, Bailey J, Eichele G, Lee CC. 1997. RIGUI, a putative mammalian ortholog of the Drosophila period gene. *Cell* 90:1003–11
115. Takahashi JS. 1993. Circadian clock regulation of gene expression. *Curr. Opin. Genet. Dev.* 3:301–9
116. Tei H, Okamura H, Sakaki Y. 1997. Circadian oscillation of a mammalian homologue of the Drosophila period gene. *Nature* 389:512–15
117. Toma DP, Robinson GE. 1996. Isolation of the *period* gene and its expression during the ontogeny of behavioral circadian rhythms in the honey bee, *Apis mellifera*. *Meet. Soc. Res. Biol. Rhythms, 5th, Jacksonville, Fl.*
118. Truman JW. 1992. The eclosion hormone system of insects. *Progr. Brain Res.* 92:361–74
119. Vafopoulou X, Steel CGH. 1996. The insect neuropeptide prothoracicotropic hormone is released with a daily rhythm: re-evaluation of its role in development. *Proc. Natl. Acad. Sci. USA* 93:3368–72
120. Vafopoulou X, Steel CGH. 1998. A photosensitive circadian oscillator in an insect endocrine gland: photic induction of rhythmic steroidogenesis in vitro. *J. Comp. Physiol. A* 182:343–49
121. Van Gelder RN, Krasnow MA. 1996. A novel circadianly expressed *Drosophila melanogaster* gene dependent on a *period* gene for its rhythmic expression. *EMBO J.* 15:1625–31
122. Vaz Nunes M, Saunders D. 1999. Photoperiodic time measurement in insects: A review of clock models. *J. Biol. Rhythms* 14:84–104
123. Vosshall LB, Price JL, Sehgal A, Saez L, Young MW. 1994. Block in nuclear localization of *period* protein by a second

clock mutation, *timeless*. *Science* 263:1606–9

124. Weber F. 1995. Cyclic layer deposition in the cockroach (*Blaberus cranifer*) endocuticle: a circadian rhythm in leg pieces cultured *in vitro*. *J. Insect Physiol.* 41:153–61
125. Welsh DK, Logothetis DE, Meister M, Reppert SM. 1995. Individual neurons dissociated from rat suprachiasmatic nucleus express independently phased circadian firing rhythms. *Neuron* 14:697–706
126. Wiedenmann G, Lukat R, Weber F. 1986. Cyclic layer deposition in the cockroach endocuticle: a circadian rhythm? *J. Insect Physiol.* 32:1019–27
127. Wilsbacher LD, Takahashi JS. 1998. Circadian rhythms: molecular basis of the clock. *Curr. Opin. Genet. Dev.* 8:595–602
128. Yang Z, Emerson M, Su HS, Sehgal A. 1998. Response of the timeless protein to light correlates with behavioral entrainment and suggests a nonvisual pathway for circadian photoreception. *Neuron* 21:215–23
129. Young MW. 1998. The molecular control of circadian behavioral rhythms and their entrainment in *Drosophila*. *Annu. Rev. Biochem.* 67:135–52
130. Zeng H, Hardin PE, Rosbash M. 1994. Constitutive overexpression of the *Drosophila period* protein inhibits *period* mRNA cycling. *EMBO J.* 13:3590–98
131. Zeng H, Qian Z, Myers MP, Rosbash M. 1996. A light-entrainment mechanism for the *Drosophila* circadian clock. *Nature* 380:129–35
132. Zerr DM, Hall JC, Rosbash M, Siwicki KK. 1990. Circadian fluctuations of period protein immunoreactivity in the CNS and the visual system of *Drosophila*. *J. Neurosci.* 10:2749–62
133. Zylka MJ, Shearman LP, Weaver DR, Reppert SM. 1998. Three period homologs in mammals: Differential light responses in the suprachiasmatic circadian clock and oscillating transcripts outside of brain. *Neuron* 20:1103–10

Annu. Rev. Entomol. 2000. 45:795–795

IMPACT OF THE INTERNET ON EXTENSION ENTOMOLOGY

J.K. VanDyk
Department of Entomology, Iowa State University, Ames Iowa 50011; e-mail: jvandyk@iastate.edu

Key Words online, World Wide Web, history, computers

■ **Abstract** Acceptance of the Internet as a common method for transferring entomological information has forced a re-evaluation of extension entomology's role. The discipline of entomology was an early adopter of Internet-based information delivery and organization on both Gopher and the World Wide Web. New opportunities exist for immediate presentation and collection of information, but organization and categorization of information for timely retrieval remains a challenge. Metadata strategies, computer literacy, and integration of the Internet with current workflow paradigms promise to change the way extension entomology is done in the near future.

This chapter, in its entirety, is available in the online version of this volume; see http://ento.annualreviews.org

0066-4170/00/0107-0795/$14.00

Subject Index

U

V

W

X

Z

Cumulative Indexes

CONTRIBUTING AUTHORS, VOLUMES 36–45

CHAPTER TITLES, VOLUMES 36–45

VECTORS OF PLANT PATHOGENS